Analytical Chemistry

INTERNATIONAL ADAPTATION

Analytical Chemistry

INTERNATIONAL ADAPTATION

Gary D. Christian
University of Washington

Purnendu K. (Sandy) Dasgupta
University of Texas at Arlington

Kevin A. Schug
University of Texas at Arlington

WILEY

Analytical Chemistry

INTERNATIONAL ADAPTATION

Contributing Subject Matter Experts: Dr. Dhanraj Masram and Dr. Sandeep Kaur, Associate Professors, Department of Chemistry, Delhi University

Founded in 1807, John Wiley & Sons, Inc. has been a valued source of knowledge and understanding for more than 200 years, helping people around the world meet their needs and fulfill their aspirations. Our company is built on a foundation of principles that include responsibility to the communities we serve and where we live and work. In 2008, we launched a Corporate Citizenship Initiative, a global effort to address the environmental, social, economic, and ethical challenges we face in our business. Among the issues we are addressing are carbon impact, paper specifications and procurement, ethical conduct within our business and among our vendors, and community and charitable support. For more information, please visit our website: www.wiley.com/go/citizenship.

ISBN: 978-1-119-77079-4

ISBN: 978-1-119-77080-0 (ePub)

ISBN: 978-1-119-77081-7 (ePdf)

Printed and bound by CPI Group (UK) Ltd, Croydon, CR0 4YY

C9781119770794_130224

Chapter 4
Stoichiometric Calculations: The Workhorse of the Analyst   133

Chapter 5
General Concepts of Chemical Equilibrium   175

Chapter 6
Acid–Base Equilibria   207

Chapter 17
Mass Spectrometry   656

Chapter 18
Thermal Methods of Analysis: Principles and Applications   692

Chapter 19
Electrochemical Cells and Electrode Potentials 711

Chapter 20
Potentiometric Electrodes and Potentiometry   731

Chapter 21
Redox and Potentiometric Titrations 771

Chapter 22
Voltammetry and Electrochemical Sensors 802

Chapter 23
Kinetic Methods of Analysis 815

Chapter 24
Automation in Measurements 832

Chapter 25
Environmental Sampling and Analysis* C1

Appendix A
Literature of Analytical Chemistry 843

Appendix B Review of Mathematical Operations: Exponents, Logarithms, and the Quadratic Formula 846

---

*Available on textbook website: www.wiley.com/college/christian.

*Available on textbook website: www.wiley.com/college/christian.

"Teachers open the door, but it is up to you to enter"
—Anonymous

This edition has two new coauthors, Purnendu (Sandy) Dasgupta and Kevin Schug, both from the University of Texas at Arlington. So the authorship now spans three generations of analytical chemists who have each brought their considerable expertise in both teaching and research interests to this book. While all chapters have ultimately been revised and updated by all authors, the three authors have spearheaded different tasks. Among the most notable changes are the following: The addition of a dedicated chapter on mass spectrometry (Chapter 17) by Kevin. Sandy provided complete rewrites of the chapters on spectrochemical methods (Chapter 11) and atomic spectrometric methods (Chapter 12), and gas and liquid chromatography (Chapters 15 and 16). Gary compiled and organized all old and new supplementary materials for the textbook companion website, and he prepared the PowerPoint presentations of figures and tables.

WHO SHOULD USE THIS TEXT?

This text is designed for college students majoring in chemistry and in fields related to chemistry. It is written for an undergraduate **quantitative analysis course**. It necessarily contains more material than normally can be covered in a one-semester or one-quarter course, so that your instructor can select those topics deemed most important. Some of the remaining sections may serve as supplemental material. Depending on how a quantitative analysis and **instrumental analysis** sequence is designed, it may serve for **both courses**. In any event, we hope you will take time to read some sections that look interesting to you that are not formally covered. They can certainly serve as a reference for the future.

WHAT IS ANALYTICAL CHEMISTRY?

Analytical chemistry is concerned with the chemical characterization of matter, both qualitative and quantitative. It is important in nearly every aspect of our lives because chemicals make up everything we use.

This text deals with the principles and techniques of quantitative analysis, that is, how to determine how much of a specific substance is contained in a sample. You will learn how to design an analytical method, based on what information is needed or requested (*it is important to know what that is, and why!*), how to obtain a laboratory sample that is representative of the whole, how to prepare it for analysis, what measurement tools are available, and the statistical significance of the analysis.

Analytical chemistry becomes meaningful when you realize that a blood analysis may provide information that saves a patient's life, or that quality control analysis assures that a manufacturer does not lose money from a defective product.

WHAT'S NEW TO THIS EDITION?

This edition is extensively rewritten, offering new and updated material. The goal was to provide the student with a foundation of the analytical process, tools, and computational methods and resources, and to illustrate with problems that bring realism to the practice and importance of analytical chemistry. We take advantage of digital technologies to provide supplementary material, including videos, website materials, spreadsheet calculations, and so forth (more on these below). We introduce the chapters with examples of representative uses of a technique, what its unique capabilities may be, and indicate what techniques may be preferred or limited in scope. The beginning of each chapter lists key learning objectives for the chapter, with page numbers for specific objectives. This will help students focus on the core concepts as they read the chapter.

Here are some of the new things:

- **Professor's Favorite Examples and Problems.** We asked professors and practicing analytical chemists from around the world to suggest new analytical examples and problems, especially as they relate to real world practice, that we could include in this new edition. It is with appreciation and pleasure that we thank the many that have generously provided interesting and valuable examples and problems. We call these Professor's Favorite Examples, and Professor's Favorite Problems, and they are annotated within the text by a margin element. We have included these in the text where appropriate and as space allows, and have placed some on the text website. We hope you find these interesting and, as appropriate, are challenged by them.

  Our special thanks go to the following colleagues who have contributed problems, analytical examples, updates, and experiments:

- Christine Blaine, Carthage College
- Andres Campiglia, University of Central Florida
- David Chen, University of British Columbia
- Christa L. Colyer, Wake Forest University
- Michael DeGrandpre, University of Montana
- Mary Kate Donais, Saint Anselm College
- Tarek Farhat, University of Memphis
- Carlos Garcia, The University of Texas at San Antonio
- Steven Goates, Brighham Young University
- Amanda Grannas, Villanova University
- Peter Griffiths, University of Idaho
- Christopher Harrison, San Diego State University
- James Harynuk, University of Alberta
- Fred Hawkridge, Virginia Commonwealth University
- Yi He, John Jay College of Criminal Justice, The City University of New York
- Charles Henry, Colorado State University
- Gary Hieftje, Indiana University
- Thomas Isenhour, Old Dominion University
- Peter Kissinger, Purdue University
- Samuel P. Kounaves, Tufts University
- Ulrich Krull, University of Toronto
- Thomas Leach, University of Washington
- Dong Soo Lee, Yonsei University, Seoul, Korea
- Milton L. Lee, Brigham Young University
- Wen-Yee Lee, University of Texas at El Paso

- Shaorong Liu, University of Oklahoma
- Fred McLafferty, Cornell University
- Michael D. Morris, University of Michigan
- Noel Motta, University of Puerto Rico, Río Piedras
- Christopher Palmer, University of Montana
- Dimitris Pappas, Texas Tech University
- Aleeta Powe, University of Louisville
- Alberto Rojas-Hernández, Universidad Autónoma Metropolitana-Iztapalapa, Mexico
- Alexander Scheeline, University of Illinois
- W. Rudolf Seitz, University of New Hampshire
- Paul S. Simone, Jr., University of Memphis
- Nicholas Snow, Seton Hall University
- Wes Steiner, Eastern Washington University
- Apryll M. Stalcup, City University of Dublin, Ireland
- Robert Synovec, University of Washington
- Galina Talanova, Howard University
- Yijun Tang, University of Wisconsin, Oshkosh
- Jon Thompson, Texas Tech University
- Kris Varazo, Francis Marion University
- Akos Vertes, George Washington University
- Bin Wang, Marshall University
- George Wilson, University of Kansas
- Richard Zare, Stanford University

- **Mass spectrometry**, especially when used as a hyphenated technique with chromatography, is increasingly a routine and powerful analytical tool, and a new chapter (Chapter 17) is dedicated to this topic. Likewise, **liquid chromatography**, including **ion chromatography** for anion determinations, is one of the most widely used techniques today, even surpassing gas chromatography. There are a wide variety of options of systems, instruments, columns, and detectors available, making selection of a suitable system or instrument a challenge for different applications. The present liquid chromatography chapter (Chapter 16) uniquely provides comprehensive coverage within the scope of an undergraduate text that not only gives the fundamentals of various techniques, how they evolved, and their operation, but also what the capabilities of different systems are and guidance for selecting a suitable system for a specific application.

- A new chapter (Chapter 18) dedicated to **thermal methods of analysis** has been introduced in the book.

- **Revised chapters**. All chapters have been revised, several extensively, especially those dealing with instrumentation to include recent technological innovations, as done for the liquid chromatography chapter. These include the **Precipitation reactions and titrations (10)**, **spectrochemical chapter (11)**, the **atomic spectrometric chapter (12)**, **extraction methods (13)**, the **gas chromatography chapter (15)** and **liquid chromatography and electrophoresis (21)**. **State-of-the-art** technologies are covered. Some of this material and that of other chapters may be appropriate to use in an Instrumental Analysis course, as well as providing the basics for the quantitative analysis course; your instructor may assign selected sections for your course.

- **Historical information** is added throughout to put into perspective how the tools we have were developed and evolved.

Thanks are due to the following students at the University of Texas at Arlington for their contributions: Barry Akhigbe, Jyoti Birjah, Rubi Gurung, Aisha Hegab, Akinde Kadjo, Karli Kirk, Heena Patel, Devika Shakya, and Mahesh Thakurathi.

OTHER MODIFICATIONS TO EXISTING CONTENT

It has been almost ten years since the last edition was published and since that time, much has changed! This regional adaptation edition-Asia of *Analytical Chemistry* is extensively revised and updated, with new materials, new problems and examples, and new references.

- **Spreadsheets.** Detailed instructions are given on how to use and take advantage of spreadsheets in analytical calculations, plotting, and data processing. But the introductory material has been moved to the end of Chapter 3 as a separate unit, so that it can be assigned independently if desired, or treated as auxiliary material. The use of Excel Goal Seek and Excel Solver is introduced for solving complex problems and constructing titration curves (see below). Mastery of these powerful tools will allow students to tackle complex problems. Several useful programs introduced in the chapters are placed on the text website and instructions are given for applying these for plotting titration curves, derivative titrations, etc. by simply inputting equilibrium constant data, concentrations, and volumes.

- **References.** There are numerous recommended references given in each chapter, and we hope you will find them interesting reading. The late Tomas Hirschfeld said you should read the very old literature and the very new to know the field. We have deleted a number of outdated references, updating them with new ones. Many references are for classical, pioneering reports, forming the basis of current methodologies, and these remain.

- **Material moved to the text website.** As detailed elsewhere, we have moved certain parts to the text website as supplemental material and to make room for updating material on the techniques to be used. This includes:

  - The **single pan balance** (Chapter 2) and **normality calculations** (Chapter 4), which may still be used, but in a limited capacity.

  - The **experiments**.

  - Auxiliary spreadsheet calculations from different chapters are posted on the website.

  - A chapter dealing with specific applications of analytical chemistry in environmental studies, **Environmental Sampling and Analysis** (Chapter 25) is moved on the text website.

SPREADSHEETS

Spreadsheets (using **Excel**) are introduced and used throughout the text for performing computations, statistical analysis, and graphing. Many titration curves are derived using spreadsheets, as are the calculations of α-values and plots of α-pH curves, and of logarithm concentration diagrams. The spreadsheet presentations are given in a "user-friendly" fashion to make it easier for you to follow how they are set up.

We provide a **list of the different types of spreadsheets** that are used throughout the text, by topic, after the Table of Contents.

GOAL SEEK

We have introduced the use of Goal Seek, a powerful Excel tool, for solving complex problems. Goal Seek performs "trial and error" or successive approximation calculations to arrive at an answer. It is useful when one parameter needs to be varied in a calculation, as is the case for most equilibrium calculations. An introduction to Goal Seek is given in Section 5.11 in Chapter 5. Example applications are given on the text website, and we list these after the Table of Contents.

SOLVER

Excel Solver is an even more versatile tool. Goal Seek can only solve one parameter in a single equation, and does not allow for incorporating constraints in the parameter we want to solve. Solver, on the other hand, can solve for more than one parameter (or more than one equation) at a time. Example applications are given on the text website, with descriptions in the text. See the list after the Table of Contents. An introduction to its use is given in Example 6.21.

REGRESSION FUNCTION IN EXCEL DATA ANALYSIS

Possibly the most powerful tool to calculate all regression related parameters for a calibration plot is the "Regression" function in Data Analysis. It not only provides the results for r, r^2, intercept, and slope (which it lists as X variable 1), it also provides their standard errors and upper and lower limits at the 95% confidence level. It also provides an option for fitting the straight line through the origin (when you know for certain that the response at zero concentration is zero by checking a box "constant is zero"). A description of how to use it is given in Chapter 11 at the end of Section 11.7, and example applications are given in Chapter 15, Section 15.5, and Chapter 23 for Example 23.1 and Example 23.2.

READY TO USE PROGRAMS

As listed above, there are numerous supplemental materials on the text website, including Excel spreadsheets for different calculations. Many of these are for specific examples and are tutorial in nature. But several are suited to apply to different applications, simply by inputting data and not having to set up the calculation program. Examples include calculating titration curves and their derivatives, or for solving either quadratic or simultaneous equations. We list here a number that you should find useful. You can find them under the particular chapter on the website.

Chapter 2
- Glassware calibration, Table 2.4

Chapter 5
- Calculate activity coefficients, equations 5.19 and 5.20 (Auxiliary data)
- Quadratic equation solution (Example 5.1) (See also Goal Seek for solving quadratic equations)

Chapter 6
- Stig Johannson pH calculator. For calculating pH of complex mixtures. Easy to use.
- CurtiPotpH calculator (Ivano Gutz) for calculating pH of complex mixtures, as well as constructing pH related curves. Learning curve higher, but very powerful.
- log C-pH Master Spreadsheet. See Section 6.16 on how to use it.

Chapter 7
- Derivative titrations—Easy method (Section 7.11)
- Universal Acid Titrator—Alex Scheeline—Easy method (Section 7.11). For polyprotic acid titration curves.
- Master Spreadsheet for titrations of weak bases—Easy method

Chapter 9
- Solving simultaneous equations (Example 9.5)

Chapter 21
- Derivative titration plots (for near the endpoint)

Chapter 11
- Calculation of unknown from calibration curve plot
- Standard deviation of sample concentration
- Two component Beer's Law solution

Chapter 12
- Standard additions plot and unknown calculation

Chapter 15
- Internal standard calibration plot and unknown calculation (Section 15.5)

EXPERIMENTS

There are 42 experiments, grouped by topic, illustrating most of the measurement techniques presented in the text, and they can be downloaded from the text website. Some of the important experiments are also part of the textbook. Each contains a description of the principles and chemical reactions involved, so the student gains an overview of what is being determined and how. Solutions and reagents to prepare in advance of the experiment are listed, so experiments can be performed efficiently. All experiments, particularly the volumetric ones, have been designed to minimize waste by preparing the minimum volumes of reagents, like titrants, required to complete the experiment.

New experiments were contributed by users and colleagues. Included are three experiments from Professor Christopher Palmer, University of Montana using a **spectrophotometric microplate reader** (Experiments 3, 24, and 29) and two experiments from Dr. Sandeep Kaur and Dr. Dhanraj Masram, University of Delhi (Experiments 12 and 22).

Experiment Video Resource. Professor Christopher Harrison from San Diego State University has a YouTube "Channel" of videos of different types of experiments, some illustrating laboratory and titration techniques: http://www.youtube.com/user/crharrison.

We would recommend that students be encouraged to look at the ones dealing with buret rinsing, pipetting, and aliquoting a sample, before they begin experiments. Also, they will find useful the examples of acid-base titrations illustrating methyl red or phenolphthalein indicator change at end points. There are a few specific experiments that may be related to ones from the textbook, for example, EDTA titration of calcium or Fajan's titration of chloride. The video of glucose analysis gives a good illustration of the starch end point, which is used in iodometric titrations.

TEXT COMPANION WEBSITE

John Wiley & Sons, Inc. maintains a companion website for your *Analytical Chemistry* textbook that contains additional valuable supplemental material.

The website may be accessed at: www.wiley.com/college/christian

Materials on the website include supplemental materials for different chapters that expand on abbreviated presentations in the text.

Following is a list of the types of materials on the website:

- Supplemental Material: WORD, PDFs, Excel, PowerPoint, JPEG

POWERPOINT SLIDES

All figures and tables in the text are posted on the text website as PowerPoint slides for each chapter, with notes on each for the instructor, and can be downloaded for preparation of PowerPoint presentations.

SOLUTIONS MANUAL

Answers for spreadsheet problems, which include the spreadsheets, are given on the text website. Answers to selected problems are given in Appendix F.

A WORD OF THANKS

The production of your text involved the assistance and expertise of numerous people. Special thanks go to the users of the text who have contributed comments and suggestions for changes and improvements; these are always welcome. A number of colleagues served as reviewers of the text and manuscript and have aided immeasurably in providing specific suggestions for revision. They, naturally, express opposing views sometimes on a subject or placement of a chapter or section, but collectively have assured a near optimum outcome that we hope you find easy and enjoyable to read and study.

First, Professors Louise Sowers, Stockton College; Gloria McGee, Xavier University; and Craig Taylor, Oakland University; and Lecturer Michelle Brooks, University of Maryland and Senior Lecturer Jill Robinson, Indiana University offered advice for revision and improvements of the 6th edition. Second, Professors Neil Barnett, Deakin University, Australia; Carlos Garcia, The University of Texas at San Antonio; Amanda Grannas, Villanova University; Gary Long, Virginia Tech; Alexander Scheeline, University of Illinois; and Mathew Wise, Condordia University, proofed the draft chapter manuscripts of this edition and offered further suggestions for enhancing the text. Dr. Ronald Majors, a leading chromatography expert from Agilent Technologies, offered advice on the liquid chromatography chapter.

The professionals at John Wiley & Sons have been responsible for producing a high quality book. Petra Recter, Vice President, Publisher, Chemistry and Physics, Global Education, shepherded the whole process from beginning to end. Her Editorial Assistants Lauren Stauber, Ashley Gayle, and Katherine Bull were key in taking care of many details, with efficiency and accuracy.

We each owe special thanks to our families for their patience during our long hours of attention to this undertaking. Gary's wife, Sue, his companion for over 50 years, has been through seven editions, and remains his strong supporter, even now. Purnendu owes his wife, Kajori, and his students much for essentially taking off from all but the absolute essentials for the last three years. He also thanks Akinde Kadjo in particular for doing many of the drawings. Kevin's wife, Dani, put up with yet another "interesting project" and lent her support in the form of keeping the kids at bay and making sure her husband was well fed while working on the text.

GARY D. CHRISTIAN
Seattle, Washington
PURNENDU K. (SANDY) DASGUPTA
KEVIN A. SCHUG
Arlington, Texas
September, 2013

"To teach is to learn twice." —**Joseph Joubert**

List of Spreadsheets Used Throughout the Text

The use of spreadsheets for plotting curves and performing calculations is introduced in different chapters. Listed in the Preface are several that are ready to use for different applications. Following is a list of the various other applications of Microsoft Excel, by category, for easy reference for different uses. All spreadsheets are given in the text website. The Problem spreadsheets are only in the website; others are in the text but also in the website.

You should always practice preparing assigned spreadsheets before referring to the website. You can save the spreadsheets in your website to your desktop for use.

Use of Spreadsheets (Section 3.20)

Filling the Cell Contents, 113

Saving the Spreadsheet, 114

Printing the Spreadsheet, 114

Relative vs. Absolute Cell References, 115

Use of Excel Statistical Functions (Paste functions), 116

Useful Syntaxes: LOG10; PRODUCT; POWER; SQRT; AVERAGE; MEDIAN; STDEV; VAR, 117

Statistics Calculations

Standard Deviation: Chapter 3, Problems 19, 20, 21, 17, 29

Confidence Limit: Chapter 3, Problems 27, 29, 30, 34

Pooled Standard Deviation: Chapter 3, Problem 39

F-Test: Chapter 3, Problems 36, 38, 40

t-Test: Chapter 3, Problems 42, 43

t-Test, multiple samples: Chapter 3, Problem 59

Propagation of Error: Chapter 3, Problems 23 (add/subtract), 24 (multiply/divide)

Using Spreadsheets for Plotting Calibration Curves

Trendline; Least squares equation; R2 (Section 3.21, Figure 3.10)

Slope, Intercept and Coefficient of Determination (without a plot) (Section 3.22; Chapter 3, Problems 53, 57, 58)

LINEST for Additional Statistics (Section 3.23, Figure 3.11)

Ten functions: slope, std. devn., R2, F, sum sq. regr., intercept, std. devn., std. error of estimate, d.f., sum sq. resid.

Plotting α vs. pH Curves (Figure 6.2, H_3PO_4), 236

use of spreadsheets for plotting curves an

Chapter 6, Problem 61 (HOAc)

Plotting log C vs. pH Curves Using Alpha Values (Section 6.16)

Chapter 6, Problem 64 (Malic acid, H_2A)

Chapter 6, Problem 68 (H_3PO_4, H_3A)

Plotting Titration Curves

HCl vs. NaOH (Figure 7.1), 269

HCl vs. NaOH, Charge Balance (Section 7.2), 271

HOAc vs. NaOH (Section 7.5), 278

Hg^{2+} vs. EDTA: Chapter 8, Problem 33

SCN^- and Cl^- vs. $AgNO_3$: Chapter 10, Problem 19

Fe^{2+} vs. Ce^{4+} (Figure 21.1): Example 21.3

Derivative Titrations
(Section 7.11), 289; Chapter 21, 792

Plotting log K' vs. pH
(Figure 8.2): Chapter 8, Problem 32

Plotting β-values vs. [ligand]($Ni(NH_3)_6^{2+}$ beta-values vs. $[NH_3]$):
Chapter 8, Problem 34

Spreadsheet Calculations/Plots
Glassware Calibration (Table 2.4), 37
Weight in Vacuum Error vs. Sample Density (Chapter 2)
Gravimetric Calculations

> Spreadsheet Examples-Grav. calcn. %Fe, 350

> Chapter 9, (Example 9.2, $\%P_2O_5$)

Solubility $BaSO_4$ vs. $[Ba^{2+}]$ Plot (Figure 9.3):
Solubility vs. Ionic Strength Plot (Figure 9.4):
Van Deemter Plot: Chapter 16, Problem 16

EXCEL SOLVER FOR PROBLEM SOLVING

This program can be used to solve several parameters or equations at a time. An introduction is given in Example 6.21.

Chapter 5 (solving quadratic equation, Example 5.1)
Example 6.21 Solver pH calculations of multiple solutions (H_3PO_4, NaH_2PO_4, Na_2HPO_4, Na_3PO_4); 242
Example 6.24 Solver calculation (buffer composition), 248
Solubility from K_{sp}: Chapter 9 (Example 9.9)

GOAL SEEK FOR PROBLEM SOLVING

The spreadsheets listed below are on the text website for the particular chapter. The page numbers refer to corresponding discussions on setting up the programs.

Excel Goal Seek for Trial and Error Problem Solving (Section 5.11):

> Practice Goal Seek—setup, answer

Goal Seek to Solve an Equation (Example 5.1—quadratic equation), 183

Solving a quadratic equation by Goal Seek—setup Goal Seek answer quadratic equation (Example 5.4); 186
Goal Seek answer Example 5.4

> Solving Example 5.13 Using Goal Seek (charge balance); 194

Example 6.7 Goal Seek solution (pH HOAc)
Example 6.8 Goal Seek solution (pH NH_3)
Example 6.9 Goal Seek solution (pH NaOAc)
Example 6.10 Goal Seek solution (pH NH_4Cl)
Example 6.19 Charge balance and Goal Seek to calc H_3PO_4 pH (See the example for details of setting up the spreadsheet)
Example 6.19b Goal Seek solution (pH H_3PO_4 + NaOAc + K_2HPO_4) (See Example 6.19 discussion for spreadsheet setup)
77PFP Goal Seek calculations—there are three tabs (Chapter 6, Problem 72). See solution on the website for a detailed description of the problem solution and appropriate equations.
Example 8.6—Goal Seek (complexation equilibria); (Section 8.6), 326 (See the example for the equation setup)
Example 10.1 Goal Seek (solubility of CaC_2O_4 in 0.001 M HCl)
Example 10.2 Goal Seek (charge balance, solubility of MA in 0.1 M HCl)
Example 10.5 Goal Seek (solubility of MX in presence of complexing ligand L)

REGRESSION FUNCTION IN EXCEL DATA ANALYSIS

This Excel tool calculates all regression related parameters for a calibration plot. It provides the results for r, r^2, intercept, and slope, and also provides their standard errors and upper and lower limits at the 95% confidence level.

Chapter 11, end of Section 11.7, Excel Exercise. Describes the use of the Excel Regression function in Data Analysis to readily calculate a calibration curve and its uncertainty, and then apply this to calculate an unknown concentration and its uncertainty from its absorbance; 403
> Section 15.5, GC internal standard determination, 554
> Chapter 15, Problem 15, GC internal standard determination
> Example 23.1, Lineweaver-Burk K_m determination
> Example 23.2, Calculating unknown concentration from reaction rate
> Problem 23.21, Lineweaver-Burk K_m determination

Gary Christian grew up Oregon, and has had a lifelong interest in teaching and research, inspired by great teachers throughout his education. He received his B.S. degree from the University of Oregon and Ph. D. degree from the University of Maryland. He began his career at Walter Reed Army Institute of Research, where he developed an interest in clinical and bioanalytical chemistry. He joined the University of Kentucky in 1967, and in 1972 moved to the University of Washington, where he is Emeritus Professor, and Divisional Dean of Sciences Emeritus.

Gary wrote the first edition of this text in 1971. He is pleased that Professors Dasgupta and Schug have joined him in this new edition. They bring expertise and experience that markedly enhance and update the book in many ways.

Gary is the recipient of numerous national and international awards in recognition of his teaching and research activities, including the American Chemical Society (ACS) Division of Analytical Chemistry Award for Excellence in Teaching and the ACS Fisher Award in Analytical Chemistry, and received an Honorary Doctorate Degree from Chiang Mai University. The University of Maryland inducted him into their distinguished alumni Circle of Discovery.

He has authored five other books, including *Instrumental Analysis*, and over 300 research papers, and has been Editor-in-Chief of the international analytical chemistry journal, *Talanta*, since 1989.

Purnendu K. (Sandy) Dasgupta is a native of India and was educated in a college founded by Irish missionaries and graduated with honors in Chemistry in 1968. After a MSc in Inorganic Chemistry in 1970 from the University of Burdwan and a brief stint as a researcher at the Indian Association for the Cultivation of Science (where Raman made his celebrated discovery), he came as a graduate student to Louisiana State University at Baton Rouge in 1973. Sandy received his PhD in Analytical Chemistry with a minor in Electrical Engineering from LSU in 1977 and managed to get a diploma as a TV mechanic while a graduate student. He joined the California Primate Research Center at the University of California at Davis as an Aerosol research Chemist in 1979 to be part of a research team studying inhalation toxicology of air pollutants. In his mother tongue, Bengali, he was once a well-published poet and a fledgling novelist but seemingly finally found his love of analytical chemistry as salvation. He joined Texas Tech in 1981 and was designated a Horn Professor in 1992, named after the first president of the University, the youngest person to be so honored at the time. He remained at Texas Tech for 25 years, joining the University of Texas at Arlington in 2007 as the Department Chair. He has stepped down as Chair, and currently holds the Jenkins Garrett Professorship.

Sandy has written more than 400 papers/book chapters, and holds 23 US patents, many of which have been commercialized. His work has been recognized by the Dow Chemical Traylor Creativity Award, the Ion Chromatography Symposium Outstanding Achievement Award (twice), the Benedetti-Pichler Memorial Award in Microchemistry, American Chemical Society Award in Chromatography, Dal Nogare Award in the Separation Sciences, Honor Proclamation of the State of Texas Senate and so on. He is the one of the Editors of *Analytica Chimica Acta*, a major international journal in analytical chemistry. He is best known for his work in atmospheric measurements, ion chromatography, the environmental occurrence of perchlorate and its effect on iodine nutrition, and complete instrumentation systems. He is a big champion of the role of spreadsheet programs in teaching analytical chemistry.

Kevin Schug was born and raised in Blacksburg, Virginia. The son of a physical chemistry Professor at Virginia Tech, he grew up running around the halls of a chemistry building and looking over his father's shoulder at chemistry texts. He pursued and received his B.S. degree in Chemistry from the College of William & Mary in 1998, and his Ph.D. degree in Chemistry under the direction of Professor Harold McNair at Virginia Tech in 2002. Following two years as a post-doctoral fellow with Professor Wolfgang Lindner at the University of Vienna (Austria), he joined the faculty in the Department of Chemistry & Biochemistry at The University of Texas at Arlington in 2005, where he is currently the Shimadzu Distinguished Professor of Analytical Chemistry.

The research in Kevin's group spans fundamental and applied aspects of sample preparation, separation science, and mass spectrometry. He also manages a second group, which focuses their efforts on chemical education research. He has been the recipient of several awards, including the Eli Lilly ACACC Young Investigator in Analytical Chemistry award, the LCGC Emerging Leader in Separation Science award, and the American Chemical Society Division of Analytical Chemistry Award for Young Investigators in Separation Science.

At present, he has authored or coauthored 65 scientific peer-reviewed manuscripts. Kevin is a member of the Editorial Advisory Boards for *Analytica Chimica Acta* and *LCGC* Magazine, and is a regular contributor to *LCGC* on-line articles. He is also Associate Editor of the *Journal of Separation Science*.

Chapter 1

ANALYTICAL OBJECTIVES, OR: WHAT ANALYTICAL CHEMISTS DO

"Unless our knowledge is measured and expressed in numbers, it does not amount to much."
—Lord Kelvin

KEY THINGS TO LEARN FROM THIS CHAPTER

- Analytical science deals with the chemical characterization of matter—what, how much?
- The analyst must know what information is really needed, and obtain a representative sample
- Few measurements are specific, so operations are performed to achieve high selectivity
- You must select the appropriate method for measurement
- Validation is important
- There are many useful websites dealing with analytical chemistry

Analytical chemistry is concerned with the chemical characterization of matter and the answer to two important questions: what is it (qualitative analysis) and how much is it (quantitative analysis). Chemicals make up everything we use or consume, and knowledge of the chemical composition of many substances is important in our daily lives. Analytical chemistry plays an important role in nearly all aspects of chemistry, for example, agricultural, clinical, environmental, forensic, manufacturing, metallurgical, and pharmaceutical chemistry. The nitrogen content of a fertilizer determines its value. Foods must be analyzed for contaminants (e.g., pesticide residues) and for essential nutrients (e.g., vitamin content). The air we breathe must be analyzed for toxic gases (e.g., carbon monoxide). Blood glucose must be monitored in diabetics (and, in fact, most diseases are diagnosed by chemical analysis). The presence of trace elements from gun powder on a perpetrator's hand will prove a gun was fired by that hand. The quality of manufactured products often depends on proper chemical proportions, and measurement of the constituents is a necessary part of **quality assurance**. The carbon content of steel will influence its quality. The purity of drugs will influence their efficacy.

> Everything is made of chemicals. Analytical chemists determine what and how much.

In this text, we will describe the tools and techniques for performing these different types of analyses. There is much useful supplemental material on the text website, including Excel programs that you can use, and videos to illustrate their use. **You should first read the Preface to learn what is available to you, and then take advantage of some of the tools.**

1.1 What Is Analytical Science?

The above description of analytical chemistry provides an overview of the discipline of analytical chemistry. There have been various attempts to more specifically define the discipline. The late Charles N. Reilley said: "Analytical chemistry is what analytical chemists do" (Reference 2). The discipline has expanded beyond the bounds of just chemistry, and many have advocated using the name *analytical science* to describe the field. This term is used in a National Science Foundation report from workshops on "Curricular Developments in the Analytical Sciences." Even this term falls short of recognition of the role of instrumentation development and application. One suggestion is that we use the term *analytical science and technology* (Reference 3).

The Federation of European Chemical Societies held a contest in 1992 to define analytical chemistry, and the following suggestion by K. Cammann was selected [*Fresenius' J. Anal. Chem.*, **343** (1992) 812–813].

> **Analytical Chemistry provides the methods and tools needed for insight into our material world . . . for answering four basic questions about a material sample:**
> - **What?**
> - **Where?**
> - **How much?**
> - **What arrangement, structure or form?**

These cover qualitative, spatial, quantitative, and speciation aspects of analytical science. The Division of Analytical Chemistry of the American Chemical Society developed a definition of analytical chemistry, reproduced in part here:

> **Analytical Chemistry seeks ever improved means of measuring the chemical composition of natural and artificial materials. The techniques of this science are used to identify the substances which may be present in a material and to determine the exact amounts of the identified substance.**
>
> **Analytical chemists serve the needs of many fields:**
> - **In *medicine*, analytical chemistry is the basis for clinical laboratory tests which help physicians diagnose disease and chart progress in recovery.**
> - **In *industry*, analytical chemistry provides the means of testing raw materials and for assuring the quality of finished products whose chemical composition is critical. Many household products, fuels, paints, pharmaceuticals, etc. are analyzed by the procedures developed by analytical chemists before being sold to the consumer.**
> - ***Environmental quality* is often evaluated by testing for suspected contaminants using the techniques of analytical chemistry.**
> - **The nutritional value of *food* is determined by chemical analysis for major components such as protein and carbohydrates and trace components such as vitamins and minerals. Indeed, even the calories in food are often calculated from its chemical analysis.**
>
> **Analytical chemists also make important contributions to fields as diverse as forensics, archaeology, and space science.**

An interesting article published by a leading analytical chemist, G. E. F. Lundell, from the National Bureau of Standards in 1935 entitled "The Analysis of Things As They Are", describes why we do analyses and the analytical process (*Industrial and Engineering Chemistry, Analytical Edition*, **5**(4) (1933) 221–225). The article is posted on the text website.

A brief overview of the importance of analytical chemistry in society, with examples that affect our lives, and the tools and capabilities, is given in the article, "What Analytical Chemists Do: A Personal Perspective," by Gary Christian, *Chiang Mai Journal of Science*, **32**(2) (2005) 81–92: http://it.science.cmu.ac.th/ejournal/journalDetail.php?journal_id=202

Reading this before beginning this course will help place in context what you are learning. A reprint of the article is posted on the text website.

1.2 Qualitative and Quantitative Analysis: What Does Each Tell Us?

The discipline of analytical chemistry consists of **qualitative analysis** and **quantitative analysis**. The former deals with the identification of elements, ions, or compounds present in a sample (we may be interested in whether only a given substance is present), while the latter deals with the determination of how much of one or more constituents is present. The sample may be solid, liquid, gas, or a mixture. The presence of gunpowder residue on a hand generally requires only qualitative knowledge, not of how much is there, but the price of coal will be determined by the percent of undesired sulfur impurity present.

Qualitative analysis tells us what chemicals are present.

Quantitative analysis tells us how much.

How Did Analytical Chemistry Originate?

That is a very good question. Actually, some tools and basic chemical measurements date back to the earliest recorded history. Fire assays for gold are referred to in Zechariah 13:9, and the King of Babylon complained to the Egyptian Pharoah, Ammenophis the Fourth (1375–1350 BC), that gold he had received from the pharaoh was "less than its weight" after putting it in a furnace. The perceived value of gold, in fact, was probably a major incentive for acquiring analytical knowledge. Archimedes (287–212 BC) did nondestructive testing of the golden wreath of King Hieron II. He placed lumps of gold and silver equal in weight to the wreath in a jar full of water and measured the amount of water displaced by all three. The wreath displaced an amount between the gold and silver, proving it was not pure gold!

The balance is of such early origin that it was ascribed to the gods in the earliest documents found. The Babylonians created standard weights in 2600 BC and considered them so important that their use was supervised by the priests.

The alchemists accumulated the chemical knowledge that formed the basis for quantitative analysis as we know it today. Robert Boyle coined the term *analyst* in his 1661 book, *The Sceptical Chymist*. Antoine Lavoisier has been considered the "father of analytical chemistry" because of the careful quantitative experiments he performed on conservation of mass (using the analytical balance). (Lavoisier was actually a tax collector and dabbled in science on the side. He was guillotined on May 8, 1793, during the French Revolution because of his activities as a tax collector.)

Gravimetry was developed in the seventeenth century, and titrimetry in the eighteenth and nineteenth centuries. The origin of titrimetry goes back to Geoffroy in 1729; he evaluated the quality of vinegar by noting the quantity of solid K_2CO_3 that could be added before effervescence ceased (Reference 4). Gay-Lussac, in 1829, assayed silver by titration with 0.05% relative accuracy and precision!

A 2000-year-old balance. Han Dynasty 10 AD. Taiwan National Museum, Taipei. From collection of G. D. Christian.

Textbooks of analytical chemistry began appearing in the 1800s. Karl Fresenius published *Anleitung zur Quantitaven Chemischen Analyse* in Germany in 1845. Wilhelm Ostwald published an influential text on the scientific fundamentals of analytical chemistry in 1894 entitled *Die wissenschaflichen Grundagen der analytischen Chemie*, and this book introduced theoretical explanations of analytical phenomena using equilibrium constants (thank him for Chapter 5 and applications in other chapters).

The twentieth century saw the evolution of instrumental techniques. Steven Popoff's second edition of *Quantitative Analysis* in 1927 included electroanalysis, conductimetric titrations, and colorimetric methods. Today, of course, analytical technology has progressed to include sophisticated and powerful computer-controlled instrumentation and the ability to perform highly complex analyses and measurements at extremely low concentrations.

This text will teach you the fundamentals and give you the tools to tackle most analytical problems. Happy journey. For more on the evolution of the field, see Reference 8.

Qualitative tests may be performed by selective chemical reactions or with the use of instrumentation. The formation of a white precipitate when adding a solution of silver nitrate in dilute nitric acid to a dissolved sample indicates the presence of a halide. Certain chemical reactions will produce colors to indicate the presence of classes of organic compounds, for example, ketones. Infrared spectra will give "fingerprints" of organic compounds or their functional groups.

A clear distinction should be made between the terms **selective** and **specific**:

Few analyses are specific. Selectivity may be achieved through proper preparation and measurement.

- A *selective* reaction or test is one that can occur with other substances but exhibits a degree of preference for the substance of interest.
- A *specific* reaction or test is one that occurs *only* with the substance of interest.

Unfortunately, very few reactions are truly specific but many exhibit selectivity. Selectivity may be also achieved by a number of strategies. Some examples are:

- Sample preparation (e.g., extractions, precipitation)
- Instrumentation (selective detectors)
- Target analyte derivatization (e.g., derivatize specific functional groups)
- Chromatography, which separates the sample constituents

For quantitative analysis, the typical sample composition will often be known (we know that blood contains glucose), or else the analyst will need to perform a qualitative test prior to performing the more difficult quantitative analysis. Modern chemical measurement systems often exhibit sufficient selectivity that a quantitative measurement can also serve as a qualitative measurement. However, simple qualitative tests are usually more rapid and less expensive than quantitative procedures. Qualitative analysis has historically been composed of two fields: inorganic and organic. The former is usually covered in introductory chemistry courses, whereas the latter is best left until after the student has had a course in organic chemistry.

In comparing qualitative versus quantitative analysis, consider, for example, the sequence of analytical procedures followed in testing for banned substances at the Olympic Games. The list of prohibited substances includes about 500 different active constituents: stimulants, steroids, beta-blockers, diuretics, narcotics, analgesics, local anesthetics, and sedatives. Some are detectable only as their metabolites. Many athletes must be tested rapidly, and it is not practical to perform a detailed quantitative analysis on each. There are three phases in the analysis: the fast-screening phase, the identification phase, and possible quantification. In the fast-screening phase, urine samples are rapidly tested for the presence of classes of compounds that will differentiate them from "normal" samples. Techniques used include immunoassays, gas chromatography–mass spectrometry, and liquid chromatography–mass spectrometry. About 5% of the samples may indicate the presence of unknown compounds that may or may not be prohibited but need to be identified. Samples showing a suspicious profile during the screening undergo a new preparation cycle (possible hydrolysis, extraction, derivatization), depending on the nature of the compounds that have been detected. The compounds are then identified using the highly selective combination of gas chromatography/mass spectrometry (GC/MS). In this technique, complex mixtures are separated by gas chromatography, and they are then detected by mass spectrometry, which provides molecular structural data on the compounds. The MS data, combined with the time of elution from the gas chromatograph, provide a high probability of the presence of a given detected compound. GC/MS is expensive and time consuming, and so it is used only when necessary. Following the identification phase, some compounds must be precisely quantified since they may normally be present at low levels, for example, from food, pharmaceutical preparations, or endogenous steroids, and elevated levels must be confirmed. This is done using quantitative techniques such as spectrophotometry or gas chromatography.

This text deals principally with quantitative analysis. In the consideration of applications of different techniques, examples are drawn from the life sciences, clinical chemistry, environmental chemistry, occupational health and safety applications, and industrial analysis.

We describe briefly in this chapter the analytical process. More details are provided in subsequent chapters.

1.3   Getting Started: The Analytical Process

See the text **website** for useful chapters from *The Encyclopedia of Analytical Chemistry* (Reference 9 at the end of the chapter) on literature searching and selection of analytical methods.

The general analytical process is shown in Figure 1.1. The analytical chemist should be involved in every step. The analyst is really a problem solver, a critical part of the team deciding what, why, and how. The unit operations of analytical chemistry that are common to most types of analyses are considered in more detail below.

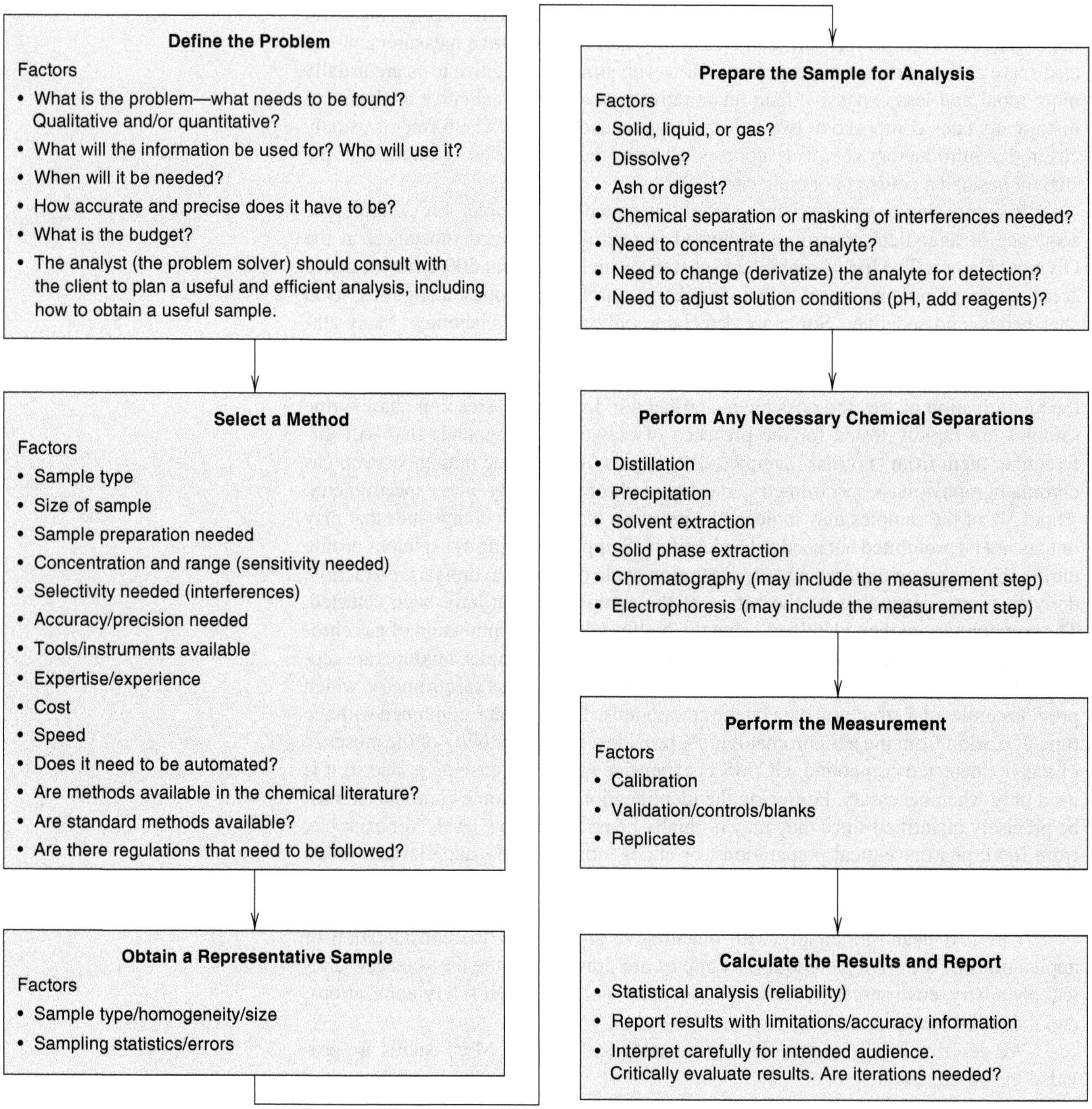

Define the Problem

Factors

- What is the problem—what needs to be found? Qualitative and/or quantitative?
- What will the information be used for? Who will use it?
- When will it be needed?
- How accurate and precise does it have to be?
- What is the budget?
- The analyst (the problem solver) should consult with the client to plan a useful and efficient analysis, including how to obtain a useful sample.

Select a Method

Factors

- Sample type
- Size of sample
- Sample preparation needed
- Concentration and range (sensitivity needed)
- Selectivity needed (interferences)
- Accuracy/precision needed
- Tools/instruments available
- Expertise/experience
- Cost
- Speed
- Does it need to be automated?
- Are methods available in the chemical literature?
- Are standard methods available?
- Are there regulations that need to be followed?

Obtain a Representative Sample

Factors

- Sample type/homogeneity/size
- Sampling statistics/errors

Prepare the Sample for Analysis

Factors

- Solid, liquid, or gas?
- Dissolve?
- Ash or digest?
- Chemical separation or masking of interferences needed?
- Need to concentrate the analyte?
- Need to change (derivatize) the analyte for detection?
- Need to adjust solution conditions (pH, add reagents)?

Perform Any Necessary Chemical Separations

- Distillation
- Precipitation
- Solvent extraction
- Solid phase extraction
- Chromatography (may include the measurement step)
- Electrophoresis (may include the measurement step)

Perform the Measurement

Factors

- Calibration
- Validation/controls/blanks
- Replicates

Calculate the Results and Report

- Statistical analysis (reliability)
- Report results with limitations/accuracy information
- Interpret carefully for intended audience. Critically evaluate results. Are iterations needed?

FIGURE 1.1   Steps in an analysis.

Defining The Problem—What Do We Really Need to Know? (Not Necessarily Everything)

Before the analyst can design an analysis procedure, he or she must know what information is needed, by whom, for what purpose, and what type of sample is to be analyzed. As the analyst, you must have good communication with the client. This stage of an analysis is perhaps the most critical. The client may be the Environmental Protection Agency (EPA), an industrial chemist, an engineer, or your grandmother—each of which will have different criteria or needs, and each having their own understanding of what a chemical analysis involves or means. It is important to communicate in language that is understandable by both sides. If someone puts a bottle on your desk and asks, "What is in here?" or "Is this safe?", you may have to explain that there are 10 million known compounds and substances. A client who says, "I want to know what elements are in here" needs to understand that at perhaps $20 per analysis for 85 elements it will cost $1700 to test for them all, when perhaps only a few elements are of interest.

Laypersons might come to analytical chemists with cosmetics they wish to "reverse engineer" so they can market them and make a fortune. When they realize it may cost a small fortune to determine the ingredients, requiring a number of sophisticated analyses, they always rethink their goals. On the other hand, a mother may come to you with a white pill that her teenage son insists is vitamin C and she fears is an illicit drug. While it is not trivial to determine what it is, it is rather straightforward to determine if it undergoes the same reactions that ascorbic acid (vitamin C) does. You may be able to greatly alleviate the concerns of an anxious mother.

The concept of "safe" or "zero/nothing" is one that many find hard to define or understand. Telling someone their water is safe is not for the analyst to say. All you can do is present the analytical data (and give an indication of its range of accuracy). The client must decide whether it is safe to drink, perhaps relying on other experts. Also, never report an answer as "zero," but as less than the detection limit, which is based on the measurement device/instrument. We are limited by our methodology and equipment, and that is all that can be reported. Some modern instruments, though, can measure extremely small amounts or concentrations, for example, parts per trillion. This presents a dilemma for policy makers (often political in nature). A law may be passed that there should be zero concentration of a chemical effluent in water. In practice, the acceptable level is defined by how low a concentration can be detected; and the very low detectability may be far below the natural occurrence of the chemical or below the levels to which it can be reasonably reduced. We analysts and chemists need to be effective communicators of what our measurements represent.

Once the problem is defined this will dictate how the sample is to be obtained, how much is needed, how sensitive the method must be, how accurate and precise[1] it must be, and what separations may be required to eliminate interferences. The determination of trace constituents will generally not have to be as precise as for major constituents, but greater care will be required to eliminate trace contamination during the analysis.

Once the required measurement is known, the analytical method to be used will depend on a number of factors, including the analyst's skills and training in different techniques and instruments; the facilities, equipment, and instrumentation available; the sensitivity and precision required; the cost and the budget available; and the time for analysis and how soon results are needed. There are often one or more standard procedures available in reference books for the determination of an **analyte** (constituent to be determined) in a given **sample type**. This does not mean that the method will

"To many . . . , the object of chemical analysis is to obtain the composition of a sample . . . . It may seem a small point . . . that the analysis of the sample is *not* the true aim of analytical chemistry . . . the real purpose of the analysis is to solve a problem . . ." H. A. Laitinen, Editorial: The Aim of Analysis, *Anal. Chem.*, **38** (1966) 1441.

The way an analysis is performed depends on the information needed.

The way you perform an analysis will depend on your experience, the equipment available, the cost, and the time involved.

The *analyte* is the substance *analyzed* for. Its concentration is *determined*.

---

[1]Accuracy is the degree of agreement between a measured value and a true value. Precision is the degree of agreement between replicate measurements of the same quantity and does not necessarily imply accuracy. These terms are discussed in more detail in Chapter 3.

Chemical Abstracts is a good source of literature.

necessarily be applicable to other sample types. For example, a standard EPA method for groundwater samples may yield erroneous results when applied to the analysis of sewage water. The chemical literature (journals) contains many specific descriptions of analyses. *Chemical Abstracts* (http://info.cas.org), published by the American Chemical Society, is a good place to begin a literature search. It contains abstracts of all articles appearing in the major chemical journals of the world. Yearly and cumulative indices are available, and many libraries have computer search facilities. If your library subscribes to Scifinder from Chemical Abstracts Service, this is the best place to start your search (www.cas.org/products/scifindr/index.html). The Web of Science, a part of the Web of Knowledge (www.isiwebofknowledge.com) is an excellent place to search the literature and provides also the information as to where a particular article has been cited and by whom. Another excellent source, available to anyone, is Google Scholar, which allows you to search articles, authors, etc. (http://scholar.google.com). The major analytical chemistry journals may be consulted separately. Some of these are: *Analytica Chimica Acta, Analytical Chemistry, Analytical and Bioanalytical Chemistry, Analytical Letters, Analyst, Applied Spectroscopy, Clinica Chimica Acta, Clinical Chemistry, Journal of the Association of Official Analytical Chemists, Journal of Chromatography, Journal of Separation Science, Spectrochimica Acta*, and *Talanta*. While the specific analysis of interest may not be described, the analyst can often use literature information on a given analyte to devise an appropriate analytical scheme. Finally, the analyst may have to rely upon experience and knowledge to develop an analytical method for a given sample. The literature references in Appendix A describe various procedures for the analysis of different substances.

Examples of the manner in which the analysis of particular types of samples are made are given in application Chapter 25 on the text's website. These chapters describe commonly performed clinical, biochemical, and environmental analyses. The various techniques described in this text are utilized for the specific analyses. Hence, it will be useful for you to read through these applications chapters both now and after completing the majority of this course to gain an appreciation of what goes into analyzing real samples and why the analyses are made.

Once the problem has been defined, the following steps can be started.

Obtaining a Representative Sample—We Can't Analyze the Whole Thing

A chemical analysis is usually performed on only a small portion of the material to be characterized. If the amount of material is very small and it is not needed for future use, then the entire sample may be used for analysis. The gunshot residue on a hand may be an example. More often, though, the characterized material is of value and must be altered as little as possible in sample collection. For example, sampling of a Rembrandt painting for authenticity would need to be done with utmost care for sample quantity, so as not to deface the artwork.

The *gross sample* consists of several portions of the material to be tested. The *laboratory sample* is a small portion of this, taken after homogenization. The *analysis sample* is that actually analyzed. See Chapter 2 for methods of sampling.

The material to be sampled may be solid, liquid, or gas. It may be homogeneous or heterogeneous in composition. In the former case, a simple "grab sample" taken at random will suffice for the analysis. In the latter, we may be interested in the variation throughout the sample, in which case several individual samples will be required. If the gross composition is needed, then special sampling techniques will be required to obtain a representative sample. For example, in analyzing for the average protein content of a shipment of grain, a small sample may be taken from each bag, or tenth bag for a large shipment, and combined to obtain a **gross sample**. Sampling is best done when the material is being moved, if it is large, in order to gain access. The larger the particle size, the larger should be the gross sample. The gross sample must be reduced in size to obtain a **laboratory sample** of several grams, from which a few grams to milligrams will be taken to be analyzed (**analysis sample**). The size reduction may require taking portions (e.g., two quarters) and mixing, in several steps, as well as crushing and sieving to obtain a uniform powder for analysis. Methods of sampling solids, liquids, and gases are discussed in Chapter 2. If one is interested in spatial structure, then homogenization must not be carried out, but spatially resolved sampling must be done.

In the case of biological fluids, the conditions under which the sample is collected can be important, for example, whether a patient has just eaten. The composition of blood varies considerably before and after meals, and for many analyses a sample is collected after the patient has fasted for a number of hours. Persons who have their blood checked for cholesterol levels are asked to fast for up to twelve hours prior to sampling. Preservatives such as sodium fluoride for glucose preservation and anticoagulants for blood samples may be added when samples are collected; these may affect a particular analysis.

Blood samples may be analyzed as whole blood, or they may be separated to yield plasma or serum according to the requirements of the particular analysis. Most commonly, the concentration of the substance external to the red cells (the extracellular concentration) will be a significant indication of physiological condition, and so serum or plasma is taken for analysis.

If whole blood is collected and allowed to stand for several minutes, the soluble protein **fibrinogen** will be converted by a complex series of chemical reactions (involving calcium ion) into the insoluble protein **fibrin**, which forms the basis of a gel, or **clot**. The red and white cells of the blood become caught in the meshes of the fibrin network and contribute to the clot, although they are not necessary for the clotting process. After the clot forms, it shrinks and squeezes out a straw-colored fluid, **serum**, which does not clot but remains fluid indefinitely. The clotting process can be prevented by adding a small amount of an **anticoagulant**, such as heparin or a citrate salt (i.e., a calcium complexor). Blood collection vials are often color-coded to provide a clear indication of the additives they contain. An aliquot of the unclotted whole blood can be taken for analysis, or the red cells can be centrifuged to the bottom, and the light pinkish-colored **plasma** remaining can be analyzed. Plasma and serum are essentially identical in chemical composition, the chief difference being that fibrinogen has been removed from the latter.

Serum is the fluid separated from clotted blood. *Plasma* is the fluid separated from unclotted blood. It is the same as serum, but contains fibrinogen, the clotting protein.

Details of sampling other materials are available in reference books on specific areas of analysis. See the references at the end of the chapter for some citations.

Certain precautions should be taken in **handling and storing samples** to prevent or minimize contamination, loss, decomposition, or matrix change. In general, one must prevent contamination or alteration of the sample by (1) the container, (2) the atmosphere, (3) heat/temperature, or (4) light. Also, a chain of custody should be established and will certainly be required for any analysis that may be involved in legal proceedings. In the O.J. Simpson case, there were television news clips of people handling samples, purportedly without proper custody, placing them in the hot trunk of a car, for example. While this may not have affected the actual analyses and correctness of samples analyzed, it provided arguments for the defense to discredit analyses.

Care must be taken not to alter or contaminate the sample.

The sample may have to be protected from the atmosphere or from light. It may be an alkaline substance, for example, which will react with carbon dioxide in the air. Blood samples to be analyzed for CO_2 must be protected from the atmosphere.

The stability of the sample must be considered. To minimize degradation of glucose, for example, a preservative such as sodium fluoride is added to blood samples. The preservative must not, of course, interfere in the analysis. Proteins and enzymes tend to denature on standing and should be analyzed without delay. Trace constituents may be lost during storage by adsorption onto the container walls.

Urine samples are unstable, and calcium phosphate precipitates out, entrapping metal ions or other substances of interest. Precipitation can be prevented by keeping the urine acidic (pH 4.5), usually by adding 1 or 2 mL glacial acetic acid per 100-mL sample and stored under refrigeration. Urine, as well as whole blood, serum, plasma, and tissue samples, can also be frozen for prolonged storage. Deproteinized blood samples are more stable than untreated ones.

Corrosive gas samples will often react with the container. Sulfur dioxide, for example, is troublesome. In automobile exhaust, SO_2 is also lost by dissolving in condensed water vapor from the exhaust. In such cases, it is best to analyze the gas by an *in situ* analyzer that operates at a temperature in which condensation does not occur.

Preparing the Sample for Analysis—It Probably Needs to be Altered

The first thing you must do is measure the size of sample to be analyzed.

The first step in analyzing a sample is to measure the amount being analyzed (e.g., volume or weight of sample). This will be needed to calculate the percent composition from the amount of analyte found. The analytical sample size must be measured to the degree of precision and accuracy required for the analysis. An analytical balance sensitive to 0.1 mg is usually used for weight measurements and balances that can weigh down to 0.01 mg are becoming increasingly common. Solid samples are often analyzed on a dry basis and must be dried in an oven at 110 to 120°C for 1 to 2 h and cooled in a dessicator before weighing, if the sample is stable at the drying temperatures. Some samples may require higher temperatures and longer heating time (e.g., overnight) because of their great affinity for moisture. The amount of sample taken will depend on the concentration of the analyte and how much is needed for isolation and measurement. Determination of a major constituent may require only 100 mg of sample, while a trace constituent may require several grams. Usually **replicate samples** are taken for analysis, in order to obtain statistical data on the precision of the analysis and thus provide more reliable results.

Solid samples usually must be put into solution.

Analyses may be nondestructive in nature, for example, in the measurement of lead in paint by X-ray fluorescence in which the sample is bombarded with an X-ray beam and the characteristic reemitted X-radiation is measured. More often, the sample must be in solution form for measurement, and solids must be dissolved. Inorganic materials may be dissolved in various acids, redox, or complexing media. Acid-resistant material may require fusion with an acidic or basic flux in the molten state to render it soluble in dilute acid or water. Fusion with sodium carbonate, for example, forms acid-soluble carbonates.

Ashing is the burning of organic matter. Digestion is the wet oxidation of organic matter.

Organic materials that are to be analyzed for inorganic constituents, for example, trace metals, may be destroyed by **dry ashing**. The sample is slowly combusted in a furnace at 400 to 700°C, leaving behind an inorganic residue that is soluble in dilute acid. Alternately, the organic matter may be destroyed by **wet digestion** by heating with oxidizing acids. A mixture of nitric and sulfuric acids is common. Perchloric acid digestion is used for complete oxidative digestion; this is a last resort as special extraction or fume hoods are required due to potential explosion hazards. Biological fluids may sometimes be analyzed directly. Often, however, proteins interfere and must be removed. Dry ashing and wet digestion accomplish such removal. Or proteins may be precipitated with various reagents and filtered or centrifuged away, to give a **protein-free filtrate** (PFF).

If the analyte is organic in nature, these oxidizing methods cannot be used. Rather, the analyte may be extracted away from the sample or dialyzed, or the sample dissolved in an appropriate solvent. It may be possible to measure the analyte nondestructively. An example is the direct determination of protein in feeds by near-infrared spectrometry.

The pH of the sample solution will usually have to be adjusted.

Once a sample is in solution, the solution conditions must be adjusted for the next stage of the analysis (separation or measurement step). For example, the pH may have to be adjusted, or a reagent added to react with and "mask" interference from other constituents. The analyte may have to be reacted with a reagent to convert it to a form suitable for measurement or separation. For example, a colored product may be formed that will be measured by spectrometry. Or the analyte will be converted to a form that can be volatilized for measurement by gas chromatography. The gravimetric analysis of iron as Fe_2O_3 requires that all the iron be present as iron(III), its usual form. A volumetric determination by reaction with dichromate ion, on the other hand, requires that all the iron be converted to iron(II) before reaction, and the reduction step will have to be included in the sample preparation.

The solvents and reagents used for dissolution and preparation of the solution should be of high purity (reagent grade). Even so, they may contain trace impurities of the analyte. Hence, it is important to prepare and analyze replicate **blanks**, particularly for trace analyses. A blank theoretically consists of all chemicals in the unknown and used in an analysis in the same amounts (including water), run through the entire analytical procedure. The blank result is subtracted from the analytical sample result to arrive at a net analyte concentration in the sample solution. If the blank is appreciable, it may invalidate the analysis. Oftentimes, it is impossible to make a perfect blank for an analysis.

Always run a blank!

Performing Necessary Chemical Separations

In order to eliminate interferences, to provide suitable selectivity in the measurement, or to preconcentrate the analyte for more sensitive or accurate measurement, the analyst must often perform one or more separation steps. It is preferable to separate the analyte away from the sample matrix, in order to minimize losses of the analyte. Separation steps may include precipitation, extraction into an immiscible solvent, chromatography, dialysis, and distillation.

Performing the Measurement—You Decide the Method

The method employed for the actual quantitative measurement of the analyte will depend on a number of factors, not the least important being the amount of analyte present and the accuracy and precision required. Many available techniques possess varying degrees of selectivity, sensitivity, accuracy and precision, cost, and rapidity. Analytical chemistry research often deals with the optimization of one or more of these parameters, as they relate to a particular analysis or analysis technique. **Gravimetric analysis** usually involves the selective separation of the analyte by precipitation, followed by the very nonselective measurement of mass (of the precipitate). In **volumetric**, or **titrimetric**, **analysis**, the analyte reacts with a measured volume of reagent of known concentration, in a process called **titration**. A change in some physical or chemical property signals the completion of the reaction. Gravimetric and volumetric analyses can provide results accurate and precise to a few parts per thousand (tenth of 1 percent) or better. However, they require relatively large (millimole or milligram) quantities of analyte and are only suited for the measurement of major constituents, although microtitrations may be performed. Volumetric analysis is more rapid than gravimetric analysis and is therefore preferred when applicable.

Instrumental techniques are used for many analyses and constitute the discipline of **instrumental analysis**. They are based on the measurement of a physical property of the sample, for example, an electrical property or the absorption of electromagnetic radiation. Examples are spectrophotometry (ultraviolet, visible, or infrared), fluorimetry, atomic spectroscopy (absorption, emission), mass spectrometry, nuclear magnetic resonance spectrometry (NMR), X-ray spectroscopy (absorption, fluorescence), electroanalytical chemistry (potentiometric, voltammetric, electrolytic), chromatography

Instruments are more selective and sensitive than volumetric and gravimetric methods. But they may be less precise.

(gas, liquid), and radiochemistry. Instrumental techniques are generally more sensitive and selective than the classical techniques but are less precise, on the order of 1 to 5% or so. These techniques are usually much more expensive, especially in terms of initial capital investment. But depending on the numbers of analyses, they may be less expensive when one factors in personnel costs. They are usually more rapid, may be automated, and may be capable of measuring more than one analyte at a time. Chromatography techniques are particularly powerful for analyzing complex mixtures. They integrate the separation and measurement steps. Constituents are separated as they are pushed through (eluted from) a column of appropriate material that interacts with the analytes to varying degrees, and these are sensed with an appropriate detector as they emerge from the column, to give a transient peak signal, proportional to the amount of each.

Table 1.1 compares various analytical methods to be described in this text with respect to sensitivity, precision, selectivity, speed, and cost. The numbers given may be exceeded in specific applications, and the methods may be applied to other uses, but these are representative of typical applications. The lower concentrations determined by titrimetry require the use of an instrumental technique for measuring the completion of the titration. The selection of a technique, when more than one is applicable, will depend, of course, on the availability of equipment, and personal experience, and preference of the analyst. As examples, you might use spectrophotometry to determine the concentration of nitrate in river water at the sub parts-per-million level, by first reducing to nitrite and then using a diazotization reaction to produce a color. Fluoride in toothpaste may be determined potentiometrically using a fluoride ion-selective electrode. A complex mixture of hydrocarbons in gasoline can be separated using gas chromatography and determined by flame ionization detection. Glucose in blood can be determined kinetically by the rate of the enzymatic reaction between glucose and oxygen, catalyzed by the enzyme glucose oxidase, with measurement of the rate of oxygen depletion or the rate of hydrogen peroxide production. The purity of a silver bar can be determined gravimetrically by dissolving a small sample in nitric acid and precipitating AgCl with chloride and weighing the purified precipitate.

TABLE 1.1 Comparison of Different Analytical Methods

Method	Approx. Range (mol/L)	Approx. Precision (%)	Selectivity	Speed	Cost	Principal Uses
Gravimetry	10^{-1}–10^{-2}	0.1	Poor–moderate	Slow	Low	Inorg.
Titrimetry	10^{-1}–10^{-4}	0.1–1	Poor–moderate	Moderate	Low	Inorg., org.
Potentiometry	10^{-1}–10^{-6}	2	Good	Fast	Low	Inorg.
Electrogravimetry, coulometry	10^{-1}–10^{-4}	0.01–2	Moderate	Slow–moderate	Moderate	Inorg., org.
Voltammetry	10^{-3}–10^{-10}	2–5	Good	Moderate	Moderate	Inorg., org.
Spectrophotometry	10^{-3}–10^{-6}	2	Good–moderate	Fast–moderate	Low–moderate	Inorg., org.
Fluorometry	10^{-6}–10^{-9}	2–5	Moderate	Moderate	Moderate	Org.
Atomic spectroscopy	10^{-3}–10^{-9}	2–10	Good	Fast	Moderate–high	Inorg., multielement
Chromatography— Mass Spectrometry	10^{-4}–10^{-9}	2–5	Good	Fast–moderate	Moderate–high	Org., multicomponent
Kinetics methods	10^{-2}–10^{-10}	2–10	Good–moderate	Fast–moderate	Moderate	Inorg., org., enzymes

The various methods of determining an analyte can be classified as either **absolute** or **relative**. Absolute methods rely upon accurately known fundamental constants for calculating the amount of analyte, for example, atomic weights. In gravimetric analysis, for example, an insoluble derivative of the analyte of known chemical composition is prepared and weighed, as in the formation of AgCl for chloride determination. The precipitate contains a known fraction of the analyte, in this case, fraction of Cl = at wt Cl/f wt AgCl = 35.453/143.32 = 0.24737.[2] Hence, it is a simple matter to obtain the amount of Cl contained in the weighed precipitate. Gravimetry, titrimetry and coulometry are examples of absolute methods. Most other methods, however, are relative in that they require comparison against some solution of known concentration (also called calibration or standardization, see below).

> Most methods require calibration with a standard.

Instrument Standardization

Most instrumental methods of analysis are relative. Instruments register a signal due to some physical property of the solution. Spectrophotometers, for example, measure the fraction of electromagnetic radiation from a light source that is absorbed by the sample. This fraction must be related to the analyte concentration by comparison against the fraction absorbed by a known concentration of the analyte. In other words, the instrumentation must be **standardized**.

Instrument response may be linearly or nonlinearly related to the analyte concentration. Calibration is accomplished by preparing a series of standard solutions of the analyte at known concentrations and measuring the instrument response to each of these (usually after treating them in the same manner as the samples) to prepare an **analytical calibration curve** of response versus concentration. Figure 1.2 shows examples of calibration curves obtained in a mass spectrometry experiment. The concentration of an unknown can then be determined from the response, using the calibration curve. With modern computer-controlled instruments, this is done electronically or digitally, and direct readout of concentration is obtained.

> A calibration curve is the instrument response as a function of concentration.

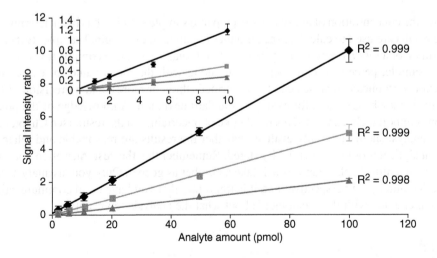

FIGURE 1.2 Calibration curves for the measurement of proteins using matrix-assisted laser desorption ionization (MALDI)—mass spectrometry and an ionic liquid matrix (Courtesy of Prof. Michael Gross, Washington University in St. Louis. Reprinted with permission).

[2]at wt = atomic weight; f wt = formula weight.

Method of Standard Additions

Standard additions calibration is used to overcome sample matrix effects.

The sample matrix may affect the instrument response to the analyte. In such cases, calibration may be accomplished by the **method of standard additions**. A portion of the sample is spiked with a known amount of standard, and the increase in signal is due to the standard. In this manner, the standard is subjected to the same environment as the analyte. These calibration techniques are discussed in more detail when describing the use of specific instruments.

See Section 12.5 and the website supplement for that section for a detailed description of the standard additions method and calculations using it. Section 15.5 illustrates its use in gas chromatography, and Example 21.8 illustrates how it is used in potentiometry. Experiments 30 (atomic spectrometry) and 32 (solid-phase extraction) on the text website employ the method of standard additions.

Internal Standard Calibration

An instrumental response is often subject to variations from one measurement to the next due to changing instrument conditions, resulting in imprecision. For example, in gas chromatography, the volume of injected sample or standard from a Hamilton microliter syringe (see Chapter 2) may vary. In atomic absorption spectrometry, fluctuations in gas flows and aspiration rates for sample introduction may occur. In order to compensate for these types of fluctuations, internal standard calibration may be used. Here, a fixed concentration of a different analyte, that is usually chemically similar to the sample analyte, is added to all solutions to be measured. Signals for both substances are recorded, and the ratio of the sample to internal standard signals is plotted versus sample analyte concentration. So, if say the volume of injected sample is 10% lower than assumed, each signal is reduced 10%, and the ratio at a given sample analyte concentration remains constant.

See Sections 12.5 (atomic spectrometry) and 15.5 (gas chromatography) for illustrations of internal standard use.

Calculating the Results and Reporting the Data

The analyst must provide expert advice on the significance of a result.

Once the concentration of analyte in the prepared sample solution has been determined, the results are used to calculate the amount of analyte in the original sample. Either an *absolute* or a *relative* amount may be reported. Usually, a relative composition is given, for example, percent or parts per million, along with the mean value for expressing accuracy. Replicate analyses can be performed (three or more), and a precision of the analysis may be reported, for example, standard deviation. A knowledge of the precision is important because it gives the degree of uncertainty in the result (see Chapter 3). The analyst should critically evaluate whether the results are reasonable and relate to the analytical problem as originally stated. Remember that the customer often does not have a scientific background so will take a number as gospel. Only you, as analyst, can put that number in perspective, and it is important that you have good communication and interaction with the "customer" about what the analysis represents.

1.4 Validation of a Method—You Have to Prove It Works!

Great care must be taken that accurate results are obtained in an analysis. Two types of error may occur: *random* and *systematic*. Every measurement has some imprecision associated with it, which results in random distribution of results, for example, a Gaussian distribution. The experiment can be designed to narrow the range of this, but it cannot be eliminated. A systematic error is one that biases a result consistently in one direction. Such errors may occur when the sample matrix suppresses the instrument signal, a weight of an analytical balance may be in error, skewed either high or low, or a sample may not be sufficiently dried.

Proper calibration of an instrument is only the first step in assuring accuracy. In developing a method, samples should be spiked with known amounts of the analyte (above and beyond what is already in the sample). The amounts determined (recovered) by the analysis procedure (after subtraction of the amount apparently present in the sample as determined by the same procedure) should be close to what was added. This is not a foolproof approach, however, and only assures that the intended analyte is measured. It cannot assure that some interferent present in the sample is not measured. A new method is better validated by comparison of sample results with those obtained with another accepted method. There are various sources of certified standards or reference materials that may be analyzed to assure accuracy by the method in use. For example, environmental quality control standards for pesticides in water or priority pollutants in soil are commercially available. The National Institute of Standards and Technology (NIST) prepares standard reference materials (SRMs) of different matrix compositions (e.g., steel, ground leaves) that have been certified for the content of specific analytes, by careful measurement by at least two independent techniques. Values are assigned with statistical ranges. Different agencies and commercial concerns can provide samples for round-robin or blind tests in which control samples are submitted to participating laboratories for analysis at random; the laboratories are not informed of the control values prior to analysis.

> The best way to validate a method is to analyze a standard reference material of known composition.

Standards should be run intermittently with samples. A *control sample* should also be run at least daily and the results plotted as a function of time to prepare a *quality control chart*, which is compared with the known standard deviation of the method. The measured quantity is assumed to be constant with time, with a Gaussian distribution, and there is a 1 in 20 chance that values will fall outside two standard deviations from the known value, and a 1 in 100 chance it will be 2.5 standard deviations away. Numbers exceeding these suggest uncompensated errors, such as instrument malfunction, reagent deterioration, or improper calibration.

Government regulations require careful established protocol and validation of methods and analyses when used for official or legal purposes. Guidelines of *good laboratory practice* (GLP) have been established to assure validation of analyses. They, of course, ideally apply to all analyses.

> Good laboratory practice (validation) is required to assure accuracy of analyses.

1.5 Analyze Versus Determine—They Are Different

The terms *analyze* and *determine* have two different meanings. We say a sample is **analyzed** for part or all of its constituents. The substances measured are called the **analytes**. The process of measuring the analyte is called a **determination**. Hence, in analyzing blood for its chloride content, we determine the chloride concentration.

> You *analyze* a sample to *determine* the amount of analyte.

The constituents in the sample may be classified as **major** (> 1% of the sample), **minor** (0.1 to 1%), or **trace** (< 0.1%). A few parts per billion or less of a constituent might be classified as **ultratrace**.

An analysis may be **complete** or **partial**; that is, either all constituents or only selected constituents may be determined. Most often, the analyst is requested to report on a specified chemical or elements or perhaps a class of chemicals or specific elements.

1.6 Some Useful Websites

In addition to the various literature and book sources we have mentioned, and those listed in Appendix A (Literature of Analytical Chemistry), there are a number of websites that are useful for supplementary resources for analytical chemists. These, of course, often change and new ones become available. But the following are good starting points for obtaining much useful information.

Textbook Companion Site

1. www.wiley.com/college/christian Select the textbook, 7th edition, and then Instructor Companion Site. This will require an assigned username and password. This site is designed to contain a variety of helpful materials to supplement this textbook, including additional problems, presentations, worksheets, and experiments.

The textbook companion site offers important supplemental materials for instructors and students.

Chemistry in General

2. www.acs.org The American Chemical Society home page. Information on journals, meetings, chemistry in the news, search databases (including *Chemical Abstracts*), and much more.

3. www.chemweb.com This is a virtual club for chemists. The site contains databases and lists relating to chemistry and incorporates discussion groups that focus on specific areas such as analytical chemistry. You must join, but it is free.

4. www.rsc.org This is the Royal Society of Chemistry website in Britain, the equivalent of the American Chemical Society.

5. http://micro.magnet.fsu.edu/primer/java/scienceopticsu/powersof10/index.html Check out this Powers of Ten visual scene, from protons to viewing the Milky Way 10 million light years from Earth.

Analytical Chemistry

6. www.analyticalsciences.org/index.php This is the home page of the Division of Analytical Chemistry of the American Chemical Society. There are a number of links and resources throughout this site that will take you all over the Internet to sites involving analytical chemistry.

7. http://wissen.science-and-fun.de/links/index.php?e-id=11&basis=(CH)Analytical Chemistry This is an analytical chemistry link center that connects to university and other sites all over the world.

8. www.nist.gov/mml/ The Materials Measurement Laboratory site of the National Institute of Standards and Technology (NIST). The national reference laboratory for measurements in the chemical, biological and materials sciences. Source of Standard Reference Materials.

9. http://www.rsc.org/Gateway/Subject/Analytical/ The Royal Society of Chemistry in Britain; the British equivalent of the ACS provides information on analytical sciences.

10. http://www.anachem.umu.se/jumpstation.htm The Analytical Chemistry Springboard. A comprehensive, annotated list of analytical chemistry resources on the Internet from Umeå University.

11. www.asdlib.org Analytical Sciences Digital Library (ASDL). A site of choice for listings of peer-reviewed websites dealing with pedagogy and techniques.

12. www.acdlabs.com Advanced Chemistry Development specializes in the support of analytical methods by providing various prediction and modeling tools. They also offer a demo version of structure drawing software (ChemSketch), as a free download.

13. www.asms.org The official website of the American Society for Mass Spectrometry. This site contains introductory and advanced information about the art of mass spectrometry, a common analytical technique.

The role of thinking logically is paramount in analytical chemistry. Here is a logical brain teaser, a favorite of **Professor Richard N. Zare, Stanford University:**

If a solution of salt (NaCl) is electrolyzed, at the negative electrode, hydrogen is generated and at the positive electrode, chlorine is generated:

$$Na^+ + e^- + H_2O = NaOH + \frac{1}{2} H_2$$

$$Cl^- - e^- = \frac{1}{2} Cl_2$$

An electrolysis cell has been set up in a small closed room and outside there are three switches—only one turns the cell on, the others are dummies. You can turn the switches on and off as many times as you want to but can enter the room only once. All switches are initially off. How will you find which switch turns the cell on? Go to the website for the answer.

1. What is analytical chemistry?
2. Differentiate between qualitative and quantitative analysis.
3. What is a problem? How does it is selected?
4. Differentiate between analyze, determine, sample and analyte.
5. What are the important factors in the preparation of the sample for analysis?
6. What is titration? Also, explain blank titration.
7. What are the methods used for the chemical separation?
8. Define gravimetric analysis.
9. Explain standard addition and when it is employed?
10. Define instrumental analysis.
11. What is calibration curve?
12. Distinguish between specific reaction and a selective reaction.
13. Suggest a method from Table 1.1 to accomplish the following analyses: (a) the purity of NaCl in the table salt, (b) the acetic acid content of vinegar, (c) the pH of swimming pool water.
14. Explain random and systematic error in validation of a method.

1. Which of the following is not a part of qualitative analysis?

    (a) Identification of elements   (b) Identification of ions

    (c) Identification of compounds   (d) Composition determination

2. Infrared spectra will give _____ of organic compounds of their functional groups.

    (a) footprints   (b) fingerprints

    (c) blueprints   (d) detailed information

3. Wet oxidation of organic matter is known as

    (a) digestion.   (b) ashing.

    (c) annealing.   (d) burning.

4. Which of the following is not an example of spectrophotometry?

    (a) Ultraviolet   (b) Infrared

    (c) Visible   (d) Fluorescence

Recommended References

General

1. J. Tyson, *Analysis. What Analytical Chemists Do*. London: Royal Society of Chemistry, 1988. A brief book that very succinctly discusses what analytical chemists do and how they do it.
2. R. W. Murray, "Analytical Chemistry Is What Analytical Chemists Do," Editorial, *Anal. Chem.*, **66** (1994) 682A.
3. D. Schatzlein and V. Thomsen, "The Chemical Analysis Process," *Spectroscopy* **23** (10), October (2008) 30. A nice summary in six pages of the analytical process, including sampling, sample preparation, and analysis of results. Examples include molten metals, water, and ores. (www.spectroscopyonline.com)
4. Ihde, A. J. *The Development of Modern Chemistry*. Harper and Row: New York, 1964. Provides an account of the development of chemistry, including early analytical chemistry.
5. T. Kuwana, Chair, *Curricular Developments in the Analytical Sciences*, A Report from Workshops held in 1996 and 1997, Sponsored by the National Science Foundation.
6. R. A. DePalma and A. H. Ullman, "Professional Analytical Chemistry in Industry. A Short Course to Encourage Students to Attend Graduate School," *J. Chem. Ed.*, **68** (1991) 383. This is a description of an industrial analytical chemistry short course for teachers and students that Proctor & Gamble Company scientists deliver at universities and colleges on invitation, to explain what analytical chemists do in industry. The emphasis is "The analytical chemist as a problem solver."
7. C. A. Lucy, "How to Succeed in Analytical Chemistry: A Bibliography of Resources from the Literature," *Talanta*, **51** (2000) 1125. Surveys the literature for advice on how to purchase equipment, how to write a manuscript, and how to get a job in analytical chemistry.
8. G. D. Christian, "Evolution and Revolution in Quantitative Analysis," *Anal. Chem.*, **67** (1995) 532A. Traces the history of analytical chemistry from the beginning of humans.

Encyclopedias and Handbooks

9. R. A. Meyers, editor-in-chief, *Encyclopedia of Analytical Chemistry*. Chichester: Wiley, 1999–2010. A 15-volume set, with volumes on theory and instrumentation and on applications, presented in 623 articles.
10. P. J. Worsfold, A. Townshend, and C. F. Poole, editors-in-chief, *Encyclopedia of Analytical Science*, 2nd ed. London: Elsevier/Academic Press: 2005. A 10-volume set with comprehensive coverage of the practice of analytical science. Covers all techniques that determine specific elements, compounds, and groups of compounds in any physical or biological matrix.
11. J. A. Dean, *Analytical Chemistry Handbook*. New York: McGraw Hill, 1995. Detailed coverage of conventional wet and instrumental techniques, preliminary operations of analysis, preliminary separation methods, and statistics in chemical analysis.

Sampling

12. F. F. Pitard, *Pierre Gy's Sampling Theory and Sampling Practice*. Vol. I: *Heterogeneity and Sampling*. Vol. II: *Sampling Correctness and Sampling Practice*. Boca Raton, FL: CRC Press, 1989.
13. P. M. Gy and A. G. Boyle, *Sampling for Analytical Purposes*. Chichester: Wiley, 1998. An abridged guide by Pierre Gy to his formula originally developed for the sampling of solid materials but equally valid for sampling liquids and multiphase media.
14. P. M. Gy, *Sampling of Heterogeneous and Dynamic Material Systems*. Amsterdam: Elsevier, 1992.
15. G. E. Baiulescu, P. Dumitrescu, and P. Gh. Zugravescu, *Sampling*. New York: Ellis Horwood, 1991.
16. J. Pawliszyn, *Sampling and Sample Preparation for Field and Laboratory*, Wilson & Wilson's Comprehensive Analytical Chemistry, D. Barceló, Ed., Vol. XXXVII. Amsterdam: Elsevier Science BV, 2002.

17. P. M. Gy, "Introduction to Theory of Sampling I. Heterogeneity of a Population of Uncorrelated Units," *Trends Anal. Chem.*, **14** (2) (1995) 67.

18. P. M. Gy, "Tutorial: Sampling or Gambling?" *Process Control and Quality*, **6** (1994) 97.

19. B. Kratochvil and J. K. Taylor, "Sampling for Chemical Analysis," *Anal. Chem.*, **53** (1981) 924A.

20. B. Kratochvil, "Sampling for Microanalysis: Theory and Strategies," *Fresenius' J. Anal. Chem.*, **337** (1990) 808.

21. J. F. Vicard and D. Fraisse, "Sampling Issues," *Fresenius' J. Anal. Chem.*, **348** (1994) 101.

22. J. M. Hungerford and G. D. Christian, "Statistical Sampling Errors as Intrinsic Limits on Detection in Dilute Solutions," *Anal. Chem.*, **58** (1986) 2567.

23. J. P. Lodge, Jr., Ed. *Methods of Air Sampling and Analysis*, 3rd ed. Boca Raton, FL: CRC Press, 1988.

24. American Public Health Association. *Standard Methods for the Examination of Water and Wastewater.* 22nd ed. http://www.standardmethods.org/

Interferences

25. W. E. Van der Linden, "Definition and Classification of Interference in Analytical Procedures," *Pure & Appl. Chem.*, **61** (1989) 91.

Standard Solutions

26. B. W. Smith and M. L. Parsons, "Preparation and Standard Solutions. Critically Selected Compounds," *J. Chem. Ed.*, **50** (1973) 679. Describes preparation of standard solutions of 72 inorganic metals and nonmetals.

27. M. H. Gabb and W. E. Latchem, *A Handbook of Laboratory Solutions*. New York: Chemical Publishing, 1968.

28. http://pubs.acs.org/reagents/, American Chemical Society Committee on Analytical Reagents.

Calibration

29. H. Mark, *Principles and Practice of Spectroscopic Calibration*. New York: Wiley, 1991.

Standard Reference Materials

30. H. Klich and R. Walker, "COMAR—The International Database for Certified Reference Materials," *Fresenius' J. Anal. Chem.*, **345** (1993) 104.

31. Office of Standard Reference Materials, Room B311, Chemistry Building, National Institute of Standards and Technology, Gaithersburg, Maryland 20899.

32. National Research Council of Canada, Division of Chemistry, Ottawa K1A OR6, Canada.

33. R. Alverez, S. D. Rasberry, and G. A. Uriano, "NBS Standard Reference Materials: Update 1982," *Anal. Chem.*, **54** (1982) 1239A.

34. R. W. Seward and R. Mavrodineanu, "Standard Reference Materials: Summary of the Clinical Laboratory Standards Issued by the National Bureau of Standards," *NBS* (NIST) *Special Publications* 260–271, Washington, DC, 1981.

35. *Standard Coal Samples.* Available from U.S. Department of Energy, Pittsburgh Mining Technology Center, P.O. Box 10940, Pittsburgh, PA 15236. Samples have been characterized for 14 properties, including major and trace elements.

Chapter 2

ANALYTICAL CHEMISTRY: BASIC TOOLS AND OPERATIONS

"Get your facts first, and then you can distort them as much as you please"
—Mark Twain

KEY THINGS TO LEARN FROM THIS CHAPTER

- Use reagent-grade chemicals
- How to use the analytical balance
- Volumetric glassware and how to use it
- How to calibrate glassware
- How to prepare standard acid and base solutions

- Common laboratory apparatus for handling and treating samples
- How to filter and prepare precipitates for gravimetric analysis
- How to sample solids, liquids, and gases
- How to prepare a solution of the analyte

Read this chapter before performing experiments.

Textbook Companion Site **www.wiley.com/college/christian.** Select the textbook, Regional Adaptation, and then Instructor Companion Site. This will require an assigned username and password. This site is designed to contain a variety of helpful materials to supplement this textbook, including additional problems, presentations, worksheets, and experiments.

Analytical chemistry requires measurements in order to get the facts. Throughout the text, specific analytical equipment and instrumentation available to the analyst are discussed as they pertain to specific measuring techniques. Several standard items, however, are common to most analyses and will be required when performing the experiments. These are described in this chapter. They include the analytical balance and volumetric glassware and items such as drying ovens and filters. Detailed explanation of the physical manipulation and use of this equipment is best done by your laboratory instructor, when you can see and practice with the actual equipment, particularly since the type and operation of equipment will vary from one laboratory to another. Some of the general procedures of good laboratory technique will be mentioned as we go along.

See the **website** for this textbook for pictures of commonly used glassware and apparatus in the analytical laboratory.

Laboratory Notebook Documentation

The laboratory notebook is a record of your job as an analytical chemist. It documents everything you do. It is the source for reports, publications, and regulatory submissions. The success or failure of a company's product or service may depend on how well you do that documentation. The notebook becomes a legal document for patent issues, government regulation issues (validation, inspections, legal actions), and the like. Remember, "if it isn't written down, it wasn't done." The notebook is where you record your original ideas that may form the basis of a patent, and so it is important to record what went into those ideas and when.

What are the features of a well-maintained notebook? They will vary with individual preferences. but here are some good rules:

- Use a hardcover notebook (no loose leafs).
- Number pages consecutively.

- Record only in ink.
- Never tear out pages. If not used, put a line through the page.
- Date each page, sign it, and have it signed and dated (soon after you complete your report) by someone else, stating "Read and Understood by."
- Record the name of the project, why it is being done, and any literature references.
- Record all data on the day you obtain it.

Modern instrument software allows the analyst to collect, store, and process data directly from the instrument signal, based on appropriate calibration. It is important that the software and calibration be validated, as for the remainder of the analysis, as a part of good laboratory practice. A variety of electronic notebooks and organizational tools are commercially available, many of which have very good functionality for storing and organizing notebook data in a variety of formats, for example, data files and spreadsheets. In addition, there are software-based laboratory information management systems (LIMS) to manage data, which ultimately aims at the complete elimination of paper notebooks. See http://en.wikipedia.org/wiki/Laboratory_information_management_system.

2.1 Laboratory Materials and Reagents

Table 2.1 lists the properties of materials used in the manufacture of common laboratory apparatus. Borosilicate glass (brand names: Pyrex, Kimax) is the most commonly used material for laboratory apparatus such as beakers, flasks, pipets, and burets. It is stable to hot solutions and to rapid changes in temperature. For more specific applications, there are several other materials employed that may possess advantage with respect to chemical resistance, thermal stability, and so forth.

The different grades of chemicals are listed on the inside *back cover* of the text. In general, only *American Chemical Society (ACS) reagent-grade* or *primary standard* chemicals should be used in the analytical laboratory.

Reagent-grade chemicals are almost always used in analyses. Primary standards are used for preparing volumetric standard solutions.

The American Chemical Society publishes a compendium of tests for evaluating the purity and quality of basic laboratory chemicals. Reagent chemicals that do not reference the ACS meet the manufacturer's own reagent specifications, which vary among suppliers.

The reagent-grade chemicals, besides meeting minimum requirements of purity, may be supplied with a report of analysis of the impurities (printed on the label). Primary standard chemicals are generally at least 99.95% pure. They are analyzed and the results are printed on the label. They are more expensive than reagent-grade chemicals and are used only for the preparation of standard solutions or for the standardization of a solution by reaction (titration) with it. Not all chemicals are available in primary standard grade. There are special grades of solvents for special purposes, for example, spectroscopy or liquid chromatography/mass spectrometry grades. These are specifically purified to remove impurities that might interfere in the particular application. Likewise, there are "semiconductor grade" acids that are specially refined and tested in greater detail for trace elemental impurities, typically in the parts per billion range.

In addition to commercial producers, the National Institute of Standards and Technology supplies primary standard chemicals. *NIST Special Publication 260* catalogs standard reference materials. (See http://ts.nist.gov/ts/htdocs/230/232/232.htm for information on the SRM program and lists of reference standards.) Reference standards are complex materials, such as alloys that have been carefully analyzed for the ingredients and are used to check or calibrate an analytical procedure.

The concentrations of commercially available acids and bases are listed on the inside *front cover*.

TABLE 2.1 Properties of Laboratory Materials

Material	Max. Working Temperature (°C)	Sensitivity to Thermal Shock	Chemical Inertness	Notes
Borosilicate glass	200	150°C change OK	Attacked somewhat by alkali solutions on heating	Trademarks: Pyrex (Corning Glass Works); Kimax (Owens-Illinois)
Soft glass		Poor	Attacked by alkali solutions	Boron-free. Trademark: Corning
Alkali-resistant glass		More sensitive than borosilicate		
Fused quartz	1050	Excellent	Resistant to most acids, halogens	Quartz crucibles used for fusions
High-silica glass	1000	Excellent	More resistant to alkalis than borosilicate	Similar to fused quartz Trademark: Vycor (Corning)
Porcelain	1100 (glazed)	Good	Excellent	
Porcelain	1400 (unglazed)			
Platinum	ca. 1500		Resistant to most acids, molten salts. Attacks by aqua regia, fused nitrates, cyanides, chlorides at > 1000°C. Alloys with gold, silver, and other metals	Usually alloyed with iridium or rhodium to increase hardness. Platinum crucibles for fusions and treatment with HF
Nickel and iron			Fused samples contaminated with the metal	Ni and Fe crucibles used for peroxide fusions
Stainless steel	400–500	Excellent	Not attacked by alkalis and acids except conc. HCl, dil. H_2SO_4, and boiling conc. HNO_3	
Polyethylene	115		Not attacked by alkali solutions or HF. Attacked by many organic solvents (acetone, ethanol OK)	Flexible plastic
Polypropylene	120	Excellent Susceptible to attack by strong oxidizing agents	Translucent. Has replaced polyethylene for many purposes	
Polystyrene	70		Not attacked by HF. Attacked by many organic solvents	Somewhat brittle
Teflon	250		Inert to most chemicals	Useful for storage of solutions and reagents for trace metal analysis. Is permeable to oxygen

2.2 The Analytical Balance—The Indispensible Tool

Weighing is a required part of almost any analysis, both for measuring the sample and for preparing standard solutions. In analytical chemistry we deal with rather small weights, on the order of a few grams to a few milligrams or less. Standard laboratory weighings are typically made to three or four significant figures, and so the weighing device must be both accurate and sensitive. There are various sophisticated ways of achieving this, but the most useful and versatile device used is the **analytical balance**.

Most analytical balances used today are electronic balances. The mechanical single-pan balance is seldom used in the modern analytical laboratory anymore. The calibration of electronic or digital balances is based on comparison of one weight against another. Factors such as zero-point drift and air buoyancy must be considered for all balance types. We really deal with masses rather than weights. The **weight** of an object is the force exerted on it by the gravitational attraction. This force will differ at different locations on Earth. **Mass**, on the other hand, is the quantity of matter of which the object is composed and is invariant.

The balance measures mass.

Modern electronic balances offer convenience in weighing and are subject to fewer errors or mechanical failures than are mechanical balances, which have become largely obsolete. The operation of dialing weights, turning and reading micrometers, and beam and pan arrest of mechanical balances are eliminated, greatly speeding the measurement. A digital-display electronic balance is shown in Figure 2.1, and the operating principle of an electronic balance is illustrated in Figure 2.2. There are no weights or knife edges as with mechanical balances. The pan sits on the arm of a movable hanger (2), and this movable system is compensated by a constant electromagnetic force. The position of the hanger is monitored by an electrical position scanner (1), which brings the weighing system back to the zero position. The compensation current is proportional to the mass placed on the pan. This is sent in digital form to a microprocessor that converts it into the corresponding weight value, which appears as a digital display. The weight of the container can be automatically subtracted.

FIGURE 2.1 Electronic analytical balance. (Courtesy of Denver Instrument Co. Denver Instrument Company owns all images.)

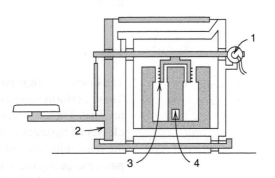

FIGURE 2.2 Operating principle of electronic balance: 1, position scanner; 2, hanger; 3, coil; 4, temperature sensor. (From K. M. Lang, *American Laboratory*, March, 1983, p. 72. Reproduced by permission of American Laboratory, Inc.)

These balances use the principle of electromagnetic force compensation first described by Angstrom in 1895. But they still use the principle of comparing one weight with another. The balance is "zeroed," or calibrated, with a known weight. When the sample is placed on the pan, its weight is electronically compared with the known. This is a form of self-calibration. Modern balances may have such features as compensating for wandering from true zero and averaging variations due to building vibrations.

A single control bar is used to switch the balance on and off, to set the display to zero, and to automatically tare a container on the pan. Since results are available as an electrical signal, they can be readily processed by a personal computer and stored. Weighing statistics can be automatically calculated.

Electronic analytical balances can be purchased with different weighing ranges and readabilities. A standard analytical balance typically has a maximum capacity of 160–300 g and a readability of 0.1 mg. Semimicro balances (readability 0.01 mg, capacity up to 200 g), microbalances (readability 1 μg, capacity up to 30 g) and ultra-micro balances (readability 0.1 μg, capacity up to 2 g) are currently commercially available.

Electrochemical quartz balances are available with 100-μg range that can detect 1 ng (10^{-9} g) changes! The balance utilizes a thin quartz crystal disk oscillating at, for example, 10 MHz. The frequency of oscillation changes with any change in mass, and the frequency change measured by the instrument is converted to mass units. A film of gold is evaporated on the quartz, and the gold substrate can be coated with the material of interest. Mass changes as small as a few percent of a monolayer coverage of atoms or molecules on the gold surface can be measured. Mass changes with time can be recorded. Such balances are incorporated in air particle mass monitors, see for example, http://www.kanomax-usa.com/dust/3521/3521.html.

> Greater precision equals greater cost. A balance readable to 0.1 mg costs $<\sim$ \$2000 whereas microbalances can cost $\sim > \$10,000$.

> Older mechanical balances used the lever principle: $M_1L_1 = M_2L_2$, where L_1 and L_2 are the lengths of the two arms of the lever and M_1 and M_2 are the corresponding masses. If L_1 and L_2 are constructed to be as nearly equal as possible then at balance, $M_1 = M_2$.

Single-Pan Mechanical Balance

Electronic balances have largely replaced mechanical balances. But they are still used some, so we have placed on the Chapter 2 text website a description of the single-pan mechanical balance.

Such balances are based on the first-class lever, like a teeter totter, that compares two masses at each end of the lever, one the unknown and the other standard weights, and the relationship $M_1L_1 = M_2L_2$ holds. The single-pan balance actually has unequal lever lengths, and the operation is based on removing weights from the lever end on which the unknown is placed, equal in value to the unknown mass. See the **website** for details.

> Burns' Hog Weighing Method: "1) Get a perfectly straight plank and balance it across a sawhorse. 2) Put the hog on one end of the plank. 3) Pile rocks on the other end until the plank is again perfectly balanced. 4) Carefully guess the weight of the rocks."
> —Robert Burns

Semimicro- and Microbalances

The discussion thus far has been limited to conventional macro or analytical balances. These perform weighings to the nearest 0.1 mg, and loads of up to 160-300 g can be handled. These are satisfactory for most routine analytical weighings. All of the above classes of balances can be made more sensitive by changing the parameters affecting the sensitivity, such as decreasing the mass of the beam (for mechanical balances) and pans, increasing the length of the beam, and changing the center of gravity of the beam. Lighter material can be used for the beam since it need not be as sturdy as the beam of a conventional balance.

> We make most quantitative weighings to 0.1 mg.

The *semimicrobalance* is sensitive to about 0.01 mg, and the *microbalance* is sensitive to about 0.001 mg (1 μg). The load limits of these balances are correspondingly less than the conventional balance, and greater care must be taken in their use.

Zero-Point Drift

The zero setting of a balance is not a constant that can be determined or set and forgotten. It will drift for a number of reasons, including temperature changes, humidity, and static electricity. The zero setting should therefore be checked at least once every half-hour during the period of using the balance.

Weight in a Vacuum—This is the Ultimate Accuracy

The weighings that are made on a balance will, of course, give the weight in air. When an object displaces its volume in air, it will be buoyed up by the weight of air displaced (**Archimedes' principle**—see the box in Chapter 1 on how analytical chemistry originated). The density of air is 0.0012 g (1.2 mg) per milliliter. If the density of the weights and the density of the object being weighed are the same, then they will be buoyed up by the same amount, and the recorded weight will be equal to the weight in a vacuum, where there is no buoyancy. If the densities are markedly different, the differences in the buoyancies will lead to a small error in the weighing: One will be buoyed up more than the other, and an unbalance will result. Such a situation arises in the weighing of very dense objects [e.g., platinum vessels (density = 21.4) or mercury (density = 13.6)] or light, bulky objects [e.g., water (density ≈ 1)]; and in very careful work, a correction should be made for this error. For comparison, the density of weights used in balances is about 8. See Reference 14 for air buoyancy corrections with a single-pan balance. (Reference 10 describes the calibration of the weights in a single-pan balance.)

> An object of 1-mL volume will be buoyed up by 1.2 mg!

Note that in most cases, a correction is not necessary because the error resulting from the buoyancy will cancel out in percent composition calculations. The same error will occur in the numerator (as the concentration of a standard solution or weight of a gravimetric precipitate) and in the denominator (as the weight of the sample). Of course, all weighings must be made with the materials in the same type of container (same density) to keep the error constant.

> The buoyancy of the weighing vessel is ignored, since it is subtracted.

An example where correction in vacuum is used is in the calibration of glassware. The mass of water or mercury delivered or contained by the glassware is measured. From a knowledge of the density of the liquid at the specified temperature, its volume can be calculated from the mass. Even in these cases, the buoyancy correction is only about one part per thousand. For most objects weighed, buoyancy errors can be neglected.

> Buoyancy corrections are usually significant in glassware calibration.

Weights of objects in air can be corrected to the weight in vacuum by

$$W_{\text{vac}} = W_{\text{air}} + W_{\text{air}} \left(\frac{0.0012}{D_o} - \frac{0.0012}{D_w} \right) \qquad (2.1)$$

where

W_{vac} = weight in vacuum, g

W_{air} = observed weight in air, g

D_o = density of object

D_w = density of standard weights

0.0012 = density of air

The density of brass weights is 8.4 and that of stainless steel weights is 7.8. A calculation with water as the object will convince you that even here the correction will amount to only about one part per thousand.

Example 2.1

The same buoyancy corrections apply for mechanical or electronic balances (which are calibrated with weights of known density).

A convenient way to calibrate pipets is to weigh water delivered from them. From the exact density of water at the given temperature, the volume delivered can then be calculated. Suppose a 20-mL pipet is to be calibrated. A stoppered flask when empty weighs 29.278 g. Water is delivered into it from the pipet, and it now weighs 49.272 g. If brass weights are used, what is the weight of water delivered, corrected to weight in vacuum?

Solution

The increase in weight is the weight of water in air:

$$49.272 - 29.278 = 19.994 \text{ g}$$

The density of water is 1.0 g/mL (to 2 significant figures from 10 to 30°C—see Table 2.4). Therefore,

$$W_{vac} = 19.994 + 19.994 \left(\frac{0.0012}{1.0} - \frac{0.0012}{8.4} \right) = 20.015 \text{ g}$$

Example 2.2

Recalculate the weight of the water delivered by the pipet in Example 2.1, using stainless steel weights at density 7.8 g/cm^3.

Solution

Do not round off until the end of the calculation. Then the same value results:

$$W_{vac} = 19.994 + 19.994 \left(\frac{0.0012}{1.0} - \frac{0.0012}{7.8} \right) = 20.015 \text{ g}$$

This illustrates that the buoyancy corrections in Table 2.4 are valid for either type of weight (see Calibration of Glassware below).

Sources of Error in Weighing

Several possible sources of error have been mentioned, including zero-point drift and buoyancy. Changes in ambient temperature or temperature of the object being weighed are probably the biggest sources of error, causing a drift in the zero or rest point due to convection-driven air currents. Hot or cold objects must be brought to ambient temperature before being weighed. Hygroscopic samples may pick up moisture, particularly in a high-humidity atmosphere. Exposure of the sample to air, prior to and during weighing, must be minimized.

General Rules for Weighing

Learn these rules!

The specific operation of your particular balance will be explained by your instructor. The main objectives are to protect all parts from dust and corrosion, avoid contamination or change in load (of sample or container), and avoid draft (air convection) errors. Some general rules you should familiarize yourself with before weighing with any type of analytical balance are:

1. Never handle objects to be weighed with the fingers. A piece of clean paper or tongs should be used.

2. Weigh at room temperature, and thereby avoid air convection currents.

3. Never place chemicals directly on the pan, but weigh them in a vessel (weighing bottle, weighing dish) or on powder paper. Always brush spilled chemicals off immediately with a soft brush.

4. Always close the balance case door before making the weighing. Air currents will cause the balance to be unsteady.

Although modern digital balances do not have user-manipulable weights, corrosion can still cause problems. Volatile corrosive substances (e.g., iodine or conc. HCl) should never be weighed in open containers in a balance.

Weighing of Solids

Solid chemical (nonmetal) materials are usually weighed and dried in a **weighing bottle**. Some of these are shown in Figure 2.3. They have standard tapered ground-glass joints, and hygroscopic samples (which take on water from the air) can be weighed with the bottle kept tightly capped. Replicate weighings can be conveniently carried out by **difference**. The sample in the weighing bottle is weighed, and then a portion is removed (e.g., by tapping) and quantitatively transferred to a vessel appropriate for dissolving the sample. The weighing bottle and sample are reweighed, and from the difference in weight, the weight of sample is calculated. The next sample is removed and the weight is repeated to get its weight by difference, and so on. This is illustrated in the Laboratory Notebook example for the soda ash experiment.

Weighing by difference is required for hygroscopic samples.

FIGURE 2.3   Weighing bottles.

It is apparent that by this technique an average of only one weighing for each sample, plus one additional weighing for the first sample, is required. However, each weight represents the difference between two weighings, so that the total experimental error is given by the combined error of both weighings. Weighing by difference *with the bottle capped* must be used if the sample is hygroscopic or cannot otherwise be exposed to the atmosphere before weighing. If there are no effects from atmospheric exposure, the bottles need not be capped.

For **direct weighing**, a **weighing dish**, weighing paper, or a weighing boat (all typically disposable) is used. The dish, paper, or boat is weighed empty and then with the added sample. This requires two weighings for each sample. The weighed sample is transferred by tapping. Direct weighing is satisfactory only if the sample is nonhygroscopic.

When making very careful weighings (e.g., to a few tenths of a milligram or less), you must take care not to contaminate the weighing vessel with extraneous material that may affect its weight. Special care should be taken not to get perspiration from the hands on the vessel because this can be quite significant. It is best to handle the vessel with a piece of paper. Alternatively, finger cots may be used. These are similar to just the fingertip region of protective gloves. Solid samples must frequently be dried to a constant weight (e.g., ±0.5 mg for a 0.5 g sample). Highly insulating material, for example laboratory ware made from fluorocarbons, easily acquire static charge that affect weigh readings. Brushes with a built in source of ionizing radiation (http://www.amstat.com/solutions/staticmaster.html) that help dissipate such charges are recommended for gently swiping such objects before weighing.

Weighing of Liquids

Weighing of liquids is usually done by direct weighing. The liquid is transferred to a weighed vessel (e.g., a weighing bottle), which is capped to prevent evaporation during weighing, and is then weighed. If a liquid sample is weighed by difference by pipetting out an aliquot from the weighing bottle, the inside of the pipet must be rinsed several times after transferring. Care should be taken not to lose any sample from the tip of the pipet during transfer.

Types of Weighing—What Accuracy Do You Need?

Only some weighings have to be done on an analytical balance, those involved in the quantitative calculations.

There are two types of weighing done in analytical chemistry, **rough** or **accurate**. Rough weighings to two or three significant figures are normally used when the amount of substance to be weighed need only be known to within a few percent. Examples are reagents to be dissolved and standardized later against a known standard, or the apportioning of reagents that are to be dried and then later weighed accurately, or simply added as is, as for adjusting solution conditions. That is, only rough weighings are needed when the weight is not involved in the computation of the analytical result. Rough weighings need not be done on analytical balances but may be completed on a top-loading balance.

Accurate weighings are reserved for obtaining the weight of a sample to be analyzed, the weight of the dried product in gravimetric procedures, or the weight of a dried reagent being used as a standard in a determination, all of which must generally be known to four significant figures or better to be used in calculating the analytical result. **These are performed only on an analytical balance**, **usually to the nearest 0.1 mg**. An exact predetermined amount of reagent is rarely weighed (e.g., 0.5000 g), but rather an approximate amount (about 0.5 g) is weighed accurately (e.g., to give 0.5129 g). Some chemicals are never weighed on an analytical balance. Sodium hydroxide pellets, for example, are so hygroscopic that they continually absorb moisture. The weight of a given amount of sodium hydroxide is not reproducible (and its purity is not known). To obtain a solution of known sodium hydroxide concentration, the sodium hydroxide is weighed on a rough balance and dissolved, and the solution is standardized against a standard acid solution.

2.3   Volumetric Glassware—Also Indispensible

Although accurate volume measurements of solutions can be avoided in gravimetric methods of analysis, they are required for almost any other type of analysis involving solutions.

Volumetric Flasks

Volumetric flasks contain an accurate volume.

Volumetric flasks are used for the dilution of solutions to a certain volume. They come in a variety of sizes, from 1 mL to 2 L or more. A typical flask is shown in Figure 2.4. These flasks are designed **to contain** an accurate volume at the specified temperature (20 or 25°C) when the bottom of the meniscus (the concave curvature of the upper surface of water in a column caused by capillary action—see Figure 2.10) just touches the etched "fill" line across the neck of the glass. The coefficient of expansion of glass is small, and for ambient temperature fluctuations the volume can be considered constant. These flasks are marked with "**TC**" to indicate "to contain." Other, less accurate containers, such as graduated cylinders, are also marked "TC." Many of these are directly

FIGURE 2.4   Volumetric flask.

marked on the face by the manufacturer as to the uncertainty of the container measurement; for example, a 250 mL volumetric flask is "±0.24 mL," or roughly a 0.1% error.

Initially, a small amount of diluent (usually distilled water) is added to the empty flask. Reagent chemicals should never be added directly to a dry glass surface, as glass is highly absorbant. When using a volumetric flask, a solution should be prepared stepwise. The desired reagent chemical (either solid or liquid) to be diluted is added to the flask, and then diluent is added to fill the flask about two-thirds (taking care to rinse down any reagent on the ground glass lip). It helps to swirl the solution before diluent is added to the neck of the flask to obtain most of the mixing (or dissolving in the case of a solid). Finally, diluent is added so that the bottom of the meniscus is even with the middle of the calibration mark (at eye level). If there are any droplets of water on the neck of the flask above the meniscus, take a piece of tissue and blot these out. Also, dry the ground-glass stopper joint.

The solution is finally thoroughly mixed as follows. Keeping the stopper on securely by using the thumb or palm of the hand, invert the flask and swirl or shake it *vigorously* for 5 to 10 s. Turn right side up and allow the solution to drain from the neck of the flask. Repeat at least 10 times.

Note. When preparing the solution of an expensive chemical, should the volume of liquid go over the calibration mark, it is still possible to save the solution as follows. Paste against the neck of the flask a thin strip of paper and mark on it with a sharp pencil the position of the meniscus, avoiding parallax error. After removing the thoroughly mixed solution from the flask, fill the flask with water to the calibration mark. Then by means of a buret or small volume graduated pipet, add water to the flask until the meniscus is raised to the mark on the strip of paper. Note and record the volume so added and use it to mathematically correct the concentration calculation. For an inexpensive chemical, start over. If the volume goes over the mark, you cannot accurately calculate concentration without determining how far over the mark you went. Be very careful and patient when filling volumetric flasks, especially when the components in the flask are irreplaceable or expensive.

FIGURE 2.5 Transfer or volumetric pipets.

Volumetric pipets deliver an accurate volume.

Pipets

The pipet is used to transfer a particular volume of solution. As such, it is often used to deliver a certain fraction (**aliquot**) of a solution. To ascertain the fraction, the original volume of solution from which the aliquot is taken must be known, but it need not all be present, so long as it has not evaporated or been diluted. There are two common types of pipets, the **volumetric**, or **transfer**, **pipet** and the **measuring** or **graduated pipet** (see Figures 2.5 and 2.6). Variations of the latter are also called **clinical**, or **serological**, **pipets**.

Pipets are designed **to deliver** a specified volume at a given temperature, and they are marked "**TD**." Again, the volume can be considered to be constant with small changes in temperature. Pipets are calibrated to account for the drainage film remaining on the glass walls. This drainage film will vary somewhat with the time taken to deliver, and usually the solution is allowed to drain under the force of gravity and the pipet is removed shortly after the solution is delivered. A uniform drainage time should be adopted.

The volumetric pipet is used for accurate measurements since it is designed to deliver only one volume and is calibrated at that volume. Accuracy to four significant figures is generally achieved, although with proper calibration, five figures may be obtained if necessary. See the table on the *back cover* for tolerances of class A transfer pipets. Measuring pipets are straight-bore pipets that are marked at different volume intervals. These are not as accurate because nonuniformity of the internal diameter of the device will have a relatively larger effect on total volume than is the case for pipets

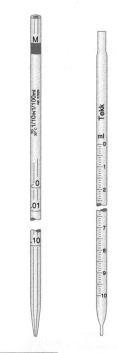

FIGURE 2.6 Measuring pipets.

with a bulb shape. Also, the drainage film will vary with the volume delivered. At best, accuracy to three significant figures can be expected from these pipets, unless you make the effort to calibrate the pipet for a given volume delivered.

Most volumetric pipets are calibrated to deliver with a certain small volume remaining in the tip. This should not be shaken or blown out. In delivering, the pipet is held vertically and the tip is touched on the side of the vessel to allow smooth delivery without splashing and so that the proper volume will be left in the tip. The forces of attraction of the liquid on the wall of the vessel will draw out a part of this.

Some pipets are **blowout** types (including measuring pipets calibrated to the entire tip volume). The final volume of solution must be blown out from the tip to deliver the calibrated amount. These pipets are easy to identify, as they will always have one or two **ground bands or rings** around the top. (These are not to be confused with a colored ring that is used only as a color coding for the volume of the pipet.) The solution is not blown out until it has been completely drained by gravity. Blowing to increase the rate of delivery will change the volume of the drainage film.

Volumetric pipets are available in sizes of 100 to 0.5 mL or less. Measuring and serological pipets range from a total capacity of 25 to 0.1 mL. Measuring pipets can be used for accurate measurements, especially for small volumes, if they are calibrated at the particular volume wanted. The larger measuring pipets usually deliver too quickly to allow drainage as fast as the delivery, and they have too large a bore for accurate reading.

In using a pipet, one should always wipe the outside of the tip dry after filling. If a solvent other than water is used, or if the solution is viscous, pipets must be recalibrated for the new solvent or solution to account for difference in drainage rate.

Pipets are filled by suction, using a rubber pipet bulb, a pipet pump, or other such pipetting device. Before using a pipet, practice filling and dispensing with water. **No solution should be pipetted by mouth**.

Syringe Pipets

Syringe pipets are useful for delivering microliter volumes.

These can be used for both macro and micro volume measurements. The calibration marks on the syringes may not be very accurate, but the reproducibility can be excellent if automatic delivery is used, such as a spring-loaded device that draws the plunger up to the same preset level each time. The volume delivered in this manner is free from drainage errors because the solution is forced out by the plunger. The volume delivered can be accurately calibrated. Microliter syringe pipets are used for introduction of samples into gas chromatographs. A typical syringe is illustrated in Figure 2.7. They are fitted with a needle tip, and the tolerances are as good as those found for other micropipets. In addition, any desired volume throughout the range of the syringe can be delivered. Syringes are available with a total volume as small as 0.5 μL, in which a wire plunger travels within the needle and so the entire syringe volume is within the needle.

The above syringe pipets are useful for accurate delivery of viscous solutions or volatile solvents; with these materials the drainage film would be a problem in conventional pipets. Syringe pipets are well suited to rapid delivery and also for thorough mixing of the delivered solution with another solution as a result of the rapid delivery.

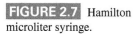 **FIGURE 2.7** Hamilton microliter syringe.

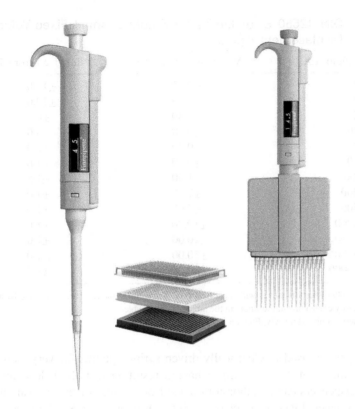

FIGURE 2.8 Single-channel and multichannel digital displacement pipets and microwell plates. (Courtesy of Thermo Fisher Scientific.)

A second type of syringe pipet is that shown in Figure 2.8. This type is convenient for rapid, one-hand dispensing of fixed (or variable) volumes in routine procedures and is widely used in the analytical chemistry laboratory. It contains a disposable non-wetting plastic tip (e.g., polypropylene) to reduce both film error and contamination. A thumb button operates a spring-loaded plunger, which stops at an intake or a discharge stop; the latter stop is beyond the former to ensure complete delivery. The sample never contacts the plunger and is contained entirely in the plastic tip. These pipets are available in volumes of 0.1 to 5000 µL and are reproducible to 1 to 2% or better, depending on the volume. Variable volume pipettes of this type typically deliver volumes across a defined range, such as 0.1 to 10 µL, 20 to 200 µL, 100 to 1000 µL, and 500 to 5000 µL. It is important to be consistent in your use of different pipettes for related procedures (e.g., always use a single draw from the 20 to 200 µL pipette to dispense 100 µL) to maintain consistency.

Sometimes, the actual volume delivered by these and other micropipets does not need to be known because they are used in relative measurements. For example, the same pipet may be used to deliver a sample and an equal volume of a standard solution for calibrating the instrument used for the measurement. Precision in delivery is usually more important than the absolute volume delivered. The European standard for pipet calibration in Europe is the German DIN 126650 (or a similar international standard ISO 8655). Calibrations are based on gravimetric testing (weighing of water). The DIN standard does not give separate limits for accuracy and precision, but rather uses a combined error limit equal to percent accuracy plus 2 times the standard deviation, that is, it gives a range in which we are 95% confident the delivery will fall (see Chapter 3 for a discussion of standard deviation and confidence limits). Table 2.2 lists the DIN error limits for single-channel displacement pipets. Table 2.3 lists accuracies and precisions for a typical variable volume single-channel pipet.

Besides the manually operated syringes, there are electronically controlled and variable-volume motor-driven syringes available for automated repetitive deliveries.

The volume may not be accurately known, but it is reproducible.

Joseph Gay-Lussac (1778–1850) designed the first buret and named the buret and pipet.

TABLE 2.2 DIN 12650 Error Limits for Single-Channel Fixed-Volume Air Displacement Pipets[a]

Nominal Volume (μL)	Maximum Error (μL)	Relative Error (%)
1	±0.15	±15.0
2	±0.20	±10.0
5	±0.30	±6.0
10	±0.30	±3.0
20	±0.40	±2.0
50	±0.80	±1.6
100	±1.50	±1.5
200	±2.00	±1.0
500	±5.00	±1.0
1000	±10.00	±1.0
2000	±20.00	±1.0
5000	±50.00	±1.0
10000	±100.00	±1.0

[a] These limits apply to manufacturers with a controlled environment. If the tests are performed by a user in a normal laboratory environment, the limits in the table may be doubled.
Courtesy of Thermo Labsystems Oy, Finland.

When syringes are used in electrically driven syringe pumps for very slow infusion of solution, there can be stick–slip behavior, resulting in pulsed flow. Inexpensive automated dispensers can be laboratory made from syringes [see for example, "Inexpensive Automated Electropneumatic Syringe Dispenser", P. K. Dasgupta and J. R. Hall, *Anal. Chim. Acta* 221 (1989) 189.] Also, you may purchase pipets with multiple syringes for simultaneous delivery, with for example, 12 or 16 channels. These are useful for delivering solutions into microwell plates used in biotechnology or clinical chemistry laboratories that process thousands of samples (Figure 2.8). You may find more information on displacement pipets from representative manufacturers, for example, www.thermoscientific.com/finnpipette or www.eppendorf.com.

Burets

FIGURE 2.9 Typical buret.

A buret is used for the accurate delivery of a variable amount of solution. Its principal use is in **titrations**, where a standard solution is added to the sample solution until the **end point** (the detection of the completion of the reaction) is reached. The conventional buret for macrotitrations is marked in 0.1-mL increments from 0 to 50 mL; one is illustrated in Figure 2.9. The volume delivered can be read to the nearest 0.01 mL by interpolation actually (good to about ±0.02 or ±0.03 mL). Burets are also obtainable in 10-, 25-, and 100-mL capacities, and microburets are available in capacities of down to 2 mL, where the volume is marked in 0.01-mL increments and can be estimated to the nearest 0.001 mL. Ultramicroburets of 0.1-mL capacity graduated in 0.001-mL (1-μL) intervals are used for microliter titrations.

Drainage film is a factor with conventional burets, as with pipets, and this can be a variable if the delivery rate is not constant. The usual practice is to deliver at a fairly slow rate, about 15 to 20 mL per minute, and then to wait several seconds after delivery to allow the drainage to "catch up." In actual practice, the rate of delivery is only a few drops per minute near the end point, and there will be no time lag between the flow rate and the drainage rate. As the end point is approached, fractions of a drop are delivered by just opening, or "cracking," the stopcock and then touching the tip of the buret to the wall of the titration vessel. The fraction of the drop is then washed down into the solution with distilled water.

TABLE 2.3 Accuracy and Precision Limits for Single-Channel Variable-Volume Finnpipettes Model F1

Range (µL)	Increment (µL)	Volume (µL)	Accuracy (µL)	Accuracy (%)	Precision[a] s.d. (µL)	Precision[a] CV (%)
0.2–2 µL	0.002 µL	2	±0.050	±2.50	0.040	2.00
		1	±0.040	±4.00	0.040	3.50
		0.2	±0.024	±12.00	0.020	10.00
0.5–5 µL	0.01 µL	5	±0.080	±1.50	0.050	1.00
		2.5	±0.0625	±2.50	0.0375	1.50
		0.5	±0.030	±6.00	0.025	5.00
1–10 µL	0.02 µL	10	±0.100	±1.00	0.050	0.50
		5	±0.080	±1.50	0.040	0.80
		1	±0.025	±2.50	0.020	2.00
1–10 µL	0.02 µL	10	±0.100	±1.00	0.080	0.80
		5	±0.080	±1.50	0.040	0.80
		1	±0.035	±3.50	0.030	3.00
2–20 µL	0.02 µL	20	±0.20	±1.00	0.08	0.40
		10	±0.15	±1.50	0.06	0.60
		2	±0.06	±3.00	0.05	2.50
2–20 µL	0.02 µL	20	±0.20	±1.00	0.08	0.40
		10	±0.15	±1.50	0.06	0.60
		2	±0.06	±3.00	0.05	2.50
5–50 µL	0.1 µL	50	±0.30	±0.60	0.15	0.30
		25	±0.25	±1.00	0.13	0.50
		5	±0.15	±3.00	0.125	2.50
5–50 µL	0.1 µL	50	±0.30	±0.60	0.15	0.30
		25	±0.25	±1.00	0.13	0.50
		5	±0.15	±3.00	0.125	2.50
10–100 µL	0.2 µL	100	±0.80	±0.80	0.20	0.20
		50	±0.60	±1.20	0.20	0.40
		10	±0.30	±3.00	0.10	1.00
20–200 µL	0.2 µL	200	±1.2	±0.60	0.4	0.20
		100	±1.0	±1.00	0.4	0.40
		20	±0.36	±1.80	0.14	0.70
30–300 µL	1 µL	300	±1.8	±0.60	0.6	0.20
		150	±1.5	±1.00	0.6	0.40
		30	±0.45	±1.50	0.18	0.60
100–1000 µL	1 µL	1000	±6.0	±0.60	2.0	0.20
		500	±4.0	±0.80	1.5	0.30
		100	±1.0	±1.00	0.6	0.60
0.5–5 mL	0.01 mL	5000	±25.0	±0.50	10.0	0.20
		2500	±17.5	±0.70	7.5	0.30
		500	±10.0	±2.00	4.0	0.80
1–10 mL	0.02 mL	10000	±50.0	±0.50	20.0	0.20
		5000	±40.0	±0.80	15.0	0.30
		1000	±20.0	±2.00	8.0	0.80

[a] s.d.= standard deviation, CV = coefficient of variation.
From https://fscimage.fishersci.com/images/D11178~.pdf

Care and Use of Volumetric Glassware

We have mentioned a few precautions in the use of volumetric flasks, pipets, and burets. Your laboratory instructor will supply you with detailed instructions in the use of each of these tools. A discussion of some general precautions and good laboratory technique follows.

Cleanliness of glassware is of the utmost importance. If films of dirt or grease are present, liquids will not drain uniformly and will leave water breaks or droplets on the walls. Under such conditions the calibration will be in error. Initial cleaning should be by repeated rinses with laboratory detergent and then water. Then try cleaning with dilute nitric acid and rinse with more water. Use of a buret or test tube brush aids the cleaning of burets and necks of volumetric flasks—but be careful of scratching the interior walls. Pipets should be rotated to coat the entire surface with detergent. There are commercial cleaning solutions available that are very effective.

Rinse pipets and burets with the solution to be measured.

Pipets and burets should be rinsed at least twice with the solution with which they are to be filled. If they are wet, they should be rinsed first with water, if they have not been already, and then a minimum of *three* times with the solution to be used; about one-fifth the volume of the pipet or buret is adequate for each rinsing. A volumetric flask, if it is wet from a previously contained solution, is rinsed with three portions of water only since later it will be filled to the mark with water. It need not be dry.

Note that analytical glassware should not be subjected to the common practice employed in organic chemistry laboratories of drying either in an oven (this can affect the volume of calibrated glassware) or by drying with a towel or by rinsing with a volatile organic solvent such as acetone (which can cause contamination). The glassware usually does not have to be dried. The preferred procedure is to rinse it with the solution that will fill it.

Avoid parallax error in reading buret or pipet volumes.

Care in reading the volume will avoid parallax error, that is, error due to incorrect alignment of the observer's eye, the meniscus, and the scale. Correct position is with your eye at the same level as the menicus. If the eye level is above the meniscus, the volume read will be smaller than that taken; the opposite will be true if the eye level is too low.

After glassware is used, it can usually be cleaned sufficiently by immediate rinsing with water. If the glassware has been allowed to dry, it should be cleaned with detergent. Volumetric flasks should be stored with the stopper inserted, and preferably filled with distilled water. Burets should be filled with distilled water and stoppered with a rubber stopper when not in use.

There are commercial glassware washing machines to automate the cleaning of glassware. These use detergents and deionized water for cleaning and rinsing. See, for example, L. Choplo, "The Benefits of Machine Washing Laboratory Glassware Versus Hand Washing," *Amer. Lab.* October (2008) 6 (http://new.american laboratory.com/914-Application-Notes/34683-The-Benefits-of-Machine-Washing-Lab oratory-Glassware-Versus-Hand-Washing/), and M. J. Felton, "Labware Washers," *Today's Chemist at Work*, November (2004) 43 (http://pubs.acs.org/subscribe/archive/ tcaw/13/i11/pdf/1104prodprofile.pdf).

General Tips for Accurate and Precise Titrating

If the buret contains a Teflon stopcock, it does not require lubrication.

Your buret probably has a Teflon stopcock, and this will not require lubrication. Make sure it is secured tightly enough to prevent leakage, but not so tight as to make rotation hard. If your buret has a ground-glass stopcock, you may have to grease the stopcock.

A thin layer of stopcock grease (not silicone lubricant) is applied uniformly to the stopcock, using very little near the hole and taking care not to get any grease in the hole. The stopcock is inserted and rotated. There should be a uniform and transparent layer of grease, and the stopcock should not leak. If there is too much lubricant, it will be forced into the barrel or may work into the buret tip and clog it. Grease can be removed from the buret tip and the hole of the stopcock by using a fine wire.

Next, we fill the buret with the solution it will deliver. The buret is filled above the zero mark and the stopcock is opened to fill the tip. Check the tip for air bubbles. If any are present, they may work out of the tip during the titration, causing an error in reading. Work air bubbles out by rapid opening and closing of the stopcock to squirt the titrant through the tip or tapping the tip while solution is flowing. No bubbles should be in the barrel of the buret. If there are, the buret is probably dirty.

The initial reading of the buret is taken by allowing it to drain slowly to the zero mark. Wait a few seconds to make certain the drainage film has caught up to the meniscus. Read the buret to the nearest 0.01 mL (for a 50-mL buret). The initial reading may be 0.00 mL or greater. The reading is best taken by placing your finger just in back of the meniscus or by using a meniscus illuminator (Figure 2.10). The meniscus illuminator has a white and a black field, and the black field is positioned just below the meniscus. Avoid parallax error by making the reading at eye level.

The titration is performed with the sample solution in an Erlenmeyer flask. The flask is placed on a white background, and the buret tip is positioned within the neck of the flask. The flask is swirled with the right hand while the stopcock is manipulated with the left hand (Figure 2.11), or whatever is comfortable. This grip on the buret maintains a slight inward pressure on the stopcock to ensure that leakage will not occur. The solution can be more efficiently stirred by means of a magnetic stirrer and stirring bar.

As the titration proceeds, the indicator changes color in the vicinity where the titrant is added, owing to local excesses; but it rapidly reverts to the original color as the titrant is dispersed through the solution to react with the sample. As the end point is approached, the return to the original color occurs more slowly, since the dilute solution must be mixed more thoroughly to consume all the titrant. At this point, the titration should be stopped and the sides of the flask washed down with distilled water from the wash bottle. A drop from the buret is about 0.02 to 0.05 mL, and the volume is read to the nearest 0.02 mL. It is therefore necessary to split drops near the end point. This can be done by slowly turning the stopcock until a fraction of a drop emerges from the buret tip and then closing it. The fraction of drop is touched off onto the wall of the flask and is washed into the flask with the wash bottle, or it is transferred with a glass stirring rod. There will be a sudden and "permanent" (lasting at least 30 s) change in the color at the end point when a fraction of a drop is added.

The titration is usually performed in triplicate. After performing the first titration, you can calculate the approximate volume for the replicate titrations from the weights of the samples and the molarity of the titrant. This will save time in the titrations. The volume should not be calculated to nearer than 0.1 mL in order to avoid bias in the reading.

After a titration is complete, unused titrant should never be returned to the original bottle but should be discarded. If the titrant isn't between pH 4 and 8 and on the short list of substances cleared to go down the drain, it should be disposed of in a recycle container.

If a physical property of the solution, such as potential, is measured to detect the end point, the titration is performed in a beaker with magnetic stirring so electrodes can be placed in the solution.

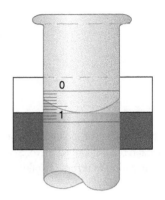

FIGURE 2.10 Meniscus illuminator for buret.

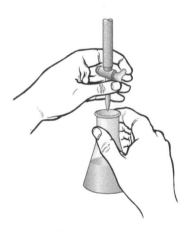

FIGURE 2.11 Proper technique for titration.

Subsequent titrations can be speeded up by using the first to *approximate* the end-point volumes.

Tolerances and Precision of Glassware

Class A glassware is accurate enough for most analyses. It can be calibrated to NIST specifications.

The National Institute of Standards and Technology (NIST) has prescribed certain *tolerances*, or absolute errors, for different volumetric glassware, and some of these are listed on the *front cover* of the text. For volumes greater than about 25 mL, the tolerance is within 1 part per thousand relative, but it is larger for smaller volumes. The letter "A" stamped on the side of a volumetric flask, buret, or pipet indicates that it complies with class A tolerances. This says nothing about the precision of delivery. Volumetric glassware that meets NIST specifications or that is certified by NIST can be purchased, but at a significantly higher price than uncertified glassware. Less expensive glassware may have tolerances double those specified by NIST. It is a simple matter, however, to calibrate this glassware to an accuracy as good as or exceeding the NIST specifications (see Experiment 2).

The variances or the uncertainties in each reading are additive. See propagation of error, Chapter 3.

The precision of reading a 50-mL buret is about ±0.02 mL. Since a buret is always read twice, the total absolute uncertainty may be as much as ±0.04 mL. The relative uncertainty will vary inversely with the total volume delivered. It becomes apparent that a titration with a 50-mL buret should involve about 40 mL titrant to achieve a precision of 1 ppt. Smaller burets can be used for increased precision at smaller volumes. Pipets will also have a certain precision of reading, but only one reading is required for volumetric pipets.

Calibration of Glassware—for the Ultimate Accuracy

Example 2.1 illustrated how Equation 2.1 may be used in the calibration of glassware, to correct for buoyancy of the water used for calibration, that is, to correct to weight in vacuum. Dividing the weight of the water in vacuum by its density at the given temperature will convert it to volume.

Table 2.4 lists the calculated volumes for a gram of water in air at atmospheric pressure for different temperatures, corrected for buoyancy with stainless steel weights of density 7.8 g/cm^3. These are used to give the volume of the glassware being calibrated, from the weight of water contained or delivered by the glassware. (The values are not significantly different for brass weights of 8.4 g/cm^3 density. See Example 2.2.) The glass volumes calculated for the standard temperature of 20°C include slight adjustments for borosilicate glass (Pyrex or Kimax) container expansion or contraction with temperature changes (volumetric glassware has a cubical coefficient of expansion of about 0.000025 per degree centigrade, resulting in changes of about 0.0025% per degree; for 1 mL, this is 0.000025 mL per degree). Water expands about 0.02% per degree around 20°C. Volume (concentration) corrections may be made using the water density data in Table 2.4, taking the ratios of the relative densities.

In the textbook **website**, the Table 2.4 spreadsheet is available, with formulas as indicated in the table. You can substitute specific weights of water in air, obtained from a flask, pipet, or buret, in cell B at the temperature of measurement to obtain the calculated calibration volume at temperature, *t*, and for 20°C. We describe the use of spreadsheets in Chapter 3. The book **website** also has a table and figure of the percent error for weight in vacuum as a function of sample density.

For those of you who live at high elevations, the density of air is slightly less than 0.0012 g/mL (at sea level), for example, about 0.0010 g/mL at 5,000 feet elevation. You may, in your downloaded spreadsheet of Table 2.4, substitute the appropriate value in the the formula in cell C14, and copy the new formula down the column.

TABLE 2.4 Glassware Calibration

	A	B	C	D	E	F	G	H
1	Glassware Calibration							
2	Weight in vacuum assuming stainless steel weights, density 7.8 g/mL.							
3	Glass expansion for borosilicate glass, 0.000025 mL/mL/ °C.							
4	The actual spreadsheet is available on the website (Table 2.4).							
5	Save it to your desktop, and use it to calculate calibrated volumes of glassware.							
6	Substitute into the appropriate Cell B, at temperature t, the weight of water							
7	in air, obtained from the glassware at the temperature of the measurement (Cell A).							
8	The calibration volume at the temperature, t (Cell D), and at 20°C (Cell F), is calculated.							
9	Round the calculated values to the appropriate number of significant figures, usually							
10	four or five.							
11								
12	t, °C	Wt. H$_2$O	Wt. in	Vol. at t,	Glass expansion,	Vol. at 20°C	Density,	
13		in air, g	vacuum, g	mL	at 20°C, mL	mL	g/mL	
14	10	1.0000	**1.0010**	**1.0013**	**-0.000250**	**1.0016**	0.9997026	
15	11	1.0000	1.0010	1.0014	-0.000225	1.0017	0.9996081	
16	12	1.0000	1.0010	1.0015	-0.000200	1.0017	0.9995004	
17	13	1.0000	1.0010	1.0017	-0.000175	1.0018	0.9993801	
18	14	1.0000	1.0010	1.0018	-0.000150	1.0020	0.9992474	
19	15	1.0000	1.0010	1.0019	-0.000125	1.0021	0.9991026	
20	16	1.0000	1.0010	1.0021	-0.000100	1.0022	0.9989460	
21	17	1.0000	1.0010	1.0023	-0.000075	1.0023	0.9987779	
22	18	1.0000	1.0010	1.0025	-0.000050	1.0025	0.9985896	
23	19	1.0000	1.0010	1.0026	-0.000025	1.0027	0.9984082	
24	20	1.0000	1.0010	1.0028	0.000000	1.0028	0.9982071	
25	21	1.0000	1.0010	1.0031	0.000025	1.0030	0.9979955	
26	22	1.0000	1.0010	1.0033	0.000050	1.0032	0.9977735	
27	23	1.0000	1.0010	1.0035	0.000075	1.0034	0.9975415	
28	24	1.0000	1.0010	1.0038	0.000100	1.0037	0.9972995	
29	25	1.0000	1.0010	1.0040	0.000126	1.0039	0.9970479	
30	26	1.0000	1.0010	1.0043	0.000151	1.0041	0.9967867	
31	27	1.0000	1.0010	1.0045	0.000176	1.0044	0.9965162	
32	28	1.0000	1.0010	1.0048	0.000201	1.0046	0.9962365	
33	29	1.0000	1.0010	1.0051	0.000226	1.0049	0.9959478	
34	30	1.0000	1.0010	1.0054	0.000251	1.0052	0.9956502	
35								
36	Formulas are entered into the boldface cells above as indicated below. They are							
37	copied down for all temperatures. See Chapter 3 for setting up a spreadsheet.							
38	**Cell C14:** $W_{vac} = W_{air} + W_{air} (0.0012/D_o - 0.0012/D_w) = W_{air} (0.0012/1.0 + 0.0012/7.8)$							
39			= B14+B14*(0.0012/1.0-0.0012/7.8)			Copy down		
40	**Cell D14:** V_t (mL)= $W_{vac,t}$ (g)/D_t (g/mL)							
41			= C14/G14			Copy down		
42	**Cell E14:** Glass expans. = (t – 20) (deg) x 0.000025 (mL/mL/deg) x Vt (mL)							
43			= (A14-20)*0.000025*D14			Copy down		
44	**Cell F14:** $V_{20o} = V_t$ – Glass$_{exp}$ =			D14-E14		Copy down		

Example 2.3

(a) Use Table 2.4 to calculate the volume of the 20-mL pipet in Example 2.2 (steel weights), from its weight in air. Assume the temperature is 23°C.

(b) Give the corresponding volume at 20°C as a result of glass contraction.

(c) Compare with the volume calculated using the weight in air with that calculated using the weight in vacuum and the density in water (Example 2.2).

Solution

(a) From Table 2.4, the volume per gram in air is 1.0035 mL at 23°C:

$$19.994 \text{ g} \times 1.0035 \text{ mL/g} = 20.064 \text{ mL}$$

(b) The glass contraction at 20°C relative to 23°C is 0.0015 mL (0.000025 mL/mL/°C × 20 mL × 3°C), so the pipet volume at 20°C is 20.062 mL.

(c) The density of water at 23°C is 0.99754 g/mL, so from the weight in vacuum:

$$20.015 \text{ g}/0.99754 \text{ g/mL} = 20.064 \text{ mL}$$

The same value is obtained.

Example 2.4

You prepared a solution of hydrochloric acid and standardized it by titration of primary standard sodium carbonate. The temperature during the standardization was 23°C, and the concentration was determined to be 0.1127_2 M. The heating system in the laboratory malfunctioned when you used the acid to titrate an unknown, and the temperature of the solution was 18°C. What was the concentration of the titrant?

Solution

$$M_{18°} = M_{23°} \times \left(D_{18°}/D_{23°}\right)$$
$$= 0.1127_2 \times (0.99859/0.99754)$$
$$= 0.1128_4 \ M$$

(See Chapter 3 for significant figures and the meaning of the subscript numbers.)

Techniques for Calibrating Glassware

You generally calibrate glassware to five significant figures, the maximum precision you are likely to attain in filling or delivering solutions. Hence, your net weight of water needs to be five figures. If the glassware exceeds 10 mL, this means weighing to 1 mg is all that is needed. This can be readily and conveniently accomplished using a top-loading balance, rather than a more sensitive analytical balance. [Note: If the volume number is large without regard to the decimal, e.g., 99, then four figures will suffice—see Chapter 3 discussion on significant figures. For example, a 10-mL pipette may be calibrated and shown to actually deliver 9.997 mL. This is as accurate as if the pipette was determined to deliver 10.003 mL (the last significant figure in both cases is one part in 10,000)].

1. **Volumetric Flask Calibration.** To calibrate a volumetric flask, first weigh the clean, dry flask and stopper. Then fill it to the mark with distilled water. There should be no droplets on the neck. If there are, blot them with tissue paper. The flask and water should be equilibrated to room temperature. Weigh the filled flask, and then record the temperature of the water to 0.1°C. The increase in weight represents the weight in air of the water contained by the flask.

2. **Pipet Calibration.** To calibrate a pipet, weigh a dry Erlenmeyer flask with a rubber stopper or a weighing bottle with a glass stopper or cap, depending on the volume of water to be weighed. Fill the pipet with distilled water (whose temperature you have recorded) and deliver the water into the flask or bottle, using proper pipetting technique, and quickly stopper the container to avoid evaporation loss. Reweigh to obtain the weight in air of water delivered by the pipet.

3. **Buret Calibration.** Calibrating a buret is similar to the procedure for a pipet, except that several volumes will be delivered. The internal bore of the buret is not perfectly cylindrical, and it will be a bit "wavy," so the actual volume delivered will vary both plus and minus from the nominal volumes marked on the buret, as increased volumes are delivered. You will ascertain the volume at 20% full-volume increments (e.g., each 10 mL, for a 50-mL buret) by filling the buret each time and then delivering the nominal volume into a dry flask. (The buret is filled each time to minimize evaporation errors. You may also make successive deliveries into the same flask, i.e., fill the buret only once. Make rapid deliveries.) Since the delivered volume does not have to be exact, but close to the nominal volume, you can make fairly fast deliveries, but wait about 10 to 20 s for film drainage. Prepare a plot of volume correction versus nominal volume and draw straight lines between each point. Interpolation is made at intermediate volumes from the lines. Typical volume corrections for a 50-mL buret may range up to ca. 0.05 mL, plus or minus.

Professor's Favorite Experiment

Contributed by Professor Alex Scheeline, University of Illinois

This illustrates a possibly more precise alternative to buret calibration, as described here, or determines the optimal approach.

Place a beaker of water in the weighing compartment of the balance so that the air becomes saturated with water vapor. Fill the buret. Place it through the top of the balance housing so the effluent will enter a beaker on the weighing table. Tare the balance and record the reading on the buret.

Now drain a few tenths of a milliliter from the buret. Record the volume, and once the balance has settled, record the mass. Continue to do this so that one has ca. 100 points from the drainage of the buret. Repeat three times. Compute the volume delivered for each reading. Plot. Now answer these questions:

- Which is more precise: the fiducial marking of the buret, as manufactured, or your attempt to calibrate the buret?
- Is there a smooth curve through the data, or is there significant variation from a smooth curve?
- Is there any indication as to what the appropriate measurement interval should be to obtain calibration to ±0.01 mL while minimizing the number of calibration points?

- How does the calibration of your buret compare with that of two other students?
- What is the best precision one can get if one ignores the results of buret calibaration?

Example 2.5

You calibrate a 50-mL buret at 10-mL increments, filling the buret each time and delivering the nominal volume, with the following results:

Buret Reading (mL)	Weight H_2O Delivered (g)
10.02	10.03
20.08	20.03
29.99	29.85
40.06	39.90
49.98	49.86

Construct a plot of volume correction versus volume delivered. The temperature of the water is 20°C and stainless steel weights are used.

Solution

From Table 2.4 (or use Table 2.4 from the website for automatic calculation of volumes):

$$W_{vac} = 10.03 + 10.03\,(0.00105) = 10.03 + 0.01 = 10.04 \text{ g}$$

$$\text{Vol.} = 10.04 \text{ g}/0.9982 \text{ g/mL} = 10.06 \text{ mL}$$

Likewise, for the others, we construct the table:

Nominal Volume (mL)	Actual Volume (mL)	Correction (mL)
10.02	10.06	+0.04
20.08	20.09	+0.01
29.99	29.93	−0.06
40.06	40.01	−0.05
49.98	50.00	+0.02

Prepare a graph of nominal volume (y axis) versus correction volume. Use 10, 20, 30, 40, and 50 mL as the nominal volumes.

Selection of Glassware—How Accurate Does It Have to Be?

Only certain volumes need to be measured accurately, those involved in the quantitative calculations.

As in weighing operations, there will be situations where you need to accurately know volumes of reagents or samples measured or transferred (accurate measurements), and others in which only approximate measurements are required (rough measurements). If you wish to prepare a standard solution of 0.1 M hydrochloric acid, it can't be done by measuring an accurate volume of concentrated acid and diluting to a known volume because the concentration of the commercial acid is not known adequately. Hence, an approximate solution is prepared that is then standardized. We see in the table on the inside *back cover* that the commercial acid is about 12.4 M. To prepare 1 L of a 0.1 M solution, about 8.1 mL needs to be taken and diluted. It would be a waste of time to measure this (or the water used for dilution) accurately. A 10-mL graduated cylinder or 10-mL measuring pipet will suffice, and the acid can be diluted in an ungraduated 1-L bottle. If, on the other hand, you wish to dilute a stock standard solution accurately, then a transfer pipet must be used and the dilution must be done in a volumetric flask. Any volumetric measurement that is a part of the actual analytical measurement must be done with the accuracy and precision required of the analytical measurement. This generally means four-significant-figure accuracy, and transfer pipets and volumetric flasks are required. This includes taking an accurate portion of a sample, preparation of a standard solution from an accurately weighed reagent, and accurate dilutions. Burets are used for accurate measurement of variable volumes, as in a titration. Preparation of reagents that are to be used in an analysis just to provide proper solution conditions (e.g., buffers for pH control) need not be prepared highly accurately, and less accurate glassware can be used, for example, graduated cylinders.

2.4 Preparation of Standard Base Solutions

Sodium hydroxide is usually used as the titrant when a base is required. It contains significant amounts of water and sodium carbonate, and so it cannot be used as a primary standard. For accurate work, the sodium carbonate must be removed from the NaOH because it reacts to form a buffer that decreases the sharpness of the end point. In addition, an error will result if the NaOH is standardized using a phenolphthalein end point (in which case the CO_3^{2-} is titrated only to HCO_3^-), and then a methyl orange end point is used in the titration of a sample (in which case the CO_3^{2-} is titrated to CO_2). In other words, the effective molarity of the base has been increased, owing to further reaction of the HCO_3^-.

Sodium carbonate is essentially insoluble in nearly saturated sodium hydroxide. It is conveniently removed by dissolving the weighed NaOH in a volume (milliliters) of water equal to its weight in grams. The insoluble Na_2CO_3 can be allowed to settle for several days, and then the clear supernatant liquid can be carefully decanted,[1] or it can be filtered in a Gooch crucible with a quartz fiber filter mat (do not wash the filtered Na_2CO_3). This procedure does not work with KOH because K_2CO_3 remains soluble.

Water dissolves CO_2 from the air. In many routine determinations not requiring the highest degree of accuracy, carbonate or CO_2 impurities in the water will result in an error that is small enough to be considered negligible. For the highest accuracy, however, CO_2 should be removed from all water used to prepare solutions for acid–base titrations, particularly the alkaline solutions. This is conveniently done by boiling the water and then cooling it under a cold-water tap.

Sodium hydroxide is usually standardized by titrating a weighed quantity of primary standard potassium acid phthalate (KHP), which is a moderately weak acid $\left(K_a = 4 \times 10^{-6}\right)$, approximately like acetic acid; a phenolphthalein end point is used. The sodium hydroxide solution should be stored in a plastic-lined glass bottle to prevent absorption of CO_2 from the air. If the bottle must be open (e.g., a siphon bottle), the opening is protected with an **Ascarite** II (fibrous silicate impregnated with NaOH) or soda-lime [$Ca(OH)_2$ and NaOH] tube.

Remove Na_2CO_3 by preparing a saturated solution of NaOH.

See Experiment 7 for preparing and standardizing sodium hydroxide.

---

[1]Concentrated alkali attacks glass whereas carbon dioxide can permeate through most organic polymers. One solution out of this dilemma is to insert a polyethylene bag inside a glass bottle and use a rubber stopper.

2.5 Preparation of Standard Acid Solutions

Hydrochloric acid is the usual titrant for the titration of bases. Most chlorides are soluble, and few possible side reactions with this acid. It is convenient to handle. It is not a primary standard (although constant-boiling HCl, which is a primary standard, can be prepared), and an approximate concentration is prepared simply by diluting the concentrated acid. For most accurate work, the water used to prepare the solution should be boiled, although use of boiled water is not so critical as with NaOH; CO_2 will have a low solubility in strongly acidic solutions and will tend to escape during shaking of the solution.

See Experiment 8 for preparing and standardizing hydrochloric acid.

Primary standard sodium carbonate is usually used to standardize HCl solutions. The disadvantage is that the end point is not sharp unless an indicator such as methyl red or methyl purple is used and the solution is boiled at the end point. A modified methyl orange end point may be used without boiling, but this is not so sharp. Another disadvantage is the low formula weight of Na_2CO_3. Tris-(hydroxymethyl)aminomethane (THAM), $(HOCH_2)_3CNH_2$, is another primary standard that is more convenient to use. It is nonhygroscopic, but it is still a fairly weak base $\left(K_b = 1.3 \times 10^{-6}\right)$ with a low molecular weight. The end point is not complicated by released CO_2, and it is recommended as the primary standard unless the HCl is being used to titrate carbonate samples.

A secondary standard is less accurate than a primary standard.

If a standardized NaOH solution is available, the HCl can be standardized by titrating an aliquot with the NaOH. The end point is sharp and the titration is more rapid. The NaOH solution is a **secondary standard**. Any error in standardizing this will be reflected in the accuracy of the HCl solution. The HCl is titrated with the base, rather than the other way around, to minimize absorption of CO_2 in the titration flask. Phenolphthalein or bromothymol blue can be used as indicator.

2.6 Other Apparatus—Handling and Treating Samples

Besides apparatus for measuring mass and volume, there are a number of other items of equipment commonly used in analytical procedures.

Desiccators

Oven-dried samples or reagents are cooled in a desiccator before weighing.

A **desiccator** is used to keep samples dry while they are cooling and before they are weighed and, in some cases, to dry a wet sample. Dried or ignited samples and vessels are cooled in the desiccator. A typical glass desiccator is shown in Figure 2.12. A desiccator is an airtight container that maintains an atmosphere of low humidity. A desiccant such as calcium chloride is placed in the bottom to absorb the moisture. This desiccant will have to be changed periodically as it becomes "spent." It will usually become wet in appearance or caked from the moisture when it is time to be changed. A porcelain plate is usually placed in the desiccator to support weighing bottles, crucibles, and other vessels. An airtight seal is made by application of stopcock grease to the ground-glass rim on the top of the desiccator. A **vacuum desiccator** has a side arm on the top for evacuation so that the contents can be kept in a vacuum rather than just an atmosphere of dry air.

The top of a desiccator should not be removed any more than necessary since the removal of moisture from the air introduced is rather slow, and continued exposure will limit the lifetime of the desiccant. A red-hot crucible or other vessel should be allowed to cool in the air about 60 s before it is placed in the desiccator. Otherwise, the air in

FIGURE 2.12 Desiccator and desiccator plate.

FIGURE 2.13 Muffle furnace. (Courtesy of Arthur H. Thomas Company.)

the desiccator will be heated appreciably before the desiccator is closed, and as the air cools, a partial vacuum will be created. This will result in a rapid inrush of air when the desiccator is opened and in possible spilling or loss of sample as a consequence. A hot weighing bottle should not be stoppered when placed in a desiccator because on cooling, a partial vacuum is created and the stopper may seize. The stopper should be placed in the desiccator with the weighing bottle.

Table 2.5 lists some commonly used desiccants and their properties. Aluminum oxide, magnesium perchlorate, calcium oxide, calcium chloride, and silica gel can be regenerated by heating at 150, 240, 500, 275, and 150°C, respectively.

Furnaces and Ovens

A **muffle furnace** (Figure 2.13) is used to ignite samples to high temperatures, either to convert precipitates to a weighable form or to burn organic materials prior to inorganic analysis. There should be some means of regulating the temperature since losses of some metals may occur at temperatures in excess of 500°C. Temperatures up to about 1200°C can be reached with muffle furnaces.

TABLE 2.5 Some Common Drying Agents

Agent	Capacity	Deliquescent[a]	Trade Name
$CaCl_2$ (anhydrous)	High	Yes	
$CaSO_4$	Moderate	No	Drierite (W. A. Hammond Drierite Co.)
CaO	Moderate	No	
$MgClO_4$ (anhydrous)	High	Yes	Anhydrone (J. T. Baker Chemical Co.); Dehydrite (Arthur H. Thomas Co.)
Silica gel	Low	No	
Al_2O_3	Low	No	
P_2O_5	Low	Yes	

[a] Becomes liquid by absorbing moisture. Take care of liquids generated. For example, P_2O_5 generates H_3PO_4.

A **drying oven** is used to dry samples prior to weighing. A typical drying oven is shown in Figure 2.14. These ovens are well ventilated for uniform heating. The usual drying temperature is about 110°C, but many laboratory ovens can be heated up to temperatures of 200 to 300°C.

Hoods

A **fume hood** is used when chemicals or solutions are to be evaporated. When perchloric acid or acid solutions of perchlorates are to be evaporated, the fumes should be collected, or the evaporation should be carried out in fume hoods specially designed for perchloric acid work (i.e., constructed from components resistant to attack by perchloric acid).

When performing trace analysis, as in trace metal analysis, care must be taken to prevent contamination. The conventional fume hood is one of the "dirtiest" areas of the laboratory since laboratory air is drawn into the hood and over the sample. **Laminar-flow hoods** or workstations are available for providing very clean work areas. Rather than drawing unfiltered laboratory air into the work area, the air is prefiltered and then flows over the work area and out into the room to create a positive pressure and prevent unfiltered air from flowing in. A typical laminar-flow workstation is shown in Figure 2.15. The high-efficiency particulate air (HEPA) filter removes all particles larger than 0.3 μm from the air. Vertical laminar-flow stations are preferred when fumes are generated that should not be blown over the operator. Facilities are available to exhaust noxious fumes. **Biohazard hoods** can also be found in many modern analytical and clinical laboratories. They are different from exhaust hoods. Their main purpose is to provide a safe place to work with potentially infectious material and particulates. They are often outfitted with high-intensity UV lamps that can be turned on prior to or after working (never during) in order to disinfect the workspace. All analysts should receive proper training prior to handling biohazard materials.

Centrifuges and Filters

A **centrifuge** has many useful applications, particularly in the clinical laboratory, where blood may have to be separated into fractions such as serum or plasma, and proteins may have to be separated by precipitation followed by centrifuging. Many laboratories also have an **ultracentrifuge**. Such an instrument has a larger capacity and can achieve higher speeds (gravitational force) in order to more readily separate sample components. Centrifugal filters are available with different molecular weight (MW) cutoff filter elements. They consist of disposable centrifuge tubes, separated by a horizontally placed filter element. A sample (typically biological) is put in the top compartment and the device is then subjected to centrifugation. Either the low MW material (often a clear filtrate) or the higher MW material retained by the filter can be of interest.

Filters for filtering precipitates (e.g., in gravimetric analysis) are of various types. The Gooch crucible, sintered-glass crucible, and porcelain filter crucible are illustrated in Figure 2.16. The **Gooch crucible** is porcelain and has holes in the bottom; a glass fiber filter disk is supported on top of it. The glass fiber filter disk will handle fine precipitates. The **sintered-glass crucible** contains a sintered-glass bottom, which is available in fine (F), medium (M), or coarse (C) porosity. The **porcelain filter crucible** contains a porous unglazed bottom. Glass filters are not recommended for concentrated alkali solutions because of the possibility of attack by these solutions. See Table 2.1 for maximum working temperatures for different types of crucible materials.

FIGURE 2.14 Drying oven. (Courtesy of Arthur H. Thomas Company.)

Laminar-flow hoods provide clean work areas.

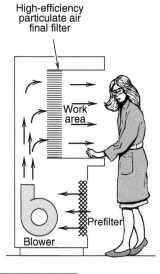

FIGURE 2.15 Laminar-flow workstation. (Courtesy of Dexion, Inc., 344 Beltine Boulevard, Minneapolis, MN.)

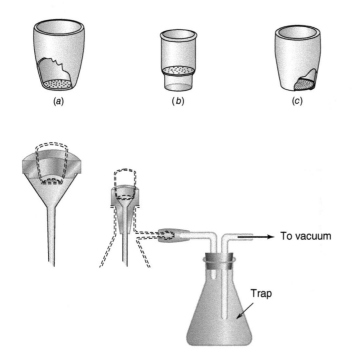

FIGURE 2.16 Filtering crucibles: (*a*) Gooch crucible; (*b*) sintered-glass crucible; (*c*) porcelain filter crucible.

To vacuum

Trap

FIGURE 2.17 Crucible holders.

Gelatinous precipitates such as hydrous iron oxide should not be filtered in filter crucibles because they clog the pores. Even with filter paper, the filtration of the precipitates can be slow.

Filter crucibles are used with a **crucible holder** mounted on a filtering flask (Figure 2.17). A safety bottle is connected between the flask and the aspirator.

Ashless filter paper is generally used for quantitative work in which the paper is ignited away and leaves a precipitate suitable for weighing (see Chapter 9). There are various grades of filter papers for different types of precipitates. These are listed in Table 2.6 for Whatman papers.

Techniques of Filtration

By proper fitting of the filter paper, the rate of filtration can be increased. A properly folded filter paper is illustrated in Figure 2.18. The filter paper is folded in the shape of a cone, with the overlapped edges of the two quarters not quite meeting (0.5 cm apart). About 1 cm is torn away from the corner of the inside edge. This will allow a good seal against the funnel to prevent air bubbles from being drawn in. After the folded paper is placed in the funnel, it is wetted with distilled water. The stem is filled with water and the top of the wet paper is pressed against the funnel to make a seal. With a proper fit, no air bubbles will be sucked into the funnel, and the suction supplied by the weight of

FIGURE 2.18 Properly folded filter paper.

TABLE 2.6 Whatman Filter Papers

Precipitate	Whatman No.
Very fine (e.g., $BaSO_4$)	50 (2.7 μm)
Small or medium (e.g., AgCl)	52 (7 μm)
Gelatinous or large crystals (e.g., $Fe_2O_3 \cdot xH_2O$)	54 (22 μm)

See http://www.whatman.com/QuantitativeFilterPapersHardenedLowAshGrades.aspx

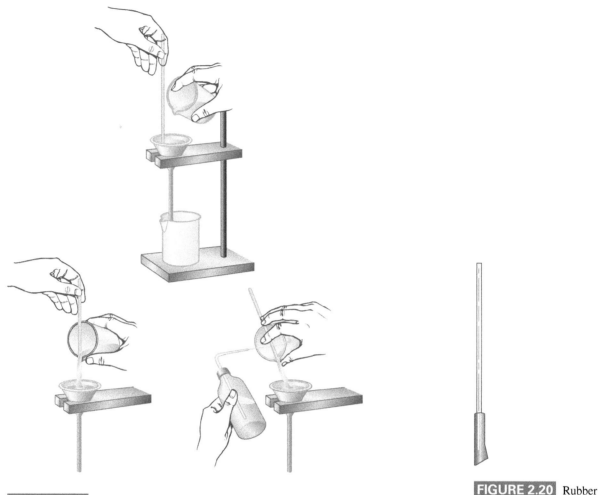

FIGURE 2.19   Proper technique for transfer of a precipitate.

FIGURE 2.20   Rubber policeman.

the water in the stem will increase the rate of filtration. The filtration should be started immediately. The precipitate should occupy not more than one-third to one-half of the filter paper in the funnel because many precipitates tend to "creep." Do not allow the water level to go over the top of the paper.

The precipitate should be allowed to settle in the beaker before filtration is begun. The bulk of the clear liquid can then be decanted and filtered at a rapid rate before the precipitate fills the pores of the filter paper.

Care must be taken in the decanting and the transferring of the precipitate to avoid losses. This is properly done by use of a stirring rod and a wash bottle, as illustrated in Figure 2.19. Note: the wash liquid is not distilled water—see Chapter 9. The solution is decanted by pouring it down the glass rod, which guides it into the filter without splashing. The precipitate is most readily washed while still in the beaker. After the mother liquor has been decanted off, wash the sides of the beaker down with several milliliters of the wash liquid, and then allow the precipitate to settle as before. Decant the wash liquid into the filter and repeat the washing operation two or three times. Then transfer the precipitate to the filter by holding the glass rod and beaker in one hand, as illustrated, and wash it out of the beaker with wash liquid from the wash bottle.

If the precipitate must be collected quantitatively, as in gravimetric analysis, the last portions of precipitate are removed by scrubbing the walls with a moistened **rubber policeman**, which contains a flexible rubber scraper attached to a glass rod (Figure 2.20). [For a description of the origin of its name, see J. W. Jensen, *J. Chem.*

Let the precipitate settle before filtering.

Wash the precipitate while it is in the beaker.

Ed., **85** (6) (2008) 776.] Wash the remainder of loosened precipitate from the beaker and from the policeman. If the precipitate is being collected in a filter paper, then instead of a rubber policeman, a small piece of the ashless filter paper can be rubbed on the beaker walls to remove the last bits of precipitate and added to the filter. This should be held with a pair of forceps.

After the precipitate is transferred to the filter, it is washed with five or six small portions of wash liquid. This is more effective than adding one large volume. Divert the liquid around the top edge of the filter to wash the precipitate down into the cone. Each portion should be allowed to drain before adding the next one. Check for completeness of washing by testing for the precipitating agent in the last few drops of the washings. Note: there will always be some precipitate in the wash due to finite solubility, i.e., finite K_{sp}, but it will be undetectable after sufficient washing.

Test for completeness of washing.

2.7 Igniting Precipitates—Gravimetric Analysis

If a precipitate is to be ignited in a porcelain filter crucible, the moisture should be driven off first at a low heat. The ignition may be done in a muffle furnace or by heating with a burner. If a burner is to be used, the filter crucible should be placed in a porcelain or platinum crucible to prevent reducing gases of the flame diffusing through the pores of the filter.

When precipitates are collected on filter paper, the cone-shaped filter containing the precipitate is removed from the funnel, the upper edge is flattened, and the corners are folded in. Then, the top is folded over and the paper and contents are placed in a crucible with the bulk of the precipitate on the bottom. The paper must now be dried and charred off. The crucible is placed at an angle on a triangle support with the crucible cover slightly ajar, as illustrated in Figure 2.21. The moisture is removed by low heat from the burner, with care taken to avoid splattering. The heating is gradually increased as moisture is evolved and the paper begins to char. Care should be taken to avoid directing the reducing portion of the flame into the crucible. A sudden increase in the volume of smoke evolved indicates that the paper is about to burst into flame, and the burner should be removed. If it does burst into flame, it should be smothered quickly by replacing the crucible cover. Carbon particles will undoubtedly appear on the cover, and these will ultimately have to be ignited. Finally, when no more smoke is detected, the charred paper is burned off by gradually increasing the flame temperature. The carbon residue should glow but should not flame. Continue heating until all the carbon and tars on the crucible and its cover are burned off. The crucible and precipitate are now ready for igniting. The ignition can be continued with the burner used at highest temperature or with the muffle furnace.

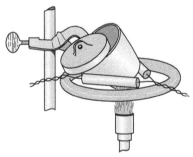

FIGURE 2.21 Crucible and cover supported on a wire triangle for charring off paper.

Before a precipitate is collected in a filter crucible or transferred to a crucible, the crucible should be dried to constant weight (e.g., 1 h of heating, followed by cooling and weighing and repeating the cycle) if the precipitate is to be dried, or it should be ignited to constant weight if the precipitate is to be ignited. Constant weight is considered to have been achieved with an analytical balance when successive weighings agree within about 0.3 or 0.4 mg. The crucible plus the precipitate are heated to constant weight in a similar manner. After the first heating, the time of heating can be reduced by half. The crucible should be allowed to cool in a desiccator for at least $\frac{1}{2}$ h before weighing. Red-hot crucibles should be allowed to cool below redness before placing them in the desiccator (use crucible tongs—usually nickel plated or stainless steel to minimize contamination from rust). Before weighing a covered crucible, check for any radiating heat by placing your hand near it (don't touch).

Do the initial ignition slowly.

Dry and weigh the crucible before adding the precipitate!

2.8   Obtaining the Sample—Is It Solid, Liquid, or Gas?

See Chapter 3 for important statistical considerations in sampling.

Collecting a representative sample is an aspect of analytical chemistry that the beginning analytical student is often not concerned with because the samples handed to him or her are assumed to be homogeneous and representative. Yet this process can be the most critical aspect of an analysis. The significance and accuracy of measurements can be limited by the sampling process. Unless sampling is done properly, it becomes the weak link in the chain of the analysis. A life could sometimes depend on the proper handling of a blood sample during and after sampling. If the analyst is given a sample and does not actively participate in the sampling process, then the results obtained can only be attributed to the sample "as it was received." And the chain of custody as mentioned earlier must be documented.

Many professional societies have specified definite instructions for sampling given materials [e.g., the American Society for Testing and Materials (ASTM: www.astm.org), the Association of Official Analytical Chemists International (AOAC International: www.aoac.org), and the American Public Health Association (APHA: www.apha.org)]. By appropriate application of experience and statistics, these materials can be sampled as accurately as the analysis can be performed. Often, however, the matter is left up to the analyst. The ease or complexity of sampling will, of course, depend on the nature of the sample.

The problem involves obtaining a sample that is representative of the whole. This sample is called the **gross sample**. Its size may vary from a few grams or less to several pounds, depending on the type of bulk material. Once a representative gross sample is obtained, it may have to be reduced to a sufficiently small size to be handled. This is called the **sample**. Once the sample is obtained, an aliquot, or portion, of it will be analyzed. This aliquot is called the **analysis sample**. Several replicate analyses on the same sample may be performed by taking separate aliquots.

Replication in sampling and analysis are key considerations.

In the clinical laboratory, the gross sample is usually satisfactory for use as the sample because it is not large and it is homogeneous (e.g., blood and urine samples). The analysis sample will usually be from a few milliliters to a fraction of a drop (a few microliters) in quantity.

Some of the problems associated with obtaining gross samples of solids, liquids, and gases are considered below.

1. **Solids.** Inhomogeneity of the material, variation in particle size, and variation within the particle make sampling of solids more difficult than other materials. The easiest but usually most unreliable way to sample a material is the **grab sample**, which is one sample taken at random and assumed to be representative. The grab sample will be satisfactory only if the material from which it is taken is homogeneous. For most reliable results, it is best to take 1/50 to 1/100 of the total bulk for the gross sample, unless the sample is fairly homogeneous. The larger the particle size, the larger the gross sample should be.

The easiest and most reliable time to sample large bodies of solid materials is while they are being moved. In this way any portion of the bulk material can usually be exposed for sampling. Thus, a systematic sampling can be performed to obtain aliquots representing all portions of the bulk. Some samples follow.

In the loading or unloading of bags of cement, a representative sample can be obtained by taking every fiftieth or so bag or by taking a sample from each bag. In the moving of grain by wheelbarrow, representative wheelbarrow loads or a shovelful from each wheelbarrow can be taken. All of these aliquots are combined to form the gross sample.

2. Liquids. Liquid samples tend to be homogeneous and representative samples are much easier to get. Liquids mix by diffusion only very slowly and must be shaken to obtain a homogeneous mixture. If the material is indeed homogeneous, a simple grab (single random) sample will suffice. For all practical purposes, this method is satisfactory for taking blood samples. The composition of some samples vary on when it is taken. This is the case for urine samples, Therefore 24-h urine sample collections are generally more representative than a single "spot sample".

The timing of sampling of biological fluids is, however, very important. The composition of blood varies considerably before and after meals, and for many analyses a sample is collected after the patient has fasted for a number of hours. Preservatives such as sodium fluoride for glucose preservation and anticoagulants may be added to blood samples when they are collected.

Blood samples may be analyzed as *whole blood*, or they may be separated to yield *plasma* or *serum* according to the requirements of the particular analysis. Most commonly, the concentration of the substance external to the red cells (the extracellular concentration) will be a significant indication of physiological condition, and so serum or plasma is taken for analysis.

If liquid samples are not homogeneous, and if they are small enough, they can be shaken and sampled immediately. For example, there may be particles in the liquid that have tended to settle. Large bodies of liquids are best sampled after a transfer or, if in a pipe, after passing through a pump when they have undergone thorough mixing. Large stationary liquids can be sampled with a *"thief" sampler*, which is a device for obtaining aliquots at different levels. It is best to take the sample at different depths at a diagonal, rather than straight down. The separate aliquots of liquids can be analyzed individually and the results combined, or the aliquots can be combined into one gross sample and replicate analyses performed. This latter procedure is probably preferred because the analyst will then have some idea of the precision of the analysis.

3. Gases. The usual method of sampling gases involves sampling into an evacuated container, often a specially treated stainless steel canister or an inert polyvinyl fluoride (Tedlar) bag is commonly used. The sample may be collected rapidly (a grab sample) or over a long period of time, using a small orifice to slowly fill the bag. A grab sample is satisfactory in many cases. To collect a breath sample, for example, the subject could blow into an evacuated bag or blow up a mylar balloon. Auto exhaust could be collected in a large evacuated plastic bag. The sample may be supersaturated with moisture relative to ambient temperature at which the sample container is. Moisture will condense in the sampling container after sample collection and the analyte of interest (e.g., ammonia in breath or nitrous acid in car exhaust) will be removed by the condensed moisture. The sample container must be heated and the sample transferred through a heated transfer line if the analyte is to be recovered.

The volume of gross gas sample collected may or may not need to be known. Often, the *concentration* of a certain analyte in the gas sample is measured, rather than the *amount*. The temperature and pressure of the sample will, of course, be important in determining the volume and hence the concentration.

Gas sampling techniques mentioned here does not concern gases dissolved in liquids, such CO_2 or O_2 in blood. These are treated as liquid samples and are then handled accordingly to measure the gas in the liquid or to release it from the liquid prior to measurement.

2.9 Operations of Drying and Preparing a Solution of the Analyte

After a sample has been collected, a solution of the analyte must usually be prepared before the analysis can be continued. Drying of the sample may be required, and it must be weighed or the volume measured. If the sample is already a solution (e.g., serum, urine, or water), then extraction, precipitation, or concentration of the analyte may be in order, and this may also be true with other samples.

In this section we describe common means for preparing solutions of inorganic and organic materials. Included are the dissolution of metals and inorganic compounds in various acids or in basic fluxes (fusion), the destruction of organic and biological materials for determination of inorganic constituents (using wet digestion or dry ashing), and the removal of proteins from biological materials so they do not interfere in the analysis of organic or inorganic constituents.

Drying The Sample

Solid samples will usually contain variable amounts of adsorbed water. With inorganic materials, the sample will generally be dried before weighing. This is accomplished by placing it in a drying oven at 105 to 110°C for 1 or 2 h. Other nonessential water, such as that entrapped within the crystals, may require higher temperatures for removal.

Decomposition or side reactions of the sample must be considered during drying. Thermally unstable samples can be dried in a desiccator; using a vacuum desiccator will hasten the drying process. A lyophilizer (freeze dryer) can also be used to remove a fairly large amount of water from a sample that contains thermally labile material; the sample must be frozen prior to placing it in a vessel within, or attached to, the apparatus. If the need to sample is weighed without drying, the results will be reported on an "as is" basis.

Plant and tissue samples can usually be dried by heating. See Chapter 1 for a discussion of the various weight bases (wet, dry, ash) used in connection with reporting analytical results for these samples.

Sample Dissolution

Before the analyte can be measured, some sort of sample manipulation is generally necessary to get the analyte into solution or, for biological samples, to rid it of interfering substances, such as proteins. Complex samples can be subjected to centrifugal filtration prior to analysis (e.g., perchlorate and iodide in milk have been chromatographically determined after centrifugal filtration). There are two types of sample preparation: those that totally destroy the sample matrix and those that are nondestructive or only partially destructive. The former type can generally be used only when the analyte is inorganic or can be converted to an inorganic derivative for measurement (e.g., Kjeldahl analysis, in which organic nitrogen is converted to ammonium ion—see below). Iodine in food is similarly determined after total oxidative digestion to HIO_3. Destructive digestion typically must be used if trace element analysis must be conducted in a largely organic matrix.

Dissolving Inorganic Solids

Strong mineral acids are good solvents for many inorganics. *Hydrochloric acid* is a good general solvent for dissolving metals that are above hydrogen in the electromotive series. *Nitric acid* is a strong oxidizing acid and will dissolve most of the common metals, nonferrous alloys, and the "acid-insoluble" sulfides.

Perchloric acid, when heated to drive off water, becomes a very strong and efficient oxidizing acid in the dehydrated state. It dissolves most common metals and destroys traces of organic matter. It must be used with extreme caution because it will react explosively with many easily oxidizable substances, especially organic matter.

Some instruments today are extraordinarily sensitive. An inductively coupled plasma mass spectrometer (ICP-MS) is essential for measuring trace impurities in semiconductor grade silicon, for example. The silicon sample is dissolved in a mixture of nitric and hydrofluoric acids before analysis. To carry out such ultra trace analysis the acids must also be ultra-pure. Such ultra-pure "semiconductor grade" acids are also very expensive.

> Fusion is used when acids do not dissolve the sample.

Some inorganic materials will not dissolve in acids, and **fusion** with an acidic or basic **flux** in the molten state must be used to solubilize them. The sample is mixed with the flux in a sample-to-flux ratio of about 1 to 10 or 20, and the combination is heated in an appropriate crucible until the flux becomes molten. When the melt becomes clear, usually in about 30 min, the reaction is complete. The cooled solid is then dissolved in dilute acid or in water. During the fusion process, insoluble materials in the sample react with the flux to form a soluble product. Sodium carbonate is one of the most useful basic fluxes, and acid-soluble carbonates are produced.

> Fusion is used when acids do not dissolve the sample.

Destruction of Organic Materials for Inorganic Analysis—Burning or Acid Oxidation

Animal and plant tissue, biological fluids, and organic compounds are usually decomposed by **wet digestion** with a boiling oxidizing acid or mixture of acids, or by **dry ashing** at a high temperature (400 to 700°C) in a muffle furnace. In wet digestion, the acids oxidize organic matter to carbon dioxide, water, and other volatile products, which are driven off, leaving behind salts or acids of the inorganic constituents. In dry ashing, atmospheric oxygen serves as the oxidant; that is, the organic matter is burned off, leaving an inorganic residue. Auxiliary oxidants (e.g., $NaNO_3$) can be used as aflux during dry ashing.

1. Dry Ashing. Although various types of dry ashing and wet digestion combinations are used with about equal frequency by analysts for organic and biological materials, simple dry ashing with no chemical aids is probably the most commonly employed technique. Lead, zinc, cobalt, antimony, chromium, molybdenum, strontium, and iron traces can be recovered with little loss by retention or volatilization. Usually a porcelain crucible can be used. Lead is volatilized at temperatures in excess of about 500°C, especially if chloride is present, as in blood or urine. Platinum crucibles are preferred for lead for minimal retention losses.

> In dry ashing, the organic matter is burned off.

If an oxidizing material is added to the sample, the ashing efficiency is enhanced. Magnesium nitrate is one of the most useful aids, and with this it is possible to recover arsenic, copper, and silver, in addition to the above-listed elements.

Liquids and wet tissues are dried on a steam bath or by gentle heat before they are placed in a muffle furnace. The heat from the furnace should be applied gradually up to full temperature to prevent rapid combustion and foaming.

After dry ashing is complete, the residue is usually leached from the vessel with 1 or 2 mL hot concentrated or 6 *M* hydrochloric acid and transferred to a flask or beaker for further treatment.

Another dry technique is that of **low-temperature ashing**. A radio-frequency discharge is used to produce activated oxygen radicals, which are very reactive and will attack organic matter at low temperatures. Temperatures of less than 100°C can be maintained, and volatility losses are minimized. Introduction of elements from the container and the atmosphere is reduced, and so are retention losses. Radiotracer studies have demonstrated that 17 representative elements are quantitatively recovered after complete oxidation of organic substrate.

Elemental analysis in the case of organic compounds (e.g., for carbon or hydrogen) is usually performed by **oxygen combustion** in a tube, followed by an absorption train. Oxygen is passed over the sample in a platinum boat, which is heated and quantitatively converts carbon to CO_2 and hydrogen to H_2O. These combustion gases pass into the absorption train, where they are absorbed in preweighed tubes containing a suitable absorbent. For example, **Ascarite** II is used to absorb the CO_2, and **Dehydrite** (magnesium perchlorate) is used to absorb the H_2O. The gain in weight of the absorption tubes is a measure of the CO_2 and H_2O liberated from the sample. Details of this technique are important, and, should you have occasion to use it, you are referred to more comprehensive texts on elemental analysis. Modern elemental analyzers are more automated and may be based on chromatographic separation of the combustion gases followed by detection with a thermal conductivity detector—Chapter 15 (see, e.g., http://www-odp.tamu.edu/publications/tnotes/tn30/tn30_10.htm).

In wet ashing, the organic matter is oxidized with an oxidizing acid.

2. Wet Digestion. Next to dry ashing, wet digestion with a mixture of nitric and sulfuric acids is the second most frequently used oxidation procedure. Usually a small amount (e.g., 5 mL) of sulfuric acid is used with larger volumes of nitric acid (20 to 30 mL). Wet digestions are usually performed in a Kjeldahl flask. The nitric acid destroys the bulk of the organic matter, but it does not get hot enough to destroy the last traces. It is boiled off during the digestion process until only sulfuric acid remains and dense, white SO_3 fumes are evolved and begin to reflux in the flask. At this point, the solution gets very hot, and the sulfuric acid acts on the remaining organic material. Charring may occur at this point if there is considerable or very resistant organic matter left. If the organic matter persists, more nitric acid may be added. Digestion is continued until the solution clears. All digestion procedures must be performed in a fume hood.

A much more efficient digestion mixture uses a mixture of nitric, perchloric, and sulfuric acids in a volume ratio of about 3:1:1. Ten milliliters of this mixture will usually suffice for 10 g fresh tissue or blood. The perchloric acid is an extremely efficient oxidizing agent when it is dehydrated and hot and will destroy the last traces of organic matter with relative ease. Samples are heated until nitric acid is boiled off and perchloric acid fumes appear; these, are less dense than SO_3 but fill the flask more readily. The hot perchloric acid is boiled, usually until fumes of SO_3 appear, signaling the evaporation of all the perchloric acid. Sufficient nitric acid must be added at the beginning to dissolve and destroy the bulk of organic matter, and there must be sulfuric acid present to prevent the sample from going to dryness, or else there is danger of explosion from the perchloric acid. A hood specially designed for perchloric acid work must be used for all digestions incorporating perchloric acid. Typically, a nitric-sulfuric acid digestion is first carried out to remove the more easily oxidizable material before digestion with the perchloric acid cocktail is carried out.

Perchloric acid digestion is even more efficient if a small amount of molybdenum(VI) catalyst is added. As soon as water and nitric acid are evaporated, oxidation proceeds vigorously with foaming, and the digestion is complete in a few seconds. The digestion time is reduced considerably.

Perchloric acid must be used with caution!

A mixture of nitric and perchloric acids is also commonly used. The nitric acid boils off first, and care must be taken to prevent evaporation of the perchloric acid to near dryness, or a violent explosion may result; this procedure *is not recommended* unless you have considerable experience in digestion procedures. **Perchloric acid**

should never be added directly to organic or biological material. Always add an excess of nitric acid first. Explosions with perchloric acid are generally associated with formation of peroxides, and the acid turns dark in color (e.g., yellowish brown) prior to explosion. Certain organic compounds such as ethanol, cellulose, and polyhydric alcohols can cause hot concentrated perchloric acid to explode violently; this is presumably due to formation of ethyl perchlorate.

A mixture of nitric, perchloric, and sulfuric acids allows zinc, selenium, arsenic, copper, cobalt, silver, cadmium, antimony, chromium, molybdenum, strontium, and iron to be quantitatively recovered. Lead is often lost if sulfuric acid is used. The mixture of nitric and perchloric acids can be used for lead and all the above elements. Perchloric acid must be present to prevent losses of selenium. It maintains strong oxidizing conditions and prevents charring that would result in formation of volatile compounds of lower oxidation states of selenium. Samples containing mercury cannot be dry ashed. Wet digestion that involves the simultaneous application of heat must be done using a reflux apparatus because of the volatile nature of mercury and its compounds. Cold or room temperature procedures are often preferred to obtain partial destruction of organic matter. For example, in urine samples, which contain a relatively small amount of organic matter compared with blood, mercury can be reduced to the element with copper(I) and hydroxylamine hydrochloride and the organic matter destroyed by potassium permanganate at room temperature. The mercury can then be dissolved and the analysis continued. Urinary mercury can be mineralized using Fenton's reagent (Fe(II) and H_2O_2) and determined as the elemental vapor after addition of sodium borohydride, a powerful reducing agent.

Many nitrogen-containing compounds can be determined by **Kjeldahl digestion** to convert the nitrogen to ammonium sulfate. The digestion mixture consists of sulfuric acid plus potassium sulfate to increase the boiling point of the acid and thus increase its efficiency. A catalyst is also added (such as copper or selenium). After destruction of the organic matter, sodium hydroxide is added to make the solution alkaline, and the ammonia is distilled into an excess of standard hydrochloric acid. The excess acid is back-titrated with standard alkali to determine the amount of ammonia collected. With a knowledge of the percent nitrogen composition in the compound of interest, the amount of the compound can be calculated from the amount of ammonia determined. This is the most accurate method for determining protein content. Protein contains a definite percentage of nitrogen, which is converted to ammonium sulfate during the digestion. See Chapter 7 for further details. Note: of course, if other nitrogen-containing species are present, the nitrogen determination will not accurately reflect the protein content. This was amply demonstrated in China when melamine, an inexpensive organic amine base containing six nitrogen atoms, was added to milk formula to boost the apparent "protein" content. This caused the death of many infants in China.

In Kjeldahl digestions, nitrogen is converted to ammonium ion, which is then distilled as ammonia and titrated.

The relative merits of various oxidation methods have been studied extensively. However, there may be no universal and generally applicable dry ashing method. Dry ashing is recommended for its simplicity and relative freedom from positive errors (contamination) since few or no reagents are added. The potential errors of dry oxidation are volatilization of elements and losses by retention on the walls of the vessel. Adsorbed metals on the vessel may in turn contaminate future samples. Wet digestion is considered superior in terms of rapidity (although it does require more operator attention), low level of temperature maintained, and freedom from loss by retention. The chief error attributed to wet digestion is the introduction of impurities from the reagents necessary for the reaction. This problem has been minimized as commercial reagent-grade acids have become available in greater purity and specially prepared high-purity acids can now be obtained commercially, albeit not inexpensively. The time

Dry and wet ashing each has advantages and limitations.

required for ashing or digestion will vary with the sample and the technique employed. Two to four hours are common for dry ashing and half to one hour is common for wet digestion.

Microwave Preparation of Samples

Microwave ovens are now widely used for rapid and efficient drying and acid decomposition of samples. Laboratory microwave ovens are specially designed to overcome limitations of household ovens, and these are discussed below. Advantages of microwave digestions include reduction of dissolution times from hours to minutes and low blank levels due to reduced amounts of reagents required.

1. How Do Microwaves Heat? The microwave region is between infrared radiation and radio waves in the electromagnetic spectrum, in the frequency range of 300 to 300,000 MHz (3×10^8 to 3×10^{11} Hz, or beginning at about 1000 μm wavelength—see Figure 11.2). Microwaves consist of an electric field and a magnetic field perpendicular to the electric field. The electric field is responsible for energy transfer between the microwave source and the irradiated sample. Microwave energy affects molecules in two ways: dipole rotation and ionic conduction. The first is the more generally important. When the microwave energy passes through the sample, molecules with non-zero dipole moments will try to align with the electric field, and the more polar ones will have the stronger interaction with the field. This molecular motion (rotation) results in heating. The energy transfer, a function of the dipole moment and the dielectric constant, is most efficient when the molecules are able to relax quickly, that is, when the relaxation time matches the microwave frequency. Large molecules such as polymers relax slowly, but once the temperature increases and they relax more rapidly, they can absorb the energy more efficiently. Small molecules such as water, though, relax more quickly than the resonating microwave energy, and they move farther away from the resonance frequency and absorb less energy as they heat up.

The ionic conduction effect arises because ionic species in the presence of an electric field will migrate in one direction or the other. Energy is transferred from the electric field, causing ionic interactions that speed up the heating of a solution. Ionic absorbers become stronger absorbers of microwave energy as they are heated since ionic conductance increases with temperature. Deionized water heats slowly, but if salt is added, it heats rapidly. Acids, of course, are good conductors and heat rapidly.

Microwaves heat by causing molecules to rotate and ions to migrate.

So microwave energy heats by causing movement of molecules due to dipole rotation and movement of ions due to ionic conductance. The microwave energy interacts with different materials in different ways. Reflective materials such as metals are good heat conductors: They do not heat and instead will reflect the microwave energy. (It is not good practice to put metals in microwave ovens due to electrical discharge occurring between one very highly charged metal segment to another.) Transparent materials are insulators because they transmit the microwave energy and also do not heat. The absorptive materials, the molecules and ions discussed above, are the ones that receive microwaves and are heated. Microwave energy is too low to break chemical bonds (a feature that has generated interest in using microwave energy to speed up chemical reactions in syntheses). The properties of reflective and insulator materials are utilized in designing microwave digestion systems.

Household microwave ovens don't work for small sample heating.

2. Design of Laboratory Microwave Ovens. Home microwave ovens were initially used for laboratory purposes, but it soon became apparent that modifications were needed. Laboratory samples are usually much smaller than food samples that are cooked and absorb only a small fraction of the energy produced by the magnetron of

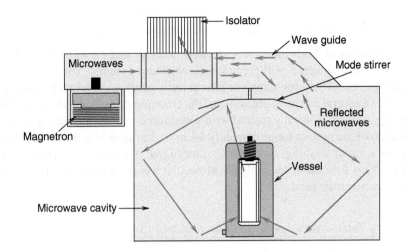

FIGURE 2.22 Schematic of a microwave system. [From G. Le Blanc, *LC/GC Suppl.*, **17**(6S) (1999) S30.] (Courtesy of *LC/GC Magazine*).

the oven. The energy not absorbed by the sample is bounced back to the magnetron, causing it to overheat and burn out. Also, arcing could occur. So laboratory ovens are designed to protect the magnetron from stray energy. The main components of these ovens (Figure 2.22) include the magnetron, an isolator, a waveguide, the cavity, and a mode stirrer. Microwave energy generated by the magnetron is propagated down the waveguide into the cavity. The stirrer distributes the energy in different directions. The isolator, made of a ferromagnetic material and placed between the magnetron and the waveguide, deflects the microwave energy returning from the cavity into a fan-cooled ceramic load, keeping it away from the magnetron.

The frequency used for cooking turns out to be good for chemistry as well, and the standard is 2450 MHz. Powers of 1200 W are typically used.

3. Acid Digestions. Digestions are normally done in closed plastic containers, either Teflon PFA (perfluoroalkoxy ethylene) or polycarbonate (insulators). This is to avoid acid fumes in the oven. It provides additional advantages. Pressure is increased and the boiling point of the acid is raised (the acid is superheated). So digestions occur more rapidly. Also, volatile metals are not lost. Modern ovens provide for control of pressure and temperature. Fiber-optic temperature probes are used that are transparent to microwave energy. Temperature control has enabled the use of the oven for microwave-assisted molecular extractions, by maintaining the temperature low enough to avoid molecular decomposition.

Partial Destruction or Nondestruction of Sample Matrix

Obviously, when the substance to be determined is organic in nature, nondestructive means of preparing the sample must be used. For the determination of metallic elements, it is also sometimes unnecessary to destroy the molecular structure of the sample, particularly with biological fluids. For example, several metals in serum or urine can be determined by atomic absorption spectroscopy by direct aspiration of the sample or a diluted sample into a flame. Constituents of solid materials such as soils can sometimes be extracted by an appropriate reagent. Thorough grinding, mixing, and refluxing are necessary to extract the analyte. Many trace metals can be extracted from soils with 1 *M* ammonium chloride or acetic acid solution. Some, such as selenium, can be distilled as the volatile chloride or bromide.

Protein-Free Filtrates

Proteins in biological fluids interfere with many analyses and must be removed nondestructively. Several reagents will precipitate (coagulate) proteins. Trichloroacetic acid (TCA), tungstic acid (sodium tungstate plus sulfuric acid), and barium hydroxide plus zinc sulfate (a neutral mixture) are some of the common ones. A measured volume of sample (e.g., serum) is usually treated with a measured volume of reagent. Following precipitation of the protein (approximately 10 min), the sample is filtered through dry filter paper without washing, or else it is centrifuged. An aliquot of the **protein-free filtrate** (PFF) is then taken for analysis. Molecular weight selective centrifugal filtration can sometimes be used.

Laboratory Techniques for Drying and Dissolving

Take care in drying or dissolving samples.

When a solid sample is to be dried in a weighing bottle, the cap is removed from the bottle and, to avoid spilling, both are placed in a beaker and covered with a ribbed watch glass. Some form of identification should be placed on the beaker.

The weighed sample may be dissolved in a beaker or Erlenmeyer flask. If there is any fizzing action, cover the vessel with a watch glass. After dissolution is complete, wash the walls of the vessel down with distilled water. Also wash the watch glass so the washings fall into the vessel. You may have to evaporate the solution to decrease the volume. This is best done by covering the beaker with a ribbed watch glass to allow space for evaporation. Low heat should be applied to prevent bumping; a steam bath or variable-temperature hot plate is satisfactory.

Use of a **Kjeldahl flask** for dissolution will avoid some of the difficulties of splattering or bumping. Kjeldahl flasks are also useful for performing digestions. They derive their name from their original use in digesting samples for Kjeldahl nitrogen analysis. They are well suited to all types of wet digestions of organic samples and acid dissolution of metals. Kjeldahl flasks come in assorted sizes from 10 to 800 mL. Some of these are shown in Figure 2.23. The sample and appropriate acids are placed in the round bottom of the flask and the flask is tilted while it is heated. In this way the acid can be boiled or refluxed with little danger of loss by "bumping." The flask may be heated with a flame or in special electrically heated Kjeldahl digestion racks, which heat several samples simultaneously.

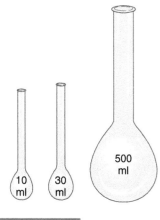

FIGURE 2.23 Kjeldahl flasks.

2.10 Laboratory Safety

You *must* familiarize yourself with laboratory safety rules and procedures before conducting experiments! Read Appendix D and the material provided by your instructor. Get a free copy of Reference 31.

Before beginning any of the experiments, you must familiarize yourself with laboratory safety procedures. You should read this material before beginning experiments. Your instructor will provide you with specific guidelines and rules for operation in the laboratory and the disposal of chemicals. For a more complete discussion of safety in the laboratory, you are referred to *Safety in Academic Chemistry Laboratories*, published by The American Chemical Society (Reference 31). This guide discusses personal protection and laboratory protocol, recommended laboratory techniques, chemical hazards, instructions on reading and understanding material safety data sheets (MSDSs), and safety equipment and emergency procedures. Rules are given for waste disposal, waste classification terminology, Occupational Safety and Health Administration (OSHA) laboratory standards for exposures to hazardous chemicals, and EPA requirements. The handling and treatment of inorganic and organic peroxides are discussed in detail, and an extensive list of incompatible chemicals is given, and maximum allowable container capacities for flammable and combustible liquids are listed. This resourceful booklet is recommended reading for students and instructors. It is available for free (one copy) from The American Chemical Society, Washington, DC (1-800-227-5558).

The Waste Management Manual for Laboratory Personnel, also published by The American Chemical Society, provides an overview of government regulations (Reference 32).

 # Professor's Favorite Example

┌Contributed by Professor
Akos Vertes, George
Washington University ┘

The Shroud of Turin Dating—A Sampling Problem

Various means have been enlisted to ascertain the validity and age of the Shroud of Turin (http://en.wikipedia.org/wiki/Shroud_of_Turin). One such study was radiocarbon dating done in 1988. Samples were obtained to be given to three separate laboratories for independent analyses. Details are given in: http://www.shroud.com/nature.htm. The paper describes in detail how these important samples were obtained and different sample preparation procedures used by the different laboratories. Three control samples of other known ancient textiles were treated and analyzed in the same way. Check the paper to see the results!

Is the controversy concluded?

Questions

1. Describe the principle and operation of the analytical balance.
2. Why is a microbalance more sensitive than an analytical balance?
3. What is zero-point drift? Explain the main reason of drift.
4. Describe the sources of errors in weighing.
5. What does TD on glassware mean? TC?
6. Describe the preparation of a standard HCl solution and a standard NaOH solution.
7. Why sodium carbonate must be removed from NaOH to achieve sharp endpoint? How it can be removed?
8. Describe the principles of dry ashing and wet digestion of organic and biological materials. List the advantages of each.
9. Perchloric acid should never be added directly to organic or biological materials in digestion procedures. Why?
10. What are the two principal means of dissolving inorganic materials?
11. Arrange the given common drying agents according to their highest to lowest capacity: $Al_2O_3, CaCl_2, CaSO_4$.
12. Match the correct pair of Whatman filter paper with the precipitate or metal.
    (a) Whatman no. 50   (i) AgCl
    (b) Whatman no. 52   (ii) $Fe_2O_3 \cdot xH_2O$
    (c) Whatman no. 54   (iii) $BaSO_4$
13. What is low temperature ashing?
14. Explain elemental analysis in organic compounds for carbon and hydrogen.
15. Differentiate among a gross sample, sample, analysis sample and grab sample.
16. What happens when microwave energy heats samples?
17. What is the frequency range of microwave range? Where does it exist in electromagnetic spectrum?
18. Why ionic conduction effect arises due to microwaves?

Problems

Glassware Calibration/Temperature Corrections

19. You calibrate a 25-mL volumetric flask by filling to the mark with distilled water, equilibrated at 22°C. The dry stoppered flask weighs 27.278 g and the filled flask and stopper is 52.127 g. The balance uses stainless steel weights. What is the volume of the flask? What is it at the standard 20°C. Also insert the weight in air at 22°C into Table 2.4 (available on the textbook website), and compare the values obtained.

20. You calibrate a 25-mL pipet at 25°C using steel weights. The weight of the delivered volume of water is 24.971 g. What is the volume of the pipet at 25 and 20°C?

21. You calibrate a 50-mL buret in the winter time at 20°C, with the following corrections:

Buret Reading (mL)	Correction (mL)
10	+0.02
20	+0.03
30	0.00
40	−0.04
50	−0.02

You use the buret on a hot summer day at 30°C. What are the corrections then?

22. You prepare a standard solution at 21°C, and use it at 29°C. If the standardized concentration is 0.05129 M, what is it when you use the solution?

PROFESSOR'S FAVORITE PROBLEMS

Contributed by Professor Bin Wang, Marshall University

23. For an electronic analytical balance with 0.1 mg readability, what is the maximum mass that can be weighed (capacity)?
    (a) 500–1000 g;
    (b) 100–300 g;
    (c) 10–20 g;
    (d) several g or less

24. The densities of air, calibration weights, and salt are 0.0012 g/mL, 7.8 g/cm³, and 2.16 g/mL, respectively. If the apparent mass of salt (i.e., NaCl) weighed in air is 15.914 g, what is its true mass?

Multiple Choice Questions

1. Choose the correct option.

Material	Working Temperature
(p) Borosilicate glass	(i) 1500
(q) Platinum	(ii) 250
(r) Teflon	(iii) 200
(s) Fused quartz	(iv) 1050

    (a) (p) – (i); (q) – (ii); (r) – (iii); (s) – (iv) (b) (p) – (iii); (q) – (i); (r) – (ii); (s) – (iv)
    (c) (p) – (iii); (q) – (ii); (r) – (i); (s) – (iv) (d) (p) – (ii); (q) – (i); (r) – (iii); (s) – (iv)

2. Which of the following is used for direct weighing?
    (a) Weighing dish
    (b) Weighing boat
    (c) Weighing paper
    (d) All of these

3. Choose which of the following is used as secondary standard during titration.
    (a) Na_2CO_3
    (b) NaOH
    (c) HCl
    (d) NH_3

Reagents and Standards

1. *Reagent Chemicals, Specifications and Procedures*, 10th ed. Washington, DC: American Chemical Society, 2005.
2. *Dictionary of Analytical Reagents*. London: Chapman & Hall/CRC, 1993. Data on over 14,000 reagents.
3. J. R. Moody and E. S. Beary, "Purified Reagents for Trace Metal Analysis," *Talanta*, **29** (1982) 1003. Describes sub-boiling preparation of high-purity acids.
4. R. C. Richter, D. Link, and H. M. Kingston, "On-Demand Production of High-Purity Acids in the Analytical Laboratory," *Spectroscopy*, **15**(1) (2000) 38. www.spectroscopyonline.com. Describes preparation using a commercial sub-boiling system.
5. www.thornsmithlabs.com. Thorn Smith Laboratories provides prepackaged materials for student analytical chemistry experiments (unknowns and standards); www.sigma-aldrich.com/analytical. A supplier of standard solutions.

Analytical Balances

6. D. F. Rohrbach and M. Pickering, "A Comparison of Mechanical and Electronic Balances," *J. Chem. Ed.*, **59** (1982) 418.
7. R. M. Schoonover, "A Look at the Electronic Analytical Balance," *Anal. Chem.*, **54** (1982) 973A.
8. J. Meija, "Solution to Precision Weighing Challenge," *Anal. Bioanal. Chem.*, **394** (2009) 11. Discusses, in detail, quantitative corrections for minor variations in the apparent mass of a stainless steel weight relative to a platinum weight as air density changes (the apparent mass of the stainless steel weight increases relative to that of the platinum weight as the air density is reduced).

Calibration of Weights

9. W. D. Abele, "Laboratory Note: Time-Saving Applications of Electronic Balances," *Am. Lab.*, **13** (1981) 154. Calibration of weights using mass standards calibrated by the National Institute of Standards and Technology is discussed.
10. D. F. Swinehart, "Calibration of Weights in a One-Pan Balance," *Anal. Lett.*, **10** (1977) 1123.

Calibration of Volumetric Ware

11. G. D. Christian, "Coulometric Calibration of Micropipets," *Microchem. J.*, **9** (1965) 16.
12. M. R. Masson, "Calibration of Plastic Laboratory Ware," *Talanta*, **28** (1981) 781. Tables for use in calibration of polypropylene vessels are presented.
13. W. Ryan, "Titrimetric and Gravimetric Calibration of Pipettors: A Survey," *Am. J. Med. Technol.*, **48** (1982) 763. Calibration of pipettors of 1 to 500 μL is described.
14. M. R. Winward, E. M. Woolley, and E. A. Butler, "Air Buoyancy Corrections for Single-Pan Balances," *Anal. Chem.*, **49** (1977) 2126.
15. R. M. Schoonover and F. E. Jones, "Air Buoyancy Corrections in High-Accuracy Weighing on Analytical Balances," *Anal. Chem.*, **53** (1981) 900.

Clean Laboratories

16. J. R. Moody, "The NBS Clean Laboratories for Trace Element Analysis," *Anal. Chem.*, **54** (1982) 1358A.

Sampling

17. J. A. Bishop, "An Experiment in Sampling," *J. Chem. Ed.*, **35** (1958) 31.
18. J. R. Moody, "The Sampling, Handling and Storage of Materials for Trace Analysis," *Phil. Trans. Roy. Soc.* London, *Ser. A*, **305** (1982) 669.
19. G. E. Baiulescu, P. Dumitrescu, and P. Gh. Zugravescu, *Sampling*. Chichester: Ellis Horwood, 1991.

Sample Preparation and Dissolution

20. G. D. Christian, "Medicine, Trace Elements, and Atomic Absorption Spectroscopy," *Anal. Chem.*, **41** (1) (1969) 24A. Describes the preparation of solutions of biological fluids and tissues.
21. G. D. Christian, E. C. Knoblock, and W. C. Purdy, "Polarographic Determination of Selenium in Biological Materials," *J. Assoc. Offic. Agric. Chemists*, **48** (1965) 877; R. K. Simon, G. D. Christian, and W. C. Purdy, "Coulometric Determination of Arsenic in Urine," *Am. J. Clin. Pathol.*, **49** (1968) 207. Describes use of Mo(VI) catalyst in digestions.
22. T. T. Gorsuch, "Radiochemical Investigations on the Recovery for Analysis of Trace Elements in Organic and Biological Materials," *Analyst*, **84** (1959) 135.
23. S. Nobel and D. Nobel, "Determination of Mercury in Urine," *Clin. Chem.*, **4** (1958) 150. Describes room temperature digestion.
24. G. Knapp, "Mechanical Techniques for Sample Decomposition and Element Preconcentration," *Mikrochim. Acta*, **II** (1991) 445. Lists acid mixtures and uses for digestions and other decomposition methods.

Microwave Digestion

25. H. M. Kingston and L. B. Jassie, eds., *Introduction to Microwave Sample Preparation: Theory and Practice*. Washington, D.C.: American Chemical Society, 1988.
26. H. M. Kingston and S. J. Haswell, eds., *Microwave-Enhanced Chemistry: Fundamentals, Sample Preparation, and Applications*. Washington, DC: American Chemical Society, 1997.
27. H. M. Kingston and L. B. Jassie, "Microwave Energy for Acid Decomposition at Elevated Temperatures and Pressures Using Biological and Botanical Samples," *Anal. Chem.*, **58** (1986) 2534.
28. R. A. Nadkarni, "Applications of Microwave Oven Sample Dissolution in Analysis," *Anal. Chem.*, **56** (1984) 2233.
29. B. D. Zehr, "Development of Inorganic Microwave Dissolutions," *Am. Lab.*, December (1992) 24. Also lists properties of useful acid mixtures.
30. www.cem.com. CEM is a manufacturer of laboratory microwave ovens.

Laboratory Safety

31. *Safety in Academic Chemistry Laboratories*, 7th ed., Vol. 1 (Student), Vol. 2 (Teacher), American Chemical Society, Committee on Chemical Safety. Washington, DC: American Chemical Society, 2003. Available online at http://portal.acs.org/portal/Public WebSite/about/governance/committees/chemicalsafety/publications/WPCP_012294 (vol. 1) and http://portal.acs.org/portal/PublicWebSite/about/governance/committees/chemicalsafety/publications/WPCP_012293 (vol. 2).
32. *The Waste Management Manual for Laboratory Personnel*, Task Force on RCRA, American Chemical Society, Department of Government Relations and Science Policy. Washington, DC: American Chemical Society, 1990.
33. A. K. Furr, ed., *CRC Handbook of Laboratory Safety*, 4th ed. Boca Raton, FL: CRC Press, 1993.
34. R. H. Hill and D. Finster, *Laboratory Safety for Chemistry Students*. Hoboken, NJ: Wiley, 2010.
35. M.-A. Armour, *Hazardous Laboratory Chemicals Disposal Guide*. Boca Raton, FL: CRC Press, 1990.

Material Safety Data Sheets

36. http://siri.org/msds. Online searchable database, with links to other MSDS and hazardous chemical sites.
37. www.env-sol.com. Solutions Software Corporation. MSDS database available on DVD or CD-ROM.

Chapter 3

ANALYTICAL CHEMISTRY: STATISTICS AND DATA HANDLING

"Facts are stubborn, but statistics are much more pliable."
—Mark Twain

"43.8% of all statistics are worthless."
—Anonymous

"Oh, people can come up with statistics to prove anything. 14% of people know that."
—Homer Simpson

KEY THINGS TO LEARN FROM THIS CHAPTER

- Accuracy and precision
- Types of errors in measurements
- Significant figures in measurements and calculations
- Absolute and relative uncertainty
- Standard deviation
- Propagation of errors
- Control charts
- Statistics: confidence limits, *t*-tests, *F*-tests

- Rejection of a result
- Least-squares plots and coefficient of determination
- Detection limits
- Statistics of sampling
- Power analysis
- How to use spreadsheets
- Using spreadsheets for plotting calibration curves

Although data handling normally follows the collection of analytical data, it is treated early in this textbook because a knowledge of statistical analysis will be required as you perform experiments in the laboratory. Also, statistical analysis is necessary to understand the significance of the data that are collected and thus sets limits on each step of the analysis. Experimental design (including required sample size, measurement accuracy, and number of analyses needed) relies on a proper understanding of what the data represent.

The availability of spreadsheets to process data has made statistical and other calculations very efficient. You will first be presented with the details of various calculations throughout the text, which are necessary for full understanding of the principles. A variety of pertinent spreadsheet calculations will also be introduced at the end of the chapter to illustrate how to take advantage of this software for routine calculations.

3.1 Accuracy and Precision: There Is a Difference

Accuracy is the degree of agreement between the measured value and the true value. An absolute true value is seldom known. A more realistic definition of accuracy, then, would assume it to be the agreement between a measured value and the *accepted* true value.

Accuracy is how close you get to the bull's-eye. *Precision* is how close the repetitive shots are to one another. It is nearly impossible to have accuracy without good precision.

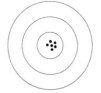  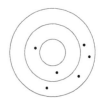

FIGURE 3.1 Accuracy versus precision.

Good precision, good accuracy   Good precision, poor accuracy   Poor precision, poor accuracy (could be accurate if shots symmetrical)

We can, by good analytical technique, such as making comparisons against a known standard sample of similar composition, arrive at a reasonable assumption about the accuracy of a method, within the limitations of the knowledge of the "known" sample (and of the measurements). The accuracy to which we know the value of the standard sample is ultimately dependent on some measurement that will have a given limit of certainty in it.

Good precision does not guarantee accuracy.

Precision is defined as the degree of agreement between replicate measurements of the same quantity. That is, it is the repeatability of a result. The precision may be expressed as the standard deviation, the coefficient of variation, the range of the data, or as a confidence interval (e.g., 95%) about the mean value. Good precision does not assure good accuracy. This would be the case, for example, if there were a systematic error in the analysis. The volume of a pipet used to dilute each of the samples may be in error. This error does not affect the precision, but it does affect the accuracy. On the other hand, the precision can be relatively poor and the accuracy may be good; admittedly, this is very rare. Since all real analyses are unknown, the higher the degree of precision, the greater the chance of obtaining the true value. It is fruitless to hope that a value is accurate despite the precision being poor; and the analytical chemist strives for repeatable results to assure the highest possible accuracy.

"To be sure of hitting the target, shoot first, and call whatever you hit the target."—Ashleigh Brilliant

These concepts can be illustrated with targets, as in Figure 3.1. Suppose you are at target practice and you shoot the series of bullets that all land in the bull's-eye (left target). You are both precise and accurate. In the middle target, you are precise (steady hand and eye), but inaccurate. Perhaps the sight on your gun is out of alignment. In the right target you are imprecise and therefore probably inaccurate. So we see that good precision is needed for good accuracy, but it does not guarantee it.

As we shall see later, reliability increases with the number of measurements made. The number of measurements required will depend on the level of uncertainty that is acceptable and on the known reproducibility of the method.

3.2 Determinate Errors—They Are Systematic

Determinate or systematic errors are nonrandom and occur when something is intrinsically wrong in the measurement.

Two main classes of errors can affect the accuracy or precision of a measured quantity. **Determinate errors** are those that, as the name implies, are determinable and that presumably can be either avoided or corrected. They may be constant, as in the case of an uncalibrated pipet that is used in all volume deliveries. Or, they may be variable but of such a nature that they can be accounted for and corrected, such as a buret whose volume readings are in error by different amounts at different volumes.

The error can be proportional to sample size or may change in a more complex manner. More often than not, the variation is unidirectional, as in the case of solubility loss of a precipitate due to its solubility (negative error). It can, however, be random in sign, i.e., a positive or negative error. Such an example is the change in solution volume and concentration occurring with changes in temperature. This can be corrected for by measuring the solution temperature. Such measurable determinate errors are classed as **systematic errors**.

Some common determinate errors are:

1. *Instrumental errors.* These include faulty equipment such as uncalibrated glassware.

2. *Operative errors.* These include personal errors and can be reduced by experience and care of the analyst in the physical manipulations involved. Operative errors can be minimized by having a checklist of operations. Operations in which these errors may occur include transfer of solutions, effervescence and "bumping" during sample dissolution, incomplete drying of samples, and so on. These are difficult to correct for. Other personal errors include mathematical errors in calculations and prejudice in estimating measurements.

3. *Errors of the method.* These are the most serious errors of an analysis. Most of the above errors can be minimized or corrected for, but errors that are inherent in the method cannot be changed unless the conditions of the determination are altered. Some sources of methodical errors include coprecipitation of impurities, slight solubility of a precipitate, side reactions, incomplete reactions, and impurities in reagents. Sometimes correction can be relatively simple, for example, by running a **reagent blank**. A blank determination is an analysis on the added reagents only. It is standard practice to run such blanks and to subtract the results from those for the sample. But a good blank analysis alone cannot guarantee correct measurements. If the method, for example, responds to an analyte present in the sample other than the intended analyte, the method must be altered. Thus, when errors become intolerable, another approach to the analysis must be made. Sometimes, however, we are forced to accept a given method in the absence of a better one.

It is always a good idea to run a blank.

Determinate errors may be *additive* or *multiplicative*, depending on the nature of the error or how it enters into the calculation. In order to detect systematic errors in an analysis, it is common practice to add a known amount of standard to a sample (a "spike") and measure its recovery (see Validation of a Method in Chapter 1) and note that good spike recovery cannot also correct for response from an unintended analyte (i.e., an interference). The analysis of reference samples helps guard against method errors or instrumental errors.

3.3 Indeterminate Errors—They Are Random

The second class of errors includes the **indeterminate errors**, often called accidental or random errors, which represent the experimental uncertainty that occurs in any measurement. These errors are revealed by small differences in successive measurements made by the same analyst under virtually identical conditions, and they cannot be predicted or estimated. These accidental errors will follow a random distribution; therefore, mathematical laws of probability can be applied to arrive at some conclusion regarding the most probable result of a series of measurements.

Indeterminate errors are random and cannot be avoided.

It is beyond the scope of this text to go into mathematical probability, but we can say that indeterminate errors should follow a **normal distribution**, or **Gaussian curve**. Such a curve is shown in Figure 3.2. The symbol σ represents the *standard deviation* of an infinite population of measurements, and this measure of precision defines the spread of the normal population distribution as shown in Figure 3.2. It is apparent that there should be few very large errors and that there should be an equal number of positive and negative errors.

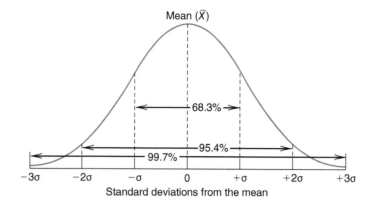

FIGURE 3.2 Normal error curve.

Indeterminate errors really originate in the limited ability of the analyst to control or make corrections for external conditions, or the inability to recognize the appearance of factors that will result in errors. Some random errors stem from the intrinsic nature of things, for example, consider that a sample of the radionuclide ^{129}I is taken. The isotope is long lived, and in a short time there will not be a perceptible change in its number. But, if a sufficient amount is taken, then based on the half-life, you can expect a decay to occur every 60 s. In reality, this may not occur every 60 s, but can fluctuate, with an average of 60 s. Sometimes, by changing conditions, some unknown error will disappear. Of course, it will be impossible to eliminate all possible random errors in an experiment, and the analyst must be content to minimize them to a tolerable or insignificant level.

3.4 Significant Figures: How Many Numbers Do You Need?

The last digit of a measurement has some uncertainty. You can't include any more digits.

The weak link in the chain of any analysis is the measurement that can be made with the least accuracy or precision. It is useless to extend an effort to make the other measurements of the analysis more accurately than this limiting measurement. The number of significant figures can be defined as **the number of digits necessary to express the results of a measurement consistent with the measured precision**. Since there is uncertainty (imprecision) in any measurement of at least ±1 in the last significant figure, the number of significant figures includes all of the digits that are known, plus the first uncertain one. In reported answers, it generally does not make sense to include additional digits beyond the first uncertain one. Each digit denotes the actual quantity it specifies. For example, in the number 237, we have 2 hundreds, 3 tens, and 7 units. If this number is reported as a final answer, it implies that uncertainty lies in the units digit (e.g., ±1).

The digit 0 can be a significant part of a measurement, or it can be used merely to place the decimal point. The number of significant figures in a measurement is independent of the placement of the decimal point. Take the number 92,067. This number has five significant figures, regardless of where the decimal point is placed. For example, $92,067$ µm, 9.2067 cm, 0.92067 dm, and 0.092067 m all have the same number of significant figures. They merely represent different ways (units) of expressing one measurement. The zero between the decimal point and the 9 in the last number is used only to place the decimal point. There is no doubt whether any zero that *follows* a decimal point is significant or is used to place the decimal point. In the number 727.0, the zero is not used to locate the decimal point but is a significant part of the figure. Ambiguity can arise if a zero *precedes* a decimal point. If it falls between two other nonzero integers, then it will be significant. Such was the case with 92,067. In the number 936,600, it is impossible to determine whether one or both or neither of the zeros is used merely to place the decimal point or whether they are a part of the measurement. It is best in cases like this to write only the significant figures you are sure about and then to locate the decimal point by scientific notation. Thus, 9.3660×10^5 has five significant figures, but 936,600 contains six digits, one to place the decimal. Sometimes, the number may be written with a period at the end to denote all digits are significant, e.g., 936,000 to avoid amiguity.

Writing a value in scientific notation avoids ambiguities about the number of significant figures it contains.

Example 3.1

State the number of significant figures in each of the following numbers and indicate which zeros are significant.

(a) 353

(b) 0.072

(c) 3.040

(d) 3000

Solution

353   three significant figures

0.072   two significant figures, leading zeroes in decimal are not significant

3.040   four significant figures, all zeroes are significant

3000   one significant figure, trailing zeroes in whole number are not significant

---

If a number is written as 500, it could represent 500 ± 100. If it is written as 5.00×10^2, then it is 500 ± 1.

The significance of the last digit of a measurement can be illustrated as follows. Assume that each member of a class measures the width of a classroom desk, using the same meter stick. Assume further that the meter stick is graduated in 1-mm increments. The measurements can be estimated to the nearest 0.1 division (0.1 mm) by interpolation, but the last digit is uncertain since it is only an estimation. A series of class readings, for example, might be

$$565.4 \text{ mm}$$
$$565.8 \text{ mm}$$
$$565.0 \text{ mm}$$
$$\underline{566.1 \text{ mm}}$$
$$565.6 \text{ mm} \ \ (\text{average})$$

Absolute and Relative Uncertainty

Whenever a measurement or the result of a calculation is numerically represented, it is implicit that the last digit of the number is uncertain to ± 1. In the above example, when we say that the average width of the desk is 565.6 mm, it is understood that the value can be between 565.5 to 565.7 mm. Here the uncertainty of the number, also called the **absolute uncertainty** is in the tenth digit, 0.1 mm. If we say the distance between Seattle, Washington, and Arlington, Texas, is 2.99×10^3 km, the uncertainty is in the tens digit, the absolute uncertainty is 10 km—this number denotes a range between 2980 to 3000 km. The **relative uncertainty** is the absolute uncertainty divided by the value of the number, in the first case, this is 0.1/565.6 or 1 part in 5656; in the second case, it is $10/2.99 \times 10^3$ or 1 part in 299.

The uncertainty of a number is often called the *absolute uncertainty*. The *relative uncertainty* is the absolute uncertainty divided by the value of the number.

Propagation of Uncertainties

We carry out arithmetic operations like addition, subtraction, division, multiplication, and so forth with numbers. Much like a chain can be no stronger than its weakest link, the result of such operations can be no more certain than the number with the greatest uncertainty involved in the calculations. However, in addition and subtraction, the greatest absolute uncertainty is preserved; the final result is no more certain than the number with the greatest absolute uncertainty. If I have 56 quarters (read: this means 55 to 57) and you have 2300 pennies (means 2299 to 2301) and we add our resources, the net result cannot have the uncertainty less than a quarter. On the other hand, when we were driving from Arlington, Texas, to Seattle, Washington, Kevin was reading the odometer and said the distance was 2990 km (read again: 2980 to 3000, relative uncertainty 1 part in 299) and Gary was keeping the time and said it took 27 hours (read: 26 to 28, relative uncertainty 1 part in 27). If we want to calculate the speed, we will have to divide the distance traveled by the time, and the final result can be no more accurate than the number with the greatest relative uncertainty, 1 part in 27.

Addition and Subtraction—Think Absolute

The answer of an addition or subtraction is known to the same number of units as the number containing the least significant unit.

In addition and subtraction, the placement of the decimal point is important in determining how many figures will be significant. Suppose you wish to calculate the formula weight of Ag_2MoO_4 from the individual atomic weights (Ag = 107.870 amu, Mo = 95.94 amu, O = 15.9994 amu); amu = atomic mass unit.

Note that the atomic weight of molybdenum is known only to the nearest 0.01 amu, while that for Ag and O are known to 0.001 and 0.0001 amu, respectively. We cannot justifiably say that we know the formula weight of a compound containing molybdenum to any closer than 0.01 atomic unit. Therefore, the most accurately known value for the atomic weight of Ag_2MoO_4 is 375.68. All numbers being added or subtracted can be rounded to the least significant unit before adding or subtracting. But for consistency in the answer, one additional figure should be carried out and then the answer rounded to one less figure.

Ag	107.87	|	0
Ag	107.87	|	0
Mo	95.94	|	
O	15.99	|	94
O	15.99	|	94
O	15.99	|	94
O	15.99	|	94
	375.67	|	76

Multiplication and Division—Think Relative

The answer of a multiplication or division can be no more accurate than the least accurately known operator.

In all measurements, the last reported digit is uncertain. This is the last significant figure in the measurement; any digits beyond it are meaningless. In multiplication and division, the uncertainty of this digit is carried through the mathematical operations, thereby limiting the number of certain digits in the answer. The final answer of a multiplication and division operation has the same relative uncertainty as the number with greatest relative uncertainty involved in the operation. The answer should contain no more significant digits than a number, involved in the operation, which has the least number of significant digits. If more than one number share the distinction of having the same least number of significant digits, the number with the lowest magnitude has the greatest relative uncertainty and it governs the final result. For example, 0.0344 and 5.39 both have the same number of significant figures; but the first has the greater relative uncertainty (1 in 344 compared to 1 in 539).

Example 3.2

Give the answer of the following computation to the correct number of significant figures:

$$\frac{\left(\dfrac{97.7}{32.42} \times 100.0\right) + 36.04}{687}$$

Solution

$$\frac{301_{.36} + 36.04}{687} = \frac{337_{.4}}{687} = 0.0491_1$$

In the first operation, the number with the greatest relative uncertainty is 97.7 and the result is $301_{.36}$. We carried an additional fifth figure until the addition step. Then we rounded the added numbers to four figures before the division since the divisor has only three significant figures. In the division step, the number with the greatest relative uncertainty is 687. If we keep the answer as 0.04911, it has more significant digits than 687. If we make the answer 0.0491, we have a greater uncertainty in the final result than in 687. Therefore, we keep all the digits and put the last one as a subscript. Note that if in the first step we had rounded to $301_{.4}$, the numerator would have become $337_{.5}$ and the final answer would be 0.491_3 (still within the experimental uncertainty).

Example 3.3

In the following pairs of numbers, pick the number with the greatest relative uncertainty that would control the result of multiplication or division. **(a)** 42.67 or 0.0967; **(b)** 100.0 or 0.4570; **(c)** 0.0067 or 0.10.

Solution

(a) 0.0967 (has three significant figures)

(b) 100.0 (both have four significant figures, but the uncertainty here is 1 part per thousand versus about 1 part in 4600)

(c) 0.10 [both have two significant figures, but the uncertainty here is 10% (1 part in 10) versus about 1 part in 70]

Example 3.4

Give the answer of the following operation to the maximum number of significant figures and indicate the number with the greatest relative uncertainty.

$$\frac{35.63 \times 0.5481 \times 0.05300}{1.1689} \times 100\% = 88.5470578\%$$

Solution

The number with the greatest relative uncertainty is 35.63. The answer is therefore 88.55%, and it is meaningless to carry the operation out to more than five figures (the fifth figure is used to round off the fourth). The 100% in this calculation is an absolute number since it is used only to move the decimal point, and it has an infinite number of significant figures. (Numerical coefficients in a formula such as the multipliers 2 and 4 or the atomic weights of Ag and O while calculating the FW of Ag_2Mo_4 also are absolute numbers, with an infinite number of significant digits.) Note that 35.63 has a relative uncertainty at best of 1 part in 3600, and so the answer has a relative uncertainty at best of 1 part in 3600; thus, the answer has a relative uncertainty at least of this magnitude (i.e., about 2.5 parts in 8900). The objective in a calculation is to express the answer to at least the precision of the least certain number, but to recognize the magnitude of its uncertainty. *The final number is determined by the measurement of significant figures.* (Similarly, in making a series of measurements, one should strive to make each measurement to about the same degree of relative uncertainty.)

A subscript number is used to indicate an added degree of uncertainty in the final result. It is used when the relative uncertainty of the result is less than that of the least certain number involved in the operation.

There can be situations, where carrying the same number of significant digits in the final answer as the number(s) with least significant digits result in a substantially smaller relative uncertainty in the final result than the number with the greatest relative uncertainty involved in the operation. In such cases, the same number of significant digits is preserved in the results, but the last digit is written as a subscript to connote uncertainty.

Example 3.5

Rationalize the following operation

$$\frac{2.301 \times 10^3 \times 97.73}{2.303 \times 10^2 \times 2030} = 4.810 \times 10^{-1}$$

Solution

All the numbers involved in the operation have the same number of significant digits, 4. The number 2.301×10^3 has the greatest relative uncertainty, 1 part in 2301. If we keep the same number of significant digits in the final result, the relative uncertainty is almost an order of magnitude smaller, 1 part in 4810. The last digit is therefore indicated as a subscript.

In multiplication and division, the answer from each step of a series of operations can statistically be rounded to the number of significant figures to be retained in the final answer. But for consistency in the final answer, it is preferable to carry one additional figure until the end and then round off.

Putting it all Together

Summarizing the importance of significant figures, there are two questions to ask. First, how accurately do you have to *know* a particular result? If you only want to learn whether there is 12 or 13% of a substance in the sample, then you need only make all required measurements to achieve a final error of ±1% (including in the case when the measurement includes multiple steps and operations). If the sample weighs about

2 g, there is no need to weigh it more precisely than to 0.1 g. The second question is, how accurately can you *make* each required measurement? Obviously, if you can read the absorbance of light by a colored solution to only three figures (e.g., $A = 0.447$), it would be useless to weigh the sample to more than three figures (e.g., 6.67 g).

When a number in a measurement is small (without regard to the decimal point) compared with those of the other measurements, there is some justification in making the measurement to one additional figure. This can be visualized as follows. Suppose you wish to weigh two objects of essentially the same mass, and you wish to weigh them with the same precision, for example, to the nearest 0.1 mg, or 1 part per thousand. The first object weighs 99.8 mg, but the second weighs 100.1 mg. You have weighed both objects with equal accuracy, but you have retained an additional significant figure in one of them to keep the relative uncertainty of the same order. This is related to keeping the last significant digit as a subscript when the final result has a smaller relative uncertainty than the number with the greatest relative uncertainty involved in the calculations.

When the number with the greatest relative uncertainty in a series of measurements is known, then the overall accuracy can be improved, if desired, either by making the number larger (e.g., by increasing the sample size) or by making the measurement to an additional figure if possible (e.g., by weighing more precisely to one additional significant figure). This would be desirable when the number has greater relative uncertainty compared to those of the other measurements in order to bring its uncertainty closer to that of the others.

In carrying out analytical operations, then, you should try to measure quantities to the same *absolute* uncertainty when adding or subtracting and to the same *relative* uncertainty when multiplying or dividing.

If a computation involves both multiplication/division and addition/subtraction, then the individual steps must be treated separately. As good practice, retain one extra figure in the intermediate calculations until the final result (unless it drops out in a subsequent step). When a calculator is used, all digits can be kept in the calculator until the end. Do not assume that a number spat out by a calculator is correct; it is common to make mistakes in entering numbers, especially when you are in a hurry. Always try to estimate the size of the answer you expect. If you expect 2% and you calculate 0.02%, you probably forgot to multiply by 100. Or if you expect 20% and the answer is 4.3, you probably made a calculation error or perhaps a measurement error. Always ask yourself: Does it make sense? Is this number reasonable?

It is good practice to keep an extra figure during stepwise calculations and then drop it in the final number.

"Check the answer you have worked out once more—before you tell it to anybody."
—Edmund C. Berkely

Logarithms—Think Mantissa

In changing from logarithms to antilogarithms, and vice versa, the number being operated on and the logarithm mantissa have the same number of significant figures. (See Appendix B for a review of the use of logarithms.) All zeros in the mantissa are significant. Suppose, for example, we wish to calculate the pH of a 2.0×10^{-3} M solution of HCl from $pH = -\log[H^+]$. Then,

$$pH = -\log 2.0 \times 10^{-3} = -(-3 + 0.30) = 2.70$$

The -3 is the characteristic (from 10^{-3}), a pure number determined by the position of the decimal. The 0.30 is the mantissa from the logarithm of 2.0 and therefore has only two digits. So, even though we know the concentration to two figures, the pH (the logarithm) has three figures. If we wish to take the antilogarithm of a mantissa, the corresponding number will likewise have the same number of digits as the mantissa. The antilogarithm of 0.072 (contains three figures in mantissa .072) is 1.18, and the logarithm of 12.1 is 1.083 (1 is the characteristic, and the mantissa has three digits, .083).

For pH = 2.70, we call "2" the characteristic and "70" the mantissa. In logarithms, it is the number of significant figures in the mantissa that determines the number of significant figures in the final value. Zeros in mantissa count as significant figures.

3.5 Rounding Off

Always round to the even number, if the last digit is a 5.

If the digit following the last significant figure is greater than 5, the number is rounded up to the next higher digit. If it is less than 5, the number is rounded to the present value of the last significant figure:

$$9.47 = 9.5$$
$$9.43 = 9.4$$

If the last digit is a 5, the number is rounded off to the nearest even digit:

$$8.65 = 8.6$$
$$8.75 = 8.8$$
$$8.55 = 8.6$$

This is based on the statistical prediction that there is an equal chance that the last significant figure before the 5 will be even or odd. That is, in a suitably large sampling, there will be an equal number of even and odd digits preceding a 5. All nonsignificant digits should be rounded off all at once. The even-number rule applies only when the digit dropped is exactly 5 (not ... 51). (For example, if we want four significant figures, then 45.365 rounds to 45.36, whereas 45.3651 rounds to 45.37.)

3.6 Ways of Expressing Accuracy

There are various ways and units in which the accuracy of a measurement can be expressed. In each case, it is assumed that a "true" value is available for comparison.

Absolute Errors

The difference between the true value and the measured value, with regard to the sign, is the **absolute error**, and it is reported in the same units as the measurement. If an analyst reports 2.62 g of copper as 2.51 g, the absolute error is −0.11 g. If the measured value is the average of several measurements, the error is called the **mean error**. The mean error can also be calculated by taking the average difference, with regard to sign, of the *individual* test results from the true value.

Relative Error

The absolute or mean error expressed as a percentage of the true value is the **relative error**. The above analysis has a relative error of $(−0.11/2.62) \times 100\% = −4.2\%$. The **relative accuracy** is the measured value or mean expressed as a percentage of the true value. The above analysis has a relative accuracy of $(2.51/2.62) \times 100\% = 95.8\%$. We should emphasize that, more often than not, neither number is known to be "true," and the relative error or accuracy is based on the mean of two sets of measurements; that the true value is "true" is an assumption, unless that value is a certified value in a reference standard.

Relative error can be expressed in units other than percentages. In very accurate work, we are usually dealing with relative errors of much less than 1%, and it is convenient to use a smaller unit. A 1% error is equivalent to 1 part in 100. It is also equivalent to 10 parts in 1000. This latter unit is commonly used for expressing small uncertainties. That is, the uncertainty is expressed in **parts per thousand**. The number 0.11 expressed as parts per thousand of the number 2.62 would be 0.11 parts per 2.62, or 42 parts per thousand. Parts per thousand is often used in expressing precision of measurement. For even smaller relative amounts, **parts per million** (1 ppm = 1 part in 1,000,000) and **parts per billion** (1 ppb = 1 part in 1,000,000,000) are commonly used.

Example 3.6

Calculate the relative error in parts per thousand, where on analysis result is 1.0148 g and the accepted value is 1.0129 g.

Solution

$$\text{Absolute error} = 1.0148 \text{ g} - 1.0129 \text{ g} = 0.0019 \text{ g}$$

$$\text{Relative error} = \frac{0.0019}{1.0129} \times 1000\%_o = 1.87 \text{ parts per thousand}$$

$\%_o$ indicates parts per thousand, just as % indicates parts per hundred.

Which is a bigger relative error: 395 ppb or 0.412 ppm? Solution: Multiply ppm by 1000 to get the value in ppb. Thus, the latter value is 412 ppb, which is obviously greater than 395 ppb and constitutes the bigger error.

3.7 Standard Deviation—The Most Important Statistic

Each set of analytical results should be accompanied by an indication of the **precision** of the analysis. Various ways of indicating precision are acceptable.

The standard deviation σ of an infinite set of experimental data is theoretically given by

$$\sigma = \sqrt{\frac{\sum (x_i - \mu)^2}{N}} \tag{3.1}$$

"If reproducibility be a problem, conduct the test only once."
—Anonymous

where x_i represents the individual measurements and μ represents the mean of an infinite number of measurements (which should represent the "true" value). This equation holds strictly only as $N \to \infty$, where N is the number of measurements. In practice, we must calculate the individual deviations from the mean of a limited number of measurements, $\bar{x}$, in which it is anticipated that $\bar{x} \to \mu$ as $N \to \infty$, although we have no assurance this will be so; $\bar{x}$ is given by $\sum (x_i/N)$.

average, $\bar{x} = \sum (x_i/N)$

For a set of N measurements, there are N (independently variable) deviations from some reference number. But if the reference number chosen is the estimated mean, $\bar{x}$, the sum of the individual deviations (retaining signs) must necessarily add up to zero, and so values of $N - 1$ deviations are adequate to define the Nth value. That is, there are only $N - 1$ independent deviations from the mean; when $N - 1$ values have been selected, the last is predetermined. We have, in effect, used one degree of freedom of the data in calculating the mean, leaving $N - 1$ **degrees of freedom** for calculating the precision.

As a result, the **estimated standard deviation s of a finite set of experimental data** (generally $N < 30$) more nearly approximates σ if $N - 1$, the number of degrees of freedom, is substituted for N ($N - 1$ adjusts for the difference between $\bar{x}$ and μ).

See Section 3.15 and Equation 3.17 for another way of estimating s for four or less numbers.

$$s = \sqrt{\frac{\sum (x_i - \bar{x})^2}{N - 1}} \tag{3.2}$$

The value of s is only an estimate of σ, and it will more nearly approach σ as the number of measurements increases. Since we deal with small numbers of measurements in an analysis, the precision is appropriately represented by s.

Example 3.7

For the set of titration values 11.52, 11.55 and 11.81 mL, calculate the mean and standard deviation.

Solution

x_i	$x_i - \bar{x}$	$(x_i - \bar{x})^2$
11.52	0.11	0.0121
11.55	0.08	0.0064
11.81	0.18	0.0324
$\Sigma = 34.88$	$\Sigma = 0.37$	$\Sigma = 0.0509$

$$\bar{x} = \frac{\sum x_i}{N} = \frac{34.88}{3} = 11.63$$

$$S = \sqrt{\frac{0.0509}{3-1}} = 0.16 \text{ mL}$$

This result would be properly displayed as 11.63 ± 0.16 mL (mean ± standard deviation) in a final report.

The standard deviation may be calculated also using the following equivalent equation:

$$s = \sqrt{\frac{\sum x_i^2 - \left(\sum x_i\right)^2 / N}{N-1}} \tag{3.3}$$

This is useful for computations with a calculator. Many calculators, in fact, have a standard deviation program that automatically calculates the standard deviation from entered individual data. All spreadsheets can calculate the mean and standard deviations of a row or column of entered data. Microsoft Excel uses the respective functions AVERAGE and STDEV. These are discussed later, in more detail, in Section 3.19.

Example 3.8

Calculate the standard deviation for the data in Example 3.7 using Equation 3.3.

Solution

x_i	x_i^2
11.52	132.71
11.55	133.40
11.81	139.47
$\Sigma = 34.88$	$\Sigma = 405.58$

$$S = \sqrt{\frac{405.58 - (34.88)^2 / 3}{3-1}} = 0.14 \text{ mL}$$

The difference of 0.02 mL from Example 3.7 is not statistically significant since the variation is at least ±0.16 mL. In applying this formula, it is important to keep an extra digit or even two in x_i^2 for the calculation.

The standard deviation calculation considered so far is an estimate of the probable error of a single measurement. The arithmetical mean of a series of N measurements taken from an infinite population will show less scatter from the "true value" than will an individual observation. The scatter will decrease as N is increased; as N gets very large the sample average will approach the population average μ, and the scatter approaches zero. The arithmetical mean derived from N measurements can be shown to be $\sqrt{N}$ times more reliable than a single measurement. Hence, the random error in the mean of a series of four observations is one-half that of a single observation. In other words, the **precision of the mean** of a series of N measurements is inversely proportional to the square root of N of the deviation of the individual values. Thus,

> The precision improves as the square root of the number of measurements.

$$\text{Standard deviation of the mean} = s_{\text{mean}} = \frac{s}{\sqrt{N}} \qquad (3.4)$$

The standard deviation of the mean is sometimes referred to as the *standard error*.

The standard deviation is sometimes expressed as the **relative standard deviation** (rsd), which is just the standard deviation expressed as a fraction of the mean; usually it is given as the *percentage* of the mean (% rsd), which is often called the **coefficient of variation**.

> $\text{rsd} = s/\bar{x}$; $\% \ \text{rsd} = (s/\bar{x}) \times 100\%$

Example 3.9

The following replicate weighings were obtained: 29.8, 30.2, 28.6, and 29.7 mg. Calculate the standard deviation of the individual values and the standard deviation of the mean. Express these as absolute (units of the measurement) and relative (% of the measurement) values.

Solution

x_i	$x_i - \bar{x}$	$(x_i - \bar{x})^2$
29.8	0.2	0.04
30.2	0.6	0.36
28.6	1.0	1.00
29.7	0.1	0.01
$\Sigma = 118.3$	$\Sigma = 1.9$	$\Sigma = 1.41$

$$\bar{x} = \frac{118.3}{4} = 29.6$$

$$s = \sqrt{\frac{1.41}{4-1}} = 0.69 \text{ mg (absolute)}; \ \frac{0.69}{29.6} \times 100\% = 2.3\% \text{ (coefficient of variation)}$$

$$s_{\text{mean}} = \frac{0.69}{\sqrt{4}} = 0.34 \text{ mg (absolute)}; \ \frac{0.34}{29.6} \times 100\% = 1.1\% \text{ (relative)}$$

The precision of a measurement can be improved by increasing the number of observations. In other words, the spread $\pm s$ of the normal curve in Figure 3.2 becomes smaller as the number of observations is increased and would approach zero as the number of observations approached infinity. However, as seen above (Equation 3.4), the deviation of the mean does not decrease in direct proportion to the number of observations, but instead it decreases as the square root of the number of observations. A point will be reached where a slight increase in precision will require an unjustifiably large increase in the number of observations. For example, to decrease the standard deviation by a factor of 10 requires 100 times as many observations.

> In Example 3.9, how many observations might you need to reduce the coefficient of variation (cv) to 1.0%? **Solution:** You presently have 4 observations and want to improve the cv by a factor of 2.3. The total number of observations needed $= 4 \times 2.3^2 \simeq 21$

"Randomness is required to make statistical calculations come out right." —Anonymous

The practical limit of useful replication is reached when the standard deviation of the random errors is comparable to the magnitude of the determinate or systematic errors (unless, of course, these can be identified and corrected for). This is because the systematic errors in a determination cannot be removed by replication.

The significance of s in relation to the normal distribution curve is shown in Figure 3.2. The mathematical treatment from which the curve was derived reveals that 68% of the individual deviations fall within one standard deviation (for an infinite population) from the mean, 95% are less than twice the standard deviation, and 99% are less than 2.5 times the standard deviation. So, a good approximation is that 68% of the individual values will fall within the range $\bar{x} \pm s$, 95% will fall within $\bar{x} \pm 2s$, 99% will fall within $\bar{x} \pm 2.5s$, and so on.

The true value will fall within $\bar{x} \pm 2s$ 95% of the time for an infinite number of measurements. See the confidence limit and Example 3.15.

Actually, these percentage ranges were derived assuming an infinite number of measurements. There are then two reasons why the analyst cannot be 95% certain that the true value falls within $\bar{x} \pm 2s$. First, one makes a limited number of measurements, and the fewer the measurements, the less certain one will be. Second, the normal distribution curve assumes no determinate errors, but only random errors. Determinate errors, in effect, shift the normal error curve from the true value. An estimate of the actual certainty a number falls within s can be obtained from a calculation of the *confidence limit* (see below).

It is apparent that there are a variety of ways in which the precision of a number can be reported. Whenever a number is reported as $\bar{x} \pm y$, you should always qualify under what conditions this holds, that is, how you arrived at $\pm y$. It may, for example, represent s, $2s$, s (mean), or the coefficient of variation.

The variance equals s^2.

A term that is sometimes useful in statistics is the **variance**. This is the square of the standard deviation, s^2. We shall use this in determining the propagation of error and in the F-test below (Section 3.12).

3.8 Propagation of Errors—Not Just Additive

When discussing significant figures earlier, we stated that the relative uncertainty in the answer to a multiplication or division operation could be no better than the relative uncertainty in the operator that had the poorest relative uncertainty. Also, the absolute uncertainty in the answer of an addition or subtraction could be no better than the absolute uncertainty in the number with the largest absolute uncertainty. Without specific knowledge of the uncertainties, we assumed an uncertainty of at least ± 1 in the last digit of each number.

From a knowledge of the uncertainties in each number, it is possible to estimate the actual uncertainty in the answer. The errors in the individual numbers will propagate throughout a series of calculations, in either a relative or an absolute fashion, depending on whether the operation is a multiplication/division or whether it is an addition/subtraction.

Addition and Subtraction—Think Absolute Variances

Consider the addition and subtraction of the following numbers:

$$(65.06 \pm 0.07) + (16.13 \pm 0.01) - (22.68 \pm 0.02) = 58.51 \ (\pm?)$$

The absolute variances of additions and subtractions are additive.

The uncertainties listed represent the random or indeterminate errors associated with each number, expressed as standard deviations of the numbers. The maximum error of the summation, expressed as a standard deviation, would be ± 0.10; that is, it could be either $+0.10$ or -0.10 if all uncertainties happened to have the same sign. The minimum uncertainty would be 0.00 if all combined by chance to cancel. Both of these extremes are not highly likely, and statistically the uncertainty will fall somewhere in between. For addition and subtraction, *absolute uncertainties* are additive. The most probable error is represented by the square root of the sum of the *absolute variances*. That is, the absolute variance of the answer is the sum of the individual variances. For $a = b + c - d$,

$$s_a^2 = s_b^2 + s_c^2 + s_d^2 \qquad (3.5)$$

$$s_a = \sqrt{s_b^2 + s_c^2 + s_d^2} \qquad (3.6)$$

In the above example,

$$s_a = \sqrt{(\pm 0.07)^2 + (\pm 0.01)^2 + (\pm 0.02)^2}$$
$$= \sqrt{(49 \times 10^{-4}) + (1 \times 10^{-4}) + (4 \times 10^{-4})}$$
$$= \sqrt{54 \times 10^{-4}} = \pm 7.3 \times 10^{-2}$$

So the answer is 58.51 ± 0.07. The number ± 0.07 represents the absolute uncertainty. If we wish to express it as a relative uncertainty, this would be

$$\frac{\pm 0.07}{58.51} \times 100\% = \pm 0.1_2\%$$

Example 3.10

You have received three shipments of Monazite sand of equal weight that contain traces of europium. Analysis of the three ores provided europium concentrations of 397.8 ± 0.4, 253.6 ± 0.3, and 368.0 ± 0.3 ppm, respectively. What is the average europium content of the ores and what are the absolute and relative uncertainties?

Solution

$$\bar{x} = \frac{(397.8 \pm 0.4) + (253.6 \pm 0.3) + (368.0 \pm 0.3)}{3}$$

The uncertainty in the summation is

$$s_a = \sqrt{(\pm 0.4)^2 + (\pm 0.3)^2 + (\pm 0.3)^2}$$
$$= \sqrt{0.16 + 0.09 + 0.09}$$
$$= \sqrt{0.34} = \pm 0.58 \text{ ppm Eu}$$

Hence, the absolute uncertainty is

$$\bar{x} = \frac{1019.4}{3} \pm \frac{0.6 \text{ ppm}}{3} = 339.8 \pm 0.2 \text{ ppm Eu}$$

Note that since there is no uncertainty in the divisor 3, the *relative* uncertainty in the europium content is

$$\frac{0.2 \text{ ppm Eu}}{339.8 \text{ ppm Eu}} = 6 \times 10^{-4} \text{ or } 0.06\%$$

Multiplication and Division — Think Relative Variances

Consider the following operation:

$$\frac{(13.67 \pm 0.02)\,(120.4 \pm 0.2)}{4.623 \pm 0.006} = 356.0 \ (\pm?)$$

The relative variances of multiplication and division are additive.

Here, the *relative uncertainties* are additive, and the most probable error is represented by the square root of the sum of the relative variances. That is, the relative variance of the answer is the sum of the individual relative variances.

For $a = bc/d$,

$$\boxed{(s_a^2)_{\text{rel}} = (s_b^2)_{\text{rel}} + (s_c^2)_{\text{rel}} + (s_d^2)_{\text{rel}}} \qquad (3.7)$$

$$\boxed{(s_a)_{\text{rel}} = \sqrt{(s_b^2)_{\text{rel}} + (s_c^2)_{\text{rel}} + (s_d^2)_{\text{rel}}}} \qquad (3.8)$$

In the above example,

$$(s_b)_{\text{rel}} = \frac{\pm 0.02}{13.67} = \pm 0.0015$$

$$(s_c)_{\text{rel}} = \frac{\pm 0.2}{120.4} = \pm 0.0017$$

$$(s_d)_{\text{rel}} = \frac{\pm 0.006}{4.623} = \pm 0.0013$$

$$(s_a)_{\text{rel}} = \sqrt{(\pm 0.0015)^2 + (\pm 0.0017)^2 + (\pm 0.0013)^2}$$

$$= \sqrt{(2.2 \times 10^{-6}) + (2.9 \times 10^{-6}) + (1.7 \times 10^{-6})}$$

$$= \sqrt{(6.8 \times 10^{-6})} = \pm 2.6 \times 10^{-3}$$

The absolute uncertainty is given by

$$s_a = a \times (s_a)_{\text{rel}}$$
$$= 356.0 \times (\pm 2.6 \times 10^{-3}) = \pm 0.93$$

So the answer is 356.0 ± 0.9.

Example 3.11

Calculate the uncertainty in the number of millimoles of chloride contained in 250.0 mL of a sample when three equal aliquots of 25.00 mL are titrated with silver nitrate with the following results: 36.78, 36.82, and 36.75 mL. The molarity of the $AgNO_3$ solution is $0.1167 \pm 0.0002 \ M$.

Solution

The mean volume is

$$\frac{36.78 + 36.82 + 36.75}{3} = 36.78 \ \text{mL}$$

The standard deviation is

x_i	$x_i - \bar{x}$	$(x_i - \bar{x})^2$
36.78	0.00	0.0000
36.82	0.04	0.016
36.75	0.03	0.0009
		$\Sigma = 0.0025$

$$s = \sqrt{\frac{0.0025}{3-1}} = 0.035 \qquad \text{Mean volume} = 36.78 \pm 0.04 \text{ mL}$$

mmol Cl^- titrated = $(0.1167 \pm 0.0002 \text{ mmol/mL})(36.78 \pm 0.04 \text{ mL}) = 4.292\ (\pm?)$

$$(s_b)_{\text{rel}} = \frac{\pm 0.0002}{0.1167} = \pm 0.0017$$

$$(s_c)_{\text{rel}} = \frac{\pm 0.035}{36.78} = \pm 0.00095$$

$$(s_a)_{\text{rel}} = \sqrt{(\pm 0.0017)^2 + (\pm 0.00095)^2}$$

$$= \sqrt{(2.9 \times 10^{-6}) + (0.90 \times 10^{-6})}$$

$$= \sqrt{3.8 \times 10^{-6}}$$

$$= \pm 1.9 \times 10^{-3}$$

The absolute uncertainty in the millimoles of Cl^- is

$$4.292 \times (\pm 0.0019) = \pm 0.0082 \text{ mmol}$$

$$\text{mmol } Cl^- \text{ in 25 mL} = 4.292 \pm 0.0082 \text{ mmol}$$

$$\text{mmol } Cl^- \text{ in 250 mL} = 10\,(4.292 \pm 0.0082) = 42.92 \pm 0.08 \text{ mmol}$$

Note that we retained one extra figure in computations until the final answer. Here, the absolute uncertainty determined is proportional to the size of the sample; it would not remain constant for twice the sample size, for example.

If there is a combination of multiplication/division and addition/subtraction in a calculation, the uncertainties of these must be combined. One type of calculation is done at a time and the uncertainties calculated at each step.

Example 3.12

You have received three shipments of iron ore of the following weights: 2852, 1578, and 1877 lb. There is an uncertainty in the weights of ± 5 lb. Analysis of the ores gives $36.28 \pm 0.04\%$, $22.68 \pm 0.03\%$, and $49.23 \pm 0.06\%$, respectively. You are to pay $300 per ton of iron. What should you pay for these three shipments and what is the uncertainty in the payment?

Solution

We need to calculate the weight of iron in each shipment, with the uncertainties, and then add these together to obtain the total weight of iron and the uncertainty in this. The relative uncertainties in the weights are

$$\frac{\pm 5}{2852} = \pm 0.0017 \quad \frac{\pm 5}{1578} = \pm 0.0032 \quad \frac{\pm 5}{1877} = \pm 0.0027$$

The relative uncertainties in the analyses are

$$\frac{\pm 0.04}{36.28} = \pm 0.0011; \qquad \frac{\pm 0.03}{22.68} = \pm 0.0013; \qquad \frac{\pm 0.06}{49.23} = \pm 0.0012;$$

The weights of iron in the shipments are

$$\frac{(2852 \pm 5 \text{ lb}) (36.28 \pm 0.04\%)}{100} = 1034.7 \, (\pm?) \text{ lb Fe}$$

We calculate the relative standard deviation of the multiplication product as before

$$(s_a)_{\text{rel}} = \sqrt{(\pm 0.0017)^2 + (\pm 0.0011)^2} = \pm 0.0020$$

$$s_a = 1034.7 \times (\pm 0.0020) = \pm 2.1 \text{ lb}$$

$$\text{lb Fe} = 1034.7 \pm 2.1 \text{ in the first shipment}$$

(We will carry an additional figure throughout.)

$$\frac{(1578 \pm 5 \text{ lb}) (22.68 \pm 0.03\%)}{100} = 357.89 \, (\pm?) \text{ lb Fe}$$

$$(s_a)_{\text{rel}} = \sqrt{(\pm 0.0032)^2 + (\pm 0.0013)^2} = \pm 0.0034$$

$$s_a = 357.89 \times (\pm 0.0034) = \pm 1.2 \text{ lb}$$

$$\text{lb Fe} = 357.9 \pm 1.2 \text{ lb in the second shipment}$$

$$\frac{(1877 \pm 5 \text{ lb}) (49.23 \pm 0.06\%)}{100} = 924.05 \, (\pm?) \text{ lb Fe}$$

$$(s_a)_{\text{rel}} = \sqrt{(\pm 0.0027)^2 + (\pm 0.0012)^2} = \pm 0.0030$$

$$s_a = 924.05 \times (\pm 0.0030) = \pm 2.8 \text{ lb}$$

$$\text{lb Fe} = 924.0 \pm 2.8 \text{ lb in the third shipment}$$

$$\text{Total Fe} = (1034.7 \pm 2.1 \text{ lb}) + (357.9 \pm 1.2 \text{ lb}) + (924.0 \pm 2.8 \text{ lb})$$

$$= 2316.6 \, (\pm?) \text{ lb}$$

Here we use absolute uncertainties:

$$s_a = \sqrt{(\pm 2.1)^2 + (\pm 1.2)^2 + (\pm 2.8)^2} = \pm 3.7 \text{ lb}$$

$$\text{Total Fe} = 2317 \pm 4 \text{ lb}$$

A price of \$300/ton is the same as \$0.15/lb since 1 ton = 2000 lbs.

$$\text{Price} = (2316.6 \pm 3.7 \text{ lb}) \left(\$0.15/\text{lb} \right) = \$347.49 \pm 0.56$$

Hence, you should pay \$347.50 ± 0.60.

Example 3.13

You determine the acetic acid (HOAc) content of vinegar by titrating with a standard (known concentration) solution of sodium hydroxide to a phenolphthalein end point. An approximately 5-mL sample of vinegar is weighed on an analytical balance in a weighing bottle (the increase in weight represents the weight of the sample) and is found to be 5.0268 g. The uncertainty in making a single weighing is ±0.2 mg. The sodium hydroxide must be accurately standardized (its concentration determined) by titrating known weights of high-purity potassium acid phthalate, and three such titrations give molarities of 0.1167, 0.1163, and 0.1164 M. A volume of 36.78 mL of sodium hydroxide is used to titrate the sample. The uncertainty in reading the buret is ±0.02 mL. What is the percent acetic acid in the vinegar? Include the standard deviation of the final result.

Solution

Two weighings are required to obtain the weight of the sample: that of the empty weighing bottle and that of the bottle plus sample. Each has an uncertainty of ±0.2 mg, and so the uncertainty of the net sample weight (the difference of the two weights) is

$$s_{wt} = \sqrt{(\pm0.2)^2 + (\pm0.2)^2} = \pm0.3 \text{ mg}$$

The mean of the molarity of the sodium hydroxide is 0.1165 M, and its standard deviation is ±0.0002 M. Similarly, two buret readings (initial and final) are required to obtain the volume of base delivered, and the total uncertainty is

$$s_{vol} = \sqrt{(\pm0.02)^2 + (\pm0.02)^2} = \pm0.03 \text{ mL}$$

	A	B
1	N1	0.1167
2	N2	0.1163
3	N3	0.1164
4	STDEV:	0.000208
5	Cell B4:	
6	STDEV(B1:B3)	

(See Section 3.20)

The moles of acetic acid are equal to the moles of sodium hydroxide used to titrate it, so the amount of acetic acid in mmol is

$$\text{mmol HOAc} = (0.1165 \pm 0.0002) \text{ mmol mL}^{-1} (36.78 \pm 0.03) \text{ mL}$$

$$= 4.284_9 \ (\pm?) \text{ mmol}$$

As before, we calculate the relative standard deviation of the product as the root mean sum of squares of the individual relative standard deviations:

$$(s_{product})_{rel} = \sqrt{[(\pm0.0002/0.1165)^2 + (\pm0.03/36.78)^2]} = \pm0.0019$$

Multiplying by 4.285, we obtain the standard deviation s:

$$s = \pm4.285 \times 0.0019 = 0.0081$$

4.285 ± 0.0081 mmol acetic acid is converted to mg HOAc by multiplying by the formula weight of 60.05 mg/mmol (we assume that there is no significant uncertainty in the formula weight), to obtain $257.3_1 \pm 0.49$ mg HOAc.

To calculate the % acetic acid content, we must divide by the sample weight, 5026.8 ± 0.3 mg:

$$\%\text{HOAc} = 257.3_1 \pm 0.49 \text{ mg}/5026.8 \pm 0.3 \text{ mg} \times 100\% = 5.119 \ (\pm?) \%\text{acetic acid}$$

Once again we calculate via the relative standard deviation:

$$(s_{product})_{rel} = \sqrt{[(\pm0.49/257.3)^2 + (\pm0.3/5026.8)^2]} = \pm0.0019$$

Multiplying by 5.119, we obtain the standard deviation to be 0.01, the final answer thus being 5.12 ± 0.01 wt% acetic acid.

The factor that increased the uncertainty the most was the variance in the molarity of the sodium hydroxide solution. This illustrates the importance of careful calibration, which is discussed in Chapter 2.

3.9 Significant Figures and Propagation of Error

We noted earlier that the total uncertainty in a computation determines how accurately we can know the answer. In other words, the uncertainty sets the number of significant figures. Take the following example:

$$(73.1 \pm 0.2)(2.245 \pm 0.008) = 164.1 \pm 0.7$$

The number of significant figures in an answer is determined by the uncertainty due to propagation of error.

We are justified in keeping four figures, even though the first number has three. Here, we don't have to carry the additional figure as a subscript since we have indicated the actual uncertainty in it. Note that the greatest relative uncertainty in the multipliers is $0.008/2.245 = 0.0036$, while that in the answer is $0.7/164.1 = 0.0043$; so, due to the propagation of error, we know the answer somewhat less accurately than the least certain number. When actual uncertainties are known, the number with the greatest uncertainty may not necessarily be the one with the smallest number of digits. For example, the relative uncertainty in 78.1 ± 0.2 is 0.003, while that in 11.21 ± 0.08 is 0.007.

Suppose we have the following calculation:

$$(73.1 \pm 0.9)(2.245 \pm 0.008) = 164.1 \pm 2.1 = 164 \pm 2$$

Now the uncertainty in the answer is the units place, and so figures beyond that are meaningless. In this instance, the uncertainty in the original number with the highest relative uncertainty (73.1) and the answer are similar (± 0.012) since the uncertainty in the other multiplier is significantly smaller.

Example 3.14

Provide the uncertainties in the following calculations to the proper number of significant figures:

(a) $(38.68 \pm 0.07) - (6.16 \pm 0.09) = 32.52$

(b) $\dfrac{(12.18 \pm 0.08)(23.04 \pm 0.07)}{3.247 \pm 0.006} = 86.43$

Solution

(a) The calculated absolute uncertainty in the answer is ± 0.11. Therefore, the answer is 32.5 ± 0.1.

(b) The calculated relative uncertainty in the answer is 0.0075, so the absolute uncertainty is $0.0075 \times 86.43 = 0.65$. Therefore, the answer is 86.4 ± 0.6, even though we know all the other numbers to four figures; there is substantial uncertainty in the fourth digit, which leads to the uncertainty in the answer. The relative uncertainty in that answer is 0.0075, and the largest relative uncertainty in the other numbers is 0.0066, very similar.

Professor's Favorite Example

⌈Contributed by Professor
Nicholas H. Snow, Seton
Hall University ⌋

Why Propagate all Those Errors?

Propagation of errors is one of the most useful and straightforward tools for determining and understanding the precision of an analytical method. In industry, precision is critical, as analysts may be called upon to perform the same procedure hundreds or even thousands of times to determine whether a product has been manufactured correctly. In consulting with industrial scientists on how to improve precision in their chemical analysis methods, I use propagation of errors as one of my first tools for evaluating the method and for leading to suggested improvements.

For example, in a method for the analysis of β-carotene, the following steps are part of the procedure. A precision of less than $\pm2\%$ is required of analysts for assays performed using this procedure.

1. Transfer about 50 mg of β-carotene into a 100-mL volumetric flask and make up to volume.
2. Pipet 5 mL of this preparation into another 100-mL volumetric flask and dilute to volume.
3. Pipet 5 mL of this preparation into a 25-mL volumetric flask and dilute to volume.

Glassware and balance uncertainties—see the inside back cover for Class A glassware (we will assume use of either Class A or Class B glassware, and compare the uncertainties):

Glassware	Uncertainty	Relative Uncertainty (%)
100-mL Vol. Flask Class A	±0.08 mL	0.08
100-mL Vol. Flask Class B	±0.16 mL	0.16
25-mL Vol. Flask Class A	±0.03 mL	0.12
25-mL Vol. Flask Class B	±0.06 mL	0.24
5-mL Vol. Pipet class A	±0.01 mL	0.2
5-mL Vol. Pipet Class B	±0.02 mL	0.4
Analytical Balance	±0.0002 g	0.4

For procedures involving glassware transfers, the relative uncertainties are additive in a similar manner as if they were multiplied in a formula. For this portion of the procedure, there are two 100-mL volumetric flasks, one 25-mL volumetric flask, two 5-mL volumetric pipets and two weighing steps (remember, weighing is always by difference).

The relative uncertainty using Class A glassware is

$$\sqrt{0.08^2 + 0.08^2 + 0.12^2 + 0.2^2 + 0.2^2 + 0.4^2 + 0.4^2} = 0.65\%$$

The relative uncertainty using Class B glassware is

$$\sqrt{0.16^2 + 0.16^2 + 0.24^2 + 0.4^2 + 0.4^2 + 0.4^2 + 0.4^2} = 0.86\%$$

The full method includes preparing all standards and samples in this manner, plus instrumental analysis and calculation steps, probably giving total uncertainty of approximately $\pm1.3\%$ for Class A glassware and $\pm1.7\%$ for Class B glassware. With a requirement of better than $\pm2\%$, this leaves little room for mistakes, poor or variable technique.

Note that a significant improvement can be made by weighing the initial sample on a microbalance, which would add one decimal place (reduce the uncertainty to

0.04%) at the cost of a more difficult to operate and expensive balance. This would give an uncertainty in the sample preparation step of 0.33% when using Class A glassware, and 0.66% for the Class B glassware (do the calculations!).

Propagation of errors tells us the best a procedure and associated calculations CAN do. Our own standard deviations generated by running the procedures with our own samples tell us what the procedure IS DOING. If there is a big difference, usually our performance is less precise than predicted using propagation of errors, then there is likely something wrong that requires further investigation. In business, propagation of errors is one of the simplest and most useful tools for evaluating procedures that must be run numerous times with cost and precision in mind.

3.10 Control Charts

A **quality control chart** is a time plot of a measured quantity that is assumed to be constant (with a Gaussian distribution) for the purpose of ascertaining that the measurement remains within a statistically acceptable range. It may be a day-to-day plot of the measured value of a standard that is run intermittently with samples. The control chart consists of a central line representing the known or assumed value of the control and either one or two pairs of limit lines, the **inner** and **outer control limits**. Usually the standard deviation of the procedure is known (a good estimate of σ), and this is used to establish the control limits.

An example of a control chart is illustrated in Figure 3.3, representing a plot of day-to-day results of the analysis of calcium in a pooled serum sample or a control sample that is run randomly and blindly with samples each day. A useful inner control limit is two standard deviations since there is only 1 chance in 20 that an individual measurement will exceed this purely by chance. This might represent a warning limit. The outer limit might be 2.5 or 3σ, in which case there is only 1 chance in 100 or 1 chance in 500 a measurement will fall outside this range in the absence of systematic error. Usually, one control is run with each batch of samples (e.g., 20 samples), so several control points may be obtained each day. The mean of these may be plotted each day. The random scatter of this would be expected to be smaller by $\sqrt{N}$, compared to individual points.

Particular attention should be paid to trends in one direction; that is, the points lie largely on one side of the central line. This would suggest that either the control is in error or there is a systematic error in the measurement. A tendency for points to lie outside the control limits would indicate the presence of one or more determinate errors in the determination, and the analyst should check for deterioration of reagents, instrument malfunction, or environmental and other effects. Trends should signal contamination of reagents, improper calibration or erroneous standards, or change in the control lot.

A control chart is constructed by periodically running a "known" control sample.

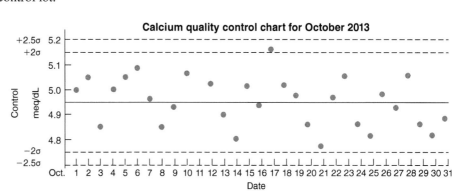

FIGURE 3.3 Typical quality control chart.

3.11 The Confidence Limit—How Sure Are You?

Calculation of the standard deviation for a set of data provides an indication of the precision inherent in a particular procedure or analysis. But unless there is a large amount of data, it does not by itself give any information about how close the experimentally determined mean $\bar{x}$ might be to the true mean value μ. Statistical theory, though, allows us to estimate the range within which the true value might fall, within a given probability, defined by the experimental mean and the standard deviation. This range is called the **confidence interval**, and the limits of this range are called the **confidence limit**. The likelihood that the true value falls within the range is called the **probability**, or **confidence level**, usually expressed as a percent. The confidence limit, in terms of the standard deviation, s (Equation 3.2), is given by

$$\text{Confidence limit} = \bar{x} \pm \frac{ts}{\sqrt{N}} \qquad (3.9)$$

The true value falls within the confidence limit, estimated using t at the desired confidence level.

where t is a statistical factor that depends on the number of degrees of freedom and the confidence level desired. The number of degrees of freedom is one less than the number of measurements. Values of t at different confidence levels and degrees of freedom v are given in Table 3.1. Note that the confidence limit is simply the product of t and the standard deviation of the mean ($s/\sqrt{N}$), also called the standard error of the mean. **(The confidence limit for a single observation ($N = 1$), x, is given by $x \pm ts$. This is larger than that of the mean by a factor $\sqrt{N}$.)**

Example 3.15

A soda ash sample is analyzed in the analytical chemistry laboratory by titration with standard hydrochloric acid. The analysis is performed in triplicate with the following results: 93.50, 93.58, and 93.43 wt% Na_2CO_3. Within what range are you 95% confident that the true value lies?

TABLE 3.1 Values of t for v Degrees of Freedom for Various Confidence Levels[a]

	Confidence Level			
v	90%	95%	99%	99.5%
1	6.314	12.706	63.657	127.32
2	2.920	4.303	9.925	14.089
3	2.353	3.182	5.841	7.453
4	2.132	2.776	4.604	5.598
5	2.015	2.571	4.032	4.773
6	1.943	2.447	3.707	4.317
7	1.895	2.365	3.500	4.029
8	1.860	2.306	3.355	3.832
9	1.833	2.262	3.250	3.690
10	1.812	2.228	3.169	3.581
15	1.753	2.131	2.947	3.252
20	1.725	2.086	2.845	3.153
25	1.708	2.060	2.787	3.078
∞	1.645	1.960	2.576	2.807

[a]$v = N - 1 = $ degrees of freedom.

Solution

Too high a confidence level will give a wide range that may encompass nonrandom numbers. Too low a confidence level will give a narrow range and exclude valid random numbers. Confidence levels of 90 to 95% are generally accepted as reasonable.

The mean is 93.50 wt%. The standard deviation s is calculated to be 0.075 wt% Na_2CO_3 (absolute—calculate it with a spreadsheet). At the 95% confidence level and two degrees of freedom, $t = 4.303$ and

$$\text{Confidence limit} = \bar{x} \pm \frac{ts}{\sqrt{N}}$$
$$= 93.50 \pm \frac{4.303 \times 0.075}{\sqrt{3}}$$
$$= 93.50 \pm 0.19 \text{ wt\%}$$

Compare with Figure 3.2 where 95% of the values fall within 2σ.

So you are 95% confident that, in the absence of a determinate error, the true value falls within 93.31 to 93.69 wt%. Note that for an infinite number of measurements, we would have predicted with 95% confidence that the true value falls within two standard deviations (Figure 3.2); we see that for $v = \infty$, t is actually 1.96 (Table 3.1), and so the confidence limit would indeed be about twice the standard deviation of the mean (which approaches σ for large N).

⌐Contributed by Professor
Dimitri Pappas, Texas Tech
University⌐

Professor's Favorite Example

How to Calculate the Minimum Number of Measurements to Get a Given Error from the Standard Error of the Mean

Note that the standard error of the mean (SE) has an inverse, square root relationship to the number of samples. The result is that there is a point in any analysis where a modest improvement in measurement statistics requires a prohibitively large sample. One of the most useful applications of the standard error of the mean is estimation of sample size number, especially if a desired confidence interval or tolerance is known or specified. One would necessarily have to make the approximation that standard deviation does not much change with the number of samples, this is not always defensible.

Example

A series of nominally identical samples are taken for analysis. The mean weight of the samples is 9.78 g, with a standard deviation being 0.09 g for very few samples having been weighed. How many samples must be weighed so that the sample mean has an error 0.02 g from the population mean?

Note that from Equation 3.9 and Table 3.1, the 95% confidence interval (CI) for an infinite population is

$$\text{CI} = \bar{x} \pm 1.96 \text{ (SE)}$$

We want

$$\text{CI} = 9.78 \pm 0.02 \text{ g at the 95\% confidence level}$$

If 1.96 (SE) is 0.02, then

$$\text{SE} = \frac{0.02 \text{ g}}{1.96} = 0.0102 \text{ g}$$

Since $\text{SE} = \frac{s}{\sqrt{N}}$, then $N - \left(\frac{s}{SE}\right)^2 = \left(\frac{0.09}{0.0102}\right)^2 \simeq 78$

In this case, a sample size of 78 measurements would yield the appropriate standard error of the mean. This simple exercise is useful for experiment planning, especially for experiments where the sample is difficult to obtain or the analysis is time consuming.

Compare this with Example 3.26 using an iterative procedure. If you use that alternate (but essentially equivalent) approach, this method gives the first iteration value. We can't do a second iteration using Table 3.1 since 78 is between $n = 25$ and $n = \infty$ in the table. You would need to find a more extensive table. For example, at www.jeremymiles.co.uk (go to the Other Stuff link on this website), t at the 95% level for 70 to 95 samples is 1.99. This would give a second (and final) iteration of 80 samples. So for a large number of samples, this approach (and the first iteration) gives a good approximation of the minimum number of samples required.

Remember from Section 3.7 and Figure 3.2 that we are 68% confident that the true value falls within $\pm 1\sigma$, 95% confident it will fall within $\pm 2\sigma$, and 99% confident it will fall within $\pm 2.5\sigma$. Note that it is possible to estimate a standard deviation from a stated confidence interval, and vice versa a confidence interval from a standard deviation. If a mean value is 27.37 ± 0.06 g at the 95% confidence interval, then since this is two standard deviations for a suitably large number of measurements, the standard deviation is 0.03 g. If we know the standard deviation is 0.03 g, then this is the confidence interval at the 68% confidence level, or it is 0.06 g at the 95% confidence level. For small numbers of measurements, t will be larger, which proportionately changes these numbers.

As the number of measurements increases, both t and $s/\sqrt{N}$ decrease, with the result that the confidence interval is narrowed. So the more measurements you make, the more confident you will be that the true value lies within a given range or, conversely, that the range will be narrowed at a given confidence level. However, t decreases exponentially with an increase in N, just as the standard deviation of the mean does (see Table 3.1), so a point of diminishing returns is eventually reached in which the increase in confidence is not justified by the increase in the multiple of samples analyses required.

3.12 Tests of Significance—Is There a Difference?

In developing a new analytical method, it is often desirable to compare the results of that method with those of an accepted (perhaps standard) method. How, though, can one tell if there is a significant difference between the new method and the accepted one? Again, we resort to statistics for the answer.

Deciding whether one set of results is significantly different from another depends not only on the difference in the means but also on the amount of data available and the spread. There are statistical tables available that show how large a difference needs to be in order to be considered not to have occurred by chance. The F-test evaluates differences between the spread of results, while the t-test looks at differences between means.

The F-Test

This is a test designed to indicate whether there is a significant difference between two methods based on their standard deviations. F is defined in terms of the variances of the two methods, where the **variance** is the square of the standard deviation:

The F-test is used to determine if two variances are statistically different.

$$F = \frac{s_1^2}{s_2^2} \tag{3.10}$$

TABLE 3.2 Values of F at the 95% Confidence Level

	$v_1 = 2$	3	4	5	6	7	8	9	10	15	20	30
$v_2 = 2$	19.0	19.2	19.2	19.3	19.3	19.4	19.4	19.4	19.4	19.4	19.4	19.5
3	9.55	9.28	9.12	9.01	8.94	8.89	8.85	8.81	8.79	8.70	8.66	8.62
4	6.94	6.59	6.39	6.26	6.16	6.09	6.04	6.00	5.96	5.86	5.80	5.75
5	5.79	5.41	5.19	5.05	4.95	4.88	4.82	4.77	4.74	4.62	4.56	4.50
6	5.14	4.76	4.53	4.39	4.28	4.21	4.15	4.10	4.06	3.94	3.87	3.81
7	4.74	4.35	4.12	3.97	3.87	3.79	3.73	3.68	3.64	3.51	3.44	3.38
8	4.46	4.07	3.84	3.69	3.58	3.50	3.44	3.39	3.35	3.22	3.15	3.08
9	4.26	3.86	3.63	3.48	3.37	3.29	3.23	3.18	3.14	3.01	2.94	2.86
10	4.10	3.71	3.48	3.33	3.22	3.14	3.07	3.02	2.98	2.85	2.77	2.70
15	3.68	3.29	3.06	2.90	2.79	2.71	2.64	2.59	2.54	2.40	2.33	2.25
20	3.49	3.10	2.87	2.71	2.60	2.51	2.45	2.39	2.35	2.20	2.12	2.04
30	3.32	2.92	2.69	2.53	2.42	2.33	2.27	2.21	2.16	2.01	1.93	1.84

where $s_1^2 > s_2^2$. There are two different degrees of freedom, v_1 and v_2, where degrees of freedom is defined as $N - 1$ for each case.

If the calculated F value from Equation 3.10 exceeds a tabulated F value at the selected confidence level, then there is a significant difference between the variances of the two methods. A list of F values at the 95% confidence level is given in Table 3.2.

The F-test is available in Excel. The preferred method is to install the "Data Analysis" Add-In. It is also discussed at the end of Section 3.24. After installation, the Data Analysis icon will appear in the top right-hand corner of the tool bar. Click on this icon. A drop-down menu will appear. Select "F-test Two samples for Variance". Another box will appear. Select your first set of data as "Variable 1 Range" and the second set as "Variable 2 Range". You can select the confidence level you want by choosing the "alpha" value. An alpha value of 0.05 means a 95% confidence level, 0.01 connotes 99% confidence level; the default is an alpha of 0.05. The F-test results will appear in a box that will occupy 3 columns and 10 rows, giving Mean, Variance, df, F, P one tail, and F Critical one tail for Variables 1 and 2 (see also 4. t-test to Compare Different Samples on page 000). Click on the output range button and specify a cell in the box that will constitute the top left corner of the output. Your F-value should always be > 1; if it is < 1, interchange the data: select your second set of data as "Variable 1 Range" and the first set as "Variable 2 Range". The output will specify the critical value of F.

The Folin-Wu method, now rarely used, uses the reaction of glucose with cupric ion (Cu^{2+}) in an alkaline medium to reduce phosphomolybdate to molybdenum blue, sometimes called heteropoly blue. Try this problem using F-test in Excel.

Example 3.16

You are developing a new colorimetric procedure for determining the glucose content of blood serum. You have chosen the standard Folin-Wu procedure with which to compare your results. From the following two sets of replicate analyses on the same sample, determine whether the variance of your method differs significantly from that of the standard method.

Your Method (mg/dL)	Folin-Wu Method (mg/dL)
127	130
125	128
123	131
130	129
131	127
126	125
129	—
mean $(\bar{x}_1)$ 127	mean $(\bar{x}_2)$ 128

Solution

$$s_1^2 = \frac{\sum (x_{i1} - \bar{x}_1)^2}{N_1 - 1} = \frac{50}{7 - 1} = 8.3$$

$$s_2^2 = \frac{\sum (x_{i2} - \bar{x}_2)^2}{N_2 - 1} = \frac{24}{6 - 1} = 4.8$$

$$F = \frac{8.3}{4.8} = 1.7_3$$

The variances are arranged so that the F value is > 1. The tabulated F value for $v_1 = 6$ and $v_2 = 5$ is 4.95. Since the calculated value of 1.7_3 is less than this, we conclude that there is no significant difference in the precision of the two methods, that is, the standard deviations are from random error alone and do not depend on the sample. In short, statistically, your method does as well as the established procedure. For a more critical look at the F-test, see the web supplements for Chapter 3, contributed by Professor Michael D. Morris, University of Michigan, on the textbook **website**.

If $F_{calc} > F_{table}$, then the variances being compared are significantly different. If $F_{calc} < F_{table}$, they are statistically the same.

The Student t-Test—Are There Differences in the Methods?

Frequently, the analyst wishes to decide whether there is a statistical difference between the results obtained using two different procedures, that is, whether they both indeed measure the same thing. The t-test is very useful for such comparisons.

In this method, comparison is made between two sets of replicate measurements made by two different methods; one of them will be the *test method*, and the other will be an *accepted* or benchmark method. In a similar manner, we may be comparing the blood concentration of a particular analyte in a population of diabetic patients vs. those measured in a control group. A statistical t value is calculated and compared with a tabulated value for the given number of tests at the desired confidence level (Table 3.1). If the calculated t value *exceeds* the tabulated t value, then there is a *significant difference* between the results by the two methods at that confidence level. If it does not exceed the tabulated value, then we can predict that there is no significant difference between the methods at the confidence level we have chosen. This in no way implies that the two results are identical.

There are several ways and several different types of situations in which a t-test can be used. Consider the following cases:

The t-test is used to determine if two sets of measurements are statistically different. If $t_{calc} > t_{table}$, then the two data sets are significantly different at the chosen confidence level.

1. You have taken a certified single sample for which the analytical result is exactly known or is known with a degree of certainty much higher than you expect from your test method. You analyze the same sample by the test method a number of times with an objective to determine if there is no difference between the certified value and the mean value obtained by your method at a specified degree of certainty.

2. The situation is the same as above except that the uncertainty of the certified value or the standard deviation of measurements by a reference method is not negligible. You compare repeated measurements of the same single sample made by the reference method with repeated measurements made by the test method. This is often referred to as *t-test by comparison of the means*. The number of measurements in the two measurement sets need not be the same.

3. Often the intent is to compare a newly developed method with another, and a certified reference standard is not available. Further, even if a reference standard is available, it can only check a method at only one concentration and it will be desirable to check the applicability of the method spanning the entire range of concentrations in which the method is to be used. You take a *number of different samples* spanning the concentration range of interest. All samples are divided in two parts; one set is analyzed by the benchmark method and the other by the test method. The pairs of analytical results thus generated are subjected to the *paired t-test* to determine if the two methods produce results that are statistically different at a specified confidence level. This test is also useful in other situations. For example, it can be used to answer the question: Does a new drug cause a statistically different change in blood pressure compared to another? A number of people are studied and each person is examined twice for a change in blood pressure: once after administration of drug 1 and another time after administration of drug 2. Necessarily, equal numbers of sampled data exist for each measurement set in a paired *t*-test.

4. You want to compare two sample populations that are unrelated to each other. Is coal from West Virginia statistically different in its sulfur content compared to coal from Pennsylvania? Are post-mortem brain tissue data for aluminum content statistically different in Alzheimer affected subjects compared to control subjects? Note that unlike the preceding example, in this case, two different sample populations are tested and the numbers in each population do not have to be equal. This type of *t*-test can be subdivided in two groups: (a) when the variance or standard deviations of the two sample sets being compared are statistically the same (as determined, e.g., by the *F*-test), the two sample sets are said to be *homoscedastic*, (b) when the variance of the two sample sets are statistically different, the sample populations are *heteroscedastic*.

Pooled standard deviation. The concept of pooled standard deviation is often used in performing *t*-tests and this is discussed first.

The pooled standard deviation is used to obtain an improved estimate of the precision of a method, and it is used for calculating the precision of the two sets of data. That is, rather than relying on a single set of data to describe the precision of a method, it is sometimes preferable to perform several sets of analyses, for example, on different days, or on different samples with slightly different compositions. If the indeterminate (random) error is assumed to be the same for each set, then the data of the different sets can be pooled. This provides a more reliable estimate of the precision of a method than is obtained from a single set. The **pooled standard deviation** s_p is given by

$$s_p = \sqrt{\frac{\sum (x_{i1} - \bar{x}_1)^2 + \sum (x_{i2} - \bar{x}_2)^2 + \cdots + \sum (x_{ik} - \bar{x}_k)^2}{N - k}} \qquad (3.11)$$

where $\bar{x}_1, \bar{x}_2, ..., \bar{x}_k$ are the means of each of k sets of analyses, and $x_{i1}, x_{i2}, ..., x_{ik}$ are the individual values in each set. N is the total number of measurements and is equal to $(N_1 + N_2 + \cdots + N_k)$. If five sets of 20 analyses each are performed, $k = 5$ and $N = 100$. (The number of samples in each set need not be equal.) $N - k$ is the degrees of freedom obtained from $(N_1 - 1) + (N_2 - 1) + \cdots + (N_k - 1)$; one degree of freedom is lost for each subset. This equation represents a combination of the equations for the standard deviations of each set of data.

The F-test can be applied to the variances of the two methods rather than assuming they are statistically equal before applying the t-test.

1. t-Test When a Reference Value Is Known. Note that Equation 3.9 is a representation of the true value μ. We can write that

$$\mu = \bar{x} \pm \frac{ts}{\sqrt{N}} \qquad (3.12)$$

Using Equation 3.12, you can calculate the confidence interval (i.e., $\bar{x} + ts/\sqrt{N}$ to $\bar{x} - ts/\sqrt{N}$) and see if the interval contains the expected value.

It follows that

$$t_{\text{calc}} = (\bar{x} - \mu)\frac{\sqrt{N}}{s} \qquad (3.13)$$

Occasionally, a good estimate of the "true" value μ for the sample or reference standard is known with a great degree of certainty, with a greater degree of relative accuracy than is possible in typical laboratory analyses. Many metals and some pure compounds are available as highly pure form, with specified purities of 99.999, even 99.9999%. If analyzing for the metal or the compound in such samples, Equation 3.12 can be used to determine whether the value obtained from a test method is statistically indistinguishable from the certified value. The same approach can also be used even when some uncertainty is specified in the certified value, provided that this uncertainty is much smaller than the standard deviation of the results of the method under test. Our goal is to see if the result from our method differs from the certified value. Another application of this method is to determine if a set of measured values exceeds a prescribed regulation limit, for example, that of fluoride in water.

If the determined confidence interval contains the expected value (e.g., μ), then the determination can be considered statistically not different from the known value.

Example 3.17

You are developing a procedure for determining traces of copper in biological materials using a wet digestion followed by measurement by atomic absorption spectrophotometry. In order to test the validity of the method, you obtain a certified reference material and analyze this material. Five replicas are sampled and analyzed, and the mean of the results is found to be 10.8 ppm with a standard deviation of ±0.7 ppm. The listed reference value is 11.7 ppm. Does your method give a statistically correct value relative to the certified value at the 95% confidence level?

Solution

$$t_{\text{calc}} = (\bar{x} - \mu)\frac{\sqrt{N}}{s}$$
$$= (10.8 - 11.7)\frac{\sqrt{5}}{0.7}$$
$$= 2_{.9}$$

The absolute value of t_{calc} is taken, since $\bar{x} - \mu$ may be negative.

There are five measurements, so there are four degrees of freedom $(N - 1)$. From Table 3.1, we see that the tabulated value of t at the 95% confidence level is 2.776.

Because $t_{calc} > t_{table}$, one is 95% sure that your procedure is producing a value that is statistically different from the true value (look at Table 3.1—we aren't quite 99% sure t_{table} at 99% is 4.6); but most work is done at the 95% confidence level). Your procedure will be considered unacceptable until you can find the source of the discrepancy and fix it. It is possible that you may not be able to do either one. For example, your method may have a negative interference from iron in the sample which is not accounted for.

Note from Equation 3.13 that as the precision is improved, that is, as s becomes smaller, the calculated t becomes larger. Thus, there is a greater chance that the tabulated t value will be less than this. That is, as the precision improves, it is easier to distinguish nonrandom differences. Looking again at Equation 3.13, this means as s decreases, so must the difference between the two methods $(\bar{x} - \mu)$ in order for the difference to be ascribed only to random error. What this means is that when one has a very large set of samples, which usually results in a reduced value of s, one is more likely to find a statistically significant difference.

2. Comparison of the Means of Two Methods. When the t-test is applied to two sets of data, μ in Equation 3.13 is replaced by the mean of the second set. The reciprocal of the standard deviation of the mean $(\sqrt{N}/s)$ is replaced by that of the differences between the two, which is readily shown to be

$$\sqrt{\frac{N_1 N_2}{N_1 + N_2}} / s_p$$

The F-test can be applied to the variances of the two methods rather than assuming they are statistically equal before applying the t-test.

where s_p is the pooled standard deviation of the individual measurements of two sets as defined in Equation 3.11 and N_1 and N_2 are the number of samples measured in each set:

$$t_{calc} = \frac{\bar{x}_1 - \bar{x}_2}{s_p} \sqrt{\frac{N_1 N_2}{N_1 + N_2}} \tag{3.14}$$

We can use this when comparing two methods, that is, when analyzing a sample by two different methods, as illustrated below.

To apply the comparison of means t-test between two methods, it is necessary that both methods have statistically the same standard deviation. This must first be verified by using the F-test.

Example 3.18

A new gravimetric method is developed for iron(III) in which the iron is precipitated in crystalline form with an organoboron "cage" compound. The accuracy of the method is checked by determining the iron in an ore sample and comparing with the results using the standard precipitation with ammonia and weighing the Fe_2O_3 formed after ignition of the $Fe(OH)_3$ precipitated. The results, reported as % Fe for each analysis, were as follows:

Test Method	Reference Method
20.10%	18.89%
20.50	19.20
18.65	19.00
19.25	19.70
19.40	19.40
19.99	$\bar{x}_2 = 19.24\%$
$\bar{x}_1 = 19.65\%$	

Is there a significant difference between the two methods?

Solution

x_{i1}	$x_{i1} - \bar{x}_1$	$(x_{i1} - \bar{x}_1)^2$	x_{i2}	$x_{i2} - \bar{x}_2$	$(x_{i2} - \bar{x}_2)^2$
20.10	0.45	0.202	18.89	0.35	0.122
20.50	0.85	0.722	19.20	0.04	0.002
18.65	1.00	1.000	19.00	0.24	0.058
19.25	0.40	0.160	19.70	0.46	0.212
19.40	0.25	0.062	19.40	0.16	0.026
19.99	0.34	0.116		$\sum (x_{i2} - \bar{x}_2)^2 = 0.420$	
		$\sum (x_{i1} - \bar{x}_1)^2 = 2.262$			

$$F = \frac{s_1^2}{s_2^2} = \frac{2.262/5}{0.420/4} = 4.31$$

This is less than the tabulated value (6.26), so the two methods have comparable standard deviations and the t-test can be applied

$$s_p = \sqrt{\frac{\sum (x_{i1} - \bar{x}_1)^2 + \sum (x_{i2} - \bar{x}_2)^2}{N_1 + N_2 - 2}}$$

$$= \sqrt{\frac{2.262 + 0.420}{6 + 5 - 2}} = 0.546$$

$$\pm t = \frac{19.65 - 19.24}{0.546} \sqrt{\frac{(6)(5)}{6 + 5}} = 1.2_3$$

The tabulated t for nine degrees of freedom ($N_1 + N_2 - 2$) at the 95% confidence level is 2.262, so there is no statistical difference in the results by the two methods.

Rather than comparing two methods using one sample, two samples could be examined for comparability using a single analysis method in a manner identical to the above examples.

The Excel Data Analysis Add-in can perform the comparison of means t-test, but requires paired data. We illustrate its use with the following problem. NIST stainless steel standard reference material (SRM) 73c certificate of analysis (https://www-s.nist.gov/srmors/certificates/73c.pdf) reports the following values for % Mo when analyzed by four analysts: 0.089, 0.092, 0.095, 0.087. You are trying to validate a new analytical method you have developed and you analyze four portions of the same SRM and get values of 0.090, 0.094, 0.098, and 0.096 % Mo, respectively. The NIST reported average is 0.091, while yours is 0.094. Is this difference significant?

Enter both sets of data in an Excel Worksheet, putting the NIST results in columns A1:A4 and your results in columns B1:B4. Go to Data/Data Analysis and scroll down the drop-down menu and select "*t*-test: Paired two sample for means". In the new data entry box we enter A1:A4 for "Variable 1 Range" and B1:B4 for "Variable 2 Range". We do not have any reason to believe that there should be a difference between the two means, so in the box "Hypothesized Mean Difference", allow the default value of zero to remain. If we had labels, e.g., "NIST Data", "My Data" put in as column headings, we would tick the labels box, so Excel will ignore the top row. Since we have no headings, we leave the labels box unchecked. Alpha has the same connotation as in the *F*-test, the default value $\alpha = 0.05$ connotes a confidence limit of 95%. You are now ready to do the test. Excel will require three columns and 14 rows to write the results. Click on the "Output Range" button and designate D1 as the cell where the output will begin. The first column has entries that are too wide to fit the default column width. Go to the column heading D and double-click on the right border. The column will expand to automatically fit the entries. The output lists mean, variance, degrees of freedom (df), the calculated *t* value (*t*-Stat), and the critical *t* value, both for a one-tailed test and for a two-tailed test. Our calculated *t* value is negative, but we really need to consider the magnitude. If we entered the data in column B as variable 1 and those in column A as variable 2, the *t* value would be calculated to be positive, with the same magnitude (try!). The one- and two-tailed test designations simply connote whether there are both halves of a Normal Distribution or just one need to be considered. In cases of comparison, where the one data set can only be higher (or lower) than the other data set, the one-tailed test is applicable. In analytical chemistry, one data set can generally either be higher or lower than the other set, there are no restrictions. So you must use the *t*-critical two-tailed value to compare with the calculated value. In the present case, t_{calc} of 2.08 is significantly smaller than the t_{crit} value of 3.18. At the 95% confidence level, your results therefore are *not* statistically different from the NIST results.

3. Paired *t*-Test. The paired *t*-test below compares the results of a series of different samples by two different methods. For the validation of a new method or intercomparison between methods, the paired *t*-test, described here, is preferred. A new method is frequently tested against an accepted method by analyzing several different samples of varying analyte content by both methods. In this case, the *t* value is calculated in a slightly different form. Often each sample has only been measured once. The difference between each of the paired measurements on each sample is computed. An average difference $\overline{D}$ is calculated and the individual deviations of each from $\overline{D}$ are used to compute a standard deviation, s_d. The *t* value is calculated from

$$ t = \frac{|\overline{D}|}{s_d} \sqrt{N} \tag{3.15} $$

$$ s_d = \sqrt{\frac{\sum (D_i - \overline{D})^2}{N - 1}} \tag{3.16} $$

where D_i is the individual difference between the two methods for each sample, with regard to sign; and $\overline{D}$ is the mean of all the individual differences.

Example 3.19

You are developing a new analytical method for the determination of blood urea nitrogen (BUN). You want to ascertain whether your method differs significantly from a standard method for determining a range of sample concentrations expected to be found in the routine laboratory. It has been ascertained that the two methods have comparable precision. Following are two sets of results for a number of individual samples:

Sample	Your Method (mg/dL)	Standard Method (mg/dL)	D_i	$D_i - \overline{D}$	$(D_i - \overline{D})^2$
A	10.2	10.5	−0.3	−0.6	0.36
B	12.7	11.9	0.8	0.5	0.25
C	8.6	8.7	−0.1	−0.4	0.16
D	17.5	16.9	0.6	0.3	0.09
E	11.2	10.9	0.3	0.0	0.00
F	11.5	11.1	0.4	0.1	0.01
			$\sum 1.7$		$\sum 0.87$
			$\overline{D} = 0.28$		

Solution

$$s_d = \sqrt{\frac{0.87}{6-1}} = 0.42$$

$$t = \frac{0.28}{0.42} \times \sqrt{6} = 1.6_3$$

The tabulated t value at the 95% confidence level for five degrees of freedom is 2.571. Therefore, $t_{calc} < t_{table}$, and there is no significant difference between the two methods at this confidence level.

Usually, a test at the 95% confidence level is considered significant, while one at the 99% level is highly significant. That is, the smaller the calculated t value, the more confident you are that there is no significant difference between the two methods. If you employ too low a confidence level (e.g., 80%), you are likely to conclude erroneously that there is a significant difference between two methods (sometimes called type I error). On the other hand, too high a confidence level will require too large a difference to detect (called type II error). See Section 3.19 on Powering a Study, for what is meant by a type I and a type II error. If a calculated t value is near the tabulated value at the 95% confidence level, more tests should be run to ascertain definitely whether the two methods are significantly different.

4. *t*-Test to Compare Different Samples. We will illustrate this using Excel. Consider the following problem. Well-water fluoride concentrations (in mg/L) have been determined in two adjoining counties, and are as follows. Does County B have a statistically different amount of fluoride in their water compared to County A?

County A	County B
0.76	1.20
0.81	1.41
0.77	1.69
0.79	0.91
0.80	0.50
0.78	1.80
0.76	1.53

We first determine by the F-test if the two sets of data have a difference in variance that is statistically significant. Do the F-test as described in Section 3.12. Enter column B data as variable 1 and column A data as variable 2 (if you do the opposite, you will find $F < 1$). The F-test output is shown below left. You clearly have very different variances in the two sets of data (if not, you would have used next: Data/Data Analysis/ t-test: Two Sample: Assuming Equal Variance). So we go to Data/Data Analysis/ t-test: Two Sample: Assuming Unequal Variances and again enter column B data as variable 1 and column A data as variable 2. The t-test output is below right. Since the t_{calc} value of 2.93 exceeds the critical two-tailed t value of 2.44, the fluoride concentrations in the wells of county B are statistically different at the 95% confidence level.

F Test: Two-Sample for Variances

	Variable 1	Variable 2
Mean	1.2914286	0.781429
Variance	0.2114476	0.000381
Observations	7	7
df	6	6
F		555.05
$P(F < f)$ one-tail		
		5.8E-08
F Critical one-tail		
		4.283866

t-Test: Two-Sample Assuming Unequal Variances

	Variable 1	Variable 2
Mean	1.291429	0.781429
Variance	0.211448	0.000381
Observations	7	7
Hypothesized Mean Difference	0	
df	6	
t Stat	2.93175	
$P(T < t)$ one-tail	0.013114	
t Critical one-tail	1.94318	
$P(T < t)$ two-tail	0.026227	
t Critical two-tail	2.446912	

3.13 Rejection of a Result: The Q Test

Finagle's third law: In any collection of data, the figure most obviously correct, beyond all checking, is the mistake.

When a series of replicate analyses is performed, it is not uncommon that one of the results will appear to differ markedly from the others. A decision will have to be made whether to reject the result or to retain it. Unfortunately, there are no uniform criteria that can be used to decide if a suspect result can be ascribed to accidental error rather than chance variation. It is tempting to delete extreme values from a data set because they alter the calculated statistics in an unfavorable way, that is, increase the standard deviation and variance (measures of spread), and they may substantially alter the reported mean. The only reliable basis for rejection occurs when it can be decided that some specific error may have been made in obtaining the doubtful result. No result should be retained in cases where a known error has occurred in its collection.

Experience and common sense may provide just as practical a basis for judging the validity of a particular observation as a statistical test would be. Frequently, the experienced analyst will gain a good idea of the precision to be expected in a particular method and will recognize when a particular result is suspect.

The Q test is used to determine if an "outlier" is due to a determinate error. If it is not, then it falls within the expected random error and should be retained.

Additionally, an analyst who knows the standard deviation expected of a method may reject a data point that falls outside $2s$ or $2.5s$ of the mean because there is about 1 chance in 20 or 1 chance in 100 this will occur.

A wide variety of statistical tests have been suggested and used to determine whether an observation should be rejected. In all of these, a range is established within which statistically significant observations should fall. The difficulty with all of them is determining what the range should be. If it is too small, then perfectly good data will be rejected; and if it is too large, then erroneous measurements will be retained too high a proportion of the time. The Q **test** is, among the several suggested tests, one of the

TABLE 3.3 Rejection Quotient, *Q*, at Different Confidence Limits[a]

No. of Observations	Confidence Level		
	Q_{90}	Q_{95}	Q_{99}
3	0.941	0.970	0.994
4	0.765	0.829	0.926
5	0.642	0.710	0.821
6	0.560	0.625	0.740
7	0.507	0.568	0.680
8	0.468	0.526	0.634
9	0.437	0.493	0.598
10	0.412	0.466	0.568
15	0.338	0.384	0.475
20	0.300	0.342	0.425
25	0.277	0.317	0.393
30	0.260	0.298	0.372

[a] Adapted from D. B. Rorabacher, *Anal. Chem.*, **63** (1991) 139.

most statistically correct for a fairly small number of observations and is recommended when a test is necessary. The ratio *Q* is calculated by arranging the data in increasing or decreasing order of numbers. If you have a large data set, you will find the Data/Sort function in Excel very helpful to arrange the data either in ascending or descending order. The difference between the suspect number and its nearest neighbor (*a*) is divided by the range (*w*), the range being the difference between the highest number and the lowest number. Referring to the figure in the margin, $Q = a/w$. This ratio is compared with tabulated values of *Q*. If it is equal to or greater than the tabulated value, the suspected observation can be rejected. The tabulated values of *Q* at the 90, 95, and 99% confidence levels are given in Table 3.3. If *Q* exceeds the tabulated value for a given number of observations and a given confidence level, the questionable measurement may be rejected with, for example, 95% confidence that some definite error is in this measurement.

If $Q_{calc} > Q_{table}$, then the data point may be an outlier and could be discarded. In practice, it is a good idea to make more measurements to be sure, especially if the decision is a close one.

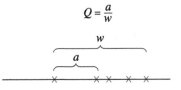

$$Q = \frac{a}{w}$$

Example 3.20

The following set of chloride determinations on separate aliquots of a pooled serum were reported: 103, 106, 107, and 114 meq/L. One value appears suspect. Determine if it can be ascribed to accidental error, at the 95% confidence level.

Solution

The suspect result is 114 meq/L. It differs from its nearest neighbor, 107 meq/L, by 7 meq/L. The range is 114 to 103, or 11 meq/L. *Q* is therefore 7/11 = 0.64. The tabulated value for four observations is 0.829. Since the calculated *Q* is less than the tabulated *Q*, the suspected number may be ascribed to random error and should not be rejected.

"And now the sequence of events in no particular order."
—Dan Rather, television news anchor

Consider reporting the median when an outlier cannot quite be rejected.

For a small number of measurements (e.g., three to five), the discrepancy of the measurement must be quite large before it can be rejected by this criterion, and it is likely that erroneous results may be retained. This would cause a significant change in the arithmetic mean because the mean is greatly influenced by a discordant value. For this reason, it has been suggested that the median rather than the mean be reported when a discordant number cannot be rejected from a small number of measurements. The **median** is the middle result of an odd number of results, or the average of the central pair for an even numbered set, when they are arranged in order of their values. The median has the advantage of not being unduly influenced by an outlying value. In the above example, the median could be taken as the average of the two middle values [$= (106 + 107)/2 = 106$]. This compares with a mean of 108, which is influenced more by the suspected number.

The following procedure is suggested for interpretation of the data of three to five measurements if the precision is considerably poorer than expected and if one of the observations is considerably different from the others of the set.

1. Estimate the precision that can reasonably be expected for the method in deciding whether a particular number actually is questionable. Note that for three measurements with two of the points very close, the Q test is likely to fail. (See the paragraph below.)

2. Check the data leading to the suspected number to see if a definite error can be identified.

3. If new data cannot be collected, run a Q test.

4. If the Q test indicates retention of the outlying number, consider reporting the median rather than the mean for a small set of data.

5. As a last resort, run another analysis. Agreement of the new result with the apparently valid data previously collected will lend support to the opinion that the suspected result should be rejected. You should avoid, however, continually running experiments until the "right" answer is obtained.

The Q test should not be applied to three data points if two are identical. In that case, the test always indicates rejection of the third value, regardless of the magnitude of the deviation, because a is equal to w and Q_{calc} is always equal to 1. The same obviously applies for three identical data points in four measurements, and so forth. A calculator/applet for performing the Q-test is available on the web (e.g., at http://asdlib.org/onlineArticles/ecourseware/Harvey/Outliers/OutlierProb1.html).

There are other useful tests for outliers. One that is recommended by ASTM International (formerly known as the American Society for Testing and Materials) is the **Grubbs Test for Outliers**. This test is often preferred for detecting outliers in a univariate data set. It assumes the data follow a normal error curve (this is difficult to determine for a limited data set and the Grubbs test is only of value when $n \geq 7$. However, $g_{critical}$ values (see below) are available for $g = 3 - 6$ and can be used as an estimate). One determines which datum is furthest away from the mean (i.e., $|X_i - \overline{X}|$ is maximum) and divides this difference by the standard deviation, s. This quotient is called the Grubbs test statistic, g. If the calculated g value exceeds the tabulated critical value of g ($g_{critical}$) at the desired confidence level, the datum can be rejected.

A description of the use of the Grubbs test, including tabulated Grubbs Critical Values for the 95% and 99% confidence levels, is given in the Chapter 3 website.

Example calculations are given along with references for further information on applicability. A useful online calculator is given at: http://www.graphpad.com/quickcalcs/Grubbs1.cfm.

3.14 Statistics for Small Data Sets

We have discussed, in previous sections, ways of estimating, for a normally distributed population, the central value (mean, $\bar{x}$), the spread of results (standard deviation, s), and the confidence limits (t-test). These statistical values hold strictly for a large population. In analytical chemistry, we typically deal with fewer than 10 results, and for a given analysis, perhaps 2 or 3. For such small sets of data, other estimates may be more appropriate.

Large population statistics do not strictly apply for small populations.

The Q test in the previous section is designed for small data sets, and we have already mentioned some rules for dealing with suspect results.

The Median may be Better than the Mean

If you have a set of N values and put them in either ascending or descending order, for an odd value of N, the center number in the series is the median value. For an even value of N, it is the average of the two central numbers. The median is denoted by M and may be used as an estimate of the central value. It has the advantage that it is not markedly influenced by extraneous (outlier) values, while the mean, $\bar{x}$ is. The efficiency of M, defined as the ratio of the variances of sampling distributions of these two estimates of the "true" mean value and denoted by E_M, is given in Table 3.4. It varies from 1 for only two observations (where the median is necessarily identical with the mean) to 0.64 for large numbers of observations. The numerical value of the efficiency implies that the median from, for example, 100 observations where the efficiency is essentially 0.64, conveys as much information about the central value of the population as does the mean calculated from 64 observations. The median of 10 observations is as efficient conveying the information as is the mean from $10 \times 0.71 = 7$ observations. It may be desirable to use the median in order to avoid deciding whether a gross error is present, that is, using the Q test. It has been shown that for three observations from a normal population, the median is better than the mean of the best two out of three (the two closest) values.

The median may be a better representative of the true value than the mean, for small numbers of measurements.

The **mean** is the arithmetic average, whereas the **median** is the middle value in a set of values. The mean can be more dramatically affected by outliers than the median in a set of data with only a few values and significant imprecision.

TABLE 3.4 Efficiencies and Conversion Factors for 2 to 10 Observations[a]

No. of Observations	Efficiency Of Median, E_M	Efficiency Of Range, E_R	Range Deviation Factor, K_R	Range Confidence Factor (t) $t_{r\,0.95}$	Range Confidence Factor (t) $t_{r\,0.99}$
2	1.00	1.00	0.89	6.4	31.83
3	0.74	0.99	0.59	1.3	3.01
4	0.84	0.98	0.49	0.72	1.32
5	0.69	0.96	0.43	0.51	0.84
6	0.78	0.93	0.40	0.40	0.63
7	0.67	0.91	0.37	0.33	0.51
8	0.74	0.89	0.35	0.29	0.43
9	0.65	0.87	0.34	0.26	0.37
10	0.71	0.85	0.33	0.23	0.33
∞	0.64	0.00	0.00	0.00	0.00

[a] Adapted from R. B. Dean and W. J. Dixon, *Anal. Chem.*, **23** (1951) 636.

Range Instead of the Standard Deviation

The range is as good a measure of the spread of results as is the standard deviation for four or fewer measurements.

The range R for a small set of measurements is highly efficient for describing the spread of results. The range is computed by subtracting the smallest value recorded from the largest value recorded. The efficiency of the range, E_R, shown in Table 3.4, is virtually identical to that of the standard deviation for four or fewer measurements. This high relative efficiency arises from the fact that the standard deviation is a poor estimate of the spread for a small number of observations, although it is still the best known estimate for a given set of data. To convert the range to a measure of spread that is independent of the number of observations, we must multiply it by the **deviation factor**, K, given in Table 3.4. This factor adjusts the range R so that on average it reflects the standard deviation of the population, which we represent by s_r:

$$s_r = RK_R \tag{3.17}$$

In Example 3.9 the standard deviation of the four weights was calculated to be 0.69 mg. The range of the data was 1.6 mg. Multiplying by K_R for four observations, $s_r = 1.6$ mg $\times 0.49 = 0.78$ mg; s and s_r are obviously quite comparable. As N increases, the efficiency of the range decreases relative to the standard deviation.

The median M may be used in computing the standard deviation, in order to minimize the influence of extraneous values. Taking Example 3.9 again, the standard deviation calculated using the median, 29.8, in place of the mean in Equation 3.2, is 0.73 mg, instead of 0.69 mg.

Confidence Limits Using the Range

Confidence limits could be calculated using s_r obtained from the range, in place of s in Equation 3.9, and a corresponding but different t table, denoted as t_r. It is more convenient, though, to calculate the limits directly from the range as

$$\boxed{\text{Confidence limit} = \bar{x} \pm Rt_r} \tag{3.18}$$

The factor for converting R to s_r has been included in the quantity, t_r, which is tabulated in Table 3.4 for 99 and 95% confidence levels. The calculated confidence limit at the 95% confidence level in Example 3.15 using Equation 3.18 is 93.50 ± 0.19 (1.3) = $93.50 \pm 0.25\%$ Na_2CO_3.

3.15 Linear Least Squares—How to Plot the Right Straight Line

"If a straight line fit is required, obtain only two data points."
—Anonymous

The analyst is frequently confronted with plotting data that fall on a straight line, as in an analytical calibration curve. Graphing, that is, curve fitting, is critically important in obtaining accurate analytical data. It is the calibration graph that is used to calculate the unknown concentration. Straight-line predictability and consistency will determine the accuracy of the unknown calculation. All measurements will have a degree of uncertainty, and so will the plotted straight line. Graphing is often done intuitively, that is, by simply "eyeballing" the best straight line by placing a ruler through the points, which invariably have some scatter. A better approach is to apply statistics to define the most probable straight-line fit of the data. The availability of statistical functions in spreadsheets today make it straightforward to prepare straight-line, or even nonlinear, fits. We will first learn the computations that are involved in curve fitting and statistical evaluation.

We should note that a straight line is a model of the relationship between observations and amount of an analyte. One can always blindly apply least squares fitting

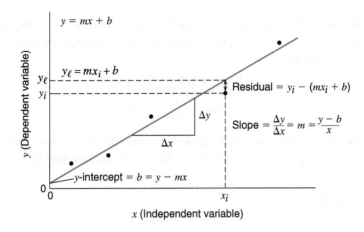

FIGURE 3.4   Straight-line plot.

(see below) to any random set of numbers. That does not necessarily mean a linear model is appropriate. Perhaps one should be fitting logarithms, or should be fitting sigmoids. We fit models to data, not data to models. To some extent, we prefer systems that are linear because they're easier to deal with. However, with facile availability of computation, we need not always avoid nonlinearity.

If a straight-line relationship is assumed, then the data fit the equation

$$y = mx + b \tag{3.19}$$

where y is the *dependent variable*, x is the *independent variable*, m is the *slope*, of the curve, and b is the *intercept* on the ordinate (y axis); y is usually the measured variable, plotted as a function of changing x (see Figure 3.4). In a spectrophotometric calibration curve, y would represent the measured absorbances and x would be the concentrations of the standards. Our problem, then, is to establish values for m and b.

Least-Squares Plots

It can be shown statistically that the best straight line through a series of experimental points is that line for which the *sum of the squares of the deviations (the residuals) of the points from the line is minimum*. This is known as the **method of least squares**. If x is the fixed variable (e.g., concentration) and y is the measured variable (absorbance in a spectrophotometric measurement, the peak area in a chromatographic measurement, etc.), then the deviation of y vertically from the line at a given value of x (x_i) is of interest. If y_l is the value *on the line*, it is equal to $mx_i + b$. The square of the sum of the differences, S, is then

$$S = \Sigma(y_i - y_l)^2 = \Sigma[y_i - (mx_i + b)]^2 \tag{3.20}$$

This equation assumes no error in x, the independent variable.

The best straight line occurs when S is minimum (S_{min}). S_{min} is obtained by use of differential calculus by setting the derivatives of S with respect to m and b equal to zero and solving for m and b. The result is

The least-squares slope and intercept define the most probable straight line.

$$m = \frac{\Sigma(x_i - \bar{x})(y_i - \bar{y})}{\Sigma(x_i - \bar{x})^2} \tag{3.21}$$

$$b = \bar{y} - m\bar{x} \tag{3.22}$$

where $\bar{x}$ is the mean of all the values of x_i and $\bar{y}$ is the mean of all the values of y_i. The use of differences in calculations is cumbersome, and Equation 3.21 can be transformed into an easier to use form:

$$m = \frac{\sum x_i y_i - \left[\left(\sum x_i \ \sum y_i\right)/n\right]}{\sum x_i^2 - \left[\left(\sum x_i\right)^2/n\right]} \tag{3.23}$$

where n is the number of data points. Aside from spreadsheets, many calculators can also perform linear regression on a set of x-y data and give the values of m and b.

Example 3.21

Riboflavin (vitamin B_2) is determined in a cereal sample by measuring its fluorescence intensity in 5% acetic acid solution. A calibration curve was prepared by measuring the fluorescence intensities of a series of standards of increasing concentrations. The following data were obtained. Use the method of least squares to obtain the best straight line for the calibration curve and to calculate the concentration of riboflavin in the sample solution. The sample fluorescence intensity was 15.4.

Riboflavin, μg/mL (x_i)	Fluorescence Intensity, Arbitrary Units (y_i)	x_i^2	$x_i y_i$
0.000	0.0	0.0000	0.00
0.100	5.8	0.0100	0.58
0.200	12.2	0.0400	2.44
0.400	22.3	0.160_0	8.92
0.800	43.3	0.640_0	34.6_4
$\sum x_i = 1.500$	$\sum y_i = 83.6$	$\sum x_i^2 = 0.850_0$	$\sum x_i y_i = 46.5_8$
$\left(\sum x_i\right)^2 = 2.250$		$n = 5$	
$\bar{x} = \dfrac{\sum x_i}{n} = 0.300_0$	$\bar{y} = \dfrac{\sum y_i}{n} = 16.7_2$		

Solution

Using Equations 3.23 and 3.22.

$$m = \frac{46.5_8 - \left[(1.500 \times 83.6)/5\right]}{0.850_0 - 2.250/5} = 53.7_5 \text{ fluor. units/ppm}$$

$$b = 16.7_2 - (53.7_5 \times 0.300_0) = 0.6_0 \text{ fluor. units}$$

Take note of the units assigned to each variable in the $y = mx + b$ linear equation. This will help you apply the equation for determination of unknowns.

We have retained the maximum number of significant figures in computation. Since the experimental values of y are obtained to only the first decimal place, we can round m and b to the first decimal. The equation of the straight line is (FU = fluorescence units; ppm = μg/mL)

$$y\,(\text{FU}) = 53.8\,(\text{FU/ppm})\,x\,(\text{ppm}) + 0.6\,(\text{Fu})$$

The sample concentration is

$$15.4 = 53.8x + 0.6$$

$$x = 0.27_5 \text{ μg/mL}$$

To prepare an actual plot of the line, take two arbitrary values of x sufficiently far apart and calculate the corresponding y values (or vice versa) and use these as points to draw

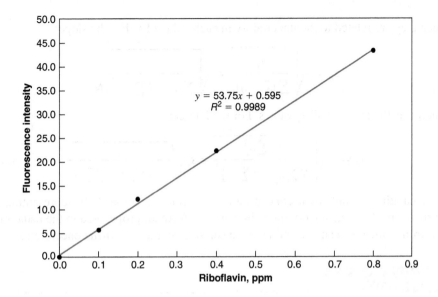

FIGURE 3.5 Least-squares plot of data from Example 3.21.

the line. The intercept $y = 0.6$ (at $x = 0$) could be used as one point. At 0.500 μg/mL, $y = 27.5$. A plot of the experimental data and the least-squares line drawn through them is shown in Figure 3.5. This was plotted using Excel, with the equation of the line and the square of the correlation coefficient (a measure of agreement between the two variables—ignore this for now, we will discuss it later). The program automatically gives additional figures, but note the agreement with our calculated values for the slope and intercept.

Note that all points are treated equally in a standard least-squares fit. For the same amount of relative deviation, such a procedure effectively gives greater weight to the higher x, y values compared to those at the low end. Weighted least-squares fitting can be used by the more adventurous. For a reference, consult, for example, Strutz, T.: *Data Fitting and Uncertainty*. Vieweg + Teubner Verlag, 2010. Also J. A. Irvin and T. I. Quickenden, "Linear Least Squares Treatment When There are Errors in Box x and y," *J. Chem. Ed.*, **60** (1983) 711–712.

Standard Deviations of the Slope and Intercept—They Determine the Unknown Uncertainty

Each data point on the least-squares line exhibits a normal (Gaussian) distribution about the line on the y axis. The deviation of each y_i from the line is $y_i - y_l = y - (mx + b)$, as in Equation 3.20. The standard deviation of each of these y-axis deviations is given by an equation analogous to Equation 3.2, except that there are two less degrees of freedom since two are used in defining the slope and the intercept:

$$s_y = \sqrt{\frac{\sum \left[y_i - (mx_i + b)\right]^2}{N - 2}}$$

$$= \sqrt{\frac{\left[\sum y_i^2 - (\sum y_i)^2/N\right] - m^2 \left[\sum x_i^2 - (\sum x_i)^2/N\right]}{N - 2}}$$

(3.24)

> The standard deviations of m and b give an equation from which the uncertainty in the unknown is calculated, using propagation of errors.

This quantity is also called the *standard deviation of regression*, s_r. The s_y value can be used to obtain uncertainties for the slope, m, and intercept, b, of the least-squares line since they are related to the uncertainty in each value of y. For the slope:

$$s_m = \sqrt{\frac{s_y^2}{\sum (\bar{x} - x_i)^2}} = \sqrt{\frac{s_y^2}{\sum x_i^2 - (\sum x_i)^2 / N}} \tag{3.25}$$

where $\bar{x}$ is the mean of all x_i values. For the intercept:

$$s_b = s_y \sqrt{\frac{s_y \sum x_i^2}{N \sum x_i^2 - (\sum x_i)^2}} = s_y \sqrt{\frac{1}{N - (\sum x_i)^2 / \sum x_i^2}} \tag{3.26}$$

In calculating an unknown concentration, x_i, from Equation 3.19, representing the least-squares line, the uncertainties in y, m, and b are all propagated in the usual manner, from which we can determine the uncertainty in the unknown concentration.

Example 3.22

Estimate the uncertainty in the slope, intercept, and y for the least-squares plot in Example 3.21, and the uncertainty in the determined riboflavin concentration for the sample with fluorescence intensity of 15.4 FU.

Solution

In order to solve for all the uncertainties, we need values for $\sum y_i^2$, $(\sum y_i)^2$, $\sum x_i^2$, $(\sum x_i)^2$, and m^2. From Example 3.21, $(\sum y_i)^2 = (83.6)^2 = 6989.0$; $\sum x_i^2 = 0.850_0$; $(\sum x_i)^2 = 2.250$, and $m^2 = (53.7_5)^2 = 2.88_9$. The $(y_i)^2$ values are $(0.0)^2$, $(5.8)^2$, $(12.2)^2$, $(22.3)^2$, and $(43.3)^2 = 0.0$, 33.6, 148.8, 497.3, and 1874.9, and $\sum y_i^2 = 2554.6$ (carrying extra figures). From Equation 3.24,

$$s_y = \sqrt{\frac{(2554.6 - 6989.0/5) - (53.7_5)^2(0.850_0 - 2.250/5)}{5 - 2}} = \pm 0.6_3 \text{ FU}$$

From Equation 3.25,

$$s_m = \sqrt{\frac{(0.6_3)^2}{0.850_0 - 2.250/5}} = \pm 1.0 \text{ FU/ppm}$$

From Equation 3.26,

$$s_b = 0.6_3 \sqrt{\frac{0.850_0}{5(0.850_0) - 2.250}} = \pm 0.4_1 \text{ FU}$$

Therefore, $m = 53._8 + 1._0$ and $b = 0.6 \pm 0.4$.

The unknown riboflavin concentration is calculated from

$$x = \frac{(y \pm s_y) - (b \pm s_b)}{m \pm s_m} = \frac{(15.4 \pm 0.6) - (0.6 \pm 0.4)}{53._8 \pm 1._0} = 0.27_5 \pm ?$$

Applying the principles of propagation of error (absolute variances in numerator additive, relative variances in the division step additive), we calculate that $x = 0.27_5 \pm 0.01_4$ ppm.

See Chapter 11 for the spreadsheet calculation of the standard deviation of regression and the standard deviation of an unknown for this.

3.16 Correlation Coefficient and Coefficient of Determination

The **correlation coefficient** is used as a measure of the correlation between two variables. When variables x and y are correlated rather than being functionally related (i.e., are not directly dependent upon one another), we do not speak of the "best" y value corresponding to a given x value, but only of the most "probable" value. The closer the observed values are to the most probable values, the more definite is the relationship between x and y. This postulate is the basis for various numerical measures of the degree of correlation.

The **Pearson correlation coefficient** is one of the most convenient to calculate. This is given by

$$r = \sum \frac{(x_i - \bar{x})(y_i - \bar{y})}{ns_x s_y} \tag{3.27}$$

where r is the correlation coefficient, n is the number of observations, s_x is the standard deviation of x, s_y is the standard deviation of y, x_i and y_i are the individual values of the variables x and y, respectively, and $\bar{x}$ and $\bar{y}$ are their means. The use of differences in the calculation is frequently cumbersome, and the equation can be transformed to a more convenient form:

$$r = \frac{\sum x_i y_i - n\bar{x}\bar{y}}{\sqrt{\left(\sum x_i^2 - n\bar{x}^2\right)\left(\sum y_i^2 - n\bar{y}^2\right)}}$$
$$= \frac{n\sum x_i y_i - \sum x_i \sum y_i}{\sqrt{\left[n\sum x_i^2 - \left(\sum x_i\right)^2\right]\left[n\sum y_i^2 - \left(\sum y_i\right)^2\right]}} \tag{3.28}$$

Despite its formidable appearance, Equation 3.28 is probably the most convenient for calculating r. Calculators that can perform linear regression calculate r directly; the formula is embedded.

The maximum value of r is 1. When this occurs, there is exact correlation between the two variables. When the value of r is zero (this occurs when $\sum x_i y_i$ is equal to zero), there is complete independence of the variables. The minimum value of r is -1. A negative correlation coefficient indicates that the assumed dependence is opposite to what exists and is therefore a positive coefficient for the reversed relation.

A correlation coefficient near 1 means there is a direct relationship between two variables, for example, absorbance and concentration.

Example 3.23

Calculate the correlation coefficient for the data in Example 3.19, taking your method as x and the standard method as y.

Solution

We calculate that $\sum x_i^2 = 903.2$, $\sum y_i^2 = 855.2$, $\bar{x} = 12.0$, $\bar{y} = 11.7$, and $\sum x_i y_i = 878.5$. Therefore, from Equation 3.28,

$$r = \frac{878.5 - (6)(12.0)(11.7)}{\sqrt{[903.2 - (6)(12.0)^2][855.2 - (6)(11.7)^2]}} = 0.991$$

A correlation coefficient can be calculated for a calibration curve to ascertain the degree of correlation between the measured instrumental variable and the sample concentration. As a general rule, $0.90 < r < 0.95$ indicates a fair curve, $0.95 < r < 0.99$ a good curve, and $r > 0.99$ indicates excellent linearity. An $r > 0.999$ can be obtained with care and is not uncommon.

The correlation coefficient gives the dependent and independent variables equal weight, which is usually not true in scientific measurements. The r value tends to give more confidence in the goodness of fit than warranted. The fit must be quite poor before r becomes smaller than about 0.98 and is really very poor when less than 0.9.

> The coefficient of determination (r^2) is a better measure of fit. The designations "r^2" and "R^2" are often used interchangeably, depending only on preference. Excel uses "R^2".

A more conservative measure of closeness of fit is the square of the correlation coefficient, r^2, and this is what most statistical programs calculate (including Excel—see Figure 3.5). An r value of 0.90 corresponds to an r^2 value of only 0.81, while an r of 0.95 is equivalent to an r^2 of 0.90. The goodness of fit is judged by the number of 9's. So three 9's (0.999) or better represents an excellent fit. We will use r^2 as a measure of fit. This is also called the **coefficient of determination**.

It should be mentioned that it is possible to have a high degree of correlation between two methods (r^2 near unity) but to have a statistically significant difference between the results of each according to the t-test. This would occur, for example, if there were a constant determinate error in one method. This would make the differences significant (not due to chance), but there would be a direct correlation between the results [r^2 would be near unity, but the slope (m) may not be near unity or the intercept (b) not near zero]. In principle, an empirical correction factor (a constant) could be applied to make the results by each method the same over the concentration range analyzed.

3.17 Detection Limits—There is No Such Thing as Zero

The previous discussions have dealt with statistical methods to estimate the reliability of analyses at specific confidence levels, these being ultimately determined by the precision of the method. All instrumental methods have a degree of noise associated with the measurement that limits the amount of analyte that can be detected. The noise is reflected in the precision of the blank or background signal, and noise may be apparent even when there is no significant blank signal. This may be due to fluctuation in the dark current of a photomultiplier tube, flame flicker in an atomic absorption instrument, and other factors.

> The concentration that gives a signal equal to three times the standard deviation of the background above the mean background is generally taken as the detection limit.

The limit of detection (LOD) is the lowest concentration level that can be determined to be statistically different from an analyte blank. There are numerous ways that detection limits have been defined. For example, the concentration that gives twice the peak-to-peak noise of a series of background signal measurements (or of a continuously recorded background signal) is sometimes taken as the detection limit (see Figure 3.6). A more generally accepted definition of the LOD is as follows. The LOD is the concentration that gives a signal that is above the background signal by three times the standard deviation of the background signal. It is instructive to compare the twice the peak-to-peak baseline noise criterion with the three times the standard deviation of the baseline criterion. Note that a continuous trace, as depicted in Figure 3.6, is tantamount to an infinite number of observations. According to the t-table (Table 3.1), 99% of the observations will fall within an interval of 2.576 standard deviations. The peak-to-peak noise is thus equivalent to ~2.6 standard deviations, and twice this value is ~5 standard deviations, a more stringent criterion than the 3 standard deviations criterion.

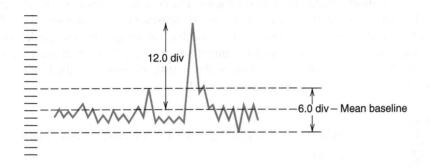

FIGURE 3.6 Peak-to-peak noise level as a basis for detection limit. The background fluctuations represent continuously recorded background signals, with the analyte measurement represented by the peak signal. A "detectable" analyte signal would be 12 divisions above a line drawn through the average of the baseline fluctuations.

Example 3.24

A series of sequential baseline absorbance measurements are made in a spectrophotometric method, for determining the purity of aspirin in tablets using a blank solution. The absorbance readings are 0.002, 0.000, 0.008, 0.006, and 0.003. A standard 1 ppm aspirin solution gives an absorbance reading of 0.051. What is the detection limit?

Solution

The standard deviation of the blank readings is ±0.0032 absorbance units, and the mean of the blank readings is 0.004 absorbance units. The detection limit is that concentration of analyte that gives a reading of $3 \times 0.0032 = 0.0096$ absorbance reading, above the blank signal. The net reading for the standard is $0.051 - 0.004 = 0.047$. The detection limit would correspond to 1 ppm $(0.0096/0.047) = 0.2$ ppm and would give a total absorbance reading of $0.0096 + 0.004 = 0.014$.

The precision at the detection limit is by definition about 33%. For quantitative measurements, the limit of quantitation (LOQ) is generally considered to be that value corresponding to 10 standard deviations above the baseline, approximately 3.3 times the LOD, about 0.7 ppm in the above example.

There have been various attempts to place the concept of detection limit on a more firm statistical ground. The International Conference on Harmonization (ICH) of Technical Requirements for Registration of Pharmaceuticals for Human Use has proposed guidelines for analytical method validation (Reference 18). The ICH Q2B guideline on validation methodology suggests calculation based on the standard deviation, s, of the response and the slope or sensitivity, S, of the calibration curve at levels approaching the limit. For the limit of detection (LOD),

$$\text{LOD} = 3.3(s/S) \tag{3.29}$$

And for limit of quantitation (LOQ)

$$\text{LOQ} = 10(s/S) \tag{3.30}$$

The standard deviation of the response can be determined based on the standard deviation of either the blank, the residual standard deviation of the least-squares regression line, or the standard deviation of the y intercept of the regression line. The Excel statistical function can be used to obtain the last two.

The International Union of Pure and Applied Chemistry (IUPAC) uses a value of 3 in Equation 3.29 (for blank measurements), derived from a confidence level of

95% for a reasonable number of measurements. The confidence level, of course, varies with the number of measurements, and 7 to 10 measurements should be taken. The bottom line is that one should regard a detection limit as an approximate guide to performance and not make efforts to determine it too precisely. The excellent article by Long and Winefordner on a closer look at the IUPAC definition of the LOD [Anal. Chem. 55:712A (1983)] is highly recommended.

3.18 Statistics of Sampling—How Many Samples, How Large?

Acquiring a valid analytical sample is perhaps the most critical part of any analysis. The physical sampling of different types of materials (solids, liquids, gases) is discussed in Chapter 2. We describe here some of the statistical considerations in sampling.

The Precision of a Result—Sampling is the Key

More often than not, the accuracy and precision of an analysis is limited by the sampling rather than the measurement step. The overall variance of an analysis is the sum of the sampling variance and the variance of the remaining analytical operations, that is,

$$s_o^2 = s_s^2 + s_a^2 \tag{3.31}$$

Little is gained by improving the analytical variance to less than one-third the sampling variance. It is better to analyze more samples using a faster, less precise method.

If the variance due to sampling is known (e.g., by having performed multiple samplings of the material of interest and analyzing it using a precise measurement technique), then there is little to be gained by reduction of s_a to less than $1/3 s_s$. For example, if the relative standard deviation for sampling is 3.0% and that of the analysis is 1.0%, then $s_o^2 = (1.0)^2 + (3.0)^2 = 10.0$, or $s_o = 3.2\%$. Here, 94% of the imprecision is due to sampling and only 6% is due to measurement (s_o is increased from 3.0 to 3.2%, so 0.2% is due to the measurement). If the sampling imprecision is relatively large, it is better to use a rapid, lower precision method and analyze more samples.

We are really interested in the value and variance of the true value. The total variance is $s_{total}^2 = s_g^2 + s_s^2 + s_a^2$, where s_g^2 describes the "true" variability of the analyte *in the system*, the value of which is the goal of the analysis. For reliable interpretation of the chemical analysis, the combined sampling and analytical variance should not exceed 20% of the total variance. [See M. H. Ramsey, "Appropriate Precision: Matching Analytical Precision Specifications to the Particular Application," *Anal. Proc.*, **30** (1993) 110.]

The "True Value"

The range in which the true value falls for the analyte content in a bulk material can be estimated from a *t*-tests at a given confidence level (Equation 3.12). Here, $\bar{x}$ is the average of the analytical results for the particular material analyzed, and s is the standard deviation that is obtained previously from analysis of similar material samples or from the present analysis if there are sufficient samples.

Minimum Sample Size

Statistical guidelines have been developed for the proper sampling of heterogeneous materials, based on the sampling variance. The minimum size of individual increments for a well-mixed population of different kinds of particles can be estimated from **Ingamell's sampling constant**, K_s:

$$wR^2 = K_s \tag{3.32}$$

where w is the weight of sample analyzed and R is the desired percent *relative* standard deviation of the determination. K_S represents the weight of sample for 1% sampling uncertainty at a 68% confidence level and is obtained by determining the standard deviation from the measurement of a series of samples of weight w. This equation, in effect, says that the sampling variance is inversely proportional to the sample weight.

> The greater the sample size, the smaller the variance.

Example 3.25

Ingamell's sampling constant for the analysis of the nitrogen content of wheat samples is 0.50 g. What weight sample should be taken to obtain a sampling precision of 0.2% rsd in the analysis?

Solution

$$w \ (0.2)^2 = 0.50 \text{ g}$$
$$w = 12.5 \text{ g}$$

Note that the entire sample is not likely to be analyzed. The 12.5-g gross sample will be finely ground, and a few hundred milligrams of the homogeneous material analyzed. If the sample were not made homogeneous, then the bulk of it would have to be analyzed.

Minimum Number of Samples

The number of individual sample increments needed to achieve a given level of confidence in the analytical results is estimated by

$$n = \frac{t^2 s_s^2}{g^2 \bar{x}^2} \tag{3.33}$$

where t is the student t value for the confidence level desired, s_s^2 is the sampling variance, g is the *acceptable* relative standard deviation of the average of the analytical results, $\bar{x}$; s_s is the *absolute* standard deviation, in the same units as $\bar{x}$, and so n is unitless. Values of s_s and $\bar{x}$ are obtained from preliminary measurements or prior knowledge.

Since g is equal to $s_x/\bar{x}$, we can write that

$$n = \frac{t^2 s_s^2}{s_x^2} \tag{3.34}$$

s_s and s_x can then be expressed in *either* absolute or relative standard deviations, so long as they are both expressed the same. Since n is initially unknown, the t value for the given confidence level is initially estimated and an iterative procedure is used to calculate n.

Example 3.26

The iron content in a blended lot of bulk ore material is about 5% (wt/wt), and the relative standard deviation of sampling, s_s, is 0.021 (2.1% rsd). How many samples should be taken in order to obtain a relative standard deviation g of 0.016 (1.6% rsd) in the results at the 95% confidence level [i.e., the standard deviation, s_x, for the 5% iron content is 0.08% (wt/wt)]?

Solution

We can use either Equation (3.33) or (3.34). We will use the latter. Set $t = 1.96$ (for $n = \infty$, Table 3.1) at the 95% confidence level. Calculate a preliminary value of n. Then use this n to select a closer t value, and recalculate n; continue iteration to a constant n.

$$n = \frac{(1.96)^2 (0.021)^2}{(0.016)^2} = 6.6$$

For $n = 7$, $t = 2.365$.

$$n = \frac{(2.365)^2 (0.021)^2}{(0.016)^2} = 9.6$$

For $n = 10$, $t = 2.23$

$$n = \frac{(2.23)^2 (0.021)^2}{(0.016)^2} = 8.6 \equiv 9$$

See if you get the same result using Equation 3.33.

Challenge: Can you apply Professor Pappas' approach (Section 3.11) to this problem? What requisite number of samples do you get?

Equation 3.33 holds for a **Gaussian distribution** of analyte concentration within the bulk material, that is, it will be centered around x with 68% of the values falling within one standard deviation, or 95% within two standard deviations. In this case, the variance of the population, σ^2, is small compared to the true value. If the concentration follows a **Poisson distribution**, that is, follows a random distribution in the bulk material such that the true or mean value $\bar{x}$ approximates the variance, s_s^2, of the population, then Equation 3.33 is somewhat simplified:

$$n = \frac{t^2}{g^2 \bar{x}} \cdot \frac{s_s^2}{\bar{x}} = \frac{t^2}{g^2 \bar{x}} \tag{3.35}$$

Note that since s_s^2 is equal to $\bar{x}$, the right-hand part of the expression becomes equal to 1, but the units do not cancel. In this case, when the concentration distribution is broad rather than narrow, many more samples are required to get a representative result from the analysis.

If the analyte occurs in clumps or patches, the sampling strategy becomes more complicated. The patches can be considered as separate strata and sampled separately. If bulk materials are segregated or stratified, and the average composition is desired, then the number of samples from each stratum should be in proportion to the size of the stratum.

3.19 Powering a Study: Power Analysis

Contributed by Professor Michael D. Morris, University of Michigan

How many samples are needed for a statistically valid result? The answer of course depends on variability of the population and confidence level needed. Chemists rarely ask this question, they tend to assume that nominally identical materials being tested *are* identical or nearly identical—only instrument accuracy and precision matter.

But these assumptions do not hold for people or animals or plants. Some claim that there is emerging iodine nutrition deficiency in the United States—how many people must we test for iodine intake to determine the veracity of this hypothesis?

We also need to introduce some new terms: Statistical tests of significance always use a null hypothesis, the *t*-test for example has the null hypothesis that the two populations are <u>not</u> statistically different. Tests are usually described as confirming or rejecting the *null hypothesis*, that is, that there is no difference between the measured parameter and some reference parameter.

A test is said to have a **Type I error**—when the null hypothesis is rejected when in fact it is true (probability $= \alpha$). **Type II error** constitutes failing to reject the null hypothesis when it is false (probability $= \beta$).

The **power** of a statistical test is the probability that it will reject a false null hypothesis (i.e., that it will not make a Type II error). The probability of a Type II error is referred to as the false negative rate (β). Therefore, power is equal to $1 - \beta$. The larger the power, the less probable a Type II error. Power analysis can be used to calculate the minimum sample size needed to accept the statistical test result at a given level of confidence.

A simple example for a null hypothesis: The average height of all 5-year-old boys in your town is the same as the average height of all 2-year-old boys in your town. An exhaustive test will involve that you measure the height of every 5-year-old boy and every 2-year-old boy, calculate the averages and calculate the difference. A more practical approach may be to measure a random sample from each age group and measure the difference. Either way, if the difference isn't zero, the null hypothesis is refuted—maybe.

Important issues are: How many 5-year-old and 2-year-old boys are needed. It is expensive and time-consuming to use too many, but the result may be meaningless with too few. What do we mean by non-zero difference—with what confidence level?

Some conventional (but arbitrary) levels are used: We want the probability of a type I error to be low, so $\alpha = 0.05$ is commonly chosen. We want the probability of a type II error to be low, i.e., the probability of correctly detecting a difference $(1 - \beta)$, to be high. The power of a study is $(1 - \beta)$. Power 0.8 is commonly chosen; it is often referred to as the study having 80% power and is the minimum that the National Institutes of Health will deem acceptable.

Typically the same number n of measurements (subjects) is used in the two groups. The mean values are X_1 and X_2 and s^2 is the mean of the variances (heights, in our example) of the two groups.

A simple example will suffice.

There are several cases, but we consider only two of the most important ones:

A given difference, $d = d_{min}$ is significant at $p = 0.05$ (95% confidence level) in both positive and negative directions, so then $2\alpha = 0.05$.

The number of standard deviations encompassed in either direction is $z_{2\alpha} = 1.96$.

In this case, n, the size of the sample needed in each of the two groups, is given by

$$n > 2(sZ_{2\alpha}/d)^2 \tag{3.36}$$

In general, we don't know what d_{min} is, but we can specify it here as, for example, 1 cm. We can use child growth data published by the Centers for Disease Control (http://www.cdc.gov/growthcharts/) to estimate the standard deviations for each of the two age groups. From the growth charts, we estimate $s \approx 2.5$ cm for 2-year-old boys and $s \approx 3.5$ cm for 5-year-old boys. We can take s to be the average standard deviation in the two populations, i.e., $1/2 (2.5 + 3.5) = 3$ cm.

$$n > 2 (3 * 1.96/1)^2 \tag{3.37}$$

In that case, using Equation 3.37, we calculate that $n > 69.1$. So, a minimum sample size is 70. Because the sample is typically divided equally between the two classes, we need at least thirty-five 2-year-old boys and at least thirty-five 5-year-old boys.

It is more common to specify the statistical power required to achieve a specified type II error rate. A given statistical power (usually 80%) has a specified true difference, $\delta > \delta_1$. Here δ is a true difference and δ_1 is the threshold true difference sought.

So, if the measured difference between the means is d, then it is significant if:

$$\delta_1 > (z_{2\alpha} + z_{2\beta})SE(d) \tag{3.38}$$

and

$$SE(d) = \frac{s}{\sqrt{n}}\sqrt{1 - \frac{n}{N}} \tag{3.39}$$

Because β is the type II error (false negative probability), $1 - \beta$ is the 95% probability of correctly reporting a difference. So for $\beta = 0.2$, $z_{2\beta} = 1.64$.

The final result is that

$$n > 2\left[\frac{(z_{2\alpha} + z_{2\beta})\,s}{\delta_1}\right]^2 \tag{3.40}$$

Using the data above and taking $\delta_1 = 2$ cm, we calculate that

$$n > 2\left[(1.96 + 1.64)\,3/2\right]^2 = 58.3 \tag{3.41}$$

or $n > 59$.

In summary, if the total population size is known and the measurement error can be estimated, and the significance (or confidence interval) and power are defined, then the number of subjects needed can be specified.

P. Armitage, G. Berry and J. N. S. Matthews, *Statistical Methods in Medical Research*, 4th Ed. Malden, MA: Blackwell Science, Inc., 2002 is a good reference on powering studies.

See the book's website for a PowerPoint™ presentation contributed by Professor Morris on diverse statistical topics including box-and-whisker plots, signal-to-noise ratio, histograms, non-Gaussian histograms and histogram equalization/normalization, sensitivity and specificity in binary tests, and receiver operating characteristics (ROC). See the website also for Dr. Morris's statistical view of the recent decision on questionable utility of mammograms in preventing breast cancer.

3.20 Use of Spreadsheets in Analytical Chemistry

A spreadsheet is a powerful software program that can be used for a variety of functions, such as data analysis and plotting. Spreadsheets are useful for organizing data, doing repetitive calculations, and displaying the calculations graphically or in chart form. They have built-in functions, for example, standard deviation and other statistical functions, for carrying out computations on data that are input by the user. Currently the most popular spreadsheet program is Microsoft Excel. We will use Excel 2010 in our illustrations.

You probably have used a spreadsheet program before and are familiar with the basic functions. But we will summarize here the most useful aspects for analytical chemistry applications. Also, the Excel Help on the tool bar provides specific information.

Use Excel Help if you are having trouble. It provides step-by-step instructions on how to carry out most of the functions.

You are referred to the excellent tutorial on using the Excel spreadsheet prepared by faculty at California State University at Stanislaus: http://science.csustan.edu/tutorial/Excel/index.htm. The basic functions in the spreadsheet are described, including entering data and formulas, formatting cells, graphing, and regression analysis. You are likely to find this helpful, even though this presentation is based on an older version of Excel. Introductory lessons on how to use Excel 2010 are available at http://www.gcflearnfree.org/excel2010/1. This will familiarize you with the contents of the following several paragraphs. The website http://www.ncsu.edu/labwrite/res/gt/gt-menu.html at North Carolina State University also provides instructions for graphing in Excel.

A spreadsheet consists of **cells** arranged in columns (labeled A, B, C, ...) and **rows** (numbered 1, 2, 3, ...). An individual cell is identified by its column letter and row number, for example, B3. Figure 3.7 has the identifiers typed into some of the cells to illustrate. When the mouse pointer (the cross) is clicked on an individual cell, it becomes the **active cell** (dark lines around it), and the active cell is indicated at the top left of the formula bar, and the contents of the cell are listed to the right of the f_x sign on the bar.

Filling the Cell Contents

You may enter *text*, *numbers*, or *formulas* in specific cells. Formulas are the key to the utility of spreadsheets, allowing the same calculation to be applied to many numbers. We will illustrate with calculations of the weights of water delivered by two different 20-mL pipets, from the difference in the weights of a flask plus water and the empty flask. Refer to Figure 3.8 as you go through the steps.

	A	B	C	D	E
1	A1	B1	C1	etc.	
2	A2	B2	C2	etc.	
3	A3	B3	C3	etc.	
4					
5					

FIGURE 3.7 Spreadsheet cells.

	A	B	C	D
1	Net weights			
2				
3	Pipet	1	2	
4	Weight of flask + water, g	47.702	49.239	
5	Weight of flask, g	27.687	29.199	
6	Weight of water, g	20.015	20.040	
7				
8	cell B6=B4-B5			
9	cell C6=C4-C5			
10				

FIGURE 3.8 Filling cell contents.

Open an Excel spreadsheet by clicking on the Excel icon (or the Microsoft Excel program under Start/All Programs/Microsoft Office/Microsoft Excel 2010). You will enter text, numbers, and formulas. Double-click on the specific cell to activate it. Enter as follows (information typed into a cell is entered by depressing the Enter key):

Cell A1: Net weights

Cell A3: Pipet

Cell A4: Weight of flask + water, g

Cell A5: Weight of flask, g

Cell A6: Weight of water, g

You may make corrections by double-clicking on a cell; then edit the text. (You can also edit the text in the formula bar.) If you single click, new text replaces the old text. You will have to widen the A cells to accommodate the lengthy text. Do so by placing the mouse pointer on the line between A and B on the row at the top, and dragging it to the right until all the text shows. You can also adjust the column width by selecting the entire column (click on the top title column marked A, B, etc.) and on the Home tab click on Format/Autofit Column Width.

Cell B3: 1

Cell C3: 2

Cell B4: 47.702

Cell C4: 49.239

Cell B5: 27.687

Cell C5: 29.199

Cell B6: =B4-B5

You can also change the number of digits shown after the decimal point, and do many other operations, by clicking on the appropriate buttons on the toolbar.

You can also enter the formula by typing =, then click on B4, then type—, and click on B5. You need to format the cells B4 to C6 to three decimal places. Highlight that block of cells by clicking on one corner and dragging to the opposite corner of the block. In the Menu bar, click on Number/number, adjust decimal places to 3, and click OK.

You need to add the formula to cell C6. You can retype it. But there is an easier way, by copying (filling) the formula in cell B6. Place the mouse pointer on the lower right corner of cell B6 and drag it to cell C6. This fills the formula into C6 (or additional cells to the right if there are more pipet columns). You may also fill formulas into highlighted cells in other ways.

Double-click on B6. This shows the formula in the cell and outlines the other cells contained in the formula. Do the same for C6. Note that when you activate the cell by either single or double-clicking on it, the formula is shown in the formula bar.

Saving the Spreadsheet

Save the spreadsheet you have just created by clicking on File:SaveAs. Give the document a File Name at the bottom, for example, Pipet Calibration. Then click Save.

Printing the Spreadsheet

Click Page Layout/Orientaion. Normally, a sheet is printed in the Portrait format, that is, vertically on the 8 1/2- × 11-inch paper. If there are many columns, you may wish to print in Landscape, that is, horizontally. If you want gridlines to print, click on Page Layout/Gridlines/Print. Now you are ready to print. Click on the MS Office icon and then Print. Just the working area of the spreadsheet will print, not the column and row identifiers. You can also set the desired print area by highlighting the cells you want to appear on a page. After you elect to Print, Excel will ask you if you want to print just the selected area or the entire worksheet.

Relative vs. Absolute Cell References

In the example above, we used *relative* cell references in copying the formula. The formula in cell B6 said subtract the cell above from the one above it. The copied formula in C6 said the same for the cells above it.

Sometimes we need to include a specific cell in each calculation, containing, say, a constant. To do this, we need to identify it in the formula as an *absolute* reference. This is accomplished by placing a $ sign in front of the column and row cell identifiers, for example, $B$2. Placing the sign in front of both assures that whether we move across columns or rows, it will remain an absolute reference.

We can illustrate this by creating a spreadsheet to calculate the means of different series of numbers. Fill in the spreadsheet as follows (refer to Figure 3.9):

A1: Titration means

A3: Titn. No.

B3: Series A, mL

C3: Series B, mL

B4: 39.27

B5: 39.18

B6: 39.30

B7: 39.20

C4: 45.59

C5: 45.55

C6: 45.63

C7: 45.66

A4: 1

We type in each of the titration numbers (1 through 4), but there are automatic ways of incrementing a string of numbers. Click on Fill/Series. Click the Columns and Linear radio buttons, and leave Step Value at 1. For Stop Value, enter 4 and click OK. The numbers 2 through 4 are inserted in the spreadsheet. You could also first highlight the cells you want filled (beginning with cell A4). Then you do not have to insert a Stop Value. Another way of incrementing a series is to do it by formula. In cell A5, type = A4 + 1. Then you can fill down by highlighting from A5 down, and clicking

	A	B	C	D
1	Titration means			
2				
3	Titn. No.	Series A, mL	Series B, mL	
4	1	39.27	45.59	
5	2	39.18	45.55	
6	3	39.30	45.63	
7	4	39.22	45.66	
8	Mean:	**39.24**	45.61	
9	Std.Dev.	0.053150729	0.047871355	
10				
11	**Cell B8=**	SUM(B4:B7)/$A$7 Copy right to Cell C8.		
12	**Cell B9=**	STDEV(B4:B7) Copy right to Cell C9.		
13	We have boldfaced the cells with formulas entered.			

FIGURE 3.9 Relative and absolute cell references.

on Fill/Down. (This is a relative reference.) Or, you can highlight cell A5, click on its lower right corner, and drag it to cell A7. This automatically copies the formula in the other cells or increments numbers in a series.

Now we wish to insert a formula in cell B8 to calculate the mean. This will be the sum divided by the number of titrations (cell A7).

$$B8 := sum(B4:B7/\$A\$7$$

We place the $ signs in the denominator because it will be an *absolute reference* that we wish to copy to the right in cell C8. Placing a $ before both the column and row addresses assures that the cell will be treated as absolute whether it is copied horizontally or vertically. The sum (B4:B7) is a **syntax** in the program for summing a series of numbers, from cell B4 through cell B7. Instead of typing in the cell addresses, you can also type "=sum(", then click on cell B4 and drag to cell B7, and type ")". We have now calculated the mean for series A. [In B8 we could also enter =AVERAGE (B4:B7) to calculate the mean, see later.] We wish to do the same for series B. Highlight cell B8, click on its lower right corner, and drag it to cell C8. *Voilà*, the next mean is calculated! Double-click on cell C8, and you will see that the formula has the same divisor (*absolute reference*), but the sum is a *relative reference*. If we had not typed in the $ signs to make the divisor absolute, the formula would have assumed it was relative, and the divisor in cell C8 would be cell B7.

Use of Excel Statistical Functions

For advanced statistical functions, you can also install the free Analysis ToolPak add-in. See Section 3.24.

Excel has a large number of mathematical and statistical functions that can be used for calculations in lieu of writing your own formulas. Let's try the statistical functions to automatically calculate the mean. Highlight an empty cell and click Formulas/Insert Function. The Insert Function window appears. Select Statistical in the select a category box. The following window appears:

Select AVERAGE for the Function name. Click OK, and then in the newly appeared windows, in the Number 1 window, type B4:B7, and click OK. The same

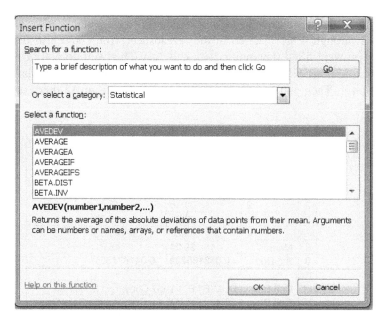

average is calculated as you obtained with your own formula. You can also type in the activated cell the syntax =average(B4:B7). Try it.

Let's calculate the standard deviation of the results. Highlight cell B9. Under the Statistical function, select STDEV for the Function name. Alternatively, you can type the syntax into cell B9, =stdev(*B*4:B7). Now copy the formula to cell C9. Perform the standard deviation calculation using Equation 3.2 and compare with the Excel values. The calculation for series A is ±0.05 mL. The value in the spreadsheet, of course, should be rounded to ±0.05 mL.

In the website of the book, you will find two video clips, titled AVERAGE and STDEV, that illustrate the use of these functions.

Useful Syntaxes

Excel has numerous mathematical and statistical functions or syntaxes that can be used to simplify setting up calculations. Peruse the Function names for the Math & Trig and the Statistical function categories under f_x in the formulas submenu. Some you will find useful for this text are:

Math and trig functions		Statistical functions	
LOG10	Calculates the base-10 logarithm of a number	AVERAGE	Calculates the mean of a series of numbers
PRODUCT	Calculates the products of a series of numbers	MEDIAN	Calculates the median of a series of numbers
POWER	Calculates the result of a number raised to a power	STDEV	Calculates the standard deviation of a series of numbers
SQRT	Calculates the square root of a number	VAR	Calculates the variance of a series of numbers

The syntaxes may be typed, followed by the range of cells in parentheses, as we did above.

This tutorial should provide you the basics for other spreadsheet applications. You can write any formula that is in this book into an active cell, and insert appropriate data for calculations. And, obviously, we can perform a variety of data analyses. We can prepare plots and charts of the data, for example, a calibration curve of instrument response versus concentration, along with statistical information. We will illustrate this later in the chapter.

Text Website Examples

Professor Steven Goates, Brigham Young University, provides tutorial examples, given on your text **website**, of using Excel for calculating confidence limits, for calculating least-squares regression lines, and for *t*-test comparisons of the means of two sets of data. Also, Professor Wen-Yee Lee, University of Texas at El Paso, gives an example for plotting and calculating least-squares regression lines for finding concentrations of unknowns using working curves.

3.21 Using Spreadsheets for Plotting Calibration Curves

While independent graphing software may provide greater flexibility, most spreadsheet programs, including Excel, allow graphing and charting, aside from statistical and least-squares calculations already mentioned. We will use the data in Example 3.21 to prepare the plot shown in Figure 3.5, using Excel.

Open a new spreadsheet and enter:

Cell A1: Riboflavin, ppm (adjust the column width to incorporate the text)

Cell B1: Fluorescence intensity

Cell A3: 0.000

Cell A4: 0.100

Cell A5: 0.200

Cell A6: 0.400

Cell A7: 0.800

Cell B3: 0.0

Cell B4: 5.8

Cell B5: 12.2

Cell B6: 22.3

Cell B7: 43.3

Format the cell numbers to have three decimal places for column A and one for column B.

If you click on the "Insert" tab on the menu bar of Excel 2010, the chart options will appear in the middle. Most often you will use scatter plots (the submenu shown on bottom left) appears upon clicking on "Scatter". The top left selection (see the figure) in the scatter plots will result in the data points being plotted without a line connecting

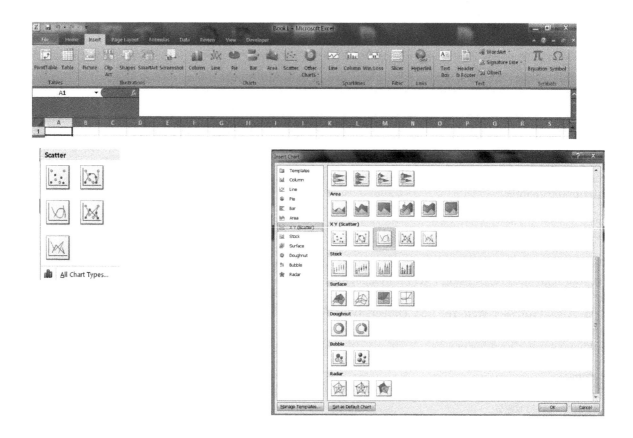

them, the top right in smooth (often called spline-fit) lines connecting them, the middle left in smooth lines but without data points, the middle right in points connected with straight lines between them, and the bottom left with straight lines between points that are invisible.

Here is how to prepare a plot using Excel 2010: Highlight the numbers in columns A and B. To the right of the Home bar, click on Insert, Column, All Chart Types. Select XY (scatter), and then Scatter with only markers. Click OK, and the plot appears. (You may also directly click on Scatter under Insert.) Add statistical data by clicking on the line in order to display Chart Tools, and then select Trendline. Select Linear Trendline, and check Display Equation on chart and Display R-squared value on chart (these are under More Trendline Options). Click Close, and the data are added to the chart. To add axis labels, go to Layout under Chart Tools and click on Axis Titles. Clicking on Primary Horizontal Axis Title and Vertical Axis Title adds these to the chart, and you can type in the actual titles. You can delete the Series notation at the right. You can place the chart on a separate sheet under Chart Tools/Design, and select Move Chart.

3.22 Slope, Intercept, and Coefficient of Determination

We can use the Excel statistical functions to calculate the slope and intercept for a series of data, and the R^2 value, without a plot. Open a new spreadsheet and enter the calibration data from Example 3.21, as in Figure 3.10, in cells A3:B7. In cell A9 type Intercept, in cell A10, Slope, and in cell A11, R^2. Highlight cell B9, click on formulas/insert function, select statistical, and scroll down to INTERCEPT under Function name, and click OK. For Known_x's, enter the array A3:A7, and for Known_y's, enter B3:B7. Click OK, and the intercept is displayed in cell B9. Now repeat, highlighting cell B10, scrolling to Slope, and entering the same arrays. The slope appears in cell B10. Repeat again, highlighting cell B11, and scrolling to RSQ. R^2 appears in cell B11. Compare with the values in Figure 3.10.

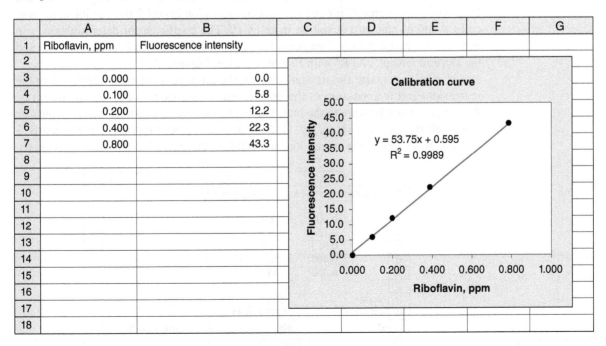

FIGURE 3.10 Calibration graph inserted in spreadsheet (Sheet 1).

3.23 LINEST for Additional Statistics

The LINEST program of Excel allows us to quickly obtain several statistical functions for a set of data, in particular, the slope and its standard deviation, the intercept and its standard deviation, the coefficient of determination, and the standard error of the estimate, besides others we will not discuss now. LINEST will automatically calculate a total of 10 functions in 2 columns of the spreadsheet.

Open a new spreadsheet, and enter the calibration data from Example 3.21 as you did above, in cells A3:B7. Refer to Figure 3.11. The statistical data will be placed in 10 cells, so let's label them now. We will place them in cells B9:C13. Type labels as follows:

> Cell A9: slope
>
> Cell A10: std. devn.
>
> Cell A11: R^2
>
> Cell A12: F
>
> Cell A13: sum sq. regr.
>
> Cell D9: intercept
>
> Cell D10: std. devn.
>
> Cell D11: std. error of estim.
>
> Cell D12: d.f.
>
> Cell D13: sum sq. resid.

Highlight cells B9:C13, and click on formulas f_x. Select Statistical function, scroll down to LINEST and click OK. For Known_y's, enter the array B3:B7, and for Known_x's, enter A3:A7. Then in each of the boxes labeled Const and Stats, type "true". Now we have to use the keyboard to execute the calculations. Depress Control, Shift, and Enter, and release. The statistical data are entered into the highlighted cells. This keystroke combination must be used whenever performing a function on an array of cells, like here. The slope is in cell B9 and its standard deviation in cell B10. The intercept is in cell C9 and its standard deviation in cell C10. The coefficient of determination is in cell B11. Compare the standard deviations with those calculated in Example 3.22, and the slope, intercept, and R^2 with Example 3.21 or Figure 3.5.

Cell C11 contains the standard error of the estimate (or standard deviation of the regression) and is a measure of the error in estimating values of y. The smaller it is, the closer the numbers are to the line. The other cells contain data we will not consider here: Cell B12 is the F value, cell C12 the degrees of freedom (used for F), cell B13 the sum of squares of the regression, and cell C13 the sum of squares of the residuals.

	A	B	C	D	E
1	Riboflavin, ppm	Fluorescence intensity			
2					
3	0.000	0.0			
4	0.100	5.8			
5	0.200	12.2			
6	0.400	22.3			
7	0.800	43.3			
8					
9	slope	53.75	0.595	intercept	
10	std. devn.	1.017759	0.419633	std. devn.	
11	R^2	0.998926	0.643687	std. error of estim.	
12	F	2789.119	3	d.f.	
13	sum sq. regr.	1155.625	1.243	sum sq. resid.	

FIGURE 3.11 Using LINEST for statistics.

How many significant figures should we keep for the least-squares line? The standard deviations give us the answer. The slope has a standard deviation of 1.0, and so we write the slope as $53.8 \pm 1._0$ at best. The intercept standard deviation is ± 0.42, so for the slope we write 0.6 ± 0.4. See also Example 3.22.

Possibly the most powerful tool to calculate all regression-related parameters is to use the "Regression" function in Data Analysis. It not only provides the results for r, r^2, intercept, and slope (which it lists as X variable 1), it also provides their standard errors and upper and lower limits at a 95% confidence level. It also provides an option for fitting the straight line through the origin (when you know for certain that the response at zero concentration is zero by checking a box "constant is zero". A description of its use is given in Chapter 11 at the end of Section 11.7, and example applications are given in Chapter 15, Section 15.5, and Chapter 23 for Example 23.1 and Example 23.2.

3.24 Statistics Software Packages

Excel offers a number of statistical functions through the Analysis ToolPak Add-in. Click on the MS Office icon on the top left corner of the screen and click through Excel Options/Add-Ins and Analysis ToolPak. Click OK. Return to the spreadsheet. Now when you go to the Data menu, you will see Data Analysis available on the top right corner. The Solver Add-In, another very useful routine, can be installed in the same way. Click on the Data Analysis link, and you will see 18 programs listed, including the F test, various types of t-tests, and Regression (this provides the best fit, along with the r^2 and uncertainties). As you experiment with the programs available under Data Analysis, you will find some very useful. The use of Solver is described in Chapter 5. See also the text **website** for a list of some commercial software packages for performing basic as well as more advanced statistical calculations.

Professor's Favorite Example

Contributed by Professor James Harynuk, University of Alberta

An excellent online Flash video tutorial is available in the Journal of Chemical Education (JCE) illustrating how to use Excel, using linear least-squares calculations as an example. It is available at: http://jchemed.chem.wisc.edu/JCEDLib/WebWare/collection/reviewed/JCE2009p0879WW/index.html.

It is available only to subscribers of JCE, and requires a username and password. If your university/college is a subscriber, you may be able to access the site on a college computer without the username/password, or may be able to get them. Otherwise, ask your instructor if she/he has access.

The specific equations used are based on those in earlier editions of another text, but the calculations can be adjusted for those in this textbook once you have learned how to create the spreadsheet. For example, the equation given for the slope is a rearrangement of Equation 3.23 (multiply both numerator and denominator by n, and you get Equation 3.23). Try setting up a spreadsheet and apply it to Example 3.21 and see if you get the same answers.

Questions

1. Distinguish between accuracy and precision.

2. What is determinate error? An indeterminate error?

3. The following is a list of common errors encountered in research laboratories. Categorize each as a determinate or an indeterminate error, and further categorize determinate errors as instrumental, operative, or methodological: (a) An unknown being weighed is hygroscopic. (b) One component of a mixture being determined quantitatively by gas chromatography reacts with the column packing. (c) A radioactive sample being counted repeatedly without any change in conditions yields a slightly different count at each trial. (d) The tip of the pipet used in the analysis is broken.

4. Explain the terms: (i) absolute error, (ii) relative error, and (iii) confidence limit.

5. What is the significance of F test and define F test?

6. From the following options, choose with high accuracy low precision:

   (a) arrows were consistently hit to the left of target.

   (b) arrows consistently miss the target and hit the tree at same spot.

   (c) arrows were consistently hit around the target but never hit the bullseye.

   (d) arrows were consistently hit at the bullseye.

7. Which of the following are used to analyze central tendency of collected data?

   (a) Standard deviation, Range and mean

   (b) Mean and normal distribution

   (c) Mean, median and mode

   (d) Median, Range and normal distribution.

Problems

For the statistical problems, do the calculations manually first and then use the Excel statistical functions and see if you get the same answers. See the textbook companion website, Problems 19–23, 25, 26, 30–35, and 42–44, 46.

Significant Figures

8. How many significant figures does each of the following numbers have?

   (a) 87009   (b) 0.0009   (c) 9.000000

9. How many significant figures does each of the following numbers have?

   (a) 5.060×10^{-3} (b) 9.00×10^{-2} (c) 2000

10. For $LiNO_3$, calculate its formula weight upto correct number of significant figures.

11. For $PdCl_2$, calculate its formula weight upto correct number of significant figures.

12. Express the result of following mathematical operations with the correct number of significant numbers.

    (a) $\dfrac{3.41\,g - 0.23\,g}{5.233\,g} \times 0.205\ mL$ (b) $\dfrac{5.556\,mL \times 2.3\,g/mL}{4.223 - 0.08\ mL}$

13. Round the following answers and show that only significant digits are retained.

    (a) $\log 4.000 \times 10^{-5} = -4.3979400$ (b) Antilog $12.5 = 3.162277 \times 10^{12}$

14. An analyst wishes to analyze spectrophotometrically the copper content in a bronze sample. If the sample weighs about 5 g and if the absorbance (A) is to be read to the nearest 0.001 absorbance unit, how accurately should the sample be weighed? Assume the volume of the measured solution will be adjusted to obtain minimum error in the absorbance, that is, so that $0.1 < A < 1$.

15. Choose the number with the greatest relative uncertainty that would control the result of multiplication or division from following pairs:

    (a) 100.0 or 0.3593,   (b) 42.77 or 0.0935 and   (c) 0.0059 or 0.10

Expressions of Results

16. A standard serum sample containing 102 meq/L chloride was analyzed by coulometric titration with silver ion. Duplicate results of 101 and 98 meq/L were obtained. Calculate (a) the mean value, (b) the absolute error of the mean, and (c) the relative error in percent.

17. A batch of nuclear fuel pellets was weighed to determine if they fell within control guidelines. The weights were 125.3, 126.2, 125.7, 127.1, and 126.8 g. Calculate (a) the mean, (b) the median, and (c) the range.

18. Calculate the absolute error and the relative error in percent and in parts per thousand in the following:

	Measured Value	Accepted Value
(a)	22.62 g	22.57 g
(b)	45.02-mL	45.31-mL
(c)	2.68%	2.71%
(d)	85.6 cm	85.0 cm

Standard Deviation

19. The tin and zinc contents of a brass sample are analyzed with the following results: (a) Zn: 33.27, 33.37, and 33.34% and (b) Sn: 0.022, 0.025, and 0.026%. Calculate the standard deviation and the coefficient of variation for each analysis.

20. Replicate water samples are analyzed for water hardness with the following results; 102.2, 102.8, 103.1, and 102.3 ppm $CaCO_3$. Calculate (a) the standard deviation, (b) the relative standard deviation, (c) the standard deviation of the mean, and (d) the relative standard deviation of the mean.

21. Replicate samples of a silver alloy are analyzed and determined to contain 95.67, 95.61, 95.71, and 95.60% Ag. Calculate (a) the standard deviation, (b) the standard deviation of the mean, and (c) the relative standard deviation of the mean (in percent) of the individual results.

PROFESSOR'S FAVORITE PROBLEMS

Contributed by Professor Wen-Yee Lee, University of Texas at El Paso

22. The mileage at which 10,000 sets of car brakes had been 80% worn through was recorded: the average was 64,700 and the standard deviation was 6,400 miles.

    (a) What fraction of brakes is expected to be 80% worn in less than 50,000 miles?

    (b) If the brake manufacturer offers free replacement for any brake that is 80% worn in less than 50,000 miles, how many extra brakes should be kept available for every 1 million product sold?

    You will need to calculate a Z-value, from normal probability distribution, and from this, an area under the Gaussian curve. See www.intmath.com/Counting-probability/14_Normal-probability-distribution.php for a discussion of this. The following website contains a Standard Normal (Z) Table that can be used to obtain the area from the Z-value: www.statsoft.com/textbook/distribution-tables.

Propogation of Error

23. Calculate the uncertainty in the answers of the following: (a) $(128 \pm 2) + (1025 \pm 8) - (636 \pm 4)$, (b) $(16.25 \pm 0.06) - (9.43 \pm 0.03)$, (c) $(46.1 \pm 0.4) + (935 \pm 1)$.

24. Calculate the absolute uncertainty in the answers of the following: (a) $(2.78 \pm 0.04)(0.00506 \pm 0.00006)$, (b) $(36.2 \pm 0.4)/(27.1 \pm 0.6)$, (c) $(50.23 \pm 0.07)(27.86 \pm 0.05)/(0.1167 \pm 0.0003)$.

25. Calculate the absolute uncertainty in the answer of the following: $[(25.0 \pm 0.1) \, (0.0215 \pm 0.0003) - (1.02 \pm 0.01) \, (0.112 \pm 0.001)] \, (17.0 \pm 0.2) / (5.87 \pm 0.01)$.

PROFESSOR'S FAVORITE PROBLEMS

Contributed by Professor Jon Thompson, Texas Tech University

26. *Climate Change and Propagation of Uncertainty.* Many factors can cause changes in Earth's climate. These range from well-known greenhouse gases such as carbon dioxide and methane to changes in ozone levels, the effective reflectivity of Earth's surface, and the presence of nm - μm sized aerosol particles that can scatter and absorb sunlight in the atmosphere. Some effects lead to a warming influence on climate, while others may cool the Earth and atmosphere. Climate scientists attempt to *keep score* and determine the *net* effect of all competing processes by assigning radiative forcing values (W/m^2) to each effect independently and then summing them. Positive radiative forcing values warm climate, while negative values cool the Earth and atmosphere. This approach is insightful since the net radiative forcing (ΔF_{net}) can be linked to expected mean temperature change $(\Delta T_{surface})$ via the climate sensitivity parameter (λ), which is often assigned values of $0.3 - 1.1 \, K/(W/m^2)$.

$$\Delta T_{surface} = \lambda \times \Delta F_{net} \qquad (3.42)$$

The Intergovernmental Panel on Climate Change (IPCC) has studied the work of many scientists to provide current best estimates of radiative forcings and associated uncertainties for each effect. These are described in the figure and table shown below.

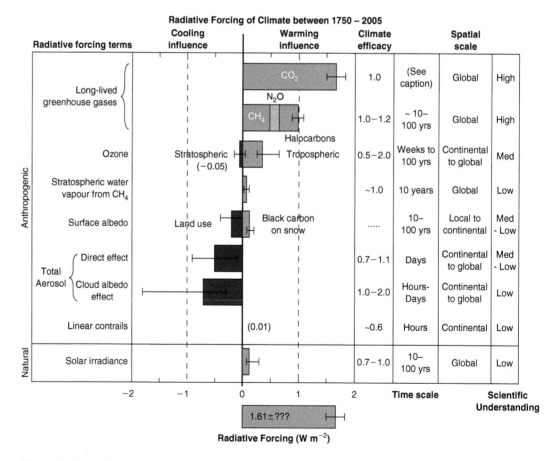

Figure Credit: "Climate Change 2007: The Physical Science Basis" Intergovernmental Panel on Climate Change (IPCC).

Best estimates of climate forcings and uncertainties

Climate Effect	Best Estimate ± Uncertainty (W/m^2)
Long-lived Greenhouse Gases	2.61 ± 0.26
Tropospheric and Stratospheric Ozone	0.30 ± 0.22
Surface Albedo	-0.10 ± 0.20
Direct Aerosol Effect	-0.50 ± 0.36
Indirect Aerosol Effect	-0.70 ± 0.7

Use the rules for propagation of uncertainty to *Ask Yourself*:

(a) The sum of the radiative forcing terms is 1.61 W/m^2. What is the uncertainty associated with this estimate? Which of the individual terms seem to dominate the magnitude of the overall uncertainty?

(b) If a value of 0.7 ± 0.4 K/(W/m^2) is used as the climate sensitivity parameter (λ), what is the expected surface temperature change? What is the associated uncertainty in this estimate?

(c) Instrumental temperature records from the 1850s to present day suggest the global mean surface temperature has increased by roughly 0.8°C. Is this estimate consistent with your calculation from b?

Confidence Limit

For an Excel-based problem related to confidence limits, courtesy of Professor Steven Goates, Brigham Young University, see the website supplement.

27. The following molarities were calculated from replicate standardization of a solution: 0.5026, 0.5029, 0.5023, 0.5031, 0.5025, 0.5032, 0.5027, and 0.5026 M. Assuming no determinate errors, within what range are you 95% certain that the true mean value of the molarity falls?

PROFESSOR'S FAVORITE PROBLEMS

Contributed by Professor Amanda Grannas, Villanova University

28. You work in an analytical testing lab and have been asked by your supervisor to purchase new pH meters to replace your old supply. You have identified four potential suppliers and now need to evaluate the performance of each pH meter.

    You obtained a solution with a known pH of 5.5. You performed 10 replicate measurements of the pH of that known solution using each of the four different pH meters and obtained the data shown below.

pH meter Brand A	pH meter Brand B	pH meter Brand C	pH meter Brand D
5.6	5.5	5.8	7.0
5.8	5.6	5.9	6.9
5.6	5.5	6.0	6.8
5.5	5.6	5.9	7.0
5.6	5.6	5.3	6.9
5.7	5.6	5.6	6.9
5.6	5.4	5.7	7.0
5.7	5.5	5.8	6.8
5.1	5.5	5.9	6.9
5.6	5.4	5.1	6.9

All other factors being equal (cost, ease of use, etc), which brand would you suggest the lab purchase? Explain your reasoning.

29. Determination of the sodium level in separate portions of a blood sample by ion-selective electrode measurement gave the following results: 139.2, 139.8, 140.1, and 139.4 meq/L. What is the range within which the true value falls, assuming no determinate error (a) at the 90% confidence level, (b) at the 95% confidence level, and (c) at the 99% confidence level?

30. Lead on leaves by a roadside was measured spectrophotometrically by reaction with dithizone. The standard deviation for a triplicate analysis was 2.3 ppm. What is the 90% confidence limit?

PROFESSOR'S FAVORITE PROBLEMS

Contributed by Professor Wen-Yee Lee, University of Texas at El Paso

31. The blood cell counts on 7 normal subjects are 5.2, 4.8, 5.4, 5.3, 5.1, 4.9, 5.5 million cells per microliter. The lab work showed that my blood cell count is 4.5 million cells per microliter. Is my blood cell count too low? How confident are you? Show the statistics that supports your answer.

32. The standard deviation established for the determination of blood chloride by coulometric titration is 0.5 meq/L. What is the 95% confidence limit for a triplicate determination?

33. A standard reference material is certified to contain 94.6 ppm of an organic pesticide in soil. Your analysis gives values of 99.3, 92.5, 96.2, and 93.6 ppm. Do your results significantly differ from the expected result at 95% confidence level? Show your work to explain your answer.

34. Estimate the range of the true molarity of the solution at the 90% confidence level from the standardization in Problem 46.

PROFESSOR'S FAVORITE PROBLEMS

Contributed by Professor Bin Wang, Marshall University

35. A repeated analysis of Cl in a given compound resulted in the following results for %Cl: 2.98, 3.16, 3.02, 2.99, and 3.07. (a) Can any of these results be rejected for statistical reasons at the 90% confidence level? (b) If the true value was 3.03%, can you be 95% confident that your results agree with the known value?

Tests of Significance

36. A study is being performed to see if there is a correlation between the concentration of chromium in the blood and a suspected disease. Blood samples from a series of volunteers with a history of the disease and other indicators of susceptibility are analyzed and compared with the results from the analysis of samples from healthy control subjects. From the following results, determine whether the differences between the two groups can be ascribed to chance or whether they are real. Control group (ppb Cr): 15, 23, 12, 18, 9, 28, 11, 10. Disease group: 25, 20, 35, 32, 15, 40, 16, 10, 22, 18.

PROFESSOR'S FAVORITE PROBLEMS

Contributed by Professor Wen-Yee Lee, The University of Texas at El Paso

37. This is a comparison of peak expiratory flow rate (PEFR) before and after a walk on a cold winter's day for a random sample of 9 asthmatics. As shown in the following table, data in one column indicate the PEFRs before the walk and the other of PEFRs after the walk. Each row represents the same subject. Is the PEFR significantly different before and after a walk at 95% confidence level?

38. An enzymatic method for determining alcohol in wine is evaluated by comparison with a gas-chromatographic (GC) method. The same sample is analyzed several times by both methods with the following results (% ethanol). Enzymatic method: 13.1, 12.7, 12.6, 13.3, 13.3. GC method: 13.5, 13.3, 13.0, 12.9. Does the enzymatic method give the same value as the GC method at the 95% confidence level?

Subject	Expiratory flow rate (L/min)	
	Before walk	After walk
1	312	300
2	242	201
3	340	232
4	388	312
5	296	220
6	254	256
7	391	328
8	402	330
9	290	231

39. Your laboratory is evaluating the precision of a colorimetric method for creatinine in serum in which the sample is reacted with alkaline picrate to produce a color. Rather than perform one set of analyses, several sets with different samples are performed over several days, in order to get a better estimate of the precision of the method. From the following absorbance data, calculate the pooled standard deviation.

Day 1 (Sample A)	Day 2 (Sample B)	Day 3 (Sample C)
0.826	0.682	0.751
0.810	0.655	0.702
0.880	0.661	0.699
0.865		0.724
$\bar{x}_A = 0.845$	$\bar{x}_B = 0.666$	$\bar{x}_C = 0.719$

40. The following replicate calcium determinations on a blood sample using atomic absorption spectrophotometry (AAS) and a new colorimetric method were reported. Is there a significant difference in the precision of the two methods?

AAS (mg/dL)	Colorimetric (mg/dL)
10.9	9.2
10.1	10.5
10.6	9.7
11.2	11.5
9.7	11.6
10.0	9.3
Mean 10.4	10.1
	11.2
	Mean 10.4

PROFESSOR'S FAVORITE PROBLEMS

Contributed by Professor Mary K. Donais, Saint Anselm College

41. This question addresses the following: Use of quality control sample to evaluate method accuracy, use of a significance test to compare two methods, sample homogeneity and data precision. The question is based on actual data collected by students for a senior research project.

  The lead content of ancient Roman bronze coins can be used to approximate their age. Lead concentrations in individual coins can vary from single percents up to over 20%. A method was developed to quantify lead in bronze coins that utilized hot plate digestion of the samples followed by flame atomic absorption spectrophotometric measurements. Hot plate temperatures can be difficult to set and are often uneven across the apparatus surface, however. As well, a quantitative transfer of the digest solution is required. A second, more efficient method was then developed that utilized a programmable temperature feedback digestion block with volumetric vessels. Data were collected using both digestion methods as provided below. Four coins (numbered 1–4) were divided in half with a sample from each half (designated as a and b) being digested and analyzed. Also, a standard reference material, NIST SRM 872, was analyzed using both methods; the certified lead content in this material is 4.13 ± 0.03%. Evaluate the data. Comment on accuracy and precision. How could the homogeneity of lead in the bronze coins affect precision? Is one method "better" or "different" than the other? How could this be shown quantitatively?

Coin	Hot Plate Method (%)	Digestion Block Method (%)
1a	14.44	15.37
1b	8.41	7.24
2a	23.77	24.14
2b	27.23	24.87
3a	6.34	6.77
3b	8.04	7.34
4a	16.16	17.20
4b	19.07	18.26
NIST SRM 872	4.21	4.12

An Excel file for the solution is in the website supplement.

42. Potassium dichromate is an oxidizing agent that is used for the volumetric determination of iron by titrating iron(II). Although potassium dichromate is a high-purity material that can be used for the direct preparation of a standard solution of known concentration, the solution is frequently standardized by titrating a known amount of iron(II) prepared from high-purity iron wire or electrolytic iron, using the same procedure as for the sample. This is because the color of the iron(III) product of the titration tends to mask the indicator color (used to detect the end of the titration), causing a slight error. A solution prepared to be 0.1012 M was standardized with the following results: 0.1017, 0.1019, 0.1016, 0.1015 M. Is the supposition that the titration values are statistically different from the actual prepared concentration valid?

43. In the nuclear industry, detailed records are kept of the quantity of plutonium received, transported, or used. Each shipment of plutonium pellets received is carefully analyzed to check that the purity and hence the total quantity is as the supplier claims. A particular shipment is analyzed with the following results: 99.93, 99.87, 99.91, and 99.86%. The listed purity as received from the supplier is 99.95%. Is the shipment acceptable?

PROFESSOR'S FAVORITE PROBLEMS

Contributed by Professor Amanda Grannas, Villanova University

44. You have developed a new method to measure cholesterol levels in blood that would be cheap, quick and patients could do tests at home (much like glucose tests for diabetics). You need to validate your method, so that you can patent it! Use the information given below and the various statistical methods of data validation you have learned to evaluate the effectiveness of your new testing method.

   (a) NIST makes a cholesterol in human serum standard that is 182.1_5 mg/dL. Your method reports values of 181.83, 182.12, 182.32 and 182.20 when taking 4 replicate measurements of this standard. Is your value the same?

   (b) You tested a sample by your method(1) and an "accepted" one(2). (1) 146.32, 146.24, 146.28, 146.36, 146.12, 146.24 (2) 146.37, 146.38, 146.28, 146.37, 146.12, 146.23. Do they agree?

   (c) You do not want to give critics an opportunity; there are many at the FDA. You compared the results you get to the accepted method when measuring *many different* samples. Using the data obtained below, compare your method of analysis to the accepted method for measuring cholesterol. Do your results agree with the accepted method?

Sample #	Your Method (mg/dL)	Accepted Method (mg/dL)
1	174.60	174.93
2	142.32	142.81
3	210.67	209.06
4	188.32	187.92
5	112.41	112.37

For an Excel-based Problem related to the *t*-test, courtesy of Professor Steven Goates, Brigham Young University, see the website supplement.

Q Test

45. Arsenic values determined after the analysis of drinking water of a city are: 5.60, 5.64, 5.70, 5.69 and 5.81 ppm. The last value appears upset. Determine whether it should be rejected at the 95% confidence level?

46. The following replicate molarities were obtained when standardizing a solution: 0.1067, 0.1071, 0.1066, and 0.1050. Can one of the results be discarded as due to accidental error at the 95% confidence level?

47. Can any of the data in Problem 19 be rejected at the 95% confidence level?

48. The precision of a method is being established, and the following data are obtained: 22.23. 22.18, 22.25, 22.09, and 22.17%. Is 22.09% a valid measurement at the 95% confidence level?

PROFESSOR'S FAVORITE PROBLEMS

Contributed by Professor Wen-Yee Lee, University of Texas at El Paso

49. Police have a hit-and-run case and need to identify the brand of red auto paint. The percentages of iron oxide, which gives paint its red color, found in paint from the victim's car are as follows: 43.15, 43.81, 45.71, 43.23, 41.99, and 43.56%.

    (a) Can we include all of the numbers above to calculate the average percentage of the iron oxide in the paint sample at 90% confidence level?

    (b) Based on your finding from subquestion (a), take only reliable numbers to calculate the average with the uncertainty, that is, average standard deviation.

    (c) The police found a suspect and collected some red paint samples from the front bumper of the suspect's car. The percentage of iron oxide was found to be (42.60 ± 0.44) % (average standard deviation) with five measurements. Do you think the police got the right person who should be charged with "hit-and-run" (at 95% confidence level)?

Statistics for Small Sets of Data

50. For Problem 20, estimate the standard deviation from the range. Compare with the standard deviation calculated in the problem.

51. For Problem 30, use the range to estimate the confidence limit at the 95% confidence level, and compare with the value calculated in the problem using the standard deviation.

52. For Problem 34, use the range to estimate the confidence limits at the 95 and 99% confidence levels, and compare with the values calculated in the problem using standard deviation.

Least Squares

53. Calculate the slope of the line in Example 3.21, using Equation 3.22. Compare with the value calculated using Equation 3.23.

54. A calibration curve for the colorimetric determination of phosphorous in urine is prepared by reacting standard solutions of phosphate with molybdenum(VI) and reducing the phospho-molybdic acid complex to produce the characteristic blue color. The measured absorbance A is

plotted against the concentration of phosphorous. From the following data, determine the linear least-squares line and calculate the phosphorous concentration in the urine sample:

ppm P	A
1.00	0.205
2.00	0.410
3.00	0.615
4.00	0.820
Urine sample	0.625

55. Calculate the uncertainties in the slope and intercept of the least-squares line in Problem 54, and the uncertainty in the phosphorus concentration in the urine sample.

For an Excel-based Problem related to least squares, courtesy of Professor Steven Goates, Brigham Young University, see the website supplement.

For an Excel-based least-squares plotting problem, courtesy of Professor Wen-Yee Lee, The University of Texas at El Paso, see the website supplement.

PROFESSOR'S FAVORITE CHALLENGES

Contributed by Professor Michael D. Morris, University of Michigan

56. Explain situations when a least-squares linear fit is not appropriate and should not be used. There are at least two common important cases.

Correlation Coefficient

57. From the data given below, determine the correlation coefficient between the amount of toxin produced by a fungus and the percent of yeast extract in the growth medium.

Sample	% Yeast Extract	Toxin (mg)
(a)	1.000	0.487
(b)	0.200	0.260
(c)	0.100	0.195
(d)	0.010	0.007
(e)	0.001	0.002

58. The cultures described in Problem 57 had the following fungal dry weights: sample (a) 116 mg, (b) 53 mg, (c) 37 mg, (d) 8 mg, and (e) 1 mg. Determine the correlation coefficient between the dry weight and the amount of toxin produced.

59. A new method for the determination of cholesterol in serum is being developed in which the rate of depletion of oxygen is measured with an oxygen electrode upon reaction of the cholesterol with oxygen, when catalyzed by the enzyme cholesterol oxidase. The results for several samples are compared with those of the standard Lieberman colorimetric method. From the following data, determine by the t-test if there is a statistically significant difference between the two methods and calculate the correlation coefficient. Assume the two methods have similar precisions.

Sample	Enzyme Method (mg/dL)	Colorimetric Method (mg/dL)
1	305	300
2	385	392
3	193	185
4	162	152
5	478	480
6	455	461
7	238	232
8	298	290
9	408	401
10	323	315

Detection Limit

60. You are determining aluminum in plants by a fluorometric procedure. Seven prepared blanks give fluorescence readings of 0.12, 0.18, 0.25, 0.11, 0.16, 0.26, and 0.16 units. A 1.0 aluminum standard solution gave a reading of 1.25. What is the detection limit? What would be the total reading at this level?

PROFESSOR'S FAVORITE PROBLEMS

Contributed by Professor Wen-Yee Lee, University of Texas at El Paso

61. In spectrophotometry, the concentration of analyte is measured by its absorbance of light. Nine reagent blanks were also measured and gave values of 0.0006, 0.0012, 0.0022, 0.0005, 0.0016, 0.0008, 0.0017, 0.0010, and 0.0009.

   (a) Find the minimum detectable signal.

   (b) A calibration curve was conducted using a series of standard solutions. The concentrations and the absorbance of the standards are listed in the following table. Absorbance is a dimensionless quantity. Find the slope of the calibration curve (including the unit).

   (c) Find the concentration detection limit.

conc. (ppb)	Absorbance
0.01	0.0078
0.10	0.0880
0.50	0.4467
1.00	0.8980
2.50	1.8770

Sampling Statistics

62. Four-tenth gram samples of paint from a bridge, analyzed for the lead content by a precise method ($< 1\%$ rsd), gives a relative sampling precision, R, of 5%. What weight sample should be taken to improve this to 2.5%?

PROFESSOR'S FAVORITE PROBLEMS

Contributed by Professor Wen-Yee Lee, University of Texas at El Paso

63. (a) You and your friends are visiting the M&M candy factory at Christmas time. Just as you walk in the door, there is a terrible spill and 200,000 red M&Ms along with 50,000 green M&Ms have just spilled on the floor. You madly start grabbing M&Ms and have managed to collect 1000 of them before management catches you. How many **red** M&Ms have you probably picked up?

(b) Continued from question above, if you repeat this scenario many times, what will be the absolute standard deviation of **green** M&Ms retrieved?

64. Copper in an ore sample is at a concentration of about 3% (wt/wt). How many samples should be analyzed to obtain a percent relative standard deviation of 5% in the analytical result at the 95% confidence level, if the sampling precision is 0.15% (wt/wt)?

Real-Life Sampling Considerations, Beyond Statistics

PROFESSOR'S FAVORITE PROBLEM

Contributed by Professor Christine Blaine, Carthage College

65. Many, if not most, environmental sampling problems go beyond mere statistical considerations. As stated, typically the greatest error occurs in the process of obtaining a representative sample of the analyte of interest. Obtaining a representative sample becomes challenging when it involves an environmental sample or a large sampling site. Let's examine the following two scenarios that focus specifically on collecting samples for analysis.

(a) State authorities in Wisconsin recently asked you to analyze the salmon in Lake Michigan for mercury contamination. Lake Michigan has a surface area of over 22,000 square miles. What factors relating to the salmon do you need to consider when collecting the samples?

(b) Poultry farmers often mix small amounts of an arsenic-containing compound, roxarsone, into chicken feed to combat parasites. As the saying goes, "what goes in, comes out." Research shows that almost all of the roxarsone is excreted by the poultry (*Environ. Sci. Technol.* **2003**, *37*, 1509–1514). Over time, the roxarsone degrades to arsenate, AsO_4^{3-}, so the concentrations of both species need to be included in your laboratory analysis. You were recently appointed to sample a 1-acre area adjacent to the coop where chickens roam free for arsenic compounds. How would you identify sites to sample? When would you sample these sites? What other types of sampling might you perform?

Obviously there are no unique solutions to these problems. The suggested solutions in the Solutions Manual provide guidelines. Consult the text **website** for more thought-provoking sample scenarios by Professor Blaine for group discussion.

Multiple Choice Questions

1. The number 0.032040 has _____ significant figures.

   (a) 3   (b) 4

   (c) 5   (d) 6

2. ±0.02 is the ± relative error determined in the density of a metal rod. If the calculated value of density is 8.6321947..., density should be reported as

   (a) 9.0   (b) 8.6

   (c) 8.63   (d) 8.632

3. Width of a confidence interval is increased by following except:

   (a) increased variability   (b) increased confidence level

   (c) increased sample size   (d) decreased sample size

4. What percent of measurements fall above median in a data set?

   (a) 49%   (b) 50%

   (c) 51%   (d) Cannot be determined.

Statistics

1. P. C. Meier and R. E. Zund, *Statistical Methods in Analytical Chemistry,* 2nd ed. New York: Wiley, 2000.
2. J. C. Miller and J. N. Miller, *Statistics and Chemometrics for Analytical Chemistry,* 4th ed. Englewood Cliffs, NJ: Prentice Hall, 2000.
3. J. C. Miller and J. N. Miller, "Basic Statistical Methods for Analytical Chemistry. A Review. Part 1. Statistics of Repeated Measurements." "Part 2. Calibration and Regression Methods," *Analyst,* **113** (1988) 1351; **116** (1991) 3.
4. A series of articles on Statistics in Analytical Chemistry by D. Coleman and L. Vanatta, *Am. Lab.*: **Part 24**—Glossary (November/December (2006) 25; **Part 26**—Detection Limits: Editorial Comments and Introduction, June/July (2007) 24; **Part 30**—Statistically Derived Detection Limits (concluded), June/July (2008) 34; **Part 32**—Detection Limits via 3-Sigma, November/December (2008) 60; **Part 34**—Detection Limit Summary, May (2009) 50; **Part 35**—Reporting Data and Significant Figures, August (2009) 34; **Part 40**—Blanks. Go to their website at www.americanlaboratory.com, click on article/archives, and you can search by title.

Q Test

5. R. B. Dean and W. J. Dixon, "Simplified Statistics for Small Numbers of Observations," *Anal. Chem.*, **23** (1951) 636.
6. W. J. Blaedel, V. W. Meloche, and J. A. Ramsay, "A Comparison of Critiera for the Rejection of Measurements," *J. Chem. Educ.*, **28** (1951) 643.
7. D. B. Rorabacher, "Statistical Treatment for Rejection of Deviant Values; Critical Values of Dixon's 'Q' Parameter and Related Subrange Ratios at the 95% Confidence Level," *Anal. Chem.*, **63** (1991) 139.
8. C. E. Efstathiou, "A Test for the Simultaneous Detection of Two Outliers Among Extreme Values of Small Data Sets," *Anal. Lett.*, **26** (1993) 379.

Quality Control

9. J. K. Taylor, "Quality Assurance of Chemical Measurements," *Anal. Chem.*, **53** (1981) 1588A.
10. J. K. Taylor, *Quality Assurance of Chemical Measurements*. Boca Raton, FL: CRC Press/Lewis, 1987.
11. J. K. Taylor, "Validation of Analytical Methods," *Anal. Chem.*, **55** (1983) 600A.
12. J. O. Westgard, P. L. Barry, and M. R. Hunt, "A Multi-Rule Shewhart Chart for Quality Control in Clinical Chemistry," *Clin. Chem.*, **27** (1981) 493.

Least Squares

13. P. Galadi and B. R. Kowalski, "Partial Least Squares Regression (PLS): A Tutorial," *Anal. Chim. Acta*, **185** (1986) 1.

Detection Limits

14. G. L. Long and J. D. Winefordner, "Limit of Detection. A Closer Look at the IUPAC Definition," *Anal. Chem.*, **55** (1983) 712A.
15. J. P. Foley and J. G. Dorsey, "Clarification of the Limit of Detection in Chromatography," *Chromatographia*, **18** (1984) 503.
16. J. E. Knoll, "Estimation of the Limit of Detection in Chromatography," *J. Chromatogr. Sci.*, **23** (1985) 422.
17. Analytical Methods Committee, "Recommendations for the Definition, Estimation and Use of the Detection Limit," *Analyst*, **112** (1987) 199.

18. *ICH-Q2B Validation of Analytical Procedures: Methodology* (International Conference on Harmonization of Technical Requirements for Registration of Pharmaceuticals for Human Use, Geneva, Switzerland, November 1996).

Sampling Statistics

19. B. Kratochvil and J. K. Taylor, "Sampling for Chemical Analysis," *Anal. Chem.*, **53** (1981) 924A.
20. M. H. Ramsey, "Sampling as a Source of Measurement Uncertainty: Techniques for Quantification and Comparison with Analytical Sources," *J. Anal. Atomic Spectrosc.*, **13** (1998) 97.
21. G. Brands, "Theory of Sampling. I. Uniform Inhomogeneous Material," *Fresenius' Z. Anal. Chem.*, 314 (1983) 6; II. "Sampling from Segregated Material," *Z. Anal. Chem.*, **314** (1983) 646.
22. N. T. Crosby and I. Patel, eds., *General Principles of Good Sampling Practice*. Cambridge, UK: Royal Society of Chemistry, 1995.
23. S. K. Thompson, *Sampling*, New York: Wiley, 1992.

Spreadsheets

24. D. Diamond and V. C. A. Hanratty, *Spreadsheet Applications in Chemistry Using Microsoft Excel*. New York: Wiley, 1997.
25. H. Freiser, *Concepts and Calculations in Analytical Chemistry: A Spreadsheet Approach*. Boca Rato, FL: CRC Press, 1992.
26. R. De Levie, *Advanced Excel For Scientific Data Analysis*. 2nd ed. Oxford University Press, 2008.
27. J. Workman and H. Mark, "Statistics and Chemometrics for Clinical Data Reporting, Part II: Using Excel for Computations," *Spectroscopy*, October, 20 (2009) Chapter 2. www.spectroscopyonline.com.
28. E. J. Billo, *Excel for Scientists and Engineers: Numerical Methods*. Hoboken, NY: Wiley, 2007.

Chapter 4

STOICHIOMETRIC CALCULATIONS: THE WORKHORSE OF THE ANALYST

KEY THINGS TO LEARN FROM THIS CHAPTER

- How to calculate molarities and moles (key Equations: 4.4, 4.5)
- How to express analytical results
- How to calculate weight and percent analyted from molarities, volumes, and reaction ratios (key Equations: 4.5, 4.17–4.20, 4.25)

- Weight relationships for gravimetric analysis (key Equation: 4.28)

Analytical chemistry deals with measurements of analytes in solids and concentrations in solution, from which we calculate masses. Thus, we prepare solutions of known concentrations that can be used to calibrate instruments or to titrate sample solutions. We calculate the mass of an analyte in a solution from its concentration and the volume. We calculate the mass of product expected from the mass of reactants. All of these calculations require a knowledge of **stoichiometry**, that is, the ratios in which chemicals react, from which we apply appropriate conversion factors to arrive at the desired calculated results.

Stoichiometry deals with the ratios in which chemicals react.

In this chapter we review the fundamental concepts of mass, moles, and equivalents; the ways in which analytical results may be expressed for solids and liquids; and the principles of volumetric analysis and how stoichiometric relationships are used in titrations to calculate the mass of analyte.

4.1 Review of the Fundamentals

Quantitative analysis is based on a few fundamental atomic and molecular concepts, which we review below. You have undoubtedly been introduced to these in your general chemistry course, but we briefly review them here since they are so fundamental to quantitative calculations.

The Basics: Atomic, Molecular, and Formula Weights

The atomic weight for any element is the weight of a specified number of atoms of that element, and that number is the same from one element to another. A gram-atomic weight of any element contains exactly the same number of atoms of that element as there are carbon atoms in exactly 12 g of carbon 12. This number is Avogadro's number, 6.022×10^{23}, the number of atoms present in 1 g-at wt of any element.[1]

---

[1] There is a proposal to change the definition of Avogadro's number in redefining the kilogram to be an invariant unit. The kilogram is the only base unit in the International System of Units (SI) that is defined by a physical artifact rather than an unvarying physical property of nature, the others being the meter (length), second (time), ampere (electric current), kelvin (temperature), mole (amount of substance), and candela (light intensity). It is currently equal to the mass of a small cylinder of platinum-iridium alloy, known as the international prototype, that was ratified as the official kilogram in 1889, and is kept in a vault at the International Bureau of Weights & Measures

Since naturally occurring elements consist of mixtures of isotopes, the chemical atomic weights will be an average of the isotope weights of each element, taking into account their relative naturally occurring abundances. For example, bromine has two isotopes: ^{79}Br with atomic weight 78.981338 at a 50.69% relative abundance, and ^{81}Br with atomic weight 80.9162921 with a 49.31% relative abundance. These average to 79.904, the natural atomic weight we use in chemical calculations. Another measurement used by chemists is **molecular weight** (mw), defined as the sum of the atomic weights of the atoms that make up a compound. The term **formula weight** (fw) is a more accurate description for substances that don't exist as molecules but exist as ionic compounds (strong electrolytes—acids, bases, salts). The term **molar mass** is sometimes used in place of formula weight.

We will use formula weight (fw) to express grams per mole.

What is a Dalton?

Biologists and biochemists sometimes use the unit **dalton** (Da) to report masses of large biomolecules and small biological entities such as chromosomes, ribosomes, viruses, and mitochondria, where the term *molecular weight* would be inappropriate. The mass of a single carbon-12 atom is equivalent to 12 daltons, and 1 dalton is therefore 1.661×10^{-24} g, the reciprocal of Avogadro's number. The number of daltons in a single molecule is numerically equivalent to the molecular weight (g/mol). Strictly speaking, it is not correct to use the dalton as a unit of molecular weight, and it should be reserved for the types of substances mentioned above. For example, the mass of an *Escherichia coli* bacterium cell is about 1×10^{-12} g, or 6×10^{11} daltons.

Moles: The Basic Unit for Equating Things

The chemist knows that atoms and molecules react in definite proportions. Unfortunately, he or she cannot conveniently count the number of atoms or molecules that participate in a reaction. But since the chemist has determined their relative masses, he or she can describe their reactions on the basis of the relative masses of atoms and molecules reacting, instead of the number of atoms and molecules reacting. For example, in the reaction

$$Ag^+ + Cl^- \rightarrow \underline{AgCl}$$

There are 6.022×10^{23} atoms in a mole of atoms.

we know that one silver ion will combine with one chloride ion. We know further, since the atomic weight of silver is 107.870 and the atomic weight of chlorine is 35.453, that 107.870 mass units of the silver will combine with 35.453 mass units of chlorine. To simplify calculations, chemists have developed the concept of the **mole**, which is Avogadro's number (6.022×10^{23}) of atoms, molecules, ions, or other species. Numerically, it is the atomic, molecular, or formula weight of a substance expressed in **grams**.[2]

---

near Paris. But it has inexplicably lost about 50 μg over time compared with copies. There are two proposals for an invariant definition of the kilogram, one based on Plank's constant and one based on Avogadro's constant, either of which would marginally change the value or definition of Avogadro's number and hence the definition of the mole. But the change would be insignificant for the units we use. Details of the proposals may be found in "Avogadro's Number Is Up ...," P. J. Karol, *Chem. & Eng. News*, March 17 (2008) 48, "Redefining the Kilogram," S. K. Ritter, *Chem. & Eng. News*, May 26 (2008) 43, and "Redefining the Kilogram and Mole," P. F. Rusch, *Chem. & Eng. News*, May 30 (2011) 58: http://pubs.acs.org/isubscribe/journals/cen/89/i22/html/8922acscomment.html.

[2] Actually, the term *gram-atomic weight* is more correct for atoms, *gram-formula weight* for ionic substances, and *gram-molecular weight* for molecules, but we will use *moles* in a broad sense to include all substances. In place of gram-formula weight we will simply use *formula weight* (fw).

Now, since a mole of any substance contains the same number of atoms or molecules as a mole of any other substance, atoms will react in the same mole ratio as their atom ratio in the reaction. In the above example, one silver ion reacts with one chloride ion, and so each mole of silver ion will react with one mole of chloride ion. (Each 107.87 g of silver will react with 35.453 g of chlorine.)

Example 4.1

Calculate the weight of one mole of $CaSO_4 \cdot 7H_2O$.

Solution

One mole is the formula weight expressed in grams. The formula weight is

Ca	40.08	
S	32.06	
11 O	176.00	
14 H	14.11	
	262.25 g/mol	

The number of moles of a substance is calculated from

$$\text{Moles} = \frac{\text{grams}}{\text{formula weight (g/mol)}} \qquad (4.1)$$

where formula weight represents the atomic or molecular weight of the substance. Thus,

$$\text{Moles Na}_2\text{SO}_4 = \frac{g}{fw} = \frac{g}{142.04 \text{ g/mol}}$$

$$\text{Moles Ag}^+ = \frac{g}{fw} = \frac{g}{107.870 \text{ g/mol}}$$

Since many experiments deal with very small quantities, a more convenient form of measurement is the **millimole**. The formula for calculating millimoles is

$$\text{Millimoles} = \frac{\text{milligrams}}{\text{formula weight (mg/mmol)}} \qquad (4.2)$$

g/mol = mg/mmol = formula weight; g/L = mg/mL; mol/L = mmol/mL = molarity.

Just as we can calculate the number of moles from the grams of material, we can likewise calculate the grams of material from the number of moles:

$$g \text{ Na}_2\text{SO}_4 = \text{moles} \times fw = \text{moles} \times 142.04 \text{ g/mol}$$
$$g \text{ Ag} = \text{moles} \times fw = \text{moles} \times 107.870 \text{ g/mol}$$

Again, we usually work with millimole quantities, so

$$\text{Milligrams} = \text{millimoles} \times \text{formula weight (mg/mmol)} \qquad (4.3)$$

Note that g/mol is the same as mg/mmol, g/L the same as mg/mL, and mol/L the same as mmol/mL.

Example 4.2

Calculate the number of moles in 500 mg Na_2WO_4 (sodium tungstate).

Solution

$$\frac{500 \text{ mg}}{293.8 \text{ mg/mmol}} \times 0.001 \text{ mol/mmol} = 0.00170 \text{ mol}$$

Example 4.3

What is the weight, in milligrams, of 0.250 mmol Fe_2O_3 (ferric oxide)?

Solution

$$0.250 \text{ mmol} \times 159.7 \text{ mg/mmol} = 39.9 \text{ mg}$$

4.2 How Do We Express Concentrations of Solutions?

Chemists express solution concentrations in a number of ways. Some are more useful than others in quantitative calculations. We will review here the common concentration units that chemists use. Their use in quantitative volumetric calculations is treated in more detail below.

Molarity—The Most Widely Used

The mole concept is useful in expressing concentrations of solutions, especially in analytical chemistry, where we need to know the volume ratios in which solutions of different materials will react. A one-**molar** solution is defined as one that contains one mole of substance in each liter of a solution. It is prepared by dissolving one mole of the substance in the solvent and diluting to a final volume of one liter in a volumetric flask; or a fraction or multiple of the mole may be dissolved and diluted to the corresponding fraction or multiple of a liter (e.g., 0.01 mol in 10 mL). More generally, the **molarity** of a solution is expressed as moles per liter or as millimoles per milliliter. Molar is abbreviated as *M*, and we talk of the *molarity* of a solution when we speak of its concentration. A one-molar solution of silver nitrate and a one-molar solution of sodium chloride will react on an equal-volume basis, since they react in a 1:1 ratio: $Ag^+ + Cl^- \rightarrow \underline{AgCl}$. We can be more general and calculate the moles of substance in any volume of the solution.

$$\boxed{\begin{aligned} \text{Moles} &= (\text{moles/liter}) \times \text{liters} \\ &= \text{molarity} \times \text{liters} \end{aligned}} \tag{4.4}$$

We often work with millimoles in analytical chemistry. Remember this formula!

The liter is an impractical unit for the relatively small quantities encountered in titrations, and we normally work with milliliters. This is what your buret reads. So,

$$\boxed{\begin{aligned} \text{Millimoles} &= \text{molarity} \times \text{milliliters} \\ (\text{or mmol} &= M \times \text{mL}) \end{aligned}} \tag{4.5}$$

Example 4.4

A solution is prepared by dissolving 1.26 g $AgNO_3$ in a 250-mL volumetric flask and diluting to volume. Calculate the molarity of the silver nitrate solution. How many millimoles $AgNO_3$ were dissolved?

Solution

$$M = \frac{1.26 \text{ g}/169.9 \text{ g/mol}}{0.250 \text{ L}} = 0.0297 \text{ mol/L} \quad (\text{or } 0.0297 \text{ mmol/mL})$$

Then,

$$\text{Millimoles} = (0.0297 \text{ mmol/mL})(250 \text{ mL}) = 7.42 \text{ mmol}$$

Always remember that *the units in a calculation must combine to give the proper units in the answer*. Thus, in this example, grams cancel to leave the proper unit, moles/liter, or molarity. Using units in the calculation to check if the final units are proper is called **dimensional analysis**. Accurate use of dimensional analysis is essential to properly setting up computations.

Always use dimensional analysis to set up a calculation properly. Don't just memorize a formula.

Example 4.5

How many grams per milliliter of NaCl are contained in a 0.250 M solution?

Solution

0.250 mol/L = 0.250 mmol/mL
0.250 mmol/mL × 58.4 mg/mmol × 0.001 g/mg = 0.0146 g/mL

Example 4.6

How many grams Na_2SO_4 should be weighed out to prepare 500 mL of a 0.100 M solution?

Solution

500 mL × 0.100 mmol/mL = 50.0 mmol
50.0 mmol × 142 mg/mmol × 0.001 g/mg = 7.10 g

Example 4.7

Calculate the concentration of potassium ion in grams per liter after mixing 100 mL of 0.250 M KCl and 200 mL of 0.100 M K_2SO_4.

Solution

$$mmol\ K^+ = mmol\ KCl + 2 \times mmol\ K_2SO_4$$
$$= 100\ mL \times 0.250\ mmol/mL$$
$$+2 \times 200\ mL \times 0.100\ mmol/mL$$
$$= 65.0\ mmol\ in\ 300\ mL$$
$$\frac{65.0\ mmol \times 39.1\ mg/mmol \times 0.001\ g/mg \times 1000\ mL/L}{300\ mL} = 8.47\ g/L$$

Normality

The equivalent weight (or the number of reacting units) depends on the chemical reaction. It may vary most often in redox reactions, when different products are obtained.

Although molarity is widely used in chemistry, some chemists use a unit of concentration in quantitative analysis called **normality** (N). A one-**normal** solution contains one equivalent per liter. An **equivalent** represents the mass of material providing Avogadro's *number of reacting units*. A reacting unit is a *proton* or an *electron*. The number of equivalents is given by the number of moles multiplied by the number of reacting units per molecule or atom; the **equivalent weight** is the formula weight divided by the number of reacting units. For acids and bases, the number of reacting units is based on the number of protons (i.e., hydrogen ions) an acid will furnish or a base will react with. For oxidation–reduction reactions it is based on the number of electrons an oxidizing or reducing agent will take on or supply. Thus, for example, sulfuric acid, H_2SO_4, has two reacting units of protons; that is, there are two equivalents of protons in each mole. Therefore,

$$\text{Equivalent weight} = \frac{98.08\ g/mol}{2\ eq/mol} = 49.04\ g/eq$$

Equivalent weight g/eq = mg/meq; eq/L = meq/mL = normality.

So, the normality of a sulfuric acid solution is twice its molarity, that is, $N =$ (g/eqwt)/L. The number of equivalents is given by

$$\boxed{\text{Number of equivalents (eq)} = \frac{wt\ (g)}{eqwt\ (g/eq)} = \text{normality (eq/L)} \times \text{volume (L)}} \quad (4.6)$$

Just as we ordinarilly use millimoles (mmol) instead of moles, we typically use milliequivalents (meq) instead of equivalents

$$\boxed{meq = \frac{mg}{eq\ wt\ (mg/meq)} = \text{normality (meq/mL)} \times mL} \quad (4.7)$$

In clinical chemistry, equivalents are frequently defined in terms of the number of charges on an ion rather than on the number of reacting units. Thus, for example, the equivalent weight of Ca^{2+} is one-half its atomic weight, and the number of equivalents is twice the number of moles. This use is convenient for electroneutrality calculations. We discuss equivalents in more detail in Section 4.3.

Formality — Instead of Molarity

Formality is numerically the same as molarity.

Chemists sometimes use the term **formality** for solutions of ionic salts that do not exist as molecules in the solid or in solution. The concentration is given as **formal** (F). Operationally, formality is identical to molarity: The former is sometimes reserved for describing makeup concentrations of solutions (i.e., total analytical concentration), and the latter for equilibrium concentrations. For convenience, we shall use molarity exclusively, a common practice.

Molality — The Temperature-Independent Concentration

In addition to molarity and normality, another useful concentration unit is **molality**, *m*. A one-**molal** solution contains one mole per 1000 g of **solvent**. The molal concentration is convenient in physicochemical measurements of the colligative properties of substances, such as freezing point depression, vapor pressure lowering, and osmotic pressure because colligative properties depend solely on the number of solute particles present in solution per mole of solvent. Molal concentrations are not temperature dependent as molar and normal concentrations are (since the solution volume in molar and normal concentrations is temperature dependent).

Molality does not change with temperature.

Density Calculations — How Do We Convert to Molarity?

The concentrations of many fairly concentrated commercial acids and bases are usually given in terms of percent by weight. It is frequently necessary to prepare solutions of a given approximate molarity from these substances. In order to do so, we must know the density in order to calculate the molarity. **Density** is the weight per unit volume at the specified temperature, usually g/mL or g/cm^3 at 20°C. (One milliliter is the volume occupied by 1 cm^3.)

Sometimes substances list **specific gravity** rather than density. Specific gravity is defined as the ratio of the mass of a body (e.g., a solution), usually at 20°C, to the mass of an equal volume of water at 4°C (or sometimes 20°C). That is, specific gravity is the *ratio of the densities of the two substances;* it is a dimensionless quantity. Since the density of water at 4°C is 1.00000 g/mL, density and specific gravity are equal when referred to water at 4°C. But normally specific gravity is referred to water at 20°C; density is equal to specific gravity × 0.99821 (the density of water is 0.99821 g/mL at 20°C).

Density of solution at 20°C = Specific gravity of solution × 0.99821 g/mL

Note that the density of the solution at a temperature other than 20°C cannot be precisely computed from the specific gravity specified for 20°C without knowing the volumetric expansion behavior of the solution, which is not the same as that of water.

Example 4.8

How many milliliters of concentrated sulfuric acid, 94.0% (g/100 g solution), density 1.831 g/cm^3, are required to prepare 1 liter of a 0.100 *M* solution?

Solution

Consider 1 cm^3 = 1 mL. The concentrated acid contains 0.940 g H_2SO_4 per gram of solution, and the solution weighs 1.831 g/mL. The product of these two numbers, then, gives the gram H_2SO_4 per milliliter of solution:

$$M = \frac{(0.940 \text{ g } H_2SO_4/\text{g solution})(1.831 \text{ g/mL})}{98.1 \text{ g/mol}} \times 1000 \text{ mL/L}$$

$$= 17.5 \text{ mol } H_2SO_4/\text{L solution}$$

Mass is conserved:

$$M_1 V_1 = M_2 V_2$$

If a solution of molarity M_1 and volume V_1 is diluted to V_2, the molarity M_2 will obey this relationship. The general equation

$$C_1 V_1 = C_2 V_2$$

will hold as long as C_1 and C_2 are in the same units, regardless of the specific unit.
Memorize this equation.

See Sections 4.5 and the text website for volumetric calculations using molarity (or normality).

The analytical concentration represents the concentration of total dissolved substance, i.e., the sum of all species of the substance in solution = C_X.

An equilibrium concentration is that of a given dissolved form of the substance = [X].

The millimoles taken for dilution will be the same as the millimoles in the diluted solution, i.e.,
$M_{stock} \times mL_{stock} = M_{diluted} \times mL_{diluted}$

We must dilute this solution to prepare 1 liter of a 0.100 M solution. The same number of millimoles of H_2SO_4 must be taken as will be contained in the final solution. Since mmol = $M \times$ mL and mmol dilute acid = mmol concentrated acid,

$$0.100 \ M \times 1000 \ \text{mL} = 17.5 \ M \times \text{mL}$$
$$x = 5.71 \ \text{mL concentrated acid to be diluted to 1000 mL}$$

Molarity and normality are the most useful concentrations in quantitative analysis. Calculations using these for volumetric analysis are discussed in more detail below.

Analytical and Equilibrium Concentrations — They are Not The Same

Analytical chemists prepare solutions of known analytical concentrations, but the dissolved substances may partially or totally dissociate to give equilibrium concentrations of different species. Acetic acid, for example, is a weak acid that dissociates a few percent depending on the concentration,

$$HOAc \rightleftharpoons H^+ + OAc^-$$

to give equilibrium amounts of the proton and the acetate ion. The more dilute the solutions, the greater the dissociation. We often use these equilibrium concentrations in calculations involving equilibrium constants (Chapter 5), usually using molarity concentrations. The **analytical molarity** is given by the notation C_X, while **equilibrium molarity** is given by [X]. A solution of 1 M $CaCl_2$ (analytical molarity) is completely ionized into constituent ions in solution and gives at equilibrium, 0 M $CaCl_2$, 1 M Ca^{2+}, and 2 M Cl^- (equilibrium molarities). Hence, we say the solution is 1 M in Ca^{2+} and 2 M in Cl^-.

Dilutions — Preparing the Right Concentration

We often must prepare dilute solutions from more concentrated stock solutions. For example, we may prepare a dilute HCl solution from concentrated HCl to be used for titrations (following standardization). Or, we may have a stock standard solution from which we wish to prepare a series of more dilute standards. The millimoles of stock solution taken for dilution will be identical to the millimoles in the final diluted solution, remember, $C_1 V_1 = C_2 V_2$.

Example 4.9

You wish to prepare a calibration curve for the spectrophotometric determination of permanganate. You have a stock 0.100 M solution of $KMnO_4$ and a series of 100-mL volumetric flasks. What volumes of the stock solution will you have to pipet into the flasks to prepare standards of 1.00, 2.00, 5.00, and 10.0×10^{-3} M $KMnO_4$ solutions?

Solution

x mL of the stock solution of 0.100 M concentration will be diluted to 100 mL of some specified concentration, say C_2. Remembering

$$C_1 V_1 = C_2 V_2$$

Let us do this for the first concentration, $C_2 = 1.00 \times 10^{-3}$. Here V_1 is x, $C_1 = 0.100\ M$ and $V_2 = 100$ mL

$$0.100\ M \times x\ \text{mL} = 1.00 \times 10^{-3}\ M \times 100\ \text{mL}$$
$$x = 1.00\ \text{mL}$$

Similarly, for the other solutions we will need 2.00, 5.00, and 10.0 mL of the stock solution, which will be diluted to 100 mL.

Example 4.10

You are analyzing for the manganese content in an ore sample by dissolving it and oxidizing the manganese to permanganate for spectrophotometric measurement. The ore contains about 5% Mn. A 5-g sample is dissolved and diluted to 100 mL, following the oxidation step. By how much must the solution be diluted to be in the range of the calibration curve prepared in Example 4.9, that is, about $3 \times 10^{-3}\ M$ permanganate?

Solution

The solution contains $0.05 \times 5\text{-g sample} = 0.25$ g Mn. This corresponds to [0.25 g/ (55 g Mn/mol)]/100mL $= 4.5 \times 10^{-3}$ mol MnO_4^-/100 mL $= 4.5 \times 10^{-2}\ M$. For $3 \times 10^{-3}\ M$, we must dilute it by $4.5 \times 10^{-2}/3 \times 10^{-3} = 15$-fold. If we have a 100-mL volumetric flask, using $C_1V_1 = C_2V_2$,

$$4.5 \times 10^{-2} M \times x\ \text{mL} = 3 \times 10^{-3} M \times 100\ \text{mL}$$
$$x = 6.7\ \text{mL needed for dilution to 100 mL}$$

Since we need to pipet accurately, we could probably take an accurate 10-mL aliquot, which would give about $4.5 \times 10^{-3} M$ permanganate for measurement.

More Dilution Calculations

We can use the relationship $C_1V_1 = C_2V_2$ to calculate the dilution required to prepare a certain concentration of a solution from a more concentrated solution. For example, if we wish to prepare 500 mL of a 0.100 M solution by diluting a more concentrated solution, we can calculate it from this relationship.

Remember, the millimoles before and after dilution are the same. See Section 4.5 (and the text's website) for volumetric calculations using molarity (and normality).

Example 4.11

You wish to prepare 500 mL of a 0.100 M $K_2Cr_2O_7$ solution from a 0.250 M solution. What volume of the 0.250 M solution must be diluted to 500 mL?

Solution

$$M_{\text{final}} \times \text{mL}_{\text{final}} = M_{\text{original}} \times \text{mL}_{\text{original}}$$
$$0.100\ \text{mmol/mL} \times 500\ \text{mL} = 0.250\ \text{mmol/mL} \times \text{mL}_{\text{original}}$$
$$\text{mL}_{\text{original}} = 200\ \text{mL}$$

Example 4.12

What volume of 0.40 M $Ba(OH)_2$ must be added to 50 mL of 0.30 M NaOH to give a solution 0.50 M in OH^-?

Solution

Volumes of dilute aqueous solutions can be assumed to be additive, i.e., if x mL of $Ba(OH)_2$ is added to 50 mL NaOH, the total volume is going to be $50 + x$ mL. Wex can use a modified form of $C_1V_1 = C_2V_2$ where all the initial solution components are added in this manner and these sum up to the final solution components:

$$\sum C_{in}V_{in} = C_{fin}V_{fin}$$

In the present case, $M_{NaOH}V_{NaOH} + 2 \times M_{Ba(OH)_2}V_{Ba(OH)_2} = M_{OH^-} \times V_{fin}$, note that 1 M $Ba(OH)_2$ is 2 M in OH^-. Thus 0.30 $M \times 50$ mL $+ 2 \times 0.40$ $M \times x$ mL $= 0.50$ $M \times (50 + x)$ mL.

Solving, $x = 33$ mL.
Alternatively,
Let $x =$ mL $Ba(OH)_2$. The final volume is $(50 + x)$ mL.

$$\text{mmol } OH^- = \text{mmol NaOH} + 2 \times \text{mmol } Ba(OH)_2$$

$$0.50 \ M \times (50 + x) \text{ mL} = 0.30 \ M \text{ NaOH} \times 50 \text{ mL} + 2 \times 0.40 \ M \text{ } Ba(OH)_2 \times x \text{ mL}$$

$$x = 33 \text{ mL } Ba(OH)_2$$

Often, the analyst is confronted with serial dilutions of a sample or standard solution. Again, obtaining the final concentration simply requires keeping track of the number of millimoles and the volumes.

Example 4.13

You are to determine the concentration of iron in a sample by spectrophotometry by reacting Fe^{2+} with 1,10-phenanthroline to form an orange-colored complex. This requires preparation of a series of standards against which to compare absorbances or color intensities (i.e., to prepare a calibration curve). A stock standard solution of $1.000 \times 10^{-3} M$ iron is prepared from ferrous ammonium sulfate. Working standards A and B are prepared by adding with pipets 2.000 and 1.000 mL, respectively, of this solution to 100-mL volumetric flasks and diluting to volume. Working standards C, D, and E are prepared by adding 20.00, 10.00, and 5.000 mL of working standard A to 100-mL volumetric flasks and diluting to volume. What are the concentrations of the prepared working solutions?

Solution

Solution A: $M_{stock} \times mL_{stock} = M_A \times mL_A$
$(1.000 \times 10^{-3} M)(2.000 \text{ mL}) = M_A \times 100.0 \text{ mL}$
$M_A = 2.000 \times 10^{-5} M$

Solution B: $(1.000 \times 10^{-3} M)(1.000 \text{ mL}) = M_B \times 100.0 \text{ mL}$
$M_B = 1.000 \times 10^{-5} M$

Solution C: $M_A \times mL_A = M_C \times mL_C$
$(2.000 \times 10^{-5} M)(20.00 \text{ mL}) = M_C \times 100.0 \text{ mL}$
$M_C = 4.000 \times 10^{-6} M$

The solute + solvent method of dilution should not be used for quantitative dilutions.

Solution D: $(2.000 \times 10^{-5}M)(10.00 \text{ mL}) = M_D \times 100.0 \text{ mL}$

$\qquad M_D = 2.000 \times 10^{-6}M$

Solution E: $(2.000 \times 10^{-5}M)(5.000 \text{ mL}) = M_E \times 100.0 \text{ mL}$

$\qquad M_E = 1.000 \times 10^{-6}M$

The above calculations apply to all types of reactions, including acid–base, redox, precipitation, and complexometric reactions. The primary requirement before making calculations is to know the ratio in which the substances react, that is, start with a balanced reaction.

Solution preparation procedures in the chemical literature often call for the dilution of concentrated stock solutions, and authors may use different terms. For example, a procedure may call for 1 + 9 dilution (solute + solvent) of sulfuric acid. In some cases, a 1:10 dilution (original volume:final volume) may be indicated. The first procedure calls for diluting a concentrated solution to 1/10th of its original concentration by adding 1 part to 9 parts of solvent; the second procedure by diluting to 10 times the original volume. The first procedure does not give an exact 10-fold dilution because volumes are not completely additive, except when all components are dilute aqueous solutions, whereas the second procedure does (e.g., adding 10 mL with a pipet to a 100-mL volumetric flask and diluting to volume—fill the flask partially with water before adding sulfuric acid!). The solute + solvent approach is fine for reagents whose concentrations need not be known accurately.

Added volumes are not completely additive, especially mixed solvents. Water and ethanol always have negative excess volumes when mixed, indicating the partial molar volume of each component is less when mixed than its molar volume when pure. That is, volume of pure alcohol plus volume of water does not equal the volume of vodka!

$Y = \text{yotta} = 10^{24}$

$Z = \text{zetta} = 10^{21}$

$E = \text{exa} = 10^{18}$

$P = \text{peta} = 10^{15}$

$T = \text{tera} = 10^{12}$

$G = \text{giga} = 10^{9}$

$M = \text{mega} = 10^{6}$

$k = \text{kilo} = 10^{3}$

$d = \text{deci} = 10^{-1}$

$c = \text{centi} = 10^{-2}$

$m = \text{milli} = 10^{-3}$

$\mu = \text{micro} = 10^{-6}$

$n = \text{nano} = 10^{-9}$

$p = \text{pico} = 10^{-12}$

$f = \text{femto} = 10^{-15}$

$a = \text{atto} = 10^{-18}$

$z = \text{zepto} = 10^{-21}$

$y = \text{yocto} = 10^{-24}$

4.3 Expressions of Analytical Results—So Many Ways

We can report the results of analysis in many ways, and the beginning analytical chemist should be familiar with some of the common expressions and units of measure employed. Results will nearly always be reported as *concentration*, on either a weight or a volume basis: the quantity of analyte per unit weight or per volume of sample. The units used for the analyte will vary.

We shall first review the common units of weight and volume in the metric system and then describe methods of expressing results. The gram (g) is the basic unit of mass and is the unit employed most often in macro analyses. For small samples or trace constituents, chemists use smaller units. The milligram (mg) is 10^{-3} g, the microgram (μg) is 10^{-6} g, and the nanogram (ng) is 10^{-9} g. The basic unit of volume is the liter (L). The milliliter (mL) is 10^{-3} L and is used commonly in volumetric analysis. The microliter (μL) is 10^{-6} L (10^{-3} mL), and the nanoliter (nL) is 10^{-9} L (10^{-6} mL). (Prefixes for even smaller quantities include pico for 10^{-12} and femto for 10^{-15}.)

Solid Samples

Calculations for solid samples are based on weight.[3] The most common way of expressing the results of macro determinations is to give the weight of analyte as a **percent** of the weight of sample (weight/weight basis). The weight units of analyte and sample are the same. For example, a limestone sample weighing 1.267 g and containing 0.3684 g iron would contain

Mass and weight are really different. See Chapter 2. We deal with masses but will use mass and weight interchangeably.

$$\frac{0.3684 \text{ g}}{1.267 \text{ g}} \times 100\% = 29.08\%\text{Fe}$$

---

[3]They are really based on mass, but the term *weight* is commonly used. See Chapter 2 for a description and determination of mass and weight.

The general formula for calculating percent on a weight/weight basis, which is the same as parts per hundred, then is

$$\% \, (\text{wt/wt}) = \left[\frac{\text{wt solute (g)}}{\text{wt sample (g)}} \right] \times 10^2 \; (\%/\text{g solute/g sample}) \tag{4.8}$$

It is important to note that in such calculations, grams of solute do *not* cancel with grams of sample solution; the fraction represents grams of solute per gram of sample. Multiplication of the above by 10^2 converts to grams of solute per 100 g of sample. Since the conversion factors for converting weight of solute and weight of sample (weights expressed in any units) to grams of solute and grams of sample are always the same, the conversion factors will always cancel. Thus, we can use any weight in the definition.

Trace concentrations are usually given in smaller units, such as **parts per thousand** (ppt, ‰), **parts per million** (ppm), or **parts per billion** (ppb). These are calculated in a manner similar to parts per hundred (%):

1 ppt (thousand) = 1000 ppm = 1,000,000 ppb; 1 ppm = 1000 ppb = 1,000,000 ppt (trillion). Usually ppt refers to parts per trillion, but in some cases it could be used as parts per thousand. Take note of the units when you see this!

$$\text{ppt} \, (\text{wt/wt}) = \left[\frac{\text{wt solute (g)}}{\text{wt sample (g)}} \right] \times 10^3 \; (\text{ppt/g solute/g sample}) \tag{4.9}$$

$$\text{ppm} \, (\text{wt/wt}) = \left[\frac{\text{wt solute (g)}}{\text{wt sample (g)}} \right] \times 10^6 \; (\text{ppm/g solute/g sample}) \tag{4.10}$$

$$\text{ppb} \, (\text{wt/wt}) = \left[\frac{\text{wt solute (g)}}{\text{wt sample (g)}} \right] \times 10^9 \; (\text{ppb/g solute/g sample}) \tag{4.11}$$

ppt = mg/g = g/kg
ppm = μg/g = mg/kg
ppb = ng/g = μg/kg

You can use any weight units in your calculations so long as both analyte and sample weights are in the same units. **Parts per trillion** (parts per 10^{12} parts) is also abbreviated ppt, so be careful to define which one you mean. Some authors like to use ppth to denote parts per thousand and pptr to denote parts per trillion. In the above example, we have 29.08 parts per hundred of iron in the sample, or 290.8 parts per thousand and 290,800 parts per million (290,800 g of iron per 1 million grams of sample, 290,800 lb of iron per 1 million pounds of sample, etc.). Working backward, 1 ppm corresponds to 0.0001 part per hundred, or $10^{-4}\%$. Table 4.1 summarizes the concentration relationships for ppm and ppb. Note that ppm is simply mg/kg or μg/g and that ppb is μg/kg, or ng/g.

Trace gas concentrations are also expressed in ppb, ppm, and so forth. In this case the ratio refers not to mass ratios, but to volume ratios (which for gases is the same as mole ratios). Thus, present atmospheric CO_2 concentration of 390 ppm means that each liter of air (this is a million microliters) contains 390 microliters of CO_2. Sometimes for this reason, these concentrations are written as ppmv or ppbv, and so on, indicating "by volume".

TABLE 4.1 Common Units for Expressing Trace Concentrations

Unit	Abbreviation	wt/wt	wt/vol	vol/vol
Parts per million	ppm	mg/kg	mg/L	μL/L
(1 ppm = $10^{-4}\%$)		μg/g	μg/mL	nL/mL
Parts per billion	ppb	μg/kg	μg/L	nL/L
(1 ppb = $10^{-7}\% = 10^{-3}$ ppm)		ng/g	ng/mL	pL/mL[a]
Milligram percent	mg%	mg/100 g	mg/100 mL	

[a]pL = picoliter = 10^{-12} L.

Example 4.14

A 2.6 g sample of plant tissue was analyzed and found to contain 3.6 μg zinc. What is the concentration of zinc in the plant in ppm? In ppb?

Solution

$$\frac{3.6\ \mu g}{2.6\ g} = 1.4\ \mu g/g \equiv 1.4\ ppm$$

$$\frac{3.6 \times 10^3\ ng}{2.6\ g} = 1.4 \times 10^3\ ng/g \equiv 1400\ ppb$$

One ppm is equal to 1000 ppb. One ppb is equal to $10^{-7}\%$.

Clinical chemists sometimes prefer to use the unit **milligram percent** (mg%) rather than ppm for small concentrations. This is defined as milligrams of analyte per 100 g of sample. The sample in Example 4.14 would then contain $(3.6 \times 10^{-3}\ mg/2.6\ g) \times 100\ mg\% = 0.14\ mg\%$ zinc.

Concentrations of Gases and Particles in Air

Example 4.15

The current National Ambient Air Quality Standards for the seven criteria pollutants listed by the U.S. Environmental Protection Agency is listed below from http://www.epa.gov/air/criteria.html. Other than lead and particulate matter, all others are gases. For particulate matter, wt/vol ($\mu g/m^3$) units are used; m^3 (equal to 1000 L) rather than $\mu g/L$. The gas concentrations are expressed by ppm(v) or ppb(v). Concentrations of CO are also given in mg/m^3 as CO largely comes from automotive exhaust and mass of CO emitted per vehicle-mile driven is often of interest.

(a) Show how 35 ppm CO is 40 mg/m^3.

(b) What is 75 ppb SO_2 at 25°C in $\mu g/m^3$?

National Ambient Air Quality Standards

Pollutant	Primary Standards		Secondary Standards	
	Level	Averaging Time	Level	Averaging Time
Carbon Monoxide	9 ppm (10 mg/m³)	8-hour [1]	None	
	35 ppm (40 mg/m³)	1-hour [1]	None	
Lead	0.15 μg/m³ [2]	Rolling 3-Month Average	Same as Primary	
	1.5 μg/m³	Quarterly Average	Same as Primary	
Nitrogen Dioxide	53 ppb [3]	Annual (Arithmetic Average)	Same as Primary	
	100 ppb	1-hour [4]	None	
Particulate Matter (PM₁₀)	150 μg/m³	24-hour [5]	Same as Primary	
Particulate Matter (PM₂.₅)	15.0 μg/m³	Annual [6] (Arithmetic Average)	Same as Primary	
	35 μg/m³	24-hour [7]	Same as Primary	
Ozone	0.075 ppm (2008 std)	8-hour [8]	Same as Primary	
	0.08 ppm (1997 std)	8-hour [9]	Same as Primary	
	0.12 ppm	1-hour [10]	Same as Primary	
Sulfur Dioxide	0.03 ppm	Annual (Arithmetic Average)	0.5 ppm	3-hour [1]
	0.14 ppm	24-hour [1]	None	
	75 ppb [11]	1-hour	None	

Solution

Because gas volumes change as a function of temperature and pressure, we must refer to some temperature and pressure. When this is not specified, we assume a temperature of 25°C and a pressure of 1 atm. The ideal gas laws ($PV = RT$, where R is the universal gas constant, 0.0821 L-atm/(mole K)) dictate that the volume of 1 mole of any gas at 1 atm pressure and 25°C (298.15 K) is 24.5 L (at 0°C, this is 22.4 L).

(a) 35 ppm is 35 μmol CO per 1 mole air. Since FW of CO is 28, we can write this as 35 μmole × 28 μg/μmole = 980 μg CO in 24.5 L air.

$$\frac{980\ \mu g}{24.5\ L} \times \frac{1\ mg}{1000\ \mu g} \times \frac{1000\ L}{1\ m^3} = 40\ mg/m^3$$

(b) 75 ppb SO_2 is 75 nmol SO_2 in 24.5 L air. The FW of SO_2 is 64, so

$$\frac{75\ nmol}{24.5\ L} \times \frac{1\ \mu mol}{1000\ nmol} \times \frac{64\ \mu g}{\mu mol} \times \frac{1000\ L}{1\ m^3} = 19_6\ \mu g/m^3;\ \text{round to } 0.20\ mg/m^3.$$

Liquid Samples

A deciliter is 0.1 L or 100 mL.

You can report results for liquid samples on a weight/weight basis, as above, or they may be reported on a **weight/volume basis**. The latter is more common, at least in the clinical laboratory. The calculations are similar to those above. Percent on a weight/volume basis is equal to grams of analyte per 100 mL of sample, while mg% is equal to milligrams of analyte per 100 mL of sample. This latter unit is often used by clinical chemists for biological fluids, and their accepted terminology is *milligrams per deciliter* (*mg/dL*) to distinguish from mg% on a weight/weight basis. Whenever a concentration is expressed as a percentage, it should be clearly specified whether this is wt/vol or wt/wt. In all but dilute aqueous solutions, this distinction is important. In dilute aqueous solutions, wt/vol and wt/wt ratios are numerically the same because the density of water is unity (1 mL = 1 g) for all practical purposes. Parts per million, parts per billion, and parts per trillion can also be expressed on a weight/volume basis; ppm is calculated from mg/L or μg/mL; ppb is calculated from μg/L or ng/mL; and ppt is calculated from pg/mL or ng/L. Alternatively, the following fundamental calculations may be used:

In dilute aqueous solution

ppm = μg/mL = mg/L
ppb = ng/mL = μg/L
ppt = pg/mL = ng/L

$$\% \text{(wt/vol)} = \left[\frac{\text{wt solute (g)}}{\text{vol sample (mL)}}\right] \times 10^2\ \text{(\%/g solute/mL sample)} \qquad (4.12)$$

$$\text{ppm (wt/vol)} = \left[\frac{\text{wt solute (g)}}{\text{vol sample (mL)}}\right] \times 10^6\ \text{(ppm/g solute/mL sample)} \qquad (4.13)$$

$$\text{ppb (wt/vol)} = \left[\frac{\text{wt solute (g)}}{\text{vol sample (mL)}}\right] \times 10^9\ \text{(ppb/g solute/mL sample)} \qquad (4.14)$$

$$\text{ppt (wt/vol)} = \left[\frac{\text{wt solute (g)}}{\text{vol sample (mL)}}\right] \times 10^{12}\ \text{(ppt/g solute/mL sample)} \qquad (4.15)$$

Note that %(wt/vol) is not pounds/100 gal of solution; the units must be expressed in grams of solute and milliliters of solution. To avoid ambiguities, increasingly it is recommended that ppm, ppb, and ppt units are not used to describe solution phase concentrations; most journals require that μg/mL or ng/mL units should be used instead.

Example 4.16

A 25.0-μL serum sample was analyzed for glucose content and found to contain 26.7 μg. Calculate the concentration of glucose in μg/mL and in mg/dL.

Solution

$$25.0 \ \mu L \times \frac{1 \ mL}{1000 \ \mu L} = 2.50 \times 10^{-2} \ mL$$

$$26.7 \ \mu g \times \frac{1 \ g}{10^5 \ \mu g} = 2.67 \times 10^{-5} \ g$$

$$\text{Glucose Concentration} = \frac{2.67 \times 10^{-5} \ g \ glucose}{2.50 \times 10^{-2} \ mL \ serum} \times 10^6 \ \mu g/g = 1.07 \times 10^3 \ \mu g/mL$$

This is numerically the same in ppm units. Also,

$$\text{Glucose Concentration} = 1.07 \times 10^3 \ \frac{\mu g}{mL} \times \frac{0.001 \ mg}{1 \ \mu g} \times \frac{100 \ mL}{1 \ dL}$$

$$= 107 \ mg/dL$$

[Note the relationship: 10 ppm(wt/vol) = 1 mg/dL]

What does 1 μg/mL, often called 1 ppm, represent in terms of moles per liter? It depends on the formula weight.

Let's do some actual conversions using real formula weights. We begin with a solution that contains 2.5 μg/mL benzene. The formula weight (C_6H_6) is 78.1. The concentration in moles per liter is $(2.5 \times 10^{-3} \ g/L)/(78 \ g/mol) = 3.8 \times 10^{-5} \ M$. Another solution contains $5.8 \times 10^{-8} M$ lead. The concentration in parts per billion is $(5.8 \times 10^{-8} \ mol/L)(207 \ g/mol) = 1.2_0 \times 10^{-5} g/L$. For parts per billion (μg/L), then $(1.2_0 \times 10^{-5} \ g/L) \times (10^6 \ \mu g/g) = 1.2_0 \times 10^1 \ \mu g/L$ or 12 ppb. A drinking water sample that contains 350 pg/L of carbon tetrachloride has a concentration in ng/L of $(350 \times 10^{-12} \ g/L)/(10^9 \ ng/g) = 350 \times 10^{-3} \ ng/L = 0.35 \ ng/L$ or 0.35 ppt. The molar concentration is $(350 \times 10^{-12} g/L)/(154 \ g/mol) = 2.3 \times 10^{-12} \ M$. (Chlorine-treated water may contain traces of chlorinated hydrocarbons—this is very low.)

A key point to remember is that solutions that have the same numerical concentrations on a weight/weight or weight/volume basis do not have the same number of molecules[4], but solutions of the same molarity do.

The relationship between μg/mL and molarity (M) units depends on the formula weight.

Example 4.17

(a) Calculate the molar concentrations of 1 mg/L (1.00 ppm) solutions each of Li^+ and Pb^{2+}.

(b) What weight of $Pb(NO_3)_2$ will have to be dissolved in 1 liter of water to prepare a 100 mg/L (100 ppm) Pb^{2+} solution?

---

[4]Unless they have the same formula weight.

Solution

(a) Li concentration = 1.00 mg/L Pb concentration = 1.00 mg/L

$$M_{Li} = \frac{1.00 \text{ mg Li/L} \times 10^{-3}\text{g/mg}}{6.94 \text{ g Li/mol}} = 1.44 \times 10^{-4} \text{ mol/L Li}$$

$$M_{Pb} = \frac{1.00 \text{ mg Pb/L} \times 10^{-3}\text{g/mg}}{207 \text{ g Pb/mol}} = 4.83 \times 10^{-6} \text{ mol/L Pb}$$

Because lead is much heavier than lithium, a given weight contains a smaller number of moles and its molar concentration is less.

(b) If 1 ppm Pb = 4.83×10^{-6} mol/L Pb
     100 ppm Pb = 4.83×10^{-4} mol/L Pb
     Therefore, we need 4.83×10^{-4} mol $Pb(NO_3)_2$.

$$4.83 \times 10^{-4} \text{ mol} \times 283.2 \text{ g } Pb(NO_3)_2/\text{mol} = 0.137 \text{ g } Pb(NO_3)_2$$

For dilute aqueous solutions, wt/wt ≈ wt/vol because the density of water is near 1.000 mg/ml.

The concentration units wt/wt and wt/vol are related through the density. They are numerically the same for dilute aqueous solutions as the density is 1 g/mL.

If the analyte is a liquid dissolved in another liquid, the results may be expressed on a **volume:volume** basis, but you will likely encounter this only in rare situations. One exception is specification of eluents in liquid chromatography, 40:60 methanol:water connotes that 40 volumes of methanol added to 60 volumes of water; the final volume is generally not available in such specifications. On the other hand, in a vol/vol designation, the first volume refers to that of the solute and the second to that of the solution; this unit is commonly used in the alcoholic beverage industry to specify ethanol content. You would handle the calculations in the same manner as those above, using the same volume units for solute and solution. As illustrated in Example 4.15, gas concentrations may be reported on a weight/volume, volume/volume, and rarely, on a weight/weight basis.

Alcohol in wine and liquor is expressed as vol/vol (200 proof = 100% vol/vol). Since the specific gravity of alcohol is 0.8, wt/vol concentration = 0.8 × (vol/vol) = 0.4 × proof

It is always best to specify clearly what is meant. In the absence of clear labels, it is best to assume that solids are being reported wt/wt, gases vol/vol, and liquids may be reported wt/wt (concentrated acid and base reagents), wt/vol (most dilute aqueous solutions), or vol/vol (the U.S. alcoholic beverage industry).

Clinical chemists frequently prefer to use a unit other than weight for expressing the amount of major electrolytes in biological fluids (Na^+, K^+, Ca^{2+}, Mg^{2+}, Cl^-, $H_2PO_4^-$, etc.). This is the unit **milliequivalent** (meq). In this context, milliequivalent is defined as the number of millimoles of analyte multiplied by the charge on the analyte ion. Results are generally reported as meq/L. This concept gives an overall view of the electrolyte balance. The physician can tell at a glance if total electrolyte concentration has increased or decreased markedly. Obviously, the milliequivalents of cations will be equal to the milliequivalents of anions. One mole of a monovalent (+1) cation (1 eq) and half a mole of a divalent (−2) anion (1 eq) have the same number of positive and negative charges (one mole each). As an example of electrolyte or charge balance, Table 4.2 summarizes the averages of major electrolyte compositions normally present in human blood plasma and urine.

The equivalents of cations and anions in any solution must be equal.

We can calculate the milliequivalents of a substance from its weight in milligrams simply as follows (similar to how we calculate millimoles):

$$\boxed{\begin{array}{l} meq = \dfrac{mg}{\text{eq wt (mg/meq)}} = \dfrac{mg}{\text{fw (mg/mmol)}/n \text{ (meq/mmol)}} \\ n = \text{charge on ion} \end{array}} \quad (4.16)$$

TABLE 4.2 Major Electrolyte Composition of Normal Human Plasma[a]

Cations	meq/L	Anions	meq/L
Na^+	143	Cl^-	104
K^+	4.5	HCO_3^-	29
Ca^{2+}	5	Protein	16
Mg^{2+}	2.5	HPO_4^-	2
		SO_4^{2-}	1
		Organic acids	3
Total	155	Total	155

[a] Reproduced from Joseph S. Annino, *Clinical Chemistry*, 3rd ed., by Boston: Little, Brown, 1964.

At blood pH of 7.4, the phosphate actually exists primarily as a mixture of mono- and dihydrogen phosphate.

The equivalent weight of Na^+ is 23.0 (mg/mmol)/1 (meq/mmol) = 23.0 mg/meq.
The equivalent weight of Ca^{2+} is 40.1 (mg/mmol)/2 (meq/mmol) = 20.0 mg/meq.

Example 4.18

The concentration of zinc ion in blood serum is about 1 mg/L. Express this as meq/L.

Solution

The equivalent weight of Zn^{2+} is 65.4 (mg/mmol)/2 (meq/mmol) = 32.7 mg/meq. Therefore,

$$\frac{1 \text{ mg Zn/L}}{32.7 \text{ mg/meq}} = 3.06 \times 10^{-2} \text{ meq/L Zn}$$

This unit is more often used for the major electrolyte constituents as in Table 4.2 rather than the trace constituents, as in the example here.

Reporting Concentrations as Different Chemical Species

Thus far, we have implied that the analyte is determined in the form it exists or for which we want to report the results. However, this is often not true. In the determination of the iron content of an ore, for example, we may measure the iron in the form of Fe_2O_3 and then report it as % Fe. Or we may determine the iron in the form of Fe^{2+} (e.g., by titration) and report it as % Fe_2O_3. This is perfectly proper so long as we know the relationship between what is really being measured and the form it is to be reported in. We may actually determine the calcium content of water, for example, but we may wish to report it as parts per million (mg/L) of $CaCO_3$ (this is the typical way of expressing *water hardness*). We know that each gram of Ca^{2+} is equivalent to (or could be converted to) grams of $CaCO_3$ by multiplying grams Ca by fw $CaCO_3$/fw Ca^{2+}. That is, multiplying the milligrams of Ca^{2+} determined by 100.09/40.08 will give us the equivalent number of milligrams of $CaCO_3$. The calcium does not have to exist in this form (we may not even know in what form it actually exists); we simply have calculated the weight that could exist and will report the result as if it did. Specific operations necessary for calculating the weight of the desired constituent will be described below.

At this point we should mention some of the different weight criteria used for expressing results with biological tissues and solids. The sample may be weighed in one of three physical forms; wet, dry, or ashed. This can apply also to fluids, although fluid volume taken is usually used for the analysis. The wet weight is taken on the fresh, untreated sample. The dry weight is taken after the sample has been dried by heating,

We may express results in any form of the analyte. This is often done to facilitate the interpretation by other professionals.

Water hardness due to calcium ion is expressed as ppm $CaCO_3$. Conveniently, the fw of $CaCO_3$ is 100. Converting from ppm $CaCO_3$ to molar units is very simple!

desiccation, or freeze-drying. If the test substance is unstable to heat, the sample should not be dried by heating. The weight of the ash residue after the organic matter has been burned off is sometimes used as the weight. This can obviously be used only for mineral (inorganic) analysis.

4.4 Volumetric Analysis: How Do We Make Stoichiometric Calculations?

Volumetric or titrimetric analyses are among the most useful and accurate analytical techniques, especially for millimole amounts of analyte. They are rapid and can be automated, and they can be applied to smaller amounts of analyte when combined with a sensitive instrumental technique for detecting the completion of the titration reaction, for example, pH measurement. Other than pedagogic purposes, manual titrations nowadays are generally used only in situations that require high accuracy for relatively small numbers of samples. They are used, for example, to calibrate or validate more routine instrumental methods. Automated titrations are useful when large numbers of samples must be processed. (A titration may be automated, for instance, by means of a color change or a pH change that activates a motor-driven buret to stop delivery. The volume delivered may be electronically registered. (Automatic titrators are discussed in Chapter 21.) Below, we describe the types of titrations that can be performed and the applicable principles, including the requirements of a titration and of standard solutions. The volumetric relationship described earlier in this chapter may be used for calculating quantitative information about the titrated analyte. Volumetric calculations are given in Section 4.5.

Titration — What are the Requirements?

We calculate the moles of analyte titrated from the moles of titrant added and the ratio in which they react.

In a **titration**, the test substance (analyte) reacts with an added reagent of known concentration, generally instantaneously. The reagent of known concentration is referred to as a **standard solution**. It is typically delivered from a buret; the solution delivered by the buret is called the **titrant**. (In some instances, the reverse may also be carried out where a known volume of the standard solution is taken and it is titrated with the analyte of unknown concentration as the titrant.) The volume of titrant required to just completely react with the analyte is measured. Since we know the reagent concentration as well as the reaction stoichiometry between the analyte and the reagent, we can calculate the amount of analyte. The requirements of a titration are as follows:

1. The reaction must be **stoichiometric**. That is, there must be a well-defined and known reaction between the analyte and the titrant. In the titration of acetic acid in vinegar with sodium hydroxide, for example, a well-defined reaction takes place:

$$CH_3COOH + NaOH \rightarrow CH_3COONa + H_2O$$

2. The reaction should be *rapid*. Most ionic reactions, as above, are very rapid.

3. There should be *no side reactions*; the reaction should be specific. If there are interfering substances, these must be removed or independently determined and their influence subtracted from the overall signal ($C_{analyte} = C_{total} - C_{interference}$). In the above example, there should be no other acids present.

4. There should be a *marked change in some property of the solution when the reaction is complete*. This may be a change in color of the solution or in

some electrical or other physical property of the solution. In the titration of acetic acid with sodium hydroxide, there is a marked increase in the pH of the solution when the reaction is complete. A color change is usually brought about by addition of an **indicator**, whose color is dependent on the properties of the solution, for example, the pH.

5. The point at which an equivalent or stoichiometric amount of titrant is added is called the **equivalence point**. The point at which the reaction is *observed* to be complete is called the **end point**, that is, when a change in some property of the solution is detected. The end point should coincide with the equivalence point or be at a reproducible interval from it.

6. The reaction should be **quantitative**. That is, the equilibrium of the reaction should be far to the right so that a sufficiently *sharp* change will occur at the end point to obtain the desired accuracy. If the equilibrium does not lie far to the right, then there will be gradual change in the property marking the end point (e.g., pH) and this will be difficult to detect precisely.

The *equivalence point* is the theoretical end of the titration where the number equivalents of the analyte exactly equals the number of equivalents of the titrant added. The *end point* is the observed end of the titration. The difference is the titration error.

Standard Solutions—There are Different Kinds

A standard solution is prepared by dissolving an accurately weighed quantity of a highly pure material called a **primary standard** and diluting to an accurately known volume in a volumetric flask. Alternatively, if the material is not sufficiently pure, a solution is prepared to give approximately the desired concentration, and this is **standardized** by titrating a weighed quantity of a primary standard. For example, sodium hydroxide is not sufficiently pure to prepare a standard solution directly. It is therefore standardized by titrating a primary standard acid, such as potassium acid phthalate (KHP). Potassium acid phthalate is a solid that can be weighed accurately. Standardization calculations are treated below.

A **primary standard** should fulfill these requirements:

A solution standardized by titrating a primary standard is itself a secondary standard. It will be less accurate than a primary standard solution due to the errors of titrations.

1. It should be *100.00% pure*, although 0.01 to 0.02% impurity is tolerable if it is accurately known.

2. It should be *stable to drying* temperatures, and it should be stable indefinitely at room temperature. The primary standard is always dried before weighing.[5]

3. It should be *readily* and relatively inexpensively *available*.

4. Although not essential, it should have a *high formula weight*. This is so that a relatively large amount of it will have to be weighed. The relative error in weighing a greater amount of material will be smaller than that for a small amount.

5. If it is to be used in titration, it should possess the *properties required for a titration* listed above. In particular, the equilibrium of the reaction should be far to the right so that a sharp end point will be obtained.

A high formula weight means a larger weight must be taken for a given molar concentration of titrant to be made. This reduces the relative error in weighing.

---

[5]There are a few exceptions when the primary standard is a hydrate.

Classification of Titration Methods—What Kinds are There?

There are four general classes of volumetric or titrimetric methods.

1. *Acid–Base.* Many compounds, both inorganic and organic, are either acids or bases and can be titrated, respectively, with a standard solution of a strong base or a strong acid. The end points of these titrations are easy to detect, either by means of an indicator or by following the change in pH with a pH meter. The acidity and basicity of many organic acids and bases can be enhanced by titrating in a *nonaqueous solvent*. The result is a sharper end point, and weaker acids and bases can be titrated in this manner.

2. *Precipitation.* In the case of precipitation, the titrant forms an insoluble product with the analyte. An example is the titration of chloride ion with silver nitrate solution to form silver chloride precipitate. Again, indicators can be used to detect the end point, or the potential of the solution can be monitored electrically.

3. *Complexometric.* In complexometric titrations, the titrant is a reagent that forms a water-soluble complex with the analyte, a metal ion. The titrant is often a **chelating agent**.[6] Ethylenediaminetetraacetic acid (EDTA) is one of the most useful chelating agents used for titration. It will react with a large number of metal ions, and the reactions can be controlled by adjustment of pH. Indicators can be used to form a highly colored complex with the metal ion.

4. *Reduction–Oxidation.* These "redox" titrations involve the titration of an oxidizing agent with a reducing agent, or vice versa. An oxidizing agent gains electrons and a reducing agent loses electrons in a reaction between them. There must be a sufficiently large difference between the oxidizing and reducing capabilities of these agents for the reaction to go to completion and give a sharp end point; that is, one should be a fairly strong oxidizing agent (strong tendency to gain electrons) and the other a fairly strong reducing agent (strong tendency to lose electrons). Appropriate indicators for these titrations are available; various electrometric means to detect the end point may also be used.

These different types of titrations and the means of detecting their end points will be treated separately in succeeding chapters.

4.5 Volumetric Calculations—Let's Use Molarity

We shall use molarity throughout the majority of the text for volumetric calculations. Some instructors prefer to introduce the concept of normality, and students are likely to encounter it in reference books. A section on normality-based calculations can be found in the text **website**.

In Equations 4.1–4.5, we previously discussed the ways of expressing in molar and millimolar units.

By rearranging these equations, we obtain the expressions for calculating other quantities.

Learn these relationships well. They are the basis of all volumetric calculations, solution preparation, and dilutions. Think units!

$$M \text{ (mol/L)} \times \text{L} = \text{mol} \qquad M \text{ (mmol/mL)} \times \text{mL} = \text{mmol} \qquad (4.17)$$

$$\text{g} = \text{mol} \times \text{fw (g/mol)} \qquad \text{mg} = \text{mmol} \times \text{fw (mg/mmol)} \qquad (4.18)$$

---

[6]A chelating agent (the term is derived from the Greek word for *clawlike*) is a type of complexing agent that contains two or more groups capable of complexing with a metal ion. EDTA has six such groups.

$$\boxed{\begin{aligned} g &= M \text{ (mol/L)} \times L \times fw \text{ (g/mol)} \\ mg &= M \text{ (mmol/mL)} \times mL \times fw \text{ (mg/mmol)} \end{aligned}} \qquad (4.19)$$

We usually work with millimole (mmol) and milliliter (mL) quantities in titrations; therefore, the right-hand equations are more useful. Note that the expression for formula weight contains the same numerical value whether it be in g/mol or mg/mmol. Note also that care must be taken in utilizing "milli" quantities (millimoles, milligrams, milliliters). Incorrect use could result in calculations errors of 1000-fold.

Assume 25.0 mL of 0.100 M $AgNO_3$ is required to titrate a sample containing sodium chloride. The reaction is

$$Cl^- \text{ (aq)} + Ag^+ \text{ (aq)} \rightarrow \underline{AgCl(s)}$$

Since Ag^+ and Cl^- react on a 1:1 molar basis, the number of millimoles of Cl^- is equal to the number of millimoles of Ag^+ needed for titration. We can calculate the milligrams of NaCl as follows:

For 1:1 reactions, $mmol_{analyte} = mmol_{titrant}$.

$$mmol_{NaCl} = mL_{AgNO_3} \times M_{AgNO_3}$$

$$= 25.0 \text{ mL} \times 0.100 \text{ (mmol/mL)} = 2.50 \text{ mmol}$$

$$mg_{NaCl} = mmol \times fw_{NaCl}$$

$$= 2.50 \text{ mmol} \times 58.44 \text{ mg/mmol} = 146 \text{ mg}$$

We can calculate the percentage of analyte A that reacts on a *1:1 mole basis* with the titrant using the following general formula:

$$\boxed{\begin{aligned} \%\text{Analyte} &= \text{fraction}_{analyte} \times 100\% = \frac{mg_{analyte}}{mg_{sample}} \times 100\% \\ &= \frac{mmol \text{ analyte} \times fw_{analyte} \text{ (mg/mol)}}{mg_{sample}} \times 100\% \\ &= \frac{M_{titrant} \text{ (mmol/mL)} \times mL_{titrant} \times fw_{analyte} \text{ (mg/mmol)}}{mg_{sample}} \times 100\% \end{aligned}} \qquad (4.20)$$

Note that this computation is a summary of the individual calculation steps taken to arrive at the fraction of analyte in the sample using proper dimensional analysis. You should use it in that sense rather than simply memorizing a formula.

Example 4.19

A 0.4671-g sample containing sodium bicarbonate was dissolved and titrated with standard 0.1067 M hydrochloric acid solution, requiring 40.72 mL. The reaction is

$$HCO_3^- + H^+ \rightarrow H_2O + CO_2$$

Calculate the percent sodium bicarbonate in the sample.

Solution

The millimoles of sodium bicarbonate are equal to the millimoles of acid used to titrate it, since they react in a 1:1 ratio.

$$mmol_{HCl} = 0.1067 \text{ mmol/mL} \times 40.72 \text{ mL} = 4.344_8 \text{ mmol}_{HCl} \equiv mmol \text{ NaHCO}_3$$

(Extra figures are carried so an identical answer is obtained when all steps are done together below.)

$$mg_{NaHCO_3} = 4.3448 \text{ mmol} \times 84.01 \text{ mg/mmol} = 365.0_1 \text{ mg NaHCO}_3$$

$$\%NaHCO_3 = \frac{365.0_1 \text{ mg NaHCO}_3}{467.1 \text{ mg}_{sample}} \times 100\% = 78.14\%NaHCO_3$$

Or, combining all the steps,

$$\%NaHCO_3 = \frac{M_{HCl} \times mL_{HCl} \times fw_{NaHCO_3}}{mg_{sample}} \times 100\%$$

$$= \frac{0.1067 \text{ mmol HCl/mL} \times 40.72 \text{ mL HCl} \times 84.01 \text{ mg NaHCO}_3/\text{mmol}}{467.1 \text{ mg}} \times 100\%$$

$$= 78.14\%NaHCO_3$$

Some Useful Things to Know For Molarity Calculations

When the reaction is not 1:1, a stoichiometric factor must be used to equate the moles of analyte and titrant.

Many substances do not react on a 1:1 mole basis, and so the simple calculation in the above example cannot be applied to all reactions. It is possible, however, to write a generalized formula for calculations applicable to *all* reactions based on the **balanced equation** for reactions.

Consider the general reaction

$$aA \times tT \rightarrow P \tag{4.21}$$

where A is the analyte, T is the titrant, and they react in the ratio a/t to give products P. Then, noting the units and using dimensional analysis,

$$\boxed{mmol_A = mmol_T \times \frac{a}{t} \text{ (mmol A/mmol T)}} \tag{4.22}$$

Still think units! We have added $mmol_{analyte}/mmol_{titrant}$.

$$\boxed{mmol_A = M_T \text{ (mmol/mL)} \times mL_T \times \frac{a}{t}\text{(mmol A/mmol T)}} \tag{4.23}$$

$$\boxed{mg_A = mmol_A \times fw_A \text{ (mg/mmol)}} \tag{4.24}$$

$$\boxed{\begin{array}{l} mg_A = M_T \text{ (mmol/mL)} \times mL_T \times \frac{a}{t} \text{ (mmol A/mmol T)} \\ \quad \times fw_A \text{ (mg/mmol)} \end{array}} \tag{4.25}$$

Note that the a/t factor serves to equate the analyte and titrant. To avoid a mistake in setting up the factor, it is helpful to remember that when you calculate the amount of analyte, you must multiply the amount of titrant by the a/t ratio (a comes first). Conversely, if you are calculating the amount of titrant (e.g., molarity) from a known amount of analyte titrated, you must multiply the amount of analyte by the t/a ratio (t comes first). The best way, of course, to ascertain the correct ratio is to always do a dimensional analysis to obtain the correct units.

In a manner similar to that used to derive Equation 4.22, we can list the steps in arriving at a general expression for calculating the percent analyte A in a sample determined by titrating a known weight of sample with a standard solution of titrant T:

$$
\begin{aligned}
\%\text{Analyte} &= \text{fraction}_{\text{analyte}} \times 100\% = \frac{\text{mg}_{\text{analyte}}}{\text{mg}_{\text{sample}}} \times 100\% \\
&= \frac{\text{mmol}_{\text{titrant}} \times (a/t)\,(\text{mmol}_{\text{analyte}}/\text{mmol}_{\text{titrant}}) \times \text{fw}_{\text{analyte}}\;(\text{mg/mmol})}{\text{mg}_{\text{sample}}} \times 100\% \\
&= \frac{M_{\text{titrant}}\,(\text{mmol/mL}) \times \text{mL}_{\text{titrant}} \times (a/t)\,(\text{mmol}_{\text{analyte}}/\text{mmol}_{\text{titrant}}) \times \text{fw}_{\text{analyte}}\;(\text{mg/mmol})}{\text{mg}_{\text{sample}}} \\
&\quad \times 100\%
\end{aligned}
$$

$$(4.26)$$

Again, note that we simply use dimensional analysis, that is, we perform stepwise calculations in which units cancel to give the desired units. In this general procedure, the dimensional analysis includes the stoichiometric factor a/t that converts millimoles of titrant to an equivalent number of millimoles of titrated analyte.

Example 4.20

A 0.2638-g soda ash sample is analyzed by titrating the sodium carbonate with the standard 0.1288 M hydrochloride solution, requiring 38.27 mL. The reaction is

$$
CO_3^{2-} + 2H^+ \rightarrow H_2O + CO_2
$$

Calculate the percent sodium carbonate in the sample.

Solution

The millimoles of sodium carbonate is equal to one-half the millimoles of acid used to titrate it, since they react in a 1:2 ratio $\left(a/t = \frac{1}{2} \right)$.

$$
\text{mmol}_{\text{HCl}} = 0.1288\;\text{mmol/mL} \times 38.27\;\text{mL} = 4.929\;\text{mmol HCl}
$$

$$
\text{mmol}_{\text{NaCO}_3} = 4.929\;\text{mmol HCl} \times \tfrac{1}{2}(\text{mmol Na}_2\text{CO}_3/\text{mmol HCl}) = 2.464_5\;\text{mmol Na}_2\text{CO}_3
$$

$$
\text{mg}_{\text{Na}_2\text{CO}_3} = 2.464_5\;\text{mmol} \times 105.99\;\text{mg Na}_2\text{CO}_3/\text{mmol} = 261.2_1\;\text{mg Na}_2\text{CO}_3
$$

$$
\%\text{Na}_2\text{CO}_3 = \frac{261.2_1\;\text{mg Na}_2\text{CO}_3}{263.8\;\text{mg}_{\text{sample}}} \times 100\% = 99.02\%\;\text{Na}_2\text{CO}_3
$$

Or, combining all the steps at once,

$$
\%\text{Na}_2\text{CO}_3 = \frac{M_{\text{HCl}} \times \text{mL}_{\text{HCl}} \times \tfrac{1}{2}(\text{mmol Na}_2\text{CO}_3/\text{mmol HCl}) \times \text{fw}_{\text{Na}_2\text{CO}_3}}{\text{mg}_{\text{sample}}} \times 100\%
$$

$$
= \frac{0.1288\;\text{mmol HCl} \times 38.27\;\text{mL HCl} \times \tfrac{1}{2}\,(\text{mmol Na}_2\text{CO}_3/\text{mmol HCl}) \times 105.99\;(\text{mg Na}_2\text{CO}_3/\text{mmol})}{263.8\;\text{mg}_{\text{sample}}} \times 100\%
$$

$$
= 99.02\%\;\text{Na}_2\text{CO}_3
$$

Example 4.21

How many milliliters of 0.25 M solution of H_2SO_4 will react with 10 mL of a 0.25 M solution of NaOH?

Solution

The reaction is

$$H_2SO_4 + 2NaOH \rightarrow Na_2SO_4 + 2H_2O$$

One-half as many millimoles of H_2SO_4 as of NaOH will react, or

$$M_{H_2SO_4} \times mL_{H_2SO_4} = M_{NaOH} \times mL_{NaOH} \times \frac{1}{2} \text{ (mmol } H_2SO_4/\text{mmol NaOH)}$$

Therefore,

$$mL_{H_2SO_4} = \frac{0.25 \text{ mmol NaOH/mL} \times 10 \text{ mL NaOH} \times \frac{1}{2}(\text{mmol } H_2SO_4/\text{mmol NaOH})}{0.25 \text{ mmol } H_2SO_4/\text{mL}}$$

$$= 5.0 \text{ mL } H_2SO_4$$

Note that, in this case, we multiplied the amount of titrant by the a/t ratio (mmol analyte/mmol titrant).

Example 4.22

A sample of impure salicylic acid, $C_6H_4(OH)COOH$ (one titratable proton), is analyzed by titration. What size sample should be taken so that the percent purity is equal to five times the milliliters of 0.0500 M NaOH used to titrate it?

Solution

Let x = mL NaOH; % salicylic acid (HA) = $5x$:

$$\%HA = \frac{M_{NaOH} \times mL_{NaOH} \times 1 \text{ (mmol HA/mmol NaOH)} \times fw_{HA} \text{ (mg/mmol)}}{mg_{sample}} \times 100\%$$

$$5x\% = \frac{0.0500 \text{ } M \times x \text{ mL NaOH} \times 1 \times 138 \text{ mg HA/mmol}}{mg_{sample}} \times 100\%$$

$$mg_{sample} = 138 \text{ mg}$$

You can apply the above examples of acid–base calculations to the titrations described in Chapter 7.

Standardization and Titration Calculations—They are the Reverse of One Another

In standardization, generally it is the concentration of the titrant that is unknown and the moles of analyte (primary standard) are known.

When a titrant material of high or known purity is not available, the concentration of the approximately prepared titrant solution must be accurately determined by **standardization**; that is, by titrating an accurately weighed quantity (a known number of millimoles) of a primary standard. From the volume of titrant used to titrate the primary standard, we can calculate the molar concentration of the titrant.

Taking the analyte A in Equation 4.21 to be the primary standard,

$$mmol_{standard} = \frac{mg_{standard}}{fw_{standard} \ (mg/mmol)}$$

$$mmol_{titrant} = M_{titrant} \ (mmol/mL) \times mL_{titrant}$$

$$= mmol_{standard} \times t/a \ (mmol_{titrant}/mmol_{standard})$$

$$M_{titrant} \ (mmol/mL) = \frac{mmol_{standard} \times t/a \ (mmol_{titrant}/mmol_{standard})}{mL_{titrant}}$$

Or, combining all steps at once,

$$M_{titrant} \ (mmol/mL) = \frac{mg_{standard}/fw_{standard} \ (mg/mmol) \times t/a \ (mmol_{titrant}/mmol_{standard})}{mL_{titrant}} \qquad (4.27) \quad \text{Units!}$$

Note once again that dimensional analysis (cancellation of units) results in the desired units of mmol/mL.

Example 4.23

An approximate 0.1 M hydrochloric acid solution is prepared by 120-fold dilution of concentrated hydrochloric acid. It is standardized by titrating 0.1876 g of dried primary standard sodium carbonate:

$$CO_3^{2-} + 2H^+ \rightarrow H_2O + CO_2$$

The titration required 35.86 mL acid. Calculate the molar concentration of the hydrochloric acid.

Solution

The millimoles of hydrochloric acid are equal to twice the millimoles of sodium carbonate titrated.

$$mmol_{Na_2CO_3} = 187.6 \ mg \ Na_2CO_3/105.99 \ (mg \ Na_2CO_3/mmol) = 1.770_0 \ mmol \ Na_2CO_3$$

$$mmol_{HCl} = M_{HCl} \ (mmol/mL) \times 35.86 \ mL \ HCl = 1.770_0 \ mmol \ Na_2CO_3$$

$$\times 2 \ (mmol \ HCl/mmol \ Na_2CO_3)$$

$$M_{HCl} = \frac{1.770_0 \ mmol \ Na_2CO_3 \times 2 \ (mmol \ HCl/mmol \ Na_2CO_3)}{35.86 \ mL \ HCl} = 0.09872 \ M$$

Or, combining all steps at once,

$$M_{HCl} = \frac{(mg_{Na_2CO_3}/fw_{Na_2CO_3}) \times (2/1) \ (mmol \ HCl/mmol \ Na_2CO_3)}{mL \ _{HCl}}$$

$$= \frac{\left[187.6 \ mg/105.99 \ (mg/mmol)\right] \times 2 \ (mmol \ HCl/mmol \ Na_2CO_3)}{35.86 \ mL}$$

$$= 0.09872 \ mmol/mL$$

Note that we multiplied the amount of analyte, Na_2CO_3, by the t/a ratio (mmol titrant/mmol analyte). Note also that although all measurements were to four significant figures, we computed the formula weight of Na_2CO_3 to five figures. This is because with four figures, it would have had an uncertainty of about one part per thousand compared to 187.6 with an uncertainty of about half that. It is not bad practice, as a matter of routine, to carry the formula weight to one additional figure.

The following examples illustrate titration calculations for different types of reactions and stoichiometry.

Example 4.24

This is a "redox" titration (see Chapter 21).

The iron(II) in an acidified solution is titrated with a 0.0206 M solution of potassium permanganate:

$$5Fe^{2+} + MnO_4^- + 8H^+ \rightarrow 5Fe^{3+} + Mn^{2+} + 4H_2O$$

If the titration required 40.2 mL, how many milligrams iron are in the solution?

Solution

There are five times as many millimoles of iron as there are of permanganate that react with it, so

$$mmol_{Fe} = \frac{mg_{Fe}}{fw_{Fe}} = M_{KMnO_4} \times mL_{KMnO_4} \times \frac{5}{1} \text{ (mmol Fe/mmol KMnO}_4)$$

$$mg_{Fe} = 0.0206 \text{ mmol KMnO}_4/mL \times 40.2 \text{ mL KMnO}_4 \times 5 \text{ (mmol Fe/mmol MnO}_4^-)$$
$$\times 55.8 \text{ mg Fe/mmol}$$
$$= 231 \text{ mg Fe}$$

Calculations of this type are used for the redox titrations described in Chapter 21.

Following is a list of typical precipitation and complexometric titration reactions and the factors for calculating the milligrams of analyte from millimoles of titrant.[7]

$$Cl^- + Ag^+ \rightarrow \underline{AgCl} \qquad mg_{Cl^-} = M_{Ag^+} \times mL_{Ag^+} \times 1 \text{ (mmol Cl}^-/\text{mmol Ag}^+) \times fw_{Cl^-}$$

$$2Cl^- + Pb^{2+} \rightarrow \underline{PbCl_2} \qquad mg_{Cl^-} = M_{Pb^{2+}} \times mL_{Pb^{2+}} \times 2 \text{ (mmol Cl}^-/\text{mmol Pb}^{2+}) \times fw_{Cl^-}$$

$$PO_4^{3-} + 3Ag^+ \rightarrow \underline{Ag_3PO_4} \qquad mg_{PO_4^{3-}} = M_{Ag^+} \times mL_{Ag^+} \times \frac{1}{3} \text{ (mmol PO}_4^{3-}/\text{mmol Ag}^+) \times fwPO_4^{3-}$$

$$2CN^- + Ag^+ \rightarrow \underline{Ag(CN)_2^-} \qquad mg_{CN^-} = M_{Ag^+} \times mL_{Ag^+} \times 2 \text{ (mmol CN}^-/\text{mmol Ag}^+) \times fw_{CN^-}$$

$$2CN^- + 2Ag^+ \rightarrow \underline{Ag\,[Ag\,(CN)_2]} \qquad mg_{CN^-} = M_{Ag^+} \times mL_{Ag^+} \times 1 \text{ (mmol CN}^-/\text{mmol Ag}^+) \times fw_{CN^-}$$

$$Ba^{2+} + SO_4^{2-} \rightarrow \underline{BaSO_4} \qquad mg_{Ba^{2+}} = M_{SO_4^{2-}} \times 1 \text{ (mmol Ba}^{2+}/\text{mmol SO}_4^{2-}) \times fw_{Ba^{2+}}$$

$$Ca^{2+} + H_2Y^{2-} \rightarrow CaY^{2-} + 2H^+ \qquad mg_{Ca^{2+}} = M_{EDTA} \times 1 \text{ (mmol Ca}^{2+}/\text{mmol EDTA}) \times fw_{Ca^{2+}}$$

These formulas are useful calculations involving the precipitation and complexometric titrations described in Chapters 7 and 10.

Example 4.25

This is a "complexometric" titration (see Chapter 8). EDTA has four protons, and in this titration, two are dissociated at the pH of the titration.

Aluminum is determined by titrating with EDTA:

$$Al^{3+} + H_2Y^{2-} \rightarrow AlY^- + 2H^+$$

A 1.00-g sample requires 20.5 mL EDTA for titration. The EDTA was standardized by titrating 20.5 mL of a 0.100 M CaCl$_2$ solution, requiring 30.0 mL EDTA. Calculate the percent Al$_2$O$_3$ in the sample.

---

[7]H_4Y = EDTA in the last equation.

Solution

Since Ca^{2+} and EDTA react on a 1:1 mole ratio,

$$M_{EDTA} = \frac{0.100 \text{ mmol CaCl}_2/\text{mL} \times 25.0 \text{ mL CaCl}_2}{30.0 \text{ mL EDTA}} = 0.0833 \text{ mmol/mL}$$

The millimoles Al^{3+} are equal to the millimoles EDTA used in the sample titration, but there are one-half this number of millimoles of Al_2O_3 (since $1Al^{3+} \rightarrow \frac{1}{2}Al_2O_3$). Therefore,

$$\%Al_2O_3 = \frac{M_{EDTA} \times mL_{EDTA} \times \frac{1}{2} (\text{mmol Al}_2O_3/\text{mmol EDTA}) \times fw_{Al_2O_3}}{mg_{sample}} \times 100\%$$

$$\%Al_2O_3 = \frac{0.0833 \text{ mmol EDTA/mL} \times 20.5 \text{ mL EDTA} \times \frac{1}{2} \times 101.96 \text{ mg Al}_2O_3/\text{mmol}}{1000\text{-mg sample}}$$

$$\times 100\% = 8.71\%Al_2O_3$$

What if the Analyte and Titrant Can React in Different Ratios?

As you might be aware from your introductory chemistry course, some substances can undergo reaction to different products. The factor used in calculating millimoles of such a substance from the millimoles of titrant reacted with it will depend on the specific reaction. Sodium carbonate, for example, can react as a diprotic or a monoprotic base:

$$CO_3^{2-} + 2H^+ \rightarrow H_2O + CO_2$$

or

$$CO_3^{2-} + H^+ \rightarrow HCO_3^-$$

In the first case, mmol Na_2CO_3 = mmol acid $\times \frac{1}{2}$ (mmol CO_3^{2-}/mmol H^+). In the second case, mmol Na_2CO_3 = mmol acid. Similarly, phosphoric acid can be titrated as a monoprotic or a diprotic acid:

$$H_3PO_4 + OH^- \rightarrow H_2PO_4^- + H_2O$$

or

$$H_3PO_4 + 2OH^- \rightarrow HPO_4^{2-} + 2H_2O$$

Example 4.26

In acid solution, potassium permanganate reacts with H_2O_2 to form Mn^{2+}:

$$5H_2O_2 + 2MnO_4^- + 6H^+ \rightarrow 5O_2 + 2Mn^{2+} + 8H_2O$$

In neutral solution, it reacts with $MnSO_4$ to form MnO_2:

$$3Mn^{2+} + 2MnO_4^- + 4OH^- \rightarrow \underline{5MnO_2} + 2H_2O$$

Calculate the number of milliliters of 0.100 M $KMnO_4$ that will react with 50.0 mL of 0.200 M H_2O_2 and with 50.0 mL of 0.200 M $MnSO_4$.

Solution

Keep track of millimoles!

The number of millimoles of MnO_4^- will be equal to two-fifths of the number of millimoles of H_2O_2 reacted:

$$M_{MnO_4^-} \times mL_{MnO_4^-} = M_{H_2O_2} \times mL_{H_2O_2} \times \frac{2}{5} \text{ (mmol } MnO_4^-/\text{mmol } H_2O_2)$$

$$mL_{MnO_4^-} = \frac{0.200 \text{ mmol } H_2O_2/\text{mL} \times 50.0 \text{ mL } H_2O_2 \times \frac{2}{5}}{0.100 \text{ mmol } MnO_4^-/\text{mL}} = 40.0 \text{ mL } KMnO_4$$

The number of millimoles of MnO_4^- reacting with Mn^{2+} will be equal to two-thirds of the number of millimoles of Mn^{2+}:

$$M_{MnO_4^-} \times mL_{MnO_4^-} = M_{Mn^{2+}} \times mL_{Mn^{2+}} \times \frac{2}{3} \text{ (mmol } MnO_4^-/\text{mmol } Mn^{2+})$$

$$mL_{MnO_4^-} = \frac{0.200 \text{ mmol } Mn^{2+}/\text{mL} \times 50.0 \text{ mL } Mn^{2+} \times \frac{2}{3}}{0.100 \text{ mmol } MnO_4^-/\text{mL}} = 66.7 \text{ mL } KMnO_4$$

Example 4.27

Oxalic acid, $H_2C_2O_4$, is a reducing agent that reacts with $KMnO_4$ as follows:

$$5H_2C_2O_4 + 2MnO_4^- + 6H^+ \rightarrow 10CO_2 + 2Mn^{2+} + 8H_2O$$

Its two protons are also titratable with a base. How many milliliters of 0.100 M NaOH and 0.100 M $KMnO_4$ will react with 500 mg $H_2C_2O_4$?

Solution

$$mmol \text{ NaOH} = 2 \times mmol \text{ } H_2C_2O_4$$

$$0.100 \text{ mmol/mL} \times x \text{ mL NaOH} = \frac{500 \text{ mgH}_2C_2O_4}{90.0 \text{ mg/mmol}} \times 2 \text{ (mmol } OH^-/\text{mmol } H_2C_2O_4)$$

$$x = 111 \text{ mL NaOH}$$

$$mmol \text{ } KMnO_4 = \frac{2}{5} \times mmol \text{ } H_2C_2O_4$$

$$0.100 \text{ mmol/mL} \times x \text{ mL } KMnO_4 = \frac{500 \text{ mg } H_2C_2O_4}{90.0 \text{ mg/mmol}} \times \frac{2}{5} \text{ (mmol } KMnO_4/\text{mmol } H_2C_2O_4)$$

$$x = 22.2 \text{ mL } KMnO_4$$

Example 4.28

Pure $Na_2C_2O_4$ plus $KHC_2O_4 \cdot H_2C_2O_4$ (three replaceable protons, KH_3A_2) are mixed in such a proportion that each gram of the mixture will react with equal volumes of 0.100 M $KMnO_4$ and 0.100 M NaOH. What is the proportion?

Solution

Assume 10.0-mL titrant, so it will react with 1.00 mmol NaOH or $KMnO_4$. The acidity is due to $KHC_2O_4 \cdot H_2C_2O_4$ denoted in the following as KH_3A_2:

$$\text{mmol } KH_3A_2 = \text{mmol NaOH} \times \frac{1}{3}(\text{mmol } KH_3A_2/\text{mmol } OH^-)$$

$$1.00 \text{ mmol NaOH} \times \frac{1}{3} = 0.333 \text{ mmol } KH_3A_2$$

From Example 4.27, each mmol $Na_2C_2O_4$ (Na_2A) reacts with $\frac{2}{5}$ mmol $KMnO_4$.

$$\text{mmol } KMnO_4 = \text{mmol } Na_2A \times \frac{2}{5}(\text{mmol } MnO_4^-/\text{mmol } Na_2A) + \text{mmol } KH_3A_2$$

$$\times \frac{4}{5}(\text{mmol } MnO_4^-/\text{mmol } KH_3A_2)$$

$$1.00 \text{ mmol } KMnO_4 = \text{mmol } Na_2A \times \frac{2}{5} + 0.333 \text{ mmol } KH_3A_2$$

$$\times \frac{4}{5} \text{ mmol } Na_2A = 1.8_3 \text{ mmol}$$

The ratio is 1.8_3 mmol $Na_2A/0.333$ mmol $KH_3A_2 = 5.5_0$ mmol $Na_2A/\text{mmol } KH_3A_2$. The weight ratio is

$$\frac{5.5_0 \text{ mmol } Na_2A \times 134 \text{ mg/mmol}}{218 \text{ mg } KH_3A_2/\text{mmol}} = 3.38 \text{ g } Na_2A/\text{g } KH_3A_2$$

If the Reaction is Slow, do a Back-Titration

Sometimes a reaction is slow to go to completion, and a sharp end point cannot be obtained. One example is the titration of antacid tablets with a strong acid such as HCl. In these cases, a **back-titration** will often yield useful results. In this technique, a measured amount of the reagent, which would normally be the titrant, is added to the sample so that there is a slight excess. After the reaction with the analyte is allowed to go to completion, the amount of excess (unreacted) reagent is determined by titration with another standard solution; the kinetics of the analyte reaction may be increased in the presence of excess reagent. So by knowing the number of millimoles of reagent taken and by measuring the number of millimoles remaining unreacted, we can calculate the number of millimoles of sample that reacted with the reagent:

In back-titrations, a known number of millimoles of reactant is taken, in excess of the analyte. The unreacted portion is titrated.

> mmol reagent reacted = mmol taken − mmol back-titrated
>
>   mg analyte = mmol reagent reacted × factor (mmol analyte/mmol reagent)
>
>   × fw analyte (mg/mmol)

Example 4.29

Chromium(III) is slow to react with EDTA (H_4Y) and is therefore determined by back-titration. Chromium(III) picolinate, $Cr(C_6H_4NO_2)_3$, is sold as a nutritional supplement for athletes with the claim that it aids muscle building. A nutraceutical preparation containing chromium(III) is analyzed by treating a 2.63-g sample with 5.00 mL of 0.0103 M EDTA. Following reaction, the unreacted EDTA is back-titrated with

1.32 mL of 0.0122 M zinc solution. What is the percent chromium picolinate in the pharmaceutical preparation?

Solution

Both Cr^{3+} and Zn^{2+} react in a 1:1 ratio with EDTA:

$$Cr^{3+} + H_4Y \rightarrow CrY^- + 4H^+$$
$$Zn^{2+} + H_4Y \rightarrow ZnY^2 + 4H^+$$

The millimoles of EDTA taken is

$$0.0103 \text{ mmol EDTA/mL} \times 5.00 \text{ mL EDTA} = 0.0515 \text{ mmol EDTA}$$

The millimoles of unreacted EDTA is

$$0.0112 \text{ mmol } Zn^{2+}/\text{mL} \times 1.32 \text{ mL } Zn^{2+} = 0.0148 \text{ mmol unreacted EDTA}$$

The millimoles of reacted EDTA is

$$0.0515 \text{ mmol taken} - 0.0148 \text{ mmol left} = 0.0367 \text{ mmol EDTA} \equiv \text{ mmol } Cr^{3+}$$

The milligrams of $Cr(C_6H_4NO_2)_3$ titrated is

$$0.0367 \text{ mmol } Cr(C_6H_4NO_2)_3 \times 418.3 \text{ mg/mmol} = 15.35 \text{ mg } Cr(C_6H_4NO_2)_3$$

$$\%Cr(C_6H_4NO_2)_3 = \frac{15.35 \text{ mg } Cr(C_6H_4NO_2)_3}{2630 \text{ mg sample}} \times 100\% = 0.584\% \; Cr(C_6H_4NO_2)_3$$

Or, combining all steps,

$$\%Cr(C_6H_4NO_2)_3 = \frac{(M_{EDTA} \times mL_{EDTA} - M_{Zn} \times mL_{Zn^{2+}}) \times 1(\text{mmol } Cr(C_6H_4NO_2)_3/ \text{mmol EDTA}) \times fw_{Cr(C_6H_4NO_2)_3}}{mg_{sample}} \times 100\%$$

$$= \frac{(0.0103 \text{ mmol EDTA/mL} \times 5.00 \text{ mL EDTA} - 0.0112 \text{ mmol } Zn^{2+}/ \text{mL} \times 1.32 \text{ mL } Zn^{2+}) \times 1 \times 418.3 \text{ mg } Cr(C_6H_4NO_2)_3/\text{mmol}}{2630 \text{ mg sample}} \times 100\%$$

$$= 0.584\% \; Cr(C_6H_4NO_2)_3$$

Example 4.30

A 0.200-g sample of pyrolusite is analyzed for manganese content as follows. Add 50.0 mL of a 0.100 M solution of ferrous ammonium sulfate to reduce the MnO_2 to Mn^{2+}. After reduction is complete, the excess ferrous ion is titrated in acid solution with 0.0200 M KMnO$_4$, requiring 15.0 mL. Calculate the percentage of manganese in the sample as Mn_3O_4 (the manganese may or may not exist in this form, but we can make the calculations on the assumption that it does).

Solution

The reaction between Fe^{2+} and MnO_4^- is

$$5Fe^{2+} + MnO_4^- + 8H^+ \rightarrow 5Fe^{3+} + Mn^{2+} + 4H_2O$$

and so there are five times as many millimoles of excess Fe^{2+} as of MnO_4^- that reacted with it.

The reactant may react in different ratios with the analyte and titrant.

The reaction between Fe^{2+} and MnO_2 is

$$MnO_2 + 2Fe^{2+} + 4H^+ \rightarrow Mn^{2+} + 2Fe^{3+} + 2H_2O$$

and there are one-half as many millimoles of MnO_2 as millimoles of Fe^{2+} that react with it. There are one-third as many millimoles of Mn_3O_4 as of MnO_2 ($1MnO_2 \rightarrow \frac{1}{3}Mn_3O_4$). Therefore,

$$\text{mmol } Fe^{2+} \text{ reacted} = 0.100 \text{ mmol } Fe^{2+}/mL \times 50.0 \text{ mL } Fe^{2+} - 0.0200 \text{ mmol } MnO_4{}^-/mL$$
$$\times 15.0 \text{ mL } MnO_4{}^- \times 5 \text{ mmol } Fe^{2+}/\text{mmol } MnO_4{}^-$$
$$= 3.5 \text{ mmol } Fe^{2+} \text{ reacted}$$

$$\text{mmol } MnO_2 = 3.5 \text{ mmol } Fe^{2+} \times \frac{1}{2} (\text{mmol } MnO_2/\text{mmol } Fe^{2+}) = 1.75 \text{ mmol } MnO_2$$
$$\text{mmol } Mn_3O_4 = 1.7_5 \text{ mmol } MnO_2 \times \frac{1}{3} (\text{mmol } Mn_3O_4)/\text{mmol } MnO_2)$$
$$= 0.58_3 \text{ mmol } Mn_3O_4$$
$$\%Mn_3O_4 = \frac{0.58_3 \text{ mmol } Mn_3O_4 \times 228.8 \text{ (mg } Mn_3O_4/\text{mmol)}}{200 \text{ mg sample}} \times 100\%$$
$$= 66.7\%Mn_3O_4$$

Or, combining all steps at once,

$$\%Mn_3O_4 = \left\{ \left[M_{Fe^{2+}} \times mL_{Fe^{2-}} - M_{MnO_4{}^-} \times mL_{MnO_4{}^-} \times 5(\text{mmol } Fe^{2+}/\text{mmol } MnO_4{}^-) \right. \right.$$
$$\times \frac{1}{2} (\text{mmol } MnO_2/\text{mmol } Fe^{2+}) \times \frac{1}{3} (\text{mmol } Mn_3O_4/\text{mmol } MnO_2)$$
$$\left. \left. \times \text{fw}_{Mn_3O_4} \right] /mg_{\text{sample}} \right\} \times 100\%$$
$$= \frac{(0.100 \times 50.0 - 0.0200 \times 15.0 \times 5) \times \frac{1}{2} \times \frac{1}{3} \times 228.8 \text{ mg/mmol}}{200 \text{ mg sample}} \times 100\%$$
$$= \%66.7Mn_3O_4$$

4.6 Titer—How to Make Rapid Routine Calculations

For routine titrations, it is often convenient to calculate the **titer** of the titrant. The titer is the weight of analyte that is chemically equivalent to 1 mL of the titrant, usually expressed in milligrams. For example, if a potassium dichromate solution has a titer of 1.267 mg Fe, each milliliter potassium dichromate will react with 1.267 mg iron, and the weight of iron titrated is obtained by simply multiplying the volume of titrant used by the titer. The titer can be expressed in terms of any form of the analyte desired, for example, milligrams FeO or Fe_2O_3.

Titer = milligrams analyte that react with 1 mL of titrant.

Example 4.31

A standard solution of potassium dichromate contains 5.442 g/L. What is its titer in terms of milligrams Fe_3O_4?

Solution

The iron is titrated as Fe^{2+} and each $Cr_2O_7{}^{2-}$ will react with $6Fe^{2+}$ (or the iron from $2Fe_3O_4$):

$$6Fe^{2+} + Cr_2O_7{}^{2-} + 14H^+ \rightarrow 6Fe^{3+} + 2Cr^{3+} + 7H_2O$$

The molarity of the $K_2Cr_2O_7$, solution is

$$M_{Cr_2O_7{}^{2-}} = \frac{g/L}{fw_{K_2Cr_2O_7}} = \frac{5.442 \text{ g/L}}{294.19 \text{ g/mol}} = 0.01850 \text{ mol/L}$$

Therefore the titer is

$$0.01850 \left(\frac{\text{mmol } K_2Cr_2O_7}{\text{mL}} \right) \times \frac{2}{1} \left(\frac{\text{mmol } Fe_3O_4}{\text{mmol } K_2Cr_2O_7} \right) \times 231.54 \left(\frac{\text{mg } Fe_3O_4}{\text{mmol } Fe_3O_4} \right)$$

$$= 8.567 \text{ mg } Fe_3O_4/\text{mL } K_2Cr_2O_7$$

4.7 Weight Relationships—You Need These for Gravimetric Calculations

In the technique of gravimetric analysis (Chapter 9), the analyte is converted to an insoluble form, which is weighed. From the weight of the precipitate formed and the weight relationship between the analyte and the precipitate, we can calculate the weight of analyte. We review here some of the calculation concepts.

The analyte is almost always weighed in a form different from what we wish to report. We must, therefore, calculate the weight of the desired substance from the weight of the gravimetric precipitate. We can do this by using a direct proportion. For example, if we are analyzing for the percentage of chloride in a sample by weighing it as AgCl, we can write

$$Cl^- \xrightarrow{\text{precipitating reagent}} AgCl(s)$$

We derive one mole AgCl from one mole Cl^-, so

$$\frac{g\ Cl^-}{g\ AgCl} = \frac{\text{at wt } Cl}{\text{fw } AgCl}$$

or

$$g\ Cl^- = g\ AgCl \times \frac{\text{at wt } Cl}{\text{fw } AgCl}$$

In gravimetric analysis, the moles of analyte is a multiple of the moles of precipitate formed (the moles of analyte contained in each mole of precipitate).

Note that when we specify at wt or fw of a substance x, it implicitly has the units g x/mol x. In other words, the weight of Cl contained in or used to create AgCl is equal to the weight of AgCl times the **fraction** of Cl in it.

Calculation of the corresponding weight of Cl_2 that would be contained in the sample would proceed thus:

$$Cl_2 \xrightarrow{\text{precipitating reagent}} 2AgCl(s)$$

We derive two moles of AgCl from each mole of Cl_2, so

$$\frac{g\ Cl_2}{g\ AgCl} = \frac{\text{fw } Cl_2}{2 \times \text{fw } AgCl}$$

and

$$g\ Cl_2 = g\ AgCl \times \frac{\text{fw } Cl_2}{2\ (\text{fw } AgCl)}$$

or

$$g\ Cl_2 = g\ AgCl \times \frac{70.906}{2 \times 143.32}$$

We may also write

$$\text{g AgCl} \times \frac{1 \text{ mol AgCl}}{143.32 \text{ g AgCl}} \times \frac{1 \text{ mol Cl}_2}{2 \text{ mol AgCl}} \times \frac{70.906 \text{ g Cl}_2}{1 \text{ mol Cl}_2} = \text{g Cl}_2$$

The **gravimetric factor** (GF) is the appropriate ratio of the formula weight of the substance *sought* to that of the substance *weighed*:

Remember to keep track of the units!

The gravimetric factor is the weight of analyte per unit weight of precipitate.

$$\boxed{\text{GF} = \text{gravimetric factor} = \frac{\text{fw of substance sought}}{\text{fw of substance weighed}} \times \frac{a}{b} \ (\text{mol sought/mol weighed})}$$

$$(4.28)$$

where a and b are integers that make the formula weights in the numerator and denominator chemically equivalent. In the above examples, the gravimetric factors were (fw Cl/fw AgCl) $\times$ 1/1, and (fw Cl$_2$/fw AgCl) $\times \frac{1}{2}$. Note that one or both of the formula weights may be multiplied by an integer in order to keep the same number of atoms of the key element in the numerator and denominator.

The weight of the substance sought is obtained by multiplying the weight of the precipitate by the gravimetric factor:

$$\boxed{\text{weight (g)} \times \frac{\text{fw of substance sought}}{\text{fw of substance weighed}} \times \frac{a}{b} = \text{sought (g)}} \qquad (4.29)$$

Note that the *species* and the *units* of the equation can be checked by dimensional analysis (canceling of like species and units).

Note also that we have calculated the amount of Cl$_2$ gas *derivable* from the sample instead of the amount of Cl$^-$ ion, the form in which it probably exists in the sample and the form in which it is weighed. If we precipitate the chloride as PbCl$_2$,

$$2\text{Cl}^- \xrightarrow{\text{precipitating agent}} \text{PbCl}_2$$

and

$$\text{Cl}_2 \rightarrow \text{PbCl}_2$$

then,

$$\text{g Cl}^- = \text{g PbCl}_2 \times \frac{2 \, (\text{fw Cl})}{\text{fw PbCl}_2} = \text{g PbCl}_2 \times \text{GF}$$

or

$$\text{g Cl}_2 = \text{g PbCl}_2 \times \frac{\text{fw Cl}_2}{\text{fw PbCl}_2} = \text{g PbCl}_2 \times \text{GF}$$

Conversion from weight of one substance to the derived there from weight of another is done using dimensional analysis of the units to arrive at the desired weight. The gravimetric factor is one step of that calculation and is useful for routine calculations. That is, if we know the gravimetric factor, we simply multiply the weight of the precipitate by the gravimetric factor to arrive at the weight of the analyte.

The grams of analyte = grams precipitate $\times$ GF.

Example 4.32

Calculate the weight of barium and the weight of Cl present in 25.0 g $BaCl_2$.

Solution

$$25.0 \text{ g BaCl}_2 \times \frac{\text{fw Ba}}{\text{fw BaCl}_2} = 25.0 \text{ g} \times \frac{137.3}{208.2} = 16.5 \text{ g Ba}$$

$$25.0 \text{ g BaCl}_2 \times \frac{2 \times \text{fw Cl}}{\text{fw BaCl}_2} = 25.0 \text{ g} \times \frac{2 \times 35.45}{208.2} = 8.51 \text{ g Cl}$$

Example 4.33

Aluminum in an ore sample is determined by dissolving it and then precipitating with base as $Al(OH)_3$ and igniting to Al_2O_3, which is weighed. What weight of aluminum was in the sample if the ignited precipitate weighed 0.2385 g?

Solution

$$\text{g Al} = \text{g Al}_2O_3 \times \frac{2 \times \text{fw Al}}{\text{fw Al}_2O_3}$$

$$= 0.2385 \text{ g} \times \frac{2 \times 26.982}{101.96} = 0.1262_3 \text{ g Al}$$

The gravimetric factor is

$$\frac{2 \times \text{fw Al}}{\text{fw Al}_2O_3} = \frac{2 \times 26.982}{101.96} = 0.52927 \text{ (g Al/g Al}_2O_3)$$

or $0.2385 \text{ g Al}_2O_3 \times 0.52927 \text{ (g Al/g Al}_2O_3) = 0.1262_3 \text{ g Al}$

Following are some other examples of gravimetric factors:

Sought	Weighed	Gravimetric Factor
SO_3	$BaSO_4$	$\dfrac{\text{fw SO}_3}{\text{fw BaSO}_4}$
Fe_3O_4	Fe_2O_3	$\dfrac{2 \times \text{fw Fe}_3O_4}{3 \times \text{fw Fe}_2O_3}$
Fe	Fe_2O_3	$\dfrac{2 \times \text{fw Fe}}{\text{fw Fe}_2O_3}$
MgO	$Mg_2P_2O_7$	$\dfrac{2 \times \text{fw MgO}}{\text{fw Mg}_2P_2O_7}$
P_2O_5	$Mg_2P_2O_7$	$\dfrac{\text{fw P}_2O_5}{\text{fw Mg}_2P_2O_7}$

The operations of gravimetric analyses are described in detail in Chapter 9.

More examples of gravimetric calculations are given in Chapter 9.

1. What is dimensional analysis? State its importance.

2. Equivalents are frequently defined in terms of the number of charges on an ion rather than on the number of reacting units in _____ chemistry.

   (a) polymer   (b) analytical

   (c) clinical   (d) pharmaceutical

3. Chemist sometimes use the term _____ for solutions of ionic salt that do not exist as molecules in the solid or in the solution.

   (a) normality   (b) formality

   (c) molality   (d) equivalent weight

4. Which of the following is a dimensionless quantity?

   (a) Density   (b) Molarity

   (c) Normality   (d) Specific gravity

5. When is the 'back titration employed? Explain with an example.

6. What is Gravimetric factor (GF)?

7. Distinguish between the expression of concentration on weight/weight, weight/volume, and volume/volume bases.

8. Express ppm and ppb on weight/weight, weight/volume, and volume/volume bases.

9. What is a standard solution? How is it prepared?

10. What are the requirements of a primary standard?

11. Why should a primary standard have a high formula weight?

Weight/Mole Calculations

12. Molecular weight is the ratio of _____ and _____.

    (a) mass, mole   (b) mole, mole

    (c) mole, mass   (d) mass, mass

13. A pseudo molecular weight obtained by dividing the mass in a mixture/solution by the number of moles in the mixture/solution is called _____.

    (a) molecular weight   (b) average molecular weight

    (c) atomic weight   (d) none of these

14. Calculate the grams of substance required to prepare the following solutions: (a) 250 mL of 5.00%(wt/vol) $NaNO_3$; (b) 500 mL of 1.00%(wt/vol) NH_4NO_3, (c) 1000 mL of 10.0%(wt/vol) $AgNO_3$.

15. What is the wt/vol% of the solute in each of the following solutions? (a) 52.3 g Na_2SO_4/L, (b) 275 g KBr in 500 mL, (c) 3.65 g SO_2 in 200 mL.

16. Calculate the formula weights of the following substances: (a) $BaCl_2 \cdot 2H_2O$, (b) $KHC_2O_4 \cdot H_2C_2O_4$, (c) $Ag_2Cr_2O_7$, (d) $Ca_3(PO_4)_2$.

17. Calculate the number of millimoles contained in 500 mg of each of the following substances: (a) $BaCrO_4$, (b) $CHCl_3$, (c) $KIO_3 \cdot HIO_3$, (d) $MgNH_4PO_4$, (e) $Mg_2P_2O_7$, (f) $FeSO_4 \cdot C_2H_4(NH_3)_2SO_4 \cdot 4H_2O$.

18. Calculate the number of grams of each of the substances in Problem 17 that would have to be dissolved and diluted to 100 mL to prepare a 0.200 M solution.

19. Calculate the number of milligrams of each of the following substances you would have to weigh out in order to prepare the listed solutions: (a) 1.00 L of 1.00 M NaCl, (b) 0.500 L of 0.200 M sucrose ($C_{12}H_{22}O_{11}$), (c) 10.0 mL of 0.500 M sucrose, (d) 0.0100 L of 0.200 M Na_2SO_4,(e) 250 mL of 0.500 M KOH, (f) 250 mL of 0.900% NaCl (g/100 mL solution).

Molarity Calculations

20. In 3.60 M sulphuric acid solution that is 29% H_2SO_4 by mass, what will be density (in g/mL) if the molar mass is 98 g/moL?

    (a) 1.45
    (b) 1.88
    (c) 1.64
    (d) 1.22

21. Aqueous urea solution is 20% by mass of solution. Calculate percent by mass of solvent.

22. Equal moles of water and urea are taken in a flask. What is the mass percent of urea in the solution?

23. Calculate the molar concentrations of all the cations and anions in a solution prepared by mixing 10.0 mL each of the following solutions: 0.100 M $Mn(NO_3)_2$, 0.100 M KNO_3, and 0.100 M K_2SO_4.

24. A solution containing 10.0 mmol $CaCl_2$ is diluted to 1 L. Calculate the number of grams of $CaCl_2 \cdot 2H_2O$ per milliliter of the final solution.

25. Calculate the molarity of each of the following solutions: (a) 10.0 g H_2SO_4 in 250 mL of solution, (b) 6.00 g NaOH in 500 mL of solution, (c) 25.0 g $AgNO_3$ in 1.00 L of solution.

26. Calculate the number of grams in 500 mL of each of the following solutions: (a) 0.100 M Na_2SO_4, (b) 0.250 M Fe $(NH_4)_2(SO_4)_2 \cdot 6H_2O$, (c) 0.667 M $Ca(C_9H_6ON)_2$.

27. Calculate the grams of each substance required to prepare the following solutions: (a) 250 mL of 0.100 M KOH, (b) 1.00 L of 0.0275 M $K_2Cr_2O_7$, (c) 500 mL of 0.0500 M $CuSO_4$.

28. How many milliliters of concentrated hydrochloric acid, 38.0% (wt/wt), specific gravity 1.19, are required to prepare 1 L of a 0.100 M solution? (Assume density and specific gravity are equal within three significant figures.)

29. Calculate the molarity of each of the following commercial acid or base solutions: (a) 70.0% $HClO_4$, specific gravity 1.668, (b) 69.0% HNO_3, specific gravity 1.409, (c) 85.0% H_3PO_4, specific gravity 1.689, (d) 99.5% CH_3COOH (acetic acid), specific gravity 1.051,(e) 28.0% NH_3, specific gravity 0.898. (Assume density and specific gravity are equal within three significant figures.)

30. 1.15 g/mL is the density of a solution which is prepared by dissolving 120 g of urea (molecular wt. = 60 u) in 1000 g of water. Molarity of solution is

    (a) 1.02 M
    (b) 2.05 M
    (c) 1.78 M
    (d) 0.50 M

31. If 23.64 mL of 0.1 M NaOH was required to neutralize 0.5632 g of acid, the molar mass of an unknown acid is

    (a) 220 g/mol
    (b) 240 g/mol
    (c) 238.24 g/mol
    (d) 236.24 g/mol

32. What will be the concentration of 0.02 M H_2SO_4, when ionization is 30%?

    (a) 0.014
    (b) 0.003
    (c) 0.002
    (d) 0.005

33. Which of the following is the most concentrated?

    (a) 20 g of NaOH in 100 mL of water
    (b) 150 g of NaOH in 200 mL of water
    (c) 180 g of NaOH in 300 mL of water
    (d) 250 g of NaOH in 600 mL of water

mg/L (PPM) Calculations

34. Which of the following best defines ppm?

    (a) For solids and liquids, it is the mole ratio.
    (b) For gases, it is the mole ratio.
    (c) It is the mole ratio for energy phase of matter.
    (d) It is the mass ratio for energy phase of matter.

35. Calculate the concentration of solution in ppm, if 0.025 g of $Pb(NO_3)_2$ is dissolved in 100 g of H_2O.

    (a) 2.5×10^{-4} ppm
    (b) 2.5 ppm
    (c) 250 ppm
    (d) 4.0×10^3 ppm

36. What is the ppm of calcium ions (mg of Ca^{2+}/L solution) in the solution prepared by dissolving 0.2500 g of $CaCO_3$ in HCl and enough water to give 500 mL solution?

    (a) 100.1 ppm
    (b) 200.2 ppm
    (c) 180.6 ppm
    (d) 210.4 ppm

37. A solution contains 6.0 μmol Na_2SO_4 in 250 mL. How many mg/L sodium does it contain? Of sulfate?

38. A solution (100 mL) containing 325 mg/L K^+ is analyzed by precipitating it as the tetraphenyl borate, $K(C_6H_5)_4B$, dissolving the precipitate in acetone solution, and measuring the concentration of tetraphenyl borate ion, $(C_6H_5)_4B^-$, in the solution. If the acetone solution volume is 250 mL, what is the concentration of the tetraphenyl borate in mg/L?

39. Calculate the molar concentrations of 1.00-mg/L solutions of each of the following. (a) $AgNO_3$, (b) $Al_2(SO_4)_3$, (c) CO_2, (d) $(NH_4)_4Ce(SO_4)_4 \cdot 2H_2O$, (e) HCl, (f) $HClO_4$.

40. Calculate the mg/L concentrations of 2.50×10^{-4} M solutions of each of the following. (a) Ca^{2+}, (b) $CaCl_2$, (c) HNO_3, (d) KCN, (e) Mn^{2+}, (f) MnO_4^-.

41. You want to prepare 1 L of a solution containing 1.00 mg/L Fe^{2+}. How many grams ferrous ammonium sulfate, $FeSO_4 \cdot (NH_4)_2SO_4 \cdot 6H_2O$, must be dissolved and diluted in 1 L? What would be the molarity of this solution?

42. A 0.456-g sample of an ore is analyzed for chromium and found to contain 0.560 mg Cr_2O_3. Express the concentration of Cr_2O_3 in the sample as (a) percent, (b) parts per thousand, and (c) parts per million.

43. How many grams NaCl should be weighed out to prepare 1 L of a 100-mg/L solution of (a) Na^+ and (b) Cl^-?

44. You have a 250-mg/L solution of K^+ as KCl. You wish to prepare from this a 0.00100 M solution of Cl^-. How many milliliters must be diluted to 1 L?

45. One liter of a 500-mg/L solution of $KClO_3$ contains how many grams K^+?

Dilution Calculations

46. What will be the concentration of the solution when solution A containing 25 M, 400 mL and solution B containing 30 M, 300 mL of same substance are mixed?

    (a) 27.14 M
    (b) 22.14 M
    (c) 14.22 M
    (d) 14.27 M

47. What will be the resulting molarity of a solution made by diluting 500 mL, 2.5 M to 800 mL?

    (a) 0.5265 M
    (b) 1.5625 M
    (c) 2.5689 M
    (d) 3.4586 M

48. 520 mL of 1.2 M solution is mixed with another 480 mL of 1.5 M solution of a substance. Calculate the molarity of final mixture?

    (a) 1.20 M
    (b) 1.344 M
    (c) 1.50 M
    (d) 2.70 M

49. A 12.5-mL portion of a solution is diluted to 500 mL, and its molarity is determined to be 0.125. What is the molarity of the original solution?

50. What volume of 0.50 M H_2SO_4 must be added to 65 mL of 0.20 M H_2SO_4 to give a final solution of 0.35 M? Assume volumes are additive.

51. How many milliliters of 0.10 M H_2SO_4 must be added to 50 mL of 0.10 M NaOH to give a solution that is 0.050 M in H_2SO_4? Assume volumes are additive.

52. You are required to prepare working standard solutions of 1.00×10^{-5}, 2.00×10^{-5}, 5.00×10^{-5}, and 1.00×10^{-4} M glucose from a 0.100 M stock solution. You have available 100-mL volumetric flasks and pipets of 1.00-, 2.00-, 5.00-, and 10.00-mL volume. Outline a procedure for preparing the working standards.

53. A 0.500-g sample is analyzed spectrophotometrically for manganese by dissolving it in acid and transferring to a 250-mL flask and diluting to volume. Three aliquots are analyzed by transferring 50-mL portions with a pipet to 500-mL Erlenmeyer flasks and reacting with an oxidizing agent, potassium peroxydisulfate, to convert the manganese to permanganate. After reaction, these are quantitatively transferred to 250-mL volumetric flasks, diluted to volume, and measured spectrophotometrically. By comparison with standards, the average concentration in the final solution is determined to be 1.25×10^{-5} *M*. What is the percent manganese in the sample?

54. A stock solution of analyte is made by dissolving 34.83 mg of copper (II) acetate hexahydrate (fw = 289.73 g/mol) in 25.00 mL of water. A second stock solution of internal standard is made by dissolving 28.43 mg of germanium (I) acetate (fw = 190.74 g/mol) into 25.00 mL of water. These solutions are used to make a series of standards for flame atomic absorption analysis calibration. The standard solutions (each 10.00 mL total volume) should have the following concentrations of copper: 10.00; 25.00; 50.00; 100.0; and 200.0 μM. Each calibration solution should also contain 50.00 μM of germanium.

    What is the analyte stock solution concentration? What is the internal standard stock solution concentration?

    Complete this table.

Concentration Cu (μM)	Volume of analyte stock (mL)	Volume of internal standard stock (mL)	Volume of diluent (mL)	Total Volume (mL)
10.00				10.00
25.00				10.00
50.00				10.00
100.0				10.00
200.0				10.00

PROFESSOR'S FAVORITE PROBLEMS

Contributed by Professor Wen-Yee Lee, The University of Texas at El Paso

55. Tom, an analytical chemist, bought a bag of decaffeinated coffee from a grocery store. However, Tom suspected that he might have received regular coffee and therefore decided to analyze his coffee for caffeine. In the lab, he took 0.5 mL of the brewed coffee and diluted it in water to make a 100.0 mL solution. He performed four analyses and found the concentrations to be 4.69, 3.99, 4.12, and 4.50 mg/L, respectively. (Assume the density of all solutions is 1.000 g/mL, 1 oz = 28.35 mL).

    (a) Report the concentration (in mg/L) of caffeine in the brewed coffee using the format as average ± standard deviation. (Note this is not the concentration in the diluted solution.)

    (b) Look up the caffeine content of regular vs. decaffeinated coffee. Do you think that Tom was given the wrong type of coffee?

    (c) Caffeine intake of 300 mg per day reportedly has no adverse effects in the vast majority of the adult population. If Tom drinks 3 cups (8 oz/cup) of this coffee daily, is his intake within this known safe zone?

Standardization Calculations

56. A 0.128 g sample of $KHC_8H_4O_4$ required 28.54 mL of NaOH to reach phenolphthalein end point. Calculate the molarity of NaOH.

57. A 0.220 *M* NaOH solution was titrated with 20 mL HCl solution. The end point required 23.72 mL of NaOH solution. What is the molarity of HCl?

58. A preparation of soda ash is known to contain 98.6% Na_2CO_3. If a 0.678-g sample requires 36.8 mL of a sulfuric acid solution for complete neutralization, what is the molarity of the sulfuric acid solution?

59. A 0.1 M sodium hydroxide solution is to be standardized by titrating primary standard sulfamic acid (NH_2SO_3H). What weight of sulfamic acid should be taken so that the volume of NaOH delivered from the buret is about 40 mL?

60. 35 mL of 0.1 M HCl solution was required to titrate given 25 mL of barium hydroxide. The molarity of barium hydroxide solution was

(a) 0.07   (b) 0.28

(c) 0.35   (d) 0.14

Analysis Calculations

61. A sample of USP-grade citric acid ($H_3C_6H_5O_7$, three titratable protons) is analyzed by titrating with 0.1087 M NaOH. If a 0.2678-g sample requires 38.31 mL for titration, what is the purity of the preparation? USP specification requires 99.5%.

62. Calcium in a 200-μL serum sample is titrated with 1.87×10^{-4} M EDTA solution, requiring 2.47 mL. What is the calcium concentration in the blood in mg/dL?

63. A 0.372-g sample of impure $BaCl_2 \cdot 2H_2O$ is titrated with 0.100 M $AgNO_3$, requiring 27.2 mL. Calculate (a) the percent Cl in the sample and (b) the percent purity of the compound.

64. An iron ore is analyzed for iron content by dissolving in acid, converting the iron to Fe^{2+}, and then titrating with standard 0.0150 M $K_2Cr_2O_7$ solution. If 35.6 mL is required to titrate the iron in a 1.68-g ore sample, how much iron is in the sample, expressed as percent Fe_2O_3? (See Example 4.31 for the titration reaction.)

65. Calcium in a 2.00-g sample is determined by precipitating CaC_2O_4, dissolving this in acid, and titrating the oxalate with 0.0200 M $KMnO_4$. What percent of CaO is in the sample if 35.6 mL $KMnO_4$ is required for titration? (The reaction is $5H_2C_2O_4 + 2MnO_4^- + 6H^+ \rightarrow 10CO_2 + 2Mn^{2+} + 8H_2O$.)

66. A potassium permanganate solution is prepared by dissolving 4.68 g $KMnO_4$ in water and diluting to 500 mL. How many milliliters of this will react with the iron in 0.500 g of an ore containing 35.6% Fe_2O_3? (See Example 4.30 for the titration reaction.)

67. A sample contains $BaCl_2$ plus inert matter. What weight must be taken so that when the solution is titrated with 0.100 $AgNO_3$, the milliliters of titrant will be equal to the percent $BaCl_2$ in the sample?

68. A 0.250-g sample of impure $AlCl_3$ is titrated with 0.100 M $AgNO_3$, requiring 48.6 mL. What volume of 0.100 M EDTA would react with a 0.350-g sample? (EDTA reacts with Al^{3+} in a 1:1 ratio.)

69. A 425.2-mg sample of a purified monoprotic organic acid is titrated with 0.1027 M NaOH, requiring 28.78 mL. What is the formula weight of the acid?

70. The purity of a 0.287-g sample of $Zn(OH)_2$ is determined by titrating with a standard HCl solution, requiring 37.8 mL. The HCl solution was standardized by precipitating AgCl in a 25.0-mL aliquot and weighing (0.462 g AgCl obtained). What is the purity of the $Zn(OH)_2$?

71. A sample of pure $KHC_2O_4 \cdot H_2C_2O_4 \cdot 2H_2O$ (three replaceable hydrogens) requires 46.2 mL of 0.100 M NaOH for titration. How many milliliters of 0.100 M $KMnO_4$ will the same-size sample react with? (See Problem 65 for reaction with $KMnO_4$.)

72. The organic matter in a 3.776 g sample of mercuric (Hg^{2+}) ointment is decomposed with HNO_3. After dilution the Hg^{2+} is titrated with 21.30 mL of 0.01144 M solution of NH_4SCN. Calculate the percent of Hg (200.6 g mol^{-1}) in the ointment.

Back-Titrations

73. A 0.500-g sample containing Na_2CO_3 plus inert matter is analyzed by adding 50.0 mL of 0.100 M HCl, a slight excess, boiling to remove CO_2, and then back-titrating the excess acid with 0.100 M NaOH. If 5.6 mL NaOH is required for the back-titration, what is the percent Na_2CO_3 in the sample?

74. A hydrogen peroxide solution is analyzed by adding a slight excess of standard $KMnO_4$ solution and back-titrating the unreacted $KMnO_4$ with standard Fe^{2+} solution. A 0.587-g sample of the H_2O_2 solution is taken, 25.0 mL of 0.0215 M $KMnO_4$ is added, and the back-titration requires 5.10 mL of 0.112 M Fe^{2+} solution. What is the percent H_2O_2 in the sample? (See Examples 4.26 and 4.30 for the reactions.)

75. The sulfur content of an iron pyrite ore sample is determined by converting it to H_2S gas, absorbing the H_2S in 10.0 mL of 0.00500 M I_2, and then back-titrating the excess I_2 with 0.00200 M $Na_2S_2O_3$. If 2.6 mL $Na_2S_2O_3$ is required for the titration, how many milligrams of sulfur are contained in the sample? Reactions:

$$H_2S + I_2 \rightarrow S + 2I^- + 2H^+$$
$$I_2 + 2S_2O_3^{2-} \rightarrow 2I^- + S_4O_6^{2-}$$

Titer

76. Express the titer of a 0.100M EDTA solution in mg BaO/mL.

77. Express the titer of a 0.0500 M $KMnO_4$ solution in mg Fe_2O_3/mL.

78. The titer of a silver nitrate solution is 22.7 mg Cl/mL. What is its titer in mg Br/mL?

Equivalent Weight Calculations

79. Calculate the equivalent weights of the following substances as acids or bases: (a) HCl, (b) $Ba(OH)_2$, (c) $KH(IO_3)_2$, (d) H_2SO_3, (e) CH_3COOH

80. Calculate the molarity of a 0.250 eq/L solution of each of the acids or bases in Problem 79.

Equivalent Weight

81. Determine the equivalent weight in the following.
    (a) The reduction half reaction for the decomposition of Br_2 in base.
    (b) The oxidation half reaction which yields bromate ion.

82. Calculate the equivalent weight of KHC_2O_4 (a) as an acid and (b) as a reducing agent in reaction with MnO_4^- ($5HC_2O_4^- + 2MnO_4^- + 11H^+ \rightarrow 10CO_2 + 2Mn^{2+} + 8H_2O$).

83. Mercuric oxide, HgO, can be analyzed by reaction with iodide and then titration with an acid: $HgO + 4I^- \rightarrow HgI_4^{2-} + 2OH^-$. What is its equivalent weight?

84. Calculate the grams of one equivalent each of the following for the indicated reaction: (a) $FeSO_4$ ($Fe^{2+} \rightarrow Fe^{3+}$), (b) H_2S ($\rightarrow S^0$), (c) H_2O_2 ($\rightarrow O_2$), (d) H_2O_2 ($\rightarrow H_2O$).

85. $BaCl_2 \cdot 2H_2O$ is to be used to titrate Ag^+ to yield AgCl. How many milliequivalents are contained in 0.5000 g $BaCl_2 \cdot 2H_2O$?

Equivalents/L (eq/L)

86. A solution is prepared by dissolving 7.82 g NaOH and 9.26 g $Ba(OH)_2$ in water and diluting to 500 mL. What is the concentration of the solution as a base in eq/L?

87. What weight of arsenic trioxide, As_2O_3, is required to prepare 1 L of 0.1000 eq/L arsenic(III) solution (arsenic As^{3+} is oxidized to As^{5+} in redox reactions)?

88. If 2.73 g $KHC_2O_4 \cdot H_2C_2O_4$ (three ionizable protons) having 2.0% inert impurities and 1.68 g $KHC_8H_4O_4$ (one ionizable proton) are dissolved in water and diluted to 250 mL, what is the concentration of the solution as an acid in eq/L, assuming complete ionization?

89. A solution of $KHC_2O_4 \cdot H_2C_2O_4 \cdot 2H_2O$ (three replaceable hydrogens) is 0.200 eq/L as an acid. What is its concentration in eq/L as reducing agent? (See Problem 65 for its reaction as a reducing agent.)

90. $Na_2C_2O_4$ and $KHC_2O_4 \cdot H_2C_2O_4$ are mixed in such a proportion by weight that the concentration of the resulting solution as a reducing agent in eq/L is 3.62 times the concentration as an acid in eq/L. What is the proportion? (See Problem 65 for its reaction as a reducing agent.)

91. What weight of $K_2Cr_2O_7$ is required to prepare 1.000 L of 0.1000 eq/L solution? (It reacts as: $Cr_2O_7^{2-} + 14H^+ + 6e^- \rightleftharpoons 2Cr^{3} + 7H_2O$.)

Charge Equivalent Calculations

92. A chloride concentration is reported as 300 mg/dL. What is the concentration in meq/L?

93. A calcium concentration is reported as 5.00 meq/L. What is the concentration in mg/dL?

94. A urine specimen has a chloride concentration of 150 meq/L. If we assume that the chloride is present in urine as sodium chloride, what is the concentration of NaCl in g/L?

Gravimetric Calculations

95. What weight of ferrosoferric oxide (Fe_3O_4) will be present in 0.5430 g of ferric oxide?

96. Zinc is determined by precipitating and weighing as $Zn_2Fe(CN)_6$.

    (a) What weight of zinc is contained in a sample that gives 0.348 g precipitate?

    (b) What weight of precipitate would be formed from 0.500 g of zinc?

97. Calculate the gravimetric factors for:

Substance Sought	Substance Weighed
Mn	Mn_3O_4
Mn_2O_3	Mn_3O_4
Ag_2S	$BaSO_4$
$CuCl_2$	$AgCl$
MgI_2	PbI_2

PROFESSOR'S FAVORITE PROBLEM

Contributed by Professor Thomas L. Isenhour, Old Dominion University

98. A 10.00 g sample contains only NaCl and KCl. The sample is dissolved and $AgNO_3$ is added to precipitate AgCl. After the precipitate is washed and dried, it weighs 21.62 g. What is the weight percent of NaCl in the original sample?

Multiple Choice Questions

1. 2 mole of Cl_2 and 3 mole of H_2 are placed in vessel and sealed. And H_2 and Cl_2 react according to the equation:

$$H_2(g) + Cl_2(g) \rightarrow 2HCl(g)$$

On reaction completion what is present in vessel?

    (a) 5 mole of HCl   **(b)** 6 mole of HCl and 1 mole of Cl_2

    (c) 4 mole of HCl and 1 mole of Cl_2 **(d)** 4 mole of HCl and 1 mole of H_2

2. For a given orthophosphoric acid solution, its molarity is 3 M then its normality is _____.

    (a) 1 N **(b)** 2 N

    (c) 0.3 N **(d)** 9 N

3. Choose the correct option.

    (a) Molality changes with temperature.

    (b) Molality does not change with temperature.

    (c) Molarity does not change with temperature.

    (d) Normality does not change with temperature.

4. 98% concentrated sulphuric acid by mass is having density of 1.80 g mL^{-1}. What volume of acid is needed to prepare 1 L of 0.1 M H_2SO_4?

    (a) 5.55 mL   **(b)** 11.10 mL

    (c) 16.65 mL   **(d)** 22.20 mL

Recommended References

1. T. P. Hadjiioannou, G. D. Christian, C. E. Efstathiou, and D. Nikolelis, *Problem Solving in Analytical Chemistry*. Oxford: Pergamon, 1988.
2. Q. Fernando and M. D. Ryan, *Calculations in Analytical Chemistry*. New York: Harcourt Brace Jovanovich, 1982.

Chapter 5

GENERAL CONCEPTS OF CHEMICAL EQUILIBRIUM

"The worst form of inequality is to try to make unequal things equal."
—Aristotle

KEY THINGS TO LEARN FROM THIS CHAPTER

- The equilibrium constant (key Equations: 5.12, 5.15)
- Calculation of equilibrium concentrations
- Using Excel Goal Seek to solve one-variable equations

- The systematic approach to equilibrium calculations: mass balance and charge balance equations
- Activity and activity coefficients (key Equation: 5.19)
- Thermodynamic equilibrium constants (key Equation: 5.23)

Even though in a chemical reaction the reactants may almost quantitatively react to form the products, reactions *never* go in only one direction. In fact, reactions reach an equilibrium in which the rates of reactions in both directions are equal. In this chapter we review the equilibrium concept and the equilibrium constant and describe general approaches for calculations using equilibrium constants. We discuss the activity of ionic species along with the calculation of activity coefficients. These values are required for calculations using thermodynamic equilibrium constants, that is, for the diverse ion effect, described at the end of the chapter. They are also used in potentiometric calculations (Chapter 20).

5.1 Chemical Reactions: The Rate Concept

In 1863 Guldberg and Waage described what we now call the law of mass action, which states that the rate of a chemical reaction is proportional to the "active masses" of the reacting substances present at any time. The active masses may be concentrations or pressures. Guldberg and Waage derived an equilibrium constant by defining equilibrium as the condition when the rates of the forward and reverse reactions are equal. Consider the chemical reaction

$$a\mathrm{A} + b\mathrm{B} \rightleftharpoons c\mathrm{C} + d\mathrm{D} \tag{5.1}$$

According to Guldberg and Waage, the rate of the forward reaction is equal to a constant times the concentration of each species raised to the power of the number of molecules participating in the reaction: that is,[1]

$$\mathrm{Rate_{fwd}} = k_{\mathrm{fwd}} [\mathrm{A}]^a [\mathrm{B}]^b \tag{5.2}$$

where rate$_{\mathrm{fwd}}$ is the rate of the forward reaction and k_{fwd} is the **rate constant**, which is dependent on such factors as the temperature and the presence of catalysts. [A] and

---

[1] represents moles/liter and here represents the effective concentration. The effective concentration will be discussed under the diverse ion effect, when we talk about activities.

[B] represent the molar concentrations of A and B. Similarly, for the reverse reaction, Guldberg and Waage wrote

$$\text{Rate}_{\text{rev}} = k_{\text{rev}} \, [\text{C}]^c \, [\text{D}]^d \qquad (5.3)$$

and for a system at equilibrium, the forward and reverse rates are equal:

$$k_{\text{fwd}} \, [\text{A}]^a \, [\text{B}]^b = k_{\text{rev}} \, [\text{C}]^c \, [\text{D}]^d \qquad (5.4)$$

> At equilibrium, the rate of the reverse reaction equals the rate of the forward reaction.

Rearranging these equations gives the **molar equilibrium constant** (which holds for dilute solutions) for the reaction, K:

$$\frac{[\text{C}]^c \, [\text{D}]^d}{[\text{A}]^a \, [\text{B}]^b} = \frac{k_{\text{fwd}}}{k_{\text{rev}}} = K \qquad (5.5)$$

The expression obtained here is the correct expression for the equilibrium constant, *but the method of derivation has no general validity*. This is because reaction rates actually depend on the *mechanism* of the reaction, determined by the number of colliding species, whereas the equilibrium constant expression depends only on the *stoichiometry* of the chemical reaction. The sum of the exponents in the rate constant gives the *order* of the reaction, and this may be entirely different from the stoichiometry of the reaction (see Chapter 17). An example is the rate of reduction of $S_2O_8^{2-}$ with I^-:

$$S_2O_8^{2-} + 3I^- \rightarrow 2SO_4^{2-} + I_3^-$$

The rate is actually given by $k_{\text{fwd}}[S_2O_8^{2-}][I^-]$ (a second-order reaction) and not $k_{\text{fwd}}[S_2O_8^{2-}][I^-]^3$, as might be expected from the balanced chemical reaction (a fourth-order reaction would be predicted). The only sound theoretical basis for the equilibrium constant comes from thermodynamic arguments. See Gibbs free energy in Section 5.3 for the thermodynamic computation of equilibrium constant values.

The value of K can be calculated empirically by measuring the concentrations of A, B, C, and D at equilibrium. Note that the more favorable the rate constant of the forward reaction relative to the backward reaction, the larger will be the equilibrium constant and the farther to the right the reaction will be at equilibrium.

> The larger the equilibrium constant, the farther to the right is the reaction at equilibrium.

When the reaction between A and B is initiated, the rate of the forward reaction is large because the concentrations of A and B are large, whereas the backward reaction is slow because the concentrations of C and D are small (that rate is initially zero). As the reaction progresses, concentrations of A and B decrease and concentrations of C and D increase, so that the rate of the forward reaction diminishes while that for the backward reaction increases (Figure 5.1). Eventually, the two rates become equal, and the system is in a state of equilibrium. At this point, the individual concentrations of A, B, C, and D remain constant (the relative values will depend on the reaction stoichiometry, the initial concentrations, and how far the equilibrium lies to the right). However, the system remains in dynamic equilibrium, with the forward and backward reactions continuing at equal rates.

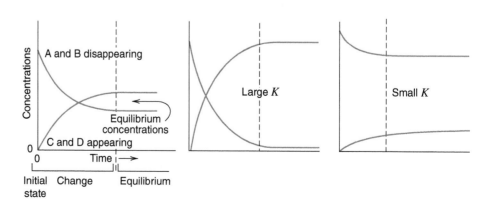

FIGURE 5.1 Progress of a chemical reaction.

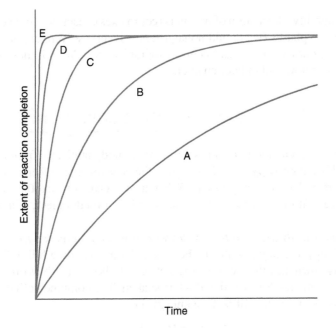

FIGURE 5.2 For the same equilibrium constant, the approach to equilibrium is controlled by the kinetics of the reaction, this cannot be predicted *a priori*. Ionic reactions are often instantaneous. Besides these and reactions involving combustion, most other reactions take measurable time. From A–E, at each step, the reaction rate increases by a factor of 3, but the position of equilibrium is not altered. An accelerated reaction rate can be brought about by an appropriate catalyst. Heating will also typically increase the reaction rate, but it may also affect the position of the equilibrium.

You will notice that the equilibrium constant expression is the ratio in which the concentrations of the products appear in the numerator and the concentrations of the reactants appear in the denominator. This is quite arbitrary, but it is the accepted convention. Hence, a large equilibrium constant indicates the equilibrium lies far to the right.

We should point out that although a particular reaction may have a rather large equilibrium constant, the reaction may proceed from *right* to *left* if sufficiently large concentrations of the *products* are initially present. Also, the equilibrium constant tells us nothing about how *fast* a reaction will proceed toward equilibrium. Some reactions, in fact, may be so slow as to be unmeasurable. The equilibrium constant merely tells us the tendency of a reaction to occur and in what direction, not whether it is fast enough to be feasible in practice. (See Chapter 17 on kinetic methods of analysis for the measurement of reaction rates and their application to analyses.)

For the reaction depicted in Equation 5.1, the rate at which equilibrium is approached will likely be different for either the forward or the reverse reaction. That is, if we start with a mixture of C and D, the rate at which equilibrium is approached may be much slower or faster than for the converse reaction.

Figure 5.2 illustrates the approach to equilibrium at different reaction rates.

A large equilibrium constant does not assure the reaction will proceed at an appreciable rate.

5.2 Types of Equilibria

We can write equilibrium constants for virtually any type of chemical process. Some common equilibria are listed in Table 5.1. The equilibria may represent dissociation

Equilibrium constants may be written for dissociations, associations, reactions, or distributions.

TABLE 5.1 Types of Equilibria

Equilibrium	Reaction	Equilibrium Constant
Acid–base dissociation	$HA + H_2O \rightleftharpoons H_3O^+ + A^-$	K_a, acid dissociation constant
Solubility	$MA \rightleftharpoons M^{n+} + A^{n-}$	K_{sp}, solubility product
Complex formation	$M^{n+} + aL^{b-} \rightleftharpoons ML_a^{(n-ab)+}$	K_f, formation constant
Reduction–oxidation	$A_{red} + B_{ox} \rightleftharpoons A_{ox} + B_{red}$	K_{eq}, reaction equilibrium constant
Phase distribution	$A_{H_2O} \rightleftharpoons A_{organic}$	K_D, distribution coefficient

(acid/base, solubility), formation of products (complexes), reactions (redox), a distribution between two phases (water and nonaqueous solvent—solvent extraction; adsorption from water onto a surface, as in chromatography, etc.). We will describe some of these equilibria below and in later chapters.

5.3 Gibbs Free Energy and the Equilibrium Constant

The tendency for a reaction to occur is defined thermodynamically from its change in **enthalpy** (ΔH) and **entropy** (ΔS). Enthalpy is the heat absorbed when an endothermic reaction occurs under constant pressure. When heat is given off (exothermic reaction), ΔH is negative. Entropy is a measure of the disorder, or randomness, of a substance or system.

A system will always tend toward lower energy and increased randomness, that is, lower enthalpy and higher entropy. For example, a stone on a hill will tend to roll spontaneously down the hill (lower energy state), and a box of marbles ordered by color will tend to become randomly ordered when shaken. The combined effect of enthalpy and entropy is given by the **Gibbs free energy**, G:

$$G = H - TS \tag{5.6}$$

where T is the absolute temperature in kelvin; G is a measure of the energy of the system, and a system spontaneously tends toward lower energy states. The change in energy of a system at a constant temperature is

$$\Delta G = \Delta H - T\Delta S \tag{5.7}$$

So a process will be *spontaneous when ΔG is negative*, will be spontaneous in the reverse direction when ΔG is positive, and will be at equilibrium when ΔG is zero. Hence, a reaction is favored by heat given off (negative ΔH), as in exothermic reactions, and by increased entropy (positive ΔS). Both ΔH and ΔS can be either positive or negative, and the relative magnitudes of each and the temperature will determine whether ΔG will be negative so that the reaction will be spontaneous.

Enthalpy or entropy change alone cannot decide if a process will be spontaneous. Many salts, NH_4Cl for example, spontaneously dissolve in water in an endothermic process (heat is absorbed, the solution gets cold). In such cases, the large positive entropy of dissolution exceeds the positive enthalpy change.

Standard enthalpy $H°$, standard entropy $S°$, and standard free energy $G°$ represent the thermodynamic quantities for one mole of a substance at standard state ($P = 1$ atm, $T = 298$ K, unit concentration). Then,

$$\Delta G° = \Delta H° - T\Delta S° \tag{5.8}$$

$\Delta G°$ is related to the equilibrium constant of a reaction by

$$\boxed{K = e^{-\Delta G°/RT}} \tag{5.9}$$

or

$$\boxed{\Delta G° = -RT \ln K = -2.303RT \log K} \tag{5.10}$$

where R is the gas constant (8.314 J K^{-1} mol^{-1}). Hence, if we know the standard free energy of a reaction, we can calculate the equilibrium constant. Obviously, the larger $\Delta G°$ (when negative), the larger will be K. Note that while $\Delta G°$ and ΔG give information about the spontaneity of a reaction, they say nothing about the *rate* at which it will

Everything in the universe tends toward increased disorder (increased entropy) and lower energy (lower enthalpy).

A spontaneous reaction results in energy given off and a lower free energy. At equilibrium, the free energy does not change.

A large equilibrium constant results from a large negative free energy change for the reaction in question.

occur. The reaction between hydrogen and oxygen to form water has a very large negative free energy change associated with it. But at room temperature, in the absence of a catalyst (or a spark!), these gases may coexist together for years without observable reaction.

5.4 Le Châtelier's Principle

The equilibrium concentrations of reactants and products can be altered by applying stress to the system, for example, by changing the temperature, the pressure, or the concentration of one of the reactants. The effects of such changes can be predicted from **Le Châtelier's principle**, which states that when stress is applied to a system at chemical equilibrium, the equilibrium will shift in a direction that tends to relieve or counteract that stress. The effects of temperature, pressure, and concentrations on chemical equilibria are considered below.

We can shift an unfavorable equilibrium by increasing the reactant concentration.

All equilibrium constants are temperature dependent, as are the rates of reactions.

5.5 Temperature Effects on Equilibrium Constants

As we have mentioned, temperature influences the individual rate constants for the forward and backward reactions and therefore the equilibrium constant (more correctly, temperature affects the free energy—see Equation 5.10). An increase in temperature will displace the equilibrium in the direction that results in absorbing heat, since this removes the source of the stress. So an endothermic forward reaction (which absorbs heat) will be displaced to the right, with an increase in the equilibrium constant. The reverse will be true for an exothermic forward reaction, which releases heat. An exothermic reaction needs to release heat to proceed, a process that will be hindered at higher temperature. The extent of the displacement will depend on the magnitude of the heat of reaction for the system.

Strictly speaking, enthalpy and entropy changes are not temperature independent. But in most cases it is a reasonable approximation that these are constant over a modest change in temperature, and the change in the equilibrium constant (K_1 at temperature T_1 and K_2 at temperature T_2) can be predicted by the Clausius-Clapeyron equation:

$$\ln \frac{K_1}{K_2} = \frac{\Delta H}{R} \left(\frac{1}{T_2} - \frac{1}{T_1} \right)$$

In addition to influencing the position of equilibrium, temperature has a pronounced effect on the rates of the forward and backward reactions involved in the equilibrium, and so it influences the *rate* at which equilibrium is approached. This is because the number and the energy of collisions among the reacting species increase with increasing temperature. The rates of many endothermic reactions increase about two- to threefold for every 10°C rise in temperature. See Figure 5.2.

5.6 Pressure Effects on Equilibria

Pressure can have a large influence on the position of chemical equilibrium for reactions occurring in the gaseous phase. An increase in pressure favors a shift in the direction that results in a reduction in the volume of the system, for example, when one volume of nitrogen reacts with three volumes of hydrogen to produce two volumes of ammonia. But for reactions occurring in solutions, normal pressure changes have a negligible effect on the equilibrium because liquids cannot be compressed the way gases can.

Note, however, that above a few atmospheres pressure, even "incompressible" fluids compress somewhat and one can learn about the electronic and mechanical changes to molecules by studying their spectroscopy under such conditions, for example, using high-pressure diamond cells.

For solutions, pressure effects are usually negligible.

5.7 Concentration Effects on Equilibria

Changes in concentration do not affect the equilibrium constant. They *do* affect the position of the equilibrium.

The value of an equilibrium constant is independent of the concentrations of the reactants and products. However, the *position* of equilibrium is very definitely influenced by the concentrations. The direction of change is readily predictable from Le Châtelier's principle. Consider the reaction of iron(III) with iodide:

$$3I^- + 2Fe^{3+} \rightleftharpoons I_3^- + 2Fe^{2+}$$

If the four components are in a state of equilibrium, as determined by the equilibrium constant, addition or removal of one of the components would cause the equilibrium to reestablish itself. For example, suppose we add more iron(II) to the solution. According to Le Châtelier's principle, the reaction will shift to the left to relieve the stress. Equilibrium will eventually be reestablished, and its position will still be defined by the same equilibrium constant.

5.8 Catalysts

Catalysts do not affect the equilibrium constant or the position at equilibrium.

See Chapter 17 for analytical uses of enzyme catalysts.

Catalysts either speed up or retard[2] the rate at which an equilibrium is attained by affecting the rates of both the forward and the backward reactions. But catalysts affect both rates to the same extent and thus have no effect on the value of an equilibrium constant. See Figure 5.2.

Catalysts are very important to the analytical chemist in a number of reactions that are normally too slow to be analytically useful. An example is the use of an osmium tetroxide catalyst to speed up the titration reaction between arsenic(III) and cerium(IV), whose equilibrium is very favorable but whose rate is normally too slow to be useful for titrations. The measurement of the change in the rate of a kinetically slow reaction in the presence of a catalyst can actually be used for determining the catalyst concentration. The same reaction between arsenic(III) and cerium(IV) is catalyzed also by iodide and the measurement of this reaction rate (through the disappearance of the yellow cerium(IV) color) constitutes the basis for what used to be the most widely used (and is still frequently used) method for measuring iodine, also called the Sandell-Kolthoff method. Modern methods used now for iodide include ion chromatography, ICP-MS, and ion-selective electrodes—see later chapters.

5.9 Completeness of Reactions

For quantitative analysis, equilibria should be at least 99.9% to the right for precise measurements. A reaction that is 75% to the right is still a "complete" reaction.

If the equilibrium of a reaction lies sufficiently to the right that the remaining amount of the substance being determined (reacted) is too small to be measured by the measurement technique, we say the reaction has gone to completion. If the equilibrium is not so favorable, then Le Châtelier's principle may be applied to make it so. We may either increase the concentration of a reactant or decrease the concentration of a product. Production of more product may be achieved by (1) allowing a gaseous product to escape, (2) precipitating the product, (3) forming a stable ionic complex of the product in solution, or (4) preferential extraction.

---

[2]Such "negative catalysts" are generally called inhibitors.

It is apparent from the above discussion that Le Châtelier's principle is the dominant concept behind most chemical reactions in the real world. It is particularly important in biochemical reactions, and external factors such as temperature can have a significant effect on biological equilibria. Catalysts (enzymes) are also key players in many biological and physiological reactions, as we shall see in Chapter 17.

5.10 Equilibrium Constants for Dissociating or Combining Species—Weak Electrolytes and Precipitates

When a substance dissolves in water, it will often partially or completely dissociate or ionize. Electrolytes that tend to dissociate only partially are called *weak electrolytes*, and those that tend to dissociate completely are *strong electrolytes*. For example, acetic acid only partially ionizes in water and is therefore a weak electrolyte. But hydrochloric acid is completely ionized and is therefore a strong electrolyte. (Acid dissociations in water are really proton transfer reactions: $HOAc + H_2O \rightleftharpoons H_3O^+ + OAc^-$.) Some substances completely ionize in water but have limited solubility; we call these *slightly soluble substances*. Substances may combine in solution to form a dissociable product, for example, a complex. An example is the reaction of copper(II) with ammonia to form the $Cu(NH_3)_4^{2+}$ species.

The dissociation of weak electrolytes or the solubility of slightly soluble substances can be quantitatively described by equilibrium constants. Equilibrium constants for completely dissolved and dissociated electrolytes are effectively infinite. Consider the dissociating species AB:

$$AB \rightleftharpoons A + B \tag{5.11}$$

The equilibrium constant for such a dissociation can be written generally as

$$\boxed{\frac{[A][B]}{[AB]} = K_{eq}} \tag{5.12}$$

The larger K_{eq}, the greater will be the dissociation. For example, the larger the equilibrium constant of an acid, the stronger will be the acid.

Some species dissociate stepwise, and an equilibrium constant can be written for each dissociation step. A compound A_2B, for example, may dissociate as follows:

$$\boxed{A_2B \rightleftharpoons A + AB \quad K_1 = \frac{[A][AB]}{[A_2B]}} \tag{5.13}$$

$$\boxed{AB \rightleftharpoons A + B \quad K_2 = \frac{[A][B]}{[AB]}} \tag{5.14}$$

The overall dissociation process of the compound is the sum of these two equilibria:

$$\boxed{A_2B \rightleftharpoons 2A + B \quad K_{eq} = \frac{[A]^2[B]}{[A_2B]}} \tag{5.15}$$

If we multiply Equations 5.13 and 5.14 together, we arrive at the overall equilibrium constant:

$$\boxed{\begin{aligned} K_{eq} = K_1 K_2 &= \frac{[A][AB]}{[A_2B]} \cdot \frac{[A][B]}{[AB]} \\ &= \frac{[A]^2[B]}{[A_2B]} \end{aligned}} \tag{5.16}$$

Equilibrium constants are finite when dissociations are less than 100%.

A weak electrolyte is only partially dissociated. Many slightly soluble substances are strong electrolytes because the portion that dissolves is totally ionized.

Successive stepwise dissociation constants become smaller.

When chemical species dissociate in a stepwise manner like this, the successive equilibrium constants generally become progressively smaller. For a diprotic acid (e.g., HOOCCOOH), the dissociation of the second proton is inhibited relative to the first ($K_2 < K_1$), because the negative charge on the monoanion makes it more difficult for the second proton to ionize. This effect is more pronounced the closer the proton sites are. Note that in equilibrium calculations we always use mol/L for solution concentrations.

If a reaction is written in the reverse, the same equilibria apply, but the equilibrium constant is inverted. Thus, in the above example, for $A + B \rightleftharpoons AB$, $K_{eq(reverse)} = [AB]/([A][B]) = 1/K_{eq(forward)}$. If K_{eq} for the forward reaction is 10^5, $K_{forward} = 1/K_{backward}$ then K_{eq} for the reverse reaction is 10^{-5}.

Similar concepts apply for combining species, except, generally, the equilibrium constant will be greater than unity rather than smaller, since the reaction is favorable for forming the product (e.g., complex). We will discuss equilibrium constants for acids, complexes, and precipitates in later chapters.

$K_{forward} = 1/K_{backward}$

5.11   Calculations Using Equilibrium Constants— Composition at Equilibrium?

Equilibrium constants are useful for calculating the concentrations of the various species in equilibrium, for example, the hydrogen ion concentration from the dissociation of a weak acid. In this section we present the general approach for calculations using equilibrium constants. The applications to specific equilibria are treated in later chapters dealing with these equilibria.

Chemical Reactions

It is sometimes useful to know the concentrations of reactants and products in equilibrium in a chemical reaction. For example, we may need to know the amount of reactant for the construction of a titration curve or for calculating the potential of an electrode in the solution. These are, in fact, applications we consider in later chapters. Some example calculations will illustrate the general approach to solving equilibrium problems.

Example 5.1

The chemicals A and B react as follows to produce C and D:

$$A + B \rightleftharpoons C + D \quad K = \frac{[C][D]}{[A][B]}$$

The equilibrium constant K has a value of 0.30. Assume 0.20 mol of A and 0.50 mol of B are dissolved in 1.00 L, and the reaction proceeds. Calculate the concentrations of reactants and products at equilibrium.

Solution

The initial concentration of A is 0.20 M and that of B is 0.50 M, while C and D are initially 0 M. After the reaction has reached equilibrium, the concentrations of A and B will be decreased and those of C and D will be increased. Let x represent the equilibrium concentration of C or the moles/liter of A and B reacting. Since we get one mole of D with each mole of C, the concentration of D will also be x. We may represent the *initial* concentration of A and B as the **analytical concentrations**, C_A and C_B. The **equilibrium concentrations** are [A] and [B]. The concentrations of A and B will each be diminished by x, that is, $[A] = C_A - x$ and $[B] = C_B - x$. So the equilibrium concentrations are

The equilibrium concentration is the initial (analytical) concentration minus the amount reacted.

	[A]	[B]	[C]	[D]
Initial	0.20	0.50	0	0
Change (x = mmol/mL reacting)	$-x$	$-x$	$+x$	$+x$
Equilibrium	0.20–x	0.50–x	x	x

We can substitute these values in the equilibrium constant expression and solve for x:

This approach to to solving such problems is often called creating an "ICE" diagram. Initial, Change, and Equilibrium conditions are charted to help construct the equilibrium expression to be solved. See for example http://www.youtube.com/user/genchemconcepts#p/a/u/5/LZtVQnILdrE.

$$\frac{(x)(x)}{(0.20 - x)(0.50 - x)} = 0.30$$

$$x^2 = (0.10 - 0.70x + x^2)0.30$$

$$0.70x^2 + 0.21x - 0.030 = 0$$

This is a quadratic equation and can be solved algebraically for x using the quadratic formula given in Appendix B (see also Example 5.1 quadratic equation solution .xlsx on the **website** supplement for a quadratic equation solution calculator,

$$x = \frac{-b \pm \sqrt{b^2 - 4ac}}{2a}$$

$$= \frac{-0.21 \pm \sqrt{(0.21)^2 - 4(0.70)(-0.030)}}{2(0.70)}$$

$$= \frac{-0.21 \pm \sqrt{0.044 + 0.084}}{1.40} = 0.11\ M$$

ICE diagram for Example 5.1

Concentration, M	[A]	[B]	[C]	[D]
Initial	0.20	0.50	0.00	0.00
Change	$-x$	$-x$	$+x$	$+x$
Equilibrium	0.20 $- x$	0.50 $- x$	x	x

$$[A] = 0.20 - x = 0.09\ M$$

$$[B] = 0.50 - x = 0.39\ M$$

$$[C] = [D] = x = 0.11\ M$$

Instead of using the quadratic equation, we could also use the **method of successive approximations**. In this procedure, we will first neglect x compared to the initial concentrations to simplify calculations, and calculate an initial value of x. Then we can use this first estimate of x to subtract from C_A and C_B to give an initial estimate of the equilibrium concentration of A and B, and calculate a new x. The process is repeated until x is essentially constant.

In successive approximations, we begin by taking the analytical concentration as the equilibrium concentration, to calculate the amount reacted. Then we repeat the calculation after subtracting the calculated reacted amount, until it is constant.

$$\text{First calculation}\ \frac{(x)(x)}{(0.20)(0.50)} = 0.30$$

$$x = 0.173$$

The calculations converge more quickly if we keep an extra digit throughout.

$$\text{Second calculation} \quad \frac{(x)(x)}{(0.20 - 0.173)(0.50 - 0.173)} = 0.30$$
$$x = 0.051$$

$$\text{Third calculation} \quad \frac{(x)(x)}{(0.20 - 0.051)(0.50 - 0.051)} = 0.30$$
$$x = 0.14_2$$

$$\text{Fourth calculation} \quad \frac{(x)(x)}{(0.20 - 0.142)(0.50 - 0.142)} = 0.30$$
$$x = 0.079$$

$$\text{Fifth calculation} \quad \frac{(x)(x)}{(0.20 - 0.79)(0.50 - 0.079)} = 0.30$$
$$x = 0.12_4$$

$$\text{Sixth calculation} \quad \frac{(x)(x)}{(0.20 - 0.124)(0.50 - 0.124)} = 0.30$$
$$x = 0.093$$

$$\text{Seventh calculation} \quad \frac{(x)(x)}{(0.20 - 0.093)(0.50 - 0.093)} = 0.30$$
$$x = 0.11_4$$

$$\text{Eighth calculation} \quad \frac{(x)(x)}{(0.20 - 0.114)(0.50 - 0.114)} = 0.30$$
$$x = 0.10_4$$

$$\text{Ninth calculation} \quad \frac{(x)(x)}{(0.20 - 0.104)(0.50 - 0.104)} = 0.30$$
$$x = 0.10_7$$

Shorten the number of iterations by taking the average of the first two for the next.

We will take 0.11 as the equilibrium value of x since it essentially repeated the value of the seventh calculation. Note that in these iterations, x oscillates above and below the equilibrium value. The larger x solved for in a particular problem is compared to C, the larger will be the oscillations and the more iterations that will be required to reach an equilibrium value (as in this example—not the best for this approach). There is a more efficient way of completing the iteration. Take the average of the first and second for the third iteration, which should be close to the final value (in this case, 0.11_2). One or two more iterations will tell us we have reached the equilibrium value. Try it!

In Example 5.1, appreciable amounts of A remained, even though B was in excess, because the equilibrium constant was not very large. In fact, the equilibrium was only about halfway to the right since C and D were about the same concentration as A. In most reactions of analytical interest, the equilibrium constants are large, and the equilibrium lies far to the right. In these cases, the equilibrium concentrations of reactants that are not in excess are generally very small compared to the other concentrations. This simplifies our calculations.

Example 5.2

Assume that in Example 5.1 the equilibrium constant was 2.0×10^{16} instead of 0.30. Calculate the equilibrium concentrations of A, B, C, and D (starting with 0.20 mol A and 0.50 mol B in 1.00 L).

If the equilibrium constant for a reaction is very large, x is very small compared to the analytical concentration, which simplifies calculations.

Solution

Since K is very large, the reaction of A with B will be virtually complete to the right, leaving only traces of A at equilibrium. Let x represent the equilibrium concentration of A. An amount of B equal to A will have reacted to form an equivalent amount of C and D (about 0.20 M for each). We can summarize the equilibrium concentrations as follows:

$$[A] = x$$

$$[B] = (0.50 - 0.20) + x = 0.30 + x$$

$$[C] = 0.20 - x$$

$$[D] = 0.20 - x$$

Or, looking at the equilibrium,

$$A + \quad B \quad \rightleftharpoons \quad C \quad + \quad D$$
$$x \quad 0.30 + x \quad 0.20 - x \quad 0.20 - x$$

Basically, we have said that all of A is converted to a like amount of C and D, except for a small amount x. Now x will be very small compared to 0.20 and 0.30 and can be neglected. So we can say

$$[A] = x$$

$$[B] \approx 0.30$$

$$[C] \approx 0.20$$

$$[D] \approx 0.20$$

The only unknown concentration is [A]. Substituting these numbers in the equilibrium constant expression, we have

$$\frac{(0.20)\,(0.20)}{(x)\,(0.30)} = 2.0 \times 10^{16}$$

$$x = [A] = 6.7 \times 10^{-18} M \text{ (usually analytically undetectable)}$$

In this case the calculation was considerably simplified by neglecting x in comparison to other concentrations. If x should turn out to be significant compared to these concentrations, then the solution should be reworked using the quadratic formula, or Goal Seek. **Generally, if the value of x is less than about 5% of the assumed concentration, it can be neglected.** In this case, the error in x itself is usually **5% or less. This simplification will generally hold if the product concentration is less than 1% at K_{eq}, that is $\leq 0.01\,K_{eq}$.** However, once you have mastered using Excel, especially its Goal Seek and Solver functions discussed later in the book, (Solver-based equilibrium calculations are also discussed in reference 8), you may find that it is just as simple or simpler to solve a problem without any approximation using Excel. This is because no judgments on what can or cannot be neglected is needed.

Neglect x compared to C (product) if $C \leq 0.01 K_{eq}$ in a reaction.

Example 5.3

A and B react as follows:

$$A + 2B \rightleftharpoons 2C \quad K = \frac{[C]^2}{[A]\,[B]^2}$$

Assume 0.10 mol of A is reacted with 0.20 mol of B in a volume of 1000 mL; $K = 1.0 \times 10^{10}$. What are the equilibrium concentrations of A, B, and C?

Solution

We have stoichiometrically equal amounts of A and B, so both are virtually all reacted, with trace amounts remaining. Let x represent the equilibrium concentration of A. At equilibrium, we have

$$A + 2B \rightleftharpoons 2C$$
$$x \quad 2x \quad 0.20 - 2x \approx 0.20$$

For each mole of A that either reacts (or is produced), we produce (or remove) two moles of C, and consume (or produce) two moles of B. Substituting into the equilibrium constant expression,

$$\frac{(0.20)^2}{(x)(2x)^2} = 1.0 \times 10^{10}$$

$$\frac{0.040}{4x^3} = 1.0 \times 10^{10}$$

$$x = [A] = \sqrt[3]{\frac{4.0 \times 10^{-2}}{4.0 \times 10^{10}}} = \sqrt[3]{1.0 \times 10^{-12}} = 1.0 \times 10^{-4} M$$

$$B = 2x = 2.0 \times 10^{-4} M$$

(analytically detectable, but not appreciable compared to the starting concentration)

Dissociation Equilibria

Calculations involving dissociating species are not much different from the example just given for chemical reactions.

Example 5.4

Calculate the equilibrium concentrations of A and B in a 0.10 M solution of a weak electrolyte AB with an equilibrium constant of 3.0×10^{-6}.

Solution

$$AB \rightleftharpoons A + B \quad K_{eq} = \frac{[A][B]}{[AB]}$$

Both [A] and [B] are unknown and equal. Let x represent their equilibrium concentrations. The concentration of AB at equilibrium is equal to its initial analytical concentration minus x.

$$AB \rightleftharpoons A + B$$
$$0.10 - x \quad x \quad x$$

In a dissociation, neglect x compared to the initial concentration C if $C \geq 100 K_{eq}$ in a dissociation.

The value of K_{eq} is quite small, so we are probably justified in neglecting x compared to 0.10. Otherwise, we will have to use a quadratic equation. Substituting into the K_{eq} expression,

$$\frac{(x)(x)}{0.10} = 3.0 \times 10^{-6}$$

$$x = [A] = [B] = \sqrt{3.0 \times 10^{-7}} = 5.5 \times 10^{-4} M$$

After the calculation is done, check if the approximation you made was valid. Here the calculated value of x can indeed be neglected compared to 0.10.

5.12 The Common Ion Effect—Shifting the Equilibrium

Equilibria can be markedly affected by adding one or more of the species present, as is predicted from Le Châtelier's principle. Example 5.5 illustrates this principle.

Example 5.5

Assume that A and B are an ion pair, which can dissociate into A (a cation) and B (an anion). Recalculate the concentration of A in Example 5.4, assuming that the solution also contains 0.20 M B.

Solution

We can represent the equilibrium concentration as follows:

	[AB]	[A]	[B]
Initial	0.10	0	0.20
Change $(x = \text{mmol/mL of AB dissociated})$	$-x$	$+x$	$+x$
Equilibrium	$0.10 - x$	x	$0.20 + x$
	≈ 0.10		≈ 0.20

The value of x will be smaller now than before because of the common ion effect of B, so we can certainly neglect it compared to the initial concentrations. Substituting in the equilibrium constant expression,

$$\frac{(x)(0.20)}{(0.10)} = 3.0 \times 10^{-6}$$
$$x = 1.5 \times 10^{-6} M$$

The concentration of A was decreased nearly 400-fold.

The common ion effect can be used to make analytical reactions more favorable or quantitative. The adjustment of acidity, for example, is frequently used to shift equilibria. Titrations with potassium dichromate, for example, are favored in acid solution, since protons are consumed in the reaction. Titrations with iodine, a weak oxidizing agent, are usually done in slightly alkaline solution to shift the equilibrium toward completion of the reaction, for example, in titrating arsenic(III):

$$H_3AsO_3 + I_2 + H_2O \rightleftharpoons H_3AsO_4 + 2I^- + 2H^+$$

Adjusting the pH is a common way of shifting the equilibrium.

5.13 Systematic Approach to Equilibrium Calculations—How to Solve Any Equilibrium Problem

Now that some familiarity has been gained with equilibrium problems, we will consider a systematic approach for calculating equilibrium concentrations that will work with all equilibria, no matter how complex. It consists of identifying the unknown concentrations involved and *writing a set of simultaneous equations equal to the number of unknowns*. Simplifying assumptions are made with respect to relative concentrations of species (not unlike the approach we have already taken) to shorten the solving of the equations. This approach involves writing expressions for **mass balance** of species and one for **charge balance** of species as part of our equations. We will first describe how to arrive at these expressions.

Mass Balance Equations

The principle of mass balance is based on the law of mass conservation, and it states that the number of atoms of an element remains constant in chemical reactions because no atoms are produced or destroyed. The principle is expressed mathematically by equating the concentrations, usually in molarities. The equations for all the pertinent chemical equilibria are written, from which appropriate relations between species concentrations are written.

Example 5.6

Write the equation of mass balance for a 0.100 M solution of acetic acid.

Solution

The equilibria are

$$HOAc \rightleftharpoons H^+ + OAc^-$$
$$H_2O \rightleftharpoons H^+ + OH^-$$

We know that the analytical concentration of acetic acid is equal to the sum of the equilibrium concentrations of all its species:

$$C_{HOAc} = [HOAc] + [OAc^-] = 0.100 \; M$$

A second mass balance expression may be written for the equilibrium concentration of H^+, which is derived from both HOAc and H_2O. We obtain one H^+ for each OAc^- and one for each OH^-:

$$[H^+] = [OAc^-] + [OH^-]$$

In a mass balance, the analytical concentration is equal to the sum of the concentrations of the equilibrium species derived from the parent compound (or an appropriate multiple).

Example 5.7

Write the equations of mass balance for a $1.00 \times 10^{-5} M$ [Ag(NH$_3$)$_2$] Cl solution.

Solution

The equilibria are

$$[Ag(NH_3)_2] Cl \rightarrow Ag(NH_3)_2^+ + Cl^-$$
$$Ag(NH_3)_2^+ \rightleftharpoons Ag(NH_3)^+ + NH_3$$
$$Ag(NH_3)^+ \rightleftharpoons Ag^+ + NH_3$$
$$NH_3 + H_2O \rightleftharpoons NH_4^+ + OH^-$$
$$H_2O \rightleftharpoons H^+ + OH^-$$

The Cl^- concentration is equal to the concentration of the salt that dissociated, that is, $1.00 \times 10^{-5} M$. Likewise, the sum of the concentrations of all silver species is equal to the concentration of Ag in the original salt that dissociated:

$$C_{Ag} = [Ag^+] + [Ag(NH_3)^+] + [Ag(NH_3)_2^+] = [Cl^-] = 1.00 \times 10^{-5} M$$

We have the following nitrogen-containing species:

$$NH_4^+ \quad NH_3 \quad Ag(NH_3)^+ \quad Ag(NH_3)_2^+$$

The concentration of N from the last species is twice the concentration of $Ag(NH_3)_2^+$. For this species, the concentration of the nitrogen is twice the concentration of the $Ag(NH_3)_2^+$, since there are two NH_3 per molecule. Hence, we can write

$$C_{NH_3} = [NH_4^+] + [NH_3] + [Ag(NH_3)^+] + 2[Ag(NH_3)_2^+] = 2.00 \times 10^{-5} M$$

Finally, we can write

$$[OH^-] = [NH_4^+] + [H^+]$$

Some of the equilibria and the concentrations derived from them may be insignificant compared to others and may not be needed in subsequent calculations.

We have seen that several mass balance expressions may be written. Some may not be needed for calculations (we may have more equations than unknowns), or some may be simplified or ignored due to the small concentrations involved compared to others. This will become apparent in the equilibrium calculations below.

Charge Balance Equations

According to the **principle of electroneutrality**, all solutions are electrically neutral; that is, there is no solution containing a detectable excess of positive or negative charge because the sum of the positive charges equals the sum of negative charges. We may write a *single* charge balance equation for a given set of equilibria.

Example 5.8

Write a charge balance equation for a solution of H_2CO_3.

Solution

The equilibria are

$$
\begin{aligned}
H_2CO_3 &\rightleftharpoons H^+ + HCO_3^- \\
HCO_3^- &\rightleftharpoons H^+ + CO_3^{2-} \\
H_2O &\rightleftharpoons H^+ + OH^-
\end{aligned}
$$

Dissociation of H_2CO_3 gives H^+ and two anionic species, HCO_3^- and CO_3^{2-}, and that of water gives H^+ and OH^-. The amount of H^+ from that portion of *completely* dissociated H_2CO_3 is equal to twice the amount of CO_3^{2-} formed, and from *partial* (first step) dissociation is equal to the amount of HCO_3^- formed. That is, for each CO_3^{2-} formed, there are 2 H^+; for each HCO_3^- formed, there is 1 H^+; and for each OH^- formed, there is 1 H^+. Now, for the singly charged species, the *charge* concentration is identical to the concentration of the *species*. But for CO_3^{2-}, the charge concentration is twice the concentration of the species, so we must multiply the CO_3^{2-} concentration by 2 to arrive at

In a charge balance, the sum of the charge concentration of cationic species equals the sum of charge concentration of the anionic species in equilibrium.

The charge concentration is equal to the molar concentration times the charge of a species.

the charge concentration from it. According to the principle of electroneutrality, positive charge concentration must equal the negative charge concentration. Hence,

$$[H^+] = 2[CO_3^{2-}] + [HCO_3^-] + [OH^-]$$

Note that while there may be more than one source for a given species (H^+ in this case), the total charge concentrations from all sources is always equal to the net equilibrium concentration of the species multiplied by its charge.

Example 5.9

Here, we neglect the dissociation of water. However, to be comprehensive, you can always include the formation of H^+ and OH^- from water in the charge balance expression for an aqueous system. Incorporating ions derived from the dissociation of water becomes more important as the concentration of other ions in the solutions decrease, i.e., it is more important in very dilute solutions. The pH of 10^{-3} M HCl can be correctly calculated, for example without taking into account the dissociation of water but the pH of 10^{-6} M HCl cannot.

Write a charge balance expression for a solution containing KCl, $Al_2(SO_4)_3$, and KNO_3. Neglect the dissociation of water.

Solution

$$[K^+] + 3[Al^{3+}] = [Cl^-] + 2[SO_4^{2-}] + [NO_3^-]$$

Example 5.10

Write a charge balance equation for a saturated solution of $CdCO_3$.

Solution

The equilibria are

$$CdCO_3 \rightleftharpoons Cd^{2+} + CO_3^{2-}$$
$$CO_3^{2-} + H_2O \rightleftharpoons HCO_3^- + OH^-$$
$$HCO_3^- + H_2O \rightleftharpoons H_2CO_3 + OH^-$$
$$H_2O \rightleftharpoons H^+ + OH^-$$

Again, the charge concentration for the singly charged species (H^+, OH^-, HCO_3^-) will be equal to the concentrations of the species. But for Cd^{2+} and CO_3^{2-} the charge concentration will be twice their concentrations. We must again equate the positive and negative charge concentrations.

$$2[Cd^{2+}] + [H^+] = 2[CO_3^{2-}] + [HCO_3^-] + [OH^-]$$

Example 5.11

Write a charge balance equation for Example 5.7.

Solution

$$[Ag^+] + [Ag(NH_3)^+] + [Ag(NH_3)_2^+] + [NH_4^+] + [H^+] = [Cl^-] + [OH^-]$$

Since all are singly charged species, the charge concentrations are equal to the molar concentrations.

Equilibrium Calculations Using the Systematic Approach—the Steps

We may now describe the systematic approach for calculating equilibrium concentrations in problems involving several equilibria. The basic steps can be summarized as follows:

1. Write the chemical reactions appropriate for the system.
2. Write the equilibrium constant expressions for these reactions.
3. Write all the mass balance expressions.
4. Write the charge balance expression.
5. Count the number of chemical species involved and the number of *independent* equations (from steps 2, 3, and 4). If the number of equations is greater than or equal to the number of chemical species, then a solution is possible. At this point, it is possible to proceed to an answer.
6. Make simplifying assumptions concerning the relative concentrations of chemical species. At this point you need to think like a chemist so that the *math* will be simplified.
7. Calculate the answer.
8. Check the validity of your assumptions!

In the systematic approach, a series of equations equal in number to the number of unknown species is written. These are simultaneously solved, using approximations to simplify.

Let us examine one of the examples worked before, but using this approach.

Example 5.12

Repeat the problem stated in Example 5.4 using the systematic approach outlined above.

Chemical reactions

$$AB = A + B$$

Equilibrium constant expressions

$$K_{eq} = \frac{[A][B]}{[AB]} = 3.0 \times 10^{-6} \qquad (1)$$

Use equilibrium constant expressions plus mass and charge balance expressions to write the equations.

Mass balance expressions

$$C_{AB} = [AB] + [A] = 0.10\,M \qquad (2)$$

$$[A] = [B] \qquad (3)$$

Remember that C represents the total analytical concentration of AB.

Charge balance expression

There is none because none of the species is charged.

Number of expressions versus number of unknowns

There are three unknowns ([AB], [A], and [B]) and three expressions (one equilibrium and two mass balance).

Simplifying assumptions: We want the equilibrium concentrations of A, B, and AB. Because K is small, very little AB will dissociate, so from (2):

$$[AB] = C_{AB} - [A] = 0.10 - [A] \approx 0.10\ M$$

Use the same rules as before for simplifying assumptions ($C_A \geq 100 K_{eq}$ for dissociations, $C \leq 0.01 K_{eq}$ for reactions).

Calculate

[AB] was found above.

[A] can be found from (1) and (3).

$$\frac{[A][B]}{0.10} = 3.0 \times 10^{-6}$$

$$[A] = \sqrt{3.0 \times 10^{-7}} = 5.5 \times 10^{-4}\ M$$

[B] can be found from (3):

$$[B] = [A] = 5.5 \times 10^{-4}\ M$$

Check

$$[AB] = 0.10 - 5.5 \times 10^{-4} = 0.10\ M \text{ (within significant figures)}$$

The systematic approach is applicable to multiple equilibria.

You see that the same answer was obtained as when the problem was worked intuitively as in Example 5.4. You may think that the systematic approach is excessively complicated and formal. For this extremely simple problem that may be a justified opinion. However, you should realize that the systematic approach will be applicable to *all* equilibrium calculations, regardless of the difficulty of the problem. You may find problems involving multiple equilibria and/or many species to be hopelessly complicated if you use only an intuitive approach. Nevertheless, you should also realize that a good intuitive "feel" for equilibrium problems is an extremely valuable asset. You should attempt to improve your intuition concerning equilibrium problems. Such intuition comes from experience gained by working a number of problems of different varieties. As you gain experience you will be able to shorten some of the formalism of the systematic approach, and you will find it easier to make appropriate simplifying assumptions. Although need for making approximations may no longer exist, the ability to set up the charge balance and mass balance equations is needed even if you use an Excel-based approach to solve the problem.

 Some Hints for Applying the Systematic Approach for Equilibrium Calculations

Mass balances:

1. One will be for the total analytical concentration of the main parent species.
2. Others will be for species of interest, e.g., H^+ and other (dissociated) species in equilibrium.

Charge balances:

1. Charge balance equations are simply adding all cationic species on one side and all anionic species on the other, each multiplied by the respective charges.

2. Both mass and charge balance equations are rarely needed for solving the equilibrium calculations; in the case of ionic equilibria, charge balance equations are often easier to write.

Solving the equilibrium concentrations:

1. Using simplifying assumptions, at least one of the equilibrium species concentrations can be estimated.
2. From (substituting) this, the other species can be calculated.

Follow the rules given after Example 5.11.

Example 5.13

Repeat the problem outlined in Example 5.5 using the systematic approach. Assume the charge on A is +1, the charge on B is −1, and that the extra B (0.20 M) comes from MB; MB is completely dissociated.

Solution

Chemical reactions

$$AB = A^+ + B^-$$

$$MB \rightarrow M^+ + B^-$$

Equilibrium expressions

$$K_{eq} = \frac{[A^+][B^-]}{[AB]} = 3.0 \times 10^{-6} \tag{1}$$

Mass balance expressions

$$C_{AB} = [AB] + [A^+] = 0.10 \, M \tag{2}$$

$$[B^-] = [A^+] + [M^+] = [A^+] + 0.20 \, M \tag{3}$$

Charge balance expression

$$[A^+] + [M^+] = [B^-] \tag{4}$$

Number of expressions versus number of unknowns

There are three unknowns ([AB], [A$^+$], and [B$^-$]; the concentration of M$^+$ is known to be 0.20 M) and three independent expressions (one equilibrium and two mass balance; the charge balance is the same as the second mass balance).

Both mass and charge balance equations are often not needed. Both are needed, however, to derive the shape of titration curve.

Simplifying assumptions

i. Because K_{eq} is small, very little AB will dissociate, so from (2).

$$[AB] = 0.10 - [A^+] \approx 0.10 \, M$$

ii. [A] ≪ [M] so from (3) or (4):

$$[B^-] = 0.20 + [A^+] \approx 0.20 \, M$$

Calculate

[A] is now found from (1):

$$\frac{[A^+](0.20)}{0.10} = 3.0 \times 10^{-6}$$

$$[A^+] = 1.5 \times 10^{-6} M$$

Check

i. $[AB] = 0.10 - 1.5 \times 10^{-6} = 0.10\ M$

ii. $[B] = 0.20 + 1.5 \times 10^{-6} = 0.20\ M$

Solving Example 5.13 Using Goal Seek

Set up the spreadsheet using individual columns for $[A^+]$, $[B^-]$, $[M^+]$, $[AB]$, and K_{eq}. The "formula view" of the spreadsheet then would look like:

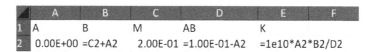

	A	B	C	D	E	F
1	A	B	M	AB	K	
2	0.00E+00	=C2+A2	2.00E-01	=1.00E-01-A2	=1e10*A2*B2/D2	

Note the 1E10 multiplier so as to avoid the aforementioned pitfall in Excel. Now Goal Seek, setting E2 to 3E4 by varying A2 (which is equal to x), you will find that A2 will readily converge to 1.50E-6, see the text website.

We will in general use the approximation approaches given in Sections 5.10 and 5.11, which actually incorporate many of the equilibria and assumptions used in the systematic approach. The use of the systematic approach for problems involving multiple equilibria is discussed in Chapter 7.

We can now write some general rules for solving chemical equilibrium problems, using the approximation approach. These rules should be applicable to acid–base dissociation, complex formation, oxidation–reduction reactions, and others. That is, all equilibria can be treated similarly.

---

1. Write down the equilibria involved.

2. Write the equilibrium constant expressions and the numerical values.

3. From a knowledge of the chemistry involved, let x represent the equilibrium concentration of the species that will be unknown and small compared to other equilibrium concentrations; other species of unknown and small concentrations will be multiples of this.

4. List the equilibrium concentrations of all species, adding or subtracting the appropriate multiple of x from the analytical concentration where needed.

5. Make suitable approximations by neglecting x compared to finite equilibrium concentrations. This is generally valid if the finite concentration is about $100 \times K_{eq}$ or more. Also, if the calculated x is less than approximately 5% of the finite concentration, the assumption is valid.

6. Substitute the approximate representation of individual concentration into the equilibrium constant expression and solve for x.

7. If the approximations in step 5 are invalid, use the quadratic formula to solve for x or use an Excel-based approach.

The application of these rules will become more apparent in subsequent chapters when we deal with specific equilibria in detail.

5.15 Heterogeneous Equilibria—Solids Don't Count

Equilibria in which all the components are in solution (a "homogeneous" medium) generally occur quite rapidly. If an equilibrium involves two phases ("heterogeneous"), the rate of approach to equilibrium will generally be substantially slower than in the case of solutions. An example is the dissolution of a poorly soluble solid or the formation of a precipitate, neither of these processes will be instantaneous.

Another way in which heterogeneous equilibria differ from homogeneous equilibria is the manner in which the different constituents offset the equilibrium. Guldberg and Waage showed that when a solid is a component of a reversible chemical process, its active mass can be considered constant, regardless of how much of the solid is present. That is, when any amount of the solid is already present, adding more solid does not bring about a shift in the equilibrium. So the expression for the equilibrium constant need not contain any concentration terms for substances present as solids. That is, the standard state of a solid is taken as that of the solid itself, or unity. Thus, for the equilibrium

$$CaF_2 \rightleftharpoons Ca^{2+} + 2F^-$$

we write that

$$K_{eq} = [Ca^{2+}][F^-]^2$$

The same is true for pure liquids (undissolved) in equilibrium, such as mercury. The standard state of water is taken as unity in dilute *aqueous solutions*, and water does not appear in equilibrium constant expressions.

5.16 Activity and Activity Coefficients—Concentration Is Not the Whole Story

Generally, the presence of diverse salts (not containing ions common to the equilibrium involved) will cause an increase in dissociation of a weak electrolyte or in the solubility of a precipitate. Cations attract anions, and vice versa, and so the cations of the analyte attract anions of the diverse electrolyte and the anions of the analyte are

> Heterogeneous equilibria are slower than solution equilibria.

> The "concentration" of a pure solid or liquid is unity.

> Consider a saturated solution of sugar with undissolved sugar remaining at the bottom. The relevant equilibrium constant is $Sugar_{solution}/Sugar_{solid}$. We know well that adding more solid sugar to a saturated solution will not increase the solution concentration further. If the equilibrium constant is indeed a constant, obviously adding more solid sugar to the system does not change the "concentration" of the solid sugar. Any amount of undissolved solid in the saturated solution system represents the same "concentration" of the solid.

> The "effective concentration" of an ion is decreased by shielding it with other "inert" ions, and it represents the activity of the ion.

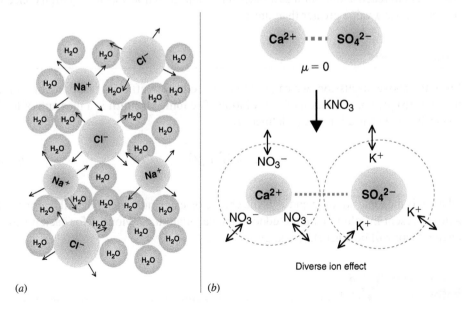

$\mu = 0$

KNO$_3$

Diverse ion effect

(a)   *(b)*

FIGURE 5.3 (a) In solution, Na$^+$ and Cl$^-$ from salt form an ion atmosphere where each ion type has more of the oppositely charged ions as nearest neighbors. The structure is very dynamic in nature (all species are rapidly shifting). (b) Addition of an inert salt, such as KNO$_3$, decreases the attraction between ion pairs by shielding and reducing the effective charge and thus increases the solubility of sparingly soluble salts such as CaSO$_4$ (see Section 5.16).

surrounded by the cations of the diverse electrolyte (Figure 5.3a). The attraction of the ions participating in the equilibrium of interest by the dissolved electrolyte effectively shields them, *decreasing their effective concentration* and shifting the equilibrium. We say that an "ion atmosphere" is formed about the cation and anion of interest. As the charge on either the diverse salt or the ions of the equilibrium reaction is increased, the diverse salt effect generally increases. This effect on the equilibrium is not predicted by Le Châtelier's principle; but it is readily understood if you think in terms of the effective concentrations being changed.

This "effective concentration" of an ion in the presence of an electrolyte is called the **activity** of the ion. To quantitatively describe the effects of salts on equilibrium constants, one must use activities, not concentrations (see the diverse salt effect below). In potentiometric measurements, it is activity that is measured, not concentration (see Chapter 20). In this section we describe how to estimate activity.

Activities are important in potentiometric measurements. See Chapter 20.

The Activity Coefficient

G. N. Lewis introduced the thermodynamic concept of activity in 1908 in a paper entitled "The Osmotic Pressure of Concentrated Solutions and the Laws of Perfect Solutions."

The **activity** of an ion a_i is defined by

$$\boxed{a_i = C_i f_i} \tag{5.17}$$

where C_i is the concentration of the ion i and f_i is its **activity coefficient**. The concentration is usually expressed as molarity, and the activity has the same units as the concentration. The activity coefficient is dimensionless, but numerical values for activity coefficients do depend on the choice of standard state. The activity coefficient varies with the total number of ions in the solution and with their charge, and it is a correction for interionic attraction. *In dilute solution, less than 10^{-4} M, the activity coefficient of a simple electrolyte is near unity*, and activity is approximately equal to the concentration. As the concentration of an electrolyte increases, or as an extraneous salt is added, the activity coefficients of ions decrease, and the activity becomes less than the concentration. Note, however, that at much higher concentrations a different effect comes into play. Ions, especially cations, are hydrated in aqueous solution and the associated water of solvation becomes unavailable to function as solvent. This causes the activity coefficient to reach a minimum as a function of concentration and at very high concentrations; it has a value greater than unity.

Ionic Strength

For ionic strengths less than 10^{-4}, activity coefficients are near unity. In 1921 Lewis and Randall first introduced the empirical concept of ionic strength and showed that in dilute solution, the logarithm of the activity coefficient is proportional to the square root of the ionic strength.

From the above discussion, we can see that the activity coefficient is a function of the total electrolyte concentration of the solution. The **ionic strength** is a measure of total electrolyte concentration and is defined by

$$\boxed{\mu = \frac{1}{2} \sum C_i Z_i^2} \tag{5.18}$$

where μ is the ionic strength and Z_i is the charge on each individual ion. All cations and anions present in solution are included in the calculation. Obviously, for each positive charge there will be a negative charge.

Example 5.14

Calculate the ionic strength of a 0.2 *M* solution of KNO_3 and a 0.2 *M* solution of K_2SO_4.

Solution

For KNO_3,

$$\mu = \frac{C_{K^+} Z_{K^+}^2 + C_{NO_3^-} Z_{NO_3^-}^2}{2}$$

$$[K^+] = 0.2\,M \qquad [NO_3^-] = 0.2\,M$$

$$\mu = \frac{0.2 \times (1)^2 + 0.2 \times (1)^2}{2} = 0.2$$

For K_2SO_4,

$$\mu = \frac{C_{K^+} Z_{K^+}^2 + C_{SO_4^{2-}} Z_{SO_4^{2-}}^2}{2}$$

$$[K^+] = 0.4\,M \qquad [SO_4^{2-}] = 0.2\,M$$

So,

$$\mu = \frac{0.4 \times (1)^2 + 0.2 \times (2)^2}{2} = 0.6$$

Note that due to the doubly charged SO_4^{2-}, the ionic strength of the same molar concentration of K_2SO_4 is three times that of the KNO_3.

> Higher charged ions contribute more to the ionic strength.

> In 1923, Dutch physicist Petrus (Peter) Debye (1884–1966), together with his assistant Erich Hückel (1896–1980), developed the Debye–Hückel theory of electrolyte solutions, an improvement of Svante Arrhenius's theory of electrical conductivity in electrolytic solutions.

If more than one salt is present, then the ionic strength is calculated from the total concentration and charges of all the different ions. For any given electrolyte, the ionic strength will be proportional to the concentration. Strong acids that are completely ionized are treated in the same manner as salts. If the acids are partially ionized, then the concentration of the ionized species must be estimated from the ionization constant before the ionic strength is computed. Very weak acids can usually be considered to be nonionized and do not contribute to the ionic strength.

Example 5.15

Calculate the ionic strength of a solution consisting of 0.30 M NaCl and 0.20 M Na_2SO_4.

Solution

$$\mu = \frac{C_{Na^+} Z_{Na^+}^2 + C_{Cl^-} Z_{Cl^-}^2 + C_{SO_4^{2-}} Z_{SO_4^{2-}}^2}{2}$$

$$= \frac{0.70 \times (1)^2 + 0.30 \times (1)^2 + 0.20 \times (2)^2}{2}$$

$$= 0.90$$

Calculation of Activity Coefficient

In 1923, Debye and Hückel derived a theoretical expression for calculating activity coefficients. The original **Debye–Hückel** equation is given as Equation 5.19a below but it is of limited use as it can be used only in extremely dilute solutions:

$$-\log f_i = AZ_i^2 \sqrt{\mu} \tag{5.19a}$$

This equation applies for ionic strengths up to 0.2.

They later provided a more useful equation, known as the **extended Debye–Hückel equation**:

$$-\log f_i = \frac{AZ_i^2 \sqrt{\mu}}{1 + Ba_i \sqrt{\mu}}$$ (5.19b)

A and B are constants; the values are, respectively, 0.51 and 0.33 for water at 25°C. At other temperatures, the values can be computed from $A = 1.82 \times 10^6 (DT)^{-3/2}$ and $B = 50.3 (DT)^{-1/2}$ where D and T are the dielectric constant and the absolute temperature, respectively; a_i is the **ion size parameter**, which is the effective diameter of the hydrated ion in angstrom units, Å. An angstrom is 100 picometers (pm, 10^{-10} meter). A limitation of the Debye–Hückel equation is the accuracy to which a_i can be evaluated. For many singly charged ions, a_i is generally about 3 Å, and for practical purposes Equation 5.19b simplifies to

The estimation of the ion size parameter places a limit on the accuracy of the calculated activity coefficient.

This equation may be used for ionic strengths less than 0.01.

$$-\log f_i = \frac{0.51 Z_i^2 \sqrt{\mu}}{1 + \sqrt{\mu}}$$ (5.20)

See Reference 10 for a tabulation of a_i values.

For common multiply charged ions, a_i may become as large as 11 Å. But at ionic strengths less than 0.01, the second term of the denominator becomes small with respect to 1, so uncertainties in a_i become relatively unimportant, and Equation 5.20 can be applied at ionic strengths of 0.01 or less. Equation 5.19b can be applied up to ionic strengths of about 0.2. Reference 10 at the end of the chapter lists values for a_i for different ions and also includes a table of calculated activity coefficients, using Equation 5.19b, at ionic strengths ranging from 0.0005 to 0.1. This paper, with the complete list of ion size parameters, is available on the text website. Excel answers using Equations 5.19b and 5.20 for the following two problems are on the website (Spreadsheets). Table 5.2 above contains a list of ion size parameters for some common ions taken from this reference.

TABLE 5.2   Ion Size Parameters for Common Ions

Ion	Ion Size Parameter Å (Angstroms)
H^+	9
$(C_3H_7)_4N^+$	8
$(C_3H_7)_3NH^+$, $\{OC_6H_2(NO_3)_3\}^-$	7
Li^+, $C_6H_5COO^-$, $(C_2H_5)_4N^+$	6
$CHCl_2COO^-$, $(C_2H_5)_3NH^+$	5
Na^+, IO_3^-, HSO_3^-, $(CH_3)_3NH^+$, $C_2H_5NH_3^+$	4–4.5
K^+, Cl^-, Br^-, I^-, CN^-, NO_2^-, NO_3^-	3
Rb^+, Cs^+, NH_4^+, Tl^+, Ag^+	2.5
Mg^{2+}, Be^{2+}	8
Ca^{2+}, Cu^{2+}, Zn^{2+}, Mn^{2+}, Ni^{2+}, Co^{2+}	6
Sr^{2+}, Ba^{2+}, Cd^{2+}, $H_2C(COO)_2^{2-}$	5
Hg_2^{2+}, SO_4^{2-}, CrO_4^{2-}	4
Al^{3+}, Fe^{3+}, Cr^{3+}, La^{3+}	9
$Citrate^{3-}$	5
PO_4^{3-}, $Fe(CN)_6^{3-}$, $\{CO(NH_3)_6\}^{3+}$	4
Th^{4+}, Zr^{4+}, Ce^{4+}	11
$Fe(CN)_6^{4-}$	5

Taken from Kielland, Reference 10 (See the text website for an arrangement of inorganic and organic ions.)

Example 5.16

Calculate the activity coefficients for K^+ and SO_4^{2-} in a 0.0020 M solution of potassium sulfate.

Solution

The ionic strength is 0.0060, so we can apply Equation 5.20:

$$-\log f_{K^+} = \frac{0.51\,(1)^2\,\sqrt{0.0060}}{1 + \sqrt{0.0060}} = 0.037$$

$$f_{K^+} = 10^{-0.037} = 10^{-1} \times 10^{0.963} = 0.918$$

$$-\log f_{SO_4^{2-}} = \frac{0.51\,(2)^2\,\sqrt{0.0060}}{1 + \sqrt{0.0060}} = 0.14_7$$

$$f_{SO_4^{2-}} = 10^{-0.14_7} = 10^{-1} \times 10^{0.85_3} = 0.71_3$$

Example 5.17

Calculate the activity coefficients for K^+ and SO_4^{2-} in a 0.020 M solution of potassium sulfate.

Solution

The ionic strength is 0.060, so we would use Equation 5.19b. From Table 5.2, we find that $a_{K^+} = 3$ Å and $a_{SO_4^{2-}} = 4.0$ Å. For K^+, we can use Equation 5.20:

$$-\log f_{K^+} = \frac{0.51\,(1)^2\,\sqrt{0.060}}{1 + \sqrt{0.060}} = 0.10_1$$

$$f_{K^+} = 10^{-0.101} = 10^{-1} \times 10^{0.899} = 0.79_4$$

For SO_4^{2-}, use Equation 5.19b:

$$-\log f_{SO_4^{2-}} = \frac{0.51\,(2)^2\,\sqrt{0.060}}{1 + 0.33 \times 4.0\sqrt{0.060}} = 0.37_8$$

$$f_{SO_4^{2-}} = 10^{-1} \times 10^{0.62_2} = 0.41_9$$

This latter compares with a calculated value of 0.39_6 using Equation 5.20. Note the decrease in the activity coefficients compared to 0.002 M K_2SO_4, especially for the SO_4^{2-} ion.

Spreadsheets for calculating activity coefficients using Equations 5.19b and 5.20 are given in the textbook's **website** for Chapter 5.

For higher ionic strengths, a number of empirical equations have been developed. One of the more useful is the **Davies modification** (see Reference 9):

$$\boxed{-\log f_i = 0.51 Z_i^2 \left(\frac{\sqrt{\mu}}{1 + \sqrt{\mu}} - 0.3\,\mu \right)} \qquad (5.21)$$

Use this equation for ionic strengths of 0.2–0.5. It gives higher activity coefficients compared to the Extended Debye–Hückel equation.

It is valid up to ionic strengths of about 0.5.

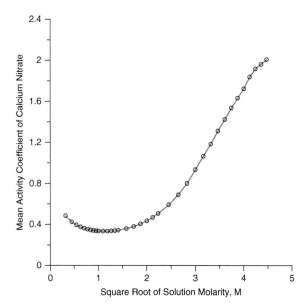

FIGURE 5.4 Mean ionic activity coefficient of calcium nitrate as a function of concentration. Data taken from R. H. Stokes and R. A. Robinson, *J. Am. Chem. Soc.* **70** (1948) 1870.

A 0.01 *M* solution of HCl prepared in 8 *M* NaCl has an activity about 100 times that in water! Its pH is actually 0.0. See F. E. Critchfield and J. B. Johnson, *Anal. Chem.*, **30**, (1958) 1247 and G. D. Christian, *CRC Crit. Rev. in Anal. Chem.*, **5**(2) (1975) 119–153.

At very high electrolyte concentrations, activity coefficients increase and become greater than unity. Note that Equation 5.21 does predict this; the last term causes an increase in the activity coefficient as μ increases. This is because the activity of the solvent, water, decreases and the water that is tied in the primary solvation shell of the ions (cations are especially hydrated) are not available to function as solvent. Consider that if the solvation number of Na^+ is 4 (meaning that 4 molecules of water are attached to each sodium ion), an 8 molal NaCl solution has 32 of the 55.5 moles of the water already tied up with the sodium, the effective amount of water remaining to function as solvent is only 43% of that in a very dilute solution. The effective concentration is thus 2.4 × higher. This change in concentration is ultimately reflected in an increased activity coefficient. Unfortunately, the real situation is more complex than this because water becomes so scarce at very high concentrations, solvation numbers are not constant and they also begin to decrease from their values in more dilute solutions. For more detailed discussion of activity coefficients in concentrated solutions, see the paper by Stokes and Robinson cited in Figure 5.4.

The Stokes–Robinson equation, which works for AB and AB_2 type electrolytes up to several molar in concentration, is given by:

$$-\log f_\pm = \frac{0.51 Z_A Z_B \sqrt{\mu}}{1 + 0.33 a_t \sqrt{\mu}} + \frac{n}{v}\log a_w + \log\ (1 - 0.018)(n - v)m \qquad (5.22)$$

where $f_\pm$ is the mean activity coefficient of the positive and the negative ion (this is a geometric mean), Z_A and Z_B are respectively the charge on ion A and ion B, a_w is the activity of water (the ratio of vapor pressure of the solution to that of pure water), m is the molality of the solution, n is the hydration number per solute molecule, and v is the number of species formed from each solute molecule, e.g., for $Ca(NO_3)_2$, this is 3.

We can draw some general conclusions about the estimation of activity coefficients.

1. The activity coefficients of ions of a given charge magnitude are approximately the same in solutions of a given ionic strength, and this activity coefficient is the same regardless of their individual concentrations.

2. The behavior of ions become less ideal as the charge magnitude increases, resulting in less confidence in calculated activity coefficients.

3. The calculated activity coefficient of an ion in a mixed electrolyte solution will be less accurate than in a single-electrolyte solution.

4. The activity coefficients of nonelectrolytes (uncharged molecules) can generally be considered equal to unity in ionic strengths up to 0.1, and deviations from this approximation are only moderate in ionic strengths as high as 1. Undissociated acids, HA, are nonelectrolytes whose activity coefficients can be taken as unity. However, in highly concentrated electrolytes, the activity coefficients of nonelectrolytes do exceed unity, again because solvent becomes unavailable. This is the basis of "salting out" a nonelectrolyte from solution, often used in organic synthesis.

The greater the charge on diverse ions, the greater their effect on the activity.

The activity of nonelectrolytes is the same as the concentration, up to ionic strengths of 1.

Salting out can be useful in analytical chemistry as well. It is possible to add sufficient $CaCl_2$ to a mixture of water and acetone containing an organic chelate of cobalt that the water will all be taken up and separate as a highly concentrated $CaCl_2$ solution layer, distinct from the acetone layer bearing the Co chelate. See C.E. Matkovich and G. D. Christian *Anal. Chem.* **45** (1973) 1915.

A final comment about activity coefficients: Kenneth S. Pitzer recast activity coefficient corrections using quantum mechanics, and provides rigorous treatment of concentrated solutions. See Reference 11.

5.17   The Diverse Ion Effect: The Thermodynamic Equilibrium Constant and Activity Coefficients

We mentioned at the beginning of the last section on activity that the presence of diverse salts will generally increase the dissociation of weak electrolytes due to a shielding (or decrease in the activity) of the ionic species produced upon dissociation. We can quantitatively predict the extent of the effect on the equilibrium by taking into account the activities of the species in the equilibrium.

In our consideration of equilibrium constants thus far, we have assumed no diverse ion effect, that is, an ionic strength of zero and an activity coefficient of 1. Equilibrium constants should more exactly be expressed in terms of activities rather than concentrations. Consider the dissociation of AB. The **thermodynamic equilibrium constant** (i.e., the equilibrium constant extrapolated to the case of infinite dilution) K_{eq}° is

Thermodynamic equilibrium constants hold at all ionic strengths.

For a qualitative picture see panel (*b*), Figure 5.3

$$K_{eq}^{\circ} = \frac{a_A \cdot a_B}{a_{AB}} = \frac{[A]f_A \cdot [B]f_B}{[AB]f_{AB}} \qquad (5.23)$$

Since the **concentration equilibrium constant** $K_{eq} = [A][B]/[AB]$, then

$$K_{eq}^{\circ} = K_{eq}\frac{f_A \cdot f_B}{f_{AB}} \qquad (5.24)$$

or

$$K_{eq} = K_{eq}^{\circ}\frac{f_{AB}}{f_A \cdot f_B} \qquad (5.25)$$

Concentration equilibrium constants must be corrected for ionic strength.

The numerical value of K_{eq}° holds for all activities. $K_{eq} = K_{eq}^{\circ}$ at zero ionic strength, but at appreciable ionic strengths, a value for K_{eq} must be calculated for each ionic strength using Equation 5.25. The equilibrium constants listed in Appendix C are for zero ionic strength; that is, they are really thermodynamic equilibrium constants. (For some reaction systems, experimental K_{eq} values are available at different ionic strengths and can be used for equilibrium calculations at the listed ionic strength, using molar concentrations without having to calculate activity coefficients. For example, chemical oceanography is of sufficient importance that all relevant equilibrium constants uniquely applicable to a seawater matrix are available.)

Example 5.18

The weak electrolyte AB dissociates to A^+ and B^-, with a thermodynamic equilibrium constant K_{eq}° of 2×10^{-8} in the presence of a diverse salt of ionic strength 0.1. If the activity coefficients of A^+ and B^- are 0.6 and 0.7, respectively, at $\mu = 0.1$:

(a) Calculate the molar equilibrium constant K_{eq} in terms of concentration.

(b) Calculate the percent dissociation of a $1.0 \times 10^{-4} M$ solution of AB in water.

Solution

(a)
$$AB \rightleftharpoons A^+ + B^-$$

$$K_{eq} = \frac{[A^+][B^-]}{[AB]}$$

$$K_{eq}^{\circ} = \frac{a_{A^+} \cdot a_{B^-}}{a_{AB}} = \frac{[A^+]f_{A^+} \cdot [B^-]f_{B^-}}{[AB]f_{AB}}$$

The activity coefficient of a neutral species is unity, so

$$K_{eq}^{\circ} = \frac{[A^+][B^-]}{[AB]} \cdot f_{A^+} \cdot f_{B^-} = K_{eq}f_{A^+} \cdot f_{B^-}$$

$$K_{eq} = \frac{K_{eq}^{\circ}}{f_{A^+} \cdot f_{B^-}} = \frac{2 \times 10^{-8}}{(0.6)(0.7)} = 5 \times 10^{-8}$$

(b)
$$\begin{array}{cccc} AB & \rightleftharpoons & A^+ & + \ B^- \\ 1 \times 10^{-4} - x & & x & x \end{array}$$

In water, $f_{A^+} = f_{B^-} \approx 1$ (since $\mu < 10^{-4}$), $x \ll 10^{-4}$

$$\frac{[A^+][B^-]}{[AB]} = 2 \times 10^{-8}$$

$$\frac{(x)(x)}{1.0 \times 10^{-4}} = 2 \times 10^{-8}$$

$$x = 1.4 \times 10^{-6} M$$

$$\% \text{ dissociated} = \frac{1.4 \times 10^{-6} M}{1.0 \times 10^{-4} M} \times 100\% = 1.4\%$$

For 0.1 M salt,

$$\frac{[A^+][B^-]}{[AB]} = 5 \times 10^{-8}$$

$$\frac{(x)(x)}{1.0 \times 10^{-4}} = 5 \times 10^{-8}$$

$$x = 2_{.2} \times 10^{-6}$$

$$\% \text{ dissociated} = \frac{2_{.2} \times 10^{-6}}{1.0 \times 10^{-4}} \times 100\% = 2_{.2}\%$$

which represents a 57% increase in dissociation.

Calculations using the diverse ion effect are illustrated in Chapter 6 for acid dissociation and in Chapter 9 for precipitate solubilities. For illustrative purposes throughout this book, we will in general neglect the diverse ion effects on equilibria. In most cases, we are interested in *relative* changes in equilibrium concentrations, and the neglect of activities will not change our arguments.

> We will generally ignore diverse salt effects.

Electrostatic effects are important in the behavior of charged molecules such as proteins, DNA and other charged biopolymers, how ions move in capillary electrophoresis, and the behavior of glass and ion selective electrodes, to name a few areas. Read Professor's Favorite Musings: Where Do Activity Coefficients Come From? The thoughts of Professor Michael D. Morris, University of Michigan, on this topic are on the text website.

Problems

Equilibrium Calculations

1. For the reaction:

$$C(s) + H_2(g) + O_2(g) \rightleftharpoons CH_3OH(l); \quad \Delta H^\circ = -238.7 \text{ kJ}$$

   From the following options, which will increase the value of K_{eq}?

   (a) Decreasing temperature
   (b) Increasing hydrogen
   (c) Increasing pressure
   (d) Decreasing carbon

2. Addition of which of the following compounds would prevent shifting of equilibrium to right as per the Le Châtelier's principle for the endothermic reaction:

$$CaCO_3(s) + CO_2(g) + H_2O(l) \rightleftharpoons Ca^{2+}(aq) + 2HCO_3^-$$

   (a) CO_2
   (b) LiOH
   (c) Na_3PO_4
   (d) $CaCO_3$

3. Among the following expressions, which is correct for equilibrium constant?

$$2H_2 + O_2 \rightleftharpoons 2H_2O$$

   (a) $K_c = \dfrac{2[H_2O]}{2[H_2][O_2]}$

   (b) $K_c = \dfrac{2[H_2][O_2]}{2[H_2O]}$

   (c) $K_c = \dfrac{[H_2O]^2}{[H_2]^2[O_2]}$

   (d) $K_c = \dfrac{[H_2]^2[O_2]}{[H_2O]^2}$

4. A and B react as follows: $A + B \rightleftharpoons C + D$. The equilibrium constant is 2.0×10^3. If 0.30 mol of A and 0.80 mol of B are mixed in 1 L, what are the concentrations of A, B, C, and D after reaction?

5. A and B react as follows: $A + B \rightleftharpoons 2C$. The equilibrium constant is 5.0×10^6. If 0.40 mol of A and 0.70 mol of B are mixed in 1 L, what are the concentrations of A, B, and C after reaction?

6. The dissociation constant for salicylic acid, $C_6H_4(OH)COOH$, is 1.0×10^{-3}. Calculate the percent dissociation of a 1.0×10^{-3} M solution. There is one dissociable proton.

7. For a 1.0×10^{-3} M solution of hydrocyanic acid, HCN calculate the percentage dissociation, where dissociation constant for HCN is 7.2×10^{-10}.

8. Calculate the percent dissociation of the salicylic acid in Problem 6 if the solution also contained 1.0×10^{-2} M sodium salicylate (the salt of salicylic acid).

9. Write the overall dissociation reactions and overall equilibrium constant for hydrogen sulfide (H_2S), which dissociates step-wise, having dissociation constant of 9.1×10^{-8} and 1.2×10^{-15}, respectively.

10. Fe^{2+} and $Cr_2O_7^{2-}$ react as follows: $6Fe^{2+} + Cr_2O_7^{2-} + 14H^+ \rightleftharpoons 6\ Fe^{3+} + 2Cr^{3+} + 7H_2O$. The equilibrium constant for the reaction is 1×10^{57}. Calculate the equilibrium concentrations of the iron and chromium species if 10 mL each of 0.02 M $K_2Cr_2O_7$ in 1.14 M HCl and 0.12 M $FeSO_4$ in 1.14 M HCl are reacted.

Systematic Approach To Equilibrium Calculations

11. Match the correct pair of types of equilibrium (chemical process) to their respective equilibrium constant.

Equilibrium	Equilibrium Constant
1. Acid base dissociation	(a) K_{sp}, solubility product
2. Solubility	(b) K_{eq}, Reaction equilibrium constant
3. Complex formation	(c) K_a, Acid dissociation constant
4. Reduction-oxidation	(d) K_D, Distribution coefficient
5. Phase distribution	(e) K_f, Formation constant

12. Write charge balance expressions for (a) a saturated solution of Bi_2S_3; (b) a solution of Na_2S.

13. Write the equations of mass balance and electroneutrality for a 0.100 M $[Cd(NH_3)_4]Cl_2$ solution.

14. Prove the following relations using the principles of electroneutrality and mass balance:

(a) $[NO_2^-] = [H^+] - [OH^-]$ for 0.2 M HNO_2 solution

(b) $[CH_3COOH] = 0.2 - [H^+] + [OH^-]$ for 0.2 M CH_3COOH solution

(c) $[H_2C_2O_4] = 0.1 - [H^+] + [OH^-] - [C_2O_4^{2-}]$ for 0.1 M $H_2C_2O_4$ solution

(d) $[HCN] = [OH^-] - [H^+]$ for 0.1 M KCN solution

(e) $[H_2PO_4^-] = \dfrac{[OH^-] - [H^+] - [HPO_4^{2-}] - 3[H_3PO_4]}{2}$ for 0.1 M Na_3PO_4 solution

(f) $[HSO_4^-] = 0.2 - [H^+] - [OH^-]$ for 0.1 M H_2SO_4 solution (assume that the dissociation of H_2SO_4 to H^+ and HSO_4^- is quantitative)

15. Write equations of mass balance for an aqueous saturated solution of BaF_2 containing the species F^-, HF, HF_2^-, and Ba^{2+}.

16. Write an equation of mass balance for an aqueous solution of $Ba_3(PO_4)_2$.

17. Calculate the pH of a 0.100 M solution of acetic acid using the charge/mass balance approach.

18. A system is formed when a 0.010 M NH_3 solution is saturated with AgBr. Write the mass-balance equation, where the equations for pertinent equilibria in the solution are:

$$AgBr\,(s) \rightleftharpoons Ag^+ + Br^-$$

$$Ag^+ + NH_3 \rightarrow AgNH_3^+$$

$$Ag\,(NH_3)^+ + NH_3 \rightleftharpoons Ag\,(NH_3)_2^{\,+}$$

$$NH_3 + H_2O \rightleftharpoons NH_4^+ + OH^-$$

$$2H_2O \rightleftharpoons H_3O^+ + OH^-$$

19. Write charge balance equation for the system in Question 18.

20. An aqueous solution that contains NaCl, $Ba(ClO_4)_2$ and $Al_2(SO_4)_3$. Write its charge balance equation.

Ionic Strength

21. Calculate the ionic strength for (a) 0.1 M solution of KNO_3 and (b) 0.1 M solution of Na_2SO_4.

22. Calculate the ionic strength for a solution of 0.05 M KNO_3 and 0.1 M Na_2SO_4.

23. Calculate the ionic strengths of the following solutions: (a) 0.30 M NaCl (b) 0.30 M Na_2SO_4 (c) 0.30 M NaCl and 0.20 M K_2SO_4; (d) 0.20 M $Al_2(SO_4)_3$ and 0.10 M Na_2SO_4.

24. Calculate the ionic strengths of the following solutions: (a) 0.20 M $ZnSO_4$; (b) 0.40 M $MgCl_2$; (c) 0.50 M $LaCl_3$; (d) 1.0 M $K_2Cr_2O_7$; (e) 1.0 M $Tl(NO_3)_3$ + 1.0 M $Pb(NO_3)_2$.

Activity

25. Using Debye-Hückel equation, calculate the activity coefficient for Hg^{2+} in a solution that has an ionic strength of 0.085 M. Use 0.5 nm for effective diameter of the ion.

See the text website, Spreadsheet Problems, for Excel answers to problems 26–29.

26. Calculate the activity coefficients of the sodium and chloride ions for a 0.00100 M solution of NaCl.

27. Calculate the activity coefficients of each ion in a solution containing 0.0020 M Na_2SO_4 and 0.0010 M $Al_2(SO_4)_3$.

28. Calculate the activity of the NO_3^- ion in a solution of 0.0020 M KNO_3.

29. Calculate the activity of the CrO_4^{2-} ion in a 0.020 M solution of Na_2CrO_4.

30. 2.5 M sulfuric acid (H_2SO_4) has a density of 1.15. The relative humidity over such a solution is 88.8%. If you assume that each proton is solvated by 4 molecules of water, what will be the mean activity coefficient according to Equation 5.22?

Thermodynamic Equilibrium Constants

31. Write thermodynamic equilibrium constant expressions for the following:

    (a) $HCN \rightleftharpoons H^+ + CN^-$

    (b) $NH_3 + H_2O \rightleftharpoons NH_4^+ + OH^-$

32. Calculate the pH of a solution of $5.0 \times 10^{-3} M$ benzoic acid (a) in water and (b) in the presence of 0.05 M K_2SO_4.

Multiple Choice Questions

1. Which of the following is true for a chemical reaction at equilibrium?

    (a) Only the forward reaction steps.   (b) Only the reverse reaction steps.

    (c) The rate constants for the forward and   (d) The rates of the forward and reverse reverse reactions are equal.   reactions are equal.

2. A chemical equilibrium may be established by starting a reaction with _____.

    (a) reactants only   (b) products only

    (c) any quantities of reactants and products (d) all of the above

3. In a container, equilibrium is established by adding 0.10 mol of A and B in 1 L flask, then which of the following is true?

    (a) [A] = [B]   (b) [A] = [B] = [C]

    (c) [B] = 2[C]   (d) [A] > [B]

Recommended References

Equilibria

1. A. J. Bard, *Chemical Equilibrium*. New York: Harper & Row, 1966.
2. T. R. Blackburn, *Equilibrium: A Chemistry of Solutions*. New York: Holt, Rinehart and Winston, 1969.
3. J. N. Butler. *Ionic Equilibrium. A Mathematical Approach*. Reading, MA: Addison-Wesley, 1964.
4. G. M. Fleck, *Ionic Equilibria in Solution*. New York: Holt, Rinehart and Winston, 1966.
5. H. Freiser and Q. Fernando, *Ionic Equilibria in Analytical Chemistry*. New York: Wiley, 1963.
6. A. E. Martell and R. J. Motekaitis, *The Determination and Use of Stability Constants*. New York: VCH, 1989.

Method of Successive Approximations

7. S. Brewer, *Solving Problems in Analytical Chemistry*. New York: Wiley, 1980.
8. J. J. Baeza-Baeza and M. C. Garcia-Alvarez-Coque, "Systematic Approach to Calculate the Concentration of Chemical Species in Multi-Equilibrium Problems," *J. Chem. Educ.* **88** (2011) 169. This article demonstrates the solution of multiple simultaneous equlibria using Excel Solver.

Activity

9. C. W. Davies, *Ion Association*. London: Butterworth, 1962.
10. J. Kielland, "Individual Activity Coefficients of Ions in Aqueous Solutions," *J. Am. Chem. Soc.*, **59** (1937) 1675.
11. K. S. Pitzer. *Activity Coefficients in Electrolyte Solutions*, 2nd ed. Boca Raton, FL: CRC Press, 1991.
12. P. C. Meier, "Two-Parameter Debye–Hückel Approximation for the Evaluation of Mean Activity Coefficients of 109 Electrolytes," *Anal. Chim. Acta*, **136** (1982) 363.

Chapter 6

ACID–BASE EQUILIBRIA

Police arrested two kids yesterday, one was drinking battery acid, the other was eating fireworks. They charged one and let the other one off.
—Tommy Cooper

Saying sulfates do not cause acid rain is the same as saying that smoking does not cause lung cancer.
—Drew Lewis

KEY THINGS TO LEARN FROM THIS CHAPTER

- Acid–base theories
- Acid–base equilibria in water (key Equations: 6.11, 6.13, 6.19)
- Weak acids and bases
- Salts of weak acids and bases (key Equations: 6.27, 6.29, 6.32, 6.36, 6.39)
- Buffers (key Equations: 6.45, 6.58)
- Polyprotic acids—α values (key Equations: 6.74–6.77)
- Using spreadsheets to prepare α vs. pH plots
- Salts of polyprotic acids (key Equations: 6.96, 6.97, 6.99, 6.100)
- Logarithmic concentration diagrams
- pH calculator programs

The acidity or basicity of a solution is frequently an important factor in chemical reactions. The use of buffers to maintain the solution pH at a desired level is very important. In addition, fundamental acid–base equilibria are important in understanding acid–base titrations and the effects of acids on chemical species and reactions, for example, the effects of complexation or precipitation. In Chapter 5, we described the fundamental concept of equilibrium constants. In this chapter, we consider in more detail various acid–base equilibrium calculations, including *weak acids* and *bases*, hydrolysis of *salts of weak acids and bases*, buffers, polyprotic acids and their salts, and *physiological buffers*. Acid–base theories and the basic pH concept are reviewed first.

6.1 The Early History of Acid—Base Concepts

The word *acid* derives from Latin *acere*, meaning sour. Bases were referred to as *alkali* in early history and that word derives from Arabic *al-qili*, the ashes of the plant saltwort, rich in sodium carbonate. In the mid seventeenth century it was recognized that acids and bases (called *alkali* in early history) tend to neutralize each other (known as the Silvio-Tachenio theory) but the concepts were vague. Acids, for example, were thought to be substances that would cause limestone to effervesce and alkalis as those that would effervesce with acids. In 1664 Robert Boyle published in *The Experimental History of Colours* that extracts of certain plants such as red roses and Brazil wood changed color reversibly as the solution was made alternately acidic and basic. Many other plant and flower extracts were shown subsequently to behave in a similar fashion. In 1675, Boyle objected to the vagueness of the Silvio–Tachenio theory and largely because of his efforts, a set of definitions emerged about acids that sought to incorporate their known properties: acids taste sour, cause limestone to effervesce, turn blue plant dyes to red, and precipitate sulfur from alkaline solutions. Alkalies are substances that are slippery to the touch and can reverse the effect of acids. Almost a hundred years elapsed before Antoine-Laurent Lavoisier formed his own opinion of how acids come to be. Based primarily on his observations on combustion and respiration, in which carbon is converted to carbon dioxide (the acidic nature of carbon dioxide dissolved in water was already obvious), he named the gas recently (1774) discovered by Joseph Priestley, so essential for combustion or respiration, as oxygen (from Greek, meaning *acid former*), since he surmised it was what created the acidic product.

Alessandro Volta announced the electric pile—an early type of battery—in 1800. Humphry Davy started playing with electricity immediately thereafter. Through electrolysis he discovered several new elements. In 1807 he electrolyzed fused potash and then soda—substances that many thought to be elements—and isolated potassium and sodium. He also similarly isolated magnesium, calcium, strontium, and barium. Davy recognized that these alkali and alkaline earth metals combine with oxygen and form already known oxides that are highly basic, which challenged Lavoisier's theory that oxygen was the acidifying element. He went on to establish that hydrochloric acid, not "oxymuriatic acid" as Lavoisier called it, was acidifying; by electrolysis he isolated hydrogen and one other element, chlorine (that he so named in 1810), which until then was believed to be a compound containing oxygen. Rather than oxygen, Davy suggested in 1815 that hydrogen may be the acidifying element. All substances that contain hydrogen, however, are not acids. It would wait for Justus von Liebig, to identify an acid in 1838 as a compound of hydrogen where the hydrogen can be replaced by a metal.

6.2 Acid–Base Theories—Not All Are Created Equal

Several acid–base theories have been proposed to explain or classify acidic and basic properties of substances. You are probably most familiar with the **Arrhenius theory**, which is applicable only to water. Other theories are more general and are applicable to other solvents or even the gas phase. We describe the common acid–base theories here.

Arrhenius Theory—H$^+$ and OH$^-$

The Arrhenius theory is restricted to aqueous solutions. See *J. Am. Chem. Soc.*, **34** (1912) 353 for his personal observations of the difficulty Arrhenius had in the acceptance of his theory.

Arrhenius, as a graduate student, introduced a dramatically new theory that an **acid** is any substance that ionizes (partially or completely) in water to give *hydrogen ions* (which associate with the solvent to give hydronium ions, H_3O^+):

$$HA + H_2O \rightleftharpoons H_3O^+ + A^-$$

A **base** ionizes in water to give *hydroxide ions*. Weak (partially ionized) bases generally ionize as follows:

$$B + H_2O \rightleftharpoons BH^+ + OH^-$$

while strong bases such as metal hydroxides (e.g., NaOH) dissociate as

$$M(OH)_n \rightarrow M^{n+} + nOH^-$$

This theory is obviously restricted to water as the solvent.

Theory of Solvent Systems—Solvent Cations and Anions

In 1905, Franklin was working in liquid NH_3 as solvent and noticed the similarity with acid-base behavior in water. In 1925, Germann, working with liquid $COCl_2$ as solvent observed the similarities as well and formulated a general solvent system concept of acids and bases. This theory recognizes the ionization of a solvent to give a cation and an anion; for example, $2H_2O \rightleftharpoons H_3O^+ + OH^-$ or $2NH_3 \rightleftharpoons NH_4^+ + NH_2^-$. An **acid** is defined as a solute that yields the characteristic *cation of the solvent* while a **base** is a solute that yields the characteristic *anion of the solvent*. Thus, NH_4Cl (which produces ammoniated NH_4^+, i.e., $[NH_4(NH_3)^+]$, and Cl^-) is a strong acid in liquid ammonia (similar to HCl in water: $HCl + H_2O \rightarrow H_3O^+ + Cl^-$) while $NaNH_2$ is a strong base in ammonia (similar to NaOH in water); both of these compounds ionize to give the characteristic solvent cation and anion, respectively. Ethanol ionizes as follows: $2C_2H_5OH \rightleftharpoons C_2H_5OH_2^+ + C_2H_5O^-$. Hence, sodium ethoxide, $NaOC_2H_5$, is a strong base in this solvent.

Franklin and Germann's theory is similar to the Arrhenius theory but is applicable also to other ionizable solvents.

Brønsted–Lowry Theory—Taking and Giving Protons

The theory of solvent systems is suitable for ionizable solvents, but it is not applicable to acid–base reactions in nonionizable solvents such as benzene or dioxane. In 1923, Brønsted and Lowry independently described what is now known as the **Brønsted–Lowry** theory. This theory states that an **acid** is any substance that can *donate a proton*, and a **base** is any substance that can *accept a proton*. Thus, we can write a "half-reaction"

The Brønsted–Lowry theory assumes a transfer of protons from an acid to a base, i.e., conjugate pairs.

$$acid = H^+ + base \qquad (6.1)$$

The acid and base of a half-reaction are called **conjugate pairs**. Free protons do not exist in solution, and there must be a proton acceptor (base) before a proton donor (acid) will release its proton. That is, there must be a combination of two half-reactions. Another way to look at it is that an acid is an acid because it can lose a proton. However, it cannot exhibit its acidic behavior unless there is a base present to accept the proton. It is like being wealthy on a deserted island with no one to accept your money. Some acid–base reactions in different solvents are illustrated in Table 6.1. In the first example, acetate ion is the conjugate base of acetic acid and ammonium ion is the conjugate acid of ammonia. The first four examples represent ionization of an acid or a base in a solvent, while the others represent a neutralization reaction between an acid and a base in the solvent.

It is apparent from the above definition that a substance cannot act as an acid unless a base is present to accept the protons. Thus, acids will undergo complete or partial ionization in basic solvents such as water, liquid ammonia, or ethanol, depending on the basicity of the solvent and the strength of the acid. But in neutral or "inert" solvents, ionization is insignificant. However, ionization in the solvent is not a prerequisite for an acid–base reaction, as in the last example in the table, where picric acid reacts with aniline.

TABLE 6.1 Brønsted Acid–Base Reactions: Conjugate acid base pairs are denoted in the same color

Solvent	$Acid_1$	+	$Base_2$	$\rightarrow$	$Acid_2$	+	$Base_1$
NH_3 (liq.)	HOAc		NH_3		NH_4^+		OAc^-
H_2O	HCl		H_2O		H_3O^+		Cl^-
H_2O	NH_4^+		H_2O		H_3O^+		NH_3
H_2O	H_2O		OAc^-		HOAc		OH^-
H_2O	HCO_3^-		OH^-		H_2O		CO_3^{2-}
C_2H_5OH	NH_4^+		$C_2H_5O^-$		C_2H_5OH		NH_3
C_6H_6	H picrate		$C_6H_5NH_2$		$C_6H_5NH_3^+$		picrate$^-$

Lewis Theory—Taking and Giving Electrons

The Lewis theory assumes a donation (sharing) of electrons from a base to an acid.

Also in 1923, G. N. Lewis introduced the electronic theory of acids and bases. In the **Lewis** theory, an acid is a substance that can accept an electron pair and a base is a substance that can donate an electron pair. The latter frequently contains an oxygen or a nitrogen as the electron donor. Thus, nonhydrogen-containing substances are included as acids. Examples of acid–base reactions in the Lewis theory are as follows:

$$H^+ \text{ (solvated)} + :NH_3 \rightarrow H:NH_3^+$$

$$AlCl_3 + :O{\overset{R}{\underset{R}{\diagup}}}{\diagdown} \rightarrow Cl_3Al:OR_2$$

$$\overset{H}{\underset{H}{\diagdown}}O: + H^+ \rightarrow H_2O:H^+$$

$$H^+ + :OH^- \rightarrow H:OH$$

In the second example, aluminum chloride is an acid and ether is a base.

6.3 Acid–Base Equilibria in Water

We see from the above that when an acid or base is dissolved in water, it will dissociate, or **ionize**, the amount of ionization being dependent on the strength of the acid or the base. A "strong" electrolyte is completely dissociated, while a "weak" electrolyte is partially dissociated. Table 6.2 lists some common electrolytes, some strong and some weak. Other weak acids and bases are listed in Appendix C.

Hydrochloric acid is a strong acid, and in water, its ionization is complete:

$$HCl + H_2O \rightarrow H_3O^+ + Cl^- \tag{6.2}$$

An equilibrium constant for Equation 6.2 would have a value of infinity. The proton H^+ exists in water as a hydrated ion, the **hydronium ion**, H_3O^+. Higher hydrates probably exist, particularly $H_9O_4^+$. The hydronium ion is written as H_3O^+ for convenience and to emphasize Brønsted behavior.

TABLE 6.2 Some Strong Electrolytes and Some Weak Electrolytes

Strong	Weak
HCl	CH_3COOH (acetic acid)
$HClO_4$	NH_3
$H_2SO_4{}^a$	C_6H_5OH (phenol)
HNO_3	$HCHO_2$ (formic acid)
NaOH	$C_6H_5NH_2$ (aniline)
	CH_3COONa

^{a} The first proton is completely ionized in dilute solution, but the second proton is partially ionized ($K_2 = 10^{-2}$).

Acetic acid[1] is a weak acid, which ionizes only partially in water (a few percent):

$$HOAc + H_2O \rightleftharpoons H_3O^+ + OAc^- \tag{6.3}$$

We can write an **equilibrium constant** for this reaction:

$$K_a^\circ = \frac{a_{H_3O^+} \cdot a_{OAc^-}}{a_{HOAc} \cdot a_{H_2O}} \tag{6.4}$$

where K_a° is the **thermodynamic acidity constant** (see Section 5.16) and a is the **activity** of the indicated species. Salt cations or anions may also partially react with water after they are dissociated. For example, acetate ion is formed from dissociated acetate salts, to give HOAc.

The activity can be thought of as representing the *effective* concentration of an ion (described in Chapter 5). The effects of protons in reactions are often governed by their activities, and it is the activity that is measured by the widely used pH meter (Chapter 20). Methods for predicting numerical values of activity coefficients were described in Chapter 5.

In dilute solutions, the activity of water remains essentially constant, and is taken as unity at standard state. Therefore, Equation 6.4 can be written as

$$K_a^\circ = \frac{a_{H_3O^+} \cdot a_{OAc^-}}{a_{HOAc}} \tag{6.5}$$

Pure water ionizes slightly, or undergoes **autoprotolysis**:

$$2H_2O \rightleftharpoons H_3O^+ + OH^- \tag{6.6}$$

The equilibrium constant for this is

$$K_w^\circ = \frac{a_{H_3O^+} \cdot a_{OH^-}}{a_{H_2O^2}} \tag{6.7}$$

> Autoprotolysis is the self-ionization of a solvent to give a characteristic cation and anion, e.g., $2CH_3OH \rightleftharpoons CH_3OH^+ + CH_3O^-$.

Again, the activity of water is constant in dilute solutions (its concentration is essentially constant at $\sim 55.5\ M$), so

$$K_w^\circ = a_{H_3O^+} \cdot a_{OH^-} \tag{6.8}$$

where K_w° is the **thermodynamic autoprotolysis**, or **self-ionization, constant**.

Calculations are simplified if we neglect activity coefficients. This simplification results in only slight errors for dilute solutions, and we shall use molar concentrations in all our calculations. This will satisfactorily illustrate the equilibria involved. Most of the solutions we will be concerned with are rather dilute, and we will frequently be interested in relative changes in pH (and large ones) in which case small errors are insignificant. We will simplify our expressions by using H^+ in place of H_3O^+. This is not inconsistent since the waters of solvation associated with other ions or molecules (e.g., metal ions) are not generally written and H_3O^+ is not an accurate representation of the actual species present; typically the proton in dilute aqueous solution has at least four water molecules in its solvation shell.

> We will use H^+ in place of H_3O^+, for simplicity. Also, molar concentrations will generally be used instead of activities.

Molar concentration will be represented by square brackets [ ] around the species. Simplified equations for the above reactions are

$$HCl \rightarrow H^+ + Cl^- \tag{6.9}$$

$$HOAc \rightleftharpoons H^+ + OAc^- \tag{6.10}$$

---

[1]We shall use the symbol OAc$^-$ to represent the acetate ion

$$CH_3-\overset{\overset{\displaystyle O}{\|}}{C}-O^-.$$

$$K_a = \frac{[H^+]\,[OAc^-]}{[HOAc]} \tag{6.11}$$

$$H_2O \rightleftharpoons H^+ + OH^- \tag{6.12}$$

$$K_w = [H^+]\,[OH^-] \tag{6.13}$$

K_a and K_w are the **molar equilibrium constants**.

K_w is exactly 1.00×10^{-14} at 24°C and even at 25°C, to a smaller number of significant figures, it is still accurately represented as 1.0×10^{-14}. The product of the hydrogen ion concentration and the hydroxide ion concentration in aqueous solution is *always* equal to 1.0×10^{-14} at room temperature:

$$[H^+]\,[OH^-] = 1.0 \times 10^{-14} \tag{6.14}$$

Chemists (and especially students!) are lucky that nature made K_w an even unit number at room temperature. Imagine doing pH calculations with a K_w like 2.39×10^{-13}. However, see Section 6.5 where you must indeed do this for other temperatures.

In pure water, then, the concentrations of these two species are equal since there are no other sources of H^+ or OH^- except H_2O dissociation:

$$[H^+] = [OH^-]$$

Therefore,

$$[H^+][H^+] = 1.0 \times 10^{-14}$$

$$[H^+] = 1.0 \times 10^{-7}\,M \equiv [OH^-]$$

If an acid is added to water, we can calculate the hydroxide ion concentration if we know the hydrogen ion concentration from the acid. *But when the hydrogen ion concentration from the acid is very small, 10^{-6} M or less, the contribution to [H^+] from the ionization of water cannot be neglected.*

Example 6.1

A 1.0×10^{-5} M solution of hydrochloric acid is prepared. What is the hydroxide ion concentration?

Solution

Since hydrochloric acid is a strong electrolyte and is completely ionized, the H^+ concentration is 1.0×10^{-5} M. Thus,

$$(1.0 \times 10^{-5})\,[OH^-] = 1.0 \times 10^{-14}$$

$$[OH^-] = 1.0 \times 10^{-9}\,M$$

6.4 The pH Scale

pScales are used to compress and more conveniently express a range of numbers that span several decades in magnitude.

The concentration of H^+ or OH^- in aqueous solution can vary over extremely wide ranges, from 1 M or greater to 10^{-14} M or less. To construct a plot of H^+ concentration against some variable would be very difficult if the concentration changed from, say, 10^{-1} M to 10^{-13} M. This range is common in a titration. It is more convenient to compress the acidity scale by placing it on a logarithm basis. The **pH** of a solution was defined by Sørenson as

$$pH = -\log[H^+] \tag{6.15}$$

The minus sign is used because most of the concentrations encountered are less than 1 M, and so this designation gives a positive number. (More strictly, **pH** is now defined as $-\log a_{H^+}$, but we will use the simpler definition of Equation 6.15.) In general, **pAnything = −log Anything**, and this method of notation will be used later for other numbers that can vary by large amounts, or are very large or small (e.g., equilibrium constants).

pH is really $-\log a_{H^+}$. This is what a pH meter (glass electrode) measures—see Chapter 20.

Example 6.2

Calculate the pH of a $2.0 \times 10^{-5} M$ solution of HCl.

Solution

HCl ion completely ionized, so

$$[H^+] = 2.0 \times 10^{-5}\, M$$
$$pH = -\log(2.0 \times 10^{-5}) = 5 - \log\, 2.0 = 5 - 0.30 = 4.70$$

A similar definition is made for the hydroxide ion concentration:

$$\boxed{pOH = -\log[OH^-]} \tag{6.16}$$

Equation 6.13 can be used to calculate the hydroxyl ion concentration if the hydrogen ion concentration is known, and vice versa. The equation in logarithm form for a more direct calculation of pH or pOH is

$$-\log K_w = -\log[H^+][OH^-] = -\log[H^+] - \log[OH^-] \tag{6.17}$$

$$\boxed{pK_w = pH + pOH} \tag{6.18}$$

At 25°C,

$$\boxed{14.00 = pH + pOH} \tag{6.19}$$

A 1 M HCl solution has a pH of 0 and pOH of 14. A 1 M NaOH solution has a pH of 14 and a pOH of 0.

Example 6.3

Calculate the pOH and the pH of a $5.0 \times 10^{-5}\, M$ solution of NaOH at 25°C.

Solution

$$[OH^-] = 5.0 \times 10^{-5}\, M$$
$$pOH = -\log(5.0 \times 10^{-5}) = 5 - \log\, 5.0 = 5 - 0.70 = 4.30$$
$$pH + 4.30 = 14.00$$
$$pH = 9.70$$

or

$$[H^+] = \frac{1.0 \times 10^{-14}}{5.0 \times 10^{-5}} = 2.0 \times 10^{-10}\, M$$
$$pH = -\log(2.0 \times 10^{-10}) = 10 - \log\, 2.0 = 10 - 0.30 = 9.70$$

Example 6.4

Keep track of millimoles!

Calculate the pH of a solution prepared by mixing 2.0 mL of a strong acid solution of pH 3.00 and 3.0 mL of a strong base of pH 10.00.

Solution

$$[H^+] \text{ of acid solution} = 1.0 \times 10^{-3} \ M$$

$$\text{mmol } H^+ = 1.0 \times 10^{-3} \ M \times 2.0 \text{ mL} = 2.0 \times 10^{-3} \text{ mmol}$$

$$\text{pOH of base solution} = 14.00 - 10.00 = 4.00$$

$$[OH^-] = 1.0 \times 10^{-4} \ M$$

$$\text{mmol } OH^- = 1.0 \times 10^{-4} \ M \times 3.0 \text{ mL} = 3.0 \times 10^{-4} \text{ mmol}$$

There is an excess of acid.

$$\text{mmol } H^+ = 0.0020 - 0.0003 = 0.0017 \text{ mmol}$$

$$\text{Total Volume} = (2.0 + 3.0) \text{ mL} = 5.0 \text{ mL}$$

$$[H^+] = 0.0017 \text{ mmol}/5.0 \text{ mL} = 3.4 \times 10^{-4} \ M$$

$$pH = -\log 3.4 \times 10^{-4} = 4 - 0.53 = 3.47$$

Example 6.5

Remember, this answer is reported to two significant figures $(2.1 \times 10^{-10}$ M) because the mantissa of the pH value (9.67) has two significant figures.

The pH of a solution is 9.67. Calculate the hydrogen ion concentration in the solution.

Solution

$$-\log[H^+] = 9.67$$

$$[H^+] = 10^{-9.67} = 10^{-10} \times 10^{0.33}$$

$$[H^+] = 2.1 \times 10^{-10} \ M$$

$[H^+] = 10^{-pH}.$

When $[H^+] = [OH^-]$, then a solution is said to be **neutral**. If $[H^+] > [OH^-]$, then the solution is **acidic**. And if $[H^+] < [OH^-]$, the solution is **alkaline**. The hydrogen ion and hydroxide ion concentrations in pure water at 25°C are each $10^{-7} \ M$, and the pH of water is 7. A pH of 7 is therefore neutral. Values of pH that are greater than this are alkaline, and pH values less than this are acidic. The reverse is true of pOH values. A pOH of 7 is also neutral. Note that the product of $[H^+]$ and $[OH^-]$ is always 10^{-14} at 25°C, and the sum of pH and pOH is always 14. If the temperature is other than 25°C, then K_w is different from 1.0×10^{-14}, and a neutral solution will have other than $10^{-7} \ M \ H^+$ and OH^- (see below).

A 10 M HCl solution should have a pH of −1 and pOH of 15.

Some mistakenly believe that it is impossible to have a **negative pH**. There is no theoretical basis for this. A negative pH only means that the hydrogen ion concentration is greater than 1 M. In actual practice, a negative pH is uncommon for two reasons. First, even strong acids may become partially undissociated at high concentrations. For example, 100% H_2SO_4 is so weakly dissociated that it can be stored in iron containers; more dilute H_2SO_4 solutions would contain sufficient protons from dissociation to attack and dissolve the iron. The second reason has to do with the *activity*, which we have chosen to neglect for dilute solutions. Since pH is really $-\log a_{H^+}$ (this is what a pH meter reading is a measure of), a solution that is 1.1 M in H^+ may actually have a positive pH because the activity of the H^+ is less than 1.0 M.[2] This is because at these high concentrations, the activity coefficient is less than unity (although at still higher concentrations the activity coefficient may become greater than unity—see Chapter 5). Nevertheless, there is mathematically no basis for not having a negative pH (or a negative pOH), although it may be rarely encountered in situations relevant to analytical chemistry.

If the concentration of an acid or base is much less than 10^{-7} M, then its contribution to the acidity or basicity will be negligible compared with the contribution from water. The pH of a 10^{-8} M sodium hydroxide solution would therefore not differ significantly from 7. If the concentration of the acid or base is around 10^{-7} M, then its contribution is not negligible and neither is that from water; hence the sum of the two contributions must be taken.

The pH of 10^{-9} M HCl is not 9!

Example 6.6

Calculate the pH and pOH of a 1.0×10^{-7} M solution of HCl.

Solution

Equilibria:

$$HCl \rightarrow H^+ + Cl^-$$
$$H_2O \rightleftharpoons H^+ + OH^-$$
$$[H^+][OH^-] = 1.0 \times 10^{-14}$$
$$[H^+]_{H_2O\ diss.} = [OH^-]_{H_2O\ diss.} = x$$

Since the hydrogen ions contributed from the ionization of water are not negligible compared to the HCl added,

$$[H^+] = C_{HCl} + [H^+]_{H_2O\ diss.}$$

Then,

$$\left([H^+]_{HCl} + x\right)(x) = 1.0 \times 10^{-14}$$
$$\left(1.00 \times 10^{-7} + x\right)(x) = 1.0 \times 10^{-14}$$
$$x^2 + 1.00 \times 10^{-7}x - 1.0 \times 10^{-14} = 0$$

Using the quadratic equation to solve [see Appendix B] or the use of Excel Goal Seek (Section 5.11),

$$x = \frac{-1.00 \times 10^{-7} \pm \sqrt{1.0 \times 10^{-14} + 4(1.0 \times 10^{-14})}}{2} = 6.2 \times 10^{-8}\ M$$

Therefore, the *total* H^+ concentration = $(1.00 \times 10^{-7} + 6.2 \times 10^{-8}) = 1.62 \times 10^{-7}$ M:

$$pH = -\log\ 1.62 \times 10^{-7} = 7 - 0.21 = 6.79$$
$$pOH = 14.00 - 6.79 = 7.21$$

---

[2] As will be seen in Chapter 20, it is also difficult to *measure* the pH of a solution having a negative pH or pOH because high concentrations of acids or bases tend to introduce an error in the measurement by adding a significant and unknown liquid-junction potential in the measurements.

or, since $[OH^-] = x$,

$$pOH = -\log(6.2 \times 10^{-8}) = 8 - 0.79 = 7.21$$

Note that, owing to the presence of the added H^+, the ionization of water is suppressed by 38% by the common ion effect (Le Châtelier's principle). At higher acid (or base) concentrations, the suppression is even greater and the contribution from the water becomes negligible. The contribution from the autoionization of water can be considered negligible if the concentration of protons or hydroxyl ions from an acid or base is 10^{-6} *M or greater.*

We usually neglect the contribution of water to the acidity in the presence of an acid since its ionization is suppressed in the presence of the acid.

The calculation in this example is more academic than practical because *carbon dioxide from the air dissolved in water substantially exceeds these concentrations,* being about 1.2×10^{-5} *M* carbonic acid. Since carbon dioxide in water forms an acid, extreme care would have to be taken to remove and keep this from the water, to have a solution of 10^{-7} *M* acid.

6.5 pH at Elevated Temperatures: Blood pH

It is a convenient fact of nature for students and chemists who deal with acidity calculations and pH scales in aqueous solutions at room temperature that pK_w is an integer number. At 100°C, for example, $K_w = 5.5 \times 10^{-13}$, and a *neutral solution* has

$$[H^+] = [OH^-] = \sqrt{5.5 \times 10^{-13}} = 7.4 \times 10^{-7} \ M$$
$$pH = pOH = 6.13$$
$$pK_w = 12.26 = pH + pOH$$

A neutral solution has pH < 7 above room temperature.

Not all measurements or interpretations are done at room temperature, however, and the temperature dependence of K_w must be taken into account (recall from Chapter 5 that equilibrium constants are temperature dependent). An important example is the pH of the body. The pH of blood at body temperature (37°C) is 7.35 to 7.45. This value represents a slightly more alkaline solution relative to neutral water than the same value would be at room temperature. At 37°C, $K_w = 2.5 \times 10^{-14}$ and $pK_w = 13.60$. The pH (and pOH) of a neutral solution is 13.60/2 = 6.80. The hydrogen ion (and hydroxide ion) concentration is $\sqrt{2.5 \times 10^{-14}} = 1.6 \times 10^{-7} \ M$. Since a neutral solution at 37°C would have pH 6.8, a blood pH of 7.4 is more alkaline at 37°C by 0.2 pH units than it would be at 25°C. This is important when one considers that a change of 0.3 pH units in the body is extreme.

The hydrochloric acid concentration in the stomach is about 0.1 to 0.02 *M*. Since $pH = -\log[H^+]$, the pH at 0.02 *M* would be 1.7. It will be the same *regardless of the temperature* since the hydrogen ion concentration is the same (neglecting solvent volume changes), and the same pH would be measured at either temperature. But, while the pOH would be $14.0 - 1.7 = 12.3$ at 25°C, it is $13.6 - 1.7 = 11.9$ at 37°C.

The pH of blood must be measured at body temperature to accurately reflect the status of blood buffers.

Not only does the temperature affect the ionization of water in the body and therefore change the pH of neutrality, it also affects the ionization constants of the acids and bases from which the buffer systems in the body are derived. As we shall see later in the chapter, this influences the pH of the buffers, and so a blood pH of 7.4 measured at 37°C will not be the same when measured at room temperature, in contrast to the stomach pH, whose value was determined by the concentration of a strong acid. For this reason, measurement of blood pH for diagnostic purposes is generally done at 37°C (see Chapter 20).

6.6 Weak Acids and Bases—What is the pH?

We have limited our calculations so far to strong acids and bases in which ionization is assumed to be complete. Since the concentration of H^+ or OH^- is determined readily from the concentration of the acid or base, the calculations are straightforward. As seen in Equation 6.3, weak acids (or bases) are only partially ionized. While mineral (inorganic) acids and bases such as HCl, $HClO_4$, HNO_3, and NaOH are strong electrolytes that are totally ionized in water; most organic acids and bases, as found in clinical applications, are weak.

The ionization constant can be used to calculate the amount ionized and, from this, the pH. The acidity constant for acetic acid at 25°C is 1.75×10^{-5}:

$$\frac{[H^+][OAc^-]}{[HOAc]} = 1.75 \times 10^{-5} \qquad (6.20)$$

When acetic acid ionizes, it dissociates to equal portions of H^+ and OAc^- by such an amount that the computation on the left side of Equation 6.20 will always be equal to 1.75×10^{-5}:

$$HOAc \rightleftharpoons H^+ + OAc^- \qquad (6.21)$$

If the original concentration of acetic acid is C and the concentration of ionized acetic acid species (H^+ and OAc^-) is x, then the final concentration for each species at equilibrium is given by

$$\begin{array}{cc} HOAc & \rightleftharpoons \quad H^+ \; + \; OAc^- \\ (C-x) & x \qquad x \end{array} \qquad (6.22)$$

Example 6.7

Calculate the pH and pOH of a 1.00×10^{-3} M solution of acetic acid.

Solution

$$HOAc \rightleftharpoons H^+ + OAc^-$$

The concentrations of the various species in the form of an ICE table are as follows:

	[HOAc]	[H^+]	[OAc^-]
Initial	1.00×10^{-3}	0	0
Change (x = mmol/mL HOAc ionized)	$-x$	$+x$	$+x$
Equilibrium	$1.00 \times 10^{-3} - x$	x	x

From Equation 6.20

$$\frac{(x)(x)}{1.00 \times 10^{-3} - x} = 1.75 \times 10^{-5}$$

The solution is that of a quadratic equation. If less than about 10 or 15% of the acid is ionized, the expression may be simplified by neglecting x compared with C (10^{-3} M in this case). This is an arbitrary (and not very demanding) criterion. The simplification applies if K_a is *smaller than about* $0.01C$, that is, smaller than 10^{-4} at $C = 0.01$ M, 10^{-3} at $C = 0.1$ M, and so forth. Under these conditions, the error in calculation is 5%

If $C_{HA} > 100K_a$, x can be neglected compared to C_{HA}.

or less (results come out too high), and within the probable accuracy of the equilibrium constant. Our calculation simplifies to

$$\frac{x^2}{1.00 \times 10^{-3}} = 1.75 \times 10^{-5}$$
$$x = 1.32 \times 10^{-4} \, M \equiv [\text{H}^+]$$

Therefore,

$$\text{pH} = -\log(1.32 \times 10^{-4}) = 4 - \log 1.32 = 4 - 0.12 = 3.88$$
$$\text{pOH} = 14.00 - 3.88 = 10.12$$

The absolute accuracy of pH measurements is no better than 0.02 pH units. See Chapter 20.

The simplification in the calculation does not lead to serious errors, particularly since equilibrium constants are often not known to a high degree of accuracy (frequently no better than ±10%). In the above example, solution of the quadratic equation results in $[\text{H}^+] = 1.26 \times 10^{-4} \, M$ (5% less) and pH = 3.91. This pH is within 0.03 unit of that calculated using the simplification, which is near the limit of accuracy to which pH measurements can be made. It is almost certainly as close a calculation as is justified in view of the experimental errors in K_a or K_b values and the fact that we are using concentrations rather than activities in the calculations. In our calculations, we also neglected the contribution of hydrogen ions from the ionization of water (which was obviously justified); this is generally permissible except for very dilute ($< 10^{-6} \, M$) or very weak ($K_a < 10^{-12}$) acids.

Similar equations and calculations hold for weak bases. It should be noted, however, a computational tool like Goal Seek or Excel Solver can solve quadratic (or higher-order) equations so easily that increasingly it is easier to solve the original equation without approximation than to reflect on whether the approximation may be valid. A Goal Seek solution of Example 6.7 can be found in the **website** section of this chapter.

Example 6.8

The basicity constant K_b for ammonia is 1.75×10^{-3} at 25°C. (It is only coincidental that this is equal to K_a for acetic acid.) Calculate the pH and pOH for a $1.00 \times 10^{-5} \, M$ solution of ammonia.

Solution

$$\text{NH}_3 \quad + \text{H}_2\text{O} \rightleftharpoons \text{NH}_4^+ + \text{OH}^-$$
$$(1.00 \times 10^{-5} - x) \qquad\qquad x \qquad x$$

$$\frac{[\text{NH}_4^+][\text{OH}^-]}{[\text{NH}_3]} = 1.75 \times 10^{-3}$$

The same rule applies for the approximation applied for a weak acid. Thus,

$$\frac{(x)(x)}{1.00 \times 10^{-5}} = 1.75 \times 10^{-3}$$
$$x = 1.32 \times 10^{-4} \, M = [\text{OH}^-]$$
$$\text{pOH} = -\log 1.32 \times 10^{-4} = 3.88$$
$$\text{pH} = 14.00 - 3.88 = 10.12$$

For an Excel Goal Seek solution of Example 6.8 without approximation, see the chapter's **website**.

Professor's Favorite Way

Contributed by Professor W. Rudolph Seitz, University of New Hampshire

Handling Bases as Their Conjugate Acids

Many tabulations of ionization constants of acids and bases list only acidity constants. That is, bases are listed in the form of their conjugate acid (see Equation 6.1). Table C.2b in Appendix C lists the corresponding acid formulas and acidity constants for the bases listed in Table C.2a. The amino groups of amine compounds are protonated (+1 charge) and can be treated as any other weak acid to give the corresponding conjugate base.

So, one can consider that there are two types of monoprotic acids, one that is uncharged (HA), e.g., HOAc, and one that has plus charge (HA^+), e.g., NH_4^+. Similarly, there are three types of diprotic acids, with charges of 0 (H_2A), e.g., oxalic acid $H_2C_2O_4$, 1 + (H_2A^+), e.g., glycinium ion $^+NH_3CH_2COOH$ and 2 + (H_2A^{2+}), e.g., the ethylenediammonium ion, $^+NH_3C_2H_4NH_3^+$. Hence, the protonated form of ammonia is NH_4^+, and the corresponding acidity constant is 5.71×10^{-10}. In order to calculate the pH of an ammonia solution (as in Example 6.8), then, the K_b of the conjugate base form of the acid NH_4^+ is calculated from $K_b = K_w/K_a (K_b = 1.00 \times 10^{-14}/5.71 \times 10^{-10} = 1.75 \times 10^{-5})$.

For a diprotic acid–base pair like ethylenediamine, $NH_2C_2H_4NH_2$, and its acid forms, the protonated forms are $^+NH_3C_2H_4NH_3^+$ and $NH_2C_2H_4NH_3^+$ (each protonated amine group has a +1 charge). The protonated amine groups dissociate stepwise to give $NH_2C_2H_4NH_3^+$ and $NH_2C_2H_4NH_2$, with $K_{a1} = 1.41 \times 10^{-7}$ and $K_{a2} = 1.18 \times 10^{-10}$. For the conjugate base forms, $K_{b1} = K_w/K_{a2}$ and $K_{b2} = K_w/K_{a1}$). Since K_{a1} is larger than K_{a2}, then K_{b2} is smaller than K_{b1}.

In this book, while dealing with α-values and in all the Excel exercises, we have followed this approach: considered all problems in terms of acid dissociation constants. It is suggested that you do so as well.

6.7 Salts of Weak Acids and Bases—They Aren't Neutral

The salt of a weak acid, for example, NaOAc, is a strong electrolyte, like (almost) all salts, and completely ionizes. In addition, the anion of the salt of a weak acid is a **Brønsted base**, which will accept protons. It partially hydrolyzes in water (a Brønsted acid) to form hydroxide ion and the corresponding undissociated acid. For example,

The hydrolysis of OAc^- is no different than the "ionization" of NH_3 in Example 6.8

$$OAc^- + H_2O \rightleftharpoons HOAc + OH^- \qquad (6.23)$$

The HOAc here is undissociated and therefore does not contribute to the pH. This ionization is also known as **hydrolysis** of the salt ion. Because it hydrolyzes, sodium acetate is a weak base (the conjugate base of acetic acid). The ionization constant for Equation 6.23 is equal to the basicity constant of the salt anion. The weaker the conjugate acid, the stronger the conjugate base, that is, the more strongly the salt will combine with a proton, as from the water, to shift the ionization in Equation 6.23 to the right. *Equilibria for these Brønsted bases are treated identically to the weak bases we have just considered.* We can write an equilibrium constant:

$$K_H = K_b = \frac{[HOAc][OH^-]}{[OAc^-]} \qquad (6.24)$$

K_H is called the **hydrolysis constant** of the salt and is the same as the basicity constant. We will use K_b to emphasize that these salts are treated the same as for any other weak base.

The value of K_b can be calculated from K_a of acetic acid and K_w if we multiply both the numerator and denominator by $[H^+]$:

$$K_b = \frac{[HOAc]\,\overline{[OH^-]}}{[OAc^-]} \cdot \frac{\overline{[H^+]}}{[H^+]} \tag{6.25}$$

The quantity inside the dashed line is K_w and the remainder is $1/K_a$. Hence,

$$K_b = \frac{K_w}{K_a} = \frac{1.0 \times 10^{-14}}{1.75 \times 10^{-5}} = 5.7 \times 10^{-10} \tag{6.26}$$

We see from the small K_b that the acetate ion is quite a weak base with only a small fraction of ionization. *The product of K_a of any weak acid and K_b of its conjugate base is always equal to K_w:*

$$\boxed{K_a K_b = K_w} \tag{6.27}$$

You will understand that this is merely a restatement of what was stated in the previous section in treating a base in terms of its conjugate acid. The product of the acid dissociation constant of any acid and the base dissociation constant of its conjugate base is K_w.

For any salt of a weak acid HA that hydrolyzes in water,

$$A^- + H_2O \rightleftharpoons HA + OH^- \tag{6.28}$$

$$\boxed{\frac{[HA][OH^-]}{[A^-]} = \frac{K_w}{K_a} = K_b} \tag{6.29}$$

The pH of such a salt (a Brønsted base) is calculated in the same manner as for any other weak base. When the salt hydrolyzes, it forms an equal amount of HA and OH^-. If the original concentration of A^- is C_{A^-}, then

$$\begin{array}{cccccc} A^- & + & H_2O & \rightleftharpoons & HA & + & OH^- \\ (C_{A^-} - x) & & & & x & & x \end{array} \tag{6.30}$$

The quantity x can be neglected compared to C_{A^-} if $C_{A^-} > 100K_b$, which will generally be the case for such weakly ionized bases.

We can solve for the OH^- concentration using Equation 6.30:

$$\frac{[OH^-][OH^-]}{C_{A^-}} = \frac{K_w}{K_a} = K_b \tag{6.31}$$

Compare this with the algebraic setup in Example 6.8. They are identical:

$$\boxed{[OH^-] = \sqrt{\frac{K_w}{K_a} \cdot C_{A^-}} = \sqrt{K_b \cdot C_{A^-}}} \tag{6.32}$$

This equation holds only if $C_{A^-} > 100K_b$, and x can be neglected compared to C_{A^-}. If this is not the case, then the quadratic formula must be solved as for other bases in this situation.

Example 6.9

Calculate the pH of a 0.10 M solution of sodium acetate.

Solution

Write the equilibria

$$NaOAc \rightarrow Na^+ + OAc^- \text{ (ionization)}$$

$$OAc^- + H_2O \rightleftharpoons HOAc + OH^- \text{ (hydrolysis)}$$

Compare this base "ionization" with that of NH_3, Example 6.8.

Write the equilibrium constant

$$\frac{[HOAc][OH^-]}{[OAc^-]} = K_b = \frac{K_w}{K_a} = \frac{1.0 \times 10^{-14}}{1.75 \times 10^{-5}} = 5.7 \times 10^{-10}$$

Let x represent the concentration of HOAc and OH^- at equilibrium. Then, at equilibrium,

$$[HOAc] = [OH^-] = x$$

$$[OAc^-] = C_{OAc^-} - x = 0.10 - x$$

Since $C_{OAc^-} \gg K_b$, neglect x compared to C_{OAc^-}. Then,

$$\frac{(x)(x)}{0.10} = 5.7 \times 10^{-10}$$

$$x = \sqrt{5.7 \times 10^{-10} \times 0.10} = 7.6 \times 10^{-6} \, M$$

Compare this last step with Equation 6.32. Also, compare the entire setup and solution with those in Example 6.8. The HOAc formed is undissociated and does not contribute to the pH:

$$[OH^-] = 7.6 \times 10^{-6} \, M$$

$$[H^+] = \frac{1.0 \times 10^{-14}}{7.6 \times 10^{-6}} = 1.3 \times 10^{-9} \, M$$

$$pH = -\log 1.3 \times 10^{-9} = 9 - 0.11 = 8.89$$

For an Excel Goal Seek solution of Example 6.9 without approximation, see the text website.

Similar equations can be derived for the cations of salts of weak bases (the salts are completely dissociated). These are **Brønsted acids** and ionize (hydrolyze) in water:

$$BH^+ + H_2O \rightleftharpoons B + H_3O^+ \tag{6.33}$$

The B is undissociated and does not contribute to the pH. The acidity constant is

$$K_H = K_a = \frac{[B][H_3O^+]}{[BH^+]} \tag{6.34}$$

The acidity constant (hydrolysis constant) can be derived by multiplying the numerator and denominator by $[OH^-]$:

$$K_a = \frac{[B]}{[BH^+]} \cdot \frac{[H_3O^+]}{[OH^-]} \cdot \frac{[OH^-]}{[OH^-]} \tag{6.35}$$

Again, the quantity inside the dashed lines is K_w, while the remainder is $1/K_b$. Therefore,

$$\frac{[B][H_3O^+]}{[BH^+]} = \frac{K_w}{K_b} = K_a \tag{6.36}$$

and for NH_4^+,

$$K_a = \frac{K_w}{K_b} = \frac{1.0 \times 10^{-14}}{1.75 \times 10^{-5}} = 5.7 \times 10^{-10} \tag{6.37}$$

We could, of course, have derived K_a from Equation 6.27. It is again coincidence that the numerical value of K_a for NH_4^+ equals K_b for OAc^-.

The salt of a weak base ionizes to form equal amounts of B and H_3O^+ (H^+ if we disregard hydronium ion formation as was done previously). We can therefore solve for the hydrogen ion concentration (by assuming $C_{BH^+} > 100K_a$:

$$\frac{[H^+][H^+]}{C_{BH^+}} = \frac{K_w}{K_b} = K_a \tag{6.38}$$

$$[H^+] = \sqrt{\frac{K_w}{K_b} \cdot C_{BH^+}} = \sqrt{K_a \cdot C_{BH^+}} \tag{6.39}$$

Again, this equation only holds if $C_{BH^+} > 100K_a$. Otherwise, the quadratic formula must be solved.

Note: One can obtain K_a directly from a list of acidity constants, as in Table C2.b in Appendix C for the acid equilibrium as given in Equation 6.33; substituting in Equation 6.34 and solving for the hydrogen ion concentration gives Equation 6.39.

Example 6.10

Calculate the pH of a 0.25 M solution of ammonium chloride.

Solution

Write the equilibria

$$NH_4Cl \rightarrow NH_4^+ + Cl^- \text{ (ionization)}$$
$$NH_4^+ + H_2O \rightleftharpoons NH_4OH + H^+ \text{ (hydrolysis)}$$
$$(NH_4^+ + H_2O \rightleftharpoons NH_3 + H_3O^+)$$

Write the equilibrium constant

$$\frac{[NH_4OH][H^+]}{[NH_4^+]} = K_a = \frac{K_w}{K_b} = \frac{1.0 \times 10^{-14}}{1.75 \times 10^{-5}} = 5.7 \times 10^{-10}$$

Let x represent the concentration of $[NH_4OH]$ and $[H^+]$ at equilibrium. Then, at equilibrium,

$$[NH_4OH] = [H^+] = x$$
$$[NH_4^+] = C_{NH_4^+} - x = 0.25 - x$$

Since $C_{NH_4^+} \gg K_a$, neglect x compared to $C_{NH_4^+}$. Then,

$$\frac{(x)(x)}{0.25} = 5.7 \times 10^{-10}$$
$$x = \sqrt{5.7 \times 10^{-10} \times 0.25} = 1.2 \times 10^{-5} \, M$$

Compare this last step with Equation 6.39. Also, compare the entire setup and solution with those in Example 6.7. The NH_4OH formed is undissociated and does not contribute to the pH:

$$[H^+] = 1.2 \times 10^{-5}\,M$$

$$pH = -\log(1.2 \times 10^{-5}) = 5 - 0.08 = 4.92$$

For an Excel Goal Seek solution of Example 6.8 without approximation, see the chapter's **website**.[3]

6.8 Buffers—Keeping the pH Constant (or Nearly So)

A **buffer** is defined as a solution that resists change in pH when a small amount of an acid or base is added or when the solution is diluted. While carrying out a reaction, this is very useful for maintaining the pH within an optimum range. A buffer solution consists of a mixture of a weak acid and its conjugate base, or a weak base, and its conjugate acid at predetermined concentrations or ratios. That is, we have a mixture of a weak acid and its salt or a weak base and its salt. Consider an acetic acid–acetate buffer. The equilibrium that governs this system is

$$HOAc \rightleftharpoons H^+ + OAc^-$$

But now, since we have added a supply of acetate ions to the system (e.g., from sodium acetate), the hydrogen ion concentration is no longer equal to the acetate ion concentration. The hydrogen ion concentration is

$$[H^+] = K_a \frac{[HOAc]}{[OAc^-]} \tag{6.40}$$

Taking the negative logarithm of each side of this equation, we have

$$-\log[H^+] = -\log K_a - \log \frac{[HOAc]}{[OAc^-]} \tag{6.41}$$

$$pH = pK_a - \log \frac{[HOAc]}{[OAc^-]} \tag{6.42}$$

Upon inverting the last log term, it becomes positive:

$$pH = pK_a + \log \frac{[OAc^-]}{[HOAc]} \tag{6.43}$$

This form of the ionization constant equation is called the **Henderson–Hasselbalch equation**. It is useful for calculating the pH of a weak acid solution containing its salt. A general form can be written for a weak acid HA that ionizes to its salt, A^-, and H^+:

$$HA \rightleftharpoons H^+ + A^- \tag{6.44}$$

$$\boxed{pH = pK_a + \log \frac{[A^-]}{[HA]}} \tag{6.45}$$

The pH of a buffer is determined by the ratio of the conjugate acid–base pair concentrations.

[3]This is an unedited student video and contains some errors of statement, e,g., it talks about K_a of NH_3 whereas it should really refer to it as K_a of NH_4^+; it mistakenly states that bases like to take up electrons whereas bases of course like to take up hydrogen ions, etc. But despite these errors of statement, it is a nicely set up example of a correctly solved problem that shows the use of Excel Goal Seek!

$$pH = pK_a + \log \frac{[\text{conjugate base}]}{[\text{acid}]} \qquad (6.46)$$

$$pH = pK_a + \log \frac{[\text{proton acceptor}]}{[\text{proton donor}]} \qquad (6.47)$$

Example 6.11

Calculate the pH of a buffer prepared by adding 10 mL of 0.10 M acetic acid to 20 mL of 0.10 M sodium acetate.

Solution

We need to calculate the concentration of the acid and salt in the solution. The final volume is 30 mL:

$$M_1 \times mL_1 = M_2 \times mL_2$$

For HOAc,

$$0.10 \text{ mmol/mL} \times 10 \text{ mL} = M_{\text{HOAc}} \times 30 \text{ mL}$$

$$M_{\text{HOAc}} = 0.033 \text{ mmol/mL}$$

For OAc$^-$,

$$0.10 \text{ mmol/mL} \times 20 \text{ mL} = M_{\text{OAc}^-} \times 30 \text{ mL}$$

$$M_{\text{OAc}^-} = 0.067 \text{ mmol/mL}$$

The ionization of the acid is suppressed by the salt and can be neglected.

Some of the HOAc dissociates to H$^+$ + OAc$^-$, and the equilibrium concentration of HOAc would be the amount added (0.033 M) minus the amount dissociated, while that of OAc$^-$ would be the amount added (0.067 M) plus the amount of HOAc dissociated. However, *the amount of acid dissociated is very small*, particularly in the presence of the added salt (ionization suppressed by the common ion effect), and can be neglected. Hence, we can assume the added concentrations to be the equilibrium concentrations:

$$pH = -\log K_a + \log \frac{[\text{proton acceptor}]}{[\text{proton donor}]}$$

$$pH = -\log(1.75 \times 10^{-5}) + \log \frac{0.067 \text{ mmol/mL}}{0.033 \text{ mmol/mL}}$$

$$= 4.76 + \log 2.0$$

$$= 5.06$$

We can use millimoles of acid and salt in place of molarity. Because the terms appear in a ratio, as long as the units are the same, they will cancel out. But it has to relate to moles or molarity, not mass.

We could have shortened the calculation by recognizing that in the log term the volumes cancel. So we can take the ratio of millimoles only:

$$\text{mmol}_{\text{HOAc}} = 0.10 \text{ mmol/mL} \times 10 \text{ mL} = 1.0 \text{ mmol}$$

$$\text{mmol}_{\text{OAc}^-} = 0.10 \text{ mmol/mL} \times 20 \text{ mL} = 2.0 \text{ mmol}$$

$$H = 4.76 + \log \frac{2.0 \text{ mmol}}{1.0 \text{ mmol}} = 5.06$$

The mixture of a weak acid and its salt may also be obtained by mixing an excess of weak acid with some strong base to produce the salt by neutralization, or by mixing an excess of salt with strong acid to produce the weak acid component of the buffer.

Example 6.12

Calculate the pH of a solution prepared by adding 25 mL of 0.10 M sodium hydroxide to 30 mL of 0.20 M acetic acid (this would actually be a step in a typical titration).

Solution

$$\text{mmol HOAc} = 0.20\ M \times 30\ \text{mL} = 6.0\ \text{mmol}$$
$$\text{mmol NaOH} = 0.10\ M \times 25\ \text{mL} = 2.5\ \text{mmol}$$

These react as follows:

$$\text{HOAc} + \text{NaOH} \rightleftharpoons \text{NaOAc} + \text{H}_2\text{O}$$

After reaction,

$$\text{mmol NaOAc} = 2.5\ \text{mmol}$$
$$\text{mmol HOAc} = 6.0 - 2.5 = 3.5\ \text{mmol}$$
$$\text{pH} = 4.76 + \log \frac{2.5}{3.5} = 4.61$$

Keep track of millimoles of reactants!

The **buffering mechanism** for a mixture of a weak acid and its salt can be explained as follows. The pH is governed by the logarithm of the ratio of the salt and acid:

$$\text{pH} = \text{constant} + \log \frac{[\text{A}^-]}{[\text{HA}]} \tag{6.48}$$

If the solution is diluted, the ratio remains constant, and so the pH of the solution does not change.[4] If a small amount of a strong acid is added, it will combine with an equal amount of the A^- to convert it to HA. That is, in the equilibrium $\text{HA} \rightleftharpoons \text{H}^+ + \text{A}^-$, Le Châtelier's principle dictates added H^+ will combine with A^- to form HA, with the equilibrium lying far to the left if there is an excess of A^-. The change in the ratio $[\text{A}^-]/[\text{HA}]$ is small and hence the change in pH is small. If the acid had been added to an unbuffered solution (e.g., a solution of NaCl), the pH would have decreased markedly. If a small amount of a strong base is added, it will combine with part of the HA to form an equivalent amount of A^-. Again, the change in the ratio is small.

The amount of acid or base that can be added without causing a large change in pH is governed by the **buffering capacity** of the solution. This is determined by the concentrations of HA and A^-. The higher their concentrations, the more acid or base the solution can tolerate. The buffer intensity or buffer index of a solution is defined as

$$\beta = dC_B/d\text{pH} = -dC_A/d\text{pH} \tag{6.49}$$

where dC_B and dC_A represent the number of moles per liter of strong base or acid, respectively, needed to bring about a pH change of dpH. Although the terms *buffer intensity* and *buffer capacity* are often used interchangeably, the buffer capacity is the integrated form of buffer intensity (e.g., the amount of strong acid/base needed to change the pH by a certain finite amount) and is always a positive number. The larger it is, the more resistant the solution is to pH change. For a simple monoprotic weak acid/conjugate base buffer solutions of concentration greater than 0.001 M, the buffer intensity is approximated by:

$$\beta = 2.303 \frac{C_{\text{HA}} C_{\text{A}^-}}{C_{\text{HA}} + C_{\text{A}^-}} \tag{6.50}$$

Dilution does not change the ratio of the buffering species.

The buffering capacity increases with the concentrations of the buffering species.

---

[4] In actuality, the pH will *increase* slightly because the activity coefficient of the salt has been increased by decreasing the ionic strength. The activity of an uncharged molecule (i.e., undissociated acid) is equal to its molarity (see Chapter 5), and so the ratio increases, causing a slight increase in pH. See the end of the chapter.

See Chapter 7, Section 7.11 for a derivation of buffer intensity.

where C_{HA} and C_{A^-} represent the analytical concentrations of the acid and its salt, respectively. Thus, if we have a mixture of 0.10 mol/L acetic acid and 0.10 mol/L sodium acetate, the buffer intensity is

$$\beta = 2.303 \frac{0.10 \times 0.10}{0.10 + 0.10} = 0.050 \text{ mol/L per pH}$$

If we add 0.0050 mol/L solid sodium hydroxide, the change in pH is

$$d\text{pH} = dC_B / \beta = 0.0050/0.050 = 0.10 = \Delta\text{pH}$$

In addition to concentration, the buffer intensity is governed by the *ratio* of HA to A⁻. It is *maximum* when the ratio is unity, that is, when the pH = pK_a:

The buffer intensity is maximum at pH = pK_a.

$$\text{pH} = pK_a + \log \frac{1}{1} = pK_a \qquad (6.51)$$

This corresponds to the midpoint of a titration of a weak acid. In general, provided the concentration is not too dilute, the buffering capacity is satisfactory over a *pH range of $pK_a \pm 1$.* We will discuss the buffering capacity in more detail in Chapter 7, when the titration curves of weak acids are discussed.

Calculating the pH of a Buffer when Strong Acid or Base is Added

Contributed by Professor Kris Varazo, Francis Marion University, Florence, South Carolina

 # Professor's Favorite Example

Example 6.13

As an example, suppose you have 100 mL of a buffer containing 0.100 *M* acetic acid and 0.0500 *M* sodium acetate. Calculate the pH of the buffer when 3.00 mL of 1.00 *M* HCl is added to it.

As a first step, calculate the pH of the buffer before adding the strong acid using the Henderson–Hasselbalch equation:

$$\text{pH} = pK_a + \log \frac{[\text{A}^-]}{[\text{HA}]}$$

All we need to determine the pH of this buffer is the pK_a of acetic acid, which is 4.76:

$$\text{pH} = 4.76 + \log \frac{0.0500}{0.100} = 4.46$$

Remember that adding acid to a solution will necessarily lower the pH, so we should expect the pH of the buffer to be lower than 4.46. The best way to solve this problem is to calculate the moles of acetic acid, moles of sodium acetate, and added moles of HCl:

$$100 \text{ mL} \times \frac{0.100 \text{ moles}}{1000 \text{ mL}} = 0.0100 \text{ moles acetic acid}$$

$$100 \text{ mL} \times \frac{0.0500 \text{ moles}}{1000 \text{ mL}} = 0.00500 \text{ moles sodium acetate}$$

$$3.00 \text{ mL} \times \frac{1.00 \text{ moles}}{1000 \text{ mL}} = 0.00300 \text{ moles hydrochloric acid}$$

Note that we expressed molarity in moles per 1000 mL instead of liters. Even though the Henderson–Hasselbalch equation uses molar concentrations, there is only one volume of buffer solution, so we can simply use the mole values we just calculated.

We also need to know the chemical reaction occurring when strong acid is added to the buffer:

$$A^- + H^+ \rightarrow HA$$

The reaction says that the acetate ion in the buffer will react with the added strong acid, and the moles of acetate will decrease and the moles of acetic acid will increase. How much will the decrease and increase be? It is equal to the amount of strong acid added. We can now write the Henderson–Hasselbalch equation and account for the decrease in moles of acetate and increase in moles of acetic acid:

$$pH = pK_a + \log \frac{(\text{moles A}^- - \text{moles H}^+ \text{ added})}{(\text{moles HA} + \text{moles H}^+ \text{ added})}$$

$$pH = 4.76 + \log \frac{(0.00500 - 0.00300)}{(0.0100 + 0.00300)} \qquad (6.52)$$

$$pH = 4.22$$

The new pH of the buffer is lower than the original value, and it makes sense because adding strong acid to a solution, even a buffer, will cause the pH to decrease. This approach also works in reverse, when you add a strong base to a buffer. In this case, the pH of the buffer will increase, and the relevant chemical reaction is:

$$HA + OH^- \rightarrow H_2O + A^-$$

This time the moles of acetic acid will decrease and the moles of acetate will increase. The Henderson–Hasselbalch equation can be written in the following form to solve for the pH:

$$pH = pKa + \log \frac{(\text{moles A}^- + \text{moles OH}^- \text{ added})}{(\text{moles HA} - \text{moles OH}^- \text{ added})} \qquad (6.53)$$

Note that a buffer can resist a pH change, even when there is added an amount of strong acid or base greater (in moles) than the equilibrium amount of H^+ or OH^- (in moles) in the buffer. For example, in Example 6.13, the pH of the buffer is 4.46 and $[H^+] = 3.5 \times 10^{-5} \, M$, and millimoles $H^+ = (3.5 \times 10^{-5} \, \text{mmol/mL}) (100 \, \text{mL}) = 3.5 \times 10^{-3}$ mmol (in equilibrium with the buffer components). We added 3.00 mmol H^+, far in excess of this. However, due to the reserve of buffer components (OAc^- to react with H^+ in this case), the added H^+ is consumed so that the pH remains relatively constant, *so long as we do not exceed the amount of buffer reserve.*

Similar calculations apply for mixtures of a weak base and its salt. We can consider the equilibrium between the base B and its conjugate acid BH^+ and write a K_a for the conjugate (Brønsted) acid:

$$BH^+ \rightleftharpoons B + H^+ \qquad (6.54)$$

$$K_a = \frac{[B][H^+]}{[BH^+]} = \frac{K_w}{K_b} \qquad (6.55)$$

The logarithmic Henderson–Hasselbalch form is derived exactly as above:

$$[H^+] = K_a \cdot \frac{[BH^+]}{[B]} = \frac{K_w}{K_b} \cdot \frac{[BH^+]}{[B]} \qquad (6.56)$$

$$-\log[H^+] = -\log K_a - \log \frac{[BH^+]}{[B]} = -\log \frac{K_w}{K_b} - \log \frac{[BH^+]}{[B]} \qquad (6.57)$$

$$\boxed{pH = pK_a + \log \frac{[B]}{[BH^+]} = (pK_w - pK_b) + \log \frac{[B]}{[BH^+]}} \qquad (6.58)$$

$$\boxed{\text{pH} = \text{p}K_a + \log \frac{[\text{proton acceptor}]}{[\text{proton donor}]} = (\text{p}K_w - \text{p}K_b) + \log \frac{[\text{proton acceptor}]}{[\text{proton donor}]}} \quad (6.59)$$

Since $\text{pOH} = \text{p}K_w - \text{pH}$, we can also write, by subtracting either Equation 6.58 or Equation 6.59 from $\text{p}K_w$,

$$\boxed{\text{pOH} = \text{p}K_b + \log \frac{[\text{BH}^+]}{[\text{B}]} = \text{p}K_b + \log \frac{[\text{proton donor}]}{[\text{proton acceptor}]}} \quad (6.60)$$

$\text{p}K_a = 14 - \text{p}K_b$ for a weak base. The alkaline buffering capacity is maximum at $\text{pOH} = \text{p}K_b(\text{pH} = \text{p}K_a).$

A mixture of a weak base and its salt acts as a buffer in the same manner as a weak acid and its salt. When a strong acid is added, it combines with some of the base B to form the salt BH^+. Conversely, a base combines with BH^+ to form B. Since the change in the ratio will be small, the change in pH will be small. Again, the buffering capacity is maximum at a pH equal to $\text{p}K_a = 14 - \text{p}K_b$ (or at $\text{pOH} = \text{p}K_b$), with a useful range of $\text{p}K_a \pm 1$. Although we show the calculations in terms of $\text{p}K_b$ as well, we recommend that you do all calculations using the $\text{p}K_a$ of the conjugate acid; consistency keeps you in the comfort zone.

When a buffer is diluted, the pH will not change appreciably because the ratio [proton donor]/[proton acceptor] will remain the same.[5]

Example 6.14

Calculate the volume of concentrated ammonia and the weight of ammonium chloride you would have to take to prepare 100 mL of a buffer at pH 10.00 if the final concentration of salt is to be 0.200 M.

Solution

We want 100 mL of 0.200 M NH_4Cl. Therefore, mmol NH_4Cl = 0.200 mmol/mL × 100 mL = 20.0 mmol

$$\text{mg NH}_4\text{Cl} = 20.0 \text{ mmol} \times 53.5 \text{ mg/mmol} = 1.07 \times 10^3 \text{mg}$$

$\text{p}K_a$ can be obtained directly from K_a given in Table C.2b in Appendix C.

Therefore, we need 1.07 g NH_4Cl. We calculate the concentration of NH_3 by

$$\text{pH} = \text{p}K_a + \log \frac{[\text{proton acceptor}]}{[\text{proton donor}]}$$

$$= (14.00 - \text{p}K_b) + \log \frac{[\text{NH}_3]}{[\text{NH}_4^+]}$$

$$10.0 = (14.00 - 4.76) + \log \frac{[\text{NH}_3]}{0.200 \text{ mmol/mL}}$$

$$\log \frac{[\text{NH}_3]}{0.200 \text{ mmol/mL}} = 0.76$$

$$\frac{[\text{NH}_3]}{0.200 \text{ mmol/mL}} = 10^{0.76} = 5.8$$

$$[\text{NH}_3] = (0.200)(5.8) = 1.1_6 \text{ mmol/mL}$$

---

[5]Buying one bottle of buffer and keeping on diluting and reusing it is not a good business plan, however. As you dilute, you lose buffering capacity. If you get it really dilute, dissolution of atmospheric CO_2 and autoionization of water will affect the buffer pH.

The molarity of concentrated ammonia is 14.8 M. Therefore, remember, $C_1V_1 = C_2V_2$,

$$100 \text{ mL} \times 1.16 \text{ mmol/mL} = 14.8 \text{ mmol/mL} \times \text{mL NH}_3$$

$$\text{mL NH}_3 = 7.8 \text{ mL}$$

Example 6.15

How many grams ammonium chloride and how many milliliters 3.0 M sodium hydroxide should be added to 200 mL water and diluted to 500 mL to prepare a buffer of pH 9.50 with a salt concentration of 0.10 M?

Solution

We need the ratio of $[NH_3]/[NH_4^+]$. From Example 6.14.

$$\text{pH} = pK_a + \log \frac{[NH_3]}{[NH_4^+]} = 9.24 + \log \frac{[NH_3]}{[NH_4^+]}$$

$$9.50 = 9.24 + \log \frac{[NH_3]}{[NH_4^+]}$$

$$\log \frac{[NH_3]}{[NH_4^+]} = 0.26$$

$$\frac{[NH_3]}{[NH_4^+]} = 10^{0.26} = 1.8$$

The final concentration of NH_4^+ is 0.10 M, so

$$[NH_3] = (1.8)(0.10) = 0.18 \ M$$

$$\text{mmol NH}_4^+ \text{ in final solution} = 0.10 \ M \times 500 \text{ mL} = 50 \text{ mmol}$$

$$\text{mmol NH}_3 \text{ in final solution} = 0.18 \ M \times 500 \text{ mL} = 90 \text{ mmol}$$

The NH_3 is formed by reacting an equal number of millimoles of NH_4Cl with NaOH. Therefore, a total of $50 + 90 = 140$ mmol NH_4Cl must be taken:

$$\text{mg NH}_4Cl = 140 \text{ mmol} \times 53.5 \text{ mg/mmol} = 7.49 \times 10^3 \text{ mg} = 7.49 \text{ g}$$

The volume of NaOH needed to react with NH_4^+ to give 90 millimoles of NH_3 is:

$$3.0 \ M \times x \text{ mL} = 90 \text{ mmol}$$
$$x = 30 \text{ mL NaOH}$$

We see that a buffer solution for a given pH is prepared by choosing a weak acid (or a weak base) and its salt, with a pK_a *value near the pH that we want*. There are a number of such acids and bases, and any pH region can be buffered by a proper choice of these. A weak acid and its salt give the best buffering in acid solution, and a weak base and its salt give the best buffering in alkaline solution. Some useful buffers for measurements in physiological solutions are described below. National Institute of Standards and Technology (NIST) buffers used for calibrating pH electrodes are described in Chapter 20.

You may have wondered why, in buffer mixtures, the salt does not react with water to hydrolyze as an acid or base. This is because the reaction is suppressed by the presence of the acid or base. In Equation 6.28, the presence of appreciable amounts of

Select a buffer with a pK_a value near the desired pH.

Buffer salts do not hydrolyze appreciably.

See Chapter 20 for a list of NIST standard buffers.

either HA or OH$^-$ will suppress the ionization almost completely. In Equation 6.33, the presence of either B or H$_3$O$^+$ will suppress the ionization.

6.9 Polyprotic Acids and Their Salts

Many acids or bases are polyfunctional, that is, have more than one ionizable proton or hydroxide ion. These substances ionize stepwise, and an equilibrium constant can be written for each step. Consider, for example, the ionization of phosphoric acid:

The stepwise K_a values of polyprotic acids get progressively smaller as the increased negative charge makes dissociation of the next proton more difficult.

$$H_3PO_4 \rightleftharpoons H^+ + H_2PO_4^- \quad K_{a1} = 1.1 \times 10^{-2} = \frac{[H^+][H_2PO_4^-]}{[H_3PO_4]} \quad (6.61)$$

$$H_2PO_4^- \rightleftharpoons H^+ + HPO_4^{2-} \quad K_{a2} = 7.5 \times 10^{-8} = \frac{[H^+][HPO_4^{2-}]}{[H_2PO_4^-]} \quad (6.62)$$

$$HPO_4^{2-} \rightleftharpoons H^+ + PO_4^{3-} \quad K_{a3} = 4.8 \times 10^{-13} = \frac{[H^+][PO_4^{3-}]}{[HPO_4^{2-}]} \quad (6.63)$$

Recall from Chapter 5 that the overall ionization is the sum of these individual steps and the overall ionization constant is the product of the individual ionization constants:

$$HPO_4^{2-} \rightleftharpoons H^+ + PO_4^{3-} \quad K_a = K_{a1}K_{a2}K_{a3} = 4.0 \times 10^{-22} = \frac{[H^+]^3[PO_4^{3-}]}{[H_3PO_4]} \quad (6.64)$$

The individual pK_a values are 1.96, 7.12, and 12.32, respectively, for pK_{a1}, pK_{a2}, and pK_{a3}. In order to make precise pH calculations, the contributions of protons from each ionization step must be taken into account. Exact calculation is difficult and requires a tedious iterative procedure since [H$^+$] is unknown in addition to the various phosphoric acid species. See, for example, References 8 and 11 for calculations. Excel or other spreadsheet-based calculations can be simple. This is illustrated later.

We can titrate the first two protons of H$_3$PO$_4$ separately. The third is too weak to titrate.

In most cases, approximations can be made so that each ionization step can be considered individually. If the difference between successive ionization constants is more than 10^3, each proton can be differentiated in a titration, that is, each is titrated separately to give stepwise pH breaks in the titration curve. (If an ionization constant is less than about 10^{-9}, then the ionization is too small for a pH break to be exhibited in the titration curve—for example, the third proton for H$_3$PO$_4$.) When the individual pK_a's are separated by three units or more, calculations are simplified because *the system can be considered as simply a mixture of three weak acids of equal concentration that largely do not interact with each other.*

Buffer Calculations for Polyprotic Acids

We can prepare phosphate buffers with pH centered around 1.96 (pK_{a1}), 7.12 (pK_{a2}), and 12.32 (pK_{a3}).

The anion on the right side in each ionization step can be considered the salt (conjugate base) of the acid from which it is derived. That is, in Equation 6.61, H$_2$PO$_4^-$ is the salt of the acid H$_3$PO$_4$. In Equation 6.62, HPO$_4^{2-}$ is the salt of the acid H$_2$PO$_4^-$, and in Equation 6.63, PO$_4^{3-}$ is the salt of the acid HPO$_4^{2-}$. So each of these pairs constitutes a buffer system, and orthophosphate buffers can be prepared over a wide pH range. The optimum buffering capacity of each pair occurs at a pH corresponding to its pK_a. The HPO$_4^{2-}$/H$_2$PO$_4^-$ couple is an effective buffer system in the blood (see next page).

Example 6.16

The pH of blood is 7.35. What is the ratio of $[HPO_4^{2-}]/[H_2PO_4^-]$ in the blood (assume 25°C)?

Solution

$$pH = pK_a + \log \frac{[\text{proton acceptor}]}{[\text{proton donor}]}$$
$$pK_{a2} = 7.12$$

$$pH = 7.12 + \log \frac{[HPO_4^{2-}]}{[H_2PO_4^-]}$$

$$7.35 = 7.12 + \log \frac{[HPO_4^{2-}]}{[H_2PO_4^-]}$$

$$\frac{[HPO_4^{2-}]}{[H_2PO_4^-]} = 10^{(7.35-7.12)} = 10^{0.23} = 1.6$$

Dissociation Calculations for Polyprotic Acids

Because the individual ionization constants are sufficiently different, *the pH of a solution of H_3PO_4 can be calculated by treating it just as we would any weak acid.* The H^+ from the first ionization step effectively suppresses the other two ionization steps, so that the H^+ contribution from them is negligible compared to the first ionization. The quadratic equation must, however, be solved because K_{a1} is relatively large.

Example 6.17

Calculate the pH of a 0.100 *M* H_3PO_4 solution.

Solution

$$H_3PO_4 \approx H^+ + H_2PO_4^-$$
$$0.100 - x \quad x \quad x$$

From Equation 6.61,

$$\frac{(x)(x)}{0.100 - x} = 1.1 \times 10^{-2}$$

In order to neglect x, C should be $\geq 100K_a$. Here, it is only 10 times as large. Therefore, use the quadratic equation to solve:

$$x^2 + 0.011x - 1.1 \times 10^{-3} = 0$$

$$x = \frac{-0.011 \pm \sqrt{(0.011)^2 - 4(-1.1 \times 10^{-3})}}{2}$$

$$x = [H^+] = 0.028 \ M$$

Treat H_3PO_4 as a monoprotic acid. But x can't be neglected compared to C.

The acid is 28% ionized:

$$pH = -\log 2.8 \times 10^{-2} = 2 - 0.45 = 1.55$$

We can determine if our assumption that the only important source of protons is H_3PO_4 was a realistic one. $H_2PO_4^-$ would be the next most likely source of protons.

From Equation 6.62, $[HPO_4^{2-}] = K_{a2}[H_2PO_4^-]/[H^+]$. Assuming the concentrations of $H_2PO_4^-$ and H^+ as a first approximation are 0.028 M as calculated, then $[HPO_4^{2-}] \approx K_{a2} = 7.5 \times 10^{-8}$ M. This is very small compared to 0.028 M $H_2PO_4^-$, and so further dissociation is indeed insignificant. We were justified in our approach.

6.10 Ladder Diagrams

⌐Contributed by Professor
Galena Talanova, Howard
University⌐

 # Professor's Favorite Example

What is the dominant species at a given pH?
A simple way to visualize what is the dominant species at a given pH is to construct what David Harvey of Depauw University (http://acad.depauw.edu/~harvey/ASDL2008/introduction.html) calls **Ladder Diagrams**. We draw a vertical axis in pH with horizontal bars for the different pK_a's present in the system, as illustrated in Figure 6.1 below.

Referring to Figure 6.1(a), the unionized acid HOAc dominates below a pH of 4.76 (the pK_a value of HOAc) while above this value the OAc$^-$ anion is dominant. Referring to Figure 6.1(b), one can consider that all three acid–base systems can be simultaneously or individually present. For the HF–F$^-$ system depicted in blue, HF dominates at pH values below its pK_a of 3.17 and above this F$^-$ dominates. Similarly H_2S and NH_4^+ dominate at pH values below 6.88 and 9.25, respectively, while HS$^-$ and NH_3 dominates at respective pH values above that. Such diagrams also indicate what species will be dominant in a mixed system at a given pH: at a pH of 6 for example, we will expect F$^-$, NH_4^+, and H_2S to be the dominant species relative to their respective conjugate acid/base. Finally, Figure 6.1(c) shows the case for EDTA, this represents a hexaprotic acid system; the free acid is actually the diprotonated form H_6Y^{2+} and

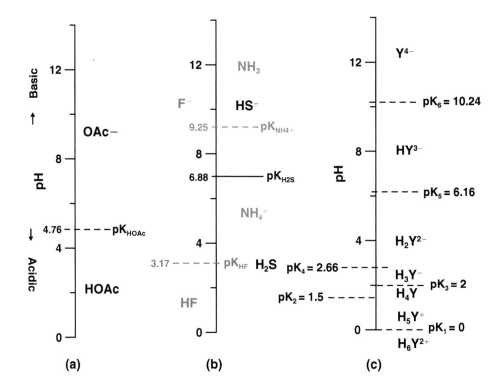

FIGURE 6.1 Ladder diagrams depicting systems containing (a) acetic acid and acetate, (b) ammonium and ammonia, hydrofluoric acid and fluoride, (di)hydrogen sulfide and hydrosulfide, and (c) diprotonated ethylenediaminetetraacetic acid (EDTA) (H_6Y^{2+}) and the other six species obtained by successive deprotonation.

it exists in dominant form only in extremely strongly acid solutions (pH < 0). The respective zones in which the other individual species dominate (e.g., Y^{4-} is dominant at pH > 10). More details and example problems are given in a PowerPoint file (ch 6 6.10 ladder diagrams.ppt) available in the text website.

6.11 Fractions of Dissociating Species at a Given pH: α Values—How Much of Each Species?

Often, it is of interest to know the distribution of the different species of a polyprotic acid as a function of pH, that is, at known hydrogen ion concentration as in a buffered solution.

Consider, for example, the dissociation of phosphoric acid. The equilibria are given in Equations 6.61 to 6.63. All the four phosphoric acid species coexist in equilibrium with one another, although the concentrations of some may be very small at a given pH. By changing the pH, the equilibria shift; the relative concentrations change. It is possible to derive general equations for calculating the fraction of the acid that exists in a given form, from the given hydrogen ion concentration.

For a given total **analytical concentration** of phosphoric acid, $C_{H_3PO_4}$, we can write

$$C_{H_3PO_4} = [PO_4^{3-}] + [HPO_4^{2-}] + [H_2PO_4^-] + [H_3PO_4] \qquad (6.65)$$

where the terms on the right-hand side of the equation represent the **equilibrium concentrations** of the individual species. We presumably know the initial total concentration $C_{H_3PO_4}$ and wish to find the fractions or concentrations of the individual species at equilibrium.

We define

$$\alpha_0 = \frac{[H_3PO_4]}{C_{H_3PO_4}} \qquad \alpha_1 = \frac{[H_2PO_4^-]}{C_{H_3PO_4}} \qquad \alpha_2 = \frac{[HPO_4^{2-}]}{C_{H_3PO_4}}$$

$$\alpha_3 = \frac{[PO_4^{3-}]}{C_{H_3PO_4}} \qquad \alpha_0 + \alpha_1 + \alpha_2 + \alpha_3 = 1$$

H_3PO_4, $H_2PO_4^-$, HPO_4^{2-}, and PO_4^{3-} all exist together in equilibrium. The pH determines the fraction of each.

where the α's are the **fractions** of each species present at equilibrium. Note that the subscripts denote the number of dissociated protons or the charge on the species. We can use Equation 6.65 and the equilibrium constant expressions 6.61 through 6.63 to obtain an expression for $C_{H_3PO_4}$ in terms of the desired species. This is substituted into the appropriate equation to obtain α in terms of $[H^+]$ and the equilibrium constants. In order to calculate α_0, for example, we can rearrange Equations 6.61 through 6.63 to solve for all the species except $[H_3PO_4]$ and substitute into Equation 6.65:

$$[PO_4^{3-}] = \frac{K_{a3}[HPO_4^{2-}]}{[H^+]} \qquad (6.66)$$

$$[HPO_4^{2-}] = \frac{K_{a2}[H_2PO_4^-]}{[H^+]} \qquad (6.67)$$

$$[H_2PO_4^-] = \frac{K_{a1}[H_3PO_4]}{[H^+]} \qquad (6.68)$$

We want all these to contain only $[H_3PO_4]$ (and $[H^+]$, the variable). We can substitute Equation 6.68 for $[H_2PO_4^-]$ in Equation 6.67:

$$[HPO_4^{2-}] = \frac{K_{a1}K_{a2}[H_3PO_4]}{[H^+]^2} \tag{6.69}$$

And we can substitute Equation 6.69 into Equation 6.66 for $[HPO_4^{2-}]$:

$$[PO_4^{3-}] = \frac{K_{a1}K_{a2}K_{a3}[H_3PO_4]}{[H^+]^3} \tag{6.70}$$

Finally, we can substitute 6.68 through 6.70 in Equation 6.65:

$$C_{H_3PO_4} = \frac{K_{a1}K_{a2}K_{a3}[H_3PO_4]}{[H^+]^3} + \frac{K_{a1}K_{a2}[H_3PO_4]}{[H^+]^2} + \frac{K_{a1}[H_3PO_4]}{[H^+]} + [H_3PO_4] \tag{6.71}$$

We can divide each side of this expression by $[H_3PO_4]$ to obtain $1/\alpha_0$:

$$\frac{C_{H_3PO_4}}{[H_3PO_4]} = \frac{1}{\alpha_0} = \frac{K_{a1}K_{a2}K_{a3}}{[H^+]^3} + \frac{K_{a1}K_{a2}}{[H^+]^2} + \frac{K_{a1}}{[H^+]} + 1 \tag{6.72}$$

Taking the reciprocal of both sides

$$\alpha_0 = \frac{1}{(K_{a1}K_{a2}K_{a3}/[H^+]^3) + (K_{a1}K_{a2}/[H^+]^2) + (K_{a1}/[H^+]) + 1} \tag{6.73}$$

multiplying both the numerator and denominator on the right with $[H^+]^3$, we have:

$$\boxed{\alpha_0 = \frac{[H^+]^3}{[H^+]^3 + K_{a1}[H^+]^2 + K_{a1}K_{a2}[H^+] + K_{a1}K_{a2}K_{a3}}} \tag{6.74}$$

Use this equation to calculate the fraction of H_3PO_4 in solution.

Similar approaches can be taken to obtain expressions for the other α's. For α_1, for example, the equilibrium constant expressions would be solved for all species in terms of $[H_2PO_4^-]$ and substituted into Equation 6.65 to obtain an expression for $C_{H_3PO_4}$ containing only $[H_2PO_4^-]$ and $[H^+]$, from which α_1 is calculated. The results for the other α's are

$$\boxed{\alpha_1 = \frac{K_{a1}[H^+]^2}{[H^+]^3 + K_{a1}[H^+]^2 + K_{a1}K_{a2}[H^+] + K_{a1}K_{a2}K_{a3}}} \tag{6.75}$$

$$\boxed{\alpha_2 = \frac{K_{a1}K_{a2}[H^+]}{[H^+]^3 + K_{a1}[H^+]^2 + K_{a1}K_{a2}[H^+] + K_{a1}K_{a2}K_{a3}}} \tag{6.76}$$

Derive these equations in Problem 57.

$$\boxed{\alpha_3 = \frac{K_{a1}K_{a2}K_{a3}}{[H^+]^3 + K_{a1}[H^+]^2 + K_{a1}K_{a2}[H^+] + K_{a1}K_{a2}K_{a3}}} \tag{6.77}$$

Note that all have the *same denominator* and that *the sum of the numerators equals the denominator*. For α_0, the first term in the denominator becomes the numerator; for α_1, the second term in the denominator becomes the numerator; for α_2, the third term becomes the numerator, and so on. See Problem 58 for a more detailed derivation of the other α's.

In general, an n-protic acid ($n = 1, 2, 3$, respectively, for HOAc, $H_2C_2O_4$, H_3PO_4, etc.) will have $n + 1$ species other than $[H^+]$ derived from the acid (e.g., $H_2C_2O_4$, $HC_2O_4^-$, and $C_2O_4^{2-}$) and thus, $n + 1$ α-values. The denominator in the α-values, Q_n, will consist in each case of $n + 1$ terms:

$$Q_n = \sum_{i=0}^{i=n} [H^+]^{n-i} K_{a0} \ldots K_{ai} \tag{6.78}$$

where K_{a0} is taken to be 1. Thus

$$Q_1 = [H^+] + K_a \tag{6.79}$$

$$Q_2 = [H^+]^2 + K_{a1}[H^+] + K_{a1}K_{a2} \tag{6.80}$$

$$Q_3 = [H^+]^3 + K_{a1}[H^+]^2 + K_{a1}K_{a2}[H^+] + K_{a1}K_{a2}K_{a3} \tag{6.81}$$

$$Q_4 = [H^+]^4 + K_{a1}[H^+]^3 + K_{a1}K_{a2}[H^+]^2 + K_{a1}K_{a2}K_{a3}[H^+] + K_{a1}K_{a2}K_{a3}\,K_{a4} \tag{6.82}$$

While Equation 6.78 may look complicated, it is easy to remember the pattern represented by Equations 6.79 through 6.82. You start with $[H^+]^n$ where n is the number of dissociable proton and then replace an H^+ with a K_a term, beginning with K_{a1} until you run out of both. For α_0 to α_n, remember that the first term in the Q expression becomes the numerator for α_0, the second term for α_1 and so on. Thus, α_1 for a monoprotic acid $\alpha_{1,m}$ is:

$$\alpha_{1,m} = K_a/Q_1 = \frac{K_a}{[H^+] + K_a} \tag{6.83}$$

and α_3 for a tetraprotic acid $\alpha_{3,te}$ is:

$$\begin{aligned}
\alpha_{3,te} &= K_{a1}K_{a2}K_{a3}[H^+]/Q_4 \\
&= K_{a1}K_{a2}K_{a3}[H^+]/([H^+]^4 + K_{a1}[H^+]^3 + K_{a1}K_{a2}[H^+]^2 \\
&\quad + K_{a1}K_{a2}K_{a3}[H^+] + K_{a1}K_{a2}K_{a3}\,K_{a4}
\end{aligned} \tag{6.84}$$

Example 6.18

Calculate the equilibrium concentration of the different species in a 0.10 M phosphoric acid solution at pH 3.00 ($[H^+] = 1.0 \times 10^{-3}\ M$).

Solution

Substituting into Equation 6.79,

$$\alpha_0 = \frac{(1.0 \times 10^{-3})^3}{(1.0 \times 10^{-3})^3 + (1.1 \times 10^{-2})(1.0 \times 10^{-3})^2 + (1.1 \times 10^{-2})}$$
$$(7.5 \times 10^{-8})(1.0 \times 10^{-3}) + (1.1 \times 10^{-2})(7.5 \times 10^{-8})(4.8 \times 10^{-13})$$
$$= \frac{1.0 \times 10^{-9}}{1.2 \times 10^{-8}} = 8.3 \times 10^{-2}$$

$$[H_3PO_4] = C_{H_3PO_4}\,\alpha_0 = 0.10 \times 8.3 \times 10^{-2} = 8.3 \times 10^{-3}\ M$$

Similarly,

$$\alpha_1 = 0.92$$

$$[H_2PO_4^-] = C_{H_3PO_4}\,\alpha_1 = 0.10 \times 0.92 = 9.2 \times 10^{-2}\ M$$

$$\alpha_2 = 6.9 \times 10^{-5}$$

$$[HPO_4^{2-}] = C_{H_3PO_4}\,\alpha_2 = 0.10 \times 6.9 \times 10^{-5} = 6.9 \times 10^{-6}\ M$$

$$\alpha_3 = 3.3 \times 10^{-14}$$

$$[PO_4^{3-}] = C_{H_3PO_4}\,\alpha_3 = 0.10 \times 3.3 \times 10^{-14} = 3.3 \times 10^{-15}\ M$$

We see that at pH 3, the majority (92%) of the phosphoric acid exists as $H_2PO_4^-$ and 8.3% exists as H_3PO_4. Only $3.3 \times 10^{-12}\%$ exists as PO_4^{3-}!

The overlapping curves represent buffer regions. The pH values where α_1 and α_2 are 1 represent the end points in titrating H_3PO_4.

See Section 6.16 for a way of representing these plots as straight lines (log–log plots).

By far the best way to understand Excel-related text is to have the relevant Excel file open on your computer while you read this text.

We can prepare a spreadsheet to calculate the fraction of each species as a function of pH. Formulas and calculations are shown in the spreadsheet[6], and Figure 6.2 shows the corresponding α versus pH plot. The K_a values are entered in cells B4, D4, and F4. The pH values are entered in column A. All the formulas needed for each cell are listed at the bottom of the spreadsheet, and they are initially entered in the boldfaced cells. The formula for calculating the corresponding hydrogen ion concentration (used in the α calculation) is entered in cell B6. The formula for the denominator used for each α calculation is entered in cell C6. Note that the constants are entered as absolute values, while the hydrogen ion concentration is computed from the specific pH values entered in column A. The formulas for the three α calculations are entered in cells D6, E6, and F6. All formulas are copied down to row 34.

Referring to Figure 6.2.xlsx that is available to you from the website (the screenshot is reproduced as Figure 6.3), we want to plot α_0, α_1, α_2, and α_3 (which appear in the columns D:G, titled a_0, a_1, a_2, and a_3) as a function of pH (column A). The most expedient way to accomplish this is to leave the x-data (pH) where it is in column A and move the three sets of y-data we want to plot in contiguous columns. We therefore select the $[H^+]$ and denominator (Q) data (B5:C34) that we do not wish to plot and cut and paste (Ctrl-X, Ctrl V) beginning in cell H5. We select B5:C34 again and delete the empty space (Alt-E D, Shift cells left). Now highlight all the data columns we wish to plot (A5:E34) and click on **Insert—Charts- Scatter**. You can choose the scatter plot shown in column 1, row 2 in the drop-down menu if you just want to see the line plot. Otherwise the scatter plot shown in column 2, row 1 if you want to see the specific points plotted as in Figure 6.2. The plot appears. You can move it to a separate sheet (by right-clicking on the frame of the chart, selecting **Move Chart**, **New Sheet**, and **OK**). Play with the Chart layout on the menu bar: You can click on **Layout 3** (the grid line and linear fit layout), the linear fits are irrelevant here can be subsequently deleted by clicking on the fit lines and deleting them, leaving the gridlines. You can also click on the axes and chart titles, and change them to what you want them to read, etc.

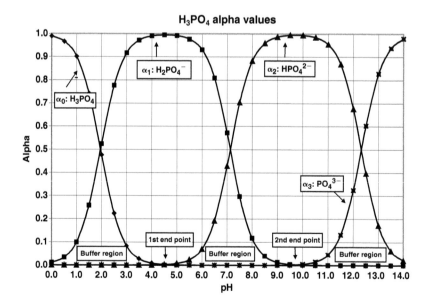

FIGURE 6.2 Fractions of H_3PO_4 species as a function of pH. (Buffer region designation suggestions courtesy of Professor Galina Talanova, Howard University)

---

[6]The "standard view" of any Excel spreadsheet consists of the numbers entered in a cell or the results of the formula entered into a cell. It does not normally show the formula contained in the cell until your cursor is on the cell and then it displays the formula in the formula bar. At any time, you can change the "standard view" into "formula view" by pressing the Ctrl and ' (accent key) together.

	A	B	C	D	E	F	G
1	Calculation of alpha values for H_3PO_4 vs. pH.						
2	Alpha (α_i) denominator = $[H^+]^3 + K_{a1}[H^+]^2 + K_{a1}K_{a2}[H^+] + K_{a1}K_{a2}K_{a3}$						
3	Numerators: $\alpha_0 = [H^+]^3$; $\alpha_1 = K_{a1}[H^+]^2$; $\alpha_2 = K_{a1}K_{a2}[H^+]$; $\alpha_3 = K_{a1}K_{a2}K_{a3}$						
4	$K_{a1}=$	1.10E-02	$K_{a2}=$	7.50E-08	$K_{a3}=$	4.80E-13	
5	pH	$[H^+]$	Denominator	α_0	α_1	α_2	α_3
6	0.0	1	1.01E+00	9.89E-01	1.09E-02	8.16E-10	3.92E-22
7	0.5	0.316228	3.27E-02	9.66E-01	3.36E-02	7.97E-09	1.21E-20
8	1.0	0.1	1.11E-03	9.01E-01	9.91E-02	7.43E-08	3.57E-19
9	1.5	0.031623	4.26E-05	7.42E-01	2.58E-01	6.12E-07	9.29E-18
10	2.0	0.01	2.10E-06	4.76E-01	5.24E-01	3.93E-06	1.89E-16
11	2.5	0.003162	1.42E-07	2.23E-01	7.77E-01	1.84E-05	2.80E-15
12	3.0	0.001	1.20E-08	8.33E-02	9.17E-01	6.87E-05	3.30E-14
13	3.5	0.000316	1.13E-09	2.79E-02	9.72E-01	2.30E-04	3.50E-13
14	4.0	0.0001	1.11E-10	9.00E-03	9.90E-01	7.43E-04	3.56E-12
15	4.5	3.16E-05	1.11E-11	2.86E-03	9.95E-01	2.36E-03	3.58E-11
16	5.0	0.00001	1.11E-12	9.02E-04	9.92E-01	7.44E-03	3.57E-10
17	5.5	3.16E-06	1.13E-13	2.81E-04	9.77E-01	2.32E-02	3.52E-09
18	6.0	0.000001	1.18E-14	8.46E-05	9.30E-01	6.98E-02	3.35E-08
19	6.5	3.16E-07	1.36E-15	2.32E-05	8.08E-01	1.92E-01	2.91E-07
20	7.0	1E-07	1.93E-16	5.19E-06	5.71E-01	4.29E-01	2.06E-06
21	7.5	3.16E-08	3.71E-17	8.53E-07	2.97E-01	7.03E-01	1.07E-05
22	8.0	1E-08	9.35E-18	1.07E-07	1.18E-01	8.82E-01	4.24E-05
23	8.5	3.16E-09	2.72E-18	1.16E-08	4.05E-02	9.59E-01	1.46E-04
24	9.0	1E-09	8.36E-19	1.20E-09	1.32E-02	9.86E-01	4.73E-04
25	9.5	3.16E-10	2.62E-19	1.21E-10	4.19E-03	9.94E-01	1.51E-03
26	10.0	1E-10	8.30E-20	1.20E-11	1.33E-03	9.94E-01	4.77E-03
27	10.5	3.16E-11	2.65E-20	1.19E-12	4.15E-04	9.85E-01	1.49E-02
28	11.0	1E-11	8.65E-21	1.16E-13	1.27E-04	9.54E-01	4.58E-02
29	11.5	3.16E-12	3.00E-21	1.05E-14	3.66E-05	8.68E-01	1.32E-01
30	12.0	1E-12	1.22E-21	8.19E-16	9.01E-06	6.76E-01	3.24E-01
31	12.5	3.16E-13	6.57E-22	4.81E-17	1.67E-06	3.97E-01	6.03E-01
32	13.0	1E-13	4.79E-22	2.09E-18	2.30E-07	1.72E-01	8.28E-01
33	13.5	3.16E-14	4.22E-22	7.49E-20	2.61E-08	6.18E-02	9.38E-01
34	14.0	1E-14	4.04E-22	2.47E-21	2.72E-09	2.04E-02	9.80E-01
35	Formulas for cells in **boldface**:						
36	Cell B6 = $[H^+]$ =	10^-A6					
37	Cell C6=denom.=	B6^3+$B$4*B6^2+$B$4*$D$4*B6+$B$4*$D$4*$F$4					
38	Cell D6 = α_0 =	B6^3/C6					
39	Cell E6 = α_1 =	($B$4*B6^2)/C6					
40	Cell F6 = α_2 =	($B$4*$D$4*B6)/C6					
41	Cell G6 = α_3 =	($B$4*$D$4*$F$4)/C6					
42	Copy each formula down through Cell 34						
43	Plot A6:A34 vs. D6:D34, E6:E34, F6:F34, and G6:G34 (series 1, 2, 3, and 4)						

FIGURE 6.3 Screenshot of the file Figure 6.2.xlsx (the spreadsheet itself is available in the text website).

The plot generated by the procedure above is given in Figure 6.2. This figure illustrates how the ratios of the four phosphoric acid species change as the pH is adjusted, for example, in titrating H_3PO_4 with NaOH. While some appear to go to zero concentration above or below certain pH values, they are not really zero, but diminishingly small. For example, we saw in Example 6.18 that at pH 3.00, the concentration of the PO_4^{3-} ion for 0.1 M H_3PO_4 is only 3.3×10^{-15} M, but it is indeed present in equilibrium. The pH regions where two curves overlap (with appreciable concentrations) represent regions in which buffers may be prepared using those two species. For example, mixtures of H_3PO_4 and $H_2PO_4^-$ can be used to prepare buffers around pH 2.0 ± 1, mixtures of $H_2PO_4^-$ and HPO_4^{2-} around pH 7.1 ± 1, and mixtures of HPO_4^{2-} and PO_4^{3-} around pH 12.3 ± 1. The pH values at which the fraction of a species is essentially 1.0 correspond to the end points in the titration of phosphoric acid with a strong base, that is, $H_2PO_4^{2-}$ at the first end point (pH 4.5), HPO_4^{2-} at the second end point (pH 9.7).

Equation 6.71 could be used for a rigorous calculation of the hydrogen ion concentration from dissociation of a phosphoric acid solution at a given H_3PO_4 concentration (no other added H^+), but this involves tedious iterations. As a first approximation, $[H^+]$ could be calculated from K_{a1} as in Example 6.17, assuming that only the first dissociation step of phosphoric acid was significant. (This is, in fact, what we did in that example.) The first calculated $[H^+]$ could then be substituted in Equation 6.71 to calculate a second approximation of $[H_3PO_4]$, which would be used for a second iterative calculation of $[H^+]$ using K_{a1}, and so forth, until the concentration was constant. A simpler way is to use Excel and Goal Seek as illustrated in Example 6.19 below.

A useful applet developed by Professor Constantinos Efstathiou at the University of Athens allows easy plotting of distribution diagrams of mono- to tetraprotic acids: http://www.chem.uoa.gr/applets/AppletAcid/Appl_Distr2.html. Check out the phosphoric acid one and compare with Figure 6.1, and the EDTA (H_4A) one with Figure 8.1 in Chapter 8. You can change the pK_a values to see how the plots change. The applet also plots log distribution diagrams for the acids (Section 6.16 below). You can also change the K_a values in the spreadsheet Figure 6.2.xlsx to see how the distribution will change.

Example 6.19

The Method of Charge Balance. Calculation of pH in a Phosphoric Acid System

Calculate the pH of 0.050 M H_3PO_4 using Excel and Goal Seek. What will be the pH if you add 0.11 mole of NaOAc and 0.02 mole of K_2HPO_4 to 1 L of this of this solution?

The charge balance method invokes that the sum of the positive charges in any solution equals the sum of negative charges. So for a solution containing only phosphoric acid, the relevant charge balance equation is:

$$[H^+] = [OH^-] + [H_2PO_4^-] + 2[HPO_4^{2-}] + 3[PO_4^{3-}] \qquad (6.85)$$

Note that the multipliers of 2 and 3 need, respectively, to be applied to the concentrations of HPO_4^{2-} and PO_4^{3-} because these ions, respectively, carry 2 and 3 units of charge. Putting all terms on one side, expressing $[OH^-]$ as $K_w/[H^+]$ and expressing the various phosphate species concentrations in terms of their α-values, we have:

$$[H^+] - K_w/[H^+] - C_p(\alpha_1 + 2\alpha_2 + 3\alpha_3) = 0 \qquad (6.86)$$

where C_p is the concentration of the total phosphate species, in this case the concentration of H_3PO_4 taken. The Example 6.19.xlsx spreadsheet is available in your text website. In cells B1:B5, we have, respectively, written down the values of K_{a1}, K_{a2}, K_{a3}, C_p, and K_w. It is unfortunate that many of the symbols we traditionally use in chemical problems aren't allowed by Excel as KA1, etc. It actually refers to the cell in column KA and row 1. Similarly C and R refer to columns and rows and are not allowed to be used for any other meaning. Thus, for our purposes, we have named K_{a1}, K_{a2}, K_{a3}, and C_p as KAA, KAB, KAC, and CP and written these names in column A next to the numeric values in column bar. Next we want to permanently ascribe these names to the specific numbers, so that every time we write KAA, Excel will know that we are referring to the value of KAA, 1.1×10^{-2}. To do this, we put our cursor on cell B2. In the top right corner on the formula bar normally it would say B2, when we put our cursor on cell B2. But notice that it says KAA. This is because we have given the number in cell B2 the name KAA. We did this (Excel 2010 only—previous versions have a different procedure) by putting the cursor on cell B2, clicking on the name box (top left corner of formula bar), typing KAA and hitting enter. Verify that cell B2 has the name KAA by moving to some other cell and coming back to cell B2 and noticing that the name box says KAA. (Practice on another spreadsheet naming cells.)

For convenience, we have named cells B3:B5 in a similar manner as KAB, KAC, CP, and KW.

Now two rows below this, we have set up column headings as pH, H^+, OH^-, Q3, Alpha1, Alpha2, Alpha3, and Equation. In cell A8, under pH is the value we will be trying to calculate. Presently you can enter any guess value for pH that you might think is reasonable, any value between 0 and 14; it is not important. For now let us enter 0. Also, go ahead and name the cell pH (we do not really need to do this, we can keep on referring to it as cell A8, but it is fun to give names to the cells to designate what they are). In cell B8, under H^+ we want to calculate the corresponding value of $[H^+]$. Excel does not *a priori* know the relationship between pH and $[H^+]$. Since we know that $[H^+]$ can be expressed as 10^{-pH}, we enter in cell B8, $= 10^{-pH}$. (If we did not name cell A8 as pH, we would have had to write in cell B8 $= 10^{-A8}$.) Again, while this is not essential, for convenience, we name cell B8 as H. In cell C8, to calculate the value of $[OH^-]$, we enter = KW/H. For good measure, we name cell C8 as OH. Next under heading Q_3, we have to put in the expression of Equation 6.81, using current names and so we enter in cell D8:

$$= H^3 + KAA*H^2 + KAA*KAB*H + KAA*KAB*KAC$$

and name the cell Q (we can name it Q or Q three, but not Q3, remember?). Similarly, we enter the alpha formulas in Equations 6.75 through 6.77 in cells E8:G8 and also name them, respectively, ALFA1, ALFA2, and ALFA3. For example, we have entered in cell F8

$$= KAA*KAB*H/Q$$

Finally, we are ready to write our Equation 6.86 in cell H8 as:

$$= (1E10) * (H-OH-CP * (ALFA1 + 2*ALFA2 + 3*ALFA3))$$

You will note that the expression following the 1E10 multiplier (remember this is to stop Excel from believing prematurely that it has found a solution) is the expression in Equation 6.86. All we have to do now is to invoke Goal Seek (Data/What-If Analysis/-Goal Seek), type in H8 in the Set Cell box, in "To Value" box type 0 (Equation 6.86), and "By changing cell" box enter pH (or A8). You instantly get your solution, the pH is 1.73. In the "Equation" cell (H8), there is merit to squaring the entire parenthetical expression because this makes it have only positive values. This is not important when we are solving a single problem, but it is important when we solve multiple problems at a time; we will discuss this in greater depth in a later section.

Now let us solve the second part of the problem in which we put into this solution in addition 0.11 *M* NaOAc, 0.02 *M* K_2HPO_4. Introducing another acid-base system would normally seem to be a formidable problem. Actually it is not. We have to do only a modest amount of additional work on the existing spreadsheet to solve this problem. The worked-out solution appears in the spreadsheet Example 6.19b.xlsx on the text website, but let us presently modify the Example 6.19.xlsx spreadsheet we have been working on.

First, we need to understand the changes. We have added sodium (0.11 *M*, we will call this CNA), potassium ($2 \times 0.02 = 0.04$ *M*, we will call this CK) and acetate species (a total of 0.11 *M*, we will call this COAC). The concentration of our total phosphate species has increased from 0.050 *M* to 0.070 *M* and we need to define the dissociation constant for acetic acid (1.75E-5; we shall call this KOAC). The addition of the new species requires that we modify Equation 6.85 to be:

$$[H^+] + [Na^+] + [K^+] = [OH^-] + [OAc^-] + [H_2PO_4^-] + 2[HPO_4^{2-}] + 3[PO_4^{3-}]$$
$$(6.87)$$

If we define Q_1 to be the relevant denominator for the acetate system (see Equation 6.79), then α_1 for the acetate system (ALFA1OAC, the name ALFA1 is already taken by the phosphate system—the two α_1 values for these acetate and phosphate systems are *not* the same—although they are in the same solution and are subject to the same pH, the K_a-values for the two systems are different) will be given by Equation 6.83 and we can write Equation 6.87 as:

$$[H^+] + [Na^+] + [K^+] - K_w/[H^+] - C_{OAc}\alpha_{1OAc} - C_p(\alpha_1 + 2\alpha_2 + 3\alpha_3) = 0 \quad (6.88)$$

Now going back to the spreadsheet, in cells D1:D4 we put in the values for KOAC, CNA, CK, and COAC, and give these cells the corresponding names. We put the cursor on column H and insert two more columns (Alt-I and C, twice in succession) so we can create headings Q1 and Alpha1OAc. In these two cells we respectively enter, for Q1, = H + KOAC in cell H8 and name it Qone. For Alpha1OAc (cell I8), we enter = KOAC/Q one and also name the cell ALFA1OAC. Now it is a matter of modifying the equation in cell J8 as:

$$= (10000000000) * (H + CK + CNA - OH - COAC * ALFA1OAC - CP$$
$$* (ALFA1 + 2*ALFA2 + 3*ALFA3))$$

Once again, invoke Goal Seek and ask it to set cell J8 to value 0 by changing the cell pH and you immediately have your answer, the pH is 5.17.

6.12   Salts of Polyprotic Acids—Acid, Base, or Both?

Salts of acids such as H_3PO_4 may be acidic or basic. The protonated salts possess both acidic and basic properties ($H_2PO_4^-$, HPO_4^{2-}), while the unprotonated salt is simply a Brønsted base that hydrolyzes (PO_4^{3-}).

1. Amphoteric Salts. $H_2PO_4^-$ possesses both acidic and basic properties. That is, it is **amphoteric**. It ionizes as a weak acid and it also is a Brønsted base that hydrolyzes:

$H_2PO_4^-$ acts as both an acid and a base. See the end of Section 6.16 for how to estimate the extent of each reaction using log–log diagrams.

$$H_2PO_4^- \rightleftharpoons H^+ + HPO_4^{2-} \quad K_{a2} = \frac{[H^+][HPO_4^{2-}]}{[H_2PO_4^-]} = 7.5 \times 10^{-8} \quad (6.89)$$

$$H_2PO_4^- + H_2O \rightleftharpoons H_3PO_4 + OH^- \quad K_b = \frac{K_w}{K_{a1}} = \frac{[H_3PO_4][OH^-]}{[H_2PO_4^-]}$$
$$= \frac{1.00 \times 10^{-14}}{1.1 \times 10^{-2}} = 9.1 \times 10^{-13} \quad (6.90)$$

The solution could, hence, be either alkaline or acidic, depending on which ionization is more extensive. Since K_{a2} for the first ionization is nearly 10^5 greater than K_b for the second ionization, the solution in this case will obviously be acidic.

An expression for the hydrogen ion concentration in a solution of an ion such as $H_2PO_4^-$ can be obtained as follows. The total hydrogen ion concentration is equal to the amounts produced from the ionization equilibrium in Equation 6.89 and the ionization of water, less the amount of OH^- produced from the hydrolysis in Equation 6.90. We can write, then,

$$C_{H^+} = [H^+]_{total} = [H^+]_{H_2O} + [H^+]_{H_2PO_4^-} - [OH^-]_{H_2PO_4^-} \quad (6.91)$$

or

$$[H^+] = [OH^-] + [HPO_4^{2-}] - [H_3PO_4] \quad (6.92)$$

We have included the contribution from water since it will not be negligible if the pH of the salt solution happens to be near 7—although in this particular case, the solution will be acid, making the water ionization negligible.

We can solve for $[H^+]$ by substituting expressions in the right-hand side of Equation 6.92 from the equilibrium constant expressions 6.61 and 6.62 and K_w to eliminate all but $[H_2PO_4^-]$, the experimental variable, and $[H^+]$:

$$[H^+] = \frac{K_w}{[H^+]} + \frac{K_{a2}[H_2PO_4^-]}{[H^+]} - \frac{[H_2PO_4^-][H^+]}{K_{a1}} \qquad (6.93)$$

from which (by multiplying each side of the equation by $[H^+]$, collecting the terms containing $[H^+]^2$ on the left side, and solving for $[H^+]^2$)

$$[H^+]^2 = \frac{K_w + K_{a2}[H_2PO_4^-]}{1 + \frac{[H_2PO_4^-]}{K_{a1}}} \qquad (6.94)$$

$$[H^+] = \sqrt{\frac{K_{a1}K_w + K_{a1}K_{a2}[H_2PO_4^-]}{K_{a1} + [H_2PO_4^-]}} \qquad (6.95)$$

That is, for the general case HA^-,

$$[H^+] = \sqrt{\frac{K_{a1}K_w + K_{a1}K_{a2}[HA^-]}{K_{a1} + [HA^-]}} \qquad (6.96)$$

For HA^{2-}, substitute $[HA^{2-}]$ for $[HA^-]$, K_{a2} for K_{a1}, and K_{a3} for K_{a2}.

This equation is valid for any salt HA^- derived from an acid H_2A (or for HA^{2-} derived from H_2A^-, etc.) where $[H_2PO_4^-$ is replaced by $[HA^-]$.

If we assume that the equilibrium concentration $[HA^-]$ is equal to the concentration of salt added, that is, that the extent of ionization and hydrolysis is fairly small, then this value along with the constants can be used for the calculation of $[H^+]$. This assumption is reasonable if the two equilibrium constants (K_{a1} and K_b) involving the salt HA^- are small and the solution is not too dilute. In many cases, $K_{a1}K_w \ll K_{a1}K_{a2}[HA^-]$ in the numerator and can be neglected. This is the equation we would have obtained if we had neglected the dissociation of water. Furthermore, if $K_{a1} \ll [HA^-]$ in the denominator, the equation simplifies to

$$[H^+] = \sqrt{K_{a1}K_{a2}} \qquad (6.97)$$

For HA^{2-}, $[H^+] = \sqrt{K_{a2}K_{a3}}$.

Therefore, if the assumptions hold, the pH of a solution of $H_2PO_4^-$ is independent of its concentration! This approximation is adequate for our purposes. The equation generally applies if there is a large difference between K_{a1} and K_{a2}. For the case of $H_2PO_4^-$, then,

$$[H^+] \approx \sqrt{K_{a1}K_{a2}} = \sqrt{1.1 \times 10^{-2} \times 7.5 \times 10^{-8}} = 2.9 \times 10^{-5}\ M \qquad (6.98)$$

and the pH is approximately independent of the salt concentration (pH $\approx$ 4.54). This would be the approximate pH of an NaH_2PO_4 solution.

Similarly, HPO_4^{2-} is both an acid and a base. The K values involved here are K_{a2} and K_{a3} of H_3PO_4 ($H_2PO_4^- \equiv H_2A$ and $HPO_4^{2-} \equiv HA^-$). Since $K_{a2} \gg K_{a3}$, the pH of a Na_2HPO_4 solution can be calculated from

$$[H^+] \approx \sqrt{K_{a2}K_{a3}} = \sqrt{7.5 \times 10^{-8} \times 4.8 \times 10^{-13}} = 1.9 \times 10^{-10} \qquad (6.99)$$

and the calculated pH is 9.72. Because the pH of amphoteric salts of this type is essentially independent of concentration, the salts are useful for preparing solutions of known pH for standardizing pH meters. For example, potassium acid phthalate, $KHC_8H_4O_2$, gives a solution of pH 4.0 at 25°C. However, these salts are poor buffers against acids or bases; their pH does not fall in the buffer region but occurs at the end point of a titration curve, where the pH can change markedly if either acid or base is added, although dilution does not affect pH as much.

2. Unprotonated Salt. Unprotonated phosphate is a fairly strong Brønsted base in solution and ionizes as follows:

$$PO_4^{3-} + H_2O \rightleftharpoons HPO_4^{2-} + OH^- \quad K_b = \frac{K_w}{K_{a3}} \tag{6.100}$$

The constant K_{a3} is very small, and so the equilibrium lies significantly to the right. Because $K_{a3} \ll K_{a2}$, hydrolysis of HPO_4^{2-} is suppressed by the OH^- from the first step, and the pH of PO_4^{3-} can be calculated just as for a salt of a monoprotic weak acid. However, because K_{a3} is so small, K_b is relatively large, and the amount of OH^- is not negligible compared with the initial concentration of PO_4^{3-}, and the quadratic equation must be solved, that is, PO_4^{3-} is quite a strong base.

Example 6.20

Calculate the pH of $0.100\ M\ Na_3PO_4$.

Solution

$$PO_4^{3-} + H_2O \rightleftharpoons HPO_4^{2-} + OH^-$$
$$0.100 - x \qquad\qquad x \qquad x$$

$$\frac{[HPO_4^{2-}][OH^-]}{[PO_4^{3-}]} = K_b = \frac{K_w}{K_{a3}} = \frac{1.0 \times 10^{-14}}{4.8 \times 10^{-13}} = 0.020$$

$$\frac{(x)(x)}{0.100 - x} = \frac{1.0 \times 10^{-14}}{4.8 \times 10^{-13}} = 0.020$$

The concentration is only five times K_b, so the quadratic equation is used:

$$x^2 + 0.020x - 2.0 \times 10^{-3} = 0$$

$$x = \frac{-0.020 \pm \sqrt{(0.020)^2 - 4(-2.0 \times 10^{-3})}}{2}$$
$$x = [OH^-] = 0.036\ M$$
$$pH = 12.56$$

The dissociation (hydrolysis) is 36% complete, and phosphate is quite a strong base. See the text **website** for a program that performs the quadratic equation calculation.

Example 6.21

Calculate the pH of 0.001, 0.002, 0.005, 0.01, 0.02, 0.05, 0.1, 0.2, 0.5, 1.0 M solutions each of H_3PO_4, NaH_2PO_4, Na_2HPO_4, and Na_3PO_4. Neglect activity corrections.

Solution

We must calculate ten separate pH values for each of the four compounds, which we will do in batches of 10 at a time using the powerful program, Solver. This problem allows us to explore the powers of Microsoft Excel Solver™, which can solve for more than one parameter (or more than one equation) at a time. At the time of this writing, Excel 2010 is the current version of Excel and Solver has many more capabilities in this version compared to previous versions, including the ability to save a scenario with up to 32 adjustable parameters and solve for up to 200 parameters at a time (although we do not recommend the latter); frequently it actually requires more time to solve say 50 one-parameter (e.g., in terms of H^+) equations four times than to solve 200 one-parameter equations at a time. In comparison, Goal Seek can only solve one parameter in a single equation, and does not allow for incorporating constraints in the parameter we want to solve (e.g., if we are solving for pH, it may be helpful for us to tell the computing algorithm that our solution lies within 0 and 14).

Solver is not automatically installed when you first install Office 2010 in your computer. After opening Excel, go to File/Options/Add-Ins, click on Go and select the Solver Add-In and click OK to install it. The next time you open Excel, if you go to the Data tab, you should see the icon for Solver in the right-hand corner of the menu bar.

Refer to the file Example 6.21.xlsx downloadable from the text **website**. But presently, just start with an empty spreadsheet. Very similar to what we had done in Example 6.19, in cells B1:B4 we have put down the numerical values of K_{a1}, K_{a2}, K_{a3}, and K_w and named them KAA, KAB, KAC, and KW, respectively. First, we are going to do phosphoric acid (H_3PO_4) and we write this down as a heading in row 5. Beginning in row 6 and starting at cell A6, we now create 10 columns and title them CP, pH, H$^+$, CNA, OH, Q3, Alpha1, Alpha2, Alpha3, and Equation. In A7:A16, we serially type in the concentrations our problem states, i.e., 0.001 to 1.0 as enumerated in the problem. In pH column (cell B7) let us enter a placeholder number, e.g., 1 for now. In the H^+ column (cell C7) we define is relationship with pH ($= 10^{-B7}$). Then in the CNA column cell D7 we type in zero. (The sodium concentration is zero in all of the pure H_3PO_4 solutions.) In the OH^- column (cell E7) we enter the relationship between OH^- and H^+ (given in cell C7) by entering = KW/C7. In the Q3 column (cell F7) we put in (much the same as in Example 6.19):

$$= C7\char`^3 + KAA*C7\char`^2 + KAA*KAB*C7 + KAA*KAB*KAC$$

Note that rather than defining and using H we are using the cell reference C7 because we will be solving a number of equations, each for the hydrogen ion concentration, in a separate row at one time. The hydrogen ion concentration in each row will be different (in C7:C16), rather than be a single value, and we cannot designate this by naming a single value called H as we did in Example 6.19. Alpha1: Alpha3 in cells G7:I16 are also put in exactly as in Example 6.19 but using C7 rather than H. Our charge balance equation is the same as Equation 6.85, except [Na$^+$] is added on the left side, for the sodium containing solutions and when expressed in terms of the equilibrium constants, is the same as Equation 6.85, again with [Na$^+$] added on the left. Finally, there is the equation in J7 to be put in; we square the charge balance expression (see why below). Although CNA is zero for H_3PO_4, we have retained this term in the charge balance expression so that we can use the same expression for all four of our cases, and multiply the whole by 10^{10}:

$$= (1E10)*(C7 + D7\text{-}E7\text{-}A7* (G7 + 2*H7 + 3*I7))\char`^2$$

We now highlight cells B7:J7 and copy and paste (Ctrl-C, Ctrl-V) beginning in cell B8; thus B8:J8 are filled. We could copy the same serially all the way down to B17:J16, but it is more expedient to highlight the two rows (B7:J8) and drag the cursor at the bottom corner of J8 down to J16 and thus fill up the remaining rows. (Note that if we had performed the drag-and-fill operation with only row 7 filled, in all the nonformula cells, by default Excel would have incremented the numerical value by 1 in each step in each of the succeeding rows, e.g., the initial assigned pH cell value would have been 2 in B8, 3 in B9, and so on, until 10 in B16; we do not necessarily want that. By having the same value in cell B8 as in B7, during drag-and-fill, Excel is being told that the increment is zero between the rows in that column and it follows that instruction.)

Next we are going to solve the set of 10 equations in J7:J16 by solving for the correct values of pH in cells B7:B16 simultaneously. First, in cell J18 we sum up the value of all of the equation expression values by typing there:

$$= SUM(J7:J16)$$

Now we are ready to invoke Solver. Go to the Data tab and then click on the Solver icon ?→ Solver on the top right-hand corner of the menu bar. Solver will open as a drop-down menu box. In the top "Set Objective" box you should already have J18 appearing and highlighted, because you opened Solver with your cursor on this cell. If not, type in J18 here. In order to solve our equations, we want J18 (the charge balance expression) to be zero. If each equation expression is zero, the sum of them must also be zero. In the digital domain, within the numerical resolution we are able to achieve, the best solution will rarely, if ever, be exactly zero but will be able to reach a very small number (especially when you remember that we have already multiplied the expression by 10^{10}). Because we solve a number of equations at one time by taking the sum of the expressions, you will understand why we use the square of the expressions rather than the expressions. From the point of view of solving an equation, one side of which is zero, obviously it does not matter if we try to solve $x = 0$ or $x^2 = 0$. However, if we have two equations $x = 0$ and $y = 0$ and try to solve them individually by trying to get $x + y = 0$, we may never reach the correct solution because for any nonzero value of x and y where $x = -y$ that condition will be made. However, for any real value of x and y, a solution that achieves $x^2 + y^2 = 0$ necessarily achieves the solutions $x = 0$ and $y = 0$.

Next in Solver, there is a choice to set our objective "To" maximize (max), minimize (min), or to a value that we have the option to specify. We can pick either the "min" button (preferred) or "to value 0," both will have the same impact. Because we have squared our expressions, the value in J18 can never be less than zero, so attempting to minimize it will have the same effect as trying to get it to be zero.

Next we click the box "By Changing Variable Cells." We need to either type in what we want to solve (B7:B16) or to use our cursor to select these cells. This box should now read $B$7:$B$16. The box "Make Unconstrained Variables Non-negative" should already be ticked. Our variable is pH and it should indeed be non-negative for most practical problems. The default method in the "Select a Solving Method" box is "GRG nonlinear" (others are "Simplex LP" and "Evolutionary") and we will use this in all our work; GRG stands for *Generalized Reduced Gradient*). You can read about the differences between these solving approaches in Excel.

In "Subject to the Constraints" box, we want to add that our pH values will be below 14. We click the "Add" button, a new box opens up. In the box "Cell Reference" we type in or select B7:B16. The condition box gives us a choice of <=, =, >=, an integer (int) among others. We are already in the default <= operator; this is what we need. So we proceed on to the next box, "Constraint," and type in 14. Then hit ENTER or click on OK. We are returned to the main Solver box. The "Options" button can open a number of adjustable options that you should explore but presently just proceed on. Move the Solver dialog Box to one corner of the screen, so that you can see the contents of J18, which you are trying to minimize. Presently it is likely a very large number. Now click on the button that says "Solve." A new box "Solver Results" will appear. Move it so that you can look at the contents of J18. You will see that it is smaller than before but it is still very large. We may need to invoke "Solver" several times so tick the box that says "Return to Solver Parameters Dialog" and click "OK." As the Solver main menu appears, click on "Solve" again. This time Solver results will likely have made a major difference in the value in J18. But we will keep doing this until there is no change. In fact, in only one more round, J18 will have gone to a value of the order of 10^{-15}, which will not change on further Solver invocations. This time when the Solver dialog box appears, click on "Close," rather than "Solve." Your solved pH values are in columns B7:B16 and range from 3.03 to 1.00.

Let us now do the same thing for NaH_2PO_4. We will write a title in row 21 as **MONOSODIUM PHOSPHATE**. Copy the entire calculation set including headings (A6:J18) beginning in cell A22. The only thing we need to change is the CNA column, noting that the CNA value for monosodium phosphate is always equal to CP, in cell D23 we write =A23 and then drag and fill that cell value though D32. Of course, your equation values will immediately change to very high values. We click on the new sum in cell J34 and invoke Solver. It will open with the previous conditions (e.g., J18 as the target cell, etc.); this you will have to change to the current needs. We set the target cell to J34, and specify that B23:B32 are the variable cells and select the entry in the constraints box and hit "Change." When the Change Constraints box appears, change B7:B16 to B23:B32 and click OK and Solve (make sure you have highlighted Cell J34 in the target cell). Again, in about four rounds you will see that the result is very small and does not change any more. The pH values will change from 5.08 at CP = 0.001 to 4.54 at CP = 1.0 M, you will also see how after about a concentration of 0.1 M, the pH hardly changes any more.

We similarly do this for Na_2HPO_4 and Na_3PO_4, the only changes we need to make is that for the Na_2HPO_4 case, we note that in the CNA cells, for the Na_2HPO_4 case, the value of the CNA cell is equal to 2* the value of the corresponding CP cell. Similarly in the Na_3PO_4 case the value of the CNA cell will be equal to 3* the value of the corresponding CP cell. When you finish solving these, notice the pattern. The pH for the Na_2HPO_4 case will converge to a value of 9.72 at high CP concentrations while the pH for the Na_3PO_4 case will continually increase with concentration.

Example 6.22

EDTA is a polyprotic acid with four protons (H_4Y); it can be further protonated to form a hexaprotic acid (H_6Y^{2+}). Calculate the hydrogen ion concentration of a 0.0100 M solution of Na_2EDTA (Na_2H_2Y).

Solution

The two equilibria involving H_2Y^{2-} are

$$H_2Y^{2-} \rightleftharpoons H^+ + HY^{3-} \quad K_{a3} = 6.9 \times 10^{-7}$$

and

$$H_2Y^{2-} + H_2O \rightleftharpoons H_3Y^- + OH^- \quad K_b = \frac{K_w}{K_{a2}} = \frac{1.0 \times 10^{-14}}{2.2 \times 10^{-3}}$$

Compared to the previously considered case of the partially neutralized salt of a diprotic acid, H_2Y^{2-} is the equivalent of HA^-, and H_3Y^- is the equivalent of H_2A. The equilibrium constants involved are K_{a2} and K_{a3} (the former for the conjugate acid H_3Y^- of the hydrolyzed salt). Thus,

$$[H^+] = \sqrt{K_{a2}K_{a3}} = \sqrt{(2.2 \times 10^{-3})\,(6.9 \times 10^{-7})}$$
$$= 3.9 \times 10^{-5}\ M$$

6.13 Physiological Buffers—They Keep You Alive

The pH of the blood in a healthy individual remains remarkably constant in a range of 7.35 to 7.45. This is because the blood contains a number of buffers that protect against pH change due to the presence of acidic or basic metabolites. From a physiological viewpoint, a change of ± 0.3 pH units is extreme. Acid metabolites are ordinarily produced in greater quantities than basic metabolites, and carbon dioxide is the principal one. The buffering capacity of blood for handling CO_2 is estimated to be distributed among various buffer systems as follows: hemoglobin and oxyhemoglobin, 62%; $H_2PO_4^-/HPO_4^{2-}$, 22%; plasma proteins, 11%; HCO_3^-, 5%. Proteins contain carboxylic and amino groups, which are weak acids and bases, respectively. They are, therefore, effective buffering agents. The combined buffering capacity of blood to neutralize acids is designated by clinicians as the "alkali reserve," and this is frequently determined in the clinical laboratory. Certain diseases cause disturbances in the acid balance of the body. For example, diabetes may give rise to "acidosis," which can be fatal.

An important diagnostic analysis is the CO_2/HCO_3^- balance in blood. This ratio is related to the pH of the blood by the Henderson–Hasselbalch Equation 6.45:

$$\boxed{pH = 6.10 + \log \frac{[HCO_3^-]}{[H_2CO_3]}} \tag{6.101}$$

where H_2CO_3 can be considered equal to the concentration of dissolved CO_2 in the blood; 6.10 is pK_{a1} of carbonic acid in blood at body temperature (37°C). Normally, the HCO_3^- concentration in blood is about 26.0 mmol/L, while the concentration of carbon dioxide is 1.3 mmol/L. Accordingly, for the blood,

$$pH = 6.10 + \log \frac{26\ \text{mmol/L}}{1.3\ \text{mmol/L}} = 7.40$$

The CO_2/HCO_3^- balance can be assessed from measuring two of the parameters in Equation 6.101.

The HCO_3^- concentration may be determined by titrimetry (Experiment 10), or the total carbon dioxide content (HCO_3^- + dissolved CO_2) can be determined by acidification and measurement of the evolved gas.[7] If both analyses are performed, the ratio of HCO_3^-/CO_2 can be calculated, and hence conclusions can be drawn concerning acidosis or alkalosis in the patient. Alternatively, if the pH is measured (at 37°C), either HCO_3^- or total CO_2 need be measured for a complete knowledge of the carbonic acid balance because the ratio of $[HCO_3^-]/[H_2CO_3]$ can be calculated from Equation 6.101.

The partial pressure, p_{CO_2} of CO_2, may also be measured (e.g., using a CO_2 electrode), in which case $[H_2CO_3] \approx 0.30 p_{CO_2}$. Then, only pH or $[HCO_3^-]$ need be determined.

---

[7] The volume of CO_2 is measured, but from the temperature and atmospheric pressure, the number of millimoles of CO_2 and hence its concentration in mmol/L in the solution it originated from can be calculated. At standard temperature and pressure (0°C and 1 atm pressure), 22.4 L contain one mole gas.

Note that these equilibria and Equation 6.101 hold although there are other buffer systems in the blood. The pH is the result of all the buffers and the $[HCO_3^-]/[H_2CO_3]$ ratio is set by this pH.

The HCO_3^-/H_2CO_3 buffer system is the most important one in buffering blood in the lung (alveolar blood). As oxygen from inhaled air combines with hemoglobin, the oxygenated hemoglobin ionizes, releasing a proton. This excess acid is removed by reacting with HCO_3^-.

$$H^+ + HCO_3^- \rightarrow H_2CO_3$$

But note that the $[HCO_3^-]/[H_2CO_3]$ ratio at pH 7.4 is $26/1.3 = 20:1$. This is not a very effective buffering ratio; and as significant amounts of HCO_3^- are converted to H_2CO_3, the pH would have to decrease to maintain the new ratio. But, fortunately, the H_2CO_3 produced is rapidly decomposed to CO_2 and H_2O by the enzyme *carbonic anhydrase*, and the CO_2 is exhaled by the lungs. Hence, the ratio of HCO_3^-/H_2CO_3 remains constant at 20:1.

Example 6.23

The total carbon dioxide content ($HCO_3^- + CO_2$) in a blood sample is determined by acidifying the sample and measuring the volume of CO_2 evolved with a Van Slyke manometric apparatus. The total concentration was determined to be 28.5 mmol/L. The blood pH at 37°C was determined to be 7.48. What are the concentrations of HCO_3^- and CO_2 in the blood?

Solution

$$pH = 6.10 + \log \frac{[HCO_3^-]}{[CO_2]}$$

$$7.48 = 6.10 + \log \frac{[HCO_3^-]}{[CO_2]}$$

$$\log \frac{[HCO_3^-]}{[CO_2]} = 1.38$$

$$\frac{[HCO_3^-]}{[CO_2]} = 10^{1.38} = 24$$

$$[HCO_3^-] = 24[CO_2]$$

But

$$[HCO_3^-] + [CO_2] = 28.5 \text{ mmol/L}$$
$$24[CO_2] + [CO_2] = 28.5$$
$$[CO_2] = 1.1_4 \text{ mmol/L}$$
$$[HCO_3^-] = 28.5 - 1.1 = 27.4 \text{ mmol/L}$$

6.14 Buffers for Biological and Clinical Measurements

Many biological reactions of interest occur in the pH range of 6 to 8. A number, particularly specific enzyme reactions that might be used for analyses (see Chapter 23), may occur in the pH range of 4 to 10 or even outside of this. The proper selection of buffers for the study of biological reactions or for use in clinical analyses can be critical in determining whether or not they influence the reaction. A buffer must have the correct pK_a, near physiological pH so the ratio of $[A^-]/[HA]$ in the Henderson–Hasselbalch equation is not too far from unity, and it must be physiologically compatible.

Phosphate Buffers

One useful series of buffers are phosphate buffers. Biological systems usually contain some phosphate already, and phosphate buffers will not interfere in many cases. By choosing appropriate mixtures of $H_3PO_4/H_2PO_4^-$, $H_2PO_4^-/HPO_4^{2-}$, or HPO_4^{2-}/PO_4^{3-}, solutions over a wide pH range can be prepared. See G. D. Christian and W. C. Purdy, *J. Electroanal. Chem.*, **3** (1962) 363 for the compositions of a series of phosphate buffers at a constant ionic strength of 0.2. Ionic strength is a measure of the total salt content of a solution (see Chapter 5), and it frequently influences reactions, particularly in kinetic studies. Hence, these buffers could be used in cases where the ionic strength must be constant. However, the buffering capacity decreases markedly as the pH approaches the values for the single salts listed, and the single salts are not buffers at all. The best buffering capacity, obtained at the half neutralization points, is within ± 1pH unit of the respective pK_a values, that is, 1.96 ± 1, 7.12 ± 1, and 12.32 ± 1.

Example 6.24

What weights of NaH_2PO_4 and Na_2HPO_4 would be required to prepare 1 L of a buffer solution of pH 7.45 that has an ionic strength of 0.100?

Solution

Let $x = [Na_2HPO_4]$ and $y = [NaH_2PO_4]$. There are two unknowns, and two equations are needed. (Remember there must be the same number of equations as unknowns to solve.) Our first equation is the ionic strength equation:

$$\mu = \tfrac{1}{2} \sum C_i Z_i^2$$

$$0.100 = \tfrac{1}{2}[Na^+](1)^2 + [HPO_4^{2-}](2)^2 + [H_2PO_4^-](1)^2$$

$$0.100 = \tfrac{1}{2}[(2x + y)(1)^2 + x(2)^2 + y(1)^2]$$

$$0.100 = 3x + y \tag{1}$$

Our second equation is the Henderson–Hasselbalch equation:

$$pH = pK_{a2} + \log \frac{[HPO_4^{2-}]}{[H_2PO_4^-]}$$

$$7.45 = 7.12 + \log \frac{x}{y} \tag{2}$$

$$\frac{x}{y} = 10^{0.33} = 2.1_4$$

$$x = 2.1_4 y \tag{3}$$

Substitute in (1):

$$0.100 = 3(2.1_4)y + y$$

$$y = 0.013_5 \ M = [\text{NaH}_2\text{PO}_4]$$

Substitute in (3):

$$x = (2.1_4)\left(0.013_5\right) = 0.028_9 \ M = [\text{Na}_2\text{HPO}_4]$$

$$g\text{NaH}_2\text{PO}_4 = 0.013_5 \ \text{mol/L} \times 120 \ \text{g/mol} = 1.6_2 \ \text{g/L}$$

$$g\text{Na}_2\text{HPO}_4 = 0.028_9 \ \text{mol/L} \times 142 \ \text{g/mol} = 4.1_0 \ \text{g/L}$$

Solving Example 6.24 Using Excel Solver

We can use Example 6.21.xlsx as a template and use only one row of the numbers. Since we know that the ionic strength is going to be 0.1, CP must be significantly less than 0.1. On an initial basis, we enter both 0.1 for CP and CNa. We enter pH (cell B7) as 7.45—this is not a variable in the present problem; this has been given to us. We insert another column before the equation, ionic strength, and in cell J7 express it as

$$= 0.5^*(C7 + D7 + E7 + A7^*(G7 + 4^*H7 + 9^*I7))$$

We need to vary CP and CNA to get the charge balance equation expression to be zero and ionic strength I to be 0.1. The latter is tantamount to saying that we want to solve the equation $I - 0.1 = 0$. In cell L7 we write this expression: $= J7\text{-}0.1$. Now in cell M7 we have the sum of the squares of the two equation expressions we want to solve, multiplied by 10^{10}. As constraints we merely put in that CP must be less than 0.1. Solver is invoked; we ask cell M7 to be minimized by changing CP (A7) and CNA (D7).

Solver comes up with the solution that CP = 0.0424 M and CNA = 0.0712 M. If I start by making 0.0424 M NaH_2PO_4, I will have left 0.0712-0.0424 = 0.0288 M Na, this will then go to making 0.0288 M Na_2HPO_4, leaving 0.0424-0.0288 = 0.0136 M NaH_2PO_4.

The solved problem is presented in the **website** as 6.24.xlsx.

The use of phosphate buffers is limited in certain applications. Besides the limited buffering capacity at certain pH values, phosphate will precipitate or complex many polyvalent cations, and it frequently will participate in or inhibit a reaction. It should not be used, for example, when calcium is present if its precipitation would affect the reaction of interest.

Tris Buffers

A buffer that is widely used in the clinical laboratory and in biochemical studies in the physiological pH range is that prepared from *tris*(hydroxymethyl)aminomethane $[(\text{HOCH}_2)_3\text{CNH}_2$ —**Tris**, or **THAM**] and its conjugate acid (the amino group is protonated). It is a primary standard and has good stability, has a high solubility in physiological fluids, is nonhygroscopic, does not absorb CO_2 appreciably, does not precipitate calcium salts, does not appear to inhibit many enzyme systems, and is compatible with biological fluids. It has a pK close to physiological pH (pK_a = 8.08 for the conjugate acid), but its buffering capacity below pH 7.5 does begin to diminish, a disadvantage. Other disadvantages are that the primary aliphatic amine has considerable potential reactivity and it is reactive with linen fiber junctions, as found in saturated calomel reference electrodes used in pH measurements (Chapter 20); a reference electrode with a ceramic, quartz, or sleeve junction should be used. These buffers are usually prepared by adding an acid such as hydrochloric acid to a solution of Tris to adjust the pH to the desired value.

Tris buffers are commonly used in clinical chemistry measurements.

Good Buffers

Norman E. Good and coworkers sought to prepare reasonably inexpensive stable optically transparent ($\lambda \geq 230$ nm) buffering components to buffer in the biologically important pH range of 6–8. They wanted these buffers to have good solubility in water but poor solubility in lipids (and hence will not permeate through cell membranes), exhibit minimum salt effects (see following Section 6.12) and temperature effects, and not interact with typical cations that are present in biological systems (at least not precipitate). Based on their experimental study came up with a list of suggested buffering compounds that also included Tris. A few are not easily commercially available, the rest are listed in Table 6.3.

TABLE 6.3 Good's Buffers[1]

Compound		pK$_a$ at 20°C
Morpholinoethane sulfonic acid (MES)		6.15
N-(2-acetamido)iminodiacetic acid (ADA)		6.6
Piperazine-N,N'-bis(ethanesulfonic acid)		6.8
N-(2-Acetamido)-2-aminoethanesulfonic acid		6.9
N,N-bis(2-hydroxyethyl)-2-aminoethanesulfonic acid (BES)		7.15
N-[Tris(hydroxymethyl)methyl] 2-aminoethanesulfonic acid (TES)		7.5
4-(2-hydroxyethyl)-1-piperazineethanesulfonic acid (HEPES)		7.55
N-[Tris(hydroxymethyl)methyl]glycine (Tricine)		8.15
Glycinamide		8.2
N-[Bis(hydroxymethyl)methyl]glycine (Bicine)		8.35

[1] N. E. Good, G. D. Winget, W. Winter, T. N. Connolly, S. Izawa, and R. M. M. Singh, *Biochemistry* 5 (1966) 467.

For a discussion of Good's buffers, see Q. Yu, A. Kardegedara, Y. Xu, and D. B. Rorabacher, "Avoiding Interferences from Good's Buffers: A Contiguous Series of Noncomplexing Tertiary Amine Buffers Covering the Entire pH Range of pH 3–11," *Anal. Biochem.*, **253(1)** (1997) 50–56.

6.15 Diverse Ion Effect on Acids and Bases: ^cK_a and ^cK_b—Salts Change the pH

In Chapter 5, we discussed the thermodynamic equilibrium constant based on activities rather than on concentrations. Diverse salts affect the activities and therefore the extent of dissociation of weak electrolytes such as weak acids or bases. The activity coefficient of the undissociated acid or base is essentially unity if it is uncharged. Then, for the acid HA,

$$K_a = \frac{a_{H^+} \cdot a_{A^-}}{a_{HA}} \approx \frac{a_{H^+} \cdot a_{A^-}}{[HA]} \tag{6.102}$$

$$K_a = \frac{[H^+]f_{H^+} \cdot [A^-]f_{A^-}}{[HA]} = {}^cK_a f_{H^+} f_{A^-} \tag{6.103}$$

$$^cK_a = \frac{K_a}{f_{H^+}f_{A^-}} \tag{6.104}$$

where K_a is the true equilibrium constant at zero ionic strength and ^cK_a is the "concentration constant" effective at a finite ionic strength.

Therefore, we would predict an increase in ^cK_a and in the dissociation with increased ionic strength, as the activity coefficients decrease. See Example 5.18, and Problem 32 in Chapter 5. A similar relationship holds for weak bases (see Problem 60 at the end of this chapter).

In the following discussions, by pH, we mean the pH measured with an electrode; this is $-\log a_{H^+}$. Since the ionic strength affects the dissociation of weak acids and bases, it will have an effect on the pH of a buffer. We can write the Henderson–Hasselbalch equation as:

$$pH = pK_a + \log \frac{a_{A^-}}{[HA]} = pK_a + \log \frac{[A^-]}{[HA]} + \log f_{A^-} \tag{6.105}$$

By adding some indifferent salt (e.g., NaCl) to a buffer, the concentration ratio $[A^-]/[HA]$ in a buffer will not change. In Equation 6.105, the only term on the right that is affected by salt addition or dilution is the log f_A term. If a buffer solution is diluted, its ionic strength will decrease, and f_{A^-} and log f_{A^-} will increase, and so will pH. See Footnote 4, earlier in this chapter.

For a $HPO_4^{2-}/H_2PO_4^-$ buffer, the ratio of $a_{HPO_4^{2-}}/a_{H_2PO_4^-}$ will also decrease with increased ionic strength because the effect is greater on the multiply charged ion.

6.16 log C—pH Diagrams

Plots of concentrations of various acid–base species vs. pH are essentially log–log plots that help visualize the status of a system over a large range of concentrations and pH. Especially when combined with the corresponding lines for [H⁺] and [OH⁻], they provide a global overview that is helpful to understand the nature of a system, even a complex one. It allows for approximate pH estimates of simple and even some reasonably complex systems. It may be argued, however, that exact pH computations of such systems may be even more straightforward by the charge balance method (see Examples 6.19 and 6.19b on the text website) relative to the effort needed to generate a log C-pH diagram. A more detailed exposition of this is therefore deferred to the text website but here we present a general description of how to create such a diagram and its usefulness. Regardless of whether exact solutions by the charge balance approach is superior, log C-pH diagrams provide a visual perspective that is unmatched by other approaches. In a way they are much fancier versions of ladder diagrams.

The first step to create such a diagram is to make a spreadsheet that has the necessary data for plotting. We could use the α-value spreadsheet that we generated previously, but let us create a spreadsheet expressly for this purpose and one that can be generically used to generate any log C-pH diagram incorporating one or more acid–base systems. The master spreadsheet is a valuable tool that is on the text website as log C-pH Master.xlsx and you should refer to that in our discussion here. Note again that for the present purposes, even if a base is involved, it is easier to treat it in terms of its conjugate acid. Referring to the spreadsheet logC-pH Master.xlsx, a partial screen shot of which appears as Figure 6.4, the first three columns (A:C) (download the spreadsheet from the website and open) has pH values (0-14, 0.1 pH units apart, A9:A149) and columns B and C have corresponding values of [H^+] and [OH^-] (calculated respectively as 10^{-pH} and $10^{-(14-pH)}$). The top of the spreadsheet has details for the constants and concentrations of up to two acid–base systems A and B. These are both set up as tetraprotic acid systems (with respective dissociation constants KAAA1, KAAA2, KAAA3, KAAA4, and KAAB1, KAAB2, etc.) and total concentrations of CONCA and CONCB, respectively. Note that mathematical considerations of a n-protic acid system *automatically* reduce to that for a $(n-1)$-protic acid system if K_n is entered as zero. Thus, if KAAA4 is entered as zero, system A will be treated as a triprotic acid, if KAAA4 and KAAA3 are both entered as zero, system A will be treated as a diprotic acid, etc. In fact, the spreadsheet will open with nonzero values for only KAAA1 and KAAB1, with all the others entered as zero, i.e., it will open with default values for two monoprotic acid–base systems (KAAA1 = 1.75×10^{-5}, the same as that for HOAc, and KAAB1 = 5.71×10^{-10}, the same as that for NH_4^+), both with default concentrations of 0.1 M. So as it opens, the spreadsheet corresponds to the situation for 0.1 M NH_4OAc.

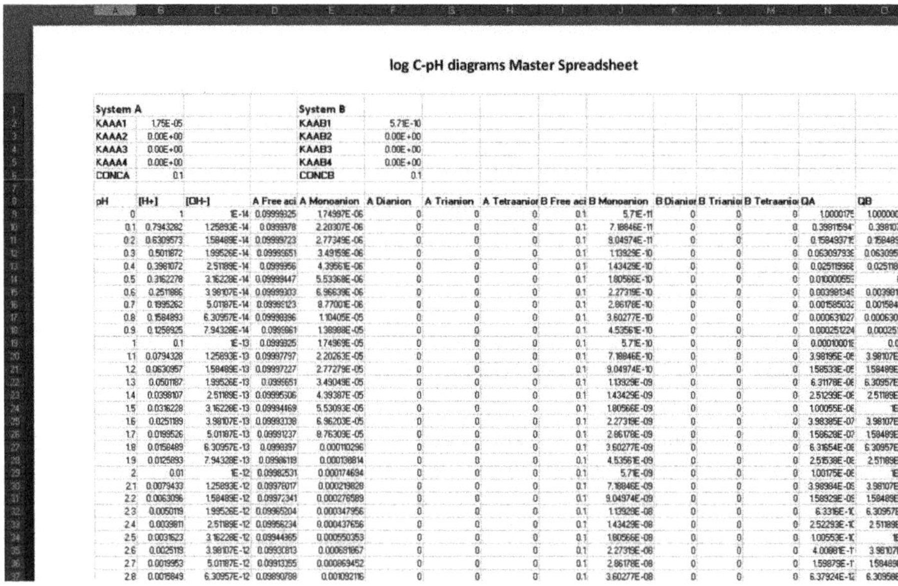

FIGURE 6.4 Screen shot of the top portion of Figure 6.5.xlsx.

Leaving 10 columns (D:M) as room for the five A-species and B-species, we enter the formulas for the alpha value denominators in columns N and O (titled QA and QB) in row 9. The concentrations of the various A-species $C\alpha_0$ through $C\alpha_4$ are then formulated in columns D:H and those for the corresponding B-species are formulated in columns I:M. Selecting D9:O9 and double-clicking on the right bottom corner of O9 will now fill up the entire spreadsheet.

For the 0.1 M NH$_4$OAc system, before we plot, let us delete (Alt-E D, Shift cells left) the A anions that don't exist for this system (F8:H149) and remove the empty space. We relabel "A Free Acid" and "A Monoanion" as "HOAc" and "OAc$^-$," respectively, and "B Free Acid" and "B Monoanion" as "NH$_4$$^+$" and "NH$_3$," respectively. We now select columns A through G and plot by selecting **Insert-Charts-Scatter**. Pick the plot with smooth lines without points; we have too many points for individual data points to be shown. Click on Layout 3 to get the grid and the linear trend lines, then delete the trend lines (you may find it more expedient to precisely click on the lines if you expand the graph by dragging outward from the corners of the frame and clicking on the lines where the plot is least crowded). Now let us label the axes, click on the **Axis Title** for the X-axis, highlight "Axis Title" and type in "pH." Similarly change the Y-axis title to read "Concentration." We do not want the pH axis to read up to 16, which it presently does. Hover your cursor around the tick mark for "16" on the X-axis until you see the label "**Horizontal (Value) Axis**"; now right-click. You will have a pop-up menu with the bottom entry being "**Format Axis**." Click on "Format Axis." A new "Format Axis" menu will appear. In the second row for the maximum value of the X-axis, click on the **Fixed** button and enter 14 for the maximum value. While you can keep your multicolor plot as is, for the purposes of the book we would like to make it in shades of gray. So we click on **Style 1** on the menu bar in **Chart Styles**.

The plot that appears is the C vs pH plot. To see the concentration changes better over a larger range, we need to make it a log C vs pH plot. Rather than computing the log of the concentrations, it is easier to make the ordinate logarithmic. Move the cursor again slowly near where the Y-axis has tick marks with the numeric labels—when you see a label pop up that says "**Vertical (Value) Axis**," you are in the right place; right-click now. Click on **Format Axis**. A new "Format Axis" menu will appear. Tick the box that says logarithmic scale. To avoid the Y-axis labels running into the Axis title, we should make the labels all appear in scientific notation. Format the Y-axis again, click on **Number** on the **Format Axis** menu (second entry on the left pane) and pick **Scientific**. This problem is gone. Now we may want to use dashed lines for some of the traces to make the traces more clearly different; let us do this for the HOAc and the NH$_3$ traces. Looking at the right edge of the graph, from the bottom, the second line is the HOAc trace. Pick this trace by right-clicking on it and in the pop-up menu, select **Format Data Series**. When this menu appears, click on **Line Style**, click on the pull-down arrow on the **Dash type** submenu and pick the short dashes (fourth entry from top). Repeat this process for the NH$_3$ trace—this is the second line from the bottom on the left edge of the graph.

You will now have the figure depicted as Figure 6.5, the log C-pH diagram for the system. This shows the distribution of all the species in the system at any pH. The diagonal line with −1 slope is the [H$^+$] line and the diagonal line with +1 slope is the [OH$^-$] line. Except at extremes of pH, these concentrations are much lower than the HOAC/OAc$^-$ or NH$_4$$^+$/NH$_3$ species.

Can we determine from Figure 6.5 what the pH of a 0.1 M NH$_4$OAc solution is? For starters, it is good to assume that the dominant species is what you are putting in. If we are putting in NH$_4$OAc, let us assume that NH$_4$$^+$ and OAc$^-$ are the dominant species and if so they must be equal. Indeed, over the entire range of pH 5.5–8.5 both of these are almost the same ~0.1 M. While this can tell us that NH$_4$OAc will

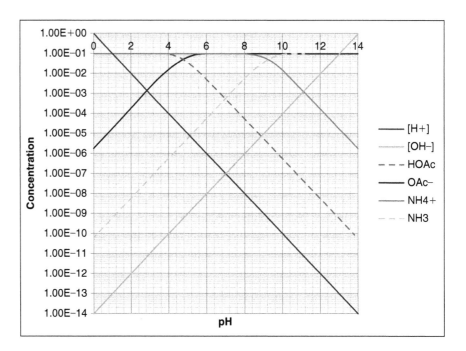

FIGURE 6.5 log C-pH diagram for 0.1 M NH$_4$OAc

have good buffer intensity across this entire pH range, it does not tell us accurately what the pH of the solution is. $[NH_4^+] = [OAc^-]$ where the two traces intersect each other. If you greatly magnify the Y-axis scale in this region by formatting the axis to have a minimum and maximum value of 0.09 and 0.1, you will be able to see that the intersection is at pH 7. If you want to magnify the X-axis by limiting its span between 6 and 8, you will be able to see that the intersection occurs at pH 7.0.

As an exercise, make a log C-pH diagram for 0.1 M (NH$_4$)$_2$HPO$_4$. You will enter the three relevant dissociation constants of H$_3$PO$_4$ in cells B2:B4, and noting that the concentration of the NH$_4^+$/NH$_3$ system will be 0.2 M, change the value in cell F6 (CONCB) to read 0.2. Once again assume that the initial species put in (NH$_4^+$ and HPO$_4^{2-}$) are the dominant species. If so, the charge balance equation will be

$$[NH_4^+] = 2\,[HPO_4^{2-}]$$

If you have generated the log C-pH diagram correctly, you will be able to verify that at a pH of ~8.4, the $[NH_4^+]$ trace has twice the value of the $[HPO_4^{2-}]$ trace and also that at this pH the concentration of all other charged species are much lower and the equality above, which neglect all other charged species, is defensible. The text website has the relevant spreadsheet and a log C-pH diagram for the system. See also the applet in Section 6.11 that provides both distribution and log plots for mono- to tetraprotic acids (the log plots show the different acid species, but not $[H^+]$ and $[OH^-]$).

6.17 Exact pH Calculators

A powerful program for calculating pH values of simple to complex mixtures of strong and weak acids and bases is described at: www.phcalculation.se. The program is accessed from the text website and detailed instructions are given therein. (See Reference for QR code access to see a description of the program.) The Newton–Raphson approach is used. Calculations can be made using concentrations (ConcpH) or activities (ActpH). Activity coefficients are automatically calculated from the input concentrations and ion size parameters for the Debye–Hückel equation (see Chapter 5). Author of this text worked with Sig Johansson in Sweden to refine this versatile program for activity calculations. Equilibrium concentrations of all species are calculated. ActpH also calculates activity coefficients of the species and the ionic strength. The two workable programs are on your text **website**.

The program requires input of pK values for ConcpH and concentrations of the reacting species, both acids and bases. For ActpH, additionally, ion size parameters are required, obtained from the Kielland table, Reference 9 in Chapter 5 (and given on the pHcalculation website). ActpH is what a pH meter measures, i.e., $-\log a_{H^+}$ (see Chapter 20).

The pH calculator website gives the following example of a complex mixture:

Calculate pH for a mixture of $0.012\ M\ K_2HAsO_4$, $0.020\ M\ NaH_2PO_4$, $0.013M\ K_2HAsO_4$, and $0.0021\ M\ NaOH$.

$H_3PO_4 : pK_1 = 2.15\quad pK_2 = 7.21\quad pK_3 = 12.36\quad H_3A = 0\quad H_2A^- = 0.020$

$HA^{2-} = 0\quad A^{3-} = 0.012\ H_3AsO_4 : pK_1 = 2.25\quad pK_2 = 7.00\quad pK_3 = 11.52$
$H_3A = 0\quad H_2A^- = 0\ HA^{2-} = 0.013\ A^{3-} = 0\quad MeOH = 0.0021$
$NH_4^+ : K_b = 4.76\quad B = 0\ BH^+ = 0.024$

Inputting the data gives (if you go to the text's website, Stig Johannson folder, 4. Products, you can more readily view and compare this and the next example – the illustrations here give you a quick overview and show the power of the calculator):

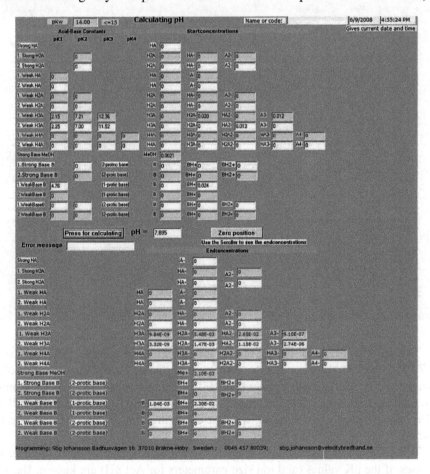

The calculated pH is 7.985. The equilibrium concentrations of all species are given.

To calculate ActpH for the same mixture, the following addition must be done. The values of ion diameters:

For all acid ions $d = 4$, $NH_4^+ d = 3$, $Li^+ d = 6$, $Na^+ d = 4$, $K^+ d = 3$

The concentrations for the nonhydrolytic ions: $Li^+ = 0.026$, $Na^+ = 0.0221$ and $K^+ = 0.026$

Inputting the parameters gives:

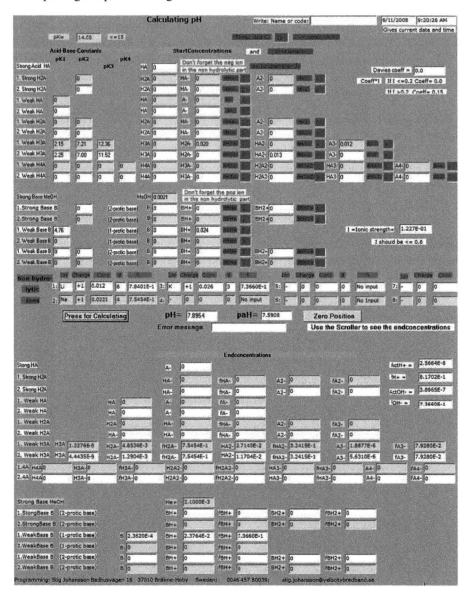

This program calculates both ConcpH and ActpH. The latter is 7.5908, compared with 7.8354 for ConcpH. The calculations should be rounded to 0.01 pH since absolute pH cannot be measured with greater accuracy, although discriminations or changes at the 0.001 pH level may be of interest. Note that the activity coefficients of all the equilibrium species are given, along with the concentrations, and so the activities are known.

This program can be used to calculate the pH of virtually any mixture of acids and bases, if the pK values (and ion size parameters for Act pH) are known. Use this example as a practice. Then try making up a mixture and calculate the pH.

A second useful program is CurTiPot by Ivano Gutz, Universidad de São Paulo, Brazil: http://www2.iq.usp.br/docente/gutz/Curtipot_.html. (See Reference 16 for QR code access.)

It is a powerful and very versatile program that performs the same pH and paH calculations, but also titration curves, alpha plots, etc. It provides:

- pH calculation of any aqueous solution of acids, bases, and salts, including buffers, zwitterionic amino acids, from single component to complex mixtures (30 or more species in equilibrium)

- buffer capacity, ionic strength, fractional distribution, activities and apparent dissociation constants of all species at equilibrium.

It has a higher learning curve, but with practices, it provides a wealth of information. You might try doing the same calculations using both programs. You should get the same result.

We can also solve the pH of mixtures of acids and bases using Goal Seek. See the **video** illustrating this for a mixture of NaOH and H_2CO_3.

 # Professor's Favorite Problems

⌐Contributed by Professor
Michael De Grandpre,
University of Montana   ⌡

Example 6.25

pH of Seawater

Did you know that ocean pH is decreasing (see the figures on the next page)? A portion of the CO_2 from fossil fuel combustion is absorbed by the oceans, forming carbonic acid. This process, named "ocean acidification," has decreased the average pH in the surface oceans by 0.1 units over the past ~100 years. Chemical oceanographers have long realized the importance of tracking and studying CO_2 in the oceans and in the 1980s began developing improved analytical tools to do this, including new methods for measuring pH. Glass pH electrodes, the workhorse for most pH applications, are not accurate enough to document these small pH changes over time. Oceanographers revitalized an old but rarely used method, the spectrophotometric measurement of pH using indicators. As you may have guessed, the function for deriving the pH is simply the Henderson–Hasselbalch equation:

$$pH = pK_a' + \log \frac{[A^-]}{[HA]}$$

where pH is defined on the total hydrogen ion scale (for a definition and description see References 17 and 19.). The pK_a' is the apparent dissociation constant, and $[A^-]$ and $[HA]$ are the unprotonated and protonated forms of the pH indicator. The improvement came in the determination of pK_a' on a pH scale consistent with the CO_2 equilibria in seawater (see References 19 and 20). The indicator concentrations $[A^-]$ and $[HA]$ are quantified on a spectrophotometer by recording 100% transmittance with pure seawater (the blank) and adding a small amount of indicator to the seawater sample at a fixed temperature. Using Beer's law you might expect the ratio $[A^-]/[HA]$ to be:

$$\frac{[A^-]}{[HA]} = \frac{A_2 \varepsilon_{1HA}}{A_1 \varepsilon_{2A}} \tag{6.106}$$

where A_i and ε are the indicator absorbances and molar absorptivities, respectively, at the absorbance maxima for the protonated (1) and unprotonated (2) forms (see the second figure on the next page). However, for the sulfonephthalein indicators used for

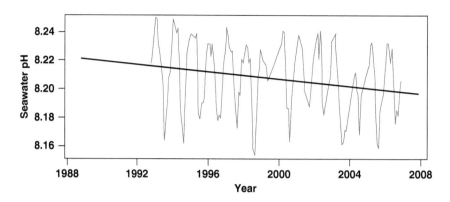

(**Source:** N. R Bates/From M.D. Degrandpre. Data for plot , courtesy of Nicholas B. Bates Reproduced with permission.)

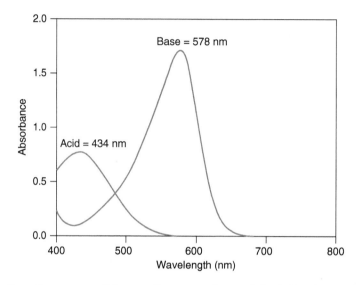

(**Source:** M. D. Degranpre/From M.D. Degranpre Reproduced with permission.)

this application, the spectra of the two forms overlap, and the following equation must be used (References 20 and 21):

$$pH = pK_a' + \log \left[\frac{(A_2/A_1) - e_1}{e_2 - e_3(A_2/A_1)} \right] \tag{6.107}$$

where $e_1 = \varepsilon_{2HA}/\varepsilon_{1\,HA}$, $e_2 = \varepsilon_{2A}/\varepsilon_{1\,HA}$, and $e_3 = \varepsilon_{1A}/\varepsilon_{1\,HA}$. Equation 6.107 gives very reproducible measurements of seawater pH at a given temperature and salinity.[8]

Example 6.26

At the Bermuda Atlantic Time-Series station (BATS), a sample taken from the ocean surface is analyzed using the indicator meta-cresol purple (mCP). The analytical wavelengths are 434 nm and 578 nm, corresponding to the absorbance maxima of the protonated and unprotonated forms, respectively. At these wavelengths e_1, e_2, and e_3 and are equal to 0.00691, 2.222, and 0.1331, respectively. The mCP $pK_a' = 1245.69/T + 3.8275 + (2.11 \times 10^{-3})(35 - S)$ where T is temperature in Kelvin and S is salinity

[8]In very recent work, to avoid any effect of the indicator addition on the sea water pH, measurements are made by adding different amounts of indicator and extrapolating the measured pH to zero indicator added. A single injection of the indicator into a flowing sea water stream, using basic flow-injection techniques can generate the necessary data.

(parts per thousand) (Reference 20). A small amount of mCP is added to the seawater sample and the absorbances, measured using a UV–VIS spectrophotometer are found to be $A_1 = 0.3892$ and $A_2 = 0.8528$. What is the pH of the seawater sample? The measurement temperature is 22.3°C and salinity = 35.16 ppt. Small effects due to perturbation of the seawater pH from the weak acid indicator and e_i temperature dependence are neglected.

Solution

The absorbance is the result of the overlapping spectra of the two species. If measured at two wavelengths, typically at the absorbance maxima of the two forms, the absorbances are the sum of the individual species absorbances:

$$A_2 = \varepsilon_{2HA}b[HA] + \varepsilon_{2A}b[A^-] \tag{1}$$

$$A_1 = \varepsilon_{1HA}b[HA] + \varepsilon_{1A}b[A^-] \tag{2}$$

Solving (1) for [HA] and substituting into (2) gives:

$$[A^-] = \frac{A_2\varepsilon_{1HA} - A_1\varepsilon_{2HA}}{\varepsilon_{1HA}\varepsilon_{2A}b - \varepsilon_{2HA}\varepsilon_{1A}b}$$

Similarly, solving (2) for [A⁻] and substituting into (1) gives:

$$[HA] = \frac{A_1\varepsilon_{2A} - A_2\varepsilon_{1A}}{\varepsilon_{1HA}\varepsilon_{2A}b - \varepsilon_{2HA}\varepsilon_{1A}b}$$

Substituting [A⁻] and [HA] into the Henderson–Hasselbalch equation gives Equation 6.107 above.

$$pH = pK'_a + \log\left[\frac{(A_2/A_1) - e_1}{e_2 - e_3(A_2/A_1)}\right]$$

Where $e_1 = \varepsilon_{2HA}/\varepsilon_{1HA}$, $e_2 = \varepsilon_{2A}/\varepsilon_{1HA}$, and $e_3 = \varepsilon_{1A}/\varepsilon_{1HA}$. When there is no overlap $e_1 = e_3 = 0$ and the equation simplifies to Equation 6.106 in the example (this can be directly derived by using Equations 6.106 and 6.107 where $\varepsilon_{2HA} = \varepsilon_{1A} = 0$):

$$pH = pK'_a + \log\left[\frac{A_2\varepsilon_{1HA}}{A_1\varepsilon_{2A}}\right]$$

Substituting the values of T (Kelvin) and S in the pK'_a equation, and of e_1, e_2, e_3, A_1, and A_2 in the pH equation, gives pK'_a to be 8.0430 at this temperature and salinity. The seawater pH from the BATS site is calculated to be 8.0967.

Questions

1. Explain the difference between a strong electrolyte and a weak electrolyte. Is an "insoluble" salt a weak or a strong electrolyte?

2. Explain the Brønsted acid–base and Lewis acid–base theory.

3. Differentiate between conjugate acid and conjugate base.

4. Write the ionization reaction of aniline, $C_6H_5NH_2$, in glacial acetic acid, and identify the conjugate acid of aniline. Write the ionization reaction of phenol, C_6H_5OH, in ethylene diamine, $NH_2CH_2CH_2NH_2$, and identify the conjugate base of phenol.

5. Write a brief note on good buffers.

Problems

Strong Acids and Bases

6. Describe the characterization of aqueous solution as acidic, basic or neutral.

7. Express hydronium and hydroxide ion concentrations on the pH and pOH scales.

8. Calculate the pH and pOH of the following strong acid solutions: (a) 0.050 M HClO$_4$, (b) $1.2 \times 10^{-2} M$ HNO$_3$, (c) 1.2 M HCl, (d) $1.2 \times 10^{-5} M$ HCl, (e) $2.0 \times 10^{-5} M$ HNO$_3$.

9. Calculate the pH and pOH of the following strong base solutions: (a) 0.030 M NaOH, (b) 0.12 M Ba(OH)$_2$, (c) 2.2 M NaOH, (d) $5.0 \times 10^{-5} M$ KOH, (e) $2.5 \times 10^{-5} M$ KOH

10. Calculate the hydroxide ion concentration of the following solutions: (a) 2.6×10^{-5} M HCl, (b) 0.020 M HNO$_3$, (c) 2.7×10^{-9} M HClO$_4$, (d) 1.9 M HClO$_4$.

11. Calculate the hydrogen ion concentration of the solutions with the following pH values: (a) 3.47, (b) 0.20, (c) 8.60, (d) −0.60, (e) 14.35, (f) −1.25.

12. Calculate the pH and pOH of a solution obtained by mixing equal volumes of 0.10 M H$_2$SO$_4$ and 0.30 M NaOH.

13. Calculate the pH of a solution obtained by mixing equal volumes of a strong acid solution of pH 3.00 and a strong base solution of pH 12.00.

PROFESSOR'S FAVORITE PROBLEMS

Contributed by Professor Noel Motta, University of Puerto Rico, Rio Piedras

14. V_a mL of a strong acid solution of pH 2.00 is mixed with V_b mL of a strong base solution of pH 11.00. Express V_a in terms of V_b if the mixture is neutral. The solution temperature is 24°C.

Temperature Effect

15. Calculate the hydrogen ion concentration and pH of a neutral solution at 50°C ($K_w = 5.5 \times 10^{-14}$ at 50°C).

16. Calculate the pOH of a blood sample whose pH is 7.40 at 37°C.

Weak Acids and Bases

17. The pH of an acetic acid solution is 3.26. What is the concentration of acetic acid and what is the percent acid ionized?

PROFESSOR'S FAVORITE PROBLEMS

Contributed by Professor Wen Yen Lee, The University of Texas at El Paso

18. K_a for acetic acid (CH$_3$COOH) is 1.75×10^{-5}. K_w is 1.00×10^{-14}. (a) Find K_b for acetate ion (CH$_3$COO$^-$). (b) When 0.1 M of sodium acetate (CH$_3$COONa) dissolves in water at 24°C, what is the pH of the solution? Assume the ions behave ideally.

19. The pH of a 0.20 M solution of a primary amine, RNH$_2$, is 8.42. What is the pK_b of the amine?

20. A monoprotic organic acid with a K_a of 6.7×10^{-4} is 3.5% ionized when 100 g of it is dissolved in 1 L. What is the formula weight of the acid?

21. Calculate the pH of a 0.25 M solution of propanoic acid.

22. Calculate the pH of a 0.10 M solution of aniline, a weak base.

23. Calculate the pH of a 0.1 M solution of iodic acid, HIO$_3$.

24. The first proton of sulfuric acid is completely ionized, but the second proton is only partially dissociated, with an acidity constant K_{a2} of 1.2×10^{-2}. Calculate the hydrogen ion concentration in a 0.0100 M H$_2$SO$_4$ solution.

25. Calculate the hydrogen ion concentration in a 0.100 M solution of trichloroacetic acid.

26. An amine, RNH_2, has a pK_b of 4.20. What is the pH of a 0.20 M solution of the base?

27. What is the concentration of a solution of acetic acid if it is 3.0% ionized?

28. By how much should a 0.100 M solution of a weak acid HA be diluted in order to double its percent ionization? Assume $C > 100K_a$.

Salts of Weak Acids and Bases

29. If 25 mL of 0.20 M NaOH is added to 20 mL of 0.25 M boric acid, what is the pH of the resulting solution?

30. Calculate the pH of a 0.010 M solution of NaCN.

31. Calculate the pH of a 0.050 M solution of sodium benzoate.

32. Calculate the pH of a 0.25 M solution of pyridinium hydrochloride (pyridine · HCl, $C_6H_5NH^+Cl$).

33. Calculate the pH of the solution obtained by adding 12.0 mL of 0.25 M H_2SO_4 to 6.0 mL of 1.0 M NH_3.

34. Calculate the pH of the solution obtained by adding 20 mL of 0.10 M HOAc to 20 mL of 0.10 M NaOH.

35. Calculate the pH of the solution prepared by adding 0.10 mol each of hydroxylamine and hydrochloric acid to 500 mL water.

36. Calculate the pH of a 0.0010 M solution of sodium salicylate, $C_6H_4(OH)COONa$.

37. Calculate the pH of a 1.0×10^{-4} M solution of NaCN.

Polyprotic Acids and Their Salts

38. Calculate the pH of the following polyprotic acids and their salts. (i) 0.0100 M phthalic acid (ii) 0.0100 M potassium phthalate (iii) 0.0100 M potassium acid phthalate (KHP) (iv) 0.050 M trisodium salt of EDTA, (Na_3HY) (v) 0.600 M solution of Na_2S (vi) 0.500 M solution of Na_3PO_4 (vii) 0.250 M solution of $NaHCO_3$ (viii) 0.600 M solution of NaHS

PROFESSOR'S FAVORITE PROBLEMS

Contributed by Professor Bin Wang, Marshall University

39. What is the dominant species in solution:

    (a) in a diprotic acid (H_2X) system if (i) pH > pK_{a2}; (ii) pK_{a1} < pH < pK_{a2}; and (c) pH < pK_{a1}?

    (b) in a triprotic acid system (H_3A) (i) if pH = $\frac{1}{2}(pK_2 + pK_3)$, (ii) pH > pK_a?

Buffers

40. What is the pH of a solution which is 0.050 M in formic acid and 0.10 M in sodium formate?

41. What is the pH of a solution prepared by mixing 5.0 mL of 0.10 M NH_3 with 10.0 mL of 0.020 M HCl?

42. An acetic acid–sodium acetate buffer of pH 5.00 is 0.100 M in NaOAc. Calculate the pH after the addition of 10 mL of 0.1 M NaOH to 100 mL of the buffer.

43. A buffer solution is prepared by adding 20 mL of 0.10 M sodium hydroxide solution to 50 mL of 0.10 M acetic acid solution. What is the pH of the buffer?

44. What is the pH of buffer prepared by adding 25 mL of 0.050 M sulfuric acid solution to 50 mL of 0.10 M ammonia solution?

45. Aspirin (acetylsalicylic acid) is absorbed from the stomach in the free (nonionized) acid form. If a patient takes an antacid that adjusts the pH of the stomach contents to 2.95 and then takes two 5-grain aspirin tablets (total 0.65 g), how many grams of aspirin are available for immediate absorption from the stomach, assuming immediate dissolution? Also assume that aspirin does not change the pH of the stomach contents. The pK_a of aspirin is 3.50, and its formula weight is 180.2.

46. *Tris*(hydroxymethyl)aminomethane [$(HOCH_2)_3CNH_2$—Tris, or THAM] is a weak base frequently used to prepare buffers in biochemistry. Its K_b is 1.2×10^{-6} and pK_b is 5.92. The corresponding pK_a is 8.08, which is near the pH of the physiological buffers, and so it exhibits good buffering capacity at physiological pH. What weight of THAM must be taken with 100 mL of 0.50 M HCl to prepare 1 L of a pH 7.40 buffer?

47. Calculate the hydrogen ion concentration for Problem 25 if the solution contains also 0.100 M sodium trichloroacetate.

PROFESSOR'S FAVORITE PROBLEMS

Contributed by Professor Bin Wang, Marshall University

48. Use the Henderson–Hasselbalch equation to find the value of $[C_6H_5COOH]/[C_6H_5COO^-]$ in a solution at (a) pH 3.00, and (b) pH 5.00. For C_6H_5COOH, pK_a is 4.20.

Buffers From Polyprotic Acids

49. What is the pH of a solution that is 0.20 M in phthalic acid (H_2P) and 0.10 M in potassium acid phthalate (KHP)?

50. What is the pH of a solution that is 0.25 M each in potassium acid phthalate (KHP) and potassium phthalate (K_2P)?

51. The total phosphate concentration in a blood sample is determined by spectrophotometry to be 3.0×10^{-3} M. If the pH of the blood sample is 7.45, what are the concentrations of $H_2PO_4^-$ and HPO_4^{2} in the blood?

PROFESSOR'S FAVORITE PROBLEMS

Contributed by Professor Bin Wang, Marshall University

52. A student weighed out 0.6529 g of anhydrous monohydrogen sodium phosphate (Na_2HPO_4) and 0.2477 g of dihydrogen sodium phosphate ($NaH_2PO_4H_2O$), then dissolved them into 100 mL of distilled water. What is the pH?

Buffer Intensity

53. A buffer solution contains 0.10 M NaH_2PO_4 and 0.070 M Na_2HPO_4. What is its buffer intensity in moles/liter per pH? By how much would the pH change if 10 μL (0.010 mL) of 1.0 M HCl or 1.0 M NaOH were added to 10 mL of the buffer?

54. You wish to prepare a pH 4.76 acetic acid–sodium acetate buffer with a buffer intensity of 1.0 M per pH. What concentrations of acetic acid and sodium acetate are needed?

Constant-Ionic-Strength Buffers

55. What weight of Na_2HPO_4 and KH_2PO_4 would be required to prepare 200 mL of a buffer solution of pH 7.40 that has an ionic strength of 0.20? (See Chapter 5 for a definition of ionic strength.)

56. What volume of 85% (wt/wt) H_3PO_4 (sp. gr. 1.69) and what weight of KH_2PO_4 are required to prepare 200 mL of a buffer of pH 3.00 that has an ionic strength of 0.20?

α Calculations

57. Calculate the equilibrium concentrations of the different species in a 0.0100 M solution of sulfurous acid, H_2SO_3, at pH 4.00 ($[H^+] = 1.0 \times 10^{-4}$ M.

58. Derive Equations 6.75, 6.76, and 6.77 for α_1, α_2, and α_3 of phosphoric acid.

Diverse Salt Effect

59. Calculate the hydrogen ion concentration for a 0.0200 M solution of HCN in 0.100 M NaCl (diverse ion effect).

60. Derive the equivalent of Equation 6.104 for the diverse salt effect on an uncharged weak base B.

Logarithmic Concentration Diagrams

You can use the HOAc spreadsheet exercise on the text website as a guide for Problem 61 (Spreadsheet Problems). Prepare a spreadsheet for Problem 64 using α-values—see the text website for HOAc log plots using α-values.

61. Construct the log–log diagram for a 10^{-3} M solution of acetic acid.

62. From the diagram in Problem 61, estimate the pH of a 10^{-3} M solution of acetic acid. What is the concentration of acetate ion in this solution?

63. For Problem 61, derive the expression for $\log[OAc^-]$ in acid solution and calculate the acetate concentration at pH 2.00 for a 10^{-3} M acetic acid solution. Compare with the value estimated from the log–log diagram.

64. Construct the log–log diagram for a 10^{-3} M solution of malic acid by preparing a spreadsheet using α values.

65. From the diagram in Problem 64, estimate the pH and concentrations of each species present in (a) 10^{-3} M malic acid and (b) 10^{-3} M sodium malate solution.

66. For Problem 64, derive the expressions for the HA^- curves in the acid and alkaline regions.

67. Derive expressions for (a) $\log[H_3PO_4]$ between pH = pK_{a1} and pK_{a2}, (b) $\log[H_2PO_4^-]$ between pH = pK_{a2} and pK_{a3}, (c) $\log[HPO_4^{2-}]$ at between pH = pK_{a2} and pK_{a1}, and (d) $\log[PO_4^{3-}]$ at between pH = pK_{a3} and pK_{a2}. Check with representative points on the curves.

68. Construct a log–log diagram for 0.001 M H_3PO_4 using α values. Start with the spreadsheet for Figure 6.2 (given in the text website). Compare the chart with Figure 6.16.3 on the text website (Addendum to Section 6.16). Vary the H_3PO_4 concentration and see how the curves change.

69. The Stig Johansson pH calculator has been shown to give pH calculations of NIST standard buffers that are within a few thousandths of a pH unit of the NIST values. The NIST buffers are given in Table 20.2 in Chapter 20. Use the calculator in Reference 15 to calculate the ActpH of the NIST phosphate buffer consisting of 0.025 M KH_2PO_4 and 0.025 M Na_2HPO_4 (footnote e) at 50°C, and compare with the NIST value of 6.833. Use $pK_w = 13.26$, $pK_1 = 2.25$, $pK_2 = 7.18$, and $pK_3 = 12.36$ for 50°C. Don't forget to enter the temperature.

70. Use the Stig Johansson pH calculator to calculate the pH in Problem 38(iv).

71. Use the Stig Johansson pH calculator to calculate the pH in Problem 38(vii).

PROFESSOR'S FAVORITE PROBLEMS

Contributed by Professor George S. Wilson, University of Kansas

72. Many geochemical processes are governed by simple chemical equilibria. One example is the formation of stalactites and stalagmites in a limestone cave, and is a good illustration of Henry's law. This is illustrated in the diagram below:

Rainwater percolates through soil. Due to microbial activity in the soil, the gaseous CO_2 concentration in the soil interstitial space (expressed as the partial pressure of CO_2, pCO_2, in atmospheres), is 3.2×10^{-2} atm, significantly higher than that in the ambient atmosphere

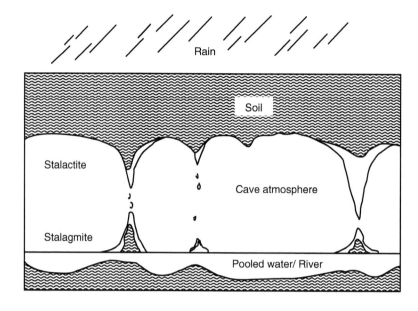

$(3.9 \times 10^{-4}$ atm; CO_2 concentration in the ambient atmosphere is presently increasing by 2×10^{-6} atm each year, see http://CO2now.org for the current atmospheric CO_2 concentration). Water percolating through the soil reaches an equilibrium (called Henry's law equilibrium) with the soil interstitial pCO_2 as given by Henry's law:

$$[H_2CO_3] = K_H pCO_2$$

where H_2CO_3 is the aqueous carbonic acid concentration and K_H is the Henry's law constant for CO_2, 4.6×10^{-2} M/atm at the soil temperature of 15°C. The CO_2-saturated water effluent from the soil layer then percolates through fractures and cracks in a limestone layer, whereupon it is saturated with $CaCO_3$. This $CaCO_3$ saturated water drips from the ceiling of the cave.

Because of the diurnal temperature variation outside the cave, the cave "breathes": the CO_2 concentration in the cave atmosphere is essentially the same as in ambient air $(3.9 \times 10^{-4}$ atm). Show that when the water dripping from the ceiling re-equilibrates with the pCO_2 concentration in the cave atmosphere, some of the calcium in the drip water will re-precipitate as $CaCO_3$, thus forming stalactites and stalagmites. Assume cave temperature to be 15°C as well. At this temperature the successive dissociation constants of H_2CO_3 are: $K_{a1} = 3.8 \times 10^{-7}$ and $K_{a2} = 3.7 \times 10^{-11}$, K_w is 4.6×10^{-15} and K_{sp} of $CaCO_3$ is 4.7×10^{-9}.

See the text website (as well as the Solutions Manual) for a detailed solution of this complex problem. Corresponding Goal Seek calculations are also given on the website.

PROFESSOR'S FAVORITE CHALLENGE

Contributed by Professor George S. Wilson, University of Kansas

73. The following 5 mathematical expressions can be used to approximately calculate H^+ concentrations in different contexts: (a) $\sqrt{K_a C_a}$ (b) $\dfrac{10^{-14}}{\sqrt{K_b C_b}}$ (c) $\sqrt{\dfrac{K_w K_a}{K_b}}$ (d) $\sqrt{\dfrac{C_a K_1 K_2}{K_1 + C_a}}$ (e) $\sqrt{\dfrac{C_a K_2 K_3}{K_2 + C_a}}$

   For each of the following salts in $0.10\ M$ aqueous solutions, match the most appropriate expression to calculate $[H^+]$: (i) K_2HPO_4, (ii) NH_4CN, (iii) CH_3NH_3Cl, (iv) Na_2CO_3, (v) $NaHSO_3$, (vi) $CH_3COONH(CH_3)_3$, (vii) Na_2H_2Y (where: $H_4Y = EDTA$, $H_4C_{10}H_{12}N_2O_8$).

Contributed by Professor Noel Motta, University of Puerto Rico

74. A given polyproticacid H_nX has the following fractional composition (alpha-values) vs. pH:

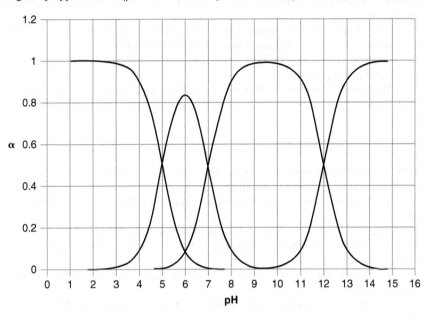

What is n?

If 15.0 mL of 0.10 M H_nX is titrated with 0.10 M NaOH, the titration curve should *clearly* show: (a) only an equivalence point at $V = 15.0$ mL; (b) an equivalence point at $V = 15.0$ mL, and another at $V = 30.0$ mL; (c) only an equivalence point at $V = 30.0$ mL; (d) an equivalence point at $V = 15.0$ mL, and another at $V = 45.0$ mL; (e) an equivalence point at $V = 15.0$ mL, another at $V = 30.0$ mL, and another at $V = 45.0$ mL.

Multiple Choice Questions

1. Choose the option that is equal to pK_a of a weak acid.
   - (a) The pK_b of its conjugate base
   - (b) Its relative molecular mass
   - (c) Equilibrium concentration of its conjugate base
   - (d) The pH of solution contains equal amount of the acid and its conjugate base.
2. A solution of 4×10^{-4} mol L^{-1} of HCl has pH _____.
   - (a) 3.21
   - (b) 3.40
   - (c) 2.67
   - (d) 4.31
3. An aqueous solution contains 2.5×10^{-8} OH$^-$ ions, calculate the pH of this solution.
   - (a) 7.40
   - (b) 6.40
   - (c) 6.42
   - (d) 7.60

Recommended References

Acid–Base Theories, Buffers

1. R. P. Buck, S. Rondini, A. K. Covington, F. G. K Baucke, C. M. A. Brett, M. F. Camoes, M. J. T. Milton, T. Mussini, R. Naumann, K. W. Pratt, and others, "Measurement of pH. Definition,

Standards, and Procedures," *Pure and Applied Chemistry*, **74** (2002) 2169. From oceanography paper as the authoritative reference on the modern definition and measurement of pH.

2. H. Kristensen, A. Salomen, and G. Kokholm, "International pH Scales and Certification of pH," *Anal. Chem.*, **63** (1991) 885A.
3. R. G. Bates, "Concept and Determination of pH," in I. M. Kolthoff and P. J. Elving, eds., *Treatise on Analytical Chemistry,* Part I, Vol. **1**. New York: Wiley-Interscience, 1959, pp. 361–401.
4. N. W. Good, G. D. Winget, W. Winter, T. N. Connally, S. Izawa, and R. M. M. Singh, "Hydrogen Ion Buffers for Biological Research," *Biochemistry*, **5** (1966) 467.
5. D. E. Gueffroy, ed., *A Guide for the Preparation and Use of Buffers in Biological Systems*. La Jolla, CA: Calbiochem, 1975.
6. D. D. Perrin and B. Dempsey, *Buffers for pH and Metal Ion Control*. New York: Chapman and Hall, 1974.

Equilibrium Calculations

7. S. Brewer, *Solving Problems in Analytical Chemistry*. New York: Wiley, 1980. Describes iterative approach for solving equilibrium calculations.
8. J. N. Butler, *Ionic Equilibria. A Mathematical Approach*. Reading, MA: Addison-Wesley, 1964.
9. W. B. Guenther, *Unified Equilibrium Calculations*. New York: Wiley, 1991.
10. D. D. DeFord, "The Reliability of Calculations Based on the Law of Chemical Equilibrium," *J. Chem. Ed.*, **31** (1954) 460.
11. E. R. Nightingale, "The Use of Exact Expressions in Calculating H⁺ Concentrations," *J. Chem. Ed.*, **34** (1957) 277.
12. R. J. Vong and R. J. Charlson, "The Equilibrium pH of a Cloud or Raindrop: A Computer-Based Solution for a Six-Component System," *J. Chem. Ed.*, **62** (1985) 141.
13. R. deLevie, *A Spreadsheet Workbook for Quantitative Chemical Analysis*. New York: McGraw-Hill, 1992.
14. H. Freiser, *Concepts and Calculations in Analytical Chemistry. A Spreadsheet Approach*. Boca Raton, FL: CRC Press, 1992.

Web pH Calculators

15. www.phcalculation.se. A program that calculates the pH of complex mixtures of strong and weak acids and bases using concentrations (ConcpH) or activities (ActpH) is described. Equilibrium concentrations of all species are calculated. The latter also calculates activity coefficients of the species and the ionic strength. The website for this programs is reproduced on the text's website and the program can be downloaded from there.
16. www2.iq.usp.br/docente/gutz/Curtipot.html. A free program that calculates the pH and paH of mixtures of acids and bases, and also titration curves and distribution curves, alpha plots, and more. The Curtipot program is available on the text's website.

Seawater pH Scales

17. I. Hansson, "A New Set of pH-Scales and Standard Buffers for Seawater". *Deep-Sea Research* **20** (1973) 479.
18. A. G. Dickon. "pH Scales and Proton-Transfer Reactions in Saline Media Such as Sea Water". *Geochim. Cosmochim. Acta* **48** (1984) 2299.
19. R. E. Zeebe and D. Wolf-Gladrow, CO$_2$ in Seawater: Equilibrium, Kinetics, Isotopes, Elsevier Science B.V., Amsterdam, Netherlands, 2001.
20. T. D. Clayton and R. H. Byrne, "Spectrophotometric Seawater pH Measurements: Total Hydrogen Ion Concentration Scale Calibration of m-Cresol Purple and At-Sea Results," *Deep-Sea Research*, **40** (1993) 2115–2129.
21. M. P. Seidel, M. D. DeGrandpre, and A. G. Dickson, "A Sensor for in situ Indicator-Based Measurements of Seawater pH," *Mar. Chem.*, **109** (2008) 18–28.

Chapter 7

ACID–BASE TITRATIONS

KEY THINGS TO LEARN FROM THIS CHAPTER

- Calculating acid–base titration curves
  - Strong acids, strong bases (Table 7.1)
  - Charge balance approach strong acid–strong base, Goal Seek, Solver
  - Spreadsheet calculations
  - Weak acids, weak bases (Table 7.2)
  - Charge balance approach weak acid–strong base, Goal Seek, Solver
  - Spreadsheet calculations, weak acid–strong base
- Indicators (key Equations: 7.12, 7.13)

- Derivative titrations, Goal Seek, Solver, website (7.11)
- Buffer intensity, buffer capacity
- Titration of Na_2CO_3
- Titration of polyprotic acids (Table 7.3)
  - Spreadsheets for deriving titration curves, 7.11 (website, Master Polyprotic Acid Titration Using Solver), Problem 66 (website Universal Acid Titrator)
- Titration of amino acids
- Kjeldahl analysis of nitrogen-containing compounds, proteins

In Chapter 6, we introduced the principles of acid–base equilibria. These are important for the construction and interpretation of titration curves in acid–base titrations. In this chapter, we discuss the various types of acid–base titrations, including the titration of strong acids or bases and of weak acids or bases. The shapes of titration curves obtained are illustrated. Through a description of the theory of indicators, we discuss the selection of a suitable indicator for detecting the completion of a particular titration reaction. The titrations of weak acids or bases with two or more titratable groups and of mixtures of acids or bases are presented. The important Kjeldahl analysis method is described for determining nitrogen in organic and biological samples.

What are some uses of acid–base titrations? They are important in a number of industries. They provide precise measurements. While manual titrations are tedious, automated titrators are often used that can perform titrations effortlessly and accurately. Acid–base titrations are used in the food industry to determine fatty acid content, acidity of fruit drinks, total acidity of wines, acid value of edible oils, and acetic acid content of vinegar. They are used in biodiesel production to determine the acidity of waste vegetable oil, one of the primary ingredients in biodiesel production; waste vegetable oil must be neutralized before a batch may be processed to remove free fatty acids that would normally react to make soap instead of biodiesel. Ammonia is very toxic to aquatic life; aquaria test for ammonia content. In the plating industry, boric acid concentration in nickel plating solutions is determined. In the metal industries, acid in etching solutions is determined. In the environmental arena, alkalinity or acidity of city and sewage water is determined. In the petroleum industry, the acid number of engine oil is determined, and acetic acid in vinyl acetate. In the pharmaceutical industry, sodium hydrogen carbonate in stomach antacids is determined. Acid–base titrations can be used to determine the percent purity of chemicals. The saponification value of carboxylic acids is used to determine the average chain length of fatty acids in fat, by measuring the mass in milligrams of KOH required to saponify the carboxylic acid in one gram of fat.

7.1 Strong Acid versus Strong Base—The Easy Titrations

An acid–base titration involves a **neutralization** reaction in which an acid is reacted with an equivalent amount of base. By constructing a **titration curve**, we can easily explain how the **end points** of these titrations can be detected. The end point signals the completion of the reaction. A titration curve is constructed by plotting the pH of the solution as a function of the volume of titrant added. *The titrant is always a strong acid or a strong base.* The analyte may be either a strong base or acid or a weak base or acid.

Only a strong acid or base is used as the titrant.

In the case of a strong acid versus a strong base, both the titrant and the analyte are completely ionized. An example is the titration of hydrochloric acid with sodium hydroxide:

$$H^+ + Cl^- + Na^+ + OH^- \rightarrow H_2O + Na^+ + Cl^- \tag{7.1}$$

The H^+ and OH^- combine to form H_2O, and the other ions (Na^+ and Cl^-) remain unchanged, so the net result of neutralization is conversion of the HCl to a neutral solution of NaCl. The titration curve for 100 mL of 0.1 M HCl titrated with 0.1 M NaOH is shown in Figure 7.1, plotted from the spreadsheet exercise setup below.

The calculations of titration curves simply involve computation of the pH from the concentration of the particular species present at the various stages of the titration, using the procedures given in Chapter 6. The volume changes during the titration must be taken into account when determining the concentration of the species.

Table 7.1 summarizes the equations governing the different portions of the titration curve. We use f to denote the fraction of analyte, which has been titrated by titrant. In Figure 7.1, at the beginning of the titration ($f = 0$), we have 0.1 M HCl, so the initial pH is 1.0. As the titration proceeds ($0 < f < 1$), part of the H^+ is removed from solution as H_2O. So the concentration of H^+ gradually decreases. At 90% neutralization ($f = 0.9$) (90 mL NaOH), only 10% of the H^+ remains. Neglecting the volume change, the H^+ concentration at this point would be $10^{-2} M$, and the pH would have risen by only one pH unit. (If we correct for volume change, it will be slightly higher—see the spreadsheet below.) However, as the **equivalence point** is approached (the point at which a stoichiometric amount of base is added), the H^+ concentration is rapidly reduced until at the equivalence point ($f = 1$), when the neutralization is complete, a neutral solution of NaCl remains and the pH is 7.0. As we continue to add NaOH ($f > 1$), the OH^- concentration rapidly increases from $10^{-7} M$ at the equivalence point and levels off between 10^{-2} and $10^{-1} M$; we then have a solution of NaOH plus NaCl. Thus, the pH remains fairly constant on either side of the equivalence point, but it changes markedly very near the equivalence point. This large change allows the determination of the completion of the reaction by measurement of either the pH or some property that changes with pH (e.g., the color of an indicator or potential of an electrode).

The equivalence point is where the reaction is theoretically complete.

TABLE 7.1 Equations Governing a Strong-Acid (HX) or Strong-Base (BOH) Titration

Fraction f Titrated	Strong Acid		Strong Base	
	Present	Equation	Present	Equation
$f = 0$	HX	$[H^+] = [HX]$	BOH	$[OH^-] = [BOH]$
$0 < f < 1$	HX/X$^-$	$[H^+] = [\text{remaining HX}]$	BOH/B$^+$	$[OH^-] = [\text{remaining BOH}]$
$f = 1$	X$^-$	$[H^+] = \sqrt{K_w}$ (Eq. 6.13)	B$^+$	$[H^+] = \sqrt{K_w}$ (Eq. 6.13)
$f > 1$	OH$^-$/X$^-$	$[OH^-] = [\text{excess titrant}]$	H$^+$/B$^+$	$[H^+] = [\text{excess titrant}]$

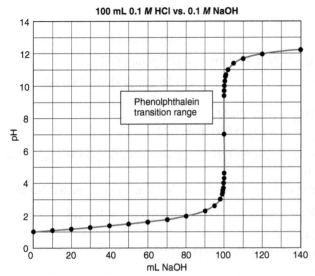

100 mL 0.1 *M* HCl vs. 0.1 *M* NaOH

	A	B	C	D	E	F	G
1	100.00 mL of 0.1000 M HCl vs. 0.1000 M NaOH						
2	mL$_{HCl}$=	100.00	M$_{HCl}$=	0.1000			
3	M$_{NaOH}$=	0.1000	K$_w$=	1.00E–14			
4	**mL$_{NaOH}$**	**[H$^+$]**	**[OH$^-$]**	**pOH**	**pH**		
5	0.00	**0.1**			**1.00**		
6	10.00	0.0818182			1.09		
7	20.00	0.0666667			1.18		
8	30.00	0.0538462			1.27		
9	40.00	0.0428571			1.37		
10	50.00	0.0333333			1.48		
11	60.00	0.025			1.60		
12	70.00	0.0176471			1.75		
13	80.00	0.0111111			1.95		
14	90.00	0.0052632			2.28		
15	95.00	0.0025641			2.59		
16	98.00	0.0010101			3.00		
17	99.00	0.0005025			3.30		
18	99.20	0.0004016			3.40		
19	99.40	0.0003009			3.52		
20	99.60	0.0002004			3.70		
21	99.80	0.0001001			4.00		
22	99.90	5.003E–05			4.30		
23	99.95	2.501E–05			4.60		
24	100.00	**0.0000001**			7.00		
25	100.05		**2.5E–05**	**4.60**	**9.40**		
26	100.10		5E–05	4.30	9.70		
27	100.20		1E–04	4.00	10.00		
28	100.40		0.0002	3.70	10.30		
29	100.80		0.0004	3.40	10.60		
30	101.00		0.0005	3.30	10.70		
31	102.00		0.00099	3.00	11.00		
32	105.00		0.00244	2.61	11.39		
33	110.00		0.00476	2.32	11.68		
34	120.00		0.00909	2.04	11.96		
35	140.00		0.01667	1.78	12.22		
36	Formulas for cells in **boldface**:						
37	**Cell B5**: [H$^+$] = (mL$_{HCl}$ x M$_{HCl}$ – mL$_{NaOH}$ x M$_{NaOH}$)/(mL$_{HCl}$ + mL$_{NaOH}$)						
38		= ($B$2*$D$2–A5*$B$3)/($B$2+A5)		Copy through Cell B23			
39	**Cell E5** = pH =	(–LOG10(B5))		Copy through Cell E24			
40	**Cell B24** = [H$^+$] = K$_w^{1/2}$ =	SQRT(D3)					
41	**Cell C25** = [OH$^-$] = (mL$_{NaOH}$ x M$_{NaOH}$ – mL$_{HCl}$ x M$_{HCl}$)/(mL$_{HCl}$ + mL$_{NaOH}$)						
42		= (A25*$B$3–$B$2*$D$2)/($B$2+A25)		Copy to end			
43	**Cell D25** = pOH = –log[OH$^-$] =	(–LOG10(C25))		Copy to end			
44	**Cell E25** = pH = 14 – pH =	14–D25		Copy to end			

FIGURE 7.1 Titration curve for 100 mL of 0.1 *M* HCl versus 0.1 *M* NaOH.

Example 7.1

Calculate the pH at 0, 10, 90, 100, and 110% titration (% of the equivalence point volume) for the titration of 50.0 mL of 0.100 *M* HCl with 0.100 *M* NaOH.

Solution

At 0% pH = – log 0.100 = 1.00

At 10%, 5.0 mL NaOH is added. We start with 0.100 *M* × 50.0 mL = 5.00 mmol H$^+$. Calculate the concentration of H$^+$ after adding the NaOH:

Keep track of millimoles reacted and remaining!

mmol H$^+$ at start		= 5.00 mmol H$^+$
mmol OH$^-$ added = 0.100 *M* × 5.0 mL		= 0.500 mmol OH$^-$
mmol H$^+$ left		= 4.50 mmol H$^+$ in 55.0 mL

$$[H^+] = 4.50 \text{ mmol}/55.0 \text{ mL} = 0.0818 \ M$$
$$pH = -\log 0.0818 = 1.09$$

At 90%

mmol H⁺ at start	= 5.00 mmol H⁺
mmol OH⁻ added = 0.100 M × 45.0 mL	= 4.50 mmol OH⁻
mmol H⁺ left	= 0.50 mmol H⁺ in 95.0 mL

$$[H^+] = 0.00526 \ M$$
$$pH = -\log 0.00526 = 2.28$$

At 100%: All the H⁺ has been reacted with OH⁻, and we have a 0.0500 M solution of NaCl. Therefore, the pH is 7.00.

At 110%: We now have a solution consisting of NaCl and excess added NaOH.

$$\text{mmol OH}^- = 0.100 \ M \times 5.00 \ \text{mL} = 0.50 \ \text{mmol OH}^- \text{ in 105 mL}$$

$$[OH^-] = 0.00476 \ M$$
$$pOH = -\log 0.00476 = 2.32; pH = 11.68$$

Note that prior to the equivalence point, when there is excess acid, the relationship is $[H^+] = (M_{acid} \times V_{acid} - M_{base} \times V_{base})/V_{total}$, where V is the volume. You can simply apply this to calculate $[H^+]$ once you understand the solution to Example 7.1. Likewise, beyond the equivalence point when there is excess base, $[OH^-] = (M_{base} \times V_{base} - M_{acid} \times V_{acid})/V_{total}$. Note that V_{total} is always $V_{acid} + V_{base}$.

The magnitude of the break will depend on both the concentration of the acid and the concentration of the base. Titration curves at different concentrations are shown in Figure 7.2. The reverse titration gives the mirror image of these curves. The titration of 0.1 M NaOH with 0.1 M HCl is shown in Figure 7.3. The selection of the indicators as presented in these figures is discussed below.

> The selection of the indicator becomes more critical as the solutions become more dilute.

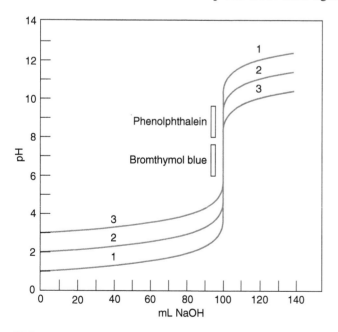

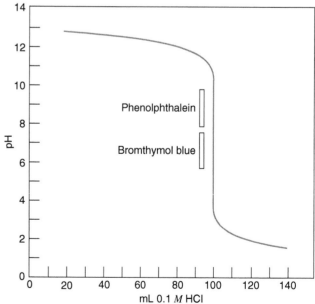

FIGURE 7.2 Dependence of the magnitude of end-point break on concentration. Curve 1: 100 mL of 0.1 M HCl versus 0.1 M NaOH. Curve 2: 100 mL of 0.01 M HCl versus 0.01 M NaOH. Curve 3: 100 mL of 0.001 M HCl versus 0.0001 M NaOH. The equivalence point pH is 7.00 in all cases.

FIGURE 7.3 Titration curve for 100 mL 0.1 M NaOH versus 0.1 M HCl. The equivalence point pH is 7.00.

Spreadsheet Exercise—Titrating a Strong Acid with a Strong Base

Let's prepare the spreadsheet for the construction of Figure 7.1. Enter the values of the HCl concentration and volume, the NaOH concentration, and K_w in specific cells (see cells B2, D2, B3, and D3). These are absolute values that will be used in the formulas. In the spreadsheet, the hydrogen ion concentration is calculated as the remaining millimoles of HCl divided by the total volume (**cell B5** formula). At the equivalence point, it is the square root of K_w (**cell B24** formula). And beyond the equivalence point, the hydroxide concentration is the millimoles of excess NaOH divided by the total volume (**cell C25** formula). Note that the concentrations of HCl and NaOH taken, the volume of HCl, and K_w are all absolute numbers in the formulas. We enter formulas to calculate pH from [H$^+$] (**cell E5**), pOH from [OH$^-$] (**cell D25**), and pH from pOH (**cell E25**).

The titration curve is plotted using Insert/Charts/Scatter by plotting A5:A35 (X-axis) versus E5:E35 (Y-axis). See the spreadsheet exercise in Chapter 6 for the α versus pH plot, Figure 6.2, for details on preparing the chart. The plot (Chart 1) is given in your textbook's **website**, Chapter 7.

7.2 The Charge Balance Method—An Excel Exercise for the Titration of a Strong Acid and a Strong Base

The charge balance method is easily implemented to compute the pH of the solution at any point during a titration using Goal Seek, or by using Excel Solver to simultaneously compute the pH of a multipoint titration plot. An important difference between the foregoing spreadsheet exercise and the charge balance approach is that we need not apply different equations, depending on where we are in the titration ($f < 1$, $f = 0$, or $f > 1$, where f is the fraction titrated) to solve for pH; rather, the same charge balance equations are (iteratively) solved.

Let us rework the problem that we have just considered: the titration of 100 mL of 0.1 M HCl with 0.1 M NaOH. The generic approach to any problem that we wish to solve by the charge balance method involves the following steps:

1. Write down all the species present in the solution at any time.

   In the present case, except the initial point where no NaOH has been added, at all other times we have present in solution: H_2O, H$^+$, OH$^-$, Na$^+$ and Cl$^-$.

2. Write down any equilibrium expression that may exist.

   Since strong acids and strong bases are fully dissociated, the only equilibrium of concern here is the ionization of water:

$$[H^+][OH^-] = K_w$$

   Realize also that the after accounting for the total volume, [Cl$^-$] at any point represents the initial amount of HCl present and similarly, after accounting for the total volume, [Na$^+$] at any point represents the total amount of NaOH added until that point.

3. Write a charge balance expression, putting the concentrations of all charged species on the left-hand side of an equation, multiplying each concentration by the magnitude of the charge on that ion (in the present case these are all 1) and setting the difference of the sum of positive charges and the sum of negative charges to be zero. In the present case, this amounts to:

$$([Na^+] + [H^+]) - ([OH^-] + [Cl^-]) = 0 \qquad (7.2)$$

4. Convert all parameters into either known values that may involve concentrations, volumes (or fraction titrated), equilibrium constants, etc., and/or a single variable, most commonly H^+. Let us call the initial volume of the acid taken V_A and the concentration of the acid taken C_A eq/L (For NaOH and HCl, there is no difference between molar units M and equivalents/liter). If V_B volume of base of concentration C_B eq/L has been added, by definition the fraction titrated will be:

$$V_B C_B = f V_A C_A \qquad (7.3)$$

Alternatively, V_B can be expressed as

$$V_B = f V_A C_A / C_B \qquad (7.4)$$

In the present case, since $C_A = C_B = 0.1$ eq/L, V_B can be written as $f V_A$. The total volume at any point $(V_A + V_B)$ can therefore be written as:

$$V_A + V_B = V_A + f V_A = V_A(1 + f) \qquad (7.5)$$

Now we can make the dilution correction for the concentration of chloride, which starts out with C_A but is diluted during the titration because of the added volume of the titrant. Thus, at any point:

$$[Cl^-] = (V_A C_A)/(V_A + V_B) = (V_A C_A)/V_A(1 + f) = C_A/(1 + f) \qquad (7.6)$$

We also must realize because of the reaction stoichiometry, $[Na^+]$, which directly reflects the amount of NaOH added, must be

$$[Na^+] = f[Cl^-] = f C_A/(1 + f) \qquad (7.7)$$

Recognizing that $[OH^-] = K_w/[H^+]$, we can now write our desired equation by putting Equations 7.6 and 7.7 into Equation 7.2:

$$f C_A/(1 + f) + [H^+] - K_w/[H^+] - C_A/(1 + f) = 0 \qquad (7.8)$$

We can put the first and the last term together:

$$[H^+] - K_w/[H^+] + (f C_A - C_A)/(1 + f)$$

This simplifies to a more pleasing form:

$$[H^+] - \left(\frac{K_w}{[H^+]} + \frac{1 - f}{1 + f} C_A \right) = 0 \qquad (7.9)$$

Given that K_w and C_A are known, $[H^+]$ or pH can be computed for each specified value of f using Goal Seek. Using Solver, multiple solutions for each point in a multipoint solution can be computed.

Construct an Excel sheet (or start with the spreadsheet 7.2.xlsx available on the book's website—Section 7.2 Charge balance and Solver for HCl vs. NaOH). In cell A1 type in CA and in B1 type in 0.1 and name this CA. Similarly type in KW in cell A2 and 1E-14 in cell B2 and define that as KW. Beginning in row 5, type in the column headings f, **pH**, H^+, K_w/H^+, and **Equation**. Enter 0 in cell A6 under f and 0.05 in cell A7. Then highlighting these two cells, drag down from the right bottom corner to fill through to cell A30. This should fill in these cells in column A with f values from 0 to 1.2 with a step of 0.05. Enter any value, say zero, in cell B6. Type in C6 the value of $[H^+]$ as $= 10^\wedge - B6$ (this should now read 1 if we entered zero in cell B6). In cell D6, type in = KW/B6. Now we are ready to write Equation 7.9. We enter: $= C6 - (D6 + CA * (1 - A6)/(1 + A6))$. Remember previous experience with numerical solutions in Excel? We square the whole thing to solve multiple problems in multiple rows at the same time (see Example 6.21). Then we multiply that with 1E10 to prevent premature solutions. Now the ultimate expression in cell E6 reads: $= 10000000000*(C6 - (D6 + CA^*(1 - A6)/(1 + A6)))^\wedge 2$.

Now highlight cell E6 and invoke Data/What-if Analysis/Goal Seek. Execute Goal Seek by seeking to set cell E6 to value 0 by changing cell B6. If you typed it correctly, Goal Seek would find the best value for cell B6 to be 1.00 and cell E6 will have a low residual, of the order of 0.0005. We can solve each row from cell B7 to cell B30 in this fashion each time solving for the pH value for the specific value of f. We can basically also try to solve all the rows at a time by using Solver. To do this, let us select cells B6:E6 and copy (Ctrl+C). Then paste the same into cells B7:E7 (Ctrl+V, beginning in B7). We then select both rows (avoid column A) B6:E7 and drag down from the right-hand corner to E30 and fill up the rest of the rows. In cell E32 the entered formula = SUM(E6:E30) sums up the values of cells E6:E30.

Now we invoke Solver (Data/Analysis/Solver). Set the objective cell as E32. The variable cells to be changed are B6:B30. We need to add constraints to permissible values of B6:B30, these will be within 0–14. Using the Add button we bring up the "Add constraint" box, type in B6:B30 in the cell reference box, leave the operator box at the default "<=" and enter 14 in the "constraint" and click on Add. In the new Add Constraint window, again enter B6:B30 but this time click on the operator box arrow to select the operator ">=" and enter 0 in the constraint box. This time click on OK and you will return to the main Solver menu. We now click on Solver Options and enter 1E-10 on the precision box, 500 seconds on the maximum time box, and 100,000 on the maximum iterations. Click OK, return to the main window, and click on "Solve."

Solver will open the Results pane and provide some preliminary (and incorrect) solutions. You will have to execute Solver several times until the value in cell E32 no longer decreases. To facilitate this, check the box on the Results pane that says "Return to Solver Parameters Dialog." Click OK. You may need to shuttle between the Solver pane and the Results pane 4–5 times before you will see the value in cell E32 no longer decreases (about 0.00578). Solver has done the best it can in this summation mode and actually all of the values except for $f = 1.00$ are correctly solved. The $f = 1.00$ value is unique because at this point $[Na^+]$ and $[Cl^-]$ are exactly equal and the charge balance is determined by $[H^+]$ and $[OH^-]$. Obviously, for Equation 7.2 to be satisfied, at this point $[H^+] = [OH^-]$ and pH = 7. However, even if Solver tries a pH value of 6, ($[H^+] = 10^{-6}$) the difference between $[H^+]$ and $[OH^-]$ will still be very small, comparable to the residuals left when a correct solution is obtained with the other rows. As a result, Solver returns a solution of pH = 6.38 for $f = 1$. This row can be examined by itself. For example, if we multiply 1E20 instead of 1E10 so that the residual remains high enough to prevent a premature solution and then invoke Goal Seek seeking to set cell E26 to zero, by changing cell B26, it will immediately reach the correct solution of pH being 7.00. In general it is advisable to solve the exact neutralization point pH by itself at the end.

You can then construct the titration curve by highlighting cells A6:B30, click on Insert/Charts/Scatter, and select the plot with "Smooth Lines and Markers" (top right). The titration curve is plotted in terms of pH as a function of the fraction titrated. You can move the chart to a separate sheet if you like by right-clicking on the frame of the chart, then Move Chart/New Sheet/OK.

7.3 Detection of the End Point: Indicators

Carrying out the titration process is of little value unless we can tell exactly when the acid has completely neutralized the base, i.e., when the equivalence point has been reached. Therefore, we wish to determine accurately when the equivalence point is reached. The point at which the reaction is *observed to be complete* is called the **end point**. A measurement is chosen such that the end point coincides with or is very close

The goal is for the end point to coincide with the equivalence point.

to the equivalence point. The difference between the equivalence point and the end point is referred to as the **titration error**; as with any measurement, we want to minimize error. The most obvious way of determining the end point is to measure the pH at different points of the titration and make a plot of this versus milliliters of titrant. This is done with a pH meter, which is discussed in Chapter 20.

It is often more convenient to add an **indicator** to the solution and visually detect a color change. An indicator for an acid–base titration is a weak acid or weak base that is highly colored. The color of the ionized form is markedly different from that of the nonionized form. One form may be colorless, but at least one form must be colored. These substances are usually composed of highly conjugated organic constituents that give rise to the color (see Chapter 11).

Assume the indicator is a weak acid, designated HIn, and assume that the nonionized form is red while the ionized form is blue:

$$\underset{\text{(red)}}{HIn} \rightleftharpoons H^+ + \underset{\text{(blue)}}{In^-} \qquad (7.10)$$

We can write a Henderson–Hasselbalch equation for this, just as for other weak acids:

$$pH = pK_{In} + \log \frac{[In^-]}{[HIn]} \qquad (7.11)$$

Your eyes can generally discern only one color if it is 10 times as intense as the other.

The indicator changes color over a pH **range**. The transition range depends on the ability of the observer to detect small color changes. With indicators in which both forms are colored, generally only one color is perceived if the ratio of the concentration of the two forms is 10:1, if the molar absorptivities (Chapter 11), i.e., color intensities, of each are not too different; only the color of the more concentrated form is visually sensed. From this information, we can calculate the pH transition range required to go from one color to the other. When only the color of the nonionized form is seen, $[In^-]/[HIn] = \frac{1}{10}$. Therefore,

$$pH = pK_a + \log \frac{1}{10} = pK_a - 1 \qquad (7.12)$$

When only the color of the ionized form is observed, $[In^-]/[HIn] = \frac{10}{1}$, and

$$pH = pK_a + \log \frac{10}{1} = pK_a + 1 \qquad (7.13)$$

So the pH in going from one color to the other has changed from $pK_a - 1$ to $pK_a + 1$. This is a pH change of 2, and *most indicators require a transition range of about two pH units*. During this transition, the observed color is a mixture of the two colors.

Choose an indicator with a pK_a near the equivalence point pH.

See the inside back cover for a comprehensive list of indicators.

Midway in the transition, the concentrations of the two forms are equal, and $pH = pK_a$. Obviously, then, *the pK_a of the indicator should be close to the pH of the equivalence point*.

Calculations similar to these can be made for weak base indicators, and they reveal the same transition range; the pOH midway in the transition is equal to pK_b, and the pH equals $14 - pK_b$. Hence, a weak base indicator should be selected such that $pH = 14 - pK_b$. Many find it convenient to treat weak base indicators in terms of their conjugate acids and use pK_a values exclusively.

Figure 7.4 illustrates the colors and transition ranges of some commonly used indicators. The range may be somewhat less in some cases, depending on the colors; some colors are easier to see than others. The transition is easier to see if one form of the indicator is colorless. For this reason, phenolphthalein is usually used as an indicator for strong acid–base titrations when applicable (see Figure 7.1, titration of 0.1 M HCl). In dilute solutions, however, phenolphthalein falls outside the steep portion of the titration curve (Figure 7.2), and an indicator such as bromothymol blue must

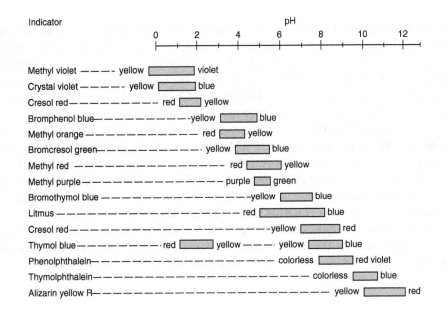

FIGURE 7.4 pH transition ranges and colors of some common indicators.

be used. A similar situation applies to the titration of NaOH with HCl (Figure 7.3). A more complete list of indicators is given on the inside back cover. See also http://en.wikipedia.org/wiki/PH_indicator, which gives pH ranges for color changes of indicators, and shows the colors for acid and base forms.

Since an indicator is a weak acid or base, the amount added should be kept minimal so that it does not contribute appreciably to the pH and so that only a small amount of titrant will be required to cause the color change. That is, the color change will be sharper when the concentration is lower because less acid or base is required to convert it from one form to the other. Of course, sufficient indicator must be added to impart an easily discernible color to the solution. Generally, a few tenths percent solution (wt/vol) of the indicator is prepared and two or three drops are added to the solution to be titrated.

Two drops (0.1 mL) of 0.004 M indicator (0.1% solution with fw = 250) is equal to 0.01 mL of 0.04 M titrant.

7.4   Standard Acid and Base Solutions

Hydrochloric acid is usually used as the strong acid titrant for the titration of bases, while sodium hydroxide is the usual titrant for acids. Most chlorides are soluble, and few side reactions are possible with HCl. It is convenient to handle.

Neither of these is a primary standard and so solutions of approximate concentrations are prepared, and then they are standardized by titrating a primary base or acid. Special precautions are required in preparing the solutions, particularly the sodium hydroxide solution. The preparation and standardization of hydrochloric acid and sodium hydroxide titrants is presented in Chapter 2.

See Chapter 2 for special procedures required to prepare and standardize acid and base solutions.

7.5   Weak Acid versus Strong Base—A Bit Less Straightforward

The titration curve for 100 mL of 0.1 M acetic acid titrated with 0.1 M sodium hydroxide is shown in Figure 7.5. The neutralization reaction is

$$HOAc + Na^+ + OH^- \rightarrow H_2O + Na^+ + OAc^- \qquad (7.14)$$

The curve is flattest and the buffering capacity the greatest at the midpoint of a weak-acid—strong base or weak base—strong acid titration, when the weak acid or base is half neutralized.

The acetic acid, which is only a few percent ionized, depending on the concentration, is neutralized to water and an equivalent amount of the salt, sodium acetate. Before the titration is started, we have 0.1 M HOAc, and the pH is calculated as described for weak acids in Chapter 6. Table 7.2 summarizes the equations governing the different portions of the titration curve, as developed in Chapter 6. As soon as the titration is started, some of the HOAc is converted to NaOAc, and a buffer system is set up. As the titration proceeds, the pH slowly increases as the ratio [OAc⁻]/ [HOAc] changes. At the *midpoint of the titration*, [OAc⁻] = [HOAc], and the *pH is equal to pK$_a$*. At the equivalence point, we have a solution of NaOAc. Since this is a Brønsted base (it hydrolyzes), *the pH at the equivalence point will be alkaline*. The pH will depend on the concentration of NaOAc (see Equation 6.32 and Figure 7.6). The greater the concentration, the higher the pH. As excess NaOH is added beyond the equivalence point, the ionization of the base OAc⁻ is suppressed to a negligible amount (see Equation 6.23), and the pH is determined only by the concentration of excess OH⁻. Therefore, *the titration curve beyond the equivalence point follows that for the titration of a strong acid*.

TABLE 7.2 Equations Governing a Weak-Acid (HA) or Weak-Base (B) Titration

Fraction f Titrated	Weak Acid		Weak Base	
	Present	Equation	Present	Equation
$f = 0$	HA	$[H^+] = \sqrt{K_a \cdot C_{HA}}$ (Eq. 6.20)	B	$[OH^-] = \sqrt{K_b \cdot C_B}$ (Example 6.8)
$0 < f < 1$	HA/A⁻	$pH = pK_a + \log \frac{c_{A^-}}{c_{HA}}$ (Eq. 6.45)	B/BH⁺	$pH = (pK_w - pK_b) + \log \frac{C_B}{C_{BH^+}}$ (Eq. 6.58)
$f = 1$	A⁻	$[OH^-] = \sqrt{\frac{K_w}{K_a} \cdot C_{A^-}}$ (Eq. 6.32)	BH⁺	$[H^+] = \sqrt{\frac{K_w}{K_b} \cdot C_{BH^+}}$ (Eq. 6.39)
$f > 1$	OH⁻/A⁻	$[OH^-] = [\text{excess titrant}]$	H⁺/BH⁺	$[H^+] = [\text{excess titrant}]$

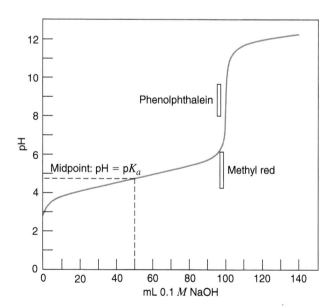

FIGURE 7.5 Titration curve for 100 mL 0.1 M HOAc versus 0.1 M NaOH. Note that the equivalence point pH is not 7, but is defined by Equation 6.32. At 0.05 M NaOAc, it is 8.73.

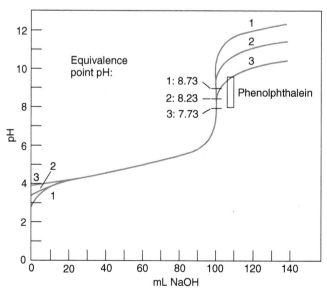

FIGURE 7.6 Dependence of the titration curve of weak acid on concentration. Curve 1: 100 mL of 0.1 M HOAc versus 0.1 M NaOH. Curve 2: 100 mL of 0.01 M HOAc versus 0.01 M NaOH. Curve 3: 100 mL of 0.001 M HOAc versus 0.001 M NaOH.

Example 7.2

Calculate the pH at 0, 10.0, 25.0, 50.0, and 60.0 mL titrant in the titration of 50.0 mL of 0.100 M acetic acid with 0.100 M NaOH.

Solution

At 0 mL, we have a solution of only 0.100 M HOAc:

$$\frac{(x)(x)}{0.100 - x} = 1.75 \times 10^{-5}$$

$$[H^+] = x = 1.32 \times 10^{-3} \, M$$

$$pH = 2.88$$

At 10.0 mL, we started with 0.100 $M \times 50.0$ mL = 5.00 mmol HOAc; part has reacted with OH$^-$ and has been converted to OAc$^-$:

$$\text{mmol HOAc at start} = 5.00 \text{ mmol HOAc}$$

$$\text{mmol OH}^- \text{ added} = 0.100 \, M \times 10.0 \text{ mL} = 1.00 \text{ mmol OH}^-$$

$$= \underline{\text{mmol OAc}^- \text{ formed in 60.0 mL}}$$

$$\text{mmol HOAc left} = 4.00 \text{ mmol HOAc in 60.0 mL}$$

Keep track of millimoles reacted and remaining.

We have a buffer. Since volumes cancel, use millimoles:

$$pH = pK_a + \log \frac{[\text{OAc}^-]}{[\text{HOAc}]}$$

$$pH = 4.76 + \log \frac{1.00}{4.00} = 4.16$$

At 25.0 mL, one-half the HOAc has been converted to OAc$^-$, so pH = pK_a:

$$\text{mmol HOAc at start} = 5.00 \text{ mmol HOAc}$$

$$\text{mmol OH}^- = 0.100 \, M \times 25.0 \text{ mL} = \underline{2.50 \text{ mmol OAc}^- \text{ formed}}$$

$$\text{mmol HOAc left} = 2.50 \text{ mmol HOAc}$$

$$pH = 4.76 + \log \frac{2.50}{2.50} = 4.76$$

At 50.0 mL, all the HOAc has been converted to OAc$^-$ (5.00 mmol in 100 mL, or 0.0500 M):

$$[\text{OH}^-] = \sqrt{\frac{K_w}{K_a}[\text{OAc}^-]}$$

$$= \sqrt{\frac{1.0 \times 10^{-14}}{1.75 \times 10^{-5}} \times 0.0500} = 5.35 \times 10^{-6} \, M$$

$$pOH = 5.27 \quad pH = 8.73$$

At 60.0 mL, we have a solution of NaOAc and excess added NaOH. The hydrolysis of the acetate is negligible in the presence of added OH$^-$. So the pH is determined by the concentration of excess OH$^-$:

$$\text{mmol OH}^- = 0.100 \, M \times 10.0 \text{ mL} = 1.00 \text{ mmol in 110 mL}$$

$$[\text{OH}^-] = 0.00909 \, M$$

$$pOH = -2.04; \, pH = 11.96$$

The slowly rising region before the equivalence point is called the **buffer region**. It is flattest at the midpoint, that is, where the ratio $[OAc^-]/[HOAc]$ is unity (see buffers and buffering capacity, Section 6.8), and so the **buffer intensity is greatest at a pH corresponding to** pK_a. The buffering capacity also depends on the concentrations of HOAc and OAc^-, and the **total buffering capacity** increases as the concentration increases. In other words, the distance of the flat portion on either side of pK_a will increase as [HOAc] and $[OAc^-]$ increase. As the pH deviates to the acid side for pK_a, the buffer will tolerate more base but less acid; the change in pH with a given small amount of added base will be greater, though, than at a pH equal to pK_a because the curve is not so flat, again, because buffer intensity is largest at pH = pK_a Conversely, on the alkaline side of pK_a, more acid but less base can be tolerated. See Chapter 6 for a discussion of buffer intensity and buffer capacity.

See Section 6.8 for a quantitative description of buffer capacity.

You may have noticed that the corresponding region for a strong acid–strong base titration (Figures 7.1 and 7.2) is much flatter than for the weak-acid case. In this respect, a solution of a strong acid or of a strong base is much more resistant to pH change upon addition of H^+ or OH^- than the buffer systems we have discussed. Indeed, high concentrations of strong acids and bases are very good buffers. The problem is that they are restricted to a very narrow pH region, either very acid or very alkaline, especially if the acid ore base concentration is to be strong enough to have any significant capacity against a pH change. These are pH regions that are rarely of much practical value. Also, solutions of strong acids and bases are less resistant to pH change upon dilution than buffers are. Therefore, we usually use mixtures of weak acids or bases with their salts as buffers. This allows the desired pH region to be selected. Often, a buffer is used only to give a specified pH, and no extraneous acids or bases are added. A desired pH can be obtained more easily and it will be less susceptible to change with conventional buffers than with a strong acid or a strong base.

Strong acids are actually good buffers, except their pH changes with dilution.

The transition range of the indicator for this titration of a weak acid must fall within a pH range of about 7 to 10 (Figure 7.5). Phenolphthalein fits this nicely. If an indicator such as methyl red were used, it would begin changing color shortly after the titration began and would gradually change to the alkaline color up to pH 6, before the equivalence point was even reached.

Weak-acid titrations require careful selection of the indicator.

The dependence of the shape of the titration curve and of the equivalence point pH on concentration is shown in Figure 7.6 for different concentrations of HOAc and NaOH. Obviously, phenolphthalein could not be used as an indicator for solutions as dilute as 10^{-3} M (curve 3). Note that the equivalence point pH decreases as the weak acid system becomes more dilute (which doesn't happen in the strong acid system).

Spreadsheet Exercise–Weak Acid–Strong Base Titration

The spreadsheet 7.2.xlsx that we used for strong acid–strong base titrations is readily modified to accommodate a titration of a weak acid with a strong base, e.g., 0.05 M HOAc with 0.05 M NaOH. We need to define K_a, the dissociation constant for acetic acid. In this problem, HOAc, unlike HCl, is not fully ionized and C_A therefore represents the combined concentration of HOAc and OAc^-.

Proceeding as before

1. The species present are: H_2O, HOAc, H^+, OH^-, Na^+, and OAc^-.
2. The equilibrium expressions are:

$$[H^+][OH^-] = K_w$$

$$\frac{[H^+][OAc^-]}{[HOAc]} = K_a$$

The expressions of the fractions of dissociating species, as described in Chapter 6 (Section 6.11) can be manipulated to give the α-values. In particular, we are interested in the concentration of the acetate ion, $[OAc^-]$:

$$[OAc^-] = \alpha_1 C_A$$

where

$$\alpha_1 = \frac{K_a}{K_a + [H^+]}$$

Realize also that the after accounting for the total volume, $[Cl^-]$ at any point represents the initial amount of HCl present and similarly, after accounting for the total volume, $[Na^+]$ at any point represents the total amount of NaOH added until that point.

3. The charge balance expression will be:

$$([Na^+] + [H^+]) - ([OH^-] + [OAc^-]) = 0 \qquad (7.15)$$

4. Accounting for dilution, we have this:

$$[OAc^-] = \alpha_1 C_A V_A / (V_A + V_B) \qquad (7.16)$$

Putting in Equation 7.4, Equation 7.16 reduces to

$$[OAc^-] = \alpha_1 C_A / (1 + f C_A / C_B) \qquad (7.17)$$

For $C_A = C_B$ as in the present case, this further simplifies to

$$[OAc^-] = \alpha_1 C_A / (1 + f) \qquad (7.18)$$

As in spreadsheet Example 7.2, Eq. 7.7

$$[Na^+] = f C_A / (1 + f)$$

Recognizing that $[OH^-] = K_w / [H^+]$, we can now write our desired equation by putting Equations 7.7 and 7.18 into Equation 7.15:

$$f C_A / (1 + f) + [H^+] - K_w / [H^+] - C_A \frac{K_a}{(K_a + [H^+])(1 + f)} = 0 \qquad (7.19)$$

Construct an Excel sheet (or modify the spreadsheet 7.2.xlsx available on the book's website). (See on the website Section 7.5. spreadsheet for HOAc vs.NaOH.) In addition to CA and KW already having been specified and the names defined, type in KA in cell A3 and 1.75E-5 in cell B3 and define this as KA. Change the value of CA to read 0.05. Beginning in row 5, type in the column headings **F, pH, H$^+$, Na$^+$, K$_w$/H$^+$, OAc$^-$**, and **Equation**. You can use the existing spreadsheet, put your cursor in column D and insert a new column by typing Alt-I, then C. Use $[Na^+]$ as the column heading and type in cell A6 the expression for Na$^+$ as given in Equation 7.7 above, = A6*CA/(1 + A6). Put the cursor in column F and again create a new column with the heading $[OAc^-]$ and enter in cell F6 the expression that is in Equation 7.18 as given in Equation 7.19 as: = KA*CA/((KA + C6) * (1 + A6)). Finally in cell G6, enter the charge balance expression, as before squared, and then multiplied by 1E10 as: = 1E10 * (C6 + D6-E6-F6^2.

Let us be more ambitious and do this titration with greater resolution in f. Delete the data in the rows below 6. Let us first check that we have set up the equations/expressions correctly. So we use Goal Seek to solve for the pH and if all is well, the pH will be 3.03 with a residual less than 0.001. However, this suggests that the residuals are in danger of becoming too small and Solver reaching premature solutions. Let us go back in cell G6 and change the multiplier to 1E15, instead of 1E10, we can do Goal Seek again but it is not needed.

We will do this problem with a resolution of 0.01 in f. So we enter 0.01 in cell A7, select cells A6 and A7 and drag the right corner down to 126 (up to a f value of 1.2). Now we copy B6:G6 and paste into B7:G7. Select B6:G7 and drag right bottom corner down to fill up the rest of the sheet. In cell G128, sum up G6:G126.

We now invoke Solver, add the constraints B6:B126 >= 0 and B6:B126 <= 14 and try to minimize the value of G128 by changing B6:B126. In short, we are trying to solve 120 equations at the same time! In 4 or 5 cycles Solver will reach a value for G128 that will not further change ($\sim$49553). If you look carefully at the values in the G column, you will see that the $f = 1$ value has a very large contribution. Once again, this particular row will need to be solved by itself to get a correct solution (the original value of 8.22 will change to 8.58 upon individual solution). Note that no change in the formula is needed when you are trying to obtain an accurate solution for $f = 1$. Rather than using the sum of all the equations together, you will ask Solver to set G106 to the minimum by changing B106. Be sure to remove the constraints that apply to B6:B126. You can have the constraints apply only to B106. In general, it will also work if you do not specify a constraint. Since you are solving for a single parameter, you can also, of course, use Goal Seek; most find that simpler.

Again, you can make a plot by selecting A6:B126, click on Insert/Charts/Scatter and select the plot with "Smooth Lines and Markers" (top right). The titration curve is plotted in terms of pH as a function of the fraction titrated.

Figure 7.7 shows the titration curves for 100 mL of 0.1 M solutions of weak acids of different K_a values titrated with 0.1 M NaOH. The sharpness of the end point decreases as K_a decreases. As in Figure 7.6, the sharpness will also decrease as the concentration decreases. Generally, for titrations at significant concentrations (ca. 0.1 M), acids with K_a values as low as 10^{-6} can be titrated accurately with a visual indicator. With suitable color comparisons, even those with K_a values approaching 10^{-8} can be titrated with reasonable accuracy. A pH meter can be used to obtain better precision for the very weak acids by plotting the titration curve.

Professor's favorite challenge: Can you modify 7.5.xlsx on the text's website with different K_a values to generate the different curves in Figure 7.7?

7.6 Weak Base versus Strong Acid

The titration of a weak base with a strong acid is completely analogous to the above case, but the titration curves are the reverse of those for a weak acid versus a strong base. The titration curve for 100 mL of 0.1 M ammonia titrated with 0.1 M hydrochloric acid is shown in Figure 7.8. The neutralization reaction is

$$NH_3 + H^+ + Cl^- \rightarrow NH_4^+ + Cl^- \tag{7.20}$$

At the beginning of the titration, we have 0.1 M NH_3, and the pH is calculated as described for weak bases in Chapter 6. See Table 7.2. As soon as some acid is added, some of the NH_3 is converted to NH_4^+, and we are in the buffer region. At the *midpoint of the titration*, [NH_4^+] equals [NH_3], and the *pH is equal to* $(14 - pK_b)$ or pK_a for NH_4^+. *At the equivalence point*, we have a solution of NH_4Cl, a weak Brønsted acid that hydrolyzes to give an **acid solution**. Again, the pH will depend on the concentration; the greater the concentration, the lower the pH (see Equation 6.39). Beyond the equivalence point, the free H^+ suppresses the ionization (see Equation 6.33), and the pH is determined by the concentration of H^+ added in excess. Therefore, the titration curve beyond the equivalence point will follow that for titration of a strong base

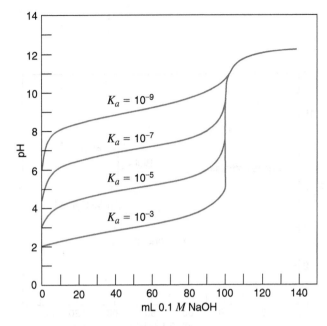

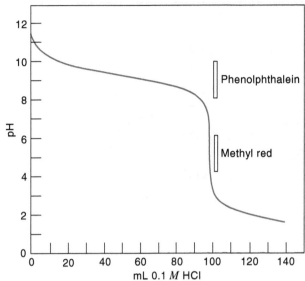

FIGURE 7.7 Titration curves for 100 mL 0.1 M weak acids of different K_a values versus 0.1 M NaOH.

FIGURE 7.8 Titration curve for 100 mL 0.1 M NH$_3$ versus 0.1 M HCl.

(Figure 7.3). Because K_b for ammonia happens to be equal to K_a for acetic acid, the titration curve for ammonia versus a strong acid is just the mirror image of the titration curve for acetic acid versus a strong base.

The indicator for the titration in Figure 7.8 must have a transition range within about pH 4 to 7. Methyl red meets this requirement, as shown in the figure. If phenolphthalein had been used as the indicator, it would have gradually lost its color between pH 10 and 8, before the equivalence point was reached.

Titration curves for different concentratioSpreadsheet Exercise–ns of NH$_3$ titrated with varying concentrations of HCl would be the mirror images of the curves in Figure 7.6. Methyl red could not be used as an indicator in dilute solutions. The titration curves for weak bases of different K_b values (100 mL, 0.1 M) versus 0.1 M HCl are shown in Figure 7.9. In titrations involving significant concentrations (ca. 0.1 M), *one can accurately titrate a base with a K_b of 10^{-6} using a visual indicator.*

Professor's Favorite Challenge: Can you generate Figure 7.9 using the spreadsheet approach of 7.5.xlsx? See the website Figure 7.9 Spreadseet plot for a solution.

7.7 Titration of Sodium Carbonate—A Diprotic Base

Sodium carbonate is a Brønsted base that is used as a primary standard for the standardization of strong acids. It hydrolyzes in two steps:

$$CO_3^{2-} + H_2O \rightleftharpoons HCO_3^- + OH^- \quad K_{H1} = K_{b1} = \frac{K_w}{K_{a2}} = 2.1 \times 10^{-4} \quad (7.21)$$

$$HCO_3^- + H_2O \rightleftharpoons CO_2 + H_2O + OH^- \quad K_{H2} = K_{b2} = \frac{K_w}{K_{a1}} = 2.3 \times 10^{-8} \quad (7.22)$$

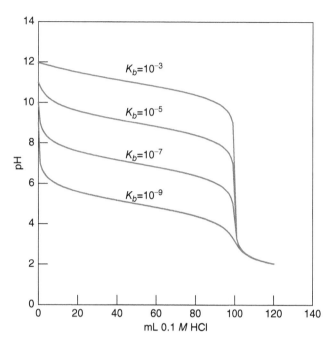

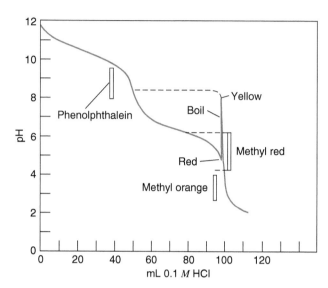

FIGURE 7.9 Titration curves for 100 mL 0.1 M weak bases of different K_b values versus 0.1 M HCl.

FIGURE 7.10 Titration curve for 50 mL 0.1 M Na$_2$CO$_3$ versus 0.1 M HCl. Dashed line represents a boiled solution with CO$_2$ removed.

where K_{a1} and K_{a2} refer to the K_a values of H$_2$CO$_3$; HCO$_3^-$ is the conjugate acid of CO$_3^{2-}$ and H$_2$CO$_3$ is the conjugate acid of HCO$_3^-$; and the K_b values are calculated as described in Chapter 6 for salts of weak acids and bases (i.e., from $K_a K_b = K_w$).

Sodium carbonate can be titrated to give end points corresponding to the stepwise additions of protons to form HCO$_3^-$ and CO$_2$. (Carbonic acid, H$_2$CO$_3$, dissociates in acid solution to CO$_2$—its acid anhydride—and H$_2$O in acid solution.) *The K$_b$ values should differ by at least* 10^4 to obtain good separation of the equivalence point breaks in a case such as this.

A titration curve for Na$_2$CO$_3$ versus HCl is shown in Figure 7.10 (solid line). Even though K_{b1} is considerably larger than the 10^{-6} required for a sharp end point, the pH break is decreased by the formation of CO$_2$ beyond the first equivalence point. The second end point is not very sharp either because K_{b2} is smaller than the 10^{-6} we would like. Fortunately, this end point can be sharpened, because the CO$_2$ produced from the neutralization of HCO$_3^-$ is volatile and can be boiled out of the solution. This is described below.

At the start of the titration, the pH is determined by the hydrolysis of the Brønsted base CO$_3^{2-}$. After the titration is begun, part of the CO$_3^{2-}$ is converted to HCO$_3^-$, and a CO$_3^{2-}$/HCO$_3^-$ buffer region is established. At the first equivalence point, there remains a solution of HCO$_3^-$, and $[H^+] \approx \sqrt{K_{a1}K_{a2}}$. Beyond the first equivalence point, the HCO$_3^-$ is partially converted to H$_2$CO$_3$(CO$_2$) and a second buffer region is established, the pH being established by $[HCO_3^-]/[CO_2]$. The pH at the second equivalence point is determined by the concentration of the weak acid CO$_2$.

Phenolphthalein is used to detect the first end point, and methyl orange is used to detect the second one. Neither end point, however, is very sharp. In actual practice, the phenolphthalein end point is used only to get an approximation of where the second end point will occur; phenolphthalein is colorless beyond the first end point and does not interfere. The second equivalence point, which is used for accurate titrations, is

The dashed line is for HCO$_3^-$ only, for which $[H^+] = \sqrt{K_{a1}K_{a2}}$ for the horizontal portion.

Boiling the solution removes the CO$_2$, raising the pH to that of HCO$_3^-$ (7.73). The second equivalence point pH without boiling is 3.92 from 0.033 M carbon dioxide, and with boiling, 7.00 (NaCl).

normally not very accurate with methyl orange indicator because of the gradual change in the color of the methyl orange. This is caused by the gradual decrease in the pH due to the HCO_3^-/CO_2 buffer system beyond the first end point.

If beyond the first equivalence point we were to boil the solution after each addition of HCl to remove the CO_2 from the solution, the buffer system of HCO_3^-/CO_2 would be removed, leaving only HCO_3^- in solution. This is both a weak acid and a weak base whose pH (≈ 8.3) is independent of concentrations ($[H^+] = \sqrt{K_{a1}K_{a2}}$ or $[OH^-] = \sqrt{K_{b1}K_{b2}}$; see Chapter 6). In effect, then, the pH would remain essentially constant until the equivalence point, when we are left with a neutral solution of water and NaCl (pH = 7). The titration curve would then follow the dashed line in Figure 7.10.

The following procedure can, therefore, be employed to sharpen the end point, as illustrated in Figure 7.10. Methyl red is used as the indicator, and the titration is continued until the solution just turns from yellow through orange to a definite red color. This will occur just before the equivalence point. The change will be very gradual because the color will start changing at about pH 6.3, well before the equivalence point. At this point, the titration is stopped and the solution is gently boiled to remove CO_2. The color should now revert to yellow because we have a dilute solution of only HCO_3^-. The solution is cooled and the titration is continued to a sharp color change to red or pink. The equivalence point here does not occur at pH 7 because there is a small amount of HCO_3^- still remaining to be titrated after boiling. That is, there will still be a slight buffering effect throughout the remainder of the titration, and dilute CO_2 will still remain at the equivalence point.

Bromcresol green can be used in a manner similar to methyl red. Its transition range is pH 3.8 to 5.4, with a color change from blue through pale green to yellow (see Experiment 8). Similarly, methyl purple can be used (see Experiment 20).

The methyl orange end point can be used (without boiling) by adding a blue dye, xylene cyanole FF, to the indicator. This mixture is called **modified methyl orange**. The blue color is complementary to the orange color of the methyl orange at about pH 2.8. This imparts a gray color at the equivalence point, the transition range of which is smaller than that of methyl orange. This results in a sharper end point. It is still not so sharp as the methyl red end point. Methyl orange can also be used by titrating to the color of the indicator in a solution of potassium acid phthalate, which has a pH close to 4.0.

7.8 Using a Spreadsheet to Perform the Sodium Carbonate—HCl Titration

We can use Solver to construct the titration curves in Figure 7.10. As was stated before, weak bases can be treated mathematically in terms of the K_a values of their conjugate acids. Sodium carbonate is a salt of carbonic acid and in Section 6.11, you have already learned to calculate the values of α_1 and α_2. For the titration of 100 mL 0.1 M Na_2CO_3 with HCl, we can consider $f = 1$ as the point where 1 mole of HCl has been added per mole of Na_2CO_3 (the solution composition being $NaHCO_3$ and NaCl) and $f = 2$ as the point where it has been fully neutralized.

Proceeding as before, a charge balance equation at any point will demand

$$[Na^+] + [H^+] - [OH^-] - [HCO_3^-] - 2[CO_3^{2-}] - [Cl^-] = 0 \qquad (7.23)$$

The initial concentration of all carbonate species (this includes H_2CO_3, HCO_3^-, and CO_3^{2-} and is typically denoted by C_T) is 0.1 M but C_T can be expressed as $C_T = 0.1/(1 + f)$ at any time. The concentration of HCO_3^- and CO_3^{2-} are respectively given by $\alpha_1 C_T$ and $\alpha_2 C_T$, where α_1 is $K_{a1}[\text{H}^+]/Q$ and α_2 is $K_{a1}K_{a2}[\text{H}^+]/Q$, Q being (see Section 6.11):

$$Q = [\text{H}^+]^2 + K_{a1}[\text{H}^+] + K_{a1}K_{a2}$$

The concentration of chloride $[\text{Cl}^-]$ at any point is simply fC_T and the concentration of sodium, $[\text{Na}^+]$ can be expressed as $2C_T$.

Create a spreadsheet 7.9.xlsx (do this on your own, but one is available on the book's website Fig. 7.10 if you need help). Define and name CTIN (initial value of CT, 0.1), KAA (K_{a1}, 4.3×10^{-7}), KAB (K_{a2}, 4.8×10^{-11}), and KW. Make the following column headings across the row beginning in cell A6: **F**, **pH**, **[H+]**, **Q**, **CT**, **[HCO3−]**, **[CO32−]**, **[Na+]**, **[Cl−]**, **[OH−]**, and **Equation**.

Type in 0 and 0.02 in cells A7 and A8, then select and drag down to cell A117 up to an f value of 2.2, you should have a resolution of 0.02 between adjacent f values. Insert some trial value for the initial pH in cell B7 (e.g., 12, sodium carbonate is going to be alkaline), and in cell C7 enter the relationship of how $[\text{H}^+]$ is to be calculated from pH: $= 10^\wedge - \text{B7}$. In cell D7, enter the expression for Q: $\text{C7}^\wedge 2 + \text{KAA}^*\text{C7} + \text{KAA}^*\text{KAB}$. In cell E7, enter the dilution corrected expression for CT: $= \text{CTIN}/(1 + \text{A7})$. In cell F7, for $[\text{HCO}_3^-]$, we enter $C_T\alpha_1$ as: $= \text{E7}^*\text{KAA}^*\text{C7}/\text{D7}$. In cell G7, for $[\text{CO}_3^{2-}]$, we enter $C_T\alpha_2$ as $= \text{E7}^*\text{KAA}^*\text{KAB}/\text{D7}$. In cell H7, $[\text{Na}^+]$ is simply $2C_T$ that we enter as: $= 2^*\text{E7}$. In cell I7, $[\text{Cl}^-]$ is fC_T and is expresses as $= \text{A7} * \text{E7}$; in cell J7, $[\text{OH}^-]$ has the usual expression, $K_w/[\text{H}^+]$ entered as $= \text{KW}/\text{C7}$. The equation (cell K7) is then simply set up as the charge balance expression (Equation 7.23) squared and multiplied by 10^{10}. Check that all is well by invoking Solver, minimizing cell K7 by changing cell B7; cell B7 should converge to a pH of 11.65. Now, as before, we select cells B7:K7 and copy and paste it into cells B8:K8. Then we select cells B7:K8 and drag from the bottom right corner of cell K8 to fill in down to row 117. Solve the $f = 2$ case (row 107) individually by itself before solving the sum of all the equation terms. Then, as before, in cell K119 sum up cells B7:B119 and minimize cell K119 by changing cells B7:B117.

Now, can we simulate by calculation what happens when past the first end point we remove CO_2 formed by boiling? Consider that up to $f = 1$, the process taking place is

$$\text{Na}_2\text{CO}_3 + \text{HCl} \rightarrow \text{NaHCO}_3 + \text{NaCl} \qquad (7.24)$$

Further addition of HCl goes to form H_2CO_3 that decomposes to H_2O and CO_2 when heated and the latter is driven off from the solution.

$$\text{NaHCO}_3 + \text{HCl} \rightarrow \text{H}_2\text{CO}_3 \uparrow + \text{NaCl} \qquad (7.25)$$

Mathematically, this loss of CO_2 is nothing more than a decrease in C_T by the same amount. Reflect on this and you will see that H_2CO_3 formed (and hence the loss in C_T according to Equation 7.25) is the chloride concentration minus half the sodium concentration. Until $f = 1$ is reached (when $[\text{Cl}^-] = \frac{1}{2}[\text{Na}^+]$), a net positive amount of H_2CO_3 will not be produced according to Equation 7.25.

To do the computation, we copy cells A57:K117 and paste into cells M57:W117. Copy the headings in cells A6:K6 in M6:W6. Remember we have the $[\text{Na}^+]$ and $[\text{Cl}^-]$ expressions in columns T and U defined in terms of C_T (column Q). If we change how we calculate C_T, then the $[\text{Na}^+]$ and $[\text{Cl}^-]$ values will change also; we don't want to have this happen. So we change the definitions of $[\text{Na}^+]$ and $[\text{Cl}^-]$ in columns T and U in terms of C_T values that we calculated before in column E and that would not be altered in any manipulation of the present C_T values in column Q. Now T57 reads $= 2^*\text{Q57}$, change it to read $= 2^*\text{E57}$. Similarly U57 reads $= \text{M57}^*\text{Q57}$, change it to read $= \text{M57}^*\text{E57}$. Then select these two cells and double-click on the right bottom corner of U57 to fill in the rows that follow. Now we can modify the expression for C_T in Q57.

It reads = CTIN/(1 + M57), make it read = CTIN/(1 + M57) − (U57 − T57/2). This is the original C_T value minus the difference between chloride and half the sodium. Alter the rest of the Q column by double-clicking on the right bottom corner of Q57. Go ahead and sum up the equation values in W119 and solve cells N57:N117 by minimizing W119. Once again, you will need to solve for the $f = 2$ value by itself, whether before or after the rest is solved.

To make a plot by selecting A7:B117, click on Insert/Charts/Scatter and select the plot with "Scatter with Smooth Lines" (middle left). The titration curve is plotted in terms of pH as a function of the fraction titrated. Right-click on the plot, click on Select Data, then select Series 1 (it will be highlighted) and click on Edit. In "Series Name" box, enter "Original." Click OK. Click on "Add." When the Edit Series Box opens, name the series "Boil Off CO_2." For the X-data, select with your cursor cells M57:M117, for the Y-data, select cells N57:N117. After clicking on OK twice, you have the plot, exactly the same as Figure 7.10. You can right-click on the "boil off" trace, choose "Shape Outline" (second from left on the drop-down menu bar), pick dashes and change the trace into a dashed style. *At any point* after the titration is half done, you are on the pH represented by the solid line (blue) trace, if you boil it and remove CO_2, you will move up to the pH in the dashed line plot, without a shift in the X-position.

7.9 Titration of Polyprotic Acids

Diprotic acids can be titrated stepwise just as sodium carbonate was. In order to obtain good end-point breaks for titration of the first proton, K_{a1} should be at least $10^4 \times K_{a2}$. If K_{a2} is in the required range of 10^{-7} to 10^{-8} for a successful titration, an end-point break is obtained for titrating the second proton. Triprotic acids (e.g., H_3PO_4) can be titrated similarly, but K_{a3} is generally too small to obtain an end-point break for titration of the third proton. Figure 7.11 illustrates the titration curve for a diprotic acid H_2A, and Table 7.3 summarizes the equations governing the different portions of the titration curve. The pH at the beginning of the titration is determined from the ionization of the first proton if the solution is not too dilute (see discussion of polyprotic acids in Chapter 6). If K_{a1} is not too large and the amount dissociated is ignored compared to the analytical concentration of the acid, the approximate equation given can be used to

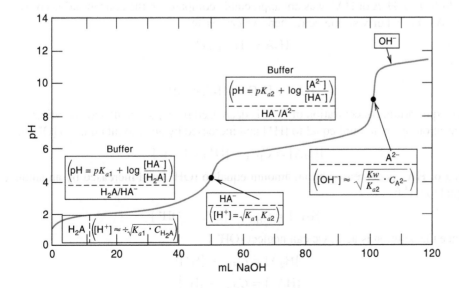

FIGURE 7.11 Titration of diprotic acid, H_2A, with sodium hydroxide.

TABLE 7.3   Equations Governing Diprotic Acid (H_2A) Titration

Fraction f Titrated	Present	Equation
$f = 0$ (0%)	H_2A	$[H^+] \approx \sqrt{K_{a1} C_{H_2A}}$ (Example 6.7) (or Eq. 6.20; Ex. 6.17, quadratic, if ~strong acid)
$0 < f < 1$ (> 0 to < 100%)	H_2A/HA^-	$pH = pK_{a1} + \log \dfrac{C_{HA^-}}{C_{H_2A}}$ (Eq. 6.45) (or $C_{HA^-} + [H^+]$ and $C_{H_2A} - [H^+]$ if a strong acid)
$f = 1$ (100%) (1st eq. pt.)	HA^-	$[H^+] \approx \sqrt{K_{a1} K_{a2}}$ (Eq. 6.84) (or Eq. 6.83 if H_2A ~ strong acid)
$1 < f < 2$ (> 100 to < 200%)	HA^-/A^{2-}	$pH = pK_{a2} + \log \dfrac{C_{A^{2-}}}{C_{HA^-}}$ (Eq. 6.45, Ex. 6.16, 6.24)
$f = 2$ (200%) (2nd eq. pt.)	A^{2-}	$[OH^-] \approx \sqrt{\dfrac{K_w}{K_{A2}} \cdot C_{A^{2-}}}$ (Eq. 6.32) (or Eq. 6.29, Ex. 6.20, quadratic if A^{2-} ~ strong base)
$f > 2$ (> 200%)	OH^-/A^{2-}	$[OH^-] = [\text{excess titrant}]$

calculate $[H^+]$. Otherwise the quadratic formula must be used to solve Equation 6.20 (see Example 6.17).

During titration up to the first equivalence point, an HA^-/H_2A buffer region is established. At the first equivalence point, a solution of HA^- exists, and $[H^+] \approx \sqrt{K_{a1} K_{a2}}$. Beyond this point, an A^{2-}/HA^- buffer exists; and finally at the second equivalence point, the pH is determined from the hydrolysis of A^{2-}. If the salt, A^{2-}, is not too strong a base, then the approximate equation given can be used to calculate $[OH^-]$. Otherwise the quadratic equation must be used to solve Equation 6.29 (see Example 6.20).

Figure 7.11 and Table 7.3 illustrate which species are present at each part of the titration curve. As noted, if either K_{a1} or K_{b1} ($= K_w/K_{a1}$) is fairly large, we can't make the simplifying assumptions, and the quadratic equation must be used for the *beginning of the titration* or at the *second equivalence point*. In practice, there are few diprotic acids where this is not the case for one or the other. Furthermore, if K_{a1} is fairly large (e.g., chromic acid, $K_{a1} = 0.18$, $K_{a2} = 3.2 \times 10^{-7}$), then the Henderson–Hasselbalch equation cannot be used for the *first buffer region* because the assumption in deriving that from the K_{a1} expression was that the amount of H^+ or OH^- from dissociation or hydrolysis of H_2A or HA^- was not appreciable compared to the concentrations of H_2A or HA^-. For a fairly strong acid, then, we can write

$$H_2A \rightleftharpoons H^+ + HA^-$$

and

$$HA^- + H_2O \rightleftharpoons H_2A + OH^-$$

The equilibrium concentration of H_2A is decreased from the calculated analytical concentration by an amount equal to $[H^+]$ and increased by an amount equal to $[OH^-]$:

$$[H_2A] = C_{H_2A} - [H^+] + [OH^-]$$

That of HA^- is decreased by an amount equal to $[OH^-]$ and increased by an amount equal to $[H^+]$:

$$[HA^-] = C_{HA^-} + [H^+] - [OH^-]$$

Since the solution is acid, we can neglect $[OH^-]$:

$$[H_2A] = C_{H_2A} - [H^+]$$
$$[HA^-] = C_{HA^-} + [H^+]$$

(For simplification, we could have written the same equations just from the H_2A dissociation above.) So, for a fairly strong acid, we must substitute in the $K_{a1} = [H^+][HA^-]/[H_2A]$ expression: $[H_2A] = C_{H_2A} - [H^+]$, and $[HA^-] = C_{HA^-} + [H^+]$, and solve a quadratic equation for the buffer region, where C_{H_2A} and C_{HA^-} are the calculated concentrations resulting from the acid–base reaction at a given point in the titration:

$$[H^+]^2 + (K_{a1} + C_{HA^-})[H^+] - K_{a1}C_{H_2A} = 0$$

Also, if K_{b1} for A^{2-} is fairly large, then just beyond the second equivalence point, OH^- from hydrolysis of A^{2-} cannot be ignored compared to the concentration of the OH^- from excess titrant. So we continue to use the K_{b1} expression, $K_{b1} = K_w/K_{a2} = [HA^-][OH^-]/[A^{2-}]$ where $[A^2] = C_{A^{2-}} - [OH^-]$, and $[HA^-] = [OH^-]$, and solve the quadratic equation. After addition of, say, 0.5 mL, of NaOH, there is enough excess OH^- to suppress the hydrolysis of A^{2-}, and we can calculate the pH just from the excess OH^- concentration. Finally, for HA^- at the *first equivalence point*, we may have to use the more exact Equation 6.96 instead of 6.97 for calculation of $[H^+]$ to get the correct pH, since K_{a1} may not be negligible compared to $[HA^-]$ ($K_{a1}K_w$ in the numerator will probably still be negligible). So,

$$[H^+] = \sqrt{\frac{K_{a1}K_{a2}[HA^-]}{K_{a1} + [HA^-]}}$$

And very near the equivalence points, even more complicated expressions may be required. You can appreciate that these types of considerations, each different for each region of a titration, can quickly get cumbersome.

In contrast, if the systematic approach using the charge balance equation is used for calculations, then no simplifying assumptions or approximations are necessary and a single charge balance equation applies at any point in the titration.

We will not construct a diprotic acid titration curve here, but you probably have realized that constructing a titration curve for a polyprotic acid by the charge balance method is quite starightforward. The website for the book in Chapter 7 has a master spreadsheet that allows you to construct a titration curve for up to a tetraprotic acid, titled Master polyprotic acid titration spreadsheet.xlsx. Note that the approach in this spreadsheet calculates pH as a constant increment of titrant is added (or f is varied with constant increments). Try with some of the polyprotic acids for which dissociation constants are given in Appendix C, Table C.1. While the Master titration spreadsheet approach mimics what one does experimentally, i.e., add a little titrant and measure the pH, computationally an alternative approach, where we ask how much titrant we need to add to get to a certain pH is much easier. This approach is discussed in Section 7.11 and is also embodied in the professor's favorite spreadsheet Exercise 7.66. For the titration of weak polyprotic bases, see Section 7.11b Supplement on the text's website. A master spreadsheet for easy calculations is given.

7.10   Mixtures of Acids or Bases

Mixtures of acids (or bases) can be titrated stepwise if there is an appreciable difference in their strengths. In general, one K_a value must be at least 10^4 times greater than the other to clearly see the individual end points separately. If one of the acids is a strong acid, a separate end point will be observed for the weak acid *only if K_a is about 10^{-5} or smaller*. See, for example, Figure 7.12, where a break is seen for the HCl. The stronger acid will titrate first and will give a pH break at its equivalence point. This will

One acid should be at least 10^4 times weaker than the other to titrate separately.

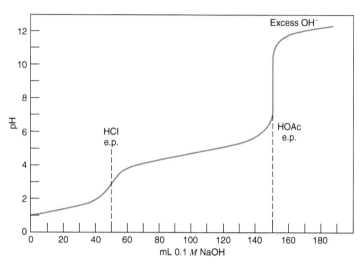

FIGURE 7.12 Titration curve for 50 mL of mixture of 0.1 M HCl and 0.2 M HOAc with 0.1 M NaOH.

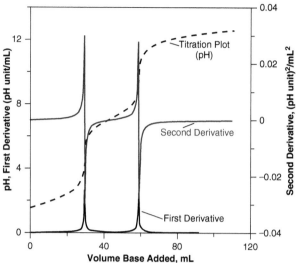

FIGURE 7.13 The titration plot and the two derivative plots for the H_3PO_4-NaOH titration generated in the spreadsheet 7.11.xlsx (text's website Sec. 7.11 Derivative titrations—Easy method).

be followed by titration of the weaker acid and a pH break at its equivalence point. The titration curve for a mixture of hydrochloric acid and acetic acid versus sodium hydroxide is shown in Figure 7.12. At the equivalence point for HCl, a solution of HOAc and NaCl remains, and so the equivalence point is acidic. Beyond the equivalence point, the OAc^-/HOAc buffer region is established, and this markedly suppresses the pH break for HCl, leading to only a small pH change in the equivalence point of HCl, compared to when HCl is titrated in the absence of HOAc. The remainder of the titration curve is identical to Figure 7.5 for the titration of HOAc.

If two strong acids are titrated together, there will be no differentiation between them, and only one equivalence point break will occur, corresponding to the titration of both acids. The same is true for two weak acids if their K_a values are not too different. For example, a mixture of acetic acid, $K_a = 1.75 \times 10^{-5}$, and propionic acid, $K_a = 1.3 \times 10^{-5}$, would titrate together to give a single equivalence point. Put the dissociation constants for citric acid, a triprotic acid, in the Master titration spreadsheet on the website and construct a titration plot to observe how many equivalence points are visible.

With H_2SO_4, the first proton is completely dissociated and the second proton has a K_a of about 10^{-2}. Therefore, the second proton is ionized sufficiently to titrate as a strong acid, and only one equivalence point break is found. The same is true for a mixture of a strong acid and a weak acid with a K_a in the neighborhood of 10^{-2}.

The first ionization constant of sulfurous acid, H_2SO_3, is 1.3×10^{-2}, and the second ionization constant is 5×10^{-6}. Therefore, in a mixture with HCl, the first proton of H_2SO_3 would titrate along with the HCl, and the pH at the equivalence point would be determined by the HSO_3^- remaining; that is, $[H^+] = \sqrt{K_{a1}K_{a2}}$, since HSO_3^- is both an acid and a base. This would be followed by titration of the second proton to give a second equivalence point. The volume of titrant required to reach the first end point would always be greater than that in going from the first to the second since the first includes the titration of both acids. The amount of H_2SO_3 could be determined from the amount of base required for the titration of the second proton. The amount of HCl could be found by subtracting from the first end point the volume of base required to titrate the second proton of H_2SO_3, which is equal to the volume required to titrate the first proton. In reality, this titration would find little practical use because H_2SO_3 is volatilized and lost from solution as SO_2 gas in a strongly acid solution.

Phosphoric acid in mixture with a strong acid acts in a manner similar to the above example. The first proton titrates with the strong acid, followed by titration of the second proton to give a second equivalence point; the third proton is too weakly ionized to be titrated. The titration of a polyprotic acid is essentially the same as that of the titration of a mixture of monoprotic acids of the corresponding K_a values, where the separate monoprotic acids all have the same concentration. Figure 7.13 shows the titration of phosphoric acid. The derivatives of titration curves are discussed in Section 7.11.

Example 7.3

A mixture of HCl and H_3PO_4 is titrated with 0.1000 M NaOH. The first end point (methyl red) occurs at 35.00 mL, and the second end point (bromthymol blue) occurs at a total of 50.00 mL (15.00 mL after the first end point). Calculate the millimoles HCl and H_3PO_4 present in the solution.

Solution

The second end point corresponds to that in the titration of one proton of H_3PO_4 $PO_4^- \rightarrow HPO_4^-$). Therefore, the millimoles H_3PO_4 is the same as the millimoles NaOH used in the 15.00 mL for titrating that proton:

$$\text{mmol}_{H_3PO_4} = M_{NaOH} \times \text{mL}_{NaOH} = 0.1000 \text{ mmol/mL} \times 15.00 \text{ mL}$$
$$= 1.500 \text{ mmol}$$

The HCl and the first proton of H_3PO_4 titrate together. A 15.00-mL portion of base was used to titrate the first proton of H_3PO_4 (same as for the second proton), leaving 20.00 mL used to titrate the HCl. Therefore,

$$\text{mmol}_{HCl} = 0.1000 \text{ mmol/mL} \times (35.00 - 15.00) \text{ mL} = 2.000 \text{ mmol}$$

Similarly, mixtures of bases can be titrated if their strengths are sufficiently different. Again, the difference in K_b values must be at least a factor of 10^4. Also, if one of them is a strong base, the weak base must have a K_b no greater than about 10^{-5} to obtain separate end points. For example, sodium hydroxide does not give a separate end point from that for the titration of CO_3^{2-} to HCO_3^- when titrated in the presence of Na_2CO_3 ($K_{b1} = 2.1 \times 10^{-4}$).

7.11 Equivalence Points from Derivatives of a Titration Curve

In instrumental monitoring of a titration as with a pH electrode (or some other potentiometric electrodes, see Chapter 20), the end point is frequently determined from a derivative of the titration curve. You will note that in all of the foregoing titration plots, the *rate at which pH changes* with added titrant is the greatest at the equivalence point. For an acid sample, titrated with a base, this rate of change is the first derivative of the pH vs. V_B plot, or dpH/dV_B. The pH increases during such a titration and dpH/dV_B

is always positive—it reaches its maximum value at the end point. Conversely, when a base is titrated with an acid, the pH decreases monotonically during the titration as acid (volume V_A) is added, and dpH/dV_A is always negative and reaches the lowest (maximum negative) value at the end point. If you have had calculus already, then you know the condition of a maximum or minimum for any function is that the derivative of that function will be zero when the maximum or minimum is reached. Although the following discussion is solely centered on titrating an acid with a base, essentially the same considerations apply when a base is titrated with an acid.

To reiterate, if we have pH data for the titration of an acid as a function of the volume of base added (pH vs. V_B), a plot of the first derivative dpH/dV_B against V_B would show a maximum at the equivalence point and a plot of the second derivative, denoted $d^2pH/dV_B{}^2$ (meaning the rate of change of dpH/dV_B with V_B), versus V_B will go through zero at the end point (or equivalence point for a calculated curve). In general, derivatives reveal subtle features that are not always observable on the original plots, although the generation of such derivatives (differentiation) from real data does make for a more noisy result. By strict definitions of calculus, differentiation pertains to increments (of volume base added for example) that are infinitesimally small. However, as long as the volume increments or the resulting pH increments are not too large, the use of $\Delta pH/\Delta V_B$ (this means the difference in two adjacent pH values, ΔpH, brought about by the addition of a finite increment of base, ΔV_B) instead of the theoretically desired dpH/dV_B, produces equally good results.

We could use any of the titration data that we have generated thus far, except that we have used a constant increment of base in this titration and near the equivalence point the pH change is too large for a good derivative plot. We have two choices: we can either generate data that will provide for higher resolution (lower volume increments) in the neighborhood of the equivalence point or we can generate titration data with high resolution throughout the titration (say, a thousand points per titration). The latter approach does not require that we select a region near the equivalence point (which requires that we already know where the equivalence point is!) to produce high-resolution data. Producing a high-resolution titration plot with 1000 points is a rather tall order if this has to be done by Goal Seek or Solver in Excel using the approach we have thus far used. On the other hand, if our primary desire is to generate a titration plot, it does not have to be done by constant volume increments of the titrant. If we know the concentration of an acid, it is rather trivial by charge balance principles to calculate how much base we need to add to get to a certain pH.

Consider the titration of V_A mL of a triprotic acid of molar concentration C_A being titrated by C_B molar NaOH, V_B mL having been added at any point. Charge balance requires that

$$[Na^+] = [OH^-] + C_A(\alpha_1 + 2\alpha_2 + 3\alpha_3) - [H^+]$$

If V_B mL of NaOH has been added with the total volume being $V_A + V_B$, the above equation becomes

$$V_B C_B/(V_A + V_B) = [OH^-] - [H^+] + V_A C_A (\alpha_1 + 2\alpha_2 + 3\alpha_3)/(V_A + V_B)$$

Multiply both sides by $(V_A + V_B)$:

$$V_B C_B = (V_A + V_B)([OH^-] - [H^+]) + V_A C_A (\alpha_1 + 2\alpha_2 + 3\alpha_3)$$

Separating the V_A and V_B multipliers on the first term on the right and transposing:

$$V_B(C_B - ([OH^-] - [H^+])) = V_A([OH^-] - [H^+]) + V_A C_A (\alpha_1 + 2\alpha_2 + 3\alpha_3)$$

or

$$V_B = V_A\{([OH^-] - [H^+]) + C_A (\alpha_1 + 2\alpha_2 + 3\alpha_3)\}/(C_B + [H^+] - [OH^-]) \quad (7.26)$$

If we know the dissociation constants and we specify pH (and hence $[H^+]$), we can easily calculate the alpha values and all terms on the right side of Equation 7.26 are readily calculated.

Let us illustrate this with 0.10 molar solution of H_3PO_4 ($K_{a1} = 1.1 \times 10^{-2}$, $K_{a2} = 7.5 \times 10^{-8}$, $K_{a3} = 4.8 \times 10^{-13}$) with 0.17 M NaOH. Refer to the spreadsheet "Sec. 7.11 derivative titrations easy method.xlsx" in the text **website** for the following discussion. In cells A1:B7, we put in the values of C_A, C_B, V_A, K_{a1}, K_{a2}, K_{a3} (these named as KAA, KAB, and KAC so Excel will accept these names) and K_w and define these names. Titles are put in cells D8:K8, for V_B, pH, $[H^+]$, Q, α_1, α_2, α_3 and $[OH^-]$. In row 9, we enter any trial value for pH in cell E9, express cell F9 as $10^\wedge - E9$ and Q, α_1, α_2, α_3 as their customary expressions, respectively, in cells G9:J9 in terms of the K_a's and $[H^+]$ (F9). Finally in cell K9, we express $[OH^-]$ as $K_w / [H^+]$. Only this first row of calculations, row 9, with no base added, is solved differently from the subsequent rows to know what the starting pH is. For this row and this row only, we enter the charge balance expression in cell A9 as $[OH^-] - [H^+] + C_A(\alpha_1 + 2\alpha_2 + 3\alpha_3)$, multiply by 1E10 and use Goal Seek to set cell A9 to zero by changing cell E9 (pH). Once this is solved, we enter the next 0.01 pH unit higher pH in (rounded to the nearest 0.01 pH unit) in cell E10 and another 0.01 higher pH unit in cell E11. We then select cells E10 and E11 and drag from the bottom right-hand corner of cell E11 down to populate the column up to the desired pH value (in the illustrated case we stopped at a pH of 12.55). We now select cells F9:K9 and double-click at the bottom right corner of cell K9 to fill the rest of the data space in columns F through K. In cell D9, the first entry for V_B, enter zero. But beginning in cell D10, enter the expression in Equation 7.26; this will now directly solve for V_B. Now double-click on the bottom right corner of cell D10 to fill up the D column. Between columns D and E (V_B and pH), you generated the titration plot!

Now we are ready to compute the first derivative. Consider that in cell M10 we have calculated ΔV_B as the difference between the V_B values (D10-D9), and in cell N10 we have similarly calculated the ΔpH value as (E10-E9). Thence, in cell O10 we have similarly calculated the ΔpH/ΔV_B value as N10/M10. We want to plot these ΔpH/ΔV_B values against V_B—but we have two V_B values—one in cell D9, one in cell D10. Since our ΔpH/ΔV_B data are derived from both V_B values, we actually use an average of the two. We have so designated column L as "Average V_B" and calculated it as the average of cells D9 and D10 ($V_{B,av}$). We are now ready to fill up column L, so we double-click on bottom right corner of cell L10 to fill up the column. Plot now the standard titration plot (pH vs. V_B) and right-click on the plot, choose Select Data and Add, and add new X and Y data corresponding to V_B and ΔpH/ΔV_B. If you wish, you can put the second plot on the right hand Y-axis by right-clicking on the trace to select it, choosing "Format data series" and choosing to plot it using a secondary axis. The chart labeled "Tit plot & 1stDeriv" shows such an overlaid plot. You can readily see that how sharp inflection points indicate the two equivalence points in this titration. In the following Professor's Favorite Example with the real example for red wine titration given in Figure 7.14, the exact location of the second equivalence point will be far more difficult to discern without a derivative plot.

Various titrations are extensively used in industrial laboratories, but few titrations are done manually. Extensive use is made of robotic sample handling and automatic titrators (often referred to as autotitrators) in which the titrant is dispensed by a motorized buret (see Chapter 21). Potentiometric sensors (see Chapter 20), of which a pH electrode is a hallmark example, do not exhibit particularly fast response times. If the titrant was delivered continuously at a fast rate, the system may go by the equivalence point too fast and would not get an accurate reading. On the other hand, if you delivered the titrant continuously at a very slow rate, it will take too long to complete a titration. Such instruments continuously calculate the first derivative and as its value begins

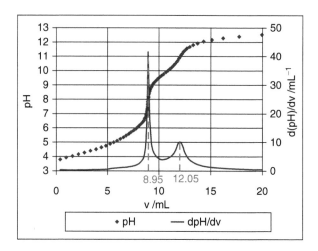

FIGURE 7.14 Titration curve for 25 mL of red wine titrated with 0.1926 M NaOH (markers) and first derivatives of the curve (solid line).

to increase, the system lowers the rate of titrant addition such that the titrant is added slower as the equivalence point is approached. This way both accuracy and analytical throughput are maintained.

Buffer Intensity

Note that the first derivative dpH/dV_B has a direct relationship to the *buffer intensity* β, first discussed in Chapter 6, Section 6.8. It was defined in Equation 6.94 as dC_B/dpH where dC_B is the infinitesimal increase in base concentration added to affect the change in pH of dpH. Note that dV_B is readily converted to the corresponding amount of base in moles by multiplying with C_B. This is then readily converted to the corresponding increment in concentration by dividing with the total volume:

$$dC_B = \left(C_B dV_B\right) / \left(V_A + V_B\right)$$

When using real data with finite increments, we will have to use $V_{B,Av}$ instead of V_B and approximate (it will be exactly equal only if it was a true differential) β as:

$$\beta \approx \Delta C_B / \Delta pH = \{(C_B V_B)/(V_A + V_{B,av})\}\Delta pH \tag{7.27}$$

β is calculated according to Equation 7.27 in column P. For a monoprotic acid base system, the strict expression for β is (see the book's **website** Sec. 7.11 Supplement–Buffer intensity derivation for a detailed derivation and discussion, some familiarity with elementary differential calculus will be needed):

$$\beta = 2.303 \left(C\alpha_0\alpha_1 + [H^+] + [OH^-]\right) \tag{7.28}$$

where C is the total concentration of the buffering species in the solution. Under conditions where $[H^+]$ and $[OH^-]$ are negligible relative to $C\alpha_0\alpha_1$, Equation 7.28 is simplified and noting that due to dilution, $C = C_A V_A/(V_A + V_B)$, we have

$$\beta = 2.303 \, C_A\alpha_0\alpha_1/(V_A + V_B) \tag{7.29}$$

Note that this is the limiting maximum value, attainable only when the increments are infinitesimally small. Refer again to the spreadsheet "Sec. 7.11 derivative titn master spreadsheet–Goal Seek.xlsx" in the text website. In the chart "Tit plot & Beta vs VB" (if you don't see the tab for this, click on the left pointing arrow to the left of the tabs to scroll to it) we have the titration plot (pH vs. V_B) and β computed both according to Equations 7.27 (column P) and 7.29 (column Q). In using Equation 7.29, we recognize rather than a monoprotic acid, here the second proton of a diprotic acid is being neutralized, so in calculating the data in column Q, we use $\alpha_1\alpha_2$ rather than $\alpha_0\alpha_1$. Also, we limit the computation to the range where only these species are important, where both α_1 and α_2 is at least 0.1 (10% of the total concentration). Observe that there is an excellent agreement between the results obtained with Equations 7.27 and 7.29. If you compare the actual numerical values in columns P and Q, you will find that maximum β is reached at a pH of 7.06 for both computations, and the finite difference computed value in column P is 99.9% of the maximum calculated in column Q. The exact pH at which β reaches a maximum value is slightly before pK_{a2} because of two reasons: (a) this is a titration system where the concentration does not remain constant but decreases continuously as the acid is titrated and the pH increases; (b) although the other dissociations are not close, they can still have a minute effect, and in this case, pK_{a1} is slightly closer to pK_{a2} than is pK_{a3}.

Buffer Intensity Computations Depend on the pH Resolution: Buffer Capacity

If we examine the "Beta vs pH" plot, in the same spreadsheet it is clear that even at its maximum where β is relatively unchanging, it is constant only over a rather small range. Although at the maximum we observe a value of ~31 meq/L per pH unit, this depends on the pH resolution of the present calculations being 0.01 pH unit. For practical purposes one is typically interested in how much base or acid is needed to cause a larger change in pH, e.g., 1 pH unit. Because β does not remain constant across a significant range, specifying β values thus calculated may not directly address the question as to how much the pH will change if a significant amount of acid or base is added. For example, if we had a buffer solution corresponding to that at the maximum β in our plot and we added ~31 meq/L base (without volume change), the pH will increase by more than 1 pH unit because the plot also shows that across 1 pH unit range β will not remain at this maximum value.

We can examine how β changes as a function of the pH resolution we use. In columns T:AF, AH:AT, and AV:BH we essentially copy and paste our previous calculations and then repeat them with pH resolutions of 0.1, 0.5, and 1 unit. The results are plotted in the "Buffer Intensity and Capacity" chart. You can readily see that while there is no significant change in β in going from a pH resolution of 0.01 to 0.1, it rapidly decreases when the pH resolution is decreased to 0.5 and 1. The apparent shift in the location of the maximum is an artifact of Excel's attempt to draw a smooth curve through too few data points (the points are shown for the two lowest resolution traces). The decrease in β over a wider range without this artifact shift in pH location can be observed by using the Moving Average function within Data/Data Analysis in Excel. (Like Solver, the "Data Analysis" Toolpak is not automatically installed. You must install it from Excel options.) The chart also shows a 200-point running average of pH and β from the 0.01 pH resolution data set; note that the position of the maximum does not actually move but β decreases.

When referring to changes over a significant pH interval, the term *buffer capacity*, rather than buffer intensity, should be used. In going from pH 7.1 to pH 8.1, the buffer capacity of our solution for example will be ~21 meq/L (see cell BH17).

The Second Derivative

The second derivative $\Delta^2 pH / \Delta V_B^2$ passes through zero at the end point and makes it electronically simpler to locate the end point in this fashion. In principle it also permits locating the end point with a precision greater than the resolution of the data points. We calculate the second derivative (column R) simply by obtaining the difference between two successive rows of the first derivative (column O). Note that to compute the second derivative, the original pH and V_B data that we ultimately use are in rows 9, 10, and 11 and for this reason the value is located centrally in row 10 and when plotted against V_B, it is plotted against the V_B data also in row 10 (and not the $V_{B,av}$ data in column L). The original titration plot and both the first and second derivative plots are shown together in the worksheet "1st and 2nd deriv." The titration plot and both the derivative plots are shown in Figure 7.13.

⌐Contributed by Professor
Alberto Rojas-Hernández,
Universidad Autónoma
Metropolitana-Iztapalapa,
Mexico ⌐

 # Professor's Favorite Example

Acid–base titration of red wine: Titrable acidity (TA), tartaric acid content (g/L) and maximum buffer capacity.

Red wines are composed of several substances with acid–base properties that may be considered among three main groups: carboxylic acids, carboxylic-polyphenolic, and polyphenolic acids. The common carboxylic acids are tartaric ($pK_{a1} = 3.2$, $pK_{a2} = 4.3$), malic ($pK_{a1} = 3.4$, $pK_{a2} = 5.1$), and citric ($pK_{a1} = 3.1$, $pK_{a2} = 4.7$, $pK_{a3} = 6.4$). The common carboxylic-polyphenolic acids are gallic ($pK_{a1} = 4.4$, $pK_{a2} = 8.6$, $pK_{a3} = 11.2$, $pK_{a4} = 12.0$) and caffeic ($pK_{a1} = 4.4$, $pK_{a2} = 8.6$, $pK_{a3} = 11.5$). Finally the common polyphenolic acids are tannic (with several pK_a values between 6 and 10), as well as flavonols and flavonoids like anthocyanines (that give color to these wines).

Figure 7.14 shows the titration of 25 mL of red wine with 0.1926 M NaOH, obtained with an automatic titrator. The titration shows two equivalence points. The first equivalence point corresponds to the titration of the carboxylic groups, often called titrable acidity (TA). The second equivalence point corresponds with the true total acidity [TAc, see for example, *Journal of Food Composition and Analysis*, Composition study of methods for determination of titrable acidity in wine, **16** (2003) 555–562] and the second step is due to titration of the phenolic groups.

If we assume that the only carboxylic acid present in the wine is tartaric acid (HOOC-CHOH-CHOH-COOH = H_2tar), the mass of tartaric acid (m_{H_2tar}) in the 25 mL red wine sample is:

$$m_{H_2tar} \approx (8.95 \text{ mL})(0.1926 \ M)(1\text{mmol } H_2\text{tar}/2\text{mmol NaOH})(150.1 \text{ mg/mmol})$$

$$= 129_{.4} \text{ mg, equivalent to } 5.18 \text{ g/L of tartaric acid.}$$

The buffer capacity of the same red wine sample is shown in Figure 7.15. The buffer capacity has a maximum value for pH ≈ 3.85, which is near the initial sample pH. The buffer capacity is 40 mmol/L per pH. This means that the amount of NaOH or HCl that should be added to change the initial pH of the wine by 1 unit is approximately (40 mmol/1000 mL)(25 mL) = 1 mmol (see Equation 6.50).

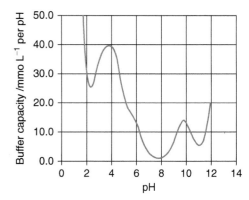

FIGURE 7.15 Buffer capacity curve as a function of pH for the sample of red wine, for which titration is shown in Figure 7.14.

7.12 Titration of Amino Acids—They Are Acids and Bases

Amino acids are important in pharmaceutical chemistry and biochemistry. These are amphoteric substances that contain both acidic and basic groups (i.e., they can act as acids or bases). The acid group is a carboxylic acid group ($-CO_2H$), and the basic group is an amine group ($-NH_2$). In aqueous solutions, these substances tend to undergo internal proton transfer from the carboxylic acid group to the amino group because the RNH_2 is a stronger base than RCO_2^-. The result is a **zwitterion**:

$$R-CH-CO_2^-$$
$$|$$
$$NH_3^+$$

Since they are amphoteric, these substances can be titrated with either a strong acid or a strong base. Many amino acids are too weak to be titrated in aqueous solutions, but some will give adequate end points, especially if a pH meter is used to construct a titration curve.

We can consider the conjugate acid of the zwitterion as a **diprotic acid**, which ionizes stepwise:

$$\underset{\substack{\text{conjugate acid} \\ \text{of zwitterion}}}{R-\underset{|}{\overset{}{C}}H-CO_2H} \underset{NH_3^+}{} \rightleftharpoons H^+ + \underset{\substack{\text{zwitterion}}}{R-\underset{|}{\overset{}{C}}H-CO_2^-} \underset{NH_3^+}{} \rightleftharpoons H^+ + \underset{\substack{\text{conjugate base} \\ \text{of zwitterion}}}{R-\underset{|}{\overset{}{C}}H-CO_2^-} \underset{NH_2}{} \quad (7.30)$$

K_{a1} and K_{a2} values are frequently tabulated for amino acids (see Table C.1 in Appendix C). The values listed represent the successive ionization of the protonated form (i.e., the conjugate acid of the zwitterion); it ionizes to give first the amphoteric zwitterion and second to give the conjugate base, which is the same as a salt of a weak acid that hydrolyzes. Acid–base equilibria of amino acids are therefore treated just as for any other diprotic acid. The hydrogen ion concentration of the zwitterion is calculated in the same way as for any amphoteric salt, such as HCO_3^-, as we described in Chapter 6; that is,

For the zwitterion, $[H^+] \approx \sqrt{K_{a1}K_{a2}}$.

$$[H^+] = \sqrt{K_{a1}K_{a2}} \quad (7.31)$$

The titration of amino acids is not unlike the titration of other amphoteric substances, such as HCO_3^-. In the latter case, titration with a base gives CO_3^{2-}, with an intermediate CO_3^{2-}/HCO_3^- buffer region, and titration with an acid gives H_2CO_3, with an intermediate HCO_3^-/H_2CO_3 buffer region.

When the zwitterion of an amino acid is titrated with strong acid, a buffer region is first established, consisting of the zwitterion (the "salt") and the conjugate acid. Halfway to the equivalence point, $pH = pK_{a1}$ (just as with HCO_3^-/H_2CO_3); and at the equivalence point, the pH is determined by the conjugate acid (and K_{a1}, as with H_2CO_3). When the zwitterion is titrated with a strong base, a buffer region of conjugate base (the "salt") and zwitterion (now the "acid") is established. Halfway to the equivalence point, $pH = pK_{a2}$ (as with CO_3^{2-}/HCO_3^-); and at the equivalence point, the pH is determined by the conjugate base (whose $K_b = K_w/K_{a2}$, as with CO_3^{2-}).

Amino acids may contain more than one carboxyl or amine group; in these cases, they may yield stepwise end points like other polyprotic acids (or bases), provided the different groups differ in K's by at least a factor of 10^4 and are still strong enough to be titrated.

7.13 Kjeldahl Analysis: Protein Determination

An important method for accurately determining nitrogen in proteins and other nitrogen-containing compounds is the **Kjeldahl analysis**. The quantity of protein can be calculated from a knowledge of the percent nitrogen contained in it. Although other more rapid methods for determining proteins exist, the Kjeldahl method is the standard on which all other methods are based.

The material is digested with sulfuric acid to decompose it and convert the nitrogen to ammonium hydrogen sulfate:

$$C_aH_bN_c \xrightarrow[\text{catalyst}]{H_2SO_4} aCO_2 \uparrow + \tfrac{1}{2}bH_2O + cNH_4HSO_4$$

The solution is cooled, concentrated alkali is added to make the solution alkaline, and the volatile ammonia is distilled into a solution of standard acid, which is in excess. Following distillation, the excess acid is back-titrated with standard base.

$$cNH_4HSO_4 \xrightarrow{OH^-} cNH_3 \uparrow + cSO_4^{2-}$$

$$cNH_3 + (c+d)HCl \rightarrow cNH_4Cl + dHCl$$

$$dHCl + dNaOH \rightarrow \tfrac{1}{2}dH_2O + dNaCl$$

---

mmol N(c) = mmol reacted HCl = mmol HCl taken
$\times (c+d)$ − mmol NaOH(d)

mmol $C_aH_bN_c$ = mmol N $\times 1/c$

The digestion is speeded up by adding potassium sulfate to increase the boiling point and by a catalyst such as a selenium or copper salt. The amount of the nitrogen-containing compound is calculated from the weight of nitrogen analyzed by multiplying it by the gravimetric factor.

Example 7.4

A 0.2000-g sample containing urea,

$$\underset{NH_2-C-NH_2}{\overset{\overset{\textstyle O}{\overset{\textstyle \|}{}}}{}}$$

is analyzed by the Kjeldahl method. The ammonia is collected in 50.00 mL of 0.05000 M H$_2$SO$_4$, and the excess acid is back-titrated with 0.05000 M NaOH, a procedure requiring 3.40 mL. Calculate the percent urea in the sample.

Solution

The titration reaction is

$$H_2SO_4 + 2NaOH \rightarrow Na_2SO_4 + 2H_2O$$

mmol NaOH consumed = 3.40 mL $\times$ 0.05 mmol/mL = 0.17 mmol

mmol H$_2$SO$_4$ neutralized = 0.17/2 mmol = 0.085 mmol

mmol H$_2$SO$_4$ originally taken = 0.05000 mmol/mL $\times$ 50.00 mL = 2.500 mmol

mmol H$_2$SO$_4$ neutralized by NH$_3$ = 2.500 − 0.085 mmol = 2.415 mmol

The reaction with the NH$_3$ is

$$2NH_3 + H_2SO_4 \rightarrow (NH_4)_2SO_4$$

Two millimoles NH_3 react with one millimole H_2SO_4 and two millimoles NH_3 come from one millimole urea. Therefore, millimoles of H_2SO_4 neutralized by ammonia, 2.415 mmol, is the same as the amount of urea. Multiplying by 60.05 mg/mmol urea, we have 2.415 mmol × 60.05 mg/mmol = 145.02 mg urea

$$\% \text{ urea by weight: } 145.02/200.0 \times 100\% = 72.51\%$$

A large number of different **proteins** contain very nearly the same percentage of nitrogen. The gravimetric factor for conversion of weight of N to weight of protein for normal mixtures of serum proteins (globulins and albumin) and protein in feeds is 6.25, (i.e., the proteins contain 16% nitrogen). When the sample is made up almost entirely of gamma globulin, the factor 6.24 is more accurate. If it contains mostly albumin, 6.27 is preferred.

Many proteins contain nearly the same amounts of nitrogen.

In the conventional Kjedahl method, two standard solutions are required, the acid for collecting the ammonia and the base for back-titration. A modification can be employed that requires only standard acid for direct titration. The ammonia is collected in a solution of boric acid. In the distillation, an equivalent amount of ammonium borate is formed:

$$NH_3 + H_3BO_3 \rightleftharpoons NH_4^+ + H_2BO_3^- \tag{7.32}$$

Boric acid is too weak to be titrated, but the borate, which is equivalent to the amount of ammonia, is a fairly strong Brønsted base that can be titrated with a standard acid to a methyl red end point. The boric acid is so weak it does not interfere, and its concentration need not be known accurately. Also, boric acid is so weak that it is not significantly ionized, it is nonconductive. In contrast, ammonium borate is ionized and is conductive. In a present-day adaptation, instead of titration, the conductivity of the ammonium borate formed is measured as a measure of ammonia absorbed.

Example 7.5

A 0.300-g feed sample is analyzed for its protein content by the modified Kjeldahl method. If 25.0 mL of 0.100 M HCl is required for titration, what is the percent protein content of the sample?

Solution

Since this is a direct titration with HCl which reacts 1 : 1 with NH_3, the millimoles of NH_3 (and therefore of N) equals the millimoles of HCl. Multiplication by 6.25 gives the milligrams of protein.

$$\% \text{ protein} = \frac{0.10 \text{ mmol/mL HCl} \times 25.0 \text{ mL HCl} \times 14.01 \text{ mg N/mmol HCl} \times 6.25 \text{ mg protein/mg N}}{300 \text{ mg}}$$

$$\times 100\% = 73.0\%$$

The boric acid method (a direct method) is simpler and is usually more accurate since it requires the standardization and accurate measurement of only one solution. However, the end point break is not so sharp, and the indirect method requiring back-titration is usually preferred for **micro-Kjeldahl analysis**. A macro-Kjeldahl analysis of blood requires about 5 mL blood, while a micro-Kjeldahl analysis requires only about 0.1 mL.

We have confined our discussion to those substances in which the nitrogen exists in the −3 valence state, as in ammonia. Such compounds include amines and amides. Compounds containing oxidized forms of nitrogen, such as organic nitro and azo compounds, must be treated with a reducing agent prior to digestion in order to obtain complete conversion to ammonium ion. Reducing agents such as iron(II) or thiosulfate are used. Inorganic nitrate and nitrite are not converted by such treatment to ammonia.

7.14 Titrations Without Measuring Volumes

Can titrations be done without measuring volumes? Yes. Professor Michael DeGrandpre at the University of Montana has developed "tracer monitored" titrations (TMT) in which the spectrophotometric monitoring of titrant dilution, rather than volumetric increment, places the burden of analytical performance solely on the spectrophotometer. Most modern titration systems are fully automated with precision pumps. The TMT method is insensitive to pump precision and reproducibility of automatic titrators, and can allow less precise and less expensive pumps.

An inert indicator tracer (e.g., dye) is added to the titrant. Dilution of a pulse of titrant in a titration vessel is tracked using the total tracer concentration, measured spectrophotometrically (an instrumental technique that measures the concentration of a colored substance from the amount of light absorbed based on linearity defined by Beer's law, see Chapter 11). The concentration of the tracer at the detected end point (e.g., from pH measurement), derived from Beer's law, is used to calculate the relative proportions of titrant and sample.

In conventional titrations, the analyte concentration in the sample, $[analyte]$, is given by

$$[analyte] = \frac{Q\,[titrant]\,V_{tit.ep}}{V_{sample}} \tag{7.33}$$

where Q is the reaction stoichiometry (moles analyte : moles titrant), $[titrant]$ is the titrant concentration, $V_{tit,ep}$ is the volume of the titrant at the end point, and V_{sample} is the original volume of the sample taken.

The TMT method is based on determining the dilution factor, D_{ep}, of the titrant at the end point, determined using the inert tracer:

$$D_{ep} = \frac{V_{tit.ep}}{(V_{tit.ep} + V_{sample})} = \frac{[tracer]_{ep}}{[tracer]_{tit}} \tag{7.34}$$

where $[tracer]_{ep}$ is the tracer concentration in the mixture at the end point, and $[tracer]_{tit}$ is the tracer concentration in the titrant. The sample analyte concentration is then given by

$$[analyte] = \frac{Q\,[titrant]}{\left(\frac{1}{D_{ep}} - 1\right)} \tag{7.35}$$

So we only need to determine D_{ep}. Any inert tracer can be used, for example, a fluorescent dye can be used in a low concentration in the titrant. The technique can be applied to other types of titrations, e.g., complexometric or redox titrations. It will also be realized, that the same principle can be applied if the sample is spiked with an indicator dye at the beginning of the titration and the titrant simply dilutes the dye level as the titration progresses. Either way, the method eliminates the need for volumetric measurements. However, Beer's law (see Chapter 11) must be obeyed through the dye concentration range of interest.

Details of the method are given in References 10 and 11.

1. What is the minimum pH change required for a sharp indicator color change at the end point? Why?
2. What criterion is used in selecting an indicator for a particular acid–base titration?
3. At what pH is the buffering capacity of a buffer the greatest?
4. Is the pH at the end point for the titration of a weak acid neutral, alkaline, or acidic? Why?
5. What would be a suitable indicator for the titration of ammonia with hydrochloric acid? Of acetic acid with sodium hydroxide?
6. Explain why boiling the solution near the end point in the titration of sodium carbonate increases the sharpness of the end point.
7. What is the approximate pK of the weakest acid or base that can be titrated in aqueous solution?
8. What must be the difference in the strengths of two acids in order to differentiate between them in titration?
9. Distinguish between a primary standard and a secondary standard.
10. What is a zwitterion?
11. What percent nitrogen is contained in a typical protein?
12. What is the preferred acid for titrating bases? Why?
13. Differentiate between end point and equivalence point.
14. Explain the terms: (a) titration error, (b) indicator and (c) modified methyl orange.
15. The concentration of NaOH is 0.5 M. If 20 mL is needed to titrate 35 mL of acid, what is the concentration of acid?

    (a) 0.875 M of acid   (b) 0.0029 M of acid

    (c) 0.29 M of acid   (d) 0.00875 M of acid

16. What is the purpose of indicator in the solution with the unknown concentration?

    (a) It will show when there is enough acid in the solution.

    (b) It will show when the equivalence point is obtained.

    (c) It will show when there is enough base in the solution.

17. The equation of neutralization reaction between solutions of potassium hydroxide and hydrochloric acid is:

    (a) $H^+ (aq) + OH^- (aq) \rightarrow H_2O$

    (b) $KOH (aq) + HCl (aq) \rightarrow H_2O (l) + KCl$

    (c) $POH (aq) + HClO (aq) \rightarrow H_2O (l) + PClO (aq)$

    (d) $KOH (aq) + HCl (aq) \rightarrow K_2Cl (aq) + H_2O (l)$

    (e) $KOH (aq) + HCl (aq) \rightarrow K_2Cl (aq) + H_2O (l) + H^+$

18. 50 mL of 0.5 M barium hydroxide is required to fully titrate a 100 mL solution of sulfuric acid. What is the initial concentration of acid?

    (a) 50 M (b) 0.5 M

    (c) 100 M (d) 0.25 M

19. In a titration of a weak acid with strong base, what is the pH of the solution at the equivalence point?

    (a) > 7   (b) < 7

    (c) 7   (d) 0

20. Which of the following describes the equivalence point on a graph of pH versus the amount of titrant added to a solution?

    (a) The point with the lowest pH.

    (b) The point with the highest pH.

    (c) The point where the magnitude of the slope of the curve is greatest

    (d) the point where the magnitude of the slope of the curve is least.

21. After adding very large volume of titrant, asymptote in curves appears, the position of asymptote is determined by which of the following experimental parameters?

    (a) pH of the initial solution

    (b) Slope of the curve of titration curve

    (c) K_a of initial solution

    (d) The pH of titrant

22. A 0.128 g sample of KHP (HKC_8C_4O4) is required for 28.54 mL of NaOH solution to reach a phenolphthalein end point. Calculate the molarity of the NaOH.

23. A 20.00 mL sample of HCl was titrated with the NaOH solution from Question 22. To reach the end point required 23.72 mL of NaOH. Calculate the molarity of the HCl.

Problems

Standardization Calculations

24. A hydrochloric acid solution is standardized by titrating 0.4541 g of primary standard tris(hydroxymethyl)aminomethane. If 35.37 mL is required for the titration, what is the molarity of the acid?

25. A hydrochloric acid solution is standardized by titrating 0.2329 g of primary standard sodium carbonate to a methyl red end point by boiling the carbonate solution near the end point to remove carbon dioxide. If 42.87 mL acid is required for the titration, what is its molarity?

26. A sodium hydroxide solution is standardized by titrating 0.8592 g of primary standard potassium acid phthalate to a phenolphthalein end point, requiring 32.67 mL. What is the molarity of the base solution?

27. A 10.00-mL aliquot of a hydrochloric acid solution is treated with excess silver nitrate, and the silver chloride precipitate formed is determined by gravimetry. If 0.1682 g precipitate is obtained, what is the molarity of the acid?

Indicators

28. In titration of weak acid with strong base, which indicator would be the best choice and why?

    (a) Methyl orange   (b) Bromocresol green   (c) Phenolphthalein

29. Write a Henderson–Hasselbalch equation for a weak-base indicator, B, and calculate the required pH change to go from one color of the indicator to the other. Around what pH is the transition?

Titration Curves

You may wish to use spreadsheets for some of these calculations.

30. Calculate the pH at 0, 10.0, 25.0, and 30.0 mL of titrant in the titration of 50.0 mL of 0.100 M NaOH with 0.200 M HCl.

31. Calculate the pH at 0, 10.0, 25.0, 50.0, and 60.0 mL of titrant in the titration of 25.0 mL of 0.200 M HA with 0.100 M NaOH. $K_a = 2.0 \times 10^{-5}$.

32. Calculate the pH at 0, 10.0, 25.0, 50.0, and 60.0 mL of titrant in the titration of 50.0 mL of 0.100 M NH_3 with 0.100 M HCl.

33. Calculate the pH at 0, 25.0, 50.0, 75.0, 100, and 125% titration in the titration of both protons of the diprotic acid H_2A with 0.100 M NaOH, starting with 100 mL of 0.100 M H_2A. $K_{a1} = 1.0 \times 10^{-3}$, $K_{a2} = 1.0 \times 10^{-7}$.

34. Calculate the pH at 0, 25.0, 50.0, 75.0, 100, and 150% titration in the titration of 100 mL of 0.100 M Na_2HPO_4 with 0.100 M HCl to $H_2PO_4^-$.

35. A titration is carried out for 25.00 mL of 0.100 M HCl with 0.100 M NaOH. Calculate the pH at the volumes of added base solution.

   (a) 0.00 mL   (b) 12.50 mL

   (c) 25.00 mL   (d) 37.50 mL

36. Calculate the pH for strong acid/strong base titration between 50.0 mL of 0.100 M HNO_3 (aq) and 0.200 M NaOH at the tested volume of added base.

   (a) 0.00 mL   (b) 15.0 mL

   (c) 25.0 mL   (d) 40.0 mL

Quantitative Determinations

37. A 0.492-g sample of KH_2PO_4 is titrated with 0.112 M NaOH, requiring 25.6 mL:

$$H_2PO_4^- + OH^- \rightarrow HPO_4^{2-} + H_2O$$

What is the percent purity of the KH_2PO_4?

38. What volume of 0.155 M H_2SO_4 is required to titrate 0.293 g of 90.0% pure LiOH?

39. An indication of the average formula weight of a fat is its saponification number, expressed as the milligrams KOH required to hydrolyze (saponify) 1 g of the fat:

$$
\begin{array}{ccc}
CH_2CO_2R & & CH_2OH \\
| & & | \\
CHCO_2R + 3KOH \rightarrow & CHOH + 3RCO_2K \\
| & & | \\
CH_2CO_2R & & CH_2OH
\end{array}
$$

where R can be variable. A 1.10-g sample of butter is treated with 25.0 mL of 0.250 M KOH solution. After the saponification is complete, the unreacted KOH is back-titrated with 0.250 M HCl, requiring 9.26 mL. What is the saponification number of the fat and what is its average formula weight (assuming the butter is all fat)?

40. A sample containing the amino acid alanine, $CH_3CH(NH_2)COOH$, plus inert matter is analyzed by the Kjeldahl method. A 2.00-g sample is digested, the NH_3 is distilled and collected in 50.0 mL of 0.150 M H_2SO_4, and a volume of 9.0 mL of 0.100 M NaOH is required for back-titration. Calculate the percent alanine in the sample.

41. A 2.00-mL serum sample is analyzed for protein by the modified Kjeldahl method. The sample is digested, the ammonia is distilled into boric acid solution, and 15.0 mL of standard HCl is required for the titration of the ammonium borate. The HCl is standardized by treating 0.330 g pure $(NH_4)_2SO_4$ in the same manner. If 33.3 mL acid is required in the standardization titration, what is the concentration of protein in the serum in g% (wt/vol)?

Quantitative Determinations of Mixtures

42. A 100-mL aliquot of a solution containing HCl and H_3PO_4 is titrated with 0.200 M NaOH. The methyl red end point occurs at 25.0 mL, and the bromthymol blue end point occurs at 10.0 mL later (total 35.0 mL). What are the concentrations of HCl and H_3PO_4 in the solution?

43. A 0.527-g sample of a mixture containing $NaCO_3$, $NaHCO_3$, and inert impurities is titrated with 0.109 M HCl, requiring 15.7 mL to reach the phenolphthalein end point and a total of 43.8 mL to reach the modified methyl orange end point. What is the percent each of Na_2CO_3 and $NaHCO_3$ in the mixture?

44. Sodium hydroxide and Na_2CO_3 will titrate together to a phenolphthalein end point ($OH^- \rightarrow H_2O$; $CO_3^{2-} \rightarrow HCO_3^-$). A mixture of NaOH and Na_2CO_3 is titrated with 0.250 M HCl, requiring 26.2 mL for the phenolphthalein end point and an additional 15.2 mL to reach the modified methyl orange end point. How many milligrams NaOH and Na_2CO_3 are in the mixture?

45. Sodium carbonate can coexist with either NaOH or $NaHCO_3$ but not with both simultaneously, since they would react to form Na_2CO_3. Sodium hydroxide and Na_2CO_3 will titrate together to a phenolphthalein end point ($OH^- \rightarrow H_2O$; $CO_3^{2-} \rightarrow HCO_3^-$). A mixture of either NaOH and Na_2CO_3 or of Na_2CO_3 and $NaHCO_3$ is titrated with HCl. The phenolphthalein end point occurs at 15.0 mL and the modified methyl orange end point occurs at 50.0 mL (35.0 mL beyond the first end point). The HCl was standardized by titrating 0.477 g Na_2CO_3, requiring 30.0 mL to reach the modified methyl orange end point. What mixture is present and how many millimoles of each constituent are present?

46. What would be the answers to Problem 45 if the second end point had occurred at 25.0 mL (10.0 mL beyond the first end point)?

47. A mixture containing only $BaCO_3$ and Li_2CO_3 weighs 0.150 g. If 25.0 mL of 0.120 M HCl is required for complete neutralization ($CO_3^{2-} \rightarrow H_2CO_3$), what is the percent $BaCO_3$ in the sample?

48. A sample of P_2O_5 contains some H_3PO_4 impurity. A 0.405-g sample is reacted with water ($P_2O_5 + 3H_2O \rightarrow 2H_3PO_4$), and the resulting solution is titrated with 0.250 M NaOH ($H_3PO_4 \rightarrow Na_2HPO_4$). If 42.5 mL is required for the titration, what is the percent of H_3PO_4 impurity?

PROFESSOR'S FAVORITE PROBLEMS

Contributed by Professor Tarek Farhat, The University of Memphis and Peter R. Griffiths, University of Idaho

49. A bubbler containing sodium hydroxide solution is used as a filter to remove CO_2 gas from an air stream that purges a glove box. The following reaction takes place:

$$2NaOH + CO_2(g) \rightarrow Na_2CO_3 + H_2O$$

If the bubbler contained 100 mL of the NaOH solution and the NaOH solution was initially at a pH of 12, calculate the drop in pH after passing an air stream (1 atm pressure, 25°C) at a flow rate of 5.0 mL/min that contains 0.5% CO_2 gas by volume at 25°C and 1 atm for (a) one hour, and (b) five hours?

50. Which curve most closely approximates that expected for the complete titration of a solution of sodium hydrogen sulfate, $NaHSO_4$ with aqueous NaOH? (Dissociation constant for $HSO_4^- = 1.2 \times 10^{-2}$)

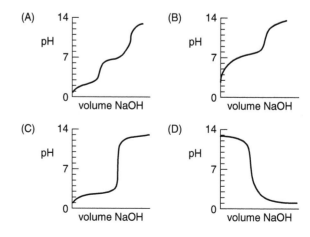

The following problems were contributed by Professor Noel Motta, University of Puerto Rico-Río Piedras

51. The following curve may correspond to the titration of

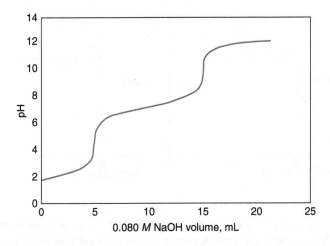

(a) 5.00 mL of a 0.080 M diprotic acid, H_2A

(b) 5.00 mL of a mixture consisting of 0.080 M H_2A and 0.080 M HA^-

(c) 5.00 mL of a mixture consisting of two acids with the same K_a, 0.080 M HA, and 0.16 M HB

(d) 5.00 mL of a mixture consisting of two acids with the same concentration (0.080 M), but different K_a

52. Changing titration conditions can have a significant influence on its feasibility as the following example illustrates. The titration of 0.010 M NH_4^+ with NaOH 0.010 M is shown at left. Select the set of conditions under which the curve at right will be observed for the titration of NH_4^+ with NaOH.

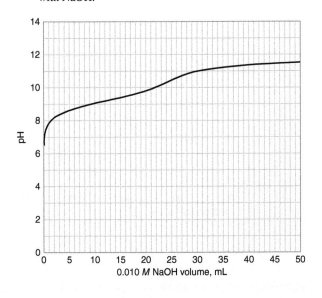

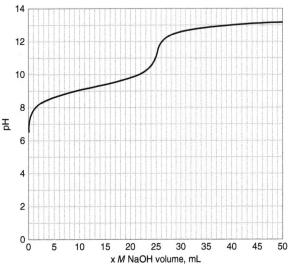

(a) Titration of 50.0 mL 0.0050 M NH_4^+ vs 0.010 M NaOH

(b) Titration of 25.0 mL 0.50 M NH_4^+ vs 0.50 M NaOH

(c) Titration of 25.0 mL 0.010 M NH_4^+ vs 1.0 M NaOH

(d) Titration of 12.5 mL 0.020 M NH_4^+ vs 0.01 M NaOH

(e) Titration of 12.5 mL 0.0050 M NH_4^+ vs 0.010 M NaOH

53. In the following example, by changing the titration conditions on a weak acid as compared to a strong acid, a difference at the pre-equivalence regions for the two is evident, while no difference at the post-equivalence regions is observed:

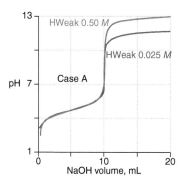
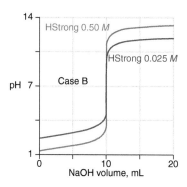

In both Case A and Case B, 0.50 M and 0.025 M acids are titrated with NaOH 0.50 M and NaOH 0.025 M, respectively. Case A corresponds to the weak acid titration (represented as HWeak) and Case B corresponds to the strong acid titration (represented as HStrong).

Explain why the pH behavior in Case A is different from Case B in the pre-equivalence region when changing conditions, while they are the same in the post-equivalence region.

54. An analyst titrated two solutions with HCl 0.0100 M. The analyst knew that one of them contained a mixture of CO_3^{2-} and OH^- and the other contained a mixture of CO_3^{2-} and HCO_3^-. However, the analyst did not know which solution corresponded to each of those two compositions. To find out, she performed a potentiometric titration on 10.00 mL of each of the two solutions and obtained the following titration curves. Determine compositions of A and B.

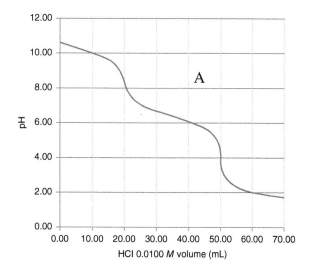
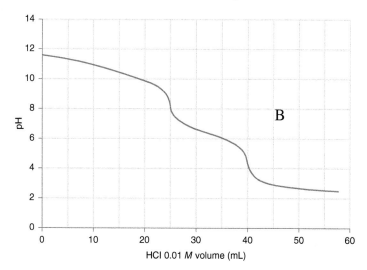

55. A newly synthesized triprotic acid showed the following general titration curve:

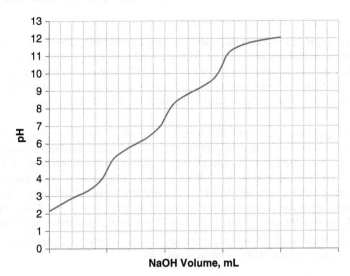

To determine the MW of the acid, a student dissolved a pure sample (400.0 mg) in a small volume of water, then transferred it to a 100.0 mL volumetric flask and made up to the mark with distilled water. The student then took a 25.00-mL aliquot of this solution and titrated it with 15.90 mL 0.0750 M NaOH to reach a phenol red end point. The MW of the acid is: (a) 111.8, (b) 167.7, (c) 218.5, (d) 286.2, (e) 335.4

56. For which of the following diprotic acids a titration curve with 0.100 M NaOH will show just one equivalence point with 2:1 stoichiometry?

   (a) Resorcinol: pK_1: 9.30; pK_2: 11.06,

   (b) Leucine: pK_1: 2.33; pK_2: 9.74,

   (c) Maleic acid: pK_1: 1.82; pK_2: 6.59,

   (d) Salicylic acid: pK_1: 2.97; pK_2: 13.74,

   (e) Tartaric acid: pK_1: 3.036; pK_2: 4.366

57. For which of the following amino acids a titration with 0.10 M NaOH will show three well-defined end points?

	pK_1	pK_2	pK_3
I. Aspartic Acid	1.99	3.90	10.00
II. Arginine	1.82	8.99	12.48
III. Histidine	1.70	6.02	9.08

   (a) I only, (b) II only, (c) III only, (d) II and III, (e) I,II, and III

The following problems were contributed by Professor Bin Wang, Marshall University

58. (a) A 0.9872 g sample of unknown potassium hydrogen phthalate (KHP) required 28.23 mL of 0.1037 M NaOH for neutralization. What is the percentage of KHP in the sample?

59. A 0.2027 g sample of finely powered limestone (mainly $CaCO_3$) was dissolved in 50.00 mL of 0.1035 M HCl. The solution was heated to expel CO_2 produced by the reaction. The remaining HCl was then titrated with 0.1018 M of NaOH and it required 16.62 mL. Calculate the percentage of $CaCO_3$ in the limestone sample.

The following problem was contributed by Professor Wen-Yee Lee, The University of Texas at El Paso

60. The Kjeldahl procedure was used to analyze 500 µL of a solution containing 50.0 mg protein/mL solution. The liberated NH_3 was collected in 5.00 mL of 0.0300 M HCl. The remaining acid required 10.00 mL 0.010 M NaOH for complete titration. What is the weight percent of nitrogen in the protein?

The following problems were contributed by Professor Christopher Harrison, San Diego State University

61. (a) What is the pH of a solution that contains 12.00 mg of H_3PO_4 in one liter of pure water? (b) How many milligrams of citric acid will you need to prepare one liter of a solution that has exactly twice the molar concentration of the H_3PO_4 solution? (c) How many milligrams of sodium hydroxide will you need to add to the citric acid solution to get to the same pH as the phosphoric acid solution? How easy will it be to do this? (See the text website for Excel calculations.)

62. A buffer is typically made from 30 millimoles of sodium hydrogensulfite ($NaHSO_3$) and 90 millimoles of disodium sulfite (Na_2SO_3) per liter. You have run out of these reagents. At your disposal you have solid sodium carbonate, solid disodium hydrogen phosphate, 1 M dimethylamine, solid sodium hydroxide, and 1 M nitric acid. Using some or all of these chemicals, prepare a buffer with the same pH and formal concentration as the sulfite-hydrogensulfite buffer. What is the simplest combination of chemicals and how many grams or milliliters of each will be needed (ignoring the water needed)? Relevant dissociation constant data are in Appendix C.

63. What volume of a stock HNO_3 solution containing 25% (w/w) nitric acid (density of 1.18 g/mL) is needed to make 500 mL HNO_3 solution of pH 2.45?

The following Professor's favorite final exam problem was contributed by Professor Michael D. Morris, University of Michigan

64. The pK_a of finalic acid is 6.50.
    (a) What is the pH of an equimolar solution of finalic acid and sodium finalate? Assume that both solutions are at 0.1 M.
    (b) Would you expect the buffer capacity to be higher or lower if you decrease the concentration of each constituent to 0.01 M? Why?
    (c) What is the effect of water dissociation on the pH of these solutions?
    (d) Only considering the pH of the solution, would it be safe to drink a 0.1 M solution of finalic acid? How about a 0.01 M solution?

PROFESSOR'S FAVORITE PUZZLE

The following puzzle was contributed by Professor Fred Hawkridge, Virginia Commonwealth University

65. You want to prepare a pH 7.0 phosphate buffer that will be exactly 0.100 M in phosphate concentration. You look up the K_a values for H_3PO_4 and decide that you will need mostly Na_2HPO_4 that may need a little bit of acid added until the pH is right. However, if you took exactly 0.100 M Na_2HPO_4, the phosphate concentration will be altered when you add the acid. How are you going to make the buffer?

PROFESSOR'S FAVORITE SPREADSHEET EXERCISES

66. Professor Alexander Scheeline, University of Illinois, has developed a spreadsheet Universal Acid Titrator (see the book's website, 7.52 Universal Acid Titrator.xlsx). You can type in the K_a value(s) for any acid (up to a decaprotic acid!) in row 11, or pK_a value(s) in row 10, and generate a titration plot (pH vs.Volume base) for titration with strong, monoprotic base like NaOH. Delight Professor Scheeline by playing with the spreadsheet and generating different titration curves. (We recommend that when the higher K's are irrelevant, directly enter 0 for these K values in row 11.) Rather than computing the pH for a given amount of base added, this approach, also based on charge balance, computes how much base you have to add to get to a given pH. Here are the following challenge questions:
    (a) Why is it so much easier to directly compute the needed base amounts without involving solver or Goal Seek rather than computing the pH for a given amount of base added?

(b) Especially at low pH values why are some of the calculated base volumes negative? (They are never plotted.)

(c) For any acid of your choice (or your creation), calculate the pH when no base has been added (V = 0) by using detailed calculation or Goal Seek or Solver. Then use the current spreadsheet to calculate how much base you will need to get to this pH. See if you indeed get zero.

(d) For the more adventurous: After computing the titration curve with this spreadsheet (that ignores activity corrections), do a successive approximations calculation at each pH that takes ionic strength and activity corrections into account, and compute a revised titration curve. Compare the activity corrected and uncorrected curves: where does activity correction matter most?

The following problems were contributed by Professor Steven Goates, Brigham Young University Excel answers are given on the text's website.

67. Develop a spreadsheet program to calculate the change of pH with volume of base added for the theoretical titration of an acid with a strong base. Calculate the titration curve for the titration of KHP with NaOH described below, and graph your results. (Even though you will plot pH on the y-axis and volume of titrant on the x-axis, as in the case of Problem 66 above you may find it much more convenient in your calculation to use pH as the *independent* variable and solve for volume of base added as the *dependent* variable.) You should calculate the volume of base required to reach increasingly higher pH values at intervals of 0.100 pH units up to a final pH of 12.500. For this problem, assume all activity coefficients to be 1, pH = − log a_{H^+}, the KHP is 100.00 % pure, the NaOH solution is carbonate-free, and volumes of solution are additive. The dissociation constant data are available in Appendix C. Specifically calculate for the following titration (but have the spreadsheet in a manner where all these values can be readily changed): A 0.994 mL aliquot of 0.4103 M KHP solution was placed in a titration vial and 5.00 mL distilled water was added. The solution was then titrated with 0.2102 M NaOH solution. What effect is there on the titration curve if 6 mL instead of 5 mL water is added?

68. Modify your calculation of Problem 67 to include activity coefficients. Go through the calculations at least two rounds. First time, as in Problem 67, calculate ionic strength μ (Equation 5.18). From this ionic strength, calculate the activity coefficients by the extended Debye-Hückel equation given in your textbook (Equation 5.19). Assume ion size parameters of 9, 3.5, 6, and 6 A for H^+, OH^-, HP^-, and P^{2-}, respectively. Convert the known activity constants K_w, K_{a1}, and K_{a2} to the corresponding values given in terms of concentrations because mass balance equations must ultimately be in terms of concentrations, and not activities. Using these constants in terms of concentration, calculate the new values of $[H^+]$, $[OH^-]$, α_1, and α_2 and ionic strength again. Then repeat the calculations and finally, plot your calculated V_b vs. pH for Problems 67 and 68 on the same graph, and discuss any differences.

Multiple Choice Questions

1. For the neutralization of strong acid (HCl) and strong base (NaOH), write the ionic reaction.

   (a) $HCl(aq) + NaOH(aq) \rightleftharpoons H_2O(l) + NaCl(aq)$

   (b) $H^+(aq) + Cl^-(aq) + Na^+(aq) + OH^-(aq) \rightleftharpoons H^+(aq) + OH^-(aq) + Na^+(aq) + Cl^-(aq)$

   (c) $H^+(aq) + Cl^-(aq) + Na^+(aq) + OH^-(aq) \rightleftharpoons H_2O(l) + Na^+(aq) + Cl^-(aq)$

   (d) None of the above are correct.

2. When the equivalence point or end point occurs between HCl and NaOH in acid/base titration,

   (a) the phenolphthalein indicator turned as joint pink color.

   (b) the acid and base neutralized each other.

   (c) the moles of H^+ = the moles of OH^-

   (d) all of these are correct.

3. A mixture was made by adding 15.0 mL of 0.500 *M* NaOH to a solution of 20 mL of 0.500 *M* HCl. Choose the correct option.

   (a) The solution will be acidic with pH < 7.

   (b) The solution will be acidic with pH > 7.

   (c) The solution will be basic with pH > 7.

   (d) None of above are correct.

4. For the above question 3, calculate the pH of the mixture.

   (a) pH = 0.3010   (b) pH = 0.9031

   (c) pH = 1.146   (d) pH = 2.602

Recommended References

Indicators

1. R. W. Sabnis, *Handbook of Acid–Base Indicators*. Boca Raton, FL: CRC Press, 2008.
2. G. Gorin, "Indicators and the Basis for Their Use," *J. Chem. Ed.*, **33** (1956) 318.
3. E. Bishop, *Indicators*. Oxford: Pergamon, 1972.

Titration Curves

4. R. K. McAlpine, "Changes in pH at the Equivalence Point," *J. Chem. Ed.*, **25** (1948) 694.
5. A. K. Covington, R. N. Goldberg, and M. Sarbar, "Computer Simulation of Titration Curves with Application to Aqueous Carbonate Solutions," *Anal. Chim. Acta*, **130** (1981) 103.

Titrations

6. Y.-S. Chen, S. V. Brayton, and C. C. Hach, "Accuracy in Kjeldahl Protein Analysis," *Am. Lab.*, June (1988) 62.
7. R. M. Archibald, "Nitrogen by the Kjeldahl Method," in D. Seligson, ed., *Standard Methods of Clinical Chemistry*, Vol. 2. New York: Academic, 1958, pp. 91–99.

Web pH Titration Calculator

8. www2.iq.usp.br/docente/gutz/Curtipot.html. A free program that calculates the pH and paH of mixtures of acids and bases, and also titration curves and distribution curves, alpha plots, and more.
9. www.phcalculation.se. A free program that calculates pH of complex mixtures of strong and weak acids and bases using concentrations (ConcpH) or activities (ActpH). Equilibrium concentrations of all species are calculated. The latter also calculates activity coefficients of the species and the ionic strength. A variation of the program allows calculation of points in a titration curve by adding different volumes of titrant to the analyte solution.

Titrations without Measuring Volumes

10. T. R. Martz, A. G. Dickson, and M. D. DeGrandpre, "Tracer Monitoring Titrations: Measurement of Total Alkalinity," *Anal. Chem.*, **78** (2006) 1817.
11. M. D. DeGrandpre, T. R. Martz, R. D. Hart, D. M. Elison, A. Zhang, and A. G. Bahnson, "Universal Tracer Monitored Titrations," *Anal. Chem.*, **83** (2011) 9217.

Chapter 8

COMPLEXOMETRIC REACTIONS AND TITRATIONS

"Simple things should be simple. Complex things should be possible."
—Alan Kay

KEY THINGS TO LEARN FROM THIS CHAPTER

- Formation constants
- EDTA equilibria (key Equations: 8.5–8.8)

- Indicators for EDTA titrations
- α_M and β values (key Equations: 8.17–8.21)

Many metal ions form slightly dissociated complexes with various ligands (complexing agents). The analytical chemist makes judicious use of complexes to mask undesired reactions. The formation of complexes can also serve as the basis of accurate and convenient titrations for metal ions in which the titrant is a complexing agent. Complexometric titrations are useful for determining a large number of metals. Selectivity can be achieved by appropriate use of *masking agents* (addition of other complexing agents that react with interfering metal ions, but not with the metal of interest) and by pH control, since most complexing agents are weak acids or weak bases whose equilibria are influenced by the pH. In this chapter, we discuss metal ions, their equilibria, and the influence of pH on these equilibria. We describe titrations of metal ions with the very useful complexing agent EDTA, the factors that affect them, and indicators for the titrations. The EDTA titration of calcium plus magnesium is commonly used to determine water hardness. In the food industry, calcium is determined in cornflakes. In the plating industry, nickel is determined in plating solutions by complexometric (also called chelometric) titration, and in the metals industry in etching solutions. In the pharmaceutical industry, aluminum hydroxide in liquid antacids is determined by similar titrations. Nearly all metals can be accurately determined by complexometric titrations. Complexing reactions are useful for gravimetry, spectrophotometry, and fluorometry, and for masking interfering ions.

8.1 Complexes and Formation Constants—How Stable Are Complexes?

Complexes play important roles in many chemical and biochemical processes. For example, the heme molecule in blood holds iron tightly because the nitrogen atoms of the heme form strong ligating or complexing bonds. In general, the nitrogen atom derived from an amino group is a good donor atom or complexer. The iron [as iron(II)] in turn bonds readily with oxygen to transport oxygen gas from the lungs to elsewhere in the body and then easily releases it because oxygen is a poor donor atom or complexer. Carbon monoxide kills because it is a strong complexer and displaces oxygen; it binds to heme 200 times more strongly than does oxygen, forming carboxyhemoglobin.

Many cations will form complexes in solution with a variety of substances that have a pair of unshared electrons (e.g., on N, O, S atoms in the molecule) capable of

Most ligands contain O, S, or N as the complexing atoms.

satisfying the coordination number of the metal cation. [The metal cation is a Lewis acid (electron pair acceptor), and the complexer is a Lewis base (electron pair donor).] The number of molecules of the complexing agent, called the **ligand**, will depend on the coordination number of the metal cation and on the number of complexing groups on the ligand molecule.

Ammonia is a simple complexing agent with one pair of unshared electrons that will complex copper ion:

$$Cu^{2+} + 4:NH_3 \rightleftharpoons \begin{bmatrix} & NH_3 & \\ H_3N : & \overset{..}{\underset{..}{Cu}} & : NH_3 \\ & NH_3 & \end{bmatrix}^{2+}$$

Here, the copper ion acts as a Lewis acid, and the ammonia is a Lewis base. The Cu^{2+} (hydrated) ion is pale blue in solution, while the ammonia (the ammine) complex is deep blue. A similar reaction occurs with the green hydrated nickel ion to form a deep blue ammine complex.

Ammonia will also complex with silver ion to form a colorless complex. Two ammonia molecules complex with each silver ion in a stepwise fashion, and we can write an equilibrium constant for each step, called the **formation constant** K_f:

$$Ag^+ + NH_3 \rightleftharpoons Ag(NH_3)^+ \quad K_{f1} = \frac{[Ag(NH_3)^+]}{[Ag^+][NH_3]}$$
$$= 2.5 \times 10^3 \tag{8.1}$$

$$Ag(NH_3)^+ + NH_3 \rightleftharpoons Ag(NH_3)_2^+ \quad K_{f2} = \frac{[Ag(NH_3)_2^+]}{[Ag(NH_3)^+][NH_3]}$$
$$= 1.0 \times 10^4 \tag{8.2}$$

The overall reaction is the sum of the two steps, and the overall formation constant is the product of the stepwise formation constants:

$$Ag^+ + 2NH_3 \rightleftharpoons Ag(NH_3)_2^+ \quad K_f = K_{f1} \cdot K_{f2} = \frac{[Ag(NH_3)_2^+]}{[Ag^+][NH_3]^2}$$
$$= 2.5 \times 10^7 \tag{8.3}$$

For the formation of a simple 1:1 complex, for example, $M + L = ML$, the formation constant is simply $K_f = [ML]/[M][L]$. The formation constant is also called the **stability constant** K_s, or K_{stab}.

We could write the equilibria in the opposite direction, as dissociations. If we do this, the concentration terms are inverted in the equilibrium constant expressions. The equilibrium constants then are simply the reciprocals of the formation constants, and they are called **instability constants** K_i, or **dissociation constants** K_d:

$K_f = K_s = 1/K_i$ or $1/K_d$.

$$Ag(NH_3)_2^+ \rightleftharpoons Ag^+ + 2NH_3 \quad K_d = \frac{1}{K_f} = \frac{[Ag^+][NH_3]^2}{[Ag(NH_3)_2^+]} = 4.0 \times 10^{-8} \tag{8.4}$$

You can use either constant in calculations, as long as you use it with the proper reaction and the correct expression. You will note that the dissociation of an acid $HA \rightleftharpoons H^+ + A^-$ is in fact very similar to the dissociation of a metal ligand complex $ML \rightleftharpoons M + L$ However, the convention is to write the first as a dissociation reaction and the second as an association reaction.

Example 8.1

A divalent metal M^{2+} reacts with a ligand L to form a 1:1 complex:

$$M^{2+} + L \rightleftharpoons ML^{2+} \quad K_f = \frac{[ML^{2+}]}{[M^{2+}][L]}$$

Calculate the concentration of M^{2+} in a solution prepared by mixing equal volumes of 0.20 M M^{2+} and 0.20 M L. $K_f = 1.0 \times 10^8$.

Solution

We have added stoichiometrically equal amounts of M^{2+} and L. The complex is sufficiently strong that their reaction is virtually complete. Since we added equal volumes, the initial concentrations were halved by dilution. Let x represent $[M^{2+}]$. At equilibrium, we have

$$\begin{array}{ccc} M^{2+} + L \rightleftharpoons & ML^{2+} \\ x \quad\quad x & 0.10 - x \approx 0.10 \end{array}$$

Essentially, all the M^{2+} (original concentration 0.20 M) was converted to an equal amount of ML^{2+}, with only a small amount of uncomplexed metal remaining. Substituting into the K_f expression,

$$\frac{0.10}{(x)(x)} = 1.0 \times 10^8$$

$$x = [M^{2+}] = 3.2 \times 10^{-5} M$$

You can also solve the equation $(0.10 - x)/x^2 = K_f$ by Goal Seek in Excel. This will apply even when K_f is not so high and you cannot really assume that x is so small that $0.10 - x$ can be approximated as 0.10.

Example 8.2

Silver ion forms a stable 1:1 complex with triethylenetetraamine, called "trien" $[NH_2(CH_2)_2NH(CH_2)_2NH(CH_2)_2NH_2]$. Calculate the silver ion concentration at equilibrium when 25 mL of 0.010 M silver nitrate is added to 50 mL of 0.015 M trien. $K_f = 5.0 \times 10^7$.

Solution

$$Ag^+ + trien \rightleftharpoons Ag(trien)^+ \quad K_f = \frac{[Ag(trien)^+]}{[Ag^+][trien]}$$

Calculate the millimoles Ag^+ and trien added:

$$mmol\ Ag^+ = 25\ mL \times 0.010\ mmol/mL = 0.25\ mmol$$

$$mmol\ trien = 50\ mL \times 0.015\ mmol/mL = 0.75\ mmol$$

The equilibrium lies far to the right, so you can assume that virtually all the Ag^+ reacts with 0.25 mmol of the trien (leaving 0.50 mmol trien in excess) to form 0.25 mmol complex. Calculate the molar concentrations:

$$[Ag^+] = x = \text{mol/L unreacted}$$

$$[trien] = (0.50 \text{ mmol/75 mL}) + x = 6.7 \times 10^{-3} + x$$

$$\approx 6.7 \times 10^{-3}$$

$$[Ag(trien)^+] = 0.25 \text{ mmol/75 mL} - x$$

$$= 3.3 \times 10^{-3} - x \approx 3.3 \times 10^{-3}$$

Try neglecting x compared to the other concentrations:

$$\frac{3.3 \times 10^{-3}}{(x)(6.7 \times 10^{-3})} = 5.0 \times 10^7$$

$$x = [Ag^+] = 9.8 \times 10^{-9} \ M$$

We were justified in neglecting x.

Try with Excel Goal Seek:

   Let us assume that in the same example instead of trien there is some other ligand L, which reacts the same way but has an association constant of only 5. What will be the concentrations of AgL^+, Ag^+, and L?

8.2 Chelates: EDTA — The Ultimate Titrating Agent for Metals

The term *chelate* is derived from the Greek term meaning "clawlike." Chelating agents literally wrap themselves around a metal ion.

Simple complexing agents such as ammonia are rarely used as titrating agents because a sharp end point corresponding to a stoichiometric complex is generally difficult to achieve. This is because the stepwise formation constants are frequently close together and are not very large, and a single stoichiometric complex cannot be observed. Certain complexing agents that have two or more complexing groups on the molecule, however, do form well-defined complexes and can be used as titrating agents. Schwarzenbach demonstrated that a remarkable increase in stability is achieved if a bidentate ligand (one with two complexing groups) is used (see Reference 4 for his many contributions). For example, he showed replacing ammonia with the bidentate ethylenediamine, $NH_2CH_2CH_2NH_2$ (en), results in a highly stable $Cu(en)_2^{2+}$ complex.

   The most generally useful titrating agents are aminocarboxylic acids, in which the amino nitrogen and carboxylate groups serve as ligands. The amino nitrogens are more basic and are protonated ($-NH_3^+$) more strongly than the carboxylate groups. When these groups bind to metal atoms, they lose their protons. The metal complexes formed with these multidentate complexing agents are often 1:1, regardless of the charge on the metal ion, because there are sufficient complexing groups on one molecule to satisfy all the coordination sites of the metal ion.

The protons in EDTA are displaced upon complexing with a metal ion. A negatively charged chelate results.

   An organic agent that has two or more groups capable of complexing with a metal ion is called a **chelating agent**. The complex formed is called a **chelate**. The chelating agent is called the *ligand*. Titration with a chelating agent is called a **chelometric titration**, perhaps the most important and practical type of complexometric titrations.

The most widely used chelating agent in titrations is **ethylenediaminetetraacetic acid (EDTA)**. The formula for EDTA is

$$
\begin{array}{ccc}
\text{HO}_2\text{CCH}_2 & & \text{CH}_2\text{CO}_2\text{H} \\
\diagdown \quad \text{H} \quad\quad \text{H} \quad\diagup & \\
\underset{+}{\text{N}}\text{CH}_2\text{CH}_2\underset{+}{\text{N}} & \\
\diagup \quad\quad\quad\quad \diagdown & \\
{}^{-}\text{O}_2\text{CCH}_2 & & \text{CH}_2\text{CO}_2{}^{-}
\end{array}
$$

Each of the two nitrogens and each of the four carboxyl groups contains a pair of unshared electrons capable of complexing with a metal ion. Thus, EDTA contains six complexing groups. We will represent EDTA by the symbol H_4Y. It is a tetraprotic acid, and the hydrogens in H_4Y refer to the four ionizable hydrogens belonging to the four carboxylic acid groups. At sufficiently low pH, the nitrogens can also be protonated and this diprotonated EDTA can be considered a hexaprotic acid. However, this occurs at a very low pH and EDTA is almost never used under such conditions. It is the unprotonated ligand Y^{4-} that forms complexes with metal ions, that is, the protons are displaced by the metal ion upon complexation.

Note that in the structure we have depicted the uncharged EDTA molecule as the double zwitterion (see Section 7.12); this is the form in which it really exists. This is why EDTA is most commonly available as the disodium salt, where the two ionized carboxylic acid groups form the salt.

The Chelon Effect—The More Complexing Groups, the Better

Multidentate chelating agents form stronger complexes with metal ions than do similar bidentate or monodentate ligands. This is the result of thermodynamic effects in complex formation. Chemical reactions are driven by decreasing enthalpy (liberation of heat, negative ΔH) and by increasing entropy (increased disorder, positive ΔS). Recall from Chapter 5, Equation 5.7, that a chemical process is spontaneous when the free energy change, ΔG, is negative, and $\Delta G = \Delta H - T\,\Delta S$. The enthalpy change for ligands with similar groups is often similar. For example, four ammonia molecules complexed to Cu^{2+} and four amino groups from two ethylenediamine molecules complexed to Cu^{2+} will result in about the same release of heat. However, more disorder or entropy is created by the dissociation of the $Cu(NH_3)_4^{2+}$ complex in which five species are formed than in the dissociation of the $Cu(H_2NCH_2CH_2NH_2)_2^{2+}$ complex, in which three species are formed. Hence, ΔS is greater for the former *dissociation*, creating a more negative ΔG and a greater tendency for dissociation. Thus, multidentate complexes are more stable (have larger K_f values), largely because of the entropy effect. This is known as the **chelon effect** or **chelate effect**. It is pronounced for chelating agents such as EDTA, which have sufficient ligand atoms to occupy up to six coordination sites on metal ions.

*For more discussion of the design of chelating agents, see C. N. Reilley, R. W. Schmid, and F. S. Sadek, "Chelon Approach to Analysis (I). Survey of Theory and Application," J. Chem. Ed., **36** (1959) 555. Illustrated experiments are given in a second article in J. Chem. Ed., **36** (1959) 619.*

The chelon effect is an entropy effect.

EDTA Equilibria

We can represent EDTA as having four K_a values corresponding to the stepwise dissociation of the four protons[1]:

$$
H_4Y \rightleftharpoons H^+ + H_3Y^- \quad K_{a1} = 1.0 \times 10^{-2} = \frac{[H^+][H_3Y^-]}{[H_4Y]} \tag{8.5}
$$

$$
H_3Y^- \rightleftharpoons H^+ + H_2Y^{2-} \quad K_{a2} = 2.2 \times 10^{-3} = \frac{[H^+][H_2Y^{2-}]}{[H_3Y^-]} \tag{8.6}
$$

---

[1] As stated before, the nitrogens on the EDTA molecule can protonate, and so there are in reality six dissociation steps and six K_a values, the first two being ~1.5 and 0.032. The two nitrogens are more basic than the carboxyl oxygens, and so protonate more readily. The nitrogen protonation does affect the solubility of EDTA in acid.

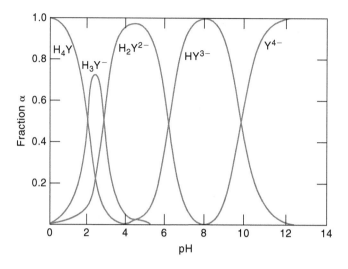

FIGURE 8.1 Fraction of EDTA species as function of pH. In this treatment we ignore further protonation of the nitrogens beyond H_4Y.

$$H_2Y^{2-} \rightleftharpoons H^+ + HY^{3-} \qquad K_{a3} = 6.9 \times 10^{-7} = \frac{[H^+][HY^{3-}]}{[H_2Y^{2-}]} \qquad (8.7)$$

$$HY^{3-} \rightleftharpoons H^+ + Y^{4-} \qquad K_{a4} = 5.5 \times 10^{-11} = \frac{[H^+][Y^{4-}]}{[HY^{3-}]} \qquad (8.8)$$

Polyprotic acid equilibria are treated in Section 6.9, and you should review this section before proceeding with the following discussion.

Figure 8.1 illustrates the fraction of each form of EDTA as a function of pH. Since the anion Y^{4-} is the ligand species in complex formation, the complexation equilibria are affected markedly by the pH. H_4Y has a very low solubility in water, and so the disodium salt $Na_2H_2Y \cdot 2H_2O$ is generally used, in which two of the acid groups are neutralized. This salt dissociates in solution to give predominantly H_2Y^{2-}; the pH of the solution is approximately 4 to 5 (theoretically 4.4 from $[H^+] = \sqrt{K_{a2}K_{a3}}$).

The following ladder diagram shows the distribution of EDTA species at varied pH. This ladder diagram was contributed by Professor Galina Talanova, Howard University. As noted above, the nitrogens on EDTA can protonate in very acidic solution, and when considering this, we can write six K_a values, in which case the $pK_3 - pK_6$ values in the ladder diagram correspond to $K_{a1} - K_{a4}$ in Equations 8.5 through 8.8 (note the protonated nitrogens occur around pH 0 and 1.5):

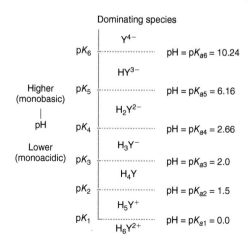

Formation Constant

Consider the formation of the EDTA chelate of Ca^{2+}. This can be represented by:

$$Ca^{2+} + Y^{4-} \rightleftharpoons CaY^{2-} \tag{8.9}$$

The formation constant for this is

$$K_f = \frac{[CaY^{2-}]}{[Ca^{2+}][Y^{4-}]} \tag{8.10}$$

The values of some representative EDTA formation constants are given in Appendix C.

Effect of pH on EDTA Equilibria—How Much is Present as Y^{4-}?

The equilibrium in Equation 8.9 is shifted to the left as the hydrogen ion concentration is increased, due to competition for the chelating anion by the hydrogen ion.

The dissociation may be represented by:

$$CaY^{2-} \rightleftharpoons Ca^{2+} + \underbrace{Y^{4-} \overset{H^+}{\rightleftharpoons} HY^{3-} \overset{H^+}{\rightleftharpoons} H_2Y^{2-} \overset{H^+}{\rightleftharpoons} H_3Y^- \overset{H^+}{\rightleftharpoons} H_4Y}_{C_{H_4Y}}$$

Note that $C_{H_4Y} = [Ca^{2+}]$. Or, from the overall equilibrium:

$$Ca^{2+} + H_4Y \rightleftharpoons CaY^{2-} + 4H^+$$

According to Le Châtelier's principle, increasing the acidity will favor the competing equilibrium, that is, the protonation of Y^{4-} (all forms of EDTA are present in equilibrium, but some are very small; see Figure 8.1). Decreasing the acidity will favor formation of the CaY^{2-} chelate.

From a knowledge of the pH and the equilibria involved, Equation 8.10 can be used to calculate the concentration of free Ca^{2+} under various solution conditions (e.g., to interpret a titration curve). The Y^{4-} concentration is calculated at different pH values as follows (see Chapter 6, polyprotic acids). If we let C_{H_4Y} represent the total concentration of all forms of uncomplexed EDTA, then

$$C_{H_4Y} = [Y^{4-}] + [HY^{3-}] + [H_2Y^{2-}] + [H_3Y^-] + [H_4Y] \tag{8.11}$$

and we can also write that

$$[Y^{4-}] = \alpha_4 C_{H_4Y} \tag{8.12a}$$

where α_4 is the fraction of C_{H_4Y} that exists as Y^{4-} and α_4 is given by (revisit Chapter 6; the discussion of how to arrive at this is in Equations 6.78 through 6.84):

$$\begin{aligned}\alpha_4 &= K_{a1}K_{a2}K_{a3}K_{a4}/Q_4 = K_{a1}K_{a2}K_{a3}K_{a4}/([H^+]^4 + K_{a1}[H^+]^3 \\ &\quad + K_{a1}K_{a2}[H^+]^2 + K_{a1}K_{a2}K_{a3}[H^+] + K_{a1}K_{a2}K_{a3}K_{a4})\end{aligned} \tag{8.12b}$$

Protons compete with the metal ion for the EDTA ion. To apply Equation 8.10, we must replace $[Y^{4-}]$ with $\alpha_4 C_{H_4Y}$ as the equilibrium concentration of Y^{4-}.

Similar equations can be derived for the fraction of each of the other EDTA species α_0, α_1, α_2, and α_3, as in Chapter 6. (This is the way Figure 8.1 was constructed.)

We can use Equation 8.12b, then, to calculate the fraction of the EDTA that exists as Y^{4-} at a given pH; and from a knowledge of the concentration of uncomplexed EDTA (C_{H_4Y}), we can calculate the free Ca^{2+} using Equation 8.10.

Example 8.3

Calculate the fraction of EDTA that exists as Y^{4-} at pH 10, and from this calculate pCa in 100 mL of solution of 0.100 M Ca^{2+} at pH 10 after adding 100 mL of 0.100 M EDTA.

Solution

Equation 6.82 states

$$Q_4 = [H^+]^4 + K_{a1}[H^+]^3 + K_{a1}K_{a2}[H^+]^2 + K_{a1}K_{a2}K_{a3}[H^+] + K_{a1}K_{a2}K_{a3}K_{a4}$$

Putting in $[H^+] = 10^{-10}$ and the K_a values as listed in Equations 8.5 through 8.8, we get

$$Q_4 = 2.3_5 \times 10^{-21}$$

The numerator in Equation 8.12b is

$$K_{a1}K_{a2}K_{a3}K_{a4} = 8.3_5 \times 10^{-22}$$

Thus, from Equation 8.12b,

$$\alpha_4 = 8.3_5 \times 10^{-22}/2.3_5 \times 10^{-21} = 0.36$$

Stoichiometric amounts of Ca^{2+} and EDTA are added to produce an equivalent amount of CaY^{2-}, less the amount dissociated:

$$\text{mmol } Ca^{2+} = 0.100 \, M \times 100 \text{ mL} = 10.0 \text{ mmol}$$

$$\text{mmol EDTA} = 0.100 \, M \times 100 \text{ mL} = 10.0 \text{ mmol}$$

We have formed 10.0 mmol CaY^{2-} in 200 mL, or 0.0500 M:

$$Ca^{2+} + EDTA \rightleftharpoons CaY^{2-}$$

$$\begin{array}{ccc} x & x & 0.0500 \, M - x \\ & & \approx 0.0500 \, M \text{ (since } K_f \text{ is large)} \end{array}$$

where x represents the total equilibrium EDTA concentration in all forms, C_{H_4Y}. $[Y^{4-}]$, needed to apply Equation 8.10, is equal to $\alpha_4 C_{H_4Y}$. Hence, we can write Equation 8.10 as:

$$K_f = \frac{[CaY^{2-}]}{[Ca^{2+}]\alpha_4[C_{H_4Y}]}$$

From Appendix C, $K_f = 5.0 \times 10^{10}$. Hence,

$$5.0 \times 10^{10} = \frac{0.0500}{(x)(0.36)(x)}$$

$$x = 1.7 \times 10^{-6} \, M$$

$$pCa = 5.77$$

Conditional Formation Constant—Use for a Fixed pH

The term *conditional formation constant* is used under specified conditions (e.g., at a certain pH) and can be convenient for calculations. It takes into account that only a fraction of the total EDTA species are present as Y^{4-}. We can substitute $\alpha_4 C_{H_4Y}$ for $[Y^{4-}]$ in Equation 8.10:

$$K_f = \frac{[CaY^{2-}]}{[Ca^{2+}]\alpha_4 C_{H_4Y}} \tag{8.13}$$

We can then rearrange the equation to yield

$$\boxed{K_f \alpha = K_f' = \frac{[CaY^{2-}]}{[Ca^{2+}]C_{H_4Y}}} \tag{8.14}$$

where K_f' is the **conditional formation constant** and is dependent on α_4 and, hence, the pH. We can use this equation to calculate the equilibrium concentrations of the different species at a given pH in place of Equation 8.10.

The conditional formation constant applies for a specified pH.

Example 8.4

The formation constant for CaY^{2-} is 5.0×10^{10}. At pH 10, α_4 is calculated (Example 8.3) to be 0.36 to give a conditional constant (from Equation 8.14) of 1.8×10^{10}. Calculate pCa in 100 mL of a solution of 0.100 M Ca^{2+} at pH 10 after addition of (a) 0 mL, (b) 50 mL, (c) 100 mL, and (d) 150 mL of 0.100 M EDTA.

Solution

(a) $pCa = -\log[Ca^{2+}] = -\log \times 1.00 \times 10^{-1} = 1.00$

(b) We started with 0.100 M × 100 mL = 10.0 mmol Ca^{2+}. The millimoles of EDTA added are 0.100 M × 50 mL = 5.0 mmol. In general, a rule of thumb for a binary association reaction (A + B ⇌ C) is that the limiting reagent can be assumed to have reacted quantitatively (> 99.9%) if the effective equilibrium constant is greater than $\sim10^6$. The conditional formation constant in this case is larger, so the reaction will be far to the right. Hence, we can neglect the amount of Ca^{2+} from the dissociation of CaY^{2-} and the number of millimoles of free Ca^{2+} is essentially equal to the number of unreacted millimoles:

$$mmol\ Ca^{2+} = 10.0 - 5.0 = 5.0\ mmol$$

$$[Ca^{2+}] = 5.0\ mmol/150\ mL = 0.033\ M$$

$$pCa = -\log\ 3.3 \times 10^{-2} = 1.48$$

(c) At the equivalence point, we have converted all the Ca^{2+} to CaY^{2-}. We must, therefore, use Equation 8.14 to calculate the equilibrium concentration of Ca^{2+}. The number of millimoles of CaY^{2-} formed is equal to the number of millimoles of Ca^{2+} started with, and $[CaY^{2-}] = 10.0\ mmol/200\ mL = 0.0500\ M$. From the dissociation of CaY^{2-}, $[Ca^{2+}] = C_{H_4Y} = x$, and the equilibrium $[CaY^{2-}] = 0.050\ M - x$. But since the dissociation is slight, we can neglect x compared to 0.050 M. Therefore, from Equation 8.14,

$$\frac{0.050}{(x)(x)} = 1.8 \times 10^{10}$$

$$x = 1.7 \times 10^{-6} M = [Ca^{2+}]$$

$$pCa = -\log\ 1.7 \times 10^{-6} = 5.77$$

Compare this value with that calculated using K_f in Example 8.3, instead of K_f'.

(d) The concentration C_{H_4Y} is equal to the concentration of excess EDTA added (neglecting the dissociation of CaY^{2-}, which will be even smaller in the presence of excess EDTA). The millimoles of CaY^{2-} will be as in (c). Hence,

$$[CaY^{2-}] = 10.0 \text{ mmol}/250 \text{ mL} = 0.0400 \ M$$

$$\text{mmol excess } C_{H_4Y} = 0.100 \ M \times 150 \text{ mL} - 0.100 \ M \times 100 \text{ mL} = 5.0 \text{ mmol}$$

$$C_{H_4Y} = 5.0 \text{ mmol}/250 \text{ mL} = 0.020 \ M$$

$$\frac{0.040}{[Ca^{2+}](0.020)} = 1.8 \times 10^{10}$$

$$[Ca^{2+}] = 1.1 \times 10^{-10} \ M$$

$$\text{pCa} = -\log 1.1 \times 10^{-10} = 9.95$$

The pH can affect stability of the complex (i.e., K'_f) by affecting not only the form of the EDTA but also that of the metal ion. For example, hydroxo species may form ($M^{2+} + OH^- \rightarrow MOH^+$). That is, OH^- competes for the metal ion just as H^+ competes for the Y^{4-}. Figure 8.2 (prepared from a spreadsheet—see Problem 32) shows how K'_f changes with pH for three metal–EDTA chelates with moderate (Ca) to strong (Hg) formation constants. The calcium chelate is obviously too weak to be titrated in acid solution ($K'_f < 1$), while the mercury chelate is strong enough to be titrated in acid. At pH 13, all K'_f values are virtually equal to the K_f values because α_4 is essentially unity; that is, the EDTA is completely dissociated to Y^{4-}. The curves are essentially parallel to one another because at each pH, each K_f is multiplied by the same α_4 value to obtain K'_f.

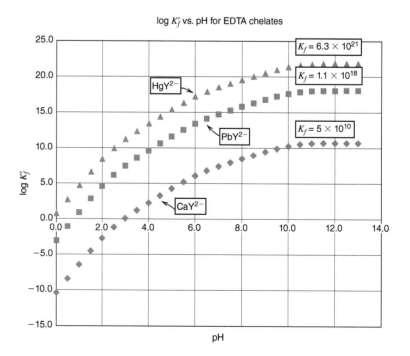

FIGURE 8.2 Effect of pH on K'_f values for EDTA chelates.

8.3 Metal–EDTA Titration Curves

A titration is performed by adding the chelating agent to the sample; the reaction occurs as in Equation 8.9. Figure 8.3 shows the titration curve for Ca^{2+} titrated with EDTA at pH 10. Before the equivalence point, the Ca^{2+} concentration is nearly equal to the amount of unchelated (unreacted) calcium since the dissociation of the chelate is slight (analogous to the amount of an unprecipitated ion). At the equivalence point and beyond, pCa is determined from the dissociation of the chelate at the given pH as described in Example 8.3 or 8.4, using K_f or K'_f. The effect of pH on the titration is apparent from the curve in Figure 8.3 for titration at pH 7.

You will recall that in pH titrations, it is simpler to calculate how much titrant is needed to get to a specified pH than to calculate the reverse, i.e., what is the pH if a certain amount of titrant is added. Revisit Section 7.11 and Problem 7.66 how to handle these. The same is true for metal ion (M)—EDTA (L) titrations. As is commonly the case, we can assume that the titration is being conducted in a buffered medium of constant pH in which the conditional formation constant is K'_f. We have taken V_M mL of a metal of analytical concentration C_M to titrate. At any given point we have added V_L mL of titrant L of concentration C_L and we wish to calculate for a given amount of free metal ion concentration (that we express as pM, as in the case of hydrogen ion concentrations being expressed as pH). In terms of pM $= -\log[M]$, what is V_L?

Based on mass balance approaches, it is readily shown (see Section 8.3's website supplement) that

$$V_L = \frac{V_M(C_M - [M])(1 + K'_f[M])}{[M](1 + K'_f(C_L + [M]))} \tag{8.15}$$

We calculate $[M]$ as 10^{-pM} and then calculate V_L. To generate the data for Figure 8.3, we took K'_f to be 5.01×10^{-10} (Appendix C, Table C4) and calculate α_4 to be 0.35_5 and 4.80×10^{-4} at pH 10 and 7, respectively, resulting in K'_f values of 1.8×10^{10} and 2.4×10^7. The calculation is illustrated in the spreadsheet Figure 8.3a.xlsx in the text website for pH 7 and 10.

It is interesting to note, however, that unlike acid–base titrations where ionization of water can complicate matters, in the present case there are mass balance equations for both M-containing species and L-containing species. This results in a quadratic

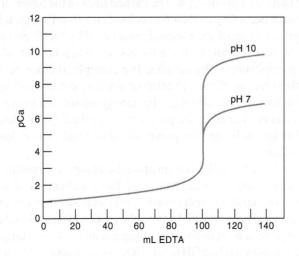

FIGURE 8.3 Titration curves for 100 mL 0.1 M Ca^{2+} versus 0.1 M Na$_2$EDTA at pH 7 and pH 10.

equation as shown below that can be explicitly solved. Consider Equation 8.14. For a generic metal ion M and a generic ligand L we can write

$$K_f' = \frac{[ML]}{[M][L_T]}$$

Where L_T indicates all L-containing species other than ML. Let us denote the total volume at any point, $V_M + V_L$ (both in units of mL), as V_T; $C_M V_M$ is the original mmol metal taken and $C_L V_L$ is the total mmol of ligand taken at any point. Mass balance for M requires that the original amount of metal taken, less the amount present as ML, divided by V_T must equal $[M]$:

$$[M] = (C_M V_M - [ML]V_T)/V_T$$

A similar mass balance for L yields:

$$[L_T] = (C_L V_L - [ML]V_T)/V_T$$

A combination of the three equations above gives

$$K_f'(C_M V_M - [ML]V_T)(C_L V_L - [ML]V_T) = [ML]V_T^2$$

This is a quadratic equation in $[ML]$, where, in the usual notations,

$$a = K_f' V_T^2, b = -V_T(K_f'(C_M V_M + C_L V_L) + V_T) \text{ and } c = K_f' C_M V_M C_L V_L$$

Generation of Figure 8.3 using the explicit equation above is illustrated in Figure 8.3b.xlsx in the website for the book. Note that of the two solutions to a quadratic equation, the negative root is meaningful in this problem; else one ends up with negative concentrations.

The more stable the chelate (the larger K_f), the farther to the right will be the equilibrium of the reaction (Equation 8.9), and the larger will be the end-point break. Also, the more stable the chelate, the lower the pH at which the titration can be performed (Figure 8.2). This is important because it allows the titration of some metals in the presence of others whose EDTA chelates are too weak to titrate at the lower pH.

Figure 8.4 shows the minimum pH at which different metals can be titrated with EDTA. The points on the curve represent the pH at which the *conditional formation constant* K_f' for each metal is 10^6 (log $K_f' = 6$), which was arbitrarily chosen as the minimum needed for a sharp end point. Note that the smaller the K_f, the more alkaline the solution must be to obtain a K_f' of 10^6 (i.e., the larger α_4 must be). Thus, Ca^{2+} with K_f only about 10^{10} requires a pH of ~≥ 8. The dashed lines in the figure divide the metals into separate groups according to their formation constants. One group is titrated in a highly acidic (pH < ~3) solution, a second group at pH ~3 to 7, and a third group at pH > 7. At the highest pH range, all the metals will react, but not all can be titrated directly due to precipitation of hydroxides. For example, titration of Fe^{3+} or Th^{4+} is not possible without the use of back-titration or auxiliary complexing agents to prevent hydrolysis. At the intermediate pH range, the third group will not titrate, and the second group of metals can be titrated in the presence of the third group. And finally, in the most acidic pH range, only the first group will titrate and can be determined in the presence of the others.

Masking can be achieved by precipitation, by complex formation, by oxidation–reduction, and kinetically. A combination of these techniques is often used. For example, Cu^{2+} can be masked by reduction to $Cu(I)$ with ascorbic acid and by complexation with I^-. Lead can be precipitated with sulfate when bismuth is to be titrated. Most masking is accomplished by selectively forming a stable, soluble complex. Hydroxide ion complexes aluminum ion [$Al(OH)_4^-$ or AlO_2^-] so calcium can be titrated. Fluoride masks $Sn(IV)$ in the titration of $Sn(II)$. Ammonia complexes copper so it is difficult to titrate $Cu(II)$ with EDTA in an ammoniacal buffer. Other metals can be titrated in the presence of $Cr(III)$ because its EDTA chelate, although very stable, forms only slowly.

Only some metal chelates are stable enough to allow titrations in acid solution; others require alkaline solution.

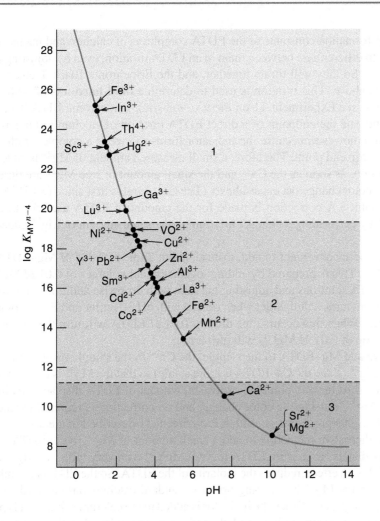

FIGURE 8.4 Minimum pH for effective titration of various metal ions with EDTA. (Reprinted with permission from C. N. Reilley and R. W. Schmid, *Anal. Chem.*, **30** (1958) 947. Copyright by the American Chemical Society.)

8.4   Detection of the End Point: Indicators—They Are Also Chelating Agents

We can measure the pM potentiometrically if a suitable electrode is available, for example, an ion-selective electrode (see Chapter 20), but it is simpler if an indicator can be used. Indicators used for complexometric titrations are themselves chelating agents. They are usually dyes of the o, o'-dihydroxy azo type.

Eriochrome Black T is a typical indicator. It contains three ionizable protons, so we will represent it by H_3In. This indicator can be used for the titration of Mg^{2+} with EDTA. A small amount of indicator is added to the sample solution, and it forms a red complex with part of the Mg^{2+}; the color of the uncomplexed indicator is blue. As soon as all the free Mg^{2+} is titrated, the EDTA displaces the indicator from the magnesium, causing a change in the color from red to blue:

$$\underset{\text{(red)}}{MgIn^-} + \underset{\text{(colorless)}}{H_2Y^{2-}} \rightarrow \underset{\text{(colorless)}}{MgY^{2-}} + \underset{\text{(blue)}}{HIn^{2-}} + H^+ \qquad (8.16)$$

This will occur over a range of pMg values, and the change will be sharper if the indicator is kept as dilute as possible but is still sufficient to give a good color change.

Of course, the metal–indicator complex must be less stable than the metal–EDTA complex, or else the EDTA will not displace it from the metal. On the other hand, it must not be too weak, or the EDTA will start replacing it at the beginning of the titration, and a diffuse end point will result. In general, *the K_f for the metal–indicator complex should be 10 to 100 times less than that for the metal–titrant complex.*

Eriochrome Black T

Water hardness is expressed as ppm CaCO₃ and represents the sum of calcium and magnesium.

The formation constants of the EDTA complexes of calcium and magnesium are too close to differentiate between them in an EDTA titration, even by adjusting pH (see Figure 8.4). So they will titrate together, and the Eriochrome Black T end point can be used as above. This titration is used to determine **total hardness of water**, (Ca^{2+} plus Mg^{2+}—see Experiment 11 on the text website). Eriochrome Black T cannot be used to indicate the endpoint of a direct EDTA titration of calcium in the absence of magnesium, however, because the indicator forms too weak a complex with calcium to give a sharp end point. Therefore, a small measured amount of Mg^{2+} is added to the Ca^{2+} solution; as soon as the Ca^{2+} and the small amount of free Mg^{2+} are titrated, the end-point color change occurs as above. (The Ca^{2+} titrates first since its EDTA chelate is more stable.) A correction is made for the amount of EDTA used for titration of the Mg^{2+} by performing a "blank" titration on the same amount of Mg^{2+} added to the buffer.

It is more convenient to add, instead, about 2 mL of 0.005 M Mg–EDTA rather than $MgCl_2$. This is prepared by adding together equal volumes of 0.01 M $MgCl_2$ and 0.01 M EDTA solutions and adjusting the ratio with dropwise additions until a portion of the reagent turns a dull violet when treated with pH 10 buffer and Eriochrome Black T indicator. When this occurs, one drop of 0.01 M EDTA will turn the solution blue, and one drop of 0.01 M $MgCl_2$ will turn it red.

If we add Mg–EDTA to the sample, the Ca^{2+} in the sample displaces the EDTA from the Mg^{2+} (since the Ca–EDTA is more stable) so that the Mg^{2+} is free to react with the indicator. At the end point, an equivalent amount of EDTA displaces the indicator from the Mg^{2+}, causing the color change, and no correction is required for the added Mg–EDTA. This procedure is used in Experiment 11 described in the text website.

An alternative method is to add a small amount of Mg^{2+} to the EDTA solution. This immediately reacts with EDTA to form MgY^{2-} with very little free Mg^{2+} in equilibrium. This, in effect, reduces the molarity of the EDTA. So the EDTA is standardized *after* adding the Mg^{2+} by titrating primary standard calcium carbonate (dissolved in HCl and pH adjusted). When the indicator is added to the calcium solution, it is pale red. But as soon as the titration is started, the indicator is complexed by the magnesium and turns wine red. At the end point, it changes to blue, as the indicator is displaced from the magnesium. No correction is required for the Mg^{2+} added because it is accounted for in the standardization. *This solution should not be used to titrate metals other than calcium.*

High-purity EDTA can be prepared from $Na_2H_2Y \cdot 2H_2O$ by drying at 80°C for 2 h. The waters of hydration remain intact; this can be used as a primary standard for preparing a standard EDTA solution.

The titration of calcium and magnesium with EDTA is done at pH 10, using an ammonia–ammonium chloride buffer. The pH must not be too high or the metal hydroxide may precipitate, causing the reaction with EDTA to be very slow. Calcium can actually be titrated in the presence of magnesium by raising the pH to 12 with strong alkali; $Mg(OH)_2$ precipitates and does not titrate.

Since Eriochrome Black T and other indicators are weak acids, their colors will depend on the pH because their ionized species have different colors. For example, with Eriochrome Black T, H_2In^- is red (pH < 6), HIn^{2-} is blue (pH 6 to 12), and In^{3-} is yellow orange (pH > 12). Thus, indicators can be used over definite pH ranges. It should be emphasized, though, that although complexometric indicators respond to pH, their mechanism of action does not involve changes in pH, as the solution is buffered. But the pH affects the stability of the complex formed between the indicator and the metal ion, as well as that formed between EDTA and the metal ion.

An indicator is useful for indication of titrations of only those metals that form a more stable complex with the titrant than with the indicator at the given pH. This may sound complex but suitable indicators are known for many titrations with several different chelating agents.

Calmagite

Xylenol orange

Calmagite gives a somewhat improved end point over Eriochrome Black T for the titration of calcium and magnesium with EDTA. It also has a longer shelf life. Xylenol orange is useful for titration of metal ions that form very strong EDTA complexes and are titrated at pH 1.5 to 3.0. Examples are the direct titration of thorium(IV) and bismuth(III), and the indirect determination of zirconium(IV) and iron(III) by back-titration with one of the former two metals. There are many other indicators for EDTA titrations. The work of Flaschka and Barnard (Reference 6) gives many examples of EDTA titrations, and you are referred to this excellent source for detailed descriptions of procedures for different metals.

There are a number of other useful reagents for complexometric titrations. A notable example is ethyleneglycol bis (β-aminoethyl ether)-N, N, N′, N′-tetraacetic acid (**EGTA**). This is an ether analog of EDTA that will selectively titrate calcium in the presence of magnesium:

EGTA allows the titration of calcium in the presence of magnesium.

Complexing agents having ether linkages have a strong tendency to complex the alkaline earths heavier than magnesium. Log K_f for calcium–EGTA is 11.0, while that for magnesium–EGTA is only 5.2. For other chelating agents and their applications, see References 4 and 6.

With the exception of the alkali metals, *nearly every metal can be determined with high precision and accuracy by complexometric titration.* These methods are much more rapid and convenient than gravimetric procedures and are therefore preferred, except in those few instances when greater accuracy is required. In more recent years, however, all of these metal determination methods are giving way to atomic and mass spectrometric measurements (see Chapter 12).

Complexometric titrations in the clinical laboratory are limited to those substances that occur in fairly high concentrations since volumetric methods are generally not too sensitive. The most important complexometric titration is the determination of calcium in blood. Chelating agents such as EDTA are used in the treatment of heavy-metal poisoning, for example, when children ingest chipped paint that contains lead. The calcium chelate (as Na_2CaY) is administered to prevent complexation and use of Ca-EDTA, rather than Na_2EDTA, prevents leaching of calcium in the bones. Heavy metals such as lead form more stable EDTA chelates than calcium does and will displace the calcium from the EDTA. The chelated lead is then excreted via the kidneys.

A table of formation constants of some EDTA chelates appears in Appendix C, Table C4.

Dithizone

8.5 Other Uses of Complexes

Analytical chemists can use the formation of complexes to advantage in ways other than in titrations. For example, metal ion chelates may be formed and extracted into a water-immiscible solvent for separation by **solvent extraction**. Complexes of metal ions with the chelating agent dithizone, for example, are useful for extractions. The chelates are often highly colored. Their formation can then serve as the basis for **spectrophotometric** or atomic spectroscopic **determination** of metal ions. Complexes that

Chelates are used in gravimetry, spectrophotometry, fluorometry, solvent extraction, and chromatography.

Dimethylglyoxime

fluoresce may also be formed. Metal chelates may sometimes precipitate. The nickel–dimethylglyoxime precipitate is an example used in gravimetric analysis. Table 9.2 lists several other metal chelate precipitates. Complexation equilibria may influence chromatographic separations, and we have mentioned the use of complexing agents as masking agents to prevent interfering reactions. For example, in the solvent extraction of vanadium with the chelating agent oxine (8-quinolinol—see Chapter 9), the extraction of copper is avoided by chelating it with EDTA, thus preventing the formation of its oxine chelate. Many metal chelates are intensely colored. It is common practice today to separate metal ions by chromatography and then introduce a chromogenic (meaning color-forming) chelating ligand that reacts with the different metals nonselectively; the product is then spectrophotometrically detected. The lack of selectivity in this case is a virtue, as the same chelating agent can be used to detect a variety of metals that have already been separated.

All of these complexing reactions are pH dependent, and pH adjustment and control (with buffers) are always necessary to optimize the desired reaction or to minimize undesired reactions.

8.6 Cumulative Formation Constants β and Concentrations of Specific Species in Stepwise Formed Complexes

While EDTA reacts with metals on a 1:1 basis, many ligands, especially those that have a more limited number of binding sites, will react with a metal ion in a stepwise fashion, one ligand being added at a time. Ammonia complexes with the nickel ion Ni^{2+}, for example, in 6 steps, forming ultimately $Ni(NH_3)_6^{2+}$. The stepwise formation constants for a metal reacting with a ligand L can be written as:

$$M + L \rightleftharpoons ML, \qquad [ML]/[M][L] = K_{f1}$$
$$ML + L \rightleftharpoons ML_2, \qquad [ML_2]/[ML][L] = K_{f2}$$

and so on to

$$ML_{n-1} + L \rightleftharpoons ML_n, \qquad [ML_n]/[ML_{n-1}][L] = K_{fn}$$

If you compare with a polyprotic acid H_nA, you will notice that H is analogous to L and likewise M to A. But we are not only writing these equilibria as associations, rather than dissociations, we also have the order of the stepwise equilibria reversed. For the dissociation of the acid, we would write as the first step the dissociation of H_nA to $H_{n-1}A^-$ and H^+ and designate that dissociation constant as K_{a1}. So, $1/K_{fn}$ corresponds to K_{a1}, $1/K_{fn-1}$ to K_{a2}, and so on until $1/K_{f1}$, which corresponds to K_{an}.

In dealing with association equilibria, the cumulative constants, designated as β, are also often used. The cumulative constant for formation of ML_n is designated β_n and it is simply the product of the K_f values until that point, thus:

$$\beta_n = K_{f1}K_{f2}K_{f3} \dots K_{fn} \tag{8.17}$$

For example, for the cumulative formation of ML_3, the equilibrium and the corresponding constant will be written as

$$M + 3L \rightleftharpoons ML_3 \qquad [ML_3]/[M][L]^3 = K_{f1}K_{f2}K_{f3} = \beta_3$$

Note that β_1 is the same as K_{f1}. Although β_0 has no physical meaning, for mathematical convenience (the reason why will be apparent in the next section), β_0 is taken to be 1.

In a metal-ligand system containing M, L, and various ML_n species, the sum of the concentrations of all the metal species, including the free metal, often called the analytical concentration of the metal, is typically designated by C_M and is given by

$$C_M = [M] + [ML] + [ML_2] + [ML_3] + \cdots + [ML_n]$$

This is readily expanded to

$$C_M = [M] + \beta_1[M][L] + \beta_2[M][L]^2 + \beta_3[M][L]^3 + \cdots + \beta_n[M][L]^n$$

$$C_M = [M](1 + \beta_1[L] + \beta_2[L]^2 + \beta_3[L]^3 + \cdots + \beta_n[L]^n)$$

(8.18)

Should we consider β_0 to be 1, Equation 8.18 can be written in a more compact form

$$C_M = [M]\sum_{i=0}^{i=n} \beta_i L^i$$

(8.19)

One fraction, that of the free metal ion, denoted α_M, is of frequent interest, and it will be readily realized from Equation 8.18 that

$$\alpha_M = [M]/C_M = \left(\sum_{i=0}^{i=n} \beta_i L^i\right)^{-1}$$

(8.20)

Since $[M]$ can thus also be expressed as

$$[M] = C_M \Big/ \left(\sum_{i=0}^{i=n} \beta_i L^i\right)$$

(8.21)

the concentration of any other species such as ML_i can also be readily calculated as $\beta_i[M]L^i$. Illustrative examples follow.

Example 8.5

Given that Cu^{2+} forms a tetrammine complex with log β_I values of 3.99, 7.33, 10.06, and 12.03 for $i = 1 - 4$, respectively, and given that from Table C4 the formation constant for the $Cu^{2+} - EDTA$ complex is 6.30×10^{18}, calculate the conditional formation constant for the $Cu^{2+} - EDTA$ complex in an $NH_3 - NH_4Cl$ buffer, which has a pH of 10 and $[NH_3] = 1.0\ M$. Comment on the feasibility of an EDTA titration of Cu^{2+} under these conditions.

Solution

The conditional formation constant depends not only on what fraction of the total EDTA exists as Y^{4-} (we have already previously calculated that $\alpha_{Y^{4-}} = 0.35$ at pH 10), it also depends on what fraction of the metal, α_{Cu}, will exist as the free Cu^{2+} not complexed by ammonia (EDTA complexation is not taken into this account). In other words, the conditional formation constant can be written as:

$$K_f = \frac{[CuY^{2-}]}{[Cu^{2+}][Y^{4-}]} = \frac{1}{\alpha_{Cu}\alpha_{Y^{4-}}}\frac{[CuY^{2-}]}{C_{Cu}C_{H_4Y}} = \frac{K_f'}{\alpha_{Cu}\alpha_{Y^4}}$$

Hence

$$K_f' = \alpha_{Cu}\alpha_{Y^{4-}}K_f$$

We calculate α_{Cu} from Equation 8.20:

$$\alpha_{Cu} = (1 + 10^{3.99} \times 1.0 + 10^{7.33} \times 1.0^2 + 10^{10.06} \times 1.0^3 + 10^{12.03} \times 1.0^4)^{-1}$$

$$= (1 + 9770 + 2.14 \times 10^7 + 1.15 \times 10^{10} + 1.07 \times 10^{12})^{-1} = 9.26 \times 10^{-13}$$

Therefore, $K_f' = 9.26 \times 10^{-13} \times 0.35 \times 6.3 \times 10^{18} = 2.0 \times 10^6$.

The minimum conditional formation constant at which a titration is feasible is 10^6. So it will be just marginally possible to do a titration under these conditions.

Example 8.6

In 100 mL of a 0.010 M NH_3 solution, 11.9 mg of $NiCl_2 \cdot 6H_2O$ was dissolved. The log β_i values for NH_3 complexation of Ni^{2+} are 2.67, 4.79, 6.40, 7.47, 8.10, 8.01 for $i = 1 - 6$, respectively. Compute the concentrations of the various $Ni(NH_3)_i^{2+}$ species.

Solution

The FW of $NiCl_2 \cdot 6H_2O$ is 237.7 g/mol, 11.9 mg is $11.9/238 = 0.0500$ mmol, dissolved in 0.1 L, $C_{Ni} = 5.00 \times 10^{-4}\ M$. A simplistic solution assumes that the equilibrium free NH_3 concentration (L) is $1.0 \times 10^{-2}\ M$; hence, we can calculate from Equation 8.19. (Note the shortened algebraic notation, e.g., for $\beta_1 L_1$: $\beta_1 L_1 = 10^{2.67} \times 10^{-2} = 10^{(2.67-2)}$):

$$\alpha_{Ni} = (1 + 10^{(2.67-2)} + 10^{(4.79-4)} + 10^{(6.40-6)} + 10^{(7.47-8)} + 10^{(8.10-10)} + 10^{(8.01-12)})^{-1}$$

$$= (1 + 4.68 + 6.17 + 2.51 + 0.30 + 0.01 + 1.00 \times 10^{-4})^{-1} = 6.8 \times 10^{-2}$$

$$[Ni^{2+}] = C_{Ni}\alpha_{Ni} = 5.00 \times 10^{-4} \times 6.8 \times 10^{-2} = 3.4 \times 10^{-5}\ M$$

$$[Ni(NH_3)^{2+}] = [Ni^{2+}]\beta_1 L = 3.4 \times 10^{-5} \times 4.68 = 1.6 \times 10^{-4}\ M$$

$$[Ni(NH_3)_2^{2+}] = [Ni^{2+}]\beta_2 L^2 = 3.4 \times 10^{-5} \times 6.17 = 2.1 \times 10^{-4}\ M$$

$$[Ni(NH_3)_3^{2+}] = [Ni^{2+}]\beta_3 L^3 = 3.4 \times 10^{-5} \times 2.51 = 8.5 \times 10^{-5}\ M$$

$$[Ni(NH_3)_4^{2+}] = [Ni^{2+}]\beta_4 L^4 = 3.4 \times 10^{-5} \times 0.30 = 1.0 \times 10^{-5}\ M$$

$$[Ni(NH_3)_5^{2+}] = [Ni^{2+}]\beta_5 L^5 = 3.4 \times 10^{-5} \times 0.01 = 3.4 \times 10^{-7}\ M$$

$$[Ni(NH_3)_6^{2+}] = [Ni^{2+}]\beta_6 L^6 = 3.4 \times 10^{-5} \times 1.0 \times 10^{-4} = 3.4 \times 10^{-9}\ M$$

This solution is unfortunately only approximate because it ignores the consumption of NH_3 in forming the $Ni(NH_3)_n$ complexes. 0.16 mM $Ni(NH_3)^{2+}$, 0.21 mM $Ni(NH_3)_2^{2+}$ 0.085 mM $Ni(NH_3)_3^{2+}$ and 0.01 mM $Ni(NH_3)_4^{2+}$ respectively ties up 0.16, 0.42, 0.26, and 0.04 mM NH_3, totaling 0.88 mM NH_3, when the original total NH_3 present was 10 mM, this consumption is not insignificant. We can solve this problem without such errors by Goal Seek.

We set up the problem by first expressing α_M and $[M]$ in the usual manner and then the total Ligand concentration ($L_T = 0.010$) as the sum of all ligand containing terms:

$$L_T - ([L] + [ML] + 2^*[ML_2] + 3^*[ML_3] + 4^*[ML_4] + 5^*[ML_5] + 6^*[ML_6]) = 0$$

The result is multiplied by 10^{10} and Goal Seek is invoked to make this result a target of zero by changing the value of L. The Excel spreadsheet example is given in the website supplement as Example 8.6.xlsx. The result for $[Ni^{2+}]$ is $3.97 \times 10^{-5}\ M$.

And additional check for the metal,

$$C_M - ([L] + [ML] + [ML_2] + [ML_3] + [ML_4] + [ML_5] + [ML_6]) = 0$$

can be put in to verify that this result also is satisfied.

Questions

1. Distinguish between a complexing agent and a chelating agent.
2. Explain the principles of indicators used in chelometric titrations.
3. Why is a small amount of magnesium salt added to the EDTA solution used for the titration of calcium with an Eriochrome Black T indicator?
4. Why would it be difficult to titrate Cu^{2+} with EDTA in a strongly ammoniacal medium?
5. Indicator for complexometric titrations are:
   (a) colmagite.
   (b) potassium permanganate.
   (c) solochrome black.
   (d) both (a) and (c).
6. What is the indicator used for EDTA titration?
7. In hard water estimation, which indicator is used for the indication of end point of direct EDTA titration of calcium in absence of magnesium?
8. Why di-sodium salt of EDTA is chosen for determination?
9. Which colour does EBT shows at pH (a) < 6 (b) 6 to 12 and (c) >12?
10. Give example of a titrant other than EDTA that can be used for selectively titrating calcium in presence of magnesium.
11. Why only pH = 10 buffer is used in titrations?

Problems

Complex Equilibrium Calculations (K_f)

12. Calcium ion forms a weak 1:1 complex with nitrate ion with a formation constant of 2.0. What would be the equilibrium concentrations of Ca^{2+} and $Ca(NO_3)^+$ in a solution prepared by adding 10 mL each of 0.010 M $CaCl_2$ and 2.0 M $NaNO_3$? Neglect diverse ion effects.
13. The formation constant of the silver–ethylenediamine complex, $Ag(NH_2CH_2CH_2NH_2)^+$, is 5.0×10^4. Calculate the concentration of Ag^+ in equilibrium with a 0.10 M solution of the complex. (Assume no higher order complexes.)
14. What would be the concentration of Ag^+ in Problem 13 if the solution contained also 0.10 M ethylenediamine, $NH_2CH_2CH_2NH_2$?
15. Silver ion forms stepwise complexes with thiosulfate ion, $S_2O_3^{2-}$, with $K_{f1} = 6.6 \times 10^8$ and $K_{f2} = 4.4 \times 10^4$. Calculate the equilibrium concentrations of all silver species for 0.0100 M $AgNO_3$ in 1.00 M $Na_2S_2O_3$. Neglect diverse ion effects.

Conditional Formation Constants

16. The formation constant for the lead–EDTA chelate (PbY^{2-}) is 1.10×10^{18}. Calculate the conditional formation constant (a) at pH 3 and (b) at pH 10.
17. Using the conditional constants calculated in Problem 16 calculate the pPb ($- \log[Pb^{2+}]$) for 50.0 mL of a solution of 0.0250 M Pb^{2+} (a) at pH 3 and (b) at pH 10 after the addition of (1) 0 mL, (2) 50 mL, (3) 125 mL, and (4) 200 mL of 0.0100 M EDTA.
18. The conditional formation constant for the calcium–EDTA chelate was calculated for pH 10 in Example 8.4 to be 1.8×10^{10}. Calculate the conditional formation constant at pH 3. Compare this with that calculated for lead at pH 3 in Problem 16. Could lead be titrated with EDTA at pH 3 in the presence of calcium?

19. Calculate the conditional formation constant for the nickel–EDTA chelate in an ammoniacal pH 10 buffer containing $[NH_3] = 0.10\ M$. See Example 8.6 for the relevant complexation constants with ammonia.

Standard Solutions

20. Calculate the weight of $Na_2H_2Y \cdot 2H_2O$ required to prepare 500.0 mL of 0.05000 M EDTA.

21. The concentration of solution of EDTA was determined by standardizing against a solution of Ca^{2+} prepared using a primary standard of $CaCO_3$. A 0.4071g sample of $CaCO_3$ was transformed to a 500 mL volumetric flask, dissolved using a minimum of 6 M HCl, and diluted to volume. After transferring a 50.0 mL portion of this solution to a 250 mL Erlenmeyer flask, the pH was adjusted by adding 5 mL of a pH 10 NH_3–NH_4Cl buffer containing a small amount of Mg^{2+}- EDTA. After adding calmagite as indicator, the solution was titrated with EDTA, requiring 42.63 mL to reach the end point. Report molar concentration of EDTA in the titrant.

22. Calculate the molar Y^{4-} concentration in a 0.0200 M EDTA solution referred to a pH of 10.00 ($\alpha_4 = 0.35$).

23. An EDTA solution is standardized against high-purity $CaCO_3$ by dissolving 0.3982 g $CaCO_3$ in hydrochloric acid, adjusting the pH to 10 with ammoniacal buffer, and titrating. If 38.26 mL was required for the titration, what is the molarity of the EDTA?

24. Calculate the titer of 0.1000 M EDTA in mg $CaCO_3$/mL.

25. If 100.0 mL of a water sample is titrated with 0.01000 M EDTA, what is the titer of the EDTA in terms of water hardness/mL?

Quantitative Complexometric Determinations

26. Calcium in powdered milk is determined by dry ashing (see Chapter 1) a 1.50-g sample and then titrating the calcium with EDTA solution, 12.1 mL being required. The EDTA was standardized by titrating 10.0 mL of a zinc solution prepared by dissolving 0.632 g zinc metal in acid and diluting to 1 L (10.8 mL EDTA required for titration). What is the concentration of calcium in the powdered milk in parts per million?

27. Calcium is determined in serum by microtitration with EDTA. A 100-μL sample is treated with two drops of 2 M KOH, Cal-Red indicator is added, and the titration is performed with 0.00122 M EDTA, using a microburet. If 0.203 mL EDTA is required for titration, what is the level of calcium in the serum in mg/dL and in meq/L?

28. In the Liebig titration of cyanide ion, a soluble complex is formed; at the equivalence point, solid silver cyanide is formed, signaling the end point:

$$2CN^- + Ag^+ \rightarrow Ag(CN)_2^- \quad \text{(titration)}$$

$$Ag(CN)_2^- + Ag^+ \rightarrow Ag[Ag(CN)_2] \quad \text{(end point)}$$

A 0.4723-g sample of KCN was titrated with 0.1025 M AgNO_3, requiring 34.95 mL. What is the percent purity of the KCN?

29. Copper in saltwater near the discharge of a sewage treatment plant is determined by first separating and concentrating it by solvent extraction of its dithizone chelate at pH 3 into methylene chloride and then evaporating the solvent, ashing the chelate to destroy the organic portion, and titrating the copper with EDTA. Three 1-L portions of the sample are each extracted with 25-mL portions of methylene chloride, and the extracts are combined in a 100-mL volumetric flask and diluted to volume. A 50-mL aliquot is evaporated, ashed, and titrated. If the EDTA solution has a $CaCO_3$ titer of 2.69 mg/mL and 2.67 mL is required for titration of the copper, what is the concentration of copper in the seawater in parts per million?

30. Chloride in serum is determined by titration with $Hg(NO_3)_2$; $2Cl^- + Hg^{2+} \rightleftharpoons HgCl_2$. The $Hg(NO_3)_2$ is standardized by titrating 2.00 mL of a 0.0108 M NaCl solution, requiring 1.12 mL to reach the diphenylcarbazone end point. A 0.500-mL serum sample is treated with 3.50 mL water, 0.50 mL 10% sodium tungstate solution, and 0.50 mL of 0.33 M H_2SO_4 solution to precipitate proteins. After the proteins are precipitated, the sample is filtered through a dry filter into a dry flask. A 2.00-mL aliquot of the filtrate is titrated with the $Hg(NO_3)_2$ solution, requiring 1.23 mL. Calculate the mg/L chloride in the serum. (Note: mercury is rarely used today due to its toxicity. The problem is illustrative.)

PROFESSOR'S FAVORITE PROBLEM

Contributed by Professor Bin Wang, Marshall University

31. A 0.1021 g sample containing ZnO was titrated using a standard EDTA solution with Erichrome Black T as indicator. It took 25.52 mL of 0.0100 M EDTA to reach the end point. What is the percentage of ZnO in the sample?

Spreadsheet Problems

See the textbook website, Chapter 8, for suggested setups.

32. Prepare a spreadsheet for Figure 8.2, log K_f' vs. pH for the EDTA chelates of calcium, lead, and mercury. This will require calculating α_4 for EDTA and the K_f values for the chelates of calcium, lead, and mercury. Calculate at 0.5 pH intervals. Compare your plot with Figure 8.2.

33. Prepare a spreadsheet for the titration of 100.00 mL of 0.1000 M Hg^{2+} with 0.1000 M Na_2EDTA at pH 6 (similar to pCa vs. mL EDTA—Figure 8.3). Start out at pHg = 1 and go up to pHg = 17 with a resolution of 0.1. See Figure 8.3.xlsx. (See the text website for the solution.)

34. Prepare a spreadsheet to plot the seven fractional values (from α_M to α_{ML6}) present in a Ni^{2+} – NH_3 system as a function of [NH_3]. Plot the results from 0.001 to 1.0 M NH_3, with a resolution of 0.001 M, use logarithmic scaling for the X-axis (ammonia concentration).

Multiple Choice Questions

1. To carry out complexometric titration, what conditions should be respected?
   (a) Presence of buffer solution
   (b) (M-EDTA) complex should be more stable than [M-Indicator] complex
   (c) Presence of a catalyst
   (d) Always solution of EDTA should be added to analyzed solution and not vice versa.

2. By direct complexometric titration method, which of the following compounds can be analyzed?
   (a) $FeSO_4$ (b) $MgSO_4$
   (c) NH_4Cl (d) $CaCl_2$

3. To determine the end point in complexometric titration, which of the following can be used?
   (a) Adsorption indicator   (b) Specific indicator
   (c) Murexide   (d) Metallochromic indicator

Recommended References

1. *Stability Constants of Metal-Ion Complexes. Part A: Inorganic Ligands*, E. Hogfeldt, ed., *Part B: Organic Ligands*, D. D. Perrin, ed. Oxford: Pergamon, 1979, 1981.
2. A. Martell and R. J. Motekaitis, *The Determination and Use of Stability Constants*. New York: VCH, 1989.
3. J. Kragten, *Atlas of Metal-Ligand Equilibria in Aqueous Solution*. London: Ellis Horwood, 1978.
4. G. Schwarzenbach, *Complexometric Titrations*. New York: Interscience, 1957.
5. H. Flaschka, *EDTA Titrations*. New York: Pergamon, 1959.
6. H. A. Flaschka and A. J. Barnard, Jr., "Titrations with EDTA and Related Compounds," in C. L. Wilson and D. W. Wilson, eds. *Comprehensive Analytical Chemistry*, Vol. 1B. New York: Elsevier, 1960.
7. F. J. Welcher, *The Analytical Uses of Ethylenediaminetetraacetic Acid*. Princeton: Van Nostrand, 1958.
8. J. Stary, ed., *Critical Evaluation of Equilibrium Constants Involving 8-Hydroxyquinoline and Its Metal Chelates*. Oxford: Pergamon, 1979.
9. H. A. Flaschka, *Chelates in Analytical Chemistry*, Vols. 1–5. New York: Dekker, 1967–1976.

Chapter 9

GRAVIMETRIC ANALYSIS AND PRECIPITATION EQUILIBRIA

"Some loads are light, some heavy. Some people prefer the light to the heavy…"

—Mao Tse-tung

KEY THINGS TO LEARN FROM THIS CHAPTER

- Steps of a gravimetric analysis: precipitation, digestion, filtration, washing, drying, weighing, calculation
- Gravimetric calculations (key Equations: 9.1, 9.3, 9.5)

- The solubility product, the common ion effect
- The diverse ion effect (key Equation: 9.10)

Gravimetric analysis is one of the most accurate and precise methods of macroquantitative analysis. In this process the analyte is selectively converted to an insoluble form. The separated precipitate is dried or ignited, possibly to another form, and is accurately weighed. From the weight of the precipitate and a knowledge of its chemical composition, we can calculate the weight of analyte in the desired form.

Gravimetric analysis is capable of exceedingly precise analysis. In fact, gravimetric analysis was used to determine the atomic masses of many elements to six figure accuracy. Theodore W. Richards at Harvard University developed highly precise and accurate gravimetric analysis of silver and chlorine. He used these methods to determine the atomic weights of 25 elements by preparing pure samples of the chlorides of the elements, decomposing known weights of the compounds, and determining the chloride content by gravimetric methods. For this work, he was the first American to receive the Nobel Prize. He was the ultimate analytical chemist!

Gravimetry does not require a series of standards for calculation of an unknown since calculations are based only on atomic or molecular weights. Only a precise analytical balance is needed for measurements. Gravimetric analysis, due to its high degree of accuracy, can also be used to calibrate other instruments in lieu of reference standards. While it is tedious and time consuming, it may find use where very precise results are needed, for example, in determining the iron content of an ore, whose price is determined by the iron content. It is used to determine the chloride content of cement. In environmental chemistry, sulfate is precipitated with barium ion, and in the petroleum field, hydrogen sulfide in desulfurization waste water is precipitated with silver ion.

This chapter describes the specific steps of gravimetric analysis, including preparing the solution in proper form for precipitation, the precipitation process and how to obtain the precipitate in pure and filterable form, the filtration and washing of the precipitate to prevent losses and impurities, and heating the precipitate to convert it to a weighable form. It gives calculation procedures for computing the quantity of analyte from the weight of precipitate, following the principles introduced in Chapter 4. It also provides some common examples of gravimetric analysis. Finally, it discusses the solubility product and associated precipitation equilibria.

9.1 How to Perform a Successful Gravimetric Analysis

A successful gravimetric analysis consists of a number of important operations designed to obtain a pure and filterable precipitate suitable for weighing. You may wish to precipitate silver chloride from a solution of chloride by adding silver nitrate. There is more to the procedure than simply pouring in silver nitrate solution and then filtering.

Accurate gravimetric analysis requires careful manipulation in forming and treating the precipitate.

What Steps are Needed?

The steps required in a gravimetric analysis, after the sample has been dissolved, can be summarized as follows:

1. Preparation of the solution	5. Washing
2. Precipitation	6. Drying or igniting
3. Digestion	7. Weighing
4. Filtration	8. Calculation

These operations and the reasons for them are described below.

First Prepare the Solution

The first step in performing gravimetric analysis is to prepare the solution. Some form of preliminary separation may be necessary to eliminate interfering materials. Also, we must adjust the solution conditions to maintain low solubility of the precipitate and to obtain it in a form suitable for filtration. Proper adjustment of the solution conditions prior to precipitation may also mask potential interferences. Factors that must be considered include the volume of the solution during precipitation, the concentration range of the test substance, the presence and concentrations of other constituents, the temperature, and the pH.

Although preliminary separations may be required, in other instances the precipitation step in gravimetric analysis is sufficiently selective that other separations are not required. The pH is important because it often influences both the solubility of the analytical precipitate and the possibility of interferences from other substances. For example, calcium oxalate is insoluble in basic medium, but at low pH the oxalate ion combines with the hydrogen ions to form a weak acid and begins to dissolve. 8-Hydroxyquinoline (oxine—also known as 8-quinolinol) can be used to precipitate a large number of elements, but by controlling pH, we can precipitate elements selectively. Aluminum ion can be precipitated at pH 4, but the concentration of the anion form of oxine is too low at this pH to precipitate magnesium ion; magnesium oxinate has a much greater solubility product, the solubility product concept is discussed later in the chapter.

Usually, the precipitation reaction is selective for the analyte.

8-Hydroxyquinoline can be used in combination with pH adjustments for selective precipitation of different metals. Al^{3+} can be selectively precipitated over Mg^{2+} at pH 4.

A higher pH is required to shift the ionization step to the right in order to precipitate magnesium. If the pH is too high, however, magnesium hydroxide will precipitate, causing interference.

The effects of the other factors mentioned above will become apparent as we discuss the precipitation step.

Then do the Precipitation—But Under the Right Conditions

After preparing the solution, the next step is to do the precipitation. Again, certain conditions are important. The precipitate should first be *sufficiently insoluble* that the amount lost due to solubility will be negligible. It should consist of *large crystals* that can be easily filtered. All precipitates tend to carry some of the other constituents of the solution with them. This contamination should be negligible. Keeping the crystals large can minimize this contamination.

We can achieve an appreciation of the proper conditions for precipitation by first looking at the **precipitation process**. When a solution of a precipitating agent is added to a test solution to form a precipitate, such as in the addition of $AgNO_3$ to a chloride solution to precipitate AgCl, the actual precipitation occurs in a series of steps. The precipitation process involves heterogeneous equilibria and, as such, is not instantaneous (see Chapter 5). The equilibrium condition is described by the solubility product, discussed at the end of the chapter. First, **supersaturation** occurs, that is, the solution phase contains more of the dissolved salt than it can carry at equilibrium. This is a metastable condition, and the driving force will be for the system to approach equilibrium (saturation). This is started by **nucleation**. For nucleation to occur, a minimum number of particles must come together to produce microscopic nuclei of the solid phase. The higher the degree of supersaturation, the greater the rate of nucleation. The formation of a greater number of nuclei per unit time will ultimately produce more total crystals of smaller size. The total crystal surface area will be larger, and there will be more danger that impurities will be adsorbed (see below).

During the precipitation process, supersaturation occurs (this should be minimized!), followed by nucleation and precipitation.

Although nucleation should theoretically occur spontaneously, it is usually induced, for example, on dust particles, scratches on the vessel surface, or added seed crystals of the precipitate (not in quantitative analysis).

Following nucleation, the initial nucleus will grow by depositing other precipitate particles to form a crystal of a certain geometric shape. Again, the greater the supersaturation, the more rapid the crystal growth rate. An increased growth rate increases the chances of imperfections in the crystal and trapping of impurities.

Von Weimarn discovered that the particle size of precipitates is inversely proportional to the relative supersaturation of the solution during the precipitation process:

$$\text{Relative supersaturation} = \frac{Q - S}{S}$$

where Q is the concentration of the mixed reagents *before* precipitation occurs, S is the **solubility** of the precipitate at equilibrium, and $Q - S$ is the **degree of supersaturation**. This ratio, $(Q - S)/S$, relative supersaturation, is also called the **von Weimarn ratio**.

As previously mentioned, when a solution is supersaturated, it is in a state of metastable equilibrium, and this favors rapid nucleation to form a large number of small particles. That is,

High relative supersaturation → many small crystals
(high surface area)
Low relative supersaturation → fewer, larger crystals
(low surface area)

Obviously, then, we want to keep Q low and S high during precipitation. Several steps are commonly taken to maintain *favorable conditions for precipitation*:

1. Precipitate from *dilute solution*. This keeps Q low.
2. Add dilute precipitating reagents *slowly*, with effective *stirring*. This also keeps Q low. Stirring prevents local excesses of the reagent.
3. Precipitate from *hot solution*. This increases S. The solubility should not be too great or the precipitation will not be quantitative (with less than 1 part per thousand remaining). The bulk of the precipitation may be performed in the hot solution, and then the solution may be cooled to make the precipitation quantitative.
4. Precipitate at as *low* a *pH* as is possible to maintain quantitative precipitation. As we have seen, many precipitates are more soluble in acid medium, and this slows the rate of precipitation. They are more soluble because the anion of the precipitate (which comes from a weak acid) combines with protons in the solution.

Here is how to minimize supersaturation and obtain larger crystals.

Most of these operations can also decrease the degree of contamination. The concentration of impurities is kept lower and their solubility is increased, and the slower rate of precipitation decreases their chance of being trapped. The larger crystals have a smaller specific surface area (i.e., a smaller surface area relative to the mass) and so have less chance of adsorbing impurities. Note that the most insoluble precipitates do not make the best candidates for pure and readily filterable precipitates. An example is hydrous iron oxide (or iron hydroxide), which forms a gelatinous precipitate of large surface area.

Very insoluble precipitates are not the best candidates for gravimetric analysis! They supersaturate too easily.

When the precipitation is performed, a slight excess of precipitating reagent is added to decrease the solubility by mass action (common ion effect) and to assure complete precipitation. A large excess of precipitating agent should be avoided because this increases chances of adsorption on the surface of the precipitate, in addition to being wasteful. If the approximate amount of analyte is known, a 10% excess of reagent is generally added. Completeness of precipitation is checked by waiting until the precipitate has settled and then adding a few drops of precipitating reagent to the clear solution above it. If no new precipitate forms, precipitation is complete.

Don't add too much excess precipitating agent. This will increase adsorption.

Check for completeness of precipitation!

Digest the Precipitate to Make Larger and More Pure Crystals

We know that very small crystals with a large specific surface area have a higher surface energy and a higher apparent solubility than large crystals. This is an initial rate phenomenon and does not represent the equilibrium condition, and it is one consequence of heterogeneous equilibria. When a precipitate is allowed to stand in the presence of the **mother liquor** (the solution from which it was precipitated), the large crystals grow at the expense of the small ones. This process is called **digestion**, or **Ostwald ripening**, and is illustrated in Figure 9.1. Small particles have greater surface energy associated with a greater surface area and display somewhat greater solubility than larger particles. The small particles tend to dissolve and reprecipitate on the surfaces of the larger crystals. In addition, individual particles **agglomerate** to effectively share a common counterion layer (see below), and the agglomerated particles finally *cement* together by forming connecting bridges. This noticeably decreases surface area.

Ostwald ripening improves the purity and crystallinity of the precipitate.

Also, imperfections of the crystals tend to disappear, and adsorbed or trapped impurities tend to go into solution. Digestion is usually done at elevated temperatures

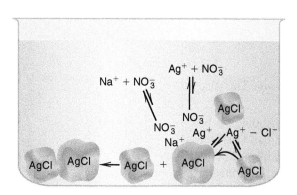

FIGURE 9.1 Ostwald ripening.

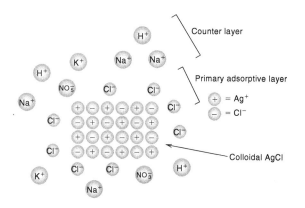

FIGURE 9.2 Representation of silver chloride colloidal particle and adsorptive layers when Cl⁻ is in excess.

to speed the process, although in some cases it is done at room temperature. It improves both the filterability of the precipitate and its purity.

Many precipitates do not give a favorable von Weimarn ratio, especially very insoluble ones. Hence, it is impossible to yield a crystalline precipitate (small number of large particles), and the precipitate is first **colloidal** (large number of small particles).

Colloidal particles are very small (1 to 100 nm) and have a very large surface-to-mass ratio, which promotes surface adsorption. They are formed by virtue of the precipitation mechanism. As a precipitate forms, the ions are arranged in a fixed pattern. In AgCl, for example, there will be alternating Ag^+ and Cl^- ions on the surface (see Figure 9.2). While there are localized + and − charges on the surface, the net surface charge is zero. However, the surface does tend to adsorb the ion of the precipitate particle that is in excess in the solution, for example, Ag^+ if precipitating Cl^- with an excess of Ag^+; this imparts a charge. (With crystalline precipitates, the degree of such adsorption will generally be small in comparison with particles that tend to form colloids.) The adsorption creates a **primary layer** that is strongly adsorbed and is an integral part of the crystal. It will attract ions of the opposite charge in a **counterlayer** or secondary layer so the particle will have an overall neutral charge. There will be solvent molecules interspersed between the layers. Normally, the counterlayer completely neutralizes the primary layer and is close to it, so the particles will collect together to form larger particles; that is, they will **coagulate**. However, if the secondary layer is loosely bound, the primary surface charge will tend to repel like particles, maintaining a colloidal state.

When coagulated particles are filtered, they retain the adsorbed primary and secondary ion layers along with solvent. Washing the particles with water increases the extent of solvent (water) molecules between the layers, causing the secondary layer to be loosely bound, and the particles revert to the colloidal state. This process is called **peptization** and is discussed in more detail below where we describe washing the precipitate. Adding an electrolyte will result in a closer secondary layer and will promote coagulation. Heating tends to decrease adsorption and the effective charge in the adsorbed layers, thereby aiding coagulation. Stirring will also help.

While all colloidal systems cause difficulties in analytical determinations, some are worse than others. Depending on the affinity of the dispsersed material for water, colloidal systems can be classified into **hydrophilic** (water loving) and **hydrophobic** (do not like water). While the former ones tend to produce stable dispersions in water, the latter ones tend to aggregate.

Peptization is the reverse of coagulation (the precipitate reverts to a colloidal state and is lost). It is avoided by washing with an electrolyte that can be volatized by heating.

Coagulation of a hydrophobic colloid is fairly easy and results in a curdy precipitate. An example is silver chloride. Coagulation of a hydrophilic colloid, such as hydrous ferric oxide, is more difficult, and it produces a gelatinous precipitate that is difficult to filter because it tends to clog the pores of the filter. In addition, gelatinous precipitates adsorb impurities readily because of their very large surface area. Sometimes a *reprecipitation* of the filtered precipitate is required. During the reprecipitation, the concentration of impurities in solution (from the original sample matrix) has been reduced to a low level, and adsorption will be very small.

Despite the colloidal nature of silver chloride, the gravimetric determination of chloride is one of the most accurate determinations compared to other techniques such as titrimetry. In fact, it was used for atomic weight determination by T. W. Richards, who used nephelometry (light scattering) to correct for colloidal silver chloride.

AgCl forms a hydrophobic colloid (a sol), which readily coagulates. $Fe_2O_3 \cdot xH_2O$ forms a hydrophilic colloid (a gel) with large surface area.

Impurities in Precipitates

Precipitates tend to carry down from the solution other constituents that are normally soluble, causing the precipitate to become contaminated. This process is called **coprecipitation**. The process may be equilibrium based or kinetically controlled. There are a number of ways in which a foreign material may be coprecipitated.

1. Occlusion and Inclusion. In the process of **occlusion**, material that is not part of the crystal structure is trapped within a crystal. For example, water may be trapped in pockets when $AgNO_3$ crystals are formed, and this can be removed to a degree by dissolution and recrystallization. If such mechanical trapping occurs during a precipitation process, the water will contain dissolved impurities. **Inclusion** occurs when ions, generally of similar size and charge, are trapped within the crystal lattice (isomorphous inclusion, as with K^+ in NH_4MgPO_4 precipitation). These are not equilibrium processes.

Occluded or included impurities are difficult to remove. Digestion may help some but is not completely effective. The impurities cannot be removed by washing. Purification by dissolving and reprecipitating is helpful.

Occlusion is the trapping of impurities inside the precipitate.

2. Surface Adsorption. As we have already mentioned, the surface of the precipitate will have a primary adsorbed layer of the lattice ions in excess. This results in **surface adsorption**, the most common form of contamination. For example, after barium sulfate is completely precipitated, the lattice ion in excess will be barium, and this will form the primary layer. The counterion will be a foreign anion, say, nitrate two for each barium. The net effect then is an adsorbed layer of barium nitrate, an equilibrium-based process. These adsorbed layers can often be removed by washing, or they can be replaced by ions that are readily volatilized. Gelatinous precipitates are especially troublesome, though. Digestion reduces the surface area and, therefore, the adsorbed amount.

Surface adsorption of impurities is the most common source of error in gravimetry. It is reduced by proper precipitation technique, digestion, and washing.

3. Isomorphous Replacement. Two compounds are said to be **isomorphous** if they have the same type of formula and crystallize in similar geometric forms. When their lattice dimensions are about the same, one ion can replace another in a crystal, resulting in a **mixed crystal**. This process is called **isomorphous replacement** or isomorphous substitution. For example, in the precipitation of Mg^{2+} as magnesium ammonium phosphate, K^+ has nearly the same ionic size as NH_4^+ and can replace it to form magnesium potassium phosphate. Isomorphous replacement, when it occurs, causes major interference, and little can be done about it. Precipitates in which it occurs are seldom used analytically. Chloride cannot be selectively determined by precipitation as AgCl, for example, in the presence of other halides and vice versa. Mixed crystal formation is a form of equilibrium precipitate formation, although it may be influenced by the rate of precipitation. Such a mixed precipitate is akin to a solid solution. The mixed crystal may be spatially homogeneous if the crystal is in equilibrium with the final solution composition (homogeneous coprecipitation) or heterogenous if it is in instaneous equilibrium with the solution as it forms (heterogenous coprecipitation), as the solution composition changes during precipitation.

4. Postprecipitation. Sometimes, when the precipitate is allowed to stand in contact with the mother liquor, a second substance will slowly form a precipitate with the precipitating reagent. This is called **postprecipitation**. For example, when calcium oxalate is precipitated in the presence of magnesium ions, magnesium oxalate does not immediately precipitate because it tends to form supersaturated solutions. But it will precipitate if the solution is allowed to stand too long before being filtered. Similarly, copper sulfide will precipitate in acid solution in the presence of zinc ions without zinc sulfide being precipitated, but eventually zinc sulfide will precipitate. Postprecipitation is a slow equilibrium process.

Washing and Filtering the Precipitates — Take Care or You May Lose Some

Coprecipitated impurities, especially those on the surface, can be removed by washing the precipitate after filtering. The precipitate will be wet with the mother liquor, which is also removed by washing. Many precipitates cannot be washed with pure water, because **peptization** occurs. This is the reverse of coagulation, as previously mentioned.

The process of coagulation discussed above is at least partially reversible. As we have seen, coagulated particles have a neutral layer of adsorbed primary ions and counterions. We also saw that the presence of another electrolyte will cause the counterions to be forced into closer contact with the primary layer, thus promoting coagulation. These foreign ions are carried along in the coagulation. Washing with water will dilute and remove foreign ions, and the counterion will occupy a larger volume, with more solvent molecules between it and the primary layer. The result is that the repulsive forces between particles become strong again, and the particles partially revert to the colloidal state and pass through the filter. This can be prevented by adding an electrolyte to the wash liquid, for example, HNO_3 or NH_4NO_3 for AgCl precipitate (but not KNO_3 since it is nonvolatile—see below).

The electrolyte must be one that is volatile at the temperature to be used for drying or ignition, and it must not dissolve the precipitate. For example, dilute nitric acid is used as the wash solution for silver chloride. The nitric acid replaces the adsorbed layer of $Ag^+|anion^-$, and it is volatilized when dried at $110°C$. Ammonium nitrate is used as the wash electrolyte for hydrous ferric oxide. It is decomposed to NH_3, HNO_3, N_2, and oxides of nitrogen when the precipitate is dried by ignition at a high temperature.

Test for completeness of washing.

When you wash a precipitate, you should conduct a test to determine when the washing is complete. This is usually done by testing the filtrate for the presence of an ion of the precipitating reagent. After several washings with small volumes of the wash liquid, a few drops of the filtrate are collected in a test tube for the testing. For example, if chloride ion is determined by precipitating with silver nitrate reagent, the filtrate is tested for silver ion by adding sodium chloride or dilute HCl. We describe the technique of filtering in Chapter 2.

Drying or Igniting the Precipitate

Drying removes the solvent and wash electrolytes.

If the collected precipitate is in a form suitable for weighing, it must be heated to remove water and to remove the adsorbed electrolyte from the wash liquid. This drying can usually be done by heating at 110 to $120°C$ for 1 to 2 h. **Ignition** at a much higher temperature is usually required if a precipitate must be converted to a more suitable form for weighing. For example, magnesium ammonium phosphate, $MgNH_4PO_4$, is decomposed to the pyrophosphate, $Mg_2P_2O_7$, by heating at $900°C$. Hydrous ferric oxide, $Fe_2O_3 \cdot xH_2O$, is ignited to the anhydrous ferric oxide. Many metals that are precipitated by organic reagents (e.g., 8-hydroxyquinoline) or by sulfide can be ignited to their oxides. The technique of igniting a precipitate is also described in Chapter 2.

9.2 Gravimetric Calculations—How Much Analyte is There?

The precipitate we weigh is usually in a different form than the analyte whose weight we wish to report. The principles of converting the weight of one substance to that of another are given in Chapter 4 (Section 4.8), using stoichiometric mole relationships. We introduced the **gravimetric factor** (GF), which represents the weight of analyte per unit weight of precipitate. It is obtained from the ratio of the formula weight of the analyte to that of the precipitate, multiplied by the moles of analyte per mole of precipitate obtained from each mole of analyte, that is,

$$\boxed{\begin{aligned} GF &= \frac{\text{fw analyte (g/mol)}}{\text{fw precipitate (g/mol)}} \times \frac{a}{b}\text{(mol analyte/mol precipitate)} \\ &= \text{g analyte/g precipitate} \end{aligned}} \tag{9.1}$$

So, if Cl_2 in a sample is converted to chloride and precipitated as AgCl, the weight of Cl_2 that gives 1 g of AgCl is

Grams precipitate × GF gives grams analyte.

$$\begin{aligned} g\,Cl_2 &= g\,AgCl \times \frac{\text{fw}\,Cl_2\,(g\,Cl_2/\text{mol}\,Cl_2)}{\text{fw}\,AgCl\,(g\,AgCl/\text{mol}\,AgCl)} \times \frac{1}{2}(\text{mol}\,Cl_2/\text{mol}\,AgCl) \\ &= g\,AgCl \times GF\,(g\,Cl_2/g\,AgCl) \\ &= g\,AgCl \times 0.2473_7\,(g\,Cl_2/g\,AgCl) \end{aligned}$$

Example 9.1

Calculate the grams of analyte per gram of precipitate for the following conversions:

Analyte	Precipitate
P	Ag_3PO_4
K_2HPO_4	Ag_3PO_4
Bi_2S_3	$BaSO_4$

Solution

$$g\,P/g\,Ag_3PO_4 = \frac{\text{at wt P (g/mol)}}{\text{fw}\,Ag_3PO_4\,(g/mol)} = \frac{1}{1}\,(\text{mol P/mol}\,Ag_3PO_4)$$

$$GF = \frac{30.97\,(g\,P/\text{mol})}{418.58\,(g\,Ag_3PO_4/\text{mol})} \times \frac{1}{1} = 0.07399\,g\,P/g\,Ag_3PO_4$$

$$\text{g } K_2HPO_4/\text{g } Ag_3PO_4 = \frac{\text{fw } K_2HPO_4 \text{ (g/mol)}}{\text{fw } Ag_3PO_4 \text{ (g/mol)}} \times \frac{1}{1} \text{ (mol } K_2HPO_4/\text{mol } Ag_3PO_4)$$

$$GF = \frac{174.18 \text{ (g } K_2HPO_4/\text{mol)}}{418.58 \text{ (g } Ag_3PO_4/\text{mol)}} \times \frac{1}{1}$$

$$= 0.41612 \text{ g } K_2HPO_4/\text{g } Ag_3PO_4$$

$$\text{g } Bi_2S_3/\text{g } BaSO_4 = \frac{\text{fw } Bi_2S_3 \text{ (g/mol)}}{\text{fw } BaSO_4 \text{ (g/mol)}} \times \frac{1}{3} \text{ (mol } Bi_2S_3/\text{mol } BaSO_4)$$

$$GF = \frac{514.15 \text{ (g } Bi_2S_3/\text{mol)}}{233.40 \text{ (g } BaSO_4/\text{mol)}} \times \frac{1}{3} = 0.73429 \text{ g } Bi_2S_3/\text{g } BaSO_4$$

In gravimetric analysis, we are generally interested in the percent composition by weight of the analyte in the sample, that is,

$$\% \text{ substance sought} = \frac{\text{weight of substance sought (g)}}{\text{weight of sample (g)}} \times 100\% \qquad (9.2)$$

We obtain the weight of substance sought from the weight of the precipitate and the corresponding weight/mole relationship (Equation 9.1):

$$\boxed{\begin{array}{l} \text{Weight of substance sought (g)} = \text{weight of precipitate (g)} \\[2mm] \qquad\qquad \times \dfrac{\text{fw substance sought (g/mol)}}{\text{fw precipitate (g/mol)}} \\[2mm] \qquad\qquad \times \dfrac{a}{b} \text{ (mol substance sought/mol precipitate)} \\[2mm] \qquad = \text{weight of precipitate (g)} \\[2mm] \qquad\qquad \times GF \text{ (g sought/g precipitate)} \end{array}}$$

$$(9.3)$$

Calculations are usually made on a percentage basis:

$$\boxed{\% \text{ A} = \frac{g_A}{g_{sample}} \times 100\%} \qquad (9.4)$$

where g_A represents the grams of analyte (the desired test substance) and g_{sample} represents the grams of sample taken for analysis.

We can write a general formula for calculating the percentage composition of the substance sought:

$$\boxed{\% \text{ sought} = \frac{\text{weight of precipitate (g)} \times GF \text{ (g sought/g precipitate)}}{\text{weight of sample (g)}} \times 100\%} \qquad (9.5)$$

Check units!

Example 9.2

Orthophosphate ($PO_4{}^{3-}$) is determined by weighing as ammonium phosphomolybdate, $(NH_4)PO_4 \cdot 12MoO_3$. Calculate the percent P in the sample and the percent P_2O_5 if 1.1682 g precipitate (ppt) were obtained from a 0.2711-g sample. Perform the % P calculation using the gravimetric factor and just using dimensional analysis.

Solution

$$\%P = \frac{1.1682 \text{ g ppt} \times \frac{P}{(NH_4)_3PO_4 \cdot 12MoO_3} \text{ (g P/g ppt)}}{0.2711 \text{ g sample}} \times 100\%$$

$$= \frac{1.1682 \text{ g} \times (30.97/1876.5)}{0.2711 \text{ g}} \times 100\% = 7.111\%$$

$$\%P_2O_5 = \frac{1.1682 \text{ g ppt} \times \frac{P_2O_5}{2(NH_4)_3PO_4 \cdot 12MoO_3} \text{(g P}_2O_5\text{/g ppt)}}{0.2711 \text{ g sample}} \times 100\%$$

$$= \frac{1.1682 \text{ g} \times [141.95/(2 \times 1876.5)]}{0.2711 \text{ g}} \times 100\%$$

$$= 16.30\%$$

Let's do the same calculation using dimensional analysis for the %P setup.

$$\%P = \frac{1.982 \text{ g } \cancel{(NH_4)_2PO_4 \cdot 12MoO_3} \times (30.97/1867.5)\text{g P/g } \cancel{(NH_4)_2PO_4 \cdot 12MoO_3}}{0.2771 \text{ g sample}}$$

$$\times 100\%$$
$$= (7.111 \text{ g P/g sample}) \times 100\% = 7.111\%P$$

Note that the $(NH_4)_2PO_4 \cdot 12MoO_3$ species cancel one another (dimensional analysis), leaving only g P in the numerator.

When we compare this approach with the gravimetric factor calculation, we see that the setups are really identical. However, this approach better shows which units cancel and which remain.

Example 9.3

An ore is analyzed for the manganese content by converting the manganese to Mn_3O_4 and weighing it. If a 1.52-g sample yields Mn_3O_4 weighing 0.126 g, what would be the percent Mn_2O_3 in the sample? The percent Mn?

Solution

$$\%Mn_2O_3 = \frac{0.126 \text{ g } Mn_3O_4 \times \frac{3Mn_2O_3}{2Mn_3O_4}(\text{g } Mn_2O_3\text{/g } Mn_3O_4)}{1.52 \text{ g sample}} \times 100\%$$

$$= \frac{0.126 \text{ g} \times [3(157.9)/2(228.8)]}{1.52 \text{ g}} \times 100\% = 8.58\%$$

$$\%Mn = \frac{0.126 \text{ g } Mn_3O_4 \times \frac{3Mn}{Mn_3O_4}(\text{g Mn/g } Mn_3O_4)}{1.52 \text{ g sample}} \times 100\%$$

$$= \frac{0.126 \text{ g} \times [3(54.94)/228.8]}{1.52 \text{ g}} \times 100\% = 5.97\%$$

The following two examples illustrate some special additional capabilities of gravimetric computations.

Example 9.4

What weight of pyrite ore (impure FeS_2) must be taken for analysis so that the $BaSO_4$ precipitate weight obtained will be equal to one-half that of the percent S in the sample?

Solution

If we have A% of S, then we will obtain $\frac{1}{2}$ A g of $BaSO_4$. Therefore,

$$A\%S = \frac{\frac{1}{2}A(g\ BaSO_4) \times \frac{S}{BaSO_4}(g\ S/g\ BaSO_4)}{g\ sample} \times 100\%$$

or

$$1\%\ S = \frac{\frac{1}{2} \times \frac{32.064}{233.40}}{g\ sample} \times 100\%$$

$$g\ sample = 6.869\ g$$

Precipitate Mixtures—We Need Two Weights

Example 9.5

A mixture containing only $FeCl_3$ and $AlCl_3$ weighs 5.95 g. The chlorides are converted to the hydrous oxides and ignited to Fe_2O_3 and Al_2O_3. The oxide mixture weighs 2.62 g. Calculate the percent Fe and Al in the original mixture.

Solution

There are two unknowns, so two simultaneous equations must be set up and solved. Let x = g Fe and y = g Al. Then, for the first equation,

$$g\ FeCl_3 + g\ AlCl_3 = 5.95\ g \tag{1}$$

$$x\left(\frac{FeCl_3}{Fe}\right) + y\left(\frac{AlCl_3}{Al}\right) = 5.95\ g \tag{2}$$

$$x\left(\frac{162.21}{55.85}\right) + y\left(\frac{133.34}{26.98}\right) = 5.95\ g \tag{3}$$

$$2.90x + 4.94y = 5.95\ g \tag{4}$$

$$g\ Fe_2O_3 + g\ Al_2O_3 = 2.62\ g \tag{5}$$

$$x\left(\frac{Fe_2O_3}{2Fe}\right) + y\left(\frac{Al_2O_3}{2Al}\right) = 2.62\ g \tag{6}$$

$$x\left(\frac{159.69}{2 \times 55.85}\right) + y\left(\frac{101.96}{2 \times 26.98}\right) = 2.62\ g \tag{7}$$

$$1.43x + 1.89y = 2.62\ g \tag{8}$$

See the **website** supplement to look up Example 9.5 Solving Simultaneous Equations that solves two-variable simultaneous equations.

Solving (4) and (8) simultaneously for x and y:

$$x = 1.07\,g$$

$$y = 0.58\,g$$

$$\%Fe = \frac{1.07\,g}{5.95\,g} \times 100\% = 18.0\%$$

$$\%Al = \frac{0.58\,g}{5.95\,g} \times 100\% = 9.8\%$$

9.3 Examples of Gravimetric Analysis

Some of the most precise and accurate analyses are based on gravimetry. There are many examples, and you should be familiar with some of the more common ones. These are summarized in Table 9.1, which lists the substance sought, the precipitate formed, the form in which it is weighed, and the common elements that will interfere and must be absent. We do not present more details because gravimetry is not currently used often (unless the much longer time and increased labor requirements can be justified by the necessity of high precision and accuracy). You should consult more advanced texts and comprehensive analytical reference books for details on these and other determinations.

9.4 Organic Precipitates

All the precipitating agents we have talked about so far, except for oxine, cupferrate, and dimethylglyoxime (Table 9.1), have been inorganic in nature. There are also a large number of organic compounds that are very useful precipitating agents for metals. Some of these are very selective, and others are very broad in the number of elements they will precipitate.

TABLE 9.1 Some Commonly Employed Gravimetric Analyses

Substance Analyzed	Precipitate Formed	Precipitate Weighed	Interferences
Fe	$Fe(OH)_3$	Fe_2O_3	Many. Al, Ti, Cr, etc.
	Fe cupferrate	Fe_2O_3	Tetravalent metals
Al	$Al(OH)_3$	Al_2O_3	Many. Fe, Ti, Cr, etc.
	$Al(ox)_3$ [a]	$Al(ox)_3$	Many. Mg does not interfere in acidic solution
Ca	CaC_2O_4	$CaCO_3$ or CaO	All metals except alkalis and Mg
Mg	$MgNH_4PO_4$	$Mg_2P_2O_7$	All metals except alkalis
Zn	$ZnNH_4PO_4$	$Zn_2P_2O_7$	All metals except Mg
Ba	$BaCrO_4$	$BaCrO_4$	Pb
SO_4^{2-}	$BaSO_4$	$BaSO_4$	NO_3^-, PO_4^{3-}, ClO_3^-
Cl^-	AgCl	AgCl	Br^-, I^-, SCN^-, CN^-, S^{2-}, $S_2O_3^{2-}$
Ag	AgCl	AgCl	Hg(I)
PO_4^{3-}	$MgNH_4PO_4$	$Mg_2P_2O_7$	MoO_4^{2-}, $C_2O_4^{2-}$, K^+
Ni	$Ni(dmg)_2$ [b]	$Ni(dmg)_2$	Pd

[a] ox = Oxine (8-hydroxyquinoline) monoanion.
[b] dmg = Dimethylglyoxime monoanion.

Chelates are described in Chapter 8.

Organic precipitating agents have the advantages of giving precipitates with very low solubility in water and a favorable gravimetric factor. Most of them are **chelating agents** that form slightly soluble, uncharged **chelates** with the metal ions. A chelating agent is a type of complexing agent that has two or more groups capable of complexing with a metal ion. The complex formed is called a chelate. See Chapter 8 for a more thorough discussion of chelates.

Since chelating agents are weak acids, the number of elements precipitated, and thus the selectivity, can usually be regulated by adjustment of the pH. The reactions can be generalized as (the underline indicates what is precipitated):

$$M^{n+} + nHX \rightleftharpoons \underline{MX_n} + nH^+$$

Metal chelate precipitates (which give selectivity) are sometimes ignited to metal oxides for improved stoichiometry.

There may be more than one ionizable proton on the organic reagent. The weaker the metal chelate, the higher the pH needed to achieve precipitation. Some of the commonly used organic precipitants are listed in Table 9.2. Some of these precipitates are not stoichiometric, and more accurate results are obtained by igniting them to form the metal oxides. Some, such as sodium diethyldithiocarbamate, can be used to perform group separations, as is done with hydrogen sulfide. You should consult specialized reference texts at the end of the chapter for applications of these and other organic precipitating reagents. The multivolume treatise by Hollingshead on the uses of oxine and its derivatives is very helpful for applications of this versatile reagent. (See Reference 4 at the end of the chapter.)

TABLE 9.2 Some Organic Precipitating Agents

Reagent	Structure	Metals Precipitated
Dimethylglyoxime	$CH_3 - C = NOH$ $\vert$ $CH_3 - C = NOH$	Ni(II) in NH_3 or buffered HOAc; Pd(II) in HCl $(M^{2+} + 2HR \rightarrow \underline{MR_2} + 2H^+)$
α-Benzoinozime (cupron)	OH NOH CH—C	Cu(II) in NH_3 and tartrate; Mo(VI) and W(VI) in H^+ $(M^{2+} + H_2R \rightarrow \underline{MR} + 2H^+;\ M^{2+} = Cu^{2+},$ $MoO_2^+, WO_2^{2+})$ Metal oxide weighed
Ammonium nitrosophenylhydroxylamine (cupferron)	N=O $\vert$ N—O—NH_4	Fe(III), V(V), Ti(IV), Zr(IV), Sn(IV), U(IV) $(M^{n+} + nNH_4R \rightarrow \underline{MR_n} + nNH_4^+)$ Metal oxide weighed
8-Hydroxyquinoline (oxine)	OH N	Many metals. Useful for Al(III) and Mg(II) $(M^{n+} + nHR \rightarrow \underline{MR_n} + nH^+)$
Sodium diethyldithiocarbamate	S $\parallel$ $(C_2H_5)_2N - C - S^-Na^+$	Many metals from acid solution $(M^{n+} + nNaR \rightarrow \underline{MR_n} + nNa^+)$
Sodium tetraphenylboron	$NaB(C_6H_5)_4$	$K^+, Rb^+, Cs^+, Tl^+, Ag^+, Hg(I), Cu(I), NH_4^+, RNH_3^+,$ $R_2NH_2^+, R_3NH^+, R_4N^+.$ Acidic solution $(M^+ + NaR \rightarrow \underline{MR} + Na^+)$
Tetraphenylarsonium chloride	$(C_6H_5)_4AsCl$	$Cr_2O_7^{2-}, MnO_4^-, ReO_4^-, MoO_4^{2-}, WO_4^{2-}\ ClO_4^-, I_3^-.$ Acidic solution $(A^{n-} + nRCl \rightarrow R_nA + nCl^-)$

9.5 Precipitation Equilibria: The Solubility Product

When substances have limited solubility and their solubility is exceeded, the ions of the dissolved portion exist in equilibrium with the solid material. So-called insoluble compounds generally exhibit this property.

When a compound is referred to as insoluble, it is actually not completely insoluble but is **slightly soluble**. For example, if solid AgCl is added to water, a small portion of it will dissolve:

$$\underline{AgCl} \rightleftharpoons (AgCl)_{aq} \rightleftharpoons Ag^+ + Cl^- \tag{9.6}$$

"Insoluble" substances still have slight solubility.

The precipitate will have a definite solubility (i.e., a definite amount that will dissolve) in g/L, or mol/L, at a given temperature (a saturated solution). A small amount of undissociated compound usually exists in equilibrium in the aqueous phase (e.g., on the order of 0.1%, although usually less for the precipitations employed for analysis, and depending on K_{sp}), and its concentration is constant. It is difficult to measure the undissociated species, and we are interested in the ionized form as a measure of a compound's solubility and chemical availability. Hence, we can generally neglect the presence of any undissociated species.

We can write an overall equilibrium constant for the above stepwise equilibrium, called the **solubility product** K_{sp}. $(AgCl)_{aq}$ cancels when the two stepwise equilibrium constants are multiplied together.

The solid does not appear in K_{sp}.

$$K_{sp} = [Ag^+][Cl^-] \tag{9.7}$$

The "concentration" of any solid such as AgCl is constant and is combined in the equilibrium constant to give K_{sp}. The above relationship holds regardless of the presence of any undissociated intermediate; that is, the concentrations of free ions are rigorously defined by Equation 9.7, and we will take these as a measure of a compound's solubility. From a knowledge of the value of the solubility product at a specified temperature, we can calculate the equilibrium solubility of the compounds. (The solubility product is determined in the reverse order, by measuring the solubility.)

The amount of a slightly soluble salt that dissolves does *not* depend on the amount of the solid in equilibrium with the solution, so long as some solid is present. Instead, the amount that dissolves depends on the *volume* of the solvent. A nonsymmetric salt (one in which the cation and anion are not in the same ratio) such as Ag_2CrO_4 would have a K_{sp} as follows:

The concentration of solute in a saturated solution is the same whether the solution fills a beaker or a swimming pool, so long as there is solid in equilibrium with it. But much more solid will dissolve in the swimming pool! The concentration in a saturated solution is also independent of how much undissolved solid is present.

$$Ag_2CrO_4 \rightleftharpoons 2\,Ag^+ + CrO_4^{2-} \tag{9.8}$$

$$K_{sp} = [Ag^+]^2[CrO_4^{2-}] \tag{9.9}$$

Such electrolytes do not dissolve or dissociate in steps because they are really strong electrolytes. That portion that dissolves ionizes completely. *Therefore, we do not have stepwise K_{sp} values.* As with any equilibrium constant, the K_{sp} product holds under all equilibrium conditions at the specified temperature. Since we are dealing with heterogeneous equilibria, the equilibrium state is achieved more slowly than with homogeneous solution equilibria.

The Saturated Solution

Example 9.6

The K_{sp} of AgCl at 25°C is 1.0×10^{-10}. Calculate the concentrations of Ag^+ and Cl^- in a saturated solution of AgCl, and the molar solubility of AgCl.

Solution

When AgCl ionizes, equal amounts of Ag^+ and Cl^- are formed; $AgCl \rightleftharpoons Ag^+ + Cl^-$ and $K_{sp} = [Ag^+][Cl^-]$. Let s represent the molar solubility of AgCl. Since each mole of AgCl that dissolves gives one mole of either Ag^+ or Cl^-, then

$$[Ag^+] = [Cl^-] = s$$
$$s^2 = 1.0 \times 10^{-10}$$
$$s = 1.0 \times 10^{-5} \, M$$

The solubility of AgCl is $1.0 \times 10^{-5} \, M$.

Decreasing the Solubility—The Common Ion Effect

If there is an excess of one ion over the other, the concentration of the other is suppressed (**common ion effect**), and the solubility of the precipitate is decreased. We can still calculate the concentration from the solubility product.

Example 9.7

Adding a common ion decreases the solubility.

Ten milliliters of $0.20 \, M$ AgNO$_3$ is added to 10 mL of $0.10 \, M$ NaCl. Calculate the concentration of Cl^- remaining in solution at equilibrium, and the solubility of the AgCl.

Solution

The final volume is 20 mL. The millimoles Ag^+ added equals $0.20 \times 10 = 2.0$ mmol. The millimoles Cl^- taken equals $0.10 \times 10 = 1.0$ mmol. Therefore, the millimoles excess Ag^+ equals $(2.0 - 1.0) = 1.0$ mmol. From Example 9.6, we see that the Ag^+ concentration contributed from the precipitate is small, that is, on the order of 10^{-5} mmol/mL in the absence of a common ion. This will be even smaller in the presence of excess Ag^+ since the solubility is suppressed. Therefore, we can neglect the amount of Ag^+ contributed from the precipitate compared to the excess Ag^+. Hence, the final concentration of Ag^+ is 1.0 mmol/20 mL = $0.050 \, M$, and

$$(0.050)\,[Cl^-] = 1.0 \times 10^{-10}$$
$$[Cl^-] = 2.0 \times 10^{-9} \, M$$

The Cl^- concentration again equals the solubility of the AgCl, and so the solubility is $2.0 \times 10^{-9} \, M$.

Because the K_{sp} product always holds, *precipitation will not take place unless the product of $[Ag^+]$ and $[Cl^-]$ exceeds the K_{sp}*. If the product is just equal to K_{sp}, all the Ag^+ and Cl^- remains in solution.

The solubility product must be exceeded for precipitation to occur.

Solubility Depends on the Stoichiometry

Table 9.3 lists some solubility products along with the corresponding calculated molar solubilities for some slightly soluble salts. The molar solubility is not necessarily directly proportional to the K_{sp} value since it depends on the stoichiometry of the salt. The K_{sp} of AgI is 5×10^{15} larger than that of $Al(OH)_3$, but its molar solubility is only twice that of $Al(OH)_3$. That is, a 1:1 salt has a lower solubility than a nonsymmetric salt for a given K_{sp}. Note that HgS has a solubility product of only 4×10^{-53}, with a molar solubility of 6×10^{-27} M! This corresponds to less than one ion each of Hg^{2+} and S^{2-} in a liter in equilibrium with the precipitate, and it would take some 280 L for two ions to exist together (can you calculate this using Avogadro's number?). So it is like two ions finding each other in a good size bathtub! (Actually, they find the precipitate.) A more complete list of solubility products appears in Appendix C.

Example 9.8

What must be the concentration of added Ag^+ to just start precipitation of AgCl in a 1.0×10^{-3} M solution of NaCl?

Solution

$$[Ag^+](1.0 \times 10^{-3}) = 1.0 \times 10^{-10}$$

$$[Ag^+] = 1.0 \times 10^{-7} M$$

The concentration of Ag^+ must, therefore, just exceed 10^{-7} M to begin precipitation. Caveat: As we have observed before, in reality supersaturation is needed before precipitation begins. In practice it is unlikely that precipitation will begin when Ag^+ just exceeds 10^{-7} M.

TABLE 9.3 Solubility Product Constants of Selected Slightly Soluble Salts

Salt	K_{sp}	Solubility, s (mol/L)
$PbSO_4$	1.6×10^{-8}	1.3×10^{-4}
AgCl	1.0×10^{-6}	1.0×10^{-5}
AgBr	4×10^{-13}	6×10^{-7}
AgI	1×10^{-16}	1×10^{-8}
$Al(OH)_3$	2×10^{-32}	5×10^{-9}
$Fe(OH)_3$	4×10^{-38}	2×10^{-10}
Ag_2S	2×10^{-49}	4×10^{-17}
HgS	4×10^{-53}	6×10^{-27}

Example 9.9

What is the solubility of PbI_2, in g/L, if the solubility product is 7.1×10^{-9}?

Solution

The equilibrium is $PbI_2 \rightleftharpoons Pb^{2+} + 2I^-$, and $K_{sp} = [Pb^{2+}] [I^-]^2 = 7.1 \times 10^{-9}$. Let s represent the molar solubility of PbI_2. Then

$$[Pb^{2+}] = s \quad \text{and} \quad [I^-] = 2s$$

$$(s) (2s)^2 = 7.1 \times 10^{-9}$$

$$s = \sqrt[3]{\frac{7.1 \times 10^{-9}}{4}} = 1.2 \times 10^{-3} \; M$$

Therefore, the solubility, in g/L, is

$$1.2 \times 10^{-3} \; \text{mol/L} \times 461.0 \; \text{g/mol} = 0.55 \; \text{g/L}$$

Note that the concentration of I^- was *not* doubled before squaring; $2s$ represented its actual equilibrium concentration, not twice its concentration. We could have let s represent the concentration of I^-, instead of the molar solubility of PbI_2, in which case $[Pb^{2+}]$ and the solubility of PbI_2 would have been $\frac{1}{2}s$. The calculated s would have been twice as great, but the concentrations of each species would have been the same. You try this calculation!

Example 9.10

A smaller K_{sp} with a nonsymmetrical precipitate does not necessarily mean a smaller solubility compared to a symmetrical one.

Calculate the molar solubility of $PbSO_4$ and compare it with that of PbI_2.

Solution

$$PbSO_4 \rightleftharpoons Pb^{2+} + SO_4{}^{2-}$$

$$[Pb^{2+}][SO_4{}^{2-}] = 1.6 \times 10^{-8}$$

$$(s)(s) = 1.6 \times 10^{-8}$$

$$s = 1.3 \times 10^{-4} \; M$$

Although the K_{sp} of PbI_2 (7.1×10^{-9}) is smaller than that of $PbSO_4$ (1.6×10^{-8}), the solubility of PbI_2 is greater (see Example 9.9), due to the nonsymmetrical nature of the precipitate.

For electrolytes of the same valence type, the order of solubility will be the same as the order of the corresponding solubility products. But when we compare salts of different valence type, the order may be different. Compound AB will have a smaller molar solubility than compound AC_2 when both have identical K_{sp} values.

We take advantage of the common ion effect to decrease the solubility of a precipitate in gravimetric analysis. For example, sulfate ion is determined by precipitating $BaSO_4$ with added barium chloride solution. Figure 9.3 illustrates the effect of excess barium ion on the solubility of $BaSO_4$.

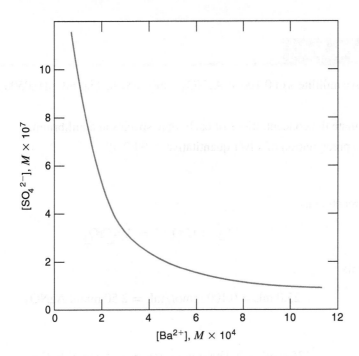

FIGURE 9.3 Predicted effect of excess barium ion on solubility of $BaSO_4$. Sulfate concentration is amount in equilibrium and is equal to $BaSO_4$ solubility. In absence of excess barium ion, solubility is 10^{-5} M.

Example 9.11

What pH is required to just precipitate iron(III) hydroxide from a 0.10 M $FeCl_3$ solution?

$Fe(OH)_3$ actually precipitates in acid solution due to the small K_{sp}!

Solution

$$Fe(OH)_3 \rightleftharpoons Fe^{3+} + 3OH^-$$

$$[Fe^{3+}][OH^-]^3 = 4 \times 10^{-38}$$

$$(0.1)[OH^-]^3 = 4 \times 10^{-38}$$

$$[OH^-] = \sqrt[3]{\frac{4 \times 10^{-38}}{0.1}} = 7 \times 10^{-13}\ M$$

$$pOH = -\log 7 \times 10^{-13} = 12.2$$

$$pH = 14.0 - 12.2 = 1.8$$

Hence, we see that iron(III) hydroxide precipitates in acid solution, when the pH just exceeds 1.8! When you prepare a solution of $FeCl_3$ in water, it will slowly hydrolyze to form iron(III) hydroxide (hydrous ferric oxide), a rust-colored gelatinous precipitate. To stabilize the iron(III) solution, you must acidify the solution with, for example, hydrochloric acid.

Again, note that precipitation generally will not begin exactly at the calculated pH, as supersaturation is needed.

Example 9.12

Twenty-five milliliters of 0.100 M AgNO$_3$ is mixed with 35.0 mL of 0.0500 M K$_2$CrO$_4$ solution.

(a) Calculate the concentrations of each ionic species at equilibrium.

(b) Is the precipitation of silver quantitative .> 99:9~/?

Solution

(a) The reaction is

$$2Ag^+ + CrO_4^{2-} \rightleftharpoons \underline{Ag_2CrO_4}$$

We mix

$$25.0 \text{ mL} = 0.100 \text{ mmol/mL} = 2.50 \text{ mmol AgNO}_3$$

and

$$35.0 \text{ mL} \times 0.0500 \text{ mmol/mL} = 1.75 \text{ mmol K}_2\text{CrO}_4$$

Hence, 1.25 mmol of CrO$_4^{2-}$ will react with 2.50 mmol Ag$^+$, leaving an excess of 0.50 mmol CrO$_4^{2-}$. The final volume is 60.0 mL. If we let s be the molar solubility of Ag$_2$CrO$_4$, then at equilibrium:

$$[CrO_4^{2-}] = 0.50 \text{ mmol/60.0 mL} + s = 0.0083 + s \approx 0.0083 \ M$$

s will be very small due to the excess CrO$_4^{2-}$ and may be neglected compared to 0.0083.

$$[Ag^+] = 2s$$
$$[K^+] = 3.50 \text{ mmol/60.0 mL} = 0.0583 \ M$$
$$[NO_3^-] = 2.50 \text{ mmol/60.0 mL} = 0.0417 \ M$$
$$[Ag^+]^2 [CrO_4^{2-}] = 1.1 \times 10^{-12}$$
$$(2s)^2 (8.3 \times 10^{-3}) = 1.1 \times 10^{-12}$$

$$s = \sqrt{\frac{1.1 \times 10^{-12}}{4 \times 8.3 \times 10^{-3}}} = 5.8 \times 10^{-6} \ M$$

$$[Ag^+] = 2(5.8 \times 10^{-6}) = 1.1_6 \times 10^{-5} \ M$$

(b) The percentage of silver precipitated is

$$\frac{2.50 \text{ mmol} - 60.0 \text{ mL} \times 1.1_6 \times 10^{-5} \text{ mmol/mL}}{2.50 \text{ mmol}} \times 100\% = 99.97\%$$

Or the percent remaining in solution is

$$\frac{60.0 \text{ mL} \times 1.1_6 \times 10^{-5} \text{ mmol/L}}{2.50 \text{ mmol}} \times 100\% = 0.028\%$$

Hence, the precipitation is quantitative.

9.6 Diverse Ion Effect on Solubility: K_{sp} and Activity Coefficients

In Chapter 5 we defined the thermodynamic equilibrium constant written in terms of activities to account for the effects of inert electrolytes on equilibria. The presence of diverse salts will generally increase the solubility of precipitates due to the shielding of the dissociated ion species. (Their activity is decreased.) Consider the solubility of AgCl. The thermodynamic solubility product K_{sp} is

$$K_{sp} = a_{Ag^+} \cdot a_{Cl^-} = [Ag^+]f_{Ag^+}[Cl^-]f_{Cl^-} \tag{9.10}$$

Since the *concentration* solubility product $^cK_{sp}$ is $[Ag^+][Cl^-]$, then

$$K_{sp} = {}^cK_{sp}f_{Ag^+}f_{Cl^-} \tag{9.11}$$

or

$$^cK_{sp} = \frac{K_{sp}}{f_{Ag^+}f_{Cl^-}} \tag{9.12}$$

K_{sp} holds at all ionic strengths. $^cK_{sp}$ must be corrected for ionic strength.

The numerical value of K_{sp} holds at all activities. $^cK_{sp}$ equals K_{sp} at zero ionic strength, but at appreciable ionic strengths, a value must be calculated for each ionic strength using Equation 9.12. Note that this equation shows, as we predicted qualitatively, that decreased activity of the ions will result in an increased $^cK_{sp}$ and, therefore, an increase in molar solubility.

Example 9.13

Calculate the solubility of silver chloride in 0.10 M NaNO$_3$.

Solution

The equilibrium constants listed in the Appendix C are for zero ionic strength; that is, they are really thermodynamic equilibrium constants.[1] Therefore, from Table C.3, $K_{sp} = 1.0 \times 10^{-10}$.

We need the activity coefficients of Ag$^+$ and Cl$^-$. The ionic strength is 0.10. From Reference 10 in Chapter 5, we find that $f_{Ag^+} = 0.75$ and $f_{Cl^-} = 0.76$. (You could also have used the values of α_{Ag^+} and α_{Cl^-} in the reference to calculate the activity coefficients using Equation 5.19.) From Equation 9.12

Diverse salts increase the solubility of precipitates and have more effect on precipitates with multiply charged ions.

$$K_{sp} = \frac{1.0 \times 10^{-10}}{(0.75)(0.76)} = 1.8 \times 10^{-10} = [Ag^+][Cl^-] = s^2$$

$$s = \sqrt{1.8 \times 10^{-10}} = 1.3 \times 10^{-5}M$$

This is 30% greater than at zero ionic strength ($s = 1.0 \times 10^{-5}$ M).

---

[1] Experimental K_{sp} values are often available at different ionic strengths and can be used to calculate molar solubilities at the listed ionic strengths without needing to calculate activity coefficients.

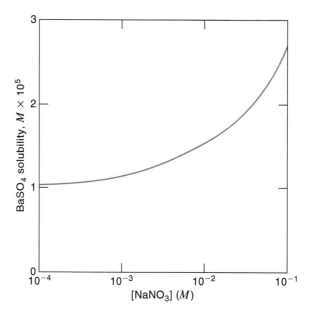

FIGURE 9.4 Predicted effect of increased ionic strength on solubility of $BaSO_4$. Solubility at zero ionic strength is 1.0×10^{-5} M.

Figure 9.4 illustrates the increase in solubility of $BaSO_4$ in the presence of $NaNO_3$ due to the diverse ion effect.

The increase in solubility is greater with precipitates containing multiply charged ions. At very high ionic strengths, where activity coefficients may become greater than unity, the solubility is decreased. In gravimetric analysis, a sufficiently large excess of precipitating agent is added so that the solubility is reduced to such a small value that we do not need to worry about this effect.

Acids frequently affect the solubility of a precipitate. As the H^+ concentration increases, it competes more effectively with the metal ion of interest for the precipitating agent (which may be the anion of a weak acid). With less free reagent available, and a constant K_{sp}, the solubility of the salt must increase:

$$M^{n+} + nR^- \rightleftharpoons \underline{MR_n} \qquad \text{(desired reaction)}$$

$$R^- + H^+ \rightleftharpoons HR \qquad \text{(competing reaction)}$$

$$\underline{MR_n} + nH^+ \rightleftharpoons M^{n+} + nHR \quad \text{(overall reaction)}$$

Similarly, a complexing agent that reacts with the metal ion of the precipitate will increase the solubility, for example, when ammonia reacts with silver chloride:

$$AgCl + 2NH_3 \rightleftharpoons Ag(NH_3)_2^+ + Cl^-$$

The quantitative treatment of these effects in solubility calculations will be covered in Chapter 10.

9.7 Electrogravimetry

Electrogravimetric analysis is similar to conventional gravimetric analysis. In this method, electrolytic deposition is used for the gravimetric determination of metals. In this, electrode is weighted before and after the process of electrolysis. For most of the metals, platinum cathode is used for the deposition, but certain cases require anodic deposition such as, determination of lead as lead dioxide on platinum and determination of chloride as silver chloride on silver.

Types of Electrogravimetric Methods

There are mainly two types of electrogravimetric methods: (i) without potential control and (ii) with potential control.

1. **Electrogravimetry Without Potential Control.** In this method, no external exertions are required to control the potential of working electrode during the electrolytic procedure. During the process of electrolysis, potential applied across the cell is sustained at a relatively constant level.

– *Apparatus.* Figure 9.5 shows apparatus used for electrogravimetric analysis. It consists of a suitable cell which is a direct current source of 6 – 12 V. For the indication of current and voltage, an ammeter and a voltmeter are used, respectively. Resistor is used for controlling the applied voltage. Cylindrical platinum gauge is usually used as cathode.

– *Physical characteristics of metal electrolytically deposited.* Preferably, metal electrolytically deposited must be strongly adherent, smooth and dense, so that without any mechanical loss, it can be washed, dried and weighed. Fine granules with metallic luster are considered as good metallic deposits, whereas spongy, powdery or flaky deposits are considered impure and less adherent.

Major factor swaying the physical characteristics of deposited metal include temperature, current density and the presence of complexing agents. Current densities below 0.1 A/cm^2 usually gives best deposits. Quality of deposits can be upgraded with mild stirring. Deposition of metal obtained from solution of metal complexes provide smoother film than metal deposits obtained using metal salts having simple ions. Best results are obtained in cases of cyanide and ammonia metal complexes.

– Applications. It is used for the separation of easily reduced cations from those that are more difficult to reduce. Consider electrolysis of copper (II) ions in an acidic medium. In its preliminary step, potential is adjusted to −2.5 V by maintaining the resistance, causing the current to be about 1.5 A (Figure 9.6*a*). It leads to copper deposition at this applied potential.

From Figure 9.6*b*, it is clear that as the reaction proceeds, *IR* declines continuously. At the cathode, concentration polarization is the main cause for the *IR* drop. It restricts the frequency of transportation of copper ions to the surface

IR drop is the electrical potential difference between the two ends of a conducting phase during a current flow. This voltage drop across any resistance is the product of current (I) passing through resistance and resistance (R).

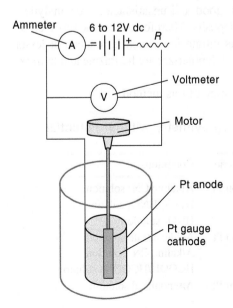

Ammeter   6 to 12V dc
A   R
Voltmeter
Motor
Pt anode
Pt gauge cathode

FIGURE 9.5 Apparatus for electrogravimetric analysis without potential control.

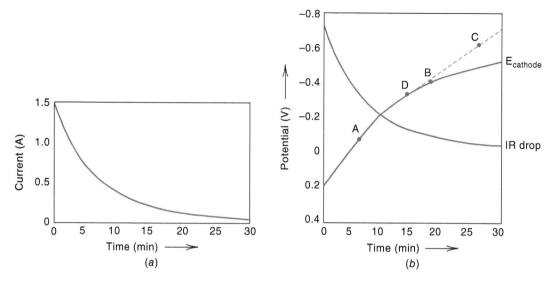

FIGURE 9.6 *(a)* Current, *(b)* IR drop, and cathode potential change during electrolytic deposition of copper at constant applied cell potential.

of the electrode and ultimately restricts the current. Compensation for decreased *IR* can be done by increasing the cathode potential as the applied cell potential remains constant.

At point B, increase in the cathode potential is decelerated due to reduction of the hydrogen ions. Because of the presence of acid in surplus within the solution, current is not restricted by concentration polarization anymore, and simultaneous co-deposition of copper and hydrogen occurs until deposition of all the copper ions is completed. It is said that cathode is depolarized by hydrogen ions under these conditions.

Consider cases, in which solution of copper (II) and lead (II) ions is taken, in that case lead (II) begins to deposit at point A. Hence, lead (II) would co-deposit well before copper deposition was complete and would interfere in copper determination. In contract, metal ion such as cobalt (II) that would deposit at cathode at point C, would not interfere because depolarization by hydrogen ions prevents the cathode from attaining that potential.

Co-deposition of hydrogen is not good and unsatisfactory for analytical purpose as it causes poor adhesion of deposits. Therefore, it can be resolved by addition of another depolarizer such as nitrate ion which is reduced at lesser negative potential. Other commonly used depolarizer are hydrazine and hydroxylamine.

Table 9.4 illustrates typical application of this method.

TABLE 9.4 Some Applications of Electrogravimetry without Potential Control

Analyte	Weighted as	Anode	Cathode	Conditions
Cd^{2+}	Cd	Pt	Cu on Pt	Alkaline CN^- solution
Cu^{2+}	Cu	Pt	Pt	H_2SO_4/HNO_3 solution
Pb^{2+}	PbO_2 (on anode)	Pt	Pt	HNO_3 solution
Zn^{2+}	Zn	Pt	Cu on Pt	Acidic citrate solution
Ag^+	Ag	Pt	Pt	Alkaline CN^- solution
Mn^{2+}	MnO_2 (on anode)	Pt dish	Pt	HCOOH/HCOONa solution
Ni^{2+}	Ni	Pt	Cu on Pt	Ammoniacal solution
Br^-	AgBr (on anode)	Ag	Pt	-

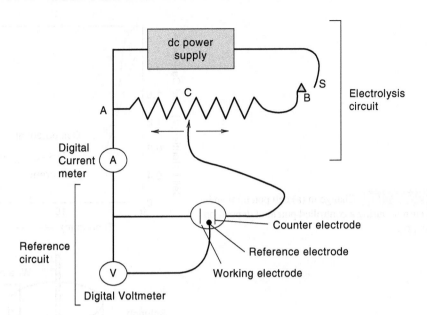

FIGURE 9.7 Apparatus for electrogravimetric analysis with potential control.

2. **Electrogravimetry with Potential Control.** By this method, species with deposition potential having difference of few tenths of volt can be separated.

- *Apparatus.* For this method, three electrode system is used shown in Figure 9.7 instead of two electrode system. This apparatus consists of two independent circuits that share a common electrode that is called as the working electrode, when deposition is done. Counter circuit consists of a DC source (12 V), a potential divider for potential variation in controlled manner, a counter electrode and an ammeter. Reference circuit have reference electrode (SCE), high resistance digital voltmeter and a working electrode. Due to large resistance of reference circuit, electrolysis circuit supplies all current for deposition. The current in reference electrode circuit is essentially zero all times. Reference circuit monitors the potential between working electrode and reference electrode.

    Figure 9.8 shows cell potential and current changes in constant potential analysis. Potential across working and counter electrode has to be decreased continuously. As the potential of counter electrode remains constant during the change, the cathode potential becomes smaller.

- *Electrodes.* Usually non-reactive platinum gauge electrodes are used. But for metals such as bismuth and zinc that cause permanent damage to platinum electrode, protective copper coating is done on platinum electrode for these metals.

    Removal of easily reducing metals is the preliminary step in the analysis. Mercury cathode is used as, shown in Figure 9.9. For example, copper, nickel, cobalt, silver and cadmium are readily separated from ions, such as aluminum, titanium, alkali metals, sulphates and phosphates. With little hydrogen evolution, precipitated elements dissolve in mercury as it has a high hydrogen over voltage.

- *Application.* It helps in successive deposition of metals from a mixture of copper, bismuth, lead, cadmium, zinc and tin. First at cathode potential, −0.2 V vs SCE copper is deposited, then at −0.4 V bismuth is deposited at copper plotted electrode. Then at −0.6 V lead is deposited over the same electrode. Thus, three elements are deposited from a neutral solution having tartrate ions that form a complex with tin (IV) to prevent its deposition. After lead deposition,

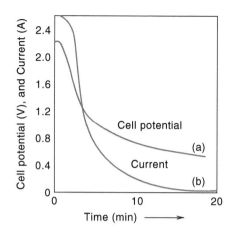

FIGURE. 9.8 Change in (a) cell potential and (b) current during a controlled potential deposition of copper.

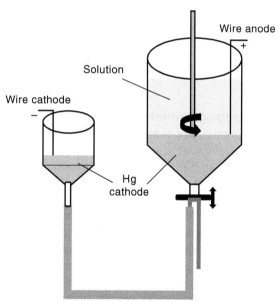

FIGURE. 9.9 A mercury cathode for electrolytical removed of metal ions from solution.

solution is made strongly ammonical and successively cadmium and zinc are deposited at −1.2 V and −1.5 V, respectively. Finally, tin-tartrate complex is redissolved under acidic condition and tin is deposited at potential of 0.65 V using fresh cathode.

Table 9.5 shows application of electrogravimetry under controlled potential.

TABLE 9.5 Some Applications of Electrogravimetry with Potential Control[a]

Metal	Potential vs. SCE	Electrolyte	Other Elements that can be present
Ag	0.10	Acetic acid/acetate buffer	Cu and heavy metals
Cu	−0.30	Tartrate + hydrazine + Cl⁻	Bi, Sb, Pb, Sn, Ni, Cd, Zn
Bi	−0.40	Tartrate + hydrazine + Cl⁻	Pb, Zn, Sb, Cd, Sn
Cd	−0.80	HCl + hydroxylamine	Zn
Ni	−1.10	Ammoniacal tartrate + sodium sulfite	Zn, Al, Fe
Sb	−0.35	HCl + hydrazine at 70°C	Pb, Sn
Sn	−0.60	HCl + hydroxylamine	Cd, Zn, Mn, Fe
Pb	−0.60	Tartrate + hydrazine	Cd, Sn, Ni, Zn, Mn, Al, Fe

[a]Reproduced from J.J. Lingane, *Electroanalytical Chemistry*, 2nd ed., p. 413. New York, Interscience. 1958.

Electrolytically metal is deposited by increasing mass of the electrode.

$$M^{2+} + 2e^- \rightarrow M(s)$$

Therefore,

$$E_{cell} = E_{cathode} - E_{anode}$$

As, per Ohm's law and Faraday's law of electrolysis, current *(I)* is directly proportional to electromotive force *(E)* and inversely proportional to resistance *(R)*.

so, $E = IR$

Therefore, $E_{applied} = E_{cell} - IR$

so,

$$E_{applied} = E_{cathode} - E_{anode} - IR$$

$$I = \frac{(E_{cathode} - E_{anode} - E_{applied})}{R}$$

$$= \frac{(E_{cell} - E_{applied})}{R}$$

Questions

1. Describe the unit operations commonly employed in gravimetric analysis, and briefly indicate the purpose of each.

2. What is the von Weimarn ratio? Define the terms in it. Also state its significance in optimum condition of precipitation.

3. What is peptization?

4. What is digestion of a precipitate, and why is it necessary?

5. Outline the optimum conditions for precipitation that will obtain a pure and filterable precipitate.

6. What is coprecipitation? List the different types of coprecipitation, and indicate how they may be minimized or treated.

7. Why must a filtered precipitate be washed?

8. Why must a wash liquid generally contain an electrolyte? What are the requirements for this electrolyte?

9. What advantages do organic precipitating agents have?

10. Explain the occlusion and the inclusion process.

11. What is isomorphous replacement?

12. Dimethylglyoxime is used as precipitating agents of _____.

    (a) Zn   (b) Mn

    (c) Ni   (d) V

13. Which organic precipitating agent is suitable for Al(III)?

    (a) Cupron   (b) DMG

    (c) Cupferron   (d) Oxine

14. What is electrogravimetric analysis? It is used for which element's gravimetric analysis?

15. Describe briefly different types of electrogravimetric method.

16. For most complete gravimetric analysis of Ca^{2+}, which reagent is best? Given, solubility product for various calcium salts.

Salt	K_{sp}
$CaSO_4$	2.4×10^{-5}
$CaCO_3$	4.5×10^{-9}
$Ca(OH)_2$	6.5×10^{-6}
$CaCl_2$	1.57×10^{-3}

    Reagents:

    (a) $0.05\ M\ Na_2SO_4$ (b) $0.05\ M\ Na_2CO_3$

    (c) $0.10\ M\ NaOH$ (d) $0.1\ M\ NaCl$

17. Which among the following analytical methods have the greatest precision?

    (a) Volumetric titration
    (b) Linear calibration
    (c) Gravimetric
    (d) Spectrophotometric

18. In gravimetric analysis, before complete drying precipitate of barium sulphate was weighed. What is impact of this error on the result?

    (a) A higher result than the correct value.

    (b) A lower result than the correct value.

    (c) Minimal, as mass of water is low.

    (d) Minimal, as the calculation assumes a hydrated sample is produced.

19. For the collection of precipitate on filtration of aqueous solution, which of the following is the best procedure to use?

    (a) Warm the solution before filtration.

    (b) Wash the precipitate with minimal amount of water.

    (c) Do not waste precipitate.

    (d) Large volume of water is required to remove solute matter from the precipitate.

20. On reaction of 20 mL of 0.6 M $MgCl_2$ with 25 mL of 0.5 M NaOH, amount of $Mg(OH)_2$ produced is

    (a) 0.364 g
    (b) 0.700 g
    (c) 0.729 g
    (d) 1.46 g

21. Which of the following properties is not required in a substance known for its use as a precipitate in gravimetric analysis?

    (a) Stable on heating at 110°C
    (b) Low solubility
    (c) Has known formula
    (d) Can be used for extended time without deterioration

22. On mixing $MgCl_2$ and $AgNO_3$, a precipitate is obtained. This precipitate is of

    (a) $AgNO_3$
    (b) AgCl
    (c) $Mg(NO_3)_2$
    (d) $MgCl_2$

23. $NaNO_3$ cannot be analyzed gravimetrically because

    (a) of very low stability of $NaNO_3$.

    (b) $NaNO_3$ is an inert substance.

    (c) compound with Na^+ and NO_3^- are soluble.

    (d) sodium nitrate is insoluble.

Problems

Gravimetric Factor

24. Calculate the weight of sodium present in 50.0 g Na_2SO_4.

25. If the salt in Problem 24 is analyzed by precipitating and weighing $BaSO_4$, what weight of precipitate would be obtained?

26. Calculate the gravimetric factors for:

Substance Sought	Substance Weighed
As_2O_3	Ag_3AsO_4
$FeSO_4$	Fe_2O_3
K_2O	$KB(C_6H_5)_4$
SiO_2	$KAlSi_3O_8$

27. How many grams CuO would 1.00 g Paris green, $Cu_3(AsO_3)_2 \cdot 2As_2O_3 \cdot Cu(C_2H_3O_2)_2$, give? Of As_2O_3?

Quantitative Calculations

28. A 523.1-mg sample of impure KBr is treated with excess $AgNO_3$ and 814.5 mg AgBr is obtained. What is the purity of the KBr?

29. What weight of Fe_2O_3 precipitate would be obtained from a 0.4823-g sample of iron wire that is 99.89% pure?

30. The aluminum content of an alloy is determined gravimetrically by precipitating it with 8-hydroxyquinoline (oxine) to give $Al(C_9H_6ON)_3$. If a 1.021-g sample yielded 0.1862 g of precipitate. What is the percent aluminum in the alloy?

31. Iron in an ore is to be analyzed gravimetrically by weighing as Fe_2O_3. It is desired that the results be obtained to four significant figures. If the iron content ranges between 11 and 15%, what is the minimum size sample that must be taken to obtain 100.0 mg of precipitate?

32. The chloride in a 0.12-g sample of 95% pure $MgCl_2$ is to be precipitated as AgCl. Calculate the volume of 0.100 M $AgNO_3$ solution required to precipitate the chloride and give a 10% excess.

33. Ammonium ions can be analyzed by precipitating with H_2PtCl_6 as $(NH_4)_2PtCl_6$ and then igniting the precipitate to platinum metal, which is weighed $[(NH_4)_2PtCl_6 \xrightarrow{heat} Pt + 2NH_4Cl(g) + 2Cl_2(g)]$. Calculate the percent ammonia in a 1.00-g sample that yields 0.100 g Pt by this method.

34. A sample is to be analyzed for its chloride content by precipitating and weighing silver chloride. What weight of sample would have to be taken so that the weight of precipitate is equal to the percent chloride in the sample?

35. Pyrite ore (impure FeS_2) is analyzed by converting the sulfur to sulfate and precipitating $BaSO_4$. What weight of ore should be taken for analysis so that the grams of precipitate will be equal to 0.1000 times the percent of FeS_2?

36. A mixture containing only BaO and CaO weighs 2.00 g. The oxides are converted to the corresponding mixed sulfates, which weigh 4.00 g. Calculate the percent Ba and Ca in the original mixture.

37. A mixture containing only $BaSO_4$ and $CaSO_4$ contains one-half as much Ba^{2+} as Ca^{2+} by weight. What is the percentage of $CaSO_4$ in the mixture?

38. A mixture containing only AgCl and AgBr weighs 2.000 g. It is quantitatively reduced to silver metal, which weighs 1.300 g. Calculate the weight of AgCl and AgBr in the original mixture.

39. 0.0715 g was weighed of a mixture of chlorides of sodium and potassium in a 0.4150 g sample of feldspar. 0.1548 g of potassium hexachloroplatinate was obtained from further treatments. Calculate the percent of Na_2O and K_2O in sample and mineral.

40. In 200 mL, a natural water sample was used for Ca determination as CaC_2O_4. Weight of empty crucible was 26.6002 g and weight of crucible with CaO (56.077 g/mol) was 26.7134 g as precipitate was filtered, washed and ignited. Determine Ca concentration (40.078 g/mol) in water in units of gram per 100 mL of water.

Solubility Product Calculations

41. Write solubility product expressions for the following: (a) AgSCN, (b) $La(IO_3)_3$, (c) Hg_2Br_2, (d) $Ag[Ag(CN)_2]$; (e) $Zn_2Fe(CN)_6$, (f) Bi_2S_3.

42. Bismuth iodide, BiI_3, has a solubility of 7.76 mg/L. What is its K_{sp}?

43. What is the concentration of Ag^+ and CrO_4^{2-} in a saturated solution of Ag_2CrO_4?

44. Calculate the concentration of barium in the solution at equilibrium when 15.0 mL of 0.200 M K_2CrO_4 is added to 25.0 mL of 0.100 M $BaCl_2$.

45. What must be the concentration of PO_4^{3-} to just start precipitation of Ag_3PO_4 in a 0.10 M $AgNO_3$ solution?

46. What must be the concentration of Ag^+ to just start precipitating 0.10 M PO_4^{3-}? 0.10 M Cl^-?

47. At what pH will $Al(OH)_3$ begin to precipitate from 0.10 M $AlCl_3$?

48. What weight of Ag_3AsO_4 will dissolve in 250 mL water?

49. What is the solubility of Ag_2CrO_4 in 0.10 M K_2CrO_4?

50. Compounds AB and AC_2 each have solubility products equal to 4×10^{-18}. Which is more soluble, as expressed in moles per liter?

51. The solubility product of Bi_2S_3 is 1×10^{-97} and that of HgS is 4×10^{-53}. Which is the least soluble?

52. A student proposes to analyze barium gravimetrically by precipitating BaF_2 with NaF. Assuming a 200-mg sample of Ba^{2+} in 100 mL is to be precipitated and that the precipitation must be 99.9% complete for quantitative results, comment on the feasibility of the analysis.

Diverse Ion Effect on Solubility

53. Write the thermodynamic solubility product expressions for the following:

    (a) $\underline{BaSO_4} \rightleftharpoons Ba^{2+} + SO_4{}^{2-}$ (b) $\underline{Ag_2CrO_4} \rightleftharpoons 2Ag^+ + CrO_4{}^{2-}$

54. Calculate the solubility of $BaSO_4$ in 0.0125 M $BaCl_2$. Take into account the diverse ion effect.

55. You are to determine fluoride ion gravimetrically by precipitating CaF_2. $Ca(NO_3)_2$ is added to give an excess of 0.015 M calcium ion after precipitation. The solution also contains 0.25 M $NaNO_3$. How many grams fluoride will be in solution at equilibrium if the volume is 250 mL?

Multiple Choice Questions

1. Formation of an insoluble form of analyte is required in which electrochemical method?
   - (a) Electrogravimetry
   - (b) Coulometry
   - (c) Potentiometry
   - (d) Voltammetry

2. Through Cu electrogravimetry, 2.0 g brass solution was analyzed, and weight of Pt-gauze changed from 14.5 to 16.0 g. The percent weight of Cu in brass is
   - (a) 55
   - (b) 50
   - (c) 75
   - (d) 60

3. 3.72 g of dolomite was used for analysis, and 1.24 g of CaO was formed during gravimetric analysis. Calculate the percent weight of $CaCO_3$ in dolomite sample.
   - (a) 33.3
   - (b) 26.9
   - (c) 59.5
   - (d) 56.0

Recommended References

General and Inorganic

1. F. E. Beamish and W. A. E. McBryde, "Inorganic Gravimetric Analysis," in C. L. Wilson and D. W. Wilson, eds., *Comprehensive Analytical Chemistry*, Vol. **1A**. New York: Elsevier, 1959, Chapter VI.
2. C. L. Wilson and D. W. Wilson, eds., *Comprehensive Analytical Chemistry*, Vol. **1C**, *Classical Analysis: Gravimetric and Titrimetric Determination of the Elements*, New York: Elsevier, 1962.

Organic Reagents

3. K. L. Cheng, K. Ueno, and T. Imamura, eds., *Handbook of Organic Analytical Reagents*. Boca Raton, FL: CRC Press, 1982.
4. R. G. W. Hollingshead, *Oxine and Its Derivatives*. London: Butterworth Scientific, 1954–56.
5. F. Holmes, "Organic Reagents in Inorganic Analysis," in C. L. Wilson and D. W. Wilson, eds., *Comprehensive Analytical Chemistry*, Vol. **1A**. New York: Elsevier, 1959, Chapter II.8.

Precipitation from Homogeneous Solution

6. L. Gordon, M. L. Salulsky, and H. H. Willard, *Precipitation from Homogeneous Solution*. New York: Wiley, 1959.

Electrogravimetry

7. A. Douglas Skoog, *Fundamental of Analytical Chemistry*, 8th Edition, Books/Cole, Thomson learning Inc.

Chapter 10

PRECIPITATION REACTIONS AND TITRATIONS

"If you're not part of the solution, then you're part of the precipitate."

—Anonymous

KEY THINGS TO LEARN FROM THIS CHAPTER

- Effects of acids on solubility (key Equations: 10.4, 10.6)
- Mass balance calculations
- Effect of complexation on solubility (key Equations: 10.10, 10.11)

- Calculating precipitation titration curves
- Indicators for precipitation titrations

A number of anions form slightly soluble precipitates with certain metal ions and can be titrated with the metal solutions; for example, chloride can be titrated with silver ion and sulfate with barium ion. The precipitation equilibrium may be affected by pH or by the presence of complexing agents. The anion of the precipitate may be derived from a weak acid and therefore combine with protons in acid solution to cause the precipitate to dissolve. On the other hand, the metal ion may complex with a ligand (the complexing agent) to shift the equilibrium toward dissolution. Silver ion will complex with ammonia and cause silver chloride to dissolve.

In this chapter, we describe the quantitative effects of acidity and complexation in precipitation equilibria and discuss precipitation titrations using silver nitrate and barium nitrate titrants with different kinds of indicators and their theory. You should review fundamental precipitation equilibria described in Chapter 9. Most ionic analytes, especially inorganic anions, are conveniently determined using ion chromatography (Chapter 16), but for high concentrations more precise determinations can be made by precipitation titration when applicable. And many of the analyses performed by gravimetry as described in Chapter 9 may be more readily performed by precipitation titrations, albeit with not as high a precision.

10.1 Effect of Acidity on Solubility of Precipitates: Conditional Solubility Product

Before describing precipitation titrations, we shall consider the effects of competing equilibria on the solubility of precipitates. Before you read any further, you might want to review the discussion of polyprotic acid equilibria and the calculation of α's, the fractions of each acid species in equilibrium at a given pH, in Chapter 6.

The solubility of a precipitate whose anion is derived from a weak acid will increase in the presence of added acid because the acid will tend to combine with the anion and thus remove the anion from solution. For example, the precipitate MA that partially dissolves to give M^+ and A^- ions will exhibit the following equilibria:

$$\left. \begin{array}{c} MA \rightleftharpoons M^+ + A^- \\ + \\ H^+ \\ \updownarrow \\ HA \end{array} \right\} A_T$$

The anion A^- can combine with protons to increase the solubility of the precipitates. The combined equilibrium concentrations of A^- and HA make up the total analytical concentration A_T, (or formal concentration, see Chapter 4) of A, which will be equal to $[M^+]$ from the dissolved precipitate (if neither M^+ or A^- is in excess). By applying the equilibrium constants for the equilibria involved, we can calculate the solubility of the precipitate at a given acidity.

Consider, for example, the solubility of CaC_2O_4 in the presence of a strong acid. The equilibria are

Protons compete with calcium ion for the oxalate ion.

$$\underline{CaC_2O_4} \rightleftharpoons Ca^{2+} + C_2O_4^{2-} \quad K_{sp} = [Ca^{2+}][C_2O_4^{2-}] = 2.6 \times 10^{-9} \tag{10.1}$$

$$C_2O_4^{2-} + H^+ \rightleftharpoons HC_2O_4^- \quad K_{a2} = \frac{[H^+][C_2O_4^{2-}]}{[HC_2O_4^-]} = 6.1 \times 10^{-5} \tag{10.2}$$

$$HC_2O_4^- + H^+ \rightleftharpoons H_2C_2O_4 \quad K_{a1} = \frac{[H^+][HC_2O_4^-]}{[H_2C_2O_4]} = 6.5 \times 10^{-2} \tag{10.3}$$

The solubility s of CaC_2O_4 is equal to $[Ca^{2+}] = Ox_T$, where Ox_T represents the concentrations of all the oxalate species in equilibrium ($= [H_2C_2O_4] + [HC_2O_4^-] + [C_2O_4^{2-}]$). We can substitute $Ox_T\alpha_2$ for $[C_2O_4^{2-}]$ in the K_{sp} expression:

$$\boxed{K_{sp} = [Ca^{2+}]Ox_T\alpha_2} \tag{10.4}$$

where α_2 is the fraction of the oxalate species present as $C_2O_4^{2-}$ ($\alpha_2 = [C_2O_4^{2-}]/Ox_T$). Using the approach described in Chapter 6 for H_3PO_4 to calculate α's, we find that

$$\alpha_2 = \frac{K_{a1}K_{a2}}{[H^+]^2 + K_{a1}[H^+] + K_{a1}K_{a2}} \tag{10.5}$$

We can write, then, that

The conditional solubility product value holds for only a specified pH.

$$\boxed{\frac{K_{sp}}{\alpha_2} = K'_{sp} = [Ca^{2+}]Ox_T = s^2} \tag{10.6}$$

where K'_{sp} is the **conditional solubility product**, similar to the conditional formation constant described in Chapter 8.

Example 10.1

Note that this is a simpler problem than asking how much CaC_2O_4 will dissolve in 0.0010 M HCl. In the problem worked out, the equilibrium H^+ concentration is assumed to be 0.0010 M. Dissolving the CaC_2O_4 in 0.0010 M HCl will consume some of the protons and an iterative calculation will be needed. See Example 10.1.xlsx in the **website** to see a Goal Seek–based solution to the problem with HCl.

Calculate the solubility of CaC_2O_4 in a solution containing 0.0010 M $[H^+]$.

Solution

$$\alpha_2 = \frac{(6.5 \times 10^{-2})(6.1 \times 10^{-5})}{(1.0 \times 10^{-3})^2 + (6.5 \times 10^{-2})(1.0 \times 10^{-3}) + (6.5 \times 10^{-2})(6.1 \times 10^{-5})}$$

$$= 5.7 \times 10^{-2}$$

$$s = \sqrt{K_{sp}/\alpha_2} = \sqrt{2.6 \times 10^{-9}/5.7 \times 10^{-2}} = 2.1 \times 10^{-4} \, M$$

This compares with a calculated solubility in water using Equation 10.1 of $5.1 \times 10^{-5} \, M$ (a 400% increase in solubility). Note that both $[Ca^{2+}]$ and $Ox_T = 2.1 \times 10^{-4} \, M$. We can obtain the concentrations of the other oxalate species in equilibrium by multiplying this number by α_0, α_1, and α_2 for oxalic acid at 0.0010 M H^+ to obtain $[H_2C_2O_4]$, $[HC_2O_4^-]$, and $[C_2O_4^{2-}]$, respectively. We will not derive α_0 and α_1

here, but the results would be $[C_2O_4^{2-}] = 1.2 \times 10^{-5}\ M$, $[HC_2O_4^-] = 2.0 \times 10^{-4}\ M$, and $[H_2C_2O_4] = 3.1 \times 10^{-6}\ M$. (Try the calculations; these have been discussed in Chapter 6.)

In the above calculations we assumed that $[H^+] = 0.0010\ M$ in the final solution. A more common situation will be to start with $[H^+] = 0.0010\ M$ and see how much CaC_2O_4 will dissolve. But this process consumes H^+. In the above calculations, see that one-fifth of it would react to form $HC_2O_4^{2-}$. The amount reacted to form $H_2C_2O_4$ is negligible. If we desire a more exact solution, then we can subtract the amount of acid reacted, as calculated above, from the initial acid concentration and then repeat the calculation using the new acid concentration. We then repeat this process until the change in the final answer is within the desired accuracy, an iterative procedure. Recalculation using $0.8 \times 10^{-3}\ M$ acid gives a calcium concentration of $1.9 \times 10^{-4}\ M$, 10% less. See the text **website** for an exact calculation using Goal Seek.

We should emphasize that, when dealing with multiple equilibria, the validity of a given equilibrium expression is in no way compromised by the existence of additional competing equilibria. Thus, in the above example, the solubility product expression for CaC_2O_4 describes the relationship between Ca^{2+} and $C_2O_4^{2-}$ ions, whether or not acid is added. In other words, the product $[Ca^{2+}][C_2O_4^{2-}]$ is always a constant as long as there is solid CaC_2O_4 present. The quantity of CaC_2O_4 that dissolves is increased, however, because part of the $C_2O_4^{2-}$ in solution is converted to $HC_2O_4^-$ and $H_2C_2O_4$.

10.2 Mass Balance Approach for Multiple Equilibria

We may solve the multiple equilibrium problem as well by using the systematic approaches described in Chapter 5, using the equilibrium constant expressions, the mass balance expressions, and the charge balance expression.

The systematic approach is well suited for competing equilibria calculations.

Example 10.2

How many moles of MA will dissolve in 1 L of 0.10 M HCl if K_{sp} for MA is 1.0×10^{-8} and K_a for HA is 1.0×10^{-6}?

Solution

The equilibria and dissociations are

$$MA \rightleftharpoons M^+ + A^-$$
$$A^- + H^+ \rightleftharpoons HA$$
$$H_2O \rightleftharpoons H^+ + OH^-$$
$$HCl \rightarrow H^+ + Cl^-$$

The equilibrium expressions are

$$K_{sp} = [M^+][A^-] = 1.0 \times 10^{-8} \tag{1}$$

$$K_a = \frac{[H^+][A^-]}{[HA]} = 1.0 \times 10^{-6} \tag{2}$$

$$K_w = [H^+][OH^-] = 1.0 \times 10^{-14} \tag{3}$$

The mass balance expressions are

$$[M^+] = [A^-] + [HA] = A_T \tag{4}$$

$$[H^+] = [Cl^-] + [OH^-] - [HA] \tag{5}$$

$$[Cl^-] = 0.10 \, M \tag{6}$$

The charge balance expression is

$$[H^+] + [M^+] = [A^-] + [Cl^-] + [OH^-] \tag{7}$$

Number of expressions versus number of unknowns:

There are six unknowns ($[H^+]$, $[OH^-]$, $[Cl^-]$, $[HA]$, $[M^+]$, and $[A^-]$) and six independent equations (the charge balance equation can be generated as a linear combination of the others, and does not count as an independent equation).

Simplifying assumptions

The number of equations must equal or exceed the number of unknowns. Make assumptions to simplify calculations.

1. In an acid solution, dissociation of HA is suppressed, making $[A^-] \ll [HA]$, so from (4):

$$[M^+] = [A^-] + [HA] \approx [HA]$$

2. In an acid solution $[OH^-]$ is very small, so from (5) and (6):

$$[H^+] = 0.10 + [OH^-] - [HA] \approx 0.10 - [HA]$$

Calculate

We need to calculate $[M^+]$ in order to obtain the moles of MA dissolved in a liter. From (1)

$$[M^+] = \frac{K_{sp}}{[A^-]} \tag{8}$$

From (2)

$$[A^-] = \frac{K_a[HA]}{[H^+]} \tag{9}$$

So, dividing (8) by (9):

$$[M^+] = \frac{K_{sp}[H^+]}{K_a[HA]} = 1.0 \times 10^{-2} \frac{[H^+]}{[HA]} \tag{10}$$

From assumption (1),

$$[M^+] \approx [HA]$$

From assumption (2),

$$[H^+] \approx 0.10 - [HA] \approx 0.10 - [M^+]$$

$$[M^+] = \frac{(1.0 \times 10^{-2})(0.10 - [M^+])}{[M^+]}$$

$$\frac{[M^+]^2}{0.10 - [M^+]} = 1.0 \times 10^{-2}$$

Use of the quadratic equation gives $[M^+] = 0.027 \, M$.

The validity of the assumptions can be checked.

So, in 1 L, 0.027 mol of MA will dissolve. This compares with 0.00010 mol in water. Check

1.

$$[HA] \approx [M^+] = 0.027\ M$$

$$[A^-] = \frac{K_{sp}}{[M^+]} = \frac{1.0 \times 10^{-8}}{0.027} = 3.7 \times 10^{-7}\ M$$

Assumption (1) is acceptable because $[A^-] \ll [HA]$.

2.

$$[H^+] \approx 0.10 - [M^+] = 0.073\ M$$

$$[OH^-] = \frac{K_w}{[H^+]} = \frac{1.0 \times 10^{-14}}{0.073} = 1.4 \times 10^{-13}$$

Assumption (2) is acceptable because $[OH^-] \ll [Cl^-]$ or $[HA]$.

Alternate Goal Seek—Based Solution

The more Excel-minded among you may find this no-approximation approach simpler and quicker.

The charge balance equation is:

$$[M^+] + [H^+] - [Cl^-] - [A^-] - [OH^-] = 0$$

This can be rewritten as:

$$s + [H^+] - 0.10 - s\alpha_1 - K_w/[H^+] = 0$$

Substituting $\sqrt{(K_{sp}/\alpha_1)}$ for s, Goal Seek readily finds a solution $s = 0.027\ M$, see Example 10.2 Goal Seek.xlsx in the text **website**.

Example 10.3

Calculate the solubility of CaC_2O_4 in a solution of 0.0010 M hydrochloric acid, using the systematic approach.

Solution

The equilibria and dissociations are

$$CaC_2O_4 \rightleftharpoons Ca^{2+} + C_2O_4{}^{2-}$$

$$C_2O_4{}^{2-} + H^+ \rightleftharpoons HC_2O_4{}^-$$

$$HC_2O_4{}^- + H^+ \rightleftharpoons H_2C_2O_4$$

$$H_2O \rightleftharpoons H^+ + OH^-$$

$$HCl \rightarrow H^+ + Cl^-$$

The equilibrium constant expressions are

$$K_{sp} = [Ca^{2+}][C_2O_4{}^{2-}] = 2.6 \times 10^{-9} \tag{1}$$

$$K_{a1} = \frac{[H^+][HC_2O_4{}^-]}{[H_2C_2O_4]} = 6.5 \times 10^{-2} \tag{2}$$

$$K_{a2} = \frac{[H^+][C_2O_4^{2-}]}{[HC_2O_4^-]} = 6.1 \times 10^{-5} \tag{3}$$

$$K_w = [H^+][OH^-] = 1.00 \times 10^{-14} \tag{4}$$

The mass balance expressions are

$$[Ca^{2+}] = [C_2O_4^{2-}] + [HC_2O_4^-] + [H_2C_2O_4] = Ox_T \tag{5}$$

$$[H^+] = [Cl^-] + [OH^-] - [HC_2O_4^-] - 2[H_2C_2O_4] \tag{6}$$

$$[Cl^-] = 0.0010 \, M \tag{7}$$

The charge balance expression is

$$[H^+] + 2[Ca^{2+}] = 2[C_2O_4^{2-}] + [HC_2O_4^-] + [Cl^-] + [OH^-] \tag{8}$$

There are seven unknowns ($[H^+]$, $[OH^-]$, $[Cl^-]$, $[Ca^{2+}]$, $[C_2O_4^{2-}]$, $[HC_2O_4^-]$, and $[H_2C_2O_4]$) and seven independent equations.

Simplifying assumptions

1. K_{a1} is rather large and K_{a2} is rather small, so assume $[HC_2O_4^-] \gg [H_2C_2O_4]$, $[C_2O_4^{2-}]$.
2. In an acid solution, $[OH^-]$ is very small, so from (6) and (7):

$$[H^+] = 0.0010 + [OH^-] - [H_2C_2O_4^-] - 2[H_2C_2O_4] \approx 0.0010 - [HC_2O_4^-] \tag{9}$$

Calculate
We need to calculate $[Ca^{2+}]$ in order to obtain the moles of CaC_2O_4 dissolved in a liter. From (1)

$$[Ca^{2+}] = \frac{K_{sp}}{[C_2O_4^{2-}]} \tag{10}$$

From (3)

$$[C_2O_4^{2-}] = \frac{K_{a2}[HC_2O_4^-]}{[H^+]} \tag{11}$$

So,

$$[Ca^{2+}] = \frac{K_{sp}[H^+]}{K_{a2}[HC_2O_4^{2-}]} \tag{12}$$

From assumption (1),

$$[Ca^{2+}] = [HC_2O_4^-] \tag{13}$$

From assumption (2),

$$[H^+] \approx 0.0010 - [HC_2O_4^-] \approx 0.0010 - [Ca^{2+}] \tag{14}$$

Substitute (13) and (14) in (12):

$$[Ca^{2+}] = \frac{K_{sp}(0.0010 - [Ca^{2+}])}{K_{a2}[Ca^{2+}]} = \frac{(2.6 \times 10^{-9})(0.0010 - [Ca^{2+}])}{(6.1 \times 10^{-5})[Ca^{2+}]}$$

$$[Ca^{2+}] = \frac{(4.6 \times 10^{-5})(0.0010 - [Ca^{2+}])}{[Ca^{2+}]}$$

Solving the quadratic equation gives $[Ca^{2+}] = 1.9 \times 10^{-4}$ M. This is the same as that calculated in Example 10.1, using the conditional solubility product approach, *after correcting for the H^+ consumed*. In the present example, we corrected for the H^+ consumed in the calculation. Note that in Example 10.1 we calculated $HC_2O_4^-$ to be 95% of the $[Ca^{2+}]$ value, so our assumption (1) was reasonable.

> The answer is the same as when using K'_{sp} (Example 10.1).

The Goal Seek solution to this problem was already presented in Example 10.1 on the text **website**.

10.3 Effect of Complexation on Solubility: Conditional Solubility Product

Complexing agents can compete for the metal ion in a precipitate, just as acids compete for the anion. A precipitate MA that dissociates to give M^+ and A^- and where the metal complexes with the ligand L to form ML^+ would have the following equilibria:

$$
\left.
\begin{array}{c}
MA \rightleftharpoons M^+ \\
+ \\
L \\
\updownarrow \\
ML^+
\end{array}
\right\}
\begin{array}{c}
+ A^- \\
\\
M_T
\end{array}
$$

The sum of $[M^+]$ and $[ML^+]$ is the analytical concentration M_T in equilibrium, which is equal to $[A^-]$. Calculations for such a situation are handled in a manner completely analogous to those for the effects of acids on solubility.

Consider the solubility of AgBr in the presence of NH_3. The equilibria are

$$AgBr \rightleftharpoons Ag^+ + Br^- \tag{10.7}$$

$$Ag^+ + NH_3 \rightleftharpoons Ag(NH_3)^+ \tag{10.8}$$

$$Ag(NH_3)^+ + NH_3 \rightleftharpoons Ag(NH_3)_2{}^+ \tag{10.9}$$

The solubility s of AgBr is equal to $[Br^-] = Ag_T$, where Ag_T represents the concentrations of all the silver species in equilibrium $(= [Ag^+] + [Ag(NH_3)^+] + [Ag(NH_3)_2{}^+])$. As before, we can substitute $Ag_T\alpha_M$ for $[Ag^+]$ in the K_{sp} expression, where α_M is the fraction of silver species present as Ag^+:

$$K_{sp} = [Ag^+][Br^-] = Ag_T\alpha_M[Br^-] = 4 \times 10^{-13} \tag{10.10}$$

Then,

$$\boxed{\frac{K_{sp}}{\alpha_M} = K'_{sp} = Ag_T[Br^-] = s^2} \tag{10.11}$$

where K'_{sp} is again the **conditional solubility product**, whose value depends on the concentration of ammonia.

> The K'_{sp} value holds for only a given NH_3 concentration.

Example 10.4

Calculate the molar solubility of silver bromide in a 0.10 M ammonia solution.

Solution

From Equation 8.20 and the K_f values (Equations 8.1 and 8.2), we calculate for 0.10 M ammonia,

$$\alpha_{Ag} = 1/(1 + K_{f1}[NH_3] + K_{f1}K_{f2}[NH_3]^2)$$
$$= 1/[1 + (2.5 \times 10^3)(0.10) + (2.5 \times 10^3)(1.0 \times 10^4)(0.10)^2] = 4.0 \times 10^{-6}$$

$$s = \sqrt{\frac{K_{sp}}{\alpha_M}} = \sqrt{4 \times 10^{-13}/4.0 \times 10^{-6}} = 3_{.2} \times 10^{-4} \ M$$

This compares with a solubility in water of $6 \times 10^{-7} \ M$ (530 times more soluble). Note again that both $[Br^-]$ and $Ag_T = 3_{.2} \times 10^{-4} \ M$. The concentrations of the other silver species in equilibrium can be obtained: $[Ag^+]$ is $Ag_T \ \alpha_M$, $[Ag(NH_3)^+]$ is then $[Ag^+]\beta_1[NH_3]$, and $[Ag(NH_3)_2^+]$ is $[Ag^+]\beta_2[NH_3]^2$, respectively. Taking the β values from Example 8.5, the results are $[Ag^+] = 1_{.3} \times 10^{-9} \ M$, $[Ag(NH_3)^+] = 3_{.2} \times 10^{-7} \ M$, and $[Ag(NH_3)_2^+] = 3_{.2} \times 10^{-4} \ M$. Note that the majority of the dissolved silver exists in the $Ag(NH_3)_2^+$ form.

Check that the equilibrium ammonia concentration assumed was correct.

We neglected the amount of ammonia consumed in the reaction with the silver. We see that it was indeed negligible compared to 0.10 M [$6 \times 10^{-4} \ M$ was used in forming $Ag(NH_3)_2^+$, even less in forming $Ag(NH_3)^+$]. Had the amount of ammonia consumed been appreciable, we could have used an iterative procedure to obtain a more exact solution; that is, we could have subtracted the amount of ammonia consumed from the original concentration and then used the new concentration to calculate new β's and a new solubility, and so on, until a constant value was reached for a solution. These problems are also amenable to solution by Goal Seek.

Example 10.5

Calculate the molar solubility of a salt MX with $K_{sp} = 1 \times 10^{-8}$ in 0.0010 M solution of ligand L, where L is a base with a K_b of 1.0×10^{-3} and binds to M with β_1 and β_2 values of 1.0×10^5 and 1.0×10^8, respectively. (See Section 8.6 for a review of β values.)

Goal Seek Solution

In Example 10.4 we ignored the base behavior of ammonia (that some of it ionizes to form NH_4^+, which does not participate in complex formation with silver); if we deal with a ligand that is not as weak a base, we cannot ignore this. Similarly, we assumed a negligible decrease in the free ligand concentration due to complex formation, this would not be the case if the salt was more soluble (K_{sp} was higher) and/or complexation constants were higher. The present problem deliberately makes all of these constraints such that approximations may not be permissible. Nevertheless, Goal Seek readily provides an iterative solution.

Consider that the base behavior of L is hydrolysis to generate OH^-:

$$L + H_2O \rightleftharpoons LH^+ + OH^-$$

And the relevant equilibrium constant expression is

$$K_b = \frac{[LH^+][OH^-]}{[L]}$$

If we neglect the autoionization of water as this will be a distinctly basic solution, the only source of the two ions in the numerator is the above equilibrium and their concentrations will be equal, i.e., $[LH^+] = [OH^-]$. It follows that

$$K_b = [LH^+]^2/[L]$$

$$[LH^+] = \sqrt{K_b[L]}$$

The mass balance for L demands

$$[L]_T = 0.001 = [L] + [LH^+] + [ML^+] + 2[ML_2{}^+]$$

If s is the solubility, we also have, as in Example 10.4, $s = \sqrt{\left(\dfrac{K_{sp}}{\alpha_M}\right)}$ and $[M^+] = s\alpha_M$, and using these, we can express the two remaining terms:

$$[ML^+] = \beta_1[M^+][L]$$

$$[ML_2{}^+] = \beta_2[M^+][L]^2$$

As long as the relationships are all defined, we can enter these already defined terms into the mass balance equation (MBE). Execution of Example 10.5 Goal Seek.xlsx on the **website** shows the solution, $s = 4.5 \times 10^{-4}\ M$.

10.4 Precipitation Titrations

Titrations with precipitating agents are useful for determining certain analytes, provided the equilibria are rapid and a suitable means of detecting the end point is available. A consideration of titration curves will increase our understanding of indicator selection, precision, and the titration of mixtures.

Titration Curves—Calculating pX

Consider the titration of Cl^- with a standard solution of $AgNO_3$. A titration curve can be prepared by plotting pCl ($-\log[Cl^-]$) against the volume of $AgNO_3$, in a manner similar to that used for acid–base titrations. A typical titration curve is illustrated in Figure 10.1; pX in the figure refers to the negative logarithm of the halide concentration. At the beginning of the titration, we have $0.1\ M\ Cl^-$, and pCl is 1. As the titration continues, part of the Cl^- is removed from solution by precipitation as $AgCl$, and the pCl is determined by the concentration of the remaining Cl^-; the contribution of Cl^- from dissociation of the precipitate is negligible, except near the equivalence point. At the equivalence point, we have a saturated solution of $AgCl$, pCl = 5, and $[Cl^-] = \sqrt{K_{sp}} = 10^{-5}\ M$ (see Chapter 9). Beyond the equivalence point, there is excess Ag^+, and the Cl^- concentration is determined from the concentration of Ag^+ and K_{sp} as in Example 9.7 in Chapter 9 ($[Cl^-] = K_{sp}/[Ag^+]$). See the text **website** for an Excel calculation of these curves.

AgI has the lowest solubility, so $[I^-]$ beyond the equivalence point is smaller and pI is larger.

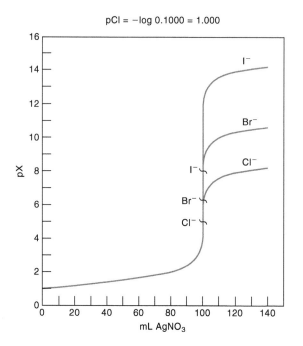

FIGURE 10.1 Titration curves for 100 mL 0.1 M chloride, bromide, and iodide solutions versus 0.1 M AgNO$_3$.

Example 10.6

Calculate pCl for the titration of 100.0 mL of 0.1000 M Cl$^-$ with 0.1000 M AgNO$_3$ for the addition of 0.00, 20.00, 99.00, 99.50, 100.00, 100.50, and 110.00 mL AgNO$_3$.

Solution

At 0.00 mL:

$$\text{pCl} = -\log\ 0.1000 = 1.000$$

At 20.00 mL:

$$\text{mmol Cl}^- = 100.0\ \text{mL} \times 0.1000\ \text{mmol/mL} = 10.00\ \text{mmol}$$

$$\text{mmol Ag}^+ = 20.00\ \text{mL} \times 0.1000\ \text{mmol/mL} = 2.000\ \text{mmol}$$

$$\text{Cl}^-\ \text{left} = 10.00 - 2.00 = 8.00\ \text{mmol}/120.0\ \text{mL} = 0.0667\ M$$
$$\text{pCl} = -\log\ 0.0667 = 1.18$$

At 99.00 mL:

$$\text{mmol Ag}^+ = 99.00\ \text{mL} \times 0.1000\ \text{mmol/mL} = 9.900\ \text{mmol}$$

$$\text{Cl}^-\ \text{left} = 10.00 - 9.90 = 0.10\ \text{mmol}/199.0\ \text{mL} = 5.0 \times 10^{-4}\ M$$

$$\text{pCl} = -\log\ 5.0 \times 10^{-4} = 3.26$$

At 99.50 mL:

$$\text{mmol Ag}^+ = 99.50\ \text{mL} \times 0.1000\ \text{mmol/mL} = 9.950\ \text{mmol}$$

$$\text{Cl}^-\ \text{left} = 10.00 - 9.95 = 0.05\ \text{mmol}/199.5\ \text{mL} = 2._5 \times 10^{-4}M$$

$$\text{pCl} = -\log\ 2._5 \times 10^{-4} = 3.60$$

At 100.00 mL, all the Cl$^-$ is reacted with Ag$^+$:

$$[\text{Cl}^-] = \sqrt{K_{sp}} = \sqrt{1.0 \times 10^{-10}} = 1.0 \times 10^{-5}\ M$$
$$\text{pCl} = -\log\ 1.0 \times 10^{-5} = 5.00$$

At 100.50 mL:

$$\text{mmol Ag}^+ = 100.50 \text{ mL} \times 0.1000 \text{ mmol/mL} = 10.05 \text{ mmol}$$

$$\text{Ag}^+ \text{ left} = 10.05 - 10.00 = 0.05 \text{ mmol}/200.5 \text{ mL} = 2._5 \times 10^{-4} \, M$$

$$[\text{Cl}^-] = K_{sp}/[\text{Ag}^+] = 1.0 \times 10^{-10}/2._5 \times 10^{-4} = 4_0 \times 10^{-7} \, M$$

$$\text{pCl} = -\log 4._0 \times 10^{-7} = 6.40$$

At 110.00 mL:

$$\text{mmol Ag}^+ = 110.00 \text{ mL} \times 0.1000 \text{ mmol/mL} = 11.00 \text{ mmol}$$

$$\text{Ag}^+ \text{ left} = 11.00 - 10.00 = 1.00 \text{ mmol}/210 \text{ mL} = 4.76 \times 10^{-3} \, M$$

$$[\text{Cl}^-] = 1.0 \times 10^{-10}/4.76 \times 10^{-3} = 2.1 \times 10^{-8} \, M$$

$$\text{pCl} = -\log 2.1 \times 10^{-8} = 7.67$$

The smaller the K_{sp}, the larger the break at the equivalence point. We can illustrate this point by comparing the titration curves for Cl^-, Br^-, and I^- versus Ag^+ in Figure 10.1. The K_{sp} values of AgCl, AgBr, and AgI are 1×10^{-10}, 4×10^{-13}, and 1×10^{-16}, respectively. The concentration of each anion has been chosen to be the same at the beginning of the titration, and so up to near the equivalence point the concentration of each remains the same, since the same fraction is removed from solution. At the equivalence point, $[\text{X}^-]$ is smaller for the smaller K_{sp} values; hence pX is larger for a saturated solution of the salt. Beyond the equivalence point, $[\text{X}^-]$ is smaller when K_{sp} is smaller; also resulting in a larger jump in pX. So the overall effect is a larger pX break at the equivalence point when the compound is more insoluble.

> The smaller the K_{sp}, the sharper the end point.

If the chloride titration were performed in reverse, that is, Ag^+ titrated with Cl^-, the titration curve would be the reverse of the curve in Figure 10.1 if pCl were plotted against the volume of Cl^-. Before the equivalence point, the Cl^- concentration would be governed by the concentration of excess Ag^+ and K_{sp}; and beyond the equivalence point, it would be governed merely by the excess Cl^-. The pAg value could be plotted, instead, against the volume of chloride solution, and the curve would look the same as that of Figure 10.1.

Note that very much like metal-ligand titrations (as discussed in Section 8.3), the present case of a precipitation titration can be explicitly solved at all points with a single quadratic equation. In fact this is simpler than the metal-ligand titration because unlike the formation constant for a metal-ligand complex ML, the expression for the corresponding equilibrium constant, the solubility product, K_{sp}, has no denominator term. The two situations are identical (except in that the metal-ligand case the equilibrium constant by convention is written as an association constant, whereas the solubility product constant is written as a dissociation constant). For an insoluble salt XY, let us assume we took V_X mL of a solution of concentration C_x molar and it is being titrated with C_Y molar solution of Y, V_Y mL having been added at any point, with the total volume being V_T (which equals $V_X + V_Y$). The implicit concentration units in the solubility product expression are in molar units. To keep all quantities in moles and molar units, we express volumes in liters.

$$K_{sp} = [\text{X}][\text{Y}] = \frac{(C_X V_X - p)}{V_T} \frac{(C_Y V_Y - p)}{V_T} \tag{10.12}$$

where p indicates the moles XY precipitated. This results in a quadratic equation in p:

$$p^2 - (C_X V_X + C_Y V_Y)p + (C_X V_X C_Y V_Y - V_T^2 K_{sp}) = 0 \tag{10.13}$$

As in the case of metal-ligand complexation, generally the negative root of this equation provides the meaningful solution for p. [X] and [Y] (and hence pX and pY) are then readily calculated as $(C_X V_X - p)/V_T$ and $(C_Y V_{Y-p})/V_T$. We demonstrate the generation of Figure 10.1 by this approach in Figure 10.1.xlsx in the **website** supplement. Note that before any Y is added, there is no precipitation of XY and the solubility product equilibrium is inapplicable. For $V_Y = 0$, $p = 0$ must be manually put in.

Stepwise Precipitation Titrations

If K_{sp} values differ sufficiently, two analytes that can be precipitated by the same reagent can be titrated stepwise. Consider a mixture of iodide and chloride that is titrated with silver ion. The K_{sp} of AgI is 10^6 times smaller than that of AgCl, and AgI is precipitated first. As soon as the first drop of titrant is added, AgI precipitates and titration proceeds exactly as in the previous section with only the solubility equilibrium of AgI being relevant. This continues until the iodide is nearly all titrated and the $[Ag^+]$ reaches a value in which the K_{sp} of AgCl is reached. Henceforth the precipitation of AgCl will begin and both AgI and AgCl precipitates will be present. We could proceed as before considering the silver precipitated, but let us this time solve this by solving for the silver remaining in the solution as the variable. Consider the following conditions:

Let

V_X be original volume of solution to be titrated, 0.1 L

I_0 be the original concentration of I^- taken, 0.1 M

Cl_0 be the original concentration of Cl^- taken, 0.1 M

C_{Ag} be the concentration of the $AgNO_3$ titrant

V_{Ag} be the titrant volume added at any point in liters

V_T be the total volume at any point $(V_x + V_{Ag})$ in liters

After AgCl precipitation begins:

moles of Ag^+ added = moles of AgI precipitated + moles of AgCl precipitated
+ moles of Ag in solution

$$V_{Ag}C_{Ag} = (\text{moles of } I^- \text{ taken} - \text{moles of } I^- \text{ in solution})$$
$$+ (\text{moles of } Cl^- \text{ taken} - \text{moles of } Cl^- \text{ in solution}) + [Ag^+]V_T$$
$$V_{Ag}C_{Ag} = (V_X I_0 - [I^-]V_T) + (V_X Cl_0 - [I^-]V_T) + [Ag^+]V_T$$

$$(10.14)$$

Because $[I^-]$ and $[Cl^-]$ can be respectively expressed as $K_{sp,AgI}/[Ag^+]$ and $K_{sp,AgCl}/[Ag^+]$, Equation 10.1 becomes

$$(V_{Ag}C_{Ag} - V_X(I_0 + Cl_0))/V_T = [Ag^+] - (K_{sp,AgI} + K_{sp,AgCl})/[Ag^+] \quad (10.15)$$

Equation 10.15 is a quadratic equation in $[Ag^+]$ where in usual notation, $a = 1$, $b = -(V_{Ag}C_{Ag} - V_X(I_0 + Cl_0))/V_T$, and $c = -(K_{sp,AgI} + K_{sp,AgCl})$. Presently the positive root gives the tenable solution. The relevant spreadsheet is given in Figure 10.2xlsx in the **website** and the plot is shown in Figure 10.2. We can assume that AgCl precipitation begins when the equivalence point of iodide has been reached. But we can find the exact volume of silver added at which this applies and how to do this is discussed in the spreadsheet. In practice, this is so close to the iodide end point, no difference will be discernible.

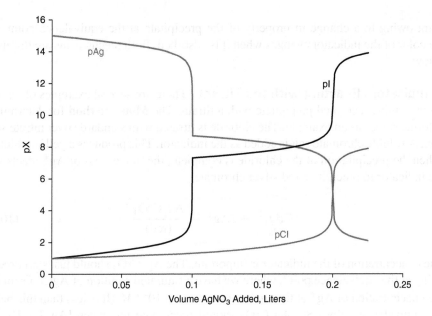

FIGURE 10.2 Titration curve for 100 mL of a mixture containing 0.1 M iodide and 0.1 M chloride titrated with 0.1 M AgNO$_3$.

Example 10.7

The iodine in United States table salt is added as sodium or potassium iodide, at a concentration of 45 mg I/kg salt. Neglecting any other components present, if you try to determine the iodide content by titration with silver nitrate, what fraction of the iodide will remain in solution when AgCl begins to precipitate? Assume that you have dissolved enough salt in water to make a 0.1 M NaCl solution.

Solution

The fw of NaCl is 58.5. The molar ratio of chloride to iodide is therefore

$$\frac{\frac{1000\ g}{58.5\ g\ NaCl/mol}}{\frac{0.045\ g}{127\ g\ iodide/mol}} = 4.82 \times 10^4$$

If we have a solution 0.1 M in Cl$^-$, the I$^-$ concentration is $0.1/(4.82 \times 10^4) = 2.07 \times 10^{-6} M$.

For a 0.1 M [Cl$^-$] solution, AgCl will begin to precipitate at [Ag$^+$] = $K_{sp,AgCl}/[Cl^-] = 10^{-10}/0.1 = 10^{-9}\ M$.

At this [Ag$^+$], iodide concentration is $K_{sp,AgI}/[Ag^+] = 10^{-16}/10^{-9} = 10^{-7}\ M$. Therefore, the iodide fraction not precipitated is $10^{-7}/2.07 \times 10^{-6} \times 100\% = 4.83\%$.

Detection of the End Point: Indicators

We can detect the end point by measuring either pCl or pAg with an appropriate electrode and a potentiometer. We discuss this in Chapter 20. It is more convenient if an indicator can be used. The indicator theory for these titrations is different from that for acid–base indicators. The properties of the indicators do not necessarily depend on the concentration of some ion in solution, that is, on pCl or pAg.

Chemists commonly use two types of indicators. The first type forms a colored precipitate with the titrant when the titrant is in excess. The second type, called an **adsorption indicator**, suddenly becomes adsorbed on the precipitate at the equivalence point owing to a change in property of the precipitate at the equivalence point, and the color of the indicator changes when it is adsorbed. Both mechanisms are discussed below.

1. **Indicators Reacting with the Titrant.** There are several examples of an indicator forming a colored precipitate with a titrant. The **Mohr method** for determining chloride serves as an example. The chloride is titrated with standard silver nitrate solution. A soluble chromate salt is added as the indicator. This produces a yellow solution. When the precipitation of the chloride is complete, the first excess of Ag^+ reacts with the indicator to precipitate red silver chromate:

$$\underset{\text{(yellow)}}{CrO_4^{2-}} + 2Ag^+ \rightarrow \underset{\text{(red)}}{\frac{Ag_2CrO_4}{}} \qquad (10.16)$$

The concentration of the indicator is important. The Ag_2CrO_4 should just start precipitating at the equivalence point, where we have a saturated solution of AgCl. From K_{sp}, the concentration of Ag^+ at the equivalence point is $10^{-5}\ M$. (It is less than this before the equivalence point.) So, Ag_2CrO_4 should precipitate just when $[Ag^+] = 10^{-5}\ M$. The solubility product of Ag_2CrO_4 is 1.1×10^{-12}. By inserting the Ag^+ concentration in the K_{sp} equation for Ag_2CrO_4, we calculate that, for this to occur, $[CrO_4^{2-}]$ should be $0.011\ M$:

$$(10^{-5})^2[CrO_4^{2-}] = 1.1 \times 10^{-12}$$

$$[CrO_4^{2-}] = 1.1 \times 10^{-2}\ M$$

If the concentration is greater than this, Ag_2CrO_4 will begin to precipitate when $[Ag^+]$ is less than $10^{-5}\ M$ (before the equivalence point). If it is less than $0.011\ M$, then the $[Ag^+]$ will have to exceed $10^{-5}\ M$ (beyond the equivalence point) before precipitation of Ag_2CrO_4 begins.

In actual practice, the indicator concentration is kept at 0.002 to 0.005 M. If it is much higher than this, the intense yellow color of the chromate ion obscures the red Ag_2CrO_4 precipitate color, and an excess of Ag^+ is required to produce enough precipitate to be seen. An indicator blank should always be determined and subtracted from the titration to correct for indicator error. Standardization of the titrant using the same titration also takes into account the indicator error.

The Mohr titration must be performed at a pH of about 8. If the solution is too acidic (pH < 6), then part of the indicator is present as $HCrO_4^-$, and more Ag^+ will be required to form the Ag_2CrO_4 precipitate. Above pH 8, silver hydroxide may be precipitated (at pH > 10). The pH is properly maintained by adding solid calcium carbonate to the solution. (While the carbonate ion is a fairly strong Brønsted base, the concentration in a saturated calcium carbonate solution is just sufficient to give a pH about 8.) The Mohr titration is useful for determining chloride in neutral or unbuffered solutions, such as drinking water.

A second example of this type of indicator is illustrated in the **Volhard titration**. This is an indirect titration procedure for determining anions that precipitate with silver (Cl^-, Br^-, SCN^-), and it is performed in acidic (HNO_3) solution. In this procedure, we add a measured excess of $AgNO_3$ to precipitate the anion and then determine the excess Ag^+ by back-titration with standard potassium thiocyanate solution:

$$X^- + Ag^+ \rightarrow \underline{AgX} + \text{excess } Ag^+$$
$$\text{excess } Ag^+ + SCN^- \rightarrow \underline{AgSCN} \qquad (10.17)$$

Titration error from the indicator providing the equivalence point too early or too late can be corrected for by a blank titration or standardizing the titrant with primary standard NaCl using the same titration.

The Mohr titration is performed in slightly alkaline solution.

The Volhard titration is performed in acid solution.

We detect the end point by adding iron(III) as a ferric alum (ferric ammonium sulfate), which forms a soluble red complex with the first excess of titrant:

$$\boxed{Fe^{3+} + SCN^- \rightarrow Fe(SCN)^{2+}} \tag{10.18}$$

If the precipitate, AgX, is less soluble than AgSCN or is equally soluble, we do not have to remove the precipitate before titrating. Such is the case with I^-, Br^-, and SCN^-. In the case of I^-, we do not add the indicator until all the I^- is precipitated, since it would be oxidized by the iron(III). If the precipitate is more soluble than AgSCN, it will react with the titrant to give a high and diffuse end point. Such is the case with AgCl:

$$\underline{AgCl} + SCN^- \rightarrow \underline{AgSCN} + Cl^- \tag{10.19}$$

Therefore, we remove the precipitate by filtration before titrating.

Obviously, these indicators must not form a precipitate with the titrant that is more stable than the titration precipitate, or the color reaction would occur when the first drop of titrant is added.

2. Adsorption Indicators. With adsorption indicators, the indicator reaction takes place on the surface of the precipitate. The indicator, which is a dye, exists in solution in the ionized form, usually an anion, In^-. To explain the mechanism of the indicator action, we must invoke the mechanism occurring during precipitation (see Chapter 9 for more detail).

Consider the titration of Cl^- with Ag^+. Before the equivalence point, Cl^- is in excess, and the *primary adsorbed layer* is Cl^-. This repels the indicator anion, and the more loosely held *secondary (counter) layer* of adsorbed ions is cations, such as Na^+:

$$AgCl: Cl^- :: Na^+$$

Beyond the equivalence point, Ag^+ is in excess, and the surface of the precipitate becomes positively charged, with the primary layer being Ag^+. This will now attract the indicator anion and adsorb it in the counterlayer:

$$AgCl: Ag^+ :: In^-$$

The color of the adsorbed indicator is different from that of the unadsorbed indicator, and this difference signals the completion of the titration. A possible explanation for this color change is that the indicator forms a colored complex with Ag^+, which is too weak to exist in solution, but whose formation is facilitated by adsorption on the surface of the precipitate (it becomes "insoluble").

The pH is important. If it is too low, the indicator, which is usually a weak acid, will dissociate too little to allow it to be adsorbed as the anion. Also, the indicator must not be too strongly adsorbed at the given pH, or it will displace the anion of the precipitate (e.g., Cl^-) in the primary layer before the equivalence point is reached. This will, of course, depend on the degree of adsorption of the anion of the precipitate. For example, Br^- forms a less soluble precipitate with Ag^+ and is more strongly adsorbed. A more strongly adsorbed indicator can therefore be used.

The degree of adsorption of the indicator can be decreased by increasing the acidity. The stronger an acid the indicator is, the wider the pH range over which it can be adsorbed. In the case of Br^-, since a more acidic (more strongly adsorbed) indicator can be used, the pH of the titration can be more acidic than with Cl^-.

Table 10.1 lists some adsorption indicators. Fluorescein can be used as an indicator for any of the halides at pH 7 because it will not displace any of them. Dichlorofluorescein will displace Cl^- at pH 7 but not at pH 4. Hence, results tend to be low when titrations are performed at pH 7. Titration of chloride using these indicators is called **Fajans' method.** Fluorescein was the original indicator described by Fajans, but

The more insoluble precipitates can be titrated in more acid solutions, using more strongly adsorbed indicators.

TABLE 10.1 Adsorption Indicators

Indicator	Titration	Solution
Fluorescein	Cl^- with Ag^+	pH 7–8
Dichlorofluorescein	Cl^- with Ag^+	pH 4
Bromcresol green	SCN^- with Ag^+	pH 4–5
Eosin	Br^-, I^-, SCN^- with Ag^+	pH 2
Methyl violet	Ag^+ with Cl^-	Acid solution
Rhodamine 6 G	Ag^+ with Br^-	HNO_3 ($\leq 0.3\ M$)
Thorin	SO_4^{2-} with Ba^{2+}	pH 1.5–3.5
Bromphenol blue	Hg^{2+} with Cl^-	$0.1\ M$ solution
Orthochrome T	Pb^{2+} with CrO_4^{2-}	Neutral, $0.02\ M$ solution

dichlorofluorescein is now preferred. Eosin cannot be used for the titration of chloride at any pH because it is too strongly adsorbed.

Because most of these end points do not coincide with the equivalence point, *the titrant should be standardized by the same titration as used for the sample.* In this way, the errors will essentially be corrected if about the same amount of titrant is used for both the standardization and analysis.

A chief source of error in titrations involving silver is photodecomposition of AgX, which is catalyzed by the adsorption indicator. By proper standardization, however, accuracies of one part per thousand can be achieved.

The precipitate is uncharged at the equivalence point (neither ion is in excess). Colloidal precipitates, such as silver chloride, therefore tend to coagulate at this point, especially if the solution is shaken. This is just what we want for gravimetry, but the opposite of what we want here. Coagulation decreases the surface area for absorption of the indicator, which in turn decreases the sharpness of the end point. We can prevent coagulation of silver chloride by adding some dextrin to the solution.

> For adsorption indicators, we want the maximum surface area for adsorption, in contrast to gravimetry.

Visual indicators are convenient for rapid precipitation titrimetry with silver ion. Potentiometric end-point detection is also widely used, particularly for dilute solutions, for example, in the millimolar concentration range (see Chapter 21).

Titration of Sulfate with Barium

Sulfate can be determined by titrating with barium ion, to precipitate $BaSO_4$. As in the gravimetric determination of sulfate by precipitation of barium sulfate, this titration is subject to errors by coprecipitation. Cations such as K^+, Na^+, and NH_4^+ (especially the first) coprecipitate as sulfates:

$$BaSO_4: SO_4^{2-} : 2M^+$$

As a result, less barium ion is required to complete the precipitation of the sulfate ion, and the calculated results are undersitimated. Some metal ions will complex the indicator and interfere. Foreign anions may coprecipitate as the barium salts to cause overestimated results. Errors from chloride, bromide, and perchlorate are small, but nitrate causes large errors and must be absent.

Cation interferences are readily removed with a strong cation exchange resin in the hydrogen form:

$$2Rz^-H^+ + (M)_2SO_4 \rightarrow 2Rz^-M^+ + H_2SO_4$$

A strong cation resin contains a $-SO_3H$ groups integral to the resin, which can exchange metal cations with protons, removing the metal ion, as in the equation. The principles of ion exchange are discussed in Chapter 16.

The titration is carried out in an aqueous–organic solvent mixture. The organic solvent decreases the dissociation of the indicator and thereby hinders formation of a barium–indicator complex. It also results in a more flocculant precipitate (formed when colloids come out of suspension in the form of flocs or flakes) with better adsorption properties for the indicator.

Questions

1. For the given salts, whose solubility depends on pH and why?
   (a) $Pb(OH)_2$ (b) $PbCl_2$

2. Compared to pure water, which salt will be more soluble in $1.0\ M\ H^+$?
   (a) NaCl (b) KI (c) $FePO_4$
   (d) AgBr (e) KNO_3

3. Which of the following indicators is commonly used in a precipitation titration?
   (a) Phenolphthalein (b) Self indicator
   (c) Adsorption indicator (d) Starch indicator

4. Differentiate between three methods that are Volhard's, Mohr's and Fajan's method used for precipitation reaction.

5. Match the adsorption indictor with proper metal ions in the precipitation titration.

Indicator	Metal Ions
1. Eosin	(a) Pb^{2+}
2. Thorin	(b) Br^-
3. Fluorescein	(c) Ba^{2+}
4. Orthochrome T	(d) Ag^+

6. Explain the Volhard titration of chloride, and the Fajan titration. Which is used for acid solution? Why?

7. Explain the principles of adsorption indicators.

Problems

Effect of Acidity on Solubility

8. Calculate the solubility of $AgIO_3$ in $0.100\ M\ HNO_3$. Also calculate the equilibrium concentrations of IO_3^- and HIO_3.

9. Calculate the solubility of CaF_2 in $0.100\ M\ HCl$. Also calculate the equilibrium concentrations of F^- and HF.

10. Calculate the solubility of PbS in $0.0100\ M\ HCl$. Also calculate the equilibrium concentrations of S^{2-}, HS^-, and H_2S.

Effect of Complexation on Solubility

11. Silver ion forms a stepwise 1 : 2 complex with ethylenediamine (en) with formation constants of $K_{f1} = 5.0 \times 10^4$ and $K_{f2} = 1.4 \times 10^3$. Calculate the solubility of silver chloride in $0.100\ M$ ethylenediamine. Also calculate the equilibrium concentrations of $Ag(en)^+$ and $Ag(en)_2^+$.

Mass Balance Calculations

12. Calculate the solubility of $AgIO_3$ in $0.100\ M\ HNO_3$, using the mass balance approach. Compare with Problem 8.

13. Calculate the solubility of PbS in $0.0100\ M\ HCl$, using the mass balance approach. Compare with Problem 10.

14. Calculate the solubility of AgCl in 0.100 M ethylenediamine. Compare with Problem 11. The formation constant is given in Problem 11.

15. In Mohr's titration, a 0.4179 g sample containing chloride and inert material is titrated with 0.1012 M AgNO$_3$, requiring 34.99 mL to reach the Ag$_2$CrO$_4$ end point. Whereas, 0.32 mL of AgNO$_3$ was required for blank sample. Calculate the percent weight chloride in the sample (Molecular weight = 35.453 g/mol).

Quantitative Precipitation Determinations

16. What CrO$_4^{2-}$ must be present to cause Ag$_2$CrO$_4$(s) to precipitate in 0.00105 M AgNO$_3$(aq)?

17. Chloride in a brine solution is determined by the Volhard method. A 10.00-mL aliquot of the solution is treated with 15.00 mL of standard 0.1182 M AgNO$_3$ solution. The excess silver is titrated with standard 0.101 M KSCN solution, requiring 2.38 mL to reach the red Fe(SCN)$^{2+}$ end point. Calculate the concentration of chloride in the brine solution, in g/L.

18. In a Mohr titration of chloride with silver nitrate, an error is made in the preparation of the indicator. Instead of 0.011 M chromate indicator in the titration flask at the end point, there is only 0.0011 M. If the flask contains 100 mL at the end point, what is the error in the titration in milliliters of 0.100 M titrant? Neglect errors due to the color of the solution.

Spreadsheet Problem

19. 50 mL of 0.1 M thiocyanate is mixed with 100 mL 0.1 M chloride and the solution is titrated with 0.1 M AgNO$_3$. Generate the titration curve, using the approach of Equation 10.15 in both titration steps. See the text website for a solution.

PROFESSOR'S FAVORITE PROBLEM

Contributed by Professor Wen-Yee Lee, University of Texas at El Paso.

20. Given K_{sp}: Mn(OH)$_2$ = 1.6 × 10^{-13}; Ca(OH)$_2$ = 6.5 × 10^{-6}. By adding NaOH to a mixture containing 0.10 M Mn^{2+} and 0.10 M Ca^{2+}, is it possible to separate 99.0% of Mn^{2+} from Ca^{2+} without precipitation of Ca(OH)$_2$? Use calculations to explain your answer.

Multiple Choice Questions

1. For Mohr's titration, the optimum pH condition is
   - (a) pH < 6
   - (b) pH = 8
   - (c) pH > 10
   - (d) None of these

2. For pH maintenance during Mohr's titration, the base added is
   - (a) NH$_4$OH
   - (b) NaOH
   - (c) Ca(OH)$_2$
   - (d) CaCO$_3$

3. Anions that can be determined by precipitation with silver in the Volhard's method of indirect titration is
   - (a) Cl$^-$
   - (b) Br$^-$
   - (c) SCN$^-$
   - (d) All of these

4. In Fajan's method, which among the following indicators cannot be used for titration of Cl$^-$ at any pH?
   - (a) Methyl violet
   - (b) Eosin
   - (c) Bromophenol blue
   - (d) Fluorescein

5. During sulfate determination with barium ion, cation causing error by co-precipitation is/are:
   - (a) K$^+$
   - (b) Na$^+$
   - (c) NH$_4^+$
   - (d) All of these

Chapter 11

SPECTROCHEMICAL METHODS

"I cannot pretend to be impartial about colours. I rejoice with the brilliant ones and am genuinely sorry for the poor browns."

—Sir Winston Churchill

KEY THINGS TO LEARN FROM THIS CHAPTER

- Wavelength, frequencies, and photon energy (key Equations: 11.1 to 11.3)
- How molecules absorb electromagnetic radiation
- UV–visible absorption and molecular structure
- IR absorption and molecular structure
- Near-IR spectrometry
- Spectral databases
- Beer's law calculations (key Equations: 11.10, 11.13)
- Mixture calculations (key Equations: 11.16 and 11.17)—use of spreadsheets for calculations

- Using spreadsheets to calculate unknown concentrations, and their standard deviations from the calibration curve
- Spectrometers (components) for UV, visible, and IR regions
- FTIR spectrometers
- Spectrometric error
- Fluorometry
- Chemiluminescence
- Fiber-optic sensors

Spectrometry, particularly in the visible region of the electromagnetic spectrum, is one of the most widely used methods of analysis. It is widely used in clinical chemistry and environmental laboratories because many substances can be selectively converted to a colored derivative. The instrumentation is also readily available; miniature UV-visible spectrometers are relatively inexpensive and easy to operate. In this chapter, we (1) describe the absorption of radiation by molecules and its relation to molecular structure; (2) make quantitative calculations, relating the amount of radiation absorbed to the concentration of an absorbing analyte; and (3) describe the instrumentation required for making measurements. Measurements can be made in the infrared, visible, and ultraviolet regions of the spectrum. The wavelength region of choice will depend upon factors such as (1) availability of instruments, (2) whether the analyte is colored or can be converted to a colored derivative, (3) whether it contains functional groups that absorb in the ultraviolet or infrared regions, and (4) whether other absorbing species are present in the solution. Infrared spectrometry is generally less suited for quantitative measurements but better suited for qualitative or fingerprinting information than are ultraviolet (UV) and visible spectrometry. However, near-infrared spectrometry is increasingly being used for quantitative analysis, especially in process control applications.

We also describe a related technique, fluorescence spectrometry, in which the amount of light emitted upon excitation is related to the concentration. This is an extremely sensitive analytical technique.

Spectrophotometric methods can be used to determine mixtures of analytes, even if their spectra may overlap, by making measurements at more than one wavelength. Modern instruments make simultaneous measurements at multiple wavelengths and

Visible spectrometry is probably the most widely used analytical technique.

include software for performing calculations. Near-infrared spectrometry can be used to determine protein content in grains with no sample preparation. Fluorescence can be used for the sensitive determination of polyaromatic hydrocarbons in environmental samples.

Who Was the First Spectroscopist?

Johannes Marcus Marci of Kronland (1595–1667) in Eastern Bohemia is likely the first spectroscopic scientist. He was interested in the phenomenon of the rainbow and performed experiments to explain it. He published a book in ca. 1648 whose title, roughly translated, is *The Book of Thaumas, about the Heavenly Rainbow and the Nature of the Colors That Appear and Also about Its Origin and the Causes Thereof*. He described the conditions responsible for the production of rainbows and wrote about the production of a spectrum by passing a beam of light through a prism. The phenomenon (and the rainbow) was correctly explained as being due to the diffraction of light. Newton, over 20 years later, performed experiments similar to Marci's and provided a more rigorous explanation of the colors of the rainbow. Although he is credited more often, Marci was the first!

11.1 Interaction of Electromagnetic Radiation with Matter

Spectrometry is based on the absorption of photons by the analyte.

In spectrometric methods, the sample solution absorbs electromagnetic radiation (EMR), i.e., light (in the rest of this chapter, we use the term light generally, this does not necessarily mean visible light) from an appropriate source, and the amount absorbed is related to the concentration of the analyte in the solution. A solution containing copper ions is blue because it absorbs the *complementary color* yellow from white light and transmits the remaining blue light (see Table 11.1). The more concentrated the copper solution, the more yellow light is absorbed and the deeper the resulting blue color of the solution. In a spectrometric method, the amount of this yellow light absorbed would be measured and related to the concentration. We can obtain a better understanding of absorption spectrometry from a consideration of the electromagnetic spectrum and how molecules absorb radiation. See http://science.csustan.edu/tutorial/color/index.htm for an excellent tutorial on the basics of light and color. It also describes the additive and subtractive primary colors. Absorption of light can also result in the emission of light; light can also be scattered in a characteristic manner by molecules; all of these phenomena have analytical utility.

TABLE 11.1 Colors of Different Wavelength Regions

Wavelength Absorbed (nm)	Absorbed Color	Transmitted Color (Complement)
380–450	Violet	Yellow-green
450–495	Blue	Yellow
495–570	Green	Violet
570–590	Yellow	Blue
590–620	Orange	Green-blue
620–750	Red	Blue-green

The Electromagnetic Spectrum

For our purposes, electromagnetic radiation can be considered any form of radiant energy that is propagated as a transverse wave. It vibrates perpendicular to the direction of propagation, and this imparts a wave motion to the radiation, as illustrated in Figure 11.1. The wave is described either in terms of its **wavelength**, the distance of one complete cycle, or in terms of the **frequency**, the number of cycles passing a fixed point per unit time. The reciprocal of the wavelength is called the **wavenumber** and is the number of waves in a unit length or distance per cycle.

When interpreting the information in Table 11.1, recognize that the color of light of the wavelength stated in column 1 appears in column 2 and when this color is absorbed by a solution, it looks the color of column 3 to us. Thus, a light-emitting diode that emits light at 660 nm is an LED emitting red light and a dye that absorbs strongly at this wavelength, methylene blue, is blue in color.

The relationship between the wavelength and frequency is

$$\lambda = \frac{c}{v} \tag{11.1}$$

where λ is the wavelength in centimeters (cm), v is the frequency in reciprocal seconds (s^{-1}), or hertz (Hz), and c is the velocity of light (3×10^{10} cm/s). The wavenumber is represented by $\bar{v}$, in cm^{-1}:

$$\bar{v} = \frac{1}{\lambda} = \frac{v}{c} \tag{11.2}$$

The wavelength of electromagnetic radiation varies from a few angstroms to several meters. The units used to describe the wavelength are as follows:

Å = angstrom = 10^{-10} meter = 10^{-8} centimeter = 10^{-4} micrometer
nm = nanometer = 10^{-9} meter = 10 angstroms = 10^{-3} micrometer
μm = micrometer = 10^{-6} meter = 10^{4} angstroms

The wavelength unit preferred for the *ultraviolet* and *visible* regions of the spectrum is nanometer, while the unit micrometer is preferred for the *infrared* region. In this last case, wavenumbers are often used in place of wavelength, and the unit is cm^{-1}. See below for a definition of the ultraviolet, visible, and infrared regions of the spectrum.

Electromagnetic radiation possesses a certain amount of energy. The energy of ultimately indivisible unit of radiation, called the **photon**, is related to the frequency or wavelength by

$$E = hv = \frac{hc}{\lambda} \tag{11.3}$$

Wavelength, frequency, and wavenumber are interrelated.

My favorite AM radio station broadcasts at 870 kHz and my favorite FM station broadcasts at 90.1 MHz. The corresponding wavelengths are 345 and 3.33 m. What are yours? My favorite TV station is Channel 33 that broadcasts over a 584–590 MHz band. What are the corresponding wavelengths?

Wavelengths in the ultraviolet and visible regions are on the order of nanometers. In the infrared region, they are micrometers, but the reciprocal of wavelength is often used (wavenumbers, in cm^{-1}).

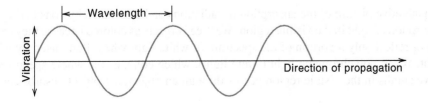

FIGURE 11.1 Wave motion of electromagnetic radiation.

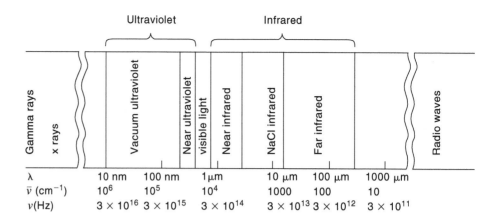

FIGURE 11.2 Electromagnetic spectrum.

Shorter wavelengths have greater energy. That is why ultraviolet radiation from the sun burns you! Sunscreen lotions contain UV absorbing compounds that prevent this radiation from reaching your skin. The sun protection factor (SPF) is a direct measure of how well the lotion absorbs the UV radiation.

Working ranges of the UV/Vis and IR spectra.

UV	190–380 nm
Vis	380–780 nm
Near-IR	0.78 – 2.5 μm
Mid-IR	2.5 – 15 μm

where E is the energy of the photon in ergs and h is Planck's constant, 6.63×10^{-34} joule-second (J-s) or 4.14×10^{-15} eV.s. It is apparent, then, that *the shorter the wavelength or the greater the frequency, the greater the energy.*

The birth of the quantum theory came about in trying to explain the electronic structure of atoms and the properties of light. It became apparent toward the end of the nineteenth century that the classical laws of physics (classical mechanics as proposed by Isaac Newton in the seventeenth century) could not be used to describe electronic structure. The new theory of quantum mechanics, developed at the beginning of the twentieth century, was a scientific breakthrough that changed the way we view atoms.

As indicated above, the electromagnetic spectrum is arbitrarily broken down into different regions according to wavelength. The various regions of the spectrum are shown in Figure 11.2. We will not be concerned with the gamma-ray and X-ray regions in this chapter, although these high-energy radiations can be used in principle in the same manner as lower-energy radiations. The *ultraviolet* region extends from about 10 to 380 nm, but the most analytically useful region is from 190 to 380 nm, called the **near-ultraviolet or quartz UV region**. Below 190 nm, air, notably oxygen, absorbs appreciably and so the instruments are operated under a vacuum; hence, this wavelength region is called the **vacuum-ultraviolet region**. The **visible (Vis) region** is actually a very small part of the electromagnetic spectrum, and it is the region of wavelengths that can be seen by the human eye, that is, where the light appears as different colors, depending on the wavelength. The visible region extends from the near-ultraviolet region (380 nm, deep violet) to about 780 nm (far red). The **infrared (IR) region** extends from about 0.78 μm (780 nm) to 300 μm, but the range from 2.5 to 15 μm is the most frequently used for analysis, corresponding to a wavenumber range of 4000 to 667 cm^{-1}. The 0.8- to 2.5-μm range is known as the **near-infrared region**, the 2.5- to 16-μm region as the **mid- or NaCl-infrared region**, and longer wavelengths as the **far-infrared region**. We shall not be concerned with lower-energy radiation (radio or microwave) in this chapter, however, they do have important applications. For example, nuclear magnetic resonance spectroscopy involves the interaction of low-energy microwave radiation with the nuclei of atoms.

Absorption of Radiation by Matter

The color of an object we see is due to the wavelengths transmitted or reflected. The other wavelengths are absorbed.

A qualitative picture of the absorption of radiation can be obtained by considering the absorption of light in the visible region. We "see" objects as colored because they transmit or reflect only a portion of the spectrum of white light when illuminated by white light. When polychromatic light (white light), which contains the whole spectrum of wavelengths in the visible region passes through an object, the object absorbs certain of the wavelengths, leaving the unabsorbed wavelengths to be transmitted. These residual transmitted wavelengths are seen as a color. This color is **complementary** to the absorbed colors. In a similar manner, opaque objects will absorb certain wavelengths, leaving the residual light to be reflected and "seen" as a color that results from the combination of the wavelengths of the residual light.

Table 11.1 summarizes the approximate colors associated with different wavelengths in the visible spectrum. As an example, a solution of potassium permanganate absorbs light in the green region of the spectrum with an absorption maximum of 525 nm, and the solution appears purple.

There are three basic processes by which a molecule can absorb radiation; in all cases, the molecule is raised to a higher internal energy level, the increase in energy being equal to the energy of the absorbed photon ($h\nu$). The three types of internal energy are **quantized**; that is, they exist at discrete levels. First, the molecule rotates about various axes, the energy of rotation being at definite energy levels, so the molecule may absorb radiation and be raised to a higher rotational energy level, in a **rotational transition**. Second, the atoms or groups of atoms within a molecule vibrate relative to each other, and the energy of this vibration occurs at definite quantized levels. The molecule may then absorb a discrete amount of energy and be raised to a higher vibrational energy level, in a **vibrational transition**. Third, the outer shell electrons of a molecule, valence electrons, may be raised to a higher electron energy, corresponding to an **electronic transition**. With absorption of even more energetic radiation, X-rays, inner shell electrons are generally ejected. An excellent tutorial on various spectroscopic techniques involving X-rays is given in http://www.p-ng.si/~arcon/xas/xas/xas.htm.

Since each of these internal energy transitions is quantized, they will occur only at *definite wavelengths* corresponding to an energy ($h\nu$) equal to the quantized jump in the internal energy. There are, however, many *different* possible energy levels for each type of transition, and several wavelengths may be absorbed. The transitions can be illustrated by an energy-level diagram like that in Figure 11.3. The relative energy levels of the three transition processes are in the order electronic > vibrational > rotational, each being about an order of magnitude different in its energy level. Rotational transitions thus can take place at very low energies (long wavelengths, that is, the microwave or far-infrared region), but vibrational transitions require higher energies in the infrared

We see only a very small portion of electromagnetic radiation.

A molecule absorbs a photon by undergoing an energy transition exactly equal to the energy of the photon. The photon must have the right energy for this quantitized transition.

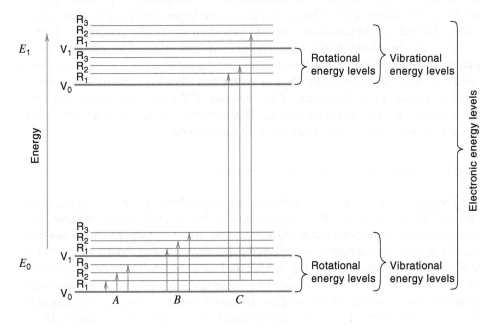

FIGURE 11.3 Energy level diagram illustrating energy changes associated with absorption of electromagnetic radiation: *A*, pure rotational changes (far infrared); *B*, rotational–vibrational changes (infrared and near infrared); *C*, rotational–vibrational–electronic transitions (visible and ultraviolet). E_0 is electronic ground state and E_1 is first electronic excited state.

to near-infrared region, while electronic transitions require still higher energies (in the visible and ultraviolet regions).

Rotational Transitions

Purely rotational transitions can occur in the **far-infrared** and **microwave** regions (ca. 100 µm to 10 cm), where the energy is insufficient to cause vibrational or electronic transitions. At or below room temperature, a molecule is usually in its lowest electronic energy state, called the **ground state** (E_0). Thus, the pure rotational transition will occur at the ground-state electronic level (*A* in Figure 11.3), although it is also possible to have an appreciable population of **excited states** of the molecule. When only rotational transitions occur, discrete absorption *lines* will occur in the spectrum, the wavelength of each line corresponding to a particular transition. Hence, fundamental information can be obtained about rotational energy levels of molecules. This region has been of little use analytically, however.

Vibrational Transitions

As the energy is increased (the wavelength decreased), vibrational transitions occur *in addition* to the rotational transitions, with different combinations of vibrational–rotational transitions. *Each* rotational level of the lowest vibrational level can be excited to different rotational levels of the excited vibrational level (*B* in Figure 11.3). In addition, there may be several different excited vibrational levels, each with a number of rotational levels. This leads to numerous discrete transitions. The result is a spectrum of *peaks* or "envelopes" of unresolved fine structure. The wavelengths at which these peaks occur can be related to vibrational modes within the molecule. These occur in the mid- and far-infrared regions. Some typical infrared spectra are shown in Figure 11.4.

Electronic Transitions

At still higher energies (visible and ultraviolet wavelengths), different levels of electronic transition take place, and rotational and vibrational transitions are superimposed on these (*C* in Figure 11.3). This results in an even larger number of possible transitions. Although all the transitions occur in quantized steps corresponding to discrete wavelengths, these individual wavelengths are too numerous and too close to resolve into the individual lines or vibrational peaks, and the net result is a spectrum of broad *bands* of absorbed wavelengths. Typical visible and ultraviolet spectra are shown in Figures 11.5 and 11.6. Due to lack of fine structure, UV-Visible spectra are not particularly useful for structure determinations. However, they can be used for confirmation by comparison with the spectra of the putative compound.

What Happens to the Absorbed Radiation?

The lifetimes of excited states of molecules are rather short, on the order of microseconds to femtoseconds, and the molecules quickly lose their energy of excitation and drop back down to the ground state. However, rather than emitting this energy as a photon of the same wavelength as absorbed, most of them get deactivated by collisional processes in which the energy is lost as heat; the heat is too small to be detected in most cases. This is the reason for a solution or a substance being colored. If the light were reemitted, then it would appear colorless. In some cases, light can be emitted, usually at longer wavelengths; this is called fluorescence, discussed later in Section 11.15.

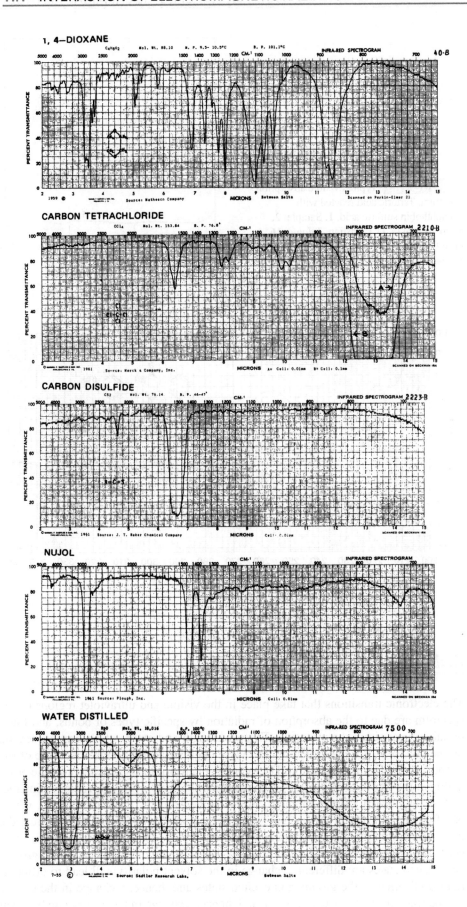

FIGURE 11.4 Typical infrared spectra. (From *26 Frequently Used Spectra for the Infrared Spectroscopist*, Standard Spectra-Midget Edition. Copyright © Sadtler Research Laboratories, Inc. Permission for the publication herein of Sadtler Standard Spectra © has been granted, and all rights are reserved by Sadtler Research Laboratories, Inc.)

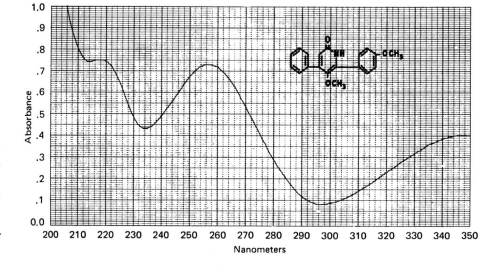

FIGURE 11.5 Typical visible absorption spectrum. Tartaric acid reacted with β-naphthol in sulfuric acid. 1, Sample; 2, blank. [From G. D. Christian, *Talanta*, **16** (1969) 255. Reproduced by permission of Pergamon Press, Ltd.]

FIGURE 11.6 Typical ultraviolet spectrum. 5-Methoxy-6-(*p*-methoxyphenyl)-4-phenyl-2(1*H*)-pyridone in methanol. (From *Sadtler Standard Spectra-u.v.* Copyright © Sadtler Research Laboratories, Inc., 1963. Permission for the publication herein of Sadtler Standard Spectra® has been granted and all rights are reserved by Sadtler Research Laboratories, Inc.)

11.2 Electronic Spectra and Molecular Structure

Radiation	Transition Type
Microwave	Rotational
Infrared	Rotational/ vibrational
Near IR	Vibrational
Visible	Outer electronic
UV	transitions

The electronic transitions that take place in the visible and ultraviolet regions of the spectrum are due to the absorption of radiation by specific types of bonds, and functional groups within the molecule. The wavelength and extent of absorption depend on the precise molecular structure. The wavelength of absorption is a measure of the energy required for the transition. The intensity of absorption is dependent on the probability of the transition occurring when the electronic system and the radiation interact and on the polarity (dipole moment) of the excited state of the chromophore, which is different from that in the ground state. The polarity of the solvent may affect the color of a chromophore, causing a hypsochromic (blue) shift or a bathochromic (red) shift. This effect is called solvatochromism. Since the dipole moments of the ground and excited states are different, a change in the solvent polarity will lead to differential stabilization of the ground and excited states and, hence, a change in the energy gap between the states, excited states, and, hence, a change in the energy gap between the states.

Kinds of Transitions

Electrons in a molecule can be classified into four different types. (1) Closed-shell electrons that are not involved in bonding. These have very high excitation energies and do not contribute to absorption in the visible or UV regions. (2) Covalent single-bond electrons (σ, or sigma, electrons). These also possess too high an excitation energy to contribute to absorption of visible or UV radiation (e.g., single-valence bonds in saturated hydrocarbons, $-CH_2-CH_2-$). (3) Paired nonbonding outer-shell electrons (n electrons), such as those on N, O, S, and halogens. These are less tightly held than σ electrons and can be excited by visible or UV radiation. (4) Electrons in π (pi) orbitals, for example, in double or triple bonds. These are the most readily excited and are responsible for the majority of visible and UV light absorption.

> π (double or triple bond) and n (outer-shell) electrons are responsible for most UV and visible electron transitions.

Electrons reside in orbitals. A molecule also possesses normally *unoccupied orbitals* called **antibonding orbitals**; these correspond to excited-state energy levels and are either σ^* or π^* orbitals. Hence, absorption of radiation results in an electronic transition to an antibonding orbital. The most common transitions are from π or n orbitals to antibonding π^* orbitals, and these are represented as $\pi \to \pi^*$ and $n \to \pi^*$ transitions, indicating a transition to an excited π^* state. The nonbonding n electron can also be promoted, at very short wavelengths, to an antibonding σ^* state: $n \to \sigma^*$. These occur at wavelengths less than 200 nm.

> Excited electrons go into antibonding (π^* or σ^*) orbitals. Most transitions above 200 nm are $\pi \to \pi^*$ or $n \to \pi^*$.

Examples of $\pi \to \pi^*$ and $n \to \pi^*$ transitions occur in ketones $\left(R-\overset{\displaystyle O}{\overset{\displaystyle \|}{C}}-R'\right)$. Representing the electronic transitions by valence bond structures, we can write

Acetone, for example, exhibits a high-intensity $\pi \to \pi^*$ transition and a low-intensity $n \to \pi^*$ transition in its absorption spectrum. An example of $n \to \pi^*$ transition occurs in ethers ($R-O-R'$). Since this occurs below 200 nm, ethers as well as thioethers ($R-S-R'$), disulfides ($R-S-S-R$), alkyl amines ($R-NH_2$), and alkyl halides ($R-X$) are transparent in the visible and UV regions; that is, they have no absorption bands in these regions.

$\pi \to \pi^*$ transitions are more probable than $n \to \pi^*$ transitions, and so the intensities of the absorption bands are greater for the former. Molar absorptivities, ϵ, at the wavelength of maximum absorption, λ_{max}, for $\pi \to \pi^*$ transitions are typically 1000 to 100,000, while for $n \to \pi^*$ transitions they are less than 1000; ϵ is a direct measure of the intensities of the bands.

Absorption by Isolated Chromophores

The absorbing groups in a molecule are called **chromophores**. A molecule containing a chromophore is called a **chromogen**. An **auxochrome** does not itself absorb radiation, but, if present in a molecule, it can enhance the absorption by a chromophore and/or shift the wavelength of absorption when attached to the chromophore. Examples are hydroxyl groups, amino groups, and halogens. These possess unshared (n) electrons that can interact with the π electrons in the chromophore (n–π conjugation).

Spectral changes can be classed as follows: (1) **bathochromic shift**—absorption maximum shifted to longer wavelength, (2) **hypsochromic shift**—absorption maximum shifted to shorter wavelength, (3) **hyperchromism**—an increase in molar absorptivity, and (4) **hypochromism**—a decrease in molar absorptivity.

In principle, the spectrum due to a chromophore is not markedly affected by minor structural changes elsewhere in the molecule. For example, acetone

$$CH_3 - \overset{\overset{\textstyle O}{\|}}{C} - CH_3$$

and 2-butanone,

$$CH_3 - \overset{\overset{\textstyle O}{\|}}{C} - CH_2CH_3$$

give spectra similar in shape and intensity. If the alteration is major or is very close to the chromophore, then changes can be expected.

Similarly, the spectral effects of two isolated chromophores in a molecule (separated by at least two single bonds) are, in principle, independent and are additive. Hence, in the molecule CH_3CH_2CNS, an absorption maximum due to the CNS group occurs at 245 nm with an ϵ of 800. In the molecule $SNCCH_2CH_2CH_2CNS$, an absorption maximum occurs at 247 nm, with approximately double the intensity ($\epsilon = 2000$). Interaction between chromophores may perturb the electronic energy levels and alter the spectrum.

Table 11.2 lists some common chromophores and their approximate wavelengths of maximum absorption. See Section 11.7 for a description of ϵ_{max}. It is a measure of

TABLE 11.2 Electronic Absorption Bands for Representative Chromophores[a]

Chromophore	System	λ_{max}	ϵ_{max}
Amine	—NH₂	195	2,800
Ethylene	—C=C—	190	8,000
Ketone	C=O	195	1,000
		270–285	18–30
Aldehyde	—CHO	210	Strong
		280–300	11–18
Nitro	—NO₂	210	Strong
Nitrite	—ONO	220–230	1,000–2,000
		300–400	10
Azo	—N=N—	285–400	3–25
Benzene		184	46,700
		202	6,900
		255	170
Naphthalene		220	112,000
		275	5,600
		312	175
Anthracene		252	199,000
		375	7,900

[a]From M. M. Willard, L. L. Merritt, and J. A. Dean, *Instrumental Methods of Analysis*, 4th ed. Copyright © 1948, 1951, 1958, 1965, by Litton Educational Publishing, Inc., by permission of Van Nostrand Reinhold Company.

how strongly light is absorbed, and depends on the wavelength and the nature of the absorbing material; the units are $cm^{-1} mol^{-1} L$ (also commonly written $M^{-1} cm^{-1}$).

It should be noted that exact wavelengths of an absorption band and the probability of absorption (intensity) cannot be calculated, and the analyst always runs standards under carefully specified conditions (temperature, solvent, concentration, instrument type, etc.). Databases of standard spectra and standard catalogs of spectra are available for reference.

Absorption by Conjugated Chromophores

Where multiple (e.g., double, triple) bonds are separated by just one single bond each, they are said to be conjugated. The π orbitals overlap, which decreases the energy gap between adjacent orbitals. The result is a bathochromic shift in the absorption spectrum and generally an increase in the intensity. The greater the degree of conjugation (i.e., several alternating double, or triple, and single bonds), the greater the shift. Conjugation of double or triple bonds with nonbonding electrons ($n - \pi$ conjugation) also results in spectral changes, for example, in $C{=}CH{-}NO_2$. This is called hyperconjugation, and electrons in a polarized sigma bond interact with an adjacent π-orbital to give an extended molecular orbital that increases the stability of the system.

Absorption by Aromatic Compounds

Aromatic systems (containing phenyl or benzene groups) exhibit conjugation. The spectra are somewhat different, however, than in other conjugated systems, being more complex. Benzene, absorbs strongly at 200 nm ($\epsilon_{max} = 6900$) with a weaker band at 230 to 270 nm ($\epsilon_{max} = 170$); see Figure 11.7. The weaker band exhibits considerable fine structure, each peak being due to the influence of vibrational sublevels on the electronic transitions.

Aromatic compounds are good absorbers of UV radiation.

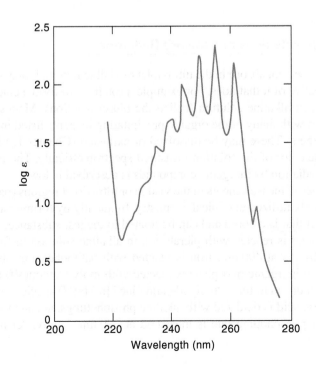

FIGURE 11.7 Ultraviolet spectrum of benzene.

As substituted groups are added to the benzene ring, a smoothing of the fine structure generally results, with a bathochromic shift and an increase in intensity. Hydroxy (—OH), methoxy (—OCH$_3$), amino (—NH$_2$), nitro (—NO$_2$), and aldehydic (—CHO) groups, for example, increase the absorption about 10-fold; this large effect is due to n–π conjugation. Halogens and methyl (—CH$_3$) groups act as auxochromes.

Polynuclear aromatic compounds (fused benzene rings), for example, naphthalene, have increased conjugation and so absorb at longer wavelengths. Naphthacene (four rings) has an absorption maximum at 470 nm (visible) and is yellow, and pentacene (five rings) has an absorption maximum at 575 nm and appears blue (see Table 11.1).

In polyphenyl compounds, para-linked molecules (1,4 positions, as shown) are capable of resonance interactions (conjugation) over the entire system, and increased numbers of para-linked rings result in bathochromic shifts (e.g., from 250 to 320 nm in going from $n = 0$ to $n = 4$). In meta-linked molecules (1,3 positions), however, such conjugation is not possible and no appreciable shift occurs up to $n = 16$. The intensity of absorption increases, however, due to the additive effects of the identical chromophores.

Many heterocyclic aromatic compounds, for example, pyridine, absorb in the UV region, and added substituents will cause spectral changes as for phenyl compounds.

Indicator dyes used for acid–base titrations and redox titrations (Chapters 7 and 21) are extensively conjugated systems and therefore absorb in the visible region. Loss or addition of a proton or an electron will markedly change the electron distribution and hence the color.

The absorption spectra of the same species can differ considerably from the vapor phase to solution. The greatest fine structure in the absorption spectra are observed in the vapor phase where intermolecular interactions are minimized. Some loss of resolution of the fine structure in the absorption spectra occurs in solution, less so in a non-interacting solvent than in one that interacts strongly with the analyte, even though the solvent itself is nonabsorbing.

What if a Molecule does not Absorb Radiation?

A nonabsorbing analyte can often be converted to an absorbing derivative.

If a compound does not absorb in the ultraviolet or visible region, it may be possible to prepare a derivative of it that does. For example, proteins will form a colored complex with copper(II) in alkaline solution (called the biuret reaction). Metals form highly colored chelates with many of the organic precipitating reagents listed in Table 9.2, as well as with others. These may be dissolved or extracted (Chapter 13) in an organic solvent, and the color of the solution measured spectrometrically. The mechanism of absorption of radiation by inorganic compounds is described below.

Spectrometric measurements in the visible or ultraviolet regions (particularly the former) are widely utilized in clinical chemistry, frequently by forming a derivative or reaction product that is colored and can be related to the test substance. For example, creatinine in blood is reacted with picrate ion in alkaline solution to form a colored product that absorbs at 490 nm. Iron is reacted with bathophenanthroline and measured at 535 nm; inorganic phosphate is reacted with molybdenum(VI) and the complex formed is reduced to form "molybdenum blue" [a Mo(V) species] that absorbs at 660 nm; and uric acid is oxidized with alkaline phosphotungstate, and the blue reduction product of phosphotungstate is measured at 680 nm. Ultraviolet measurements include the determination of barbiturates in alkaline solution at 252 nm, and the monitoring of many enzyme reactions by following the change in absorbance at 340 nm due to changes in the reduced form of nicotinamide adenine dinucleotide (NADH), a common reactant or product in enzyme reactions. Clinical measurements are discussed in more detail in Chapter 24 on the text website.

How do Inorganic Chelates Absorb So Intensely?

The absorption of ultraviolet or visible radiation by a metal complex can be ascribed to one or more of the following transitions: (1) *excitation of the metal ion*, (2) *excitation of the ligand*, or (3) *charge transfer transition*. Excitation of the metal ion in a complex usually has a very low molar absorptivity (ϵ), on the order of 1 to 100 cm^{-1} mol^{-1} L, and is not useful for low-level quantitative analysis. Most ligands used are organic chelating agents that exhibit the absorption properties discussed above, that is, can undergo $\pi \rightarrow \pi^*$ and $n \rightarrow \pi^*$ transitions. Complexation with a metal ion is similar to protonation of the molecule and will result in a change in the wavelength and intensity of absorption. These changes are slight in most cases.

    The intense color of metal chelates is frequently due to charge transfer transitions. This is simply the movement of electrons from the metal ion to the ligand, or vice versa. Such transitions include promotion of electrons from π levels in the ligand or from σ bonding orbitals to the unoccupied orbitals of the metal ion, or promotion of σ-bonded electrons to unoccupied π orbitals of the ligand.

    When such transitions occur, a redox reaction actually occurs between the metal ion and the ligand. Usually, the metal ion is reduced and the ligand is oxidized, and the wavelength (energy) of maximum absorption is related to the ease with which the exchange occurs. A metal ion in a lower oxidation state, complexed with a high electron affinity ligand, may be oxidized without destroying the complex. An important example is the 1,10-phenanthroline chelate of iron(II).

    Molar absorptivities of charge transfer complexes are very high, with ϵ values typically 10,000 to 100,000 cm^{-1} mol^{-1} L; they occur in either the visible or UV regions. The intensity (ease of charge transfer) is increased by increasing the extent of conjugation in the ligand. Metal complexes of this type are intensely colored due to their high absorption and are well suited for the detection and measurement of trace concentration of metals.

Charge transfer transitions between a metal ion and complexing ligand lead to highly absorbing complexes.

11.3   Infrared Absorption and Molecular Structure

Infrared spectroscopy is very useful for obtaining qualitative information about molecules. But molecules must possess certain properties in order to undergo absorption.

Absorption of Infrared Radiation

Not all molecules can absorb in the infrared region. For absorption to occur, there must be a *change in the dipole moment (polarity) of the molecule*. The alternating electrical field of the radiation (electromagnetic radiation consists of an oscillating electrical field and an oscillating magnetic field, perpendicular to each other) interacts with fluctuations in the dipole moment of the molecule. If the frequency of the radiation matches the vibrational frequency of the molecule, then radiation will be absorbed, causing a change in the amplitude of molecular vibration. A diatomic molecule must have a permanent dipole (polar covalent bond in which a pair of electrons is shared unequally) in order

The molecule must undergo a change in dipole moment in order to absorb infrared radiation.

to absorb, but larger molecules do not. For example, nitrogen, $N \equiv N$, cannot exhibit a dipole and will not absorb in the infrared region. An unsymmetrical diatomic molecule such as carbon monoxide does have a permanent dipole and hence will absorb. Carbon dioxide, $O = C = O$, does not have a permanent dipole, but by vibration it may exhibit a dipole moment. Thus, in the vibration mode $O \Rightarrow C \Leftarrow O$, there is symmetry and no dipole moment. But in the mode $O \Leftarrow C \Leftarrow O$, there is a dipole moment and the molecule can absorb infrared radiation, that is, via an induced dipole. The types of absorbing groups and molecules for the infrared and other wavelength regions will be discussed below.

Our discussions have been confined to molecules since nearly all absorbing species in solution are molecular in nature. In the case of single atoms (which occur in a flame or an electric arc) that do not vibrate or rotate, only electronic transitions occur. These occur as sharp lines corresponding to definite transitions and will be the subject of discussion in the next chapter.

Single atoms only undergo electronic transitions. So the spectra are sharp lines.

Infrared Spectra

The IR region is the "fingerprint" region.

Absorbing (vibrating) groups in the infrared region absorb within a certain wavelength region, and the exact wavelength will be influenced by neighboring groups. The absorption peaks are much sharper than in the ultraviolet or visible regions and thus easier to identify. In addition, each molecule will have a complete absorption spectrum unique to that molecule, and so a "fingerprint" of the molecule is obtained. See, for example, the top spectrum in Figure 11.4. Catalogs of infrared spectra are available for a large number of compounds for comparison purposes. See the references at the end of the chapter and the Web address at the end of this section. Mixtures of absorbing compounds will, of course, exhibit the combined spectra of compounds. Even so, it is often possible to identify the individual compounds from the absorption peaks of specific groups on the molecules. Typical functional groups that can be identified include alcohol, hydroxyl, ester carbonyl, olefin, and aromatic unsaturated hydrocarbon groups. Figure 11.8 summarizes regions where certain types of groups absorb. Absorption in the 6- to 15-μm region is very dependent on the molecular environment, and this is called the **fingerprint region**. A molecule can be identified by a comparison of its unique absorption in this region with cataloged known spectra.

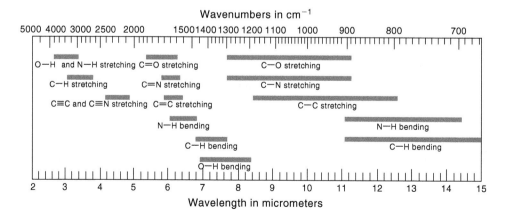

FIGURE 11.8   Simple correlations of group vibrations to regions of infrared absorption. (From R. T. Conley, *Infrared Spectroscopy,* 2nd ed. Boston: Allyn and Bacon, Inc., 1972. Reproduced by permission of Allyn and Bacon, Inc.)

Although the most important use of infrared spectroscopy is in identification and structure analysis, sometimes it is useful for quantitative analysis of complex mixtures of similar compounds because some absorption peaks for each compound will occur at a definite and selective wavelength, with intensities proportional to the concentration of absorbing species. See http://science.csustan.edu/tutorial/ir/index.htm for a quick summary of common absorption bands and compare with Figure 11.8. This gives a nice brief description of how to identify different types of compounds from the occurrence of one or more absorption bands. Equally importantly, certain classes of compounds can be excluded when their characteristic absorption bands are absent.

Usefulness of Infrared Spectrometry in Everyday Life

Infrared spectrometry is used for quantitative analysis in many applications, such as industrial hygiene and air quality monitoring. When you have your car emission tested (in the United States, this is presently required in all states except Kentucky and Minnesota), an infrared probe is inserted in the exhaust tailpipe to measure CO, CO_2, and hydrocarbons (based on an average absorptivity for hydrocarbons).

If a person is arrested for driving under the influence of alcohol, chances are he or she will be tested for blood alcohol content by measurement of the alcohol in breath using a "breathalyzer" infrared instrument. You are directed to blow into a tube to collect alveolar (deep lung) air in a sample chamber (the air in equilibrium with the capillary blood in the lungs). Alcohol is measured using its absorption band at 3.44 µm. However, this absorption is not specific for alcohol; acetone in the breath is the most likely substance to interfere. Significant concentrations of acetone can occur in breath in cases of acidosis (ketosis) if a person is diabetic or has not eaten for a period of time. To correct for this, the absorbance is also measured at 3.37 µm where acetone absorbs much more strongly than ethanol. The two absorbances are used to calculate a corrected breath alcohol content. Calibration is done by blowing air through a standard aqueous alcohol solution, at a set temperature, into the sample cell. Breath alcohol content is converted to (and read out as) percent blood alcohol, using an average conversion factor of 2100:1 (the concentration of ethanol in the breath at body temperature 37°C, in grams/cc, is about 1/2100th that of blood.) This conversion factor varies some in reality, but is the accepted value in the United States. For measurement, one instrument collects 55.2 cc of breath at 50°C (to prevent water condensation), which is equivalent to 52.5 cc at mouth temperature of 34°C, which is 1/40th of 2,100. So multiplying the result by 40 gives the equivalent blood alcohol content. Many states have adopted a blood alcohol content of under 0.08% (0.08 g/100 cc) as the legal limit, that is, when your blood alcohol level reaches or exceeds this, you are presumed to be under the influence. In many European countries, it is even less. For an underage individual, any amount of breath alcohol may lead to loss of license until official adulthood. A blood alcohol content of 0.08 g/100 cc is equal to 0.0000381 g/100 cc breath.

11.4   Near-Infrared Spectrometry for Nondestructive Testing

The mid-infrared region (mid-IR) (1.5 to 25 µm) is widely used for identification purposes because of the extensive and unique fine structure contained in the spectra. Quantitative analysis is more limited because of the necessity of diluting samples to make measurements and the difficulty in finding solvents that do not absorb in the regions of interest. The region of the spectrum just beyond the visible end of the electromagnetic spectrum, from 0.75 to 2.5 µm (750 to 2500 nm), is called the **near-infrared** region (NIR region). Absorption bands in this region are weak and rather featureless but are useful for nondestructive quantitative measurements, for example, for analysis of solid samples.

Overtones and Bands—The Basis of NIR Absorption

NIR absorption is due to vibrational **overtones** and **combination bands**, which are *forbidden transitions* of low probability and hence the reason they are weak. These are related to fundamental vibrations in the mid-IR. Excitation of a molecule from the ground vibrational state to a higher vibrational state, where the vibrational quantum number v is ≥ 2, results in overtone absorptions. Thus, the first overtone band results from a $v = 0$ to $v = 2$ transition, while the second and third overtones result from a $v = 0$ to $v = 3$ and a $v = 0$ to $v = 4$ transition, respectively. Combination absorption bands arise when two different molecular vibrations are excited simultaneously. The intensity of overtone bands decreases by approximately one order of magnitude for each successive overtone. Absorption in the NIR is due mainly to C—H, O—H, and N—H bond stretching and bending motions.

Short- and Long-Wavelength NIR

The NIR region is often subdivided into the short-wavelength NIR (750 to 1100 nm) and the long-wavelength NIR (1100 to 2500 nm). This is based solely on the types of detectors used for the two regions (silicon detectors for the former and PbS, germanium and especially InGaAs detectors for the latter). Absorbances are generally weaker in the short-wavelength NIR region. So a 1- to 10-cm pathlength is typical, while a shorter 1- to 10-mm cell may be required for the long-wavelength NIR. NIR absorption, in general, is 10 to 1000 times less intense than in the mid-IR region, and so samples are usually run "neat" as powders, slurries, or liquids, with no dilution. In the mid-IR, samples are usually diluted, in the form of KBr pellets, thin films, mulls, or solutions, and cell pathlengths are limited to between 15 µm and 1 mm.

Calibration of NIR for Nondestructive Testing

NIR absorption is useful for nondestructive quantitative measurements. For example, the protein content of grains can be rapidly measured.

While near-IR spectra are rather featureless and have low absorption, the signal-to-noise ratio is high due to intense radiation sources, high radiation throughput, and sensitive detectors in the near-IR. The operating noise range for the mid-IR is typically in the milliabsorbance range, while near-IR detectors operate at microabsorbance noise levels, 1000 times lower (the definition of absorbance follows). Hence, with proper calibration, excellent quantitative results can be achieved.

Because of its penetration of undiluted samples and the ability to use relatively long pathlengths, NIR is useful for nondestructive and rapid measurement of samples that are not necessarily homogeneous; the long pathlength permits more representative samples. However, the low resolution of the technique limited its use for many years until the advent of inexpensive computing power and the development of statistical (chemometric) techniques to "train" software to recognize and resolve analyte spectra in a complex sample matrix, for example, using principal components regression analysis. Chemometric techniques utilize multivariate mathematical procedures for multicomponent measurements, measuring all at once, rather than one parameter or component at a time. Sophisticated software is available for automatic calibration and determination. In essence, calibrating standards containing the analyte at different concentrations are prepared in the sample matrix. These are used as training spectra from which the software is able to extract the analyte spectrum and prepare a calibration curve. Generally, the entire spectrum is measured simultaneously (see Section 11.8) and hundreds or thousands of wavelength absorbance data are used to extract the spectrum. For quantitative analysis (the main use of NIR spectroscopy), the composition of standards must be known or determined by an accepted method.

NIR Spectrometry: Uses

A major use of NIR is the bulk determination of nutrient values in milled grains such as wheat, corn, rice, and oats. The classical ways for analyzing these samples include Kjeldahl analysis for protein, Soxhlet extraction for fat, air drying for moisture, and refractometry for sugars. With proper calibration using mixtures of these constituents, it is possible to put a powdered grain sample in a cup and obtain a complete analysis in a few minutes. But each sample matrix (wheat, corn, etc.) requires its own set of calibrants since matrix matching is required for accurate analysis. Even different geographical sources of grains may require a different calibration model for each source. And usually hundreds of standard mixtures are needed. So the speed and flexibility of the technique is balanced by more time and effort needed to prepare standards and calibrate the instrument, and the technique is really limited to measurements in which thousands of samples are routinely analyzed. Another example of the use of NIR is the petrochemical industry, for measuring octane number, vapor pressure, aromatic content, and the like in refineries. These properties are linked to the composition of the hydrocarbons whose spectra are measured. Small portable analyzers for ethanol content of petroleum-based fuels rely on NIR measurements. One of the major uses of NIR spectroscopy in the process industry is moisture analysis. Such analyzers are capable of determining the moisture content continuously as the sample rapidly goes by, e.g., on a conveyor belt; see for example, http://www.processsensors.com/products/ProductsCat.asp.

11.5 Spectral Databases—Identifying Unknowns

The references at the end of the chapter list a number of useful catalogued spectral sources for UV–Vis and IR spectra, for compound identification. Bourassa et al. have compiled an excellent bibliography on spectral interpretation [*Spectroscopy*, **12**(1) (January) (1997) 10].

Powerful and versatile commercial spectral databases are available. Also, there are some basic free databases. We list some here. Details of each are given on the text website.

1. http://webbook.nist.gov. Gateway to NIST data collection of UV–Vis and IR spectra.
2. www2.chemie.uni-erlangen.de/services/telespec/. Simulation of infrared spectra.
3. www.ftirsearch.com. Pay-per-use spectral libraries for smaller laboratories, 87,000 spectra (Jan, 2012).
4. http://www.bio-rad.com/evportal/evolutionPortal.portal?_nfpb=true&_pageLabel =verticalLandingPage&catID=1300. Large collection, subscription-based access.

11.6 Solvents for Spectrometry

Obviously, the solvent used to prepare the sample must not absorb appreciably in the wavelength region where the measurement is being made. In the visible region, this limitation typically does not represent a big problem. There are many colorless solvents and, of course, water is used for inorganic substances. Water can also be used in the ultraviolet

TABLE 11.3 Lower Wavelength Usability Limit of Solvents in the Ultraviolet Region

Solvent	Cutoff Point (nm)	Solvent	Cutoff Point (nm)[a]
Water[b]	200	Dichloromethane	233
Ethanol (95%)	205	Butyl ether	235
Acetonitrile	210	Chloroform	245
Cyclohexane	210	Ethyl proprionate	255
Cyclopentane	210	Methyl formate	260
Heptane	210	Carbon tetrachloride	265
Hexane	210	N,N-Dimethylformamide	270
Methanol	210	Benzene	280
Pentane	210	Toluene	285
Isopropyl alcohol	210	m-Xylene	290
Isooctane	215	Pyridine	305
Dioxane	220	Acetone	330
Diethyl ether	220	Bromoform	360
Glycerol	220	Carbon disulfide	380
1,2-Dichloroethane	230	Nitromethane	380

[a]Wavelength at which the absorbance is unity (10% light is transmitted) for a 1-cm cell, with water as the reference.
[b]with respect to air

region. Many substances measured in the ultraviolet region are organic compounds that are insoluble in water, requiring the use of an organic solvent. Table 11.3 lists a number of solvents for use in the ultraviolet region. The cutoff point is the lowest wavelength at which the absorbance (see below) is unity, using a 1-cm cell with water as the reference. These solvents can all be used at least through to the visible region.

The choice of solvent can sometimes affect the spectrum in the ultraviolet region due to solvent–solute interactions. In going from a nonpolar to a polar solvent, loss of fine structure may occur and the wavelength of maximum absorption may shift (either bathochromic or hypsochromic, depending on the nature of the transition and the type of solute–solvent interactions).

The problem of finding a suitable solvent is more serious in the infrared region, where it is difficult to find one that is completely transparent. The use of either carbon tetrachloride or carbon disulfide (both of which, unfortunately, are volatile and toxic) will cover the most widely used region of 2.5 to 15 μm (see Figure 11.4). Water exhibits strong absorption bands in the infrared region, and it can be used only for certain limited regions of the spectrum. Also, special cell materials compatible with water must be used; cells for infrared measurements are usually made of NaCl because glass absorbs the radiation, but NaCl would dissolve in water. Any solvent used with cells made of NaCl must obviously be moisture-free.

Transparent solvents in the IR region are limited. Rather concentrated solutions of the sample must often be used.

11.7 Quantitative Calculations

The fraction of radiation absorbed by a solution of an absorbing analyte can be quantitatively related to its concentration. Here, we consider both pure compounds and mixtures.

This is not a beverage law, although it applies to the absorption of radiation by beer (and to measure the intensity of its color)!

Beer's Law—Relating the Amount of Radiation Absorbed to Concentration

The amount of monochromatic radiation absorbed by a sample is described by the Bouguer–Lambert-Beer's law, commonly called **Beer's law**. Consider the absorption of monochromatic radiation as in Figure 11.9[1]. Incident radiation of radiant power P_0 passes through a homogeneous medium, such as a solution of an absorbing species at concentrations c and pathlength b, and the emergent (transmitted) radiation has radiant power P. This radiant power is the quantity measured by spectrometric detectors. Bouguer in 1729 (P. Bouguer, *Essai d'otique sur la gradation de la lumier,* Paris, 1729) and Lambert in 1760 (J. H. Lambert, *Photometria,* Ausburg, 1760) recognized that when EMR is absorbed, the power of the transmitted radiation decreases exponentially. Assume, for example, that 25% of the incident radiation in Figure 11.9 is absorbed in a pathlength of b. Twenty-five percent of the remaining energy (25% of 0.75 P_0) will be absorbed in the next pathlength b, leaving 56.25% as the emergent radiation. Twenty-five percent of this would be absorbed in another pathlength of b, and so on, so that an infinite pathlength would be required to absorb all the radiant energy. Since the fraction of radiant energy transmitted decays exponentially with pathlength, we can write it in exponential form:

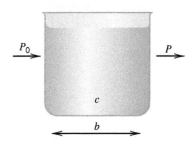

FIGURE 11.9 Absorption of radiation. P_0 = power of incident radiation, P = power of transmitted radiation, c = concentration, b = pathlength.

$$T = \frac{P}{P_0} = 10^{-kb} \qquad (11.4)$$

where k is a constant and T is called the **transmittance**, the fraction of radiant energy transmitted. Putting this in logarithmic form,

$$\log T = \log \frac{P}{P_0} = -kb \qquad (11.5)$$

In retrospect, it may seem obvious that, after all, the length of the path alone through which the light travels cannot be the only factor governing light attenuation. Rather, molecules are responsible for the absorption of light. The population density of such absorbing molecules per unit pathlength, tantamount to concentration, must play a role. However, nearly a century elapsed before this additional important relationship was explored. In 1852, Beer [A. Beer, *Ann. Physik Chem.*, **86** (1852) 78] and in 1853 Bernard [F. Bernard, *Ann. Chim. et Phys.*, **35** (1853) 385] each stated that the transmittance T depends on the concentration, c, just as it depends on the pathlength. Although they did not state it in this form, mathematically this will be similar to Equation 11.4:

$$T = \frac{P}{P_0} = 10^{-k'c} \qquad (11.6)$$

where k' is a new constant, or

$$\log T = \log \frac{P}{P_0} = -k'c \qquad (11.7)$$

The combination of these two relationships, represented by Equations 11.4 and 11.6, describes the dependence of T on both the pathlength and the concentration:

$$T = \frac{P}{P_0} = 10^{-abc} \qquad (11.8)$$

where a is a combined constant of k and k', and

$$\log T = \log \frac{P}{P_0} = -abc \qquad (11.9)$$

It is a more convenient to omit the negative sign on the right-hand side of the equation and to define a new term, **absorbance**:

$$A = -\log T = \log \frac{1}{T} = \log \frac{P_0}{P} = abc \qquad (11.10)$$

---

[1] Instead of P and P_0, the symbols I and I_0 are also commonly used.

Beer's law is as simple as abc!

where A is the absorbance. The relationships in Equations 11.8 through 11.10 are collectively called Beer's law, Equation 11.10 being the most common form. Note that it is the *absorbance* that is directly proportional to the concentration.

The **percent transmittance** is given by

$$\%T = \frac{P}{P_0} \times 100 \tag{11.11}$$

Equation 11.10 can be rearranged. Since $T = \%T/100$,

$$A = \log \frac{100}{\%T} = \log 100 - \log \%T$$

Or

$$\boxed{\begin{array}{l} A = 2.00 - \log \%T \\ \%T = 10^{(2.00-A)} \end{array}}$$

and

$$T = 10^{-A} \tag{11.12}$$

The following spreadsheet calculation and plot of absorbance and transmittance illustrates the exponential change of absorbance as a function of a linear change in transmittance.

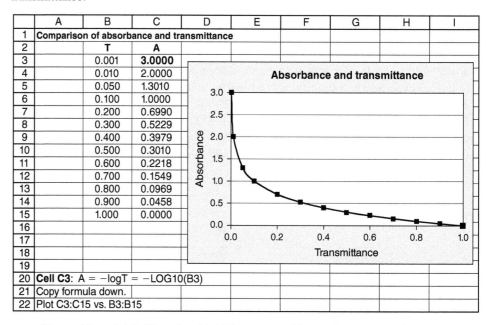

	A	B	C	D	E	F	G	H	I
1	Comparison of absorbance and transmittance								
2		T	A						
3		0.001	**3.0000**						
4		0.010	2.0000						
5		0.050	1.3010						
6		0.100	1.0000						
7		0.200	0.6990						
8		0.300	0.5229						
9		0.400	0.3979						
10		0.500	0.3010						
11		0.600	0.2218						
12		0.700	0.1549						
13		0.800	0.0969						
14		0.900	0.0458						
15		1.000	0.0000						
16									
17									
18									
19									
20	**Cell C3**: A = −logT = −LOG10(B3)								
21	Copy formula down.								
22	Plot C3:C15 vs. B3:B15								

The absorptivity varies with wavelength and represents the absorption spectrum.

The pathlength b in Equation 11.10 is expressed in centimeters and the concentration c in grams per liter. The constant a is called the **absorptivity** and is dependent on the wavelength and the nature of the absorbing material. In an absorption spectrum, the absorbance varies with wavelength in direct proportion to a (b and c are held constant). The product of the absorptivity and the molecular weight of the absorbing species is called the **molar absorptivity** ϵ. Thus,

$$\boxed{A = \epsilon b c} \tag{11.13}$$

$a = \text{cm}^{-1}\,\text{g}^{-1}\,\text{L}$

$\epsilon = \text{a} \times \text{f.wt.} = \text{cm}^{-1}\,\text{mol}^{-1}\,\text{L}$

where c is now in *moles per liter*. Most commonly, the cell pathlength in UV–Vis spectrometry is 1 cm; ϵ has the units $\text{cm}^{-1}\,\text{mol}^{-1}\,\text{L}$, while a has the units $\text{cm}^{-1}\,\text{g}^{-1}\,\text{L}$. Absorptivity varies with wavelength; thus, Beer's law holds strictly for monochromatic radiation. Table 11.4 lists recommended and older Beer's law symbols you will find in the literature.

TABLE 11.4 Spectrometry Nomenclature[2]

Recommended Name	Older Names or Symbols
Absorbance (A)	Optical density (OD), extinction, absorbancy
Absorptivity (a)	Extinction coefficient, absorbancy index, absorbing index
Pathlength (b)	l or d
Transmittance (T)	Transmittancy, transmission
Wavelength (nm)	mμ (millicron)

There are many Beer's law related symbols and terms in the literature. Here are some of them.

Example 11.1

A sample in a 1.0-cm cell transmits 80% light at a certain wavelength. If the absorptivity of this substance at this wavelength is 2.0, what is its concentration?

T is unitless. Check dimensional units.

Solution

The percent transmittance is 80%, and so $T = 0.80$:

$$\log \frac{1}{0.80} = 2.0 \text{ cm}^{-1} \text{ g}^{-1} \text{ L} \times 1.0 \text{ cm} \times c$$

$$\log 1.2_5 = 2.0 \text{ g}^{-1} \text{ L} \times c$$

$$c = \frac{0.10}{2.0} = 0.050 \text{ g/L}$$

Example 11.2

A solution containing 1.00 mg iron (as the thiocyanate complex) in 100 mL was observed to transmit 70.0% of the incident light compared to an appropriate blank.

(a) What is the absorbance of the solution at this wavelength?

(b) What fraction of light would be transmitted by a solution of iron four times as concentrated?

Solution

(a)

$$T = 0.700$$

$$A = \log \frac{1}{0.700} = \log 1.43 = 0.155$$

(b) According to Beer's law, absorbance is linearly related to concentration. If the original solution has an absorbance of 0.155, a four times more concentrated solution will have an absorbance four times as much or $4 \times 0.155 = 0.620$. The transmittance T will be: $T = 10^{-0.620} = 0.240$.

---

[2]A term commonly used in older literature is "Sandell's Sensitivity" for which there is no present counterpart. It is the concentration in μg/cm^3 that provides an absorbance of 0.001 in a 1-cm pathlength cell, it is expressed in units of μg/cm^2.

Example 11.3

Amines, RNH_2, react with picric acid to form amine picrates, which absorb strongly at 359 nm ($\epsilon = 1.25 \times 10^4$). An unknown amine (0.1155 g) is dissolved in water and diluted to 100 mL. A 1-mL aliquot of this is diluted to 250 mL for measurement. If this final solution exhibits an absorbance of 0.454 at 359 nm using a 1.00-cm cell, what is the formula weight of the amine?

Solution

$$A = \epsilon bc$$

$$0.454 = 1.25 \times 10^4 \text{ cm}^{-1} \text{ mol}^{-1} \text{ L} \times 1.00 \text{ cm} \times c$$

$$c = 3.63 \times 10^{-5} \text{ mol/L}$$

$$\frac{(3.63 \times 10^{-5} \text{ mol/L})(0.250 \text{ L})}{1.00 \text{ mL}} \times 100 \text{ mL} = 9.08 \times 10^{-4} \text{ mol in original flask}$$

$$\frac{0.1155 \text{ g}}{9.08 \times 10^{-4} \text{ mol}} = 127_{.2} \text{ g/mol}$$

Example 11.4

Chloroaniline in a sample is determined as the amine picrate as described in Example 11.3. A 0.0265-g sample is reacted with picric acid and diluted to 1 L. The solution exhibits an absorbance of 0.368 in a 1-cm cell. What is the percentage chloroaniline in the sample?

Solution

$$A = \epsilon bc$$

$$0.368 = 1.25 \times 10^4 \text{ cm}^{-1} \text{ mol}^{-1} \text{ L} \times 1.00 \text{ cm} \times c$$

$$c = 2.94 \times 10^{-5} \text{ mol/L}$$

$$(2.94 \times 10^{-5} \text{ mol/L}) (127.6 \text{ g/mol} = 3.75 \times 10^{-3} \text{ g chloroaniline}$$

$$\frac{3.75 \times 10^{-3} \text{ g chloroaniline}}{2.65 \times 10^{-2} \text{ g sample}} \times 100\% = 14.2\%$$

Mixtures of Absorbing Species

It is possible to make quantitative calculations when two absorbing species in solution have overlapping spectra. It is apparent from Beer's law that the total absorbance A at a given wavelength will be equal to the sum of the absorbances of all absorbing species. For two absorbing species, then, if c is in grams per liter,

$$\boxed{A = a_x bc_x + a_y bc_y} \tag{11.14}$$

The absorbances of individual absorbing species are additive.

or if c is in moles per liter,

$$\boxed{A = \epsilon_x bc_x + \epsilon_y bc_y} \tag{11.15}$$

where the subscripts refer to substances x and y, respectively.

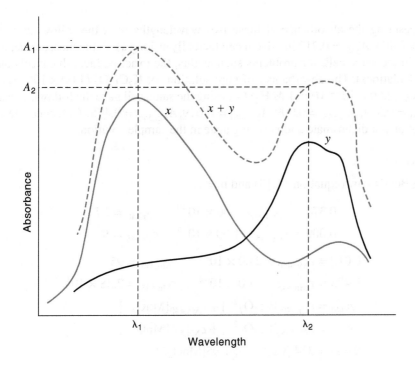

FIGURE 11.10 Absorption spectra of pure substances x and y and of mixture of x and y at same concentrations.

Consider, for example, the determination of substances x and y whose individual absorption spectra at their given concentration would appear as the solid curves in Figure 11.10, and the combined spectrum of the mixture is the dashed curve. Since there are two unknowns, two measurements will have to be made. The technique is to choose two wavelengths for measurement, one occurring at the absorption maximum for x (λ_1 in the figure) and the other at the maximum for y (λ_2 in the figure). We can write, then,

$$A_1 = A_{x1} + A_{y1} = \epsilon_{x1}bc_x + \epsilon_{y1}bc_y \qquad (11.16)$$

$$A_2 = A_{x2} + A_{y2} = \epsilon_{x2}bc_x + \epsilon_{y2}bc_y \qquad (11.17)$$

where A_1 and A_2 are the absorbances at wavelengths 1 and 2, respectively (for the mixture); A_{x1} and A_{y1} are the absorbances contributed by x and y, respectively, at wavelength 1; and A_{x2} and A_{y2} are the absorbances contributed by x and y, respectively, at wavelength 2. Similarly, ϵ_{x1} and ϵ_{y1} are the molar absorptivities of x and y, respectively, at wavelength 1; while ϵ_{x2} and ϵ_{y2} are the molar absorptivities of x and y, respectively, at wavelength 2. These molar absorptivities are determined by making absorbance measurements on pure solutions (known molar concentrations) of x and y at wavelengths 1 and 2. So c_x and c_y become the only two unknowns in Equations 11.16 and 11.17, and they can be calculated from the solution of the two simultaneous equations.

We have two unknowns (c_x and c_y). We need to write two equations that can be solved simultaneously.

Example 11.5

Potassium dichromate and potassium permanganate have overlapping absorption spectra in 1 M H_2SO_4. $K_2Cr_2O_7$ has an absorption maximum at 440 nm, and $KMnO_4$ has a band at 545 nm (the maximum is actually at 525 nm, but using the longer wavelength is advantageous because interference from $K_2Cr_2O_7$ is less). A mixture is analyzed by measuring the absorbance at these two wavelengths with the following results: $A_{440} = 0.405, A_{545} = 0.712$ in a 1-cm cell (actually as long as all the measurements are made in the same cell, for problems such as this, the exact pathlength cancels out in the calculations). The absorbances of pure solutions of $K_2Cr_2O_7$ (1.00×10^{-3} M) and $KMnO_4$ (2.00×10^{-4} M) in 1 M H_2SO_4, using the same cell gave the following results: $A_{Cr,440} = 0.374, A_{Cr,545} = 0.009, A_{Mn,440} = 0.019, A_{Mn,545} = 0.475$. Calculate the concentrations of dichromate and permanganate in the sample solution.

Solution

Using Beer's Law (equation 11.13) and b = 1,

$$0.374 = \varepsilon_{Cr,440} \times 1.00 \times 10^{-3} \quad \varepsilon_{Cr,440} = 374$$
$$0.009 = \varepsilon_{Cr,545} \times 1.00 \times 10^{-3} \quad \varepsilon_{Cr,545} = 9$$

$$0.019 = \varepsilon_{Mn,440} \times 2.00 \times 10^{-4} \quad \varepsilon_{Mn,440} = 95$$
$$0.475 = \varepsilon_{Mn,545} \times 2.00 \times 10^{-4} \quad \varepsilon_{Mn,545} = 2.38 \times 10^3$$

$$A_{440} = \varepsilon_{Cr,440}\,[Cr_2O_7^{2-}] + \varepsilon_{Mn,440}[MnO_4^-]$$
$$A_{545} = \varepsilon_{Cr,545}\,[Cr_2O_7^{2-}] + \varepsilon_{Mn,545}[MnO_4^-]$$
$$0.405 = 374\,[Cr_2O_7^{2-}] + 95[MnO_4^-]$$
$$0.712 = 9\,[Cr_2O_7^{2-}] + 2.38 \times 10^3[MnO_4^-]$$

Solving simultaneously,

$$[Cr_2O_7^{2-}] = 1.01 \times 10^{-3}\ M \quad [MnO_4^-] = 2.95 \times 10^{-4}\ M$$

Note that for Cr at 545 nm, where it overlaps the main Mn peak, the absorbance was measured to only one significant figure since it was so small. This is fine. The smaller the necessary correction, the better. Ideally, it should be zero.

General Solution to the Determination of two Components in a Mixture

Consider that at wavelengths λ_1 and λ_2 analyte A has molar absorptivities of $\varepsilon_{A,1}$ and $\varepsilon_{A,2}$, respectively. Similarly consider that analyte B has molar absorptivities of $\varepsilon_{B,1}$ and $\varepsilon_{B,2}$, respectively. In a mixture in which A and B are present at molar concentrations [A] and [B], respectively, the absorbance at wavelengths λ_1 and λ_2, A_1 and A_2 in a cell of pathlength b, must respectively be given by:

$$A_1 = \varepsilon_{A,1}b[A] + \varepsilon_{B,1}b[B] \tag{1}$$

and

$$A_2 = \varepsilon_{A,2}b[A] + \varepsilon_{B,2}b[B] \tag{2}$$

Multiplying equation 1 by $\varepsilon_{A,2}$ and equation 2 by $\varepsilon_{A,1}$ gives, respectively:

$$\varepsilon_{A,2}A_1 = \varepsilon_{A,1}\,\varepsilon_{A,2}b[A] + \varepsilon_{A,2}\,\varepsilon_{B,1}b[B] \tag{3}$$

$$\varepsilon_{A,1}A_2 = \varepsilon_{A,1}\,\varepsilon_{A,2}b[A] + \varepsilon_{A,1}\,\varepsilon_{B,2}b[B] \tag{4}$$

Subtracting (3) from (4) gives:

$$\varepsilon_{A,1}\,A_2 - \varepsilon_{A,2}\,A_1 = b[B](\varepsilon_{A,1}\,\varepsilon_{B,2} - \varepsilon_{A,2}\,\varepsilon_{B,1}) \tag{5}$$

[B] is thence

$$[B] = (\varepsilon_{A,1} A_2 - \varepsilon_{A,2} A_1)/(b(\varepsilon_{A,1} \varepsilon_{B,2} - \varepsilon_{A,2} \varepsilon_{B,1})) \tag{6}$$

Similarly by multiplying Equation (1) with $\varepsilon_{B,2}$ and Equation (2) with $\varepsilon_{B,1}$, subtracting and transposing,

$$[A] = (\varepsilon_{B,1} A_2 - \varepsilon_{B,2} A_1)/(b(\varepsilon_{A,2} \varepsilon_{B,1} - \varepsilon_{A,1} \varepsilon_{B,2})) \tag{7}$$

An Excel-based calculator for the solution of such simultaneous equations was previously presented in Equation 9.5. A similar calculator, specifically formulated to solve two-component Beer's law–based problems is presented in the web-based supplement of this chapter as Two-component Beer's Law Solution and used to solve Example 11.5.

While two-component compositions can be exactly determined in this fashion, this approach does not facilitate the determination of the component concentrations when the number of components exceeds two. However, n components are readily determined from n equations that result from measurement at n different wavelengths by matrix inversion approaches that are readily performed in Excel. See Matrix Approach to Solving Simultaneous Equations in the **website** supplement to this chapter. Software available from spectrophotometer manufacturers can in principle determine a large number of components at a time. However, it should be recognized the success of any such approach depends on the orthogonality of the spectral information among the different components, i.e., sufficient difference must exist in the spectra of the different components to perform accurate quantitation in this manner. Even in a two-component mixture, if the spectra of the two components were identical or one was an exact multiple of the other ($\varepsilon_{A,\lambda}/\varepsilon_{B,\lambda}$ values are the same for all values of λ), it will not be possible to determine either concentration with any certainty. Thus, although there is no mathematical barrier to determining n components with absorbance measurement at n or more wavelengths, in practice this is rarely carried out beyond three or four components.

As we have seen before, no measurement is exact. But if we solve for n components with n equations, we will have the apparent (but misleading!) satisfaction of having obtained some exact answers. One common practice is therefore to use more measurements than necessary (N equations to find n unknowns, N > n) to provide the best-fit value for each of the unknowns. It is possible to take the many possible ($\frac{N!}{(N-n)!n!}$) combination of n out of N equations and calculate the average and standard deviation of the best-fit parameters. Excel Solver can also readily provide the best-fit solutions. The following expanded example based on Example 11.5 illustrates this.

Example 11.6

In addition the data provided in Example 11.5, the following data were made available:

$$\varepsilon_{Mn,350} = 1850; \varepsilon_{Mn,400} = 215; \varepsilon_{Cr,350} = 2620; \varepsilon_{Cr,400} = 1860$$

The same mixture as in Example 11.5, diluted 10-fold in $1\,M\,H_2SO_4$, had measured absorbances of 0.320 and 0.194 at 350 and 400 nm, respectively. Use the data presented in both problems together to calculate the concentrations of $Cr_2O_7{}^{2-}$ and $MnO_4{}^-$ and the associated uncertainties.

Solution

We have four equations, two already stated in Example 11.5:

$$\text{At 440 nm} \qquad 374[Cr_2O_7^{2-}] + 95[MnO_4^-] = 0.405 \qquad (8)$$

$$\text{At 545 nm} \quad 9[Cr_2O_7^{2-}] + 2.38 \times 10^3 [MnO_4^-] = 0.712 \qquad (9)$$

And two new ones as given in this problem:

$$2620 \times 0.1 \times [Cr_2O_7^{2-}] + 1850 \times 0.1 \times [MnO_4^-] = 0.320$$

Simplifying,

$$\text{At 350 nm} \quad 262[Cr_2O_7^{2-}] + 185[MnO_4^-] = 0.320 \qquad (10)$$

Similarly

$$\text{At 400 nm} \quad 186[Cr_2O_7^{2-}] + 21.5[MnO_4^-] = 0.197 \qquad (11)$$

We have two different possibilities to approach this problem. The long-handed approach is as follows. Only a pair of equations is needed to produce a solution; four equations make six possible pairs: (8)–(9), (8)–(10), (8)–(11), (9)–(10), (9)–(11), and (10)–(11). We can solve each pair at a time and take the mean and standard deviations of the results. These will respectively lead to (retaining excess significant figures until the final step):

$[Cr_2O_7^{2-}]$, mM = 1.008, 1.005, 1.039, 1.013, 1.025, and 1.027
Mean ($\pm$ standard deviation) = 1.02 $\pm$ 0.01 mM

$[MnO_4^-]$, mM = 0.295, 0.306, 0.172, 0.295, 0.295, and 0.275
Mean ($\pm$ standard deviation) = 0.273 $\pm$ 0.051 mM

Obviously this approach is already tedious with only four measurements and will be untenable as many more measurements are made. Solver readily provides a best-fit solution as illustrated in Example 11.6.xlsx in the **website** supplement. The values generated by Solver, $[Cr_2O_7^{2-}] = 1.01$ mM and $[MnO_4^-] = 0.295$ mM are close but not identical to those obtained in the above approach. This is because while all solutions are considered individually and given equal weight in the first approach (this means that a single errant measurement can affect the overall final value more), Solver takes the least-squares minimization approach, which considers the entire data set holistically.

Determining the uncertainty of the best-fit values from Solver is a bit less straightforward, however. Robert De Levie provides a "Macrobundle" for Excel for downloading (http://www.bowdoin.edu/~rdelevie/excellaneous/#downloads) and it contains a "Solveraid" Macro that can provide the uncertainties of the fit parameters. E. J. Billo's book (Excel for Chemists, 2nd ed., Wiley: New York, 2001) also has a "solverstat" macro that can similarly determine the uncertainties. For those not ready to jump into "Macros" in Excel, the "Jackknife" approach (see, e.g., M. S. Caceci, *Anal. Chem.* **61** (1989) 2324) provides a reliable measure of the uncertainties; it is convenient for small data sets. In this approach, the best-fit parameter values are calculated n times omitting each of the n equations at a time. The standard deviation of these n sets of parameters then gives us the desired results, $[Cr_2O_7^{2-}] = 1.01 \pm 0.01$ mM and $[MnO_4^-] = 0.295 \pm 0.010$ mM as illustrated in Example 11.6.xlsx.

With multiple wavelength measurements, we may analyze for multiple components at a time. See Section 11.10 and Figure 11.26.

In making these difference measurements, we have assumed that Beer's law holds over the concentration ranges encountered. This may not always be true, further, if the absorbance due to one substance is much greater than that due to the other at both wavelengths, the determination of the other substance will not be very accurate.

Excel Exercise: Calculation of Unknown Sample Concentration from Calibration Curve, and Uncertainty—Application of Data Analysis Regression Function

We can use the Excel Regression function in Data Analysis to readily calculate a calibration curve and its uncertainty, and then apply this to calculate an unknown concentration and its uncertainty from its absorbance. See a description of the information it provides in Chapter 3 at the end of Section 3.23, and the video in that chapter illustrating its use. All we need to do is enter the standard concentration and absorbance values in a spreadsheet, and apply the function.

Iron is determined spectrophotometrically by reacting with 1,10-phenanthroline to produce a complex that absorbs at 510 nm. A series of standard solutions give the following absorbances: 0.100 ppm: 0.081; 0.200 ppm: 0.171; 0.500 ppm: 0.432; and 1.0 ppm: 0.857. The sample solution gives an absorbance of 0.463. Prepare a spreadsheet to calculate the calibration equation and calculate the concentration of the sample solution, including uncertainties.

Enter the calibration concentration values in cells A1:A4 and the corresponding absorbance values in cells B1:B4 (consult the text **website** spreadsheet: Uncertainty in calculating unknown concentration from regression.xlsx). Use the Data/Data Analysis/Regression tool in Excel that we discussed earlier in Chapter 3. Because our intention is to calculate the concentration corresponding to a known absorbance value when we bring up Data/Data Analysis/Regression, we use the absorbance values B1:B4 as the x-array and the concentration values A1:A4 as the y-array. The resulting regression equation with 95% uncertainty (note that you can choose some other value, e.g., 99%; in that case, the uncertainties will be higher) is (from the X Variable 1 and Intercept Coefficients and Intercepts):

$$\text{Concentration, ppm} = 1.161_5 \pm 0.0070 \times \text{absorbance} + 0.0025 \pm 0.0034$$

Note that as calculated, the equation is in the form $x = my + b$, the opposite of the usual least-squares representation of $y = mx + b$.

The r^2 value for this equation is calculated to be > 0.9999, so it should be an excellent fit to a straight line (make a chart to fit the data and see!). For an absorbance of 0.463, simple propagation of error calculations (see the **website** spreadsheet) for the expression can be applied:

$$(1.1615 \pm 0.0070 \times 0.463 + 0.0025 \pm 0.0034) \text{ ppm} = 0.540 \pm 0.005 \text{ ppm}$$

Note that if we had made multiple measurements of the sample and had obtained an absorbance of 0.463 ± 0.003, this could be put in just as well in the above expression and the final result would have been 0.540 ± 0.006 ppm.

See Chapter 15, Section 15, and Chapter 23, Examples 23.1 and 23.2, for examples of application of the regression program.

For an alternative rigorous approach to calculating the standard deviation of the sample concentration, see the relevant sections in the **website** supplement (Standard Deviation of the Sample Concentration, and the spreadsheet Std Devn of Sample Concn.xlsx), illustrated for this problem. See that the results are the same.

Quantitative Measurements from Infrared Spectra

IR spectrometry is normally used to identify functional groups. However, quantifcation can also be performed. Infrared instruments usually record the percent transmittance as a function of wavelength. The presence of scattered radiation, especially at higher concentrations in infrared work, makes direct application of Beer's law difficult. Also, due to rather weak sources, it is necessary to use relatively wide slits (which give rise to apparent deviations from Beer's law—see below). Therefore, empirical methods are often employed in quantitative infrared analysis, keeping experimental conditions constant. The **baseline** or **ratio method** is often used, and this is illustrated in Figure 11.11. A peak is chosen that does not fall too close to others of the test substance or of other substances. A straight line is drawn at the base of the band, and P and P_0 are measured at the absorption peak. (The curve is upside down from the usual absorption spectrum because transmittance is recorded against wavelength.) Log P_0/P is plotted against concentration in the usual manner. Unknowns are compared against standards run under the same instrumental conditions. This technique minimizes relative errors that are in proportion to the sample size, but it does not eliminate simple additive errors, such as factors that offset the baseline.

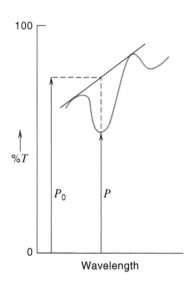

FIGURE 11.11 Baseline method for quantitative determination in infrared region of spectrum.

The types of instrument components will depend on the wavelength region.

11.8 Spectrometric Instrumentation

A **spectrometer** or **spectrophotometer** is an instrument that will resolve polychromatic radiation into different wavelengths and measure the light intensitry at one or more wavelengths. A block diagram of a spectrometer is shown in Figure 11.12. All spectrometers require (1) a **source** of continuous radiation over the wavelengths of interest, (2) a **monochromator** for dispersing the light into its component wavelengths and frequently, choosing a narrow band of wavelengths from the source spectrum, (3) a **sample cell**, (4) a **detector**, or transducer, for converting radiant energy into electrical energy, and (5) a device to read out the response of the detector. The sample may precede or follow the monochromator. Each of these, except the readout device, will vary depending on the wavelength region.

Sources

Sources for:
Vis—incandescent lamp, white LED
UV–H$_2$ or D$_2$ discharge lamp
IR—rare-earth oxide or silicon carbide glowers

The source should have a readily detectable output of radiation over the wavelength region for which the instrument is designed to operate. No source, however, has a constant spectral output. The most commonly used source for the **visible** region is a quartz tungsten–halogen (QTH) *lamp*. The spectral output of a typical QTH lamp is illustrated in Figure 11.13. The useful wavelength range is from about 325 or 350 nm to 25 μm, so it can also be used in the near-ultraviolet and near-infrared regions. A stable, regulated power supply is required to power a spectrometer light source. LEDs are very power-efficient light sources. They provide approximately monochromatic light; a white LED is composed of a blue LED and a phosphor that emits primarily in the green but extends into red. The usable wavelength for a white LED is 425–700 nm

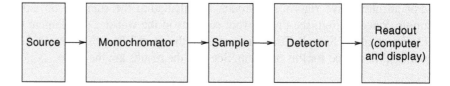

FIGURE 11.12 Block diagram of spectrometer.

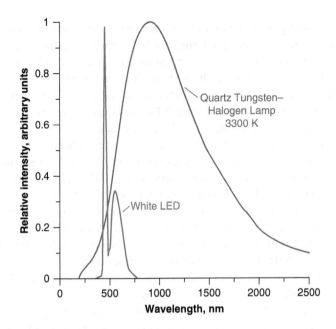

FIGURE 11.13 Intensity of radiation as function of wavelength for two typical light sources: Quartz tungsten–halogen lamp operating at 3300 K exhibits useful light intensity from 350–2500 nm. A white LED actually contains a blue InGaN LED emitting at 450 nm that is coated with a broadband phosphor that has peak emission in the green, at 550 nm; this source is useful from 425–700 nm.

(Figure 11.13). Such a low power consumption source is particularly well suited for battery-operated spectrometers. Sometimes, a 6-V storage battery is used as the voltage source.

For the **ultraviolet** region, a low-pressure *deuterium discharge lamp* is generally used as the source. The deuterium continuum emission extends from 185 to 370 nm but the lamp has useful spectral output out to 600 nm. Ultraviolet sources must have a quartz window, because glass is not transparent to ultraviolet radiation.

Infrared radiation is essentially heat, and so hot wires, light bulbs, or glowing ceramics are used as sources. The energy distribution from the black-body sources tends to peak at about 100 to 2000 nm (near-IR) and then tails off in the mid-IR. Infrared spectrometers usually operate from about 2 to 15 μm, and because of the relatively low-intensity radiation in this region, relatively large slits are used to increase the light throughput. But this degrades the wavelength resolution. For this reason, an interferometer is preferred for its increased throughput (see discussion of Fourier transform infrared instrument in Section 11.11). A typical infrared source is the *Nernst glower*. This is a rod consisting of a mixture of rare-earth oxides. It has a negative temperature coefficient of resistance and is nonconducting at room temperature. Therefore, it must be heated to excite the element to emit radiation, but once in operation it becomes conducting and furnishes maximum radiation at about 1.4 μm, or 7100 cm^{-1} (1500 to 2000°C). Another infrared source is the *Globar*. This is a rod of sintered silicon carbide heated to about 1300 to 1700°C. Its maximum radiation occurs at about 1.9 μm (5200 cm^{-1}), and it must be water-cooled. The Globar is a less intense source than the Nernst glower, but it is more satisfactory for wavelengths longer than 15 μm because its intensity decreases less rapidly. IR sources have no protection from the atmosphere, as no satisfactory envelope material exists.

In **fluorescence spectrometry**, the intensity of fluorescence is proportional to the intensity of the radiation source (see Section 11.15). Various continuum high-intensity UV-visible sources are used to excite fluorescence (see below). The use of lasers has also gained in importance because these monochromatic radiation sources can provide very high intensities. The wavelength and power characteristics of commercially available lasers are listed in many sources on the web; http://upload.wikimedia.org/wikipedia/commons/4/48/Commercial_laser_lines.svg

Working ranges of common UV/Vis sources:
Pulsed xenon arc: 180–2500 nm
dc deuterium: 185–2500 nm
dc arc: 200–2500 nm
quartz tungsten–halogen filament: 320–2200 nm
LEDs are available at discrete wavelengths from 240 to 4450 nm.

Lasers are intense monochromatic sources, good for fluorescence excitation.

provides an extensive graphical description. From the fluorine excimer laser at 157 nm to the methanol chemical laser at 0.7 mm, the lasers range in continuous power from fractions of a milliwatt as used in laser pointers, to > 1 kW for CO_2 lasers used for machining. Only those that lase in the visible-ultraviolet region are generally useful for exciting fluorescence. There is also increased interest in dyes that fluoresce in the NIR, and hence in the deep red, and NIR lasers are inexpensively available as solid-state diode lasers. Solid-state diode lasers are now available from the near-UV to NIR and with rare exceptions, such lasers are used whenever possible. Mass-produced, some of these devices are very inexpensive. The current version of a SONY Playstation™, for example, sports three lasers at 405, 640, and 780 nm, respectively, for Blu-Ray, DVD, and CDs, within a single package. Tunable lasers allow variation of the laser wavelength. Originally these were based on fluorescent organic dyes; each dye emission can generally be tuned over several tens of nm. Through choice of dyes that fluoresce at different wavelengths, the laser emission wavelength can be chosen to vary from UV to NIR. Present use of tunable lasers is dominated by Ti-Sapphire lasers (650–1100 nm), tunable fiber lasers (high-power fiber lasers have led to "super-continuum" sources that span from UV to IR) and semiconductor-based tunable diode lasers. Variable wavelength lasers are also useful as sources in absorption spectrometry because they provide good resolution and high intensity to interrogate highly absorbing samples.

We shall see below how spectrometric instruments can be adjusted to account for the variations in source intensity with wavelength as well as for the variation in detector sensitivity with wavelength.

Monochromators

A monochromator consists chiefly of a dispersing element to "separate" the wavelengths of the polychromatic radiation from the source. It additionally uses lenses or mirrors to focus the radiation, entrance and exit slits to reject unwanted radiation and help control the spectral purity of the radiation emitted from the monochromator. There are mainly two types of dispersing elements, the prism and the diffraction grating. Various types of optical filters may also be used to select specific wavelengths.

Dispersion by prisms is good at short wavelengths, poor at long wavelengths (IR).

1. Prisms.   When electromagnetic radiation passes through a prism, it is refracted because the index of refraction of the prism material is different from that of air. The index of refraction depends on the wavelength and, therefore, so does the degree of refraction. Shorter wavelengths are refracted more than longer wavelengths. The effect of refraction is to "spread" the radiation apart into different wavelengths (Figure 11.14). By rotation of the prism, different wavelengths of the spectrum can be made to pass through an exit slit and through the sample. A prism works satisfactorily in the ultraviolet and visible regions and can also be used in the infrared region. However, because of its **nonlinear dispersion**, it works more effectively for the shorter wavelengths. Glass prisms and lenses can be used in the visible region, but quartz or

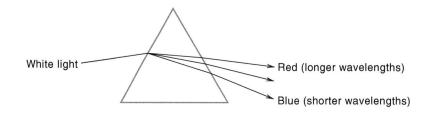

FIGURE 11.14 Dispersion of polychromatic light by prism.

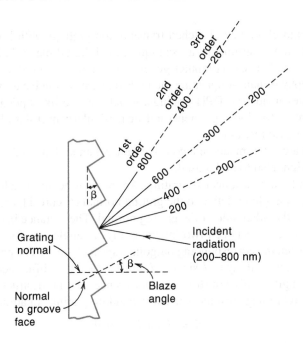

FIGURE 11.15 Diffraction of radiation from grating.

fused silica must be used in the ultraviolet region. The latter can also be used in the visible region.

In the infrared region, glass and fused silica transmit very little, and the prisms and other optics must be made from large crystals of alkali or alkaline earth halides, which are transparent to infrared radiation. Sodium chloride (rock salt) is used in most instruments and is useful from 2.5 to 15.4 μm (4000 to 650 cm^{-1}). For longer wavelengths, KBr (10 to 25 μm) or CsI (10 to 38 μm) can be used. These (and the entire monochromator compartment) must be kept dry.

2. Diffraction Gratings. These consist of a large number of parallel lines (grooves) ruled on a highly polished surface such as aluminum, about 6000 to 12000 per millimeter for the ultraviolet and visible regions and 1800 to 2400 per millimeter for the infrared region. The grooves act as scattering centers for rays impinging on the grating. The result is equal dispersion of all wavelengths of a given order, that is, **linear dispersion** (Figure 11.15). The resolving power depends on the number of ruled grooves, but generally the resolving power of gratings is better than that of prisms, and they can be used in all regions of the spectrum. They are particularly well suited for the infrared region because of their equal dispersion of the long wavelengths. Virtually all current instruments rely on gratings rather than prisms.

Originally, gratings were made with "ruling engines" that were complex instruments. Photolithographic techniques allowed gratings to be created from a holographic interference pattern. These have sinusoidal grooves and are preferred over ruled gratings because of their low stray light characteristics. High-quality replicas can be made of a master grating to lower fabrication costs. Holographic gratings can also be made from a photosensitive gel sandwiched between two plates by a technique called volume phase holography (VPH). They have no grooves but the refractive index within the gel varies in a periodic manner. These gratings are highly efficient, and complicated patterns can be incorporated into a single grating. Losses and aberrations from surface scattering and influence of surface defects are largely eliminated in such a grating.

Semiconductor/microfabrication technologies have brought about holographically patterned gratings made by deep reactive ion etching of both silicon and fused silica. These high-performance devices can be mass manufactured at low cost and

Dispersion by gratings is independent of wavelength, but the reflection efficiency varies with wavelength.

combine the high efficiency of etched transmission gratings with low stray light. In integrated miniature instruments, a technique called digital planar holography (DPH) is increasingly used. These computer-generated patterns are produced with standard micro-lithography or nano-imprinting methods that are conducive to mass production. Light propagates inside the DPH gratings, which behave like optical fibers (light is confined by a refractive index gradient) and are particularly useful in chip-scale or integrated spectrometric instruments.

Gratings are often made on a concave rather than a planar substrate; this allows the diffracted light also to be focused.

We consider here a reflective grating. An incident beam of radiation strikes the grating face at an angle i relative to the grating normal (Figure 11.15) and is reflected at an angle θ on the other side of the grating normal. The distance between grooves is d. The path difference between two incoming rays at angle i is $d \sin i$, and the path difference between the corresponding outgoing rays is $d \sin \theta$. The path difference for an incident and reflected ray is $d \sin i - d \sin \theta$. When this difference is equal to one or more wavelengths, fully constructive interference, and no destructive interference, occurs and a bright image results. The corresponding grating equation is

$$n\lambda = d(\sin i - \sin \theta) \tag{11.18}$$

where n is the *diffraction order*, and is an integer. It is apparent that if n is increased and the wavelength decreased by the same multiple, these shorter (higher-order) wavelengths will be reflected at the same angle, θ. These have to be filtered before they reach the detector (see below). To access light of different wavelengths, the grating is rotated so that the angle i changes.

The *dispersion* of a grating for a given incident angle, i, is given by

$$\frac{d\theta}{d\lambda} = \frac{n}{d \cos \theta} \tag{11.19}$$

that is, it equals the order divided by the product of the grating spacing and cosine of the angle of reflection. The *resolving power* of a grating is the product of the number of rulings and the order. So a large grating has greater resolving power than a small one.

The intensity of radiation reflected by a grating varies with the wavelength, the wavelength of maximum intensity being dependent on the angle from which the radiation is reflected from the surface of the groove in the grating: this angle is called the blazing angle (see Figure 11.15). Hence, gratings are blazed at specific angles for specific wavelength regions, and one blazed for the blue region would be poor for an infrared spectrometer. As mentioned, gratings will diffract radiation also at wavelength *multiples* of the diffracted wavelength (see Figure 11.15). These multiples are called **higher orders** of the radiation. The primary order is called the first order, twice the wavelength is the second order, three times the wavelength is the third order, and so on. So a grating produces first-order spectra, second-order spectra, and so on. The higher-order spectra are more greatly dispersed and the resolution increased. Because of the occurrence of higher orders, radiation at wavelengths less than that of the desired spectral region must be filtered out, or else its higher orders will overlap the radiation of interest. This can be accomplished with various types of optical filters (see below) that pass radiation only above a certain wavelength. For example, imagine a source is radiating over a broad range between 300 and 700 nm and after diffraction we wish to measure the intensity at 650 nm. Unless removed, light at 325 nm would overlap first-order radiation at 650 nm. The 325-nm light can be eliminated by placing a filter that blocks radiation ≤ 400 nm between the light source and the grating, preventing the 325-nm radiation from reaching the grating.

Higher orders are better dispersed.

In fluorescence, higher-order radiation from a shorter emitting (primary) wavelength may overlap a longer primary wavelength that is being measured. The shorter primary radiation must be filtered before reaching the grating. See also Section 11.9, single-beam spectrometers.

3. **Optical Filters.**   Traditional optical filters are basically of two types. Glass or plastic filters that contain various inorganic or organic colored compounds rely on absorbing some wavelengths while transmitting others. These are absorptive filters. Dichroic filters (also called "interference" or "thin film" or "reflective" filters) are made by coating a glass substrate with a series of different optical coatings. These filters rely on the principle of interference. The different layers function as a series of reflective cavities that resonate at the desired wavelengths. Other wavelengths are either reflected or eliminated by destructive interference as the wave peaks and troughs overlap.

Functionally, optical filters are generally classified into three types: short-pass, long-pass, and bandpass. The first type only transmits light that is below the stated wavelength, the second type only transmits light above the stated wavelength. At the stated wavelength, the transmittance is 50% of the maximum transmittance. Typically $\leq 1\%$ light is transmitted in the blocked portion and $> 80\%$, often close to 100%, of the light is transmitted in the permitted wavelengths. The third type, the bandpass filter, is a combination of a short-pass filter with a 50% cut point at λ_s and a long-pass filter with 50% of maximum transmittance at λ_L, with $\lambda_s > \lambda_L$. Thus, only the light between these wavelengths is transmitted. Bandpass filters are usually characterized by three parameters: (a) the center wavelength of the transmitted band [typically $\frac{1}{2}(\lambda_s + \lambda_L)$] (b) the width of the pass band, $\lambda_s - \lambda_L$, often called the half-width or full-width half maximum (FWHM); this can be fraction of a nm to more than 100 nm, 10–50 nm being common, and (c) maximum transmittance at the peak transmission wavelength; this generally decreases as the FWHM decreases and ranges from 10 to 90%. A "notch filter" is a special filter that transmits all wavelengths except a very narrow band at a certain wavelength, and at this wavelength the light is attenuated by as much as a millionfold or more. Such devices, centered on the laser wavelength, are used to filter any residual excitation laser light from the emitted light in laser-induced fluorescence or Raman spectroscopy. Notch filters can be regarded as a special bandpass filter where λ_L is just very slightly greater than λ_s.

An Acousto-Optic Tunable Filter (AOTF) is a solid-state, electronically tunable bandpass filter that utilizes the acousto-optic interaction inside an anisotropic medium to control what wavelength of light will be transmitted. The filters can be used with multi-line lasers or broadband light sources. They provide wavelength tunability over a broad range, good resolution (down to 0.4 nm), and intensity control. But they are not inexpensive.

Sample Cells

The cell holding the sample (usually a solution) must, of course, be transparent in the wavelength region being measured. The materials described above for the optics are used for the cell material in instruments designed for the various regions of the spectrum.

The cells for use in **visible** and **ultraviolet** spectrometers are usually cuvettes 1 cm in pathlength (*internal* distance between parallel walls, the walls are typically 1 mm thick, resulting in a 12×12 mm external cross section), although cells of different pathlengths and volumes can be used. These are illustrated in Figure 11.16. Flow-through spectrometers with integrated flow-through cells are commonly used in liquid chromatography, flow-injection analysis and other flow applications. Conventional spectrophotometers can also be adapted for such use with small volume flow-through cells as illustrated in the right image in Figure 11.16. In such applications, it is important to know the "z-dimension" of the spectrophotometer, this is the distance from the floor of the cell compartment to the center of the collimated light beam that passes through the cell. If the z-dimension of the cell does not match that of the spectrophotometer, the beam will hit the opaque portion of the cell.

Cells for:
UV—quartz
Vis—glass, quartz
IR—salt crystals

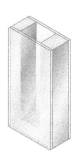

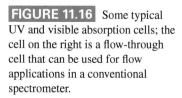

FIGURE 11.16 Some typical UV and visible absorption cells; the cell on the right is a flow-through cell that can be used for flow applications in a conventional spectrometer.

For **infrared** instruments, various types of cells are used. The most common is a cell with sodium chloride windows. Fixed-thickness cells are available for these purposes and are the most commonly used. The solvent, of course, must not attack the windows of the cell. Sodium chloride cells must be protected from atmospheric moisture (stored in desiccators) and moist solvents. They require periodic polishing to remove "fogging" due to moisture contamination. Silver chloride windows are often used for wet samples or aqueous solutions. These are soft, easily scratched, and gradually darken because of photoreduction to silver by visible light.

Table 11.5 lists the properties of several infrared transmitting materials. Especially when the windows must be polished, the pathlengths are difficult to reproduce, making quantitative analysis difficult. Use of an internal standard helps. The pathlength of the empty cell can be measured from the interference fringe patterns. Variable pathlength cells are also available in thicknesses from about 0.002 to 3 mm.

TABLE 11.5 Properties of Infrared Materials

Material	Useful Range (cm^{-1})	General Properties
NaCl	40,000–625	Hygroscopic, water soluble, low cost, most commonly used material.
KCl	40,000–500	Hygroscopic, water soluble.
KBr	40,000–400	Hygroscopic, water soluble, slightly higher in cost than NaCl and more hygroscopic.
CsBr	40,000–250	Hygroscopic, water soluble.
CsI	40,000–200	Very hygroscopic, water soluble, good for lower wavenumber studies.
LiF	83,333–1425	Slightly soluble in water, good UV material.
CaF$_2$	77,000–1110	Insoluble in water, resists most acids and alkalis.
BaF$_2$	67,000–870	Insoluble in water, brittle, soluble in acids and NH$_4$Cl.
AgCl	10,000–400	Insoluble in water, corrosive to metals. Darkens upon exposure to short-wavelength visible light. Store in dark.
AgBr	22,000–333	Insoluble in water, corrosive to metals. Darkens upon exposure to short-wavelength visible light. Store in dark.
40% TlBr, 60%TlI (KRS-5)	16,600–285	Insoluble in water, highly toxic, soluble in bases, soft, good for ATR work.
ZnS	50,000–760	Insoluble in water, common acids and bases, brittle.
ZnSe	20,000–500	Insoluble in water, common acids and bases, brittle.
Ge	5000–560	Brittle, high index of refraction.
Si	83,333–1430 400–30	Insoluble in most acids and bases.
UV quartz	56,800–3700	Unaffected by water and most solvents.
IR quartz	40,000–3000	Unaffected by water and most solvents.
Polyethylene	625–10	Low-cost material for far-IR work.

Adapted from McCarthy Scientific Co. Catalogue 489, with permission.

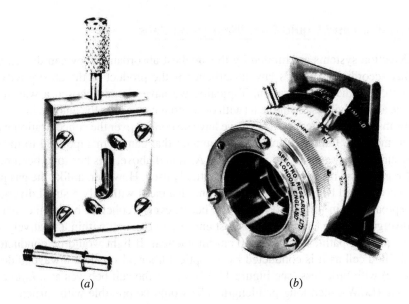

FIGURE 11.17 Typical infrared cells. (*a*) Fixed-path cell. (Courtesy of Barnes Engineering Co.) (*b*) Variable-pathlength cell. (Courtesy of Wilks Scientific Corporation.)

Pure liquid samples are usually run without dilution ("neat") in the infrared region, when a chemist is trying to identify or confirm the structure of an unknown or new compound. For this purpose, the cell length must be short in order to keep the absorbance within the optimum region, generally pathlengths of 0.01 to 0.05 mm are needed. If a solution of the sample is to be prepared, a high sample concentration is used, to keep the solvent absorbance minimal (no solvent is completely transparent in the infrared). So again, short pathlengths are required, generally 0.1 mm or less.

However, samples may not be sufficiently soluble in available solvents to give a high enough concentration to measure in the infrared region. Powders may be run as a suspension or thick slurry (mull) in a viscous liquid having about the same index of refraction in order to reduce light scattering. The sample is ground in the liquid, which is often Nujol, a mineral oil (see Figure 11.4). If Nujol masks any C—H bands present, chlorofluorocarbon greases are useful. The mull technique is useful for qualitative analysis, but it is difficult to reproduce for quantitative work. Samples may also be ground with KBr (which is transparent in the infrared region) and pressed into a pellet that is mounted for measurement.

Gases may be analyzed by infrared spectrometry, and for this purpose a long-path cell is used, usually 10 cm in length. Some typical infrared cells are shown in Figure 11.17. For workplace measurement of toxic gases in low concentrations, much longer pathlengths are needed. "White" cells, named after the inventor, use (concave) mirrors on both sides of the cell. The beam enters through an aperture in the entrance mirror and is reflected multiple times before it exits through the aperture in the exit mirror. With a 0.5 or 1 m distance between the mirrors (base path)., pathlengths of tens of meters are readily attainable. Some white cells permit angular adjustment of the mirror(s) to vary the number of reflections before the beam exits and hence the pathlength. Even longer pathlengths are needed for measurements in ambient air. A focused light source and a detector can be placed hundreds of meters even kilometers apart, utilizing the open atmosphere as the cell. Commercial versions of such differential optical absorption spectrometers (DOAS) typically have the light source and the detector at the same location and use a distantly located mirror to reflect the source light back to the detector; they can span the wavelength range from UV to IR.

Optical Fibers and Liquid Core Waveguide Cells

Most detection systems are limited by the smallest absorbance they can detect. Consider that according to Beer's law, absorbance is the product of the absorptivity, the pathlength, and the concentration. Typically, we have already chosen a wavelength where the absorptivity is the highest (with or without converting the analyte into a more strongly absorbing species), and since we have no control over the concentration we are trying to measure, the only remaining parameter that can be manipulated to increase the measured absorbance is the pathlength. As noted above, this has long been recognized for gas phase measurements and is routinely used. However, unlike the gas phase where a collimated beam can travel over long distances with only a small divergence (beam spreading), in the liquid phase, the beam, even a coherent beam like that of a laser, diverges rapidly and the light is lost entirely to the wall within a relatively short distance, making pathlengths over 10 cm impractical. If light would be conducted in a liquid-filled cell as it is conducted by an optical fiber (also called a waveguide or a fiber optic) with little loss (see Figure 11.18), i.e., if the cell behaves as a *liquid core waveguide* (LCW), then long pathlength cells would be possible with sufficient light transmission through the cell.

The angle of acceptance, θ_a, is the greatest angle of incidence that will be totally reflected for a given core–cladding refractive index difference. Any light entering at an angle greater than θ_a will not be transmitted. The numerical aperture (NA) of the fiber is a measure of the light collection ability of the fiber and is given by:

$$\text{NA} = n_{\text{ext}} \sin \theta_a = \sqrt{(n_1^2 - n_2^2)} \tag{11.20}$$

where n_{ext} is the refractive index of the external medium, typically air for which $n \simeq 1$. The acceptance angle θ_a, measured with respect to the core axis, is given by:

$$\theta_a = \cos^{-1} \left(\frac{n_2}{n_1} \right) \tag{11.21}$$

Manufacturers typically provide numerical aperture data for different fibers. Another property usually provided is the light loss per unit length for different wavelengths. A spectral curve is given that shows attenuation versus wavelength. Attenuation is usually expressed in decibels per kilometer (dB/km), and is given by

$$\text{dB} = 10 \log \frac{P_0}{P} \tag{11.22}$$

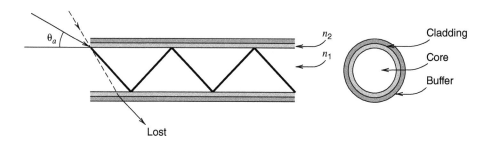

FIGURE 11.18 An optical fiber or "waveguide" transmits light with little loss. It consists of two regions, the core and the clad and typically an outer protective "buffer." The core has a higher refractive index (RI) n_1 than that of the clad (n_2). Light entering the optical fiber within the acceptance angle undergoes total internal reflection (TIR) at the core–clad interface and is transmitted with little loss.

where P_0 is the input intensity and P is the output intensity. Thus, the attenuation for silica-based fibers at 850 nm is in the order of 10 dB/km. Note that dB = 10 × absorbance. So a 10-m (0.01-km) fiber would exhibit an absorbance of 0.01 (0.1 dB attenuation), corresponding to 97.7% transmittance.

Fiber optics may be purchased that transmit radiation from the ultraviolet (190 nm) to the infrared (≥ 5 μm), but each has a limited range. Plastic and compound glass materials are used for short distances in the visible region, while silica fibers can be used from the UV through the near-IR (2.3 μm) regions, but they are expensive. Fluoride and calcogenide glasses extend farther into the infrared.

In coupling fiber optics to spectrometers, there is a trade-off between increased numerical aperture to collect more light and the collection angle of the spectrometer itself, which is usually limiting. That is, light collected with a fiber of numerical aperture greater than that of the spectrometer itself will not be seen by the spectrometer. See Reference 21 for a discussion of design considerations for fiber-optic/spectrometer coupling.

Fiber optics may be used as probes for conventional spectrophotometric and fluorescence measurements. Light must be transmitted from a radiation source to the sample and back to the spectrometer. While there are couplers and designs that allow light to be both transmitted and received by a single fiber, usually a **bifurcated fiber** cable is used. This consists of two fibers in one casing, split at one end, where one goes to the radiation source and the other to the spectrometer. Often, the cables consist of a bundle of several dozen small fibers, and half are randomly separated from the other at one end. For absorbance measurements, a small mirror is mounted (attached to the cable) a few millimeters from the end of the fiber. The source radiation penetrates the sample solution and is reflected back to the fiber for collection and transmission to the spectrometer. The radiation path length is twice the distance between the fiber and the mirror.

> With bifurcated cables, one is used to transmit the source radiation and the other is used to receive the absorbed or fluorescent radiation.

Fluorescence measurements are made in a similar fashion, but without the mirror. Radiation emitting from the end of the fiber in the shape of a cone excites fluorescence in the sample solution, which is collected by the return cable (the amount depends on the numerical aperture) and sent to the spectrometer. Often, a laser radiation source is used to provide good fluorescence intensity.

For a liquid-filled tube to behave as an optical fiber, the clad must be optically transparent and have an RI less than that of the liquid. For example, a glass (RI ~ 1.45) tube filled with CS_2 (refractive index 1.63) behaves as an LCW and conducts visible light with little loss. Such examples are of little practical use, however, as most liquid-phase measurements are done in dilute aqueous solutions (RI of water is 1.33). For the construction of a tubular LCW cell with an aqueous solution, one thus requires a tube constructed of a material that has an RI < 1.33 and is transparent in the wavelength region of intended use. This had proven elusive until the 1990s when a new fluoropolymer Teflon AF, which can be formulated with an RI as low as 1.29, first became commercially available. For $n_2 = 1.29$ and $n_1 = 1.33$, θ_a is 14.1°. While this may be a relatively poor optical fiber, in recent years LCW cells have been extensively used for trace measurements with pathlengths as long as 5 m. For a review of LCW cells see T. Dallas, and P. K. Dasgupta, *Trends Anal. Chem.* **23** (2004) 385.

Detectors

The choice of detectors depend on the wavelength of interest.

UV–Vis Detectors. A **phototube** was once common for measurement in the *ultraviolet* and *visible regions*. This consists of a photoemissive cathode and an anode.

> Detectors for:
> UV—phototube, PM tube, silicon diode array, CCD array
> Vis—phototube, PM tube, silicon diode array, CCD array
> IR—thermocouples, bolometers, thermistors, InGaAs diode array

A potential, anywhere from tens of volts to several thousand, depending on the tube, is impressed between the anode and cathode. When a photon strikes the cathode, an electron is emitted and is attracted to the anode, causing current to flow that is measured. (Albert Einstein received the 1921 Nobel Prize in Physics for its discovery in 1905, not for the special theory of relativity which he also introduced in 1905—this was still controversial–see http://en.wikipedia.org/wiki/Albert_Einstein). The response of the photoemissive material is wavelength dependent, and different phototubes are available for different regions of the spectrum. For example, one may be used for the blue and ultraviolet portions and a second for the red portion of the spectrum. "Solar blind" phototubes respond only to UV radiation, typically in the <320 nm range. In most other applications, phototubes have largely been replaced by photodiodes.

Different photocathode materials are listed below:

Ag-O-Cs: This is one of the oldest photocathode materials (often referred to as S-1) and responds over a 300 to 1200 nm range. It exhibits relatively high thermionic dark emission (referred to as dark current); present use is limited to the NIR range with a cooled photocathode to reduce the dark current.

GaAs(Cs): Cesium-activated GaAs responds over a broad range from 300 to 930 nm, with relatively flat response over 300 to 850 nm.

InGaAs(Cs): It has greater extended sensitivity in the infrared range than GaAs(Cs). In 900 to 1000 nm, this photocathode exhibits much higher S/N than S-1.

Sb-Cs: A widely used photocathode with a spectral response in the ultraviolet to visible range.

Bialkali (Sb-Rb-Cs, Sb-K-Cs): A spectral response range similar to the Sb-Cs photocathode, but with higher sensitivity and lower noise.

High temperature bialkali or low noise bialkali (Na-K-Sb): This is particularly useful at higher operating temperatures (175°C limit). A major application is in the oil well logging industry. At room temperatures, the dark current is very low, making it ideally suited for photon counting.

Multialkali (Na-K-Sb-Cs): This has a wide spectral response with high sensitivity from the UV to NIR. It is widely used for broadband spectrophotometers. The long wavelength response can be extended out to 930 nm by specially processing the photocathode. Possibly the most widely used photocathode.

Cs-Te, Cs-I: "Solar blind" photocathodes with no response to visible light. Cs-Te responds only to $\lambda < 320$ nm, and Cs-I to $\lambda < 200$ nm.

A **photomultiplier tube** (PMT) is more sensitive than a phototube and is widely used for sensitive detection in *visible* and *ultraviolet regions*. It consists of a photoemissive cathode, which the photon strikes, and a series of electrodes (dynodes), each at a more positive potential (50 to 90 V) than the one before it. When an electron strikes the photoemissive surface, a primary electron is emitted (this is the photoelectric effect). The primary electron released from the photoemissive surface is accelerated toward the first dynode. The impact of the electron on the dynode surface causes the release of many secondary electrons, which in turn are accelerated to the next electrode where each secondary electron releases more electrons, and so on; up to about 10 stages of amplication is common. The electrons are finally collected by the anode. The final output of the photomultiplier tube may, in turn, be further electronically amplified.

Again, different photomultiplier tubes utilize different photocathode materials that have different wavelength response characteristics. Figure 11.19 illustrates the response characteristics of some typical photomultiplier tubes with different photoemissive cathode surfaces. The high sensitivity of a PMT permits detection of very low levels of light and narrower slit widths can be used for better wavelength resolution.

Albert Einstein received the 1921 Nobel Prize in Physics for his elucidation of the photoelectric effect in 1905. Contrary to a popular misconception, he did not receive the Nobel Award for his discovery of the special theory of relativity, which he also introduced in 1905. His theory on relativity was still too controversial in the early 1920's. (see http://en.wikipedia.org/wiki/Albert_Einstein).

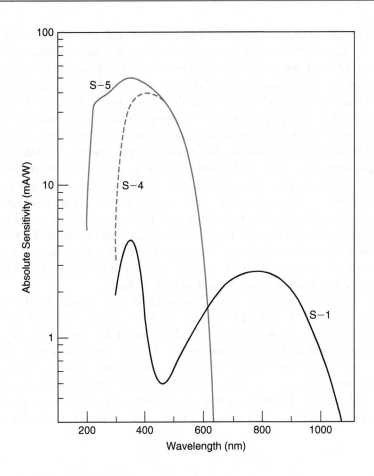

FIGURE 11.19 Some spectral responses of different photocathode materials. S-1:Ag-O-Cs, S-4, and S-5: two different types of bialkali photocathodes. (Adapted from G. D. Christian and J. E. O'Reilly, *Instrumental Analysis*, 2nd ed. Boston: Allyn and Bacon, Inc., 1986. Reproduced by permission of Allyn and Bacon, Inc.)

The most widely used photodetector for general light detection applications is a semiconductor diode with a junction that is accessible to light exposure, called a photodiode. A p-type semiconductor has a deficit of lattice electrons, as obtained with doping silicon (with 4 valence electrons) with Al, Ga, or In (with 3 valence electrons); a n-type semiconductor has a surplus of lattice electrons, as obtained with doping Si with N, P, or As (with 5 valence electrons). The simplest diode consists of the junction of a p-doped and an n-doped semiconductor. The short-circuit current is linearly related to the light intensity falling on the photodiode. Solar cells (photovoltaic cells) are essentially large area photodiodes. Applying a reverse potential (n-side positive) to the diode greatly improves response speed at the expense of increased noise. PIN diodes have an insulator layer between the p and n-type semiconductors and are particularly well suited for such reverse-biased detection, with very high response speed as may be needed to detect fast laser pulses. Avalanche photodiodes (APDs) use a very high reverse bias (often several hundred volts); like a PMT, the initially generated photoelectrons create an avalanche of secondary electrons, leading to very high sensitivity. In a typical application, the photodiode output is processed by an external current-to-voltage converter. Integrated photodiode—operational amplifier combinations provide current-to-voltage conversion, and substantial (fixed or variable) amplification; they are widely available and are routinely used as photodetectors.

The applicable wavelength range of photodiodes depends on the specific semiconductor used. Typical silicon photodiodes normally respond in the 400 to 900 nm range, but special fabrication can provide response from 170 to 1100 nm. No single phototube/PMT can span such a wide range. SiC detectors have an intrinsic response in the 200 to 400 nm range. GaP photodiodes have a peak response near 440 nm and a response range of 190 to 550 nm; in the visible range, their response mimics the responsivity of the human eye. The response of GaAsP photodiodes varies greatly depending on the nature and extent of doping; the response range can extend from 190 to 760 nm. GaN and AlGaN photodiodes respond over 200 to 370 nm and 200 to 320 nm, respectively. Depending on the doping, InGaN photodiodes are available in UV versions (200–400 nm) or in near-UV to visible (300–510 nm). Great strides have been made in InGaAs photodiodes in recent years, driven by the interest in night vision equipment that "sees" in the NIR. Depending on doping, the response spans from 850 to 2500 nm. To reduce thermal noise, thermoelectric cooling is commonly used to cool these and other detectors. An attractive (clickable) pictorial summary of wavelength response range of some of these and other photodetectors can be found in http://jp.hamamatsu.com/en/product_info/wave/index.html . Photodiodes with built in colored filters or interference filters that restrict their response to a desired wavelength range are available. Figure 11.20 illustrates some photodiode arrays, and Figure 11.21 the spectral response of a UV-sensitive silicon photodiode.

The charge coupled device (CCD) photosensor was invented by Willard Boyle and George Smith in AT&T Bell laboratories in 1969; they shared the Nobel Prize in Physics 40 years later for this discovery with Charles Kao, who was credited for his contributions to optical fiber communications. An individual CCD element is rarely used; they are nearly always used as arrays. A linear array is used in facsimile (fax) machines and digital scanners, while a two-dimensional array is used for image sensing in digital cameras. Individual sensors are referred to as elements or pixels, hence the designation of cameras as having a 5 megapixel sensor, etc. In CCDs, the light

FIGURE 11.20 Photo of 1024-element diode arrays. (Courtesy of Hamatsu Photonics, K. K.)

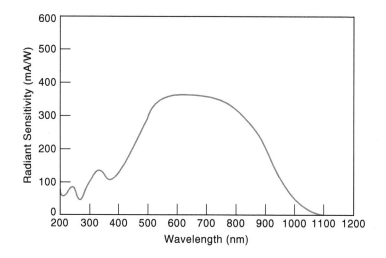

FIGURE 11.21 Typical spectral response of a UV-sensitive silicon photodiode. (From M. Kendall-Tobias, *Am Lab.,* March 1989, p. 102. Reproduced by permission of International Scientific Communications, Inc.)

intensity falling on a sensor is converted into quantized (digital) amounts of electrical charge. The charges are read, along a row of sensor pixels, in discrete time intervals. The CCD is, therefore, a discrete-time device, i.e., the light signal is sampled at discrete time intervals; however, the time interval can be very short. The use of linear CCD arrays as spectrometric detectors ranges from inexpensive miniature UV–Vis spectrometers to moderately expensive miniature thermoelectrically cooled back-thinned back-illuminated CCD-based spectrometers that provide sensitivities rivaling those of PMTs.

Silicon and InGaAs diodes are also made in the form of multi-pixel (from 2 to 2048 pixels) arrays. Silicon photodiode array (PDA) based modestly priced spectrometers that cover the entire UV–Vis range 190 to 900 (often to 1100) nm are popular. Both desktop and miniature fiber-optic InGaAs diode array–based NIR spectrometers are available but the cost is significant. An advantage of array detector-based spectrometers (whether CCD or PDA) is that they do not need a mechanically scanned monochromator; a typical arrangement is shown later in Figure 11.23. Array detector elements are sequentially read, and depending on the electronics, the number of pixels and the overall arrangement, this can take < 1 ms, but tens of milliseconds are more typical in low-cost spectrometers.

The journal *Analytical Chemistry* defines a spectrophotometer as a spectrometer that measures the *ratio* of the radiant power of two beams, that is, P/P_0, and so it can record absorbance. The two beams may be measured simultaneously or separately, as in a double-beam or a single-beam instrument—see below. An exception is when the radiation source is replaced by a radiating sample whose spectrum and intensity are to be measured, as in fluorescence spectrometry—see below. However, in a high-end fluorescence spectrophotometer, the measured fluorescence intensity is often ratioed to the excitation light intensity. If the prism or grating monochromator in a spectrophotometer is replaced by an optical filter that passes a narrow band of wavelengths, the instrument is often called a photometer or colorimeter.

IR Detectors. Infrared detectors fall in two categories, one relies on the quantum properties of EMR and the other relies on the fact that infrared radiation is essentially heat and these detectors are heat sensors. Among the first type are **photovoltaic** detectors that are completely equivalent to Si photodiodes used in the UV–Vis range. These include Ge (0.8 to 1.8 μm), InGaAs (0.8 to 2.5 μm), InSb (1 to 5.5 μm), InAsSb (1 to 5.8 μm), and HgCdTe (2 to 16 μm, often called MCT, for mercury cadmium telluride). The second type consists of quantum effect **photoconductive** detectors. The resistance of these devices decreases exponentially with the light intensity falling on them. Such detectors are available with response maxima in the near UV (ZnS) or visible (CdS), but are rarely used in analytical instrumentation due to the superiority of photovoltaic silicon detectors. Photoconductive IR detectors include PbS (1–3.6 μm), PbSe (1.5-5.8 μm), and PbSnTe (3–14 μm); InGaAs, InSb, and HgCdTe can also be used in the photoconductive mode and generally result in extended response to higher wavelengths compared to the photovoltaic mode; InSb and HgCdTe detectors can be used over 1–6.7 and 2–25 μm, respectively. For long wavelength IR detection, the detector must be cooled to liquid nitrogen temperatures. The archtypical thermal detector is a **thermocouple** that consists typically of two dissimilar metal wires. A **thermopile** consists of thermocouples connected in series, or less commonly, in parallel. A typical thermocouple may consist of a pair of antimony and bismuth wires connected at two points. When a temperature difference exists between the two points, a potential difference is developed, which can be measured. One of the junctions, then, is placed in the path of the light from the monochromator. A thermopile consists of up to six thermocouples in series, mounted in a vacuum to minimize heat loss by conduction. Half are sensing and half are bonded to a substrate. Thermopiles have response times of about

In an array detector, an individual detector element is called a pixel.

Common detectors:

Photomultiplier tubes: 160–1100 nm

Silicon-photodiode (arrays): 170–1100 nm

Charge-coupled devices (CCDs): 180–1100 nm

Indium gallium arsenide (InGaAs) photodiodes: 850–2550 nm

Lead sulfide (PbS): 1000–3300 nm

30 ms. **Bolometers** and **thermistors** rely on the temperature dependence of electrical *resistance* (typically negative). Thermistors are made of sintered oxides of cobalt, manganese, and nickel. Their change in resistance is measured in a Wheatstone bridge circuit. The advantage of these over thermocouples is the more rapid response time (4 ms, compared with 30–60 ms), and thus improved resolution and faster scanning rates can be accomplished, but sensitivity is compromised. The response of thermal detectors is essentially independent of the wavelengths measured. A bolometer consists of a thin layer of an absorber, which acts as a resistance thermometer connected to a large capacity thermal reservoir at a constant temperature (often cooled to liquid nitrogen or even lower temperatures). The response time is proportional to the ratio of the heat capacity of the absorptive element to the thermal conductance between the absorptive element and the reservoir. Typically semiconductor or superconductor absorptive elements are used.

For rapid measurements required with FTIR instruments, and for high-sensitivity measurements, photon detectors are used. Examples are the solid-state PbS, PbSe, InGaAs, or InSb photoconductive detectors. Photovoltaic detectors are even faster (InGaAs detectors, often used in optical communications, can respond in a sub-ps timescale) and more sensitive, but typically require cooling. InGaAs offers the highest sensitivity in the near-IR and has become the detector of choice.

So-called two-color photodetectors use an IR-transparent silicon detector atop a Pbs, PbSe, or InGaAs detector in the same optical axis, resulting, respectively, in effective response ranges of 0.2–3, 0.2–4.85, and 0.32–2.55 µm.

Present high-end spectrophotometers cover the entire UV–NIR range and use multiple monochromators and multiple detectors. For example, the Perkin–Elmer Lambda 1050 spans a 175–3300 nm range, using a PMT to cover the 175–860 nm range and cooled InGaAs/PbS detectors to cover 860–1800/1800–3300 or 860–2500/2500–3300 nm ranges, depending on the choice of the InGaAs detector. The wavelength resolution can be as good as 0.05 nm, absorbance noise as low as 2×10^{-5}, and upper dynamic range as high as 8 absorbance units (1 in 10^8 incident photons are transmitted through the sample!).

Slit Width—Physical vs. Spectral

The radiation passed by a slit is not monochromatic.

We previously mentioned that it is impossible to obtain spectrally pure wavelengths from a monochromator. Instead, a **band** of wavelengths emanates from the monochromator and the width of this band will depend on both the dispersion of the grating or prism and the exit slit width. The dispersive power of a prism depends on the wavelength and on the material from which it is made, as well as on its geometrical design, while that of a grating depends on the number of grooves per inch. Dispersion is also increased as the distance to the slit is increased.

After the radiation has been dispersed, a certain portion of it will fall on the exit slit, and the width of this slit determines how broad a band of wavelengths the sample and detector will see. Figure 11.22 depicts the distribution of wavelengths leaving the slit. The **nominal wavelength** is that set on the instrument and is the wavelength of maximum intensity passed by the slit. The intensity of radiation at wavelengths on each side of this decreases, and the width of the band of wavelengths passed at one-half the intensity of the nominal wavelength is the **spectral bandwidth**, or **bandpass**. The **spectral slit width** is approximately twice the spectral bandwidth (approximating the Gaussian band shape in Figure 11.22 by an isosceles triangle), and this is a measure of the total wavelength spread that is passed by the slit. Note that the spectral slit width is not the same as the mechanical slit width, which may vary from a few micrometers to a millimeter or more (the spectral slit width is the band of radiation passed by the

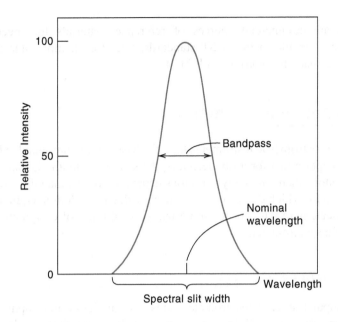

FIGURE 11.22 Distribution of wavelengths leaving the slit of monochromator.

mechanical slit and is measured in units of wavelength). For a Gaussian band shape, ~76% of the radiation intensity is contained within the wavelengths of the spectral bandwidth.

If the intensity of the source and the sensitivity of the detector allow, the spectral purity can be improved (the bandpass decreased) by decreasing the slit width. The decrease is typically not linear, however, and a limit is reached due to aberrations in the optics and diffraction effects caused by the slit at very narrow widths. The diffraction effectively increases the spectral slit width. In actual practice, the sensitivity limit of the instrument is usually reached before diffraction effects become significant.

The bandwidth or the spectral slit width is essentially constant with a grating dispersing element for all wavelengths of a given spectral order at a constant slit width setting. This is not so with a prism because of the variation of dispersion with changing wavelength. The bandwidth will be smaller at shorter wavelengths and larger at longer wavelengths.

The bandwidth varies with wavelength with a prism, but is constant with a grating.

Instrumental Wavelength and Absorbance Calibration

The wavelength reading of a spectrophotometer can be checked using solutions of known absorbance maxima and minima. Potassium dichromate at pH 2.9 has maximum absorbances at 257 and 350 nm, and minima at 235 and 313 nm. A holmium oxide glass filter absorbs sharply at 279.2, 222.8, 385.8, 446.0, 536.4, and 637.5 nm.

The National Institute of Standards and Technology (NIST) provides standard reference materials (SRMs) to verify the wavelength accuracy and accuracy of absorbance (transmittance) readings. SRM 930E for UV–Vis analysis consists of a set of three neutral density glass filters of standard thickness with nominal transmittances of 10, 20, and 30%. Other SRMs consist of standard solutions of, for example, potassium dichromate or potassium acid phthalate in perchloric acid. See the NIST Web page, http://srdata.nist.gov/gateway/gateway?keyword=absorbance for absorbance standards. SRM 1921A is a polystyrene film for infrared calibration. See R. A. Spragg and M. Billingham, *Spectroscopy* **10**(1) (1995) 41 for corrections to apply in routine use (correcting for effects of resolution, peak-finding algorithm, and temperature on the band positions).

There are commercial sources of reference materials for spectral calibration that are traceable to the NIST standards. See, for example Starna Cells, Inc. http://www.starnacells.com/d_ref/stds.html .

Types of Instruments

Although all spectrometric instruments have the basic design presented in Figure 11.12 (with the exception that for array detectors, the sample generally is placed before the monochromator), there are many variations depending on the manufacturer, the wavelength regions for which the instrument is designed, the resolution required, and so on. We will indicate here a few of the important general types of design and the general operation of a spectrometer.

Single-Beam Spectrometers

The most popular student spectrometers today are miniature fiber-optic single-beam spectrometers. Broadband light from a source such as that shown in Figure 11.13 is incident on the sample cell using a fiber optic or focusing optics. The transmitted light is brought to the spectrometer by an optical fiber. There are many optical configurations of a miniature spectrometer. Figure 11.23 illustrates an asymmetric crossed Czerny–Turner design. Light enters the spectrometer through a slit and is incident on a collimating concave mirror that reflects the light (gray solid lines) onto a grating. Depending on the wavelength range and the resolution, the grating will have a specific blaze angle and groove density. The grating then diffracts the beam (blue dashed lines) to a second mirror, which images onto the detector array, usually a CCD array. Some manufacturers (e.g., Stellarnet Inc.) use a concave grating that serves the combined

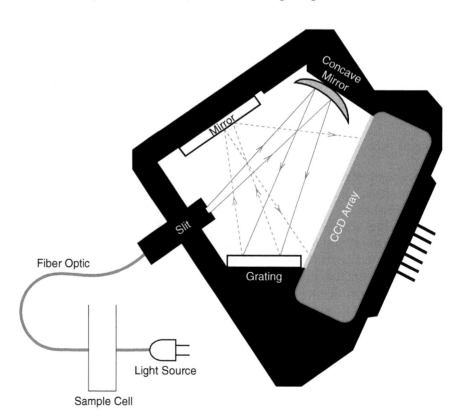

FIGURE 11.23 Schematic diagram of a miniature fiber-optic spectrometer-based measurement system. See Figure 11.24 for a photograph of a similar system.

FIGURE 11.24 Miniature fiber-optic spectrometer. Box is the spectrometer. Light source is to right, and fiber-optic cable guides light to cuvet. Second cable takes transmitted radiation to spectrometer. (Photo courtesy of Ocean Optics, Inc.)

function of a concave mirror and a planar grating and permits greater light throughput. In a calibrated system, each pixel of the detector will correspond to a specific wavelength of light.

Depending on the light source, an order sorting filter may be necessary. A white LED source emits over a limited wavelength range and since the ratio of the maximum/minimum usable wavelength is <2, an order sorting filter is not needed. With a quartz-tungsten–halogen lamp, however, an order sorting filter will be needed.

The selection of the filter depends on what radiation must be restricted. For most applications, a short-pass or long-pass filter may be used.

Any radiation not absorbed by the sample falls on the detector, where the intensity is converted to an electrical signal that is acquired on a computer and displayed.

We have illustrated that the spectral intensity of the sources and the spectral response of the detectors are dependent on the wavelength. Therefore, some means must be used to provide sufficient but not excessive light to the detector. This can be accomplished by one of two ways: by adjusting the slit width to allow more or less light to fall on the detector, but this is not typically user-adjustable in inexpensive fiber-optic spectrometers. An adequate amount of light is typically adjusted by adjusting the exposure time of the detector before the signal is read, much like a camera.

All detectors display some **dark response** in the absence of light, owing to thermal effects. This is usually small but can be taken into account by acquiring the dark response across the spectrum over the same integration time that is to be used with the blank and the sample. Now, the cell filled with solvent is placed in the beam path and the detector reading taken. We have the P_0 value at each wavelength when we subtract the corresponding dark response. The instrument scale is now ready to read the sample spectrum. Sample is put in the cell and readings taken. After subtracting the dark response at each wavelength, we have the P values. The transmittance (P/P_0) or ($-\log T$) are easily calculated and displayed.

The detector has a finite response even when no radiation falls on it. This is the dark response.

In an analytical determination, instead of the solvent, the blank solution[3] is often used for taking the 100% transmittance (P_0) reading. Any blank absorbance is then automatically corrected for (subtracted). This method should only be used if the blank reading has been demonstrated to be constant. A large blank reading would be more likely to be variable. An advantage of zeroing the instrument with the blank is that one reading, which always contains some experimental error, is eliminated. If this technique is used, it would be a good practice to check repeated blank solutions to make sure the blank is constant.

Duble-Beam Spectrometers

Double-beam spectrometers use a more complex design than single-beam instruments, but they have the advantage that any fluctuation in the source intensity can be readily compensated for. The instrument has two light paths, one goes through the sample and the other constitutes the reference beam, it may directly go to a detector, or be directed through a reference or blank. In a typical setup, the beam from the source strikes a vibrating or rotating mirror that alternately passes the beam through the reference cell and the sample cell and, from each, to the detector. In effect, the detector alternately sees the reference and the sample beam and the output of the detector is proportional to the ratio of the intensities of the two beams (P/P_0). In another arrangement, a stationary beam splitter is used to divide the light into two separate portions that go separately to two matched detectors, the sample beam travels through the sample.

In the first arrangement, the output is an alternating signal whose frequency is equal to that of the vibrating or rotating mirror. An ac amplifier is used to amplify this signal, and stray dc signals are not recorded. The wavelength is changed by a motor that drives the dispersing element at a constant rate, and the slit is continually adjusted by a servomotor to keep the energy from the reference beam at a constant value; that is, it automatically adjusts to 100% transmittance through the reference cell (which usually contains the blank or the solvent).

This is a simplified discussion of a double-beam instrument. There are variations on this design and operation, but it illustrates the utility of these instruments. They are very useful for qualitative work in which the entire spectrum is required, and they can automatically compensate for absorbance by the blank, as well as for drifts in source intensity.

Single Beam or Double Beam?

Early UV–Vis and IR spectrophotometers, back in the 1950s, were very large instruments that usually had double-beam monochromators to compensate for optical drift and electronic noise. They were slow and not very sensitive. Improvements in optical and electronic technology have reduced the necessity for double-beam optical systems that reduce the energy of the transmitted beam. Modern single-beam instruments are smaller, faster, more sensitive, and more economical than the older versions. However, double-beam instruments still provide the optimal stability, and the choice depends on your needs. All modern dispersive IR instruments are single beam. Array detectors allow the acquisition of the entire spectrum at once and are presently popular and inexpensive. Since they allow referencing at a wavelength where the sample does not absorb, this allows for another way to compensate for source fluctuations.

---

[3]This contains all reagents used in the sample, but no analyte.

The choice of resolution of instruments ranges from low-resolution student instruments, such as the Spectronic 20 with 20 nm resolution to better than 0.05 nm in research instruments. The typical instrument will have built-in software that allows calibration with multiple standards, polynomial curves, and statistical calculations.

11.10 Array Spectrometers—Getting the Entire Spectrum at Once

As shown in Figure 11.23, single-beam spectrometers today typically use an array detector. Instead of a CCD array, a photodiode array (PDA) is also commonly used. Either can record an entire spectrum in a few milliseconds. The basic design of an array detector-based spectrometer is shown in Figure 11.25 (a simplified version of Figure 11.23). Broadband (polychromatic) light passes through the sample, and the dispersing element is placed after the sample. Because a single wavelength of light is not isolated, the dispersing element, typically a grating, is often referred to as the polychromator, instead of a monochromator. The use of an exit slit to isolate a given wavelength is eliminated, and the dispersed light is allowed to fall on the face of the array detector. Resolution is limited by the number of idividual pixels in the array. Recognize also that there is dead isolation space between each element/pixel.

Array spectrometers are very useful for the analysis of mixtures of absorbing species with overlapping spectra. With an array spectrometer, the absorbance at many points can be simultaneously measured, using data on the sides of absorption bands as well as at absorption maxima. This method of "overdetermination," in which more measurement points than analytes are obtained, improves the reliability of quantitative measurements, allowing six or more constituents to be determined in simple

In array spectrometers, there is no exit slit, and all dispersed wavelengths that fall on the array are recorded simultaneously.

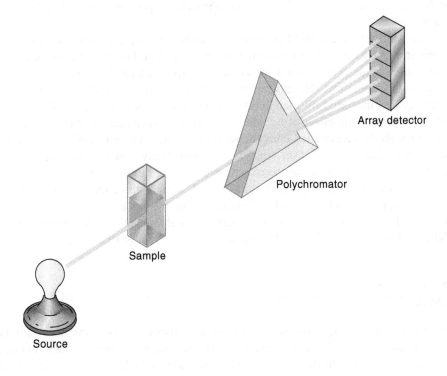

Array detector

Polychromator

Sample

Source

FIGURE 11.25 Schematic of array spectrometer.

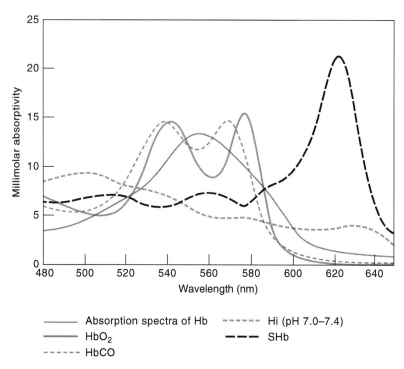

FIGURE 11.26 Absorptivities in mmol^{-1} L cm^{-1}. [From A. Zwart, A. Buursma, E. J. van Kampen, and W. G. Zijlstra, *Clin. Chem.*, **30** (1984) 373. Reproduced by permission.]

Absorption spectra of Hb —— Hi (pH 7.0–7.4) -------

HbO$_2$ —— SHb ---

HbCO -------

Note: Hb = hemoglobin, HbO$_2$ = oxyhemoglobin, HbCO = carboxyhemoglobin, Hi = methemoglobin, SHb = sulfhemoglobin.

mixtures of components with similar but not identical spectra. An example of a multicomponent analysis is shown in Figure 11.26 for the simultaneous measurement of five hemoglobins. The five spectra were quantitatively resolved by comparing against standard spectra of each compound stored in the computer memory. Full-spectrum analysis can be performed in a variety of software packages. Mixtures of standards may be used for calibration, and this can compensate for possible interactions between components.

The measurement precision is improved by averaging many measurements.

The ability of array spectrometers to acquire data rapidly also allows the use of measurement statistics to improve the quantitative data. For example, 10 measurements can be made at each point in one second, from which the standard deviation of each point is obtained. The instrument's computer then weights the data points in a least-squares fit, based on their individual precision. This "maximum-likelihood" method minimizes the effect of bad data points on the quantitative calculations. The ability to make fast automated computer-controlled measurements and computerized data interpretation also makes such instruments particularly suitable for making kinetic measurements where spectra are taken repeatedly at fixed time intervals.

A CCD array, especially a back-thinned CCD array, is more sensitive than a PDA. It provides superior sensitivity in low light detection - as in various forms of luminescence spectrometry. On the other hand, if sufficient light is available, the response from a PDA is more reproducible; it is the preferred detector in absorbance measurements.

11.11 Fourier Transform Infrared Spectrometers

Conventional infrared spectrometers are known as **dispersive instruments**. With the advent of computer- and microprocessor-based instruments, these have been replaced by Fourier transform infrared (FTIR) spectrometers, which possess a number of

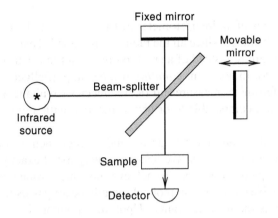

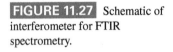

FIGURE 11.27 Schematic of interferometer for FTIR spectrometry.

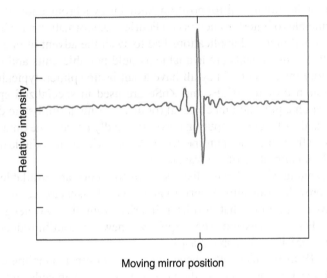

FIGURE 11.28 Typical interferogram. Point marked "0" is where both mirrors of the interferometer are the same distance from beam splitter. [From D. W. Ball, *Spectroscopy*, **9**(8) (1994) 24. Reproduced by permission.]

advantages. Rather than a grating monochromator, an FTIR instrument employs an interferometer to obtain a spectrum.

The basis of an interferometer instrument is illustrated in Figure 11.27. Radiation from a conventional IR source is split into two paths by a beam splitter, one path going to a fixed position mirror, and the other to a moving mirror. When the beams are reflected, one is slightly displaced (out of phase) from the other since it travels a smaller (or greater) distance due to the moving mirror, and they recombine to produce an interference pattern (of all wavelengths in the beam) before passing through the sample. The sample sees all wavelengths simultaneously, and the interference pattern changes with time as the mirror is continuously scanned at some chosen linear velocity. The result of absorption of the radiation by the sample is a spectrum in the **time domain**, called an **interferogram**, that is, absorption intensity as a function of the optical path difference between the two beams.

A typical interferogram is shown in Figure 11.28. The tall part of the signal corresponds to when the two mirrors are equidistant from the beam splitter, when destructive interference between the two beams is zero, and is called the centerburst. The intensity drops off rapidly away from this, due to destructive interference. This is converted, using a computer, into the frequency domain via a mathematical operation known as a **Fourier transformation** (hence the name **Fourier transform infrared spectrometer**). A conventional appearing infrared spectrum is obtained by the transformation.

FTIR spectrometers have largely replaced dispersive IR spectrometers.

An interferogram is a spectrum in the time domain. Fourier transformation converts it to the frequency domain.

Advantages of FTIR spectrometers: greater light throughput, increased signal-to-noise ratio, simultaneous measurement of all wavelengths.

The advantages of an interferometer instrument is that there is greater throughput (Jacquinot's advantage) since all the radiation is passed. That is, the sample sees all wavelengths at all times, instead of a small portion at a time. This results in increased signal-to-noise ratio. In addition, a *multiplex advantage* (Fellget's advantage) results because the interferometer measures all IR frequencies simultaneously, and so a spectrum with resolution comparable to or better than that with a grating is obtained in a few seconds.

In order to take many interferograms and average them to increase the signal-to-noise level, the computer must average the centerburst at exactly the same position along the mirror's path every time. To achieve this, interferometers have a small red helium–neon (He–Ne) laser whose monochromatic beam passes through the interferometer, the same as the infrared source. Upon recombining, it produces interference fringes separated by the exact wavelength of the laser, 632.8 nm. These fringes serve as a calibration for the moving mirror position, allowing synchronization of all the spectra.

The principles of interferometers and Fourier transformation have been known for over a century, but practical applications had to await the advent of high-speed digital computers. FTIR instruments are available as field portable units and more sophisticated laboratory instruments. They all have a salt beam splitter, typically germanium coated potassium bromide (CaF_2 and ZnSe are used in specialized applications), a moving mirror on a precision mechanical (or air) bearing, a solid-state detector (often cryocooled), and sufficient computing power to rapidly process the time-domain interferogram into a frequency-domain spectrum. A laser with a photodiode detector is also incorporated for calibration of the wavelength.

Fast scanning single-beam dispersive instruments are available in the NIR, where they provide competitive performance. The FT-spectroscopic technique has a lower wavelength limit that results primarily from the wavelength of the reference laser. For a discussion, see http://www.newport.com/Introduction-to-FT-IR-Spectroscopy/405840/1033/content.aspx.

Modern IR instruments often have reflectance or other sampling capabilities for obtaining IR spectra that eliminate the necessity of salt plate cells and simplify sample handling. The most useful is an internal reflectance method called attenuated total reflectance (ATR). The sample is pressed on a diamond substrate and the infrared radiation penetrates the sample, being reflected internally, and then exits for detection.

11.12 Near-IR Instruments

NIR sources are more intense and detectors more sensitive than for the mid-IR region, so noise levels are 1000-fold lower.

Radiation sources for near-IR instruments are operated at typically 2500 to 3000 K, compared to 1700 K in the mid-IR region, resulting in about 10 times more intense radiation and improved signal-to-noise ratios. This is possible because the IR radiation of typical sources tails off in the mid-IR region and the maximum intensity shifts further into the near-IR region as the temperature is increased. The higher temperature results in weaker mid-IR radiation, but is beneficial in the near-IR region. A quartz tungsten–halogen lamp provides intense radiation in the 750- to 1750-nm range.

An indium gallium arsenide (InGaAs) detector is most commonly used in the near-IR and is roughly 100 times more sensitive than mid-IR detectors. The combination of intense radiation sources and sensitive detectors results in very low noise levels, on the order of microabsorbance units. Glass and quartz are transparent to near-IR radiation, and so the optics and cells are easier to design and use than for the mid-IR region. Near-IR radiation can be sent for long distances over fiber optics, high-speed optical communications are typically carried out in the 1270–1625 nm range, 1310 and 1550 nm being the two most common wavelengths used. Commercial instruments for process or field (portable) testing often use fiber-optic probes (see below) for nondestructive sample testing.

> ### Spectroscopy for Sleuthing for Counterfeit Drugs
>
> A serious problem facing pharmaceutical companies and end users is the lucrative market of counterfeit drugs, some worthless, without the active pharmaceutical ingredient (API), some dangerous, containing toxic ingredients, and some being only partially effective, containing much less of the API than labeled. While efforts are made to make it easier to detect fakes, for example, through sophisticated packaging, counterfeiters are clever at avoiding detection. Analytical chemistry comes to the rescue. Pharmaceuticals can be sent to a laboratory for analysis using sophisticated techniques to measure small differences in formulations or amounts of APIs. But this takes time, allowing counterfeiters to continue until shut down. So field-based instruments, usable by laypersons, have been developed to quickly identify counterfeit drugs before they reach users. These are generally based on infrared measurements, providing nondestructive testing with no sample preparation or reagents. Other approaches for rapid testing include portable laboratories that can be rapidly deployed. See the following articles in C&ENews for details of the problems and the analytical approaches: http://cen.acs.org/articles/90/i33/Instrumentation-Firms-Develop-Portable-Technology.html and http://cen.acs.org/articles/90/i33/REMOTE-TESTING.html

11.13 Spectrometric Error in Measurements

There will always be a certain amount of error or irreproducibility in reading an absorbance or transmittance scale. Uncertainty in the reading will depend on a number of instrumental factors and on the region of the scale being read, and hence on the concentration.

Because of the logarithmic relationship between transmittance and concentration, small errors in measuring transmittance cause large relative errors in the calculated concentration at low and high transmittances. It is probably obvious to you that if the sample absorbs only a very small amount of the light, an appreciable *relative* error may result in reading the small decrease in transmittance. At the other extreme, if the sample absorbs nearly all the light, an extremely stable instrument would be required to read the very small amount of light transmitted by the sample precisely. There is, therefore, some optimum transmittance or absorbance where the relative error in making the reading will be minimal.

> It is difficult to precisely measure either very small or very large values of absorbance.

The transmittance for minimum relative error can be derived from Beer's law by calculus, assuming that the error results essentially from the uncertainty in reading the instrument scale (or in digitizing the data and displaying it) and also that the *absolute* error in reading the transmittance is constant, independent of the value of the transmittance. The result is the prediction that the minimum relative error in the concentration theoretically occurs when $T = 0.368$ or $A = 0.434$. See Considerations on Optimum Absorbance for Minimum Error in the web supplement.

Figure 11.29 illustrates the dependence of the relative error on the transmittance, calculated for a small but constant error in T (e.g., 0.01). It is evident from the figure that, while the minimum occurs at 36.8% T, a nearly constant minimum error occurs over the range of 20 to 65% T (0.7 to 0.2 A). The percent transmittance should fall

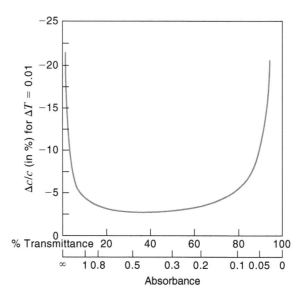

Relative concentration error as function of transmittance for 1% uncertainty in % T.

within 10 to 80% T ($A = 1$ to 0.1) in order to prevent large errors in spectrophotometric readings. Hence, samples should be diluted (or concentrated), and standard solutions prepared, so that the absorbance falls within the optimal range.

For many present-day instruments, however, the above guidelines are likely overly restrictive. Figure 11.29 in practice approximates the error only for instruments with **Johnson** or **thermal noise-limited detectors**, such as photoconductive detectors or thermocouples, bolometers, and Golay detectors in the infrared region. Johnson noise is produced by random thermal motion in resistance circuit elements. Other sources of noise, such as shot noise, inversely related to the square root of the light intensity received by the detector, is often the limiting factor in the UV–Vis range.

For minimum error, the absorbance should fall in the 0.1 to 1 range.

11.14 Deviation from Beer's Law

Deviations from Beer's law result in nonlinear calibration curves, especially at higher concentrations.

It cannot always be assumed that Beer's law will apply, that is, that a linear plot of absorbance versus concentration will occur. Deviations from Beer's law occur both as a result of chemical and instrumental factors. Most "deviations" from Beer's law are really only "apparent" deviations because if the factors causing nonlinearity are accounted for, the true or corrected absorbance-versus-concentration curve will be linear. True deviations from Beer's law will occur when the concentration is so high that the index of refraction of the solution measurably changes from that of the blank. Anytime light is incident from a medium of one refractive index to another, there is a reflection loss from the mismatch of refractive indices. Called Fresnel loss, the extent of this loss is proportional to

$$\frac{(n_1 - n_2)^2}{(n_1 + n_2)^2}$$

where n_1 and n_2 are the refractive indices of the two different media. For this reason, light transmission changes as the refractive index of the medium changes. An air-filled cuvette ($n_{air} \simeq 1$, $n_{glass} \sim 1.45$) has a greater Fresnel loss than a water-filled cuvette ($n_{water} = 1.33$); with an air-filled cuvette as reference, the absorbance of a water-filled cuvette registers negative (typically -0.03 to -0.04).

A similar situation would apply for mixtures of organic solvents with water, and so the blank solvent composition should closely match that of the sample. The solvent may also have an effect on the absorptivity of the analyte.

Chemical Deviations

Chemical causes for nonlinearity occur when an asymmetric chemical equilibrium exists. An example is a weak acid that absorbs at a particular wavelength but has an anion that does not:

$$\underset{\text{(absorbs)}}{HA} \rightleftharpoons H^+ + \underset{\text{(transparent)}}{A^-}$$

The ratio of the acid form to the salt form will, of course, depend on the pH (Chapter 6). So long as the solution is buffered or is very acidic, this ratio will remain constant at all concentrations of the acid. However, in unbuffered solution, the degree of ionization will increase as the acid is diluted, that is, the above equilibrium will shift to the right. Thus, a smaller fraction of the species exists in the acid form available for absorption for dilute solutions of the acid, causing apparent deviations from Beer's law. The result will be a positive deviation from linearity at higher concentrations (where the fraction dissociated is smaller). If the anion form were the absorbing species, then the deviation would be negative. Actually, in by far the majority of systems, the anion form absorbs at a higher wavelength and with a greater epsilon. So in practice, more commonly, the anion is the more absorbing species.

Similar arguments apply to colored (absorbing) metal ion complexes or chelates in the absence of a large excess of the complexing agent. That is, in the absence of excess complexing agent, the degree of dissociation of the complex will increase as the complex is diluted. Here, the situation may be extremely complicated because the complex may dissociate stepwise into successive complexes that may or may not absorb at the wavelength of measurement. pH also becomes a consideration in these equilibria.

Apparent deviations may also occur when the substance can exist as a dimer as well as a monomer. Again, the equilibrium depends on the concentration. An example is the absorbance by methylene blue, which exhibits a negative deviation at higher concentrations due to association of the methylene blue. In some systems, both pH and concentration play a role; the equilibrium $2CrO_4^{2-} + 2H^+ \rightleftharpoons Cr_2O_7^{2-} + H_2O$ is an example.

The best way to minimize chemical deviations from Beer's law is by adequate buffering of the pH, adding a large excess of complexing agent, ionic strength adjustment, and so forth. Preparation of a calibration curve over the measurement range will correct for most deviations.

If both species of a chemical equilibrium absorb, and if there is some overlap of their absorption curves, the wavelength at which this occurs is called the **isosbestic point**, and the molar absorptivity of both species is the same. Such a point is illustrated in Figure 11.30. The spectra are plotted at different pH values since the pH generally causes the shift in the equilibrium. Obviously, the effect of pH could be eliminated by making measurements at the isosbestic point, but the sensitivity is decreased. By making the solution either very acid or very alkaline, one species predominates and the sensitivity is increased by measuring at this condition.

For a two-component system in which the two absorbing species are in equilibrium, all curves intersect at the isosbestic point where they have the same ϵ value. The existence of an isosbestic point is a necessary (although not sufficient) condition to prove that there are only two absorbing substances in equilibrium with overlapping absorption bands. If both of the absorbing species follow Beer's law, then the

The absorptivity of all species is the same at the isosbestic point.

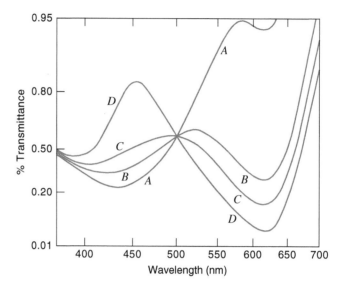

FIGURE 11.30 Illustration of isosbestic point of bromothymol blue (501 nm): (*A*) pH 5.45, (*B*) pH 6.95, (*C*) pH 7.50, (*D*) pH 11.60. *A* represents the spectrum of the acid form, and *D* of the base form, of bromothymol blue. *B* and *C* represent the sums of the spectra of the two species in equilibrium as the pH shifts. They both have the same absorbance at 500 nm.

absorption spectra of all equilibrium mixtures will intersect at a fixed wavelength. For example, the different colored forms of indicators in equilibrium (e.g., the red and yellow forms of methyl orange) often exhibit an isosbestic point, supporting evidence that two and only two colored species participate in the equilibrium.

The existence of an isosbestic point is not proof of the presence of only two components. There may be a third component with $\epsilon = 0$ at this particular wavelength. The absence of an isosbestic point, however, is definite proof of the presence of a third component, provided the possibility of deviation from Beer's law in the two-component system can be dismissed. For a two-component system, the isosbestic point is a unique wavelength for quantitative determination of the total amount of two absorbing species in mutual equilibrium.

If an acid–base indicator dissolved in the solution of a salt of a weak acid (NaA) is injected into a flowing stream of an acid HA, as the indicator disperses, the dispersed portions of the indicator are distributed in fluid elements that have different ratios of NaA and HA, i.e., they are at different pH. This results in qualitatively different spectra at different time points as the indicator bolus passes through an array detector. However, the indicator is also diluted to different degrees, causing the magnitude of the absorption to be dependent on the local concentration. Vithanage and Dasgupta (*Anal. Chem.* **58** (1986) 326) showed how the absorbance at the isosbestic wavelength can be used to correct for (normalize) the change in concentration and generate results such as in Figure 11.30 by a single injection in a flowing stream and also calculate the acid dissociation constant or a metal-ligand association constant.

Instrumental Deviations

The basic assumption in applying Beer's law is that monochromatic light is used. We have seen in the discussions above that it is impossible to extract monochromatic radiation from a continuum source. Instead, a band of radiation is passed, the width of which depends on the dispersing element and the slit width or the size of or the spacing between array detector elements. In an absorption spectrum, different wavelengths are absorbed to a different degree; that is, the absorptivity changes with wavelength. At a wavelength corresponding to a fairly broad maximum on the spectrum, the band of wavelengths will all be absorbed to nearly the same extent. However, on a steep portion of the spectrum, they will be absorbed to different degrees. The slope of the spectrum increases as the concentration is increased, with the result that the fractions of the amounts of each wavelength absorbed may change. This becomes worse if the instrument setting drifts over the period of the measurements made. So a negative deviation in the absorbance-versus-concentration plot will be observed. The greater the slope of the spectrum, the greater is the deviation.

Obviously, it is advantageous to make the measurement on an absorption peak whenever possible, in order to minimize this curvature, as well as to obtain maximum sensitivity. Because a band of wavelengths is passed, the absorptivity at a given wavelength may vary somewhat from one instrument to another, depending on the resolution, slit width, and sharpness of the absorption maximum. Therefore, you should check the absorptivity and linearity on your instrument rather than relying on reported absorptivities. It is common practice to prepare calibration curves of absorbance versus concentration rather than to rely on direct calculations of concentration from Beer's law.

> The absorptivity at a given wavelength may vary from instrument to instrument. Therefore, always run a standard.

If there is a second (interfering) absorbing species whose spectrum overlaps with that of the test substance, nonlinearity of the total absorbance as a function of the test substance concentration will result. It may be possible to account for this in preparation of the calibration curve by adding the interfering compound to standards at the same concentration as in the samples. This will obviously work only if the concentration of the interfering compound is essentially constant, and its contribution to the absorbance at the measurement wavelength relatively small. Otherwise, simultaneous two-component analysis as described earlier will be required.

Other instrumental factors that may contribute to deviations from Beer's law include stray radiation (also called **stray light**, radiation at wavelengths other than the intended wavelength band reaching the detector), internal reflections of the radiation within the monochromator, and mismatched cells (in terms of pathlength) used for different analyte solutions or used in double-beam instruments (when there is appreciable absorbance by the blank or solvent in the reference cell). Stray light becomes especially limiting at high absorbances and eventually causes deviation from linearity. Consider, for example, an instrument with 0.1% stray light. If 100.0 units of light proceeds through the sample, 0.1 units of light goes directly to the detector without going through the sample. If the sample has a true transmittance of 1%, instead of 1 unit of light reaching the detector, 1.1 unit is registered and the transmittance is read to be $1.1/100.1 \simeq 0.011$ instead 0.010. The absorbance registered is 1.959 instead of 2.000. It is for this reason that most low-end spectrometers do not produce reliable data at absorbance values over 2 (although present-day high-end, ultra-low stray light spectrometers can operate up to 6–8 absorbance units). The effects of large amounts of stray light on the transmittance and absorbance are shown in the text **website**: Figure 11.a. Stray light. Noise resulting from stray light also becomes a major contributor to the spectrometric error or imprecision at high absorbances. Radiation that does not interact with the sample can originate from light leaks in the instrument, from scattering of light from the optical components, or scattered light through the sample itself. A stray light component equivalent to 0.1% transmittance results in an error of 0.4% for a sample with an absorbance of 1.0.

> Stray light is the most common cause of negative deviation from Beer's law. For Beer's law, the light falling on the detector goes to zero at infinite concentration (all the light is absorbed). But this is impossible when stray light falls on the detector. The effects of a large bandwidth used to measure a narrow absorption band or making a measurement on the shoulder of a steeply rising absorption profile is akin to the effects of stray light. For an attractive applet, see http://www.chem.uoa.gr/applets/AppletBeerLaw/Appl_Beer2.html.

Other chemical and instrumental sources of nonlinearity in absorbance measurements may include hydrogen bonding, interaction with the solvent, nonlinear detector response or nonlinear electronic amplification, noncollimated radiation, and signal saturation.

Nonuniform cell thickness can affect a quantitative analysis. This is potentially a problem, especially in infrared spectrometry, where cell spacers are used. Air bubbles can affect the pathlength and stray light, and it is important to eliminate these bubbles, especially in infrared cells.

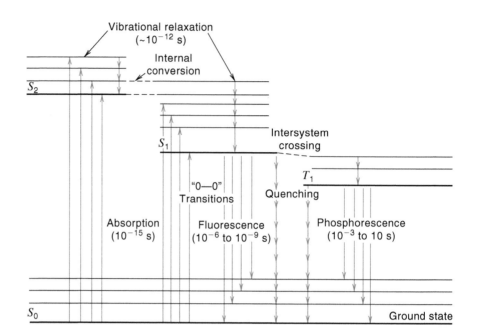

FIGURE 11.31 Energy level diagram (sometimes referred to as a Jablonski diagram) showing absorption processes, relaxation processes, and their rates.

11.15 Fluorometry

Fluorometric analysis is extremely sensitive and is used widely in many disciplines.

Principles of Fluorescence

Some molecules that absorb UV radiation lose only part of the absorbed energy by collisions. The rest is reemitted as radiation at longer wavelengths.

When a molecule absorbs electromagnetic energy, this energy is usually lost as heat, as the molecule is deactivated via collisional processes. Some molecules lose only part of the energy via collisions, and then the electron drops back to the ground state by emitting a photon of lower energy (longer wavelength) than was absorbed. This phenomenon is called fluorescence; an estimated 5 to 10% of all molecules fluoresce, especially when excited by energetic UV radiation. Refer to Figure 11.31.

A molecule at room temperature normally resides in the ground state. The ground state is usually a **singlet state** (S_0), with all electrons paired. Electrons that occupy the same molecular orbital must be "paired," that is, have opposite spins. In a singlet state, the electrons are paired. If electrons have the same spin, they are "unpaired" and the molecule is in a **triplet state**. Singlet and triplet states refer to the **multiplicity** of the molecule. The process leading to the emission of a photon begins with the absorption of a photon (a process that takes 10^{-15} s) by the fluorophore, resulting in an electronic transition to a higher-energy (excited) state. In most organic molecules at room temperature, this absorption corresponds to a transition from the lowest vibrational level of the ground state to one of the vibrational levels of the first or second electronic excited state of the same multiplicity (S_1, S_2). The spacing of the vibrational levels and rotational levels in these higher electronic states gives rise to the absorption spectrum of the molecule.

If the transition is to an electronic state higher than S_1, a process of **internal conversion** rapidly takes place. It is envisioned that the excited molecule passes from the vibrational level of this higher electronic state to a high vibrational level of S_1 that is isoenergetic with the original excited state. Collision with solvent molecules at this point rapidly removes the excess energy from the higher vibrational level of S_1; this process is called **vibrational relaxation**. These energy degradation processes (internal conversion and vibrational relaxation) occur rapidly ($\sim 10^{-12}$ s). Because of this rapid energy loss, fluorescence emission from states higher in energy than the first excited state is rare.

Once the molecule reaches the first excited singlet state, internal conversion to the ground state is a comparatively slow process. Thus, decay of the first excited state by emission of a photon can effectively compete with other decay processes. This emission process is **fluorescence**. Generally, fluorescence emission occurs very rapidly after excitation (10^{-6} to 10^{-9} s). Consequently, it is not possible for the eye to perceive fluorescence emission after removal of the excitation source. Because fluorescence occurs from the lowest excited state, the fluorescence spectrum, that is, the wavelengths of emitted radiation, is independent of the wavelength of excitation. The intensity of emitted radiation, however, will be proportional to the intensity of incident radiation (i.e., the number of photons absorbed).

Another feature of excitation and emission transitions is that the longest wavelength of excitation corresponds to the shortest wavelength of emission. This is the "0–0" band, corresponding to the transitions between the 0 vibrational level of S_0 and the 0 vibrational level of S_1 (Figure 11.31).

While the molecule is in the excited state, it is possible for one electron to reverse its spin, and the molecule transfers to a lower-energy triplet state by a process called **intersystem crossing**. Through the processes of internal conversion and vibrational relaxation, the molecule then rapidly attains the lowest vibrational level of the first excited triplet (T_1). From here, the molecule can return to the ground state S_0 by emission of a photon. This emission is referred to as **phosphorescence**. Since transitions between states of different multiplicity are "forbidden," this process is slow and T_1 has a much longer lifetime than S_1 and phosphorence is much longer-lived than fluorescence ($> 10^{-4}$ s). Consequently, one can quite often perceive an "afterglow" in phosphorescence when the excitation source is removed. In addition, because of its relatively long life, radiationless processes can effectively compete with phosphorescence. For this reason, phosphorescence is not normally observed from solutions due to collisions with the solvent or with oxygen. Phosphorescence measurements are made by cooling samples to liquid nitrogen temperature ($-196°C$) to freeze them and minimize collision with other molecules. Solid samples will also phosphoresce, and many inorganic minerals exhibit long-lived phosphorescence. Studies have been made in which molecules in solution are adsorbed on a solid support from which they can phosphoresce. Phosphorescence may be observed with minerals. "Fluorescent" lamps contain mercury and a phosphor-coated glass envelope. Mercury is excided by electrical discharge and emits in the UV, which excites the phosphors that emit visible light. Notice that the glow from such lamps persists for a while after the lamp is turned off. So the glow comes from the coatings on the wall of the glass tube. These absorb the UV emissions of the Hg lamp, and fluoresce at a longer (lower-energy) wavelength, resulting in a bathochromic shift. Modern fluorescent lamps use a mixture of phosphors, e.g., from barium aluminate that emits in the blue to lanthanum phosphate that emits in the green, to yttrium oxide that emits in the orange-red. The precise composition of the phosphor mixture can generate "cool" to "warm" white light. For an informative article on fluorescent lamp phosphors, see http://www.lamptech.co.uk/Documents/FL%2520Phosphors.htm.

A typical excitation and emission spectrum of a fluorescing molecule is shown in Figure 11.32. The excitation spectrum usually corresponds closely in shape to the absorption spectrum of the molecule. There is frequently (but not necessarily) a close relationship between the structure of the excitation spectrum and the structure of the emission spectrum. In many relatively large molecules, the vibrational spacings of the excited states, especially S_1, are very similar to those in S_0. Thus, the form of the emission spectrum resulting from decay to the various S_0 vibrational levels tends to be a

The wavelengths of emitted radiation are independent of the wavelength of excitation. The intensity of the emitted radiation is, however, dependent both on the intensity and the wavelength of the exciting radiation.

Phosphorescence is longer lived than fluorescence, and it may continue after the excitation source is turned off.

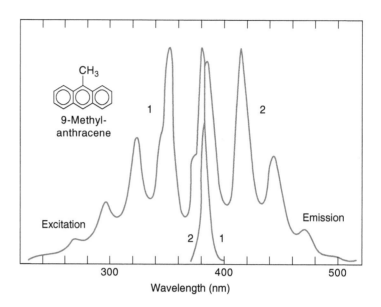

FIGURE 11.32 Excitation and emission spectra of fluorescing molecule.

"mirror image" of the excitation spectrum arising from excitation to the various vibrational levels in the excited state, such as S_1. Substructure, of course, results also from different rotational levels at each vibrational level.

The longest wavelength of absorption and the shortest wavelength of fluorescence tend to be the same (the 0–0 transition in Figure 11.31). More typically, however, this is not the case due to solvation differences between the excited molecule and the ground-state molecule. The heats of solvation of each are different, which results in a decrease in the energy of the emitted photon by an amount equal to the difference in these two heats of solvation.

A minority of all molecules fluoresce; even fewer phosphoresce. This is an advantage with respect to selective detection or measurement. The emitted radiation may be in the ultraviolet region, especially if the compound absorbs at less than 300 nm, but is often in the visible or the NIR region. It is the emitted radiation that is measured and related to concentration.

Chemical Structure and Fluorescence

In principle, any molecule that absorbs radiation and is promoted to an electronically excited state could fluoresce. Most, however, do not for a number of different reasons. We point out below what types of substances may be expected to fluoresce.

First of all, the greater the absorption by a molecule, the greater its fluorescence intensity. Many aromatic and heterocyclic compounds fluoresce, particularly if they contain certain substituted groups. Multiple conjugated double bonds favor fluorescence. One or more electron-donating groups such as $-OH$, $-NH_2$, and $-OCH_3$ enhances the fluorescence. Polycyclic compounds such as vitamin K, purines, and nucleosides and conjugated polyenes such as vitamin A are fluorescent. Groups such as $-NO_2$, $-COOH$, $-CH_2COOH$, $-Br$, $-I$, and azo groups tend to *inhibit* fluorescence. The nature of other substituents may alter the degree of fluorescence. The fluorescence of many molecules is greatly pH dependent because only the ionized or un-ionized form may be fluorescent. For example, phenol, C_6H_5OH, is fluorescent but its anion, $C_6H_5O^-$, is not. The amino acid tryptophan fluoresces with optimum excitation at ~280 nm and maximum emission at ~360 nm. All proteins show some fluorescence from the tryptophan moiety, albeit this fluorescence is not strong. The lifetime of this fluorescence, however, is very sensitive to the specific environment of the amino acid and is often used to determine structural or conformational changes.

If a compound is nonfluorescent, it may be converted to a fluorescent derivative. For example, nonfluorescent steroids may be converted to fluorescent compounds by dehydration with concentrated sulfuric acid. These cyclic alcohols are converted to phenols. Similarly, dibasic acids, such as malic acid, may be reacted with β-naphthol in concentrated sulfuric acid to form a fluorescing derivative. White and Argauer have developed fluorometric methods for many metals by forming chelates with organic compounds (see Reference 21). A variety of metals form highly fluorescent chelates with 8-hydroxyquinoline-5-sulfonic acid (sulfoxine) or quench the fluorescence of other metal-sulfoxine chelates (see Reference 22). Antibodies may be made to fluoresce by condensing them with fluorescein isocyanate, which reacts with the free amino groups of the proteins. NADH, the reduced form of nicotinamide adenine dinucleotide, fluoresces. It is a product or reactant (cofactor) in many enzyme reactions (see Chapter 25 on the text website), and its fluorescence serves as the basis of many sensitive assays for enzymes or their substrates. Amino acids other than tryptophan do not fluoresce, but intensely fluorescent derivatives are formed by reaction with dansyl chloride.

Fluorescence Quenching

One difficulty frequently encountered in fluorescence is that of **fluorescence quenching** by many substances. These are substances that, in effect, compete for the electronic excitation energy and decrease the quantum yield (the efficiency of conversion of absorbed radiation to fluorescent radiation—see below). Iodide ion is an extremely effective quencher. Iodine and bromine substituents decrease the quantum yield. Quenchers themselves may be determined indirectly by measuring the extent of fluorescence quenching by adding the quencher analyte to a constant concentration of a fluorophore. Some molecules do not fluoresce because they may have a bond whose dissociation energy is less than that of the exciting radiation. In other words, a molecular bond may be broken, preventing fluorescence.

A colored species in solution with a fluorescing analyte may interfere by absorbing the excitation radiation or the emitted fluorescent radiation or both. This is the so-called **inner-filter effect**. For example, in sodium carbonate solution, potassium dichromate exhibits absorption peaks at 245 and 348 nm. These overlap with the excitation and emission peaks for tryptophan and would interfere. The inner-filter effect can also arise from too high a concentration of the fluorophore itself. Some of the analyte molecules will reabsorb the emitted radiation of others (see the discussion of fluorescence intensity and concentration below).

Quenching of fluorescence is often a problem in quantitative measurements.

Relationship between Concentration and Fluorescence Intensity

It can be readily derived from Beer's law (Problem 57) that the fluorescence intensity F is given by

$$\boxed{F = \phi P_0(1 - 10^{-abc})} \qquad (11.23)$$

where ϕ is the **quantum yield**, a proportionality constant and a measure of the fraction of absorbed photons that are converted into fluorescent photons. The quantum yield is, therefore, less than or equal to unity. The other terms in the equation are the same as for Beer's law. It is evident from the equation that if the product abc is large, the term 10^{-abc}, the same as the fractional transmittance T, becomes negligible compared to 1, and F becomes constant:

$$F = \phi P_0 \qquad (11.24)$$

Fluorescence intensity is proportional to the intensity of the source. Absorbance, on the other hand, is independent of it.

For highly efficient fluorophores, the quantum yield can be close to unity: For ethanolic acidic rhodamine, it is 1.0, while for aqueous alkaline fluorescein, it is 0.79. For aqueous tryptophan buffered at pH 7.2, it is 0.14.

On the other hand, if abc is small ($\leq 0.01, T \geq 98\%$), it can be shown[4] by expanding Equation 11.23 that to a good approximation,

$$F = 2.303\phi P_0 abc \tag{11.25}$$

For low concentrations, fluorescence intensity becomes directly proportional to the concentration.

Thus, for low concentrations, the fluorescence intensity is directly proportional to the concentration. Also, it is proportional to the intensity of the incident radiation.

This equation generally holds for concentrations up to a few parts per million, depending on the substance. At higher concentrations, the fluorescence intensity may decrease with increasing concentration. Consider that in dilute solutions, the absorbed radiation is distributed equally through the entire depth of the solution. But at higher concentrations, the first part of the solution in the path will absorb more of the radiation. So the equation holds only when most of the radiation goes through the solution, when more than about 92% is transmitted [this is to say that at low absorbance values (A $\leq$ 0.04), absorbance and transmittance are linearly related, see the spreadsheet figure following Equation 11.12].

Fluorescence Instrumentation

For fluorescence measurements, it is necessary to separate the emitted radiation from the incident radiation. This is most easily done by measuring the fluorescence at right angles to the incident radiation. The fluorescence radiation is emitted in all directions, but the incident radiation passes straight through the solution.

A simple fluorometer design is illustrated in Figure 11.33. An ultraviolet source is required. Most fluorescing molecules absorb ultraviolet radiation over a band of wavelengths, and so a simple line source is sufficient for many applications. Such a source is a medium-pressure mercury vapor lamp. A spark is passed through mercury vapor at low pressure, and principal lines are emitted at 253.7, 365.0, 520.0 (green), 580.0 (yellow), and 780.0 (red) nm. *Wavelengths shorter than 300 nm are harmful to the eyes*, and one must never look directly at a short-wavelength UV source. The mercury vapor itself absorbs most of the 253.7-nm radiation (self-absorption), and a blue filter in the envelope of the lamp may be added to remove most of the visible light.[5] The 365-nm

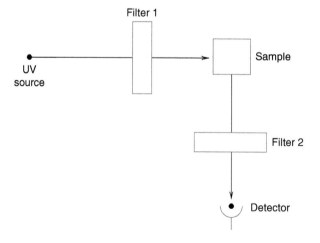

FIGURE 11.33 Simple fluorometer design.

[4]It is known that $e^{-x} = 1 - x + x^2/2!\ldots$ and that $10^{-x} = e^{-2.303x}$. Therefore, $1 - e^{-2.303abc} = 1 - [1 - 2.303abc + (2.303abc)^2/2!\ldots]$. The squared term and higher-order terms can be neglected if $abc \leq 0.01$, and so the expanded term reduces to $2.303abc$. This is a Taylor expansion series.

[5]The intensity of the 253.7 nm line depends on the pressure of Hg in the lamp. In low pressure lamps, it is the most intense line.

line is thus the one used primarily for the activation. A high-pressure xenon arc (a continuum source) is usually used as the source in more sophisticated instruments that will scan the spectrum (spectrofluorometers) because it has a more uniform disribution of energy throughout the UV-Visible spectrum. The lamp pressure is 7 atm at 25°C and 35 atm at operating temperatures and is typically housed in a protective but well-vented housing.

In the simple filter fluorometer instrument in Figure 11.33, a excitation filter (filter 1) is used to select wavelengths that efficiently excite the analyte fluorescence. This filter is typically a short-pass or bandpass filter with a long wavelength cutoff that is shorter than the cut-on wavelength of the emission filter (filter 2), usually a long-pass filter. Thus filter 1 allows the passage of only the wavelength of excitation while filter 2 passes the wavelength of emission but not the wavelength of excitation, which may find its way to the detector by scattering. Depending on the application, glass or nonfluorescing grade quartz cells are suitable.

LCW-based fluorescence detectors provide a very simple approach to flow-through fluorescence detection in dedicated applications. The principle is illustrated in Figure 11.34(a). Light is incident transversely on an LCW tube through which the analyte-bearing solution is flowing. Any unabsorbed light passes out in the radial direction. When a fluorescent molecule comes in the light path, it absorbs the light and the resulting fluorescence is emitted in all directions. The portion of this fluorescence that is within the acceptance angle of the fiber proceeds in both axial directions and is measured at one end of the LCW. In such an arrangement, the detected fluorescence emission is largely free of the exciting radiation. Actual implementation is carried out in the arrangement shown in Figure 11.34(b) where one end of the LCW is connected to a tee and the fluorescence emission is carried via a fiber optic to a detector. To

Filter 1 passes the wavelengths desired for excitation (or all wavelengths shorter than the desired excitation wavelength) while filter 2 passes the fluorescence emission (and longer wavelengths) and effectively rejects the scattered excitation light.

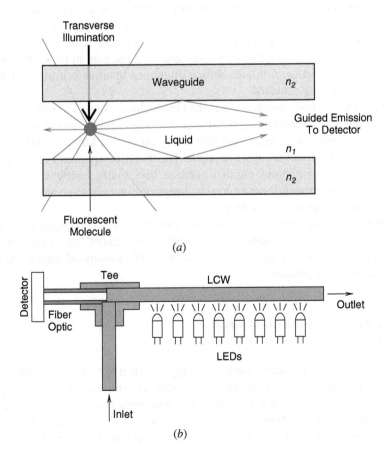

FIGURE 11.34 (*a*) Principle of operation of a liquid core waveguide (LCW) based fluorescence detector, (*b*) A typical use configuration.

enhance sensitive detection, an essentially monochromatic light source such as one or more LEDs or a miniature fluorescent black light (365 nm) is used. An emission filter that further filters out any excitation light can be optionally placed in front of the detector to remove any stray excitation light. Very good detection limits can be obtained for many analytes with such an approach (see Reference 23). Typically the most expensive part of such a system is the photomultiplier tube that is used as the detector. The use of LEDs as light sources allow rapid on/off switching without sacrificing reproducibility of the light intensity and the use of optical fibers to carry the emitted light allows multiple fibers from multiple detection cells to be connected to the same detector, with the excitation lights in only one detection cell being turned on at a time. For an account of the fluorometric measurement of atmospheric hydrogen peroxide and organic peroxides using such a multiplexed detector, see Z. Genfa, P. K. Dasgupta, and G. A. Tarver (*Anal. Chem.* **75** (2003) 1203.

In a spectrofluorometer, the filters are replaced with scanning monochromators. Either the excitation spectrum (similar to the absorbance spectrum) or the emission spectrum may be recorded.

In a **spectrofluorometer**, the measurement is also made at right angles to the direction of the incident radiation. But instead of using filters, the instrument incorporates two monochromators, one to select the wavelength of excitation and one to select the wavelength of fluorescence. The wavelength of excitation from a continuum source can be scanned and the fluorescence measured at a set wavelength to give a spectrum of the excitation wavelengths. This allows the establishment of the wavelength of maximum excitation. Then, by setting the excitation wavelength for maximum excitation, the emission wavelength can be scanned to establish the wavelength of maximum emission. When this spectrum is scanned, there is usually a "scatter peak" corresponding to the wavelength of excitation. High-end spectrofluorometers often use a double grating monochromator at the excitation or emission stage or both, to reduce stray light to a minimum.

In typical spectrofluorometers, variations in intensity from the source or response of the detector at different wavelengths are not corrected for, and calibration curves are generally prepared under a given set of conditions. Since the source intensity or detector response may vary from day to day, the instrument is usually calibrated by measuring the fluorescence of a standard solution and adjusting the gain to bring the instrument reading to the same value. A dilute solution of quinine in dilute sulfuric acid is usually used as the calibrating standard.

Sophisticated instruments such as the Horiba Jobin-Yvon Fluorolog can provide "corrected spectra" by monitoring the source intensity continuously at all wavelengths by a calibrated photodiode array, and also correcting for the known response behavior of the detector as a function of wavelength. The emission spectrum can be presented directly in quanta of photons emitted per unit bandwidth. Instruments such as the Hamamatsu Quantaurus-QY are expressly designed to measure quantum efficiency as a function of the excitation wavelength.

There are many instruments that can simultaneously (synchronously) scan the excitation and emission monochromators. While this is beyond the scope of the present discussion, "Synchronous fluorescence scanning" offers many advantages, especially for multicomponent analysis.

Fluorescence Lifetime and Gated Fluorescence/Phosphorescence Measurement

For most common fluorescent compounds, typical excited state decay times for photon emissions with energies from the UV to near-infrared are within the range of 0.5 to 20 ns. These "fluorescence lifetimes" are often highly sensitive to the immediate environment of the fluorophore and hence can provide information about structure and configuration. With the advent of LEDs that can be pulsed at high speeds, lifetime measurements are almost exclusively done by phase-resolved fluorometry. Consider that we excite a sample once with a single flash of light that is of a duration much smaller compared to the fluorescence lifetime. While light absorption is essentially instantaneous, the peak in the fluorescence intensity will occur after a finite time interval equal to the mean lifetime of the excited state. Now consider that instead of a single flash, the LED is turning on and off with a short on period and at an overall frequency of 10 MHz. One whole cycle is thus 100 ns long and a whole cycle is 360 degrees. If the fluorescence emission waveform has a peak that occurs 10 ns after the excitation peak, we would observe that the emission is $10/100*360° = 36°$ phase-shifted relative to the excitation wave form. The phase difference between two such signals can be determined with a high degree of resolution and precision and this constitutes the basis of lifetime measurements by phase-resolved fluorometry. For example, the Hamamatsu Quantaurus Tau instrument uses user-selectable LEDs emitting at 280, 340, 365, 405, 470, 590, and 630 nm, allows cooling of samples down to liquid nitrogen temperatures, and can measure fluorescence lifetimes within a minute. Lifetimes can be resolved to better than 0.1 ns.

As previously mentioned, fluorescence lifetimes become longer when intersystem crossing is involved and generates "phosphorescence." A major analytical application involves the unique behavior of certain lanthanide metal ions, notably europium (Eu^{3+}) and terbium (Tb^{3+}). In aqueous solution, these ions both absorb and fluoresce weakly, with a short lifetime. When chelated with a suitable of organic complexing agent, the absorption maximum increases greatly and shifts to the UV. The ligand absorbs the UV radiation and is promoted to the excited state, but then the energy is transferred to the metal center by intersystem crossing. The metal complex fluoresces in the visible with intensities as much as 10,000 times greater than the bare metal ion and with a fluorescence lifetime that is as much as several hundred microseconds. This type of system is ideal for performing what is called (time)-gated fluorescence detection where the detector views the fluorescence a finite time after the excitation flash/pulse illuminates the sample. The great advantage of this type of technique is that detection takes place truly in an otherwise totally dark environment. Because the excitation source is off at the time measurement is made, there is no scattered excitation light. Photomultiplier tubes are specially made for time-gated operation. Note that even when power is not applied to a sensitive photo detector, exposure to bright illumination leads to memory effects. The view of the detector is therefore physically blocked with choppers, tuning forks/vanes, and other mechanical means during the excitation pulse.

Fluorescence-based fingerprint detection/imaging is highly sensitive but is often hindered by native fluorescence of the substrate bearing the fingerprint. Using europium chelate-based fingerprint developing reagents and time-resolved imaging, it is possible to visualize fingerprints on otherwise very difficult surfaces (see Reference 24). A large portion of the mass of bacterial spores, including anthrax, consists of calcium dipicolinate, which it apparently uses as an energy source during germination. The terbium complex of dipicolinic acid has a long fluorescence lifetime and can be used as a very sensitive measure of the presence of airborne spores. For an account of such spore measurement using gated fluorescence detection, see Reference 25.

Fluorescence vs. Absorbance

The reason that fluorometric methods are more sensitive than absorption spectrometry lies in the following. In trying to measure a very small absorbance, we are trying to measure a very small difference in a large amount of transmitted light. The best flow-through absorbance detectors today approach a noise level of 10^{-6} absorbance units,

Fluorescence measurements are 1000-fold more sensitive than conventional absorbance measurements. Very long path absorbance measurements and stable solid-state light sources are gradually making absorbance measurements competitive, however.

equivalent to a light intensity stability of ~2 parts in a million. In fluorescence, in principle, we measure the difference between no light and a small amount of light, and so the limit of detection is governed by the intensity of the source and the sensitivity and stability of the detector (the "shot noise"). In both absorbance and fluorescence, the signal depends linearly on concentration, and a wide dynamic range of response is observed; a dynamic range of 10^3 to 10^4 is not uncommon. In practice, the detection limit in fluorescence measurements is controlled by scattered light, source stability, and detector dark noise. In a favorable case, such as laser-induced fluorescence (LIF), detection of a fluorescent analyte of high quantum efficiency using a notch filter and a PMT that is cooled to reduce dark noise, it is possible to detect a single molecule, a feat not matched by any other technique.

11.16 Chemiluminescence

Fluorescence and phosphorescence are both subclasses of the general phenomena of luminescence; specifically they are classified as *photoluminescence*, emission of light following excitation by photons. **Radioluminescence** involves excitation by energetic radiation (γ-rays, etc.); **electroluminescence** involves direct electrical excitation; **piezoluminescence** involves luminescence created by pressure. **Thermoluminescence** is somewhat of a misnomer; it does not really involve excitation by heat. Rather, exposure to omnipresent energetic radiation (cosmic rays) creates lattice defects in solids that are released in the form of luminescence when the material is heated. Heating a solid to high temperatures releases all the latent luminescence; solids that have previously been fired to high temperatures (e.g., pottery) have their thermoluminescence clock thus reset to zero at that time. Cosmic ray dosage is essentially constant over time; ancient pottery can thus be dated by thermoluminescence measurements. **Triboluminescence** is generated when material is pulled apart, ripped, scratched, crushed, or rubbed from energy released due to the breaking of chemical bonds. Crushing sugar crystals in one's mouth actually generates luminescence, but because this is mostly in the near UV, it is not easily visible. Wintergreen (methyl salicylate) flavored candies, on the other hand, visibly glow when crushed in the dark (try Wint-O-Green flavored lifesavers!); in this case, the near **UV** triboluminescence from sugar excites the methyl salicylate (fluorescent); this emits in the visible, facilitating visual detection. **Chemiluminescence** provides electronic excitation by energy derived from a chemical reaction; such reactions may be brought about in natural biological systems (**bioluminescence**, as in fireflies). **Electrogenerated chemiluminescence** (ECL) is typically also considered a special subclass. For general succinct reviews on luminescence and chemiluminescence, see References 27 and 28.

Chemiluminescence (CL) reactions generally involve those that are highly energetic, often involving strong oxidants. Ozone reacts with many substances, producing light. In particular, gas-phase reactions respectively with ethylene and NO are the basis of standard methods for measuring ozone and oxides of nitrogen. Many other substances are indirectly determined by the $NO-O_3$ reaction. Sulfur-selective CL detectors rely on a hydrogen-rich flame to generate SO; this is oxidized by O_3 to excited SO_2, denoted SO_2^*; this emits in the blue to near-UV. Virtually all alkenes undergo a CL reaction with ozone. Many metal hydrides react with ozone and emit light; the reaction with AsH_3 is the basis for sensitive assays for arsenic.

One important gas-phase CL reaction that does not involve ozone is the production of monoatomic sulfur in a hydrogen flame, two of these combine to make excited diatomic S_2^*; this emits in the deep violet (394 nm). Phosphorus compounds similarly form excited HPO* in a H_2-flame; this emits at 526 nm. These form the basis of the sulfur and phosphorus flame photometric detectors used in gas chromatography.

In the liquid phase, the best-known CL reaction involves luminol, an oxidant is needed and the reaction is catalyzed by various species. Either the catalyst (many metal ions and many other species, including enzymes such as peroxidase) or the oxidant (H_2O_2, hypochlorite, etc.) can be determined by this reaction. Other important solution-phase CL reactions involve acidic potassium permanganate (CL is produced by a variety of substances it oxidizes), hypohalites (hypochlorite (OCl^-), hypobromite (OBr^-), both are energetic oxidants), diaryloxalates, luminol, and tris(2,2'-bipyridyl) ruthenium(III), abbreviated as Ru(bipy)$^{3+}$. In the last case, as Ru(bipy)$^{3+}$ oxidizes some analyte, it is reduced to excited Ru(bipy)$^{2+}$, which returns to the ground state with the emission of orange light. With rare exceptions, these reagents react with so many substances in a similar manner, producing similar CL, that direct CL reactions are rarely of value as a selective or specific assay. If the desired analyte can be isolated, e.g., by selective preconcentration or chromatography, CL reactions can be profitably used with great sensitivity. Atmospheric hydrogen peroxide, for example, has been determined at the parts per trillion levels using the luminol reaction (see Reference 29).

For most CL measurements, unless characterizing the CL spectrum is of importance, a spectrometer is not necessary, just a sensitive detector is needed; photomultiplier tubes are commonly used. Many fluorometers can measure CL (excitation source turned off) or can be modified to do so. "Luminometers" are often used for bioassays and are available both in the single tube/cell or the microplate format. Most of the traditional CL reactions listed in the above paragraph occur too fast, however, for reagents to be manually mixed and put in front of the detector to read, as the reproducibility of the time between the mixing of the reagents and reading the light emission is critical. For this reason, most practical CL assays with the above reactions are conducted in a continuous flow mode where the reagents are mixed directly in front of the detector in a cell (two confluent entrance ports) with at least one transparent window facing the detector; the mixed liquid typically flows out through a spiral-shaped path (to maximize residence time in front of the detector) before exiting. The Global FIA Firefly CL detector uses an LCW cell within which the liquids are mixed; the light emitted can be collected from both ends of the tube and sent to the detector.

11.17 Fiber-Optic Sensors

There has been a great deal of interest in recent years in developing optical sensors that function much as electrochemical sensors (Chapter 22) do. Optical fibers have proven valuable as platforms for sensitive absorbance- and fluorescence-based sensors. They can be assembled to have multiple separate or collocated (coaxial) termini—while one receives the source light, another can be connected to a detector/spectrometer and a third can address an analyte.

The simplest fiber-optic sensor is a spectrometric dip-probe that measures the spectrum of a solution by dipping the probe in it. A bifurcated fiber optic, made of a multitude of small individual fibers, is used (Figure 11.35). Each individual bundle combines to form a common leg. This common terminus can be randomized (the position of the fibers from each separate bundle appear randomly in the common face (the most common arrangement) or has a deliberate geometry (e.g., the fibers from one leg forms the central core of the common phase, and the fibers from the other leg are arranged around this circle). The fibers can be made of glass or silica, depending on the application. Custom fiber bundles where visible fluorescence is to be excited

Optical sensors do not have the requirement and associated difficulties of a reference electrode.

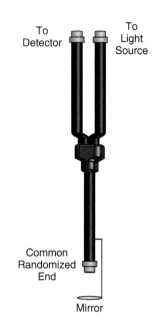

To Detector   To Light Source

Common Randomized End

Mirror

FIGURE 11.35 Bifurcated fiber-optic-based spectrophotometric probe.

by UV excitation benefit from a combination of glass and silica fiber legs, the latter brings in the excitation light while the fluorescence is sent back through the former to the detector, as the glass intrinsically blocks the transmission of the exciting UV radiation.

In the dip-probe shown in Figure 11.35, source light is brought in through one leg and is transmitted through the solution to the attached mirror and the reflected light proceeds through the other leg to the detector. The effective path length is twice the physical distance from the common leg face to the mirror. With a broadband source and an array detector, a complete spectrum can be obtained, while an LED and a photodiode is sufficient to provide transmittance measurements at a given wavelength for a specific application.

Instead of a mirror, an immobilized reagent that selectively reacts with a desired analyte on a substrate can be used. Such a reagent can also be directly chemically bonded to the face of the common leg. A major advantage of a fiber-optic sensor over electrochemical sensors is that a reference electrode (and salt bridge) is (are) not needed, and electromagnetic fields do not influence the response. For example, a fluorometric pH sensor may be prepared by chemically immobilizing the indicator fluorescein isothiocyanate (FITC) on a porous glass bead and attaching this to the end of the fiber with transparent epoxy adhesive. The fluorescence spectrum of FITC changes with pH over the range of about pH 3 to 7, centered around the pK of the indicator. The fluorescence intensity measured at the fluorescence maximum is related to the pH via a calibration curve. See References 32 and 33 for a discussion of the limitations of fiber-optic sensors for measuring pH and ionic activity.

If an enzyme, for example, penicillinase, is immobilized along with an appropriate indicator, then the sensor is converted into a biosensor for measuring penicillin. The enzyme catalyzes the hydrolysis of penicillin to produce penicilloic acid, which causes a pH decrease. Fiber-optic sensors have been developed for alkali metals, O_2, CO_2, moisture, and many other analytes. In order for such sensors to be attractive, the indicator chemistry must be reversible and bound in a robust manner to the sensing surface. Fiber-optic sensors for pH, O_2, CO_2, and relative humidity have become commercially available.

A different type of fiber-optic sensor uses evanescent waves for sensing. In this sensor, the fiber has no clad region or the clad is removed. The sensing indicator is directly chemically bonded to the surface of the fiber. When such a glass or silica fiber is immersed in an aqueous medium, it still behaves as a light guide, as the refractive index of glass or silica is significantly higher than the surrounding medium. Consider as light travels through one medium (such as the core of a fiber), it must penetrate some to the other medium: Otherwise how would a photon "sense" that the refractive index on the other side is lower? Light travelling through the core actually penetrates a small depth (to an extent of one-fourth the wavelength) into the interface, and this is termed evanescent interaction. As a result, any change in optical properties of the surface of the core is reflected in the transmitted light at the other end. But because the evanescent wave penetrates to such a small depth, many reflections at the core-interface wall are necessary; the length of the optical fiber used in an evanescent probe is always significant. Polyaniline is a polyprotic base; its absorption at 1428 nm changes continuously over a pH range of (at least) 3–14. One of the early demonstrations of an evanescent wave fiber-optic pH sensor used a thin coating of polyaniline on a silica core, probed by an NIR spectrometer and a quartz tungsten–halogen lamp source (see Reference 34). The fluorescence of many fluorescent molecules is quenched by oxygen, and such molecules have been used as the basis of fiber-optic oxygen sensors in both standard and the evanescent wave configurations (see Reference 35).

11.18 Photoacoustic Spectroscopy

Photoacoustic (PAS) or optoacoustic spectroscopy is a new method and is used less frequently due to the paucity of instruments. PAS is a measurement of the effect of absorbed electromagnetic energy (particularly of light) on matter by means of acoustic detection and is based on thermal effects and changes in the velocity of sound. The photoacoustic effect based on the light absorption effect was first investigated by Alexander Graham Bell in 1880 but PAS spectroscopy was developed only in the early 1970s and has taken several decades from that time until its application as a measurement technique. The PAS effect is observed on irradiating a gas in a closed cell with a chopped beam of radiation of a wavelength that can be absorbed by the gas. Periodic heating of the gas is caused by the absorbed radiation, that results in regular pressure fluctuation within the chamber. The pulses of pressure can be detected by a sensitive microphone if the chopping rate is within the acoustical frequency range. Analysis of absorbing gases has been done using the PAS effect. The photoacoustic effect can be used to obtain ultraviolet, visible, and infrared absorption spectra of solids, semisolids or turbid liquids. Obtaining spectra for these types of samples is not easy or often impossible because of light scattering and reflection by these samples. The major advantage of PAS is that it is suitable for highly absorbing samples. The PAS effect has gained new importance with the advent of tunable infrared lasers as sources.

The application of the PAS effect in the IR region is a new technique. With the design and development of highly-sensitive FTIR instruments, this technique is being widely researched for analyzing solid samples.

Principle

The energy can go on many paths on absorption of light by a material. The light hitting the sample either has to be absorbed, transmitted through the material, or reflected off of the material or in other words, should be conserved. The focus of PAS is on the absorbed light as that is responsible for the release of heat. When light strikes the sample and photons ar e absorbed, it leads to excitation of electrons because of the energy that is created. Acoustic waves are formed when this energy is released as heat and the heat expands (Figure 11.36).

Electrons can be excited either electronically or vibrationally when light is absorbed, as a result, electrons jump to a higher energy level. When the electrons drop back to its ground state, the extra energy is given off as heat. Heat can also be generated on collision of atoms. Also, energy can be dissipated through chemical reactions or radiative emissions for electronic excitation (Figure 11.36). These processes reduce the amount of heat formed since energy is spent somewhere else. Thermal expansion occurs with the formation of heat. The heat expansion creates localized pressure waves and in turn, can be measured as an acoustic wave. Apart from the formation of energy, heat can also be lost through the surroundings by diffusion which lowers the temperature around the emitted energy source and decreases the pressure fields. A sensor is then used to measure the acoustic waves that is sent after every pulse of light. A spectrum can be plotted by measuring the corresponding acoustic wave obtained by adjusting the wavelength of each pulse of light.

A theory explaining the PAS effect for all types of samples is yet to be developed. The theory that is prevalent is based on the photoacoustic signal and three parameters; sample thickness, light absorption length (related to absorbance coefficient), and the thermal diffusion length (related to thermal diffusion coefficient).

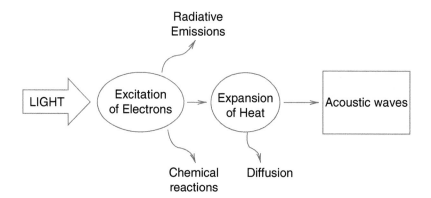

FIGURE 11.36 Scheme for absorption of light leading to acoustic waves.

Depending on these parameters, samples are classified as optically transparent/opaque; thermally thick/thin. The PAS signal reflects the sample optical properties, regardless of the sample thermal properties for an optically transparent sample. The photoacoustic signal is proportional to the sample absorbance coefficient and the thermal diffusion length for a sample that is optically opaque and thermally thick and hence, is said to be photo acoustically transparent.

Further, the depth from which the photoacoustic signal is emitted can be changed by changing the modulation frequency because the photoacoustic signal is proportional to the thermal diffusion length for a thermally thick sample.

Measurement and Instrumentation

The solid sample to be measured is placed in a sealed vessel containing air or some other non-absorbing gas with a small, sensitive microphone attached to it (Figure 11.37 and 11.38). The sample is then irradiated with a chopped beam from a monochromator. The chopper modulates the intensity of the radiation and is not required for a pulsed

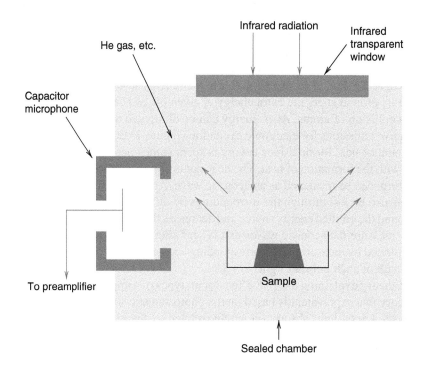

FIGURE 11.37 Measurement of a sample in photoacoustic spectroscopy.

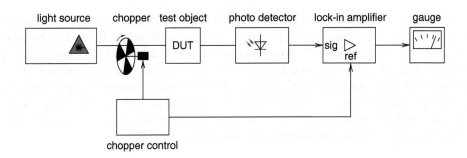

FIGURE 11.38 Components of a photoacoustic spectrometer.

source. When the radiation is absorbed by the sample, heat is generated due to the incident light, which in turn causes pressure changes in the surrounding gaseous layer and can be detected by the highly sensitive microphone (acoustic detector). For liquids a piezoelectric detector is used. The detector is not a photomultiplier tube. The detector is used in the form of a tube or disk and is a solid material like $PbZrO_2$, $PbTe:O_3$. The extent of absorption determines the power of the resulting sound. The microphone is not affected by the reflection or scattering of the radiation by the sample. The photoacoustic cell consists of a small, well-sealed vessel fitted with a highly sensitive microphone. The shape and size of the cell depends on the physical state of the sample. The sensitivity of the instrument increases by making the cell volume as small as possible. Also, the cell is mounted on a vibration-isolation stand in order to eliminate vibrations from the floor that can produce noise. The construction material of the cell has thermal mass (for example, steel, aluminum, silver or gold) so that external signals do not interfere with the sample.

The most commonly used **source** for PAS spectroscopy is a xenon arc lamp (pressure of 50-70 atmospheres, operating at wavelengths from 250-2500 nm). For lower wavelength hydrogen, tungsten or krypton lamps are used. **Lasers** (argon ion laser or a tunable dye laser) are also used for other regions. Amplifiers, light sources, and sensors have significantly improved with the advancement in technology.

Photoacoustic Spectra

Photoacoustic spectroscopy detects the light absorbed by the sample as sound waves. The signal is zero (0) for light that is not absorbed. So, the peaks in photoacoustic spectra point upwards, and are similar to absorption spectra (Figure 11.39).

Importance and Applications

PAS has unique aspects in spite its similarity with other infrared techniques such as Fourier transform infrared spectroscopy (FTIR). PAS is extremely accurate as it measures directly from the sample and not from the effect relative to the background. Tiny amounts of samples (as a few milliliters) and samples with multiples gases can be measured. The amount of sample that can be measured is however, small compared to other techniques used. The PAS method is very useful in the absence of fluorescence or phosphorescence. Small PAS signals result when samples lose energy by luminescence.

Samples can be measured without pretreatment regardless of its form; optically neutral samples that do not transmit or reflect incident light can also be measured. Information at different depths can be obtained by changing the velocity of the movable mirror in the interferometer. In addition, by controlling the intensity of the source, the method can be made more sensitive.

PAS finds applications in gas detectors because of the high sensitivity and accuracy, thus providing details on any dangers of rising toxic gases. It can also be used to determine the materials of unknown samples. This is because each material has a unique spectrum and the acoustic waves produced can be matched with specific profiles of materials. PAS also finds applications in high resolution imaging for analyzing the topography of a sample. Overall, PAS can be exploited to analyze irregularly shaped samples, perform measurement of the surface treatments on fabrics, films, and powders, and can be used for measurement in the depth direction.

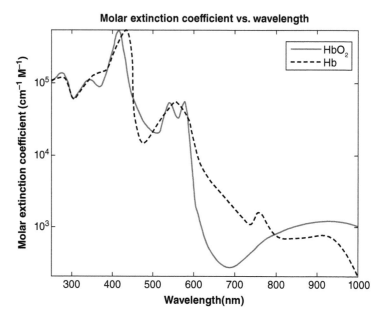

FIGURE 11.39 PAS spectra of hemoglobin.

Questions

Absorption of Radiation

1. Describe the absorption phenomena taking place in the far-infrared, mid-infrared, and visible–ultraviolet regions of the spectrum.

2. What types of electrons in a molecule are generally involved in the absorption of UV or visible radiation?

3. What are the most frequent electronic transitions during absorption of electromagnetic radiation? Which results in more intense absorption? Give examples of compounds that exhibit each.

4. What is a necessary criterion for absorption to occur in the infrared region?

5. What types of molecular vibration are associated with infrared absorption?

6. What is the fingerprint region in the infrared spectra?

7. Define anti-bonding orbitals.

8. What distinguishes near-infrared absorption from mid-infrared absorption? What are its primary advantages?

9. Define the following terms: chromophore, chromogen, auxochrome, bathochromic shift, hypsochromic shift, hyperchromism, and hypochromism and hyperconjugation.

10. Which of the following pairs of compounds is likely to absorb radiation at the longer wavelength and with greater intensity?

(a) $CH_3CH_2CO_2H$ or $CH_2 = CHCO_2H$

(b) $CH_3CH = CHCH = CHCH_3$ or $CH_3C \equiv C - C \equiv CCH_3$

(c)

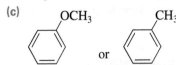
OCH$_3$ CH$_3$

or

11. In the following pairs of compounds, describe whether there should be an increase in the wavelength of maximum absorption and whether there should be an increase in absorption intensity in going from the first compound to the second:

(a)

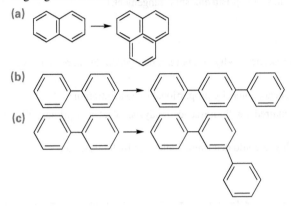

(b)

(c)

12. Why do acid–base base indicators change color in going from acid to alkaline solution?

13. What are the mechanisms by which a metal complex can absorb radiation?

Quantitative Relationships

14. Define absorption, absorbance, percent transmittance, and transmittance.

15. Define absorptivity and molar absorptivity.

16. Why is a calibration curve likely to be linear over a wider range of concentrations at the wavelength of maximum absorption compared to a wavelength on a shoulder of the absorption curve?

17. List some solvents that can be used in the ultraviolet, visible, and infrared regions, respectively. Give any wavelength restrictions.

18. What is an isosbestic point?

19. Describe and compare different causes for deviations from Beer's law. Distinguish between real and apparent deviations.

20. Why does a deuterium lamp produce a continuous rather than a line spectrum in the ultraviolet?

21. Define scattered radiation (in a monochromator).

22. What are the differences between a photon detector and a heat detector?

23. Define "effective bandwidth of a filter".

24. Why do quantitative and qualitative analyses often require different slit widths for monochromators?

25. What are the advantages of Fourier transform IR spectrometer in comparison to a dispersive instrument?

26. Why do IR spectra seldom show regions at which transmittance is 100%?

Instrumentation

27. Describe radiation sources and detectors for the ultraviolet, visible, and infrared regions of the spectrum.

28. Discuss the effect of the slit width on the resolution of a spectrophotometer and the adherence to Beer's law. Compare it with the spectral slit width.

29. Compare the operations of a single-beam spectrophotometer and a double-beam spectrophotometer.

30. Given the weak absorption in the near-infrared region, why do near-infrared instruments function with reasonable sensitivity?

31. Describe the operation of an array spectrometer.

32. Describe the operation of an interferometer. What are its advantages?

33. Referring to Figure 11.30, what would be the color of an acid solution and an alkaline solution at maximum absorption? What color filter would be most applicable for the analysis of each in a filter colorimeter? (A filter replaces the prism and slit arrangement.)

Fluorescence

34. Describe the principles of fluorescence. Why is fluorescence generally more sensitive than absorption measurements?

35. Under what conditions is fluorescence intensity proportional to concentration?

36. Describe the instrumentation required for fluorescence analysis. What is a primary filter? A secondary filter?

37. Suggest an experiment by which you could determine iodide ion by fluorescence.

Problems

Wavelength/Frequency/Energy

38. Express the wavelength 2500 Å in micrometers and nanometers.

39. Convert the wavelength 400 nm into frequency (Hz) and into wavenumbers (cm^{-1}).

40. The most widely used wavelength region for infrared analysis is about 2 to 15 μm. Express this range in angstroms and in wavenumbers.

41. One mole of photons (Avogadro's number of photons) is called an *Einstein* of radiation. Calculate the energy, in calories, of one Einstein of radiation at 300 nm.

Beer's Law

42. Most spectrophotometers can display either absorbance or in percent transmittance. What would be the absorbance reading at 20% T? At 80% T? What would the transmittance reading be at 0.25 absorbance? At 1.00 absorbance?

43. A 20-ppm solution of a DNA molecule (unknown molecular weight) isolated from *Escherichia coli* was found to give an absorbance of 0.80 in a 2-cm cell. Calculate the absorptivity of the molecule.

44. A compound of formula weight 280 absorbed 65.0% of the radiation at a certain wavelength in a 2-cm cell at a concentration of 15.0 μg/mL. Calculate its molar absorptivity at the wavelength.

45. Titanium is reacted with hydrogen peroxide in 1 M sulfuric acid to form a colored complex. If a 2.00×10^{-5} M solution absorbs 31.5% of the radiation at 415 nm, what would be (a) the absorbance and (b) the transmittance and percent absorption for a $6.00 \times 10^{-5} M$ solution?

46. A compound of formula weight 180 has an absorptivity of 290 cm^{-1} g^{-1} L. What is its molar absorptivity'?

47. Aniline, $C_6H_5NH_2$, when reacted with picric acid gives a derivative with an absorptivity of 144 cm^{-1} g^{-1} L at 379 nm. What would be the absorbance of a 1.50×10^{-4} M solution of reacted aniline in a 1.00-cm cell?

Quantitative Measurements

48. The drug tolbutamine (fw = 270) has a molar absorptivity of 703 at 262 nm. One tablet is dissolved in water and diluted to a volume of 2 L. If the solution exhibits an absorbance in the UV region at 262 nm equal to 0.687 in a 1-cm cell, how many grams tolbutamine are contained in the tablet?

49. Amines (weak base) form salts with picric acid (trinitrophenol), and all amine picrates exhibit an absorption maximum at 359 nm with a molar absorptivity of 1.25×10^4. A 0.200-g sample of aniline, $C_6H_5NH_2$, is dissolved in 500 mL of water. A 25.0-mL aliquot is reacted with picric acid in a 250-mL volumetric flask and diluted to volume. A 10.0-mL aliquot of this is diluted to 100 mL and the absorbance read at 359 nm in a 1-cm cell. If the absorbance is 0.425, what is the percent purity of the aniline?

50. Phosphorus in urine can be determined by treating with molybdenum(VI) and then reducing the phosphomolybdate with aminonaphtholsulfonic acid to give the characteristic molybdenum blue color. This absorbs at 690 nm. A patient excreted 1270 mL urine in 24 h, and the pH of the urine was 6.5. A 1.00-mL aliquot of the urine was treated with molybdate reagent and aminon-aphtholsulfonic acid and was diluted to a volume of 50.0 mL. A series of phosphate standards was similarly treated. The absorbance of the solutions at 690 nm, measured against a blank, were as follows:

Solution	Absorbance
1.00 ppm P	0.205
2.00 ppm P	0.410
3.00 ppm P	0.615
4.00 ppm P	0.820
Urine sample	0.625

(a) Calculate the number of grams of phosphorus excreted per day.

(b) Calculate the phosphate concentration in the urine as millimoles per liter.

(c) Calculate the ratio of HPO_4^{2-} to $H_2PO_4^-$ in the sample:

$$K_1 = 1.1 \times 10^{-2} \quad K_2 = 7.5 \times 10^{-8} \quad K_3 = 4.8 \times 10^{-13}$$

51. Iron(II) is determined spectrophotometrically by reacting with 1,10-phenanthroline to produce a complex that absorbs strongly at 510 nm. A stock standard iron(II) solution is prepared by dissolving 0.0702 g ferrous ammonium sulfate, $Fe(NH_4)_2SO_4 \cdot 6H_2O$, in water in a 1-L volu-metric flask, adding 2.5 mL H_2SO_4, and diluting to volume. A series of working standards is prepared by transferring 1.00-, 2.00-, 5.00-, and 10.00-mL aliquots of the stock solution to sep-arate 100-mL volumetric flasks and adding hydroxylammonium chloride solution to reduce any iron(III) to iron(II), followed by phenanthroline solution and then dilution to volume with water. A sample is added to a 100-mL volumetric flask and treated in the same way. A blank is prepared by adding the same amount of reagents to a 100-mL volumetric flask and diluting to volume. If the following absorbance readings measured against the blank are obtained at 510 nm, how many milligrams iron are in the sample?

Solution	A
Standard 1	0.081
Standard 2	0.171
Standard 3	0.432
Standard 4	0.857
Sample	0.463

52. Nitrate nitrogen in water is determined by reacting with phenoldisulfonic acid to give a yellow color with an absorption maximum at 410 nm. A 100-mL sample that has been stabilized by adding 0.8 mL H_2SO_4/L is treated with silver sulfate to precipitate chloride ion, which inter-feres. The precipitate is filtered and washed (washings added to filtered sample). The sample solution is adjusted to pH 7 with dilute NaOH and evaporated just to dryness. The residue is treated with 2.0 mL phenol disulfonic acid solution and heated in a hot-water bath to aid disso-lution. Twenty milliliters distilled water and 6 mL ammonia are added to develop the maximum color, and the clear solution is transferred to a 50-mL volumetric flask and diluted to volume with distilled water. A blank is prepared using the same volume of reagents, starting with the disulfonic acid step. A standard nitrate solution is prepared by dissolving 0.722 g anhydrous KNO_3 and diluting to 1 L. A standard addition calibration is performed by spiking a separate 100-mL portion of sample with 1.00 mL of the standard solution and carrying through the entire procedure. The following absorbance readings were obtained: blank, 0.032; sample, 0.270; sam-ple plus standard, 0.854. What is the concentration of nitrate nitrogen in the sample in parts per million?

53. Two colorless species, A and B, react to form a colored complex AB that absorbs at 550 nm with a molar absorptivity of 450. The dissociation constant for the complex is 6.00×10^{-4}. What would the absorbance of a solution, prepared by mixing equal volumes of 0.0100 M solutions of A and B in a 1.00-cm cell, be at 550 nm?

Fluorescence

54. Derive Equation 11.27 relating fluorescence intensity to concentration.

Photoacoustic Spectroscopy

55. Does vibrational or electronic excitation of electrons produce more heat? Why is that?

56. Why does PAS require pulses of light instead of a continuous steady source of light to hit the sample?

PROFESSOR'S FAVORITE PROBLEMS

Contributed by Professor Robert E. Synovec, University of Washington, Seattle
(This problem is repeated in Chapter 13 with an alternative approach for solution in the Solutions Manual.)

57. Lead (Pb) concentration in polluted water is determined using solid-phase extraction (SPE—see Chapter 13 for how it works) for preconcentration, combining preconcentration and the standard addition method (SAM), followed by absorbance measurement of a suitable Pb-complex. Assume only the Pb-complex absorbs light at the wavelength used. For the first step of SPE, a 1000-mL volume of either the *original* sample or *spiked* sample was preconcentrated (extracted). Then for the second step, a 5-mL volume of elution solvent was used to remove the preconcentrated Pb-complex from the SPE cartridge for each sample.

   For the *preconcentrated, original* polluted water sample, the absorbance was $A = 0.32$. Meanwhile, the *spiked* water sample contained an additional 5.0×10^{-8} mol Pb-complex per 1000 mL of *original* sample. The absorbance of the *preconcentrated, spiked* solution was 0.44. The pathlength was 1 cm. The molar absorptivity for the Pb-complex is 2.0×10^4 L mol^{-1} cm^{-1}.

   (a) What is the concentration of the Pb-complex in the *original* polluted water sample (prior to preconcentration)?

   (b) If we define P_{ideal} as the theoretical or ideal preconcentration factor that will be attained if all of the analyte were captured by the SPE cartridge and all of the captured material was eluted in the eluting solvent, what is P_{ideal} in the present case?

   (c) What is the experimental preconcentration factor, P_{expt}?

Contributed by Professor James Harynuk, University of Alberta, Edmonton, Canada
58. The core of a simple single-cavity bandpass-type interference filter consists of a thin layer of a dielectric material with thin reflective metal layers on each face. Only the wavelength λ that meets the constructive interference criterion

$$n\lambda = 2tn$$

is transmitted, where n is the order of interference, and t and n are, respectively, the thickness and the refractive index of the dielectric layer, and light is incident perpendicularly on the filter. For a core dielectric of 200-nm thickness and a refractive index of 1.377, what is the transmitted wavelength through this filter if the emission from a white LED (Figure 11.13) is focused on it?

Contributed By Professor Gary M. Hieftje, Indiana University, Bloomington
59. Consider the absorption cell of L geometry as pictured below:

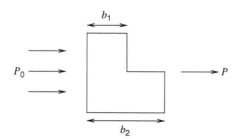

(a) If equal amount of radiant power is incident on the top half as the lower half, show that the following general relation is valid for the observed absorbance A as a function of b_1 and b_2, the molar absorptivity ε of the absorbing substance, and the molar concentration c of the absorbing substance:

$$A = \log 2 - \log \left(10^{-\varepsilon b_1 c} + 10^{-\varepsilon b_2 c}\right)$$

(b) What are the limiting slopes in an A versus c plot, at low and high c values, respectively?

Contributed by Professor Paul S. Simone, Jr., University of Memphis

60. An analyst would like to measure the concentrations of benzene in ethanol with a spectrophotometer using 1-cm pathlength cells. Benzene has two characteristic absorption wavelengths at 204 nm and 256 nm, and ethanol has a minimum usable (cutoff) wavelength at 220 nm.

   (a) Which absorption wavelength of benzene should be used for the analysis?

   (b) Which kinds of light source would be ideal for this measurement?

   (c) What cell material would be used in the spectrophotometer for this analysis?

Multiple Choice Questions

1. A 0.1 M solution of compound A shows 50% transmittance when a cell of 1 cm width is used at λ_1 nm. Another 0.1 M solution of compound B gives the optical density value of 0.1761 using 1 cm cell at λ_1 nm.
   What will be the transmittance of a solution that is simultaneously 0.1 M in A and 0.1 M in B using the same cell and at the same wavelength?
   ($\log 20 = 1.301$; $\log 30 = 1.4771$; $\log 50 = 1.699$)

   (a) 33.3% (b) 50%
   (c) 66.7% (d) 70%

2. Which of the following will result in deviation from Beer's law:
   P. Change in refractive index of medium.
   Q. Dissociation of analyte on dilution.
   R. Polychromatic light
   S. Path length of cuvette

   (a) P, Q and R (b) Q, R and S
   (c) P, R and S (d) P, Q and S

3. The spectroscopic technique that can distinguish unambiguously between *trans*-1,2-dichloroethylene and *cis*-1,2-dichoroethylene without any numerical calculation is

   (a) Microwave spectroscopy. (b) UV-visible spectroscopy.
   (c) X-ray photoelectron spectroscopy. (d) γ-ray spectroscopy.

4. Spectrophotometric monitoring is not suitable to determine the end point of titration of

   (a) oxalic acid *vs* potassium permanganate.

   (b) iron (II) *vs* 1,10-phenanthroline.

   (c) cobalt (II) *vs* eriochrome black T.

   (d) nickel (II) *vs* dimethylglyoxime.

5. The intensity of a light beam decreases by 50% when it passes through a sample of 1.0 cm path length. The percentage of transmission of the light passing through the same sample, but of 3.0 cm path length, would be

   (a) 50.0 (b) 25.0
   (c) 16.67 (d) 12.5

6. Considering the following parameters with reference to the fluorescence of a solution:
   P. Molar absorptivity of fluorescent molecule.
   Q. Intensity of light source used for excitation.
   R. Dissolved oxygen

The correct answer for the enhancement of fluorescence with the increase in these parameters is/are

(a) P and Q   (b) R and S

(c) P and R   (d) S only

7. Identify the correct statement(s) for phosphorimetric measurement from the following:
   P. It is done after a time delay when florescence, if present becomes negligible.
   Q. Immobilization of analyte increases phosphorescence.
   R. Phosphorescence decreases in the presence of heavy atoms.
   Answer(s) is/are:

(a) P only   (b) P and Q

(c) P and R   (d) Q and R

8. The first electronic absorption band maximum of a polar and relatively rigid aromatic molecule appears at 310 nm but its fluorescence maximum in acetonitrile solution appears with a large stroke shift at 450 nm. The most likely reason for the strokes shift is

(a) large change in molecular geometry in the excited state.

(b) increase in dipole moment of the molecule in the excited state.

(c) decrease in polarizability of the molecule in the excited state.

(d) lowered interaction of the excited state with polar solvent.

9. Using a double beam UV-visible spectrophotometer, Beer's law fails for $K_2Cr_2O_7$ solution when

(a) intensity of light source is changed.

(b) detector is not a photomultiplier tube.

(c) cuvette of 2 cm size is used.

(d) pH is not kept same in all measurements.

10. The molar extinction coefficient of B (molecular weight = 180) is 4×10^3 L /mol cm. One-liter solution of C which contains 0.1358 g pharmaceutical preparation of B, shows an absorbance of 0.411 in a 1 cm quartz cell. The percentage (*w/w*) of B in the pharmaceutical preparation is

(a) 10.20   (b) 14.60

(c) 20.40   (d) 29.12

11. Using cuvettes of 0.5 cm path length, a $10^{-4}M$ solution of a chromatophore shows 50 % transmittance at certain wavelength. The molar extinction coefficient of the chromophore at this wavelength is (log 2 = 0.301)

(a) 1500 M^{-1} cm^{-1} (b) 3010 M^{-1} cm^{-1}

(c) 5000 M^{-1} cm^{-1} (d) 6020 M^{-1} cm^{-1}

12. An aqueous solution of hemoglobin has a molar absorptivity value of 18,600 L / mol cm for an absorbance value of 0.1 at 540 nm (Given: cell thickness = 1 cm). The concentration (in *M*) of hemoglobin solution is

(a) 0.537   (b) 5.37

(c) 53.7   (d) 537.0

13. Which among the following is not a reason for laser not being generally used as a source of radiation for UV-visible spectroscopy?

(a) High cost   (b) Limited range of wavelength

(c) Less intensity   (d) Complex to work with

14. Which among the following is a source used in spectroscopy?

(a) LASER   (b) Tube light

(c) Sodium vapor lamp   (d) Tungsten lamp

15. The wave number of near infrared spectrometer is _____.

(a) 4000 – 200 cm^{-1} (b) 200 – 10 cm^{-1}

(c) 12500 – 4000 cm^{-1} (d) 50 – 1000 cm^{-1}

16. In case of Fourier Transform Infrared (FTIR) spectrometer which statement is not true?

(a) It is of non-dispersive type.

(b) It is useful where repetitive analysis is required.

(c) Size has been reduced over the years.

(d) Size has increased over the years.

17. The phenomenon when the absorption of electromagnetic radiation by matter leads to emission of radiation of the longer or same wavelengths for a short or a long time, is termed as:

(a) luminescence.

(b) fluorescence.

(c) phosphorescence.

(d) spontaneous emission.

18. The following needs to be computed to find transmittance and absorbance at various frequencies:

(a) Ratio of signal and noise

(b) Ratio of sample and reference spectra

(c) Sample spectra

(d) Reference spectra

19. The reference that is generally used in FTIR interferometer is:

(a) Air

(b) NaCl solution

(c) Alcohol

(d) Base solution

20. The unit of absorbance which can be derived from Beer-Lambert's law is _____.

(a) $L\ mol^{-1}\ cm^{-1}$

(b) $L\ g^{-1}\ cm^{-1}$

(c) cm

(d) No unit

21. In fluorescence spectroscopy the purpose of secondary filter is:

(a) allows only excitation radiation.

(b) allows only emission radiation.

(c) allows both excitation and emission radiations.

(d) allows transmitted radiation.

22. The absorption spectrum of a certain organic compound has its λ_{max} at 240 nm. This is in which region of the electromagnetic spectrum?

(a) Ultraviolet

(b) Visible

(c) Infrared

(d) Radio wave

23. When monochromatic light is passed through a $1\ M$ solution of a compound for distance of 1 cm, the observed absorbance is known as

(a) molar emission.

(b) molar absorption.

(c) radiance.

(d) molar extinction coefficient.

24. Which of the following transitions require the least energy?

(a) $n \rightarrow \pi^*$

(b) $\pi \rightarrow \pi^*$

(c) $\sigma- \rightarrow \sigma^*$

(d) $n- \rightarrow \sigma^*$

25. Spectroscopic methods are advantages over the conventional methods because

(a) these are very quick and extremely sensitive.

(b) information is obtained in the form of a permanent record and is highly reliable.

(c) very small quantities of the substance are required for the determination of spectra.

(d) all of the above.

26. The value of λ_{max} of $R-NO_2$ is equal to

(a) 271 nm

(b) 250 nm

(c) 230 nm

(d) 200 nm

27. Coronene shows absorption at

(a) 305 mμ

(b) 342 mμ

(c) 428 mμ

(d) None of these

28. The wavelength associated with an ultraviolet radiation is 285 nm. Determine the energy associated with it in k cal mole^{-1}.

    (a) 100.49 kcal mole^{-1} (b) 410.00 kcal mole^{-1}

    (c) 210.00 kcal mole^{-1} (d) 375.58 kcal mole^{-1}

29. Which one of the following statements regarding the UV absorption of the carbonyl chromophore present acetophenone if correct?

    (a) Absorption due to $n \to \pi^*$ is of much high intensity than that due to $\pi \to \pi^*$ transitions.

    (b) The $\pi \to \pi^*$ and $n \to \pi^*$ transition occur with equal intensity of absorption.

    (c) The $\pi \to \pi^*$ transition occurs at a longer wavelength than the $n \to \pi^*$ transition does.

    (d) The $\pi \to \pi^*$ transition occurs at a shorter wavelength than the $n \to \pi^*$ transition does.

30. Which of the following bands appear in butadiene due to $\pi \to \pi^*$ transition in UV spectroscopy?

    (a) K* – Band   (b) R* –Band

    (c) B– Band   (d) E –Band

31. The dark purple colour of $KMnO_4$ is due to

    (a) d–d transitions   (b) ligand filled transition

    (c) charge transfer transition   (d) $\sigma \to \pi^*$ transition

32. In the UV spectrum of cyclohexanone, the absorption at $\lambda_{max} \approx 215$ nm is due to the transition

    (a) $\sigma \to \pi^*$ (b) $\sigma \to \eta$

    (c) $\pi \to n$ (d) $\pi \to \pi^*$

33. A water solution of a coloured compound has a motor absorptivity (ε) of 3200 at 525 nm. The absorbance and % transmittance of a 3.40×10^{-4} solution respectively

    (a) 1.09, 0.91 %   (b) 1.09, 8.1%

    (c) 1.09, 1.8 %   (d) 1.09, 2 %

34. Which one of the statements is not true?

    (a) The spectrum obtained from molecular species appears to be broad and termed as band spectra.

    (b) Electronic transitions of atoms and molecules are observed in ultraviolet and visible region.

    (c) Molecular vibrations are observed in infrared regions.

    (d) Infrared rays are below the lower limits of visible region and produce low thermal effects.

35. Which group of compounds does not involve the $\pi \to \Pi^*$ transitions in UV spectroscopy?

    (a) Alkene   (b) Azo compound

    (c) Alcohols   (d) Cyanides

36. Infrared spectra C − H stretching peaks of acetylenic, ethylenic and paraffinic species occur respectively at

    (a) 2800 cm^{-1}, 3080 cm^{-1}, 3300 cm^{-1}

    (b) 3300 cm^{-1}, 3080 cm^{-1}, 2800 cm^{-1}

    (c) 3080 cm^{-1}, 2800 cm^{-1}, 3300 cm^{-1}

    (d) 3080 cm^{-1}, 3300 cm^{-1}, 2800 cm^{-1}

Recommended References

General

1. D. F. Swinehart, "The Beer-Lambert Law," *J. Chem. Ed.*, **39** (1962) 333.
2. D. W. Ball, "The Electromagnetic Spectrum: A History," *Spectroscopy*, **21**(3), March 2007, 14. www.spectroscopyonline.com.
3. D. W. Ball, "Light: Particle or Wave?" *Spectroscopy*, **21**(6), June 2008, 30. www.spectroscopyonline.com.
4. D. W. Ball, "Prisms," *Spectroscopy*, **23**(9), September 2008, 30. www.spectroscopyonline.com.

5. D. W. Ball, "Lenses," *Spectroscopy*, **23**(12), December 2008, 74.
6. R. M. Silverstein and F. X. Webster, *Spectrometric Identification of Organic Compounds, 6th ed.* New York: Wiley, 1997.
7. G. D. Christian and J. B. Callis, eds., *Trace Analysis: Spectroscopic Methods for Molecules.* New York: Wiley, 1986.
8. G. H. Morrison and H. Freiser, *Solvent Extraction in Analytical Chemistry.* New York: *Wiley,* 1975, pp. 189–247. Colorimetric determination of metals.

Cataloged Spectra

9. *Catalogue of Infrared Spectral Data.* Washington, DC: American Petroleum Institute, Research Project 44. Multivolume series started in 1943.
10. *Catalogue of Ultraviolet Spectral Data.* Washington, DC: American Petroleum Institute, Research Project 44. Multivolume series started in 1945.
11. "Infrared Prism Spectra," in *The Sadtler Standard Spectra,* Vols. 1–36; "Standard Infrared Grating Spectra," in *The Sadtler Standard Spectra*, Vols. 1–16. Philadelphia: Sadtler Research Laboratories.
12. L. Lang, ed., *Absorption Spectra in the Ultraviolet and Visible Regions*, Vols. 1–23. New York: Academic, 1961–1979.
13. *U.V. Atlas of Organic Compounds*, Vols. I–V. London: Butterworths, 1966–1971.
14. "Ultraviolet Spectra," in *The Sadtler Standard Spectra*, Vol. 1–62. Philadelphia: Sadtler Research Laboratories. A comprehensive catalog of ultraviolet spectra of organic compounds.
15. D. L. Hansen, *The Spouse Collection of Spectra. I. Polymers, II. Solvents by Cylindrical Internal Reflectance, III. Surface Active Agents, IV. Common Solvents—Condensed Phase, Vapor Phase and Mass Spectra.* Amsterdam: Elsevier Science, 1987–1988. Peak table search software available for each.

Infrared

16. P. R. Griffiths, *Fourier Transform Infrared Spectrometry,* 2nd ed. New York: Wiley, 1986.
17. J. Workman and L. Weyer, *Practical Guide to Interpretive Near-Infrared Spectroscopy.* Boca Raton, FL: CRC Press, 2008
18. D. A. Burns and E. W. Ciurczak, eds. *Handbook of Near-Infrared Analysis,* 3rd ed. Atlanta: CRC Press, 2007.
19. R. Raghavachari, ed., *Near-Infrared Applications in Biotechnology.* New York: Marcel Dekker, 2000.

Fluorometry

20. A. Sharma and S. G. Schulman, *An Introduction to Fluorescence Spectroscopy.* New York: Wiley, 1999.
21. C. E. White and R. J. Argauer, *Fluorescence Analysis: A Practical Approach.* New York: Marcel Dekker, 1970.
22. K. Soroka, R. S. Vithanage, D. A. Phillips, B. Walker, and P. K. Dasgupta, "Fluorescence Properties of Metal Complexes of 8-Hydroxyquinoline-5-Sulfonic Acid and Chromatographic Application." *Anal. Chem.*, **59** (1987) 629.
23. P. K. Dasgupta, Z. Genfa, J. Z. Li, C. B. Boring, S. Jambunathan, and R. S. Al-Horr, "Luminescence Detection with a Liquid Core Waveguide." *Anal. Chem.* **71** (1999) 1400.
24. E. R. Menzel, "Recent Advances in Photoluminescence Detection of Fingerprints." *The Scientific World* **1** (2001) 498.
25. Q. Y. Li, P. K. Dasgupta, and H. Temkin, "Airborne Bacterial Spore Counts. Terbium Enhanced Luminescence Detection: Pitfalls and Real Values." *Environ. Sci. Technol.* **42** (2008) 2799.
26. R. J. Hurtubise, *Phosphorimetry. Theory, Instrumentation and Applications.* New York: VCH, 1990.

Chemiluminescence

27. N. W. Barnett and P. S. Francis. *Encyclopedia of Analytical Sciences.* 2nd ed. Elsevier, 2005. Luminescence Overview. p. 305; Chemiluminescence Overview p. 506, Chemiluminescence-Liquid Phase, p. 511.

28. N. W. Barnett, P. S. Francis, and J. S. Lancaster. *Encyclopedia of Analytical Sciences.* 2nd ed. Elsevier, 2005. Luminescence-Gas Phase, p. 521.

29. G. Zhang, P. K. Dasgupta, and A. Sigg, "Determination of Gaseous Hydrogen Peroxide at Parts per Trillion Levels with a Nafion Membrane Diffusion Scrubber and a Single-Line Flow-Injection System." *Anal. Chim. Acta* **260** (1992) 57.

30. O. S. Wolfbeis, ed., *Fiber Optic Chemical Sensors and Biosensors*, Vols. 1 and 2. Boca Raton, FL: CRC Press, 1991.

31. L. W. Burgess, M.-R. S. Fuh, and G. D. Christian, "Use of Analytical Fluorescence with Fiber Optics," in P. Eastwood and L. J. Cline-Love, eds. *Progress in Analytical Luminescence, ASTM STP 1009.* Philadelphia: American Society for Testing and Materials, 1988.

32. J. Janata, "Do Optical Fibers Really Measure pH?" *Anal. Chem.,* **59** (1987) 1351.

33. J. Janata, "Ion Optrodes," *Anal. Chem.,* **64** (1992) 921A.

34. Z. Ge, C. W. Brown, L. Sun, and S. C. Yang, "Fiber-optic pH Sensor Based on Evanescent Wave Absorption Spectroscopy." *Anal. Chem.* **65** (1993) 2335.

35. J. N. Demas, B. A. DeGraff, and P. B. Coleman, "Peer Reviewed: Oxygen Sensors Based on Luminescence Quenching." *Anal. Chem.* **71** (1999) 793A.

Photoacoustic Spectroscopy

36. A. Rosencwaig, "Photoacoustics and Photoacoustic Spectroscopy", Wiley, New York (1980).

37. R. H. Williams, (ed.) " Spectroscopy-New Uses and Implications", Academic Press (2011).

Chapter 12

ATOMIC SPECTROMETRIC METHODS

KEY THINGS TO LEARN FROM THIS CHAPTER

- Distribution of atoms as a function of temperature (key Equation: 12.1)
- Flame emission spectrometry
- Atomic absorption spectrometry (AAS)
- Flame AAS
- Electrothermal AAS
- Cold vapor and hydride generation
- Interferences in AAS

- Sample preparation
- Internal standard and standard addition calibration
- Atomic emission spectrometry the induction-coupled plasma (ICP)
- Laser ablation ICP–optical emission spectrometry/mass spectrometry
- Atomic fluorescence spectrometry

Chapter 11 dealt with the spectrometric determination of substances in solution, that is, the absorption of energy by molecules, either organic or inorganic. This chapter deals with the spectroscopy of atoms. **Note:** Spectrometry and spectroscopy are often used interchangeably. Strictly speaking, spectroscopy is the study of the interaction between matter and radiated energy, while spectrometry connotes quantitative measurement of light intensity as a function of wavelength. Thus, a *spectroscope* disperses the light to see the components of visible light, while a *spectrometer* permits the quantitative measurement of that dispersed light. The term *spectrometry* is generally used here; keep in mind that both terms are used in the literature. The usage of these terms is not intuitive: Experts in the field are almost always referred to as *spectroscopists* and not *spectrometrists*; this hardly indicates their studies are only qualitative!

Since atoms are the simplest and purest form of matter and do not have different rotational and vibrational energy states like a molecule, both absorption and emission spectra of atoms consist of sharp lines, corresponding to various wavelengths of light. Imaging of the solar spectrum with high resolution readily reveals the presence of narrow dark lines in the solar continuum. William Hyde Wollaston (otherwise best known for his discovery of palladium and rhodium in 1803) first observed these dark lines in 1802. But it would remain for Fraunhofer to independently discover, characterize, and catalog them in detail in 1813–1814. To date, these absorption lines are called Fraunhofer lines and most are still referred with Fraunhofer's original nomenclature. One often refers to the ability of a monochromator to resolve the sodium D_1 line at 589.592 nm from the sodium D_2 line at 588.995 nm as a benchmark, while lesser monochromators see these collectively as the "sodium D-line" at an average 589.3 nm. The refractive indices of most materials vary with the wavelength of light (as noted in Chapter 11, this is why prisms can disperse light), but we often refer to the refractive index of a material with a single value. Invariably, this is the value measured at 589.3 nm, the sodium D-line. Various atoms are present in the photosphere of the sun; Fraunhofer's lines in the continuum spectra are caused by the sharp, well-defined absorption by the atoms present in the photosphere (the absorption lines from helium were discovered before the element was: Helium is indeed named after the sun (Helios). Although Brewster in 1820 expressed the view that Fraunhofer's lines are caused by absorption in the solar atmosphere, the fact that they are atomic absorption lines was recognized nearly 50 years after their original discovery, when Kirchhoff and Bunsen noted that the emission spectra of some elements (produced on heating) contain the same lines and they went on to correctly deduce the origin of Fraunhofer lines.

As in molecular spectroscopy, atomic spectroscopy is divided broadly into absorption and emission spectroscopy. The notable differences are that atomic spectrometry is always carried out in the gas phase. The measurement conditions require elevated temperatures; with the exception of Hg, Cd and the inert gases, elements are not present as a monoatomic gas at room temperature. Also, as the name implies, we measure atoms; atomic spectroscopy is then a form of elemental analysis.

In **atomic absorption spectrometry (AAS)**, the half width (FWHM) of the absorption line is < 0.01 nm; this brings constraints not encountered in molecular spectroscopy. You will recall that the typical FWHM of a monochromator commonly used in molecular spectroscopy is 2 nm. Even in the fortunate instance that only one absorption line (and that being due to the desired analyte) is located within this chosen wavelength window, *complete* absorption of the radiation at the line (very high analyte concentration) will result in a difference of < 0.5% in the transmitted light. This is akin to carrying out molecular spectroscopy with > 99.5% stray light, and quantitation would be comparable to trying to judge if the color of a thread strung across a sunlit window is white, gray, or black, by measuring differences in the total light transmitted through the window. We describe below how to get around this problem by using a sharp line source (or else, a very high resolution spectrometer).

The second issue is that to convert the desired analyte into the atomic state, we typically need to introduce the sample into a flame or a plasma, or put it into a mini-furnace. The sample may not be completely converted into the atomic state. However, especially in flames, a low-energy source, substances present in the sample (or components of the flame itself) may absorb at the wavelength of interest. This **molecular absorption background** causes further complications in AAS. AAS is typically carried out using a hot flame for atom formation **(flame AAS)**, or an electrically heated mini-furnace, most commonly consisting of graphite **[electrothermal AA, graphite furnace AA (GFAA)]**, and rather uniquely for Hg (and occasionally Cd), it is possible to do this without heating **[cold vapor AA (CVAA)]**.

Similar to molecular luminescence, atomic luminescence can be produced by excitation of atoms by photons; this represents the technique of **atomic fluorescence spectrometry (AFS)**. Also, unlike molecules that thermally decompose long before they can be thermally excited sufficiently to emit light, atoms can be heated to high enough temperatures to emit their characteristic radiation. While atoms undergo true thermoluminescence, this term is not used to describe the relevant measurement techniques, rather they are always described with reference to the specific means of exciting the atoms. Examples are the techniques of **flame emission spectrometry** (in practical use today, the desired wavelengths are selected using interference filters—hence, the technique is more accurately called **flame photometry**), arc/spark source emission spectrometry, direct current plasma (DCP) emission spectrometry, and **induction coupled plasma** (also called **inductively coupled plasma**) **atomic** (or **optical**) **emission spectrometry (ICP-AES or ICP-OES)**. The ICP is sufficiently energetic to strip off one (or more) outer shell electron(s) to generate positive ions from the analyte atoms; they can be analyzed by a mass spectrometer after being sorted by their m/z values (see Chapter 17 for more on mass spectrometry); this technique is hence called **induction coupled plasma mass spectrometry (ICP-MS)**. The ICP as an atom/ion source is discussed in this chapter; ICP-MS is discussed in Chapter 17 along with other mass spectrometric techniques.

Atomic spectrometry is widely used in many laboratories for trace element analyses. Environmental samples are analyzed for heavy-metal contamination, and

Atomic absorption/emission lines are very narrow (half-width < 0.01 nm); concurrently present broad absorption from molecular species make it impossible to simply use equipment designated for molecular spectrometry to perform atomic absorption spectrometry.

pharmaceutical samples may be analyzed for metal impurities. The semiconductor industry requires precise doping of some elements into others and composition must be accurately known. The steel industry needs to determine minor components, as well as major ones. The composition of a batch of steel frequently must be confirmed (and amended as needed) before the molten mass is poured out. The particular technique used in any given application will depend on the sensitivity required, the number of samples to be analyzed, and whether multielement measurements are needed. Individual techniques are discussed in more detail below.

12.1   Principles: Distribution between Ground and Excited States—Most Atoms Are in the Ground State

The relative populations of ground-state (N_0) and excited-state (N_e) populations at a given flame temperature can be estimated from the **Maxwell–Boltzmann expression**:

$$\frac{N_e}{N_0} = \frac{g_e}{g_0} e^{-(E_e - E_0)/kT} \tag{12.1}$$

Nearly all the gaseous atoms are in the ground state. Atomic emission is still sensitive, for the same reason that fluorescence spectrometry is. We do not have to measure a small decrease in a signal (which has some noise) as in absorption.

where g_e and g_0 are the *degeneracies* of the excited and ground states, respectively; E_e and E_0 are the energies of the two states ($E_e = h\nu$, where ν is the frequency of the photon that is emitted when transition occurs between the excited and the ground states; E_0 is usually zero); k is the Boltzmann constant (1.3805×10^{-16} erg K^{-1}); and T is the absolute temperature. *Degeneracy* is the term used to represent the number of equivalent energy states an electron can reside in a given energy level, and they are available from quantum mechanical calculations for particular energy states. Many physical chemistry texts (see, e.g., Chapter 17 in T. Engel, P. Reid, *Physical Chemistry*, 3rd ed., 2012) discuss this in more detail. In essence, for any given energy state, if the total angular momentum operator resulting from spin-orbit coupling is J, then the degeneracy g is ($2J + 1$). See Example 12.1 for an illustrative calculation. If the wavelength of the photon that is absorbed or emitted in the transition between the two energy states is λ, Equation 12.1 is written in more practical terms as:

$$\frac{N_e}{N_0} = \frac{g_e}{g_0} e^{-(hc)/\lambda kT} \tag{12.2}$$

where h is Planck's constant and c is the velocity of light (see Equation 11.3). Figure 12.1 illustrates graphically the implications of this equation.

Table 12.1 summarizes the relative population ratios for a few elements at 2000, 3000, and 10,000 K.

TABLE 12.1   Values of N_e/N_0 (excited-state to ground-state population ratio) for Different Transition Lines

Line (nm)	N_e/N_0		
	2000 K	3000 K	10,000 K
Na 589.0	9.9×10^{-6}	5.9×10^{-4}	2.6×10^{-1}
Ca 422.7	1.2×10^{-7}	3.7×10^{-5}	1.0×10^{-1}
Zn 213.8	7.3×10^{-15}	5.4×10^{-10}	3.6×10^{-3}

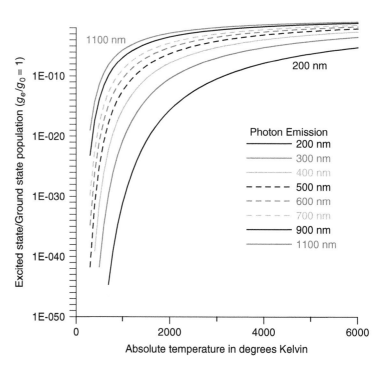

FIGURE 12.1 Excited-state to ground-state population ratio for photon energies of 200–1100 nm at temperatures from 300–6000 K. Commonly, the excited-state degeneracy is higher than the ground state and the absolute values of $\frac{N_e}{N_0}$ are likely to be higher. Note that a longer wavelength photon means less energy gap between the two states and the excited state is more easily populated.

We see that even for a relatively easily excited element such as sodium, the excited-state population is small except at 10,000 K, as obtained in an ICP. Emission spectrometry is not very useful in lower temperature excitation sources like flames (typically < 3000 K) for elements that have low wavelength transition lines; e.g., arsenic with its most prominent transition line at 193.7 nm. Those with long-wavelength emissions will exhibit acceptable sensitivity in a flame. Arc and spark sources were once used in commercially available emission spectrometers, but have long given way to the ICP. All elements are predominantly in the ground state at temperatures < 3000 K and the ground-state population varies very little with temperature. Measurement of ground-state atoms, as in AAS, is less dependent on the transition line wavelength. On the other hand, remember that to carry out atomic spectrometry, we must generate the element in the atomic form in the first place. Forming such atoms from molecules requires breaking of bonds; this is typically accomplished thermally; only Hg and Cd can be generated in the atomic form by room temperature as "cold vapors." For all others, atomic absorption is nearly exclusively carried out with a thermal excitation source such as a flame or a furnace.

Example 12.1

The NIST reference data tables http://physics.nist.gov/PhysRefData/Handbook/element_name.htm lists the energy states and their relative energy level (given in terms of the equivalent photon energy in wavenumber units) and *J* values, for each element. For the neutral iodine atom, what are the degeneracies of the ground state and the lowest excited state? What would be the wavelength of the transition line between these two states and their population ratio at 3000 K? 10,000 K? This line is difficult to use in practice, why? The transition between the lowest excited state and the energy state at 65669.99 cm^{-1} produces an analytically useful transition line. What is the corresponding wavelength and what is the relative population of this state relative to the lowest excited state at 10,000 K?

Solution

The website lists the element names. We click on iodine and then the "Energy Levels" button under "Neutral Atom." The ground state (the first entry listed with an energy level of 0.00) has a listed J value of $\frac{3}{2}$, and therefore, $g_0 = 2 * \frac{3}{2} + 1 = 4$. The lowest excited state is the one at 54633.46 cm^{-1} and it has a J value of $\frac{5}{2}$ and hence, $g_e = 6$. The corresponding photon wavelength is

$$\lambda = (10^7 \text{ nm/cm})/(54633.46 \text{ cm}^{-1}) = 183 \text{ nm}$$

This is in the vacuum UV and will be difficult to use. Using Equation 12.2 and values for h, c, and k given in the back cover, at 3000 K,

$$\frac{N_e}{N_0} = \frac{6}{4} \exp\left(\frac{-4.1357 \times 10^{-15} \text{ eV} \times \text{s} \times 2.99792 \times 10^8 \text{ m/s}}{183 \times 10^{-9} \text{ m} \times 8.6173 \times 10^{-5} \text{ eV/K} \times 3000 \text{ K}}\right) = 6.23 \times 10^{-12}$$

(Compare this result with the 200-nm trace at 3000 K in Figure 12.1.)

And at 10,000 K, we similarly calculate the population ratio to be 5.78×10^{-5}.

When energy-state values are expressed in cm^{-1}, the difference in energy between the two states can be obtained by simple subtraction. The transition energy between the two states at 65669.99 and 54633.46 cm^{-1} is therefore 11036.53 cm^{-1}. The corresponding photon wavelength is:

$$\lambda = (10^7 \text{ nm/cm})/(11036.53 \text{ cm}^{-1}) = 906 \text{ nm}$$

This NIR wavelength is easily measured. In the NIST table, we also see that the J value for this higher excited state is $\frac{7}{2}$, hence degeneracy of this state, $g'_e = 8$. The population ratio with respect to the lowest excited state can now be calculated from Equation 12.2, using g'_e/g_e instead of g_e/g_0 and $\lambda = 906$ nm, to be 0.272 at 10,000 K. Although the 183-nm line is stronger, a substantial amount of emission will also occur at this NIR wavelength. The NIST tables also contain data for the relative intensities of such emission lines (click the persistent lines button—in this table the wavelengths are given in Angstrom units, 1 nm = 10 Å), the conditions are unspecified but the 906-nm line is listed as having ~20% of the intensity of the 183-nm line.

All of an element present in the measurement path may not, however, be in an atomic form. Especially in flames, some elements, even though originally atomized, form stable oxides and hydroxides and thus are not present as atoms. This problem is not present in high temperature plasmas, but ionization of some easily ionizable elements becomes increasingly important in high temperature sources.

In atomic emission spectrometry, we measure the excited-state population; and in atomic absorption and atomic fluorescence methods, we measure the ground-state population. Except for elements that are easily excited (long wavelength photon emission), emission spectrometry benefits from a very high-temperature source such as a plasma. Sensitivities and detection limits cannot generally be predicted simply from the ground-state to excited-state population ratios. In AES we measure a narrow line signal against a low background making it intrinsically sensitive. In contrast, in absorption measurements the attainable limit of detection (LOD) depends on the ability to detect a small difference in a large signal. In cases where AFS is applicable, it can provide the best of both worlds, in measuring an emission signal against a low background and coming from a large atom population. Unlike molecular spectroscopy, however, in AFS the excitation and emission wavelengths are identical, making it vital to eliminate or minimize any scattered radiation.

12.2 Flame Emission Spectrometry[1]

In this technique, the source of excitation energy is a flame. The sample is introduced into the flame in the form of a solution. A flame is a low-energy source, and so the emission spectrum is simple and there are few emission lines. In practice, the technique is inexpensive and attractive for only a few elements. As a direct result of the work of Kirchoff and Bunsen in the early 1860s, the analytical utility of measuring the characteristic radiation emitted by specific elements excited in flames was realized. The earliest such instrument was used for the measurement of sodium in plant ash using a Bunsen flame. The difficult part in such an instrument was how best to introduce the sample into the flame. It wasn't until 1929, when Lundegardh utilized a nebulizer to introduce a significant fraction of the sample reproducibly into the flame, that a breakthrough was made. Characteristic atomic emission lines were dispersed by a quartz prism spectrograph and recorded photographically. Use of optical filters and electrical photodetectors improved convenience and precision in later years, and such instruments became widely useful for measuring Na, K, Li, and Ca. Work on many other elements then became possible with the use of grating spectrometers equipped with more sensitive photomultiplier detectors. But the commercial development of the more broadly applicable AAS technique in the 1960s essentially restricted the scope of flame emission spectrometry and arrested its further development.

Most commercial instruments in use today are dedicated to the measurement of Li, Na, K, Ca, and Ba, using interference filters centered at 670, 589, 766, 622, and 515 nm, respectively, and thus they are commonly called **flame photometers**. Calcium is actually measured by the emission of molecular CaOH, as this emission at 622 nm is more intense than the emission by atomic Ca at 423 nm. A propane-air flame (temperature 1900–2000°C) is commonly used. Butane-air or natural gas-air flames can also be used; these flames are not as hot and the measurement is less sensitive. With a propane-air flame, the best of current instruments claim LODs of 0.02 mg/L for Na and K to 10 mg/L for Ba. Most analyze one element at a time; the filter is manually changed to switch the target element, but some, such as the BWBTech XP Flame Photometer, are capable of providing simultaneous multiple readouts using multiple detection channels. The formation of CaOH is much affected by the presence of Ba, however, and the two generally cannot be simultaneously determined. The atomic emission line of Ca at 423 nm is less susceptible to interference by Ba but is not commonly used because of lower sensitivity. Many such shortcomings of flame emission measurement can be overcome with higher temperature and more reducing flames such as air-acetylene and higher-resolution spectrometric detection. In many ways, flame emission as practiced today in the form of flame photometry (FP) is a step backward from the heights previously attained because incorporation of high temperature flames and high-resolution spectrometry is not cost competitive relative to the more broadly applicable flame AAS technique.

Burners Used for Flame Spectrometry and Desirable Flame Characteristics

In early years, there was considerable debate on the superiority of the two main types of burners. In one type, the fuel (propane/acetylene) and the oxidant (air/oxygen/nitrous oxide) are premixed and the solution is nebulized by the flow of the premixed gas before the flame. This is the type of arrangement pioneered by Lundegardh. Much of the liquid that is aspirated by the nebulizer actually tends to form large droplets and simply drains

---

[1]See Chapter 11 for the distinction between spectrometry and spectrophotometry.

from the burner. In the second type of burner, often called a **total consumption burner**, the fuel and oxidant gases are not premixed. The design is basically that consisting of three concentric tubes that terminate in a nozzle. The outermost typically carried the fuel, the next one the oxidant, and the central tube, a capillary, was the sample inlet. The Venturi suction created by the fuel-oxidant flow aspirates the sample into the flame; the name derives from the fact that the entire aspirated sample enters the flame.

Notwithstanding the superiority the "total consumption" name may imply, the **premix burner**, sometimes called the **laminar flow burner**, ultimately produces better results. It is the only type of burner in present use. The primary reason for the difference is the sample droplet size generated. Whereas the direct sample injection process in the total consumption burner results in larger droplets of size ~7 nL, the premix burner nebulizer produces droplets of size as small as 0.05 nL. The smaller droplets evaporate and eventually atomize much more easily. A typical design is shown in Figure 12.2. A three-slotted burner head, originally due to Boling, as shown in the figure, is the most common. It produces a wide flame, sufficient to accommodate the widest probe beam cross section. Atmospheric oxygen enters only the edges of the flame, permitting optimum reducing conditions in the central portion of the flame. The atom population is uniform with height over a significant portion of the flame, making adjustments simple. Three-slot burners have been shown to exhibit less noise and clog less easily with samples of significant dissolved solute content than most other designs. Past the nebulization point, most burner/nebulizer assemblies have baffles in their path that effectively remove the large droplets; these do not atomize as efficiently and result in local cooling of the flame.

A flame should meet several desirable characteristics for use in flame emission applications: (a) it should provide sufficient energy for the sample to be atomized; for the metals determined presently by flame photometry, the propane-air flame is sufficient, (b) the flame should be nonturbulent so that atom population shows the least

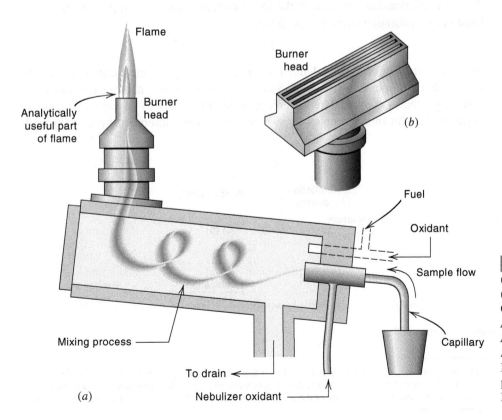

(a)

FIGURE 12.2 Premix burner. (a) Nebulizer, chamber, and burner. (b) Burner head. (Adapted from G. D. Christian and F. J. Feldman, *Atomic Absorption Spectroscopy. Applications in Agriculture, Biology, and Medicine.* New York: Interscience, 1970. Reproduced by permission of John Wiley & Sons, Inc.)

spatiotemporal variation, (c) by itself the flame should have minimum emission and absorption at the wavelengths of interest, and (d) the flame should be able to operate at low gas velocities so the emitting atoms remain in the view volume as long as possible. In addition, the flame operation should be safe and inexpensive.

Flame Processes Occurring in Typical Flame Photometry

Figure 12.3 schematically depicts the processes that go on when a solution of KCl is aspirated into the flame. The solvent evaporates from the fine droplets, leaving the dehydrated salt. The salt is dissociated into free gaseous atoms in the ground state. A certain fraction of these atoms can absorb energy from the flame and be raised to an excited electronic state. The excited levels have a short lifetime (1–10 ns) and drop back to the ground state, emitting photons[2] of characteristic wavelengths, with energy equal to $h\nu$. Only electronic transitions are involved and very narrow emission lines are observed. Some of the atoms also ionize by thermal excitation; a fraction of them is excited by thermal energy to form excited ions. In the case of measuring calcium by flame photometry, side reactions in the relatively low-temperature propane-air flame result in largely CaOH formed in the flame, which is excited and the emission from this excited molecule (much broader than atomic emission lines) is measured in typical filter flame photometers. For all the other metals, the emission from the excited atom is measured.

In flame emission, we measure $K^{\circ*}$. In atomic absorption, we measure K°.

As indicated in the figure, side reactions in the flame may decrease the population of the desired emitting species and hence the emission signal. The intensity of emission is linearly proportional to the concentration of the analyte in the solution being aspirated only at the low end of the calibration curve. For a number of reasons (including the fact that there is a much greater population of the atoms in the ground state that can reabsorb the emitted radiation) the signal at higher analyte concentrations is less than the linear relationship observed at the lower concentrations would predict. A typical large range calibration curve has the form:

$$I = kC^{n} \tag{12.3}$$

where I is the emission signal intensity, C is the analyte concentration and k and n are calibration constants ($n < 1$). Some present-day flame photometers can automatically process multianalyte multipoint calibration data to fit equations akin to Equation 12.3 (or more complex forms) and perform measurements up to 1000 mg/L

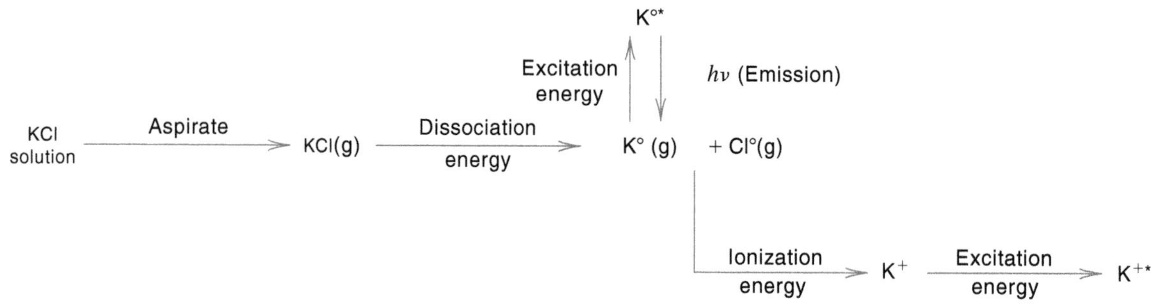

FIGURE 12.3   Processes occurring in flame.

---

[2]This is in contrast to excited molecules in solution, where there is much greater probability for collisions with solvent and other molecules. In the flame, there is less probability for collision because there are much fewer flame molecules and, therefore, many of the atoms lose their energy of excitation as electromagnetic radiation rather than as heat.

without dilution. Although few metals are determined by FP, it is widely used in niche areas. Sodium, potassium, and lithium are routinely determined by FP in the clinical laboratory. Residual alkali metals in biodiesel and the measurement of Na, K, and Ca in cement are among some of the more important applications. In the heyday of flame emission spectrometry, up to 60 elements have been measured in hot nitrous oxide (N_2O)–acetylene (C_2H_2) flames. But today, measurements of all but the stated metals are conducted by AAS.

12.3 Atomic Absorption Spectrometry

Since the number of atoms in the ground state far outweighs those in the excited state, one would normally think that the study of absorption by atoms in a flame will parallel or even precede flame emission spectroscopy studies. This in fact has not been the case because in flame AAS we encounter a problem that has no parallel in molecular absorption spectrometry. This is that the flame itself emits some (often a lot) of light at the wavelength(s) where we are trying to measure light absorption. Moreover, this background emission from the flame is not particularly constant over time; a simple subtraction of its contribution is not possible. Imagine a flashlight being aimed at a detector in an otherwise dark room. It is not difficult to tell when some of this light is being absorbed by species present in the light path; this is the situation in molecular spectroscopy.

On the other hand, if the same experiment is conducted in a brightly sunlit room (and moreover, clouds are occasionally passing by, changing the background light intensity), it would not be easy for the detector to tell minor changes in the light it is receiving from the flashlight because of the presence of absorbing species. This was the problem facing Australian spectroscopist Alan Walsh (see sidebar) as his colleague John Shelton reminded him. Walsh's solution was equivalent to turning the flashlight on and off at a constant and relatively high frequency and then filter the detector signal so we look only at this frequency. This process eliminates all other signals and looks only at the flashlight signal. This is no more complicated than being immersed in radio waves of all different frequencies but still being able to tune to the desired frequency and listen to/watch one's favorite radio or TV station. Pulsing on and off light sources at high speeds reproducibly is only practical for solid-state sources that did not exist until recently. Walsh accomplished this by putting a motorized mechanical chopper in front of the light source. As the chopper rotates, it alternately blocks the beam and allows it to pass through.

The rotational speed basically controls the frequency with which the light is "turned on and off." (Walsh also proved the concept by using a sodium lamp powered at 50 Hz; this did not need a chopper.) The detector is tuned to this frequency. Walsh also recognized that he did not have monochromators that can provide resolution comparable to those of atomic absorption lines (he estimated the line width to be 2 pm). Such resolution will be needed to essentially exclude stray light. He chose therefore to use a light source that contained the analyte element of interest. A discharge is created; this results in the emission of line(s) characteristic of the element. (If more than one line is emitted, they are far enough apart for a simple monochromator, usually placed after the sample, to isolate the principal emission line of interest.) This line is used as the source light. In effect, the source itself functions as a very high-resolution monochromator, providing an extremely narrow source line. Note that a different lamp is needed for the analysis of each element. Note also that the temperature of the light source should not

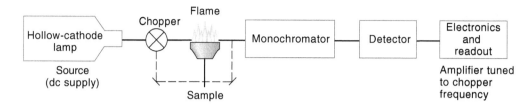

FIGURE 12.4 Schematic diagram of atomic absorption instrument. (Adapted from G. D. Christian and F. J. Feldman, *Atomic Absorption Spectroscopy. Applications in Agriculture, Biology, and Medicine.* New York: Interscience, 1970. Reproduced by permission of John Wiley & Sons, Inc.)

be very high, as temperature also causes broadening of the emitted line. A line much broader than the absorption line width will also effectively be stray light.

Principles of Flame AAS

A simplified functional diagram of the first commercial flame AAS instrument (Perkin Elmer Model 303) is shown in Figure 12.4. The light source is a hollow cathode lamp that emits at the atomic absorption line of interest (more on this later). A mirror-equipped chopper alternately directs the beam through the flame and bypassing it, effectively providing a double-beam arrangement and compensating for any source drift. Modern AAS instruments mostly do not use such a double-beam arrangement, however. The drift of present-day light sources following a warm-up time is much smaller than drifts in the atomizer. Rather, the reference reading is often taken just before and after the sample is measured, thus effectively providing a double beam in time arrangement. The sample solution is aspirated into a flame as in flame emission spectrometry, and the analyte element is converted to atomic vapor. The flame then contains atoms of that element. Some are thermally excited by the flame, but most remain in the ground state, as shown in Table 12.1. These ground-state atoms can absorb radiation emitted by the source that is deliberately composed of that element so its characteristic lines are emitted.

Atomic absorption spectrophotometry is identical in principle to absorption spectrophotometry described in the previous chapter. The absorption follows Beer's law. That is, the *absorbance* is directly proportional to the pathlength in the flame and to the concentration of atomic vapor in the flame. Both of these variables are difficult to determine, but the pathlength is essentially held constant in a given burner and flame conditions and the concentration of atomic vapor is directly proportional to the concentration of the analyte in the solution being aspirated. In practice, one calibrates the instrument response by aspirating samples of different concentration.

Types of AAS Instrumentation

Broadly there are two types of AAS instrumentation in present use:

(a) Line source AAS (LS-AAS) instruments, and

(b) Continuum source AAS (CS-AAS) instruments

Either type can utilize flame or a graphite furnace (also called electrothermal) atom source.

Walsh originally believed CS-AAS instruments would not be possible, not only because single-digit pm resolution is needed, but also because even if such resolution could be attained, there would not be enough light from a continuum source through such a narrow bandwidth. Technology has advanced to the point, however, that such instruments have become commercially available since 2004, equipped with a very high-power water-cooled Xenon lamp operated in a "hot-spot" mode, a very high-resolution double monochromator, and a CCD-array detector. They have the advantage that very many different lamp sources are not needed for the analysis of different elements. Bernhard Welz, perhaps the most ardent champion of AAS (see Reference 8), simply calls CS-AAS "The better way to do AAS" (Reference 9).

As in molecular absorption spectrophotometry, the requirements for AAS include a light source, a beam path through the analyte (the flame or furnace), a monochromator, and a detector. As in molecular absorption spectroscopy with array detectors, the monochromator follows the sample.

The various components of an atomic absorption spectrophotometer are described as follows.

Light Sources for AAS

Two types of light sources are presently used in LS-AAS. The more common source is a **hollow-cathode lamp** (HCL). The basic construction of an HCL is illustrated in Figure 12.5.

It consists of a cylindrical hollow cathode made of the element to be determined or an alloy of it, and an anode, usually of W or Zr. Alloying is used where an electrode cannot be made easily from the element (e.g., Na, As, etc.), in the case of precious elements where significant cost savings are possible without sacrificing performance, and where a superior long-term performance is observed from an alloy rather than the pure element (e.g., Cr, Cd, etc.). The alloying elements are carefully chosen not to interfere spectrally with the element of interest. The functioning cathode elements are typically pressed into a cathode cup, commonly made from steel but other metals may be used. The electrodes are enclosed in a borosilicate glass tube with the window material being borosilicate glass for wavelengths over 400 nm, special UV transparent glass for wavelengths between 240 (the exact low-end cut-off varies with the manufacturer) to 400 nm, and quartz for shorter wavelengths. The tube is under reduced pressure and filled with an inert gas, usually neon. Argon is used for those elements where interference from neon lines is possible. A high voltage is impressed across the electrodes, causing the gas atoms to be ionized at the anode. These positive ions are accelerated toward the negative cathode. When they bombard the cathode, they cause some of the metal to "sputter" and become vaporized. The vaporized metal is excited to higher electronic levels by continued collision with the high-energy gas ions. When the electrons

A sharp-line source is used in LS-AAS. The source emits the lines of the element to be measured. These possess the precise energies required for absorption by the analyte atoms.

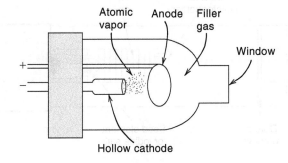

FIGURE 12.5 Design of a hollow-cathode lamp.

return to the ground state, the characteristic lines of that metal are emitted. The filler gas lines are also emitted.

"Boosted" HCLs are offered by some manufacturers. A secondary discharge, electrically isolated from the sputtering current, excites the already vaporized atoms. This provides a sharper and more intense line emission and is discussed in more detail in Section 12.7 as a source for atomic fluorescence spectrometry. Single-element HCLs are available for some 70 elements. Elements whose emission lines do not interfere with each other can be combined to make multielement lamps. Lamps with two to as many as seven selected elements are available. In some cases, intensities comparable to single-element HCLs are attained, but in many cases significant intensity compromises have to be made to provide the convenience of a multielement source. They may also have shorter lifetimes than single-element HCLs due to selective volatilization ("distillation") of one of the elements from the cathode with condensation on the walls. Most LS-AAS instruments come with a multi-lamp turret so that a different lamp can be brought online rapidly, whether in manual or automated fashion.

The second type of line source is an **electrodeless discharge lamp** (EDL). An EDL contains a small quantity of the analyte as either the element or often as the iodide salt in a quartz bulb (Figure 12.6). The bulb is filled with an inert gas, typically Ar, at low pressure. The bulb is surrounded by a radio frequency coil.

When the coil is powered to generate an intense electromagnetic field, an inductively coupled discharge occurs in the low-pressure lamp with characteristic line emission of the element. The EDL is a more intense source than an HCL and often has a narrower line width. EDLs require a different type of power supply. It is particularly for the more volatile elements that the EDL can be attractive: lamps are available for As, Bi, Cd, Cs, Ge, Hg, P, Pb, Rb, Sb, Se, Sn, Te, Tl, and Zn.

Solid-state diode lasers can be tuned over a (limited) wavelength range and have been used in LS-AAS. Many advantages can be seen: The source is readily modulated at any desired frequency. With a properly designed reflective geometry, the coherent nature of a laser beam can allow multiple passes through the atom source, greatly increasing the sensitivity. Isotopic differences in an element make a small difference in the exact position of a transition line; the 670.8-nm line shifts by 0.015 nm from ^{6}Li to ^{7}Li, for example. The ability of a diode laser to be scanned over such ranges in a highly precise manner thus provides the opportunity to perform isotopic analysis. Despite these advantages, until such lasers become readily available in the deep UV, they are unlikely to be of major importance in the general practice of AAS or AFS. Especially since many laser sources will still be needed to cover the entire wavelength range of interest, the ability to use a single high-power continuum source as discussed below is likely to be increasingly attractive.

For CS-AAS, the extremely high resolution needed also means that only a very small fraction of the total radiant power emitted by the continuum source will reach

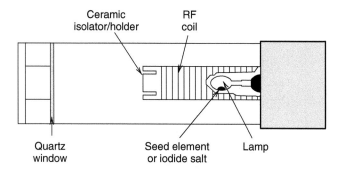

FIGURE 12.6 Design of electrodeless discharge lamp.

the detector. The lamp source must be very intense. This problem has been solved with high-power (300 W) high-pressure (17 bar at room temperature) short-arc (electrode distance < 1 mm) Xe lamps made with special electrode materials that can be operated in the "**hot-spot**" mode; in this mode, rather than the diffuse emission from a typical Xe arc lamps, the light originates from a small (0.2 mm diameter) intense spot close to the cathode. The housing is water-cooled. The lamp output is reflected by a focusing mirror through the analyte atoms produced by the atom source.

The position of the hot spot varies with time; this hot spot jitter is compensated for by monitoring the position of the spot with a four-quadrant position detector and using this information to move the micro actuator-mounted mirror appropriately. Because an array detector is used, any fluctuations in the lamp intensity due to spatiotemporal variations can be compensated for by monitoring the pixels surrounding the pixel corresponding to the absorption line. The lamp lifetime is at least 1000 h, over which the output drops by 4x with a two-fold deterioration of S/N. Because such lamps are under high pressure, it is essential to use proper protection to remove/install or handle such lamps.

Supercontinuum white-light coherent sources are presently available from 400 nm out to NIR (see http://www.rp-photonics.com/supercontinuum_generation.html). Intense broadband white light is generated by launching high-power ultrafast (tens of femtoseconds) laser pulses into nonlinear media such as optical fibers. Such an extremely intense coherent broadband source is foreseen to be ideal for CS-AAS, once the lower wavelength limit can be extended into the UV.

Analyte Atom Sources

There are several ways to generate atomic vapor for AAS measurements, including various types of flames, electrothermal atomizers, and cold-vapor and hydride generation for atomization.

1. Flame Atomization. The burners and nebulizers used in flame AAS are the same as those in FP but air- C_2H_2 (2250°C) and N_2O–C_2H_2 (2960°C) are essentially exclusively used. In flame AAS, there are additional criteria that the flame should meet beyond those required in FP. The flame conditions/fuel-oxidant ratio should permit controlled variation to create either oxidizing or reducing conditions in the flame. Different elements exhibit optimum sensitivity under different flame conditions; both of the above flame types permit such variation and permit low gas velocities (maximum velocity ~160 cm/s) so the residence time of atoms in the flame can be prolonged. Note that under a given flame condition, the best sensitivity for a given element may be at different observational heights in the flame.

Air-propane flames are not hot enough for general use in AAS because most elements are not efficiently atomized. However, note that although the N_2O–C_2H_2 flame is much hotter than the air- C_2H_2 flame, it is not necessarily better for all elements because if the temperature is too high, some elements begin to significantly ionize. Indeed, this is the reason that in FP the elements that are measured are measured with a flame no hotter than propane-air. Air- C_2H_2 is the preferred flame in AAS for the majority of elements except for about 30 elements that tend to form heat-stable (refractory) oxides. For these, the N_2O–C_2H_2 flame is preferred. The burner must be suitable for supporting the very high temperature N_2O–C_2H_2 flame. A special, thick, stainless steel burner head is used. The length of the flame is equivalent to the cell pathlength in molecular spectroscopy. The Boling design provides such a long path.

The air–acetylene flame is the most popular for AAS. The nitrous oxide–acetylene flame is best for refractory elements.

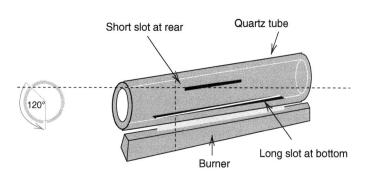

FIGURE 12.7 Slotted quartz tube put above the flame has a shorter exit slit at an angle, prolonging the residence time of atoms.

As we go down in wavelength, by 200 nm hydrocarbon-fueled flames absorb ~ > 50% of the radiation. As such, these flames are not suitable for monitoring absorption below this wavelength. In an argon–hydrogen-entrained air flame (also called argon–hydrogen diffusion flame), hydrogen is used as the fuel and argon is the nebulizer or auxiliary gas. As the flame is lit, the hydrogen burns in the surrounding air. The outer part of the flame reaches a temperature of 850°C while in the middle the temperature is much lower (300–500°C, depending on the flame height). Even at 190 nm, the flame transmittance in a typical burner is 80%. Elements, such as arsenic (193.5 nm) and selenium (197.0 nm), easily form hydrides (AsH_3, H_2Se, discussed in more detail later) and these gases are directly introduced into such a flame.

2. Getting a Little More from the Atoms. The reason that we are interested in low-flame velocities is that once atoms are formed, they stay around in the observation zone longer than at high-flame velocities; a longer time to observe the signal improves the signal-to-noise ratio. One way to prolong the residence time of the atoms with an oxygen–acetylene flame is to use a slotted quartz tube atop the flame (Figure 12.7). Such a tube is readily put in, but the gain in S/N is modest (only 2–3 times).

Before GFAA became widely used, the use of a tantalum/nickel boat/cup with a handle was popular; this was often used in conjunction with the quartz tube shown above. A desired volume of the sample (e.g., blood, to measure lead in blood) was put in the boat and the sample was then evaporated to dryness. The boat with the sample residue was then put in the flame, and the atoms formed travel into the quartz tube placed above and were measured. Obviously, the boat/cup cannot reach a temperature any greater than that of the flame; the technique was useful only for the relatively easily atomized elements, but many of these elements are of particular interest in biosamples, e.g., Cd, Pb, Hg, Zn, Se, etc.

3. Electrothermal Atomization. Although aspiration into a flame is the most convenient and reproducible means of obtaining atomic vapor, it is not a particularly efficient means of converting all the analyte into atomic vapor and have it present in the optical path for a long enough time to measure the absorption. From the dissolved molecular/ionic entities in solution, as little as 0.1% of the aspirated analyte may actually be atomized and measured. The volume of solution required for flame atomization is minimally of the order of a milliliter or more.

Electrothermal atomization uses some type of a mini-furnace (typically less than 1 cm³ in volume) in which an aliquot of the sample is put in and dried. The furnace is made of an electrically conductive material. While tantalum and other substances have had utility in specific applications (e.g., for elements that form refractory carbides), presently graphite furnaces are nearly exclusively used.

The furnace is then rapidly electrically heated (currents in excess of 100 amperes and heating rates in excess of 1000°C/s are common) to a very high temperature to produce an atomic vapor cloud.

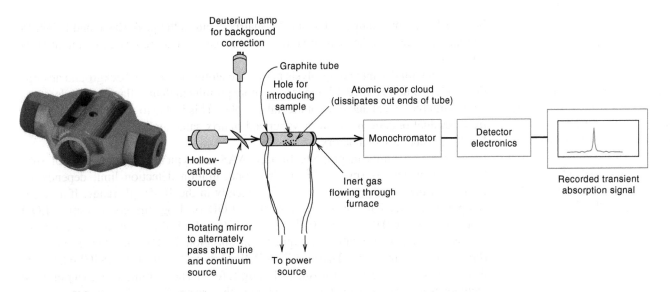

FIGURE 12.8 Left: Photograph of a transversely heated graphite tube with an integrated platform. The light beam passes from front to back; electrical contact is between the left and right. The sample is introduced through the hole at the top directly onto the platform. (Courtesy PerkinElmer Inc.) Right: A typical electrothermal AAS instrument using an axially heated graphite tube shown schematically.

Electrothermal atomizers have conversion efficiencies approaching 100%, so absolute detection limits are often 100 to 1000 times improved over those of flame aspiration methods. Our discussion will center on resistively heated atomizers (graphite furnace). Although these are not generally useful for emission measurements, they are well suited for atomic absorption measurements. A schematic of a typical electrothermal atomizer is shown in Figure 12.8.

In a typical electrothermal atomization assay, a few microliters of sample are placed inside the graphite tube. Ordinary graphite is porous and leads to problems; to address this, the furnace is made of or coated with nonporous pyrolytic graphite. Original furnaces were simply a cylindrical tube (often referred to as the Massman furnace after Hans Massman) but much better results are obtained when the sample is put on a platform (often called the L'vov platform) resting within the tube. The furnace is heated resistively by passing electrical current through it. The furnace can be heated axially (i.e., the contact electrodes are placed on two sides of the long axis of the tube along which the light beam passes) or transversely, at right angles to the beam path. The latter arrangement leads to a more uniform axial temperature distribution; this is used in the high-end instruments. The sample is first dried at a low temperature for a few seconds (~100 to 200°C), followed by pyrolysis at 500 to 1400°C to destroy organic matter that produces smoke and scatters the source light during measurement; the smoke from pyrolysis is flushed out by flowing argon gas. Finally, the sample is rapidly thermally atomized at a high temperature, up to 3000°C. The light path passes over the atomizer (or through the tube). A sharp peak of absorbance versus time is recorded as the atomic cloud forms through the light path (Figure 12.8). Most commonly, the area of the observed peak, rather than its height, is used for quantitation. The heating, of course, is done in an inert atmosphere (e.g., argon) to prevent combustion of the graphite; refractory metal oxides can also form if oxygen is present. Electrothermal atomization can be nearly 100% efficient and typically only a few microliters of sample are required.

Inter-element effects are generally more pronounced in electrothermal AAS than in flame AAS. The inter-element effects are typically addressed by using a standard addition method in which the sample is remeasured after a known amount of a standard is added to the sample. This is discussed in more detail later in

"Imagine my wonder," he writes, "when the bright sodium lines from the hollow cathode tube started to weaken and disappeared completely…." (*Spectrochim. Acta B.* 39 (1984) 149). L'vov's first paper on total vaporization of a sample in a graphite cuvette (as he called it) appeared in Russian in 1959 and was translated into English 25 years later.

Section 12.5. Any change in the sample matrix changes the peak shape and height to a greater degree than the peak area; it is for this reason that area-based quantitation is preferred.

Even with drying and pyrolysis before the atomization step, background absorption in electrothermal methods tends to be more prominent than in flame methods, especially with biological and environmental samples. This is due to molecular absorption from residual organic material or vaporized matrix salts, and background correction for this is generally mandatory.

Detection limits in graphite furnace AAS are typically on the order of sub- to several picograms (pg, 10^{-12} g). The concentration detection limit depends, of course, on the sample volume, which is typically in the 10–50 μL range. If a 10-μL sample is analyzed for an element with an LOD of 1 pg, the concentration LOD would be 10^{-12} g/0.01 mL or 10^{-10} g/mL or 100 ng/L (100 part per trillion). A web-accessible comparison table of typical concentration LODs for various elements (http://www.thermo.com/eThermo/CMA/PDFs/Articles/articlesFile_18407.pdf) lists detection limits of 7.5 ng/L for zinc to 100 μg/L for P. The extreme sensitivity of these techniques is impressive even when dealing with very small sample volumes.

Electrothermal methods are complementary to flame methods. The latter are better suited when the analyte element is at a sufficiently high concentration to measure and adequate solution volume is available. They provide excellent reproducibility, and interferences are usually easier to deal with. On the other hand, electrothermal atomization excels when either the concentrations or available sample amounts are very small. Additionally, in many cases it is possible to analyze solid samples directly by electrothermal AAS. Method development and calibration of electrothermal methods require more care, however.

4. Cold Vapor and Hydride Generation for Atomization. The best-known example in this class is that of mercury. Mercury is easily reduced to the element by reducing agents such as $SnCl_2$. When a suitable reductant is added to a sample solution containing mercury, it is reduced to Hg(0); simultaneous purging of the solution (typically with argon) flushes the liberated mercury into an AAS measurement cell. Sodium borohydride ($NaBH_4$) is a much more powerful reducing agent, and it can reduce organically bound Hg to Hg(0). The use of $SnCl_2$ and $NaBH_4$ as respective reducing agents allows the measurement of inorganic and total Hg, respectively. Although the vapor pressure of cadmium is very low at room temperature, Cd can also be successfully generated as Cd(0) vapor from solution by $NaBH_4$. $NaBH_4$ also generates the hydrides of As, Bi, Ge, Pb, Sb, S, Se, Sn, Te, In, and Tl as the gases—these are easily decomposed to the element by introducing the generated gas into a flame, a hot quartz measurement cell, or even a graphite furnace. The great advantage of cold vapor and hydride generation is **matrix isolation**, meaning that the analyte of interest is totally isolated as a vapor from the original sample matrix. There are reports that while many other metals such as Au, Ag, Co, Cr, Cu, Fe, Ir, Mn, Ni, Os, Pd, Pt, Rh, Ru, and Zn do not form hydrides, they do form nanometer-size elemental aerosols, which can be then readily atomized.

Continuum Source AAS Instrumentation

A continuum source can be used for AAS measurements if a suitably high-resolution spectrometer is used to isolate a narrow wavelength span. A block diagram of such an instrument is shown in Figure 12.9. Either a flame or a graphite furnace can be used as an atom source. The very high wavelength resolution is provided by an *Echelle* grating (from French *eschelle*, meaning steps or stairs), usually ruled in a coarse fashion, operated at a high incidence angle to create higher order diffraction, almost always greater

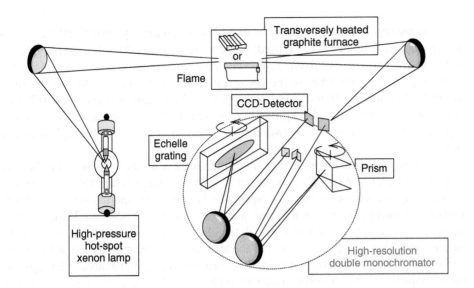

FIGURE 12.9 Continuum source–high-resolution spectrometer setup for AAS. The instrument uses an intense high-pressure Xe lamp, operated in the hot-spot mode, and an *Echelle* polychromator with a prism for order sorting. The beam then passes through the atom source (flame or graphite furnace) to a CCD array detector.

than 5). (See Chapter 11 for a description of diffraction gratings and diffraction order.) Such gratings can lead to very high dispersion and hence wavelength resolution, but multiple orders of diffraction usually overlap and must be sorted out, e.g., by a secondary dispersing device such as a second grating instrument of modest resolution, or more commonly, a prism.

Interferences in AAS

These fall under three classes, spectral, chemical, and physical. We will discuss these briefly and point out their relative effects in emission and absorption measurements.

1. Spectral Interferences. Atomic absorption lines are very narrow and absorption lines from different elements almost never overlap; a much more common problem is molecular absorption. Such absorption can be caused by flame gases/combustion products, undissociated sample-derived molecules and also scattering by particles that may be generated from the sample, especially in electrothermal atomization or in flame AAS when the aspirated sample has much dissolved solids, making it difficult to atomize all of the sample. Unless corrected for, this erroneously measured high absorbance can lead to a gross overestimation of the analyte concentration.

The effects of such **background absorption** can be corrected for by making sequential measurements of the absorbance due to the background and that due to the analyte + background and determining that due to the analyte by difference. The most common way of doing this is to use an additional light source, a continuum source such as a D_2 lamp, as shown Figure 12.8. Such a **continuum source background correction method** takes advantage of the fact that the background absorption is broad across the entire wavelength span that passes through the slit-monochromator combination (the *passband*). When a mirror switches the light source from the HCL to the D_2 lamp, the average absorbance is measured across the wavelength region. The atomic absorption line is so narrow compared to this passband that the absorbance due to the analyte atoms is considered essentially negligible. On the other hand, when the absorbance is measured with the HCL, it measures the absorbance due to the analyte atoms as well as the background absorbance at that wavelength. Typically a rotating mirror automatically switches between the two light sources and the difference signal is electronically generated; separate manual measurement or correction is not necessary. A D_2 lamp does not provide enough energy above 330 nm, and a quartz-halogen lamp needs to be used; low-end instruments do not often provide the ability to use another switchable lamp. The continuum source correction method is far from perfect; mismatches in the correction beam geometry with that of the HCL and lack of uniformity in its spatial and spectral energy distribution and contribution of atomic absorption in the background measurement process, all cause difficulties in high-accuracy measurements. However, it is relatively inexpensive to implement and convenient to use.

The **Smith–Hieftje correction method**, named after the inventors, involves an elegant principle: when an HCL is pulsed at high currents, the emitted line broadens substantially and for many elements, especially those that are volatile, undergoes a **self-reversal**, meaning that the original emission line largely disappears while there is now strong emission on both sides of the analytical line wavelength that is absorbed by the background. Thus, the measurement of the absorption under normal and pulsed high-current conditions respectively provides the measurement of the analyte + background and background only. This correction method has the advantage of using a single light source. However, the method has fallen out of favor as sensitivity is sacrificed when the line does not undergo significant self-reversal and recovery from pulsed condition is not fast enough to perform adequate correction under the rapidly changing conditions of electrothermal atomization.

The **Zeeman correction method** is likely the most sophisticated and the most accurate of the background correction methods used in line source AAS. This principle depends on the splitting of the line source into three principal polarized components in a magnetic field, applied either to the light source or the atomizer (the latter is more common, especially for heated graphite atomizers). The π component(s) remain at or close to the original wavelength corresponding to the atomic absorption line, and there are two σ components, one moves to a wavelength slightly greater, the other to a wavelength slightly less relative to the original; the exact shifts depend on the strength of the magnetic field. Fairly strong magnetic fields are needed to attain useful splitting; some high-end instruments allow tuning of the field strength (e.g., from 0.1 to 1.1 Tesla in 0.1 Tesla steps). For reasons that relate to the dimensions of the atomizer, it is much more practical to attain high magnetic fields with the much smaller graphite furnace atomizer than a flame or a cold vapor source, especially with the preferred configuration of the magnetic field being longitudinal. The transversely heated graphite tube is thus particularly well suited for applying the magnetic field, although specially designed longitudinally heated tubes can be used. In a typical instrument equipped with a Zeeman corrector, an alternating magnetic field that turns on and off is applied. The analyte + background absorbance is measured when the magnetic field is off, the background alone is measured with the field on, via the absorption measured with the two off-resonance σ lines while the π component(s) are taken out by a rotating polarizer.

The great advantage of **high-resolution continuum source AAS** is that background correction can be made within a single measurement; sequential measurements of absorbance of analyte and analyte + background are not needed. This allows for correcting accurately even for the fastest changing background, as may be encountered in electrothermal atomization. Because a very high-resolution *Echelle* monochromator is used along with an array detector, it is possible to simultaneously measure the absorbance at the resonance line as well as on either side of it, as the light intensity on each pixel is available. If the background contains a fine structure, a more complex correction method is used that involves appropriately matching, with a least-squares

Broadband absorption due to molecules or loss of light by scattering from particles is a common problem in AAS. This can be corrected for by measuring the background absorption separately.

Background correction is often more critical in electrothermal atomization.

algorithm, the background spectra with stored spectra of known diatomic molecules that can exist under the measurement conditions.

2. **AAS Ionization Interference.** An appreciable fraction of alkali and alkaline earth elements and several other elements may be ionized in very hot flames. Since we are measuring the un-ionized atoms, both emission and absorption signals can be decreased. In the present practice of flame photometry, only flames of modest temperatures are used, so ionization interference is not a problem. Much hotter flames are used in flame AAS, however, and ionization of easily ionizable elements do occur. If a fixed fraction of the element of interest is ionized, this would largely be included in its calibration behavior, except that the sensitivity and the linear dynamic range may suffer some. The main problem comes from the effects of one ionizable element upon another. Consider that we wish to measure calcium by flame AAS and a calibration curve was constructed with pure calcium standards. There is some ionization of the calcium, but this is accounted for in the calibration. The samples, on the other hand, contain significant and variable amounts of sodium. The sodium ionizes easily, increasing the free electron density in the flame and in turn suppressing the ionization of calcium, thus increasing the calcium atom/ion ratio and causing a positive measurement error. Ionization interference can usually be overcome either by adding large amounts of the interfering, easily ionizable, element to both the samples and the standards, to make the enhancement constant and ionization minimal. Ionization can usually be detected by noting that the calibration curve has a positive deviation or curvature upward at higher concentrations because free electron density in the flame increases sufficiently at high analyte concentrations so that a lesser fraction of the atoms are ionized compared to that at lower concentrations.

> Ionization can be suppressed by adding a solution of a more easily ionized element, for example, potassium or cesium.

3. **Refractory Compound Formation.** The sample may contain constituents that form a refractory (heat-stable) compound with the analyte element of interest. For example, when determining calcium, the presence of phosphate in the sample can cause the formation of calcium pyrophosphate, $Ca_2P_2O_7$ that is not easily broken down to the atoms at flame temperatures. The lack of complete atomization of calcium thus causes a negative error in its determination. The best way to handle such interference is chemical intervention; in the above example, a high concentration (ca. 1%) of strontium chloride or lanthanum nitrate can be added to the solution. Called a releasing agent, the strontium or lanthanum will preferentially combine with the phosphate and prevent its reaction with the calcium. Alternatively, a high concentration of EDTA can be added to the solution to form a chelate with the calcium; this prevents its reaction with phosphate. The calcium–EDTA chelate is decomposed in the flame to give free calcium atoms. A hotter flame such as the N_2O–C_2H_2 flame also successfully atomizes many compounds that are not easily decomposed at lower temperatures. In flame AAS, some elements, e.g., Al, Ti, V, Mo, etc., react with flame species, notably O and OH, forming refractory oxides and hydroxides that can only be decomposed in a N_2O–C_2H_2 flame. The flame is usually operated in the reducing (fuel-rich) condition in which a large red feather-like secondary-reaction zone is observed; this zone arises from the presence of CN, NH, and other highly reducing radicals. These (or the lack of oxygen-containing species), combined with the high temperature of the flame, decompose and/or prevent the formation of refractory oxides.

> Refractory compound formation is avoided by addition of a chemical competitor or use of very high temperatures.

Electrothermal atomization of some elements can suffer from interferences as well. Chemical modifiers, for specific elements or specific types of samples, have been developed. Volatilization of NaCl from highly saline samples such as seawater is a problem, and addition of NH_4NO_3 solves this by forming volatile NH_4Cl and

$NaNO_3$, which are easily thermally decomposed. Specific recommended additives for the analysis of different elements have been widely tabulated, see, e.g., http://www. chem.agilent.com/library/usermanuals/public/gta_analytical_methods_0848.pdf. The formation of refractory carbides can cause serious problems in GFAA. Elements such as Ba, Mo, Ti, and V readily form carbides but do not readily react with pyrolytically coated graphite. However, Ta, W, and Zr form carbides nevertheless and are difficult to sensitively determine by GFAA.

4. **Physical Interferences.** Most parameters that affect the rate of sample uptake in the burner and the atomization efficiency can be considered physical interferences. This includes such things as variations in the gas flow rates, variation in sample viscosity due to temperature or solvent variation, high and variable solids content, and changes in the flame temperature. Sample surface tension changes can affect the size of the nebulized droplets. These can generally be accounted for by frequent calibration, use of internal standards, or standard addition.

12.4   Sample Preparation—Sometimes Minimal

Sample preparation in flame AAS can often be kept to a minimum. As long as chemical or spectral interferences are absent, essentially all that is required is to obtain the sample in the form of a diluted solution, filtered to be free from particles. Atomic spectrometry measures the elements; the specific chemical form of the element is immaterial; it will be dissociated to the free elemental atomic vapor in the measurement process. Thus, several elements can be determined in blood, urine, cerebrospinal fluid, and various biological fluids by direct aspiration of the diluted sample. GFAA allows the analysis of not only dilute solutions, but liquid samples containing suspended solids and even solids. Specialized atomizers have been developed that facilitate the accurate analysis of solids and are used if solids are frequently analyzed.

In preparing standards, the analyte matrix must be matched as closely as possible. Methylcyclopentadienyl manganese tricarbonyl (MMT) is added to make high-octane number gasoline. If gasoline is to be analyzed for Mn, an appropriate hydrocarbon solvent matrix must be used for standards, not water. Reference 7 reviews the applications of AAS to biological samples. AAS is widely used for metal analysis in biological fluids and tissues, in environmental samples such as air and water, and in occupational health and safety areas. In the clinical laboratory, alkali and alkaline earth metals were measured by flame photometry, but to a large extent these have been replaced by ion-selective electrode measurements (Chapter 20); the latter have the advantage that ISE-based bedside point-of-care monitors are available.

Many factors determine the choice between flame AAS and GFAAS. Flame AAS is easier to use, takes less time per sample, but is less sensitive. Perkin-Elmer provides a very useful guide that compares the relative cost, measurement range, and limits of detection by the various measurement techniques (flame AAS, GFAAS, ICP-AES and ICP-MS) http://www.perkinelmer.com/PDFs/Downloads/BRO_World LeaderAAICPMSICPMS.pdf.

12.5   Internal Standard and Standard Addition Calibration

An internal standard undergoes similar interferences as the analyte. Measurement of the ratio of the analyte to internal standard signals cancels the interferences.

In atomic spectrometric methods, signals can frequently vary with time due to factors like fluctuations in gas flow rates and aspiration rates. Precision can be improved by the technique of **internal standards**. As an example, a multichannel flame photometer such as that discussed in Section 12.2 can be used for the measurement of sodium and potassium in serum. The accuracy can be significantly improved by adding a fixed concentration of lithium to all standards and samples and interpreting the data in terms of the *ratios* of the K/Li and Na/Li signals. If the aspiration rate, for example, fluctuates, each signal is affected to the same extent and the ratio, at a given K or Na concentration, remains constant. See the spreadsheet exercise in Section 15.5 for an example of how to perform internal standard calibrations from the signals of the analyte and the internal standard. Ideally, the internal standard element should be chemically similar to the analyte element, and their wavelengths should not be too different.

Another difficulty often encountered in atomic-spectrometric methods (and in fact, many other analytical methods) is suppression or enhancement of the signal by the sample matrix. This can be due to physical or chemical reasons that arise from a mismatch of the sample matrix and the standards used for calibration. For example, if the sample has higher viscosity than the standards, the aspiration rate will not be the same; if the sample contains a combustible organic solvent, the effective temperature in the flame will be different compared to a purely aqueous standard. Numerous causes of chemical interferences have already been stated. The technique of **standard addition** can be utilized to minimize matrix induced errors. The technique was previously discussed in Section 21.9 in the context of potentiometric titrations, one principal difference of atomic spectrometry (and most other analytical techniques) with potentiometry is the linear dependence of the analytical response on concentration rather than on the logarithm of the concentration as is encountered in potentiometry. A linear dependence generally simplifies the interpretation of the data.

Consider the blank-corrected absorbance A_s arising from a sample of unknown concentration (C_{unk}). We take V_s mL of a sample and spike it with V_{std} mL volume of a standard of known concentration C_{std} ($V_{std} \ll V_s$ to minimally affect matrix composition. C_{std} is so chosen that the change in concentration of the spiked sample from the original sample concentration is of the same order as the anticipated sample concentration). We measure the blank corrected absorbance of the spiked sample to be A_{spk}, representing C_{spk}, which is the sum of the sample plus spiked concentrations.

Assuming a linear analytical response,

In standard addition calibration, the standard is added to the sample, and so it experiences the same matrix effects as the analyte.

$$A_s = kC_{unk} \tag{12.4}$$

$$A_{spk} = kC_{spk} \tag{12.5}$$

where

$$C_{spk} = \frac{(V_s C_{unk} + V_{std} C_{std})}{V_s + V_{std}} \tag{12.6}$$

These combine to

$$C_{unk} = \frac{A_s V_{std} C_{std}}{A_{spk}(V_s + V_{std}) - A_s V_s} \tag{12.7}$$

In practice, at least two standard addition readings resulting in two different final concentrations should be taken in addition to the original sample to ensure that one is still operating within the linear range. Further, as blanks can often be significant, it is important to perform blank corrections. Multiple standard additions are often carried out for high accuracy. It is convenient to add multiple aliquots of the same standard. To have a constant matrix, if n standard additions are to be made, rather than analyzing the pure sample, as the sample we analyze V_s mL sample + nV_{std} mL water (or whatever the matrix for the standard is). The first spiked sample consists of V_s mL sample + $(n-1)V_{std}$ mL water + V_{std} mL standard, the second spiked sample consists of V_s mL sample $+ (n-2)V_{std}$ mL water $+ 2V_{std}$ mL standard and so on until the nth spiked sample consisting of V_s mL sample $+ nV_{std}$ mL standard. The total volume V_t in all cases remains constant as $V_s + nV_{std}$. (If the volume is not held constant, in each case the signal must be multiplied by V_t/V_s to account for dilution, V_t being calculated for each case. That is, if any amount of standard is added, it dilutes the sample and changes the matrix—unless such a small volume is added that it is negligible with respect to the total volume. For the most accurate results, we want that dilution to be the same for all measured samples. Therefore, the total volume added to the sample is kept constant, equal to the largest volume of the standard added. When a smaller volume of standard is added, the difference is made up by the solvent, often water.) Normally, a calibration curve is constructed, similar to Figure 21.6, where the Y axis would be the atomic emission or absorbance signal and V_{std} the volume of the standard added in each case as the X-value. It will be apparent that for any of the measurements, the observed absorbance, A_{obs}, will be given by

$$A_{obs} = \frac{kC_{unk}V_s}{V_t} + \frac{kC_{std}nV_{std}}{V_t} \qquad (12.8)$$

A plot of A_{obs} as a function of nV_{std} thus yields a straight line with a slope of kC_{std}/V_t and an intercept of $\frac{kC_{unk}V_s}{V_t}$. The intercept to slope ratio gives $\frac{C_{unk}V_s}{C_{std}}$, from which C_{unk} is readily computed.

Calculation of uncertainties and use of spreadsheet calculations for standard addition experiments are discussed in the website supplement for this chapter. Here we look at a simple standard addition problem.

Example 12.2

A serum sample is analyzed for potassium by flame emission spectrometry using the method of standard additions. Two 0.500-mL aliquots are added to 5.00-mL portions of water. To one portion is added 10.0 μL of 0.0500 M KCl solution. The net emission signals in arbitrary units are 32.1 and 58.6. What is the concentration of potassium in the serum?

Solution

The amount of standard added is

$$0.0100 \text{ mL} \times 0.0500 \, M = 5.00 \times 10^{-4} \text{ mmol}$$

This produces a signal of

$$58.6 - 32.1 = 26.5 \text{ arbitrary units}$$

The millimoles potassium in the sample, then, is

$$5.00 \times 10^{-4} \text{ mmol} \times \frac{32.1 \text{ units}}{26.5 \text{ units}} = 6.06 \times 10^{-4} \text{ mmol}$$

This is contained in 0.500 mL serum, so the concentration is

$$\frac{6.06 \times 10^{-4} \text{ mmol}}{0.500 \text{ mL}} = 1.21 \times 10^{-3} \text{ mmol/mL serum}$$

12.6 Atomic Emission Spectrometry: The Induction Coupled Plasma (ICP)

Excited-state atoms are necessary to observe characteristic emission from the excited atoms. The emission intensity is thus directly proportional to the excited-state population. As we have seen in Table 12.1, the relative population of the atoms in the excited state is very small until very high temperatures are reached. Except for very easily excitable elements, such as the alkali and alkaline earth metals (for which flame photometry still enjoys some utility, see Section 12.2), over the last two decades all other excitation sources (arc, spark, high temperature flame, etc.) have given way to the inductively coupled plasma source which reaches temperatures up to 10,000 K. As such, present-day atomic emission spectrometry (AES) exclusively uses the induction coupled plasma (ICP) as the excitation source and is thus called ICP-AES or ICP-OES (optical emission spectrometry). The ICP was first introduced for atomic spectrometry by Velmer Fassel at Iowa State University and Stanley Greenfield in the UK in the 1960s, and commercial instruments were first introduced in 1974. A chronological history of ICP development is given by Greenfield, from his perspective: S. Greenfield, "Invention of the Annular Inductively Coupled Plasma as a Spectroscopic Source," *J. Chem. Ed.*, **27** (2000) 584–591.

An ICP is schematically shown in Figure 12.10. A radio frequency coil around a quartz tube excites argon gas flowing through it; the frequency range is typically 5–75 MHz with typically 1–2 kW power dissipated. The plasma is initiated by a pilot discharge from a Tesla coil or the like, and the ionized argon then begins to conduct. The alternating magnetic field generates an eddy current within the conducting gas and the dissipated energy heats it to plasma temperatures.

The plasma can be viewed from a radial direction as indicated in Figure 12.10. Most of the early instruments utilized this view configuration. An axial view configuration was introduced much later (in this case, in the perspective of Figure 12.10, one is looking down on the plasma from the top). In the axial flow configuration, a further flow of gas is typically necessary for thermal management to protect the optics for axial view. Air can be used as a "shear flow" but some manufacturers choose to use

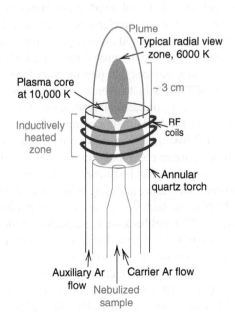

FIGURE 12.10 An induction-coupled plasma. The sample is introduced as a nebulized aerosol through the central tube in the torch, much like sample introduction in a flame. An argon carrier flow is used to nebulize the solution; high-end instruments often use a thermoelectrically cooled nebulizer to minimize introduction of water vapor that cools the plasma. A much larger flow of argon goes through the annulus of the torch, confining the plasma. Total argon consumption of a typical ICP torch is significant, between 10–20 L/ min. Total power consumption of the instrument often ranges between 3–4 kW. Some instruments permit viewing the plasma image through a suitably placed camera to adjust conditions for optimum measurement.

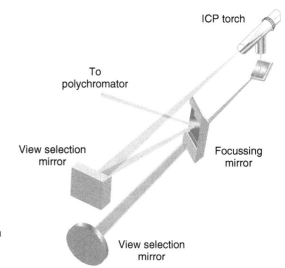

FIGURE 12.11 The optical layout of the Thermo iCAP 6000, a dual-view ICP-OES instrument. The position of the two easily switchable view selection mirrors dictate which view is imaged. (Courtesy of Thermo Fisher Scientific.)

argon or nitrogen purging of this region to extend the ability to view down to the deep UV, as low as 166 nm, such wavelengths would otherwise be completely absorbed if air or oxygen was present. The long wavelength limit is typically > 800 nm. Neither the radial view nor the axial view configuration is superior to the other in all respects. The radial view configuration is more robust, but it observes a more limited volume of the plasma than the axial view, so the LODs are significantly better in the axial view, but the interferences also tend to be greater. In axial view, it becomes particularly difficult to accurately determine a weak emission line from an element present in trace concentrations in the presence of an intense neighboring emission of an element present at larger concentrations. Several top-end instruments permit operating an ICP-OES instrument in either configuration. The optical layout of the Thermo Fisher Scientific iCAP 6000 instrument is shown in Figure 12.11 as an illustration.

Polychromators and Detectors in ICP-OES

An ICP is hot enough such that chemical interference is essentially eliminated. However, one needs to be concerned about interference from spectral overlaps. Aside from background emission from the plasma and plasma gases, there are a multitude of emission lines of each element, increasing the probability of interference. Reliable performance is attained by dispersing devices that provide very high wavelength resolution, usually *Echelle* gratings. ICP-OES instruments boasting a wavelength resolution of 7 pm at 200 nm and capable, e.g., of baseline resolution of the thallium doublet emission lines at 190.856 nm and 190.870 nm, are commercially available.

While early ICP-OES instruments used one or more photomultiplier tubes as detectors, array detectors, or more commonly, two-dimensional (2-D) imaging detectors are exclusively used in present-day instruments. The use of a 2-D imaging detector allows imaging the plasma as a whole, allowing different elements to be determined at different locations in the plasma, such that each can be determined at its optimum temperature. While the use of CCD and back-thinned, back-illuminated CCD (these offer significantly superior sensitivity) imaging detectors are common, at least one manufacturer offers 2-D charge injection detectors (CIDs).

CIDs were originally developed at General Electric; the first CID imaging camera goes back to 1972. CIDs differ from CCDs in that every pixel is isolated from others and every pixel is individually addressable. Saturation of a CCD pixel causes "blooming," i.e., the charge spills over to neighboring pixels. CIDs are greatly resistant to such blooming. Moreover, present-day CIDs allow different integration times on different pixels, allowing the rapid readout of a strong emission line from a major constituent at its corresponding pixel, while allowing more integration time on another pixel corresponding to the weak emission line of another element present in trace concentrations. Current ICP-OES instruments boast LODs below 0.01 μg/L for at least 10 elements, below 0.1 μg/L for another 17 and $\leq$ 1 μg/L for others, for a total of 66 elements (see, e.g., https://static.thermoscientific.com/images/D01567~.pdf). While ICP-MS (see below) can in general provide even better LODs, ICP-MS LODs for low atomic weight elements are often not as good. The measurement of trace silicon in particular, an important measurement in the semiconductor industry, is better conducted by ICP-OES than ICP-MS. Samples containing silicon are usually dissolved in HF, extremely corrosive to metals and glass/quartz. Specialized instruments that utilize sample handling lines entirely built of HF-inert material are available for such analyses.

Ionization and ICP-MS

A plasma is an ionized state of matter. If the temperature is high enough, one or more outer electrons comes off, and the plasma consists of the resulting positive ions and free electrons. The ease with which the first electron comes off is called the first ionization potential (IP) and is typically given in electron volts (1 eV = 1.60×10^{-19} J). Tables of ionization potentials are widely available on the web, see for example http://images.tutorvista.com/content/periodic-classification-elements/periodic-table.jpeg. Alkali metals are easiest to ionize, and Cs being the largest atom among Li-Cs, Cs is the easiest to ionize with an IP of 3.9 eV. The inert gases and highly electronegative elements are the most reluctant to ionize, with IPs of 24.6, 21.6, and 17.4 eV for He, Ne, and F. The Maxwell–Boltzmann distribution represented by Equation 12.1 applies here as well and we can easily estimate the population ratio of the first ionized state to the ground-state atoms for argon. With the Boltzmann constant k being 1.38×10^{-23} J/K, at 10,000 K, kT is 1.38×10^{-19} J, and with an IP of 15.8 eV for Ar and assuming a degeneracy ratio of unity, Equation 12.1 suggests that the ratio of Ar^+ to Ar even at 10,000 K is 10^{-8}. Most other elements are much easier to ionize, however, with IPs of 3.9, 7.5, and 10.4 eV for Cs, Ru, and I, and the ion/atom ratios are 1.1×10^{-2}, 1.7×10^{-4}, and 5.8×10^{-6}, assuming again the degeneracy ratio is unity. Although the ratio may seem small, charged species can be accelerated in an electric field and detected with exquisite sensitivity using an electron multiplier, a channel multiplier or a PMT (after ions are converted to photons by a scintillator) as detector. Such detectors are used in mass spectrometry and are discussed in more detail in Chapter 17, as is the MS portion of ICP-MS instrumentation.

While most commonly a mass spectrometer is used to characterize organic molecules after ionizing them by various means, the ICP serves as a unique ion source for a mass spectrometer. At plasma temperatures, no organic molecules survive but elements present in the sample ionize to various degrees depending on their IP and can be introduced into a mass spectrometer where they are sorted based on their mass to charge ratio and then detected with great sensitivity. There are very few inter-element interferences per se, however, interferences through ionization suppression, very similar to that described in Section 12.3 is common. To wit, if an instrument is calibrated with a standard solution of the element of interest but the sample contains an abundance of an easily ionizable element, e.g., Na, the ionization of the Na will increase the free electron density of the plasma, which in turn will reduce the efficiency with which the analyte element is ionized. Using an internal standard is essential for good quantitation. The best possible standard is an isotopic tracer of the same element that is not naturally present in the sample, e.g., [129]I is commonly used as the isotopic tracer for iodine analysis. In many cases, this is not possible, however. Another element that is not present in the sample is used as the tracer; a tracer with an IP close to the element of interest is preferred. Ideally two tracers, bracketing the element of interest in their IP, should be used.

Both ICP-MS and ICP-OES allow essentially simultaneous measurement of a large number of elements with very high sensitivity. Evans Analytical Group provides a useful comparison for the two techniques of limits of detection in solution and after typical dissolution of a solid sample, see http://www.eaglabs.com/documents/icp-oes-ms-detection-limit-guidance-BR023.pdf.

Laser Ablation ICP-OES/MS

A laser ablation (LA) tool is equivalent to a microprobe for an atomic spectrometrist. The basic principle involves directing a high-energy UV laser pulse on to a (solid) specimen. The deposited energy vaporizes/aerosolizes the sample spot. The released aerosol/vapor is carried directly into an ICP by an inert carrier gas, and elemental analysis is carried out by OES or MS. Commonly neodymium doped yttrium aluminum garnet (Nd-YAG) lasers, optically pumped by a flash lamp, are used in a pulsed mode, and typically the fundamental NIR emission of these lasers is frequency multiplied by a nonlinear optical crystal to 266 or 213 nm. ArF excimer lasers operating at 193 nm are also used.

In most cases, shorter wavelengths provide better results, but shorter wavelength LA systems of equal power are more expensive. Typical pulse duration is in the several nanosecond scale, although femtosecond pulse duration laser ablation systems are also commercially available. The spot diameter can range from 2 μm to $\sim > 1$ mm, with maximum energy deposition of $5 - 15$ J/cm^2. However, the quality of results is not dependent on the energy deposited per se (very high-power infrared lasers that melt the specimen produce very poor results), but on the reproducibility of the laser spot from pulse to pulse and its spatial uniformity. Imaging modes can involve repeatedly ablating the same spot to create an elementally mapped depth profile (depth of the ablation is a function of the laser repetition rate, the scanning speed and the nature of the specimen itself; a layer of 100–500 nm thickness is typically removed per pulse). An entire area can also be mapped in a raster fashion. A visible light image (either in the transmitted or in the reflected light mode, depending on the specimen) is simultaneously obtained by optical microscopy to delineate the area of examination.

A single spot depth profile to modest depths can be generated in 2–3 min while a 25 mm length can be mapped with 50 μm resolution in 20–30 min. See Figure 12.12 for an example of a laser ablated crater. The best case reproducibility in LA-ICP experiments is 2%; more often it ranges up to 8%. The great advantage of LA-ICP elemental measurements is that no sample preparation is necessary. Compared to other elemental mapping techniques such as the electron microprobe or secondary ion mass spectrometry, LA-ICP techniques are more sensitive, quantitatively accurate and reproducible. Suitably uniform solid standards and appropriate calibration for quantitative accuracy are the most difficult aspects of this technique. It is invaluable for the examination of archaeologic, forensic and geologic specimens, and for the examination of the distribution of metals (often originating from catalysts used during manufacturing) in polymers. For a more detailed discussion see Reference 13. As with solutions, in most cases LA-ICP-MS is more sensitive than LA-ICP-OES. However, for several elements, LA-ICP-OES can reach LODs of ≤ 1 μg/g and some find that in cases where it has sufficient sensitivity to make measurements, it provides better quantitative accuracy than LA-ICP-MS (see, e.g., A. Stankova, N. Gilon, L. Dutruch and V. Kanicky, *J. Anal. At. Spectrom.*, **26** (2011) 443).

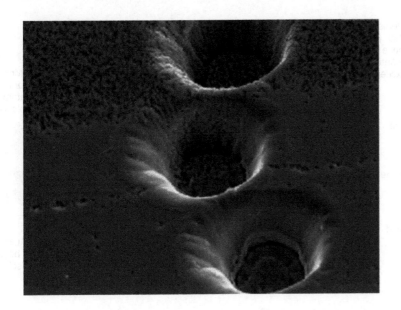

FIGURE 12.12 Laser ablation craters created in a ZnS specimen (mineral Sphalerite). Each laser ablated crater is approximately 30 μm in diameter and 20 μm deep. (Courtesy of Alan E. Konig, US Geological Survey, Denver, CO)

12.7 Atomic Fluorescence Spectrometry

Atomic fluorescence spectrometry (AFS) differs from atomic emission spectrometry only in that the atoms are excited by photons, rather than thermally. However, atoms can be excited efficiently only at the same resonance line where they emit; as such, unlike molecular fluorescence where excitation and emission maxima differ, there is no way to distinguish scattered excitation light from emitted fluorescence; they are at the same wavelength. As such, in atomic fluorescence spectrometry scattered light must be avoided; note that pulsing the excitation source and detecting the fluorescence signal only at that frequency provides some immunity against background radiation from the atom source, but not against scattered radiation. This puts considerable limitations on the atom source.

Normal flames result in poor detection limits. Although the graphite furnace has been used in AFS and record detection limits at the attogram (10^{-18} g) level have been achieved with tunable laser sources, this is largely limited to the research laboratory—no commercial AFS instrument uses electrothermal atomization or a laser excitation source. The air-entrained argon-hydrogen flame (see Section 12.3, Analyte Atom Sources) has high transparency in the UV region and is nearly nonluminous; this has often been used as an atom source. Note that this flame is relatively cool (does not even reach 1000°C) and can effectively atomize only the hydride forming elements when they are introduced into the flame as the hydrides. These hydrides are also efficiently decomposed to the atom by a heated quartz tube. In effect, the air–Ar–H$_2$ flame and the heated quartz tube are the only atomizers that are used in commercially available instruments. The use of AFS is thus limited to the hydride-forming elements and Hg (which is readily generated at room temperature as the atomic vapor and no separate atomizer is needed), but for these elements, AFS can provide exquisite sensitivity.

Because it is so readily generated as a cold atomic vapor, and inexpensive mercury lamps provide an intense excitation source at 253.7 nm, mercury is by far the most common element measured by AFS. Without enrichment, commercial instruments can achieve an LOD of 0.2–1 ng/L; with prior enrichment on a gold trap and thermal desorption, the LOD can be an order of magnitude better. Because a separate flame or hot tube atomizer is not needed, AFS based dedicated mercury analyzers, typically with an option to perform preconcentration on gold, are marketed by several manufacturers.

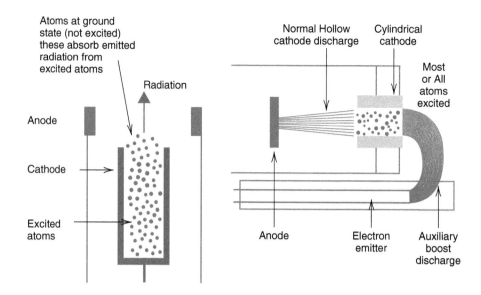

FIGURE 12.13 Hollow-cathode vs. boosted discharge hollow-cathode lamp. Boosted discharge HCL (right), standard hollow-cathode lamp (note closed bottom) on the left. An additional boost current induces a second discharge within the lamp. The atom plume formed above the cathode is re-excited by the secondary discharge, reenergizing the atoms and boosting the emission output 3–5x, compared to a standard HCL. Self-absorption and resulting line broadening are reduced, resulting in a narrower emission profile. (Courtesy of Photron Pty Ltd.)

Due to the high toxicity of this element, most nations have regulations on the maximum permissible levels of mercury on a variety of matrices ranging from drinking water to industrial discharge, and there is much need for low level determination of Hg. Some instruments actually permit both an AAS or an AFS configuration.

General AFS instruments use a changeable light source specific for the element being analyzed. Recall that the fluorescence signal is directly proportional to the exciting light intensity; standard hollow cathode lamps are generally considered unsuitable for AFS. Boosted discharge hollow cathode lamps (Figure 12.13) provide much greater intensity, are more stable than EDLs and hence are the preferred sources. (The use of such lamps in AAS also provides superior results, particularly in the deep UV region where transmission efficiency and detector sensitivity may be poor; for example, in both AAS and AFS, As and Se are measured in this wavelength domain, at 193.7 and 196.0 nm, respectively.)

The emitted fluorescence radiation is collected at 90° to the excitation beam and focused on to the detector. Because line sources are used and background radiation is minimal, the demands on the monochromator placed before the PMT detector are modest, and it serves to take out other lines that may be emitted by the source (especially so for an Hg lamp). Low-resolution high-throughput monochromators or changeable interference filters are commonly used.

Questions

Principles

1. Why do ion lines predominate in spark spectra and atom lines in arc and inductively coupled plasma spectra?

2. (a) Describe the principles of atomic emission, atomic absorption and atomic fluorescence spectrometry.

   (b) Define chemical interference, self absorption and ionization suppressor.

3. Compare flame photometry and flame atomic absorption spectrometry with respect to applicability of instrumentation, sensitivity, and interferences.

4. Why is source modulation employed for atomic absorption spectrometry?

5. Name a continuous and a non-continuous type of atomizer that is used in atomic spectrometry. How do the output signals from a spectrometer differ?

6. What causes the red feather in a reducing nitrous oxide–acetylene flame?

7. (a) Why is an electrothermal atomizer more sensitive than a flame ionizer?

    (b) Why are sources often blanketed with a stream of inert gas?

8. What is the purpose of an internal standard in flame emission methods?

9. Gary M. Hieftje, one of the pioneers in atomic spectrometry, asked the following question in an undergraduate textbook on chemical analysis that he coauthored in 1974: Why would the simultaneous analysis of several elements by means of atomic absorption spectrometry be more difficult than by either atomic fluorescence or flame emission spectrometry? What developments have made possible the (near) simultaneous measurement of several elements by AAS? How would you design an AFS instrument that can measure several elements simultaneously? Why would this not be as sensitive as measuring one element at a time?

PROFESSOR'S FAVORITE PROBLEM

Contributed by Professor Gary M. Hieftje, Indiana University.

10. (a) If the hollow-cathode lamp primarily determines the spectral bandwidth of the detected radiation in AAS, why is a monochromator needed at all?

    (b) For the alkali metals such as Na, K, Li, etc., many of the atomic species present in a flame are ionized. Why do we not choose to measure these ionic emissions and instead measure the atomic emission?

Instrumentation

11. Explain the mechanism of operation of a hollow-cathode lamp.

12. Describe the premix chamber burner. What flames can be used with it?

13. Explain why the radiation source in atomic absorption instruments is usually modulated.

Interferences

14. Lead in seawater was determined by atomic absorption spectrometry. The APDC (ammonium pyrrolidinedithiocarbamate) chelate was extracted into methylisobutyl ketone and the organic solvent was aspirated. A standard and reagent blank were treated in a similar manner. The blank reading was essentially zero. Measurements were made at the 283.3-nm line. An independent determination using anodic-stripping voltammetry revealed the atomic absorption results to be high by nearly 100%. Assuming the anodic-stripping voltammetry results are correct, suggest a reason for the erroneous results and how they might be avoided in future analyses.

15. Why is a high-temperature nitrous oxide–acetylene flame sometimes required in atomic absorption spectrometry?

16. Why is a high concentration of a potassium salt sometimes added to standards and samples in flame absorption or emission methods?

17. Chemical interferences are more prevalent in "cool" flames such as air–propane, but this flame is preferred for the determination of the alkali metals. Suggest why.

18. An analyst notes that a 1-ppm solution of sodium gives a flame emission signal of 110, while the same solution containing also 20-ppm potassium gives a reading of 125. It was determined that a 20-ppm solution of potassium exhibited no blank reading at the sodium emission wavelength. Explain the results.

19. Mercury can be determined by cold vapor AAS. How would you distinguish between inorganic mercury and organically bound mercury?

20. The hydrides of certain elements can be generated by reacting with $NaBH_4$, and flushed into a flame or other atomizer. What is the advantage of this method?

21. (a) What are the advantages of plasma sources compared with flame sources for emission spectrometry?

    (b) Why are ionization interferences less severe in induction coupled plasma (ICP) than in flame emission spectroscopy?

22. Describe methods for introducing a sample into an ICP.

23. Laser ablation is an effective tool for probing micro areas of solid samples. How does it work?

24. Atomic fluorescence spectrometry is an attractive alternative to atomic absorption spectrometry for certain elements. For which elements is it best suited? Why?

Problems

Sensitivity

25. In AAS, sensitivity is often defined as that analyte concentration that results in 1% absorption of the incident light. A 15-ppm solution of lead gives an atomic absorption signal of 8.0% absorption. What is the atomic absorption sensitivity?

26. Silver exhibits an atomic absorption sensitivity of 0.050 ppm under a given set of conditions. What would be the expected absorption for a 0.70-ppm solution?

Boltzmann Distribution

27. The transition for the cadmium 228.8-nm line is a $^1S_0 - {}^1P_1$ transition. Calculate the ratio of N_e/N_0 in an air–acetylene flame. What percent of the atoms is in the excited state? The velocity of light is 3.00×10^{10} cm/s, Planck's constant is 6.62×10^{-27} erg-s, and the Boltzmann constant is 1.380×10^{-16} erg K^{-1}.

Quantitative Calculations

28. Calcium in a sample solution is determined by atomic absorption spectrometry. A stock solution of calcium is prepared by dissolving 1.834 g $CaCl_2, 2H_2O$ in water and diluting to 1 L. This is diluted 1:10. Working standards are prepared by diluting the second solution, respectively, 1:20, 1:10, and 1:5. The sample is diluted 1:25. Strontium chloride is added to all solutions before dilution, sufficient to give 1% (wt/vol) to avoid phosphate interference. A blank is prepared, to give 1% $SrCl_2$. Absorbance signals are acquired by a data system as the solutions are aspirated into an air–acetylene flame, as follows: blank, 1.5 units; standards, 10.6, 20.1, and 38.5 units; sample, 29.6 units. What is the concentration of calcium in the sample in parts per million?

29. Lithium in the blood serum of a manic-depressive patient treated with lithium carbonate is determined by flame emission spectrometry, using a standard additions calibration. One hundred microliters of serum diluted to 1 mL gives an emission signal of 6.7 units. A similar solution to which 10 μL of a 0.010 M solution of $LiNO_3$ has been added gives a signal of 14.6 units. Assuming linearity between the emission signal and the lithium concentration, what is the concentration of lithium in the serum, in parts per million?

30. Chloride in a water sample is determined indirectly by atomic absorption spectrometry by precipitating it as AgCl with a measured amount of $AgNO_3$ in excess, filtering, and measuring the concentration of silver remaining in the filtrate. Ten-milliliter aliquots each of the sample and a 100-ppm chloride standard are added to separate dry 100-mL Erlenmeyer flasks. Twenty-five milliliters of a silver nitrate solution is added to each with a pipet. After allowing time for the precipitate to form, the mixtures are transferred partially to dry centrifuge tubes and are centrifuged. Each filtrate is aspirated for atomic absorption measurement of silver concentration. A blank is similarly treated in which 10-mL deionized distilled water is substituted for the sample. If the following absorbance signals are recorded for each solution, what is the concentration of chloride in the water sample?

    Blank: 12.8 units, Standard: 5.7 units, Sample: 6.8 units

Standard Addition

PROFESSOR'S FAVORITE PROBLEM

Contributed by Professor Wen-Yee Lee, The University of El Paso

Total # of M&Ms added	0	10	20	30	40
Total mass (g)	52.7	58.1	63.6	68.5	74.4

31. This is a generic standard additions problem, to illustrate the concept here for atomic spectrometry. You have a magical paper bag that contains some M&Ms. The empty bag has a mass of 10.0 g and you cannot see into it. Your task is to find out the number of M&Ms originally placed in the bag. To help you figure out the answer, a known number of M&Ms are added into the bag and the whole bag is weighed. The number of M&Ms and the mass of the whole bag after each addition are included in the following table. Find the number of M&Ms originally present in the bag.

Multiple Choice Questions

1. The principle of atomic absorption spectroscopy is:

    (a) Radiation is absorbed by non-excited atoms in vapor state and are excited to higher states

    (b) Medium absorbs radiation and transmitted radiation is measured.

    (c) Color is measured.

    (d) Color is simply observed.

2. The following is the function of the flame or emission system in atomic absorption spectroscopy

    (a) Splitting the beam into two

    (b) Breaking the steady light into pulsating light

    (c) Filtering unwanted components

    (d) Reducing the sample into atomic state

3. Atomic absorption spectroscopy is also called as absorption flame photometry. State true or false.

4. The following is not used as a fuel in flame photometry

    (a) Acetylene   (b) Propane

    (c) Hydrogen   (d) Camphor oil

5. In flame emission photometers, the measurement of the following is used for qualitative analysis:

    (a) Color   (b) Intensity

    (c) Velocity   (d) Frequency

6. Which among the following is not an advantage of Laminar flow burner used in flame photometry?

    (a) Noiseless

    (b) Stable flame for analysis

    (c) Efficient atomization of sample

    (d) Sample containing two or more solvents can be burned efficiently

7. In flame emission photometers the following is not used as a detector

    (a) Photronic cell   (b) Photovoltaic cell

    (c) Photoemissive tube   (d) Chromatogram

8. The correct statements for hollow cathode lamp (HCL) from the following are
   P. HCL is suitable for atomic absorption spectroscopy (AAS).
   Q. lines emitted from HCL are very narrow.
   R. the hardening of lamp makes it unsuitable for AAS.
   S. transition elements used in the lamps have short life.

    (a) P, Q and R   (b) Q, R and S

    (c) R, S and P   (d) S, P and Q

9. How does continuous wedge filter differ from normal interference filter used in absorption spectroscopy?

    (a) It permits continuous selection of different wavelength.

    (b) It allows a narrow band of wavelengths to pass.

    (c) It has two semi-transparent layers of silver.

    (d) Space layer is made of a substance having low refractive index.

10. The following can be used as the layer of dielectric in interference filters in absorption spectroscopy?

    (a) Graphite (b) MgF_2

    (c) Fe (d) $AgNO_3$

11. The gas commonly used in generating plasma in inductively coupled plasma-atomic emission spectroscopy (ICP-AES) is

    (a) argon. (b) carbon dioxide.

    (c) nitrous oxide. (d) hydrogen.

12. The following is not an application of flame emission photometers?

    (a) Analysis of biological fluids

    (b) Determination of sodium, potassium in soil.

    (c) Determination of metals such as Mn and Cu

    (d) Analysis of complex mixtures

Recommended References

General

1. V. Thomsen, "A Timeline of Atomic Spectroscopy," *Spectroscopy*, **21**(10), October 2006, p. 32. A short history of the experimental and theoretical development of atomic spectroscopy for elemental spectrochemical analysis. www.spectroscopyonline.com.
   Contains pictures of pioneers: Isaac Newton, Gustav Kirchoff, Robert Bunsen, Anders Ångström, Henry Rowland, Johann Balmer, Wilhelm Röntgen, Hans Geiger, Niels Bohr, Henry Moseley, Louis de Broglie, Werner Heisenberg, Lise Meitner, Alan Walsh, and others.
2. L. Ebdon and E. H. Evans, eds., *An Introduction to Analytical Spectrometry*, 2nd. ed. Chichester: Wiley, 1998.
3. J. W. Robinson, *Atomic Spectroscopy*, 2nd. ed. New York: Marcel Dekker, 1996.
4. G. M. Hieftje, "Atomic Spectroscopy—A Primer," http://www.spectroscopynow.com/coi/cda/detail.cda?id=1905&type=EducationFeature&chId=1&page=1 (search G. M. Hieftje).

Flame Emission and Atomic Absorption Spectrometry

5. J. A. Dean, *Flame Photometry*. New York: McGraw-Hill, 1960. A classic on the fundamentals.
6. G. D. Christian and F. J. Feldman, *Atomic Absorption Spectroscopy. Applications in Agriculture, Biology, and Medicine*. New York: Wiley-Interscience, 1970. Describes sample preparation procedures.
7. G. D. Christian, "Medicine, Trace Elements, and Atomic Absorption Spectroscopy," *Anal. Chem.*, **41**(1) (1969) 24A. Lists levels of trace elements in biological samples.
8. B. Welz and M. Sperling, *Atomic Absorption Spectroscopy*, 3rd ed., Wiley, 1999.
9. B. Welz, H. Becker-Ross, S. Florek, and U. Heitmann. *High Resolution Continuum Source AAS: The Better Way to Do Atomic Absorption Spectrometry*. Wiley, 2005.
10. M. T. C. de Loos-Vollebregt. *Background Correction Methods in Atomic Absorption Spectroscopy. Wiley On-line Encyclopedia of Analytical Chemistry*. Wiley, 2006. http://onlinelibrary.wiley.com/doi/10.1002/9780470027318.a5104/abstract. Only the abstract is freely accessible; institutional access is required for the full article. Reference 8 also covers this topic.
11. X. Hou and B. T. Jones. *Inductively Coupled Plasma/Optical Emission Spectrometry*. 2000. http://www.wfu.edu/chemistry/courses/jonesbt/334/icpreprint.pdf.
12. Cai, Y. *Atomic Fluorescence in Environmental Analysis*. 2000. http://www2.fiu.edu/~cai/index_files/Cai_Encyclopedia%20Anal%20Chem.pdf.
13. A. E. Koenig. *Laser Ablation ICP-MS: Performance, Problems, Pitfalls and Potential*. http://www.armi.com/News/Presentations/2005/02%20-%20ICP-MS%20analysis%20and%20where%20it%20fits%20in%20the%20Analytical%20Lab_files/frame.htm#slide0112.htm.

Chapter 13

SAMPLE PREPARATION: SOLVENT AND SOLID-PHASE EXTRACTION

KEY THINGS TO LEARN FROM THIS CHAPTER

- Distribution coefficient, distribution ratio (key Equations: 13.1, 13.3, 13.8)
- Percent extracted (key Equation: 13.10)
- Solvent extraction of metal ions–complexes, chelates
- Accelerated and microwave-assisted solvent extraction
- Solid-phase extraction
- Solid-phase microextraction

The next chapter introduces chromatographic techniques for analyzing complex samples, whereby multiple analytes are separated on a column and detected as they emerge from the column. But very often, samples need to be "cleaned up" prior to introduction into the chromatographic column. The techniques of solvent extraction and solid-phase extraction and related techniques are very useful for isolating analytes from complex sample matrices prior to chromatographic analysis. Solvent extraction is also useful for spectrophotometric determination.

Solvent extraction involves the distribution of a solute between two immiscible liquid phases. This technique is extremely useful for very rapid and "clean" separations of both organic and inorganic substances. In this chapter, we discuss the distribution of substances between two phases and how this can be used to form analytical separations. The solvent extraction of metal ions into organic solvents is described.

Solid-phase extraction is a technique in which hydrophobic functional groups are bonded to solid particle surfaces and act as the extracting phase. They reduce the need for large volumes of organic solvents.

13.1 Distribution Coefficient

A solute S will distribute itself between two phases (after shaking and allowing the phases to separate) and, within limits, the ratio of the concentrations of the solute in the two phases will be a constant:

$$K_D = \frac{[S]_1}{[S]_2} \tag{13.1}$$

where K_D is the **distribution coefficient** and the subscripts represent solvent 1 (e.g., an organic solvent) and solvent 2 (e.g., water). If the distribution coefficient is large, the solute will tend to be quantitatively partitioned in solvent 1.

The apparatus used for solvent extraction is the **separatory funnel**, illustrated in Figure 13.1. Most often, a solute is extracted from an aqueous solution into an immiscible organic solvent. After the mixture is shaken for about a minute, the phases are allowed to separate and the bottom layer (the denser solvent) is drawn off in a completion of the separation.

Many substances are partially ionized in the aqueous layer as weak acids. The extraction now becomes dependent on the pH of the solution. Consider, for example, the extraction of benzoic acid from an aqueous solution into either. Benzoic acid (HBz)

Neutral organics distribute from water into organic solvents. "Like dissolves like."

FIGURE 13.1 Separatory funnel.

is a weak acid in water with a particular ionization constant K_a (given by Equation 13.4). The distribution coefficient is given by

$$K_D = \frac{[HBz]_e}{[HBz]_a} \qquad (13.2)$$

where e represents ether and a represents the aqueous solvent. However, part of the benzoic acid in the aqueous layer will exist as Bz^- that does not transfer into ether, depending on the magnitude of K_a and on the pH of the aqueous layer; hence, quantitative separation may not be achieved when a significant fraction exists as Bz^-.

13.2 Distribution Ratio

It is more meaningful to describe a different term, the **distribution ratio**, which is the ratio of the concentrations of *all* the species of the solute in each phase. In this example, it is given by

$$D = \frac{[HBz]_e}{[HBz]_a + [Bz^-]_a} \qquad (13.3)$$

We can readily derive the relationship between D and K_D from the equilibria involved. The acidity constant K_a for the ionization of the acid in the aqueous phase is given by

$$K_a = \frac{[H^+]_a[Bz^-]_a}{[HBz]_a} \qquad (13.4)$$

Hence,

$$[Bz^-]_a = \frac{K_a[HBz]_a}{[H^+]_a} \qquad (13.5)$$

From Equation 13.2,

$$[HBz]_e = K_D[HBz]_a \qquad (13.6)$$

Substitution of Equations 13.5 and 13.6 into Equation 13.3 gives

$$D = \frac{K_D[HBz]_a}{[HBz]_a + K_a[HBz]_a/[H^+]_a} \qquad (13.7)$$

$$D = \frac{K_D}{1 + K_a/[H^+]_a} \qquad (13.8)$$

This equation predicts that when $[H^+]_a \gg K_a, D$ is nearly equal to K_D, and if K_D is large, the benzoic acid will be quantitatively extracted into the ether layer; D is maximum under these conditions. If, on the other hand, $[H^+] \ll K_a$, then D reduces to $K_D[H^+]_a/K_a$, which will be small, and the benzoic acid will remain in the aqueous layer. That is, in alkaline solution, the benzoic acid is ionized and cannot be extracted, while in acid solution, it is largely undissociated. These conclusions are what we would intuitively expect from inspection of the chemical equilibria.

In solvent extraction, the separation efficiency is usually independent of the concentration.

Equation 13.8, like Equation 13.1, predicts that the *extraction efficiency will be independent of the original concentration of the solute.* This is one of the attractive features of solvent extraction; it is applicable to tracer (e.g., radioactive) levels and to macrolevels alike, a condition that applies only so long as the solubility of the solute in one of the phrases is not exceeded and there are no side reactions such as dimerization of the extracted solute.

Of course, if the hydrogen ion concentration changes, the extraction efficiency (D) will change. In this example, the hydrogen ion concentration will increase with increasing benzoic acid concentration, unless an acid–base buffer is added to maintain the hydrogen ion concentration constant (see Chapter 6 for a discussion of buffers).

In deriving Equation 13.8, we actually neglected to include in the numerator of Equation 13.3 a term for a portion of the benzoic acid that exists as the dimer in the organic phase. The extent of dimerization tends to increase with increased concentration, and by Le Châtelier's principle, this will cause the equilibrium to shift in favor of the organic phase with increased concentration. So, in cases such as this, the efficiency of extraction will actually increase at higher concentrations. As an exercise, derivation of the more complete equation is presented in Problem 12.

13.3 Percent Extracted

The distribution ratio D is a constant independent of the volume ratio. However, the fraction of the solute extracted will depend on the volume ratio of the two solvents. If a larger volume of organic solvent is used, more solute must dissolve in this layer to keep the concentration ratio constant and to satisfy the distribution ratio.

The fraction of solute extracted is equal to the millimoles of solute in the organic layer divided by the total number of millimoles of solute. The millimoles are given by the molarity times the milliliters. Thus, the percent extracted it given by

$$\%E = \frac{[S]_o V_o}{[S]_o V_o + [S]_a V_a} \times 100\% \qquad (13.9)$$

where V_o and V_a are the volumes of the organic and aqueous phases, respectively. It can be shown from this equation (see Problem 11) that the percent extracted is related to the distribution ratio by

$$\boxed{\%E = \frac{100D}{D + (V_a/V_o)}} \text{ or, } \boxed{D = \frac{(V_a/V_o) \cdot E}{100 - E}} \qquad (13.10)$$

If $V_a = V_o$, then

Extraction will be quantitative (99.9%) for D values of 1000.

$$\boxed{\%E = \frac{100D}{D + 1}} \qquad (13.11)$$

In the case of equal volumes, the solute can be considered quantitatively retained in the aqueous phase if D is less than 0.001. It is essentially quantitatively extracted if D is greater than 1000. The percent extracted changes only from 99.5 to 99.9% when D is increased from 200 to 1000.

Example 13.1

Twenty milliliters of an aqueous solution of 0.10 M butyric acid is shaken with 10 mL ether. After the layers are separated, it is determined by titration that 0.5 mmol butyric acid remains in the aqueous layer. What is the distribution ratio, and what is the percent extracted?

Solution

We started with 2.0 mmol butyric acid, and so 1.5 mmol was extracted. The concentration in the ether layer is 1.5 mmol/10 mL = 0.15 M. The concentration in the aqueous layer is 0.5 mmol/20 mL = 0.025 M. Therefore,

$$D = \frac{0.15}{0.025} = 6.0$$

Since 1.5 mmol was extracted, the percent extracted is $(1.5/2.0) \times 100\% = 75\%$. Or

$$\%E = \frac{100 \times 6.0}{6.0 + (20/10)} = 75\%$$

Equation 13.10 shows that the fraction extracted can be increased by decreasing the ratio of V_a/V_o, for example, by increasing the organic phase volume. However, a more efficient way of increasing the amount extracted using the same volume of organic solvent is to perform successive extractions with smaller individual volumes of organic solvent. For example, with a D of 10 and $V_a/V_o = 1$, the percent extracted is about 91%. Decreasing V_a/V_o to 0.5 (doubling V_o) would result in an increase of $\%E$ to 95%. But performing two successive extractions with $V_a/V_o = 1$ would give an overall extraction of 99%.

13.4 Types of Solvent Extraction

1. **Batch Extraction.** In this method, solute dissolved in solvent is extracted to other solvent which is immiscible, and both form different layers after shaking until an equilibrium is attained. This is simplest method of extraction, and is mostly used for analytical separations. Extent of extraction's completeness depends upon the times of extraction carried out. For best result, large number of extractions are done with small volume of solvent. Apparatus used includes a separating funnel.

It can be expressed as 'W' gram of solute present in 'V' mL of phase I was extracted with 'S' mL of phase II which is immiscible. W_1 is amount unextracted solute remaining in phase I, so,

$$\text{Phase I concentration, } C_1 = \frac{W_1}{V} \text{g/mL} \tag{13.12}$$

$$\text{Phase II concentration, } C_2 = \left(\frac{W - W_1}{S}\right) \text{g/mL} \tag{13.13}$$

$$\text{Distribution ratio, } D = C_2/C_1$$

Using Eq. (13.12) and (13.13)

$$D = \frac{(W - W_1)/S}{(W_1)/V}$$

$$W_1 = W\left[\frac{V}{DS + V}\right] \tag{13.14}$$

as second round, extraction is carried out with 'S' mL, then extracted solute 'W_2' is

$$W_2 = W_1\left[\frac{V}{DS + V}\right]$$
$$= W\left[\frac{V}{DS + V}\right]\left[\frac{V}{DS + V}\right] = W\left[\frac{V}{DS + V}\right]^2$$

So, therefore,

$$W_n = W\left[\frac{V}{DS + V}\right]^n \tag{13.15}$$

So, for best results, (n) should be large and (S) should be small.

This method is best in cases where value of D is large.

A large number of batch extractions are required when the distribution ratio is small, this method is called continuous extraction. Although efficiency in this method depends on various factors including viscosity, relative volumes of the phases and the D values. Normally a large area of contact will result in improved efficiency. The volume required to decrease the extracted amount to half of its original volume is called half extraction volume. Using this volume, the distribution factor can be obtained as mathematically,

$$K = \frac{0.693\,W}{V} \tag{13.16}$$

where, K is ratio of concentration of solute in extraction solvent to concertation of original solution, V is the half extraction volume and W is the volume of the original solution.

2. **Continuous Counter Current Extraction.** It is an improved method for separating analytes with similar distribution ratios. This method utilizes a serial extraction of both the same along with the extracting phases.

3. **Discontinuous Method.** In this method, the material is first treated with fresh solvent and extracted. After recovering the solvent layer, fresh solvent is added, extracted and this process is repeated many times. Although this method offers simple and robust apparatus construction, but on the other hand suffers from the limited capacity and discontinuous product output.

 # Professor's Favorite Example

⌐Contributed by Professor
Galina Talanova, Howard
University ⌐

We can illustrate the value of multiple extractions from equations for the fraction of analyte that will be extracted into the organic phase upon n consecutive extractions, or even better (due to simplicity of the final mathematical expression), the fraction of analyte remaining in the aqueous phase after n consecutive extractions (see the text **website** for the derivations):

$$E \text{ (fraction extracted)} = DV_o\left[\frac{1}{DV_o + V_a} + \frac{V_a}{(DV_o + V_a)^2} + \frac{V_a^2}{(DV_o + V_a)^3} + \cdots \right.$$
$$\left. + \frac{V_a^{(n-1)}}{(DV_o + V_a)^n}\right] \tag{13.17}$$

or

$$\text{Fraction remaining} = \left[\frac{V_a}{(DV_o + V_a)}\right]^n \tag{13.18}$$

where n is the number of consecutive extractions, D is the distribution ratio, and V_a and V_o are the volumes of the aqueous and organic phases, respectively.

Example 13.2

Assume that in extraction from water into toluene, analyte A has the distribution ratio $D = 10$. A 20-mL portion of an aqueous solution of A is extracted with toluene. Which of the following procedures will result in the most efficient removal of A from the aqueous phase into toluene?

(a) one extraction with 40 mL of toluene;

(b) two extractions with 20 mL of toluene each;

(c) four extractions with 10 mL of toluene each?

Solution

(a) Fraction remaining $= [V_a/(DV_o + V_a)]^n = [20/(10 \times 40 + 20)]^1 = 0.048 = 4.8\%$; Extracted approximately 95% of A.

(b) Fraction remaining $= [20/(10 \times 20 + 20)]^2 = 0.0083 = 0.83\%$
Extracted approximately 99% of A.

(c) Fraction remaining $= [20/(20/10 \times 20 + 20)]^4 = 7.7 \times 10^{-4} = 0.077\%$
Extracted approximately 100% of A.

Conclusion: Several extractions using small portions of organic solvent allows one to separate the analyte more efficiently than one large-portion extraction.

Solvent Extraction of Metals

To extract a metal ion into an organic solvent, its charge must be neutralized, and it must be associated with an organic agent.

Solvent extraction has one of its most important applications in the separation of metal cations. In this technique, the metal ion, through appropriate chemistry, distributes from an aqueous phase into a water-immiscible organic phase. Solvent extraction of metal ions is useful for removing them from an interfering matrix, or for selectively (with the right chemistry) separating one or a group of metals from others. The technique is widely used for the spectrophotometric determination of metal ions since the reagents used to accomplish the extraction often form colored complexes with the metal ion. It is also used in flame atomic absorption spectrophotometry for introducing the sample in a nonaqueous solvent into the flame for enhanced sensitivity, and removal of matrix effects.

The separation can be accomplished in several ways. You have noted above that the uncharged organic molecules tend to dissolve in the organic layer while the charged anion from the ionized molecules remains in the polar aqueous layer. This is an example of "like dissolves like." Metal ions do not tend to dissolve appreciably in the organic layer. For them to become soluble, their charge must be neutralized and something must be added to make them organic-like. There are two principal ways of doing this.

Extraction of Ion – Association Complexes

In one method, the metal ion is incorporated into a bulky molecule and then associates with another ion of the opposite charge to form an **ion pair**, or the metal ion associates with another ion of great size (organic-like). For example, it is well known that iron(III) can be quantitatively extracted from hydrochloric acid medium into diethyl ether. The mechanism is not completely understood, but evidence exists that the chloro complex of the iron is coordinated with the oxygen atom of the solvent (the solvent displaces the coordinated water), and this ion associates with a solvent molecule that is coordinated with a proton:

$$\{(C_2H_5)_2O : H^+, FeCl_4[(C_2H_5)_2O]_2^-\}$$

Similarly, the uranyl ion UO_2^{2+} is extracted from aqueous nitrate solution into isobutanol by associating with two nitrate ions $(UO_2^{2+}, 2NO_3^-)$, with the uranium probably being solvated by the solvent to make it solvent-like. Permanganate forms an ion pair with tetraphenylarsonium ion $[(C_6H_5)_4As^+, MnO_4^-]$, which makes it organic-like, and it is extracted into methylene chloride. There are numerous other examples of ion-association extractions. Unsightly foam often appears in the shoreline of natural waters, sometimes because of the presence of natural surfactants but also often because of contamination from man-made detergents. These are mostly anionic, and anionic surfactants can be extracted into an organic solvent such as chloroform as the ion pair with the cationic dye methylene blue.

Halogenated organic solvents have often been used in the past as extraction solvents of choice. There is considerable concern about their toxicity, however, and in most laboratories the use of chlorinated organic solvents are being phased out. These solvents tend to be more polar than hydrocarbon solvents and can sometimes be replaced by water-insoluble ionic liquids.

Extraction of Metal Chelates

The most widely used method of extracting metal ions is formation of a chelate molecule with an organic chelating agent.

As mentioned in Chapter 8, a chelating agent contains two or more complexing groups. Many of these reagents form colored chelates with metal ions and form the basis of spectrophotometric methods for determining the metals. The chelates are often insoluble in water and will precipitate. They are, however, usually soluble in organic solvents such as methylene chloride. Many of the organic precipitating agents listed in Chapter 9 are used as extracting agents.

Extraction Process for Metal Chelates

Most chelating agents are weak acids that ionize in water; the ionizable proton is displaced by the metal ion when the chelate is formed, and the charge on the organic compound neutralizes the charge on the metal ion. An example is *diphenylthiocarbazone (dithizone)*, which forms a chelate with lead ion:

(Courtesy Professor Galina Talanova, Howard University)

The usual practice is to add the chelating agent, HR, to the organic phase. It distributes between the two phases, and in the aqueous phase it dissociates as a weak acid.

The metal ion, M^{n+}, reacts with nR^- to form the chelate MR_n, which then distributes into the organic phase. The distribution ratio is given by the ratio of the metal chelate concentration in the organic phase to the metal ion concentration in the aqueous phase. The following equation can be derived:

$$D = \frac{[MR_n]_o}{[M^{n+}]_a} = K\frac{[HR]_o^n}{[H^+]_a^n} \tag{13.19}$$

where K is a constant that includes K_a of HR, K_f of MR_n, and K_D of HR and MR_n. Note that the distribution ratio is independent of the concentration of the metal ion, provided the solubility of the metal chelate in the organic phase is not exceeded. HR is often in large excess and is considered constant. The extraction efficiency can be affected only by changing the pH or the reagent concentration. A 10-fold increase in the reagent concentration will increase the extraction efficiency the same as an increase in the pH of one unit (10-fold decrease in $[H^+]$). Each effect is greater as n becomes greater. By using a high concentration of reagent, extraction can be performed in more acid solution.

Chelates of different metals extract at different pH values. Some can be extracted over a broad range of pH, whereas others may only be extracted from alkaline solution. Once a metal has been extracted into an organic solvent and the solvent separated, it may be extracted back into a aqueous layer, if desired, at an appropriate lower pH. Such a process is often referred to as back extraction. By appropriate adjustment of pH, selectivity can be achieved in the extraction. Also, judicious use of masking agents, complexing agents that prevent one metal ion from reacting with the chelating agent, can enhance the selectivity.

For reasons of toxicity and disposal, most organic solvent extraction methods for metals have been replaced by solid-phase resins bearing chelating functionalities. Alternatively, ICP-based methods are used; these have high sensitivities and few interferences and do not require extraction, as described in Chapter 12. The principles of solvent extraction, especially sequential extraction, are key to understanding how chromatography (next chapter) works, essentially by the same continuous sequential partitioning process.

13.5 Accelerated and Microwave-Assisted Extraction

Accelerated solvent extraction is a technique for the efficient extraction of analytes from a solid sample matrix into a solvent. Accelerated solvent extraction instrument is depicted in Figure 13.2. The sample and solvent are placed in a closed vessel and heated to 50 to 200°C. The high pressure allows heating above the boiling point, and the high temperature accelerates the dissolution of analytes in the solvent. Both time of extraction and the volume of solvent needed are greatly reduced over atmospheric extraction.

In **microwave-assisted extraction** (MAE), the solvent is heated by microwave energy. The analyte compounds are again partitioned from the sample matrix into the solvent. This approach is an extension of closed-vessel acid digestion described in Chapter 2. A closed vessel containing the sample and solvent is placed in a microwave oven similar to the one described in Figure 2.23. The kinetics of extraction is affected by the temperature and the choice of solvent or solvent mixture. Atmospheric heating for extraction is limited to the boiling point of the solvent. Closed-vessel temperatures at 175 psig typically reach on the order of 150°C, compared with boiling points of 50 to 80°C for commonly used solvents. Solvent mixtures may be used so long as one of them absorbs microwave energy. Some solvents are microwave transparent, for example, hexane, and do not heat, but a mixture of hexane and acetone heats rapidly.

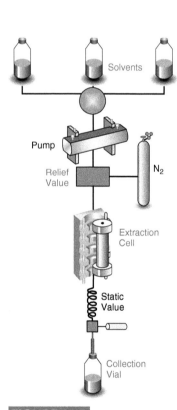

FIGURE 13.2 Schematic depiction of operation of an Accelerated Solvent Extraction Instrument. (Courtesy Thermo Fisher Inc.) Typical temporal sequence: Load cell and fill solvent (0.5–1 min), heat and pressurize (5 min), static extraction (5 min), may cycle between this and previous step, flush with fresh solvent (0.5 min), purge with nitrogen (1–2 min)–extract is ready in 12–14 min. (Courtesy Thermo Fisher Scientific Inc.)

The closed vessels must be inert to solvents and be microwave transparent. The body is made of polyetherimide (PEI), with a perfluoroalkoxy fluorocarbon (PFA, Teflon) liner. Several sample vessels may be placed in the oven at the same time for multiple extractions.

Microwave extractions may also be performed at atmospheric pressure, without the need for pressurized vessels (see Reference 9). Heating and cooling cycles are utilized to prevent boiling of the solvent. This technique also reduces extraction times substantially. For information on commercial MAE systems, see www.cem.com.

13.6 Solid-Phase Extraction

Liquid–liquid extractions are very useful but have certain limitations. The extracting solvents are limited to those that are water immiscible (for aqueous samples). Emulsions tend to form when the solvents are shaken, and relatively large volumes of solvents are used that generate a substantial waste disposal problem. The operations are often manually performed and may require a back extraction. Many solvents used for extraction in yesteryears are presently considered hazardous.

Many of these difficulties are avoided by the use of **solid-phase extraction** (SPE), which has become a widely used technique for sample cleanup and concentration prior to chromatographic analysis (next chapter) in particular. In this technique, a particular type of organic functional group is chemically bonded to a solid surface, for example, powdered silica. A common example is the bonding of C_{18} chains on silica, with particle sizes on the order of 40 μm. The bonded hydrocarbon creates a virtual liquid phase, into which hydrophobic organic analytes in an aqueous sample can partition and be extracted. There are a variety of different types of phases commercially available with different polarities. The same solid phases used in high-performance liquid chromatography (Chapter 16) are used for solid-phase extraction except in a larger particle size.

The powdered phase is generally placed in a small cartridge, similar to a plastic syringe. Sample is placed in the cartridge and forced through by means of a plunger (positive pressure) or a vacuum (negative pressure), or by centrifugation (see Figure 13.3). Trace organic molecules can be extracted, preconcentrated on the column, and separated away from the sample matrix. Then they can be eluted with a solvent such as methanol and finally analyzed, for example, by chromatography (Chapters 14–16). They may be further concentrated prior to analysis by evaporating the solvent.

The nature of the extracting phase, particularly the type of bonded functional group, can be varied to allow extraction of different classes of compounds. Figure 13.4 illustrates bonded phases based on van der Waals forces, hydrogen bonding (dipolar attraction), and electrostatic attraction.

When silica particles are bonded with a hydrophobic phase, they become "waterproof" and must be conditioned in order to interact with aqueous samples. This is accomplished by passing methanol or a similar solvent through the sorbent bed. The solvent penetrates into the bonded layer and permits water molecules and analyte to diffuse into the bonded phase. After conditioning, water is passed to remove the excess solvent prior to adding the sample. Polystyrenedivinylbenzene or other polymer-based supports are also commonly used, especially for ion exchanger based SPE phases. Chelating groups (iminodiacetate, 8-hydroxyquinoline) based SPEs are widely used for online or off-line extraction and preconcentration of trace metals, e.g., from seawater.

Figure 13.5 illustrates a typical sequence in a solid-phase extraction. Following conditioning, the analyte and other sample constituents are adsorbed on the sorbent

In solid-phase extraction, the bonded C_{18} chains take the place of the organic solvent.

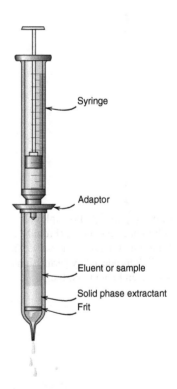

Syringe

Adaptor

Eluent or sample

Solid phase extractant
Frit

FIGURE 13.3 Solid-phase cartridge and syringe for positive pressure elution.

Silica base

NH₂   SO₃⁻

↔ van der Waals   Bonded functional
forces   group (C8)

Silica base

↔ Dipolar attraction or hydrogen
bonding

Silica base

SO₃⁻ → NH₃⁺   OH   CO₂H

↔ Electrostatic
attraction

CO₂⁻   N⁺   OCON(CH₃)₂

FIGURE 13.4 Solid-phase extractants utilizing nonpolar, polar, and electrostatic interactions. (Adapted from N. Simpson, *Am. Lab.*, August, 1992, p. 37. Reproduced by permission of American Laboratory, Inc.)

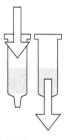

CONDITIONING
Conditioning the sorbent prior to sample
application ensures reproducible retention
of the compound of interest (the isolate).

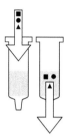

RETENTION
■ Adsorbed isolate
● Undesired matrix constituents
▲ Other undesired matrix components

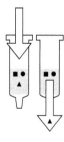

RINSE
▲ Rinse the columns to remove other
undesired matrix components

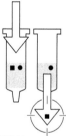

ELUTION
■ Purified and concentrated isolate ready
for analysis
● Undesired components remain

FIGURE 13.5 Principles of solid-phase extraction. (From N. Simpson, *Am. Lab.*, August, 1992, p. 37. Reproduced by permission of American Laboratory, Inc.)

extraction bed. A rinsing step removes some of the undesired constituents, while elution removes the desired analyte, perhaps leaving other constituents behind, depending on the relative strengths of interaction with the solid phase or solubility in the eluting solvent. Such a procedure is used for the determination of organic compounds in drinking water in the official Environmental Protection Agency (EPA) method.

SPE Cartridges

A good rule of thumb is that a typical SPE packing can retain approximately 5% of its mass without significant breakthrough (e.g., 500 mg of packing can be expected to provide capacity for ~25mg of retained compounds).

The SPE sorbent is prepackaged in polypropylene syringe barrels, with typically 500 mg of packing in 3- or 5-mL syringe barrels. Smaller 1-mL syringes packed with 100 mg are becoming more popular because of reduced sample and solvent requirement and faster cleanup times, and even smaller packed beds down to 10 mg are available. These smaller packings, of course, have smaller capacity. Larger ones may be required for preparing large volumes of environmental samples such as polluted water that has large amounts of contaminants to be removed.

The SPE cartridges are used for the isolation and concentration of drugs from biological samples and are typically processed in batches of 12 to 24 using vacuum manifolds (Figure 13.6). There are automated liquid handling systems to improve the

FIGURE 13.6 16-Port vacuum manifold for use with solid-phase extraction tubes. (Courtesy of Alltech.)

efficiency. SPE cartridges are generally considered to be for a single use and disposable. However, cost can be a significant issue if a large number of samples are to be processed.

SPE Pipet Tips

SPE pipet tips provide a convenient means of processing a small amount of sample and can be used to either retain undesired material, or more commonly, to selectively retain the analyte and then elute it with a solvent. These were originally introduced with a packed sorbent bed held with a bed support. The reproducibility of these devices was greatly dependent on operator skill and this original design has largely disappeared. Present pipet tip SPEs (a) contain either a porous monolithic sorbent bed (no bed support is needed, e.g., the OMIX series from Agilent, capable of handling as little as 500-nL sample and used primarily for protein and peptide purification), or (b) are very loosely packed; the sample, often with a solvent or other reagent added, is aspirated through the loose bed whereupon it forms a slurry; air is then aspirated through the slurry to mix it thoroughly; the liquid is then dispensed and the sorbent retains the undesired constituents (for a video demonstration of producing a water-clear eluate from a milk-acetonitrile:water mixture (proteins and sugars are retained), see www.dpxlabs.com), or (c) have the sorbent embedded in the walls of the bottom portion of the tip, permitting as little as 100-nL sample to be processed, NuTip™ (http://www.zirchrom.com/glygen1.asp) and TopTip™ from the same manufacturer that uses a small amount of 20–30 μm particles with the pipette tip having a 2 μm wide slit to retain the particles; these can also handle as little as a 100-nL sample volume.

SPE Disks

The small cross-sectional area SPE pipet tips are prone to plugging by protein samples. So solid-phase extractants are also available in filter form (Empore™ extraction disks) in which 8-μm silica particles are enmeshed into a web of PTFE [poly(tetrafluoroethylene)] fibrils. Fiberglass-based disks, which are more rigid, are also available. The greater cross-sectional area disks with shorter bed depths allow higher flow rates for large-volume samples with low concentrations of analyte, typically encountered in environmental analysis. The disks are less prone to channeling found with packed cartridges. They tend to plug if samples contain particulate matter, and so a prefilter may have to be used. Disk cartridges (disks preloaded in a plastic barrel) are also available that operate like a regular cartridge.

96-Well SPE Plates

Liquid chromatography (Chapter 16) combined with mass spectrometry (Chapter 17) is widely used for rapid and selective drug analysis, and samples can be run in 1 to 3 min. So, faster ways of sample cleanup are needed for processing large numbers of samples. 96-Well plates with small wells (so-called microtiter plates) are widely used for processing large numbers of samples in automated instruments.

Solid-phase extraction systems have been designed in a 96-well microtiter plate format, so they can be processed automatically. Single-block plates with 96 wells contain either packed beds or disks of sorbent particles, in an 8-row × 12-column rectangular format (Figure 13.7). The plates sit on top of a 96-well plate collection system. The chemistry is the same as in the above formats. Samples are processed using a vacuum manifold or centrifuge using a microplate carrier. The SPE columns are 1 to 2 mL, with 10 to 100 mg packing of sorbent particles. The bed mass loading determines the solvent and elution volumes, as well as the capacity for analyte and sample matrix constituents. The smallest bed that provides adequate capacity should be used. This minimizes extraction times and the smaller elution volumes require less time for evaporation prior to reconstitution and analysis.

The optimum use of SPE procedures requires investigating different stationary phases, their masses, the volumes of conditioning, sample load, wash and elution solvents, and the sample size. These variables are readily studied in a column format. But it is expensive and inconvenient to use only a fraction of the 96 wells to perform all the studies. Hence, modular well plates have been developed that have small removable plastic SPE cartridges that fit tightly in the 96-hole base plate, and only a portion needs to be used to develop a method.

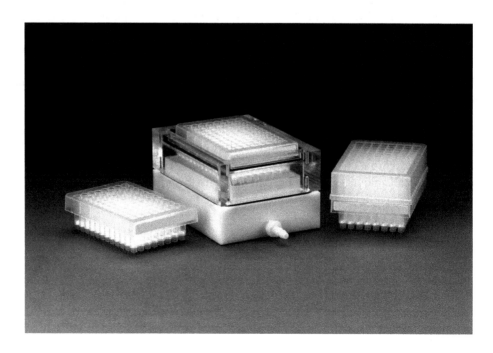

FIGURE 13.7 96-Well extraction plates and vacuum manifold with collection plate. (Courtesy of Thermo Labsystems.)

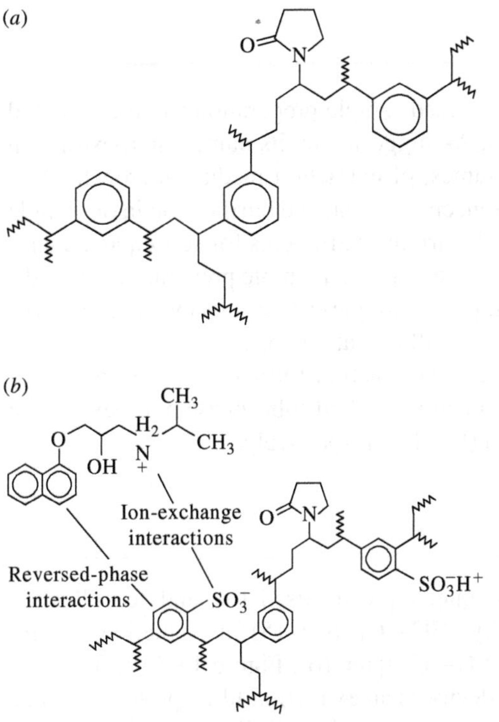

(a)

(b)

FIGURE 13.8 "Universal sorbents": Chemical structures of Waters' Oasis (a) HLB and (b) MXC polymer sorbents. The top structure in (b) is the basic drug propanolol demonstrating drug–sorbent interaction. [From D. A. Wells, *LC·GC*, **17**(7) (1999) 600. Reproduced by permission of LC.GC.]

Other Sorbents for Solid-Phase Extraction

Sorbents are available in long chain lengths (C_{20} and C_{30}) for isolation of hydrophobic molecules. "Universal sorbents" have been developed that will sorb a group of structurally similar compounds. An example in Figure 13.8a is a synthetic polymer of N-vinylpyrrolidone (top half of molecule) and divinylbenzene (bottom half). It provides hydrophilicity for wetting and hydrophobicity for analyte retention. A sulfonated version (Figure 13.8b) is a mixed-mode sorbent that has both ion exchange and solvent extraction properties and will retain a range of acidic, neutral, and basic drugs. These wettable sorbents do not require conditioning.

Polymeric Phases

Besides the common silica-based SPE particles, polymer-based supports are also available. These have advantages of being stable over a wide pH range, and they do not possess residual silanol groups (hydroxyl groups residing on the surface of unmodified silica gel) that can interact with, for example, metal ions or other cationic species. The particles are spherical, while silica-based SPE particles are irregular in shape, and the polymeric particles have been designed to be wettable. They typically have higher capacity than silica-based particles.

Dual Phases

The use of two different phases can extend the range of compounds extracted. Three types are used, mixed mode, layered, and stacked phases. In the mixed mode, two different types of chemically bonded phases are mixed together in the cartridge. An example is a mixture of C_8 and cation exchange particles. In the layered mode, the two different phases are packed one on top of the other. Stacked phases use two cartridges in series to provide enhanced separations. The first two modes are more readily adapted to automation since only a single cartridge is used.

13.7 Microextraction

(Contributed by Professor Yi He, The City University of New York.)

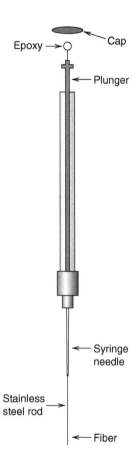

FIGURE 13.9 Schematic of a solid-phase microextraction device. [From C. L. Arthur, D. W. Potter, K. D. Buchholz, S. Motlagh, and J. Pawliszn, *LC·GC*, **10**(9) (1992) 656. Reproduced by permission of LC.GC.]

Microextraction is a simplified and miniaturized sample preparation technique aimed at zero or minimum solvent consumption. As suggested by its name, microextraction uses only microliter, or even smaller volumes, of extraction media. Microextraction methods usually integrate sample extraction, enrichment, and purification in one single step, and can be conveniently coupled with various instruments for subsequent analysis. Compared with conventional, or even some improved sample preparation methods, microextraction methods significantly improve sample-use efficiency, reduce errors from multistep procedures, and lower the overall operating cost.

Although there are many different microextraction formats, they are generally classified as solid-based and liquid-based methods, called solid-phase microextraction (SPME) and liquid-phase microextraction (LPME), respectively.

Solid-Phase Microextraction (SPME)

SPME is a solvent-free extraction technique, typically used for analyte collection for determination by gas chromatography (GC—Chapter 15) or, sometimes, high-performance liquid chromatography (HPLC—Chapter 16). Figure 13.9 illustrates an SPME assembly. The key feature of this device is an extraction fiber, protected inside the needle of a syringe. A typical SPME fiber is made of fused silica coated with a thin layer (7 μm to 100 μm thick) of immobilized polymer or a solid adsorbent, or a combination. In a solution or headspace (vapor in equilibrium with the solution in a closed system) analytes are exposed to the fiber and distribute between the sample matrix and the fiber coating during extraction.

Solid, liquid, or gaseous matrices can be sampled by SPME. The fiber is exposed to a gaseous or liquid sample, or the headspace above a solid or liquid sample, for a fixed time and temperature; samples are often agitated to increase the extraction efficiency. After extraction, the fiber is withdrawn back into the needle and directly transferred to the GC injector port, where the analyte is thermally desorbed and introduced into the GC column for separation. For LC analysis, the fiber can be introduced into a specially designed chamber for solvent desorption. Importantly, the fiber-in-needle assembly provides an excellent means for field sampling; once used for an extraction, the apparatus can be easily transported from the field to the laboratory for subsequent analysis.

Various SPME fibers and coatings are commercially available and can be used to target different analytes (Table 13.1). Many coatings are similar to available commercial gas chromatography stationary phases. The principle of extraction is based on "like dissolves like" (i.e., good extraction efficiencies can be obtained by a fiber with matching polarity to that of the target analytes). For example, a widely used fiber is based on a poly(dimethylsiloxane) (PDMS) coating, which is relatively nonpolar due to the presence of methyl groups. This fiber is useful for sampling nonpolar volatile or semivolatile compounds and can be used to determine flavor components from beverages, foods, and the like. Another example is a fiber with 85-μm coating of polyacrylate. It is more polar due to the presence of carboxyl groups, and thus, it is used to extract polar compounds such as phenols.

Liquid-Phase Microextraction (LPME)

LPME is a miniaturized form of liquid–liquid extraction (LLE), usually using less than 10 μL of solvent. However, in contrast with LLE, which aims to extract exhaustively all

TABLE 13.1 Commercial SPME fiber coatings and their applications[a]

Fiber coating	Analytes
Polydimethylsiloxane (PDMS)	Nonpolar analytes
Polydimethylsiloxane/Divinylbenzene (PDMS/DVB)	Many polar compounds (esp. amines)
Polyacrylate	Highly polar (ideal for phenols)
Carboxen/Polydimethylsiloxane (CAR/PDMS)	Gaseous/volatile analytes
Carbowax/Divinylbenzene (CW/DVB)	Polar analytes (esp. alcohols)
DVB/CAR/PDMS	Broad range of polarities (good for C3–C20 range)
Carbowax/Templated resin (CW/TPR)	For HPLC applications

[a]Information adapted from Supelco application note

target analytes, LPME only extracts a small representative fraction of the analytes from the sample matrix. LPME can be easily performed using a medical or chromatographic syringe, a capillary column, or just a sample tube. A typical LPME setup involves the placement of a solvent droplet on the tip of a microsyringe (a representative setup is shown in Figure 13.10, without the separate organic layer—the drop is the organic solvent). After extraction, the solvent drop is collected, usually withdrawn back into the syringe, and analyzed using chromatographic or spectroscopic instruments. In order to improve the stability of the solvent droplet, the solvent can be filled in a small segment of a hollow fiber e.g., of porous polypropylene, and then attached to a syringe for extraction. Water-immiscible solvents such as toluene, octanol, octane, heptane, and n-dodecane are frequently used. Solvent mixtures can also be used to better match the analyte polarity and thus enhance extraction efficiency.

LPME can be used to extract ionizable compounds through a three-phase setup, which includes a sample solution (donor phase), an organic solvent, and an aqueous phase drop (acceptor phase) dipping in the organic phase. The thin layer of organic phase separates the aqueous sample and acceptor phases (see Figure 13.10). For compounds with acidic functional group(s), the pH of sample solution is adjusted before

Dispersive liquid-liquid microextraction (DLLME): A new powerful liquid-liquid extraction technique. In this relatively new technique an aqueous sample is extracted by rapidly injecting into it a few microliters of a binary solvent mixture, mutually miscible with each other, but one component is miscible with water and the other immiscible. Further, the latter component is preferably denser than water, e.g., chlorobenzene, chloroform or carbon disulfide, whereas the water-miscible component may be methanol, acetonitrile or acetone. With rapid injection into the sample, very small droplets of the immiscible extractant solvent are produced. The large interfacial surface area leads to very rapid extraction. The mixture can then be centrifuged to recover the heavier extractant (the small volume used leads to very high enrichment) and subjected to analysis by a suitable technique. See M. Rezaee, Y. Yamini and M. Faraji, "Evolution of Liquid-Liquid Microextraction Method" *Journal of Chromatography A* **1217** (2010) 2342.

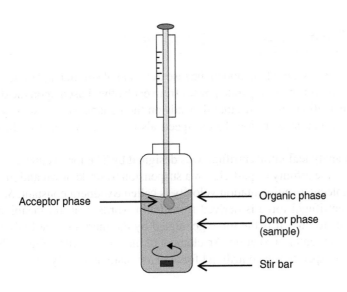

FIGURE 13.10 Liquid-phase microextraction setup.

Acceptor phase
Organic phase
Donor phase (sample)
Stir bar

extraction to a value at least three units lower than the analyte pKa. Under this condition, an acidic analyte will be in protonated form and can be easily extracted by the organic solvent. For the acceptor phase, pH must be basic, preferably three units higher than the pKa value so that the analyte is totally ionized and exhibits higher solubility in the aqueous acceptor phase. Similarly, for compounds with basic functional group(s), the pH of donor solution is adjusted to basic and the acceptor phase to acidic. This process integrates extraction (from sample to organic solvent) and back-extraction (from organic solvent to aqueous acceptor phase) into one single step. The recovered aqueous drop can then be analyzed using instruments such as HPLC or capillary electrophoresis. Three-phase LPME has been used to determine phenols, aromatic amines, and ionizable pharmaceutical residues from environmental water samples, as well as illicit drug such as methamphetamine from urine samples.

13.8 Solid-Phase Nanoextraction (SPNE)

Contributed by Professor Andres Campiglia, University of Central Florida

 # Professor's Favorite Example

SPNE is an emerging preconcentration technique recently developed for the water extraction and preconcentration of polycyclic aromatic hydrocarbons (PAHs). PAH extraction is based on the adsorption of PAHs on the surface of Au nanoparticles. The extraction procedure consists of mixing microliters of water sample with microliters of an aqueous solution of Au NPs with 20-nm average particle diameters.

Although SPNE was originally developed for PAHs, several features make this approach an attractive alternative for the generic extraction of organic pollutants in water samples. These include low analysis cost and rather low usage of organic solvents via an environmentally friendly experimental procedure. The small water volume required for quantitative extraction of pollutants facilitates simultaneous centrifugation for routine monitoring of numerous samples.

This technique is described in more detail on the text website.

13.9 Centrifugation Methods

Centrifuge is an essential method in the research field of cellular and molecular biology. Centrifugation is used to separate phases from each other based upon the differences in their sedimentation properties that depends on their shape, size, viscosity and density using centrifugal acceleration. Rotor speed also plays a critical role in defining the separation.

First analytical ultracentrifuge was designed by Thedor Svedberg in 1925. Sedimentation is the affinity of particles in a suspension to settle down and in a centrifugal field the molecular sedimentation velocity is called Svedberg constant, S.

As centrifugal force is derived from Latin words centrum meaning center and figure means to flux. It is the force generated by the inertia of the body that draws it away from the center of rotation. An element of mass m in a centrifuge, tube filled with a liquid will experience a centrifugal force given mathematically by:

$$F_e = m \cdot r \cdot \omega^2 \tag{13.20}$$

Where m is the mass of the element, ω is the angular velocity and r is the distance from the axis.

And,

$$S = \frac{V}{\omega^2 r} = \frac{M(1 - \overline{V}\rho_{sol})}{N_A f} \qquad (13.21)$$

where M is the Molecular weight, S is Svedberg coefficient, N_A is Avogadro's number, f is frictional coefficient and $\overline{V}\rho_{sol}$ is the partial specific volume of the molecule.

Types of Centrifugation

1. **Differential Centrifugation.** Separation depends on the differences in particle's size, shape and mass, so in this method, successive centrifugation is done using increasing centrifugal force. When subjected to centrifugal force, the heavier particles settle first at lower g values, while lighter particles require high centrifugal force to settle down. This method produces enriched fraction rather than purified fractions.

2. **Rate Zonal Centrifugation.** This method separates particles based on their size and conformation. Sample is kept on the topmost position of the tube and it starts sedimentation over the gradient in distinct zones. The particle having more frictional coefficient (f) will transfer slower (rod shaped will move slower than globular).

3. **Isopycnic or Density Gradient Centrifugation.** In this method, particles are separated purely on the basis of their density and size. Particles are subjected to constant agitation with the gradient material and each particle will reside as the gradient density becomes equal to its specific density.

4. **Moving Boundary or Zone Centrifugation.** In this method, entire tube is filled and centrifuged. The slowest sedimenting component is purified. The two fractions are recovered by decanting the supernatant solution from the pellet and this solution can be recentrifuged at higher speed with the formation of pellet and supernatant again.

Because of so much diversity, centrifugation is very useful for separation, interpretation, de-gritting and concentration of various components in a mixture.

Questions

1. Define the distribution coefficient and the distribution ratio. List the limitations of the Nernst distribution law.
2. Suggest a method for the separation of aniline, $C_6H_5NH_2$, an organic base, from nitrobenzene, $C_6H_5NO_2$ (extremely toxic!).
3. Describe two principal solvent extraction systems for metal ions. Give examples of each.
4. What is the basic approach to achieve selectivity in the metal ion extraction by chelating agents?
5. What is the largest concentration of a metal chelate that can be extracted into an organic solvent? The smallest concentration?
6. Discuss the effect of the pH and of the reagent concentration on the solvent extraction of metal chelates.
7. What is the basis of accelerated solvent extraction?
8. What is the basis of microwave-assisted extraction?
9. How does solid-phase extraction differ from solvent extraction?
10. What is solid-phase microextraction? How does dispersive liquid-liquid microextraction work?

Problems

Extraction Efficiencies

11. Derive Equation 13.10 from Equation 13.9.

12. In deriving Equation 13.8, we neglected the fact that benzoic acid partially forms a dimer in the organic phase $(2HBz \rightleftharpoons (HBz)_2; K_p = [(HBz)_2/[HBz]^2$, where K_p is the dimerization constant). Derive an expression for the distribution ratio taking this into account.

13. The volumes of the aqueous and organic phase in solvent extraction of uranium with 8-hydroxyquinoline in chloroform were 30 mL, when the percentage extraction was 99.8%. Calculate the distribution ratio.

14. An aqueous solution has an iodine concentration of 2.5×10^{-3} M. Calculate the percentage of iodine remaining in the aqueous phase after extraction of 0.150 L of this aqueous solution with 0.075 L of CCl_4 at 25°C. The distribution coefficient for the partition of molecular iodine between water and CCl_4 is 85.

15. Ninety-six percent of a solute is removed from 100 mL of an aqueous solution by extraction with two 70-mL portions of an organic solvent. What is the distribution ratio of the solute?

16. The distribution ratio between 3 M HCl and tri-n-butylphosphate for $PdCl_2$ is 2.6. What percent $PdCl_2$ will be extracted from 30.0 mL of a $7.0 \times 10^{-4} M$ solution into 15.0 mL tri-n-butylphosphate?

17. Ninety five percent of a metal chelate is extracted when equal volumes of aqueous and organic phases are used. What will be the percent extracted if the volume of the organic phase is doubled?

PROFESSOR'S FAVORITE PROBLEMS

Contributed by Professor Shaorong Liu, University of Oklahoma.

18. 10 mL of an aqueous solution containing 0.020 M RCOOH ($pK_a = 6.00$) is mixed with 10 mL of CCl_4. The partition coefficient is 3.0. When the pH of the aqueous phase is adjusted to 6.00, 0.012 M of RCOOH is measured in the CCL_4 phase. What will be the formal concentration of RCOOH in the aqueous solution if its pH was adjusted to 7.00 before extraction?

Multiple Extractions

19. For a solute with a distribution ratio of 30.0, show by calculation which is more effective, extraction of 15 mL of an aqueous solution with 15-mL organic solvent or extraction with two separate.

20. Arsenic(III) is 75% extracted from 7 M HCl into an equal volume of toluene. What percentage will remain unextracted after three individual extractions with toluene?

Solid-phase Extraction/Standard Addition Method

PROFESSOR'S FAVORITE PROBLEMS

Contributed by Professor Robert Synovec, University of Washington.

21. **Absorbance Spectrometry and the Standard Addition Method Using Complexation Chemistry, SPE with Preconcentration**

Lead (Pb) concentration in polluted water is determined using solid-phase extraction (SPE), combining preconcentration and the standard addition method (SAM), followed by absorbance measurement of a suitable Pb-complex. Assume only the Pb-complex absorbs light at the wavelength used. For the first step of SPE, a 1000-mL volume of either the *original* sample or *spiked* sample was preconcentrated (extracted). Then for the second step, a 5-mL volume of elution solvent was used to remove the preconcentrated Pb-complex from the SPE cartridge for each sample.

For the *preconcentrated, original* polluted water sample the absorbance was 0.32. Meanwhile, the *spiked* water sample contained an additional 5.0×10^{-8} mol Pb-complex per 1000 mL of *original* sample. The absorbance of the ***preconcentrated, spiked*** solution was 0.44. The pathlength was 1 cm. The molar absorptivity for the Pb-complex is $\epsilon = 2.0 \times 10^4 \text{Lmol}^{-1}\text{cm}^{-1}$.

(a) What is the concentration of the Pb-complex in the ***original*** polluted water sample (prior to preconcentration)? Either use an equation or graphical approach.

(b) What is the theoretical (ideal) preconcentration factor, P_{ideal}, for this analysis?

(c) What is the experimental preconcentration factor, P_{expt}, and how does it quantitatively compare to the theoretical P_{ideal}, from part b?

Multiple Choice Questions

1. Distribution coefficient is written as
   (a) K
   (b) S
   (c) H
   (d) G

2. The following additives can be used to extract perchlorate anion from an aqueous phase to an organic phase
   (a) No additive is needed
   (b) Use of a crown ether
   (c) Use of an anionic ligand
   (d) Use of an ion pair reagent (e.g. tetrabutyl ammonium chloride)

3. Calculate the fraction of the solute that remains in the aqueous phase after four extractions using 25 mL of organic solvent for each extraction, assume that $V_{aq} = 50$ mL and partition Coefficient of $K = 4$.
   (a) 0.33
   (b) 1.20%
   (c) 0.012
   (d) 0.215

4. Distribution ratio of "A" between $CHCl_3$ and water is 9.0. It is extracted with several, 5 mL aliquots of $CHCl_3$. The number of aliquots needed to extract 99.9% of "A" from its 5 mL aqueous solution are
   (a) 2
   (b) 3
   (c) 4
   (d) 5

5. What fraction of I_2 initially present in 50 mL of water is expected to partition into the organic phase after a single extraction with 50 mL of chloroform?
   (a) 0.01
   (b) 0.03
   (c) 0.50
   (d) 0.99
   (e) 0.97

6. What aqueous concentration of I^- should be used to remove roughly 99% of the I_2 initially present in 50 mL of chloroform using 5 extractions with 50 mL volumes of water?
   (a) 0.174
   (b) 0.075
   (c) 0.288
   (d) 11.48
   (e) None of the above

Recommended References

1. C. C. Akoh and D. B. Min, *Food Lipids*; *Chemistry, Nutrition and Biotechnology*. CRC press: Taylor and Francis Group
2. S. M. Khopkar, *Basic Concepts of Analytical Chemistry*. New Age International (P) Ltd.

Solvent Extraction

3. G. H. Morrison and H. Freiser, *Solvent Extraction in Analytical Chemistry*. New York: Wiley, 1957. A classic. Detailed coverage of extraction of metals.

4. J. Stary, *The Solvent Extraction of Metal Chelates*. New York: Macmillan, 1964.

5. J. R. Dean, *Extraction Techniques in Analytical Sciences*. New York: Wiley, 2009.

6. J. M. Kokosa, A. Przyjazny, and M. Jeannot. *Solvent Extraction: Theory and Practice*. New York: Wiley, 2009.

7. R. E. Majors, "Practical Aspects of Solvent Extraction," *LCGC North America*, **26** (12), December 2008, p. 1158. www.chromatographyonline.com.

Accelerated and Microwave-Assisted Solvent Extraction

8. B. E. Richter, B. A. Jones, J. L. Ezzell, N. L. Porter, N. Avdalovic, and C. Pohl, Jr., "Accelerated Solvent Extraction: A Technique for Sample Preparation," *Anal. Chem.*, **68** (1996) 1033.

9. K. Ganzler, A. Salgo, and K. Valco, "Microwave Extraction. A Novel Sample Preparation Method for Chromatography," *J. Chromatogr.*, **371** (1986) 371.

Solid-Phase Extraction

10. N. J. K. Simpson, ed., *Solid-Phase Extraction, Principles, Techniques, and Applications*. New York: Marcel Dekker, 2000.

11. J. S. Fritz, *Analytical Solid-Phase Extraction*. New York: Wiley-VCH, 1999.

12. R. E. Majors, "Advanced Topics in Solid-Phase Extraction," *LCGC North America*, **25**(1), January, 2007, p. 1. www.chromatographyonline.com.

13. SPME Application Guide, Supelco (www.sigma-aldrich.com). References categorized according to application, analyte/matrix.

14. J. Pawiliszyn and R. M. Smith, eds., *Applications of Solid Phase Microextraction*. Berlin: Springer, 1999.

15. S. A. S. Wercinski, ed., *Solid Phase Microextraction. A Practical Guide*. New York: Marcel Dekker, 1999.

16. J. Pawliszyn and L. L. Lord, *Handbook of Sample Preparation*. Hoboken, NJ: Wiley, 2010.

17. Y. He and H. K. Lee, "Liquid Phase Microextraction in a Single Drop of Organic Solvent by Using a Conventional Microsyringe," *Anal. Chem.*, **69** (1997) 4634.

18. S. Pedersen-Bjergaard and K. E. Rasmussen, "Liquid Phase Microextraction with Porous Hollow Fibers, a Miniaturized and Highly Flexible Format for Liquid-Liquid Extraction," *J. Chromatogr. A*, **1184** (2008) 132.

19. G. Shen and H. K. Lee, "Hollow Fiber-Protected Liquid-Phase Microextraction of Triazine Herbicides," *Anal. Chem.*, **74**(2002) 648.

20. G. Shen and H. K. Lee, "Headspace Liquid-Phase Microextraction of Chlorobenzenes in Soil with Gas Chromatography-Electron Capture Detection," *Anal. Chem.*, **75** (2003) 98.

21. L. Hou and H. K. Lee, "Dynamic Three-Phase Microextraction as a Sample Preparation Technique Prior to Capillary Electrophoresis," *Anal. Chem.*, **75** (2003) 2784.

22. Y. He and Y.-J. Kang, "Single Drop Liquid-Liquid-Liquid Microextraction of Methamphetamine and Amphetamine in Urine," *J. Chromatogr. A*, **1133** (2006) 35.

Centrifugation Methods

23. David Rickwood, "Centrifugation Techniques," *Encyclopedia of Life Sciences*, John Wiley & Sons Ltd., 2001, *www.els.net.*

Chapter 14

CHROMATOGRAPHY: PRINCIPLES AND THEORY

KEY THINGS TO LEARN FROM THIS CHAPTER

- Countercurrent extraction
- How chemicals are separated on a column
- Types of chromatography: adsorption, partition, ion exchange, size exclusion
- Chromatographic nomenclature (see tables of terms and key equations below)
- Theory of column efficiency

  - Plate number (key Equations: 14.7, 14.8)
  - van Deemter equation for packed GC columns (key Equation: 14.13)

  - Golay equation for open tubular GC columns (key Equation 14.24)
  - Huber and Knox equations for HPLC (key Equations: 14.25, 14.27)

- Retention factor (key Equation: 14.12)
- Chromatographic resolution (key Equations: 14.31, 14.33)
- Separation factor (key Equation: 14.32)
- Chromatography simulation software and databases

Key Equations:

Plate height	$H = \dfrac{L}{N}$	(14.5)
Plate number	$N = 5.545 \left(\dfrac{t_R}{w_{1/2}} \right)^2$	(14.7)
Adjusted retention time	$t_R' = t_R - t_M$	(14.10)
Retention factor (older term: Capacity factor)	$k = \dfrac{t_R'}{t_M}$	(14.12)
van Deemter equation Packed GC column	$H = A + \dfrac{B}{\bar{u}} + C\,\bar{u}$	(14.13)
Golay equation Capillary GC column	$H = \dfrac{B}{\bar{u}} + C\,\bar{u}$	(14.24)
Huber equation Liquid chromatography	$H = A + \dfrac{B}{\bar{u}} + C_s\bar{u} + C_m\bar{u}$ A and B are negligible for small particles except at very slow velocities	(14.25)
Knox equation Liquid chromatography	$h = A\nu^{1/3} + \dfrac{B}{\nu} + C\nu$	(14.27)
Resolution	$R_s = \dfrac{t_{R2} - t_{R1}}{(w_{b1} + w_{b2})^2}$	(14.31)

Separation factor (older term: Selectivity)	$\alpha = \dfrac{t'_{R2}}{t'_{R1}} = \dfrac{k_2}{k_1}$	(14.32)
Resolution	$R_s = \dfrac{1}{4}\sqrt{N}\left(\dfrac{\alpha - 1}{\alpha}\right)\left(\dfrac{k_2}{k_{ave} + 1}\right)$	(14.33)

In 1901, the Russian botanist, Mikhail Tswett, invented adsorption chromatography during his research on plant pigments. He separated different colored chlorophyll and carotenoid pigments of leaves by passing an extract of the leaves through a column of calcium carbonate, alumina, and sucrose, eluting them with petroleum ether/ethanol mixtures. He coined the term **chromatography** in a 1906 publication, from the Greek words *chroma* meaning "color" and *graphos* meaning "to write." Tswett's original experiments went virtually unnoticed in the literature for decades, but eventually others adopted it. Today there are several different types of chromatography. Chromatography is taken now to refer generally to the separation of components in a sample by distribution of the components between two phases—one that is stationary and one that moves, usually (but not necessarily) in a column.

The International Union of Pure and Applied Chemistry (IUPAC) has drafted a recommended definition of chromatography: "Chromatography is a physical method of separation in which the components to be separated are distributed between two phases, one of which is stationary (stationary phase), while the other (the mobile phase) moves in a definite direction" [L. S. Ettre, "Nomenclature for Chromatography," *Pure & Appl. Chem.*, **65** (1993) 819–872]. The stationary phase is usually in a column, but may take other forms, such as a planar phase (flat sheet). Chromatographic techniques have been in valuable in the separation and analysis of highly complex mixtures and revolutionized the capabilities of analytical chemistry. In this chapter, we introduce the concepts and principles of chromatography, including the different types, and describe the theory of the chromatographic process.

GC and HPLC are the most widely used forms of chromatography.

The two principal types of chromatography are gas chromatography (GC) and liquid chromatography (LC). Gas chromatography separates gaseous substances based on partitioning in a stationary phase from a gas phase and is described in Chapter 15. Liquid chromatography includes techniques such as size exclusion (separation based on molecular size), ion exchange (separation based on charge), and high-performance liquid chromatography (HPLC—separation based on partitioning from a liquid phase). These are presented in Chapter 16, along with thin-layer chromatography (TLC), a planar form of LC, electrophoresis where separation in an electrical field is based on the sign and magnitude of solute charge, and ion chromatography (IC), designed to separate and uniquely detect ionic analytes.

Birth of Modern Liquid and Gas Chromatography

In June 1941, the British chemists A. J. P. Martin and R. L. M. Synge presented a paper at the Biochemical Society meeting in London on the separation of monoamino monocarboxylic acids in wool using a new liquid–liquid chromatography technique called partition chromatography. The details are published in *Biochem. J.*, **35** (1941) 91. For this work, they received the 1952 Nobel Prize in Chemistry. In a second paper, they stated "The mobile phase need not be a liquid but may be a vapour..." and "Very refined separations of volatile substances should therefore be possible in columns in which permanent gas is made to flow over gel impregnated with a nonvolatile solvent." But this was largely missed during World War II, when many libraries did not receive journals, and it was not until 1950 that Martin, along with a young colleague A. T. James, successfully demonstrated "liquid–gas partition chromatography" at the October meeting of the Biochemical Society [A. T. James and A. J. P. Martin, *Biochem. J. Proc.*, **48**(1) (1950) vii.]. Thus were born two of the most powerful analytical techniques in use today. For a fascinating historical account of these developments, see L. S. Ettre, "The Birth of Partition Chromatography," *LC-GC*, **19**(5) (2001) 506.

14.1   Countercurrent Extraction: The Predecessor to Modern Liquid Chromatography

Let us imagine that we have individual extraction tubes (numbered $0, 1, 2, 3, 4 \ldots, r$) in which we are going to carry out solvent extraction. Let us imagine that all of the tubes have V mL of water in them. We add a unit amount of solute to tube 0, followed by V mL of an organic solvent and shake up tube 0 to perform an extraction. After the extraction, let the fraction of the total solute mass s in the aqueous phase be a and that in the organic phase be b ($a + b = 1$). In the chromatographic context, we typically refer to the distribution constant (K_D in Equation 13.1), as the partition constant and denote it by K. Casting Equation 13.1 in these terms,

$$K = \frac{c_o}{c_{aq}} = \frac{b/V}{a/V} = \frac{b}{a} \tag{14.1}$$

It is readily seen that the fraction a that remains in the aqueous phase (sometimes called the *raffinate*) is:

$$a = \frac{1}{K+1} \tag{14.2}$$

Conversely, the extracted fraction b is

$$b = \frac{K}{K+1} \tag{14.3}$$

Let us assume that the organic phase is lighter and we quantitatively transfer the top organic layer in tube 0 to tube 1 and put fresh organic solvent in tube 0 and shake up both tubes. Then we take the top layer from tube 1 and transfer it to tube 2, and take the top layer from tube 0 and transfer it to tube 1, and again replenish tube 0 with fresh organic solvent. This process is continued. The results are depicted in Figure 14.1.

Although in the scenario that we have envisioned, the aqueous phase remains stationary and the organic phase is moving from the left to right, we could have just as well started the process with all the tubes filled with an organic phase heavier than water, and begin by adding the solute as an aqueous solution and move the aqueous phase after extraction from right to left. The point is not that one phase remains stationary but that one liquid phase is moving with respect to the other. In actual industrial extraction processes, the lighter phase may be pumped from the bottom and the heavier extractant phase may be pumped from the top as fine dispersed droplets to perform the extraction and be collected at the bottom, while the extracted raffinate goes out at the top. Hence, the process is aptly called countercurrent extraction (CCE).

Note that in each extraction stage, we multiply the *total* content of each tube (both phases after transfer) by b, the fraction that goes to the organic phase, and by a, the fraction that remains in the aqueous phase. The ratio of the organic to aqueous phase after extraction is thus always $b/a = K$. We took a unit amount of solute; therefore, the sum total of the contents of all tubes must add up to 1. Note that after the nth extraction, the sum of the content of all the tubes can simply be given by expanding the binomial $(a + b)^n$. As $a + b = 1$, unity sum is indeed maintained. Moreover as the total content of both phases in each tube is simply the corresponding sequential term of the binomial expansion, the total contents of tuber r after n extractions $f_{n,r}$ is given by:

$$f_{n,r} = \frac{n!}{r!(n-r)!}a^{n-r}b^r$$

Based on Equations 14.1–14.3, the above is readily transformed to:

$$f_{n,r} = \frac{n!}{r!(n-r)!}\frac{K^r}{(1+K)^n}$$

Solutes with different distribution constants will move at different rates through the tubes, as in chromatography, and separation is achieved.

For more details, see E. W. Berg, *Physical and Chemical Methods of Separation*, McGraw-Hill, 1963.

FIGURE 14.1 Distribution of an analyte between the two phases in countercurrent extraction after sequential extraction steps.

Extraction — Distribution of an analyte between the two phases (Tube # 0–6):

Extraction	Phase	0	1	2	3	4	5	6
	Org							
	Aq	1						
EXTRACT↓								
1	Org	b						
	Aq	a						
TRANSFER↓								
	Org		b					
	Aq	a						
EXTRACT↓								
2	Org	ab	b^2					
	Aq	a^2	ab					
TRANSFER↓								
	Org		ab	b^2				
	Aq	a^2	ab					
EXTRACT↓								
3	Org	a^2b	$2ab^2$	b^3				
	Aq	a^3	$2a^2b$	ab^2				
TRANSFER↓								
	Org		a^2b	$2ab^2$	b^3			
	Aq	a^3	$2a^2b$	ab^2				
EXTRACT↓								
4	Org	a^3b	$3a^2b^2$	$3ab^3$	b^4			
	Aq	a^4	$3a^3b$	$3a^2b^2$	ab^3			
TRANSFER↓								
	Org		a^3b	$3a^2b^2$	$3ab^3$	b^4		
	Aq	a^4	$3a^3b$	$3a^2b^2$	ab^3			
EXTRACT↓								
5	Org	a^4b	$4a^3b^2$	$6a^2b^3$	$4ab^4$	b^5		
	Aq	a^5	$4a^4b$	$6a^3b^2$	$4a^2b^3$	ab^4		
TRANSFER↓								
	Org		a^4b	$4a^3b^2$	$6a^2b^3$	$4ab^4$	b^5	
	Aq	a^5	$4a^4b$	$6a^3b^2$	$4a^2b^3$	ab^4		
EXTRACT↓								
6	Org	a^5b	$5a^4b^2$	$10a^3b^3$	$10a^2b^4$	$5ab^5$	b^6	
	Aq	a^6	$5a^5b$	$10a^4b^2$	$10a^3b^3$	$5a^2b^4$	ab^5	
TRANSFER↓								
	Org		a^5b	$5a^4b^2$	$10a^3b^3$	$10a^2b^4$	$5ab^5$	b^6
	Aq	a^6	$5a^5b$	$10a^4b^2$	$10a^3b^3$	$5a^2b^4$	ab^5	
EXTRACT↓								
7	Org	a^6b	$6a^5b^2$	$15a^4b^3$	$20a^3b^4$	$15a^2b^5$	$6ab^6$	b^7
	Aq	a^7	$6a^6b$	$15a^5b^2$	$20a^4b^3$	$15a^3b^4$	$6a^2b^5$	ab^6

Martin and Synge at the Wool Research Laboratory in the United Kingdom were utilizing CCE to separate and purify different *N*-acetyl amino acids derived from wool. They manually transferred one phase from one separatory funnel to another. The tedium of this led them to reinvent a modern version of partition chromatography four decades after Tswett. At the Rockefeller Institute of Medical Research in New York, as WWII approached, Lyman C. Craig was working with a war-relevant problem: the separation and purification of the anthelmintic and antimalarial drug Atabrine and its metabolites in urine and blood. He was using CCE as well. The tedium of the manual process led this gifted instrument designer to devise a clever multistage countercurrent extraction apparatus that could be operated in a semi-automated manner. Craig's CCE instrument consists of a series of glass tubes that function as separatory funnels. Extraction takes place in all the tubes when they are rocked back and forth. The tubes are designed and arranged in a manner such that after extraction, a full rotation results in the top phase being transferred from one tube to the next. Professor Constantinos E. Efstathiou of the University of Athens maintains a web page about Craig and his apparatus. He presents an animated demonstration how the Craig tubes work (http://www.chem.uoa.gr/applets/AppletCraig/Appl_Craig2.html). This site also contains an applet where you can select the partition constants of two solutes and repeatedly extract and transfer in the manner of Figure 14.1 and observe how each solute begins to distribute itself among the tubes in a Gaussian fashion, and if the distribution coefficients are sufficiently different, how separation occurs as the process is repeated. Craig's original paper describing his apparatus appeared in the *Journal of Biological Chemistry* in 1944; its novelty and originality are reflected in this solo author paper containing no cited references! Otto Post helped Craig to design and make available a commercial version of the CCE apparatus (they generally referred the technique as Countercurrent Distribution) and over the next two decades more than a thousand papers were published that used the technique.

Craig's automated CCE apparatus evolved into droplet countercurrent chromatography (droplet CCC) where droplets of the mobile phase move by gravitational force through discrete elements of a liquid stationary phase. Invented by Yoichiro Ito at the National Institutes of Health, this was versatile but very slow. Centrifugal partition chromatography was the next evolution; in this, the equivalents of tubes and channels are created in a disk-shaped rotor. As the disk rotates, centrifugal force holds the stationary phase in place. At high rotation speeds, this force can be many times the force of gravity; mobile phase could be pumped through the system much faster than in the droplet case, permitting faster separations. It would be apparent that such an instrument will require rotating seals at the inlet and outlet; while initial designs were problematic, current generations of such instruments have highly reliable rotating seals. More recently, Ito again invented a type of planetary centrifuge that spins spools of tubing (usually stainless steel or polytetrafluoroethylene–PTFE) at high speeds. Alternating zones of mixing and phase separation progress along the coil. The forces retain the stationary phase while separation occurs. The first generation operated at 80 g and was termed high-speed CCC; high mobile phase flow rates and separations in a few hours were possible for the first time. The current generation of such instruments operate at 240 g, can complete separations in less than an hour and are generally referred to as high-performance CCC. While CCC is alive and well, it is used primarily in limited niche applications (often preparative separations of natural products); aficionados maintain an informative website (http://theliquidphase.org). An in-depth account is provided in the monograph by Berthod (Reference 1).

Chromatography and Numerical Simulation

In CCE, stepwise equilibration and transfer between two phases are involved. Each step is discrete and each funnel or stage is discrete. Although the separation stages in chromatography are not discrete, much of the conceptual framework is based on such theoretical stages or plates, as will become implicit later in this chapter. Also, the mobile phase moves continuously through the column without necessarily reaching equilibrium at each plate before reaching the next one. While complete stage-to-stage transfer following complete equilibration is decidedly unrealistic, this does not make the conceptual equilibrium model totally wrong. If, for example, we assume that at all stages we reach 90% of complete equilibrium before the mobile phase is transferred, it just means that the effective value of K is accordingly lower. In the case of chromatography, Equation 14.1 will normally be expressed as

$$K = \frac{c_s}{c_m} \tag{14.4}$$

where C_s and C_m are the stationary and the mobile phase concentrations (note that this is really opposite to what would typically be considered the mobile phase in CCC). As long as the sorption isotherm is linear, i.e., C_s is linearly dependent on C_m, CCC-based models predict chromatographic behavior well. Accordingly, an attractive applet (http://www.chem.uoa.gr/applets/appletchrom/appl_chrom2.html) using a Gaussian approximation of the binomial expression of $f_{n,r}$ given in Figure 14.1 simulates the separation of five solutes. You can pick the partition constants for each solute. It also simulates the peaks leaving the column and entering the detector where you can pick different response factors for different analytes.

Few simulations, however, demonstrate the chromatographic principles as well as a numerical simulation that you can do yourself in Excel, as in the website file Figure 14.2.xlsx. You can use this file as is to see how separations differ, simply by changing the Input K value. We discuss here the implementation of the file. Referring to Figure 14.2, we first define our partition constant as K, in cell B2. Imagine that Excel column A is the stationary phase (S) of a real chromatographic column and each cell is an equilibration plate. Column B is the mobile phase (M); each cell here is to be equilibrated with the corresponding stationary phase cell. In our first snapshot (step) of this process we input 100 units of the sample in the mobile phase (B4). In the next snapshot (step), we simply equilibrate the sum of the contents of A4 and B4 and write the results in C4 and D4, respectively, as (A4 + B4)*K/(K+1) and (A4 + B4)/(K+1) according to Equations 14.2 and 14.3. Next, we copy these values and paste them (Paste values: Alt + E − S − V) in cells E4 and F4. Next we will simulate the movement of the mobile phase by copying the contents (in F4) one row below (in F5). We equilibrate again, the contents of E4:F4 and E5:F5 in the same above fashion, and write the results, respectively, in G4:H4 and G5:H5. We repeat these steps of copying/moving and equilibrating. By columns AE and AF, we are getting an idea of how the solute and its accompanying mobile phase are moving down the column. Excel 2010 allows conditional formatting of the color of a particular cell in a column to be rendered according to the magnitude of its contents—this gives us a clear idea of how the solute is distributed in the column, both in the stationary phase and in the mobile phase. The actual Excel file Figure 14.2.xlsx is in the web supplement; change the partition constant and examine its behavior.

Admittedly, the above may be interesting for 10 plates, but it quickly gets tedious. Fortunately, Excel allows you to write a subroutine, called a "Macro," that allows you

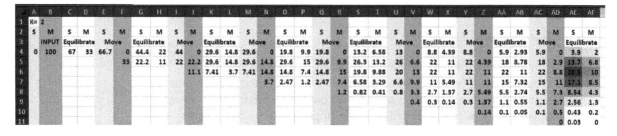

FIGURE 14.2 Excel simulation involving repeated equilibration and mobile-phase movement steps. See text for detailed discussion.

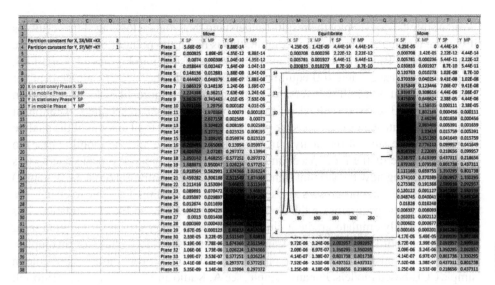

FIGURE 14.3 Column after 50 equilibrations.

do this automatically and keep things in the same set of columns. We have written such a subprogram, and placed it on the **website** with directions on how to write such a program (see "Writing the chromatographic Excel simulation Figure 14.3 – 4.docx"), such that each time it executes, 50 equilibration/mobile phase movement steps occur sequentially and the data are plotted as the calculations progress. Figures 14.3 and 14.4, showing chromatographic peak separation, were generated using this program.

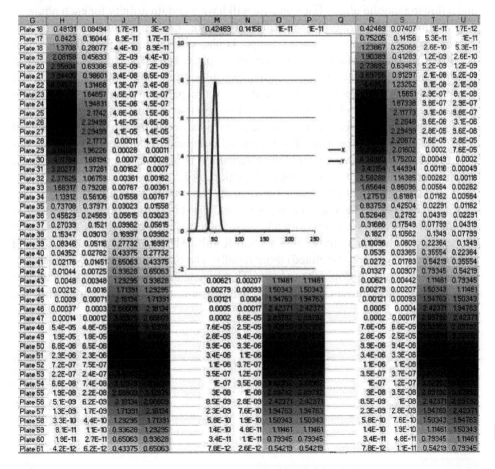

FIGURE 14.4 Column after 100 equilibrations.

We can simultaneously plot the results (understand that you are looking at the results as if you are looking at a transparent column; this is not at the point of the detector). The behaviors of a "Blue" (X) and a "Gray"(Y) solute are shown in Figures 14.3 and 14.4, respectively, after 50 and 100 equilibrations (plates). In the **website** supplement for this chapter you will find two versions of the program: Figure 14.3–4 autoscale.xlsm and Figure 14.3–4 fixed scale.xlsm; the first makes it easier to see the separation (it continuously autoscales the y-axis the plot is in terms of the total amount of the analyte in the plate, the sum of the analyte in the stationary phase and the mobile phase) to make better use of the plot space; and the second holds the y-axis scale fixed—this helps visualizing how the amount (tantamount to the concentration or the peak height) decreases as the band progresses and spreads along the column. Feel free to change the parameters in the programs on the **website** and run them! In an Excel simulation, it is possible to easily emulate gradient elution, peak asymmetry arising from a nonlinear isotherm, or poor kinetics of mass transfer out of the stationary phase, or incomplete plate-to-plate transfer of the mobile phase (e.g., due to poor packing), not possible in simple CCE-based models. For an example, see A. Kadjo and P. K. Dasgupta, "Tutorial: Simulating Chromatography with Microsoft Excel Macros," *Anal. Chim. Acta*, **778** (2013) 1.

14.2 Principles of Chromatographic Separations

Chromatographic retention of a compound is often associated with the establishment of an equilibrium between the stationary phase and the mobile phase, but in reality, chromatography is a dynamic process, and true equilibrium may not be reached.

While the mechanisms of retention for various types of chromatography differ, they are all based on the dynamic distribution of an analyte between a fixed stationary phase and a flowing mobile phase. Each analyte will have a certain affinity for each phase.

Figure 14.5 illustrates the separation of components in a mixture on a chromatographic column. A small volume of sample is placed at the top of the column, which is filled with particles constituting the stationary phase and the solvent. Rather than an equilibrium-based "plate view" of chromatography, many hold that a "rate view" of chromatography to be more rigorous: in this view, the partition ratio is simply the ratio of the time a solute spends in the stationary phase to that it spends in the mobile phase.

More solvent, functioning as mobile phase, is added to the top of the column and percolates through the column. The individual components interact with the stationary phase to different degrees, and the distribution is given in terms of the idealized equilibrium relationship represented by Equation 14.4.

The distribution of the analyte between the two phases is governed by many factors: the temperature, the type of compound, and the stationary and mobile phases. As Equation 14.4 implies, solutes with a large K value will be retained more strongly by the stationary phase than those with a small K value. The result is that the latter will move along the column (be eluted) more rapidly. The band broadens and decreases in amplitude as it travels down the column. This broadening of the injected rectangular wave sample pulse into a Gaussian peak is intrinsic to the chromatographic process,

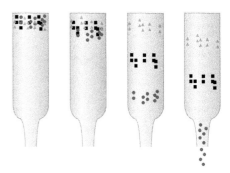

FIGURE 14.5 Principle of chromatographic separations.

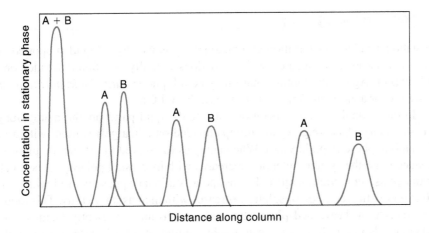

FIGURE 14.6 Distribution of two substances, A and B, along a chromatographic column in a typical chromatographic separation.

and is not due to the lack of attaining equilibrium, parabolic profile of laminar flow (see Chapter 16) or any other nonideal characteristics. The areas under the respective peaks, proportional to the analyte masses, remain the same. Band-broadening effects are treated below.

Figure 14.6 illustrates the distribution of two species A and B along a column as they move down the column, a time-lapse snapshot of the process described in Figures 14.3 and 14.4, and one that can be dynamically visualized by executing the fixed-scale program given in the **website** supplement. A plot of the concentrations of the analytes as they emerge from the column and as a function of time (or, less commonly, as a function of the volume of the mobile phase passed through the column) is called a chromatogram. A flow-through detector is placed at the end of the column to automatically measure the eluted compounds and print out a chromatogram of the peaks for the separated substances.

Although there are several different forms of chromatography, this simplified model typifies the mechanism of each. That is, *there is nominally an equilibrium between two phases, one mobile and one stationary.* (True equilibrium is never really achieved.) By continually adding mobile phase, the analytes will distribute between the two phases and eventually be eluted, and if the distribution is sufficiently different for the different substances, they will be separated.

14.3 Classification of Chromatographic Techniques

Chromatographic processes can be classified according to the type of equilibration process involved, which is governed by the type of stationary phase. Various bases of equilibration are: (1) adsorption, (2) partition, (3) ion exchange, and (4) size dependent pore penetration. More often than not, solute stationary-phase–mobile-phase interactions are governed by a combination of such processes.

Adsorption Chromatography

The stationary phase is a solid on which the sample components are adsorbed. The mobile phase may be a liquid (*liquid–solid chromatography*) or a gas (*gas–solid chromatography*); the components distribute between the two phases through a combination of sorption and desorption processes. Thin-layer chromatography (TLC) is a special example of adsorption chromatography in which the stationary phase is planar, in the form of a solid supported on an inert plate, and the mobile phase is a liquid.

Partition Chromatography

The stationary phase of partition chromatography is usually a liquid supported on a solid or a network of molecules, which functions virtually as a liquid, bonded on the solid support. Again, the mobile phase may be a liquid (*liquid–liquid partition chromatograpy*) or a gas (*gas–liquid chromatography*, GLC).

Compounds in a mixture must be soluble in the mobile phase in order to be separated by chromatography. Normal-phase chromatography is used to separate compounds that are not water soluble. Reversed-phase chromatography is used to separate water-soluble compounds based on their differing degrees of hydrophobicity and is more common.

In the normal mode of operations of liquid–liquid partition chromatography, a polar stationary phase (e.g., cyano groups bonded on silica gel) is used, with a nonpolar mobile phase (e.g., hexane). When analytes (dissolved in the mobile phase) are introduced into the system, retention increases with increasing polarity. This is called **normal-phase chromatography**. If a nonpolar stationary phase is used with a polar mobile phase, the retention of solutes decreases with increasing polarity. This mode of operation is termed **reversed-phase chromatography** and is presently the most widely used mode. "Normal-phase" chromatography significantly predates the reversed-phase mode, and was originally called liquid chromatography. Only after "reversed-phase" chromatography came along, the need arose to distinguish between the two, and the older version, still more prevalent then, was termed "normal-phase."

Ion Exchange and Size Exclusion Chromatography

Ion exchange chromatography uses supports with ion exchange functionalities as the stationary phase. The mechanism of separation is based on ion exchange equilibria. Hydrophobic interactions play a strong role in most ion exchange separations nevertheless, particularly in anion exchange chromatography. In size exclusion chromatography, solvated molecules are separated according to their size by their ability to penetrate into porous pockets and passages in the stationary phase.

Some types of chromatography are considered together as a separate technique, such as *gas chromatography* for gas–solid and gas–liquid chromatography. In every case, successive equilibria determine to what extent the analyte stays behind in the stationary phase or moves along with the eluent (mobile phase). In column chromatography, the column may be packed with small particles that act as the stationary phase (adsorption chromatography) or are coated with a thin layer of liquid phase (partition chromatography). In gas chromatography, the most common form today is a capillary column in which a virtual liquid phase, often a polymer, is coated or bonded on the wall of the capillary tube. We will see in Chapter 15 that this results in greatly increased separation efficiency.

| Chromatography Nomenclature and Terms |

In the fundamental discussions that follow, we use the IUPAC recommended symbols and terms, published in 1993 (Reference 6). The listing is very extensive, filling 54 pages. L. S. Ettre, who chaired the IUPAC committee, has published an abbreviated list of symbols and the most significant changes from traditional use [L. S. Ettre, "The New IUPAC Nomenclature for Chromatography," *LC.GC*, **11**(7) (July) (1993) 502]. Majors and Carr published a very useful updated "Glossary of Liquid-Phase Separation Terms," R. E. Majors and P. W. Carr, *LC.GC*, **19**(2) (February) (2001) 124, www.lcgcmag.com/lcgc/data/articlestandard/lcgc/482001/2936/article.pdf. They incorporate the recommended IUPAC terms.

Some of the older terms and the corresponding recommended terms are given in the following table:

Old		New	
Symbol	Term	Symbol	Term
α	Selectivity factor	α	Separation factor
HETP	Height equivalent to a theoretical plate	H	Plate height
k'	Capacity factor	k	Retention factor
n	Number of theoretical plates	N	Efficiency, number of plates
n_{eff}	Effective number of theoretical plates	N_{eff}	Effective theoretical plates; effective plate number
t_m	Mobile-phase holdup time	t_M	Mobile-phase holdup time
t_r	Retention time	t_R	Retention time
t'_r	Adjusted retention time	t'_R	Adjusted retention time
w	Base peak width	w_b	Bandwidth of peak

In addition to these terms, we will use a number of others throughout the chapter in describing the properties of gas and liquid chromatography. These are summarized here for easy reference.

A = eddy diffusion term = $2\lambda d_p$

λ = packing factor

d_p = average particle diameter

B = longitudinal diffusion term = $2\gamma D_m$

γ = obstruction factor

D_m = diffusion coefficient of solute in the mobile phase

C = interphase mass transfer term = $\dfrac{d_p^2}{6\,D_m}$

C_m = mobile-phase mass transfer term

C_s = stationary-phase mass transfer term

L = column length

u = mobile-phase linear velocity, cm/s

$\bar{u}$ = average mobile-phase linear velocity, cm/s

v = reduced velocity

h = reduced plate height

R_s = resolution

14.4   Theory of Column Efficiency in Chromatography

Band broadening in chromatography is the result of several factors, which influence the efficiency of separations. Mathematically, column chromatography is the easiest to treat; we can quantitatively describe the efficiency of a column and evaluate the factors that contribute to it.

A theoretical plate represents a single equilibrium step. The more theoretical plates, the greater the resolving power (the greater the number of equilibrium steps).

Theoretical Plates

The separation efficiency of a column can be expressed in terms of the number of theoretical plates in the column. A theoretical plate is a concept derived from distillation theory, whereby each theoretical plate in chromatography can be thought of as representing a single equilibrium step, such as in our Excel simulations. They are a measure of the efficiency or resolving power of a column; the more the number of plates, the more efficient is the column. The **plate height**, H, is the length of a column, L, divided by the number of theoretical plates, N:

$$H = \frac{L}{N} \tag{14.5}$$

To avoid a long column, H should be as small as possible. These concepts apply to all forms of column chromatography.

Experimentally, the plate height is a function of the variance, σ^2, of the chromatographic band and the distance, x, it has traveled through the column, and is given by σ^2/x ; σ is the standard deviation of the Gaussian chromatographic peak. The width at half-height, $w_{1/2}$, corresponds to 2.355 σ, and the base width, w_b, corresponds to 4 σ (Figure 14.7). The **number of plates**, N, for a solute eluting from a column is given by:

$$N = \left(\frac{t_R}{\sigma}\right)^2 \tag{14.6}$$

Putting in $w_{1/2} = 2.355 \, \sigma$, we have

$$N = 5.545 \left(\frac{t_R}{w_{1/2}}\right)^2 \tag{14.7}$$

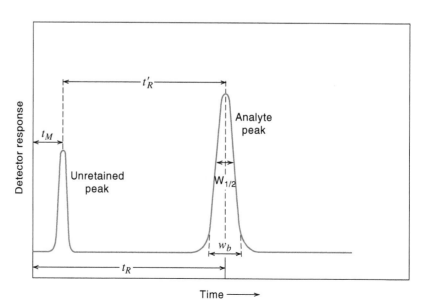

FIGURE 14.7   Characteristics of a chromatographic peak. $w_{1/2} = 2.355$, $w_b = 1.70 \, w_{1/2}$ $w_b = 4\sigma$.

where N, the number of plates of a column, is strictly applicable for only that specific analyte, t_R is the retention time, and $w_{1/2}$ is the peak width at half-height in the same units as t_R. These are illustrated in Figure 14.7. *Retention volume* V_R may be used in place of t_R. It should be noted that w_b is not the base width of the peak but the width obtained from the intersection of the baseline with tangents drawn through the inflection points at each side of the peak. N can also be expressed in terms of w_b.

$$N = 16 \left(\frac{t_R}{w_b} \right)^2 \tag{14.8}$$

The narrower the peak, the greater the number of plates.

Example 14.1

Calculate the number of plates in the column resulting in the chromatographic peak in Figure 14.7.

Solution

Measuring with a ruler, $t_R = 52.3$ mm and $w_{1/2} = 5.3$ mm

$$N = 5.545 \left(\frac{52.3}{5.3} \right)^2 = 5.4 \times 10^2$$

This is not a very efficient column, as we will see below.

The *effective plate number* corrects theoretical plates for dead (void) volume and hence is a measure of the true number of useful plates in a column:

$$N_{\text{eff}} = 5.545 \left(\frac{t'_R}{w_{1/2}} \right)^2 \tag{14.9}$$

where t'_R is the *adjusted retention time*.

$$t'_R = t_R - t_M \tag{14.10}$$

and t_M is the time required for the mobile phase to traverse the column and is the time it would take an unretained solute to appear. In GC, an air peak often appears from unretained air injected with the sample, and the time for this to appear is taken as t_M. The most common detector in GC is the flame ionization detector (FID) that responds better to carbonaceous compounds. It is common to inject butane, drawn from a lighter, to determine t_M in GC-FID systems.

The above equations assume a Gaussian-shaped peak, as in Figure 14.7, and the position of the peak maximum is taken for calculations. For asymmetric (tailing) peaks, the efficiency is better determined by the peak centroid and variance by mathematical analysis as described by the *Foley–Dorsey equation* [J. P. Foley and J. G. Dorsey, "Equations for Calculation of Figures of Merit for Ideal and Skewed Peaks," *Anal. Chem.*, **55**(1983) 730]. They derived empirical equations based solely on the graphically measurable retention time, t_R, peak width at 10% of peak height, $w_{0.1}$, and the empirical *asymmetry factor* at 10% peak height, $A_{s,0.1}$, B/A. $A + B = w_{0.1}$, and are the widths from t_R to the left and right sides (the direction of time progression is from left to right), respectively, of the asymmetric peak. (When the peak is symmetrical, $A = B = 1/2$ the peak width at 10% height.)

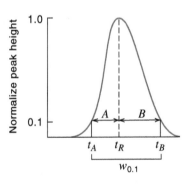

Asymmetric peak and Foley–Dorsey equation terms.

For the most efficient operation, we want H to be minimum. However, the critical issue is to accomplish the desired separation in the minimum possible time. More often than not, this operational optimum is reached at a velocity greater than the optimum velocity $\bar{u}_{\text{opt}}$ that generates the most efficient separation (minimum H).

Foley and Dorsey derived the number of theoretical plates as:

$$N_{\text{sys}} = \frac{41.7(t_R/w_{0.1})^2}{B/A + 1.25} \qquad (14.11)$$

This equation corrects the retention time and plate count for peak tailing and extracolumn sources of broadening.

For a symmetric peak $(B/A = 1)$, this becomes $N_{\text{sys}} = 18.53(t_R/w_{0.1})^2$, which is within 0.6% of the theoretical equation of $N_{0.1} = 18.42(t_R/w_{0.1})^2$, that is, the equation holds for ideal as well as asymmetric peaks.

Once the number of plates is known, H can be obtained from Equation 14.5. The width of the peak, then, is related to H, being narrower with smaller H. H has dimensions of length, and is typically expressed in micrometers. The *effective* plate height, H_{eff}, is L/N_{eff}.

The term H is usually determined for the last eluting compound, as it is usually the most conservative specification. For a well-packed high-performance liquid chromatography (HPLC) column of 5-μm particles, H should ideally be about 2 to 3 times the particle diameter. Values of 10 to 30 μm are typical.

Rate Theory of Chromatography—The van Deemter Equation

The solvent extraction derived term, distribution constant or partition constant, implies equilibrium and to some extent also a mechanism by which the solute distributes itself between two phases, as in chromatography (in adsorption or ion exchange chromatography, a solvent extraction model will at best be stretched). Chromatography is rarely an equilibrium process, and we use the term *retention factor*, k, instead; it is the ratio of the time the solute spends in the stationary phase to the time it spends in the mobile phase:

$$k = \frac{t'_R}{t_M} \qquad (14.12)$$

Note that the IUPAC recommended as far back as 1993 (see the Chromatography Nomenclature and Terms box in Section 14.2) that the older term used for this purpose, capacity factor, denoted by k', be replaced; it has not been widely adopted. While we will follow the IUPAC recommended notation and terminology here, you are likely to encounter the term capacity factor and the notation k' in the literature as well.

The plate theory of chromatography and equilibrium models cannot explain the dynamics of separation in a quantitative fashion. How might the separation efficiency change, for example, if we double the mobile phase flow? The van Deemter equation is the best known and most used to explain and determine conditions for efficient separations.

For a packed GC column, van Deemter showed that the broadening of a peak is the summation of interdependent effects from several sources. The **van Deemter equation** expresses the plate height, H as:

$$\boxed{H = A + \frac{B}{\bar{u}} + C\bar{u}} \qquad (14.13)$$

where A, B, and C are constants for a given system and are related to the three major factors affecting H, and $\bar{u}$ is the average linear velocity of the carrier gas in cm/s. While the van Deemter equation was developed for GC, the general principles hold for LC as well, although the diffusion term becomes less important while the equilibration term becomes more critical (see below).

The value of $\bar{u}$ is equal to the column length, L, divided by the time for an unretained substance to elute, t_M (Figure 14.7):

$$\bar{u} = \frac{L}{t_M} \tag{14.14}$$

The general flow term for chromatography is the *mobile-phase velocity*, u. However, in GC, the linear velocity will be different at different positions along the column (lower in the beginning and greater towards the exit) due to the compressibility of gases. So we use the *average linear velocity*, $\bar{u}$. In LC, compressibility is negligible, and $\bar{u} = u$. Below we will generally use the term u with the understanding that in GC we are really referring to $\bar{u}$.

The significance of the three terms A, B, and C in packed-column gas chromatography is illustrated in Figure 14.8, a plot of H as a function of carrier gas velocity. Here, A represents *eddy diffusion* and is due to the variety of tortuous (variable-length) pathways available between the particles in the column and is independent of the gas- or mobile-phase velocity. The heterogeneity in axial velocities (eddy diffusion) is related to particle size and geometry of packing by:

$$A = 2\lambda d_p \tag{14.15}$$

where λ is an empirical constant that is dependent on how well the column is packed. As a first approximation, λ is dependent on u; the typical values range from 0.8 to 1.0 for a well-packed column; d_p is the average particle diameter. It is minimized by using small and uniform particles, packing them well. However, gas chromatography is used at modest pressures, and very fine tightly packed supports are not used.

The term B represents **longitudinal** (axial) or **molecular diffusion** of the sample components in the carrier gas or mobile phase, due to concentration gradients within the column. That is, there is a gradient at the interface of the sample zone and the mobile phase, and molecules tend to diffuse to where the concentration is smaller. The diffusion in the mobile phase is represented by:

$$B = 2\gamma D_m \tag{14.16}$$

where γ is an obstruction factor, typically equal to 0.6 to 0.8 in a packed GC column, and D_m is the diffusion coefficient. Molecular diffusion is a function of both the sample and the carrier gas (D_m is high in the gas phase at typical GC temperatures and is therefore important). In a given analysis, the sample components are fixed, and the only way to change B or $B/\bar{u}$ is by varying the type, pressure, or flow rate of the carrier gas. High flow rates reduce the contribution of molecular diffusion, as this is a time-dependent contribution, and high flow rates reduce the total analysis time. Denser gases, such as nitrogen or carbon dioxide, also reduce B, compared to helium or hydrogen. In liquid chromatography, molecular diffusion is very small

Peaks are broadened by eddy diffusion, molecular diffusion, and limitations in mass transfer rates. Small, uniform particles minimize eddy diffusion. Faster flow decreases molecular diffusion but increases mass transfer effects. There will be an optimum flow.

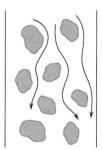

Eddy diffusion.

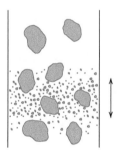

Molecular diffusion.

The effects of molecular diffusion are nearly negligible in LC but important in GC.

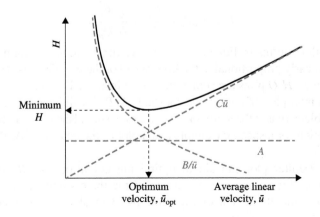

FIGURE 14.8 Illustrative contributions from each term in the van Deemter equation.

compared to that in gases. In GC, it dominates at flow rates less than $\bar{u}_{opt}$, and for LC it is often negligible under normal operating conditions. Almost always one operates at flow rates greater than $\bar{u}_{opt}$; as Figure 14.8 indicates, the H-u curve is not steep near $\bar{u}_{opt}$, and separations are faster when operating at flow rates greater than $\bar{u}_{opt}$. Demonstrative videos that illustrate how the A- and B-terms originate have been presented by McNair, one of the best-known educators of chromatography, in http://chromedia.org/chromedia?waxtrap=xqegzCsHqnOxmO1IecCbC&subNav=wnj edDsHqnOxmO1EcCzBkF.

Mass transfer dominates in LC.

The constant C is the **interphase mass transfer** term and is due to the finite time required for solute distribution equilibrium to be established between the two phases as it moves between the mobile and stationary phases. The C-term has two separate components, C_m and C_s, respectively, representing mass transfer limitations in the mobile and the stationary phases. The C_m term originates from nonuniform velocities across the column cross section: the flow velocity in the immediate vicinity of the column wall or packed particles is lower relative to that away from them. This difference in velocities causes peak broadening and the differences increase with increasing velocity. On the other hand, diffusive mixing, proportional to the diffusion coefficient of the analyte in the mobile phase, ameliorates this difference. For uniformly packed columns with spherical packing, C_m is given by:

$$C_m = \frac{C_1\,\omega\,d_p^2}{D_m}u \tag{14.17}$$

where C_1 is a constant, ω is related to the total volume of mobile phase in the column, and the other terms have been explained previously.

The stationary phase mass transfer term, C_s, is proportional to the amount of stationary phase, and increases with the retention factor for the analyte, and the thickness of the stationary phase film d_f (through which the solute must diffuse; $\frac{d_f^2}{D_s}$ represents the characteristic time for the solute to diffuse in and out of the stationary phase. C_s also increases directly with u, as less time is available for the diffusive equilibrium to be achieved; note that C_s will be zero for an unretained solute ($k = 0$). C_s is given by:

$$C_s = C_2\frac{k}{(1+k)^2}\frac{d_f^2}{D_s}u \tag{14.18}$$

where C_2 is a constant and other terms have already been defined.

Packed column GC, to which much of the above discussion pertains, is presently infrequently used. It is the only type of analytical chromatography where irregular, rather than spherical, particles are still commonly used as packing. For analytes with a low retention factor, the C_s term becomes small and the plate height can be estimated from the approximate equation:

$$H = 1.5d_p + \frac{D_m}{\bar{u}} + \frac{1}{6}\frac{d_p^2}{D_m}\bar{u} \tag{14.19}$$

Note that all three terms in Equation 14.19 have dimensions of length. Because the velocity $\bar{u}$ is linearly proportional to the flow rate Q of the mobile phase (which is more readily available), H-Q plots are sometimes made instead. They have the same profile as the van Deemter plot; of course, the constants A, B, C will be numerically different.

See Problem 16 and the textbook **website** for a spreadsheet calculation of the van Deemter equation for a specific example and a plot of the change of A, $B/\bar{u}$, and $C\bar{u}$ as a function of $\bar{u}$.

A van Deemter plot can aid in optimizing conditions. A, B, and C can be determined from three points and a solution of the three simultaneous van Deemter equations. Theoretically, a plot of Equation 14.13 results in a minimum, H_{min}, of $A + 2\sqrt{BC}$ at $\bar{u}_{opt} = \sqrt{B/C}$. Note the importance of the slope beyond $\bar{u}_{opt}$. The smaller the slope, the better, since the efficiency will then suffer little at velocities in excess of $\bar{u}_{opt}$. As previously stated, practical chromatography is rarely done at $\bar{u}_{opt}$, it takes too long. One is far more likely to use a longer or a more efficient column that has more plates.

An efficient packed gas chromatography column will have several thousand theoretical plates, and capillary columns have plate counts that depend on the column internal diameter, ranging from, e.g., 3,800 plates/m for a 0.32 mm i.d. column with a film thickness of 0.32 μm to 6,700 plates/m for a 0.18 mm i.d. column with 0.18 μm film thickness (for a solute of $k = 5$). The columns are typically 20–30 m long and total plate counts can be well in excess of 100,000. In a high-performance liquid chromatograph (below), efficiency on the order of 200–5000 theoretical plates per centimeter is typically achieved, the highest plate numbers being for sub-2 μm, superficially porous core-shell particles; columns are 5 to 25 cm in length, although the longer columns are impractical with the very small particles.

Reduced Plate Height

For comparing the performance of different columns, Knox introduced a dimensionless parameter called the *reduced plate height*, h, obtained by dividing by the particle diameter:

$$h = \frac{H}{d_p} \tag{14.20}$$

A well-packed column should have an h value at the optimum flow of 2 or less. For open tubular columns

$$h = \frac{H}{d_c} \tag{14.21}$$

where d_c is the inner diameter of the column.

The reduced plate height is used with the *reduced velocity*, v, for comparing different packed columns over a broad range of conditions; v relates the diffusion coefficient in the mobile phase and the particle size of the column packing:

$$v = \bar{u}\frac{d_p}{D_m} \tag{14.22}$$

(For open-tubular columns, d_p is replaced by d_c.) The *reduced form of the van Deemter equation* is

$$h = A + \frac{B}{v} + Cv \tag{14.23}$$

Open Tubular Columns

As we will see in Chapter 15, open tubular capillary columns, ranging in inner diameter from 180 to 560 μm, are the most widely used in GC; they provide a large number of plates. The use of open tubular columns is not yet widespread in LC: the characteristic time for a solute molecule located at the center of the tube to diffuse to the active coating on the column wall is given by $0.25d_c^2/D_m$. Considering that typical values of D in an LC setting is 1000–10,000 times smaller than in GC, to maintain the same characteristic time one would need d_c values 30–100 times smaller, in the range of 2–20 μm. Injection and detection volumes would have to be proportionally reduced. The technical challenges of such small injections in the pL volume scale and achieving sensitive detection have not yet been overcome, although efforts are continually being made. Open tubular columns have no packing, and so the eddy diffusion term in the

There is no eddy diffusion in open tubular columns.

van Deemter equation disappears. For open tubular columns, the modification of the van Deemter equation, called the *Golay equation*, applies:

$$H = \frac{B}{\bar{u}} + C\bar{u} \qquad (14.24)$$

Marcel Golay was a pioneer in the development of capillary columns for GC and correctly predicted that they would be much more efficient than packed columns then in vogue. Fortunately, the technology for drawing long lengths of glass and silica capillaries was also developed shortly thereafter.

High-Performance Liquid Chromatography: The Huber and Knox Equations

That the *C*-term in the van Deemter equation contains both mobile-phase and stationary phase contributions (C_m and C_s, see Equations 14.17 and 14.18) was first noted by Josef F. K. Huber in the context of LC. Unlike in packed column GC, the C_s term cannot be ignored and the van Deemter equation takes the form

$$H = A + B/\bar{u} + C_m\bar{u} + C_s\bar{u} \qquad (14.25)$$

This is known as the *Huber equation*. Except at very low mobile-phase velocities, the longitudinal diffusion term *B* is negligible because D_m is very small in the liquid phase (see Equation 14.16). With present uniform spherical particles of very small particle size (< 5 μm) that are tightly and uniformly packed, the contribution of the *A*-term is also very small and can be neglected. So Equation 14.25 reduces further to:

$$H = C_m\bar{u} + C_s\bar{u} \qquad (14.26)$$

The term C_s is relatively constant for a given *k* value; C_m, in this case, includes stagnant mobile-phase transfer (in the pores of the particles). Representative *H* versus *u* plots for HPLC are shown below in Figure 14.9 for different size particles.

At very slow velocities for small particles, molecular diffusion does become appreciable and *H* increases again. At high flow velocities, the increase in *H* with *u* is much smaller compared to gas chromatography, especially for the smaller particles. For well-packed columns of typical 5-μm particles, *H* values are usually in the range of 0.01 to 0.03 mm (10 to 20 μm). Note the scale in Figure 14.9 is in that range.

Knox developed an empirical equation for liquid chromatography that better fits observed experimental behavior, containing the third root of the velocity. The *Knox equation* is usually expressed in the dimensionless form as a function of reduced velocity:

$$h = Av^{1/3} + \frac{B}{v} + Cv \qquad (14.27)$$

FIGURE 14.9 van Deemter plots in HPLC as a function of particle size. The smaller particle sizes are more efficient, especially at higher flow rates. Column i.d.: 4.6 mm; mobile phase: 65% acetonitrile/35% water; sample: *t*-butylbenzene. [From M. W. Dong and M. R. Gant, *LC.GC*, **2** (1984) 294. Reprinted with permission.] See also Figure 16.25 for behavior of even smaller particles.

The term A is typically 1 to 2, B is about 1.5, and C about 0.1. So a typical good column follows:

$$h = v^{1/3} + \frac{1.5}{v} + 0.1v \qquad (14.28)$$

Efficiency and Particle Size in Hplc

Column efficiency is related to particle size. It turns out that for well-packed HPLC columns, H is about two to three times the mean particle diameter, that is,

$$H = (2 \text{ to } 3) \times d_p \qquad (14.29)$$

For core-shell particles, H is often lower than what Equation 14.29 suggests. These are particles with a solid core and a thin porous shell, so diffusion distance is limited. They are also called fused core particles; similiar but larger size particles, referred to as pellicular packing, had been advocated actually since the early days of LC but suffered from poor capacity, except for specialized ion exchange packings. Current generation of particles with a 1.25-μm solid core and a 0.23-μm porous shell can achieve minimum plate heights down to 1.5 μm.

Retention Factor Efficiency and Resolution

The retention factor k (Equation 14.12) is a direct measure of how strongly an analyte is retained by the column under the given conditions. If a pair of analytes are poorly separated (often referred to as poorly resolved), separation (resolution) improves if chromatographic conditions (temperature in GC, eluent strength in LC) are altered to increase k. While a large retention factor favors good separation, large retention factors mean increased elution time, so there is a compromise between separation efficiency and separation time. The retention factor can be increased by increasing the stationary-phase volume. A gradual decrease in the retention factor is an indication of degradation of the stationary phase.

Retention factor is a measure of retention time. Too small retention factors make good separation difficult but too high retention factors require too much time for elution.

The effective plate number is related to the retention factor and plate number via:

$$N_{\text{eff}} = N \left(\frac{k}{k+1} \right)^2 \qquad (14.30)$$

The volume of a chromatographic column consists of the stationary-phase volume and the **void volume**, the volume occupied by the mobile phase. The latter can be determined from t_M and the flow rate. One void volume of the mobile phase is required to flush the column once; however, flushing out a previous solvent and re-equilibrating with a new solvent minimally requires 5–10 column volumes.

Example 14.2

Calculate the retention factor for the chromatographic peak in Figure 14.7.

Solution

Measuring with a ruler, $t_R = 52.3$ mm and $t_M = 8.0$ mm

$$k = \frac{52.3 - 8.0}{8.0} = 5.5_4$$

The preferred retention factor values are 1 to 5. If too low, the compounds pass rapidly through the column and the degree of separation may be insufficient. Further, the detection of a weakly retained analyte will be subject to interferences from other nonretained species. Too large k values mean long retention time and long analysis times.

Resolution in Chromatography

The resolution of two chromatographic peaks is defined by:

You should strive for a resolution of at least 1.0.

$$R_s = \frac{t_{R2} - t_{R1}}{(w_{b1} + w_{b2})/2} \tag{14.31}$$

where t_{R1} and t_{R2} are the retention times of the two peaks (peak 1 elutes first), and w_b is the baseline width of the peaks. The R_s value indicates the quality of separation between two adjacent peaks. A resolution of 0.6 is needed to discern a valley between two peaks of equal heights. A value of 1.0 results in 2.3% overlap of two peaks of equal width and is considered the minimum separation to allow good quantitation. A resolution of 1.5 is an overlap of only 0.1% for equal width peaks and is considered sufficient for baseline resolution of equal height peaks.

We can describe resolution in thermodynamic terms, without regard to peak width. The **separation factor**, α, also widely referred to as the selectivity, is a thermodynamic quantity that is a measure of the relative retention of analytes, and is given by:

$$\alpha = \frac{t'_{R2}}{t'_{R1}} = \frac{k_2}{k_1} \tag{14.32}$$

The master resolution equation has three distinct terms: (1) the efficiency term; (2) the selectivity term; and (3) the retention or capacity term.

where t'_{R1} and t'_{R2} are the adjusted retention times (Equation 14.10) and k_1 and k_2 are the corresponding retention factors (Equation 14.12). This describes how well the chromatographic conditions discriminate between the two analytes. It is also the ratio of the amount of time each analyte spends in the stationary phase. The resolution can, then, be written as:

$$R_s = \frac{1}{4}\sqrt{N} \left(\frac{\alpha - 1}{\alpha}\right)\left(\frac{k_2}{k_{ave} + 1}\right) \tag{14.33}$$

where k_{ave} is the mean of the two peak capacity factors. This form relates resolution to efficiency, that is, band broadening and retention time (Equation 14.7), and is known as the *master resolution equation* or the *Purnell equation*. Note that since N is proportional to L, the resolution is proportional to the square root of the column length, $\sqrt{L}$. So doubling the length of the column increases the resolution by $\sqrt{2}$ or 1.4. A fourfold increase would double the resolution. Retention times, of course, would be increased in direct proportion to the length of the column. For asymmetric peaks, the centroids of the peaks should be used in calculating retention times for calculating α values.

The number of plates required for a given degree of resolution is given by:

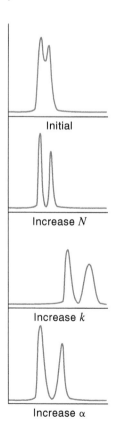

Initial

Increase N

Increase k

Increase α

$$N_{req} = 16R_s^2 \left(\frac{\alpha}{\alpha - 1}\right)^2 \left(\frac{k_{ave} + 1}{k_2}\right)^2 \tag{14.34}$$

Substituting from Equation 14.30, the number of effective plates required is

$$N_{eff(req)} = 16R_s^2 \left(\frac{\alpha}{\alpha - 1}\right)^2 \tag{14.35}$$

The figure in the margin illustrates how resolution increases differently with increasing values of N, k, or α. Note that increasing k increases the retention time for both peaks and broadens them. In uniformly packed columns, the widths of bands increase with the square root of the distance migrated, while the distance between centers of peaks increases in direct proportion to the distance traveled. Since the bands or peaks move faster than the broadening, separation improves.

While it is desirable to increase the number of plates, the resolution in a packed column, as noted above, increases only with the square root of N (e.g., by increasing L), and the pressure drop increases essentially linearly with L. It is more effective to try to increase the selectivity (α) or retention factor (k) by varying the stationary and mobile phases. Increasing the retention time, of course, lengthens the analysis time, and a compromise is generally chosen between speed and resolution.

Example 14.3

Methanol and ethanol are separated in a capillary GC column with retention times of 375 and 390 s, respectively, and half widths ($w_{1/2}$) of 9.42 and 10.0 s. An unretained butane peak occurs at 10.0 s. Calculate the separation factor and the resolution.

Solution

Use the longest eluting peak to calculate N (Equation 14.7):

$$N = 5.54 \left(\frac{390}{10.0} \right)^2 = 8.43 \times 10^3 \text{ plates}$$

From Equation 14.32,

$$\alpha = \frac{390 - 10}{375 - 10} = 1.04_1$$

From Equation 14.12,

$$k_1 = \frac{375 - 10}{10.0} = 36.5$$

$$k_2 = \frac{390 - 10}{10.0} = 38$$

$$k_{ave} = (36.5 + 38.0)/2 = 37.25$$

From Equation 14.33,

$$R_s = \frac{1}{4} \sqrt{8.43 \times 10^3} \left(\frac{1.041 - 1}{1.04_1} \right) \left(\frac{38.0}{37.25 + 1} \right) = 0.90$$

We can calculate w_b values from $w_{1/2}$ from the relationship $w_b = 1.70\, w_{1/2}$, to be 16.0 and 17.0 s for methanol and ethanol, respectively.

Then, from Equation 14.31, we can also verify

$$R_s = \frac{390 - 375}{(17.0 + 16.0)/2} = 0.91$$

This resolution is insufficient ($R_s < 1.5$) for baseline separation of the compounds. The peaks for methanol and ethanol would be seen to partly overlap in the chromatogram.

14.5 Chromatography Simulation Software

You are in charge of developing a new chromatographic separation. This involves selecting the proper column (stationary phase) and dimensions, mobile phase, and optimizing variables such as percent organic solvent, solvent or temperature gradient, and so forth. Optimization normally will require many repetitive chromatographic runs. But help is here! There are commercial software packages that assist the analyst in method development and optimization. Some of these are posted on the text website, with detailed descriptions of their capabilities. They are listed here: *Dry-Lab* (LC Resources): www.lcresources.com; *ACD/GC Simulator, ACD/LC Simulator,* and *ACD/ChromManager* (ACD/Labs): www.acdlabs.com; and *ChromSword*® (Merck KGaA): www.chromsword.com; *Virtual Column*: Software for ion chromatography simulation, University of Tasmania, www.virtualcolumn.com, basic version freely downloadable.

Questions

1. What is the description of chromatography?
2. Define: Plate height, Retention factor, Eddy diffusion, Column resolution and Stationary phase.
3. Describe a method for determining the number of plates in a column.
4. List the variables that lead to:
   (a) Zone broadening
   (b) Zone seperation
5. Classify the different chromatographic techniques, and give examples of principal types of applications.
6. What is the van Deemter equation? Define terms.
7. How does the Golay equation differ from the van Deemter equation?
8. How do the Huber and Knox equations differ from the van Deemter equation?

PROFESSOR'S FAVORITE PROBLEM

Contributed by Professor Michael D. Morris, University of Michigan

9. In chromatography the governing phase equilibrium constant is defined as larger when the equilibrium favors the stationary phase. In liquid chromatography, for example, $K = C_s/C_m$, where the subscripts refer to stationary and mobile phases, respectively. Explain what the effect of an increase in K will be on the time required for elution of an analyte.

Problems

Chromatography Resolution

10. A 125 cm long column is operated at 160°C. Retention time (in minutes): 0.80 (air peak), 1.25 (heptane) and 1.50 (octane) were obtained. The base widths of the bands were 0.14 min for heptane and 0.20 min for octane. Find the relative retention and the resolution for these bands.

11. A gas-chromatographic peak had a retention time of 70 s. The base width obtained from intersection of the baseline with the extrapolated sides of the peak was 5.8 s. If the column was 4 ft in length, what was H in cm/plate?

12. It is desired to baseline resolve two gas-chromatographic peaks with retention times of 85 and 100 s, respectively, using a column that has an H value of 1.5 cm/plate under the operating conditions. What length columns is required? Assume the two peaks have the same base width.

13. The following gas-chromatographic data were obtained for individual 2-μL injections of *n*-hexane in a gas chromatograph with a 3-m column. Calculate the number of plates and H at each flow rate, and plot H versus the flow rate to determine the optimum flow rate. Use the *adjusted* retention time t'_R.

Flow rate (mL / min)	t_M(Air Peak)(min)	t_R'(min)	Peak Width (min)
120.2	1.18	5.49	0.35
90.3	1.49	6.37	0.39
71.8	1.74	7.17	0.43
62.7	1.89	7.62	0.47
50.2	2.24	8.62	0.54
39.9	2.58	9.83	0.68
31.7	3.10	11.31	0.81
26.4	3.54	12.69	0.95

14. Three compounds, A, B, and C, exhibit retention factors on a column having only 500 plates of $k_A = 1.40$, $k_B = 1.85$, and $k_C = 2.65$. Can they be separated with a minimum resolution of 1.05?

15. Describe how the following situations could occur. In one case, two compounds are well resolved ($R > 1.5$) despite the separation being characterized by only a selectivity of 1.02. In another, two compounds are unresolved ($R < 1.5$) in spite of having a selectivity of 1.8 between them.

Spreadsheet Problem

16. Prepare a spreadsheet for a van Deemter plot for the following hypothetical A, B, and C terms: $A = 0.5$ mm, $B = 30$ mm · mL/ min, and $C = 0.05$ mm · min /mL. Plot H vs. $\bar{u}$ at linear velocities of 4, 8, 12, 20, 28, 40, 80, and 120 mL/ min. Also, on the same chart, plot A vs. $\bar{u}$, $B/\bar{u}$ vs. $\bar{u}$, and $C\bar{u}$ vs. $\bar{u}$, and note how they change with the linear velocity, that is, how their contributions to H change. Calculate the hypothetical H_{min} and $\bar{u}_{opt}$ and compare with the H_{min} on the chart. Also calculate $B/\bar{u}_{opt}$ and $C\bar{u}$. Look on the chart and see where the $B/\bar{u}$ and $C\bar{u}$ lines cross. Check your results with those in the text **website**, Chapter 14.

Multiple Choice Questions

1. The stationary phase in chromatography can be _____ supported on a solid.

   (a) solid or liquid
   (b) liquid or gas
   (c) solid only
   (d) liquid only

2. Which among the following cannot be used as an adsorbent in column adsorption chromatography?

   (a) Magnesium oxide
   (b) Silica gel
   (c) Activated alumina
   (d) Potassium permanganate

3. The remaining retention time after subtracting _____ from _____ is the adjusted retention time.

   (a) solute migration rate; retention time
   (b) retention time; solute migration rate
   (c) dead time; retention time
   (d) retention time; dead time

4. The ratio of moles of solute in stationary phase to the moles of solute in the mobile phase is defined as

   (a) distribution constant.
   (b) volumetric phase ratio.
   (c) retention factor.
   (d) total porosity.

5. What should be the value of the selectivity factor?

   (a) Equal to 1
   (b) Less than 1
   (c) Greater than 1
   (d) Greater than 0

6. The expression for plate number, N when 't_R' is the adjusted retention time and 'w_b' is the width at the base of the peak which is equal to 4 standard deviations is given as

   (a) $16\, t_R^2/w_b$
   (b) $4\, t_R^2/w_b$
   (c) $(4t_R/w_b)^2$
   (d) $4\, t_R/w_b^2$

7. Which of the following values is taken for the obstruction factor in an unpacked coated capillary column

   (a) 0

   (b) 0.6

   (c) 1

   (d) 1.6

8. The expression for longitudinal diffusion in the column, if 'γ' represents obstruction factor, 'd_m' represents particle diameter, 'D_m' represents solute diffusion coefficient and 'λ' represents function for packing uniformity is given as:

   (a) $\lambda\, d_m$

   (b) $2\,\gamma\, D_m$

   (c) $D_m\,\gamma$

   (d) λ/d_m

9. In the conditions given below which one will cause the efficiency of the column to increase?

   (a) Plate number becomes greater, and plate height becomes smaller.

   (b) Plate number becomes smaller, and plate height becomes smaller.

   (c) Plate number becomes greater, and plate height becomes larger.

   (d) Plate number becomes greater, and plate height becomes larger.

Recommended References

General

1. A. Berthod, Countercurrent Chromatography, *The Support-Free Liquid Stationary Phase. Comprehensive Analytical Chemistry* (D. Barceló, Ed.), Volume **XXXVIII**, Elsevier, 2002.
2. J. C. Giddings, *Unified Separation Science*. New York: Wiley-Interscience, 1991.
3. C. E. Meloan, *Chemical Separations: Principles, Techniques, and Experiments*. New York, Wiley, 1999.
4. J. Cazes, ed., *Encyclopedia of Chromatography*. New York: Marcel Dekker, 2001.
5. R. E. Majors and P. W. Carr, "Glossary of Liquid-Phase Separation Terms," *LC.GC*, **19**(2) (2001) 124.
6. L. S. Ettre, "Nomenclature for Chromatography," *Pure Appl. Chem.*, **65** (1993) 819.

Historical Accounts

These references give a good review of how separation techniques have evolved, particularly the stationary phases used and why we use the ones we do today.

7. H. J. Isaaq, ed., *A Century of Separation Science*, New York: Marcel Dekker, 2002.
8. L. S. Ettre, "The Rebirth of Chromatography 75 Years Ago," *LC.GC North America*, **25**(7), July 2007, p. 640. An excellent tracing of the birth and evolution of chromatography. www.chromatographyonline.com.
9. W. Jennings, The History of Chromatography. "From Academician to Entrepreneur—A Convoluted Trek," *LC.GC North America*, **26**(7), July 2008, 626. An interesting account of the life of capillary chromatography pioneer Walt Jennings and the establishment of his column preparation company. www.chromatographyonline.com.

Computer Simulation

10. R. G. Wolcott, J.W. Dolan, and L. R. Snyder, "Computer Simulation for the Convenient Optimization of Isocratic Reversed-Phase Liquid Chromatography Separations by Varying Temperature and Mobile Phase Strength," *J. Chromatogr. A*, **869** (2000) 3.
11. J. W. Dolan, L. R. Snyder, R. G. Wolcott, P. Haber, T. Baczek, R. Kaliszan, and L. C. Sander, "Reversed-phase liquid chromatographic separation of complex samples by optimizing temperature and gradient time: III. Improving the accuracy of computer simulation," *J. Chromatogr. A*, **857** (1999) 41.

Chapter 15

GAS CHROMATOGRAPHY

KEY THINGS TO LEARN FROM THIS CHAPTER

- Gas chromatograph
- GC columns—packed, capillary
- Stationary phases—polar to nonpolar
- GC detectors (see Table 15.2)
- Temperature programming

- Quantitative measurements—internal standards, spreadsheets for calculation
- Headspace analysis, thermal desorption, purging, and trapping
- Small columns for fast separations
- 2-D gas chromatography

Gas chromatography (GC) is one of the most versatile and ubiquitous analytical techniques in the laboratory. It is widely used for the determination of organic compounds. The separation of benzene and cyclohexane (bp 80.1 and 80.8°C) is extremely simple by gas chromatography, but it is virtually impossible by conventional distillation. Although Martin and Synge invented liquid–liquid chromatography in 1941, the introduction of gas–liquid partition chromatography by James and Martin a decade later had a more immediate and larger impact for two reasons. First, as opposed to manually operated liquid–liquid column chromatography, GC required instrumentation for application, which was developed by collaboration among chemists, engineers, and physicists; and analyses were much more rapid and done on a small scale. Second, at the time of its development, the petroleum industry badly needed improved analytical monitoring and immediately adopted GC. Within a few short years, GC was used for the analysis of almost every type of organic compound.

Very complex mixtures can be separated by this technique. The recent technique of two-dimensional GC (also called GC-GC) has further improved these capabilities. When coupled with mass spectrometry as a detection system, positive identification of virtually all of the eluted compounds is possible at very high sensitivity, creating a very powerful analytical system.

There are two types of GC: **gas–solid (adsorption) chromatography** and **gas–liquid (partition) chromatography**. The more important of the two is gas–liquid chromatography (GLC), used in the form of a capillary column. In this chapter, we describe the principles of operation of gas chromatography, the types of GC columns, and GC detectors. The principles of mass spectrometry (MS) are described in Chapter 17, along with coupling of the gas chromatograph with a mass spectrometer (GC-MS).

The volume "75 Years of Chromatography, A Historical Dialogue" (L. S. Ettre and A. Zlatkis, Eds., Journal of Chromatography Library Volume 17, Elsevier Science, 1979) published on the occasion of the 75th anniversary of the discovery of chromatography by Tswett contains short individual accounts of all prominent scientists involved in the development of chromatography in the early years. As many accounts are very personal, this volume is a treasure that contains the excitement and energy of these years perhaps like no other.

15.1 Performing GC Separations

In gas chromatography, the sample is converted to the vapor state (if it is not already a gas) by injection into a heated port, and the eluent is a gas (the **carrier gas**). The stationary phase is generally a nonvolatile liquid or a liquid-like phase supported on or bonded to a capillary wall or inert solid particles such as diatomaceous earth (kieselguhr—derived from skeletal remains of microscopic marine single-celled algae, consisting mainly of silica); the kieselguhr is usually calcined to increase particle size, creating what is known as firebrick, sold as Chromosorb P or W, for example. There are a large number of liquid phases available, and it is by changing the liquid phase, rather

Analyte in the vapor state distributes between the stationary phase and the carrier gas. Gas-phase equilibria are rapid, so resolution (and the number of plates) can be high.

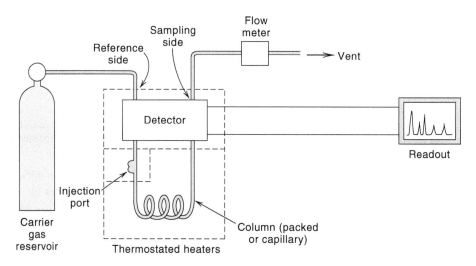

FIGURE 15.1 Schematic diagram of gas chromatograph.

FIGURE 15.2 Modern automated gas chromatography system. (Courtesy of Shimadzu North America.)

than the mobile phase, that different separations are accomplished. The most important factor in gas chromatography is the selection of the proper column (stationary phase) for the particular separation to be attempted. The nature of the liquid or solid phase will determine the exchange equilibrium with the sample components; and this will depend on the solubility or adsorbability of the analytes, the polarity of the stationary phase and sample molecules, the degree of hydrogen bonding, and specific chemical interactions. Most separation protocols have been developed empirically, however, theoretical approaches as well as suitable software are now available.

A schematic diagram of a gas chromatograph is given in Figure 15.1, and a picture of a modern GC system is shown in Figure 15.2. The sample is rapidly injected by means of a hypodermic syringe through a septum or from a gas sampling valve. Typically, the injected sample first goes into the inlet/inlet liner and then the carrier gas carries it (or if split, a portion of it—often done with capillary columns to avoid overloading) to the column. The sample injection port, column, and detector are heated to temperatures at which the sample has a vapor pressure of at least 10 torr, usually about 50°C above the boiling point of the highest boiling solute. The injection port and detector are usually kept somewhat warmer than the column to promote rapid vaporization of the injected sample and prevent sample condensation in the detector. For packed columns, liquid samples of 0.1 to 10 μL are injected, while gas samples of 1 to 10 mL are injected. Gases may be injected by means of a gastight syringe or through a special gas inlet chamber of constant volume (gas sampling valve). For capillary columns, volumes of only about 1/100 these sizes must be injected because of the lower capacity (albeit greater resolution) of the columns. Sample splitters are included on chromatographs designed for use with capillary columns that deliver a small fixed fraction of the sample to the column, with the remainder going to waste. They usually also allow splitless injection when packed columns are used (split/splitless injectors).

Separation occurs as the vapor constituents equilibrate between carrier gas and the stationary phase. The carrier gas is a chemically inert gas available in pure form such as argon, helium, or nitrogen. A highly dense gas gives best efficiency since diffusivity is lower, but a low-density gas gives faster speed. The choice of gas is often dictated by the type of detector. Gas chromatography always uses flow-through detectors that automatically detect the analytes as they elute from the column; the majority of GC detectors are destructive.

The sample emerges from the column at a constant flow rate. A variety of detectors are used, the specific response is dependent upon the analyte (see below). Some detectors contain a **reference side** and a **sampling side**. The carrier gas is passed through the reference side before entering the column and emerges from the column through the sampling. The difference in response of the sampling side relative to the reference side is processed as the analytical signal. The signal, representing the chromatographic peaks is acquired and displayed by a data system as a function of time. By measuring the **retention time** (the minutes between the time the sample is injected and the time the chromatographic peak appears) and comparing this time with that of a standard of the pure substance, it may be possible to identify the peak (agreement of retention times of two compounds does not guarantee the compounds are identical). The area under the peak is proportional to the concentration, and so the amount of substance can be quantitatively determined. The peaks are often very sharp (narrow in their temporal width, this requires fast detectors) and, if so, the peak height can be compared with a calibration curve prepared in the same manner. Chromatography data handling systems usually have automatic detection of peaks, readout of the peak area and/or peak height, as well as the retention time.

The separation ability of this technique is illustrated in a chromatogram in Figure 15.3. The entire analysis time is amazingly short for a highly complex sample. This, coupled with the very small sample required, explains the popularity of the technique. This is not to exclude the more important reason that many of the analyses performed simply cannot be done by other methods.

With complex mixtures, it is not a simple task to identify the many peaks. Instruments are commercially available in which the gaseous effluent is fed into a mass spectrometer where they are ionized, sorted on the basis of their mass-to-charge ratio, and identified based on mass/charge ratio mass (as well as and fragmentation pattern). This important analytical technique is called **gas chromatography–mass spectrometry** (GC–MS—see Chapter 17). The mass spectrometer is a sensitive and selective detector, and when a capillary GC column (very high resolution—see Section 15.2) is used (capillary GC–MS), this technique is capable of identifying and quantifying very complex mixtures of trace substances. For example, hundreds of compounds may be identified in sewage effluents, and traces of complex drugs in urine or blood or pollutants in water can be determined. GC-GC offers yet an order of magnitude greater peak

Detection of the analytes as they emerge from the column makes measurements rapid and convenient. Retention times are used for qualitative identification. Peak areas are used for quantitative measurements.

See more on GC–MS in Chapter 17.

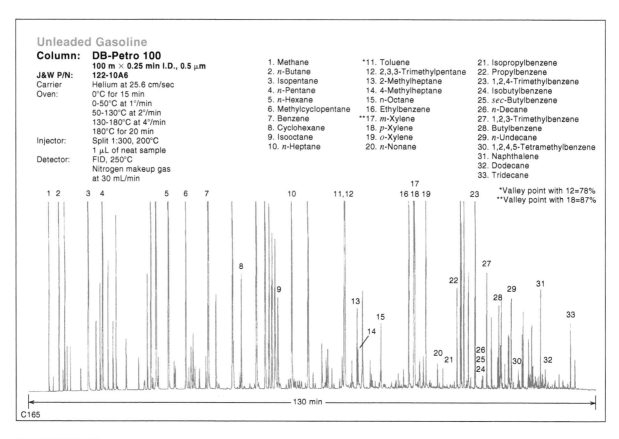

Unleaded Gasoline

Column:	**DB-Petro 100**
	100 m × 0.25 min I.D., 0.5 μm
J&W P/N:	**122-10A6**
Carrier	Helium at 25.6 cm/sec
Oven:	0°C for 15 min
	0-50°C at 1°/min
	50-130°C at 2°/min
	130-180°C at 4°/min
	180°C for 20 min
Injector:	Split 1:300, 200°C
	1 μL of neat sample
Detector:	FID, 250°C
	Nitrogen makeup gas
	at 30 mL/min

1. Methane
2. *n*-Butane
3. Isopentane
4. *n*-Pentane
5. *n*-Hexane
6. Methylcyclopentane
7. Benzene
8. Cyclohexane
9. Isooctane
10. *n*-Heptane

*11. Toluene
12. 2,3,3-Trimethylpentane
13. 2-Methylheptane
14. 4-Methylheptane
15. n-Octane
16. Ethylbenzene
**17. *m*-Xylene
18. *p*-Xylene
19. *o*-Xylene
20. *n*-Nonane

21. Isopropylbenzene
22. Propylbenzene
23. 1,2,4-Trimethylbenzene
24. Isobutylbenzene
25. *sec*-Butylbenzene
26. *n*-Decane
27. 1,2,3-Trimethylbenzene
28. Butylbenzene
29. *n*-Undecane
30. 1,2,4,5-Tetramethylbenzene
31. Naphthalene
32. Dodecane
33. Tridecane

*Valley point with 12=78%
**Valley point with 18=87%

C165

FIGURE 15.3 Typical gas chromatogram of unleaded gasoline, a complex mixture, using a capillary column. (Courtesy of Agilent Technologies.)

capacity, some 4000 compounds have been identified in cigarette smoke. For best sensitivity, though, some of the element or compound-type specific detectors listed later offer exquisitely low detection limits.

What Compounds Can Be Determined by GC?

Many, many compounds may be determined by gas chromatography, but there are limitations. They must be volatile and stable at operational temperatures, typically from 50 to 300°C. GC is useful for:

- All gases
- Most nonionized organic molecules, solid or liquid, containing up to about 25 carbons
- Many organometallic compounds (volatile derivatives of metal ions may be prepared)

If compounds are not volatile or stable, often they can be derivatized to make them amenable to analysis by GC. GC cannot be used for macromolecules nor salts, but these can be determined by HPLC and ion chromatography.

15.2 Gas Chromatography Columns

The two types of columns used in GC are **packed columns** and **capillary columns**. Packed columns came first and were used for many years. Capillary columns are more commonly used today, but packed columns are still used for applications that do not require high resolution or when increased capacity is needed.

Packed Columns

Columns can be in any shape that will fill the heating oven. Column forms include coiled tubes, U-shaped tubes, and W-shaped tubes, but coils are most commonly used. Typical packed columns are 1 to 10 m long and 0.2 to 0.6 cm in diameter. Well-packed columns may have 1000 plates/m, and so a representative 3-m column would have 3000 plates. Short columns can be made of glass or glass/silica-lined stainless steel, but longer columns may be made of stainless steel or nickel so they can be straightened for filling and packing. Columns are also made of Teflon. For inertness, glass is still preferred for longer columns. The resolution for packed columns increases only with the square root of the length of the column. Long columns require high pressure and longer analysis times and are used only when necessary (e.g., analytes that are poorly retained require more stationary phase to achieve adequate retention). Separations are generally attempted by selecting columns in lengths of multiples of 3, such as 1 or 3 m. If a separation isn't complete in the shorter column, then the next longer one is tried.

The column is packed with small particles that may themselves serve as the stationary phase (adsorption chromatography) or more commonly are coated with a nonvolatile liquid phase of varying polarity (partition chromatography). Gas–solid chromatography (GSC) is useful for the separation of small gaseous species such as H_2, N_2, CO_2, CO, O_2, NH_3, and CH_4 and volatile hydrocarbons, using high surface area inorganic packings such as alumina (Al_2O_3) or porous polymers (e.g., Porapak Q—a polyaromatic cross-linked resin with a rigid structure and a distinct pore size). The gases are separated by their size due to retention by adsorption on the particles. Gas–solid chromatography is preferred for aqueous samples.

The solid support for a liquid phase should have a high specific surface area that is chemically inert but wettable by the liquid phase. It must be thermally stable and available in uniform sizes. The most commonly used supports are prepared from diatomaceous earth, a spongy siliceous material. They are sold under many different trade names. Chromosorb P is a pink-colored diatomaceous earth prepared from crushed firebrick. Chromosorb W is diatomaceous earth that has been heated with an alkaline flux to decrease its acidity; it is lighter in color. Chromosorb G was the first support expressly developed for GC, combining the good efficiency and handling characteristics of Chromosorb G while having the low adsorptive properties of Chromosorb W. Generally, all of the above are available in non-acid washed, acid washed, and silanized with dimethylchlorosilane (DMCS, this greatly reduces polarity) and in high-performance versions (HP, controlled uniform fine particles). Chromosorb 750 is a very inert and efficient support that is acid washed and DMCS treated. Chromosorb T is useful for separating permanent gases and small molecules, it is largely based on fluorocarbon (Teflon) particles. Chromosorb P is much more acidic than Chromosorb W, and it tends to react with polar solutes, especially those with basic functional groups.

Column-packing support material is coated by mixing with the correct amount of liquid phase dissolved in a low-boiling solvent such as acetone or pentane. About a 5 to 10% coating (wt/wt) will give a thin layer. After coating, the solvent is evaporated by heating and stirring; the last traces may be removed in a vacuum. A newly prepared column should be conditioned at elevated temperature by passing carrier gas through

Packed columns can be used with large sample sizes and are convenient to use.

Capillary columns can provide very high resolution, compared with packed columns.

it for several hours, preferably before connecting detectors or other downstream components. The selection of liquid phases is discussed below.

Particles should be uniform in size for good packing and have diameters in the range of 60 to 80 mesh (0.25 to 0.18 mm), 80 to 100 mesh (0.18 to 0.15 mm), or 100 to 120 mesh (0.15 to 0.12 mm). Smaller particles are impractical due to high pressure drops generated.

Capillary Columns—The Most Widely Used

In 1957 Marcel Golay published a paper entitled "Vapor Phase Chromatography and the Telegrapher's Equation" [*Anal. Chem.*, **29** (1957) 928]. His equation predicted increased number of plates in a narrow open-tubular column with the stationary phase supported on the inner wall. Band broadening due to multiple paths (eddy diffusion) would be eliminated. And in narrow columns, the rate of mass transfer is increased since molecules have small distances to diffuse. Higher flow rates can be used due to decreased pressure drop, which decreases molecular diffusion. Golay's work led to the development of various **open-tubular columns** that today provide extremely high resolution and have become the mainstay for gas-chromatographic analyses. These columns are made of thin fused silica (SiO_2) coated on the outside with a polyimide polymer for support and protection of the fragile silica capillary, allowing them to be coiled. The polyimide layer is what imparts a brownish color to the columns, and it often darkens on use. The inner surface of the capillary is chemically treated to minimize interaction of the sample with the silanol groups (Si–OH) on the tubing surface, by reacting the Si–OH group with a silane-type reagent (e.g., DMCS).

Capillaries can also be made of stainless steel or nickel. Stainless steel interacts with many compounds and so is deactivated by treatment with DMCS, producing a thin siloxane layer to which stationary phases can be bonded. Stainless steel columns, though less common, are more robust than fused silica columns and are used for applications requiring very high temperatures.

The capillaries are 0.10 to 0.53 mm internal diameter, with lengths of 15 to 100 m and can have several hundred thousand plates, even a million. They are sold as coils of about 0.2 m diameter (Figure 15.4). Capillary columns offer advantages of high resolution with narrow peaks, short analysis time, and high sensitivity (with detectors designed for capillary GC) but are more easily overloaded by too much sample. Split injectors by and large alleviate the overloading problem.

Figure 15.5 illustrates the improvements in separation power in going from a packed column (6.4 × 1.8 m) to a very long but fairly wide-bore stainless steel capillary column (0.76 mm × 150 m), to a narrow but shorter glass capillary column (0.25 mm × 50 m). Note that the resolution increases as the column becomes narrower, even when the capillary column is shortened.

Increasing the film thickness increases capacity but increases plate height and retention time.

There are three types of open-tubular columns. **Wall-coated open-tubular** (WCOT) columns have a thin liquid film coated on and supported by the walls of the capillary. The walls are coated by slowly passing a dilute solution of the liquid phase through the columns. The solvent is evaporated by passing carrier gas through the columns. Following coating, the liquid phase is cross-linked to the wall. The resultant stationary liquid phase is 0.1 to 0.5 μm thick. Wall-coated open-tubular columns typically have 5000 plates/m. So a 50-m column will have 250,000 plates.

In **support coated open-tubular** (SCOT) columns, solid microparticles coated with the stationary phase (much like in packed columns) are attached to the walls of the capillary. These have higher surface area and have greater capacity than WCOT columns. The tubing diameter of these columns is 0.5 to 1.5 mm, larger than WCOT columns. The advantages of low pressure drop and long columns is maintained, but

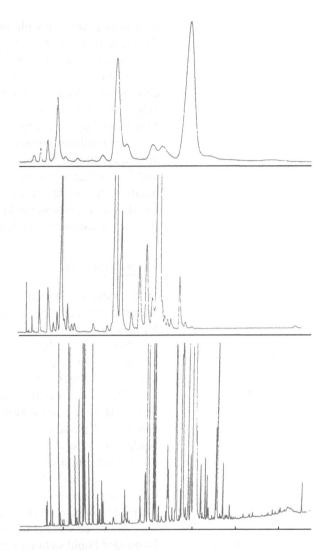

FIGURE 15.5 Three generations in gas chromatography. Peppermint oil separation on (top) $\frac{1}{4}$-in. × 6-ft packed column; (center) 0.03-in. × 500-ft stainless steel capillary column; (bottom) 0.25-mm × 50-m glass capillary column. [From W. Jennings, *J. Chromatogr. Sci.*, **17** (1979) 363. Reproduced from the *Journal of Chromatographic Science* by permission of Preston Publications, A Division of Preston Industries, Inc.]

capacity of the columns approaches that of packed columns. Flow rates are faster and dead volume connections at the inlet and detector are less critical. Sample splitting is not required in many cases, so long as the sample volume is 0.5 μL or less. If a separation requires more than 10,000 plates, then a SCOT column should be considered instead of a packed column.

The third type, **porous layer open-tubular** (PLOT) columns, have solid-phase particles attached to the column wall, for adsorption chromatography. Particles of alumina or porous polymers (molecular sieves) are typically used. These columns, like packed GSC columns, are useful for separating permanent gases, as well as volatile hydrocarbons. The resolution efficiency of open-tubular columns is generally in the order: WCOT > SCOT > PLOT. Wide-bore (0.5-mm) open-tubular columns

The resolution for open-tubular columns is WCOT > SCOT > PLOT. SCOT columns have capacities approaching those of packed columns.

have been developed with thicker stationary liquid phases, up to 5 μm, that approach the capacity of SCOT and packed columns, but their resolution is decreased. Many wide-bore columns are only available with thicker films.

Columns can tolerate a limited amount of analyte before becoming overloaded, causing peak distortion and broadening, and shifts in retention time. Sample capacity ranges are from approximately 100 ng for a 0.25-mm-i.d. column with 0.25-μm-thick film, up to 5 μg for a 0.53-mm-i.d. column with a 5-μm-thick stationary phase.

Open-tubular columns, being the mainstay for gas-chromatographic separation, are manufactured by numerous companies. Some of the major ones include Agilent Technologies (www.chem.agilent.com/Scripts/PHome.asp), Perkin-Elmer Instruments (http://instruments.perkinelmer.com), Phenomenex (www.phenomenex.com), Quadrex Corp. (www.quadrexcorp.com), Restek Corp. (www.restekcorp.com), SGE, Inc. (www.sge.com), and Sigma-Aldrich (Supelco) (www.sigmaaldrich.com).

Stationary Phases—The Key to Different Separations

Over a thousand stationary phases have been proposed for gas chromatography, and numerous phases are commercially available. Hundreds of phases have been used for packed columns, necessitated by their low overall efficiency, and stationary-phase selection is critical for achieving selectivity. Several attempts have been made to predict the proper selection of liquid immobile phase without resorting exclusively to trial-and-error techniques (see below).

Phases are selected based on their polarity, keeping in mind that "like dissolve like." That is, a polar stationary phase will interact more with polar compounds, and vice versa. A phase should be selected in which the solute has some solubility. Nonpolar liquid phases are generally nonselective because there are few forces between the solute and the solvent, and so separations tend to follow the order of the boiling points of the solutes, with the low-boiling ones eluting first. Polar liquid phases exhibit several interactions with solutes such as dipole interactions, hydrogen bonds, and induction forces, and there is often no correlation between the retention factor and volatility.

> *Liquid stationary phases are selected based on polarity, determined by the relative polarities of the solutes.*

For fused silica columns, the majority of separations can be done with fewer than 10 bonded liquid stationary phases of varying polarity. This is because with their very high resolving power; selectivity of the stationary phase is less critical. The stationary phases are high-molecular-weight, thermally stable polymers that are liquids or gums. The most common phases are polysiloxanes and polyethylene glycols (Carbowax), with the former the most widely used. The polysiloxanes have the backbone:

> *Polysiloxanes are the most common stationary phases for capillary GC.*

$$\left[O - \underset{\underset{R_2}{|}}{\overset{\overset{R_1}{|}}{Si}} \right]_n$$

The R functional groups determine the polarity, and include methyl (CH_3), phenyl $\left(\bigcirc\!\!\!\bigcirc \right)$, cyanopropyl ($CH_2CH_2CN$), and trifluoropropyl ($CH_2CH_2F_3$). Table 15.1 lists several commonly used stationary phases. Those with cyano functions are susceptible to attack by water and by oxygen. The carbowaxes must be liquid at operating temperatures. Incorporating either phenyl or carborane groups in the siloxane polymer backbone strengthens and stiffens the polymer backbone, which inhibits stationary-phase degradation at higher temperatures, and results in lower column bleed (loss of stationary phase by vaporization). These columns are important when coupling to a highly sensitive mass spectrometer for detection (see Chapter 17), where bleeding must be minimized. Ionic liquid (IL) stationary phases are a group of recently introduced

TABLE 15.1 Capillary Fused Silica Stationary Phases

Phase	Polarity	Use	Max.Temp. (°C)
100% Dimethyl polysiloxane	Nonpolar	Basic general-purpose phase for routine use. Hydrocarbons, polynuclear aromatics, PCBs.	320
Diphenyl, dimethyl polysiloxane	Low ($x = 5\%$) Intermediate ($x = 35\%$) Intermediate ($x = 65\%$)	General-purpose, good high-temperature characteristics. Pesticides.	320 300 370
14% Cyanopropylphenyl–86% dimethylsiloxane	Intermediate	Separation of organochlorine pesticides listed in EPA 608 and 8081 methods. Susceptible to damage by moisture and oxygen.	280
80% Biscyanopropyl–20% cyanopropylphenyl polysiloxane	Very polar	Free acids, polysaturated fatty acids, alcohols. Avoid polar solvents such as water and methanol.	275
Arylenes	Vary R as above to vary polarity	High temperature, low bleed	300–350
Carboranes open circles = boron filled circles = carbon	Vary R as above to vary polarity	High temperature, low bleed	430

(continued)

TABLE 15.1 (*Continued*)

Phase	Polarity	Use	Max.Temp. (°C)
Poly(ethyleneglycol) (Carbowax) $-[O-CH_2CH_2]_n-$	Very polar	Alcohols, aldehydes, ketones, and separation of aromatic isomers, e.g., xylenes	250
Ionic Liquids (structure with $(CH_2)_n$, 2 TfO$^-$)	Polar	Good general purpose phase. Excellent for alcohols, fatty acid methyl esters (FAMEs), aromatics, pesticides, etc. Generally they show less retention for nonpolar compounds (compared to most other phases) and more retention for polar compounds.	360
$HO-(H_2C)_n-N$...(PEG)...$N-(CH_2)_n-OH$, 2 NTF$_2$	Highly polar	Very high polarity permits separation of water from most organic solvents	300
$TfO^- \equiv O=\overset{CF_3}{\underset{O}{\overset{|}{S}}}-O^-$ $NTf_2^- \equiv$ (bis(trifluoromethylsulfonyl)imide structure)		Alcohols, aldehydes, ketones, and separation of aromatic isomers, e.g., xylenes	250

stationary phases that provide selectivities orthogonal to the majority of other phases. Uniquely, they are inert to O_2 and H_2O. They range in polarity from moderate to very highly polar; the last phase in Table 15.1 is by far the most polar. Next to carboranes, IL phases have the highest thermal stability and their low bleed makes them well suited for MS detection, arguably the most common detector in use by major users of GC.

A number of manufacturers provide very useful guides for capillary GC column selection (References 9–11).

Retention Indices for Liquid Stationary Phases

Selecting the proper packed-column stationary phase from the myriad of possible phases is challenging. Methods have been developed that group phases according to their retention properties, for example, according to polarity. The **Kovats indices** and **Rohrschneider constant** are two approaches used to group different materials. Supina and Rose (Reference 8) tabulated the Rohrschneider constants for 80 common liquid phases used in packed-column GC, which enables one to decide, almost by inspection, if it is worth trying a particular liquid phase. Equally important, it is easy to identify phases that are very similar and differ only in trade name. McReynolds described a similar approach, defining phases by their **McReynolds constants** (Reference 7). McReynolds used a standard set of test compounds for measuring retention times at 120°C on columns with 20% liquid phase loading to classify stationary phases.

Although the Kovats retention index (KRI) was originally developed for classifying liquid phases and comparing retention behavior of different analytes, the general scheme is just as applicable to capillary GC. The KRI system is designed primarily to provide a means to identify unknowns by comparing their retention with the retention of a set of known standards on the same column. *n*-Alkanes (paraffins) are used as the standards on the Kovats scale. The KRI for an alkane C_nH_{2n+2} is $100n$, being denoted by I, and is simply 100x the number of carbon atoms. The log of the adjusted retention time for the paraffins (and in general typically for any homologous series of compounds) under isothermal conditions is linearly related to the KRI. The KRI for any other compound is then simply assigned based on its log t'_R: the KRI of the unknown corresponds to where the log $t'_{R,\text{unk}}$ falls when plotted on the same line that represents the behavior of the paraffins. Mathematically, if $t'_{R,\text{unk}}$ falls between the adjusted retention times of paraffins bearing n and $n+1$ carbon atoms, the KRI of the unknown will be given as

$$I_{\text{unk}} = 100\,n + \frac{\log t'_{R,\text{unk}} - \log t'_{R,C_n}}{\log t'_{R,C_{n+1}} - \log t'_{R,C_n}} \times 100 \tag{15.1}$$

An analyte can be identified by its KRI on different columns, as these indices are tabulated for many analytes on many columns. Isothermal operation is not practical for compounds spanning a significant molecular weight range; some tabulated data are also available for specified programmed temperature conditions.

Rohrschneider envisioned a system to quantify the polarities and selectivities of stationary phases for specific compound classes. This was further developed by McReynolds, using ten selected compounds (benzene, n-butanol, 2-pentanone, nitropropanone, pyridine—these first five are often considered to provide enough information; the original ten also include 2-methyl-2-pentanol, 1-iodobutane, 2-octyne 1,4-dioxane, and cis-hydrindane) with which to characterize a phase. The McReynold's constants (MRC) of a given probe compound on a given column is the difference between the KRI of that probe and that of a standard (often squalene, a C_{30} hydrocarbon with six evenly spaced double bonds)—this is a measure of the relative polarity of the column. This suite of standards serves to measure the extent of intermolecular interaction between the stationary phase and the probe molecule. The higher the McReynolds constant, the more polar is the analyte; with the same analyte, the higher the MRC on a test stationary phase compared to 20% squalene as the stationary phase, the more polar is the test phase. As an example on 20% squalene as a stationary phase, benzene has a KRI of 649, while n-hexane and n-heptane has by definition KRI values of 600 and 700. For the same number of carbon atoms, benzene has a greater KRI than n-hexane. Under the same conditions KRI values of benzene on a dinonylphthalate and an SP-2340 phase are 733 and 1169, respectively. Therefore, dinonylphthalate is a more polar stationary phase than squalene and SP-2340 is far more polar still.

Practical applications of KRI and McReynold's constant values include comparison of different phases for similarity, or ranking them by polarity and predicting the retention order of analytes. An excellent reference that discusses KRI and the Rohrschneider/McReynold's constants and how to use them is freely accessible on the Web (Reference 12) and tabulates an extensive series of data. A more scholarly account is given in an official (International Union of Pure and Applied Chemistry) IUPAC publication (Reference 13).

Analyte Volatility

In the above discussions, we have emphasized the role of the polarity of the stationary phase (and of the analyte) in providing effective separations. The other important factor is the relative volatility of the analyte species. More volatile species will tend to migrate down the column more rapidly than lower volatility species. Gaseous species, especially small molecules such as CO, will migrate rapidly. The retention factor, k (see Equation 14.12), is related to volatility by

$$\ln k = \Delta H_v/RT - \ln \gamma + C$$

The torr is a traditional unit of pressure, now defined as exactly 1/760 of a standard atmosphere, while 1 atmosphere is 101,325 pascals. Thus, one torr is about 133.3 pascals, it is approximately equal to 1 mm of Hg.

where ΔH_v is the analyte heat of vaporization, so a higher value (higher boiling point) results in lower volatility and a larger k. Increasing the temperature T decreases this contribution to retention. The $\ln \gamma$ term is a function of the analyte-stationary-phase interaction (polarity, etc.), and is an activity term that decreases from unity for the pure undiluted analyte: interaction increases for the diluted analyte causing k to increase; C is a constant (and R is the gas constant). Quite a bit of boiling point selectivity and separation tuning capability is provided by the T-dependent term in the equation. This is why people do temperature programming (see below). The volatility of an analyte can be markedly changed by derivatization. Silylation, specifically to introduce a trimethylsilyl group, is widely practiced to increase volatility and also detectability. With the same number of cabon atoms, volatility decreases with increased extent of oxidation. n-Tetradecane at 25°C has a vapor pressure of 20 mtorr; the corresponding terminals aldehyde, alcohol, and nitrate reduce the vapor pressures to 3, 0.8, and 0.2 mtorr, going to the carboxylic acid, this drops to 7 μtorr. For refererence, n-pentadecane has a vapor pressure of 4 mtorr.

So the selection of chromatographic conditions (column, temperature, carrier flow rate) will be influenced by the compound volatility, molecular weight, and polarity.

15.3 Gas Chromatography Detectors

Since the initial experiments with gas chromatography were begun, a large number of detectors have been developed. Some are designed to respond to most compounds in general, while others are designed to be selective for particular types of substances. We describe some of the more widely used detectors. Table 15.2 lists and compares some commonly used detectors with respect to application, sensitivity, and linearity.

Thermal conductivity detectors are inexpensive and exhibit universal response, but they are not very sensitive.

The original GC detector was the **thermal conductivity**, or **hot wire**, **detector** (TCD). As a gas is passed over a heated filament wire, the temperature and thus the resistance of the wire will vary according to the thermal conductivity of the gas. Typically it is deployed in a referenced configuration: The pure carrier is passed over one filament, and the effluent gas containing the sample constituents is passed over another. These filaments are in opposite arms of a Wheatstone bridge circuit (Reference 14) that generates a voltage as the resistance of the sensing filament changes. So long as there is only carrier gas in the effluent, the resistance of the wires will be the same. But whenever a sample component elutes, a small resistance change will occur in the effluent arm. The change in the resistance, which is proportional to the concentration of the sample component in the carrier gas, is registered by the data system. The TCD is particularly useful for the analysis of gaseous mixtures, and of permanent gases such as CO_2.

Hydrogen and helium carrier gases are preferred with thermal conductivity detectors because they have a very high thermal conductivity compared with most other gases, and so the largest change in the resistance occurs in the presence of sample component gases (helium is preferred for safety reasons). The thermal conductivity of hydrogen is 53.4×10^{-5} and that of helium is 41.6×10^{-5} cal/°C-mol at 100°C, while those of argon, nitrogen, carbon dioxide, and most organic vapors are typically one-tenth of these values. The advantages of thermal conductivity detectors are their simplicity and approximately equal response for most substances. Also, their response is very reproducible. They are not the most sensitive detectors, however.

Most organic compounds form ions in a flame, generally cations such as CHO^+. This forms the basis of an extremely sensitive detector, the **flame ionization detector** (FID) (Figure 15.6). The ions are measured (collected) by a pair of oppositely charged

TABLE 15.2 Comparison of Selected Gas-Chromatographic Detectors

Detector	Application	Sensitivity Range	Linearity	Remarks
Thermal conductivity	General, responds to all substances	Fair, 5–100 ng, 10 ppm–100%	Good, except thermistors at higher temperatures	Sensitive to temperature and flow changes; concentration sensitive
Catalytic combustion	Very similar to the FID	Fair, very similar to TCD	Good	Prone to burnout at high sample concentrations
Flame ionization	All organic substances; some oxygenated products respond poorly. Good for hydrocarbons	Very good, 10–100 pg, 10 ppb–99%	Excellent, up to 10^6	Requires very stable gas flow; response for water is 10^4–10^6 times weaker than for hydrocarbons; mass sensitive
Flame photometric	Sulfur compounds (393 nm), phosphorus compounds (526 nm)	Very good, 10 pg S, 1 pg P	Excellent	
Flame thermionic	All nitrogen- and phosphorus-containing substances	Excellent, 0.1–10 pg, 100 ppt–0.1%	Excellent	Needs recoating of sodium salts on screen; mass sensitive
Rubidium silicate bead	Specific for nitrogen- and phosphorus-containing substances	Excellent		Mass sensitive
Argon ionization (β-ray)	All organic substances; with ultrapure He carrier gas, also for inorganic and permanent gases	Very good; 0.1–100 ng, 0.1–100 ppm	Good	Very sensitive to impurities and water; needs very pure carrier gas; concentration sensitive
Electron capture	All substances that have affinity to capture electrons; no resonse for aliphatic and naphthenic hydrocarbons	Excellent for halogen containing substancs, 0.05–1 pg, 50 ppt—1 ppm	Poor	Very sensitive to impurities and temperature changes; quantitative analysis complicated; concentration sensitive
Vacuum UV absorption	Nearly all substances but inert gases and nitrogen	Excellent down to pg levels	Good up to 10^4	Very recently introduced detector, expected to have a wide and universal application area, provides some structural confirmation based on spectral match
Mass spectrometry	Nearly all substances. Depends on ionization method	Excellent	Excellent	Can provide structural and molecular weight information

The flame ionization detection is both general and sensitive. It is the most commonly used general detector.

electrodes. The response (number of ions collected) depends on the number of carbon atoms in the sample and on the oxidation state of the carbon. Those atoms that are completely oxidized do not ionize, and the compounds with the greatest number of low oxidation state carbons produce the largest signals. This detector gives excellent sensitivity, permitting measurement of components in the ppb concentration range. The FID is about 1000 times more sensitive than the TCD. However, the dynamic range is more limited, and samples of pure liquids are generally restricted to 0.1 µL or less. The carrier gas is relatively unimportant. Helium, nitrogen, and argon are most frequently employed. The flame ionization detector is insensitive to most inorganic compounds, including water, and so aqueous solutions can be injected (but only if you have a compatible column). If oxygen is used as the flame support gas in place of air, then many inorganic compounds can be detected because a hotter flame is produced that can ionize them.

When sulfur and phosphorus compounds are burned in an FID-type flame, chemiluminescent species are produced that produce light at 393 nm (sulfur) and 526 nm (phosphorous). An optical interference filter passes the appropriate light to a photomultiplier tube, a sensitive photon detector. These detectors are known as flame photometric detectors (FPD).

In the FPD for sulfur, light emission takes place from the excited diatomic sulfur (S_2^*) species. As a result, the emission signal is proportional not to the sulfur concentration but to the square of it as the probability of two S atoms combining to form S_2 is proportional to the square of the concentration. This unusual dependence of the signal on a power function of the concentration leads to some unusual procedures to increase sensitivity. The sensitivity in detecting sulfur compounds *actually increases* if the carrier gas is deliberately doped with a small concentration of a sulfur compound, e.g., 1 ppb of sulfur hexafluoride (SF_6). Consider that without any sulfur doping, with zero sulfur in the background, the background would be zero and the response from a sample containing 1 and 3 ppb S will be 1 and 9 arbitrary units. If I have 1 ppb S in the carrier gas, the background will be 1 unit and when the 1 and 3 ppb samples elute, the total concentrations of 2 and 4 ppb S will produce signals of 4 and 16 units, the net signals being 3 and 15 units, much greater than those obtained with no doping. Such a beneficial effect of deliberately contaminating the background is unique.

The **catalytic combustion detector** (CCD) responds much like an FID in regard to the type of compounds it responds to and has the sensitivity of a TCD. The detector is very small (typically 1 cm diameter). The sensor element consists of a Pt wire coil embedded in an alumina ceramic containing noble metal catalysts. It is well suited for use with an air-carrier GC application; with other carrier gases, air is added prior to passage over the detector. During operation, a separate heater heats the bead to 500 °C, sufficiently hot for hydrocarbons to be rapidly oxidized (burn) in the presence of the air and the catalyst. This heat of combustion increases the temperature of the Pt filament that is sensed through its change in resistance. One needs to be careful with such a detector to not inject too much sample; too much heat will destroy the sensing filament.

The **flame thermionic detector** is essentially a two-stage flame ionization detector designed to give an increased specific response for nitrogen- and phosphorus-containing substances. A second flame ionization detector is mounted above the first, with the flame gases from the first passing into the second flame. The two stages are divided by a wire mesh screen coated with an alkali salt or base such as sodium hydroxide. This detector is also known as a nitrogen–phosphorous detector (NPD).

The column effluent enters the lower flame, which acts as a conventional FID whose response may be recorded. A small current normally flows in the second flame due to evaporation and ionization of sodium from the screen. However, if a substance

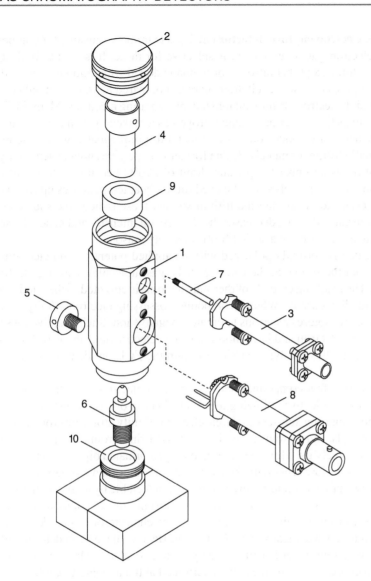

FIGURE 15.6 The anatomy of a flame ionization detector.
1: Detector body, 2: Chimney, 3: Electrical collection assembly, 4: Collecting electrode, 5: Retaining screw, 6: Jet (flame emanates from here), 7: Collecting electrode, 8: Flame igniter, 9: Ceramic insulator, 10: Detector base. (Courtesy ThermoFisher Scientific.)

containing nitrogen or phosphorus is burned in the lower flame, the ions resulting from these greatly increase the volatilization of the alkali metal from the screen. This results in a response that is much greater (at least 100 times) than the response of the lower flame to the nitrogen or phosphorus. By recording the signals from both flames, one can obtain the usual chromatogram of an FID from the lower flame; a prominent response from the top channel (relative to the bottom channel) appears only when nitrogen- and phosphorus-containing compounds elute.

In the β-**ray**, or **argon ionization, detector**, the sample is ionized by bombardment with β rays from a radioactive source (e.g., strontium-90). The carrier gas is argon, and the argon is excited to a metastable state by the β particles. Argon has an excitation energy of 11.5 eV, which is greater than the ionization potential of most organic compounds, and the sample molecules are ionized when they collide with the excited argon atoms. The ions are detected in the same manner as in the flame ionization detector. This detector is about 300 times more sensitive than is the TCD. The **helium ionization detector** (HID) operates on the same principle except that He, instead of Ar is used. In the **discharge ionization detector** (DID), excited helium atoms are created by a electrical discharge, these then ionize the analyte molecules.

The ECD is very sensitive for halogen-containing compounds, for example, pesticides.

The **electron capture detector** (ECD) is extremely sensitive for compounds that contain electronegative atoms and is selective for these. It is similar in design to the β-ray detector, except that nitrogen or methane doped with argon is used as the carrier gas. These gases have low excitation energies compared to argon and only compounds that have high electron affinity are ionized, by capturing electrons. Many ECDs operate with helium as the carrier gas, using nitrogen as a makeup gas in the detector.

The detector cathode consists of a metal foil impregnated with a β-emitting element, usually tritium or nickel-63. The former isotope gives greater sensitivity than the latter, but it has an upper temperature limit of 220°C because of losses of tritium at high temperatures; nickel-63 can be used routinely at temperatures up to 350°C. Also, nickel is easier to clean than the tritium source; these radioactive sources inevitably acquire a surface film that decreases the β-emission intensity and hence the sensitivity. The β sources are used in a sealed form for safety reasons.

The cell is normally polarized with an applied potential, and electrons (β rays) emitted from the source at the cathode strike gas molecules, causing electrons to be released. The resulting cascade of thermal electrons is attracted to the anode, and establishes a standing current. When a compound possessing electron affinity is introduced into the cell, it captures electrons to create a negative ion. Such negative ions are much larger than an electron and have mobilities in an electric field about 100,000 times less than electrons Thus, analytes passing through the ECD are detected by a decrease in standing current.

Chlorofluorocarbons (CFCs) were first made by Thomas Midgley at General Motors in 1928. They became the choice refrigerants as they had low boiling point, low toxicity, and were largely nonreactive. In a demonstration for the American Chemical Society in 1930, Midgley flamboyantly demonstrated all these properties by inhaling a breath of the gas and using it to blow out a candle. They became the aerosol propellants of choice. By the 1970s, more than a million tons were annually being used and released into the air. Because they are unreactive, CFCs would persist and they should have been measurable in air but through the 1960s there were no detectors sensitive enough to detect the minuscule atmospheric concentrations of these compounds.

Relatively few compounds show significant electronegativities, and so electron capture is quite selective, allowing the determination of trace constituents in the presence of noncapturing substances. High-electron-affinity atoms or groups include halogens, carbonyls, nitro groups, certain condensed ring aromatics, and certain metals. The ECD is widely used for pesticides and polychlorinated biphenyls (PCBs). Electron capture has very low sensitivity for hydrocarbons other than aromatics.

Many analytes of interest are not directly detectable by an ECD but may be determined by preparing appropriate derivatives. Most biological compounds, for example, possess low electron affinities. Steroids such as cholesterol can be determined as the chloroacetates. Trace metals, e.g., Al, Cr, Cu, Be, etc., have been determined at pg-ng levels by preparing volatile trifluoroacetylacetone chelates. Methylmercuric chloride, present in contaminated fish, can be determined at the nanogram level.

The gas chromatograph may be interfaced with atomic spectroscopic instruments for specific element detection. Chromatography is used to separate different forms of an element, and atomic spectroscopy detection identifies the element. This powerful combination is useful for speciation of different forms of toxic elements in the environment. For example, a helium microwave induced plasma **atomic emission detector** (AED) can be used to detect volatile methyl and ethyl derivatives of mercury in fish, separated by GC. The AED can simultaneously determine the atomic emissions of many elements The emitted light passes through a monochromator and is detected by an array detector. Also, gas chromatographs are interfaced to inductively coupled plasma–mass spectrometers (ICP–MS) in which atomic species from the plasma are introduced into a mass spectrometer (see Chapters 12 and 17), for very sensitive simultaneous detection of species of several elements and even to differentiate between different isotopes of the same element.

There are other specific detectors that enjoy considerable use in niche applications. The **sulfur chemiluminescence detector** (SCD) is among the most sensitive and selective chromatographic detectors available for the analysis of sulfur compounds, of interest to several industries. High-temperature combustion of sulfur-containing compounds is used to form sulfur monoxide (SO). This is then reacted with ozone; the highly energetic reaction results in light emission, detected by a photomultiplier tube. A linear and equimolar response to the sulfur present in different compounds is observed, generally without interference from the sample matrix. The **nitrogen chemiluminescence detector** (NCD) works similarly; NO is produced from nitrogen-containing compounds and then reacts with ozone to produce chemiluminescence.

The **photoionization detector** (PID) provides a different type of selectivity. Sufficiently energetic UV radiation can dislodge an electron from many compounds, producing an electron and a positive ion, which can then be detected the same way as in an FID. The column effluent is exposed to intense UV radiation. What is detected depends on the lamp used, and current instruments mostly use inert gas-filled electrodeless discharge lamps (EDLs, see Chapter 12). Each analyte has its own unique ionization potential (IP). The photon energy must exceed the IP for the analyte to be detected. The 10.6 eV lamp (116.9 nm, krypton gas, MgF_2 window) is the most widely used because it detects most volatile organic compounds and the lamp is easy to clean. The Kr lamp also produces 10.0 eV (123.9 nm) radiation. An argon lamp can provide 11.7 eV radiation (105.9 nm) and will ionize even methanol. However, it requires an expensive and hygroscopic LiF window and is not commonly used. All nonflame ionization detectors ionize only a very small portion of the sample and can generally be considered nondestructive. Other nondestructive detectors used in GC include absorption spectroscopic detectors; both IR and UV, and even nuclear magnetic resonance (NMR) spectroscopy is used as a GC detector, but none of these are common.

While GC detectors have often been used for detection in liquid chromatography, the opposite is rare. The one exception is a solution phase detector developed in the 1970s by Randall C. Hall, a Purdue entomologist, for detecting pesticides. The **Hall electrolytic conductivity detector** (HECD) is a solution phase electrical conductivity detector (see Chapter 16) that can detect compounds of certain elements with very high sensitivity. It consists of two distinctly different components, the furnace (500–1000°C) and the conductivity cell, which uses a scrubbing solvent, typically 1-propanol (See Figure 15.7). The detector can selectively detect halogen, nitrogen, or sulfur. In the halogen mode, an inert carrier gas is used and passage through the hot nickel reaction tube decomposes any halogenated hydrocarbon to the corresponding acid (HCl, HBr, etc.); sulfur compounds also produce H_2S under these conditions. The gases are transferred through a PTFE conduit where it meets slightly acidic 1-propanol as the scrubbing solvent at a tee and proceeds through the conductivity cell. The solvent is recycled through a disposable ion exchange bed and reused. The use of slightly acidic 1-propanol prevents H_2S from dissolving in the solvent and provides selectivity over S compounds. As HCl, HBr, etc. dissolve in the alcohol, they ionize fully and raise the conductivity. Single digit pg levels of halogenated organics can be detected with a discrimination ratio over hydrocarbons as large as 10^9:1. In the nitrogen mode, the furnace is operated at a much higher temperature (at the furnace temperatures used for halogenated compounds, NH_3 is not significantly formed) and the evolved gases pass through a $Sr(OH)_2$-filled cartridge to remove acidic gases—the evolved ammonia is scrubbed into slightly alkaline aqueous ~15% 1-propanol. The detection limits are in the pg range and discrimination over hydrocarbons is 10^5:1, but because ammonia is a weak base (the scrubbing solvent is mostly water because ionization in pure propanol is even poorer), the linear range spans three orders of magnitude, compared to six orders

Lovelock developed the ECD in 1970 and became the first to detect CFCs in the air and reported that the dominant CFC in use had an atmospheric concentration of about 60 parts per trillion. To put this in perspective, the atmospheric concentration of methane is ~1.5 ppm, merely detecting methane was considered a major feat in 1950. The discovery that CFCs can even be detected in unpolluted air prompted Lovelock to ask the UK government for modest funds to place his apparatus on board a ship bound from England to Antarctica. His request was summarily rejected; one reviewer commented that it probably could not be measured, and even if it could be, there could not be a more useless bit of knowledge. Lovelock persisted, however. With his own money, he put his experiment aboard the research vessel Shackleton in 1971. Two years later he reported the detection of CFC-11 in essentially all air samples collected in the North and South Atlantic. Later it would be discovered that CFCs photochemically decompose to form chlorine atoms in the stratosphere where they destroy ozone in a chain reaction. Because of their threat to the ozone layer (that protects us from UV radiation), the production of CFCs has been phased out. Lovelock is also the father of the "Gaia" hypothesis that posits that organisms interact with their inorganic surroundings on Earth to form a self-regulating, complex system that contributes to maintaining the conditions for life on the planet; his book by that name is considered a classic.

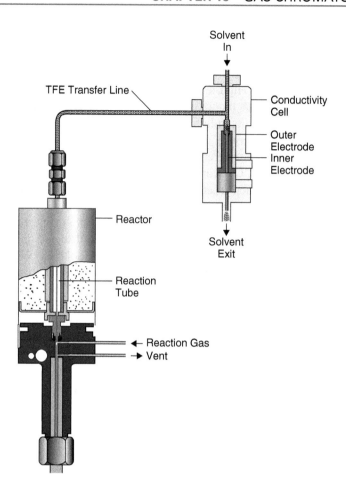

FIGURE 15.7 Cross-sectional view of a Hall electrolytic conductivity detector. (Courtesy ThermoFisher Scientific.)

of magnitude for halogen detection. In the sulfur mode, air or oxygen is added and sulfur is converted to SO_2 and/or SO_3. Lower temperatures favor the production of the latter. 1-Propanol or other alcohols are used as the scrubbing solvent—performance and selectivity are not as good as in the other modes.

Another halogen selective detector for GC, called a **dry electrolytic conductivity detector** (DELCD) is somewhat of a misnomer: it does not as much measure conductivity as it measures the current resulting from the reduction of halogen oxides (e.g., ClO_2 or BrO_2). Air is added to the column effluent and then passed through a ceramic tube heated to 1000°C, where the halogenated compounds are significantly converted to the dioxides. These reduce back to the elemental halogen when passing between a Pt cathode and a nichrome anode located towards the exit of the tube and this reduction current (incorrectly referred to as conductivity) is measured. Halogenated compounds can be detected at the ppb level. The DELCD can also be used after an FID, but most of the halogen has already been converted to HCl or HBr, which does not respond to the DELCD. Only ~0.1% of the halogen is converted to the oxides, allowing detection of halogenated organics only at the ppm level.

One of the most versatile GC detectors that can operate both in the universal and selective mode is the **pulsed discharge (ionization) detector** (PDD) (Reference 15). The PDD, developed by William Wentworth at the University of Houston in 1992 (Reference 16), is based on a pulsed high voltage discharge between platinum electrodes, typically used with a helium carrier gas. The transition of excited diatomic He_2 to the He ground state results in UV radiation (60–100 nm, ~13.5–17.5 eV) of sufficient energy to ionize all elements and compounds except for neon. The PDD can be configured as a universal, selective or an atomic/polyatomic emission detector. In the universal mode, called the helium pulsed discharge photoionization detector (He-PDPID), the eluting analyte is photoionized and the resulting electrons produce a current across suitably placed downstream electrodes. Response to permanent gases is positive (increase in standing current), with LODs in the low ppb range. The He-PDPID is an attractive replacement for an FID in situations where the flame and use of hydrogen may pose a risk. The PDPID can have the He discharge gas doped with Ar, Kr, or Xe, and the excited atoms of these gases are then exclusively produced, resulting in their characteristic photon emissions; what classes of compounds will be ionized and detected depend on the dopant gas (see the PID above).

For operation in the electron capture mode (PDECD), He and CH_4 are introduced just before the detector. The presence of methane results in a significant standing current between the sensing electrodes, and high electron affinity compounds such as freons, chlorinated pesticides, and other halogen compounds are very sensitively detected as they capture electrons generated by methane ionization. In this case, the LODs are comparable to the conventional ECD, in the fg-pg range. The detector can be operated temperatures up to 400°C.

The pulsed discharge emission detection mode (PDED) is presently the least commonly used. In this case the PDD has a terminal quartz window and the emission lines resulting from the analytes excited by the high-energy photons are imaged with a monochromator and a detector. The original work was done with a single photomultiplier tube detector, but current availability of array detectors may bring much to this detector configuration.

Detectors are either concentration sensitive or mass flow sensitive. The signal from a concentration-sensitive detector is related to the concentration of the solute in the detector and is decreased by dilution with a makeup gas. The sample is usually not destroyed. Thermal conductivity, argon-ionization, and electron capture detectors are concentration sensitive. In mass-flow-sensitive detectors, the signal is related to the rate at which solute molecules enter the detector and is not affected by the makeup gas. These detectors usually destroy the sample, such as flame ionization and flame thermionic detectors. Earlier, when two-column GC was used to increase resolution, cuts of the effluent from a first column were taken and directed to a second column for secondary separation. The first detector must be nondestructive or else the eluent split prior to detection, with a portion going to the second column. In current practice of GC-GC (see below), the detector is almost always a mass spectrometer and generally the effluent from the first column passes on to the second dimension without going through a detector.

GC-VUV

An emerging technology for gas chromatography detection is vacuum ultraviolet (VUV) spectroscopy. Involving the absorption of light ranging in wavelength from approximately 120–200 nm, VUV spectroscopy has historically been limited to investigations at specialized synchrotron facilities, where these short-wavelength photons could be harnessed from electrons accelerated to extremely high kinetic energies. Very few accounts can be found of VUV spectroscopy in textbooks and in the literature. Only recently, have manufacturers devised bench-top devices with this capability (see VUV Analytics, Inc., www.vuvanalytics.com) that utilize a standard deuterium light source. The windows of the lamp and flow cell are made from MgF_2, which allows for the transmission of low wavelength VUV radiation (for more on such materials, see http://www.esourceoptics.com/vuv_material_properties.html). GC provides an ideal means for separating and introducing vapor phase analytes into the spectrophotometer; as a detector, VUV provides exceptional quantitative and qualitative capabilities, potentially poised to meet or exceed the performance of many modern GC detectors.

The biggest advantages of vapor phase VUV spectroscopy are that all molecules absorb in this region of the electromagnetic spectrum and when measurements are made in the gas phase, absorption features are sharp (in solution, interactions with solvent blur spectral absorption features); every molecule has a unique spectrum, which can be used to unequivocally assign its identity. Shown are normalized absorption spectra for two common metabolites of naphthalene, 1- and 2-naphthol. Perhaps surprisingly, standard GC-MS with electron ionization cannot distinguish these two compounds based on their fragmentation patterns; however, their VUV absorption patterns are very different. Additionally, absorption in the VUV still follows Beer's Law principles; therefore, quantitative analysis can be performed. Accounts place detection limits for most molecules in the low picogram levels for GC-VUV. This rivals the capabilities of mass spectrometry and flame ionization detection capabilities.

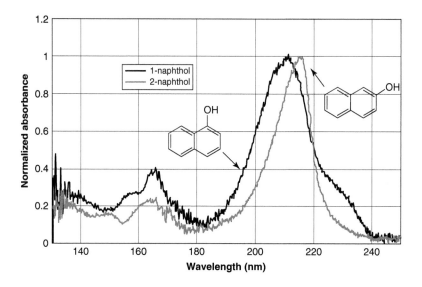

15.4 Temperature Selection

Chromatographic conditions represent a compromise between speed, resolution, and sensitivity.

The proper temperature selection in gas chromatography is a compromise between several factors. The **injector or injection port temperature** should be relatively high, consistent with thermal stability of the sample, to give the fastest rate of vaporization to get the sample into the column in a small volume to get decreased band broadening and increased resolution. Too high an injection temperature, though, will tend to degrade the injection septa and foul the injection port. The **column temperature** is a compromise between *speed*, *sensitivity*, and *resolution*. At high column temperatures, the sample components spend most of their time in the gas phase and so they are eluted quickly, but resolution is poor. At low temperatures, they spend more time in the stationary phase and elute slowly; resolution is increased but sensitivity is decreased due to increased broadening of the peaks. The **detector temperature** must be high enough to prevent condensation of the sample components. The sensitivity of the thermal conductivity detector decreases as the temperature is increased and so its temperature is kept at the minimum required.

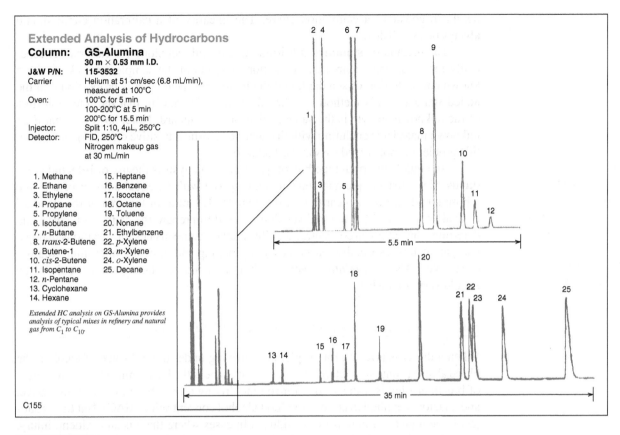

FIGURE 15.8 Temperature-programmed analysis. (Courtesy of Agilent Technologies.)

Separations can be facilitated by **temperature programming**, and most gas chromatographs have temperature programming capabilities. The temperature is automatically increased at a preselected rate during the running of the chromatogram; this may be linear, exponential, steplike, and so on. In this way, the compounds eluted with more difficulty can be eluted in a reasonable time without forcing the others from the column too quickly.

Figure 15.8 shows a temperature programmed separation of a complex hydrocarbon mixture with stepwise linear temperature programming. The first 12 gaseous or light compounds are readily eluted and resolved at a low fixed (100°C) temperature for 5.5 min , while the others require higher temperatures. After 5.5 min , the temperature is linearly increased at 5°C/ min for 20 min to 200°C, and then the temperature is held at that value until the last two compounds are eluted.

As stated before, if the constituent to be determined is not volatile at the accessible temperatures, it may be converted to a **volatile derivative**. For example, nonvolatile fatty acids are converted to their volatile methyl esters. Some inorganic halides are sufficiently volatile that at high temperatures they can be determined by gas chromatography. Metals may be made volatile by complexation, for example, with trifluoroacetylacetone.

> Temperature programming from low to higher temperatures speeds up separations. The more difficult to elute solutes are made to elute faster at the higher temperatures. The more easily eluted ones are better resolved at the lower temperatures.

15.5 Quantitative Measurements

The concentrations of eluted solutes are proportional to the areas under the recorded peaks. Electronic integrations in GC instruments report the areas of peaks, and the retention times of peaks are also generally given. It is also possible to measure peak height to construct a calibration curve. The linearity of a calibration curve should always be established.

The method of **standard additions** is a useful technique for calibrating, especially for occasional samples. One or more aliquots of the sample are spiked with a known concentration of standard, and the increase in peak area is proportional to the added standard. This method has the advantage of verifying that the retention time of the unknown analyte is the same as that of the standard. Note, however, that if an unknown material is coeluting with the analyte causing a positive error in quantitation, this problem is not solved by standard addition.

An equally important method of quantitative analysis is the use of **internal standards**. Here, the sample and standards are spiked with an equal amount of a solute whose retention time is near that of the analyte. The ratio of the area of the standard or analyte to that of the internal standard is used to prepare the calibration curve and determine the unknown concentration. This method compensates for variations in physical parameters, especially inaccuracies in pipetting and injecting microliter volumes of samples. Also, the *relative* retention should remain constant, even if the flow rate should vary somewhat.

An internal standard is usually added to standard and sample solutions. The ratio of the analyte peak area to internal standard peak area is measured and will remain unaffected by slight variations in injected volume and chromatographic conditions.

Spreadsheet Exercise: Internal Standard Calibration

When a driver is arrested for suspicion of driving under the influence of alcohol, the blood alcohol content is determined to see if it exceeds the legal limit. The measurement of breath alcohol is usually done for routine driving arrests because it is noninvasive, and a factor is applied to convert to blood alcohol concentration (BAC). But this is subject to biological variations in individuals. In cases where there is an accident, injury, or death, the blood alcohol is usually determined directly, by analyzing a blood sample by gas chromatography.

A 5.00-mL blood sample from a suspect is spiked with 0.500 mL of aqueous 1% propanol internal standard. A 10-µL portion of the mixture is injected into the GC, and the peak areas are recorded. Standards are treated in the same way. The following results were obtained:

% EtOH (wt/vol)	Peak Area EtOH	Peak Area PrOH
0.020	114	457
0.050	278	449
0.100	561	471
0.150	845	453
0.200	1070	447
Unknown	782	455

Use Excel Data Analysis-based regression analysis for the ratio of the EtOH/PrOH areas vs. EtOH concentration, and calculate the unknown concentration with the associated uncertainty. In this State, a BAC level of 0.08% or higher is considered the legal definition of intoxication; was the person legally intoxicated at the 95% confidence level?

We use the regression function as described in Chapter 11, Section 11.7. See the spreadsheet Section 15.5 on the text website for the calculations. Click on the calculated numbers to see the formulas. We calculate the peak area ratio (EtOH/PrOH) and plot this as the *x*-variable and % w/v EtOH as the *y*-variable, as this will facilitate the calculation of the unknown y-value. The output of the regression analysis gives us the best fit equation, based on the 5 calibration points ($n = 5$):

$$\%EtOH \ (w/v) = (8.30 \pm 0.16) \times 10^{-2} * \text{area ratio EtOH/PrOH} - (9.40 \pm 24.3) \times 10^{-4}$$

Note that the uncertainties given in the program output are stated to be the standard errors, that is, they are equal to $s/\sqrt{n}$. Should we want to write the above equation in terms of the standard deviation, we need then to multiply the uncertainties by $\sqrt{n}$ or 2.236:

$$\%EtOH \ (w/v) = (8.30 \pm 0.36_4) \times 10^{-2} * \text{area ratio EtOH/PrOH} - (9.40 \pm 54.3) \times 10^{-4}$$

Noting that for the sample, the area ratio of EtOH/PrOH is 1.719,

$$\%EtOH(w/v) \text{ in unknown} = (8.30 \pm 0.36_4) \times 10^{-2} * 1.719 - (9.40 - 54.3) \times 10^{-4}$$

$$\%EtOH(w/v) \text{ in unknown} = (14.3 \pm 0.63) \times 10^{-2} - (9.40 \pm 54.3) \times 10^{-4}$$

Standard propagation of error calculations gives

$$\%EtOH(w/v) \text{ in unknown} = 0.142 \pm 0.008$$

As the uncertainty here is based on one standard deviation and a 95% confidence limit requires an uncertainty of ± 2 standard deviations, we can say with 95% confidence that the BAC falls between $0.142 \pm 0.016\%(w/v)$ or $0.126 - 0.158\%(w/v)$. We could have also calculated the 95% certainty lower bound from the 95% certainty lower limit of the slope (7.78×10^{-2}) and the intercept (-8.68×10^{-3}) as in the first equation above (remembering the sign of the intercept) to be 0.125% (w/v) and the 95% certainty upper bound from the 95% certainty upper limit of the slope (8.82×10^{-2}) and the intercept (6.80×10^{-3}) to be 0.158% (w/v). This individual will be considered legally intoxicated.

The ability to drive exponentially decreases with alcohol concentration and the level of intoxication roughly doubles for every 0.05% BAC, so someone with 0.20% blood alcohol is 8 times as intoxicated as one with 0.05%!

15.6 Headspace Analysis

In Chapter 13, we described solvent extraction and solid-phase extraction sample preparation methods, which are applicable to GC analyses as well as others. A convenient way of sampling volatile samples for GC analysis is the technique of **headspace analysis**. A sample in a sealed vial is equilibrated at a fixed temperature, for example, for 10 min, and the vapor in equilibrium above the sample is sampled and injected into the gas chromatograph. A typical 20-mL glass vial is capped with a silicone rubber septum lined with polytetrafluoroethylene (PTFE). A syringe needle can be inserted to withdraw a 1-mL portion. Or the pressurized vapor is allowed to expand into a 1-mL sample loop at atmospheric pressure, and then an auxiliary carrier gas carries the loop contents to the GC loop injector. Volatile compounds in solid or liquid samples can be determined at parts per million or less. Pharmaceutical tablets can be dissolved in a water–sodium sulfate solution for headspace analysis The sodium sulfate is added to "salt out" the volatile analyte more effectively. Figure 1 in your text **website**, Chapter 15, shows a headspace chromatogram of volatile compounds in a blood sample. For this type of analysis, the prevalent present practice is to use an SPME fiber to sample the headspace (see Chapter 13).

Headspace analysis avoids the need for solvent extraction for volatile analytes.

15.7 | Thermal Desorption

In thermal desorption, the volatile analyte is desorbed from the sample by heating and introduced directly into the GC.

Thermal desorption (TD) is a technique in which solid or semisolid samples are heated under a flow of inert gas. Volatile and semivolatile organic compounds are extracted from the sample matrix into the gas stream and introduced into a gas chromatograph. Samples are typically weighed into a replaceable PTFE tube liner, which is inserted into a stainless steel tube for heating.

The thermal desorption must take place at a temperature below the decomposition point of other materials in the sample matrix. Solid materials should have a high surface area (e.g., powders, granules, fibers). Bulk materials are ground with a coolant such as solid carbon dioxide prior to weighing. This technique simplifies sample preparation and avoids the necessity of dissolving samples or solvent extraction. Thermal desorption is well suited for dry or homogeneous samples such as polymers, waxes, powders, pharmaceutical preparations, solid foods, cosmetics, ointments, and creams. There is essentially no sample preparation required.

See www.markes.com for literature on sorbent selection for thermal desorption.

An example of the use of TD is for the analysis of water-based paints for organic volatiles. The TD tube is used in combination with a second tube containing a sorbent that removes the water, which cannot be introduced into most capillary GC columns, except for ionic liquid phases. A small aliquot of paint (e.g., 5 μL) is placed on glass wool in the TD tube. Solids from the paint remain behind.

A somewhat related technique is pyrolysis-GC, most commonly in conjunction with an MS detector. The sample, typically a polymer or a composite sample like paint, is heated under controlled conditions to decomposition—the vapors from this pyrolysis process are analyzed by GC-MS. The "fingerprints" thus generated can be remarkably characteristic and surprisingly reproducible, provided that pyrolysis conditions are exactly reproduced. See Reference 17 for a review and Reference 18 for practical aspects. There are many pyrolysis-GC instruments available and are invaluable for characterizing many types of materials. The *Journal of Analytical and Applied Pyrolysis* is a scholarly publication solely devoted to this topic.

15.8 | Purging and Trapping

Purge-and-trap is a form of headspace analysis in which the volatile analyte is trapped on a sorbent and then thermally desorbed.

The **purge-and-trap** technique is a variation of thermal desorption analysis in which volatiles are purged from a liquid sample placed in a vessel by bubbling a gas (e.g., air) through the sample and collecting the volatiles in a sorbent tube containing a suitable sorbent. The trapped volatiles are then analyzed by thermally desorbing them from the sorbent. This can be considered a form of "headspace" analysis in which analytes are concentrated prior to introduction into the GC. A typical example involves the determination of chloroform and other halogenated organics in water. Generally a hydrophobic sorbent that can collect organics as volatile as hexane is used. Examples are Tenax TA or graphitized carbon. These sorbents allow bulk polar solvents such as water or ethanol to pass through unretained. Whiskey is analyzed for C4 to C6 ethyl esters, which are markers of maturity, but the alcohol is not significantly collected by the sorbent and therefore does not interfere in the chromatogram.

Purge-and-trap is suitable for samples that are far from homogeneous since fairly large samples can be taken, including samples of high water-content. Examples include foods such as pizza or fruits. The measurement of malodorous organic volatiles in the headspace vapor above a sample of aged food is used to determine whether it still meets the "freshness" requirements. The food sample, placed in a large purge vessel, is heated under a flow of air, and the effluent air is collected on the sorbent.

Sorbents more selective than Tenax-type sorbents may be used, and two or more may be used in series for measurement of different classes of compounds. Other sorbents include many materials used for chromatography such as alumina, silica gel, Florisil (for PCBs), coconut charcoal, Porapak, and Chromosorb. Some may be coated for specific applications, for example, silica gel coated with sulfuric acid or sodium hydroxide for collecting bases or acids.

Another important use of trapping is for the direct analysis of gaseous samples such as air. The sample is passed directly through the sorbent tube, and the trapped volatiles are subsequently desorbed or removed by extraction. This technique is widely used for indoor and outdoor air monitoring.

For more information about the sorbents and thermal desorption, see SKC (www.skcinc.com) and Markes International (www.markes.com). SKC lists sorbents for the determination of specific analytes.

Volatile organic compounds (VOCs) in ambient air are also analyzed by sampling in evacuated stainless steel canisters that have the interior specially electropolished or passivated. The evacuated containers range in volume from 0.4 to 6 L and are equipped with an inlet valve, 2-μm particle filter, an orifice to control flow in and a vacuum gauge. The orifice can be selected to collect "grab samples" in less than 1 min or collect time-integrated samples over 24–48 h. Canister sampling is generally applicable to VOCs present in concentrations > 0.5 ppb. One great advantage of canister sampling is stability of the collected sample for periods up to 30 days such that samples can be collected in the field and shipped to a central laboratory for analysis. They allow analysis of many VOCs that cannot be thermally desorbed from sorbents without loss or decomposition. The USEPA has several prescribed methods for both indoor and outdoor air that rely on canister sampling. Table S15.1 on the text **website** provides a list of compounds in air that are amenable to determination by EPA-prescribed methods using canister sampling and GC-MS. For low-level ambient analysis, a 250- to 500-ml sample is withdrawn from the canister and concentrated using a cryogenically cooled trap. After the sample is adequately concentrated, the VOCs are thermally desorbed, placed onto a cryofocuser, and introduced onto a gas chromatographic (GC) column for separation.

15.9   Small and Fast

The combination of narrow-bore chromatographic columns with very thin liquid-phase films, and fast, small volume detectors and data systems capable of acquiring and processing data at kHz rates, has made possible very rapid GC analyses, for example, in a few seconds. Small diameter and short capillary columns are used with hydrogen as the carrier gas (which allows more rapid mass transfer) and fast temperature program rates. Faster flow rates and higher pressures are used. Fast elution of analytes requires fast detector response time and data acquisition rates because peak widths are on the order of only 0.5 s, compared to 0.5 to 2 s or more for conventional capillary GC, improving analysis times 5- to 10-fold. High-speed chromatography with short columns makes column selection (selectivity) more critical than in conventional capillary GC. Figure 15.9 shows a fast GC chromatographic separation of hydrocarbons in less than 10 s using a 0.32-mm-diameter column of 5 m length, and 0.25-μm-thick stationary phase, compared to 10 min using conventional GC.

Shorter columns and faster temperature programming causes some loss in resolution, but this is partially offset by the smaller internal diameter and thinner liquid film. The second dimension in a two-dimensional GC separation (see Section 15.10 below) is always operated in such a fast GC mode.

Thin-film stationary phases, small diameter and short columns, and a light carrier gas gives fast separations.

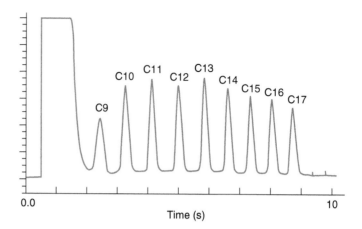

FIGURE 15.9 Separation of C9 to C17 hydrocarbons by fast chromatography, with rapid temperature programming. Conditions: 5 m × 0.32 mm, 0.25 μm d_f, 60°C, 19.2°C/s. [From G. L. Reed, K. Clark-Baker, and H. M. McNair. *J. Chromatogr. Sci.*, **37** (1999) 300. Reproduced from the *Journal of Chromatographic Science* by permission of Preston Publications, A Division of Preston Industries, Inc.]

There are small portable GCs available from a number of manufacturers. Many advances have been made in portable GC and GC-MS instruments (see C. M. Harris "GC to Go," Today's Chemist at Work, March, 2003, pp. 33–38). Very recently, the combination of a high-speed, high-resolution gas chromatograph with a miniature toroidal ion trap mass spectrometer has been introduced to identify volatile and semivolatile organic compounds in the field. In some versions, the sampling is done by an SPME fiber. In others, greater sensitivity is attained by sampling through a "needle-trap," a small sorbent-packed needle, introduced by Janusz Pawliszyn of the University of Waterloo, who invented SPME as well. Many applications of a portable GC-MS are available on the Web as presentations (www.torion.com) and an extensive listing of portable GC/MS and miniature MS instrumentation is available in www.gcms.de.

15.10 Separation of Chiral Compounds

Many naturally occurring compounds are chiral (there is asymmetry at least around one atom—this results in structural mirror images not being superimposable). Properties of such optical isomers can vary markedly, especially in regard to their physiological effects. Many pharmaceuticals are chiral, and more often than not, only one isomer provides therapeutic benefits while the other only contributes to undesirable side effects. Efforts are made therefore to make such drugs in chirally pure form, often starting with chirally pure ingredients and stereoselective synthetic routes. At all steps in such processes, the exact ratio of each optical isomer (enantiomer) must be ascertained; given that the world market for chiral drugs exceeds $350 billion as of 2012, this is a very important application area.

Achiral or conventional stationary phases cannot separate chiral enantiomers. A basic tenet of chiral separations is that on a molecular scale, the analyte must interact with the binding site(s) on the stationary phase at a minimum of three independent locations, otherwise stereoselective recognition does not result. Daniel Armstrong (presently at the University of Texas at Arlington) pioneered GC stationary phases that contained bonded cyclodextrin molecules and/or their derivatives (he also invented the ionic liquid-based GC stationary phases). Cyclodextrins are basket-shaped molecules (Figure 15.10) where chiral analytes can interact with the mouth of the basket that provides a 3-point interaction reaction. Presently, chiral GC stationary phases consist mainly of various derivatized cyclodextrins. An estimated > 95% of all reported GC enantiomeric separations utilize some type of cyclodextrin-based chiral selector. It should be noted, however, that different derivatives can have very different selectivities.

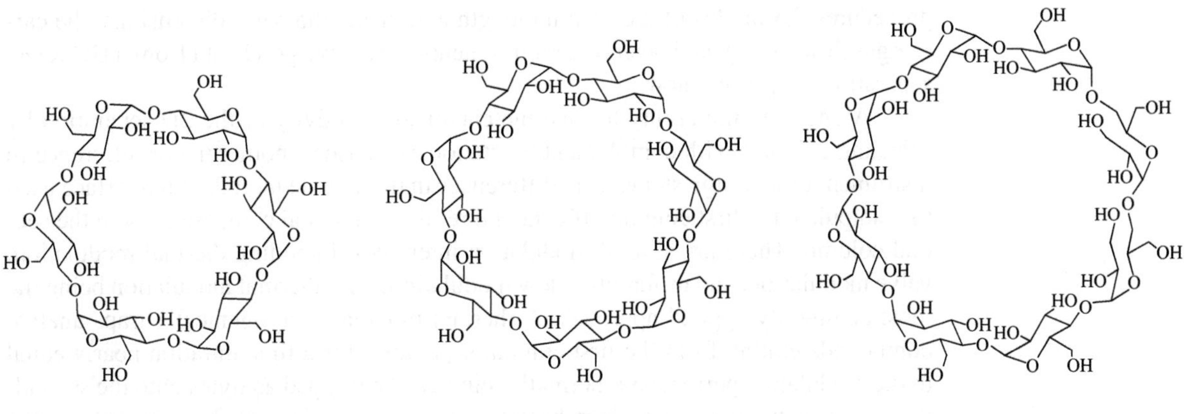

α-Cyclodextrin   β-Cyclodextrin   γ-Cyclodextrin

FIGURE 15.10 Structures of α- β-, and γ-cyclodextrin molecules—the cavity size increases in that order.

Not only do different cyclodextrin derivatives separate different classes of molecules, but they sometimes reverse the enantiomeric elution order. For an account of the historical development of enantiomeric separations, see Reference 19.

15.11 Two-Dimensional GC

Comprehensive two-dimensional (2D) gas chromatography (GC × GC), pioneered in 1991 by John Phillips at the University of Southern Illinois (Reference 20), has evolved into a powerful separation technique that is well suited for the analysis of complex chemical samples. A GC × GC instrument utilizes two gas chromatographic capillary columns interconnected by a sample modulation interface. The sample modulation interface injects portions of the first column effluent onto the second column at rapid intervals. The first column is the longer of the two columns, and often has a nonpolar stationary phase and provides a separation time of ~30 to 60 minutes. The second dimension separation is performed in real time as effluent (carrier gas and separated analyte compounds) leaves the first column, with the second column separation run time taking ~1 to 5 s for each modulation (referred to as the modulation period). Analyte peaks leaving the first column should be sampled by ~3 to 4 modulation periods to sufficiently preserve the resolving power of the first-dimension separation. For example, for an analyte peak eluting from the first column that is 8 s wide (4 standard deviation peak width at the base), would suggest use of a 2 s modulation period (second column run time) to provide a sufficiently comprehensive separation.

In the second-dimension, a polar stationary phase column is commonly used, providing a complementary separation relative to the first column, so species not separated on the first-dimension at a given first-dimension retention time have the opportunity to be separated on the second-dimension column. Utilization of two columns providing complementary separations greatly enhances peak capacity, making GC × GC more powerful than one-dimensional GC. In recent applications, researchers are using a polar column in the first dimension and a nonpolar column in the second-dimension to leverage the longer polar column more effectively. Other stationary phases are gaining popularity such as ionic liquids, chiral, and so on.

Peak widths on the second-dimension separation are on the order of 100 to 200 ms, so the detector must have a fast enough duty cycle to measure an adequate number of points (e.g., 10 to 20) across such narrow peaks, to facilitate data analysis

Contributed by Professor Rob Synovec, University of Washington

procedures. Examples of the column length and inside diameter dimensions, the carrier gas flow rate(s), and detection requirements to achieve good data from a GC × GC separation are given below.

Along with the ability to tune the chemical selectivity of the 2D separation by selecting columns with complementary stationary phases, another major difference in instrumentation design stems from differences in the sample modulation interface used to manipulate the first column effluent and reinject chemical compounds onto the second column. There are several modulation interfaces, including thermal modulation, valve modulation, and differential flow modulation, with thermal modulation being the most commonly applied approach. A thermal modulator cryogenically traps analyte compounds eluting from the first column separation for a time duration nearly equal to the modulation period, and thermally reinjects the trapped analytes onto the second-dimension column using a short heating period (e.g., ~30 ms). For example, if the modulation period (second-dimension separation run time) is exactly 2 s, then analytes are trapped for 1,970 ms, and then thermally reinjected onto the second column with a 30-ms pulse width. Other modulators provide the same basic function as the thermal modulator, albeit with different hardware, and some trade-offs in performance.

Detection is another aspect of the instrument design that must be considered. One commonly applied detector for GC × GC is the FID. Since the FID is a workhorse detector for GC due to the excellent sensitivity for all C-H containing compounds, it is not surprising it is often the detector of choice for GC × GC. Similarly, since GC coupled with mass spectrometry (MS) is extremely popular, it is not surprising that MS is also a popular detector with GC × GC. (See Chapter 17 for a description of MS detectors.) Since the peaks ultimately detected in GC × GC are those from the second-dimension, and they are on the order of 100 to 200 ms wide, the MS must be able to provide a fast scan rate duty cycle. For this purpose, the typical quadrupole MS (qMS) for one-dimensional GC is not fast enough, and a time-of-flight MS (TOFMS) is used. Both qMS and TOFMS typically use electron impact ionization with unit-mass resolution. The mass spectral scan rate for a typical TOFMS collects spectra at a rate of 500 Hz (2 ms/spectrum) over a m/z range of 5 to 1000, or about 50 to 100 mass spectra per second-dimension peak width prior to averaging ~10 to 20 spectra per peak width, for S/N optimization. In contrast, the qMS scan rate is about 3 scans per second. With this high MS scan rate, the GC × GC − TOFMS provides a considerable amount of data for a given sample run (eg., typically ~300 to 500 MB per sample run). In order to readily glean useful information from this significant volume of data, powerful data analysis software methods must be used.

As an example, a portion of a GC × GC − TOFMS chromatogram of a derivatized metabolite sample is shown in Figure 15.11, in which the total ion current (TIC) is plotted in a gray scale image [in the TIC monitoring mode of a mass spectrometer, the instrument sums the currents from all fragment ions as a molecule (or molecules) in a GC peak passes through the detector, to provide a conventional looking gas chromatogram (see Chapter 17)]. Each metabolite peak, indicated as a dark spot in the separation, is characterized and identified by retention times on both separation axes, combined with mass spectral library matching. The total run time was 38 minutes (we show about one-half of that) and the second-dimension run time was 1.5 seconds (we show two-thirds of that). The first-dimension column was a 250-μm inside diameter 20-m Rtx-5MS (0.5-μm film), and the second-dimension column was a 180-μm inside diameter 2-m Rtx-200 (0.2-μm film). The carrier gas was helium at a 1-mL/min flow rate. The portion of a 2D separation shown provides ample evidence of the powerful separation ability of this technology for many important applications, including metabolomics, forensics, environmental studies, and fuel characterization.

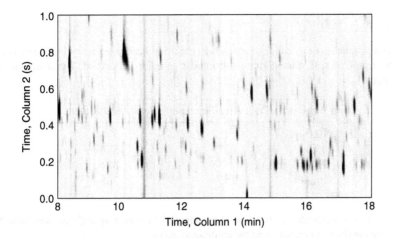

FIGURE 15.11 Total ion current GC × GC – TOFMS chromatogram of a derivatized metabolite sample. Overlapped analyte peaks at a given time from column 1 are often resolved into several peaks on column 2. Darker spots are indicative of more concentrated analyte peaks. Further identification can be achieved by looking at the mass spectrum fragmentation of a given peak.

GC × GC, in general, and GC × GC – TOFMS in particular, are gaining rapidly in popularity, due to the ability to analyze very complex samples. For any complex sample difficult to completely separate by one-dimensional GC, the analyst should consider the option of applying GC × GC. The added chemical selectivity of having the second separation dimension is often essential to more effectively analyze complex samples.

Questions

1. Describe the principles of gas chromatography.
2. What compounds can be determined by gas chromatography?
3. How do gas-liquid and gas-solid chromatography differ?
4. What are the main types of gas chromatography?
5. Compare packed and capillary columns in number of plates.
6. Compare WCOT, SCOT, and PLOT columns.
7. Describe the principles of the following gas chromatography detectors:

    (a) thermal conductivity   (b) flame ionization   (c) electron capture

8. Compare the detectors in Question 7 with respect to sensitivity and types of compounds that can be detected.
9. What is temperature programing as used in the gas chromatography?
10. What is required for fast GC analysis?
11. What is the difference between mass flow-sensitive and a concentration-sensitive detector?
12. What is the effect of thickness of stationary phase film on gas chromatograms?
13. Why are gas chromatographic stationary phases often bonded and cross-linked?

Problems

14. Gas reduction valves used on gas tanks in gas chromatography usually give the pressure in psig (pounds per square inch above atmospheric pressure). Given that atmospheric pressure (760 torr) is 14.7 psi, calculate the inlet pressure to the gas chromatograph in torr, for 50.0 psig, if the ambient pressure is 730 torr.

Spreadsheet Problem

15. A water sample is analyzed for traces of benzene using headspace analysis. Samples and standards are spiked with a fixed amount of toluene as internal standard. The following data are obtained:

ppb Benzene	Peak Area Benzene	Peak Area Toluene
10.0	252	376
15.0	373	370
25.0	636	371
Sample	533	368

What is the concentration of benzene in the sample? Prepare a spreadsheet similar to the one described in the chapter, and print the calibration curve.

Literature Search

16. Using Chemical Abstracts or SciFinder Scholar (the online access to Chemical Abstracts—see Appendix A) if your library subscribes to it, find at least one article on the gas chromatography determination of ethanol in blood. Read the journal article and write a summary of the principle of the method and prepare a synopsis of the procedure used including any sample preparation. Is there information on the accuracy and precision of the method?

Multiple Choice Questions

1. Capillary columns are open tubular columns constructed from the following materials.
   - (a) Glass
   - (b) Metal
   - (c) Stainless steel
   - (d) Fused silica

2. Height equivalent to theoretical plate (HETP) in gas-liquid chromatography depends significantly on the following factors.
   - P. Temperature of column
   - Q. Velocity of carrier gas
   - R. Packing of column
   - S. Column material

   Correct answer is:
   - (a) P, Q and R
   - (b) R and S
   - (c) Q, R and S
   - (d) P and R

3. Liquid samples are injected into the column in gas chromatography in which of the following methods
   - (a) Gas tight syringe
   - (b) Micro-syringe
   - (c) Rotary sample valve
   - (d) Solid injection syringes

4. The following is the disadvantage of nitrogen, that is sometimes used as carrier gas in gas chromatography.
   - (a) Dangerous to use
   - (b) Expensive
   - (c) Reduced sensitivity
   - (d) High density

5. The following detector is widely used to detect environmental samples like chlorinated pesticides and polychlorinated biphenyls.
   - (a) Flame ionization detector
   - (b) Thermal conductivity detector
   - (c) Argon ionization detector
   - (d) Electron capture detector

6. Filter photometer detector is primarily responsive to the following compounds/elements?
   - (a) Volatile sulphur or phosphorous compounds
   - (b) Nitrogen
   - (c) Halogen
   - (d) Potassium

7. The following is not an advantage of thermal conductivity detector used in gas chromatography.

   (a) Simple in construction   (b) High sensitivity

   (c) Large linear dynamic range   (d) Non-destructive character

8. The following components cannot be retained by gas-liquid columns but can be separated by using gas-solid chromatography.

   (a) Formaldehyde   (b) Hydrogen sulphide

   (c) Benzene   (d) Carbon dioxide

9. The distribution coefficients of Gas-solid chromatography are greater than that of Gas-liquid chromatography. State true or false.

10. The following is most often used in the chromatograph in gas chromatograph MS.

    (a) Curvette   (b) Paper support

    (c) Capillary tube   (d) Flask

11. The system for measurement of ion intensity in GS-MS system consists of _____.

    (a) Electrometer   (b) Ion meter

    (c) Ion transducer   (d) Intensity meter

12. The system for measurement of ion intensity in GS-MS system consists of _____.

    (a) Band pass amplifier   (b) Narrow band amplifier

    (c) Wide band amplifier   (d) Low pass amplifier

Recommended References

General

1. L. S. Ettre, "The Beginnings of Gas Adsorption Chromatography 60 Years Ago," *LCGC North America*, **26** (1), January 2008, p. 48. www.chromatographyonline.com. Traces the start of GC in Austria.

2. P. J. T. Morris and L. S. Ettre, "The Saga of the Electron-Capture Detector," *LCGC North America*, **25** (2), February 2007, p. 164. www.chromatographyonline.com. Chronicles the work of James Lovelock which led to the invention of the EC detector.

3. P. J. Marriott and P. D. Morrison, "Nonclassical Methods and Opportunities in Comprehensive Two-Dimensional Gas Chromatography," *LCGC North America*, **24** (10), October 2006, p. 1067. www.chromatographyonline.com. Describes the GC × GC method and its advantages of increased separation power.

4. H. M. McNair and J. M. Miller, *Basic Gas Chromatography*, 2nd ed. Hoboken, NJ: Wiley, 2009.

5. W. Jennings, E. Mittlefehldt, and P. Stremple, eds., *Analytical Gas Chromatography*, 2nd ed. San Diego: Academic, 1997.

6. R. Kolb and L. S. Ettre, *Static Headspace-Gas Chromatography*. New York: Wiley, 1997.

7. W. O. McReynolds, "Characterization of Some Liquid Phases," *J. Chromatogr. Sci.*, **8** (1970) 685.

8. W. R. Supina and L. P. Rose, "The Use of Rohrschneider Constants for Classification of GLC Columns," *J. Chromatogr. Sci.*, **8** (1970) 214.

9. Sigma-Aldrich, Supelco Analytical. GC column Selection guide. http://www.sigmaaldrich.com/content/dam/sigma-aldrich/docs/Supelco/General_Information/t407133.pdf.

10. Agilent J&W GC Column Selection Guide. http://www.crawfordscientific.com/downloads/pdf_new/GC/Agilent_J&W_GC_Column_Selection_Guide.pdf.

11. SGE Analytical Science. GC Capillary Column Selection Guide. http://www.sge.com/support/training/columns/capillary-column-selection-guide.

12. Sigma-Aldrich, Supelco Analytical. The Retention Index System in Gas Chromatography: McReynolds Constants. www.sigmaaldrich.com/Graphics/Supelco/objects/7800/7741.pdf.

13. D. T. Burns, "Characteristics of Liquid Stationary Phases and Column Evaluation for Gas Chromatography," *Pure & AppL Chem.* **58** (1986) 1291–1306.

14. Calculator Edge. Wheatstone's Bridge calculator. http://www.calculatoredge.com/new/Wheatstone%20Bridge%20Calculator.htm.

15. D. S. Forsyth, "Pulsed Discharge Detector: Theory and Applications," *J. Chromatography A* **1050** (2004) 63–68.

16. W.E. Wentworth, S. V. Vasnin, S.D. Stearns, and C. J. Meyer, "Pulsed Discharge Helium Ionization Detector," *Chromatographia* **34** (1992) 219–225.

17. K. L. Sobeih, M. Baron, and J. Gonzalez-Rodriguez. "Recent Trends and Developments in Pyrolysis–Gas Chromatography," *J. Chromatogr A* **1186** (2008) 51–66.

18. CDS Analytical. Pyrolyzer Overview. http://www.cdsanalytical.com/instruments/pyrolysis/pyrolysis_overview.html.

19. D. W. Armstrong, "Direct Enantiomeric Separations in Liquid Chromatography and Gas Chromatography." In: *A Century of Separation Science*. Ed. H. J. Issaq, Marcel Dekker, Inc. New York, 2002, Ch. 33, pp. 555–578.

20. Z. Liu, and J. B. Phillips, "Comprehensive Two-Dimensional Gas Chromatography using an On-Column Thermal Modulator Interface," *J. Chromatogr. Sci.*, **29** (1991), 227–231.

Chapter 16

LIQUID CHROMATOGRAPHY AND ELECTROPHORESIS

KEY THINGS TO LEARN FROM THIS CHAPTER

LC SYSTEMS, COLUMNS
- **HPLC**
- **Separation modes:** NPC, RPC, IEC, IC, HILIC, SEC (GFC), ICE, chiral LC, affinity LC
- **Stationary phases** and particles
- Hydrolytically and pH stable silica particles
- Macro-meso/microporous (perfusion) particles
- Superficially porous or superficially active particles
- Ion exchange and ion exchange resins
- Modern ion exchanger phases
- Monolithic columns
- HILIC stationary phases
- Chiral stationary phases: Pirkle-type, cavity, helical polymer, ligand exchange
- Other supports: alumina, zirconia, titania, carbon

EQUIPMENT
- HPLC equipment
- Solvent delivery
- Sample injection
- Separation columns

DETECTORS
- What do we need in a detector?
- Universal detectors, refractive index
- Viscosity and light scattering detectors
- Conductivity detector
- Key equations: Stokes-Einstein equation and Einstein relation, ionic (electrophoretic) mobility and limiting equivalent conductance
- Electrical resistance, conductance, specific conductance, and cell constant
- Key equations: Nernst-Einstein equation and limiting molar conductivity
- Four-electrode, bipolar pulse, and contactless conductance measurement
- Aerosol detectors: ELSD, ACD, and CNLSD
- Power function linearization of detector response
- UV-Visible detectors, PDA detectors, absorbance ratio to determine peak coelution
- UV-Vis cell design, RI effects, LCW cells
- Fluorescence detection, (confocal) LIF
- Chiral detectors: polarimetric ORD, CD detectors
- Electrochemical detection: coulometric and amperometric detectors
- Pulsed amperometric detector
- Chemiluminescence detector
- Radioactivity detector and coincidence counting
- Postcolumn reaction detection, reagent introduction and reactor geometries, photochemical reactors

ION CHROMATOGRAPHY
- Ion exchange separation of amino acids and PCR detection
- Ion chromatography: how suppressed IC works
- Single column ion chromatography
- Suppressors in IC
- Membrane suppressors in IC
- Electrodialytic eluent generation and carbonate removal
- Carbon dioxide removal device
- Ion pair/ion interaction chromatography
- **HPLC method development**, gradient elution
- **UHPLC and Fast LC**
- **Open tubular liquid chromatography (OTLC)**

Although modern chromatography was born with Martin and Synge's introduction of liquid–liquid partition chromatography, Martin's introduction of gas chromatography (GC) in 1950 using immobilized liquid stationary phases proved to be far more popular. The speed and sensitivity of GC and its reasonably broad applicability (especially in the then rapidly growing petrochemical industry) led to its rapid adoption and development. By now, however, liquid chromatography (LC), especially in the form of **high-performance liquid chromatography (HPLC),** has more than caught on. It is potentially broader because approximately 80% of known compounds are not sufficiently volatile or stable to be separated by gas chromatography. Although initially LC was far behind GC in performance, the wealth of accumulated chromatographic knowledge (primarily from GC) has led to HPLC techniques today that rival GC in performance and even allows separations to be made in seconds. As the petrochemical industry championed GC in the 1950s, the pharmaceutical industry made HPLC the workhorse beginning in the 1970s. Today the HPLC market is not just larger than the GC market, it is the single largest segment of the analytical instrumentation industry, estimated to exceed US $4 billion by 2017. There is a vast array of systems and techniques to choose from for particular applications. For this reason, we provide here extensive coverage of modern instrumentation and columns, which will serve to guide in selection and inform the user of the principles of the columns and detectors used.

HPLC is the liquid-phase analog of GC. The secret to its success is small uniform particles to give small eddy diffusion and rapid mass transfer.

In this chapter, we describe the principles of HPLC and its development. Thin-layer chromatography (TLC), LC in a planar form, is also considered. In addition, we discuss electrophoresis, in which separations are based on differing electrical mobilities of charged species. A supercritical fluid is a fluid at a temperature and pressure above its critical point. Under these conditions, neither a distinct liquid nor a distinct gas phase exists. A supercritical fluid thus has properties between a gas and liquid. **Supercritical fluid chromatography (SFC)** is an important type of chromatography that has attributes both common to and different from GC and HPLC and stands somewhere in between. SFC is not discussed in this book.

16.1 High-Performance Liquid Chromatography

Early LC was carried out in large columns with relatively large particles under gravity feed, with manual collection of effluent fractions for off-line measurement; a technique that is still practiced in some synthetic organic chemistry or preparative biochemistry laboratories. In 1964, in a benchmark paper [*Anal. Chem.*, **36** (1964) 1890] J. Calvin Giddings (University of Utah) predicted much improved efficiency if small particles are used with the concomitant use of high pressure to overcome the flow resistance. Shortly thereafter, Horvath and Lipsky at Yale University built the first high-pressure liquid chromatograph. The technology of producing small particles that will allow high efficiency and performance came in the 1970s. While HPLC today largely connotes "high-performance liquid chromatography" rather than high pressure, the continued move to smaller particles does necessitate the use of higher and higher pressures. Some commercially available systems are capable of pumping at 15,000–19,000 psi pressures and differentiate themselves by the term **ultra-high-pressure liquid chromatography (UHPLC)** systems.

Principles

Figure 16.1 shows the basic components of an HPLC system and Figure 16.2 illustrates a modern HPLC instrument. These instruments tend to be assembled in modular form, unlike most GC instruments, allowing the user to readily change different components.

In HPLC, analytes are separated based on their differential affinity between a solid stationary phase and a liquid mobile phase. The kinetics of distribution of solutes between the stationary and the mobile phase is largely diffusion-controlled. Compared to gases, the diffusion coefficient of analytes in liquids is 1000 to 10,000 times slower. To minimize the time required for the interaction of the analytes between the mobile and the stationary phase, two criteria should be met. First, the packing particles should be small and as uniformly and densely packed as possible. This criterion is met by uniformly sized spherical particles and results in a smaller A value in the van Deemter equation (smaller eddy diffusion). Second, the stationary phase should be effectively a thin uniform film with no stagnant pools and provide a small C value (more rapid mass transport between the phases—necessary for high flow rates). Because molecular diffusion in liquids is small, the B term in Equation 14.13 is small. Hence, the detrimental increase in H at slow flow rates is much less pronounced than what was generically shown in Figure 14.8. This is illustrated in Figure 16.3 and is expressed both by the Huber equation (Equation 14.26) and the Knox equation (Equation 14.27).

Jorgenson and his students first described UHPLC. Read the classic paper: J. E. McNair, K. C. Lewis and J. W. Jorgenson, *Anal. Chem.*, **69** (1997) 983. This paper described packing columns at 60,000 psi (4100 bar) and operating at 2000 psi (~1400 bar). Two years later they described operation at 72,000 psi (5000 bar) and a pump capable of pressures up to 130,000 psi (9000 bar): *Anal. Chem.*, **71** (1999) 700. Still later they will go on to do chromatography at pressures exceeding 100,000 psi (6800 bar), generating in excess of 730,000 plates per meter. Current commercial UHPLC systems obviously fall rather short of the Jorgenson "ultra" standard!

Molecular or longitudinal diffusion in liquids is slow and can be neglected.

Mass transfer is the primary determinant of H in HPLC.

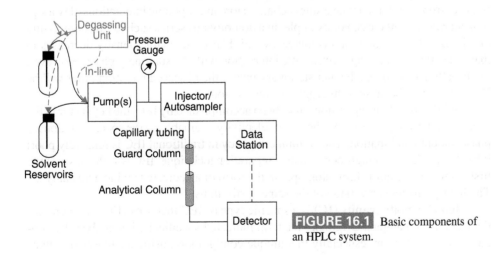

FIGURE 16.1 Basic components of an HPLC system.

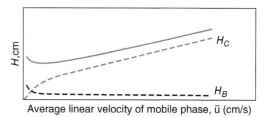

FIGURE 16.3 van Deemter plot for HPLC. Theoretical plate height is plotted versus average mobile-phase velocity. The contributions of each term to the shape of the van Deemter curve are as given in Figure 14.4. See Figure 16.25 for the behavior of some actual particles in present use.

H_C

H_B

Average linear velocity of mobile phase, $\bar{u}$ (cm/s)

FIGURE 16.2 A modular (U)HPLC system. From top: solvent tray and degasser, gradient pump (binary or quaternary), thermostated autosampler, thermostated column compartment with automated switching valve(s), UV–Vis absorbance detector. (Courtesy of Thermo Fisher Scientific Inc.)

HPLC Subclasses

Normal phase chromatography (NPC) utilizes a polar stationary phase and relatively nonpolar to intermediate polarity solvents such as hexane, tetrahydrofuran (THF), etc. Early HPLC was mostly conducted with bare silica particles, and partitioning into water sorbed on the silica provided for the mechanism of separation. Later, as bonded nonpolar phases such as octadecyl silica (ODS or C18, more on this later) were introduced, polar hydroorganic solvents, mostly acetonitrile:water, methanol:water, etc. were used as the mobile phase. Since the polarity of the stationary and the mobile phases were reversed from the major type of chromatography then in vogue, this was termed **reverse phase chromatography (RPC)**. Over time, RPC became far more popular; today RPC is used at least ten times as often as NPC, but the nomenclature remains. Normal phase chromatography does not mean that this is the mode of chromatography that is normally used.

   Ion exchange chromatography (IEC) is one of the forms of aqueous chromatography that has been practiced since ion exchange resins were first made in the 1930s. Ion exchange particles carry fixed positive or negative charges; a sulfonic acid type resin, for example, has $SO_3^-H^+$ groups where the H^+ groups can be exchanged for other cations. Such a resin is therefore called a cation exchange resin. Different cations, e.g., metal ions, or species that can form positive ions, e.g., amines, can be separated on such a column, based on their differing affinities for the stationary phase. Ion exchange-based separations played an important early role in the enrichment of Uranium in the Manhattan Project (the making of the first atomic bomb).

   One of the important aspects of an ion exchange separation is that electroneutrality must always be maintained. To separate cations in the above example, for the analyte cation(s) to move down the column, another cation must take its place—the eluent must therefore be ionic. However, it should not be construed that electrostatic interaction is the only governing factor in ion exchange affinities. While it is generally true that a triply charged ion is more strongly retained than a doubly charged ion, which is more strongly retained than a singly charged ion, hydrophobic interactions play a significant role, nonetheless. For example, in a homologous series such as that comprising of Cl^-, Br^-, I^-, the ionic radius increases and thence the charge density decreases in that order, thus decreasing electrostatic interaction with the stationary phase. However, in virtually any anion exchanger stationary phase, the observed retention order will be $I^- > Br^- > Cl^-$ because of hydrophobic interactions.

   While important separations have been accomplished by IEC, the key factor in the success of these separations has been the selectivity attainable by the specific stationary phase–eluent combination. The column efficiency in traditional IEC is relatively poor; it hardly qualifies as a high-performance separation technique. IEC was the basis of the first commercial liquid chromatograph, in the form of a dedicated amino acid analyzer. The high-performance version of the same is still in use.

   Ion chromatography (IC) is a specific type of IEC that uses efficient microparticulate ion exchangers. Originally the term indicated specifically ion analysis by conductometric detection, especially in a unique configuration using a suppressor device (discussed later). Presently, the term is commonly used to indicate efficient IEC with a variety of detection methods.

In **hydrophilic interaction chromatography (HILIC)**, one of the most recent entries, water is adsorbed on a hydrophilic surface to provide the partitioning process. This mode of separation is well suited for highly polar water soluble analytes that includes many drug molecules of interest. The basic mechanism is the same as that in NPC. But in practice, different types of column-eluent combinations are used. Acetonitrile-water is commonly used as the eluent system, in which water is the strong eluent. In the exact reverse order as RPC, to accomplish gradient elution, one starts with a high acetonitrile content, and the water content is increased with time. Not surprisingly, HILIC elution order is often reverse of that of RPC. To establish both its similarity and differences with NPC, early allusions to HILIC were made as **aqueous normal phase (ANP)** chromatography. The practice and importance of HILIC is presently growing very fast.

In **size exclusion chromatography (SEC)**, molecules are separated based on their size. The stationary phase volume is largely occupied by pores. Molecules that are larger than the largest pores cannot enter any pores and hence are "excluded" from the pores and come out in the void volume. Molecules that are smaller than the smallest pores, in contrast, can explore the entire pore space in the stationary phase and come out last. As a result, all analytes in SEC elute within this finite retention volume window, the molecules above a certain size/MW (exact value depending on the pore size distribution in the stationary phase) coming out first, and molecules below a certain size coming out together at the end. Molecules of intermediate size are separated in the retention window.

Specific SEC phases have different pore size distributions and will effectively separate molecules that are within the corresponding size range. Molecules larger or smaller than the range will elute, but without separation within the larger or smaller groups. Although in principle there are no analyte specific interactions between the analyte(s) and the stationary-phase matrix, in some cases such interactions do occur and modify the expected retention behavior.

When SEC is conducted for the separation of proteins and other biomolecules, aqueous eluents are used and the technique is often referred to as **gel filtration chromatography (GFC)**. The earliest and still a dominant application area of SEC is in determining the molecular weight distribution of polymers. This technique typically uses porous polymer stationary phases with organic solvents as eluents, often at an elevated temperature. It is widely referred to as **gel permeation chromatography (GPC)**. Referring to polystyrene standards, GPC columns with high-end MW **exclusion limits** as low as 1500 to as high as 2×10^8 are readily available

An animated cartoon depicting a SEC separation can be seen at http://www.separations.eu. tosohbioscience.com/ Products/HPLCColumns/ ByMode/SizeExclusion/.

Although it cannot be classified as HPLC, SEC is widely used in biochemistry for low-pressure/gravity-fed preparative separation of biomolecules. Pharmacia (later a part of GE Healthcare) made cross-linked dextran gels called Sephadex (the name derives from <u>Se</u>paration <u>pharma</u>cia <u>dex</u>tran) with the fractionation range (in terms of globular proteins) ranging from ≤ 700 (type G-10) to 5,000–600,000 (type G-200). Bio-Gel P from Bio-Rad is a similar polyacrylamide gel available in particle sizes $< 45 - 180 \, \mu m$) with fractionation ranges of 100–1800 (Bio-Gel P-2) to 5,000–100,000 (Bio-Gel P100). Such techniques are also useful for the desalting of proteins that may have been partially purifed by salting out with a high concentration of some salt. A gel with a low exclusion limit, such as Sephadex 25, allows the protein to pass through the column much faster than the salts.

Ion exclusion chromatography / ion chromatography exclusion (ICE), like SEC, depends on principles of exclusion to accomplish separation and like SEC, all analytes elute within a finite retention window. Weak electrolytes can be separated by this technique, the dominant application area being the separation of organic acids that can be separated from strong acids and further separated according to their pK_a.

Consider the column to be consisting of a –SO$_3$H type cation exchange resin. The sulfonate group is fully ionized and the resin matrix is therefore negatively charged. Electrostatic forces (often loosely termed the *Donnan Potential*) thus inhibit the penetration of an anion into the interior of the resin, but no such penetration barrier exists towards a neutral molecule. If we consider a weak acid HA, the anion A$^-$ will be excluded from the interior of the resin, but not the neutral unionized acid HA, hence the terminology **ion exclusion**. The consequence is that fully ionized acids elute in the void volume and others elute in the order of increasing pK_a; acids that are largely unionized elute last. An acidic eluent is used to maintain the pH and partial ionization; it is also possible to run a gradient by decreasing the concentration of the eluent acid during the run. Weak bases can also be similarly separated on gel-type strong anion exchanger phases, but this is not as commonly used.

The importance of chiral separations, both in an analytical and preparative scale, especially in the context of pharmaceuticals, many of which contain chiral centers, has been mentioned in the previous chapter. HPLC plays a much larger role in **chiral chromatography** than GC, as many of the analytes of interest are not amenable to GC separations. To achieve differentiation between chiral enantiomers, the stationary phase must act differently with the two enantiomers and must therefore itself be chiral. Occasionally, it is possible to add chiral additives to the mobile phase and perform chiral separations on achiral stationary phases.

Affinity chromatography is widely used for the separation/purification of specific biomolecules. This relies on the highly specific binding between an analyte and its counterpart, as in antibody-antigen binding. The counterpart is immobilized on the stationary phase; this is called the affinity column. When the desired analyte, along with any variety of other substances is passed through the column, only the analyte is retained on the column, everything else passes through. The analyte is then eluted as a sharp band by some eluent that is capable of dislodging the analyte from its counterpart. While affinity purification is straightforward in principle, having an agent that binds the analyte specifically and with high affinity and then devising another agent that releases the analyte without denaturation is not always a trivial task.

16.2 Stationary Phases in HPLC

Early microparticles were irregularly shaped porous silica gel or alumina of equivalent diameter $\leq$ 10 μm. Spherical particles (Figure 16.4) since developed can be packed more homogeneously and provide improved efficiency. Presently **high-purity silica** particles, low in trace metal content, < 10 μm, even < 2 μm diameter particles, generally functionalized in some fashion, are the mainstay of HPLC. Smaller diameter particles create higher back pressure, but they also exhibit very little loss of efficiency at higher flow rates (see Figure 16.3), permitting faster separations.

For the analysis of small molecules, polypeptides and many proteins, and for very high molecular weight proteins, particles with respective pore sizes of 60–150, 200–300, and 1,000–4,000 Å are used to allow the analyte to penetrate the pores. Most HPLC is performed in the liquid–liquid partition mode, albeit the stationary liquid phase is bonded to the support particles rather than adsorbed or coated as in GC packings. Adsorption chromatography is occasionally useful for some applications.

See, e.g., http://www.bio-rad.com/ webroot/web/pdf/lsr/literature/ LIT-174B.pdf for the different Bio-Gel fractionation ranges.

An animated cartoon depicting an affinity chromatography separation can be seen at http://www.separations.eu. tosohbioscience.com/ ServiceSupport/TechSupport/ ResourceCenter/Principlesof Chromatography/Affinity/.

FIGURE 16.4 Spherical microporous silica particles, 10 μm, 800x magnification. Particles are fully porous with 100-Å pores. They are available as base silica for adsorption chromatography, or with bonded phases. (AstroSil® from Stellar Phases, Inc. Courtesy of Stellar Phases, Inc.)

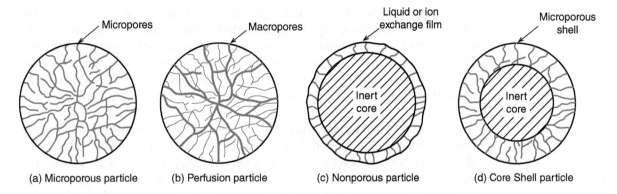

FIGURE 16.5 Structural types of particles used in high-performance liquid chromatography. (Adapted in part from D. C. Scott in *Modern Practice of Liquid Chromatography*, J. J. Kirkland ed. New York, Wiley-Interscience, 1971, with permission.)

The landscape of preferred HPLC stationary phase support particles is continuously changing, but particles most commonly used until now are **microporous particles**; the pores being permeable to the analytes and the eluting solvent (Figure 16.5a). The majority of the surface area (the active region) is within the pores.

Most of the mobile phase moves around the particles. The solute diffuses into the stagnant mobile phase within the pores to interact with the stationary phase, and then diffuses out into the bulk mobile phase. The use of small particles minimizes the pathlength for diffusion and hence band broadening. Exact methods of manufacturing HPLC particles are mostly proprietary. However, rather than endowing a single particle with pores, high-purity ultrafine colloiodal silica can be agglomerated to form effectively microporous spherical particles (Figure 16.6a). Acidification of soluble silicates, including organic alkyl silicates, can be used to generate an amorphous, high surface area, high-porosity, rigid particle, often called Xerogels (Figure 16.6b).

Silica particles have surface silanol groups, SiOH. Silanol groups provide polar interaction sites; this may be undesirable or can be used to advantage. It can be used as the site for functionalizing the silica particle. For example, a reaction with the monochlorosilane $R(CH_3)_2SiCl$ (where $R = CH_3(CH_2)_{16}CH_2-$) will lead to the most commonly used "C18 silica" (also called ODS for octadecyl silica) stationary phase:

Endcapping. The free –SiOH group in the middle has been endcapped. The solid line on the left represents the silica surface.

$$-Si-OH + Cl-\underset{\underset{CH_3}{|}}{\overset{\overset{CH_3}{|}}{Si}}-R \rightarrow -Si-O-\underset{\underset{CH_2}{|}}{\overset{\overset{CH_3}{|}}{Si}}-R + HCl$$

The extent to which silanol groups are functionalized depends on the chain length of the functionalizing agent. In the above example, with R being a 18 carbon chain, it is difficult to functionalize more than ~30% of the silanol groups. The unreacted silanol groups can be "endcapped" with the smallest trialkylsilane, trimethylchlorosilane ($R = CH_3-$, see margin figure). Even so, only about 50% of the residual silanols are usually reacted. There are, however, proprietary techniques that claim to accomplish a greater degree of encapping.

Similary functionalized are: reverse phases, e.g., phenyl ($R = -C_6H_5$, provides π-π interaction, often spaced with one or more methylene bridges, biphenyl and diphenyl are also common), C8 [$R = -(CH_2)_7CH_3$, less hydrophobic than C18]; normal phases (in the order of increasing polarity) cyanopropyl [$R = (CH_2)_3CN$], diol [$R = -(CH_2)_2OCH_2CH(OH)CH_2OH$], amino and dimethylamino, both with a 2–3 carbon methylene bridge [$R = -(CH_2)_3NH_2$], and [$R = -(CH_2)_3N(CH_3)_2$].

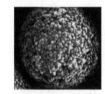

(a) ZORBAX Rx-SIL (Silica Sol)

(b) Xerogel Silica

FIGURE 16.6 *(a)* Zorbax porous silica microsphere particle, 50% porosity, 100 Å pores. *(b)* Xerogel silica particle, 70% porosity, 100 Å pores. (Courtesy of Agilent Technologies.)

The extent of functionalization, especially for C18 phases, is often expressed in terms of wt %C (as obtained by elemental analysis). When a bonded phase is prepared from a monochlorotrialkylsilane as above, the reagent can only form a single bond with a surface –OH. This forms a monomeric bonded phase, resembling a brush-type structure, in which each of the bristles represents the same chemical entity. For a long hydrocarbon chain like C18, this would very much represent a bonded molecular film of oil.

Instead of a monochlorotrialkyl silane (this is said to lead to a "monomeric phase"), imagine that a dichlorodialkylsilane or a trichloroalkylsilane is reacted with silica without complete exclusion of water. On an average, between one and two chlorine atoms from each reagent will react with the silica particle, all the rest remain unreacted. These hydrolyze, forming –SiOH groups which then react with more reagent, thus essentially forming a "polymeric phase." This would resemble a three-dimensional network rather than a brush-like phase and may provide greater shape- and size-selectivity. This may be beneficial in affecting separations between geometrical isomers, polynuclear aromatic hydrocarbons, etc.

Percent carbon content of a polymeric phase can be much higher than that of a monomeric phase. The carbon content is an index of the column capacity, although increasing carbon content of a C18 phase is sometimes wrongly interpreted as increasing nonpolar nature of the phase. Endcapping adds little to the carbon load but reduction of the free –SiOH groups greatly alters the polarity of the phase. Both monomeric and polymeric stationary phases are available in endcapped and non-endcapped versions, however, free residual –SiOH groups inside a 3-D polymeric network cannot be capped as easily. As a result, a particular polymeric phase may have a greater carbon load but its polarity may not necessarily be lower than a well-endcapped monomeric C18 phase.

Improving hydrolytic stability at pH and temperature extremes. Standard silica based columns have limited lifetime at pH levels below 2 or above 8. In the silylation reaction used for endcapping / functionalization, if the two CH_3– groups in the chlorodimethylalkyl silane are replaced by the larger isopropyl ($n = 1$) or isobutyl ($n = 2$) groups, $[CH(CH_3)_nCH_2^-]$, access of H^+ to the Si–O–Si bond becomes more difficult in the sterically hindered hydrophobic environment, and stability to acidic pH is significantly improved.

Cross-linked polymeric particles, for example, poly(methacrylate)s, and especially cross-linked poly(styrene), can withstand the full range of pH. Chromatographic columns based on graphitic carbon, alumina, titania, and zirconia are also available and all exhibit greater pH tolerance than silica. However, at comparable particle sizes, no other support particle exhibits chromatographic efficiencies provided by silica or are capable of being functionalized by the varieties of well-characterized silane chemistries that have been developed.

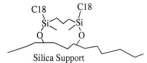
Silica Support

Different approaches to improve alkaline pH stability of a C18 silica packing have been developed. Scientists at Agilent developed a technique where a propylene bridged bidentate C18 silane anchors the silica at two points (see margin figure). Because the bonded layer is sterically fixed in position, this makes it difficult for OH^- to attack the underlying silica. At room temperature, with organic base buffers such a phase may be used for extended periods even up to a pH of 11.5. Another approach, developed by Waters Corp., towards making an organic–inorganic hybrid silica particle with extended stability at alkaline pH involves the synthesis of particles from two high-purity monomers, tetraethylorthosilicate and bis(triethoxysilylethane), which incorporates the preformed ethylene bridge (see margin figure) to synthesize a polyethoxysilane that upon hydrolysis gives the desired silica particle containing covalent hydrolytically stable –Si–CH_2–CH_2–Si– linkages.

Macroporous-micro/mesoporous structures. Perfusion packings. The dividing boundaries of micro-, meso-, and macropores are not strictly defined. But in the general context of HPLC packings, pores smaller than ~100 Å are referred to as micropores while those larger than ~1000 Å are regarded as macropores. Pores of intermediate size are sometimes referred to as mesopores. Macroporous structures are obviously necessary for the analysis of very large molecules; extensive macroporosity also leads to much greater surface area and thus provides greater capacity. Increased capacity of a macroporous packing is often advantageously used in ion exchange chromatography. Most often, some micro/mesoporosity will also be associated with macropores. At Purdue University, Regnier deliberately designed polymeric packings with large macropores (6000–8000 Å) connected to a great number of mesoporous (~800 Å) channels (where most of the surface area lies, Figure 16.5b); and some microporosity was also present. The accessible surface area is then functionalized in the desired manner. This type of particle is particularly well suited for the separation of very large molecules, especially large proteins, which have very small diffusion coefficients. Consider that you have to deliver and collect mail from a large number of houses, which are all situated in small lanes. Your task will be a lot faster if the lanes emanate from fewer large avenues. The optimum flow rate will be faster and the solute is brought more quickly to the mesoporous channels where separation occurs. While these packings were originally made as 10–20 μm diameter beads, the current offering spans particle diameters of 10–50 μm and is targeted primarily at large-scale separation tasks in the biopharmaceutical industry. Affinity ligands such as recombinant protein A or protein G bound to perfusion packings are popular for rapid analytical scale purifications of the corresponding antibodies. Free reactive aldehyde or epoxide groups in an affinity column are used similarly to bind proteins with free –NH$_2$ groups.

Nonporous packings (Figure 16.5c) primarily of silica in very small particle size (as low as 1.5 μm) have had a transient popularity in HPLC. Mass transfer in porous silica is limited by the rate of intraparticle diffusion; in addition, uncapped active sites in the pores may lead to undesirable interactions. If there are no pores, the pore diffusion and longitudinal diffusion limitations disappear. With very small particles, the diffusion distance from the mobile to the stationary phase is very short; column efficiency is virtually flow rate independent. However, the pressure needed to attain a certain flow rate through a given column varies inversely as the square of the particle diameter, the pressure needed to pump through a column at a given flow rate is 1100% higher when the column is packed with 1.5 μm particles than with 5 μm particles, the particles being otherwise identical. This limits the lowest practical particle diameters and/or the highest flow rate that can be attained, because of pressure limitations of current pumping technologies. In addition, although high-strength silica particles that can withstand 19000 psi (1300 bar) are now commercially available, most available particles are deformed/crushed at pressures much less than this.

Nonporous packings have a much lower surface area, limiting how much sample can be injected on column (low column capacity) and thus exhibit much lower retention times under otherwise identical elution conditions. Because of high pressure required for the very small particles, column lengths are obligatorily short. It is also more difficult to pack a column densely and homogenously with nonporous particles. Superficially porous particles discussed below have essentially replaced nonporous packings.

Superficially porous particles. From the beginning days of liquid chromatography, it has been realized that a solid-core particle with a thin superficial active layer should have good mass transfer characteristics. As early as 1967 Horvath et al. described HPLC separation of nucleosides on glass beads coated with ion exchange resins and other material [*Anal. Chem.* **39** (1967) 1422]. When Small, Stevens, and Bauman et al. from Dow Chemical introduced ion chromatography in 1975, to make an efficient ion exchanger phase, they used solid-core poly(styrenedivinylbenzene)

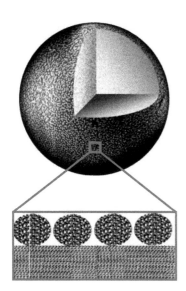

Surface agglomerated ion exchangers constitute early examples of core-shell particles.

(PSDVB) particles and very lightly sulfonated them on the surface so that the surface contained negatively charged $-SO_3^-$ groups. Upon passing a suspension of colloidal size positively charged anion exchange nanoparticles (commonly called latex; (this term is discussed in more detail later) through a column packed with such PSDVB particles, the positively charged latex binds so tightly to the surface of the negatively charged core particles that an efficient stable ion exchanger results (see margin picture).

In a similar manner, **superficially porous** particles (SPP, also called **core-shell** or **fused core** particles of small diameter (Figure 16.5d) have been introduced more recently for general use. The packing material is composed of an inner fused or non-porous particle core, and a porous outer particle shell. Thus, analytes only interact with the outer shell, reducing resistance to mass transfer and providing superior separation efficiency. As stagnant regions in the middle of the particles are removed, analytes and mobile phase can transfer more efficiently from the stationary phase back to the mobile phase. Columns packed with SPP with particle size as small as 1.3 μm are now available.

Chromatography on classical ion exchange resins. Although many chromatographic separations have been carried out on classical ion exchange resins that were typically **gel-type ion exchangers** (meaning a solid polymer bead with no porosity other than on a molecular scale in the polymer lattice) and were fairly large particles (at least 25–37 μm diameter), such resins are no longer used for analytical separations. They are, however, extensively used in water softening, water purification, high-purity water production, certain large-scale separation of metals (including radionuclides), as catalysts, and the production of many pharmaceuticals, sugar, and beverages (including purifying fruit juice). They are also widely used in the laboratory to convert a compound in one ionic form to another. It is important to have an appreciation of the classical ion exchange resins and how they work before modern stationary phases for ion exchange are discussed.

While a few commercial ion exchange resins are based on an acrylate skeleton, by far the large majority are made of a polystyrene polymer, cross-linked with divinylbenzene. The aromatic skeleton of the cross-linked polymer resin is easily functionalized to contain permanent ionic functional groups (e.g., $-SO_3^-$, $-NR_3^+$) or protolytically ionizable groups (e.g., $-COOH$, $-NH_2$). There are chemically four basic types of ion exchange resins used in analytical chemistry: strongly acidic, strongly basic, weakly acidic, and weakly basic, as summarized in Table 16.1.

TABLE 16.1 Types of Ion Exchange Resins

Type of Exchanger	Functional Exchanger Group	Trade Name
Cation		
Strong acid	Sulfonic acid	Dowex[a]50; Amberlite[b]200C; Ionac[c] C249; Rexyn[d] 101(H)
Weak acid	Carboxylic acid	Amberlite IRC-50; Rexyn 102; Amberlite CG-50
Anion		
Strong base	Quaternary ammonium ion	Dowex 1; Amberlite IRA 400; Ionac A544; Rexyn 201; Amberlite IRA-900*
Weak base	Amine group	Dowex M43; Dowex 22*; Amberlite IR-45; Ionac A365

[a] Dow Chemical Company.   [b] Rohm and Haas, now part of Dow Chemical.
[c] Sybron Chemical.   [d] Fisher Scientific.
* Macroporous structure, others are gel-type.

Aside from the original solid-bead gel-type resin, resins are also made now with "Porogens," solvents that are present in the structure during the formation of the beads but are later washed out, leaving macro/meso pores. The latter are termed **macroporous** or **macroreticular** resins. These of course have much greater surface area than gel-type resins.

Cation exchange resins contain functional groups where the cation is mobile and is replaceable by another cation. They are typically sold in the H^+ or the Na^+ form. The strong-acid type exchangers have $-SO_3H$ groups that are fully ionized. Weak-acid cation exchangers have $-COOH$ groups (or in special resins $-PO_3H$ (phosphonic acid) groups), these are only partially ionized.

In a cation exchange resin, cations are replaceable by another cation; this is an equilibrium process.

$$n \text{ Resin-SO}_3{}^- H^+ + M^{n+} \rightleftharpoons (\text{Resin-SO}_3)_n M + n H^+ \qquad (16.1)$$

and

$$n \text{ Resin-COOH}^- H^+ + M^{n+} \rightleftharpoons (\text{Resin-COO})_n M + n H^+ \qquad (16.2)$$

The equilibrium can be shifted to the left or the right by respectively increasing $[H^+]$ or $[M^{n+}]$ or by changing the amount of resin present at constant $[H^+]$ or $[M^{n+}]$.

The **exchange capacity** of a resin is the total number of equivalents of replaceable hydrogen per unit volume or per unit weight of resin, and it is determined by the number and strength of fixed ionic groups on the resin. Typical ion exchange capacities are of the order of 1–4 milliequivalents/gram. The greater the ion exchange capacity of a column, the greater is the solute retention. Whereas the behavior of strong-acid type resins is largely independent of pH, retention by weak-acid cation exchangers is highly pH dependent. Below ~pH 4, the resins "hold on" to the protons too strongly for exchange to occur. But this pH-dependent tunability of affinity for non-H^+ cations provides a dimension for separation control that is not available with strong-acid type resins. Stationary phases used today for cation separation in ion chromatography are nearly always based on weak-acid exchangers. These type of exchangers do not, however, interact well with weak bases—these are better separated on strong-acid exchangers.

Anion exchange resins contain permanently cationic ($-NR_3{}^+$) or protolytically generated cationic groups ($-NR_2 + H^+ \rightleftharpoons -NR_2H^+$) that are permanently bonded to the resin; the resin is typically supplied in the Cl^- or OH^- form; the anions are replaceable. The exchange reactions can be represented by

$$n \text{ Resin NR}_3{}^+ Cl^- + A^{n-} \rightleftharpoons (\text{Resin} - NR_3)_n A + n Cl^- \qquad (16.3)$$

where R is an alkyl group (most commonly methyl, also benzyl, hydroxyethylbenzyl) for strong-base resins, and for a weak-base resin, one or more of the R groups can be H. In present-day hydroxide eluent IC discussed later, alkanolamine groups, e.g., ethanolamine ($-CH(NH_2)CH_2OH$), etc. type functionalities make the phase more selective for OH^- and are often used.

Strong-base type exchangers retain their ability to exchange anions up to a pH of 12, but weak-base resins are not effective exchangers at alkaline pH. They also do not bind weak acids effectively, but may be good for separating strong acids like sulfonates.

Styrene has only one unsaturated group; polymerization results in a soft, easily deformable polymer. Incorporating a monomer that has more than one unsaturation, e.g., divinylbenzene (DVB) or its ethyl derivative, results in the linear chains of polystyrene being braced by the DVB bridge. This **cross-linking** makes the material more rigid, reduces swelling by a solvent and hence makes the resin more pressure tolerant. With increasing **degree of cross-linking** (typically specified as % DVB), the increasing rigidity also increases the differences in selectivity towards different exchangeable ions. Gel-type resins are available with cross-links as low as 2% DVB to as high as 16% DVB, although 4 to 8% cross-linked resins are most common. The resin

For an illustration of a cross-linked structure, see http://courses.chem.psu.edu /chem112/materials/crosslink.jpg

name often indicates the degree of cross-linking, Dowex 50WX4 and Dowex 50WX8 are the same strong-acid type resin that are, respectively, 4 and 8% cross-linked.

Although *prima facie*, a cation has no affinity for anion exchangers and cations per se cannot be separated on anion exchangers, separations of the majority of transition metals, heavy metals or rare earth metals have long been carried out on anion exchangers. The trick here is that in the presence of complexing anions, the metals actually form anionic complexes, which then bind to anion exchangers. Concentrated hydrochloric acid forms anionic chlorocomplexes with all but a few metals; a variety of metals have thus been separated on a strong-base type anion exchanger with an HCl gradient. The highly corrosive nature of HCl has led to alternatives: weak complexing agents [hydroxyisobuytyric acid (HIBA) or pyridine, 2–6 dicarboxylic acid (PDCA)] are often used in modern IC as the eluent with anion exchangers and even mixed cation-anion exchanger phases. In fact, the difference in selectivity among different transition metals or different rare earth metals on a standard cation exchanger is so little that a straight cation exchange separation will simply not be possible. It is the differences in the complexation constant with the complexing eluent that ultimately brings about the separation. As the complexing agents are either weak acids or bases, their complexing ability depends on pH as well, and a pH gradient can be a further tool to control the separation. Uncharged complexing agents can also affect the partitioning equilibrium, either directly by affecting the concentration of the exchangeable form of the metal ion, or indirectly because the neutral metal-ligand complex has a different (often greater, due to hydrophobic interactions) affinity for the ion exchanger.

Modern ion exchanger phases. Figure 16.7 depicts the various types of ion exchanger phases in current use, following the classifications proposed by Christopher Pohl of Thermo Fisher Dionex; several of these packing classes were pioneered by him. As previously discussed in the context of core-shell particles (see superficially porous

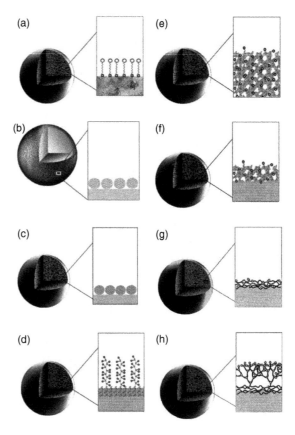

FIGURE 16.7 Types of ion exchanger stationary phases: (a) silane-modified porous silica substrate, (b) electrostatically agglomerated latex on nonporous substrate, (c) electrostatically agglomerated latex on macroporous substrate, (d) polymer grafted porous substrate, (e) proprietary chemically modified substrate, (f) polymer encapsulated substrate, (g) adsorbed polymer-coated substrate, and (h) step-growth electrostatic graft porous polymer substrate. Courtesy Christopher A. Pohl, Thermo Fisher Scientific Inc.

particles), from early days of the introduction of IC, ion exchanger nanoparticles (colloidal organic ion exchanger particles, typically made by **emulsion polymerization** and often called latex) of 50–200 nm size, agglomerated on a substrate particle have been used as the stationary phase. Even earlier, porous silica-based ion exchangers, with ion exchange groups bonded through silane chemistry, as shown in Figure 16.7a, have been and continue to be used in HPLC-type applications. Significant ion exchange capacity is possible with this type of packing. It is not used in IC applications because of poor stability of the base material with purely aqueous eluents, with either extreme of pH.

Figure 16.7b shows the earliest core-shell type of IC packing discussed above under superficially porous particles and illustrated in the margin note. This type of packing has a relatively low capacity. The original substrates used were of low cross-link; these have gradually been replaced by higher cross-link substrates to allow compatibility with organic solvents and high pressure. In principle, it is possible to use pH-stable inorganic substrates or inorganic nanoparticle ion exchangers for agglomeration, but this has not yet been realized in practice. This structure is used today only in preconcentrator or guard columns.

Figure 16.7c shows a macroporous substrate-based packing that uses the same basic strategy as in Figure 16.7b but uses a substrate with 100–300 nm pores and a latex size small enough to attach inside the pores, permitting nearly an order of magnitude greater exchange capacity compared to nonporous substrates of similar size. This is widely used in present-day IC columns.

Figure 16.7d shows polymer grafted films on a porous substrate, a strategy that leads to high-capacity packings but where cross-linking cannot be controlled (and thence manipulation of selectivity through cross-linking is not possible). The substrate is either prepared with polymerizable groups on the surface, or it is modified to introduce polymerizable groups. Monomer(s) and initiator are then allowed to react to produce the grafted particle. Either polymer or inorganic substrates can be used in theory; in practice only polymeric substrates are in commercial use.

Figure 16.7e depicts a general class of IC packings in which a porous polymer is directly chemically derivatized to form the functional groups to make high-capacity packings. Of many chemistries used, the most common approach is to form a copolymer with a cross-linking monomer and a reactive monomer such as vinyl benzyl chloride or glycidyl methacrylate, which are subsequently allowed to react with a tertiary amine to produce quaternary ion exchange sites. This architecture is popular among Japanese column manufacturers.

Polymer encapsulated substrates (Figure 16.7f) were pioneered by Gerard Schomburg of Max Planck Institute für Kohlenforschung. A preformed polymer with residual double bonds, and a suitable free-radical initiator dissolved in solvent are added to the substrate particles. The solvent is then stripped off to leave a polymer film that is cured at an elevated temperature to yield a cross-linked film permanently encapsulating the substrate. The most successful example of this technique is poly(butadiene-maleic acid) copolymer coated on porous silica. This makes an efficient weak-acid exchanger for cation separations.

Figure 16.7g shows a stationary phase made with the substrate in contact with a low molecular weight ionic polymer or a long chain ionic surfactant. Such coatings are stable in purely aqueous solvents but are not solvent compatible. Figure 16.7h shows a type of stationary phase that is now widely used by Thermo Fisher Dionex; more than 10 different column types are offered that rely on this type of chemistry. The stationary phase is prepared using a series of alternating reaction steps, beginning with a particle with a negatively charged surface. First, a 1:1 mixture of a primary amine and a diepoxide makes a "basement coating." Then diepoxide and primary amine reaction steps are alternately applied. This reaction sequence doubles the column capacity with each diepoxide and primary amine reaction cycle. Note that an already packed column is chemically modified *in situ*, a rather rare practice.

A monolithic column, as the name suggests, constitutes essentially of a single solid rod that is thoroughly permeated by interconnecting pores. The general notion that a chromatographic column must be made of discrete fine particles was challenged by Stellan Hjertén of the University of Uppsala in Sweden in the late 80s; he asserted in 1989 that "A continuous gel plug with channels sufficiently large to permit an hydrodynamic flow might be the ideal chromatographic column" and demonstrated separation of proteins on a "continuous polymer bed" of a cation exchanger where the separation efficiency was flow rate independent over an order of magnitude flow rate range. Two years later he and his students described the detailed preparation and use of such columns. Polymeric monoliths were later developed by Tennikova (Russian Academy of Sciences) and Svec (University of California, Berkeley) in a collaborative effort and first commercialized as disks. Large monoliths (up to several liters in column volume) based on this technology has since been developed for large-scale bioseparations. Very short (5 mm long, 5 mm dia) columns are also now commercially available for both analytical and semipreparative use for high resolution and rapid separations of antibodies (IgG, IgM), plasmid DNA, viruses, phages and other macro biomolecules.

Although chronologically silica monoliths became commercially available subsequent to the polymer versions, the basic synthetic strategies were extensively developed earlier by Tanaka et al. at the Kyoto Institute of Technology. As with perfusion packings, they have a bimodal pore structure (Figure 16.8). Macropores, which act as flow-through pores, are about 2 μm in diameter. The silica skeleton contains mesopores with diameters of about 13 nm (130 Å). It can be surface modified with stationary phases like C18. The rod is shrink-wrapped in a polyetheretherketone (PEEK) plastic holder to prevent "wall effects" of solution flowing along the walls. The surface area of the mesopores is about 300 m²/g, and the total porosity is 80%, compared with 65% for packed particles. The column exhibits a van Deemter curve close to that for 3.5-μm packed particles, but with a pressure drop ~40% of the packed column run at the same linear velocity. Standard diameter (4.6 mm) columns can be made up to a maximum length of 10 cm. The shorter lengths and high porosity allows flow rates up to 9 mL/min, permitting fast separations. Mass transport is facilitated by convection in addition to diffusion, making these columns well suited for efficient separations of both large and small molecules. For greater plate counts, more than one column can be connected in series.

More recently, the mesopore/macropore ratio of such silica monoliths were further optimized. A 50% gain in efficiency compared to the first-generation monolith was attained at the price of a slightly higher pressure. The separations achieved with

Read the classic papers on monolithic columns: S. Hjertén, J. L. Liao, and R. Zhang, *J. Chromatogr.* **473** (1989) 273; J. L. Liao, R. Zhang, and S. J. Hjertén, *J. Chromatogr.* **586** (1991) 21.

FIGURE 16.8 Electron micrograph of (*a*) mesoporous and (*b*) macroporous structures in a Merck Chromolith® monolithic rod.Chromolith [From D. Lubda, K. Cabrera, W. Krass, C. Schaefer, and D. Cunningham, *LC·GC* **19** (12) (2001) 1186. Reproduced by permission.]

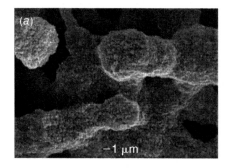
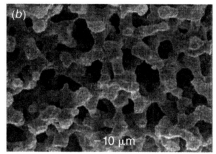

these columns are comparable to 2.6 μm superficially porous particles under otherwise comparable conditions, but at a lower pressure drop.

Until recently, silica-based and polymer-based monoliths were observed to work best for small and large molecules, respectively. Newer generation of high surface area monoliths are useful for separations across the size range. The polymeric monolith phases have shown high mechanical robustness and low swelling in organic solvents because a very high level of cross-linking is used. The utility of agglomerating ion exchange latex nanoparticles on the surface of an oppositely charged monolith rod has also been demonstrated. Polymeric monoliths are available in capillary formats from 0.1 to 1 mm in diameter and up to 250 mm in length, ideal for a detector such as a mass spectrometer that requires only a small liquid input flow. Such phases can separate small ions as well as large biomolecules. An impressive 200,000 plates have been demonstrated for four 25-cm columns connected in series.

Stationary phases for hydrophilic interaction chromatography (HILIC). All hydrophilic surfaces are not optimally suitable for HILIC; phases where water sorption is strongly dependent on pH are susceptible to changes in retention when the pH changes. There are three different types of HILIC phases: neutral, charged, and zwitterionic. A typical neutral HILIC phase contains amide or diol functionalities bonded to porous silica. A charged phase exhibits strong electrostatic interactions and may consist of bare silica or more commonly, amino, aminoalkyl, or sulfonate functionalities bonded to porous silica. The separation selectivity between different analytes can be favored by the electrostatic (ionic) interactions that are contributing to the retention with charged HILIC stationary phases. However, if the electrostatic interactions are too strong, a highly saline eluent is needed to affect elution in a reasonable period. Nonvolatile saline eluents are not compatible with the electrospray ionization mass spectrometer, commonly used as a detector (see Chapter 17); this complicates matters.

Balancing the charges with a zwitterionic bonded phase, a concept originally developed by Knut Irgum at University of Umeå, Sweden, is one of the most successful approaches used in HILIC. A ^-O_3S $(CH_2)_3$—$N^+(CH_3)_2CH_2$— zwitterionic group may be the functional entity, bonded to a porous silica support for better efficiency, or a polymeric support for greater pH range. It is interesting that for analytes that may exhibit significant electrostatic interaction, it makes a difference as to which end of the zwitterion is bonded to the support. In the above example, the + end of the zwitterion is closer to the support. Columns are available also with the reverse orientation and in many cases, selectivity is reversed.

A zwitterionic HILIC (ZIC-HILIC™) column with the negative end outside. In the other form, the exterior group is $-N(CH_3)_3^+$ and the bonded end is $-CH(SO_3^-)-CH_2$-silica.

Chiral stationary phases. There are four basic types of chiral stationary phases (CSPs) for HPLC.

1. Pirkle phases. William H. Pirkle first developed the π-acceptor CSPs at the University of Illinois. These CSPs resolve enantiomers containing π-donor groups, the π-electrons being supplied by the aromatic groups within the enantiomer. A chiral π-electron acceptor molecule, the active component of the CSP, is covalently bonded to porous silica particles. The interaction sites for chiral recognition on the CSPs are classified as π-basic or π-acidic aromatic rings, acidic sites, basic sites, or steric interaction sites. π-π interactions occur between the aromatic rings in the analyte and those in the CSP. Acidic sites supply protons for potential hydrogen bonding, and basic sites supply π-electrons. In addition, steric interaction can occur between large bulky groups.

Altogether, as with chiral GC stationary phases, minimally there must be interaction at three different points between the enantiomeric analyte and the CSP to achieve a separation. Common π-acceptor CSPs are based on dinitrobenzoyl derivatives of L- or D-phenylglycine, L- or D-leucine, aminophenylalkyl esters and aminoalkyl phosphonates, etc.; some are based on β-lactams. The one remarkable attribute of such phases is that a given CSP is often available in either of the two chiral configurations. The order of elution of an enantiomer pair can thus be chosen by choosing a column of the appropriate configuration. This is very valuable when one enantiomer must be separated from a very large excess of the other; invariably separation and accurate quantitation is facilitated when the trace constituent elutes first.

π-electron donating CSPs are designed to separate amines, amino acids, alcohols, and thiols. The chiral recognition mechanisms are the reciprocal of the π-acceptor CSPs. Naphthylleucine is a common CSP of this type. A third type of Pirkle CSP has attributes of both a π-donor and a π-acceptor. Examples include dinitrobenzoyl derivatives of aminotetrahydrophenanthrene, diphenylethylenediamine, diaminocyclohexane, etc. Most Pirkle-type CSPs can be used either in the reverse or the normal phase mode.

2. Chiral cavity phases. Cavity type phases such as cyclodextrins (CDs, see Section 15.10 and Figure 15.10) for gas chromatography have been previously discussed. α, β, and γ-cyclodextrins are, respectively, 6, 7, and 8 α-D-glucopyranose units linked in a α-1,4 fashion. In the basket-shaped molecule, the central cavity is somewhat hydrophobic and the outer surface is hydrophilic. For HPLC, β- and γ-cyclodextrins are bonded to porous silica, typically through a hydrolytically stable ether linkage. Stationary phases based on various derivatized CDs are also available. Commercially available since 1983, CD-based CSPs presently remain the workhorses of chiral HPLC. These are typically used in the reverse-phase mode. Chiral differentiation occurs when an enantiomer can enter in the cavity and interacts with the interior while one or more functional groups attached to it interacts with the mouth of the cavity.

Cyclofructans (CFs) contain D-fructofuranose units, linked in the β-2,1 fashion. CF6 and CF7 contain 6 and 7 units, respectively. In contrast to CDs, CFs have a polar crown ether core. Like CDs, CFs are also derivatized easily. CF-based stationary phases were first introduced in 2011—these CSPs are commonly used in the normal separation mode of HPLC; they have the unique ability to separate chiral primary amine enantiomers; these compounds are of great importance as precursor molecules for various pharmaceuticals. Sulfonated CF derivatives can be used in the HILIC mode.

Like CD and CF stationary phases, CSPs based on macrocyclic glycopeptide antibiotics were introduced also by Daniel Armstrong. Vancomycin, which contains 18 chiral centers surrounding three cavities (five aromatic ring structures bridge these strategic cavities), is the basis for one such CSP. Hydrogen donor acceptor sites are readily available close to the ring structures. Another is based on the amphoteric glycopeptide, Teicoplanin, which contains 23 chiral centers surrounding four cavities. Yet another CSP is based on Ristocetin A, the largest and most complex of the above ligands. Ristocetin A has 38 chiral centers surrounding four cavities. Six sugar moieties, a peptide chain, and additional ionizable groups give this CSP the complexity and diversity to separate a wide variety of analytes.

18-Crown-6 tetracarboxylic acid, bonded to porous silica, is also commercially available in either (+) or (−) configuration, and is particularly well suited for the separation of amino acid enantiomers.

3. Helical polymer phases. Polymers such as cellulose esters have a helical structure. Right-handed and left-handed helices are not superimposable—they are inherently chiral and can be used for enantiomer separations. Both cavity inclusion and H-bonding and/or hydrophilic/hydrophobic interactions are involved in the separation. Popular phases are based on tris-(dimethylphenyl or chloromethylphenyl) derivatives of carbamoyl cellulose or amylose.

4. Ligand exchange columns. The substrate, typically porous silica, contains one enantiomer (L- or D-) of an amino acid, e.g., proline or a derivative of the amino acid (to provide an optimum length for the spacer). When such a column is treated with a copper salt, Cu^{2+} reversibly coordinates to the bonded amino acid moieties in a multidentate fashion, i.e., at more than one binding site. Any chiral molecule that can bind to Cu^{2+} can displace one of the bound enantiomeric amino acids, but the strength of this interaction depends on the specific chiral configuration of the analyte—the analyte enantiomers are thus separated. As with the Pirkle π-acceptor CSPs, the columns are available in either L- or D-configuration, and this choice will dictate which analyte enantiomer will elute first. In general, if the column is in L-configuration, analyte D-enantiomers elute first, but there are exceptions; for one popular ligand exchange column from Astec, in a separation of lactic, malic, tartaric, and mandelic acids, only the tartaric acid enantiomers elute opposite to the expected order.

Other supports. Alumina played an important role as support material in early normal phase HPLC. Presently alumina columns (typically with 5 µm diameter microporous particles that are coated, e.g., with polybutadiene or polysiloxane with bonded alkyl functionalities) are available and can withstand a 1.3–12 pH range.

Based on the original work of Carr at the University of Minnesota, **zirconia** (ZrO_2) based column packings are available with particle sizes down to less than 2 µm (sub-two micron is often abbreviated as STM in the current HPLC literature) and coated with polybutadiene, polystyrene, various ion exchange groups, elemental carbon, C18 bonded to the carbon layer, chiral selectors, etc. Amazingly, some of these phases are chemically stable over the pH range 1 to 14 and are also thermally stable up to 200°C. Unusual GC-like separations, with GC-style detectors, become possible in LC with such phases; an example of the elution of aromatics with pure water is shown in the margin figure. Pressure drop and vapor pressure calculations can be used to show that water remains in the liquid state through most of the column. The flow rate used is equivalent to 5.6 mL/min through a conventional 4.6-mm diameter column. It would be prohibitive to attain such a flow rate through a column packed with 3 µm particles at room temperature. It is only possible because of the greatly reduced viscosity of superheated water.

Titania (TiO_2) is also available in particle sizes down to 3 µm and in a large variety of pore sizes and can also tolerate a large range of pH and temperature. Finally, **porous graphitic carbon** withstands the full pH range of 0–14 and is available in 3, 5, and 7 µm particle sizes; this stationary phase has a highly crystalline homogeneous surface and is uniquely useful for the separation of highly polar compounds and geometric isomers/diastereomers.

16.3 Equipment for HPLC

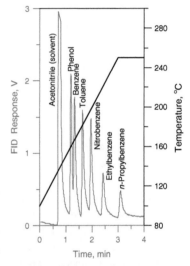

Temperature ramped (100–250°C) separation of aromatics with pure water as eluent, eluent flow rate 8.6 µL/min ., 0.18 × 130 mm stainless steel capillary packed with 3 µm Zirchrom-CARB particles. Flame ionization detector. The expansion of liquid to gas is mostly responsible for the observed tailing. Adapted from T. S. Kephart and P. K. Dasgupta, Talanta, **56** (2002) 977.

Any HPLC system must have a minimum of four components: a pump, an injector, a separation column, and a detector. Commonly a computer system is used to acquire the data and to control the other components; an autosampler is also a commonly used accessory in busy laboratories (see Figure 16.2). Not only is there a considerable difference in scale between analytical and preparative HPLC (not covered here), there is an equally great span between the small end of analytical scale HPLC (0.1 mm diameter capillaries) and the original scale of 4.6 mm diameter. As the linear velocity through the column is the governing parameter in scaling the flow, the optimum flow rate between columns of two different diameters must change in proportion to the square of the diameter. A flow rate range of 1–2 mL/min on a 4.6 mm dia. column will translate to a flow rate of ∼0.48 − 0.96 µL/min for a 0.1 mm dia. column, a ∼2000-fold difference. Optimum injection volumes also vary proportionally. No single system can span this large a range optimally for any of the components. Although some systems claim to span a 3.7 orders of magnitude range (e.g., 1 − 5000 µL/min), it is rare that more than two orders of magnitude range can be spanned with equivalent performance. It is fortunate that open tubular LC, where the column is an open (wall-modified) tube, optimally

< 20 μm in diameter and where the optimum flow rates will be in sub to double digit nL/min, is still only in research laboratories; available pumps do not need to claim to be operable in this range as well!

The great variety of column diameter, length, and particle size naturally necessitates somewhat different types of equipment. Nevertheless, the majority of separations at the time of this writing is still conducted by standard HPLC (pressures < 400 bar or 5800 psi), largely performed on short columns packed with 3- and 5-μm silica-based particles. While STM particle packed columns may require pressures over 5800 psi, especially in longer columns (> 10 cm), and UHPLC pumps of pressure capability up to 19,000 psi (1300 bar) are available, these are still used by a minority. The recent availability of SPP packings that permit efficiencies similar to STM columns but at lower pressures, is slowing the migration to UHPLC, which, at least as yet, is not in wide use. However, an estimated ~25% of the HPLC column market has been taken over by STM and STM-SPP packings, which are largely used with UHPLC systems, even though the majority of chromatographers are still not using UHPLC systems. Many analyses in the pharmaceutical/biotech industry also must use prescribed regulatory methods that are based on specified columns packed with 5-μm particles that do not require pressures > 6000 psi.

1. Solvent Delivery System. While the principal component of the solvent delivery system is the pump, other important ancillary components are an inlet filter, a solvent degassing system, and a pulse dampener.

If particles enter into the pump chamber, they may score the piston or the wall, or cause check valve malfunction, plugging of the column frit, etc.; filtration of the eluent is therefore a must. The inlet filer is typically a 0.2 μm pore size fritted stainless steel or polymeric filter that constitutes the terminal end of a PTFE or PEEK tube that is immersed in the solvent reservoir, the other end connects to the pump. In addition, the majority of HPLC users prefer to use highly pure HPLC-grade solvents. UV absorption is one of the most common modes of detection in HPLC, and aside from already being filtered, HPLC-grade solvents have low levels of UV absorbing impurities.

Unless anything else is done, the eluent reservoir will be in contact with the air and the eluent will contain dissolved air. In some isocratic elution arrangements, it is sufficient to locate the solvent reservoir above the pump so that gravity aids the solvent aspiration by the pump without forming air bubbles. Application of a modest amount of backpressure after the detector then prevents the formation of gas bubbles in the detector and attendant problems. The large majority of HPLC applications, however, require gradient elution. Solubility of air (and most common gases, other than gases that dissolve to form ions, e.g., CO_2) is significantly greater in organic solvents than in water or mixed hydroorganic solvents. As such, gas bubbles may form when an organic solvent is added to an aqueous eluent, i.e., when the solvent composition of the eluent is blended/changed online. This is further promoted by the heat generated by pumping and mixing; gas solubility decreases with increased temperature. The resulting gas bubbles will create problems with proper check valve function and/or pump and detector operation. In many cases, the detector cell is thermostated by maintaining it at a slightly elevated temperature to reduce detector noise, especially with refractive index (RI), viscosity, or conductivity detectors that are particularly sensitive to temperature changes. Again, gas bubbles will form in the absence of solution degassing.

There are a number of different ways to remove dissolved gases, by off-line procedures such as (a) heated stirring and (b) vacuum degassing, and online procedures such as (c) heated stirring, (d) vacuum degassing across a gas-permeable membrane (membranes of Teflon AF, some other fluorocarbons, and supported silicone are highly gas permeable), and (e) helium sparging. Strategies (d) and (e) are commonly used.

An excellent discussion of why gas bubbles form, the different problems they can cause and different methods of degassing and their advantages/ disadvantages appear in a Shimadzu corporation website. http://www.shimadzu.com/an /hplc/support/lib/lctalk/s5/051.html

The in-line degassers are incorporated between the solvent reservoir and the pump. A membrane-based vacuum degasser can remove 70 to 95% of the dissolved gas, decreasing with increasing flow rate. Because oxygen absorbs at low wavelengths, a flow rate gradient will result in a rising baseline at 190–220 nm. In the best-performing vacuum degassers, the vacuum pump turns on and off at short fixed intervals. Multiple degassing lines are often provided and greater degassing is possible by connecting them serially, i.e., running the eluent through the degasser multiple times.

Helium degassing (e) is most commonly used and can provide better degassing than vacuum units; for optimum operation, they do, however, require more care. Bubbling helium into the mobile phase causes the helium to displace the dissolved air. This helps to obtain a stable baseline with high sensitivity, not only for UV and fluorescence detection (fluorescence is often highly affected by dissolved oxygen), but also for RI detectors. To prevent reexposure to air, a good helium purging system is sealed, and to facilitate solvent delivery to the pump, the eluent reservoir is modestly pressurized. In cases where the highest degree of degassing is needed, helium sparging is used in conjunction with a vacuum degasser. But even this may not be enough for sensitive reductive electrochemical detection where the applied potential is sufficient to reduce dissolved O_2, which must be removed as completely as possible. In this case, any residual O_2 is removed by an electrochemical reduction unit prior to the pump. Metal, rather than polymer, tubing is used in all areas where reintrusion of O_2 may occur.

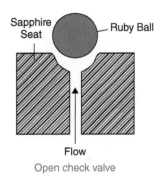

Open check valve

Pumps for solvent delivery. All HPLC pumps are positive displacement pumps that rely on the incompressibility of liquids. Within limits of their pressure capabilities, they ideally provide a constant flow rate, independent of solvent viscosity or column backpressure.

The most commonly used pump for HPLC is the reciprocating pump. In its simplest configuration, this comprises of a small cylindrical piston chamber equipped with two one-way check valves. The inlet check valve is connected to the eluent reservoir and opens during the aspiration stroke, allowing the pump chamber to refill, and during this time the outlet check valve is closed. During the pumping stroke, the outlet check valve opens and the inlet check valve closes (so flow back to the eluent reservoir is eliminated) and the solvent flows to the downstream components. The seat of the check valve is made of sapphire, the ball of ruby, and the piston of sapphire (this may seem like a veritable jewel box, but both ruby and sapphire are a very hard form of Al_2O_3 that is mass produced industrially for such purposes). Although the leakage rate through such a check valve is very low, at very low flow rates and/or high pressures, the leakage rate can be appreciable. It is common to use two or more check valves in series, although they may be contained in a single housing. A spring-loaded design that has a spring pressing down on the ball to secure it to the seat is also common. In this case, the pressure of the flowing liquid must exceed the pressure exerted by the spring for the check valve to open and permit flow.

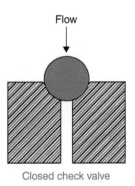

Closed check valve

In a single piston pump, the piston driving cam is so designed that the aspiration stroke is much faster than the dispensing stroke, to minimize the duration of flow stoppage during the aspiration period. Nevertheless, pulsation from a single piston pump is severe and must be dampened. Pulse dampeners are obligatorily used with such pumps. The pump output passes through a device with a flexible component (stainless steel is often used, sufficiently thin to flex but not so thin as to rupture) prior to downstream components. In suitable cases, e.g., with purely aqueous eluents used in ion chromatography, long lengths of Tefzel® are effective.

Tefzel® is ethylenetetrafluoroethylene (ETFE), a copolymer of ethylene and tetrafluoroethylene. Tefzel is nearly as inert as PTFE but it has much greater burst strength than PTFE.

Single piston pumps like this are rarely used today as the principal HPLC pump. Typically a dual piston pump head is used; the eluent reservoir feeds both heads through individual check valves, and the output of each head through individual check valves are combined by a tee. The two heads are driven out of phase with each other so that

The freely available web-based book on HPLC by Yuri Kazakevich (Seton Hall University) provides drawings for the various HPLC pump designs (http://hplc.chem.shu.edu/HPLC/index.html) discussed here. It is a good resource on HPLC in other aspects as well.

the output flow remains constant. A computer-designed cam drives both heads and pulsation is vastly reduced relative to single-head pumps.

In another design, the flow from one chamber, called the low-pressure head, feeds the serially placed high-pressure head that delivers the eluent downstream—this arrangement requires two inlet and one outlet check valves. During the high-pressure chamber refill period, the low-pressure piston can fill the high-pressure chamber very rapidly, in $<\sim 1\%$ of the pumping cycle, minimizing the pulse duration. In yet another serial dual piston design, while the smaller piston is dispensing eluent downstream, the bigger piston chamber is filling up with eluent. The bigger piston chamber is equipped with an inlet and an outlet check valve, the latter is directly connected to the second piston chamber. No other check valves are needed. When the pistons change their direction, the bigger piston refills the smaller chamber, which simultaneously dispenses eluent into the system.

The piston chamber is commonly made of stainless steel. This, however, is incompatible with a variety of eluents, halides at low pH and many chelating agents can attack stainless steel, especially when the eluent is not completely deoxygenated. In ion chromatography, as well as in many metal susceptible biosystems, PEEK is used. This is inert to aqueous solutions (except for high concentrations of HNO_3 and H_2SO_4) and many organic solvents including methanol and acetonitrile, but there are limitations on maximum pressure and it may swell in CH_2Cl_2, tetrahydrofuran (THF) and dimethyl sulfoxide (DMSO). PEEK is also limited to pressures < 400 bar. In high-end UHPLC systems, medical grade titanium is used for the piston chamber.

Gradient elution is vital to most HPLC separations. Pumping systems capable of running a gradient containing up to four components (a quaternary gradient) are commercially available. Gradient pumping systems fundamentally differ in whether the solvent composition changes on the feed (low-pressure) side or the output (high-pressure) side of the pump. To form a binary high-pressure gradient between, e.g., methanol and a phosphate buffer, two separate independent pumps, pump A pumping methanol and pump B pumping the aqueous buffer, are used. The outputs of the two pumps are combined at a mixing tee. An actively stirred mixing chamber can be used; presently efficient passive mixers are also common (UHPLC systems often use such static mixers). During the chromatographic run, a gradient programmer that controls both the pumps, slowly raises the flow rate of pump A while simultaneously decreasing the flow rate of pump B. The total flow rate is maintained constant. It is possible to use a flow rate gradient also but this is rather rarely used. Because two identical pumps are needed, a high-pressure gradient pumping system is expensive; beyond a binary gradient system, it is rarely configured this way. They do have the advantage that mixing volume in the high-pressure side can be very small, and very sharp step gradients can be effectively attained.

In low-pressure gradient systems, to form a n-component gradient, n independent 2-port solenoid valves configured in a single housing, is used. The inlet port of each valve is connected to independent modestly pressurized eluent reservoirs each modestly pressurized), and the outlet port of all the valves are connected to a common port that in turn feeds the pump. These valves can be turned on and off very fast (actuation time is $\sim$10-15 milliseconds), and during the intake stroke of the pump, individual valves open and close rapidly. During this total intake stroke, if the valve connected to solvent A and that connected to solvent B are each open a total of 50% of the time (each valve may open and close multiple times), the intake feed will nominally be composed of 50% A and 50% B and so on. Note that during a constant flow rate solvent gradient, the pump pressure may change in a marked manner, as the viscosity of the mixed solvent can be higher than either of the pure components; this is especially so for a water:methanol gradient system.

Syringe pumps, as the name implies, constitute of a motor-driven syringe (typically of stainless steel, with a highly polished bore), of a large enough volume to complete a desired chromatographic run without having to refill. The open end is connected to the common port of a 3-way valve.[1] The two other ports are respectively connected to downstream components, e.g., an injector, and the solvent reservoir. As needed, the valve switches to the solvent reservoir for aspiration and refill. One major manufacturer offers pumps with capacities of 65–1000 mL; flow rate ranges vary according to the pump capacity (10nL/ min -25mL/ min and 0.1–410 mL/ min, respectively), while the maximum pressure capability varies inversely (20,000 and 2000 psi, respectively). Aside from realistically attainable flow rates over a very large span, the great virtue of a syringe pump is nearly pulseless operation, of particular value when detection systems very sensitive to pump pulsations are used. Syringe pumps for HPLC are expensive, however, and gradient operation will require multiple pumps.

Constant pressure pumps, often a pneumatically pressurized reservoir, do not see much use in present-day LC, but this may change if open tubular LC becomes practical.

2. Sample Injection System. Most HPLC injection systems consist of 6-port loop-type injection valves or variations thereof. An injector of this type consists of a "stator," to which external connections are made and an internal disk-shaped "rotor", and is shown schematically in Figure 16.9 with the perspective from the rear of the injector. We will label the ports 1 through 6 clockwise from the top of the injector. The ports are threaded and all connections with tubes to the external world are made with appropriately threaded nuts and compression ferrules (not shown).

A fixed-volume injection loop is connected between ports 1 and 4. Port 2 connects to the sample injector (this can be a manually operated syringe, or an autosampler syringe, or a friction-fit needle port, often located at the front center of the valve). Any excess sample drains through port 3. Ports 5 and 6 connect to the column and the pump, respectively. The rotor disk does not have any ports; the black circles are drawn on the rotor only to indicate that when it is placed against the stator, how it will be positioned. Instead, three alternate grooves are engraved on the rotor between what would have been the port locations. Consider the left half of Figure 16.9; the rotor is pressed tightly against the stator. When a sample is introduced through port 2, it travels along the slot in the rotor to port 1, and as we continue introducing the sample,

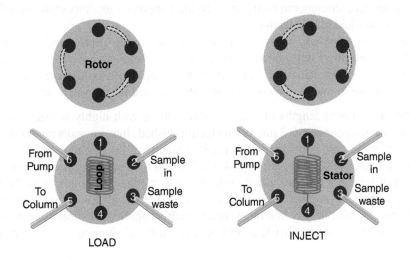

FIGURE 16.9 A 6-port sample injector, shown in the "LOAD" position on the left and in the "INJECT" position on the right.

---

[1]A "3-way valve" constitutes of three ports, a common port and two others, as in a Y-structure. Liquid can flow one of two ways: from the common leg to either of the other two branches. The term "3-way" valve can therefore be potentially misleading, but this nomenclature is too entrenched.

it fills the injection loop; the liquid travels through port 4 and the corresponding rotor slot to port 3 to waste. Meanwhile the pump output, coming in through port 6, travels through the corresponding rotor slot through port 5 directly to the column. The ports are distributed symmetrically at 60° intervals. The rotor has two positions, with travel stops at the end of each position to prevent further rotation. When the rotor is rotated by 60° manually, by a pneumatic actuator, or most commonly, by an electric motor, the valve now enters the inject position (Figure 16.9 right). Notice that in this position, the pump output now proceeds through the corresponding rotor slots to flush the contents of the loop to the column; an injection is made. Meanwhile, the sample input port connects directly to the waste port. This type of injector is called an external loop injector and the injection volume can be changed by changing to a different volume injection loop. However, at small loop volumes, internal volumes within the ports/slots also become relatively significant. The rotor is generally made of a softer material than the stator. PEEK and reinforced PTFE composites are common rotor materials as they provide a good seal and low friction. In some ingenious valve designs, the flow through the valve is configured in a manner that the fluid pressure is used to seal the rotor against the stator.

Both because of minimum length of the tubing needed to externally connect the loop ports, as well as the small but finite volumes internal to the valve ports, it is not practical to have sub-µL injection loops. However, instead of the external loop, it is possible to have a (replaceable) internal disk with a small engraved slot that serves as the injection loop. Injectors with internal loops as small as 4 nL with 100 µm bore connecting passages are commercially available for "nanoLC." Internal loop injectors of up to 5 µL volume are also available; there is of course no upper limit to the volume of external loops. Injection volumes scale with column cross section (so the injected sample occupies the same length of the column, regardless of the column dimension); a 10 µL injection, typical with a 4.6 × 250 mm column with $d_p = 5$ µm, will translate to a ~5 nL injection on a 0.1-mm bore packed capillary.

Injection volume reproducibility of filled loop injections are generally better than the reproducibility attainable by the rest of the equipment; so it is not usually a factor in the overall analytical reproducibility. Partial loop injections are possible by partially loading the loop with a microliter syringe for front loading needle-port injection valves or for automated valves, keeping the valve in the inject position for a period too short for the entire loop contents to be injected. Neither approach exhibits reproducibilities comparable to filled loop injection.

Similar to the 6 port valves with the rotor connecting alternate adjacent ports, 8 to 14 (even numbers only) port valves are available and can be used for variety of functions, e.g., to simultaneously inject the same sample to two different columns, to inject the sample to one of two columns already plumbed in the system, etc.

3. **Column.** Straight lengths of stainless steel tubing with highly polished (typical RMS surface roughness < 0.2 µm, often electropolished) interior walls are most commonly used as column housing (Figure 16.10). "Standard" bore columns usually mean an inner diameter of 4.6 mm, although 4 and 2.1 mm bore columns are also common. Columns of bore ≤ 1 mm are usually referred to as capillary columns, with the term "nanoLC" generally applied to columns of bore ≤ 0.1 mm. PEEK column housing is used in ion chromatography and other metal sensitive applications; capillary columns generally use fused silica, PEEK or PEEK-encased fused silica tubes as the column housing.

To filter out particles, to evenly distribute the flow across the column cross section, and to provide a support for the column bed, frits are used both at the inlet and the outlet of the column. Sintered stainless steel disks (typically of 0.5–2 µm pore size)

For a variety of applications of multiport valves and the many different configurations they can be used in, see http://www.vici.com/support/app/2p_japp.php.

FIGURE 16.10 Typical HPLC columns. (Courtesy of Waters Corporation.)

surrounded by a polymer ring (PEEK, PTFE, ETFE) to provide sealing, constitute a typical frit. A flow distribution disk is sometimes used below the inlet frit, especially in larger bore columns such as those used for preparative chromatography. Note that a typical stainless steel frit has far greater surface area than the rest of the column and such frits cannot be used in iron-sensitive applications, as in many bioseparations. Titanium and polymer frits are available for such applications.

Column lengths vary from 5 mm for affinity columns based on perfusion particles to 250 mm for a "standard" 4.6-mm bore column packed with 5 μm diameter particles. Recall that for SEC separations, the total retention window depends on the available pore volume within the column; this in turn is dependent on the column dimensions. Columns that use the exclusion mode (SEC/ICE) are often larger in dimension; SEC columns as large as 7.8×300 mm are not uncommon.

While it is generally true that for a given particle type, column efficiency increases approximately with the reciprocal of the particle diameter (d_p), there is a great deal of variation from manufacturer to manufacturer. In addition, no column is equally efficient for all analytes. In the study referred to in the margin notes, five of the columns used $d_p = 10$ μm, three used $d_p = 4$ μm, and the rest used $d_p = 5$ μm. With the neutral solute toluene, the highest and lowest observed efficiency (~115,000 and 32,000 plates/m, respectively) were both observed with $d_p = 5$ μm columns. The top performing column above was among the worst performers with a basic solute, pyridine; the top performer in this case was also a $d_p = 5$ μm column and attained ~65,000 plates/m.

To determine the effect of d_p on plate counts, it is more meaningful to compare similar particles of different sizes made by the same manufacturer. For a particular type of their porous silica-based C18 columns, Agilent scientists reported ~85,000, ~140,000, and ~240,000 plates/m for $d_p = 5, 3.5$ and 1.8 μm, respectively. In comparison, a brand X column ($d_p = 2$ μm), a brand Y column ($d_p = 2.5$ μm) and a monolithic column all provided almost the same efficiency under comparable conditions and were in fact marginally worse than the column with $d_p = 3.5$ μm. In comparison, present SPP-based Kinetex columns are stated by the manufacturer Phenomenex to exhibit efficiencies of $> 400,000, 320,000, 280,000$, and 180,000 plates/m, respectively, for their $d_p = 1.3, 1.7, 2.6$, and 5 μm packing.

See http://www.mac-mod.com/pdf/technical-report/036-ColumnComparisonGuide.pdf for a very detailed comparison of 60 different C18 phases. They are categorized by relative hydrophobicity and polarity, and column efficiencies for neutral and basic compounds are given.

Columns packed with $d_p \leq 2$ μm particles are typically available in 3–5 cm lengths due to the very large pressure drops; their use is practical only with UHPLC equipment. For particles with $2.5 \leq d_p < 5$ μm, the maximum available column length is typically 150 mm; for $d_p = 5$ μm, standard 25 cm length columns are available and are routinely used.

With $d_p = 5$ μm columns, a short (1–5 cm long) **guard column** is usually placed between the injector and the analytical column, generally containing the same packing as the analytical column. It is placed there for two reasons. First, it will retain debris (e.g., pump-seal fragments) and sample particulate matter that would otherwise get on the analytical column and foul it, changing column efficiency and selectivity. (Some researchers like to place an in-line filter between the pump and the injector as an additional safeguard.) Second, it retains highly sorbed compounds that would be caught on and not be eluted from the analytical column. The guard column thus extends the life of the analytical column. It must be regenerated/cleaned or more commonly, replaced periodically. Some analytical columns are designed with an integrated guard column; this minimizes band broadening through additional tubing/connections. However, with highly efficient columns packed with small particles, considerations of extra-column dispersion become all important and guard columns are generally not used.

Column oven. Temperature control of the column is not essential but greatly improves reproducibility. Most high-end HPLC systems include a column oven and the typical operational temperature is a little higher than the ambient temperature, e.g., 30°C, as this can be easily maintained. Diffusivity increases with temperature and the column efficiency increases. There is a shift in the plate height minimum in the van Deemter plot towards higher velocities, making higher flow rates more desirable and leading to faster separations. Higher flow rates are also easier to attain because the necessary pressure is reduced as the eluent viscosity decreases at a higher temperature. The penalty paid for elevated temperature operation is accelerated column degradation, especially for silica-based columns and anion exchangers. In addition, unlike GC, an increase in column temperature does not necessarily mean a decreased retention factor. Temperature programming, although possible, is rarely used in LC. Some detectors, especially RI and conductivity detectors, are very sensitive to temperature changes. While better detectors of this type already have some means of temperature equilibration/thermostating, it is advisable to operate the column and the detector at the same temperature. For a conductivity detector, the measurement cell is often detachable; this can easily be placed in the column oven.

Even in a basic HPLC setup, when a column oven is not available, rapid temperature changes can be avoided by wrapping the column in a bubble wrap or insulation foam.

4. Detectors. Generally Desirable Criteria for HPLC detectors and Data Acquisition. Low noise and high sensitivity promotes good detection limits and are thus desirable in all detectors. As a single chromatographic run on a complex sample can last a significant time (≥ 1 hour), it is highly desirable that the detector baseline does not drift substantially. As column efficiency increases, peaks become narrower and span a smaller range of time; this necessitates that the detector must respond fast and the resulting data must be acquired with a sufficiently fast time resolution. This also requires that the detector cell volume be sufficiently small and of a geometry that the entire passage to and through the cell contributes little to overall band dispersion. If any of these criteria are not met, the acquired chromatogram will display a poorer separation than what was actually achieved at the column exit.

As a rule of thumb, to faithfully represent a chromatographic peak, at least 20 digitized measurements must be made across the peak. If the entire peak occurs within 1 s, the detector response time should be no greater than 50 ms (milliseconds). Quite

A guard column extends the life of the analytical column and improves separations by retaining strongly sorbed compounds and debris. It is not typically used with columns packed with STM particles because of added dispersion from connections, etc.

often a single response time is quoted for a detector; it is typically the "e-folding" time.[2] Note also, the rise time and fall time of a detector signal may not always be the same. At least 3, more conservatively 5, e-folding times (of a rising or falling signal, whichever is limiting), should be considered the response time. In this example, we are looking then for a detector with an e-folding response time of 10 ms. If the detector output is internally digitized, then data will be directly transmitted to the acquisition system. This is typically the case for present-day detectors. In the above example, a minimum data transfer rate of 100 Hz is needed. Detectors used for fast LC measurements today can typically transfer data up to 200 Hz (but see the sidebar).

Other desirable characteristics in an HPLC detector is that it should not be sensitive to small changes in flow rate (and thence, backpressure) and temperature (not met for many detectors), be simple to operate, and be nondestructive with low dispersion so (an)other detector(s) can be serially deployed. A wide linear dynamic range is also a sought-after detector characteristic, as it makes quantitation over a broad range of concentrations easier. In an age of inexpensive data storage (where a high-resolution calibration curve can be stored and retrieved) and abundant computational power, if the response slope (change in response per unit change in concentration) remains sufficiently high, this emphasis on linearity is questionable. Indeed, some detector responses (e.g., most evaporation-based detectors that rely on forming aerosol particles) are inherently nonlinear over a large range, and manufacturers frequently incorporate signal processing in the instrument firmware to "linearize" the output, a process to which the user is blind. In a more enlightened future, this would be transparent to the user and the user should be able to choose one of many ways available to utilize a nonlinear calibration plot for quantitation.

Whether a detector selectively responds to certain species or nearly universally responds to all species can be regarded as a blessing or a curse. Universal response may be of interest in a scouting, exploratory venture. But if one is interested in a specific compound, e.g., a particular impurity of interest in a pharmaceutical sample, a detector with selective response to that analyte will be much better, especially when your analyte peak has a coeluting interferent but the detector only responds to the analyte of interest. Short of a mass spectrometer operating in selected ion fragmentation and monitoring mode (see Chapter 17), there is no detector that is likely to respond only to a specific analyte. Mass spectrometry and LC-MS are discussed in Chapter 17, but even in this case, chromatographic separation of the analyte from the matrix is needed, else quantitation errors will occur from the effect of the matrix on the ionization efficiency of the analyte.

Universal and quasi-universal detectors. A mass spectrometer and a refractive index (RI) detector are both truly universal detectors. **A refractive index (RI) detector** is a bulk property detector; any change in its composition is reflected in the RI. The RI detector has two major shortcomings: (1) RI changes considerably with temperature; as a result all good RI detectors use a thermostated optical block, and the entrance liquid is thermally equilibrated before it enters the detection cell. This is typically accomplished by embedding a significant length of a narrow-bore thermally conductive entrance tubing in the same block. While temperature stabilities of ≤ 0.01 °C are achieved, the entrance tubing does contribute to significant band broadening. (2) RI detection is generally incompatible with gradient elution. Consider that during a solvent gradient, the eluent composition changes at the percent level, changing the refractive index proportionally. In contrast, one may be trying to detect the elution of a solute a few ppm in

If the analog signal from the detector is to be externally digitized and acquired, further considerations are necessary. The rule of thumb for acquiring data is that to faithfully represent an event that is occurring at n Hz, it must be sampled and data ideally acquired at a rate at least 5 times greater, i.e., at $5n$ Hz. Thus, if a detector e-folding time of 10 ms (corresponding to a frequency of 100 Hz) is to be faithfully recorded, the data acquisition frequency should be ≥ 500 Hz. However, if the detector itself has a slow response time, fast data acquisition will be of little value.

---

[2]If a step function is applied, the detector output reaches the final value in a sigmoidal manner, with $1/e = 0.37$ of the span remaining to be traversed in each e-folding time. If the e-folding response time is, e.g., 10 ms, in 10, 20, 30, 40, and 50 ms, the response reaches 63, 85, 95, 98, and 99.3% of its final value. The e-folding time is the same as the "time constant" for an RC filter.

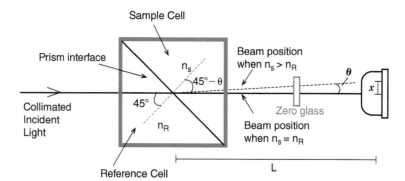

FIGURE 16.11 Simplified schematic of a differential refractometer.

concentration. This will be akin to detecting the presence of a pebble on a steep hillside from an elevation map. Further, RI may not be a single-valued function of the solvent composition. For the commonly used water methanol solvent system, going from pure water to pure methanol results in the RI (as well as the viscosity) increasing with methanol addition, reaching a maximum and then decreasing again.

Although other designs have been investigated and even been commercialized, the vast majority of HPLC RI detectors are based on the differential refractometer design shown in Figure 16.11. The heart of the differential refractometer is a two-compartmented quartz flow cell, each half of which contains a hollow 45° prism with flow in/out connections. In the figure, the interface is shown as the thick solid black line, while the perpendicular to the interface is shown as the thin dashed blue line. The reference cell is filled with the intended eluent (reference, RI n_R), this generally remains stationary during operation. The column effluent (sample, RI n_S) flows through the other cell. Collimated monochromatic light (RI is dependent on the wavelength, an essentially monochromatic light source, e.g., an LED, must be used) is incident on the reference cell. While it can be shown that the same principles apply to all other angles of incidence, presently for simplicity we assume a 45° angle of incidence. The light beam passes through the reference cell and then along the same path through the sample cell when $n_S = n_R$, as no refraction occurs (in reality refraction occurs at each quartz-liquid interface, but the effect cancels out) and falls on the photodetector.

The photodetector constitutes of two identical adjacent photodiodes, often with a common cathode and two separate anodes (such a detector is often called a one-dimensional position sensitive detector, PSD). The signal processing circuitry measures the difference in current generated by the two photodiodes. When $n_S = n_R$, an equal amount of light falls on the two photodiodes, the PSD generates zero signal. Minor offsets to achieve zero can be conducted electronically, while the major offset is carried out by translating/rotating the "zero glass" piece, which can bend the beam. When $n_S \neq n_R$ (typically the presence of a solute will increase the RI, as indicated in the figure), the beam will bend due to refraction and the PSD will generate a finite signal.

Consider that according to Snell's law of refraction

$$n_R \sin(45) = n_S \sin(45 - \theta) \tag{16.4}$$

or

$$\frac{n_R}{n_S} \sin(45) = \sin(45 - \theta) \tag{16.5}$$

It is readily shown (see the text website for derivation) that for small values of θ,

$$\sin(45 - \theta) = \sin(45) * (1 - \tan(\theta)) \tag{16.6}$$

Combining Equations 16.5 and 16.6,

$$\frac{n_R}{n_S} = 1 - \tan(\theta) \tag{16.7}$$

If the distance from the liquid–liquid interface to the detector surface is L and the beam center on the detector is shifted by the distance x,

$$\tan(\theta) = \frac{x}{L} \tag{16.8}$$

The refractive index difference between the reference and the sample is usually very small ($n_R \simeq n_S$). Expressing the difference $n_S - n_R$ as Δn, and subtracting both sides of Equation 16.7 from 1 and incorporating Equation 16.8 gives us:

$$\Delta n = n_R \frac{x}{L} \tag{16.9}$$

To increase the beam shift for a given Δn, commonly a mirror is placed behind the cell and the detection is made on the same side of the cell as the beam entrance. Effectively, L is doubled, necessitating that the deflection distance x be doubled as well. In any case, since L and n_R are constants for a given operating condition, x is linearly proportional to Δn. The PSD output signal is linearly proportional to the shift x, so the detector output signal is linearly proportional to Δn. Over a small solute concentration range as is typically of interest in chromatography, the refractive index difference between a solution and the pure solvent is linearly related to the solute concentration, so the detector output signal is also linearly related to the analyte concentration.

For a typical detector, the measurement range in a given setting is ~ 0.0005–0.0006 RI units (RIU) with a noise level of $1.5 - 3 \times 10^{-9}$ RIU. As a frame of reference, the slope of a plot of the RI of an aqueous sucrose solution vs. sucrose concentration is 0.15 RIU g^{-1} mL. An LOD of 5×10^{-9} RIU will thus connote a concentration LOD of $\sim 3 \times 10^{-8}$ g/mL. Instead of using a pair of photodiodes, it is possible to use a linear photodiode array detector. The use of a 512-element photodiode array as the detector gives the Wyatt T-rEX™ a vastly increased range of $\pm 4.7 \times 10^{-3}$ RIU, and thanks to temperature stabilization to $\leq 5 \times 10^{-3}$ °C, a noise level of 7.5×10^{-10} RIU. Because of the considerable range of beam shift that a detector array of significant length can accommodate, from the absolute beam position on the array, the detector can read out the absolute RI within a 1.2–1.8 RI range with a precision of 2×10^{-3} RIU. A high concentration version of this detector can accommodate an $\sim 8\times$ greater ΔRI range, with only $\sim 2\times$ increase in the noise level.

While RI detectors are, in general, less sensitive than UV absorbance detectors, they have been markedly improved in recent years. In particular, statements such as RI detectors are 1000-fold less sensitive than UV detectors have little meaning; many solutes, e.g., carbohydrates, have little or no useful UV absorption.

A number of other detectors are in principle universal, but in practice the sensitivity for general applications are so low that they are used only in niche applications. Various polymers are extensively used in present-day society. Most synthetic polymers do not constitute a polymer of a given chain length or molecular weight. Rather, they are a mixture of polymers of different molecular weight. The characterization of polymers and the determination of their molecular weight distribution, which is a prime determinant of certain polymer characteristics, are most commonly carried out by SEC. In SEC analysis of soluble polymers (many polymers dissolve at high temperatures in suitable organic solvents, and many SEC instruments allow operation at elevated temperatures, e.g., 150 °C), the use of **viscosity detectors** are common.

In its simplest form, the viscosity of a chromatographic effluent can be readily measured by connecting a capillary to the column exit by a tee. The backpressure caused by the capillary is given by the Hagen–Poiseuille equation:

$$\Delta P = \frac{8F\eta L}{\pi r^4} \tag{16.10}$$

where ΔP is the pressure drop, L is the length of the capillary of internal radius r, F is the flow rate, and η is the viscosity of the liquid. For a given capillary and at constant flow rate, the pressure drop is therefore a linear measure of the viscosity. Like RI, viscosity is also a strong function of temperature. A single capillary viscosity detector like this, or one using a differential pressure transducer where the differential pressure is brought to zero by pressurizing the other side of the transducer with a static pressure source, is very inexpensive but rather insensitive.

A **differential viscometer** is configured as in a Wheatstone's bridge circuit using four fluidic resistors through which the column effluent flows. With a thermal pre-equilibration coil and a measurement block thermostated to ≤ 0.01 °C, the sensitivity of a differential viscometer is comparable to a refractometer in SEC applications. In SEC, after a molecular weight (MW) vs. retention time calibration is established using known MW standards under some given chromatographic conditions, a concentration detector (e.g., an RI or similar detector) can provide the MW distribution of a polymer sample. The viscometer, on the other hand, responds more to a higher MW polymer at the same mass concentration and magnifies differences in the higher MW region compared to a refractometer. Further, if an RI and a viscosity detector are used together, both the MW distribution and degree of branching of the polymer can be established. In this context, in conjunction with a concentration detector, a **low-angle (laser) light scattering (LALS, LALLS) detector** that measures light scattering at a small angle and interprets the results in terms of zero angle (intrinsic) scattering, can provide both number-average molecular weight (M_n) and weight-average molecular weight (M_w). A **multiangle light scattering (MALS)** detector is the most powerful in this regard; the generated data can yield the molecular weight and the size (specifically, the radius of gyration, R_g, sometimes called the root mean square radius, a shape sensitive parameter) without reference to any calibration standards, in an absolute manner. A log–log plot of R_g vs MW can indicate if the conformation/shape of the molecule changes with MW.

An **electrical conductivity detector**, like a refractometer, is a bulk property detector. However, it can be considered a **quasi-universal detector** in the sense that all dissolved substances are either charged or uncharged. All charged substances, including charged colloids, have a finite electrical mobility in solution—they will move under an electrical field. The charge transported per unit time is electrical current, and the current resulting from a given electrical field is thus a measure of the electrical conductivity of the solution.

The electrical mobility μ (also interchangeably called ionic mobility μ^0, or the electrophoretic mobility, μ_{ep}, the latter is especially used in the context of electrophoresis) of an ion depends on its charge and its size, expressed as its Stokes radius, r (units of length):

$$r = \frac{kT}{6\pi\eta D} \tag{16.11}$$

where η is the viscosity of the solution and D is the diffusion coefficient of the charged species (the term ion often has a more restrictive use for species with small r—not generally applied for example to colloids or charged proteins). Equation 16.11 is generally referred to as the **Stokes-Einstein equation**. The ionic mobililty, which has units of velocity per unit field strength $\left[\frac{\text{m/s}}{\text{V/m}}\right] = \text{m}^2 \text{ V}^{-1}\text{s}^{-1}$ (the unit $\text{cm}^2 \text{ V}^{-1}\text{s}^{-1}$ is also often used) is given by:

$$\mu = \frac{DzF}{RT} \tag{16.12}$$

In a differential viscometer, the distinction between the baseline liquid vs. the sample in the same stream is accomplished by incorporating a delay coil. For the configuration, see, e.g., http://www.wyatt.com/files/events/news/LCGC3-61-05e.pdf.

Albert Einstein in 1905 and Marian Smoluchowski in 1906 independently discovered in their papers on Brownian motion the previously unexpected connection; in more general form

$$D = \mu k_B T$$

where k_B is Boltzmann's constant

z being the charge magnitude on the ion (+1, −2, etc.) and F the Faraday constant. Equation 16.12 is one form of the **Einstein–relation** or **Einstein–Smoluchowski relation**. Note that at a constant electrical field, the current will not only depend on how fast the ions are moving but how many current carriers are there, i.e., the ionic concentration. The concentration dependence is incorporated in the quantity **limiting equivalent conductance** (λ^0) where

$$\lambda^0 = \mu F \qquad (16.13)$$

The superscript ^{0} is used on the equivalent conductance λ (the conductance of a solution 1 equivalent/L in concentration, in a measurement cell containing two electrodes of unit area, unit distance apart) to indicate that this value is the limiting value reached in very dilute solutions. Note that λ or λ^0 already takes ionic charge into account; some texts denote λ on a molar basis; in that case, the charge magnitude z will also appear on the right side of Equation 16.13. In addition, while λ has no sign, μ, especially when used in the electrophoretic context, μ_{ep}, is considered a vector quantity; the μ_{ep} values of cations and anions are respectively positive and negative. Any given ion in solution is surrounded by ions of opposite charge. As ionic concentration in a solution increases, this oppositely charged "ion atmosphere" shields the central ion from fully experiencing the electric field. As a result, z effectively decreases. As we lower the concentration, z (and hence λ), asymptotically reaches a limiting value. This limiting λ value, λ^0, is also called **equivalent conductance at infinite dilution**.

A table of limiting equivalent conductance values of various ions appears in Table 16.2. Up to a concentration of 1 meq/L, the error in using these infinite dilution limiting values is small, $\sim< 2\%$.

Electrical conductance is the reciprocal of electrical resistance [unit: Ohm (Ω)], the unit of conductance was earlier called Mho (Ohms spelled backwards!) but now is termed Siemens (S). The **conductance G** (S) is related to the **specific conductance** σ_c (S/cm), a characteristic property of any ionic solution by the **cell constant** κ:

$$G = \frac{\sigma}{\kappa}\sigma_c \qquad (16.14)$$

where κ is given by:

$$\kappa = \frac{L}{A} \qquad (16.15)$$

in which two parallel electrodes, each of area A, are separated by the distance L. Note that κ has the dimensions of reciprocal length. The specific conductance of a solution is the sum of the contribution of all the ions that contribute to this conductance:

$$\sigma_c = \Sigma \, \lambda_i C_i \qquad (16.16)$$

where λ_i is the equivalent conductance of each ion and C_i is the concentration of the ion in eq/L.

For a general discussion of diffusion in liquids and the interrelationships of limiting equivalent conductance, diffusion coefficient and ionic mobilities see the notes of Professor Daniel Autrey on the web: http://faculty.uncfsu.edu/dautrey/ CHEM%20324/CHEM%20324 %20Notes%20-%20Chapter% 2024f.pdf.

TABLE 16.2 Limiting Equivalent Ionic Conductance of Various Ions, at 25 °C

Inorganic Cation		Inorganic Anion		Organic Cation		Organic Anion	
Ion	Limiting Equivalent Ionic Conductance, S cm² equivalent⁻¹	Ion	Limiting Equivalent Ionic Conductance, S cm² equivalent⁻¹	Ion	Limiting Equivalent Ionic Conductance, S cm² equivalent⁻¹	Ion	Limiting Equivalent Ionic Conductance, S cm² equivalent⁻¹
Ag^+	61.9	Br^-	78.1	i-Butylammonium	38	Acetate	40.9
Ba^{2+}	63.9	Cl^-	76.35	n-Decylpyridinium	29.5	p-Anisate	29
Be^{2+}	45	ClO_2^-	52	Diethylammonium	42	Azelate^{2-}	40.6
Ca^{2+}	59.5	ClO_3^-	64.6	Dimethylammonium	51.5	Benzoate	32.4
Cd^{2+}	54	ClO_4^-	67.9	Dipropylammonium	30.1	n-Butyrate	32.6
Ce^{3+}	70	CN^-	78	n-Dodecylammonium	23.8	Chloroacetate	39.7
Co^{2+}	53	CO_3^{2-}	72	Ethylammonium	47.2	Chlorobenzoate	33
Cs^+	77.3	CrO_4^{2-}	85	Ethyltrimethylammonium	40.5	Citrate^{3-}	70.2
Cu^{2+}	55	F^-	54.4	Methylammonium	58.3	α-Crotonate	33.2
D^+ (deuterium)	213.7(18°)	$Fe(CN)_6^{4-}$	111	Histidyl	23	Cyanoacetate	41.8
Eu^{3+}	67.9	$Fe(CN)_6^{3-}$	101	Piperidinium	37.2	Decyl sulfonate	26
Fe^{2+}	54	$H_2AsO_4^-$	34	Propylammonium	40.8	Dichloroacetate	38.3
Fe^{3+}	68	HCO_3^-	44.5	Pyridylammonium	24.3	Diethyl barbiturate^{2-}	26.3
Gd^{3+}	67.4	HF_2^-	75	Tetra-n-butylammonium	19.1	Dihydrogen citrate	30
H^+	349.82	HPO_4^{2-}	57	Tetraethylammonium	33	Dimethyl malonate^{2-}	49.4
Hg^{2+}	53	$H_2PO_4^-$	33	Tetramethylammonium	45.3	3,5-Dinitrobenzoate	28.3
K^+	73.5	HS^-	65	Tetra-n-propylammonium	23.5	Dodecyl sulfonate	24
Li^+	38.69	HSO_3^-	50	Triethylammonium	34.3	Formate	54.6
Mg^{2+}	53.06	HSO_4^-	50	Triethylsulfonium	36.1	$HC_2O_4^-$	40.2
Mn^{2+}	53.5	I^-	76.8	Trimethylammonium	46.6	Lactate	38.8
NH_4^+	73.5	IO_3^-	40.5	Trimethylsulfonium	51.4	Malonate^{2-}	63.5
Na^+	50.11	NO_2^-	71.8	Tripropylammonium	26.1	Methyl sulfonate	48.8
Nd^{3+}	69.6	NO_3^-	71.4			$C_2O_4^{2-}$	74.2
Ni^{2+}	50	$NH_2SO_3^-$	48.6			Octyl sulfonate	29
Pb^{2+}	71	N_3^-	69			Phenyl acetate	30.6
Ra^{2+}	66.8	OCN^-	64.6			Picrate	30.2
Rb^+	77.8	OH^-	198.6			Propionate	35.8
Sr^{2+}	59.46	PO_4^{3-}	69			Propyl sulfonate	37.1
UO_2^{2+}	32	SCN^-	66			Salicylate	36
Zn^{2+}	52.8	SO_3^{2-}	79.9			Succinate^{2-}	58.8
		SO_4^{2-}	80			Tartrate^{2-}	64
		$S_2O_3^{2-}$	85			Trichloroacetate	36.6

Example 16.1

Given the data in Table 16.2, calculate the specific conductance and specific resistance of pure water and the conductance of pure water between the tips of two 10 μm diameter electrodes placed 1 mm apart.

Solution

Pure water only contains H^+ and OH^-, each at a concentration of 10^{-7} eq/L, that we can express as 10^{-10} eq/cm^3. Using Equation 16.16,

$$\sigma_c = \lambda_{H^+}[H^+] + \lambda_{OH^-}[OH^-] = 349.82 \text{ S cm}^2 \text{ eq}^{-1} \times 10^{-10} \text{ eq cm}^{-3}$$
$$+198.6 \text{ S cm}^2 \text{ eq}^{-1} \times 10^{-10} \text{eq cm}^{-3}$$
$$= 548.4 \times 10^{-10} \text{ S/cm} = 54.84 \text{ nS/cm}$$

The specific resistance or **specific resistivity** ρ of pure water is simply the reciprocal of the specific conductance (the reciprocal of nS will be 10^9 Ω), thus:

$$\rho = \frac{1}{54.84} \times 10^9 \text{ } \Omega \text{ cm} = 1.823 \times 10^7 \text{ } \Omega \text{ cm} = 18.23 \text{ M}\Omega \text{ cm}$$

The quality of high-purity water is often expressed by its specific resistance. Water of the above specific resistance contains no ions other than the H^+ and OH^- resulting from autodissociation.

The area A of a 10 μm diameter disk electrode is

$$A = \pi(10^{-3} \text{ cm})^2/4 = 7.85 \times 10^{-7} \text{ cm}^2$$

If $L = 0.1$ cm, then according to Equation 16.15,

$$\kappa = 0.1 \text{ cm}/7.85 \times 10^{-7} \text{ cm}^2 = 1.27 \times 10^5 \text{ cm}^{-1}$$

From Equation 16.14,

$$G = 54.84 \text{ nS cm}^{-1}/1.27 \times 10^5 \text{ cm}^{-1} = 430 \text{ fS}$$

The **Nernst–Einstein equation** relates the limiting equivalent conductance to the diffusion coefficient:

$$\lambda° = \frac{DzF^2}{RT} \tag{16.17}$$

At least in a dilute solution, ionic migration is assumed to be independent of the presence of other ions, and the quantity, **limiting molar conductivity**, $\Lambda_m°$, is sometimes used:

$$\Lambda_m° = v_+\lambda_+° + v_-\lambda_-° \tag{16.18}$$

where λ_+ and λ_- are the limiting molar conductivities of the cation and anion, respectively, and where v_+ and v_- are the number of cations and anions, respectively, per formula unit of electrolyte. The combination of the various identities above leads to the general expression of the Nernst–Einstein equation:

$$\Lambda_m° = (v_+D_+z_+ + v_-D_-z_-)\left(\frac{F^2}{RT}\right) \tag{16.19}$$

where D_+ and D_- are the diffusion coefficients for the cation and anion, respectively. The Nernst–Einstein equation is used to determine ionic diffusion coefficients from conductivity measurements and to predict conductivities using ionic diffusion models.

Example 16.2

Given the mobility of the silver ion in aqueous solution at 298 K is 6.40×10^{-8} m^2 V^{-1} s^{-1}, and the viscosity of water at 298 K of 8.94×10^{-4} kg m^{-1} s^{-1}: Calculate the diffusion coefficient of the silver ion, its effective radius, the limiting equivalent conductance and the limiting molar conductivity for AgNO$_3$, using the λ^0 value of NO$_3^-$ from Table 16.2.

Solution

Using Equation 16.12, recognizing that at 298 K, $\dfrac{RT}{F} = \dfrac{0.0592}{2.303}$, and V $= 0.0257$ V and $z = 1$:

$$D = \frac{\mu}{z}\frac{RT}{F} = \frac{6.4 \times 10^{-8} m^2 \times 0.0257\ V}{Vs} = 1.64 \times 10^{-9}\ m^2\ s^{-1}$$

The Stokes radius can now be calculated from Equation 16.11:

$$r = \frac{kT}{6\pi\eta D} = \frac{1.38\ x\ 10^{-23}\ J\ K^{-1} \times 298\ K}{6\pi \times 0.894\ g\ m^{-1}s^{-1} \times 1.64 \times 10^{-9}\ m^2\ s^{-1}} = 1.49 \times 10^{-13}\ \frac{Js^2}{g\ m}$$

$$= 1.49 \times 10^{-15}\ \frac{Js^2}{g\ cm} = 1.49 \times 10^{-15}\ \frac{J\ cm}{erg} \times \frac{10^7}{J}\ erg = 0.149\ nm$$

Using Equation 16.13,

$$\lambda^0_{Ag^+} = \mu F = 1 \times 6.40 \times 10^{-8}\ m^2\ V^{-1}\ s^{-1} \times 96485\ Coulomb\ equiv^{-1}$$

$$= 61.8\ cm^2\ V^{-1}\ s^{-1} \times V\ \Omega^{-1}\ s\ equiv^{-1} = 61.8\ S\ cm^2\ equiv^{-1}$$

(Table 16.2 lists 61.9 S cm^2 equiv^{-1})

$$\lambda^0_{NO_3^-} = 71.4\ (from\ Table\ 16.2)$$

Limiting molar conductivity for AgNO$_3$ is now computed from Equation 16.18:

$$\Lambda^\circ_m = v_+ \lambda^\circ_+ + v_- \lambda^\circ_- = (1 \times 61.8 + 1 \times 71.4)\ S\ cm^{-2}\ mol^{-1}$$

$$= 133.2\ S\ cm^{-2}\ mol^{-1}$$

A common geometry for a flow-through electrical conductivity detector consists of two disk-shaped stainless steel electrodes (~1–1.5 mm in dia.) that are spaced ~1 mm apart, resulting in an effective cell volume of 0.8-1.8 μL and a cell constant of 12.7-8.3 cm^{-1}. A temperature sensor, usually a low thermal mass thermistor, is also placed in contact with the flowing liquid, typically at right angles to the electrodes, or simply imbedded in the cell block, close to the flow path to correct for the temperature dependence of the measured conductance. A typical correction coefficient assumes that conductivity increases 1.7%/°C but most detectors allow user-selectable temperature compensation values to be input. In another geometry, two ring-shaped electrodes are separated by a ring-shaped insulating spacer.

In the simplest arrangement, an alternating voltage, 1–15 kHz in frequency is applied across the electrodes and the resulting current is measured, rectified, and converted to a voltage signal. The ratio of the current to applied voltage is the conductance. Application of DC potential is generally not permissible, as this will lead to undesirable electrochemical processes at the electrodes. Conductometry is really not an electrochemical technique—no electrochemistry occurs at the electrodes. However, when the background resistance is high, as in suppressed ion chromatography, DC measurements can provide a very simple inexpensive alternative.

In a **four electrode conductivity measurement** arrangement, a constant current is applied across the outer pair of electrodes and the voltage drop across the inner pair of electrodes is measured; this is directly proportional to the reciprocal of the conductance between the electrodes. Both the above measurements are, however, affected by the capacitance at the electrode-solution interface. A **bipolar pulse conductance measurement** technique is not affected by the presence of capacitance, either serially or in parallel to the resistive element. In this technique, two successive voltage pulses of equal magnitude but opposite polarity are applied to the measurement cell, and the current passing through the cell at the end of the second pulse is measured.

In micro or nanoscale systems such as open tubular liquid chromatography or capillary electrophoresis that uses small diameter capillaries, it is not practical to make measurements in a separate cell, external to the separation system, due to excessive dispersion from connecting tubing, etc. The measurement must be made directly on the capillary. Although fine wire-electrodes in galvanic contact with the solution inserted through holes drilled through the capillary walls have been demonstrated, a more elegant generally applicable solution is **capacitively coupled contactless conductivity detection (C^4D)**. In this technique, a pair of ring-shaped electrodes is put on the separation / measurement capillary ~1 mm apart. An excitation voltage, typically with a frequency of several hundred kHz, is applied to one electrode. This frequency is sufficiently high to penetrate (be capacitively coupled) through the capillary walls to the solution inside. There are many different approaches to measure the conductance. In the simplest approach, a current to voltage converter is connected between the second electrode and ground. The resulting signal is rectified and is directly proportional to the solution conductance.

Another class of quasi-universal detectors relies on measurement of aerosol particles. The technique is applicable to all nonvolatile solutes that form aerosol particles when the column effluent is nebulized with the help of a dry, particle-free, gas. Obviously, volatile solutes that leave no residue cannot be detected by such methods. The oldest and the most widely used detector of this type is the **evaporative light scattering detector (ELSD)**, offered by a number of manufacturers. The first stage of all aerosol detectors is largely the same as the nebulizer used in atomic spectroscopy (see Chapter 12) except that an effort is made to keep the flow-through liquid passage streamlined and of minimum volume to minimize dispersion. Minute particles form from residual nonvolatile analytes in the effluent. The aerosol bearing gas stream passes through a focused light beam, commonly from a diode laser, perpendicular to the gas flow direction. The light scattered by the particles is then registered on a photodiode placed at right angles to the light beam as well as the direction of flow.

The **aerosol charge detector (ACD)** was introduced by Roy Dixon of California State University at Sacramento. It has become popular among this genre of detectors because of its greater sensitivity and greater uniformity of response among various analytes compared to the ELSD. In the ACD, the aerosol stream passes through a corona discharge; the particles acquire charge in proportion to their size; the gas molecules are not charged. Much like various ionization detectors used in GC (see Chapter 15), the charged particles are then detected and the total charge carried by them measured in a continuous manner.

The most recent detector of this type is the **condensation nucleation light scattering detector (CNLSD)**. In this device, supersaturated water vapor is allowed to condense on the aerosol particles making them grow, prior to measuring light scattering. This allows more sensitive measurements compared to the ELSD while achieving fast response and relatively low baseline noise.

Aerosol detectors are popular in the pharmaceutical industry where both trace and major constituents in a formulation are of interest and the analytes are largely nonvolatile. Over a large concentration range, the intrinsic response of these detectors is

An interesting consequence of applying such power functions with chromatographic detectors is that for values of $m > 1$, an artifact increase in chromatographic efficiency is observed. For details of consequences of power functions applied to chromatographic signals, see P. K. Dasgupta, Y. Chen, C. A. Serrano, G. Guiochon, H. Liu, J. N. Fairchild, and R. A. Shalliker, *Anal. Chem.* **82** (2010) 10,143.

not linear. Rather, the baseline corrected response Y is a power function of the concentration C:

$$Y = kC^n \tag{16.20}$$

where k is a proportionality constant and n is typically slightly less than 1. The response appears linear over a limited range of concentrations, but at higher concentrations the response slope becomes smaller than that at lower concentrations. To linearize the response behavior, often the detector signal is raised to the power m, where ideally $m = 1/n$ and the processed signal Y^m is output to the external world. The response behavior does not, however, completely correspond to Equation 16.20; as a result, linearization is often far from perfect.

From the beginning days of HPLC, the most commonly used detection mode has been UV absorbance measurement. In early years, a fixed wavelength absorbance detector that utilized an Hg lamp emitting at 254 nm and equipped with a band pass filter that allowed the transmission of only this line was much used. Especially with the advent of near-monochromatic LEDs with emission wavelengths into deep UV, it will be even easier today to fabricate such dedicated single wavelength detectors with high performance, and they will be useful in niche applications. In practice, all HPLC systems are presently sold with **variable wavelength UV-visible absorbance detectors**. The simplest of these detectors requires manual selection of wavelength and the absorbance at a single wavelength is monitored. At the other extreme, **photodiode array (PDA) detectors** (also called a **diode array detector** or **DAD**) can monitor a large range of wavelengths, even the entire spectrum, on a near-continuous basis. Photodiode array spectrometers have been discussed in detail in Chapter 11—the HPLC detector version is equipped with an appropriate flow cell (some are based on a LCW) and optimized for fast data throughput. Because of its versatility, the PDA is presently the most used HPLC detector.

Typical noise level of a present-day HPLC absorbance detector with a single wavelength being monitored is 5×10^{-6} absorbance units (AU), and baseline drift is 10^{-4} AU/h, while for a PDA both noise ($2\text{-}3 \times 10^{-5}$ AU) and drift ($5\text{-}10 \times 10^{-4}$ AU/h) are somewhat higher. An intermediate class of detectors that can continuously monitor 2–4 different wavelengths have noise and drift specifications only slightly worse than detectors that monitor a single wavelength. These may represent the best of both worlds. They cannot, of course, take the entire spectrum of the eluite that is useful for the purposes of confirmation. Mass or molar concentration LODs depend of course on the absorptivity of the analyte and its molecular weight. For a strongly absorbing analyte (e.g., molar absorptivity $\varepsilon = 10^4$), a noise level of 10^{-5} AU will translate to an LOD of 3×10^{-5} AU or 3×10^{-9} M at the detector. A typical chromatographic dilution factor (the ratio of the injected concentration to the peak concentration at the detector) is minimally ~10, and the detection limit for the injected solute will be 3×10^{-8} M. For a MW of 200, the concentration LOD will then be 6 μg/L. Beyond changes in light transmission due to temperature-induced RI changes, absorbance detectors have relatively low-temperature sensitivity. However, the most sensitive detectors typically do include a thermal stabilization arrangement prior to the measurement cell. Common HPLC grade solvents used in RPC have minimal absorption down to 190 nm, and water : methanol and water : acetonitrile gradients exhibit negligible baseline change with UV detection. Although there are vast differences in sensitivities among different analytes, the vast majority of substances have measurable absorption in the deep UV (190–220 nm), and at these wavelengths, UV detection is nearly universally applicable. It cannot, of course, be used with solvents that have significant absorption in the UV or with sample components that do not absorb in the UV.

A table of commonly used HPLC solvent properties including the UV cutoff (defined as the wavelength where the absorbance is 1.0 in a 1 cm cell) is available in http://www51.honeywell.com/sm/rlss/bandj/common/documents/honeywell-burdick-jackson-product-guide.pdf.

Absorbance ratiograms reveal peak purity and peak identity.

For a detector that can monitor the absorbance at two different wavelengths, the ratio of the absorbance values at two different wavelengths provides valuable information about both the purity and the identity of a chromatographic peak. Consider that an analyte has a molar absorptivity of ε_1 and ε_2 at wavelength λ_1 and λ_2, respectively. At any point in the peak, if the analyte concentration is c and the cell path length is b, the two absorbance values A_1 and A_2 will be given by $\varepsilon_1 bc$ and $\varepsilon_2 bc$, respectively. The noteworthy item is that the ratio A_1/A_2 is the constant $\varepsilon_1/\varepsilon_2$, which is both a characteristic of the particular analyte and it is independent of concentration. Thus, this ratio will be flat across the peak.[3] The peak identity can be confirmed by matching with a previously determined value of $\varepsilon_1/\varepsilon_2$.

Partial or complete coelution of two analytes in the same peak is hardly uncommon in chromatography—and frequently this is not readily apparent from the chromatogram. Except when the two analytes elute at the same retention time with 100% overlap, the composition of the peak in the leading half and the trailing half will vary. Because it is extraordinarily rare that the $\varepsilon_1/\varepsilon_2$ values will be the same for both analytes, the ratio will change across the peak, showing a consistent increase or decrease and indicating that there is coelution.

Cell design in UV–V is absorption measurements is important and depends in part on the scale. For capillary scale measurements, only direct on-capillary measurements are possible to avoid dispersion. The typical capillary detection arrangement involves just a radial pass: A focused beam is passed through a small (~1 mm long) window in the capillary where a portion of the protective sheath is removed and the transmitted light is detected; the use of a ball lens and/or an optical fiber for coupling to the capillary is common. The radial pathlength is the same as the inner diameter of the capillary and is very small. Multiple reflections in an externally mirrored capillary where the beam enters at an angle and exits after multiple reflections has been advocated but is not in common practice. Capillaries containing a "bubble" (where the capillary is expanded to a larger diameter and detection is made) are available from one manufacturer. In some cases, it is possible to connect an external Z-path capillary cell with a path length ≥ 1 mm, as pictured in the margin figure.

The Z-geometry flow path provides good washout, and the same design is common in macroscale flow cells. These are typically 6–8 mm in path length and ≤ 1 mm in diameter. Flow-through optical absorbance measurement is always subject to some refractive index effects. As the analyte band moves into the cell with a parabolic front, if the analyte has a refractive index different from that of the eluent, a virtual lens is formed. Depending on whether the analyte RI is higher or lower than that of the eluent, the light is better focused on the exit window, or is defocused, respectively causing more or less light to reach the detector. As the analyte peak leaves the detector and the analyte band is replaced by the chasing parabolic front of the eluent, precisely the reverse occurs. As a result, even when there is no light absorption, a refractive index induced artifact baseline disturbance (a peak followed by a dip or vice versa) is observed. For example, when the sample is dissolved in pure methanol or acetonitrile and a hydroorganic eluent is being used, the solvent associated with the sample elutes without retention. Even though the solvent has no significant absorbance at the monitoring wavelength, a bidirectional response as described above is observed (and in fact provides a convenient void volume marker). The superposition of such RI effects on the true absorbance signal is of course undesirable. For solutes that have a low ε and hence must be injected at high concentrations to be detected, this can cause misshapen peaks and errors in quantitation. The exact extent of the RI effect depends on the reflectivity of the cell walls and the exact optical system. One manufacturer offers a tapered cell where the light exit window is larger than the entrance window: this design does minimize RI effects. A cell design where a bifurcated fiber optic is used on the same

Similar ends as a bubble cell can be attained with the separation capillary being inserted inside a larger capillary whose inner diameter equals the outer diameter of the separation capillary (see, e. g., S. Liu and P. K. Dasgupta, *Anal. Chim. Acta* **283** (1993) 747).

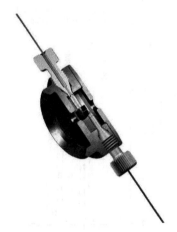

A Z-path capillary flow cell (Courtesy of Agilent Technology).

---

[3] A caveat is needed. At the edges of the peak, as the signal approaches the baseline noise value, the ratio will be very noisy and eventually will approach the indeterminate value of 0/0. In practice, the ratio is only meaningful when the S/N is at least 10. Typically, the ratio is computed within the bounds of 10% of the peak height.

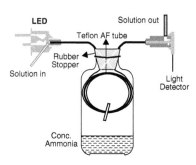

LCW-based postcolumn reactor and absorbance measurement cell. A 0.15 × 1420 mm Teflon AF tube is suspended over concentrated ammonia. Light and liquid in/out connections are provided at each end, a 400 nm LED is used as the light source and a photodiode-operational amplifier combination as the detector.

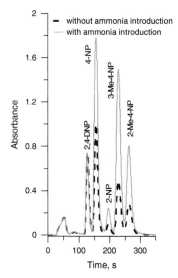

Chromatogram of a standard mixture of nitrophenols (NPs) found in the atmosphere with and without ammonia introduction. 1 mL of the indicated NPs (4, 8, 4, 8, and 4 ng/mL, respectively) were preconcentrated and separated on a 2 × 50 mm C-18 column packed with 2.2 μm particles. Both above figures adapted from Ganranoo et al., *Anal. Chem.* **82** (2010), 5838.

window (the other "window" is a mirror that reflects the light back, through the solution to the entrance window), to both bring light to the cell and take the transmitted light back is > 10x less susceptible to RI effects as the light travels the same path in opposite directions (see e.g., P. K. Dasgupta, H. S. Bellamy, H. Liu, *Talanta* **40** (1993) 341). It also has the advantage that the pathlength is doubled for the same cell volume.

Liquid core waveguide (LCW) based cells are finding increased use in HPLC. Light loss in all cells is exponential with the cell pathlength and dramatically increases with decreasing inner diameter of the cell. Long pathlength very narrow bore conventional cells are impractical, as very little light reaches the detector, and this becomes the primary determinant of detector noise. The principles of LCW operation have been previously discussed in Chapter 11. The light loss per unit length of an LCW (most commonly a Teflon AF clad silica tube) is much less than that of a standard cylindrical tube of the same inner diameter. As such, narrow-bore long-path cells become practical. Teflon AF itself can also be used as a waveguide but it is highly gas-permeable, and this is generally not desirable. One ingenious deliberate use of this permeability is the simultaneous use of a Teflon AF tube both as an optical absorbance cell and a postcolumn reactor. The arrangement is depicted in the margin figure. The device was used to measure various nitrophenols in rain and ambient air—nitrophenols behave as acid–base indicators that are yellow in basic solution (absorb at ∼400 nm). Because the separation was conducted on a C-18 silica column that could not be used at high pH, gaseous ammonia was directly introduced through the walls of the Teflon AF tube to increase the pH and permit long path detection.

Fluorescence detectors are among the most sensitive detectors used in HPLC. A small minority of compounds exhibit native fluorescence and thus this is a rather selective mode of detection. However, it is also common to derivatize specific class of compounds, either before separation or in a postcolumn fashion after separation. Common reactions for making fluorescent derivatives of amino acids/amines and alcohols are listed in Figure 16.S1 in the web supplement.

A spectrofluorometric detector uses both excitation and emission monochromators. A broad band Xe lamp or a Xe flash lamp is used as the source. The latter provides greater light intensity and somewhat better S/N. For dedicated applications, a combination of specialized lamp source, excitation band pass filter and emission long pass filter can provide excellent performance at low cost. Some detectors use an excitation monochromator, but a long pass filter on the emission side.

The signal-to-noise ratio for a fluorescence detector is typically specified in terms of the Raman emission of water when excited at 350 nm. The energy loss is 3382 cm^{-1}, thus the emission occurs at 397 nm. Since pure water is easily available and the weak Raman band mimics weak fluorescence, this is an easy parameter to measure. The typical S/N for the Raman band of water is 300–550 for 1–1.5 s integration time in a typical HPLC-type fluorescence detector, compared to > 10,000 for a state-of-the-art benchtop spectrofluorometer.

When a number of compounds are to be determined, it is often possible to find a wavelength usually in the deep UV where all the analytes of interest can be excited. But they fluoresce at different wavelengths. Some detectors allow the simultaneous monitoring of up to four different emission wavelengths. This is very helpful in such cases and provides another dimension of selectivity. Some detectors allow not only excitation and emission scanning but also synchronous scanning in which the excitation and emission monochromators are simultaneously scanned, providing several dimensions for selectivity. Some programmable detectors permit different excitation and emission wavelengths at different times, so that each analyte can be optimally detected. For a compound that has a good quantum efficiency ($\Phi \geq 0.1$), limits of detection in the low-sub femtogram range are attainable.

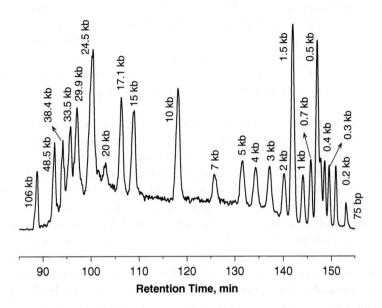

FIGURE 16.12 Hydrodynamic chromatographic separation of a DNA mixture from 75 base pairs (bp) to 106 kbp. The sample is a mixture of different types of DNA. The column is a 4.4 m long, 5 μm i.d. nonfunctionalized open bare silica capillary, with a radius of 2.5 μm and a total length of 445 cm (440 cm effective). The sample was injected at 100 psi for 10 s, and the separation was carried out at 360 psi. The fluorescent dye YOYO was intercalated in the DNA and confocal LIF was used for detection. Reprinted with permission from X. Wang, et al., *J. Am. Chem. Soc.* **132** (2010) 40. Copyright 2010, American Chemical Society.

For capillary scale fluorescence detection, a laser source is ideal as it can be focused into a very small spot and because it provides very high fluence (light intensity per unit area). A confocal arrangement, shown in the margin figure, is popular. Collimated laser light is incident on a dichroic mirror (this reflects the laser wavelength efficiently but will pass other wavelengths) and is reflected and focused by a microscope objective to the window made on the detection capillary. The emitted fluorescence passes through the dichroic mirror, a band pass filter to reject any residual laser light, and then focuses through a spatial filter to a detector. A capillary separation using confocal **laser-induced fluorescence (LIF)** detection is shown in Figure 16.12.

LCW-based fluorescence detectors can be uniquely simple and powerful; the general arrangement was shown in Figure 11.34 and a detailed description appears therein.

Chiral detectors offer unique information in chiral chromatography. Chiral compounds rotate plane polarized light; the direction of optical rotation identifies an isomer. The degree of rotation is dependent on both the concentration of a chiral compound and its **specific rotation** (a characteristic of a specific compound) as defined by **Biot's law**:

$$[\alpha]_\lambda^T = \frac{a_\lambda^T}{cl} \tag{16.21}$$

where $[\alpha]_\lambda^T$ is the specific rotation at temperature T and wavelength λ; l is the optical path length in dm; α = observed optical rotation, c = concentration in g/mL. In a **polarimetric detector** the signal strength depends on the specific rotation of the molecule, not its absorption characteristics. Polarimetric detectors typically use a red diode laser (670 nm) or an LED at 400–450 nm as the light source. **Drude's equation** states:

$$[\alpha]_\lambda = \frac{\sum A_i}{(\lambda_i - \lambda_R)} \tag{16.22}$$

where A_i is a molecular constant at wavelength λ_i and λ_R is a constant reference wavelength. The equation shows the normal behavior of **optical rotatory dispersion (ORD)**, which describes the dependence of rotational strength of optically active molecules on the wavelength of light in the absence of chromophores or in spectral regions that are distant from absorption bands. The equation also points out that the degree of rotation, as a function of wavelength, increases with decreasing wavelength.

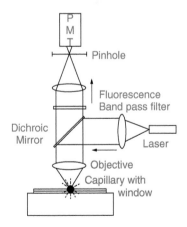

Confocal laser-induced fluorescence detection arrangement.

Lower probe wavelengths thus yield higher response. However, at lower wavelengths right- and left-handed circularly polarized light propagate at different velocities and are not absorbed by molecules to the same extent. This is called **circular dichroism** (**CD**, see below). When CD is present, it causes a deviation from Drude's equation, known as the "**Cotton effect**." Specific rotation ($[\alpha]$) is maximized and Cotton effect is minimized in the range of 400–460 nm. The use of a 430 nm light source optimizes these effects.

Faraday's scheme is used in HPLC polarimeters, to compensate for the rotation from the sample and relate the compensation to a measurement of the degree of rotation and its direction. Michael Faraday demonstrated that plane polarized light could also be rotated by an electric field as it passed through a transparent medium; the rotation is proportional to the strength of the field. A nonmechanical polarimetric detector is thus constructed, in which the light is rotated by a chiral compound in the flow cell, but a Faraday stage that follows (this simply has a coil around the beam path) rotates it back to the baseline condition. The light detector output is monitored, and using a feedback circuit a current is applied of the correct sign and magnitude to the Faraday stage so as to rotate the beam back to the original position. It is this current that is taken as the magnitude of rotation. Good polarimetric detectors have noise levels of 20 $\mu°$. To put matters in perspective, this corresponds to an arc of 7 ft for a circle the size of the earth. As Biot's law indicates, a longer pathlength permits more sensitive detection, and cells with pathlengths up to 25 mm are common in polarimetric and other chiral detectors. Some manufacturers use multiple reflections to effectively increase the pathlength.

The **ORD detector** is very similar to the polarimeter except that it uses a broadband Hg-Xe source along with a monochromator, thus permitting any wavelength between 350–900 nm. Relative to a polarimeter, the dependence of the rotation on wavelength provides a further identifying characteristic of the molecule. The typical noise level is the same as that in a polarimeter, with a drift of 50 $\mu°$/hour.

The signal from a **CD detector** is the difference between the absorbance measured with right and left circularly polarized light, $[A_r - A_l]$, often a very small value. Like UV absorbance, the CD signal is wavelength dependent. A chiral molecule should have absorption in the 200–420 nm range to have a strong CD signal. A typical CD detector has a wavelength range of 220–420 nm, a noise level of 30 $\mu°$ and a drift of 50 $\mu°$/hour. Both CD and UV absorbance output are available from a CD detector, but the UV detector performance is generally substantially below that of a standard UV detector.

The **coulometric** and the **amperometric detector** are highly selective and sensitive electrochemical detectors (see Chapter 22) that can detect substances that readily undergo redox conversion or substances that form a product with an ion derived from electrode oxidation (e.g., amino acids may form a chelate with Ag^+ derived from an Ag electrode). The cell design typically involves a thin layer cell (solution flows between closely spaced planar electrodes) or a wall-jet cell (solution impinges on the working electrode).

Coulometry differs from amperometry in that the analyte is quantitatively converted in a coulometric detector (which necessarily uses larger electrodes and a longer residence time). If the electron stoichiometry of the reaction is known, coulometry can be used for absolute quantitation without reference to any calibration. If the redox process in a coulometric detector produces a product that has a novel detection property, following the coulometric detector with another detector provides substantial new information. Consider that atmospheric nitro polynuclear aromatics (NPAH's) are of considerable concern because they are direct mutagens. The $-NO_2$ group can be easily reduced to $-NHOH$ or further to the $-NH_2$ functionality. Both the latter functionalities result in highly fluorescent products. A direct extract of atmospheric particles may

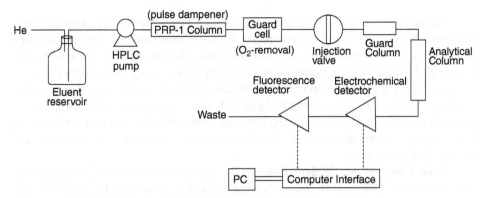

FIGURE 16.13 An electrochemical coulometric detector and a fluorescence detector setup in tandem.

contain many electroreducible compounds other than nitroaromatics, and interpretation of the chromatograms to identify the compounds is difficult. If on the other hand a fluorescence detector follows the coulometric detector, peaks that appear in the second detector *only when the first detectors is in the active reduction mode*, vastly limit compounds that will respond under these circumstances. Figure 16.13 shows the general diagram of such a setup; and Figure 16.14 shows chromatograms illustrating its application; note the use of a packed column as a pulse dampener and the use of a pre-injector "guard cell" (often integral to a coulometric detector) that reductively removes any O_2 present in the eluent. The main limitation of a tandem detector setup is the significant dispersion caused by the coulometric cell.

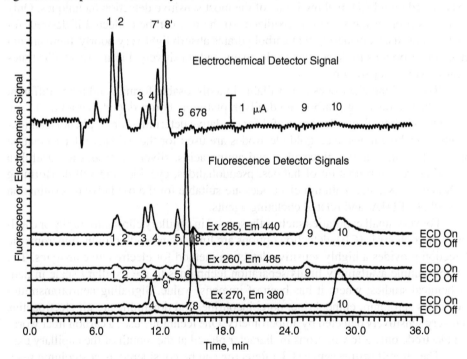

FIGURE 16.14 Chromatogram of an NPAH test mixture. The peaks are due to (amounts injected in nmols given in parentheses):**1**,1-amino-4-nitronaphthalene (10); **2**: 9-hydroxy-3-nitrofluorene (10); **3**:1-nitronaphthalene (1); **4**:2-nitronaphthalene (2); **5**: 2-nitronaphthol (0.05); **6**: 9-nitroanthracene (0.05); **7**: 2-nitrofluorene (0.05); **8**: 2-nitro-9-fluorenone (2); **9**: 1-nitropyrene (0.01); **10**: 1-amino-7-nitrofluorene (0.2). Peaks 7' and 8' are due to unknown impurities that are present in compounds 7 and 8. The traces show the coulometric detector signal and the fluorescence detector signals under three different excitation/emission conditions. In each case, the chromatofluorogram is accompanied by a trace immediately below it that shows the fluorescence detector signal with the coulometric detector off. Note the sensitivity of the electrochemical detector (ECD) to pump pulsations. Figures 16.13 and 16.14 reprinted with permission from M. Murayama and P. K. Dasgupta, *Anal. Chem.* **68** (1996) 1226. Copyright 2006, American Chemical Society.

Microfiber electrode amperometry coupled to capillary scale measurement is not new; neurotransmitters have been measured in single neuronal cells both by open tubular LC and CE more than two decades ago [R. T. Kennedy and J. W. Jorgenson, *Anal. Chem.* **61** (1989) 436; T. M. Olefirowicz, and A. G. Ewing, *J. Neurosci. Met* **34** (1990) 11].

Although coulometric detectors have a much larger signal than amperometric detectors, the noise tends to be even greater. As such, amperometric detectors are significantly superior to coulometric detectors in terms of LODs, etc., and are analytically more useful. Amperometric detection in a capillary format using microfiber electrodes has proven very useful in neurochemical studies, especially for the detection of trace quantities of catecholamines in the brain in both chromatographic and electrophoretic separation modes. In routine analyses, however, amperometric detection was difficult to use because minor alteration of the electrode surface changed the response, and reproducibility was poor.

Pulsed amperometric detection (PAD), introduced by Dennis Johnson (Iowa State University) largely solved this problem by cleaning/renewing the electrode surface in a near continuous fashion. PAD involves the application of various potentials to a working electrode over a specific time period, usually well below 1 second, that is continuously repeated. Carbohydrates (sugars), for example, contain –OH groups that are easily oxidized at a specific potential on a gold electrode and results in a current that is measured, ensuring selective and sensitive detection.

After oxidation, a reduction and reoxidation step (triple pulse waveform, a four-potential waveform is also commonly used) is applied to remove the bound analyte and renew the electrode surface. The waveform is typically repeated at frequencies greater than 1 Hz, each step lasting a few hundred milliseconds, thus allowing chromatographic data points to be recorded at least every second. Pico- and femtomol sensitivity can be achieved with PAD, making it one of the most sensitive detection techniques. Only a small portion of the sample is oxidized, so the rest can be collected if desired, for further analysis. Considering that carbohydrates absorb light very poorly, there are not many other options to detect them selectively and sensitively. Figure 16.15 illustrates a carbohydrate separation.

The working electrodes are available in a disposable format, and many different electrode materials, e.g., carbon, gold, silver, platinum, boron-doped-diamond, etc. are available. Carbon electrodes are useful for oxidative determination of phenols, antioxidants, catecholamines, etc., gold electrodes are used for the oxidative determination of carbohydrates, amino glucosides, and amino acids; silver electrodes are useful in the oxidative determination of halides, pseudohalides, cyanide, and sulfide (forming AgX and Ag_2X), and platinum electrodes are suitable for the oxidative determination of alcohols, EDTA, and related chelating agents.

For very small capillaries such as those used in capillary chromatography or capillary electrophoresis (CE), detection options are limited and end-column amperometric detection provides a highly sensitive detection method for electroactive analytes. The limited sample availability limits separation techniques to the capillary scale in neurochemical studies, where it has been of immense value. Signaling neurotransmitter molecules serotonin, epinephrine, norepinephrine, and dopamine are all electroactive and very sensitively detected by amperometry. One technique uses a carbon fiber working electrode only a few microns in diameter placed at the mouth of the capillary exit.

The **chemiluminescence (CL) detector** can be considered as postcolumn reaction (PCR) type detectors because one or more reagents must be added to the column effluent before detection. Flow-through chemiluminescence detectors were discussed in Section 11.16. Most CL reactions are very rapid. As illustrated in the margin figure, together with the column effluent, CL reagent(s) are introduced right in the CL cell, which is fixed atop a PMT window in a dark enclosure. Figure 16.16 shows a CL chromatogram. CL detection is adaptable to the microchip scale; an example is given later in Figure 16.33.

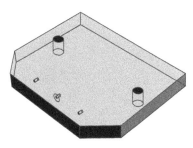

A spiral flow channel inscribed inside a transparent block constitutes the CL cell. The column effluent and the CL reagent flow in through the two left ports and after traveling through the spiral exits through the right port. From Mohr et al., *Analyst* **134** (2009) 2233. Reproduced by permission of the Royal Society of Chemistry.

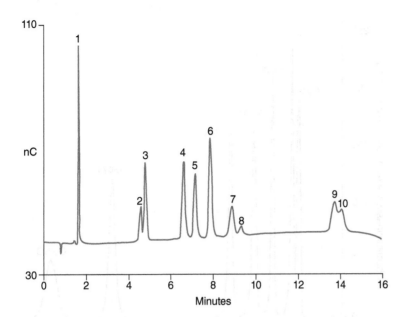

FIGURE 16.15 Carbohydrates are very weak acids, the pK_a values range for example from 12.31 for fructose to 13.60 for sorbitol. This separation was accomplished on a 3×150 mm anion exchange column with a gradient KOH eluent and pulsed amperometric detection using a quadruple pulse waveform repeated at 2 Hz. Analytes: (1) Mannitol, (2) 3-*O*-Methylglucose, (3) Rhamnose, (4) Galactose, (5) Glucose, (6) Xylose, (7) Sucrose, (8) Ribose, (9) Lactose, and (10) Lactulose. Injection: 10 μL; all concentrations except Ribose (0.32 μg/mL): 1.67 μg/mL. Courtesy Thermo Fisher Scientific Inc.

The **radioactivity detector** is another specialized detector that is used to detect radiolabeled analytes. In one design, the flow cell is positioned between the photocathodes of two highly sensitive photomultiplier tubes operating in coincidence. **Coincidence counting** of this type distinguishes from random noise (counts) in each PMT by only recognizing a real count when both PMTs register a count simultaneously. ("Simultaneous" is a relative word: in practical circuitry this is adjustable and the interval is often hundreds of nanoseconds wide.) When secondary scintillators are used between the flow cell and the PMTs, the coincidence time window for each type of scintillator needs to be adjusted. In some cases, a liquid scintillation cocktail is added postcolumn prior to entry to the detector. Radioactivity detectors provide unique capabilities to chromatographic experiments with radiotracers.

5. Postcolumn Reaction (PCR) Detection. In postcolumn reaction detection, a chemical reaction occurs before detection. Although most commonly a reagent is introduced, analyte conversion can take place without the physical addition of a reagent but brought about by UV irradiation or just heating. Conversion can also be carried out by passing through a solid reactor, e.g., an immobilized enzyme or an ion exchange column. Indeed, as we will shortly see, "suppression" in conductometric ion chromatography is an example of a unique PCR detection method. John P. Ivey of the Australian Government Analytical Laboratories in Tasmania, who contributed much to chromatographic trace analysis, especially ionic analysis, demonstrated a unique way to carry out suppressed conductometric ion chromatography—use acetoacetic acid as eluent and heat the column effluent to thermally decarboxylate it to a solution of poorly conductive acetone and CO_2:

$$CH_3COCH_2COOH \xrightarrow{\text{Heat}} CH_3COCH_3 + CO_2$$

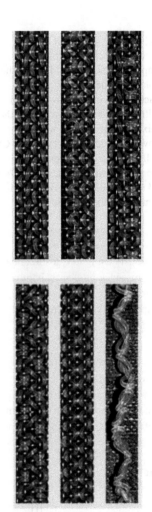

PCR Reactors, Top, from left to right: 3-D (Superserpentine) I, II, III, Bottom, from left to right: Superserpentine IV, 2-D Serpentine II, knotted reactor. The Superserpentine I-IV designations for the designs are according to its manufacturer, www.globalfia.com.

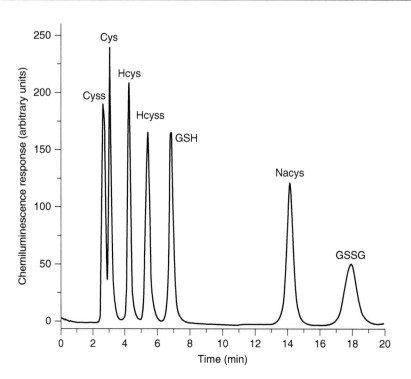

FIGURE 16.16 CL chromatogram of biologically important thiols, 10 μM each. The column effluent (5 μm particle packed C18 column) was mixed with formaldehyde at a tee and the mixed stream merged with the powerful oxidant Mn(IV) at the CL cell. Reprinted with permission from McDermott et al., *Anal. Chem.* **83** (2011) 6034. Copyright 2011, American Chemical Society.

Numerous examples of photochemical PCR detection variously using UV absorbance, fluorescence or electrochemical detection are given in Liu et al., http://www.currentseparations. com/issues/16-2/cs16-2a.pdf.

For a general account of PCR detection, see http://www.sigmaaldrich.com/etc/ medialib/docs/Supelco/Bulletin/ 4500.Par.0001.File.tmp/4500.pdf.

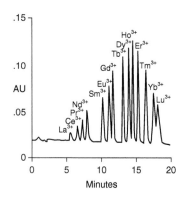

Separation of lanthanides by anion exchange, followed by reaction with ammoniacal 4(2-pyridylazo) resorcinol (PAR) and visible detection at 530 nm.Courtesy Thermo Fisher Scientific Inc.

Photochemically used PCR examples are more common. Teflon, especially FEP (fluorinated ethylene propylene copolymer) Teflon is transparent to UV radiation, and knitted Teflon tubes that are wrapped around a UV pen lamp and function as PHotochemical Reactors for Enhanced Detection (PHRED) are commercially available. There are numerous analytes that are converted to more sensitively detectable forms upon irradiation. As an example, barbiturates are an important class of soporific drugs for which the sensitivity for detection by absorbance measurement at 270 nm increases (10–30 times from barbital to pentobarbital) upon photochemical conversion. The photoirradiation products can also be determined with high sensitivity by oxidative amperometry.

The simplest chemical reaction is pH adjustment. While this can be done by adding an acid or base stream (as described in the margin figure in discussing Liquid Core Waveguides on page 684), this can be also accomplished without volumetric dilution by introducing a gaseous acid or base (e.g., HCl or NH_3) through a suitable gas-permeable tube such as Teflon AF or silicone rubber. The absorption maxima for a variety of compounds are red-shifted in alkaline medium, with an attendant increase in the ε_{max}; detection in a basic medium thus permits better LODs. However, the incompatibility of many silica-based columns does not permit the use of alkaline eluents; this problem can be solved by postcolumn alkalization.

Examples of postcolumn reaction carried out by reagent introduction are far too many to be listed here. Two classic examples are the determination of amino acids by visible absorbance detection at 530 nm after reaction with ninhydrin. This reaction requires an elevated temperature to be completed quickly; the reaction involves a series of steps that are shown the web supplement in Figure 16.S2. The other example, shown in the margin figure, is the PCR detection of lanthanide metal ions following reaction with a chromogenic ligand.

There are two important parameters in PCR detection methods. The first is mixing efficiency: while the traditional tee may be most commonly used to introduce reagent(s), an arrow mixer (input streams through two arms, output through the shaft, which may be packed with beads, see margin figure) is superior. It has long been known that a "screen-tee" reactor or the introduction of a pressurized reagent through a porous membrane (equivalent to millions of apertures/tees through which the liquid is introduced) provides superior mixing (see R. M. Cassidy, S. Elchuk, and P. K. Dasgupta, *Anal. Chem.* **59** (1987) 85). Active mixing, as with a microscale magnetic stirrer, is generally not possible because of significant dispersion.

The second aspect is the time needed for the reaction. Few reactions are instantaneous. This means that some reaction time must be allowed and the residence time needed is directly controlled by the residence volume divided by flow rate. Further, in many cases, heating is needed to accelerate the reaction. A tubular reactor, typically made of PTFE, is used. The second important aspect is the geometry of the reactor; this should minimize dispersion. A narrower bore tube of the same residence volume leads to less dispersion, but according to Equation 16.6, the pressure drop, proportional to the fourth power of the diameter, soon becomes prohibitively high. A straight tube produces the largest dispersion, and tubes knotted/knitted in particular fashions provide much lower dispersion. Knitted tubes that are serpentine in two or three dimensions are commercially available. A "pearl string reactor" (a tube filled with beads of diameter ~60% of the inner diameter of the tube) also provides reasonably low dispersion with a low pressure drop. Schemes that require more than 2 minutes of residence time generally produce too much dispersion to be useful in HPLC. In such cases, gas-segmentation can be used to minimize dispersion. Before detection, the gas and liquid phases can be separated by a gas-liquid separator. In some instances, the analyte may in fact be converted to a gaseous product that can be detected with a suitable reactor. There are cases where solid-phase reactors, such as immobilized enzyme reactors, are also used in a PCR system. Figure 16.17 embodies many of the above concepts.

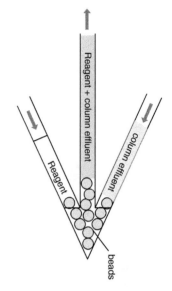

Bead-packed arrow mixer

An Important PCR Detection Application: Ion Exchange Separation of Amino Acids

The great power of varying the pH of the eluent to bring about complex separations even with low efficiency stationary phases was aptly demonstrated by Moore and Stein [*J. Biol. Chem.* **192** (1951) 663], work that eventually led the 1972 Nobel laurel to be awarded to them. They were able to separate all 50 amino acids in 175 hours (slightly more than a week) using a 0.9×100 cm column packed with 25–37 μm size Dowex 50WX8, at a flow rate of 67 μL/min using a pH gradient from 4.25 to 11.0, and using different temperatures of 25–75 °C along the way (the temperature affects the ion exchange equilibria). The ionic forms of many substances are affected by the eluent pH. Hydrolysis of metal ions and ionization of weak acids and bases can be controlled by adjusting the pH. Weak acids will not dissociate at low pH and will not ion exchange; the same is true for weak bases at high pH. Amino acids are **amphoteric** (they can act both as an acid and a base). There are three possible forms:

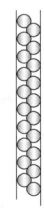

Pearl string reactor; optimum: $d_{\text{bead}} = 0.60\, d_{\text{tube}}$

$$R-\underset{\underset{A}{\overset{|}{NH_3^+}}}{CH}-CO_2H \underset{+H^+}{\rightleftharpoons} R-\underset{\underset{B}{\overset{|}{NH_3^+}}}{CH}-CO_2^- \underset{-H^+}{\rightleftharpoons} R-\underset{\underset{C}{\overset{|}{NH_2}}}{CH}-CO_2^-$$

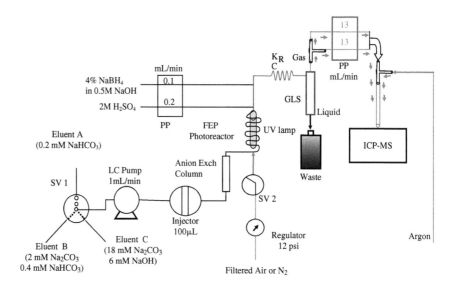

FIGURE 16.17 System for separation of arsenic compounds. An isocratic pump is used with a 3-liquid selector solenoid valve SV1 to run a 3-step gradient on an anion exchange column. Arsenic is converted to AsH_3 upon reaction with $NaBH_4$ and acid, but this conversion is very poor for organic As. Therefore, all forms of As are photooxidized to As(V) in a UV photoreactor before adding the reagents. For quantitative photooxidation, ~4 min is required; so gas segments are introduced by introducing air or N_2 by pulsing solenoid valve SV2. The gaseous AsH_3 is separated by a gas–liquid separator and sent to the ICP-MS. The chromatogram is shown in the margin figure. Reprinted with permission from M. K. Sengupta and P. K. Dasgupta, *Anal. Chem.* **81** (2009) 9737. Copyright, 2009, American Chemical Society.

Form B, called a **zwitterion**, is the dominant form at the pH corresponding to the **isoelectric point** (often called **pI**) of the amino acid. The isoelectric point is the pH at which the net charge on the molecule is zero. In more acid solutions than this, the –COO⁻ group is protonated to form –COOH, the molecule (form A) thus has an overall positive charge. At pH > pI, the NH_2^+ group loses a proton to become $-NH_2$, the molecule (form C) is now an anion. The isoelectric point will vary from one amino acid to another, depending on the relative acidity and basicity of the carboxylic acid and amino groups. Thus, group separations based on the isoelectric points are possible by pH control. In principle, at a given pH, the amino acids can be separated into three groups by being passed successively through an anion and a cation exchange column. The uncharged zwitterions (amino acid at its isoelectric point) will pass through both columns, while the positively and negatively charged amino acids will each be retained by one of the columns. (In practice, the buffering ions accompanying the amino acids may also undergo exchange, resulting in a change in pH and thus complicating such a class separation.) In fact, as Moore and Stein demonstrated, a complete separation can be carried out with only one type of an ion exchanger if the pH is slowly varied.

While LC-MS and proteomics-based methods have reduced the use of amino acid analyzers for characterizing proteins, the results that can be obtained with present-day high-performance amino acid analyzers remain most impressive (see Figure 16.18).

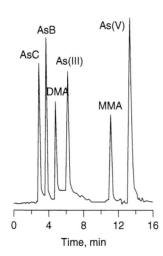

Separation of arsenocholine (AsC), arsenobetaine (AsB), dimethylarsinate (DMA), arsenic (III), monomethylarsonate (MMA), arsenic (V), 25 µg As/L each.

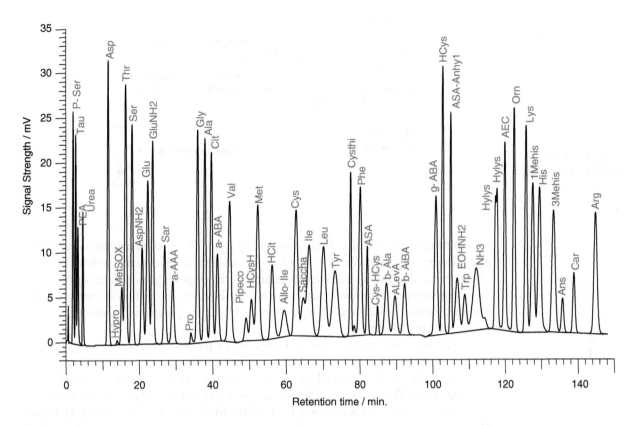

FIGURE 16.18 Gradient chromatogram of 53 amino acids and ammonia (2 nmol ea) on a 4.6 × 60 mm cation exchange column (3 μm particles) on a Hitachi-L8900 Amino Acid Analyzer. Postcolumn reaction with ninhydrin. Courtesy Hitachi High-Technologies Corp.

16.4 Ion Chromatography

Ion chromatography is a unique application of postcolumn conversion that is used for ionic analysis, especially for small ions. It takes advantage of the separating power of high-efficiency ion exchangers with the ability of the conductivity detector to detect ions (and only ions) with great sensitivity. The ability of ion exchangers to accomplish remarkable separations, e.g., the amino acid analysis application discussed above, has long been known. However, the problem has been finding an appropriate means to detect them. Unlike amino acids, no general reaction is known for many analyte ions that will render them detectable by optical absorption by PCR. Many ions of interest do not have sufficient optical absorption to be directly determined sensitively, and unlike heavy metal ions, alkali metal ions or ammonium do not easily form a complex in aqueous solution that can be optically detected. Let alone detection after chromatographic separation, the measurement of one of the most ubiquitous ions, sulfate, at low to sub-ppm levels was a major analytical challenge. While conductivity does sensitively detect ions, we have previously discussed that electroneutrality requires an ionic eluent to displace analyte ions and elute them from an ion exchange column. It would be difficult, if not impossible, to see trace ionic analytes eluting in a high conductivity, high ionic concentration (relative to analyte) eluent background by monitoring conductivity changes.

In 1975, this problem was solved by Hamish Small, Timothy Stevens, and William Bauman at Dow Chemical Company. They demonstrated a way to remove the background eluent using a second ion exchange column for post column conversion of the eluent to a less conductive form and the analyte to a more conductive form,

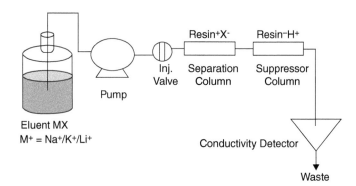

FIGURE 16.19 Schematic depiction of an ion chromatograph performing anion separation.

Read the classic paper on ion chromatography: H. Small, T. S. Stevens, and W. C. Bauman, *Anal. Chem.* **47** (1975) 1801.

thus allowing very sensitive detection of the analyte. This second column is called the **suppressor (column)** because it suppresses (reduces) the eluent conductivity and also enhances the analyte conductivity compared to the original form. For anion analysis, the suppressor is a cation exchange column in the H^+-form. For cation analysis, the separation column is a cation exchange column and the suppressor is an anion exchange column in the OH^--form.

The principle is illustrated in Figure 16.19 for anion chromatography. Let us imagine that the eluent MX is NaOH, such that the first column, the separator, is in OH^--form. As the eluent NaOH enters the suppressor, the Na^+ is replaced by H^+, thus forming water. This enters the detector, the poor conductivity of pure water providing for a very low detector background. If salts MA, M_2B, and M_3C (where HA, H_2B, and H_3C are <u>not</u> *very* weak acids) are injected into the system, A^-, B^{2-}, and C^{3-} are retained by the separator, respectively releasing 1, 2, and 3 OH^- ions previously occupying the ion exchange sites. M^+ is not retained by the separator and immediately passes through in the void volume. It enters the suppressor as MOH, where the M^+ is exchanged for H^+ and water is formed again—there is thus no void response (water in water). Imagine that A^-, B^{2-}, and C^{3-}, elute from the separator in that order and enter the suppressor as NaA, Na_2B, and Na_3C and after exchange of Na^+ for H^+, exit the suppressor respectively as HA, H_2B, and H_3C. Because these acids are fully or at least partially ionized, this registers as an increased conductivity signal on a pure water background.

A hydroxide eluent as described above leads to pure water as background and therefore has the lowest background conductivity, baseline noise and thence the best LODs compared to other eluents. If gradient elution is executed where the hydroxide eluent concentration is increased during the run, in principle, the background still remains water and there is no baseline change. However, for a variety of reasons discussed later, in the early stages of the development of IC, a hydroxide eluent could not be used. Salts of weak acids, most notably sodium carbonate and/or bicarbonate, were used as eluent. Sodium tetraborate and sodium *p*-cyanophenolate were also used. The different suppressor reactions for the different eluents would be:

$$NaOH \xrightarrow{\text{Cation exchange for } H^+} H_2O$$

$$Na_2CO_3 \xrightarrow{\text{Cation exchange for } H^+} H_2CO_3$$

$$Na_2B_4O_7 \xrightarrow{\text{Cation exchange for } H^+} 4 \, B(OH)_3$$

$$CNC_6H_4ONa \xrightarrow{\text{Cation exchange for } H^+} CNC_6H_4OH$$

In all these cases, the suppressed background is water or a very weak acid that has a low conductivity background on which enhanced conductivity signals from a stronger acid analyte can be seen.

In a similar fashion, strong acid eluents, e.g., CH_3SO_3H, are used in suppressed cation chromatography. The separation column is therefore in the H^+-form; alkali, alkaline earth metals and ammonium are the cationic analytes that can be determined by suppressed cation chromatography: the need for subsequent passage through an OH^--form anion exchanger suppressor means that only cations that form soluble hydroxides can be measured. In the suppressor $CH_3SO_3^-$ is replaced by OH^-, thus forming water from CH_3SO_3H (other strong acids can also be used as eluent) as the background, while Li^+, Na^+, Ca^{2+} leave the suppressor as $LiOH$, $NaOH$, $Ca(OH)_2$ and are sensitively detected. Unlike suppressed anion chromatography where analytes form the corresponding acids and very few acids are insoluble, the number of metals/cationic analytes that form soluble hydroxides are far more limited. Coupled to the fact that many atomic spectrometric alternatives are available for the determination of metals, suppressed anion chromatography is used far more commonly than its cation counterpart.

Single column (or nonsuppressed) ion chromatography (SCIC). It is possible to carry out conductometric ion chromatography without suppression. The considerable success of suppressed ion chromatography and the limitations of original packed column suppressors (discussed later) led others to seek alternatives to this patented technique. The key to this is that in any type of ion exchange, electroneutrality is always obligatorily preserved. Therefore, the effluent concentration from an ion exchange concentration in ionic equivalents/L is also always constant. If one is working with an x meq/L $NaOH$ eluent in an anion chromatography system, the $[Na^+]$ concentration in the column effluent remains constant at x meq/L. When, say, SO_4^{2-} elutes from the column, if the concentration of SO_4^{2-} at the peak is y meq/L, the concentration of the $[OH^-]$ at that point must be x-y meq/L as the total anion equivalent concentration must be x meq/L to maintain electroneutrality.

Consider that the background specific conductance of the eluent, σ_{bgnd}, will be $x(\lambda_{Na^+} + \lambda_{OH^-})/1000$, whereas at the sulfate peak apex the specific conductance, σ_{peak}, will be $(x\lambda_{Na^+} + y\lambda_{SO_4^{2-}} + (x-y)\lambda_{OH^-})/1000$. The net difference, or the peak height in terms of the specific conductance, will then be:

$$\sigma_{response} = \sigma_{peak} - \sigma_{bgnd} = y(\lambda_{SO_4^{2-}} - \lambda_{OH^-})/1000.$$

Noting that the limiting equivalent conductance of OH^- is higher than that of any other anion, the second term in parentheses is negative and the response is actually negative. The sensitivity of such a method is directly related to the difference in equivalent conductance of the eluent ion and the analyte ion, whereas the attainable LOD is also a function of the baseline noise; the LOD is generally reciprocally related to the background conductance. In the case of suppressed ion chromatography in a comparable situation, the response will be proportional to the sum of λ_{H^+} and $\lambda_{SO_4^{2-}}$ with a background of water. The second situation is obviously much more favorable.

Although we discussed the above in terms of a hydroxide eluent (and SCIC with such an eluent has been reported in the literature), hydroxide as an eluent has relatively poor eluting power. This means that a high eluent concentration is needed, and given the high equivalent conductance of OH^-, the background conductance, and thence the baseline noise is very high, such an approach has never been used in practice. Rather, organic eluent ions such as phthalate or benzoate, which have much greater eluting power than hydroxide and also have a low equivalent conductance have been used in conjunction with low capacity columns (so low eluent concentrations can be practical). Most small inorganic ions have a greater equivalent conductance and thus produce a positive signal. While this method is simpler in not requiring a suppressor, the overall performance is not competitive with the suppressed approach and it is no longer used.

The situation is not completely the same for cation chromatography with H^+ as the eluent ion. For any analyte cation M^+, the sensitivity index in the SCIC case, $(\lambda_{H^+} - \lambda_{M^+})$, is always greater than the sensitivity index $(\lambda_{M^+} + \lambda_{OH^-})$ for the suppressed case. But a greater calibration slope (sensitivity) does not equate to a better LOD. The background conductance in the SCIC case (and hence noise) is always higher. However, relatively low concentrations of complexing acids, e.g., tartaric acid, that help elute the metal by complexation, can be successfully used with weak-acid cation exchangers in the SCIC mode to keep the background conductance at a manageable level and get attractive detection limits. Note that since precipitation as hydroxide is not an issue, this approach can in principle measure a number of metal ions that are not measurable in the suppressed format. Many weak bases will also be detectable.

Evolution of suppressors. Consider that in suppressed anion chromatography, the suppressor column initially in the H^+-form is gradually converted to the Na^+-form as NaOH passes through it. Obviously, the column will eventually be completely converted to the Na^+-form and will no longer work. The original implementation of suppressed IC used two suppressor columns. While one was being used, the other was being regenerated to the H^+-form by regeneration with acid and then washed thoroughly with water, using a second pump.

When a water-washed suppressor is switched to the eluent being used, it takes significant time for the baseline to stabilize. Second, to get significant operational time, the suppressor column must contain some significant volume of resin. The suppressor column was generally significantly larger than the separator column. The dispersion in the suppressor thus limited the overall separation efficiency. Third, the hydration extent of an H^+-form and a Na^+-form cation exchange resin are not the same; the former internally holds more water. While strong acid anions are excluded from the interior of a cation exchange resin, there is no such restriction against the molecular form of a partially ionized acid, i.e., CH_3COO^- may be excluded from the interior of a cation exchanger but unionized CH_3COOH is not (cf. ion exclusion chromatography, under HPLC SUBCLASSES). Thus, the columnar suppressor does not cause additional retention of strong acid anions, but it does of the weak acid anions. Moreover, this additional retention is less for the Na^+-form of the suppressor compared to the H^+-form, as the latter has more internal water. Consequently, as the suppressor is used and it is converted from the H^+- to the Na^+- form, there is a minor decrease in the retention of the weak acid anions, complicating retention time-based identification. Fourth, reactive and easily decomposed acids like nitrous acid or thiosulfuric acid can undergo significant decomposition during the long passage through a significant suppressor column volume.

One solution is not to use a large-volume suppressor that will last for several hours of operation, but to use a very small-volume columnar suppressor that has enough capacity to carry out a single chromatogram. The system then can switch to another identical suppressor for the next chromatogram. In the current commercial incarnation of this concept, there are three identical microscale columnar suppressors (each configured in a U-geometry) incorporated within a programmable rotating valve. While one processes the separation column effluent and directs it to the detector, the second and third are washed and regenerated (both with low-pressure pumps), respectively. When the chromatogram is completed, #2 rotates into the position of operation, #3 starts being washed, and #1 is regenerated, and so on.

Membrane devices in ion chromatography. The above design has the advantage of being able to tolerate high pressure, but an arguably superior design is based on the use of ion exchange membranes that permit suppressors that can be operated

An excellent video describes the operation of the rotating suppressor from Metrohm Corporation: http://www.youtube.com/watch?v=8KQktLNIHf8&noredirect=1.

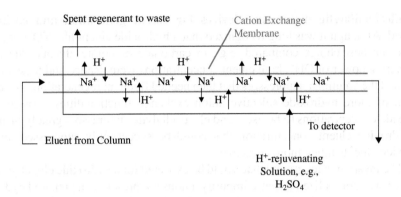

Spent regenerant to waste

Cation Exchange
Membrane

Eluent from Column

To detector

H$^+$-rejuvenating
Solution, e.g.,
H$_2$SO$_4$

FIGURE 16.20 Schematic depiction of a membrane suppressor used in anion chromatography exchanging all influent Na$^+$ for H$^+$.

continuously, without such intermittent operation. The membrane-based suppressor actually predates the above rotating columnar suppressor; a membrane suppressor for anion chromatography is shown schematically in Figure 16.20.

In a typical embodiment, the anion exchanger column effluent (alkaline) flows through a cation exchanger (either a tube or two planar cation exchange membranes), the exterior of which is bathed by an acid. Na$^+$ is exchanged for H$^+$, and barring mass transfer limitations, the exchange is virtually quantitative. The transmembrane passage of analyte or the regenerant anion is inhibited by the negative charge of the cation exchange membrane. Note that the completion of the exchange is driven by the proton gradient across the membrane. If one used a continuous stream of NaCl instead of NaOH, the exchange will not be quantitative: no net exchange will occur when the internal [H$^+$] equals the external [H$^+$]. Continuous regeneration of membrane suppressors keeps the state of the suppressor the same at all times, permitting results reproducible over time. Although the exact value depends on the design, membrane suppressors have a lower backpressure tolerance than packed column suppressors.

The membrane suppressor allows electrodialytic regeneration where the eluent cation is removed by an electric field gradient and electrolytically generated H$^+$ takes its place. This is schematically shown in Figure 16.21.

Note that potentials above the electrolytic threshold must be applied, but the electrodes are isolated by the two membranes from the central channel so no gas is formed in the central channel. Such electrodialytic suppressors have substantial capacity; 200 mM KOH flowing at 1 mL/min can be completely converted to water, for example. Convenient recycling of the detector waste in the outer channels makes such recycled electrodialytic suppression the most commonly used suppression mode. However, chemical regeneration, as shown in Figure 16.20, leads to the lowest attainable detector noise.

Whether in a tubular or a planar membrane suppressor, the flow geometry is optimized to enhance mass transport to the membrane and minimize dispersion. Tubular suppressors contain an inserted filament to reduce internal volume and are then knit to reduce dispersion (http://www.sequant.com/default.asp?ML=11515), while planar membranes are placed very close to each other and contain screens to enhance transport to the membrane (http://www.dionex.com/en-us/webdocs/4366-Man-031727-05-MMS-300-Sep09.pdf). The regenerant anion penetration into the inner stream and hence the suppressed conductivity is the lowest when the regenerant anion is a large multiply charged cation, e.g., a polymeric ion like polystyrenesulfonate. In order to avoid generating large amounts of liquid waste, the spent regenerant itself can be recirculated through a significant size cation exchange resin bed in H$^+$-form. When the latter is spent, it can be regenerated offline or disposed.

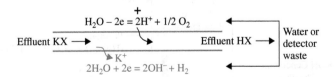

$$H_2O - 2e = 2H^+ + 1/2\ O_2$$

Effluent KX

Effluent HX

Water or detector waste

K$^+$

$$2H_2O + 2e = 2OH^- + H_2$$

FIGURE 16.21 Schematic depiction of an electrodialytic membrane suppressor used in anion chromatography. The membranes shown as the horizontal lines are cation exchangers. Eluent KX (typically KOH) flows between the membranes, while water (detector waste can be used as any anions present therein cannot pass through the membranes) flows countercurrent through the outer (regenerant) channels.

Electrodialytic membrane devices for eluent generation and carbonate removal. Although it was long recognized that a hydroxide eluent like KOH or NaOH will be the best choice compared, e.g., to carbonate or borate eluents, due to the poor eluting power of OH^-, high eluent concentrations were needed and these quickly exhausted the columnar suppressors that then needed frequent regeneration. The development of more hydroxide selective anion exchange functionalities allowed lower hydroxide concentrations to be used, and electrodialytic suppressors greatly increased the hydroxide eluent concentration that could be successfully suppressed, making hydroxide eluents much more attractive.

The invariant background that would be expected for a hydroxide eluent gradient, however, was not realized because impurity anions are present in a prepared hydroxide eluent—these accumulate on the column during the initial portion of the gradient and then elute as a large hump as the eluent strength increases. The highest purity alkali hydroxides available still contain detectable levels of impurity ions. Because of the way it is manufactured, NaOH always contains NaCl and $NaClO_3$, for example. Further, no matter how pure the eluent initially may be, it manages to form carbonate, as we live in a virtual ocean of CO_2.

These problems were solved by electrodialytically generating the hydroxide eluent. Referring to Figure 16.22, to generate KOH, a low-pressure reservoir contains KOH as a source for K^+ and a platinum anode is disposed therein. A stack of cation exchange membranes (functioning a single thick cation exchange membrane—the thickness provides for pressure tolerance and reduces any impurity anion leakage) connects it to the high-pressure cathode side though which water is pumped by the chromatographic pump. When a current is applied, K^+ is transported to the cathode side where OH^- and H_2 are produced. The stream of KOH and H_2 then proceeds downstream to the rest of the system (injector / column / suppressor / detector) via a gas permeable Teflon AF tube. The suppressor regenerant waste flows outside the AF tube. The generated H_2 is transported through the walls of the tube due to the pressure differential.

An electrical eluent generator like this has several attractive attributes. The first and most important is that any anion impurity in the low-pressure reservoir (chloride, etc.) is not transported to the generated eluent in the cathode chamber, both because of the cation exchange membranes and the direction of the electrical field. Second, the amount of KOH formed on the cathode side is directly proportional to the applied current, as governed by Faraday's laws. At a constant water flow rate through the cathode chamber, the eluent concentration is thus directly proportional to the current. To generate an eluent gradient, multiple pumps or solenoid selection valves on the low-pressure side are not necessary—the eluent concentration is programmed by programming the

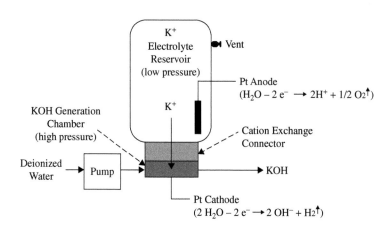

FIGURE 16.22 Schematic depiction of an electrodialytic eluent generator for generating KOH. A methanesulfonic acid (MSA) generator for cation chromatography will contain MSA in the reservoir (cathode), an anion exchange connector, and an anode disposed in the high-pressure flow-through chamber. Courtesy Thermo Fisher Scientific Inc.

current. Third, the use of an eluent generator requires the chromatographic pump to pump only pure water, reducing pump maintenance needs and prolonging piston seal life.

Electrolytic generators to generate LiOH, NaOH, KOH, K_2CO_3, and MSA are commercially available. The KOH generator is the most commonly used because K^+ has the highest mobility of the alkali metals listed above and thus leads to the lowest voltage drop and least Joule heating across the generator membrane. The greater mobility of K^+ also allows more facile removal in an electrodialytic suppressor, allowing the largest eluent concentration to be suppressed. The K_2CO_3 generator makes pure carbonate, but part of the potassium can be taken out in a controlled fashion by a cation exchange membrane based electrodialytic device to make a carbonate-bicarbonate eluent.

The purity of an electrodialytically generated hydroxide eluent is primarily governed by the purity of the water pumped into the cathode chamber. If the water contains dissolved CO_2 (this is difficult to avoid if the system is not integrally connected with a freshly deionized water source), it will be converted to carbonate when the hydroxide eluent is produced. To remove any non-hydroxide anions from the generated eluent, a secondary purifier called a continuously regenerated anion trap column (CR-ATC) is often used after the eluent generator.

The CR-ATC is basically an anion exchange resin bed containing a cathode that is separated from a flowing water receptor (suppressor waste effluent is often used) containing the anode by an anion exchange membrane (Figure 16.23). The impurity anions (as well as some hydroxide ions) are removed by the electrical field to the anode compartment, and fresh electrogenerated hydroxide at the cathode takes their place.

Postsuppressor removal of CO_2 through a membrane device. Carbonate-bicarbonate eluents were the first successful eluents in anion chromatography, and they are still widely used. The H_2CO_3 formed after suppression is the major contributor to the background conductivity and hence the background noise. The H_2CO_3 background can be removed if the suppressor effluent is passed through a CO_2-permeable membrane tube bathed outside with an alkaline absorber (e.g., suppressor regenerant waste), thus leading to lower background and better detection limits. Some other anions that form volatile acid gases will also be lost to various degrees. H_2S, HCN, etc. are lost significantly, but these acids are too weak to be detected by suppressed conductometric IC anyway. Organic acids like acetic or formic acids are lost only to a minor extent and this loss is taken into account in the overall calibration procedure. Such a carbon dioxide removal device (CRD) thus permits carbonate-based eluents to have a performance closer to that attained by a hydroxide eluent.

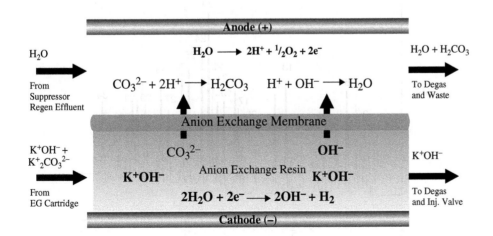

FIGURE 16.23 A continuously regenerated anion trap column removes carbonate and other anionic impurities from the generated eluent. Courtesy Thermo Fisher Scientific.

Peak identification for figure on right. Concentrations in mg/L in parentheses. 1: Isopropyl methylphosphonate (5), 2: Quinate (5), 3: Fluoride (1), 4: Acetate (5), 5: Propionate (5), 6: Formate (5), 7: Methanesulfonate (5), 8: Pyruvate (5), 9: Chlorite (5), 10: Valerate (5), 11: Monochloroacetate (5), 12: Bromate (5), 13: Chloride (2), 14: Nitrite (5), 15: Trifluoroacetate (5), 16: Bromide (3), 17: Nitrate (3), 18: Chlorate (3), 19: Selenite (5), 20: Carbonate (5), 21: Malonate (5), 22: Maleate (5), 23: Sulfate (5), 24: Oxalate (5), all the following 10 mg/L: 25: Ketomalonate, 26: Tungstate, 27: Phthalate, 28: Phosphate, 29: Chromate, 30: Citrate, 31: Tricarballylate, 32: Isocitrate, 33: cis-Aconitate, 34: trans-Aconitate

PEEK [poly (ether-ether-ketone)] tubing is commonly used to make plumbing connections between different components of an HPLC system. PEEK is relatively inert and chemically resistant to most aqueous-organic mobile-phase compositions. It is not appropriate for normal phase separations where a pure organic solvent is used.

Electrodialytic eluent generation, current programmable eluent gradients, and postcolumn electrodialytic eluent modification (suppression) are unique to ion chromatography among other HPLC techniques, and permit superb detection limits, even without sample preconcentration. But preconcentration is also particularly facile in IC. A low-pressure drop short "concentrator" column is substituted for the injection loop and the desired amount of sample is passed through this concentrator prior to it being switched to the inject position. Measurement of trace ionic impurities in water are of great importance in the nuclear power industry as even small levels of ions like chloride can cause corrosion problems. Parts per trillion levels of impurities are routinely measured in reactor cooling water after the preconcentration of $\geq$ 10 mL of sample.

Ion exchange columns are also commonly used in two-dimensional HPLC experiments, since aqueous ion chromatography eluents (as a first-dimension separation) are very compatible with an RPC separation in the second dimension. The nitrophenol separation shown earlier on page 684 using an LCW-based postcolumn reactor utilized analyte preconcentration on a ion exchange column and separation on a C-18 silica column.

The power of modern electrodialytically generated eluent electrodialytically suppressed conductometric IC on high-efficiency columns is aptly shown in Figure 16.24. Such analyses would be difficult to perform by any other method.

Finally, **ion pair chromatography** (perhaps more appropriately called **ion interaction chromatography** as ion pairs are rarely formed in the mobile phase) relies on the use of the salt of a hydrophobic cation or anion in the eluent in conjunction with a reverse-phase column. For example, when a salt like tetrabutylammonium chloride (NBu_4Cl) is put in the eluent used with a C-18 column, a significant amount of the NBu_4^+ cation adsorbs on the column, generating virtual anion exchange sites. Both strong and weak acid anions can be separated. Although suppressed conductometric ion interaction chromatography has been demonstrated [see, e.g., P. K. Dasgupta, *Anal. Chem.* **56** (1984) 769], typically optical detection methods are used.

Putting it all together. In putting the components of an HPLC system together, it is vital that there be a minimum of connection volume. The shortest connecting tube with narrowest possible bore should be used. Stainless steel or PEEK capillary tubes are typically used to connect system components. Tubing of 0.005 inch (125 μm) inner diameter is normally recommended for modern high-efficiency HPLC/UHPLC systems.

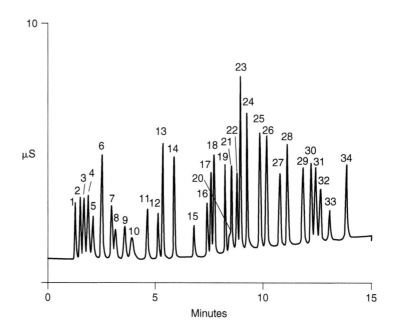

FIGURE 16.24 Hydroxide gradient separation of various organic and inorganic ions at 1–10 mg/L level. 10 μL injection; 13 μm particle size AS-11 column. Courtesy Thermo Fisher Scientific Inc.

16.5 HPLC Method Development

Analyte sizes and their polarities are the key factors for choosing a separation mode. As previously discussed, macromolecules, whether biomolecules or polymers, are generally separated in the exclusion mode. Affinity chromatography is particularly good for protein purification. For chiral separations, one must of course choose a chiral stationary phase. For separation and detection of low MW ions, ion chromatography is the method of choice. While weak acids and weak bases do not respond well to suppressed conductometry, other detection methods following an ion exchange / ion interaction separation are often applicable. For all other low MW analytes, polarity is the key to choosing a separation mode, as it influences both retention on a particular column and solubility in the mobile phase. Compound polarity follows the approximate order of: hydrocarbons and derivatives < oxygenated hydrocarbons < proton donors < ionic compounds; that is, $RH < RX < RNO_2 < ROR$ (ethers) $< RCOOR$ (esters) $< RCOR$ (ketones) $< RCHO$ (aldehydes) $< RCONHR$ (amides) $< RNH_2, R_2NH, R_3N$ (amines) $< ROH$ (alcohols) $< H_2O < ArOH$ (phenols) $< RCOOH$ (acids) $<$ nucleotides $< {}^+NH_3RCO_2{}^-$ (amino acids).

In both NPC and HILIC, the stationary phase is polar. In the former case, a nonpolar mobile phase is used, such as n-hexane, dichloromethane, or chloroform. Predominantly nonpolar or low polarity analytes, which are soluble in the above mobile phases or the like, are separated based on their degree of polarity. More polar compounds are retained longer.

Very hydrophilic analytes, however, do not dissolve in NPC mobile phases. They cannot be separated on a reverse phase, either. Many drugs and biological metabolites fall in this class. HILIC in such cases is an attractive separation mode. Both the stationary and the mobile (typically acetonitrile:water) phases are polar. More hydrophilic compounds are retained longer. Water is the strong solvent in HILIC; separations are typically performed with a high initial concentration of acetonitrile to achieve retention and then increasing the water content. HILIC method development can be challenging for the novice because of the wide variety of stationary phases available, and the variety of interaction modes between the analytes and these phases. A successful HILIC separation is critically dependent on the injected sample being introduced in a solution that is equivalent to or weaker than the initial mobile-phase composition (e.g., high acetonitrile content); otherwise, poor peak shapes result. HILIC columns are available today both on silica and polymer substrates and a variety of functionalities, zwitterionic, strong cation exchanger, strong anion exchanger, diol, etc.

Analytes in RPC are separated based on their degree of hydrophobicity. A wide range of organic compounds can dissolve in the mixed water–organic solvent phases and be separated, so RPC is the most popular form of HPLC. The mobile phase is typically methanol or acetonitrile, containing various amounts of water. Tetrahydrofuran is also used as the organic solvent, but less commonly. All of these solvents are sufficiently transparent in the UV to permit UV detection.

In contrast to GC, where the carrier gas has little influence on the separation, in HPLC, the eluent greatly influences the separation, and this is much easier to change than changing columns. Optimizing the mobile phase composition is part of developing an HPLC separation method. In most applications, a pure solvent is not useful; a blend of two or more solvents is typically used. The weak and the strong solvents are designated the A and B solvents typically. Trial-and error (or available software programs that predict retention on given columns with given eluent composition) can be used to obtain an optimum solvent composition for isocratic elution or the changes in %B with time in a gradient elution protocol.

For a good discussion of HILIC-based separations and available columns, see http://www.sequant.com/default.asp?ml=11625.

DryLab (http://www.molnar-inst itut.com/HP/Software/DryLab.php) and Virtual Column (http://www.virtualcolumn.com/) are examples of software that allow simulation and retention prediction for HPLC and ion chromatography, respectively.

The two parameters of interest are the retention factor, k, a *capacity* term, and the separation factor, α, a *selectivity* term (see Equation 14.32). Generally, one tries to adjust the solvent strength to position all solute bands within a k range of about 1 to 10. Adjusting for retention in this window may provide adequate resolution for many compounds. With protolyzable analytes, the chromatographic behavior of the protonated and unprotonated forms varies considerably. One must consider the addition of acids, bases, and buffers to the mobile phase for pH adjustment, in order obtain good peak shapes and accomplish the desired resolution. For many molecules of biological interest, buffering at a compatible pH is often mandatory. The buffer composition or pH may also be varied during a gradient run as part of a gradient elution protocol using a ternary or quaternary gradient.

Gradient elution in HPLC is one of the most effective ways of separating several analytes with vastly differing retention times under isocratic conditions. In isocratic elution, there is generally poor resolution of peaks early in the chromatogram and broadening of peaks at the longer retention times. In RPC, the gradient is accomplished by increasing the fraction of the strong solvent (%B) during the run. This allows weakly retained compounds to be eluted later and be better resolved than they would in isocratic elution, and longer retained ones to be eluted more quickly, giving a chromatogram with well-spaced peaks, with improved peak shape and lower detection limits since band broadening is lessened. As in GC temperature programming, the gradient profile may be stepwise or be continuously changing. Typically, the total flow rate is held constant. The starting eluent should have a composition that rapidly elutes and resolves the first components. The eluent is then gradually altered to a composition that resolves the last peaks in a reasonable period.

A price paid for gradient elution is that the column has to be re-equilibrated with the beginning solvent between runs. This typically requires flushing with 15 to 20 column volumes of the initial mobile phase. Ultimately, the practicing analyst is interested in accomplishing the desired separation with sufficient resolution in the shortest possible time. From this perspective, gradient elution may or may not always be better. The pharmaceutical and biotech industry account for the major share of the HPLC market (followed by food products and the chemical/petrochemical industries). In pharmaceutical quality control, analysts generally need to look for a few active ingredients in a tablet or formulation.

16.6 UHPLC and Fast LC

Many analyses are of the same or similar samples of limited complexity; isocratic separation can be successfully used. Available equipment may dictate whether small particle size, high-efficiency STM columns (often very short ones and/or columns of smaller bore) can be used at high linear velocities to achieve a fast separation. In addition, the total amount of solvent consumed can also be a factor, especially if an expensive eluent is used. From this standpoint, if the equipment permits, narrower bore columns are to be preferred. Where time is of the essence, UHPLC systems are of value. As particles get smaller, the van Deemter H vs. u plot becomes flat at high flow rates (see Figure 16.25) because smaller particles are less resistant to mass transfer (the C term in the van Deemter equation decreases).

Note that SPP particles of 2.7 μm d_p give similar efficiency as STM particles but at half the pressure—a big advantage. Longer columns can be used, and faster flow rates can be used to shorten analysis time. In many cases, older generation HPLC systems (6000 psi maximum) can be used without having to invest in UHPLC instruments.

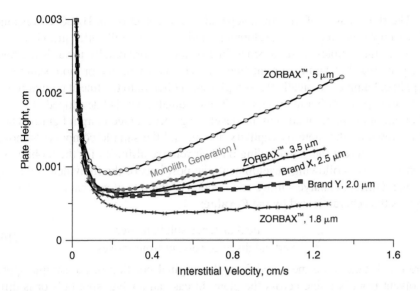

FIGURE 16.25 *H* vs. *u* plots for various columns of different particle size and a monolith. Column dimensions : $4.6 \times 50/30/20$ mm, Eluent: 85:15 ACN:Water, Flow rates: 0.05–5.0 mL/ min, temperature: 20°C, Sample: 1.0 μL Octanophenone in eluent. Note that as particle size decreases, there is very little increase in the plate height with an increase in the linear velocity. This suggests that at high flow rates fast separations can be performed with little or no loss in efficiency on the small particle columns, as long as the equipment and the column can work at the necessary pressure, and the detector/data systems have a sufficiently fast response/data acquisition time. Digitized and replotted from data presented in http://www.chem.agilent.com/Library/posters/ Public/Performance%20Characterizations%20of%20HPLC%20for%20High%20Res%20RRLC% 20separations.pdf.

16.7 Open Tubular Liquid Chromatography (OTLC)

While GC today is dominantly carried out in wall-coated/functionalized open capillaries, this is not the case for LC. Diffusion in the liquid phase is $10^3 - 10^4$ times slower, and the tube diameters need to be smaller by the square root of this factor. Efforts have been made for the last 30 years, but the difficulties in creating an active phase, injection and detection in such small capillaries have made progress difficult. Nevertheless, OTLC in capillaries as small as 800 nm radius has been successfully carried out; a favorite example is provided by Professor Shaorong Liu (University of Oklahoma) in the **website** supplement of the text for a charge-based separation in a bare silica capillary.

16.8 Thin-Layer Chromatography

Thin-layer chromatography (TLC) is a planar form of chromatography widely used for rapid qualitative analysis. It can also be used in a high-performance mode (HPTLC). Quantitative analysis is also possible, although the technique is most widely used for rapid screening, e.g., in a synthetic organic chemistry laboratory to check if the desired compound is being made and how many impurities there are. The stationary phase is a thin layer of finely divided sorbent supported on a glass, metal (most commonly aluminum) or plastic sheet. Virtually any stationary phase used in HPLC can be used, provided a suitable binder can make it adhere well to the substrate. It differs with HPLC in that multiple samples can be simultaneously analyzed.

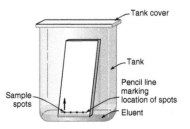

Thin-layer chromatography setup.

The three stages of injection, separation and detection in HPLC correspond in TLC to sample spotting, plate development, and detection (quite often just visual examination). In the simplest case, a pencil line is drawn horizontally towards the bottom of the plate (ca. 5–10 mm from the bottom). This is where one or more sample spots are applied. Samples (typically $0.5 - 5 \mu L$) are spotted onto the line at regular intervals (ca. 20 mm apart) with a micropipette. The chromatogram is "developed" by placing the bottom of the plate or strip in the developing solvent (see margin figure). The solvent is drawn up the plate by capillary action, and the sample components move up the plate at different rates, depending on their relative affinities for the mobile vs. the stationary phase. Following development, the positions of the individual solute spots are noted. Different analytes move at a fraction of the rate of solvent movement; each analyte is thus characterized by the R_f **value**:

$$R_f = \frac{\text{vertical distance solute moves}}{\text{vertical distance solvent front moves}} \tag{16.23}$$

where the distances are measured from the original position of the sample spot and the solvent front is a line across the plate. In case an analyte spot tails or is diffuse, the position of maximum density is taken to be the post-development solute position. As with the retention factor in HPLC, the R_f value is characteristic for a given stationary phase–solvent combination. Few users or researchers today make their own TLC plates. Plate uniformity of commercial plates is generally good. However, when doing qualitative screening for a particular compound (or a set of compounds), it is always a good idea to spot the particular analytes on the same plate (singly or in a mixture) to compare the R_f values of the components in the test sample.

The principles of chromatography are illustrated by using TLC as an example in http://www.chemguide.co.uk/analysis/chromatography/thinlayer.html. This is a simple, short, and very readable account.

Stationary Phases for TLC/HPTLC

Although essentially any HPLC stationary phase can be used in TLC, some 80% of TLC applications involve an unmodified 60 Å pore size silica stationary phase, according to Merck-Millipore, a leading vendor of TLC/HPTLC plates. Such an unmodified phase sees very limited use in modern HPLC. The remainder 20% comprises of CN-, diol-, $-NH_2$- or C18-modified silica and Al_2O_3, as well as cellulose. A typical TLC plate is 20×20 cm, although often narrower sections may be cut and used. Commonly, the sample can be applied on any edge of the plate and the plate can be accordingly developed with that edge at the bottom but some manufacturers also make plates with one area designated as the concentration zone. In such a case, the sample must be applied in this zone. Plates coded for good laboratory practice (GLP) also have specified direction of development. The mean particle size in TLC plates is 10–12 μm with a range of 5–20 μm (in HPTLC plates, mean size 5–6 μm with a range of 4–8 μm). The sorbent layer thickness of TLC plates is typically 250 μm on glass and 200 μm on other substrates, while that on HPTLC plates is 100–200 μm. Preparative plates have layer thicknesses of 0.5–2 mm and permit preparative separations in the gram scale. The typical migration distance on a TLC plate is 10–15 cm, while it is significantly smaller in HPTLC plates at 3–6 cm. However, the latter permits more efficient separations with typical plate heights of 12 μm (typical separation time 3–20 min) compared to 30 μm for a standard TLC plate (typical separation time 20–200 min). Although HPTLC plates are significantly more expensive, if a large number of analyses are to be conducted at a time, considering savings in time, the use of HPTLC plates may be more attractive. In addition, since band broadening is significantly less, nearly four times as many samples can be analyzed on an HPTLC plate than on a TLC plate and with better mass detection limits.

Merck-Millipore has an excellent instructional video about TLC and their TLC/HPTLC offerings at http://www.merckmillipore.com/chemicals/tlc-video/c_DIab.s1O39YAAAEqF1ck445w.

Mobile Phases for TLC

In adsorption chromatography, as with pure silica or alumina stationary phases, the eluting power of solvents increases in the order of their polarities (e.g., from hexane to acetone to alcohol to water). The developing solvent should contain no more than three components because mixed solvents tend to themselves migrate differently and themselves separate by chromatography as they move up the thin layer, causing a continual change in the solvent composition with distance on the plate. This may result in varying R_f values, depending on how far the spots are allowed to move up the plate. Minor variations in solvent composition and temporal/spatial variations in temperature may then cause major reproducibility problems in R_f. However, with appropriate controlled development chambers discussed later, reproducible solvent gradients may be deliberately exploited. The developing solvent must be of high purity. The presence of small amounts of water or other impurities can produce irreproducible chromatograms.

Sample Application

Manual sample application with capillary micropipettes is usually performed for simple analyses. Sample volumes of 0.5 to 5 µL can be applied as spots onto conventional layers without intermediate drying. HPTLC plates can handle less sample; up to 1 µL per spot is acceptable. If samples are very dilute and more is needed for detection/visualization, the sample may be repeatedly applied on the same spot after drying.

More demanding qualitative, quantitative, and preparative analyses or separations are made possible only by automated instrumental sample application using the spray-on technique. To take full advantage of the separation power and reproducibility of HPTLC, precise automated positioning and volume dosage is mandatory.

At the high end, autosamplers/applicators are available, e.g., from CAMAG, that require no operator presence and can apply sample as spots by contact transfer or as a rectangular band by a spray-on technique, essentially using technology similar to that used in ink-jet printers. The spray-on technique permits sample application as bands or rectangles with volumes as little as 0.5 µL to > 50 µL. Starting zones sprayed on as narrow bands offer the best separation attainable. Application in the form of rectangles also allows precise application of large volumes without damaging the layer. Prior to chromatography, these rectangles are focused into narrow bands with a solvent of high elution strength.

Developing the Chromatogram

A TLC development chamber can be as simple as a beaker covered with a watch glass. Detailed considerations, however, reveal a much more complex system: a TLC system in fact is the only chromatographic system where a solid, liquid, and gas phase all play a role in the overall chromatographic process. While the role of the gas phase is typically ignored, it can significantly influence the result of the separation.

The standard TLC chromatogram development involves placing the plate in a chamber, which contains a sufficient amount of developing solvent. The lower end of the plate is immersed several mm into the solvent, but not to the level of the sample application point. The solvent moves up the layer due to capillary forces until the desired distance is reached. The establishment of equilibrium between the component(s) of the developing solvent and the vapor phase is called chamber saturation. The composition of the gas phase can differ significantly from that of the developing solvent and depends on the relative vapor pressures of the different solvent components. In adsorption chromatography, in a closed chamber, the following processes occur:

www.camag.com describes the many aspects of TLC and available equipment at different levels of sophistication to carry out the process.

(a) The dry stationary phase adsorbs molecules from the gas phase. This adsorptive saturation processes also approaches an equilibrium: for silica and similar phases, in this process the more polar gas-phase components selectively transfer to the solid sorbent phase.

(b) Simultaneously the part of the sorbent already wetted by the mobile phase interacts with the gas phase. This results in selective transfer of the less polar components of the mobile phase to the gas phase. Unlike (a), this process is more governed by sorption equilibria than by vapor–liquid equilibria.

(c) During migration, the components of the mobile phase can themselves be potentially separated by the stationary phase, causing the formation of secondary fronts.

With the exception of a one component developing solvent, the terms "developing solvent" and "mobile phase," although used interchangeably, are really not the same. The mobile phase composition changes as it moves up the plate, while the liquid originally placed into the chamber is the developing solvent; only the composition of the latter is known. Processes (a) and (b) can both be experimentally manipulated by: putting developing solvent soaked material (e.g., paper towels) generously within the development chamber, waiting long enough for the developing solvent to saturate the chamber with vapors before beginning chromatography, and allowing the sample spotted plate to interact with the developing solvent vapors without contact with the liquid solvent (also called preconditioning). Processes (b) and (c), on the other hand, can be effectively stopped by placing a second TLC plate at a distance of one or a few mm facing the chromatographic layer of the analytical plate; this is called the "sandwich" mode of operation. The further an equilibrium according to (a) and/or (b) has been established and the less different the components of the mobile phase in their sorption behavior, the less important will be (c). In well-saturated chambers and with preconditioning, secondary fronts are rare. But secondary fronts are prominent in the sandwich mode.

With the exception of very polar components like water, methanol, etc., constituents of the developing solvent that are preferentially sorbed from gas phase may be pushed ahead of the real solvent front during chromatography. This results in R_f values being lower in saturated chambers and particularly on preconditioned layers, than in unsaturated chambers and sandwich configurations. Because of these possible complications from process (c), development in the sandwich mode or in an unsaturated horizontal developing chamber (see below) works best with single component solvents or mixtures where the behavior of the different components are very similar to each other.

It is important to understand that TLC development proceeds in the majority of cases under non-equilibrium conditions and all parameters must be maintained the same to obtain reproducible results. Development chambers are available in a number of different geometries. Chamber type and saturation play a dominant role; this also means that the precise R_f values may be different in each chamber even with the same developing solvents. No chamber is uniquely good but in some chambers, the relevant parameters can be better controlled.

The flat-bottom classical TLC development chamber shown previously in the margin figure is used under conditions of partial or complete saturation of the tank atmosphere with solvent vapors. The degree of sorbent presaturation cannot generally be controlled in such a chamber. In a twin-trough chamber, the floor of the classical chamber is modified with a partition of low height. Only one of the troughs may be filled with solvent to reduce solvent consumption. In addition, with one trough filled and the plate put in the other trough, the plate may be preconditioned. The development does not begin until solvent is put in the trough the plate is in.

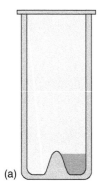

(a)

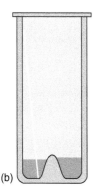

(b)

A twin-trough chamber (a) top: used in the preconditioning mode, (b) bottom: development mode.

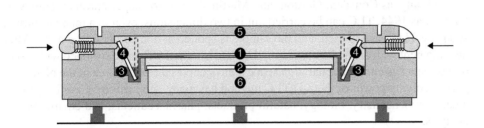

FIGURE 16.26 A horizontal developing chamber. (1) An HPTLC plate (face down), (2) glass plate (location for second plate for "sandwich" mode operation), (3) solvent reservoir, (4) glass strip (provides wicking), (5) cover plate, (6) conditioning tray (solvent-soaked pad can be put here). This and margin figures courtesy of CAMAG Scientific.

Development must be stopped when the maximum the solvent front can reach has been reached or earlier. If not stopped, the analytes will continue to move and the resulting R_f values will be erroneously high. Optical monitors that alert the operator when the solvent front has reached some preselected height are available and prevent such overdevelopment.

More sophisticated development chambers are frequently used with HPTLC. The horizontal developing chamber (Figure 16.26) is a versatile development tank that provides reproducible results. In this device, it is possible to spot the sample(s) on opposite ends of a plate. The chamber contains a solvent trough at both ends. Development and sample application can be from one end or both ends. In the latter case, after sample spotting at both ends, the plate is laid face down on edges of the solvent troughs, and development begun simultaneously from both ends. This doubles sample throughput, provided the migration distance in double-ended development is enough for the intended separation. The horizontal chamber is suitable for developments in unsaturated, saturated, and sandwich modes and can be used for preconditioning of plates.

Fully automatic developing chambers are available and provide superb reproducibility. Not only operator intervention to begin chromatography after chamber saturation or plate preconditioning is not needed, the activity of the layer prior to start of chromatography can be set. After chromatography, the plate is rapidly and completely dried. With a twin-trough chamber, a salt solution that provides for some constant preselected humidity during development can then be put in the second trough.

In HPLC, gradient elution with solvents of increasing or decreasing polarity is rarely carried out on a silica gel stationary phase. Once a polar solvent is used, time to recondition the stationary phase can be prohibitively long; with certain eluents, the stationary phase may also degrade irreversibly. In TLC, the situation is different; the stationary phase is not used multiple times with multiple samples. In addition, while whole column imaging has been occasionally demonstrated, separation in HPLC is visualized by an external end-column detector; there are no opportunities, e.g., to stop chromatography when constituent A has reached a certain position in the column. This is not the case in TLC.

Especially if the sample contains components that range widely in polarity, the principle of automated multiple development (AMD) is of great benefit. The plate is developed repeatedly in the *same direction* and *stopped* when the solvent front moves a preset distance. The solvent is then completely removed and the plate dried *in situ* under vacuum. The next development is done with a *less polar (weaker)* solvent, but each successive run progressively extends the distance (in a preprogrammed manner) the solvent front is allowed to move before development is stopped. In this way, a stepwise gradient elution is achieved. The combination of focusing effect and gradient elution results in extremely narrow bands with widths of ~1 mm on an

Whole column imaging literally means looking at the column in its entirety and being able to tell where the analyte bands are and their concentrations. When Tswett first separated plant pigments on a glass column containing a white sorbent, the progress of the separation was visible to him, ironically it is not so simple in the modern practice of HPLC.

Read the classic paper on 2-D paper chromatography: R. Consden, A. H. Gordon, and A. J. P. Martin, *Biochemical Journal,* **38** (1944) 224.

HPTLC plate: for a separation distance of 80 mm, this means that up to 40 components can be baseline resolved!

Finally, as Consden, Gordon, and Martin showed for paper chromatography as far back as 1944, TLC can be carried out in two dimensions, even in a manual format. A square TLC plate is used, and the sample is spotted at one corner of the plate. After development with one solvent system, the plate is turned 90° and developed with a second solvent system. The separation space then occupies the entire area of the plate. The great resolving power of the 2-D TLC method has many applications, especially in the areas of biochemistry, biology, natural products, pharmaceuticals, and environmental analysis.

Visualization. Detection of the Spots

A visibly colored substance can obviously be readily seen on a white TLC plate. Indeed one of the most popular demonstrations of a TLC separation involves spotting a TLC plate with a black marker pen and separating the components of the "black" ink, which invariably turns out to be the mixture of a number of different, easily distinguishable, dyes. A fluorescent analyte can also be readily visualized by examining the developed plate under long (365 nm) or short (254 nm) wavelength UV light. The reverse technique is sometimes used—TLC plates are available with a green fluorescent marker bound to the stationary phase; in this case, under UV illumination the nonfluorescent analytes are visualized as dark spots.

Instrumental "Visualizers" that provide uniform short-wave UV, long-wave UV, or white light illumination and are equipped with a digital camera to take either transmitted or reflected light images are commercially available. Software is supplied that can perform blank image corrections and can decode the images in terms of spot intensities.

In TLC, analytes remain stored on the plate post separation and can be chemically treated prior to visualization/quantitation. The objective of such treatment may be to change otherwise invisible substances to visible, or to improve detectability (e.g., form a fluorescent derivative), or to detect only a certain class of substances by a selective derivatizing reagent. In contrast, in other instances, it may be desirable to detect all sample components with comparable intensity.

The mode in which the plate is exposed to the derivatization reagent controls the uniformity of the derivatization process and hence the accuracy of any quantitation carried out post-derivatization. Gas-phase derivatization, such as exposure to iodine vapor, can be carried out uniformly: iodine vapor interacts with the sample components, either chemically or by dissolving in them, to produce a color. Liquid-phase reagents are more commonly used. For example, amino acids and amines are detected by treating the plate with a solution of ninhydrin; blue-purple spots are ultimately produced after heating. Another common but destructive technique for organic compounds is spraying the plate with a sulfuric acid solution and then heating it to char the compounds and develop black spots. The two common modes for exposing the plate to a liquid reagent are spraying the reagent on the plate or immersion in the reagent. Reagent sprayers (and spray cabinets that confine the aerosol droplets within to prevent inhalation toxicity), immersion devices and specialized planar heaters for such applications are commercially available. Whenever possible, direct immersion is to be preferred because greater homogeneity can be attained. However, immersion and withdrawal must be performed smoothly; dedicated immersion devices can do this reproducibly at controlled rates.

Quantitative Measurements

Once it was common to scrape the spots to be quantitated, extract the material with a suitable extraction solvent, and measure the extract spectrophotometrically. Those trying to make quantitative TLC measurements today are invariably using HPTLC and making the measurement instrumentally. Photographic imaging and software based densitometry is straightforward and has the advantage that the entire plate can be imaged at once. But the limitations of present cameras limit this to the visible wavelengths. Variable wavelength reflectance-based densitometric scanners are available commercially and covers the entire 190–900 nm range. Broadband light source(s) are used with a monochromator; a slit of adjustable length and width then controls the spatial resolution of the scan. Each chromatographic track is scanned at a time, measuring the diffusely reflected light. Background corrected absorption spectra for any desired spot can also be acquired for identification of the analyte, as well as selecting the best measurement wavelength.

In principle, diode-array-based TLC scanners as well as fluorescence scanners should be feasible. These, as well as transmission-based TLC densitometry, have been repeatedly demonstrated in research publications. However, such instruments thus far have not been commercially successful. With the increasing popularity of matrix-assisted laser desorption ionization mass spectrometry (MALDI-MS, see Chapter 17), TLC-MS interfaces are now available. Similarly, bioluminescence is now a popular assay to determine, for example, toxicity of substances, detection systems specifically to measure bioluminescence on TLC plates are also available.

16.9 Electrophoresis

The word electrophoresis connotes the movement of a charged particle [Greek: *elektron* + *pherein* (to carry)]. Electrophoretic methods of separation thus rely on differences in the mobility of different charged substances in an electric field. Electrophoretic methods of separation can be classified in two broad groups, depending on the medium that it is conducted in. The first group involves separations in free solution, conducted primarily in small diameter capillaries or microfabricated chips. The sample is injected as a finite zone or band, and then further subdivides into individual zones. This is thus referred to as **capillary zone electrophoresis (CZE)** or simply **capillary electrophoresis (CE)**. Electrophoresis conducted on a microchip is very similar. In the second case, the separation medium is a hydrogel, e.g., of agarose. With some large biomolecules, e.g., DNA, the charge/size ratio remains almost constant regardless of the molecular weight, thus there is very little difference in electrophoretic mobility in free solution. So DNA fragments cannot be separated by CZE. In a high viscosity gel medium, however, the mobility of these large molecules is very much a function of their tertiary structure permitting a separation. Gel-based electrophoresis can be carried out in a capillary (**capillary gel electrophoresis, CGE**) for high resolution analytical separations of large biomolecules (Figure 16.27).

Preparative gel electrophoresis can be carried out in a tube (column) with electrodes at the top and bottom (with a polarity that causes the bands to migrate down the column). It can also be run in a planar form on a thin gel layer (much like TLC), often called **slab gel electrophoresis**. For tube gels, little specialized equipment is required other than an appropriate power supply. Such tubular gel separations have the advantage that the molecules move through the gels without much lateral movement. This leads to somewhat better resolution of the bands compared to a slab gel, particularly for proteins. However, a slab-gel platform, much like TLC, can be turned around and run with an electric field perpendicular to the previous field direction, after the gel has

The word *electrophoresis* (and the science) is of relatively recent origin. According to Merriam-Webster, the first known use was in 1911. Today it is so widely used that a major journal, titled *Electrophoresis*, is solely devoted to this topic and a scientific society (http://www.aesociety.org/) has electrophoresis as its sole focus.

Large molecules, such as proteins, migrate in an electric field based on their mobilities, but their movement is affected by how their three-dimensional structure interacts with a gel.

A highly interactive virtual lab allows you to run a slab gel electrophoresis experiment to separate DNA fragments of different lengths, beginning at the beginning, making the gel! http://learn.genetics.utah.edu/content/labs/gel/ Check it out!

FIGURE 16.27 (a) CGE separation of *E. coli* extract (6 µg/µL) on a 75 µm i.d. 30 cm active length capillary with inner wall coated with polyacrylamide gel. Field strength 290 V/cm, absorbance detection at 220 nm. (b) Calibration mixture run under same conditions: a Aprotinin (6.5 kD), b Lysozyme (14.4 KD), c Trypsin inhibitor (21.5 kD), d Carbonic anhydrase (31 kD), e Ovalbumin (45 kD), f Serum albumin (66.2 kD), g Phosphorylase b (97.4 kD), h *β*-galactosidase (116.25 kD) and i Myosin (200 kD). Reprinted with permission from J. J. Lu, S. Liu, and Q. Pu, *J. Proteome Res.* **4** (2005) 1012. Copyright 2005, American Chemical Society.

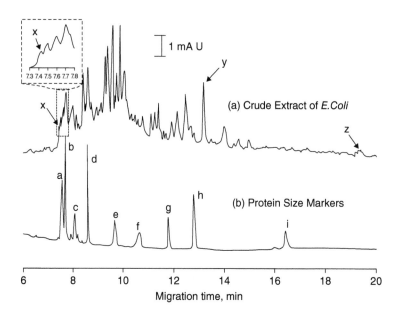

Preparation of polyacrylamide gels with various degrees of crosslinking and the effects thereof are discussed in an animated interactive slideshow from the National Forensic Science Technology Center: http://www.nfstc.org/pdi/Subject 05/images/pdi_s05_m01_01 b.swf.

Agar is a term derived from the Malay word agar-agar, meaning jelly. It is typically derived from the *Gelidium* genera of seaweed and is composed of agarose and agaropectin. Agarose is readily separated from the mixture.

Velocity is the product of the mobility and the field strength; the latter is directly proportional to the applied voltage. The mobility can be affected by the pH (which influences the charge on the analyte).

been run in one direction. Such a 2-D gel electrophoresis system has great separation power and is routinely used in biological laboratories.

Commonly used in gel electrophoresis are gels of agarose, polyacrylamide, starch, etc. Proteins are often treated with the detergent sodium dodecyl sulfate (SDS), an anionic detergent that linearizes the protein molecule and imparts a negative charge to it. Generally the charge is evenly distributed per unit mass, thereby resulting in an approximately uniform separation by size during polyacrylamide gel electrophoresis (PAGE). The resulting technique SDS-PAGE is widely used for protein separations.

Depending on the sample preparation and separation conditions, different analytes in the sample can be both positively and negatively charged, or more commonly, as in SDS-PAGE, only negatively charged. In the first case, the sample will be applied to the middle of the gel plate so the oppositely charged analytes can migrate in opposite directions. With only negatively charged analytes, the sample is applied near the negative electrode end of the gel.

The migration rate of a molecule at a given pH is controlled by the applied field strength, expressed in volts per centimeter (V/cm). The electric field controls the current flowing through the system. The product of the current and the field strength gives the heat generated per unit length and hence is proportional to the temperature rise. The greater the field strength, the faster is the migration rate and quicker is the separation. However, if the heat buildup is high, increased temperature will compromise the achieved separation by creating local temperature gradients and enhancing diffusive mixing. In planar gel electrophoresis, field strengths of 4–6 V/cm are common. In pulsed field gel electrophoresis, discussed a bit later, the optimum field strength decreases with increasing size of the molecule to be separated—for separating Mb size DNA, an applied field of 2 V/cm is optimum. Given that the gel length in planar gels rarely exceeds 20 cm, the maximum voltages supplied by power supplies for this purpose range from ~125–300 V.

In small-bore capillaries or microchips, heat dissipation is much better than in planar or large-diameter tubular gels. Also, because of very small cross section, the electrical resistance is high; this limits the current. Hence, much higher field strengths can be applied. Typical field strength in a capillary electrophoretic separation is 300 V/cm for 50–75 µm bore capillaries; 10–60 cm length capillaries are used. The power supplies can typically supply up to 30 kV but maximum current capability is usually ~300 μA. However, much greater field strengths, up to several kV/cm, can be used over short separation distances and in very narrow-bore capillaries/microchips.

Another important technique for separation of proteins takes advantage of the change in net charge of individual proteins as a function of pH and the fact that at the isoelectric point, they have no net charge and have no electrical mobility. In this technique, called **isoelectric focusing (IEF)**, sometimes simply called electrofocusing, the separation medium is an immobilized pH gradient (IPG) gel. IPGs constitute of an acrylamide gel matrix co-polymerized with a pH gradient formed by immobilizing weak acids and bases (and their salts) in varying ratios across the gel. An ampholyte solution is used as the running buffer. The negative electrode is placed at the high pH end of the gel. A protein that is located at a pH below its pI will be positively charged and so will migrate towards the negative electrode. As it does so, it travels through media of continually increasing pH, eventually reaching a pH region equal to its pI. Now it has no net charge and stops moving. Thus, the proteins focus into individual well-defined bands with each protein positioned at a point in the gel where the pH equals its pI. IEF can provide very high resolution with proteins differing by a single protolyzable group being separable. IEF is also used in a capillary format; **capillary isoelectric focusing (CIEF)** is also carried out in a microchip-based variant.

One separation technique may be followed by a second separation method. Bands separated by agarose gel electrophoresis may be cut out and separated by SDS-PAGE.

At the pH of their isoelectric point, proteins have no net charge and do not migrate.

Visualization and Quantitation

The visualization/quantitation of bands separated by planar gel electrophoresis is carried out very similar to that done for a TLC plate. The most commonly used post separation visualization agent for DNA is ethidium bromide. It intercalates in DNA strands and in that form fluoresces reddish orange/brown. The visible fluorescence can be digitally imaged and quantitated using image analysis tools.

However, nucleic acids can be readily destroyed by UV exposure and the ability to do any further work with the separated DNA is compromised. A blue light–excitable stain (e.g., SYBR Green I, blue excitation leads to green fluorescence) along with a blue light excitation source is much preferred. Ethidium bromide is highly mutagenic. Further, without appropriate protection, UV light can cause eye damage. With a blue light excited fluorescent tag, both are avoided. In addition, the blue excitation light is not significantly attenuated by passage through clear glass or plastic, permitting greater latitude in materials and illumination and imaging.

To retrieve a separated DNA band, it can be cut out of the gel, and extracted by dissolution. Quantitation in solution then also becomes possible. The use of gel electrophoresis for separation of nucleic acids in DNA sequencing is significant.

For quantitation of proteins, Coomassie Brilliant blue is often used as the stain; it binds strongly to all proteins. After removing the unbound dye by washing, the stained protein bands are easily located and quantified; the amount of bound dye is proportional to protein content. Stained gels can also be dried and preserved. Quantitation by a scanning densitometer, as in TLC, is also possible. Occasionally, the proteins may be radioactively labeled; they can be detected by autoradiography. This involves contact exposure of the gel to a photographic film; presently technology also exists to directly image this on a solid-state imager, but such equipment is expensive.

Structure of ethidium bromide

Image analysis software is widely used for quantitation of planar gels. ImageJ, freely available from the National Institutes of health, is particularly powerful and is popular in the gel electrophoresis community (http://rsbweb.nih.gov/ij/).

Structure of SYBR Green I

Related Techniques

DNA fragment lengths are often specified by the number of base pairs they contain. A 20 kb DNA fragment contains 20,000 base pairs.

In a typical unidirectional field agarose gel, DNA fragments above 10–15 kb may migrate in a nonpredictable manner unrelated to size. In **pulsed field gel electrophoresis (PFGE)**, the voltage is applied in one direction for several seconds, and then the direction is reversed for several seconds (the forward step is longer than the reverse step, providing a net forward field direction). Reversing the field direction back and forth shuttles the DNA back and forth in the gel. This slows down the large DNA fragments relative to the smaller ones because the smaller molecules can change direction faster than the larger ones. Separation with greater resolution thus becomes possible. In practice, a fixed alternating time program is useful in separating DNA within a fixed size range. To achieve a much greater separation range, the duration of the forward step, the backward step, or more commonly both, are increased during the separation, which can be as long as 24 h. DNA fragments up to several Mb in length can be separated by PFGE; it is the gold standard for identifying particular strains of pathogens.

There are several configurations in which PFGE is used. The above description is termed **field inversion gel electrophoresis (FIGE)**. Separation is actually possible even when the forward and reverse time steps are equal. This is because of the asymmetric rate at which the molecules reverse their movement. Such **zero integrated field electrophoresis (ZIFE)** is much slower than FIGE, but capable of separating extremely large DNA fragments. The other major category of PFGE reorients the DNA at oblique angles, causing the DNA to move forward in a zigzag pattern down the gel. For a similar size range under optimal conditions, these separations are faster, resolve a wider size range, and utilize a larger useful portion of the gel compared to FIGE. An extreme variant of this is **rotating field gel electrophoresis (RGE)** where the field direction continually changes (present instruments rotate the electrodes, e.g., the Rotaphor 6.0 from Analytik Jena), allowing a large separation space and excellent separation.

Read the classic papers in Free Solution Electrophoresis and CZE: S. Hjertén, *Chromatographic Reviews*, **9** (1967) 122; J. W. Jorgenson and K. D. Lukacs, *Anal. Chem.* **53** (1981) 1298.

16.10 Capillary Electrophoresis

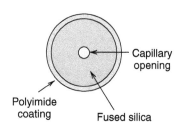

Capillary opening

Polyimide coating

Fused silica

CE fused silica capillary

Fused silica capillaries are very fragile and must be coated with an outer protective coating. The most common coating is polyimide. Polyimide is not optically clear. Capillaries coated with Teflon AF are transparent down to 200 nm and are also available.

Stellan Hjertén is justly considered a luminary in the separation sciences. He pioneered the use of agarose gels in electrophoresis (and chromatography) and made seminal contributions to monolithic columns. While Tiselius first showed that free solution electrophoresis can be used to separate proteins without the need for a gel medium, it was Hjertén who laid out the theoretical principles and demonstrated free solution zone electrophoresis in small (1–3 mm bore) quartz tubes in 1967. Detectors were not available to monitor separation in tubes any smaller; he minimized the effects of thermal convection by slowly rotating the tubes around their long axis. The potential of CE would not be fully realized until James Jorgenson (University of North Carolina at Chapel Hill, another luminary in the separation sciences, he is credited also with the invention of UHPLC), along with his student Lukacs, would demonstrate the immense power of this technique. Using 75 µm bore silica capillaries (fused silica is an excellent thermal conductor) and voltages up to 30 kV, they obtained plate counts in excess of 400,000 and performed separations in 10–30 minutes that were deemed unattainable at that time. In small capillaries, aside from electrophoretic movement, electroendosmosis (often just called electroosmosis) generates **electroosmotic flow (EOF)** of the bulk fluid in the capillary. The EOF can be large compared to electrophoretic movement, permitting both positively and negatively charged analytes to be detected by the same stationary detector.

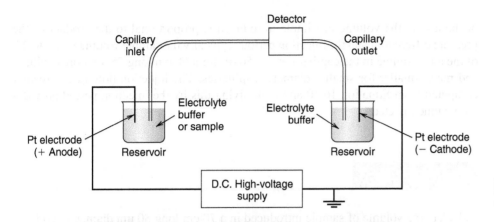

FIGURE 16.28 Capillary electrophoresis system setup.

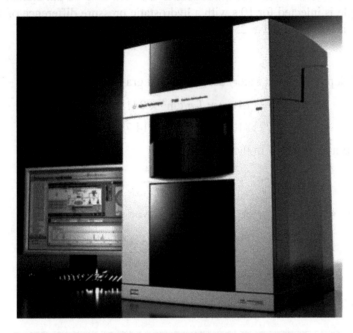

FIGURE 16.29 Agilent 7100 Capillary electrophoresis system. The high voltage inside a CE instrument is potentially hazardous. All instruments provide an interlock: whenever the instrument cover is opened, the HV source is disconnected. Courtesy Agilent Technologies.

Operation

The basic setup for the technique is illustrated in Figure 16.28. A commercial instrument is shown in Figure 16.29.

The instrumentation requirements for CE, with the possible exception of the detector, are modest, and using CE is straightforward. Electrophoresis is conducted in a fused silica capillary tube, typically 25–75 μm i.d. ≤ 375 μm o.d., 25–75 cm long, filled with a background electrolyte (BGE) solution, often with some buffering capacity. The two ends of the capillary are dipped in small reservoirs containing the BGE and platinum electrodes that are connected to a high-voltage power supply (HVPS, typically up to 30 kV and 300 μA). For all HV power supplies, one end is grounded (in Figure 16.28, this would be the negative cathode end). The capillary passes through the on-column detector to the grounded destination reservoir.

With the HV disconnected, the HV end of the capillary is lifted and put into a sample vial. A small amount of sample is then put into the capillary. This can be done by **pressure-based injection** via either (a) by lifting the sample vial to a preset height and holding it there for a programmed amount of time or (b) by pneumatic pressure applied on to a sealed sample vial for a preset period (in some instruments a vacuum may be applied to the destination reservoir instead). Sample introduction following (a) is also referred to as **hydrodynamic injection** or **gravity-based injection**. In all

Electroosmosis is the bulk flow of the fluid in the capillary induced by an electric field. Professor David Chen of the University of British Columbia in Canada likens sample injection to a group of swimmers jumping into a river. Some of the swimmers are trying to swim against the current and some with it. As long as the river flows faster than any of the countercurrent swimmers can swim, all swimmers will have a net downstream movement. A detector, akin to an observer, placed downstream from the jump-off point, will eventually see all the analytes injected, the fastest concurrent swimmer first and the fastest countercurrent swimmer last.

of the above, the volume of sample introduced is proportional to the product of the pressure differential and the injection period. Typically the sample volume is < 1 to 2% of the total volume in the capillary, < 25–50 nL for a 60 cm long 75 μm bore capillary and much smaller for smaller diameter capillaries. The injection flow rate is readily computed from Equation 16.10 and multiplying this by the injection period provides the volume injected.

Example 16.3

Calculate the volume of sample introduced in a 70 cm long 80 μm diameter capillary, if the sample is injected for 10 s with a hydrostatic pressure difference of 15 cm. The instrument is thermostated at 25 °C.

Solution

From fluid depth, h, fluid (water) density, ρ, and gravity acceleration, g:

$$\Delta P = h\rho g = 15 \text{ cm} \times 1 \text{ g/cm}^3 \times 981 \text{ cm/s}^2$$
$$= 14.72 \times 10^3 \text{ g cm}^{-1}\text{s}^{-2} = 14.72 \times 10^3 \text{ dyne/cm}^2$$

Rewriting Equation 16.10 in the form

$$F = \frac{\Delta P \pi r^4}{8\eta L}$$

where η, the viscosity of water, is 8.90×10^{-3} dyne s/cm^2 at 25°C.

$$F = \frac{14.72 \times 10^3 \text{ dyne cm}^{-2} \times 3.14 \times \left(4.00 \times 10^{-3}\right)^4 \text{ cm}^4}{8 \times 8.90 \times 10^{-3} \text{ dynes s/cm}^2 \times 70 \text{ cm}}$$
$$= 2.37 \times 10^{-6} \text{ cm}^3/\text{s} = 2.37 \text{ nL/s}$$

In 10 s, 23.7 nL will be injected.

Alternatively, a sample can be introduced by electrokinetic means, a technique unique to CE. In **electrokinetic injection** (also called **electromigration injection**), with the capillary inlet in the sample vial, a chosen voltage (injection voltage) of appropriate polarity is applied to the sample vial for a desired period of time, usually for a few seconds. In addition to bulk EOF that introduces the sample without preference for different ions, ions from the sample will electromigrate into the capillary based on their mobility and the polarity of the potential. For example, if a positive injection potential is used, cations from the sample will migrate into the capillary in preference to anions and the relative number of cations that are injected into the capillary will depend on their mobility: higher mobility cations will be preferentially injected relative to the lower mobility ones. In other words, in the absence of EOF, for sample ions that are moving into the capillary due to the applied field, if the concentration of ions i and j in the sample are C_i and C_j, respectively, the ratio of these ions injected into the capillary will be $(\lambda_i C_i)/(\lambda_j C_j)$ where λ_i is the equivalent conductance of ion i. In contrast, pressure-based injection results in some volume of the sample being injected into the capillary without preference for any particular ion.

Example 16.4

Deproteinized plasma of a patient contains 4.2 mM K^+ and 140 mM Na^+. The sample was diluted 50-fold with water and injected into a 50 cm long 75 µm i.d. capillary electrokinetically by applying 3 kV for 5 s. What is the ratio of $[K^+]$ to $[Na^+]$ in the injected sample?

Solution

The problem gives much extraneous information not needed for the solution. Using the equivalent conductance data in Table 16.2, and the concentrations of K^+ and Na^+ in the sample, the actual ratio in the injected aliquot will be

$$\frac{[K^+]}{[Na^+]} = \frac{73.5 \times 4.2}{50.1 \times 140} = 4.4 \times 10^{-2}$$

Note that this is significantly higher than the $[K^+]/[Na^+]$ ratio of 3.0×10^{-2} in the original sample, due to differences in ion mobilities. If determining potassium is the objective, this will ease the determination.

Electrokinetic injection can be advantageous as noted above. Extended electrokinetic injection can also be used to actually concentrate sample ions via such injection relative to the sample itself, but the problem of bias (different ions injected to different degrees) remains. Indeed, the exact amount of an ion injected depends also on the sample conductivity. Consider that the path between the electrode in the sample reservoir and the capillary tip and the BGE-filled capillary are electrically in series. The field strength across the sample solution path is thus larger when its conductivity is smaller. The numbers of a given ion introduced per unit time is dependent on the electrical field strength across this path. The exact amount of an analyte ion introduced is not only proportional to the analyte ion mobility and its concentration, but also the overall conductivity of the solution. All of these factors can complicate quantitation in electrokinetic injection. If the same sample type (i.e., when the matrix remains more or less the same) is repeatedly analyzed, and calibration is conducted by varying the analyte concentration in such a matrix, electrokinetic sample injection can be used. It can also be used if standard addition is used for sample injection. It has been shown that if a small finite sample volume is used for electrokinetic injection, it is possible to essentially exhaustively electromigrate the sample ions to reduce mobility based bias (P. K. Dasgupta and K. Surowiec, *Anal. Chem.* 68 (1996) 4291), but the strategy has not been widely adopted.

In IC, a significant volume of a sample can be preconcentrated on a preconcentrator column prior to analysis. A large volume of sample introduced into the capillary by a pressure-based injection does not accomplish the same ends. The separated band merely gets broader unless the sample conductivity is significantly less than the BGE conductivity. Under these conditions **electrostacking** occurs; this phenomenon makes it particularly facile to perform large volume injections of low conductivity samples in CE.

Consider that in a 50 cm capillary length, a 1 cm length of a very dilute sample solution (with conductivity much less than the BGE) is hydrodynamically introduced. We also assume that positive ions are of interest to us and +HV, as in Figure 16.28, is applied. After sample introduction, whether pressure or electromigration based, the capillary is lifted from the sample vial and transferred back into the source BGE vial. Then voltage is applied to begin electrophoresis. In the present case, the electric field is not uniform along the capillary. The resistance per unit length in the sample zone is much greater in the dilute sample zone than the rest of the capillary filled with the BGE. Consequently, the field strength in this region is proportionally higher (if the current flowing through the entire system is i, the voltage drop across any length of resistance R is iR). The ions in this region are therefore moving forward much faster than ions from the BGE are going to move into what was originally the sample zone. Sample cations thus run up to the boundary where the BGE section begins and "stack" against this boundary (electroneutrality is maintained by BGE anions moving in the opposite direction). This process continues until the conductivity of the "stacked" sample zone becomes comparable to that of the BGE and further stacking does not occur. Electrostacking is vitally important in achieving attractive sensitivity and sharp peaks in CE.

The fact that electrostacking does not occur unless the sample is of significantly lower conductivity than the BGE will suggest that the BGE should be as concentrated as possible. However, as the BGE concentration and thus its conductance is increased, more current flows. When the applied voltage is constant, the power dissipated (and thence the temperature rise) linearly increases with conductance. The temperature increase is not radially uniform but is maximum at the center. This compromises the separation efficiency. Thus, the upper limit of the BGE concentration is dictated by Joule heating. Except for ampholytes, zwitterions, and poorly ionized secies, BGE concentrations rarely exceed 10–20 mM in 50–75 μm bore capillaries, although higher concentrations can be used in smaller capillaries. There are some reports of operating with a BGE of concentration as high as 1.5 M NaCl [see e.g., J. S. Fritz, *J. Chromatogr. A* **884** (2000) 261], but this is rarely possible. Going to smaller capillaries may allow greater BGE concentrations but what is gained in better electrostacking is generally lost in difficulties in detection.

Detectors in CE

Indirect photometric detection was first proposed for ion chromatography. Read the classic paper: H. Small and T. E. Miller, Jr. *Anal. Chem.* **54** (1982) 462.

The detectors used in the capillary scale for LC and for CE are largely the same and have already been discussed. UV–Vis absorbance detection has remained the mainstay. The use of indirect photometric detection is much more common in CE than in ion chromatography. It is the exact optical absorbance analog of conductivity based non-suppressed ion chromatography. The BGE contains a strongly absorbing ion (often in the visible) of the same charge sign as the analyte ions. Accompanying analyte elution, there is a concomitant decrease in the BGE ion concentration, and the difference in absorptivity (on an equivalents basis) between the analyte ion and the BGE ion then gives rise to the signal, which appears as negative excursions from the baseline. However, the signal is often inverted to correspond to the appearance of other **electropherograms** obtained by direct detection. Figure 16.30 illustrates the separation of 27 different metal ions in ~6 min by indirect detection using methylbenzylammonium ion as the BGE cation that is monitored. This electropherogram illustrates both the utility of indirect photometric detection and the great separation power of CE.

For compounds that have native fluorescence or when fluorescent derivatives are made, fluorescence detection is very attractive. The use of focused laser sources to perform LIF detection has pushed detection limits to the zeptomol (10^{-21} mol) level. With liquid core waveguide capillaries or microchips, radial excitation by LEDs and end-column detection with very simple and inexpensive setup (no focusing optics), low- to sub-femtomol detection of amino acids is readily possible [Dasgupta et al. *Anal. Chem.* **71** (1999) 1400; Wang et al., *Anal. Chem.* **73** (2001) 4545].

Galvanic conductivity detection can be used in CE. Even suppressed conductivity detection has been demonstrated. The separation capillary is connected to the detection

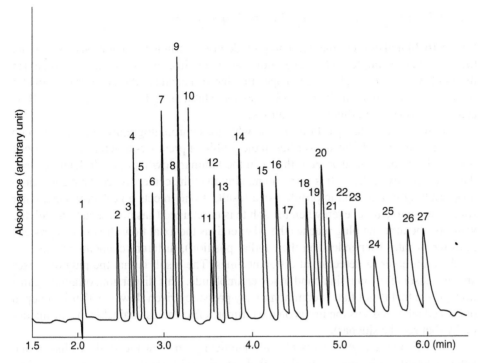

FIGURE 16.30 Separation of metal ions by capillary electrophoresis with complexing electrolyte. 1: K+; 2: Ba2+; 3: Sr2+; 4: Na+; 5: Ca2+; 6: Mg2+; 7: Mn2+; 8: Cd2+; 9: Li+; 10: Co2+; 11: Pb2+; 12: Ni2+; 13: Zn2+; 14: La3+; 15: Ho3+; 16: Pr3+; 17: Nd3+; 18: Sm3+; 19: Gd3+; 20: Cu2+; 21: Tb3+; 22: Tb3+; 23: Ho3+; 24: Er3+; 25: Tm3+; 26: Yb3+; 27: Lu3+. BGE: 8 mM 4 methylbenzylamine, 15 mM lactic acid, 5% methanol, pH 4.25. Applied voltage: 30 kV. *J. Chromatogr. A*, **640**, Y. Shi and J. S. Fritz, p. 473–479. Copyright 1993, with permission from Elsevier.

capillary (containing electrodes in contact with the liquid), by an ion exchange membrane capillary acting as the suppressor. The membrane is immersed in the regenerant solution; this is grounded to isolate the detector from the HV (Reference 24). A similar configuration is used for in-capillary electrochemical detection (see below). However, contactless conductivity detection is more common and typically displays LODs of the order of 0.3 μM. With a 10 nL injection, this also connotes a low fmol LOD.

As previously noted, amperometric detection can also be attractive in CE. Previously, end-column detection with very small carbon fiber working electrodes (WEs) was discussed; detection can also be conducted by inserting the fiber within the capillary. However, some means of grounding the field prior to the sensing electrode is essential, otherwise it will be impossible to detect very low current levels in the presence of a large electric field. Such grounding has been achieved by connecting the separation capillary to a small detection capillary segment containing the WE by (a) a porous glass joint or (b) an ionically conductive ion exchange membrane tubing and placing the HV ground outside the porous glass sleeve or ion exchange tube while liquid continues to flow from the separation to the detection capillary. Other grounding methods have also been used.

Electrospray ionization mass spectrometry (see Chapter 17) is a natural complement to CE. The HV ground is usually placed at the mass spectrometer inlet. For a typical electrospray arrangement, a double annular geometry (tube in tube in tube: CE capillary at the center, additional sheath liquid outside it and nebulizing gas outside that) is used to create a drop of charged droplets that are sampled by the MS inlet. Coupling to ICP-MS or ICP-OES instruments also uses a similar arrangement.

With advent of "nanospray" devices that can operate with sub-μL/min liquid flow, the need for auxiliary liquid and gas flow is obviated. Typically, a second power supply is used. This and the CE HVPS share a common ground that is connected to the MS inlet. The CE capillary terminates in the nanospray needle; this is connected to the second power supply that provides the electrospray voltage. The CE separation voltage is then the difference potential between the CE HVPS and that provided by the second power supply.

How CE can Provide Highly Efficient Separations

Will chromatography be ever as efficient as CE?

In very small dimensions if a fluid does not have any attractive interaction with the walls, the boundary layer velocity is <u>not</u> zero, even in pressure driven flow. Professor Mary Wirth (Purdue University) and her students have recently shown that under these conditions, **slip flow** can occur (B. Wei, B. J. Rogers, M. J. Wirth, *J. Am. Chem.* **134** (2012) 10780; see also http://www.chromatography online.com/lcgc/article/article Detail.jsp?id=790681) The velocity profile is no longer parabolic. Using a packed bed of sub-200 nm crystalline colloidal silica to effectively create very small channels, they were able to get plate heights as small as 15 nm(!) and over a million plates in a very short column. Indeed, number of plates is independent of the column length. Compare with CE: The efficiency does not depend on the length of the capillary, only the voltage!

Figure 16.30 makes clear that CE can provide highly efficient separations, and in fact there are many examples of CE separations far greater in efficiency than that depicted in this book. Chromatography, even in open tubular format and in the same inner diameter or even smaller bore than CE capillaries, do not exhibit as high efficiencies as CE. How can CE provide such efficient separations?

In reality, a better question may be why does chromatography not provide such efficient separations? The answer lies in the profile of pressure-driven flow (in an open bore capillary of small diameter, this is in the laminar flow regime). In laminar flow, the fluid velocity at the wall is zero and maximum at the center (twice the mean velocity), resulting in a parabolic profile. As a result, if one injects a perfectly rectangular solute plug into a tube, even when the solute is unretained (has no interaction with the wall), by the time the plug reaches the other end, as the axial flow streamlines proceed at different rates (dependent on their radial position), different elements of the plug reach an observer (a detector) at different times. The result is that the plug disperses and one observes a Gaussian band of lower amplitude. This dispersion, originating in a flow profile that is not flat, is common to any pressure driven flow, whether in an open or packed tube, and sets the limits on the ultimate efficiencies attainable in chromatography (but see the sidebar).

In CE, the analyte moves electrophoretically and via the EOF. Unlike radial velocity profile in laminar flow, the radial electric field profile is flat—the ions electrophoretically move much like a plug.

To discuss the velocity profile of EOF, we need to first understand how it originates. Helmholtz carried out experiments in the 1800s with the passage of current through a horizontally placed glass tube filled with a salt solution, and first observed a net flow of the fluid from the positive to the negative electrode. From the observed current and Faraday's laws, it is easy to compute the amount of water that will be associated with the transport of the cations (recall that cations are far more extensively hydrated than anions) but the overall bulk water transport is much greater than that. Where then does this flow originate? Similar to glass tubes, in fused silica capillaries there are ionizable $-SiOH$ groups on the surface (analogous conditions exist on most polymeric tubes as well: these inevitably have $-OH$ or $-COOH$ groups on the surface from surface oxidation). At all but very acidic pH, the $-SiOH$ groups are ionized to SiO^-. The stationary negative charge, measured by the **zeta potential** of the surface, has to be compensated. This role is played by the BGE counter ion. If, for example, a borate buffer (consisting of Na^+ and $B(OH)_4^-$ ions) is used as BGE, the Na^+ ions form a static layer around the wall to provide electroneutrality, this is variously called the **double layer** or the **Stern layer**. The locus of the center of the solvated ions is called the **outer Helmholtz plane**. The remaining Na^+ is present as a diffuse layer, their concentration approaching the bulk concentration as we move out to the center of the capillary. The anions are present throughout the diffuse layer and the bulk solution and provide electroneutrality. Note that the cations closest to the Stern layer are the least shielded by negative ions. They effectively experience the greatest force when an electric field is applied. Thus, when a voltage is applied, the cations in the rigid Stern layer do not move. But the ones in the diffuse layer and those in the bulk solution are free to move. They move towards the negative electrode. The ions are extensively solvated, and the water of solvation of them is also hydrogen bonded to each other. In a small capillary, as movement begins in the diffuse layer, the associated solvent front, held together by viscous coupling, moves together as a front. To a limited degree, the more tightly water is bound to the cation, the greater is the EOF. EOF is quantitatively reported in terms of the electroosmotic mobility, μ_{eo}, the electroosmotic velocity (cm/s) per unit field strength (V/cm), with a unit of $cm^2\ V^{-1}\ s^{-1}$. Reference 24 reports that 2 mM K, Na and Li tetraborate

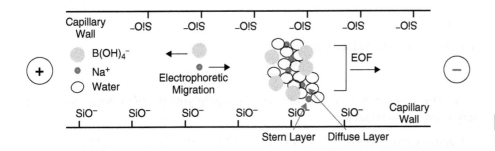

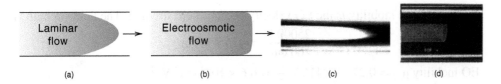

FIGURE 16.31 Electroosmotic flow origin, schematically shown.

FIGURE 16.32 Laminar flow (a, c) vs. Electroosmotic flow (b, d). (a, b) conceptual, (c, d) experimental. (c) Pressure-driven flow profile in a 500 μm wide 100 μm tall glass microchannel. (Courtesy P. K. Dasgupta and K. Surowiec, The University of Texas at Arlington and Texas Tech University), (d) EOF profile in a 75 μm diameter fused silica capillary (Reprinted with permission from Fujimoto et al., *Anal. Chem.* **68** (1996) 2753. Copyright American Chemical Society).

electrolytes respectively show μ_{eo} values of 9.0, 9.1, and 9.5 × (10^{-4}) cm^2 V^{-1} s^{-1} in a bare silica capillary. This situation is depicted as cartoon in Figure 16.31. Figure 16.32 shows the flow profiles in laminar flow and EOF, both as a depiction and as observed experimentally.

The direction and magnitude of the EOF is dependent on the sign and the magnitude of the zeta potential. In silica and polymeric capillaries at neutral to alkaline pH, the surface has a net negative charge (zeta potential negative) and as indicated in Figure 16.31, EOF proceeds from the positive to the negative electrode. At modestly alkaline pH (e.g., pH 9) and with a relatively low ionic strength BGE, a fused silica surface produces among the highest EOF. For a given BGE and a given capillary surface, this quantity is independent of the capillary inner diameter (for $d \leq 100$ μm), but it is dependent on the ionic strength of the BGE and decreases with increasing ionic strength. For a 10 mM Na$_2$B$_4$O$_7$ (pH 9) BGE, fused silica capillaries typically exhibit a μ_{eo} value of 6–7 × 10^{-4} cm^2 V^{-1}s^{-1}, but can often be higher with surface pretreatment. By convention, a positive value of μ_{eo} indicates flow from the positive to the negative electrode; a negative value will indicate that flow proceeds from the negative to the positive electrode.

Example 16.5

Acetone, a UV absorbing uncharged solute, is often used as a "neutral marker" to measure EOF. A 75 μm bore silica capillary is used with a 10 mM Na$_2$B$_4$O$_7$ (pH 9) BGE. A small amount of 10% acetone in the BGE is hydrodynamically injected. The capillary tip is then returned to the BGE vial and a voltage of +25 kV is applied across the capillary of length 60 cm. The active length of the capillary (distance from the injection end to the detector) is 50 cm. If the migration time (the time from the application of the voltage to the peak registered by the detector) is 181.5 s, what is the EOF, the electroosmotic velocity and the electroosmotic mobility?

When the EOF is very small (or even near zero), approaches as in Example 16.5 are not practical to determine the precise value of the EOF. See the review by J. Horvath and V. Dolnik [*Electrophoresis* **22** (2001) 644] on polymer coatings used in CE capillaries on different types of polymer coatings and techniques to determine EOF. Huhn et al. [*Anal. Bioanal. Chem.* **396** (2010) 297] have also reviewed CE coatings with particular attention to CE-MS.

Solution

The electroosmotic velocity is

$$\frac{50 \text{ cm}}{181.5 \text{ s}} = 0.275 \text{ cm/s}$$

EOF is the volumetric flow rate, obtained by multiplying the EO velocity by the capillary cross-sectional area:

$$\text{Capillary cross section} = \pi d^2/4 = 3.14 \times (7.5 \times 10^{-3})^2/4 \text{ cm}^2 = 4.42 \times 10^{-5} \text{ cm}^2$$

$$\text{EOF} = 0.275 \text{ cm/s} * 4.42 \times 10^{-5} \text{ cm}^2 = 1.22 \times 10^{-5} \text{ cm}^3/\text{s} = 12.2 \text{ nL/s}$$

Electroosmotic mobility is the EO velocity per unit field strength. The field strength is

$$\frac{25000 \text{ V}}{60 \text{ cm}} = 416.7 \text{ V/cm}$$

EO mobility $\mu_{eo} = 0.275 \frac{\text{cm}}{\text{s}}/416.7 \frac{\text{V}}{\text{cm}} = 6.6 \times 10^{-4} \text{ cm}^2 \text{ V}^{-1} \text{ s}^{-1}$

The EOF may be too fast or too slow or too pH dependent for an optimum separation; in many cases it may be best to have no EOF at all. A capillary can be coated with a sulfonated polymer to have a pH-independent EOF (flow from + to − electrode), with the absolute value of the EOF being controlled by the degree of sulfonation of the polymer. A capillary can be coated with an immobilized neutral material like polyvinyl alcohol or polyacrylamide to obtain a near zero zeta potential and hence a zero EOF. A capillary can be coated with a cationic polymer like polyethyleneimine or polybrene to have a positive zeta potential and hence reversed EOF (the flow is from the − to the + electrode). Capillary wall coatings can also be dynamically applied, for example, incorporating a cationic surfactant in the BGE can lower or even reverse the EOF, depending on the surfactant concentration.

It is commonly misstated that CE displays high separation efficiency because the EOF profile is plug-like. In reality, some of the most efficient CE separations are conducted with coated capillaries that have zero EOF. CE provides efficient separations because the electric field is radially uniform; this causes the electrophoretic movement also to be plug-like. The role that EOF plays is that being also plug-like, it does not deteriorate this separation.

Another elegant concept is the control of the zeta potential on a dynamic basis by applying an independent potential on the capillary wall from the outside [K. Ghowsi and R. J. Gale, *J. Chromatogr.* **559** (1991) **95**; Wu et al., *Anal. Chem.* **64** (1992) 2310]. This has not seen much use.

Note that the plug-like profile of EOF is only maintained below about a diameter of 100 μm. By the time a diameter of 200 μm is reached, the flat profile is no longer maintained, and the viscous coupling fails to keep the velocity at the center the same as that at the wall; quite opposite to laminar flow, the velocity at the center is less than that near the walls. At even larger diameters, the flow generated at the wall essentially recirculates into the center; there is very little overall flow.

Joule heating. The power dissipated in the capillary manifests itself as heat. We had previously discussed that while high BGE concentrations are preferred for electrostacking, the increased heat generated with increased BGE concentrations sets the upper limit of the usable BGE concentration. As the diameter increases, with the same BGE, the conductance increases with the square of the diameter. At the same field strength, the power dissipated and the heat generated thus increases with the square of the diameter. On the other hand, whether the capillary is externally air cooled or liquid cooled or simply remains in ambient air, the heat is dissipated through the outer surface. A high surface area-to-volume ratio, which increases with decreasing inner diameter of the capillary, is therefore preferred. On the other hand, in most cases detection limits get poorer as the capillary diameter is reduced, so this cannot be indefinitely reduced. Most CE experiments are done with fused silica capillaries 25–75 μm in inner diameter. Fused silica is not only optically transparent down to at least 190 nm, it is also an excellent thermal conductor that helps dissipate the heat.

In understanding the consequences of Joule heating, recognize that the main problems come from not just an increase in temperature but a radially (and longitudinally) *nonuniform* rise in temperature. Because the capillary dissipates the heat on the outer surface, the center of the fluid is the hottest and there is a radial temperature gradient that is proportional to the total power dissipated. It is also proportional to the square of the capillary inner diameter and inversely as the thermal conductivity of the fluid. Temperature affects viscosity and changes both electrophoretic mobility and EOF. The electrophoretic mobilities of common ions typically change about 1.7%/°C. If the ions or the bulk fluid is moving faster at the center of the capillary than at the walls, we approach the same situation that we have in laminar flow. In addition, the fluid obviously leaves hotter than it enters. So there must be an axial temperature gradient as well. While there is only one bulk flow that can be observed for the whole capillary, if the generated EOF is less where the capillary is cooler in the entrance region and higher towards the hotter exit end, obviously the EOF cannot have the same flat profile throughout.

Electrophoretic Mobility and Separation

Unlike an uncharged analyte, a charged species is not only carried by the bulk EOF, it is also moves electrophoretically. The electrophoretic mobility, μ_{ep}, can be calculated from the equivalent conductance λ from Equation 16.13; the λ values are listed for common ions in Table 16.2 (the table actually lists the limiting equivalent conductance (λ^0) values, but most CE experiments are conducted at sufficiently low BGE concentrations where significant error is not incurred in assuming $\lambda \simeq \lambda^0$). The sole determinant of whether two different charged species can be separated depends solely on the difference in their electrophoretic mobilities, whereas the applied voltage and the distance from injection to detection determine the time and efficiency for such a separation. It is important to note that because EOF is bidirectional (it is a vector quantity), the net mobility, μ_{net}, a charged species exhibits, will be the vector sum of the electroosmotic and the electrophoretic mobilities.

$$\mu_{net} = \mu_{ep} + \mu_{eo} \tag{16.24}$$

The migration time, t_m, is simply given by the length to traverse divided by the net velocity:

$$t_m = L/(\mu_{net} \times E) = L^2/(\mu_{net} \times V) \tag{16.25}$$

Equation 16.25 assumes that the distance over which the voltage is applied and the migration distance are approximately the same. If not, it is understood that L in the first right-hand term implies L_m, the migration distance, while E is V/L_E where L_E is the capillary length between the electrodes. If $L_m \neq L_E$, then L^2 in Equation 16.25 should be replaced by $L_m L_E$.

Example 16.6

In a suppressed conductometric CE system (Reference 24), the authors observe an average μ_{eo} value of 9.1 cm^2 V^{-1} s^{-1} with a 2 mM Na$_2$B$_4$O$_7$ electrolyte of pH 9. With +20 kV applied across 60 cm, which in this particular system is also approximately the distance to the detector, predict the migration times of (a) benzoate, (b) iodate, (c) sulfamate, (d) fluoride, (e) sulfate, and (f) bromide. They were observed to elute in this order. Also predict the migration time of phosphate.

Solution

The values of λ^0 listed in Table 16.2 for the six anions and $HPO_4{}^{2-}$ and $PO_4{}^{3-}$ are, respectively, in S cm^2 equiv^{-1}:32.4, 40.5, 48.6, 54.4, 80, 78.1, 57, and 69. We compute, using Equation 16.13, for benzoate for example (for anions note the −ve sign in μ_{ep}),

$$\mu_{ep,Bz-} = \lambda_{Bz-}/(zF) = \frac{32.4 \text{ S cm}^2\text{equiv}^{-1}}{96485 \text{ coulombs equiv}^{-1}}$$

Noting the equality S/coulombs = (ampere/volt)/(ampere*s)

$$\mu_{ep,Bz-} = -3.36 \times 10^{-4} \text{ cm}^2 \text{ V}^{-1} \text{ s}^{-1}$$

For iodate, sulfamate, fluoride, sulfate, and bromide, we similarly calculate respective μ_{ep} values of $-(4.20, 5.04, 5.64, 8.29, \text{ and } 8.9) \times 10^{-4}$ cm^2 V^{-1} s^{-1}.

At pH 9 phosphate is present primarily as $HPO_4{}^{2-}$ with a trace of $PO_4{}^{3-}$. The K_3 value is 4.8×10^{-13}, thus $\alpha_{HPO_4{}^{2-}} = [H^+]/(K_3 + [H^+]) = 0.9995$. The weighted average λ for phosphate at pH 9 is thus $57 * 0.9995 + 69 * 0.0005 = 57.01$. The corresponding μ_{ep} value is 5.91×10^{-4} cm^2 V^{-1} s^{-1}.

Using Equation 16.24, we thus calculate for benzoate:

$$\mu_{net,Bz-} = (9.1 - 3.4) \times 10^{-4} \text{ cm}^2 \text{ V}^{-1} \text{ s}^{-1} = 5.7 \times 10^{-4} \text{ cm}^2 \text{ V}^{-1} \text{ s}^{-1}$$

From Equation 16.25,

$$t_{m,Bz-} = (60)^2/(5.7 \times 10^{-4} \times 2 \times 10^4) \text{ s} = 316 \text{ s}$$

The migration time of all of the other anions can be similarly computed, and they actually elute (as reported in Reference 24) in the predicted order. The one exception is that bromide is predicted to be eluted before sulfate, but the reverse is observed. In addition, the predicted migration times are close to those observed for the early eluting ions, but the late eluting ions elute much sooner than the predicted values. Consider that the uncertainty in μ_{net} increases as the magnitude of μ_{ep} increases. If μ_{ep} is smaller than those calculated from limiting equivalent conductance values, observed t_m will be smaller than the calculated values. In addition, this discrepancy is likely to be higher for doubly charged ions like $SO_4{}^{2-}$ or $HPO_4{}^{2-}$.

Example 16.7

The authors of Reference 24 prepared a capillary with positively charged walls by coating the walls with a composite polymer of polyvinyl alcohol, phosphoric acid, and poly(diallyldimethylammonium) chloride. With −20 kV applied across the active distance of 60 cm, they observed the migration time for $NO_3{}^-$ (λ^0 listed as 71.4 S cm^2 equiv^{-1}) to be 150 s. Estimate μ_{eo}:

Solution

Based on Equation 16.13, $\mu_{ep,NO_3{}^-} = -7.40 \times 10^{-4}$ cm^2 V^{-1} s^{-1}
According to Equation 16.25,

$$\mu_{net,NO_3{}^-} = (60 \text{ cm})^2/(150 \text{ s} \times -2 \times 10^4\text{V}) = -1.2 \times 10^{-3} \text{ cm}^2 \text{ V}^{-1} \text{ s}^{-1}$$

Note that the negative sign in the net mobility indicates that the overall movement is taking place towards the positive electrode.

$$\mu_{eo} = \mu_{net,NO_3{}^-} - \mu_{ep,NO_3{}^-} = \left(-1.2 \times 10^{-3} + 7.4 \times 10^{-4}\right) \text{ cm}^2 \text{ V}^{-1} \text{ s}^{-1}$$
$$= -4.6 \times 10^{-4} \text{ cm}^2 \text{ V}^{-1} \text{ s}^{-1}$$

Note that the minus sign indicates that flow is towards the positive electrode; because of the positive charge on the wall, the flow is towards the positive electrode.

Plate Numbers and Resolution in CE

Recall that the dispersion of a band can be expressed in terms of its variance σ, (see Equation 14.6). In CE, none of the chromatographic reasons for dispersion apply, except for diffusion. While Equation 14.6 expresses the variance in terms of time, it can also be written in terms of the length the analyte traverses. In this case, Equation 14.6 will be written in length terms as

$$N = \left(\frac{L}{\sigma_{\text{length}}}\right)^2$$

The variance σ_{length} can be expressed in terms of the characteristic distance any species with a diffusion coefficient D diffuses in the migration time t_m as:

$$\sigma_{\text{length}} = \sqrt{(2Dt_m)} \tag{16.26}$$

Or, after incorporating the value of t_m from Equation 16.25,

$$\sigma_{\text{length}}^2 = 2Dt_m = 2DL^2/(\mu_{net} \times V) \tag{16.27}$$

Putting the value of σ_{length} in Equation 16.26, we obtain the expression for N:

$$N = (\mu_{net} \times V)/2D \tag{16.28}$$

Remarkably, the efficiency in CE depends only on the applied voltage and the net mobility of the analyte. The latter is reciprocally related to the time it takes for the analyte to migrate and longer the time, the more dispersed the band and the lower is the efficiency. Note that both μ_{net} and D are dependent on the analyte. As such, in CE the efficiency is analyte dependent.

Example 16.8

Consider the capillary of Example 16.6, where +20 kV is applied across a 60 cm capillary and the 2 mM $Na_2B_4O_7$ BGE gives a μ_{eo} of 9.1 cm^2 V^{-1} s^{-1}. Calculate the number of plates for a small ion, Ag$^+$ and for a 100 kD protein with $D = 3 \times 10^{-7}$ cm^2 s^{-1}. Assume that the BGE pH is very close to the isoelectric point of the protein and hence μ_{ep} for the protein is near zero.

Solution

According to Example 16.2, μ_{ep} for Ag$^+$ is 6.4×10^{-4} cm^2 V^{-1} s^{-1} (calculable also from limiting equivalence data in Table 16.2) from which D was computed to be 1.64×10^{-5} cm^2 s^{-1}.

Therefore, $\mu_{net} = \mu_{eo} + \mu_{ep} = (9.1 + 6.4) \times 10^{-4}$ cm^2 V^{-1} s$^{-1} = 1.55 \times 10^{-3}$ cm^2 V^{-1} s^{-1}

$$N = \frac{(1.55 \times 10^{-3} \text{ cm}^2 \text{ V}^{-1}\text{s}^{-1}) * 20000 \text{ } V}{2(1.64 \times 10^{-5} \text{ cm}^2 \text{ s}^{-1})} = 945{,}000 \text{ plates}$$

For the protein:

$$N = \frac{(9.1 \times 10^{-4} \text{ cm}^2 \text{ V}^{-1} \text{ s}^{-1}) * 20000 \ V}{2(3.0 \times 10^{-7} \text{ cm}^2 \text{ s}^{-1})} = 30,000,00 \text{ plates}$$

Both plate counts above are very high, and the number of plates for the protein is over 30-fold greater! There are many practical factors that limit plate counts: operating with very high μ_{net} values may provide very high efficiencies for a given peak, but as we will shortly observe, does not lead to a good separation. In addition, the injected sample has a finite width, some heating in the capillary is unavoidable, and the detector interrogates a zone of finite width. For small ions, plate counts of 100,000 would be considered very good; plate counts over a million are not uncommon in protein separations.

Similar to chromatographic resolution (see Equations 14.31 and 14.32), resolution R_s between two analytes 1 and 2 in CE is given by:

$$R_s = \frac{1}{4} (\Delta\mu_{net}/\mu_{net,Av}) \sqrt{N} \tag{16.29}$$

Noting that $\Delta\mu_{net} = \Delta\mu_{ep}$ and incorporating the expression of N from Equation 16.28,

$$R_s = 0.177 \, \Delta\mu_{ep} \sqrt{\frac{V}{\mu_{net,Av} D_{Av}}} \tag{16.30}$$

Equation 16.30 indicates that increasing the difference between the electrophoretic mobilities of the two analytes, whether by pH, BGE composition, or chemical derivatization, is much more effective to increase the resolution than increasing the voltage. The voltage has to be quadrupled to double the resolution. Increasing μ_{net} actually decreases the resolution. It is also implicit in Equation 16.30 that if EOF can be exactly controlled with high resolution, a pair of analytes with very close μ_{ep} values can still be resolved. Consider that we have two analytes with close μ_{ep} values, both are electrophoretically moving towards the detector. If we have opposing EOF and can adjust it such that it exactly balances the electrophoretic movement of the slower moving analyte, μ_{net} for this analyte will become zero and it will not move at all. The other one, slow as its net movement may be, it will eventually reach the detector.

Equations 16.28–16.30 show no dependence on capillary length and diameter. However, lower limit on the length and an upper limit on the diameter are set by Joule heating considerations. A short capillary length promotes faster analysis, while a larger capillary bore facilitates lower LODs. Practical analysis therefore requires a compromise. It is facile to carry out electrophoretic analysis on a microchip, and many different techniques have been developed for appropriate sample introduction. Figure 16.33 shows a chip-scale electropherogram.

In the late 1980s to early 1990s, CE was considered to have enormous promise for small ion analysis. Many felt that it will displace ion chromatography in such applications. Indeed, many, including major vendors of IC instruments, introduced CE instruments for such purposes. None remain in the market today. CE did not provide the robustness, the reproducibility, the ability to perform true trace analysis in real samples (especially in the presence of an overwhelming amount of another component) to be ultimately successful. EOF strongly influences CE separations; it is generated by the surface of the capillary. Any substance that adsorbs on the wall will change the EOF. This is a major issue with many real-world samples. However, there may be niche applications where CE, especially in the microchip format, may be successfully applied. This would be particularly true when the samples are very dilute (such that the ionic constituents can readily undergo electrostacking) and also when a surfactant can

See how gated injection in a microchip works by manipulating applied voltages: http://www.youtube.com/watch?v=PYbXEoxKkV4.

Small cross-section short-length channels on a chip can provide extremely fast separations. For a video demonstration of the separation of two compounds that takes less time than it takes you to read this sentence, see http://www.youtube.com/watch?v=famss0gHwM0.

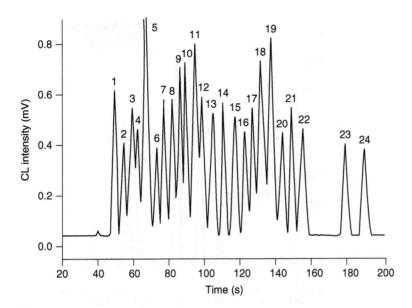

FIGURE 16.33 Microchip separation and chemiluminescence detection after labeling with N-(4-aminobutyl)-N-ethylisoluminol (ABEI). The BGE is 15 mM pH 9.8 borate buffer containing 1.0 mM Co^{2+} as catalyst, 1.0 mM adenine as enhancer and 35 mM SDS. Separation takes place in a channel 65 µm wide, 25 µm deep and ~6 cm long under a field gradient of ~350 V/cm, obtained from the separation of a mixture containing dopamine (DA), noradrenaline (NE), γ-amino n-butyric acid (GABA), Agmatine (Agm), Octopamine(Oct), Carnosine (Car), Homocarmosine (Hcar), Anserine (Ans), and 20 common amino acids. The concentration of each analyte was 1 µM. Peak identification: 1, ABEI; 2, Agmatine + Octopamine; 3, Arginine; 4, Lysine; 5, excess ABEI; 6, Noradrenaline; 7, Dopamine + Valine; 8, Proline; 9, Asparagine; 10, Glutamine + Methionine; 11, Leucine + Isoleucine; 12, Histidine; 13, Carnosine; 14, Homocarnosine; 15, Anserine; 16, Phenylalanine; 17, γ-amino n-butyric acid; 18, Alanine + Tryptophan; 19, Threonine + Tyrosine; 20, Serine; 21, Cystine; 22, Glycine; 23, Glutamate; 24, Aspartate. *Journal of Chromatography A*, **1216**, S. Zhao, Y. Huang, M. Shi, and Y-M. Liu, Quantification of biogenic amines by microchip electrophoresis with chemiluminescence detection, pp. 5155–5159. Copyright 2009, with permission from Elsevier.

be used in the BGE such that it will be the substance that will most strongly adsorb to the wall and thus maintain a constant EOF. Figure 16.34 shows the electropherogram of a sample of power plant steam condensate (cooling water in power plants is maintained to a high degree of purity to inhibit corrosion of metal conduits) obtained on a microchip CE instrument.

16.11 Electrophoresis Related Techniques

Separation of Neutral Molecules: MEKC

In CE separation takes place only on the basis of differences in μ_{ep}. All neutral molecules have $\mu_{ep} = 0$; as such, while a charged species can be separated from a neutral molecule, neutral molecules cannot be separated from each other. Shigeru Terabe and students at the Himeji Institute of Technology in Japan realized that this situation can be altered if a charged surfactant (e.g., sodium dodecylsulfate) above its critical micelle concentration is the BGE or is present in the BGE. Now a micellar phase will be present. Such a micelle has a hydrophobic core in which neutral solutes can partition. The less hydrophilic the solute, the more it will prefer to partition to the interior of the micelle than remain in the bulk BGE. If we have traces of acetone and toluene in

Read the classic paper on MEKC: S. Terabe, K. Otsuka, K. Ichikawa, A. Tsuchiya and T. Ando, *Anal. Chem.* **56** (1984) 111.

FIGURE 16.34 Electropherogram of a power plant steam condensate on a microchip. Electromigration injection for 1.5 s (−125 V/cm). The separation channel 50 × 50 μm, 7 cm long and separation field strength also −125 V/cm (negative HV is applied). The detection side (+) is grounded; $\mu_{eo} < \mu_{ep}$ for all ions, the net movement is against EOF. Detection is by galvanic conductivity. Courtesy www.advancedmicrolabs.com

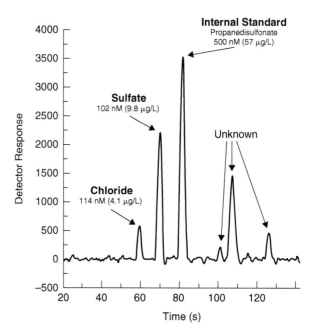

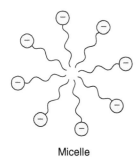

Micelle

The word **surfactant** is short for Surface Active Agent, so-called because a surfactant lowers the surface tension. A surfactant has a hydrocarbon tail, often 12 or more carbon atoms long and a polar head group which can be ionic. Detergents are typically anionic surfactants, bearing a sulfonate or sulfate head group, e.g., sodium dodecylsulfate. Conditioners applied to hair are typically cationic surfactants, most often bearing quaternary ammonium groups. Above a certain concentration, called the **critical micelle concentration** (CMC), the surfactant molecules will self-aggregate, forming micelles in which the hydrophilic tails are turned inward, to give a nonpolar core to which neutral solutes can partition. The polar head groups form an outer shell. A micelle from an anionic surfactant is shown above.

our sample, for example, the toluene molecules, being more hydrophobic, will partition more effectively into an SDS micelle interior than acetone.

An ionic micelle, as shown in the margin figure, is charged, it is electrophoretically moving as well as being transported by bulk EOF. In contrast, the neutral molecules outside the micellar core in the bulk BGE is moving only by the bulk EOF. In our illustrative case with SDS present in a borate buffer and +HV applied, μ_{eo} is positive and is greater in magnitude than the μ_{ep} (of negative sign) of the negatively charged micelle. Thus, the BGE moves towards the detector with a greater net velocity than the micelles do. Therefore, an analyte that partitions into the micelle to a greater extent arrives later at the detector.

The separation in this case occurs because of different partitioning behavior of different analytes into the micellar phase, i.e., the principle of this separation is chromatographic in nature even though movement may take place by electrically induced forces. This technique is thus aptly called **micellar electrokinetic chromatography** (MEKC); it is a form of chromatography in that the micelles act as a pseudo-stationary phase. Note that in the above example, the movement of cations is also affected because of electrostatic interactions with the charged micelles. MEKC is useful for the separation of water-insoluble neutral compounds; many small molecules as well as biological substances, e.g., steroids, fall in this class. There is no difference in hardware needed between CE and MEKC whether in standard or microchip formats. The separation in Figure 16.33 is actually being conducted by MEKC.

The retention time of an analyte depends on the retention factor k, which is given by the ratio of the total moles of the solute in the micelle (n_{MC}) to that in the bulk BGE (n_{BGE}), or

$$k = \frac{n_{MC}}{n_{BGE}} \tag{16.31}$$

The retention time t_R should appear in the range $t_0 < t_R < t_{MC}$, where t_0 is the observed migration time of a neutral marker that is not partitioned to the micelle and t_{MC} is the observed migration time of the micelle. The R value, the fraction of the solute in the aqueous phase, is given by

$$R = \frac{\mu_i + \mu_{MC}}{\mu_{eo} + \mu_{MC}} \tag{16.32}$$

where μ_i is the migration velocity of the analyte i, note that μ_{MC} is a vector quantity and in the negatively charged SDS micelle example we were considering, it will have a negative sign. R can obviously be expressed as

$$R = \frac{1}{1+k} \tag{16.33}$$

Recognizing that $\mu_{eo} = L/t_0$, $\mu_{MC} = L/t_{MC}$, and $\mu_i = L/t_R$, where L is the length from the injection end to the detection point, putting into Equation 16.32 and combining with Equation 16.33 gives

$$k = \frac{t_R - t_0}{t_0 \left(1 - \frac{t_R}{t_{MC}}\right)} \tag{16.34}$$

The term $(1 - (t_R/t_{MC}))$ comes from the retention behavior characteristic of electrokinetic separations. Recognize that when t_{MC} becomes infinite (i.e., the micellar pseudophase is stationary), Equation 16.34 reduces to the familiar equation for conventional chromatography (see Equation 14.12).

MEKC is arguably the most versatile form of electrodriven capillary separations. Chiral separations can be carried out with a chiral surfactant. The presence of a surfactant also typically prevents irreversible adsorption of analyte components on capillary walls and permanently alter EOF.

Capillary Electrochromatography

The features of capillary electrophoresis and HPLC can be combined to form a hybrid technique that has some of the features of each, called **capillary electrochromatography** (CEC), for the separation of neutral molecules as done in HPLC. Typically the capillary is packed with reversed-phase chromatographic stationary-phase particles, usually $d_p \leq 3$ μm particles, but other stationary phases are also used. A voltage is applied across the column; this generates electroosmotic flow. The flow through the column is thus driven by electroosmosis rather than pressure. However, it is not uncommon to have pressure applied by a pump as well (this is called pressure-assisted CEC or p-CEC). The general operation of CEC is similar to that in HPLC, but efficiencies are typically 2–3 times better.

The jury is still out whether CEC combines the best or worst features of CE and HPLC. While there is clearly an efficiency advantage, the flow is generated by the column bed and therefore any fouling of the bed (irreversible adsorption of a sample component) can permanently alter the flow rate. The flow velocity generated by electroosmosis alone is also often not as much as optimally desired; this is the reason that pressure assistance is used. Further, the EOF generated must be the same throughout the system in the electric field. Because of a greater surface area, if the exit frit generates a greater EOF than the column that precedes it, bubbles will form. To many, CEC is a solution looking for a problem.

Isotachophoresis and Capillary Isotachophoresis

Up until the work of Jorgenson and Lukacs, isotachophoresis (ITP) was the most widely used tubular electrophoretic technique, carried out in tubes of 0.25–0.50 mm inner diameter. Commercial instruments were available since the 1970s to carry out ITP. Today ITP is largely carried out in the capillary format. Capillary isotachophoresis (CITP) is carried out in a zero EOF capillary and for one type of ions (cations or anions) at a time. Let us consider the case for an anion ITP. First, an electrolyte, called the leading electrolyte, is chosen where the electrolyte anion has a mobility higher than

The CMC decreases with increasing carbon chain length; that for SDS is 8.2 mM. For a table of CMC values, see www.nist.gov/data/nsrds/NSRDS-NBS361.pdf.

Unimicro Technologies (www.unimicrotech.com) is one of the few vendors specializing in CEC. Visit this website for a listing of the CEC literature.

The Greek word *Tachos* means speed. Isotachophoresis thus depicts a situation where the different species of interest are all moving at the same speed. The principles of isotachophoresis were established by Kohlrausch in 1897: F. Kohlrausch, *Ann. Phys. Chem.* **62** (1897) 209.

FIGURE 16.35 Isotachophoretic separation. The ions stack according to their mobilities, the most mobile towards the leading electrolyte. The top panel is shown as the sample is first introduced; the bottom panel shows the situation after a steady state has been reached. The zone widths are related to the concentration of the analyte ions.

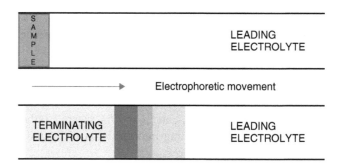

any of the anionic constituents in the sample. After filling the capillary with the leading electrolyte, the sample is injected. The + electrode end of the capillary is now put in a reservoir of the leading electrolyte and the—electrode end in a reservoir of the terminating electrolyte. The anion of the terminating electrolyte has lower mobility than any of the sample anions. Separation then occurs in the space between the leading and the terminating electrolytes.

The different anionic analytes stack up according to their mobilities. Stable zone boundaries form between the individual anionic constituents as shown in Figure 16.35.

The unique characteristic of isotachophoresis is that once the steady state is reached, all bands move at the same velocity. Since velocity is the product of the ionic mobility μ and the electric field E, the steady-state **isotachophoretic condition** is

$$\mu_L E_L = \mu_A E_A = \mu_B E_B = \mu_C E_C = \mu_T E_T \tag{16.35}$$

where L, A, B, C, T, respectively, indicate the leading electrolyte anion, the analyte anions A, B, and C, and the terminating electrolyte anion and E_i is the electric field in the zone containing ion i. In a series circuit, the same current is flowing through each of the zones; the electric field is therefore directly related to the resistance per unit zone length in any zone. Considering that conductance per unit length is approximately given by $\lambda_i C_i$, where C_i is the concentration of ion i in equivalents/L and assuming that the same counterion q is present throughout the capillary, the resistance is readily calculated as $1/C_i(\lambda_i + \lambda_q)$. Taking only the first equality in Equation 16.35, it follows that

$$C_i = \frac{C_L \mu_i (\lambda_L + \lambda_q)}{\mu_L (\lambda_i + \lambda_q)} \tag{16.36}$$

Note that C_i denotes the concentration of ion i in the separated zone, not in the original sample. The total number of ions i present in the original sample (related to its concentration in the sample) is proportional to the product of C_i and the zone length containing ion i. Thus the zone length is a measure of the concentration of ion i in the original sample, but it is also a function of its mobility and the counterion mobility as indicated in Equation 16.36. Figure 16.36 shows an isotachopherogram as observed with a resistance detector trace at the bottom and with a differential conductometer, top.

In ITP, the sharp focusing of the bands is self-correcting. Imagine that an ion from a particular band moves into the band of the next faster moving ion. Because it will now encounter a smaller electric field than when it was in its own zone, it will slow down and rejoin the original band. The obverse will occur if the ion slows down and falls into the zone originally behind it. ITP is increasingly being used as a preconcentration technique for CZE.

Because of the broad coverage and length of this chapter, we provide in the text's Solutions Manual guidances for searching for information in the chapter to answer questions and problems. Your instructor may give suggestions for assigned questions or problems. See also the Learning Objectives list for further guidance.

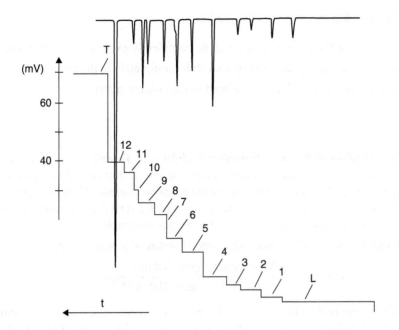

FIGURE 16.36 Isotachophoretic separation of anions in a 105 μm i.d. fluorocarbon capillary. L: 10 mM HCl titrated to pH 6 with histidine, T: 2 mM hydroxyethylcellulose, 5 mM MES. The bottom trace is in terms of increasing resistance; 1: chloride, 2: sulfate, 3: chlorate, 4: chromate, 5: malonate, 6: adipate, 7: benzoate, 8: unknown impurity, 9: acetate, 10: ß-bromopropionate, 11: naphthalene-2-sulfonate, 12: glutamate, 13: heptanoate. Adapted from *Journal of Chromatography A*, **267**, D. Kaniansky, M. Koval, and S. Stankoviansly, Simple cell for conductimetric detection in capillary isotachophoresis, pp. 67–73. Copyright 1983, with permission from Elsevier.

16.12 Gel Filtration Chromatography

This term gel filtration was coined by Porath and Floidin. Gel filtration chromatography separates the analyte based on their molecular size and by passing through porous beads. It was first reported in 1955 to separate proteins based on their size with the swollen gel of maize starch.

This technique utilizes various beads made up of cross-linked materials to form a 3-D mesh, that swells in the mobile phase to develop pores of different sizes. The pore size within the gel beads are controlled by the extent of cross linking.

Principle

The column is packed with the beads containing pores to allow entry of molecules on the basis of their sizes. The smaller size molecules are at the inner part and larger molecules are excluded from entering into the gel. The molecules are separated based on the travel time to come out from the pores, when the mobile phase is passed through the column, then the smaller molecules present in the inner part of the gel will take longer time. As a result, the larger molecules and smaller molecules get separated in this chromatography.

Classification of Analyses

The volume of mobile phase to elute a column is known as elution volume (V_e). The distribution coefficient is given by

$$K_d = \frac{V_e - V_0}{V_i} \tag{16.37}$$

where V_0 is the void volume, V_i is the pore volume and K_d is the distribution coefficient.

Based on K_d, analytes are classified as

1. $K_d = 0$ or $V_e = V_0$, these analytes will be completely excluded from the column.
2. $K_d = 1$ or $V_e = V_0 + V_i$, these analytes will be completely in the pores of the column.
3. $K_d > 1$, in this case analyte will adsorb to the column matrix.

Applications

1. **Determination of Native Molecular Weight of a Protein.** First a set of known molecular weight standard proteins are run to get the elution volume. In a separate run, we can get the elution volume for the unknown sample. Then, K_d can be calculated using Equation 16.37 and finally a plot of K_d versus log of molecular weight will provide the molecular weight of the unknown sample.

2. **Oligomeric Status of Protein**. Oligomeric status of a protein is given by

$$\text{Oligomeric method} = \frac{\text{Molecular weight (Gel filtration)}}{\text{Molecular weight (SDS-PAGE)}}$$

3. **Studying Protein Folding.** The different protein conformations during unfolding pathway has distinct hydrodynamic surface areas and it can be used to follow protein folding-unfolding stages with the gel filtration.

4. **Studying Protein-Ligand Interaction**

5. **Desalting**

Questions

Basis of HPLC

1. How does HPLC differ from GC, instrumentally and van Deemter behavior?
2. Draw a block diagram of a basic HPLC system and that of your dream system (with an unlimited budget!).
3. Describe the HPLC subclasses based (a) on the mode of separation, and (b) physical nature of the column.
4. There are two forms of HPLC – normal phase HPLC and reversed phase HPLC. In normal phase HPLC, the stationary phase is silica and the mobile phase is a non-polar solvent like hexane. What are the stationary and mobile phases in reversed phase HPLC?
5. HPLC is an advance on column chromatography where the solvent is forced through the column at a very high pressure of up to 400 atmospheres. This makes the whole process much faster, but also allows you to use a much smaller particle size in the column. Why is that an advantage?

Ion Exchange Chromatography

6. What are ion exchange resins? What is a must-have constituent in the eluent to elute an ion off an ion exchange resin?
7. Distinguish between gel-type and macroreticular ion exchange resins.
8. Distinguish between weak-acid vs. strong-acid and weak-base vs. strong-base resins.
9. An intimate mixture of an H^+-form strong-acid type cation exchanger and an OH^--form strong-base type anion exchanger is called a mixed-bed resin. If you slowly percolate some tap water through a bed of such resin and measure its conductivity, what will you expect the specific conductance of the effluent to be?
10. Describe the principle involved in the separation of proteins by ion exchange chromatography. Indicate the importance of:
    (a) pH of the buffer.
    (b) increasing salt concentration.

11. Why is it important to dialyze a protein sample obtained by ammonium sulfate precipitation before subjecting it to ion exchange chromatography?

HPLC Phases

12. What is a monolithic column? How does it differ from a standard packed bed in architecture and performance? Who invented monolithic columns? What other contributions to separation sciences did (s)he make?

13. The elution order of analytes with significant electrostatic interactions with a zwitterion-bonded HILIC stationary phase is often reversed depending on which way the zwitterion is bonded. Explain.

14. List the major types of chiral stationary phases. Describe in more detail the mechanism by which at least one of these differentiate between chiral isomers.

15. What stationary phases can withstand extremes of pH and temperature? How is this accomplished?

HPLC Equipment

16. What component(s) other than the pump(s) constitute a HPLC solvent delivery system?

17. Describe the two basic pump types used in HPLC.

18. What is meant by a low-pressure vs. high-pressure gradient system? What would be meant by a quaternary pump?

19. "An HPLC pump is a veritable jewel box." Explain.

20. Explain how a 6-port external loop injector works. What is an internal loop injector? How many external ports does it have?

Other Phases, Applications

21. In what order would the following compounds be eluted from a C-18 reverse phase column using acetonitrile:water as the eluting solvent? Benzene, acetone, and benzoic acid?

22. What are some commonly used nonpolar bonded phases for reversed-phase HPLC? What are some bonded phases for HILIC?

23. Why are silica particles endcapped in bonded reversed-phase particles?

24. Briefly describe the differences between microporous particles, perfusion particles, nonporous and superficially porous particles. What are their unique features/uses?

25. What is a guard column and why is it used? When is it difficult to use a guard column?

26. How do HPLC and UHPLC differ?

PROFESSOR'S FAVORITE QUESTIONS

Contributed by Professor Michael D. Morris, University of Michigan

27. Gradient elution is used in liquid chromatography to separate compounds of widely differing partition coefficients in relatively short times and with reasonably similar efficiencies. How and why does it work?

    Briefly explain the methodology used to achieve a similar result in gas chromatography.

28. What advantage do narrow-bore columns have in HPLC?

Size Exclusion Chromatography

29. Describe the principle of size exclusion chromatography. What is the exclusion limit?

30. Compare size exclusion chromatography and ion exclusion chromatography. What are the principal application areas of each?

31. Compare and contrast: micellar electrokinetic chromatography vs. size exclusion chromatography.

32. What features of protein structure allow their separation by:

    (a) Size exclusion ("gel-filtration") chromatography

    (b) Ion exchange chromatography

33. Describe the principle involved in the separation of proteins by size exclusion ("gel filtration") chromatography.

Detectors

34. What detectors are considered universal detectors in HPLC? What are their limitations?

35. What is the principal application area of refractive index, viscosity and light scattering detectors? Describe the operating principle of a refractive index detector.

36. How can the designation of an aerosol detector or a solution conductivity detector as a quasi-universal detector be justified?

37. Who established AC measurements of conductivity? What problems may be encountered with DC measurement of conductivity?

38. What are the generally desirable criteria for a HPLC detector?

39. You are pumping a liquid of known viscosity through a tube of known diameter and length at a certain flow rate. State the equation from which you will calculate the pressure needed. By what factor would the necessary pressure change if you reduce the diameter of the tube by a factor of 2?

40. How is it possible to measure the conductivity of a solution without contacting the solution with the electrode? What is this technique called? Do you see any advantages of measuring conductivity in this fashion?

41. Describe the principles of operation of aerosol detectors used in HPLC.

Ion Chromatography

42. Describe the principles of suppressed conductometric ion chromatography and compare and contrast with nonsuppressed ion chromatography.

43. Compare conductometric detection in nonsuppressed IC with indirect photometric detection.

44. Compare the advantages and disadvantages of a packed-column suppressor vs. a membrane suppressor.

45. Compare the advantages and disadvantages of a chemically regenerated vs. a electrodialytically regenerated membrane suppressor.

46. Why would a hydroxide eluent be the one of choice in a gradient anion IC? How is a pure hydroxide eluent generated? How is any carbonate impurity in the generated eluent removed?

47. For electrodialytically generated eluents, why is KOH more commonly used than NaOH or LiOH?

Other Detection Methods

48. What would be the advantage of using an LCW cell in absorbance detection? In fluorescence detection?

49. What is the most common detector used in HPLC? Can you draw a block diagram?

50. How can absorbance ratio plotting help determine if coelution of two analytes is occurring?

51. Draw the block diagram of a photodiode array detector used in HPLC.

52. List some analytes amenable to electrochemical detection (coulometry/(pulsed) amperometry?

53. What is postcolumn reaction detection? What are the two important criteria in a postcolumn reaction detection scheme? Can PCR detection be done without introduction of reagents?

54. Discuss different designs of postcolumn reactors. What design provides the longest reaction times?

Applications

55. Ion chromatography is unique among HPLC techniques in using a number of electrodialysis techniques to generate and manipulate the eluent. Explain.

56. Gas removal devices can be used in two different places in ion chromatography. One is before the injector; the other is before the detector. Explain.

57. Discuss when gradient elution might not be desirable.

58. What are the advantages of operating with smaller bore columns?

59. In what type of chromatography the separation space is increased by using a larger column *volume*?

Thin-Layer Chromatography

60. How is the R_f value for a spot on a TLC plate calculated? What can the R_f value be used for?

61. Compare retention factor in HPLC with the R_f value in TLC. What are the largest and smallest R_f values possible?

62. How does TLC and HPTLC differ? When is the latter worthwhile? What kind of stationary phase is used by the vast majority of (HP)TLC applications?

63. What is the best way to apply a sample to an HPTLC plate? If a reaction with a solution phase reagent is needed for visualizing the spot, what is the best way to carry it out? What is an analog of indirect photometric detection in TLC?

64. Explain the statements: Thin layer chromatography is typically a nonequilibrium process. The composition of the developing solvent put in the tank may be known but the effective solvent is different and its composition is hard to characterize and may even vary with the location on the plate.

65. How is gradient elution accomplished in HPTLC?

66. Describe the visualization and quantitation methods used in (HP)TLC.

Electrophoresis

67. Describe the basis of separation in CZE and contrast with that in chromatography

68. Why is it not possible to separate all analytes by free solution electrophoresis? Why do gels need to be used?

69. How does pulsed field gel electrophoresis differ from standard slab gel electrophoresis?

70. List the different parameters influencing electrophoresis.

71. What is isoelectric focusing? What kind of resolution is possible in capillary isoelectric focusing (CIEF)?

72. Compare the types of information we obtain about proteins using molecular exclusion chromatography and SDS-PAGE.

73. Compare the advantages or disadvantages of CE vs. HPLC.

74. Ethidium bromide is still the most common DNA intercalating fluorescence tag. But there is increasingly a move to other tags. Why?

75. What is ZIFE? How is possible to achieve a separation with a net zero field?

76. What safety precaution is adapted in all commercial CE instruments?

77. Describe what two fundamental injection modes are used in CE and the advantage and disadvantage of each.

78. Under what conditions does electrostacking occur? Is electrostacking beneficial or detrimental in CE?

79. Who first observed EOF? List any of his other contributions to science. How does EOF originate?

80. Your friend's book asserts that the EOF profile is the primary reason for the high efficiencies observed in CE. Do you agree? Defend your position.

81. Although the number of theoretical plates increases with applied voltage in CE, too much voltage leads to too much power dissipated, and past a certain point the efficiency deteriorates. Describe *exactly* how this happens (it gets heated is not a good answer).

82. What is the zeta potential? How can the zeta potential be changed?

83. The core of a typical micelle is hydrocarbon-like. The number of molecules that aggregate to form a micelle can be ~2-200; that for SDS is ~60. If we were to compare MEKC with HPLC, what HPLC subclass will it be closest to? What classes of molecules may be well suited for MEKC separations and what classes of molecules are unlikely to be separated by MEKC?

84. Explain what happens in isotachophoresis. What is the isotachophoretic condition?

Problems

Revisit Basic Chromatography One Last Time

PROFESSOR'S FAVORITE PROBLEM

Contributed by Professor Christa L. Colyer, Wake Forest University

85. Referring to Figure 16.16, arrange the partition constant C_s/C_m in order (largest to smallest) for GSH, GSSG, and HCys and explain to a friend, how did you determine the order?

PROFESSOR'S FAVORITE PROBLEMS

Contributed by Professor Milton L. Lee, Brigham Young University

86. A student was asked to separate two substances, A and B, on a 30.0 cm column. She obtained a chromatogram that gave retention times of 15.80 and 17.23 min for A and B, respectively, and an elution time of an unretained compound of 1.60 min. The base peak widths for A and B were 1.25 and 1.38 min, respectively. Please calculate (a) the average number of theoretical plates for the column, (b) the plate height, (c) the resolution of A and B (d) the length of column that would be required to achieve a resolution of 1.5, and (e) the time required to elute compounds A and B on the longer column.

Ion Exchange Chromatography

87. Alkali metal ions can be determined volumetrically by passing a solution of them through a cation exchange column in the hydrogen form. They displace an equivalent amount of hydrogen ions that appear in the effluent and can be titrated. How many millimoles of potassium ion are contained in a liter of solution if the effluent obtained from a 5.00-mL aliquot run through a cation exchange column requires 26.7 mL of 0.0506 *M* NaOH for titration?

88. The sodium ion in 200 mL of a solution containing 10 g/L NaCl is to be removed by passing through a cation exchange column in the hydrogen form. If the exchange capacity of the resin is 5.1 meq/g of dry resin, what is the minimum weight of dry resin required?

89. What will be the composition of the effluent when a dilute solution of each of the following is passed through a cation exchange column in the hydrogen form? (a) NaCl; (b) Na_2SO_4; (c) $HClO_4$; (d) $FeSO_4$; $(NH_4)_2SO_4$.

Detectors, Data Rates, and Detection Limits

PROFESSOR'S FAVORITE PROBLEM

Contributed by Professor Apryll M. Stalcup, Dublin City University, Ireland

90. Given the information content in IR, compared to UV, why isn't IR used much for detection in HPLC?

Contributed by Professor Michael D. Morris, University of Michigan

91. Two intensely absorbing fluorescent dyes were separated by CE and by conventional scale HPLC. The concentration detection limits by fluorescence are about the same, but the detection limits are much worse for CE in absorbance detection. Why?

92. You have a detector providing digitized output to the data system at selectable rates of 1 Hz, 5 Hz, 10 Hz, 20 Hz, 50 Hz, 100 Hz, and 200 Hz. Higher data rates also produce higher noise. If peaks are perfectly Gaussian and the sharpest peak has a half-width of 3 s, what data transfer rate should you use?

93. Many biological compounds are chiral and exhibit optical rotation. Taxol, a naturally occurring compound isolated from the Pacific Yew tree, is sold under the trade name Paclitaxel. It has a specific rotation of $-49°$ at 589 nm. During chromatography, it is difficult to avoid coelution of Taxol from another compound. While an MS detector can be used, the coeluting compound is achiral and a polarimetric detector should be able to measure Taxol selectively. Given that the available polarimetric detector uses a 589 nm LED, and utilizes a 20 mm path length cell, and given the detector noise level is the same as that of a typical such detector described in the text, what would be the LOD of Taxol in a sample given a chromatographic dilution factor of 10?

94. The S/N ratio of a fluorescence detector is measured by the Raman emission of water. You are using a low-pressure Hg lamp with a fixed excitation wavelength of 253.7 nm. At what wavelength should you observe the Raman emission? The energy loss is 3382 cm^{-1}.

95. In indirect photometric detection in CE, a visibly absorbing additive, e.g., chromate, is often used in the BGE and the absorbance in the visible is measured. None of the analytes have visible absorption. It has been stated that if the detector is calibrated with a known chromate concentration, no further calibration is necessary for quantitating a fully dissociated analyte, at least in eq/L units. Defend or contest this statement in a quantitative basis.

Injection in HPLC and CE; Pump Pressure

96. In a 50 cm capillary operated at +25 kV, pressure injection is used to introduce a dilute aqueous sample. How much sample is introduced by a 5 psi, 2 s injection?

    In the same capillary, a sample containing chloride and iodate is electrokinetically introduced at +2 kV for 5 s. If equal amounts of the two ions were introduced, what was their concentration ratio in the original sample?

97. In the previous example, actually indirect photometric detection was used for quantitation and the peak areas for iodate and chloride were the same. From this, the analyst erroneously concluded that the analyte amounts were the same. But you know better. You ascertained that the neutral marker indicated a μ_{eo} of 9×10^{-4} cm^2 V^{-1} s^{-1}. What was the actual concentration ratio of the two anions in the actual sample?

98. Can you construct a low pressure injector with a number of three-way solenoid valves?

99. Describe how you would plumb a single 10-port valve to inject the same sample into two different columns, each connected to a separate pump and detector.

100. A table is given in the web supplement in the book regarding the viscosity of various compositions of methanol:water and acetonitrile:water, as well as pure water as a function of temperature. Calculate the pressure needed to pump 50% methanol at 30 °C through a 100 μm inner diameter silica capillary, 1 m in length, at a flow rate of 100 μL/min. Do the same for 50% acetonitrile and pure water.

Conductivity, Specific Conductance

PROFESSOR'S FAVORITE PROBLEMS

Contributed by Professor Dong-Soo Lee, Yonsei University, South Korea

101. Your friend asserts that in a purely aqueous medium (i.e., not containing any organic solvent), it is impossible to have a specific conductance less than that of pure water. You, on the other hand, know that in very dilute strong acid and base solutions, the conductivity can indeed be less than that of pure water. Compute and plot the specific conductance of a 0.05–5 μM solution of LiOH.

102. 1 mM KCl is often used as a standard for calibrating conductivity cells. Given the data in Table 16.2, calculate the anticipated specific conductance. How does it compare with the actual value of 147.2 μS/cm.

103. Calculate the diffusion coefficients of K^+ and Cl^- ions at 25°C. What are the Stokes radii of these ions?

104. Extremely small detectors are necessary for operating with open tubular chromatography or CE systems. Imagine the 10 μm diameter platinum wires are sealed into holes drilled into a 50 μm inner diameter capillary. The electrode faces are flush with the capillary inner surface. What is the cell constant of such a cell? What is the resistance if pure water is filled into the cell? What is the conductance if the capillary is filled with 1 mM KCl?

105. Of common electrolytes, KCl is used most often in many electrochemical devices like reference electrodes, etc. What property leads to this choice?

Electrophoresis

PROFESSOR'S FAVORITE PROBLEMS

Contributed by Professor Bin Wang, Marshall University, 106–109

106. Referring to Figure 16.33, why is arginine the first amino acid to elute?

107. Three molecules of similar size are bearing one, two, and three negative charges, respectively.

   (a) Which molecule has the highest electrophoretic mobility?

   (b) Is the electrophoretic mobility a positive or negative value?

108. A molecule with an electrophoretic mobility of 0.8×10^{-8} m^2 V^{-1} s^{-1} migrated through a surface-modified silica capillary. At pH 9.00, the molecule's apparent mobility was 3.1×10^{-8} m^2 V^{-1} s^{-1}. At pH 3.00, its apparent mobility was -1.2×10^{-8} m^2 V^{-1} s^{-1}. Calculate this molecule's electroosmotic mobilities and identify the direction of flow under each of these two pH conditions

109. For Example 16.6, calculate the theoretical resolution you would expect to observe between benzoate and iodate.

110. You are having a difficult time separating potassium and ammonium ion by CE using a certain BGE at pH 7 under +HV applied. It has been suggested that you could (a) add some 18-crown-6 to the BGE, (b) increase the pH, (c) increase the voltage, (d) use a longer capillary, (e) reverse the voltage polarity, and (f) use a smaller capillary. What would you expect to happen if you take these steps individually, without altering other parameters?

111. In an SDS micelle MEKC experiment, with +20 kV applied in a 50 cm capillary, the neutral marker, the micelle and the analyte respectively appear at the detector 2.5 min, 12.0, and 7.5 min after applying voltage. What is the retention factor of the analyte in the micellar pseudophase?

112. In an anion isotachophoretic experiment, with 500 V applied on a 0.1 mm inner diameter PTFE capillary, the counterion was sodium. If the zone length for sulfamate was twice the zone length for nitrate and the sulfamate concentration in the sample was 0.2 mM, calculate the sample nitrate concentration.

Spreadsheet Problems

113. Power functions are sometimes used to linearize aerosol detector outputs. In the website supplement of this text, a data file (DS_with Degas) is provided, with time in column A and the detector output (in µS) in column B. Plot the chromatogram. Now square the data in column B and list these in column C. Plot A vs. C. What differences do you see between the chromatogram and this (chromatogram)^2?

PROFESSOR'S FAVORITE PROBLEM

Contributed by Prof. Wen-Yee Lee, The University of Texas at El Paso

114. Caffeine in coffee is determined by LC-MS by using a ^{13}C-labeled caffeine internal isotopic standard via standard addition. See the website supplement of the text for the detailed problem and the spreadsheet.

PROFESSOR'S FAVORITE QUESTION AND CASE STUDY

115. Professor's Favorite Questions and Case Studies on the Website Supplement:
See the text website supplement for a favorite question of Professor Michael D Morris about van Deemter behavior and another about electroosmotic flow and the Boltzmann distribution.
Also, see the website supplement for two case study questions by Professor Christa L. Colyer, one on chromatographic resolution, the other on the van Deemter equation.

Multiple Choice Questions

1. The method of passing a mobile phase through a chromatography column is called
   (a) Flushing
   (b) Washing
   (c) Elution
   (d) Partitioning

2. The following is not a factor in determining column efficiency in liquid chromatography:
   (a) Flow rate
   (b) column length
   (c) Detector response
   (d) Packing particle size
   (e) Diffusion coefficient of the analyte

3. In Thin Layer Chromatography, the sample is initially
   (a) in contact with mobile phase.
   (b) not in contact with mobile phase.
   (c) coated at the level of mobile phase.
   (d) coated below the level of mobile phase.

4. The following method is not used for identification of spots on the TLC plate
   (a) Spraying with reagents
   (b) Under microscope
   (c) Fluorescence
   (d) fluorescent adsorbent

5. The electrophoresis is not used for the following:
   (a) Separation of proteins
   (b) Separation of amino acids
   (c) Separation of lipids
   (d) Separation of nucleic acids

6. The acidic solution containing trimethylamine (A), dimethylamine (B) and methyl amine (C) (pK_a of cations 9.8, 10.8 and 10.6, respectively) was loaded on a cation exchange column. The order of their elution with a gradient of increasing pH>7 is
   (a) $A < C < B$
   (b) $B < C < A$
   (c) $B < A < C$
   (d) $C < B < A$

7. When considering the speed, efficiency, sensitivity and ease of operation, the following technique is much superior:

    (a) Adsorption Chromatography

    (b) Ion–exchange Chromatography

    (c) High Performance Liquid Chromatography

    (d) All of Above

8. Which among the following molecules will be eluted in the end in Gel Permeation Chromatography?

    (a) Small molecules   (b) Intermediate molecules

    (c) Larger molecules   (d) All of these

9. The most appropriate conditions for eluting target proteins from an affinity chromatography matrix are:

    (a) Low salt concentrations

    (b) High salt concentrations

    (c) Adding a soluble ligand which competes with the affinity tagged protein for binding to the column

    (d) Just keep washing buffer through the column, isocratic elution

10. Diffusion of the molecule through the matrix of the exchanger in the ion-exchange chromatography depends upon:

    (a) the degree of cross-linkages of the exchanger.

    (b) the ionic strength of the buffer.

    (c) both of above.

    (d) none of above.

Recommended References

High-Performance Liquid Chromatography

1. Snyder, Lloyd R., Joseph J. Kirkland, and John W. Dolan. *Introduction to Modern Liquid Chromatography*. New York: Wiley, 2011.
2. V. R. Meyer, *Practical High-Performance Chromatography*, 3rd ed. New York: Wiley, 1999.
3. L. R. Snyder, J. J. Kirkland, and J. L. Glajch, *Practical HPLC Method Development*, 2nd ed. New York: Wiley, 1997.
4. A. Weston and P. R. Brown, *HPLC and CE: Principles and Practice*. San Diego: Academic, 1997.
5. U. D. Neue and M. Zoubair, *HPLC Columns: Theory, Technology, and Practice*. New York: Wiley, 1997.
6. R. P. W. Scott, *Liquid Chromatography for the Analyst*. New York: Dekker, 1994.
7. P. C. Sadek, *The HPLC Solvent Guide*. New York: Wiley, 1996.
8. L. Huber and S. A. George, eds., *Diode-Array Detection in HPLC*. New York: Dekker, 1993.
9. R. L. Cunico, K. M. Gooding, and T. Wehr, *Basic HPLC and CE of Biomolecules*. Hercules, CA: Bay Analytical Laboratory, 1998.

Size Exclusion Chromatography

10. Chi-San Wu, ed. *Handbook of Size Exclusion Chromatography and Related Techniques*: Vol. **91**. Boca Raton, FL: CRC Press, 2003.
11. A. Striegel, W. W. Yau, J. J. Kirkland, D. D. Bly, *Modern Size-exclusion Liquid Chromatography: Practice of Gel Permeation and Gel Filtration Chromatography*. New York: Wiley, 2009.

Ion Exchange and Ion Chromatography

12. J. Inczedy, *Analytical Applications of Ion Exchangers*. Oxford: Pergamon, 1996.
13. O. Samuelson, *Ion Exchange Separations in Analytical Chemistry*. New York: Wiley, 1963.
14. J. S. Fritz and D. T. Gjerde, *Ion Chromatography*. Wiley-VCH, 2009.
15. H. Small, *Ion Chromatography*. New York: Plenum, 1989.
16. H. Small and B. Bowman, "Ion Chromatography: A Historical Perspective," *Am. Lab.*, October 1998, 56C. By a pioneer in ion chromatography.
17. Special supplement to *LC.GC*, devoted entirely to ion chromatography, April, 2013 https://www.pixelmags.com/awrvl/#magazines/1361/issues/68479/pages/1

Thin-Layer Chromatography

18. M. Srivastava, ed. "High-performance thin-layer chromatography (HPTLC)." *Anal. Bioanal. Chem.* **401**, no. 8 (2011): 2331–2332.
19. Spangenberg, Bernd, Colin F. Poole, and Christel Weins. *Quantitative Thin-Layer Chromatography: a Practical Survey*. Springer, 2011.
20. B. Fried and J. Sherma, *Thin-Layer Chromatography*, 4th ed. New York: Marcel Dekker, 1999.

Capillary Electrophoresis

21. P. G. Righetti and A. Guttman, *Capillary Electrophoresis*. New York: Wiley, 2012.
22. G. Lunn, *Capillary Electrophoresis Methods for Pharmaceutical Analysis*. New York: Wiley, 1999.
23. J. P. Landers, ed. *Handbook of Capillary and Microchip Electrophoresis and Associated Microtechniques*. CRC Press, Boca Raton, FL, 2007.
24. P. K. Dasgupta and L. Bao, "Suppressed Conductometric Capillary Electrophoresis Separation Systems," *Anal. Chem.* **65** (1993) 1003.

Capillary Electrochromatography

25. I. S. Krull, R. L. Stevenson, K. Mistry, and M. E. Schwartz, *Capillary Electrochromatography and Pressurized Flow Capillary Electrochromatography: An Introduction*. New York: HNB, 2000.
26. Z. Deyl and F. Švec, eds. *Capillary Electrochromatography*. Vol. **62**. Chromatography Library, Elsevier Science, 2001.
27. D. B. Gordon, G. A. Lord, G. P. Rozing and M. G. Cikalo, *Capillary Electrochromatography*. Royal Society of Chemistry, London, 2001.
28. Website source: www.ceandcec.com. Very basic coverage of both capillary electrophoresis and capillary electrochromatography, but has links to many other useful websites.

Gel Filtration Chromatography

29. Ciarán Ó'Fágáin, M. Philip Cummins, and F. Brendan O'Connor, *Gel Filtration Chromatography*, Methods in Molecular Biology, January, 2011.

Chapter 17

MASS SPECTROMETRY

KEY THINGS TO LEARN FROM THIS CHAPTER

- About mass spectrometer and its basic components
- Use exact mass and monoisotopic masses, instead of average masses
- Resolution and mass accuracy are key performance parameters for mass spectrometers
- Inlets and ion sources vary, depending on application (e.g., GC–MS vs. LC–MS)
- Mechanisms of ion generation vary, depending on source (e.g., EI vs. ESI)

- Atomic mass spectrometry (e.g., ICP-MS)
- Mass analyzers, held under high vacuum, are used to separate ions
- Both electric and magnetic fields are used in different ways to separate ions
- Basic mass analyzers can be combined to reduce limitations in hybrid instruments
- Tandem mass spectrometry can be used to aid quantitative and qualitative analysis

Mass spectrometry (MS) has evolved and matured over the past several decades to become one of the most powerful analytical techniques available for quantitative and qualitative analysis. It is now commonplace for MS to be used in support of research and routine analysis in wide-ranging fields of science and engineering, too numerous to mention. In the end of the nineteenth century, Eugene Goldstein, J. J. Thomson, and Wilhelm Wien used electric and magnetic fields to discover the proton and the electron. In the early twentieth century, Francis Aston used similar means to discover isotopes of elements (see Reference 4 for an excellent historical account of the development of mass spectrometry). Today, researchers routinely apply MS techniques to study and unravel complex biological systems; even whole virus particles have been given flight as charged species in mass spectrometer instruments. A field that began and developed firmly in the grasp of experimental physicists has now shifted into the realm of the analytical chemist. We are currently in the midst of an era where the power of MS, and the availability of more robust, more sensitive, and faster systems, continues to increase at an astonishing pace. A well-known contemporary analytical chemist once said, "If you can't do it by mass spectrometry, it's probably not worth doing." Such a statement is telling of the vast capabilities MS offers.

17.1 Mass Spectrometry: Principles

Mass spectrometry is a sophisticated instrumental technique that produces, separates, and detects ions in the gas phase. The basic components of a mass spectrometer are shown in Figure 17.1. Samples are introduced into an ionization source through some type of inlet system. Depending on the phase and nature of the sample and the analytes, different inlets and ionization sources will be more or less optimal for producing ions. Analyte molecules are typically neutral and must be ionized. This can be accomplished in many ways. Analytes can be subjected to high-energy sources, such as electron bombardment, a laser, or an electrical discharge. Lower energy, or "softer," ionization sources also exist, such as those encountered in atmospheric pressure ionization. We describe common ionization sources in the sections below.

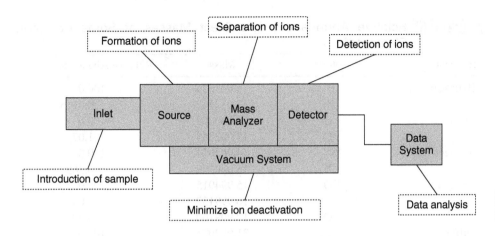

FIGURE 17.1 Components of a typical mass spectrometer.

Ions are either created or rapidly transferred into a high-vacuum (low-pressure; 10^{-4}–10^{-7} Torr) environment and can be separated based on their mass-to-charge (m/z) ratio through the application of electric and magnetic fields, in a variety of configurations. Different mass analyzers have different advantages, disadvantages, costs, and applications. Once the ions are separated, they are made to impinge on a detector, which converts the ion into an electrical signal that can be read by a data system. A variety of detectors exist, but some, such as the electron multiplier, are more common than others.

One of the biggest places where MS instrumentation has developed and matured in the past two decades is in the software platforms that collect and process data. Depending on the application, different software packages are available that specialize in qualitative and quantitative analysis, including sophisticated operations to support elemental formula determination, comparative analysis, bioinformatics, and many others. Specific software forms and functions are often manufacturer dependent, and are not discussed in detail. That said, once you understand how a mass spectrometer works, the largest obstacle to being productive with one is often learning how to use the software effectively.

Mass Spectrometry: Types of Masses

In most analytical techniques, it suffices to calculate the formula weight of a compound based on the values you obtain for elements from a standard periodic table. In MS, you have to do this a little differently, because mass spectrometers can distinguish the masses contributed by different isotopes. The formula weight that you calculate from masses on the periodic table will give you the **average mass**. The average mass is weighted by the natural abundances of the different isotopes present. In MS, we use **exact mass**. The exact mass, also called **monoisotopic mass**, is calculated based solely on the contribution to the mass by individual (often, the most abundant) isotopes of an element. We still use the convention of atomic mass units, where carbon-12 is set to exactly 12.0000 Daltons (Da). Thus, the exact mass of methane composed from the most abundant isotope of the elements ($^{12}C^{1}H_4$) would be calculated as $12.0000 + (1.007825 \times 4) = 16.0313$ Da. A signal for this exact mass would be observed if methane ions were detected by mass spectrometry. Since 1% of naturally occurring carbon also contains a carbon-13 isotope, a separate signal for an isotope of methane, $^{13}C^{1}H_4$ ($13.00335 + (1.007825 \times 4) = 17.03465$ Da), would also be observed in lower abundance. Table 17.1 lists masses and relative abundances of different isotopes for common elements.

The output of an MS experiment is called a **mass spectrum** (mass spectra, for plural; a mass spectrum is shown later in the chapter in Figure 17.4). The mass spectrum is a plot of ion abundance vs. mass-to-charge ratio. The ion abundance reflects

1 Da = 1 atomic mass unit (amu)

TABLE 17.1　Relative Abundances and Exact Masses of Some Common Elements[a]

Element	Isotope	Mass	Relative Abundance (%)
Hydrogen	^{1}H	1.007825	100.0
	^{2}H	2.014102	0.0115
Carbon	^{12}C	12.000000	100.0
	^{13}C	13.003355	1.07
Nitrogen	^{14}N	14.003074	100.0
	^{15}N	15.000109	0.369
Oxygen	^{16}O	15.994915	100.0
	^{17}O	16.999132	0.038
	^{18}O	17.999160	0.205
Sulfur	^{32}S	31.972071	100.0
	^{33}S	32.971450	0.803
	^{34}S	33.967867	4.522
Chlorine	^{35}Cl	34.968852	100.0
	^{37}Cl	36.965903	31.96
Bromine	^{79}Br	78.918338	100.0
	^{81}Br	80.916291	97.28

[a] The most abundant isotope is assigned an abundance of 100%, and the others are listed relative to it.

the amount of a particular ion that hits the detector, and it can be correlated with concentrations of the original analyte to perform quantitative analysis. The m/z is a qualitative parameter that can be used to help identify an analyte. In the example above, if we ionize methane to form CH_4^+ on a "unit resolution" instrument (resolution is discussed below), we would expect to see two signals. One higher abundance signal at m/z 16 (often denoted as M^+, because it represents the molecular ion) would be accompanied by a much smaller signal of approximately 1% relative intensity at m/z 17 (often denoted as M + 1 to indicate an isotope of the molecular ion). Ion abundance in a mass spectrum (on the y-axis) is often presented as relative intensity, where the signals of all ions recorded are normalized relative to the **base peak**, or the most abundant ion.

Resolution

In mass spectrometry, the resolving power, that is, the ability to differentiate two masses, is given by the resolution R, defined as the nominal mass divided by the difference between two masses that can be separated:

$$R = \frac{m}{\Delta m} \qquad (17.1)$$

Resolution tells you how well you can differentiate two closely spaced masses. Unit resolution tells you that you can differentiate one mass unit difference.

where Δm is the mass difference between two resolved peaks and m is the nominal mass at which the peak occurs. The mass difference is usually measured at the mean of some fixed fraction of the peak heights. For example, resolution at full-width half maximum (FWHM) is commonly encountered in manufacturer's specifications. Given any mass spectral signal, resolution at FWHM can be calculated by dividing the m/z at the apex of the signal by the width of the signal at one-half its height. Based on Equation 17.1, a resolution of 1000 means that a molecule of m/z 1000 would be resolved from m/z 1001 (or m/z 10.00 from 10.01; or m/z 500.0 is resolved from m/z 500.5). The term **unit resolution** is sometimes used to indicate the ability to differentiate next integer masses. An instrument that provides unit resolution would allow one to (at least) distinguish

m/z 50 from m/z 51, or m/z 100 from m/z 101, or m/z 500 from m/z 501. Obviously, the higher the molecular weight, the higher R must be to achieve unit resolution. A resolution of 1500 is generally suitable for routine MS detection, but some applications require higher values.

Example 17.1

Calculate the mass spectral resolution associated with an ion signal recorded with a peak apex at 465.1 m/z and an FWHM value of 0.35 m/z.

Solution

$$R_{FWHM} = m/\Delta m = 465.1/0.35 = \mathbf{1330}$$

The concept of resolution is closely tied with **mass accuracy**. Mass accuracy is often reported as a parts-per-million (ppm) relative error value, which can be calculated as

$$\frac{m_{measured} - m_{true}}{m_{true}} \times 10^6 \qquad (17.2)$$

where $m_{measured}$ is the measured ion mass and m_{true} is the expected monoisotopic mass of the ion. In organic synthesis, to report a new chemical compound in the scientific literature, many journals require that the exact mass of the compound be determined and reported to a value having less than 5 ppm error in mass accuracy. To achieve this level of accuracy, a mass analyzer capable of high-resolution measurements, such as a time-of-flight or an ion cyclotron resonance instrument, is necessary. If a newly synthesized compound is determined to have an exact mass of 634.45792 amu, to achieve less than 5 ppm relative error, the actual measurement can not deviate from this value by more than ±0.0032 amu. An instrument capable of providing a resolution close to 200,000 (634.45792/0.0032 ≈ 198000) would be needed to fully resolve the desired ion signal from a nearest neighbor only 3 mDa away. This level of resolution is possible with modern instruments, but they are extremely expensive (> $1M). In practice, however, one can achieve the necessary mass accuracy without extremely high resolution ($R \sim 30{,}000$ can be sufficient), if suitable calibration is performed to tune the instrument and minimize systematic error. Internal (calibrant standard present in the same mass spectrum as the sample) and external (calibrant standard measured separately) standard calibration can be used to carefully characterize and correct for systematic errors in mass determination by the mass spectrometer to obtain the necessary mass accuracy.

> In mass spectrometry, mass accuracy refers to the nearness of a measured m/z ratio with the "true" m/z value for an ion of interest.

Example 17.2

Ion clusters formed by sodium trifluoroacetate during electrospray ionization can be used to calibrate mass analyzers. What is the error in mass accuracy (ppm) if the $[Na_3(CF_3COO)_2]^+$ ion was recorded to have an m/z ratio of 294.9357?

Solution

The monoisotopic mass of the $[Na_3(CF_3COO)_2]^+$ ion is 294.938839.

Element	Monoisotopic mass	Quantity	Total monoisotopic mass
^{23}Na	22.989221	3	68.967664
^{12}C	12.00000	4	48.00000
^{19}F	18.998403	6	113.990419
^{16}O	15.994915	4	63.980756

$$\text{ppm error} = (294.9357 - 294.938839)/294.938839 \times 10^6 = \mathbf{-10.6 \ ppm}$$

A negative error in mass accuracy indicates that the instrument is measuring consistently below the "true" value.

Figure 17.2 shows a simulation of what mass spectral signals at different m/z values would look like if they were recorded at different resolutions. Many commercial software packages allow such simulations to be performed. We can see that a modest resolution ($R = 1000$) for a low molecular weight compound, valine, is sufficient to separate the molecular ion from its nearest isotope ion signal. However, for a larger compound, such as prymnesin-2 (monoisotopic mass = 1967.7958 Da), a resolution of 1000 is not sufficient to resolve the envelope of expected isotope ions. Instead, a resolution of 5000 would be needed. The simulated spectrum for the protonated molecular ion of prymnesin-2 also illustrates the contributions of isotopes. Since the compound contains three chlorine atoms, a collection of these molecules will contain about 25% chlorine-37 isotopes and 75% chlorine-35 isotopes. In this case, the observable isotope pattern becomes quite complex. More information on isotope distributions and access to a downloadable molecular weight and isotope distribution calculator can be found at: http://www.alchemistmatt.com/MwtHelp/IsotopicDistribution.htm.

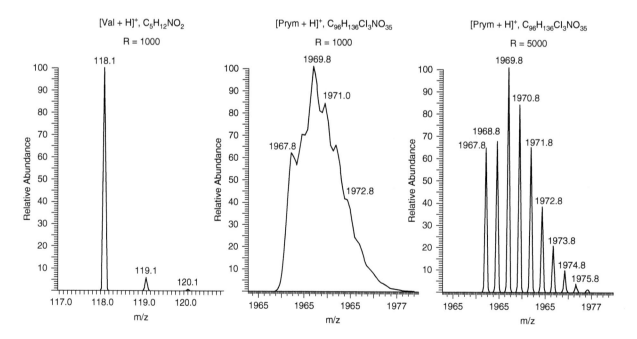

FIGURE 17.2   Simulated mass spectra for protonated ion forms of the amino acid valine (Val) and the algal toxin prymnesin-2 (Prym) at different levels of resolution.

Determination of Elemental Formula

Determination or confirmation of the molecular weight of an atom or molecule is one of the fundamentally most useful applications of MS. However, to assign the identity (and structure) of a true unknown compound, the use of MS alone can be limited. Some instruments, in combination with appropriate software tools, can be used to determine the elemental formula of an unknown. Such measurements rely on high-resolution and high mass accuracy to delineate the m/z ratio and intensity of a molecular ion and its accompanying isotopes. With the help of some well-established rules regarding allowable ratios of atoms, which can be combined to form a molecule (one would not expect an elemental formula of C_2H_{18}!), and some input by the user on expected amounts of particular atoms present (C, H, O, and N are common, but might the compound be expected to contain S or other atoms?), it is possible to discern potential elemental formulae, which have a formula weight close to the measured mass of the observed ion. The more accurately and precisely an ion mass can be measured, the fewer the number of elemental formulae that would be expected to fit. For example, there are many more combinations of C, H, N, and O that can be assembled to get a formula weight close to m/z 823.1, but the possibilities decrease significantly if the instrument can specify the mass of the ion to be m/z 823.1342, especially if we know the latter to be accurate to $< \pm 2$ ppm error in mass accuracy.

A variety of rules can be used to guide elemental formula generation from MS data (see Reference 5). A common one is the **nitrogen rule**. In applying this, a molecular mass will be an even integer number if it contains an even number $(0, 2, 4 \ldots)$ of nitrogen atoms, and will otherwise be an odd number, that is, when it contains an odd number of nitrogens. Another rule is called the "rule of 13." This approach takes a measured mass and divides by 13 to generate a base formula (containing only carbon and hydrogen). From the base formula, an index of hydrogen deficiency can be determined and used to consider whether other atoms types, besides carbon and hydrogen, are present. An excellent tutorial on this approach can be found from the website of Dr. Neil Glagovich, Department of Chemistry, Central Connecticut State University: http://www.chemistry.ccsu.edu/glagovich/teaching/316/ms/formula.html. Of course, as the molecular mass in question becomes larger and contains an increasing number of heteroatoms, solving for elemental formulae by hand can become virtually impossible. For this reason, manufacturers and researchers have developed their own computer-aided tools. The Fiehn group at University of California–Davis has developed a very comprehensive approach to determination of unknowns, especially those encountered in metabolomics research. The "Seven Golden Rules" approach (http://fiehnlab.ucdavis.edu/projects/Seven_Golden_Rules/) incorporates a framework for using a variety of mass spectral data to generate elemental formulae in combination with searching multiple databases of known chemical compounds. Even so, for a brand-new molecule, or an unknown not present in the database, one cannot rely on MS data alone. A good analytical chemist will always use other qualitative tools, such as NMR, IR, and elemental analysis (e.g., CHN analysis) to determine and confirm the identity of a new molecule.

17.2 Inlets and Ionization Sources

The eventual goal of the inlet and ionization source is to convert molecules into gas-phase ions in a high vacuum environment. The analysis of ions under vacuum is necessary in order to increase the **mean free path** of the ion, or the average distance a particle will travel before it encounters a collision. Collisions can deactivate (neutralize) ions and reduce ion yield, or sensitivity. The mean free path is inversely correlated with pressure.

There are many different inlet types, depending on the sample and analyte form. If an analyte is thermally stable and volatile, the application of heat is a relatively simple way to convert it into a gas-phase molecule. If such analytes are present in a mixture, then gas chromatography can be used to resolve the analyte from interferences, and the outflow of the capillary column will act as the inlet into the ionization source of the mass spectrometer. Gas chromatography–mass spectrometry is a very common laboratory technique, and is discussed more fully below. If the analyte is in a more pure state, a direct vapor inlet or direct insertion probe can be used. Both use heat to vaporize the analyte, so that it can be transferred to the ionization source. If the analyte is not volatile or is thermally labile (unstable), then other strategies are needed to convert analytes into gas-phase ions. Polar and ionic compounds, especially biomolecules, typically fall under this category. Mixtures of these compounds in solution will generally be resolved by liquid chromatography. The use of liquid chromatography–mass spectrometry has increased in popularity in the past 20 years, especially with the advent of atmospheric pressure ionization (API) sources, such as electrospray ionization (ESI), which are necessary to both remove excess solvent and convert nonvolatile compounds into gas-phase ions. These techniques are also suitable for introduction of solutions for analysis by direct infusion, direct injection, or flow injection analysis. Flow injection analysis is discussed in Chapter 24. Solid samples can also be analyzed by mass spectrometry. These can be introduced in various formats and subjected to ionizing radiation in the form of laser radiation, atom or ion bombardment, or electrical discharge. Laser radiation (for laser desorption/ionization) and ion bombardment (for secondary ion mass spectrometry) are currently the most readily applied techniques for sampling solid materials for mass spectrometry analysis.

17.3   Gas Chromatography–Mass Spectrometry

When the retention time of an unknown compound in a chromatogram is the same as a pure standard (using equivalent separation methods), this suggests but does not absolutely confirm the identity of the unknown. The probability of positive identification will depend on factors such as the type and complexity of the sample and sample preparation procedures used. Gas chromatography–mass spectrometry (GC–MS) is a powerful technique to confirm the presence of various compounds. GC–MS is used routinely in forensics and environmental laboratories for the analysis of complex mixtures of volatile organic compounds. It can provide very precise quantitative analysis, supported with the high specificity of mass spectrometry for qualitative analysis. GC–MS systems used to fill a room and cost several hundred thousand dollars, but today, relatively inexpensive compact benchtop systems are widely available and commonplace in the analytical laboratory. Figure 17.3 shows a picture of a modern benchtop GC–MS system.

FIGURE 17.3 A modern gas chromatograph–mass spectrometer instrument. (Reproduced with permission from Shimadzu Scientific Instruments, Inc.).

TABLE 17.2 Comparison of Ionization Methods[a]

Ionization Method	Typical Analytes	Sample Introduction	Mass Range	Method Highlights
Electron ionization (EI)	Relatively small volatile	GC or liquid–solid probe	To 1000 Daltons	Hard method. Versatile, provides structure information
Chemical ionization (CI)	Relatively small, volatile	GC or liquid–solid probe	To 1000 Daltons	Soft method. Molecular ion peak $[M + H]^+$
Electrospray ionization (ESI)	Peptides, proteins, nonvolatile	Liquid chromatography or syringe	To 200,000 Daltons	Soft method. Ions often multiply charged
Matrix-assisted laser desorption ionization (MALDI)	Peptides, proteins, nucleotides	Sample mixed in solid matrix	To 500,000 Daltons	Soft method, very high mass

[a] From web page of Professor Vicki Wysocki, The Ohio State University. Reproduced by permission.

Table 17.2 lists common ionization methods used for GC–MS, as well as for LC–MS. The most commonly used ionization source in GC–MS is **electron ionization (EI)**. The gaseous molecules eluted from the GC column are bombarded by a high-energy beam of electrons, usually 70 eV, generated from a tungsten filament under high vacuum. An electron that collides with a neutral molecule may impart sufficient energy to remove an electron from the molecule, resulting in a singly charged ion:

$$M + e^- \rightarrow M^+ + 2e^-$$

where M is the analyte molecule and M^+ is the molecular ion or parent ion. The M^+ ions are produced in different energy states and the internal energy (rotational, vibrational, and electronic) is dissipated by fragmentation reactions, producing fragments of lower mass, which are themselves ionized or converted to ions by further electron bombardment. The fragmentation pattern is fairly consistent for given conditions (electron beam energy). Only a small amount or none of the molecular ion may remain. If it does appear, it will be the highest mass in an EI spectrum, if there are not multiple isotopes. Compounds with aromatic rings, cyclic structures, or double bonds are more likely to show higher abundance molecular ion peaks because of delocalization effects that decrease fragmentation. Figure 17.4 shows a simple EI spectrum of a small molecule, methanol. The molecular ion peak is at $m/z = 32$, the formula weight of CH_3OH. The small peak at $m/z = 33$ is the molecular ion for the ^{13}C isotope of methanol, which has a relative abundance of 1.11% compared to ^{12}C at 100%. The base peak at $m/z = 31$ is from the CH_2OH^+ fragment.

A major advantage of GC–MS for qualitative analysis is the availability of libraries of EI mass spectra for hundreds of thousands of different chemical compounds. Instruments that are used to routinely perform EI are operated with a source energy of 70 eV. This energy is high enough so that significant fragmentation occurs for all molecules, and molecules fragment reproducibly via EI on different instruments. Thus, very consistent spectra can be generated on different instruments, and instrumental software allows these spectra to be searched against libraries (e.g., the NIST Mass Spectral Library, which can be found at http://www.nist.gov/srd/nist1a.cfm) to find possible matches and determine the identity of an unknown signal. Even if the signal of interest represents a compound for which no library entry exists, fragmentation patterns can be tracked and matched with those of compounds that have similar functional units. For example, if a molecule contains a substituted benzene ring, there is a good chance that a signal at m/z 77 will be observed; this represents the signal for a $C_6H_5^+$ radical cation. Table 17.3 lists ion forms and neutral losses that are commonly produced by EI for chemical compounds, which contain specific functional units. With this information, as well as some of the rules mentioned previously (e.g., the Nitrogen rule, the rule of 13, etc.) and an elemental formula, it is possible to manually elucidate

Electron ionization is considered to be a "hard" ionization process that produces many fragments.

the general structural components of an unknown compound from its EI spectrum; but, of course, library matching is a far simpler task.

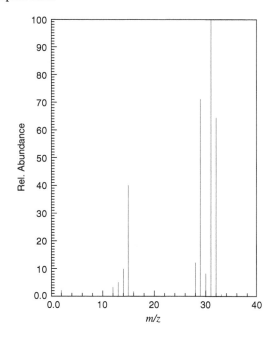

FIGURE 17.4 Electron ionization mass spectrum of methanol. From NIST MassSpec Data Center. Reproduced by permission.

TABLE 17.3 Common chemical compound neutral losses and ions formed using EI sources. Adapted from: Chhabil Dass, *Fundamentals of Contemporary Mass Spectrometry*. John Wiley & Sons, Inc. Hoboken, NJ. 2007. ISBN: 978-0-471-68229-5

Mass	Neutral losses	Possible Functionality	Mass	Neutral losses	Possible Functionality
14	Impurity, homologue		29	C_2H_5	Alkyl
15	CH_3	Methyl	30	CH_2O	Methoxy
16	CH_4	Methyl		NO	Aromatic nitro
	O (rarely)	Amine oxide		C_2H_6	Alkyl (Cl)
	NH_2	Amide	31	CH_3O	Methoxy
17	NH_3	Amine (Cl)	32	CH_3OH	Methyl ester
	OH	Acid, tertiary alcohol	33	$H_2O^+ CH_3$	Alcohol
18	H_2O	Alcohol, aldehyde, acid (Cl)		HS	Mercaptan
19	F	Fluoride	35	Cl	Chloro compound
20	HF	Fluoride	36	HCl	Chloro compound
26	C_2H_2	Aromatic	42	CH_2CO	Acetate
27	HCN	Nitrile, hetero-aromatic	43	C_3H_7	Propyl
28	CO	Phenol	44	CO_2	Anhydride
	C_2H_4	Ether	46	NO_2	Aromatic nitro
	N_2	Azo	50	CF_2	Fluoride

Mass	Ion	Possible Functionality	Mass	Ion	Possible Functionality
15	CH_3^+	Methyl, alkane	50	$C_4H_2^+$	Aryl
29	$C_2H_5^+, HCO^+$	Alkane, aldehyde	51	$C_4H_3^+$	Aryl
30	$CH_2 = NH_2^+$	Amine	77	$C_6H_5^+$	Phenyl
31	$CH_2 = OH^+$	Ether or alcohol	83	$C_6H_{11}^+$	Cyclohexyl
39	$C_3H_3^+$	Aryl	91	$C_7H_7^+$	Benzyl
43	$C_3H_7^+, CH_3CO^+$	Alkane, ketone	105	$C_6H_5C_2H_4^+$	Substituted benzene
45	CO_2H^+, CHS^+	Carboxylic acid, thiophene		$CH_3C_6H_4CH_2^+$	Disubstituted benzene
47	CH_3S^+	Thioether		$C_6H_5CO^+$	Benzoyl

Professor's Favorite Example

⌐Contributed by Professor
Fred McLafferty, Cornell
University   ⌐

Example 17.3

Professor Fred McLafferty, Cornell University, started his career in industry, and remembers how his expertise in mass spectrometry solved a serious problem in the production of polystryrene. He was in charge of the mass spectrometry lab in 1953. A 1 million pound/day polystyrene plant had to be shut down because of black specks in the polymer. He ran a mass spectrum of the styrene monomer and found an impurity, identified as the culprit. From the mass spectrum of this compound below, can you identify the impurity? A world expert in infrared spectroscopy did not find this. Can you say why?

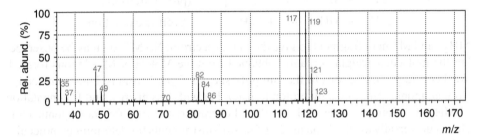

(Mass spectrum courtesy of the Shimadzu Center for Advanced Analytical Chemistry at the University of Texas at Arlington)

Solution

The mass spectrum of the styrene monomer showed about 300 ppm of CCl_4, with dominant peaks at 117, 119, 121, and 123, whose isotopic intensity ratio can only be CCl_3^+. The plant manager originally believed the IR expert who did not report CCl_4. (The reason is that CCl_4 is not IR active since it is a symmetrical molecule—see Chapter 11.) When asked where the styrene came from (Dow Texas tank cars) and what did they carry last, it was CCl_4. Dr. McLafferty's application for an expensive new MS instrument was magically approved!

EI is a hard ionization technique that rarely leaves a significant abundance of the molecular ion. In some cases, it might be desirable to produce a higher abundance molecular ion for molecular weight determination. In theory, this can be accomplished by reducing the EI filament voltage, but it is far more common to use a technique called **chemical ionization (CI)**. CI is a "softer" technique that does not cause much fragmentation, and the molecular ion is often the dominant one in CI mass spectra. In CI, a reagent gas such as methane, isobutane, or ammonia is introduced into the EI ionization chamber at a high pressure (large excess, 1 to 10 torr) to react with analyte molecules to form ions by either proton or hydride transfer (anion abstraction and charge exchange are other possible charging mechanisms). The CI process begins by electron ionization of the reagent gas in an EI source. With methane, electron collisions produce CH_4^+ and CH_3^+, which further react with CH_4 to form CH_5^+ and $C_2H_5^+$:

$$CH_4^+ + CH_4 \rightarrow CH_5^+ + CH_3$$
$$CH_3^+ + CH_4 \rightarrow C_2H_5^+ + H_2$$

TABLE 17.4 Chemical Ionization Characteristics of Different CI Reagents

Reagent	Adducts Produced	Uses/Limitations
Methane	$M-H^+$, $M-CH_3^+$	Most organic compounds. Adducts not always abundant. Extreme fragmentation.
Isobutane	$M-H^+$, $M-C_4H_9^+$	Less universal. Adducts more abundant. Some fragmentation.
Ammonia	$M-H^+$(basic compounds) $M-NH_4^+$(polar compounds)	Polar and basic compounds. Others not ionized. Virtually no fragmentation.

These react with the sample by transferring a proton (H^+) or extracting a hydride (H^-), which imparts a +1 charge on the sample molecule:

$$CH_5^+ + MH \rightarrow MH_2^+ + CH_4 \qquad \text{(proton transfer)}$$
$$C_2H_5^+ + MH \rightarrow M^+ + C_2H_6 \qquad \text{(hydride extraction)}$$

MH_2^+ and M^+ may fragment to give the mass spectrum. No M^+ ion may be observed, but the molecular weight is readily obtained from the $M + H$ or $M - H$ ions formed. Weaker acid gas-phase ions further simplify spectra. $C_4H_9^+$ from isobutane and NH_4^+ from ammonia also ionize by proton transfer, but with less energy, and fragmentation of MH_2^+ is minimal. Table 17.4 lists the CI characteristics of different reagents. CI is almost universally available on modern GC–MS instruments for determining molecular weights of eluting compounds otherwise difficult to obtain due to extreme fragmentation with "harder" sources.

In general, EI produces only positive ions; and the discussion of CI above centers on the formation of positive ions. We often refer to the radical-based ions formed by EI as "odd-electron" ions (there is one unpaired electron), whereas those produced by proton transfer and hydride extraction in CI are predominantly "even-electron" ions (all electrons are paired). It is also possible to produce negative ions in the gas phase by CI, a process referred to as **negative chemical ionization (NCI)**. In order to be observed as a negative ion, the compound must be able to stabilize a negative charge, often following deprotonation or electron capture. Because only some compounds can effectively do this (e.g., those that contain acidic groups or halogens), NCI provides a level of selectivity in ion formation that can be useful when analyzing complex samples. Halogenated compounds, such as flame retardants, pesticides, and polychlorinated biphenyl (PCB) molecules, have a high electronegativity, which enables formation of stable negative ions. These compounds are common analytes in environmental analysis, which makes NCI a popular technique among environmental analysts. Other compounds can be made amenable for NCI through derivatization; simply appending a perfluoroalkyl group on a chemical compound will allow it to stabilize a negative ion. See Reference 6 for more about NCI.

One of the biggest problems in the early development of GC–MS was interfacing the column outlet to the mass spectrometer. Packed columns were used in the early days, and the high volumes of both sample and carrier gas overwhelmed the MS system, which operates under low pressure; special interfaces had to be built. The advent of fused silica capillary columns meant that the GC–MS interface could be dispensed with, and the column eluent could be introduced directly into the ion source. It is essential that column bleeding be minimized since the mass spectrometer will detect the stationary-phase materials. Bleeding is prevented by chemically bonding alkylsiloxanes to the column wall and by chemical cross-linking of the stationary phase. Low bleed stationary phases have been discussed in Chapter 15. Capillary GC, with thousands of theoretical plates, can resolve hundreds of molecules into separate peaks, and mass spectrometry can provide identification. Even if a peak contains two or more compounds, identifying peaks through manual assignment and library matching can still provide positive identification, especially when combined with retention data. Of course, monitoring GC peaks (several times per second) generates enormous amounts of data. The evolution of fast and high-capacity computers is the other technology advancement that made GC–MS routine. The separated ions are detected by means of an electron multiplier, which is similar in design to photomultiplier tubes described in Chapter 11 (ion detectors are also discussed briefly, later in this chapter). Detection limits at the picogram level are common with modern GC–MS instruments.

17.4 Liquid Chromatography–Mass Spectrometry

The combination of MS with liquid chromatography, like with gas chromatography, has become a powerful analysis tool for sensitive and selective mass detection to characterize complex samples. It is more difficult to interface a liquid chromatograph to a mass spectrometer because of the necessity to remove the solvent. Also, the analytes are generally nonvolatile and may be thermally labile, but they must be presented in gaseous form. Hence, the combination of LC and MS was termed an "unlikely marriage," and it took several years to develop interfaces to the stage of reliable and easy use. Today, there are several types of atmospheric pressure ionization (API) interfaces that make LC–MS a routine technique. Compact and benchtop systems of enormous variety are now commercially available (Figure 17.5).

The most commonly used API interfaces are electrospray ionization (ESI) and atmospheric pressure chemical ionization (APCI) sources. The choice depends largely on the polarity and thermal stability of the analyte. ESI is preferred for polar and ionic molecules, and can accommodate a wide range of analyte sizes from small molecules to very large biomolecules, such as proteins and peptides. APCI is more suited for small molecules and for compounds that are relatively less polar. Most commercial instruments provide the user with both sources, and they can be quickly interchanged in a matter of minutes. Some manufacturers also provide an atmospheric pressure photoionization (APPI) source, which can be useful if ions for a particular analyte cannot be efficiently generated by either ESI or APCI.

Electrospray ionization (ESI) is a soft ionization technique. The sample solution is sprayed across a high potential difference of a few thousand volts from a needle into an orifice in the interface (Figure 17.6). Heat and gas flow desolvate the charged droplets, causing the droplets to shrink, subdivide, and eventually emit charged analyte

Electrospray ionization is so soft that even noncovalent complexes can be transferred intact from solution to the gas phase.

FIGURE 17.5 A modern liquid chromatograph–mass spectrometer instrument. (Reproduced with permission from Shimadzu Scientific Instruments, Inc.)

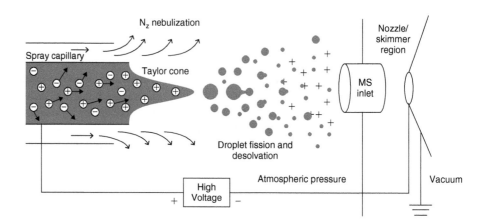

FIGURE 17.6 Schematic of an electrospray ionization interface.

molecules (and analyte ion–solvent clusters) that enter the mass spectrometer. Because very little internal energy is imparted onto the formed ions, they remain intact; ESI spectra are largely dominated by molecular and quasimolecular (or "adduct") even-electron ions (e.g., $[M + H]^+$, $[M + Na]^+$, $[M + NH_4]^+$, etc. in the positive ionization mode and $[M − H]^-$, $[M + Cl]^-$, etc. in the negative ionization mode). Even complexes in solution held together by noncovalent forces can be converted intact into gas-phase ionic complexes. The polarity of ions produced can be changed from positive to negative simply by changing the polarity of the potential imparted on the spray capillary. Generally, basic compounds more easily form positive ions by protonation, whereas acidic compounds more easily form negative ions by deprotonation. Some instrument manufacturers even allow for rapid polarity switching during the course of a single analytical run, so that ions of both polarity can be monitored simultaneously.

The detailed mechanism of ESI has been studied for many years, and depending on the analyte system, analytes may be released from the droplets in different ways (see References 7 and 8). While certainly a topic for interesting fundamental research, in practice, it is important to realize that ESI is a competitive ionization process—meaning that the composition of each droplet controls the relative yield of gas-phase ions. For ions to be generated, they need to be able to migrate to the surface of the droplet and assimilate a charge. Highly surface active species will generate the most abundant signals. In cases where it is necessary to analyze compounds in the presence of lipids or detergents, these highly surface active species can significantly reduce the response of the analytes of interest, because they effectively compete for a limited number of sites on the droplet surface. Such matrix effects are very common in ESI-MS analysis from complex mixtures. Thus, it becomes extremely important to incorporate a **stable isotopically labeled internal standard (SIL-IS)** into the analysis, and to calibrate response using a matrix that is very similar, if not identical, in composition with the sample containing the unknowns to be quantified. Use of a SIL-IS, such as a deuterated or ^{13}C-substituted version of the analyte of interest, is critical and will ensure that the internal standard is present in the same ESI droplets as the analyte. It will elute from a liquid chromatograph at the same time as the analyte and will appear a few m/z units higher in the mass spectrum. Thus, the internal standard will experience (and can appropriately correct for) the same competitive ionization effects experienced by the analyte. Even so, cost and commercial availability of appropriate SIL-IS compounds can be a problem. If these constraints are carefully addressed, the HPLC-ESI-MS can be effectively used to quantify parts-per-trillion (or femtogram) levels of analytes in samples.

A major advantage is that ESI can also be used to analyze very large molecules, such as proteins, because it can produce multiply-charged ions, with the number of charges increasing with molecular weight of the molecule. The discovery of this phenomenon earned Professor John B. Fenn a portion of the 2002 Nobel Prize in Chemistry. Proteins and peptides (and other similarly highly functionalized molecules) have multiple sites along their backbones where protonation or deprotonation can take place. As such, a large protein can be observed as a distribution of multiply-charged ions, which can be deconvoluted through a simple algorithm to determine its molecular weight. Therefore, relatively large molecular weights, well beyond the normal 2000-to 3000-dalton m/z ratio limit of a typical mass analyzer, can be measured due to the increase of z in the m/z ratio. For example, a 50,000 Da protein that has acquired 50 excess protons ($m = 50,050$; $z = 50$) will be observed as a signal at m/z 1005. This signal will almost certainly be accompanied by signals for other charge states of the same protein. Figure 17.7 shows an ESI spectrum generated for the protein cytochrome c from bovine heart. Observed are two distributions of multiply-charged ions. The more highly charged ions (+8 to +16) represent the denatured form of the protein, whereas the lower charge states (+5 to +7) represent the folded form. The denatured, or unfolded, form of the protein has more basic sites accessible for protonation, relative to the folded (native) form; and thus, higher charge states can be accommodated. The generation of extremely highly charged (or "supercharged") proteins is of significant interest to those who perform *de novo* sequencing of proteins, because more highly charged ions can be fragmented more efficiently by some tandem mass spectrometry fragmentation techniques [e.g., electron capture dissociation and electron transfer dissociation (see tandem mass spectrometry below)] for qualitative analysis.

Atmospheric pressure chemical ionization (APCI) is another common atmospheric pressure ionization interface for coupling liquid chromatography to mass spectrometry. The interface is similar to that used for ESI, but a corona discharge is used to ionize the analyte present in a vaporized state under ambient pressure. (Note: This is different than the chemical ionization source described for GC–MS that operated under vacuum.) The effluent from the liquid chromatograph is sprayed into a heater

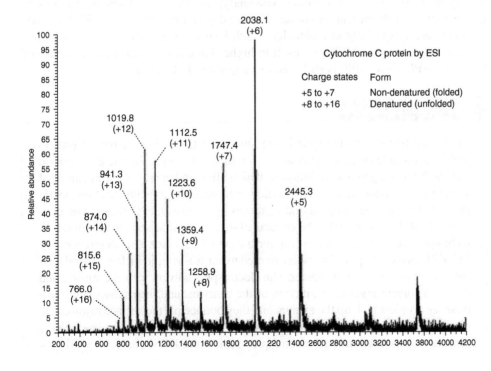

FIGURE 17.7 Electrospray ionization mass spectrum of cytochrome c from bovine heart. Adapted from Reference 10.

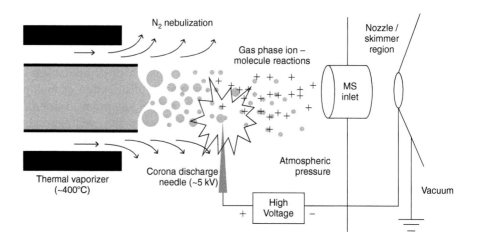

FIGURE 17.8 Schematic of an atmospheric pressure chemical ionization (APCI) interface.

jacket (~400°C) to create a vapor, which is then passed over the corona discharge needle, held at a potential of ~5 kV (Figure 17.8). Whereas ESI relies predominantly on solution-phase ionization (acid–base chemistry), APCI generates ions through gas-phase ion-molecule reactions. The gas-phase ionization of less polar analytes is more efficient than with ESI, but the mass range is limited to about 2000 Daltons. ESI and APCI are complementary; where one fails to generate ions, the other can often be used. When APCI is used, it is common to have methanol in the mobile phase, since the corona discharge can convert CH_3OH into $CH_3OH_2^+$ (among other species), which can then initiate proton transfer to create a positively charged analyte ion. Overall, it is the bulk solvent that is ionized at the corona needle, and then these ionic species collide with the analyte of interest and initiate charge transfer.

In **atmospheric pressure photoionization (APPI)**, the corona discharge needle from the APCI source is replaced with a high-intensity UV lamp. The lamp supplies energy (~10 eV) sufficient to ionize some molecules. Similar to APCI, methanol is often used to create reagent ions (its ionization potential is low enough to enable ionization by the lamp), which then interact with analytes of interest to form analyte ions by charge transfer. Toluene can also be an effective dopant for increasing APPI efficiency. In some cases, especially in molecules which bear electron-deficient aromatic units (e.g., nitro-aromatics), APPI can result in higher-intensity ions than can be achieved by ESI or APCI. Yet, APPI is still used to a significantly lesser degree.

Ambient Ionization

In the past few years, atmospheric pressure ionization sources have matured to yield a multitude of new technologies for in situ analysis. Ambient ionization (AI) is the name given to techniques, which allow independent optimization of sample introduction and ionizing radiation to generate ions under atmospheric pressure that can be sampled and analyzed by a mass spectrometer. The concept is best described under the context of one of the first and most popular AI techniques, desorption electrospray ionization (DESI) (see also Reference 11). In DESI, an electrospray source is oriented towards a sample surface placed near the inlet to the mass spectrometer. The electrospray solvent wets the surface and creates a solvent layer for extraction of the analytes, and then subsequent electrospray droplets pick up the analyte and they bounce off the surface towards the mass spectrometer inlet. Analyte ions are generated from the droplets in much the same way as in conventional electrospray ionization. DESI enables a workflow where samples (e.g., pharmaceutical tablets) can be cycled through the interface and sampled.

Other types of ionizing radiation can also be used. In desorption analysis in real time (DART) (see Reference 12), a contemporary of DESI, metastable ions produced in a glow discharge are accelerated towards a sample. These energetic species are able to desorb and ionize analyte molecules present in the sample. The setup is also highly amenable for continuous process monitoring.

Since DESI and DART have been introduced, a very large number of other AI techniques have been developed. A variety of literature reviews can be found on this topic (see Reference 13). Like DESI, some make use of electrically generated nebulized solvent sprays; others utilize chemical ionization, thermal ionization, photoionization, or other energetic sources to accomplish sampling of a variety of sample types. One of the basic tenets of AI techniques is that minimal sample preparation should be necessary. Some sources, such as extractive electrospray ionization (EESI) (Reference 14), where the electrospray plume is mixed with a nebulized sample aerosol in front of the MS inlet, have been shown to be viable for direct analysis of complex samples, such as urine.

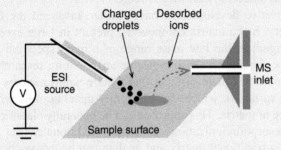

Schematic of desorption electrospray ionization.

17.5 Laser Desorption/Ionization

Ions can be formed from molecules in many different ways. One very effective means is through the application of laser light. Lasers provide highly intense monochromatic radiation, are available in many different formats, and are very useful for the ionization of species in virtually any physical state. In modern mass spectrometry, lasers are used: (a) to directly desorb and ionize solid materials [termed **laser desorption ionization (LDI)**]; (b) to desorb and indirectly ionize molecules following excitation of a matrix, which was co-crystallized with the analyte [termed **matrix-assisted laser desorption/ionization (MALDI)**]; (c) to ablate and sample materials for subsequent analysis by other techniques, such as inductively coupled plasma mass spectrometry (discussed later in this chapter, as well as previously in Chapter 12); and (d) to probe ions trapped in a mass spectrometer and perform gas-phase photochemistry experiments (this is a more advanced topic, which is not covered in this text).

LDI and MALDI are very similar in their instrumental setup on commercial systems. In most, a solid-state or nitrogen UV laser is focused onto a stainless steel plate containing the sample of interest. Ions generated upon irradiation by the laser light are accelerated into the mass analyzer—usually a time-of-flight mass analyzer (see discussion on mass analyzers below). The main difference between LDI and MALDI is the addition of a matrix compound to the sample in MALDI. This makes it a soft ionization technique, capable of analyzing a wide range of small and large molecules of variable polarity. In contrast, LDI exposes analytes to the full intensity of the laser and is often considered a hard ionization technique that can generate fragments of ions.

Demonstration of MALDI for the analysis of large biomolecules earned Professor Koichi Tanaka a share of the 2002 Nobel Prize in Chemistry (he shared this prize with Professor John Fenn, mentioned earlier in connection with ESI). In the past 15–20 years, the use and development of MALDI has grown substantially to include capabilities for analyzing a wide array of different analytes, including small molecules, lipids, high and low molecular weight polymers, peptides, and, of course, proteins. A key ingredient for successful ion generation in MALDI is the selection of an appropriate matrix compound. Matrices are typically small organic compounds mixed with the analyte of interest in a small sample spot (~1 μL of liquid) applied to the surface of a polished stainless steel plate. The primary purpose of a matrix compound [present in significant excess relative to the analyte(s) of interest] is to: (a) co-crystallize with the analyte upon drying of the sample spot; (b) absorb radiation supplied by the laser; (c) be desorbed and excited into the gas phase upon irradiation; and (d) initiate proton transfer to, or extraction from, the analyte to create the ion of interest for subsequent mass analysis. Different matrices perform better for different applications, and a small subset of common matrices is shown in Figure 17.9.

Many innovative developments have further advanced the utility of MALDI mass spectrometry. Since matrix compounds, present in large excess, often produce an intense ion signal in the low mass range of a mass spectrum and can suppress the signals of low abundance, small molecule analyte ions, researchers have successfully developed modified plate surfaces that feature nanostructures (carbon nanotube and silica-based surfaces), which can efficiently generate ions with minimal background in absence of matrix. This approach can be generally classified as nanoparticle-assisted laser desorption/ionization (NALDI). Additionally, the co-crystallization process between a traditional matrix and analyte can be irreproducible from spot-to-spot. Because of this, quantitative work is hampered significantly, and MALDI is not generally considered to be a good quantitative analysis technique. In an effort to create more homogeneous sample spots, researchers have developed ionic liquid-based MALDI matrices. Thus, analytes are literally dissolved in a molten salt, generally comprised of an anionic matrix compound mixed with a direct stoichiometric ratio of a

α-cyano-4-hydroxycinnamic acid
"α-cyano" or "CHCA"

Peptides, small molecules (< 10 kDa)

2,5-dihydroxybenzoic acid
"DHB"

Peptides, carbohydrates, polymers

3,5-dimethoxy-4-hydroxycinnamic acid
"Sinapinic Acid"

Peptides, proteins (> 10kDa)

2-(4-hydroxy-phenylazo)benzoic acid
"HABA"

Proteins, oligosaccharides, polymers

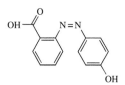

FIGURE 17.9 Some common MALDI matrices and their applications.

bulky counter-cation (to inhibit crystallization), and this homogeneous mixture generates a consistent ion signal regardless of where the laser is targeted on the sample spot. Because MALDI can be used to support a wide range of analytical determinations, from metabolomics, to lipidomics, to peptidomics and protein analysis, further development of methods to improve the technique is a rich area of research in modern analytical chemistry. Even so, many still regard sample preparation and plate spotting for MALDI analysis as an art that must be carefully practiced in order to obtain optimal results.

Mass Spectrometry Imaging

The ability to image an intact sample provides information about its content in context that might not be gleaned from a typical extraction experiment. The recent development of mass spectrometry imaging (MSI) has provided a whole new way to look at materials. In traditional MALDI, sample solutions containing analytes are mixed with a matrix and spotted on a stainless steel plate. In a similar fashion, a thin slice ($\sim$10 μm thickness) of tissue can be placed on the plate and coated with a thin layer of matrix. Then, the laser can be systematically moved (rastered) across the surface to collect mass spectra at different locations. If a mass spectrum is collected at each spatially resolved point on the surface, then a map can be generated to evaluate the relative abundance of different ion signals across the surface. An experiment such as this creates an enormous amount of data, but the ability to project a map that shows the abundance of any one or a number of ions can provide valuable information about the spatial distribution of proteins, lipids, or drug compounds in the tissue slice. An illustrative demonstration of the usefulness of this technique might be to profile and compare the expression of proteins outside, on the border, and within a cancerous portion of a tissue slice.

While very powerful, increasing spatial resolution across the sample surface is still an active area of research. Resolution of "pixels" on the sample may be limited by the size of the laser beam ($\sim$50–100 μm), or by how the sample is prepared for analysis. For the latter, the use of solvents to extract interfering species or the application of matrix can "blur" the features of interest. Some migration of chemical compounds in the sample could occur, and this would further worsen spatial resolution. As such, instrument manufacturers have developed a wide array of means for optimal slicing, fixing, and applying matrix to tissue samples prior to MSI. Besides MALDI, MSI can be performed using various ambient ionization techniques, as well as secondary ion mass spectrometry.

17.6 Secondary Ion Mass Spectrometry

Secondary ion mass spectrometry (SIMS) was developed through the middle of the twentieth century as a premier tool for analyzing the composition of solid surfaces and materials. Primarily elemental determinations are performed, but SIMS can also generate molecular ion fragments. A beam of primary ions (e.g., Xe^+, Ar^+, O^-, O_2^+, SF_5^+, or C_{60}^+ generated by an ion gun) is focused onto a surface or thin film under high vacuum ($< 10^{-4}$ Pa) in order to generate secondary ions from the sample that can be mass analyzed. The ion bombardment can create neutrals, positive ions, and negative ions simultaneously, and appropriate ion optics can be used to transfer the generated ions into the mass analyzer region. SIMS is usually used in conjunction with sector field, quadrupole, or time-of-flight mass analyzers. The low detection limits (picomoles of atoms per cubic centimeter), small size of the incident ion beam (spatial resolution as low as 10 µm), and ability to raster the surface for imaging applications, makes SIMS useful for direct elemental analysis from a wide variety of surface types, including circuit boards, meteorites, individual pollen particles, and microfossils. Depth profiling is also possible. One important consideration is that samples should be highly polished before analysis and nonconducting samples should be coated with a conducting pure metal to avoid charge build-up and to facilitate reproducible results. An excellent SIMS theory and training tutorial can be found from the Evans Analytical Group at http://www.eaglabs.com/mc/sims-theory.html. A book by Benninghoven et al. (Reference 1) is also commonly cited as one of the most authoritative works on the subject.

17.7   Inductively Coupled Plasma–Mass Spectrometry

The invention of the inductively coupled plasma (ICP) in the 1960s, and its subsequent commercialization in the 1970s, significantly expanded the capabilities of elemental analysis. ICP used as an atomization source for atomic emission spectroscopy, as well as some aspects of its use as a source for mass spectrometric analysis (ICP-MS) was covered in Chapter 12. Its main utility in elemental analysis arises from the circa 10,000 K environment in the argon plasma torch; this ensures complete destruction of organic compounds, and as shown through calculation of Maxwell–Boltzmann distributions (see Equation 12.1 in Chapter 12), promotes a sufficient proportion of elements exposed to the plasma from ground into excited or ionized states. ICP-MS yields many of the same advantages touted for ICP-AES, but often supplies superior sensitivity, especially for transition metal elements. In ICP-AES, discrete emission lines for different elements can be detected sequentially (with a scanning monochromator) or simultaneously (with a diode array detector). Similarly, in ICP-MS, atomic ions are efficiently transferred to a mass analyzer, where they can be separated based on their m/z ratio, and detected. Detection limits in the parts-per-trillion range are commonly reached, and linear ranges can extend as many as seven orders of magnitude.

Ions are typically extracted into the mass spectrometer from the torch region through a sampling cone and skimmer arrangement. This arrangement preserves and isolates the high vacuum region of the mass spectrometer from the ion source. While the ability to measure atomic masses of species adds to the specificity of analysis, some considerations have to be given to potential interferences. For example, measurement of signals at m/z 40, 56, and 80 are significantly hampered by the high abundance of Ar^+, ArO^+, and Ar_2^+ ions generated from the plasma. Immediately, the difficulty with measuring signals for ^{40}Ca and ^{56}Fe, the most abundant isotope of each of these elements, can be imagined based on these isobaric interferences. A comprehensive table of isobaric interferences, compiled by May and Wiedmeyer, can be found at http://shop.perkinelmer.com.cn/content/RelatedMaterials/Articles/ATL_TableOfPolya tomicInterferences.pdf. Many of these interferences are polyatomic ions, including oxides of various elements. Because of this, ions are typically made to traverse a collision or reaction cell, before they reach the mass analyzer. Different manufacturers have designated different trade names for their own arrangements, but typically a collisional (He) or reactive (H_2, CH_4, NH_3) gas, or a mixture of multiple gases, is introduced into the path of the ions in order to reduce the abundance of polyatomic interferences. Other radiofrequency ion guides of various configurations (quadrupoles, hexapoles, octapoles) or ion lenses are also integrated to stabilize the path of the ions to the mass analyzer, while allowing stray neutral species and solid particles to be pulled off by the vacuum. Most commercial ICP-MS instruments use a quadrupole mass analyzer for separation and selection of the ions of interest. This and other mass analyzer configurations are discussed below.

17.8 Mass Analyzers and Detectors

The mass analyzer is the heart of the mass spectrometer. Ions generated in the source, which may (e.g., electron ionization) or may not (e.g., electrospray ionization) be a high vacuum environment, are transferred to the mass analyzer region along a potential gradient. Positive ions will be repelled by regions of positive potential and attracted to regions of negative potential. Ion lenses and ion guides are used to stabilize the ion beam their path from source to detector. Neutrals are removed by a combination of physical barriers (i.e., ions can be made to turn corners whereas neutral species will not) and turbomolecular pumps used to achieve very low-pressure operating conditions. Once the ion packet reaches the mass analyzer, it can be manipulated and separated based on the application of precise electric and/or magnetic fields. The ion forms of interest are then directed toward an ion detector to measure their abundance.

Because ions have to be moved over significant distances to be mass analyzed, a high or ultrahigh vacuum environment is needed to maintain a reasonable **mean free path** for typical instruments. The mean free path is a measure of collision probability; expressed in units of distance, it is an indication of how far a species can travel at a given pressure without running into another species. Table 17.5 provides some general relationships between the number of molecules present and the mean free path at various pressures. In mass spectrometry, it is desirable to maintain a mean free path of ten to one hundred times the distance that an ion must travel to be analyzed. Incidentally, the resolution that can be achieved by a given mass analyzer is related to a considerable extent with the distance the ion is made to travel. For example, a time-of-flight instrument with a two-meter flight tube would provide higher resolution than an instrument with only a one-meter tube. An ion cyclotron resonance instrument achieves extremely high resolution by inducing the continual orbiting of ions under the influence of electric and magnetic fields. But, as you can see in Table 17.5, to achieve an adequate mean free path, vacuum pumps are needed to operate in the high vacuum to ultra-high vacuum regime. The cost and power associated with vacuum technology adds significant cost to high-performance instruments, and likewise, has been a major barrier to the production of portable mass spectrometers. Even so, researchers have made significant progress in miniaturized mass spectrometry in recent years. When microchip or even smaller-sized instruments can be fabricated, the requirements for vacuum technology becomes more modest, since a sufficient mean free path can be reached near ambient or sub-ambient pressures.

Following on the work of Eugene Goldstein and Wilhelm Wien in the late nineteenth century, Joseph John ("J.J.") Thomson, in 1905 at the Cavendish Laboratory in Cambridge, performed the necessary experiments to separate "blurry" projections of positive rays that had been produced when gas discharges were manipulated by electric

TABLE 17.5 Vacuum pressures and mean free paths for mass spectrometry

Vacuum range	Pressure (mbar)	Molecules/cm^3	Mean free path (m)	Mass analyzer
Ambient pressure	$\sim 10^3$	$10^{20} - 10^{19}$	$10^{-8} - 10^{-7}$	Nanoscale
Low vacuum	$10^2 - 1$	$10^{19} - 10^{16}$	$10^{-7} - 10^{-4}$	Microchip
Medium vacuum	$1 - 10^{-3}$	$10^{16} - 10^{13}$	$10^{-4} - 10^{-1}$	Miniature
High vacuum	$10^{-3} - 10^{-7}$	$10^{13} - 10^9$	$10^{-1} - 10^3$	Quadrupole; Ion trap
Ultra high vacuum	$10^{-7} - 10^{-12}$	$10^9 - 10^4$	$10^3 - 10^8$	Time-of-flight; Cyclotron

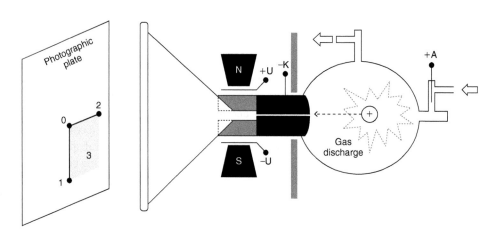

FIGURE 17.10 J.J. Thomson's parabola mass spectrograph. Ions accelerated only by a potential between $+A \rightarrow -K$ will strike the plate at 0. Application of an electric field only between $+U \rightarrow -U$ will direct positive ions to strike the plate between 0 and 1 (based on charge, q). Application of a magnetic field only will deflect ions to strike the plate between 0 and 2 (based on momentum, mv). Application of both an electric and magnetic field will cause particles to strike the plate in region 3 and be separated based on both q and mv.

and magnetic fields. Thomson is often referred to as the "Father of Mass Spectrometry." By reducing the pressure in the discharge tube, he reduced the contributions of collisions, which had blurred the output. His instrument is referred to as the parabola mass spectrograph, a diagram for which is shown in Figure 17.10. In this device, positively charged species produced during a discharge could be repelled toward a photographic plate. By passing the ions through a narrow slit, and then subjecting them to electric and magnetic fields, the ions could be separated based on their size and charge. This was such a seminal contribution to the field of mass spectrometry, it was suggested that the mass-to-charge unit be named in his honor. While the unit of Thomson (Th) has been used in many research papers to date (1 Th = 1 m/z unit), it is not an *SI* unit, nor has it been accepted by IUPAC.

  In the parabola mass spectrograph shown in Figure 17.10, ions generated on the right side of the apparatus are repelled through a transfer tube (to the left), after which they can be subjected to either an electric field, a magnetic field, or both. If no fields are applied, ions will strike the point indicated as "0" on the photoplate projection (far left). If an electric field is applied perpendicular to the path of ion travel, with the positive electrode on top and the negative electrode on the bottom, positive ions will be deflected along the line between 0 and 1 on the plate. The force ($\mathbf{F} = q\mathbf{V}$) felt by the ion is equal to the magnitude of the electric field ($\mathbf{V}$) multiplied by the charge on the ion [q; where $q = ze$, the magnitude of the charge (z) multiplied by the charge of an elementary particle ($e = 1.602 \times 10^{-19}$ Coulombs)]. Thus, a doubly charged ion would feel a force twice as large as a singly charged ion, and would be deflected twice as much. When a homogeneous magnetic field is applied perpendicular to both the electric field and the path of the ion, the effect on ion motion can be described by the Lorentz force ($\mathbf{F} = qv\mathbf{B}$, where q is the charge on the ion, v is the velocity of the ion, and $\mathbf{B}$ is the magnitude on the magnetic field). With a magnetic field only, the ions will strike on the line between 0 and 2. The orientation of the magnetic field as shown in Figure 17.10 induces the positive ions to make a right turn. The radius R of the turn, as depicted schematically in Figure 17.11, is dependent on the momentum of the ion and the strength of the applied magnetic field ($R = mv/q\mathbf{B}$). Thus, mass spectrometers that use solely magnetic fields for separations of ions are often referred to as momentum separators. When both electric and magnetic fields are applied in the parabola mass spectrograph, ions would strike somewhere in the region between 0 and 3, depending on the ion's charge, mass, and momentum, as well as on the magnitude of the fields applied. From gas discharge experiments, different ionic species produced could be

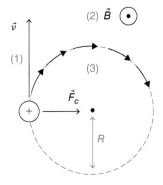

FIGURE 17.11 Motion of a charged particle in a magnetic field. A charged particle, initially accelerated to velocity v (1) enters a homogeneous magnetic field $\mathbf{B}$ (2) (oriented out from the page). The field changes the direction of the particle, and causes it to take a circular path (3) characterized by radius R.

clearly separated. Through this technology, Thomson and his student Francis Aston are given attribution for the discovery of isotopes of elements.

Example 17.4

A singly protonated ion having $m/z = 380.0$ is initially accelerated by an electric potential of 5000 V. After it is accelerated, it enters a homogeneous magnetic field with a strength of 4 T, applied perpendicular to the path of the ion's travel. What is the resultant radius of curvature for this ion in the magnetic field?

Solution

First, the velocity of the ion must be calculated.

Potential energy ($q\mathbf{V}$) can be set equal to kinetic energy ($0.5\,mv^2$) to solve for velocity.

$$q\mathbf{V} = (+1)(1.602 \times 10^{-19}\ \text{C})(5000\ \text{V}) = 8.01 \times 10^{-16}\ \text{J}$$

$$8.01 \times 10^{-16}\ \text{J} = 0.5\,((380.0\ \text{u})(1.6605 \times 10^{-27}\ \text{kg/u}))\,(v^2)$$

$$v = 50{,}387\ \text{m/s}$$

Next, the radius of curvature can be calculated

$$R = mv/q\mathbf{B}$$

$$R = \frac{(380.0\ \text{u})(1.6605 \times 10^{-27}\ \text{kg/u})\,(50{,}387\ \text{m/s})}{(+1)(1.602 \times 10^{-19}\ \text{C})\,(4\ \text{T})}$$

$$R = 0.0496\ \text{m} = \mathbf{4.96\ cm}$$

The units in these calculations are a bit tricky. Use the following conversions to help you reconcile the units and follow the solution:

1 V	= 1 J C^{-1}
1 J	= 1 kg m^2 s^{-2}
1 C	= 1 A s
q	= ze
e	= 1.602 × 10^{-19} C
1 u	= 1 amu
	= 1.6605 × 10^{-27} kg
1 T	= 1 kg C^{-1} s^{-1}
	= 1 kg A^{-1} s^{-2}
v (velocity)	= m s^{-1}

Discoveries at the Cavendish Laboratory quickly gave rise to the construction of various electric, magnetic, and double-focusing (both electric and magnetic) sector instruments through the middle of the twentieth century. **Double-focusing sector instruments** have been utilized in various configurations. A C-type arrangement (i.e., both sectors push ions in the same direction), also known as a Nier–Johnson geometry, passes ion through an electric and then a magnetic field ("forward focusing EB") or vice versa ("reverse focusing BE") in sequence in a consistent orbit to separate ions with very high resolution. Alternatively, the Mattauch–Herzog S-type (i.e., sectors push ions in opposite directions) EB arrangement has also been used. An excellent tutorial, which explains interactively the effect of applied electric and magnetic fields on ion separations in a double-focusing sector instrument, can be found through the website of the National High Magnetic Field Laboratory at Florida State University (http://www.magnet.fsu.edu/education/tutorials/java/dualsector/index.html). Today, double-focusing instruments are less common, given alternatives that are more affordable, faster, and smaller; however, they are still available for purchase (e.g., www.jeol.com) and are used by some specialized laboratories for high-resolution mass measurements in combination with a variety of different ion sources.

Quadrupole Mass Filter

In the 1950s and 1960s, German physicist Wolfgang Paul developed the **quadrupole mass analyzer** as a new simplified and affordable means for selective passage and separation of gas-phase ions. He was awarded the Nobel Prize in Physics in 1989 for this work. Figure 17.12 shows the basic design of the quadrupole analyzer. It consists

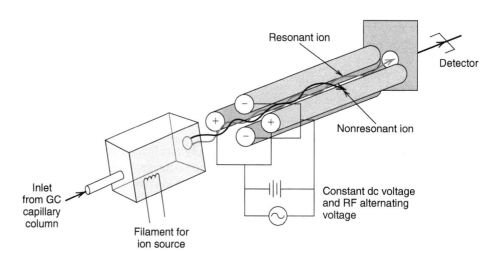

FIGURE 17.12 Quadrupole mass spectrometer.

of four parallel hyperbolic metal rods to which both a dc voltage (U) and an oscillating radiofrequency voltage (V cos ωt, where ω is the frequency and t the time) is applied. Two opposite poles are positively charged and the other two negatively charged, and their polarities change throughout the experiment. The applied voltages are U + V cos ωt and −(U + V cos ωt). The applied voltages determine the trajectory of the ions down the flight path between the four poles. As ions from the ionization source enter the RF field along the z-axis of the electrodes, they oscillate along the z-axis. Only those with a specific mass-to-charge ratio will resonate along the field and have a stable path through to the detector. Other nonresonant ions will be deflected (unstable path) and collide with the electrodes and be lost (they are filtered out). By rapidly varying the voltages, ions of one mass after another will take the stable path and be collected by the detector. Either ω is varied while holding U and V constant or U and V are varied, keeping U/V constant.

> The quadrupole analyzer is the most commonly used for GC–MS.

    The quadrupole analyzer has a number of advantages that make it ideally suited for mass analysis. The path does not depend on the kinetic energy (e.g., velocity) or angular deflection of the incoming ions. So the transmission rate is high. Since only a change in voltage is required, a complete scan can be very fast. As many as twenty spectra per second can be recorded over a range of about 800 mass units. Rapid scanning is needed to monitor chromatographic peaks that may be a fraction of a second wide or when multiple co-eluting signals are to be monitored. A resolution of about 1500 can be achieved, providing approximately unit resolution for small molecules. Additionally, quadrupoles generally provide higher dynamic ranges than other mass analyzers; this makes them attractive for quantitative analysis applications. When a quadrupole (or similarly, a hexapole or an octapole) is operated in RF-only mode (without the application of a DC voltage), then they can be used as efficient ion guides to transfer ions from one part of an instrument to another. In the RF-only mode, the motions of all ions are stabilized along the z-axis of the quadrupole.

Ion Trap

Wolfgang Paul also developed the **ion trap** mass analyzer. The three-dimensional ion trap shown in Figure 17.13 is often referred to as the quadrupole ion trap or Paul trap, but since then, many other ion trap configurations have also been developed (e.g., linear (2-D), rectilinear, Orbitrap, etc.). Similar to quadrupole analyzers, ion traps rely on oscillating electric fields. Using a three-electrode (entrance, ring, and endcap) arrangement, the applied fields allow for ions to be trapped for extended periods of time and

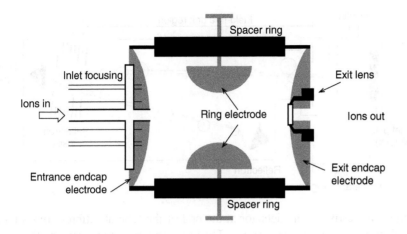

FIGURE 17.13 Cross-section view of a quadrupole ion trap mass analyzer. Adapted from the Mass Spectrometry Resource of the University of Bristol, School of Chemistry (http://www.chm.bris.ac.uk/ms/theory/qit-massspec.html).

manipulated during the course of an extended time domain. As a consequence, consecutive tandem mass spectrometry (excitation, fragmentation, and detection of ions) experiments can be performed.

Beginning in the mass-selective instability mode, the trap behaves as a total ion storage device when a potential of $U - V \cos \omega t$ is applied to the ring electrode; the endcap and entrance electrodes are held at ground. Ions precess in the trapping field with a frequency dependent on their mass-to-charge ratio. A minimal helium background ($\sim 10^{-3}$ Torr) is present to collisionally cool the ions; the collisions are not energetic enough to induce fragmentation. By increasing the magnitude of the DC and RF potential, as well as the frequency of the RF, the trap forces ions to become unstable and leave the trap to be detected. During the course of experiments, resonant excitation can be performed to isolate a particular ion desired to be fragmented. **Collision activated dissociation** in an ion trap involves isolating the ion of interest and subjecting it to multiple collisions over an extended timeframe (~ 10 msec) to induce fragmentation. A much more detailed treatment of ion trap theory can be found in an article by Raymond March (see Reference 17).

Like the quadrupole, the Paul trap has limited resolution, but is compact and inexpensive. Unlike the quadrupole, the ion trap does not sacrifice sensitivity when a wide m/z range is scanned; however, there are limitations on the number of ions that can be stored at any given time; space-charge effects can occur when too many ions are trapped together, and the result will be significant decreases in resolution, mass accuracy, and sensitivity. While scan rates are fast, extended duty cycles for more complicated ion manipulations can limit application to fast chromatography or for highly complex samples. Some limitations in sensitivity and resolution can be addressed with alternate geometries and specialized operation modes. Linear ion traps (see Reference 18) are typically more sensitive than quadrupole ion traps and the Orbitrap (see Reference 19) can achieve very high resolution by virtue of the extended orbiting of ions around a carefully machined spindle electrode.

Time-of-Flight

The **time-of-flight (TOF)** analyzer was developed in the 1970s, but did not become a popular tool for mainstream mass spectrometry practitioners until the 1990s. Figure 17.14 indicates the basic principle for separation of ions by time-of-flight. Ions formed are accelerated as a pulsed packet using repeller or accelerator plates, which supply a potential of thousands to ten-thousands of volts. The ion packets are pulsed into the flight tube at rates greater than 20,000 times per second, with the goal of supplying a constant kinetic energy to each ion. Ions of different m/z will travel at different

Ion traps have limited dynamic ranges because when too many ions are loaded into the trap, space-charge effects distort normal ion motions, leading to poor mass accuracy and decreased resolution.

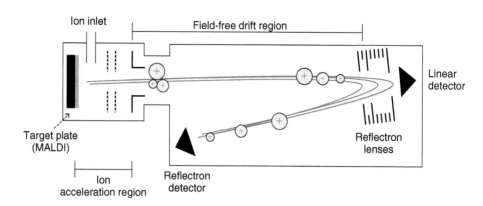

FIGURE 17.14 Time-of-flight mass analyzer.

velocities and arrive at the detector at the end of the tube at different times (smaller ions will travel faster than larger ions). The kinetic energy of ions leaving the source is given by:

$$\frac{mv^2}{2} = \mathbf{V}q \qquad (17.3)$$

where m = mass of the ion (kg), v = velocity of the ion (m s^{-1}), V = accelerating voltage (V), and q = ion charge (ze). Rearranging,

$$v = \sqrt{\frac{2\mathbf{V}q}{m}} \qquad (17.4)$$

So the ion velocity is inversely proportional to the square root of the mass-to-charge ratio of the ion. The time t to reach the detector is $t = L/v$, where L is the length of the flight tube. The difference, Δt, in arrival time that separates two ions is

$$\Delta t = L\frac{\sqrt{m_1} - \sqrt{m_2}}{\sqrt{2\mathbf{V}q}} \qquad (17.5)$$

depending, then, on the square root of the masses. Separation times between ions typically occur in the low microsecond to nanosecond timeframe.

Example 17.5

What is the flight time associated with a 455.67 m/z ion accelerated into a 2.250 m flight tube by an acceleration voltage of 15.0 kV?

Solution

From before (Example 17.4), we can reconcile the units as $(\text{V C kg}^{-1})^{0.5} =$ $(\text{J C}^{-1}\text{ C kg}^{-1})^{0.5} =$ $(\text{kg m}^2\text{ s}^{-2}\text{ kg}^{-1})^{0.5} = \text{m s}^{-1}$

$$v = \sqrt{\frac{2\mathbf{V}q}{m}}$$

$q = ze = (+1)\,(1.602 \times 10^{-19}\text{ C}) = 1.602 \times 10^{-19}\text{ C}$

$m = (455.67\text{ u})\,(1.6605 \times 10^{-27}\text{ kg/u}) = 7.5664 \times 10^{-25}\text{ kg}$

$v = ((2)\,(15000\text{ V})\,(1.602 \times 10^{-19}\text{ C})/7.5664 \times 10^{-25}\text{ kg})^{0.5} = 79698\text{ m/s}$

$t = L/v = 2.250\text{ m}/79698\text{ m/s} = 2.823 \times 10^{-5}\text{s} = \mathbf{28.23\ \mu\ sec}$

Pulsed-mode acceleration is needed since continuous ionization and acceleration would lead to a continuous stream of all ions with overlapping masses. The sequence of events for pulsed operation is to turn on the electron source for 10^{-9} s to form a packet of ions, and then the accelerating voltage for 10^{-4} s to accelerate the ions into the drift tube. Then the power is turned off for the remainder of the pulse interval as the ion packets drift down the tube to the detector (a process that takes microseconds to occur). TOF analyzers, like quadrupoles, scan the mass spectrum rapidly, but they are more amenable for high mass ion detection since they have no real upper mass limit. Even so, several limitations needed to be overcome before TOF analyzers became as popular as they currently are today.

The resolution in a TOF instrument is controlled by the temporal width of iso-mass ions when they reach the detector. When ions are initially formed, there is inherent dispersion in time, space, and velocity for a packet of ions. The ion formation process (e.g., through the input of laser energy) creates ions with a distribution of kinetic energies. Since a uniform kinetic energy profile is essential to achieving high resolution, some significant developments in technology were needed to control dispersion. Current TOF instrumentation, which can now routinely reach resolution exceeding 20,000, makes use of (a) longer flight tubes, (b) reflectron lenses, and (c) delayed extraction technology. Longer flight tubes could be accommodated through the incorporation of improved vacuum systems. Additionally, the use of reflectron lenses, which redirect and refocus ion packets back to a second detector inherently increases the flight tube length. Those isomass ions with higher kinetic energies fly further into the reflectron field (lower kinetic energies penetrate to a lesser degree), so that when the ion packet is redirected, dispersions in kinetic energy are corrected. While reflectron TOF instruments provide higher resolution than linear TOF systems, a limit is imposed on the size of ions (typically less than 20,000 m/z) that can be redirected by the lenses. Further, the use of delayed extraction, where a time delay is incorporated between the ionization and ion acceleration events, gives time for high kinetic energy ions to move away from the acceleration region and for especially small (neutral) species to be removed by the vacuum. Thus, the packet of ions that is accelerated after the delay has a narrower kinetic energy profile and it is not accelerated through a plume of other species that could collide with the ions and alter their trajectory. With these improvements, TOF mass analyzers have become widely used in coupling with a wide array of inlet and ionization techniques.

The linear TO analyzer is good for large molecules.

 # Professor's Favorite Example

Contributed by Professor
Michael Morris, University
of Michigan

Example 17.6

The dominant mass analyzer in proteomics, especially if MALDI is used, is the time-of-flight system (Figure 17.14). However, the dominant mass analyzer in GC–MS and LC–MS remains the quadrupole (Figure 17.12). Explain what it is about these applications and mass analyzers that make one or the other preferred.

Solution

The TOF system works with samples introduced in a pulse, as is the case with MALDI. Separation usually takes less than about 100 microseconds. A laser pulse of duration 5–20 nanoseconds is close to instantaneous compared to drift times.

The quadrupole doesn't handle pulsed introduction very gracefully. It can handle continuously introduced samples, because the ions exiting the quadrupole are selected by the ratio of DC/AC voltage. That makes the quadrupole well suited to GC–MS or LC–MS, because the separated constituents pass the detector region in times of a few seconds to a minute or two. That is much longer than the minimum scan time of about 0.1 second.

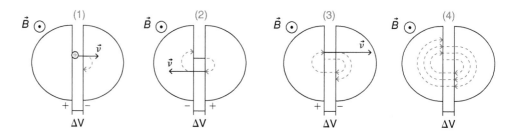

FIGURE 17.15 Manipulation of charged particle in a cyclotron. (1) Ion trapped in the cyclotron is accelerated by a potential ΔV. The ion enters a cyclotron orbit under influence of the magnetic field. (2) The ion orbits back to the acceleration region and its velocity is increased; the resulting orbit radius increases. (3) Again, the ion is accelerated to a higher kinetic energy and a higher radius. (4) The ion continues to be accelerated, and the cyclotron radius increases.

Ion Cyclotron Resonance

In 1941, Professor Ernest Lawrence at the University of California–Berkeley modified a cyclotron particle accelerator to act as a 180° magnetic sector instrument for the enrichment of radioactive uranium. The resulting "Calutron" (short name for California University Cyclotron) was soon mass-produced by the U.S. government as an integral component for preparative-scale uranium enrichment in the Manhattan Project. A **cyclotron** features a conducting cylindrical chamber segregated into two "D"s and held under high vacuum to allow acceleration of ions to very high speeds. As shown in Figure 17.15, ions that are introduced into the chamber experience a perpendicular homogeneous magnetic field, which confines the particle. As the ion returns to the space between the "D"s, it is further accelerated by an electric field. The ions are continually accelerated and confined in an orbit as they move across the gaps and feel the influence of the magnetic field, respectively. The time t it takes for an ion to orbit the device is dependent on its mass-to-charge ratio and the strength of the magnetic field, as shown by:

$$t = \frac{2\pi m}{qB} \tag{17.6}$$

For example, the time that it takes for one rotation of a proton that has been accelerated to a kinetic energy of 50 MeV in a 1 T magnetic field is 66 nsec (very fast!). This could be realistically achieved in a total of 80 μsec through 1225 rotations where the proton gains 40 keV on each trip around the cyclotron, but would require a switching frequency as high as 30 MHz. In reality, when very high energies are reached (and velocity begins to approach the speed of light), the time for orbit does not remain constant, and the frequency must be adjusted to correct for relativistic effects; the type of instrument which can do this is called a synchrotron.

Development of the Calutron set the stage for–modern **Fourier transform–ion cyclotron resonance (FT–ICR)** mass spectrometry. In FT–ICR, ions are injected into a modified cyclotron cell, commonly referred to as a Penning trap, as depicted in Figure 17.16. Once injected, ions are confined laterally by a strong electric field and axially by a strong magnetic field. Thus, it is essentially a different ion trap configuration. In the trap, ions are projected into a high radius cyclotron orbit through the application of a broadband RF pulse (called a "chirp") from the excite plates. As the ions pass the detector plates, they create an image current, which is recorded as a time domain signal. In order to visualize the recorded image current as a mass spectrum, a fast Fourier transform is used to convert the time domain signal into the frequency domain, which can then be converted to a mass spectrum.

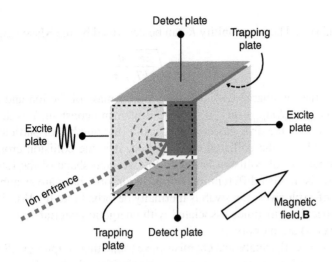

FIGURE 17.16 Ion cyclotron resonance mass analyzer.

Analogous to Equation 17.5, the cyclotron frequency (v_c) for an ion,

$$v_c = \frac{1.536 \times 10^7 B_0}{(m/z)} \qquad (17.7)$$

is dependent on its m/z ratio and the magnitude of the applied magnetic field; it is independent of the velocity of the ion. Typical magnetic field strengths available in modern commercial instruments range from 4–12 T, but some instruments with fields as high as 25 T have been developed in research facilities, such as the National High Magnetic Field Laboratory in Florida. Higher magnetic fields allow larger ions accelerated to higher kinetic energies to be contained, and they provide increased resolving power (in many cases greater than $R = 100,000$) to distinguish between ions having very similar masses. An instrument with a 9.4 T magnet has an upper mass limit of 10,000 m/z. Extremely high resolution provides the capability for very accurate mass measurements (< 1 ppm error in mass accuracy is commonly achieved).

However, this performance does not come without a price. To achieve high-resolution measurements on modest-size ions, they must be cycled for significant lengths of time in the instrument. Achieving $R > 100,000$ on a 1000 m/z ion in a 9.4 T instrument requires more than one second of analysis. On a chromatographic time scale, particularly in the analysis of complex samples, such as a cell lysate digest, it would be intractable to perform high-resolution analysis on every compound. Instead, FT-ICR analyzers are often coupled to faster mass analyzers to create a hybrid instrument, so that only those ions that are of greatest interest are sent to the FT–ICR instrument for high mass accuracy measurement. Additionally, FT–ICR instruments are among the most expensive on the market. The initial purchase of an instrument exceeds $1M, and the annual upkeep, especially for cryogenic liquids (nitrogen and helium) to maintain the superconducting magnet, can exceed $20,000 per year.

Ion Mobility Spectrometry

Ion mobility spectrometry (IMS) was originally developed in the 1950s and 1960s, and through the years it has enjoyed significant use for military and security purposes. In IMS, ionized molecules in the gas phase are separated based on their differential mobility in a carrier buffer gas. The buffer gas retards the motion of the ions as they travel the length of a tube under the influence of an electric potential. The migration time of an ion is dependent on its mass, charge, shape, and size. The ion mobility K can be described by the Mason equation as:

$$K = \frac{3}{16} \sqrt{\frac{2\pi}{\mu k T}} \frac{q}{n\sigma}$$

where q is the ion charge (ze), μ is the reduced mass of the ion and the buffer gas ($\mu = m_{ion}m_{gas}/(m_{ion} + m_{gas})$), k is the Boltzmann constant, T is the drift gas temperature, n is the drift gas number density (the number of particles per unit volume), and σ is the collision cross section. It is the collision cross section, or the probability of collision based on the size and shape of the ion, that can be used to distinguish different ion forms. In airports, bags are swabbed for the presence of explosives; the swab is thermally desorbed into the IMS instrument and the specific drift times associated with energetic material or their markers (e.g., solvents) are monitored.

More recently, analytical chemists have found that coupling IMS with mass spectrometry offers a vast array of new capabilities. With IMS separations on the millisecond timescale, and MS analysis on the microsecond timescale, the hyphenation of IMS and MS in tandem makes for a good marriage. This combination was initially pioneered in Bell Labs in the early 1960s, but only recently have commercial manufacturers begun to offer packaged instrument solutions. IMS-MS has been shown to be highly valuable in discerning different isoforms of biomolecules that might not otherwise be resolved by MS alone. In the end, addition of IMS provides another dimension of separation, which can be extremely useful for speciating complex mixtures.

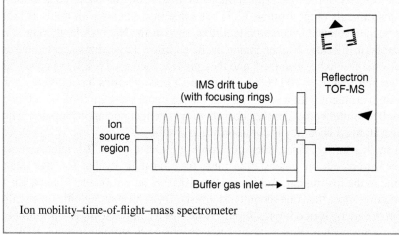

Ion mobility–time-of-flight–mass spectrometer

Ion Detectors

Once an ion is separated, isolated, or fragmented by the mass spectrometer, it is necessary to convert the abundance of that ion into an electric signal, which can be read by a data station. To be honest, mass spectrometry can be an inherently inefficient technique; only a small fraction of the molecules are converted to ions and ever reach a detector. High gain detectors are imperative in order to measure small ion currents reliably, but the fact that femtomole (10^{-15} mole) and attomole (10^{-18} mole; this is only 10,000 molecules!) amounts of analytes can be detected gives a good indication of the exceptional state-of-the-art of such devices.

J.J. Thomson originally used photographic plates for ion detection (Figure 17.10), but shortly thereafter, electrometers were introduced. Today, the field is dominated by the use of Faraday cups and electron multipliers. In a Faraday cup, the ions are made to enter a concave device in which the current required to neutralize the charge from the impinging ions is measured by an electrometer. However, these devices have no inherent gain and are only applicable for use with high-energy or high-abundance ions. Most modern-day instruments rely on electron multiplier detectors. These can be in discrete or continuous dynode arrangements, but the main concept is that an ion is accelerated toward a conversion dynode. Once struck by an ion, the conversion dynode releases electrons into the electron multiplier, which is a series of dynodes that sequentially increase the number of electrons produced. In the end, gains of $10^6 - 10^8$ can be achieved by electron multipliers. The gain is dependent on the number of dynodes and the intensity of impinging ions. Because signals can also be recorded for neutral molecules that reach the conversion dynode, detectors are often kept off-axis from the mass spectrometer inlet.

Other detector systems are currently under development. Microchannel plate electron multiplier arrays are being developed to reduce the distance that electrons must travel; this in turn reduces the response time and the power requirements for the detector. As researchers develop smaller mass spectrometer systems, power consumption becomes a primary consideration. In some time-of-flight systems, cryogenic detectors have been developed. While these can provide near 100% efficiency, which makes them ideal for slow-moving high-mass ions, the detector must be operated at near 2 K temperature, and current response times pale in comparison with modern microchannel plate detectors.

17.9 Hybrid Instruments and Tandem Mass Spectrometry

Every mass analyzer has its own advantages and limitations. Table 17.6 provides a comparison of general performance characteristics for stand-alone mass analyzers. In order to minimize the limitations and maximize advantages, various hybrid instruments, where multiple mass analyzers are connected in series, have been created. For example, quadrupole mass analyzers alone cannot be used to perform experiments, which involve the fragmentation of ions for various qualitative and quantitative applications. Instead, the quadrupole mass filter can be placed in series with other quadrupole analyzers [to create a triple quadrupole (QQQ) instrument] or with a time-of-flight instrument, to create a quadrupole–time-of-flight (Q–TOF) instrument. The QQQ is commonly used for very sensitive and selective quantitative analysis; the Q–TOF provides the ability for high mass accuracy measurement of filtered or fragmented ions.

TABLE 17.6 Some basic characteristics of common mass analyzers

Mass analyzer	Typical m/z range	Resolution	Error in mass accuracy (at 1000 m/z)	Dynamic range	Tandem MS?	Cost
Quadrupole	10^3	10^3	0.1%	$10^5 - 10^6$	No	$
Quadrupole ion trap	10^3	$10^3 - 10^4$	0.1%	$10^3 - 10^4$	Yes, MS^n	$
Triple quadrupole	10^3	10^3	0.1%	$10^5 - 10^6$	Yes, MS/MS	$$
Time-of-flight (linear)	10^6	$10^3 - 10^4$	0.1–0.01%	10^4	No	$$
Time-of-flight (reflectron)	10^4	10^4	5–10 ppm	10^4	Yes, PSD only	$$
Dual-sector	10^4	10^5	< 5 ppm	10^7	Yes, MS/MS	$$$$
Fourier transform–ion cyclotron resonance	$10^4 - 10^5$	10^6	< 5 ppm	10^4	Yes	$$$$

A variety of hybrid instruments continue to evolve. An ion trap–time-of-flight (IT–TOF) instrument allows for multistage tandem mass spectrometry in the ion trap followed by high mass accuracy measurements in the reflectron TOF mass analyzer. An ion trap–Fourier transform–ion cyclotron resonance mass analyzer provides greater versatility for highly complex determinations, such as those performed in proteomics measurements. Because the FT–ICR mass analyzer is relatively slow to obtain very high mass accuracy measurements, the addition of a linear ion trap allows for components of the analysis workflow to be performed with or without the FT–ICR part. For example, precursor peptide ions can be sent to the FT–ICR for high-resolution analysis, while the ion trap is used to concurrently perform tandem mass spectrometry measurements. The system is outfitted with multiple detector systems, so that ions analyzed by each mass analyzer can be recorded at the same time.

Another example of a common hybrid instrument is a TOF–TOF mass spectrometer. A TOF alone is not capable of performing typical collisional fragmentation measurements. In the reflectron mode, it can perform post-source decay (PSD) measurements, where the dissociation of meta-stable ions, those ions formed with excess internal energy, is monitored. In a TOF–TOF arrangement, a collision cell is placed between the two TOF mass analyzers. This provides two main advantages. The high mass range of the mass analyzer is maintained and ions can be accelerated to very high kinetic energies to enable high-energy collisional fragmentation. Afterward, even high mass fragment ions can still be monitored in the second TOF analyzer. Thus, the TOF–TOF is well suited for studying the fragmentation of proteins and polymers. High-energy collisional dissociation, as compared to the relatively low-energy collisions induced in an ion trap or a QQQ instrument, provides a greater wealth of fragmentation information. With high-energy collisions, large molecules cannot effectively redistribute the energy through bond vibrations and rotations (a situation that leads to poor fragmentation efficiency with low-energy collisions); instead, bonds are more easily broken throughout the molecule to provide rich fragmentation information.

Tandem Mass Spectrometry

One of the main reasons that hybrid instruments are created is so that **tandem mass spectrometry (MS/MS or MS^n)** can be performed. In MS/MS, selected ions are made to decompose and fragment through the application of additional energy. This energy can be supplied by collisions, photons, or electrons, and gives rise to product ions that are diagnostic of the precursor ion, which was targeted. MS/MS and MS^n can be very useful for qualitative analysis. MS^n, performed in ion traps, refers to the repeated isolation of ions following multiple fragmentation events. For example, MS^3 refers to the fragmentation of a precursor ion to obtain product ions, followed by isolation of one of the product ions for further fragmentation. In MS^4, another round of fragmentation is performed on the MS^3 products, and so on. The combinations of fragment ions (observed) and neutral losses (not observed) can be pieced together like a jigsaw puzzle to better ascertain the identity of the precursor ion. Additionally, tandem mass spectrometry can be useful for increasing sensitivity and selectivity in quantitative analysis experiments.

The usefulness of tandem mass spectrometry is best illustrated by the range of experiments, which can be performed in a QQQ mass spectrometer. Figure 17.17 describes four operation modes in which the first (Q1) and third (Q3) quadrupole are varied between selected ion monitoring (SIM) mode, where the quadrupole is set to let ions of only one m/z pass, and scan mode, where a range of ions are progressively passed through the rods by changing the magnitude of applied DC and RF voltages. The middle quadrupole (Q2) is used as a collision cell. It is maintained as an RF-only ion guide and collision gas (e.g., helium or argon) is introduced to induce fragmentation of ions as they pass through it.

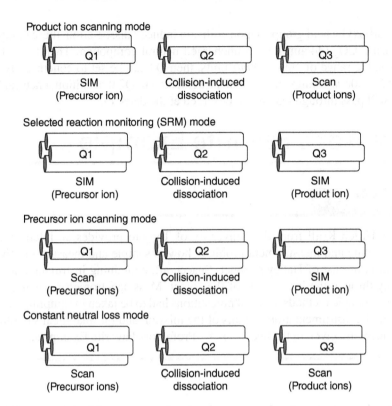

Product ion scanning mode

Q1 — SIM (Precursor ion)

Q2 — Collision-induced dissociation

Q3 — Scan (Product ions)

Selected reaction monitoring (SRM) mode

Q1 — SIM (Precursor ion)

Q2 — Collision-induced dissociation

Q3 — SIM (Product ion)

Precursor ion scanning mode

Q1 — Scan (Precursor ions)

Q2 — Collision-induced dissociation

Q3 — SIM (Product ion)

Constant neutral loss mode

Q1 — Scan (Precursor ions)

Q2 — Collision-induced dissociation

Q3 — Scan (Product ions)

FIGURE 17.17 Modes of tandem mass spectrometry in a triple quadrupole mass spectrometer.

Product ion scanning mode refers to the use of tandem mass spectrometry for qualitative analysis. In the QQQ arrangement, a specific precursor ion is selected in Q1 (SIM mode), is fragmented in Q2, and the product ions are scanned in Q3 to observe all fragment ions formed. Another common mode of QQQ operation is called **selected reaction monitoring (SRM)**. In SRM, both Q1 and Q3 are operation in SIM mode. Q1 selects for the precursor ion of interest. The ion is fragmented in Q2. Q3 is also operated in SIM mode, set specifically to monitor the m/z of a particular product ion that is unique to the precursor. This increases sensitivity, because the background noise is greatly reduced as a result of the SIM mode operation. Quadrupoles also provide the highest sensitivity when they are set to only monitor one specific m/z ratio. SRM mode enhances specificity, because the product ion monitored after Q3 should only result from the precursor ion of interest selected in Q1. Therefore, the chance of obtaining a signal from interfering species is substantially diminished. Because QQQ mass spectrometers also have fast scanning speeds, it is common to set multiple reaction events if it is desired to quantify multiple analytes in a single run. This is referred to as **multiple reaction monitoring (MRM)**; multiple precursor/product ion pairs are programmed into the software to be simultaneously monitored in a given time frame. Some manufacturers have touted the ability to efficiently monitor as many 500 different compounds in a single analytical run using the MRM mode in a QQQ instrument.

A QQQ can also be used in **precursor ion scanning** mode. In this mode, Q1 is scanned over a range of m/z values, and Q3 is operated in SIM mode to monitor for a specific fragment ion. Thus, all ions that pass through Q1 are fragmented in Q2. A signal will be observed at the detector for all ions that release a particular fragment. For example, phosphorylated molecules will often release a phosphate ion (PO_3^-, $m/z = 79$) following collision-induced dissociation. In precursor ion scanning mode, if Q3 is set to only allow ions of $m/z = 79$ to pass, then all ions that pass through Q1 and fragment to create this product ion can be identified. In a similar manner, ions that give rise to loss of a specific neutral compound can be monitored using the **constant neutral loss** (CNL) or neutral loss monitoring mode. In this mode, both Q1 and Q3 are scanned, but Q3 is scanned a certain number of m/z units lower than Q1. To illustrate the usefulness of this technique, consider that one is interested in all compounds that contain carboxylic acid groups. Upon collision induced dissociation, these compounds often release CO_2 (44 amu) as an uncharged neutral compound. Therefore, if Q1 is set to scan the range of 100–350 m/z units, then Q3 can be set to concurrently scan a range of 56–306 m/z units; thus, all ions fragmented in Q2 that exhibit a neutral loss of 44 amu will pass through Q3 and be observed at the detector.

⌐Contributed by Professor
Ulrich Krull, University of
Toronto⌐

 # Professor's Favorite Example

Example 17.7

Professor Ulrich Krull from the University of Toronto provides an interesting real-world sampling and measurement problem, how to sample and measure the chemical composition of dust of Halley's comet. Spacecraft, containing instrumentation, were sent to fly through the inner coma of the comet. Mass spectrometry was used, along with spectroscopic methods. Special precautions had to be taken in sampling to prevent damage to the instrumentation. Details of the mission and the analytical sampling and measurement are given in the text website, Professors Favorite Examples.

Questions

1. What are the six main components of a mass spectrometer instrument?
2. What is the difference between monoisotopic mass and average mass?
3. How are resolution and mass accuracy connected to one another?
4. What is the molecular ion?
5. Why is there low pressure in a mass spectrometer?
6. What is the nitrogen rule?
7. What ion sources are commonly used for GC–MS?
8. What ion sources are commonly used for LC–MS?
9. How abundance is measured in a mass spectrometer?
10. Why do samples need to be ionized in a mass spectrometer?
11. Describe how ions are generated by electrospray ionization.
12. The discovery of what aspect of electrospray ionization won John Fenn a portion of the 2002 Nobel Prize in Chemistry?
13. What is the difference between atmospheric pressure chemical ionization (APCI) and conventional chemical ionization (CI)?
14. What is mean free path, and why is it an important consideration for different mass analyzers?
15. Why do you think the "time-of-flight" mass analyzer is called so?
16. What is the Lorentz force?
17. Compare and contrast the capabilities offered by single quadrupole vs. ion trap instruments.
18. Why is the upper limit of an FT–ICR instrument dependent on the strength of its magnet?
19. How can a triple quadrupole be used to perform quantitative and qualitative analysis experiments?

Problems

Masses and Isotopes

20. Calculate the monoisotopic mass for the protonated and sodiated ion forms of caffeine.
21. What ion signals (m/z), and at what relative intensities, would you expect to observe the deprotonated ion form of 2-chloro-benzoic acid?

Resolution and Mass Accuracy

22. What resolution is needed to resolve an ion of m/z 432.1124 from an ion of m/z 432.1186?

23. Calculate the resolution (FWHM) exhibited by the ion signal below. What mass analyzers are capable of achieving this resolution?

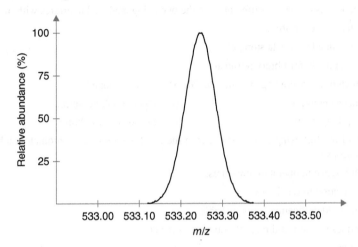

24. A 34,525 Da protein is ionized by electrospray ionization. What is the resolution needed for the signals of the +34 and +35 charge states (assuming all charges are obtained by protonation) to be completely resolved. Consider only the signal for the most abundant isotopes of each ion.

25. A standard when protonated should have a monoisotopic mass of 1234.1223 Da. When the compound is measured on a time-of-flight mass analyzer, the recorded mass of the ion is 1234.1198 Da. What is the error in mass accuracy for this measurement?

Ions in Electric and Magnetic Fields

26. What is the velocity (m/s) of a singly protonated ion with a m/z ratio of 324.9, which has been accelerated to a kinetic energy of 0.75 MeV? What is the radius of curvature for this ion, when it experiences a homogeneous magnetic field of 7 T?

27. What would be the difference in arrival time at the detector for two singly charged ions in a time-of-flight mass analyzer; one with $m/z = 1252.054$ and one with $m/z = 1253.138$, accelerated by a potential of 20 kV in a 1.750 m flight tube?

Multiple Choice Questions

1. Which of the following is determined using mass spectrometers?

    (a) Composition in sample   (b) Concentration of elements in sample

    (c) Relative mass of atoms   (d) Properties of sample

2. Mass fragment of $[IrCl]^+$ in mass spectrometry shows three mass peaks at $m/z = 226, 228$ and 230. Given that natural abundances of ^{191}Ir, ^{193}Ir, ^{35}Cl, and ^{37}Cl are 37%, 63%, 76% and 24% respectively, the intensities of the mass peaks are in the order:

    (a) 49.5 : 100 : 26.6   (b) 100 : 49.5 : 26.6

    (c) 26.6 : 100 : 49.5   (d) 26.6 : 49.5 : 100

3. Mass spectrum of a compound shows an $[M^{2+}]$ ion peak that is about 4% of M^+. This indicates that the compound has one

    (a) fluorine.   (b) sulphur.

    (c) bromine.   (d) chlorine.

4. In mass spectrometer, the sample to be analyzed is bombarded with _____.

   (a) Protons
   (b) Electrons
   (c) Neutrons
   (d) Alpha particles

5. Which of the following statements is false about mass spectrometry?

   (a) Impurities of masses different from the one being analyzed interferes with the result.
   (b) It has great sensitivity.
   (c) It is suitable for data storage.
   (d) It is suitable for library retrieval.

6. For which of the following systems has GC-MS been developed?

   (a) Packed column
   (b) Open tubular column
   (c) Capillary column
   (d) Porous layer column

7. Which of the following problems occur in the combination of gas chromatography and mass spectroscopy?

   (a) Difference in operating pressures.
   (b) Reduction in sensitivity.
   (c) Direct identification is not possible.
   (d) It does not permit direct introduction of the effluent.

8. The percent of the effluent of the liquid chromatography that must be introduced in the mass spectrometer is _____.

   (a) 1–2
   (b) 1–5
   (c) 1–20
   (d) 1–15

9. The gas burden from conventional LC flow rates is _____.

   (a) 1 mL/min of water produces 1.2 L/mm of gas
   (b) 1 mL/min of water produces 2.4 L/mm of gas
   (c) 2 mL/min of water produces 3.2 L/mm of gas
   (d) 1 mL/min of water produces 4.2 L/mm of gas

10. Which of the following devices can be combined with tandem mass spectroscopy?

    (a) Mass spectrometer and gas-solid chromatograph
    (b) Mass spectrometer and gas-liquid chromatograph
    (c) Mass spectrometer and gas chromatograph
    (d) Mass spectrometer and mass spectrometer

11. The analyzer used in a tandem mass spectrometer is _____.

    (a) time of flight mass analyzer.
    (b) magnetic deflection analyzer.
    (c) radiofrequency analyzer.
    (d) quadrupole analyzer.

12. The ions in the time of flight mass spectrometer are formed by which of the following methods?

    (a) Pulsed ionization method
    (b) Acceleration method
    (c) Dynamic method
    (d) Ion excitation method

13. The heart of quadrupole instrument is _____.

    (a) electrodes
    (b) choke
    (c) DC potential
    (d) detector

14. The least sensitive ion detector is _____.

    (a) faraday cup collector
    (b) channeltron
    (c) micro-channel plate
    (d) electron multiplier transducer

15. ICP's principle is similar to _____.

    (a) flame emission spectroscopy
    (b) fourier transforms spectroscopy
    (c) atomic emission spectroscopy
    (d) absorption spectroscopy

16. Liquid samples are introduced into the ICP spectrometer using _____.

    (a) nebulizer   (b) cuvette having glass windows

    (c) probe   (d) laser ablation system

17. The spectral range of SIMS is _____.

    (a) 0–10 amu   (b) 0–100 amu

    (c) 0–500 amu   (d) 0–1000 amu

18. _____ is the amount of matrix effect that occurs in SIMS.

    (a) Very low   (b) Low

    (c) Some   (d) Severe

Recommended References

General

1. F. W. McLafferty and F. Turecek, *Interpretation of Mass Spectra*, 4th ed. Sausalito, CA: University Science Books, 1993.
2. C. Dass, *Fundamentals of Contemporary Mass Spectrometry*. Hoboken, NJ: John Wiley & Sons, Inc., 2007.
3. J. H. Gross, *Mass Spectrometry: A Textbook*, 2nd ed. Heidelberg, Germany: Springer-Verlag, 2011.
4. M. A. Grayson, (Ed.) *Measuring Mass: From Positive Rays to Proteins*. American Society for Mass Spectrometry. Santa Fe, NM. 2002.
5. T. Kind; O. Fiehn. "Seven Golden Rules for heuristic filtering of molecular formulas obtained by accurate mass spectrometry," *BMC Bioinformatics*. **8**, (2007), 105.

Specific Techniques

6. H. Budzikiewicz, "Negative Chemical Ionization (NCI) of Organic Compounds," *Mass Spectrom. Rev.* **5**, (1986), 345–380.
7. R. B. Cole, "Some Tenets Pertaining to Electrospray Ionization Mass Spectrometry," *J. Mass Spectrom* **35**, (2000), 763–772.
8. N. B. Cech and C. G. Enke, "Practical Implications of some Recent Studies in ESI Fundamentals," *Mass Spectrom. Rev.* **20**, (2001) 362–387.
9. J. B. Fenn, "Electrospray Ionization Mass Spectrometry: How It All Began," *Journal of Biomolecular Techniques*. **13**, (2002), 101–118.
10. A. Lo and K. A. Schug, "A Birds-Eye View of Modern Proteomics," *Separation & Purification Reviews*. **38**, (2009), 148–172.
11. Z. Takats, J. M. Wiseman, B. Gologan, and R. G. Cooks "Mass Spectrometry Sampling Under Ambient Conditions with Desorption Electrospray Ionization," *Science*. **306**, (2001), 471–473.
12. R. B. Cody, J. A. Larance, and H. D. Durst, "Versatile New Ion Source for the Analysis of Materials in Open Air Under Ambient Conditions," *Anal. Chem.* **77**, (2005), 2297–2302.
13. H. Chen, G. Gamez, and R. Zenobi "What Can We Learn from Ambient Ionization Techniques?" *J. Am. Soc. Mass Spectrom.* **20**, (2009), 1947–1963.
14. H. Chen, A. Venter, and R. G. Cooks, "Extractive electrospray ionization for direct analysis of undiluted urine, milk and other complex mixtures without sample preparation," *Chem. Commun.* (2006) 2042–2044.
15. A. Benninghoven, F. G. Rüdenauer, and H. W. Werner, Secondary Ion Mass Spectrometry: Basic Concepts, Instrumental Aspects, Applications, and Trends. New York: Wiley, 1987.
16. Z. Ouyang and R. G. Cooks, "Miniature Mass Spectrometers," *Ann. Rev. Anal. Chem.*, **2**, (2009) 187–214.
17. R. E. March, "An Introduction to Quadrupole Ion Trap Mass Spectrometry," *J. Mass Spectrom.* **32**, (1997), 351–369.
18. D. J. Douglas, A. J. Frank, and D. Mao, "Linear Ion Traps in Mass Spectrometry," *Mass Spectrom.* **24**, (2005), 1–29.
19. M. Scigelova and A. Makarov "Orbitrap Mass Analyzer—Overview and Applications in Proteomics," *Proteomics*. **6**, (2006), 16–21.

Chapter 18

THERMAL METHODS OF ANALYSIS: PRINCIPLES AND APPLICATIONS

"Heat, like gravity, penetrates every substance of the universe, its rays
occupy all parts of space."

—Baron Jean-Baptiste-Joseph Fourier

KEY THINGS TO LEARN FROM THIS CHAPTER

- Principles and applications
- Thermogravimetric analysis
- Differential thermal analysis

- Differential scanning calorimetry, (key Equations: 18.1–18.8)
- Thermometric titrations, (key Equations: 18.9–18.10)

What is analytical chemistry? It is the study of the separation, identification, and quantification of the chemical components of natural and artificial materials. The basis is qualitative analysis (identity of the chemical species in the sample) and quantitative analysis (amount of one or more of the components). The separation of components is often performed prior to analysis. Analytical methods can be separated into classical and instrumental. In the previous chapters, a number of analytical methods have been discussed.

For quantitative analysis, an old and one of the most important methods is gravimetric analysis. Gravimetric analysis is a class of laboratory techniques used to determine the mass or concentration of a substance by measuring a change in mass. The chemical that is being quantified is called the analyte. In gravimetry, there is very little room for instrumental error and this method does not require a series of standards for the calculation of an unknown. The disadvantage is that this method is meticulously time consuming, however, this method is simple and anyone can carry out the estimation without any specific training. The basic instruments required are only a simple crucible, burner, and a good balance. There were however, questions on the reproducibility, limits of detection and sensitivity of this method, which were overcome to an extent with the development of thermogravimetric analysis (by ascertaining temperature at which constant weight is obtained).

In principle, there are four fundamental types of gravimetric analysis: physical gravimetry, *thermogravimetry*, precipitative gravimetric analysis, and electrodeposition. These differ in the preparation of the sample before weighing of the analyte. Physical gravimetry is mostly used in environmental engineering.

This chapter will focus mainly on thermo-analytical methods. The shortcomings of thermogravimetry (TGA) have been overcome by the development of various other methods like differential thermal analysis (DTA), differential scanning calorimetry (DSC), thermomechanical analysis (TMA), dynamic mechanical analysis (DMA), microthermal analysis and thermometric titration analysis (TTA). Over the years, a great deal of progress has been made in this field, as can be seen with the availability of accurate single pan balance, improved ovens (for drying), devices (thermocouples, analog devices) that can be used to easily raise the temperature of ovens to a fixed temperature. In this chapter mainly TGA, DTA, DSC and TTA will be discussed.

Thermal analysis can provide information regarding-variation of density, thermal stability; and free water; bound water; purity, melting point, boiling point, heats of transition, specific heats, phase diagrams, reaction kinetics studies of catalysts, glass transitions, and so on.

18.1 Principles and Applications

In the thermal method of analysis, a physical property of a substance and/or its reaction products is measured as a function of temperature or time while the substance is subjected to a controlled temperature program, in a specified atmosphere. The measurements are usually as a dynamic function of temperature. This determination method was proposed by Mckenzie in 1979 and is accepted by the International Confederation for Thermal Analysis (ICTA). The different types of thermal methods include—TGA, DTA, DSC, TMA, DMA, microthermal analysis and thermometric titrations (Table 18.1) (References 1–5). For an ideal thermal technique, the three criteria that need to be followed are: measurement of physical property; expression (directly or indirectly) of the measurement as a function of temperature; measurement under a controlled-temperature program with increase or decrease of temperature under isothermal, aerobic or anaerobic conditions.

The mass changes by evaporation, dehydration, decomposition, and oxidation reaction are measured by TGA. The exothermic and the endothermic phenomena can be measured from DTA, by simultaneous measurement with TGA in combination with DTA.

DSC is used mainly for the determination of melting, glass transition, crystallization, chemical reactions, thermal history, and specific heat capacity. Due to the mass change by the decomposition and the corrosion of the sensor by decomposed gas, sublimation, evaporation, and thermal decomposition are usually not measured using this technique.

The information for the phenomena resulting from deformation including the thermal expansion, thermal shrinkage, glass transition, and curing reaction are obtained using TMA. Measurement of the melting and crystallization is avoided with TMA as there is possibility of adhesion of the sample to the probe on melting.

Changes in the molecular motion and the molecular structures such as relaxation, crystallization, and the curing reaction are measured by DMA. This method can also be used to measure the melting at the early stage until the melt flow occurs.

In thermometric (thermal) titration analysis (TTA), change in temperature is measured against the volume of the titrant. In TTA the titrant is added at a known constant rate to a titrand till the reaction completion is indicated by a change in temperature.

The gradual and reversible transition in amorphous materials from a hard and relatively brittle "glassy" state into a viscous or rubbery state with increase in temperature is known as the glass–liquid transition, or glass transition. Amorphous solids that show a glass transition can be called as glass. The transformation of the viscous liquid into the glass state by supercooling is known as vitrification (reverse transition).

The glass-transition temperature, T_g of a sample determines the range of temperatures during which the glass transition takes place. T_g is always lower than T_m (the melting temperature) of the crystalline state of the sample, incase it exists.

Hard plastics like polystyrene, poly(methyl methacrylate) and rubber elastomers like polyisoprene and polyisobutylene are known to show glass transition temperatures.

TMA is a technique in which deformation of the sample under non-oscillating stress (compression, tension, flexure or torsion) is monitored against time or temperature while programming the temperature of the sample in a specified atmosphere.

DMA is a technique used to study the viscoelastic behavior of polymers.

TABLE 18.1 Thermal Measurement Technique and Property Measured

Method	Property	Unit
Thermogravimetric analysis (TGA)	Mass	gram
Differential Thermal Analysis (DTA)	Temperature difference	°C or μV[a]
Differential Scanning Calorimetry (DSC)	Enthalpy	W = J/sec
Thermometric titration analysis (TTA)	Change of temperature	°C or μV[a]
Thermomechanical Analysis (TMA)	Deformation	meter
Dynamic MechanicalAnalysis (DMA)	Elasticity	Pa = N/m^2

[a]Thermocouple electromotive force.

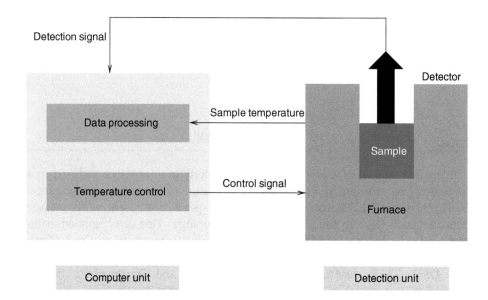

FIGURE 18.1 Block diagram of a thermal analysis instrument.

The data obtained from all the above methods are used to plot curves i.e. thermal spectra and are known as *thermograms*. The method requires little sample for tests and gives a variety of results on a single graph.

The methods find applications for quality control and research applications on industrial products like polymers, clays and minerals, pharmaceuticals, metals, alloys and several areas in food, catalysis, ceramics, civil engineering, inorganic, organic, petrochemical and glass. The disadvantage in the use of thermal analysis is the relatively high cost of the equipment.

Instrumentation

The thermal analysis instrument consists of the following parts (Figure 18.1):

1. *Detection Unit* consists of a furnace, sample and reference holder; and sensor that can heat and cool the sample in the furnace in addition to detecting the sample temperature and property.
2. *Temperature Control Unit* that can control the furnace temperature.
3. *Data Recording Unit* records the signals of sensor and sample temperature, and analyzes the data obtained.

Temperature control, data recording and *analy*sis are all computer-controlled. The combination of the furnace and sensor (thermocouple) enables to measure in various types of techniques. Thermocouple is an example of a *transducer*. Transducer can convert variation in a physical quantity (pressure, brightness) into an electrical signal and *vice versa*. Hence, the thermocouple functions as the sensor that measures the temperature and gives an output in the form of voltage (e.m.f, electromotive force). The computer can be connected to several other instruments (other types of measurement techniques) thus, enabling the simultaneous measurement and analysis.

18.2 Thermogravimetric Analysis (TGA)

In *TGA*, sample mass/weight is recorded as a function of temperature or time in a controlled atmosphere (helium, nitrogen, air, argon) (environment is heated or cooled at a controlled rate), with increase in the temperature (usually linearly with time) of the sample. The *thermogram* or *thermal decomposition curve* is obtained from the plot

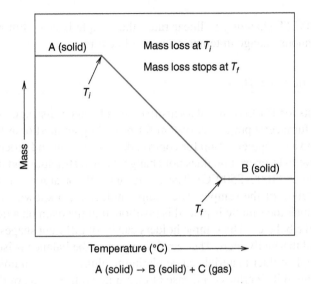

FIGURE 18.2 General example of a thermogram obtained from TGA.

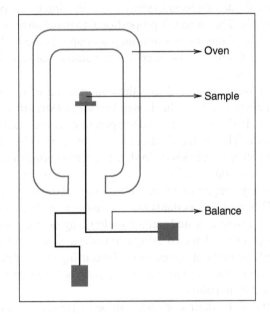

FIGURE 18.3 Instrument set-up for TGA.

of mass or mass percent as a function of time (Figure 18.2). The TGA curve is not a "*fingerprint curve.*" Mass is lost if the substance contains a volatile fraction. There are a large number of chemical substances that invariably decompose upon heating which forms the underlying principle of TGA i.e. heating a sample leads to observable weight changes (Figure 18.3). TGA can only help to develop the mass loss at a particular temperature, it cannot identify the species involved in the mass loss. In order to obtain information about the species, the output for a TG analyzer needs to be connected to FTIR or Mass spectrometer (TGA/FTIR or TGA/MS). Other ways could be to use X-ray diffraction and scanning electron microscope (SEM) techniques or using gas detecting devices.

For analysis the following types of techniques are generally adopted:

1. Isothermal or static TG (Constant T), constant temperature is maintained for a given time during which the weight is being recorded.

2. Quasi-static TG (Constant weight), the sample is heated to a constant mass at each series of constant temperature.

3. Dynamic TG (T changing in linear rate), the sample is heated in an environment with continuous change in temperature at a linear rate.

Instrumentation for TGA

The instruments for TGA consist of a sensitive microbalance, also known as thermobalance; oven or furnace; a purge-gas system for providing an inert or at times a reactive atmosphere and a computer set-up for control, data collection and processing. The furnace is sensitive to temperature, does not change its own structure and usually sustains high temperatures (1000–2500°C). The carrier gas (N_2 or argon) in the reactor prevents the oxidation of the sample. The sample holder (also known as pan, basket or crucible) in which the sample is placed is positioned in the oven on a quartz beam that is attached to the balance. The sample holders can be of different shapes and need to be chemically and thermally inert. The remaining part of the balance is isolated from the oven. The sample holder material can be platinum, aluminium or alumina; platinum is mostly used due to its inertness and ease of cleaning. Null position of the quartz beam is maintained by a current flowing through the transducer of the electromagnetic balance. Photosensitive diodes are used to sense the deflection of the beam due to change in weight of the sample. The current is proportional to the change in weight of the sample. In TGA, mostly a platinum-rhodium thermocouple is used. Common heating or cooling rates of 5–10°C/min are followed and the amount of sample that can be measured varies from 1–300 ng.

The computer software also allows the computation of rate of change of mass i.e. dm/dt versus the temperature. This is also known as *Derivative Thermogravimetry* (DTG) (Figure 18.4). In this case, time and temperature are included in the same spectra in the derivative mode. This method is more accessible than TGA, though the information obtained from both the methods is similar. The temperature changes are easily read out from DTG, the area under the DTG curve gives the mass loss value and the height of the DTG peak at any temperature gives information about the rate of mass change at that temperature. Other related methods are *evolved gas analysis* (EGA) that is used to study the gas evolved from a heated sample undergoing decomposition or desorption. Either the evolved gases are detected using *evolved gas detection* (EGD) or analysis is carried out explicitly to find out the gases evolved using *evolved gas analysis* (EGA). Another method, *emanation thermal analysis* (ETA) is based on the measurement of the inert gas released from solids.

The sensitivity of the thermobalance can be increased by using a good recorder to record the mass loss or gain as a function of temperature or time; using a furnace that sustains high temperatures; by allowing the accuracy of mass loss to be within ± 0.001 % and temperature accuracy to be ± 1 to ± 0.1 %. Moreover, the balance should not be affected by physical effects like radiation, convection and magnetic effects because of heating. The sample should be maintained at uniform temperature

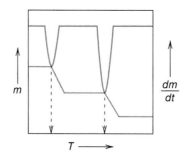

FIGURE 18.4 TG and DTG thermograms.

by fixing the position of the crucible so that it is uniformly heated. A small scan rate (5–10°C/min) should be maintained at lower temperatures while a higher scan rate (20–30°C/min) can be considered for higher temperatures. The decomposition temperature of the sample can be controlled through heating rate and sample size. Calibration of the balance should be possible with a standard certified substance (e.g.,alloy) so that the accuracy of the method can be preserved. Further, TGA is affected by particle size of the sample, packing in the sample, the crucible shape and the gas flow rate from the purge-gas system.

Applications of TGA

TGA can be applied in thermal decomposition of substances (calcination and heat treatment and polymer stability), corrosion of metals, determination of moisture, volatiles, and ash content, measuring evaporation rates and sublimation, distillation and evaporation of liquids, reaction kinetics studies, compound identification, heats of vaporization and vapor pressure determination and so on.

Example 18.1

Explain the decomposition (in N_2) profile of calcium oxalate monohydrate following a TGA experiment.

Solution

The thermogram for decomposition (in N_2) of calcium oxalate monohydrate is shown below.

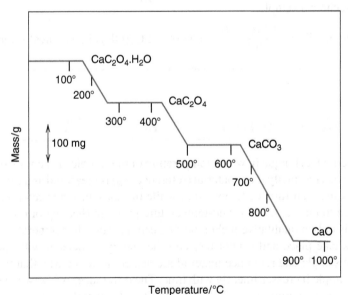

The clearly defined horizontal regions correspond to temperature ranges in which the indicated calcium compounds are stable. The equations for the decomposition are as follows:

$$CaC_2O_4 \cdot H_2O \, (s) \rightarrow CaC_2O_4 \, (s) + H_2O \, (g)$$

$$CaC_2O_4 \, (s) \rightarrow CaCO_3 \, (s) + CO \, (g) \, (in \, N_2)$$

$$CaCO_3 \, (s) \rightarrow CaO + CO_2 \, (g)$$

Contributed by Dr. Sandeep Kaur-Ghumaan, University of Delhi, India

 # Professor's Favorite Example

Example 18.2

In the TGA of 0.250 g of $Ca(OH)_2$, the loss in weight at different temperatures is as follows:

(i) 0.021 g at 100–150°C (loss of hygroscopic water)
(ii) 0.035 g at 500–560°C (dehydration)
(iii) 0.0233 g at 900–950°C (dissociation)

What is the composition of calcium hydroxide?

Solution

The composition can be determined assuming the reaction:

$$Ca(OH)_2 \cdot H_2O \xrightarrow{\Delta,\,500°C} Ca(OH)_2 \xrightarrow{\Delta,\,900°C} CaO + H_2O$$

Amount of hygroscopic water $= \dfrac{0.021 \times 100}{0.250} = 8.4\%$

From mol. wt., $Ca(OH)_2 = 74$, $H_2O = 18$

Amount of $Ca(OH)_2$ dehydrated $= \dfrac{74 \times 0.035}{18} = 0.144$ grams of $Ca(OH)_2$ in 0.250 g of technical sample.

% weight will be $\dfrac{0.144 \times 100}{0.250} = 57.6\%$ of $Ca(OH)_2$ in technical sample.

18.3 Differential Thermal Analysis (DTA)

DTA is a thermal technique in which temperature of the sample (T_S) compared with the temperature of a thermally inert material (reference, T_R) is measured as a function of the sample/inert material/furnace temperature while the substance and reference material are subjected to the same controlled-temperature program (heating or cooling). However, DTA is only a qualitative technique that can measure the temperatures at which the changes take place and cannot measure the energy associated with the changes. Usually, the sample and reference material are heated in such a way that the temperature of the sample increases linearly with time. The difference, ΔT between the sample and reference temperatures (i.e., T_S–T_R), is monitored and plotted against the temperature of the sample resulting in a differential thermogram (Figure 18.5). The area under the DTA peaks is proportional to the heat of the reaction and the mass loss and is also, inversely proportional to the thermal diffusivity of the sample.

The temperature changes in the sample can lead to either physical changes or chemical reactions which can be either exothermic or endothermic that may or may not be accompanied by a weight change. When T_S–T_R decreases the process is endothermic and when T_S–T_R increases the process is exothermic. Vaporization, fusion, melting, sublimation, absorption and desorption are endothermic physical processes while physical changes like adsorption and crystallization can be exothermic. Endothermic chemical changes include phase transitions, dehydration, decomposition, reduction

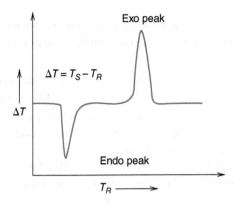

FIGURE 18.5 DTA thermogram.

whereas exothermic chemical changes may result from oxidation, chemisorption, polymerization or catalytic reactions. The above-mentioned changes cannot be measured using TGA.

DTA experiments give information that something is happening at a specific temperature. They usually do not tell us, what is happening. Combination with other methods like X-ray diffraction, spectroscopy, microscopic investigation and composition analysis (e.g., Electron probe microanalysis) are required for the interpretation of the results.

Instrumentation for DTA

The main components in a DTA instrument are—*sample holders, furnace, temperature controller, recorder, thermocouple* and *a cooling device* (Figure 18.6).

1. The *sample holder* can be made of metallic (nickel, stainless steel (up to 1000°C), platinum and its alloys) or non-metallic (glass, vitreous silica or sintered alumina) materials. Sharp exotherms and flat endotherms are obtained in the case of metallic sample holders and vice versa for non-metallic holders.

2. The tubular shape of the *furnace* is mostly preferred. Also, the dimension of the furnace depends upon the length of the uniform temperature zone desired and is calculated by taking into account the size of the sample holder, length of uniform temperature zone required, rate of heating and cost.

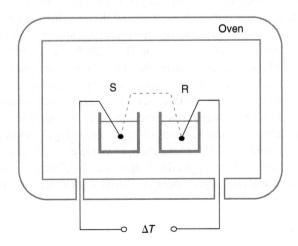

FIGURE 18.6 Instrument set-up for DTA.

3. The *temperature controller* comprises a sensor, control element and heater. The rate of heat-input required to match heat loss from the system is governed by the control element.

4. The *recorder* can be deflection or null type. The recorded signals are produced on paper or film by ink (using heating style, electric writing or optical beam).

5. The *thermocouples* in DTA are the temperature sensors. The thermocouple is selected based on temperature interval, thermoelectric coefficient, chemical compatibility with the sample, chemical gaseous environment, its price and availability. Depending on the temperature (< 1100°C or > 1100°C) chromel P and alumel or pure platinum and platinum-rhodium alloy wires are used.

6. The *cooling device* used is automatic and simple and is designed to be separated from the temperature programmer.

The difference in temperature between the sample and reference, thermocouples connected in series is continuously measured. The data is collected with amplification of high signal and low noise. Because the thermocouples are placed in direct with the sample, DTA provides the highest thermometric accuracy of all thermal methods.

Factors that Affect the DTA Curve

DTA is a dynamic temperature technique and the DTA curve is affected by environmental and instrumental factors and also depends on sample characteristics.

1. The *environmental factors* are related to the gaseous environment around the sample, which may react with the sample resulting in extra peaks in the DTA curve (oxidation in the presence of air, etc.).

2. The *instrumentals factors* include size and shape of the furnace; material and geometry of the sample holder; rate of heating; characteristics of thermocouple and its location in the sample; scan rate; speed and response of the recording instrument.

The DTA curve can be used only as a finger print for identification purposes.

3. The third i.e., *sample characteristics* concern the particle size of the sample, amount taken, its nature (hygroscopic or not), the thermal conductivity, heating capacity of material, packing density and degree of crystallinity. The size of holders and the amount of sample should be as small as possible for better resolution.

Applications of DTA

The advantages of DTA are: instruments can be used at very high temperatures, instruments are highly sensitive, characteristic transition or reaction temperatures can be accurately determined.

The disadvantages of DTA are: precision is not good, sharp thermal changes cannot be predicted and the uncertainty of heats of fusion, transition, or reaction estimations is 20-50%.

DTA finds *applications* in physical and analytical chemistry. In physical chemistry, it helps to find the heat of reaction, specific heat and thermal diffusivity of the sample.

In analytical chemistry DTA is used for the identification of substances, products as the DTA curve for two substances is not identical; for measuring melting points to check the purity of substances, finding glass transition temperature of polymers. It also finds applications in quantitative analysis, quality control of a large number of substances like cement, glass, soil, catalysts, textiles, explosives, resins, etc.

The thermal stability of many inorganic compounds/complexes (oxalates, metal amine complexes, carbonates and oxides) has been studied using DTA technique as it helps in distinguishing between reversible phase changes and irreversible decompositions.

DTA method has also been applied to help in the identification, purity determination and quantitative analysis of polymers, explosives, pharmaceuticals, oils, fats and other organic chemicals.

Example 18.3

Write the equations for the thermal decomposition of manganese phosphinate monohydrate.

Solution

$$Mn\left(PH_2O_2\right)_2 \cdot H_2O\,(s) \rightarrow Mn\left(PH_2O_2\right)_2 + H_2O\,(g)$$

$$Mn\left(PH_2O_2\right)_2 \rightarrow MnHPO_4\,(s) + PH_3\,(g)$$

$$\alpha\text{-}MnHPO_4\,(s) \rightarrow \beta\text{-}MnHPO_4\,(s)$$

$$2MnHPO_4\,(s) \rightarrow Mn_2P_2O_7\,(s) + H_2O\,(g)\ (\text{and recrystallization})$$

$$Mn_2P_2O_7\,(s) \rightarrow Mn_2P_2O_7\,(l)$$

18.4 Differential Scanning Calorimetry (DSC)

Differential scanning calorimetry (DSC) is a technique of thermal analysis that looks into the heat effects associated with phase transitions and chemical reactions for a sample. In DSC, the difference in heat flow rate (mW = mJ/sec) between a sample and reference at the same temperature is recorded as a function of time and temperature. In other words, DSC measures the energy necessary to establish a nearly zero temperature difference between a substance and an inert reference material. The temperature of both the sample and reference are increased at a constant rate.

For example, when a solid sample melts to a liquid (endothermic phase transition from solid to liquid), more heat flowing to the sample is needed to increase the sample temperature at the same rate as the reference. Likewise, when the sample is involved in an exothermic process (eg. crystallization) less heat is required to raise the sample temperature. Differential scanning calorimeters can measure the amount of heat absorbed or released during such transitions by observing the difference in heat flow between the sample and reference.

The result of a DSC experiment is a curve of heat flux versus temperature or versus time. This curve is used to calculate the enthalpy of transition by integrating the peak corresponding to a given transition. The enthalpy of transition can be expressed using the equation:

$$\Delta H = \frac{KA}{m} \tag{18.1}$$

where, ΔH is the enthalpy of transition, K is the calorimetric/calibration constant, m is mass of the sample and A is the area under the peak.

DSC is at constant pressure and the heat flow that is equivalent to enthalpy changes can be written as:

$$\left(\frac{dq}{dt}\right)_p = \left(\frac{dH}{dt}\right) \quad \left(\frac{dH}{dt} = \text{heat flow in mcal/sec}\right) \tag{18.2}$$

The heat flow difference between the sample and the reference is given as:

$$\Delta\left(\frac{dH}{dt}\right)_p = \left(\frac{dH}{dt}\right)_{sample} - \left(\frac{dH}{dt}\right)_{reference} \tag{18.3}$$

DSC was developed by E. S. Watson and M. J. O'Neill in 1962. It was commercially introduced at the 1963 Pittsburgh Conference on Analytical Chemistry and Applied Spectroscopy.

P. L. Privalov and D. R. Monaselidze developed the first adiabatic differential scanning calorimeter in 1964 at Institute of Physics in Tbilisi, Georgia. This could be used in biochemistry and allowed to measure energy directly alongwith the precise measurement of heat capacity. The term DSC was coined to describe this instrument.

The measurement of changes in the state variables of a body derived from the heat transfer associated with changes of its state i.e., chemical reactions, physical changes, or phase transitions under specified conditions is known as calorimetry.

For an endothermic process, the heat flow to the sample is higher than that to the reference such as in most phase transitions, heat is absorbed and, so $\Delta\left(\dfrac{dH}{dt}\right)_p$ is positive. Hence, DSC is a calorimetric quantitative method in which the differences in energy are measured.

The change in enthalpy,

$$\Delta H = Q_S - Q_R \tag{18.4}$$

where, Q_S is the rate of heat flow to or from the sample and Q_R is the rate of heat flow to or from the reference.

$$Q = \frac{T_2 - T_1}{R_{th}} \tag{18.5}$$

It is the thermal analog of Ohm's law.

Using Equations 18.4 and 18.5:

$$\Delta H = \frac{T_C - T_S}{R_{th}} - \frac{T_C - T_R}{R_{th}}$$

where, T_C is the constant temperature, T_R = reference temperature and T_S = sample temperature)

$$\Delta H = -\frac{T_C - T_R}{R_{th}} \tag{18.6}$$

The calorimetric constant varies from instrument to instrument and can be determined by analyzing a well-characterized material of known enthalpies of transition. The area under the peak is directly proportional to heat absorbed or evolved by the reaction, height of the peak is directly proportional to the rate of the reaction.

A typical DSC response for carbon tetrachloride is shown in Figure 18.7. The DSC curve of carbon tetrachloride exhibits three solid-state phase transformations before melting.

Types of DSC

Depending on the mechanism of operation, DSC can be classified into two types: Heat flux DSC and Power Compensated DSC (Figure 18.8) (Reference 5).

1. *Heat-flux DSC:* In this method, the temperature difference is allowed to vary, and the signal is converted to heat flow and can be written as:

$$q = \left(\frac{dT}{dR}\right) \tag{18.7}$$

where, R is well-defined thermal resistance of the transducer.

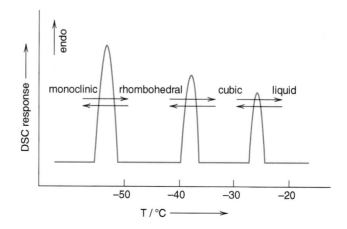

FIGURE 18.7 An example of DSC curve for CCl_4.

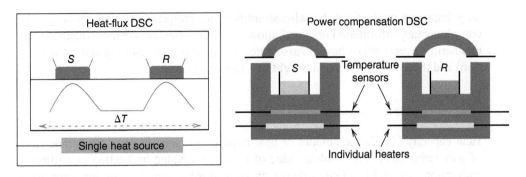

FIGURE 18.8 Set-up for Heat-flux and Power-compensation DSC.

This is a quantitative DTA method and measures the temperature difference between the sample and reference. The sample and the reference are heated by a single heating unit to provide a low-resistance heat flow path.

2. *Power-compensation DSC*: In this type, the temperature difference between the sample and reference is maintained constant as the sample is scanned with a linear increase or decrease of temperature. Here, the power needed to maintain the sample temperature equal to the reference temperature is measured and the resulting power difference is proportional to the heat flow. Also, two-independent sensors and heating units are employed for the sample and reference. This method measures enthalpy changes and can compensate the heat release or gained during a thermal event.

The response time is more rapid for this method; however, it has lower sensitivity than heat-flux DSC. Hence, power-compensated DSC is well-suited for kinetic measurements. Also, power-compensated DSC gives higher resolution than heat-flux DSC.

Instrumentation for DSC

The instrumentation for the two DSC methods comprises thermo-balance; sample holder; sensors; furnace; temperature controller. The thermo-balance is the most important part of the instrument which governs the accuracy, sensitivity and reproducibility in DSC. The balance can be of two types: *Deflection thermo-balance* like beam type or helical type and *Null thermo-balance*. The *Deflection balance*s monitor the change in the sample mass or deflection of the beam or the amplified photodiode current. On the other, the *Null point balance* detects the deviation of the balance beam from its null position with the help of a sensor and the deviation is directly proportional to the weight change. Generally, the balances are used in a way that they can cover wide range of temperature, have high degree of mechanical rigidity and electronic stability, the temperature recording is within $\pm 1°C$, have an adequate range of automatic weight adjustment, are unaffected by any kind of vibration and give a quick response to weight change.

Aluminium, stainless steel and platinum sample holders are used in DSC. In heat-flux DSC thermocouples are usually used as the temperature sensors and the furnace is one block for both the sample (**S**) and reference (**R**) cells temperature while Pt resistance thermocouples are used for power-compensated DSC with the furnace as separate block for S and R cells temperature. The furnaces are made from high quality metal and may vary in size and shape for temperatures from 1000–2000°C (tungsten or rhenium are used for higher temperatures).

Simultaneous DSC-TGA (SDT) can allow the measurement of both heat flow and weight changes in a material as a function of temperature or time in a controlled atmosphere. Simultaneous measurement of these two material properties not only improves productivity but also simplifies the interpretation of the results. The complimentary information obtained allows differentiation between endothermic and exothermic events which have no associated weight loss (e.g., melting and crystallization) and those which involve a weight loss (e.g., degradation).

Measuring Heat Capacity with DSC

Heat capacity (C_p) is the amount of heat required to raise or lower the temperature of a material and is the absolute value of heat flow divided by heating rate (times a calibration factor). Most DSC's do not measure absolute heat flow or heat capacity, a baseline subtraction is required when measuring C_p on most DSC's. Heat capacity is important as it is the thermodynamic property of a material while heat flow is not; it is a measure of molecular motion, increasing with increase in molecular motion and it provides useful information about physical properties of the material as a function of temperature.

Using DSC, C_p is measured as the absolute value of the heat flow $\left(\dfrac{dH}{dt}\right)$ divided by the heating rate $\left(\dfrac{dT}{dt}\right)$, and multiplied by a calibration constant (K).

$$\frac{dH}{dt} = C_p \frac{dT}{dt}$$

Sample heat capacity,

$$C_p = \left[\frac{\frac{dH}{dt}}{\frac{dT}{dt}}\right] \times K \tag{18.8}$$

$$C_p \left(\frac{J}{g.°C}\right) = \frac{\text{Heat flow} \left(\frac{mJ}{sec}\right)}{\text{Heat rate} \left(\frac{°C}{min}\right) \times \text{wt(mg)}} \times 60 \left(\frac{sec}{min}\right)$$

Whenever the heating rate of the sample and reference calorimeters are not identical, the measured heat flow is not the actual sample heat flow rate and hence, resolution and sensitivity suffer due to this. The sample and reference calorimeter heat capacities do not match thereby giving rise to non-zero empty DSC heat flow baseline. Therefore, heat capacity should be measured using the dedicated DSC set up of thermal analysis Instrument which can measure the capacitance and resistance of each DSC cell without the assumption that they are identical.

DSC Calibration

In DSC, calibration as close to the desired temperature range as possible is very important. It can be done by matching the melting onset temperatures to the known melting points of standards analyzed. The heat flow is maintained by using calibration standards of known heat capacity, applying slow accurate heating rates (0.5–2.0°C/min), and using similar sample and reference pan weights. The calibrants should have high purity, accurately known enthalpies, should be thermally stable and light stable, should not be hygroscopic, should not react with the pan or atmosphere (calibrants can be indium, zinc, KNO_3, $KClO_4$, benzoic acid, etc.). Baseline calibration is also important. Several pan configurations, e.g., open, pinhole, or hermetically sealed pan with same material and configuration for the sample and the reference should be used. The pan should not be overfilled in order to minimize thermal lag from the bulk material to the sensor. Moreover, the material should completely cover the bottom of the pan for good thermal contact.

The purge gas with a certain flow rate (maybe inert-argon or nitrogen) should be selected based on the reactivity of the sample being studied. The moisture content of the environment around the sample should be controlled. Reactive gas can be deliberately chosen in some cases, e.g. hydrogen to reduce oxide to metal, carbon dioxide which affects the decomposition of metal carbonate. Additionally, the purge gas system helps to remove waste products from sublimation or decomposition.

Applications of DSC

DSC can help to measure glass transition (T_g), melting (for sample purity)/boiling temperature and crystallization temperature and time; percentage crystallinity; heat of fusion and reaction; specific heat capacity; oxidative/thermal stability; reaction kinetics and purity of samples.

DSC is useful for quantitative analysis finger printing of clays, metal alloys, polymers (purity and studying the curing process) and minerals and in the pharmaceutical industry to obtain well-characterized and pure drug compounds. In DSC, the cross-linking of polymer molecules taking place during the curing process is exothermic and hence, gives a positive peak.

DSC can also be used in the study of liquid crystals. During the transitions from solid to liquid, there is often goes a third state known as the liquid (anisotropic liquid) crystalline or mesomorphous state. DSC can be used to observe the small energy changes that occur during transitions from a solid to a liquid crystal or from a liquid crystal to an isotropic liquid.

In food science area, water dynamics and determination of oils and fats are studied by using DSC in combination with other thermal analytical methods.

18.5 Thermometric Titration Analysis (TTA)

In a simple *thermometric titration analysis*, the titrant is added at a particular rate to the analyte until the completion of the reaction is indicated by a change in temperature. The endpoint is determined by an inflection in the curve generated by the output of temperature measuring device. The temperature is plotted against the volume of the titrant (Figure 18.9). The entire enthalpy change that includes entropy, as well as free energy involved, can be measured from the thermometric titration curve. This method is also known as *enthalpimetric titrations*.

The first possibly recognizable thermometric titration method was reported early in the 20th century by Bell and Cowell in 1913.

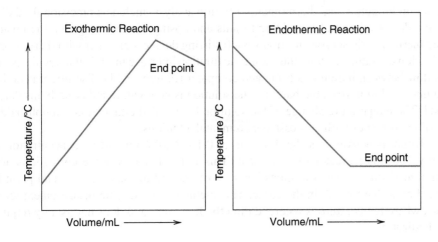

FIGURE 18.9 TTA curves for exothermic and endothermic reactions.

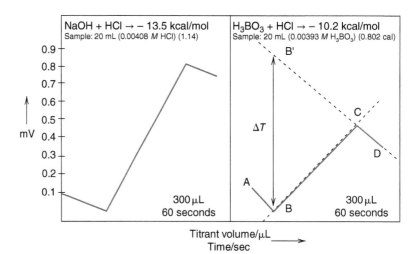

FIGURE 18.10 Examples of thermometric titrations.

Heat changes (loss or gain) accompany all chemical reactions. This can be observed from the resulting change in temperature (T). In TTA, the heat content of the reaction, i.e. enthalpy (H) is measured and not the free energy, G while potentiometric titrations ($\Delta G = -RT$ lnK) depend solely on the equilibrium constant 'K'.

Enthalpy is related to free energy and entropy (S) change as shown in Equation 18.9.

$$\Delta H = \Delta G - T\Delta S \qquad (18.9)$$

For example, when G is zero and $T\Delta S$ is large and negative as in the case of a weak acid like boric acid, a well-defined end point is observed. The titration curves of hydrochloric acid against sodium hydroxide and boric acid against sodium hydroxide are shown in Figure 18.10, respectively.

The change in temperature during the titration curve depends on the heat of reaction of the system (Equation 18.10).

$$\Delta T = \left(\frac{N\Delta H}{Q} \right) \qquad (18.10)$$

where, N is number of moles of water formed in the neutralization, ΔH is the molar enthalpy of neutralization and Q is heat capacity of the system. And hence, the change in temperature is proportional to the number of moles of the analyte.

Instrumentation for TTA

Thermistor has been found to be the most practical sensor for measuring the temperature change in titrating solutions. Thermistors are small solid state devices that exhibit relatively large changes in electrical resistance or small changes in temperature.

A thermistor is capable of resolving temperature changes as low as 10^{-5} K.

The set-up for automated thermometric titration analysis includes: precision fluid dispensing devices (burette) for adding titrants and dosing of other reagents; thermistor based thermometric sensor; titration vessel; stirring device; computer with thermometric titration operating system; thermometric titration interface module that regulates the data flow between burettes, sensors and computer (Figure 18.11). The titrant is delivered into a solution from the burette. The solution is kept within a thermally insulated vessel. The temperature change of the solution is recorded either by continuous addition of the titrant or with successive incremental additions.

Titrant is delivered at the flow rate of 0.1 – 1 mL/min and its concentration is usually 100 times greater than reactant. This is to obviate volume corrections and to minimize temperature variations between titrant and the sample. The end point is marked by a sharp break in the curve. This method is a one-time measurement technique, where temperature is recorded directly. It is rapid and does not usually require standardization.

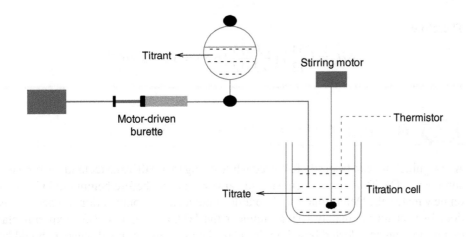

FIGURE 18.11 Set-up device used in TTA.

Applications of TTA

TTA is used for the determination of concentration of unknown substances, thermo-dynamic quantities; titration of boric acid; titrating acetic anhydride in acetic acid; study of acetylating baths; enzyme assay and immunological determinations; analysis of alkaloid drugs such as codeine phosphate or morphine sulphate, study of proteins and lipids and titrations of Ag^+, Ca^{2+}, Fe^{3+}, Tl^{3+} can be carried out. This method can also be used for non-aqueous systems, gases and molten salts. Also, all types of titra-tions viz. acid-base, redox, complexometric or precipitation and reactions of acids or bases with strong counterparts can be studied. Several examples of thermometric titra-tion are: Ti(III) with Ce(IV); $[Fe(CN)_6]^{4-}$ with Ce(IV); Fe(II) with $Cr_2O_7^{2-}$; Ca with $H_2C_2O_4$ or Ag with HCl, Ca^{2+} with EDTA.

 # Professor's Favorite Example

Contributed by Dr Dhanraj T. Masram, University of Delhi, India

Example 18.4

Heat of fusion for compound X is 6.85 kJ/mol and its relative molar mass is 98.4. Whereas, compound Y having relative molar mass of 64.3 melts at same tememperature as of X. DTA peak areas for X and Y were found to be 72 and 54 cm², respectively on using about 600 g sample of each. Calculate heat of fusion for compound Y.

Solution

On using equation:

$$\Delta H = \frac{AK}{m}$$

We get,

$$\Delta H_y = \left(\frac{Ay}{Ax}\right)\left(\frac{mx}{my}\right)\Delta H_x$$

Covert m_x and m_y into molar quantities,

$$m_x = \frac{600}{98.4} = 6.0975 \text{ moles}$$

$$m_y = \frac{600}{64.3} = 9.3312 \text{ moles}$$

Therefore,

$$\Delta H_y = \left(\frac{54}{72}\right)\left(\frac{6.0975}{9.3312}\right)6.85 = 3.36\,\text{kJ mol}^{-1}$$

18.6 Thermal Analysis: Best Practices

A few guidelines can be easily followed while trying to use thermal techniques for measurements. The TGA experiment should "always" be run before beginning DSC tests on new materials. Determination of volatile content and decomposition temperature is done by heat approximately 10 mg sample in the TGA at 10°C/min. DSC experimental conditions can be selected based on the TGA data set. Fine grained powder should be used to achieve greater contact area and better equilibrium conditions. One should keep in mind that chemical nature and flow of purge gas affect TA data. The time at any temperature must be sufficiently long in order to permit completeness of reactions. Larger mass and larger heating rate produce larger peak, but make detection of closely spaced thermal events more difficult. Powder samples increase oxidation, reduce heat flow. Evaporation can reduce sample mass, lead to incorrect measurement of enthalpy and contamination of instrument. Regular calibration while performing the measurements is extremely important. The instrument should be isolated from mechanical vibrations. Sample with unknown decomposition products must be carefully studied so that all evolved gases can be removed safely. One should make sure that crucible size, shape and material do not hamper the measurements. Artifacts can be identified by using empty pan runs as the baseline.

Problems

Principle

1. In case of TGA, DTA and DSC thermal methods, describe the quantity that is measured and how the measurement is performed?

2. Describe, with examples, the various types of curves obtained from thermogravimetric (TG) experiments.

3. Thermogravimetry and Differential Thermal Analysis are complementary techniques. Explain giving suitable examples.

4. Describe the quantity measured in thermomechanical analysis (TMA) and dynamic mechanical analysis (DMA).

5. Explain thermometric titration with suitable examples.

6. What are the physical and chemical changes that can lead to endothermic and exothermic peaks in DSC and DTA methods?

7. Are sublimation and fusion endothermic or exothermic. Explain.

8. Describe in detail the differences between differential scanning calorimetry (DSC) and differential thermal analysis (DTA). Also discuss the relative advantages and disadvantages of the two techniques.

9. Why is the thermocouple for measuring temperature in the thermal methods of analysis not immersed directly in the sample?

10. What kind of reference materials is used in DTA?

11. Estimation of the purity of a sample that melts can be made from analysis of a DSC melting endotherm. Describe the method.

Instrumentation

12. Explain what is meant by a transducer and describe the transducers used in the different thermal methods of analysis.

13. Discuss the common features for all the thermal analysis instruments and describe how the individual techniques differ from the generalized instrument.

Multiple Choice Questions

1. A solid sample of $Na[Fe(EDTA)(H_2O)_n](X)$ showed 5.6% weight loss at 120°C in a thermogravimetric experiment. Identify the complex left after this weight loss.

    (a) $Na[Fe(EDTA)(H_2O)]$
    (b) $Na[Fe(EDTA)]$
    (c) $Na[Fe(EDTA)(H_2O)_2]$
    (d) $Na[Fe(EDTA)(H_2O)_3]$

2. Which of the following option appropriately describes TGA and DTA?

    (a) TGA and DTA measures only weight

    (b) TGA measures only weight while DTA measures other effects

    (c) TGA and DTA measures only temperature

    (d) TGA measures only temperature while DTA measures other effects

3. For which of the following process can DTA be used?

    (a) Line positions of the crystals.

    (b) Mechanical properties of the crystals.

    (c) Phase diagrams

    (d) Catalytic properties of enzymes

4. Which of the following steps occur, for the decomposition of anhydrous calcium oxalate?

    (a) Intermediates, transition state, product

    (b) Intermediates, anhydrous oxalate, calcium oxysalts

    (c) Intermediates, aqueous hydrates, calcium hydroxides

    (d) Intermediates, anhydrous calcium oxalate, calcium carbonate

5. Thermal analysis can be defined as _____.

    (a) measurement of concentration of materials as a function of temperature

    (b) measurement of solubility of materials as a function of temperature

    (c) measurement of physical properties as a function of temperature

    (d) measurement of line positions of crystals as a function of temperature

6. What kind of reference material is used for DTA experiments?

    (a) Chemically active
    (b) Physically active
    (c) Inert
    (d) Having catalytic property

7. Differential scanning calorimetry can be used for determining the following.

    (a) melting temperature, glass transition temperature, heat of fusion etc.

    (b) volatilities of plasticizers and other additives

    (c) quantitative determination of additives in polymers

    (d) structural imperfections

8. The main factor that differentiates DTA and DSC cells is _____.

    (a) sensitivity
    (b) thermal conductivity
    (c) nature of the cell
    (d) designing of the cell

9. Which of the given factor determines the initial, T_i and final, T_f temperatures?

   (a) Cooling rate

   (b) Mechanical property of the material

   (c) Thermal expansion coefficient

   (d) Atmosphere above the sample

Recommended References

1. W.W. Wendlandt, *Thermal Methods of Analysis*, John Wiley & Sons, New York, 1974.
2. W.F. Hemminger, H.K. Cammenga, *Methoden der Thermischen Analyse*, Springer-Verlag, Berlin, 1989.
3. T. H. Gouw, *Guide to Modern Methods of Instrumental Analysis*, Wiley-Interscience, New York, 1972.
4. H.H. Willard, L.L. Merritt, Jr., J.A. Dean, F.A. Settle, Jr. *Instrumental Methonds of Analysis, 7th edition*, Wadsworth, Belmont, 1988.
5. G.W.H. Höhne, W.F. Hemminger, H.-J. Flammersheim, *Differential Scanning Calorimetry*, Second edition, Springer, Berlin, 2003.

Chapter 19

ELECTROCHEMICAL CELLS AND ELECTRODE POTENTIALS

"I know of nothing sublime which is not some modification of power."
— Edmund Burke

KEY THINGS TO LEARN FROM THIS CHAPTER

- Voltaic cells
- Using standard potentials to predict reactions
- Anodes, cathodes, and cell voltages (key Equation: 19.19)

- The Nernst equation (key Equation: 19.22)
- Calculating electrode potentials before and after reaction
- The formal potential

An important class of titrations is reduction–oxidation or "redox" titrations, in which an oxidizing agent and a reducing agent react (see Equation 19.1). We define an **oxidation** as a loss of electrons to an oxidizing agent (which itself gets reduced) to give a higher or more positive oxidation state, and we define **reduction** as a gain of electrons from a reducing agent (which itself gets oxidized) to give a lower or more negative oxidation state. We can gain an understanding of these reactions from a knowledge of electrochemical cells and electrode potentials. In this chapter, we discuss electrochemical cells, standard electrode potentials, the Nernst equation (which describes electrode potentials), and limitations of those potentials. Chapter 20 discusses potentiometry, the use of potential measurements for determining concentration, including the glass pH electrode and ion-selective electrodes. In Chapter 21, we describe redox titrations and potentiometric titrations in which potentiometric measurements are used to detect the end point. We review in that chapter the balancing of redox reactions since this is required for volumetric calculations. You may wish to review that material now.

Electrochemical methods, with the exception of the nearly universal use of the potentiometric pH meter, are generally not as widely used as spectrochemical or chromatographic methods. Spectrophotometric methods are widely applicable and are more readily amenable to automation than electrochemical methods, and are particularly applied in high throughput clinical chemistry laboratories, for example. There are, however, automatic clinical chemistry instruments that use potentiometry or amperometry to measure pH and electrolyte ions or CO_2 and O_2 gases. Electrochemical methods typically require more routine calibrations than other techniques.

Electrochemical methods do possess certain advantages for some applications. The instrumentation is comparatively inexpensive. Electrochemical methods can be selective for a particular chemical form of an element. For example, a mixture of Fe^{2+} and Fe^{3+} can be analyzed electrochemically. Potentiometric methods are unique in that they measure the activity of a chemical species rather than concentration, and can detect the free form of a metal ion, some of which may be tied up in complexation. In the next four chapters, we will explore these different techniques and their applicability.

Oxidation is a loss of electrons. Reduction is a gain of electrons. "OIL RIG" is a good mnemonic to help remember this.

Before beginning our discussions, it is helpful to describe some fundamental electrochemical terms, some of which will be discussed later on:

Potential (volts) originates from separation of charge

Ohm's law: $E = iR$, where E potential in volts, i is current in amperes, and R is resistance in ohms

$P = Ei$, where P is power in volt amperes

The first law of thermodynamics says that there is a conservation of energy in a chemical reaction. Work is done by the system or on the system to the extent of energy evolved or absorbed, work being equal to qE, where q is charge, in coulombs transferred across a potential difference E. In the case of electrochemical cells, free energy change, ΔG, is related to the electrical work done in the cell. If E_{cell} is the electromotive force (emf) of the cell and n moles of electrons are involved in the reaction, the electrical work done will be $\Delta G = -nFE_{cell}$, where F is the Faraday constant, 96,487 coulombs/equivalent. If reactants and products are in their standard states, then $\Delta G° = -nFE°$, where $E°$ is the standard cell potential.

While our emphasis in these chapters is on analytical applications of redox reactions and electrochemistry, such reactions are important in biochemistry, energy conversion, batteries, environmental chemistry, and other aspects of our lives, and your knowledge of basic redox chemistry and electrochemistry will be helpful in your understanding of these processes.

19.1 What Are Redox Reactions?

Note the similarity with the Brønsted-Lowry acid-base concept. Whereas acid-base behavior is centered on proton transfer, redox behavior is centered on electron transfer. Like conjugate acid-base pairs there are redox couples. But redox couples may differ by more than just the gain or loss of electron(s).

The oxidizing agent is reduced. The reducing agent is oxidized.

A reduction–oxidation reaction—commonly called a **redox** reaction—is one that occurs between a reducing and an oxidizing agent:

$$\boxed{Ox_1 + Red_2 \rightleftharpoons Red_1 + Ox_2} \qquad (19.1)$$

Ox_1 is reduced to Red_1, and Red_2 is oxidized to Ox_2. Ox_1 is the oxidizing agent, and Red_2 is the reducing agent. The reducing or oxidizing tendency of a substance will depend on its reduction potential, described below. An oxidizing substance will tend to take on an electron or electrons and be reduced to a lower oxidation state:

$$M^{a+} + ne^- \rightarrow M^{(a-n)+} \qquad (19.2)$$

for example, $Fe^{3+} + e^- \rightarrow Fe^{2+}$. Conversely, a reducing substance will tend to give up an electron or electrons and be oxidized:

$$M^{a+} \rightarrow M^{(a+n)+} + ne^- \qquad (19.3)$$

for example, $2I^- \rightarrow I_2 + 2e^-$. If the oxidized form of a metal ion is complexed, it is more stable and will be more difficult to reduce; so its tendency to take on electrons will be decreased if the reduced form is not also complexed to make it more stable and easier to form.

We can better understand the oxidizing or reducing tendencies of substances by studying electrochemical cells and electrode potentials.

19.2 Electrochemical Cells—What Electroanalytical Chemists Use

There are two kinds of electrochemical cells, **voltaic** (galvanic) and **electrolytic**. In voltaic cells, a chemical reaction spontaneously occurs to *produce electrical energy*. The lead storage battery and the ordinary flashlight cell are common examples of voltaic cells. In electrolytic cells, on the other hand, *electrical energy is used* to force a nonspontaneous chemical reaction to occur, that is, to go in the reverse direction it would in a voltaic cell. An example is the electrolysis of water. In both types of these cells, the electrode at which oxidation occurs is the **anode**, and that at which reduction occurs is the **cathode**. Voltaic cells will be of importance in our discussions in the next two chapters, dealing with potentiometry. Electrolytic cells are important in electrochemical methods such as voltammetry, in which electroactive substances like metal ions are reduced at an electrode to produce a measurable current by applying an appropriate potential to get the nonspontaneous reaction to occur (Chapter 22). The current that results from the forced electrolysis is proportional to the concentration of the electroactive substance.

In a voltaic cell, a spontaneous chemical reaction produces electricity. This occurs only when the cell circuit is closed, as when you turn on a flashlight. The cell voltage (e.g., in a battery) is determined by the potential difference of the two half reactions. When the reaction has gone to completion, the cell runs down, and the voltage is zero (the battery is "dead"). In an electrolytic cell, the reaction is forced the other way by applying an external voltage greater than and opposite to the spontaneous voltage. A battery is a combination of cells in series. A C-cell is a cell but if you break open a 9V battery you will find six individual cells.

Voltaic Cell and Spontaneous Reactions—What is the Cell Potential?

Consider the following redox reaction in a voltaic cell:

$$Fe^{2+} + Ce^{4+} \rightleftharpoons Fe^{3+} + Ce^{3+} \tag{19.4}$$

If we mix a solution containing Fe^{2+} with one containing Ce^{4+}, there is a certain tendency for the ions to transfer electrons. Assume the Fe^{2+} and Ce^{4+} are in separate beakers connected by a **salt bridge**, as shown in Figure 19.1. (A salt bridge allows ion transfer into the solutions and prevents mixing of the solutions.) No reaction can occur since the solutions do not make contact. A salt bridge is not always needed—only when the reactants or products at the anode or cathode react with each other so that it is necessary to keep them from mixing freely. Now put an inert platinum wire in each solution and connect the two wires. The setup now constitutes a voltaic cell. If a microammeter is connected in series, it indicates that a current is flowing. The Fe^{2+} is being oxidized at the platinum wire (the anode):

$$Fe^{2+} \rightarrow Fe^{3+} + e^- \tag{19.5}$$

The released electrons flow through the wire to the other beaker where the Ce^{4+} is reduced (at the cathode):

$$Ce^{4+} + e^- \rightarrow Ce^{3+} \tag{19.6}$$

The salt bridge typically contains a gel containing KCl. KCl is the preferred ingredient since K^+ and Cl^- move at equal speeds. As current flows, Cl^- comes into the anode compartment and K^+ comes into the cathode compartment to maintain charge balance.

This process occurs because of the tendency of these ions to transfer electrons. The net result is the reaction written in Equation 19.4, which would occur if Fe^{2+} and Ce^{4+} were added together in a single beaker (in that case, the electrical energy that could be harvested from the voltaic cell is simply liberated as heat). The platinum wires can

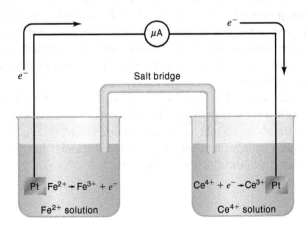

FIGURE 19.1 Voltaic cell.

be considered **electrodes**. Each will adopt an electrical **potential** that is determined by the tendency of the ions to give off or take on electrons, and this is called the **electrode potential**. A voltmeter placed between the electrodes will indicate the *difference* in the potentials between the two electrodes. The larger the potential difference, the greater the tendency for the reaction between Fe^{2+} and Ce^{4+}. The driving force of the chemical reaction (the potential difference) can be used to perform work such as lighting a light bulb or running a motor, as is done with a battery.

Half-Reactions—Giving and Accepting Electrons

Equations 19.5 and 19.6 are **half-reactions**. No half-reaction can occur by itself in much the same way as in the world of finance, there has to be a lender for a borrower to exist and without both, no transaction can occur. There must be an **electron donor** (a reducing agent) and an **electron acceptor** (an oxidizing agent). In this case, Fe^{2+} is the reducing agent and Ce^{4+} is the oxidizing agent. Each half-reaction will generate a definite potential that would be adopted by an inert electrode dipped in the solution.

Half-Reaction Potentials—They are Measured Relative to Each Other

If the potentials of all half-reactions could be measured, then we could determine which oxidizing and reducing agents will react. Unfortunately, there is no way to measure individual electrode potentials. But, as we just saw, the *difference* between two electrode potentials can be measured. The electrode potential of the half-reaction[1]

$$2H^+ + 2e^- \rightleftharpoons H_2$$ (19.7)

has arbitrarily been assigned a value of 0.000 V. This is called the **normal hydrogen electrode** (NHE), or the **standard hydrogen electrode** (SHE). This consists of a platinized platinum electrode (one coated with fine "platinum black" by electroplating platinum on the electrode) contained in a glass tube, immersed in an acid solution in which $a_{H_+} = 1$ and where hydrogen gas ($P_{H_2} = 1$ atm) is bubbled at the electrode/solution interface. The platinum black catalyzes Reaction 19.7. The potential differences between this half-reaction and other half-reactions have been measured using voltaic cells and arranged in decreasing order. Some of these are listed in Table 19.1. Potentials are dependent on concentrations, and all standard potentials refer to conditions of unit activity for all species (or 1 atmosphere partial pressure in the case of gases, as for hydrogen in the NHE). The effects of concentrations on potentials are described below. A more complete listing of potentials appears in Appendix C.

We arbitrarily define the potential of this half-reaction as zero (at standard conditions). All others are measured relative to this.

The potentials are for the half-reaction written as a *reduction*, and so they represent **reduction potentials**. We will use the Gibbs–Stockholm electrode potential convention, adopted at the 17th Conference of the International Union of Pure and Applied Chemistry in Stockholm, 1953. In this convention, the half-reaction is written as a reduction, and the potential increases as the tendency for reduction (of the oxidized form to be reduced) increases.

In the Gibbs–Stockholm convention, we always write the half-reaction as a reduction.

The electrode potential for $Sn^{4+} + 2e^- \rightleftharpoons Sn^{2+}$ is +0.15 V. In other words, the potential of this half-reaction relative to the NHE in a cell like that in Figure 19.1 would be 0.15 V. Since the above couple has a larger (more positive) reduction potential than the NHE, Sn^{4+} has a stronger tendency to be reduced than H^+ has. We can draw some general conclusions from the electrode potentials:

---

[1]The reaction could have been written $H^+ + e^- \rightleftharpoons \frac{1}{2}H_2$. The way it is written does not affect its potential.

TABLE 19.1 Some Standard Potentials

Half-Reaction	$E^0 (V)$
$H_2O_2 + 2H^+ + 2e^- \rightleftharpoons 2H_2O$	1.77
$MnO_4^- + 4H^+ + 3e^- \rightleftharpoons MnO_2 + 2H_2O$	1.695
$Ce^{4+} + e^- \rightleftharpoons Ce^{3+}$	1.61
$MnO_4^- + 8H^+ + 5e^- \rightleftharpoons Mn^{2+} + 4H_2O$	1.51
$Cr_2O_7^{2-} + 14H^+ + 6e^- \rightleftharpoons 2Cr^{3+} + 7H_2O$	1.33
$MnO_2 + 4H^+ + 2e^- \rightleftharpoons Mn^{2+} + 2H_2O$	1.23
$2IO_3^- + 12H^+ + 10e^- \rightleftharpoons I_2 + 6H_2O$	1.20
$H_2O_2 + 2e^- \rightleftharpoons 2OH^-$	0.88
$Cu^2 + I^- + e^- \rightleftharpoons CuI$	0.86
$Fe^{3+} + e^- \rightleftharpoons Fe^{2+}$	0.771
$O_2 + 2H^+ + 2e^- \rightleftharpoons H_2O_2$	0.682
$I_2(aq) + 2e^- \rightleftharpoons 2I^-$	0.6197
$H_3AsO_4 + 2H^+ + 2e^- \rightleftharpoons H_3AsO_3 + H_2O$	0.559
$I_3^- + 2e^- \rightleftharpoons 3I^-$	0.5355
$Sn^{4+} + 2e^- \rightleftharpoons Sn^{2+}$	0.154
$S_4O_6^{2-} + 2e^- \rightleftharpoons 2S_2O_3^{2-}$	0.08
$2H^+ + 2e^- \rightleftharpoons H_2$	0.000
$Zn^{2+} + 2e^- \rightleftharpoons Zn$	−0.763
$2H_2O + 2e^- \rightleftharpoons H_2 + 2OH^-$	−0.828

1. The more *positive* the electrode potential, the greater the tendency of the oxidized form to be reduced. In other words, **the more positive the electrode potential, the stronger an oxidizing agent the oxidized form is and the weaker a reducing agent the reduced form is**.

2. The more *negative* the electrode potential, the greater the tendency of the reduced form to be oxidized. In other words, **the more negative the reduction potential, the weaker an oxidizing agent is the oxidized form is and the stronger a reducing agent the reduced form is**.

The reduction potential for $Ce^{4+} + e^- \rightleftharpoons Ce^{3+}$ is very positive, so Ce^{4+} is a strong oxidizing agent, while Ce^{3+} is a very weak reducing agent. On the other hand, the potential for $Zn^{2+} + 2e^- \rightleftharpoons Zn$ is very negative, and so Zn^{2+} is a very weak oxidizing agent, while metallic zinc is a very strong reducing agent.

Ce^{4+} is a good oxidizing agent because of the high reduction potential. (But Ce^{3+} is a poor reducing agent.) Zn is a good reducing agent because of the low reduction potential. (But Zn^{2+} is a poor oxidizing agent.) Note again the similarity with conjugate acid-base pairs: The stronger the acid, the weaker is the conjugate base and *vice-versa*.

What Substances React?

The oxidized form of a species in a half-reaction is capable of oxidizing the reduced form of a species in a half-reaction whose reduction potential is more *negative* than its own, and vice versa: The reduced form in a half-reaction is capable of reducing the oxidized form in a half-reaction with a more *positive* potential.

For example, consider the two half-reactions

$$Fe^{3+} + e^- \rightleftharpoons Fe^{2+} \quad E^0 = 0.771 \text{ V} \qquad (19.8)$$

$$Sn^{4+} + 2e^- \rightleftharpoons Sn^{2+} \quad E^0 = 0.154 \text{ V} \qquad (19.9)$$

Note again that the E^0 value is a measure of the tendency of the reaction to occur. Multiplying a half-reaction such as 19.8 by 2 does not alter the tendency of the reaction to occur. E^0 remains the same.

There are two combinations for possible reaction between an oxidizing and a reducing agent in these two half-reactions, which we arrive at by *subtracting* one from the other (multiplying the first half-reaction by 2 so the electrons cancel):

$$2Fe^{3+} + Sn^{2+} \rightleftharpoons 2Fe^{2+} + Sn^{4+} \tag{19.10}$$

and

$$Sn^{4+} + 2Fe^{2+} \rightleftharpoons Sn^{2+} + 2Fe^{3+} \tag{19.11}$$

[There is no possibility of reaction between Fe^{3+} and Sn^{4+} (both oxidizing agents) or between Fe^{2+} and Sn^{2+} (both reducing agents).] Perusal of the potentials tells us that Reaction 19.10 will take place; that is, the reduced form Sn^{2+} of Reaction 19.9 (with the more negative potential) will react with the oxidized form of Reaction 19.8 (with the more positive potential). Note that the number of electrons donated and accepted must be equal (see Chapter 21 on balancing redox reactions).

Another way to combine two half-reactions is to take one of the half-reactions and write it as oxidation. The E^0 value will change sign. If the number of electrons are different in the two half-reactions, one or both must be multiplied by appropriate multipliers first so that the number of electrons involved in half-reactions are the same. Then when we add them up (the E^0 values will also add algebraically), *the reaction will proceed from left to right, if the summed E^0 value is positive.* Thus, for example, we multiply Equation 19.8 by 2 and reverse Equation 19.9 and then add them:

$$2Fe^{3+} + 2e^- \rightleftharpoons 2Fe^{2+} \qquad E^0 \rightleftharpoons 0.771 \text{ V} \tag{19.12}$$

$$Sn^{2+} \rightleftharpoons Sn^{4+} + 2e^- \qquad E^0 \rightleftharpoons -0.154 \text{ V} \tag{19.13}$$

$$\overline{2Fe^{3+} + Sn^{2+} \rightleftharpoons 2Fe^{2+} + Sn^{4+} \qquad E^0 \rightleftharpoons 0.617 \text{ V}} \tag{19.14}$$

The reaction in Equation 19.14 will proceed to the right because the net potential is positive. The electrons will appear on both sides after the addition and are therefore canceled. If we reversed Equation 19.8 (after multiplication with 2) and added to Equation 19.9, the net potential will be negative and the reaction will not proceed as written.

Example 19.1

For the following substances, list the oxidizing agents in decreasing order of oxidizing capability, and the reducing agents in decreasing order of reducing capability: MnO_4^-, Ce^{3+}, Cr^{3+}, IO_3^-, Fe^{3+}, I^-, H^+, Zn^{2+}.

Solution

Looking at Table 19.1, the following must be oxidizing agents (are in the oxidized forms) and are listed from the most positive E^0 to the least positive: MnO_4^-, IO_3^-, Fe^{3+}, H^+, Zn^{2+}. MnO_4^- is a very good oxidizing agent, Zn^{2+} is very poor. The remainder are in the reduced form, and their reducing power is in the order I^-, Cr^{3+}, and Ce^{3+}. I^- is a reasonably good reducing agent; Ce^{3+} is poor.

The net potential as represented in Equation 19.14 is called the cell voltage.[2]

---

[2]We refer to electrode **potentials** and cell **voltages** to distinguish between half-reactions and complete reactions.

To reiterate, if this calculated cell voltage is positive, the reaction goes as written. If it is negative, the reaction will occur in the reverse direction. This is the result of the convention that, for a spontaneous reaction, the free energy is negative. The free energy at standard conditions is given by

The spontaneous cell reaction is the one that gives a positive cell voltage when subtracting one half-reaction from the other.

$$\Delta G° = -nFE^0 \tag{19.15}$$

where n is the number of moles of electrons involved in the balanced reaction and F is the Faraday constant ($96,487$ C mol^{-1}); and so a positive potential difference provides the necessary negative free energy. Hence, we can tell from the relative standard potentials for two reactions, and from their signs, which reaction combination will produce a negative free-energy change and thus be spontaneous. For example, for the Ce^{4+}/Ce^{3+} half-reaction, E^0 is $+1.61$ V (Table 19.1); and for the Fe^{3+}/Fe^{2+} half-reaction, E^0 is $+0.771$ V. $\Delta G°$ for the former is more negative than for the latter, and subtraction of the iron half-reaction from the cerium one will provide the spontaneous reaction that would occur to give a negative free energy. That is, Ce^{4+} would spontaneously oxidize Fe^{2+}.

1 Coulomb (C) = 1 Ampere-second (A · s), the quantity of electricity carried by a current of 1 Ampere in 1 second.

Which is the Anode? And which is the Cathode?

By convention, a cell is written with the anode on the left:

$$\boxed{\text{anode/solution/cathode}} \tag{19.16}$$

The single lines represent a boundary between either an electrode phase and a solution phase or two solution phases. In Figure 19.1, the cell would be written as

$$Pt/Fe^{2+}(C_1), Fe^{3+}(C_2)//Ce^{4+}(C_3), Ce^{3+}(C_4)/Pt \tag{19.17}$$

where C_1, C_2, C_3, and C_4 represent the concentrations of the different species. The double line represents the salt bridge. If a voltaic cell were constructed for the above iron and tin half-reactions with platinum electrodes, it would be written as

$$Pt/Sn^{2+}(C_1), Sn^{4+}(C_2)//Fe^{3+}(C_3), Fe^{2+}(C_4)/Pt \tag{19.18}$$

Since oxidation occurs at the anode and reduction occurs at the cathode, *the stronger reducing agent is placed on the left* and *the stronger oxidizing agent is placed on the right*. The potential of the voltaic cell is given by

The anode is the electrode where oxidation occurs, i.e., the more negative or less positive half-reaction occurs in the anode compartment.

$$\boxed{E_{\text{cell}} = E_{\text{right}} - E_{\text{left}} = E_{\text{cathode}} - E_{\text{anode}} = E_+ - E_-} \tag{19.19}$$

where E_+ is the more positive electrode potential and E_- is the more negative of the two electrodes.

When the cell is set up properly, *the calculated voltage will always be positive*, and the cell reaction is written correctly, that is, the correct cathode half-reaction is written as a reduction and the correct anode half-reaction is written as an oxidation. In cell represented by Equation 19.18, we would have at standard conditions

$$E^0_{\text{cell}} = E^0_{Fe^{3+},Fe^{2+}} - E^0_{Sn^{4+},Sn^{2+}} = 0.771 - 0.154 = 0.617 \text{ V}$$

To take some more examples of possible redox reactions, Fe^{3+} will not oxidize Mn^{2+}. Quite the contrary, MnO_4^- will oxidize Fe^{2+}. I_2 is a moderate oxidizing agent and will oxidize Sn^{2+}. On the other hand, I^- is a fairly good reducing agent and will reduce Fe^{3+}, $Cr_2O_7^{2-}$, and so on. *To obtain a sharp end point in a redox titration, the reaction will need to be nearly quantitative, and there should be at least 0.2 to 0.3 V difference between the two electrode potentials.*

In a redox titration (Chapter 21), the potential difference between the titrant and the analyte half-reaction should be 0.2–0.3 V for a sharp end point.

Combining Equation 19.15 with 5.10 leads to an expression that relates difference in standard potentials to the equilibrium constant:

$$\Delta G° = -2.303RT \, \log \, K = -nFE^0$$

$$\log \, K = \frac{nE^0}{2.303RT/F} \tag{19.20}$$

At 298 K, 2.303RT/F is equal to 0.05915 V (see Section 19.3). For $n = 1$ and $E^0 = 0.2$ V, one calculates from Equation 19.20 that K ≈ 2400.

Example 19.2

From the potentials listed in Table 19.1, determine the reaction between the following half-reactions, and calculate the corresponding cell voltage:

$$Fe^{3+} + e^- \rightleftharpoons Fe^{2+} \quad E^0 \rightleftharpoons 0.771 \text{ V}$$

$$I_3^- + 2e^- \rightleftharpoons 3I^- \quad E^0 \rightleftharpoons 0.5355 \text{ V}$$

Solution

Since the Fe^{3+}/Fe^{2+} potential is the more positive, Fe^{3+} is a better oxidizing agent than I_3^-. Hence, Fe^{3+} will oxidize I^- and $E^0_{cell} = E_{cathode} - E_{anode} = E^0_{Fe^{3+},Fe^{2+}} - E^0_{I_3^-,I^-}$. In the same fashion, the second half-reaction must be subtracted from the first (multiplied by 2) to give the overall cell reaction:

$$2Fe^{3+} + 3I^- = 2Fe^{2+} + I_3^- \quad E^0_{cell} = 0.771 \text{ V} - 0.536 \text{ V} = +0.235 \text{ V}$$

Note again that multiplying a half-reaction by any number does not change its potential.

19.3 Nernst Equation—Effects of Concentrations on Potentials

Activities should be used in the Nernst equation. We will use concentrations here because we are primarily concerned here with titrations. Titrations end points involve large potential changes, and the errors are small by doing so.

The potentials listed in Table 19.1 were determined for the case when the concentrations of both the oxidized and reduced forms (and all other species) were at **unit activity**, and they are called the **standard potentials**, designated by E^0. Volta originally set up empirical E^0 tables under very controlled and defined conditions. Nernst made them practical by establishing quantitative relationships between potential and concentrations. This potential is dependent on the concentrations of the species and varies from the standard potential. This potential dependence is described by the **Nernst equation**[3]:

$$\boxed{a\text{Ox} + ne^- \rightleftharpoons b\text{Red}} \tag{19.21}$$

$$\boxed{E = E^0 - \frac{2.3026RT}{nF} \log \frac{[\text{Red}]^b}{[\text{Ox}]^a}} \tag{19.22}$$

where a and b are the numbers of Ox and Red species in the half-reaction, E is the reduction potential at the specific concentrations, n is the number of electrons involved in the

---

[3]More correctly, activities, rather than concentrations, should be used; but we will use concentrations for this discussion. In the next chapter, involving potential measurements for direct calculation of concentrations, we will use activities.

half-reaction (equivalents per mole), R is the gas constant (8.3143 V C K^{-1} mol^{-1}), T is the absolute temperature in Kelvin, and F is the Faraday constant (96,487 C eq^{-1}). At 25°C (298.16 K), the value of $2.3026RT/F$ is 0.05916 V, or 1.9842×10^{-4} (°C + 273.16 V). *The concentration of pure substances such as precipitates and liquids (H_2O) is taken as unity. Note that the log term of the reduction half-reaction is the ratio of the concentrations of the right-side product(s) over the left-side reactants(s).*

Example 19.3

A solution is 10^{-3} M in $Cr_2O_7{}^{2-}$ and 10^{-2} M in Cr^{3+}. If the pH is 2.0, what is the potential of the half-reaction at 298 K?

Solution

$$Cr_2O_7{}^{2-} + 14H^+ + 6e^- \rightleftharpoons 2Cr^{3+} + 7H_2O$$

$$E = E^0_{Cr_2O_7{}^{2-},Cr^{3+}} - \frac{0.05916}{6} \log \frac{[Cr^{3+}]^2}{[Cr_2O_7{}^{2-}][H^+]^{14}}$$

$$= 1.33 - \frac{0.05916}{6} \log \frac{(10^{-2})^2}{(10^{-3})(10^{-2})^{14}}$$

$$= 1.33 - \frac{0.05916}{6} \log 10^{27} = 1.33 - 27\left(\frac{0.05916}{6}\right)$$

$$= 1.06 \text{ V}$$

This calculated potential is the potential an electrode would adopt, relative to the NHE, if it were placed in the solution, and it is a measure of the oxidizing or reducing power of that solution. Theoretically, the potential would be infinite if there were no Cr^{3+} at all in solution. In actual practice, the potential is always finite (but impossible to calculate from the simple Nernst equation). Either there will be a small amount of impurity of the oxidized or reduced form present or, more probably, the potential will be limited by another half-reaction, such as the oxidation or reduction of water, that prevents it from going to infinity.

Equilibrium Potential—After the Reaction has Occurred

Imagine that I have a solution of Fe^{2+} containing a small amount of Fe^{3+}. The potential of the solution can be readily calculated from Equations 19.8 and 19.22. Now a small amount of Ce^{4+} is added to the solution. This results in a new equilibrium being reached, with some of the Fe^{2+} being oxidized to Fe^{3+}, with the concomitant production of a corresponding amount of Ce^{3+}. Based on the new composition, a new potential can be calculated, either from the same two equations as before or from the Ce^{3+}–Ce^{4+} half reaction potential and Equation 19.22. The potential of an inert electrode in a solution containing the ions of two half-reactions at equilibrium (e.g., at different points in a titration) can be calculated relative to the NHE using the Nernst equation for *either* half-reaction. This is because when the reaction comes to equilibrium after each aliquot of titrant addition, the potentials for the two half-reactions become identical; otherwise, the reaction would still be going on. An electrode dipped in the solution will adopt the **equilibrium potential**. The equilibrium potential is dictated by the equilibrium concentrations of either half-reaction and the Nernst equation.

To construct a titration curve, we are interested in the equilibrium *electrode* potential (i.e., when the *cell* potential is zero—after the titrant and analyte have reacted). The two electrodes have identical potentials then, as determined by the Nernst equation for each half-reaction.

Example 19.4

A 5.0 mL portion of 0.10 M Ce^{4+} solution is added to 5.0 mL of 0.30 M Fe^{2+} solution. Calculate the potential at 298 K of a platinum electrode dipping in the solution (relative to the NHE).

Solution

We start with $0.30 \, \text{mmol mL}^{-1} \times 5.0 \, \text{mL} = 1.5 \, \text{mmol} \, Fe^{2+}$ and add $0.10 \, \text{mmol}$ $\text{mL}^{-1} \times 5.0 \, \text{mL} = 0.50 \, \text{mmol} \, Ce^{4+}$. So we form $0.50 \, \text{mmol}$ each of Fe^{3+} and Ce^{3+} and have 1.0 mmol Fe^{2+} remaining. The reaction lies far to the right at equilibrium if there is at least 0.2 V difference between the standard electrode potentials of two half-reactions. But a small amount of Ce^{4+} $(= x)$ will exist at equilibrium, and an equal amount of Fe^{2+} will be formed:

$$Fe^{2+} \; + \; Ce^{4+} \; \rightleftharpoons \; Fe^{3+} \; + \; Ce^{3+}$$
$$1.0 + x \qquad x \qquad 0.50 - x \quad 0.50 - x$$

These are the equilibrium concentrations, following reaction.

where the numbers and x represent millimoles. To calculate the concentration of each species, the amounts in millimoles will need to be divided by the total volume of 10 mL, but since this same divisor appears in both the numerator and denominator, it cancels out when calculating concentration ratios. At equilibrium, the potential of the Fe^{3+}/Fe^{2+} half-reaction must be the same as that of the Ce^{4+}/Ce^{3+} half-reaction:

$$0.771 - 0.059 \; \log \frac{(1.0 + x)}{(0.50 - x)} = 1.61 - 0.059 \; \log \frac{(0.50 - x)}{x}$$

$$1.61 - 0.771 = 0.839 = 0.059 \; \log \frac{(0.50 - x)^2}{x(0.10 + x)}$$

$$\frac{0.839}{0.059} = 14.22 = \log \frac{(0.50 - x)^2}{x(0.10 + x)}$$

A solution to this quadratic equation that is also readily solved by Goal Seek will result in $x = 1.51 \times 10^{-15} \, M$. Putting this value of x in either half reaction will produce $E = 0.753$ V.

We could do this, however, a lot simpler. Consider that this reaction is analogous to "ionization" of the product in precipitation or acid–base reactions written as association reactions; a slight shift of the equilibrium here to the left would be the "ionization." The quantity x is very small compared with 0.50 or 1.0 and can be neglected. Either half-reaction can be used to calculate the potential. Since the concentrations of both species in the Fe^{3+}/Fe^{2+} couple are known, we will use this:

$$Fe^{3+} \; + \; e^- \; \rightleftharpoons \; Fe^{2+}$$
$$0.50 \qquad\qquad 1.0$$

$$E = 0.771 - 0.05916 \; \log \frac{[Fe^{2+}]}{[Fe^{3+}]}$$

$$E = 0.771 - 0.05916 \; \log \frac{1.0 \, \text{mmol}/10 \, \text{mL}}{0.50 \, \text{mmol}/10 \, \text{mL}} = 0.771 - 0.05916 \; \log 2.0$$

$$= 0.771 - 0.05916(0.30)$$
$$= 0.753 \, \text{V}$$

Note that this approach can only succeed where the standard potentials of the two half-reaction are sufficiently far apart such that the addition of Ce^{4+} will result in essentially quantitative conversion of a corresponding amount of Fe^{2+} to Fe^{3+} (assuming sufficient Fe^{2+} was present).

Cell Voltage—Before Reaction

The voltage of a cell can be calculated by taking the difference in potentials of the two half-reactions, to give a positive potential, calculated using the Nernst equation,

$$E_{cell} = E_+ - E_- \qquad (19.23)$$

as given in Equation 19.19.

In Example 19.2 for $2Fe^{3+} + 3I^- \rightleftharpoons 2Fe^{2+} + I_3^-$ at 298 K,

$$E_{cell} = E_{Fe^{3+},Fe^{2+}} - E_{I_3^-,I^-}$$

$$= \left(E^0_{Fe^{3+},Fe^{2+}} - \frac{0.05916}{2}\log\frac{[Fe^{2+}]^2}{[Fe^{3+}]^2} \right) - \left(E^0_{I_3^-,I^-} - \frac{0.05916}{2}\log\frac{[I^-]^3}{[I_3^-]} \right)$$

$$= E^0_{Fe^{3+},Fe^{2+}} - E^0_{I_3^-,I^-} - \frac{0.05916}{2}\log\frac{[Fe^{2+}]^2[I_3^-]}{[Fe^{3+}]^2[I^-]^3}$$

$$(19.24)$$

Note that the log term for the cell potential of a spontaneous reaction is always the *ratio of the product concentration(s) over the reactant concentration(s), that is, right side over left side* (as for a reduction half-reaction). Notice it was necessary to multiply the Fe^{3+}/Fe^{2+} half-reaction by 2 (as when subtracting the two half-reactions) in order to combine the two log terms (with $n = 2$), and the final equation is the same as we would have written from the cell reaction. Note also that $E^0_{Fe^{3+},Fe^{2+}} - E^0_{I_3^-,I^-}$ is the cell standard potential, E^0_{cell}.

The term on the right of the log sign is the **equilibrium constant expression** for the reaction:

$$2Fe^{3+} + 3I^- \rightleftharpoons 2Fe^{2+} + I_3^- \qquad (19.25)$$

> The cell voltage represents the tendency of a reaction to occur when the reacting species are put together (just as it does in a battery; that is, it represents the potential for work). After the reaction has reached equilibrium, the cell voltage necessarily becomes zero and the reaction is complete (i.e., no more work can be derived from the cell). That is, the potentials of the two half-reactions are equal at equilibrium. This is what happens when a battery runs down.

Example 19.5

One beaker contains a solution of 0.0200 M $KMnO_4$, 0.00500 M $MnSO_4$, and 0.500 M H_2SO_4; and a second beaker contains 0.150 M $FeSO_4$ and 0.00150 M $Fe_2(SO_4)_3$. The two beakers are connected by a salt bridge, and platinum electrodes are placed in each. The electrodes are connected via a wire with a voltmeter in between. What would be the potential of each half-cell **(a)** before reaction and

(b) after reaction? What would be the measured cell voltage (c) at the start of the reaction and (d) after the reaction reaches equilibrium? Assume H_2SO_4 to be completely ionized and in equal volumes in each beaker.

Solution

The cell reaction is

$$5Fe^{2+} + MnO_4^- + 8H^+ \rightleftharpoons 5Fe^{3+} + Mn^{2+} + 4H_2O$$

and the cell is

$$Pt/Fe^{2+}\ (0.150\ M\,),\, Fe^{3+}\ (0.00300\ M\,)//MnO_4^-(0.0200\ M\,),$$

$$Mn^{2+}(0.00500\ M\,),\ H^+(1.00\ M\,)/Pt$$

(a)

$$E_{Fe} = E^0_{Fe^{3+},Fe^{2+}} - 0.05916\ \log \frac{[Fe^{2+}]}{[Fe^{3+}]}$$

$$= 0.771 - 0.05916\ \log \frac{0.150}{0.00300} = 0.671\ V$$

$$E_{Mn} = E^0_{MnO_4^-,Mn^{2+}} - \frac{0.05916}{5}\ \log \frac{[Mn^{2+}]}{[MnO_4^-]\,[H^+]^8}$$

$$= 1.51 - \frac{0.05916}{5}\ \log \frac{0.00500}{(0.0200)(1.00)^8} = 1.52\ V$$

(b) At equilibrium, $E_{Fe} = E_{Mn}$. We can calculate E from either half-reaction. First, calculate the equilibrium concentrations. Five moles of Fe^{2+} will react with each mole of MnO_4^-. The Fe^{2+} is in excess. It will be decreased by $5 \times 0.0200 = 0.100\ M$, so $0.050\ M\ Fe^{2+}$ remains and $0.100\ M\ Fe^{3+}$ is formed (total now is $0.100 + 0.003 = 0.103\ M$). Virtually all the MnO_4^- is converted to $Mn^{2+}(0.0200\ M)$ to give a total of $0.0250\ M$. A small unknown amount of MnO_4^- remains at equilibrium, and we would need the equilibrium constant to calculate it; this can be obtained from $E_{cell} = 0$ at equilibrium—as in Equation 19.24—and as carried out in Example 19.4, this is treated in more detail in Chapter 21. But we need not go to this trouble since $[Fe^{2+}]$ and $[Fe^{3+}]$ are known:

$$E_{Mn} = E_{Fe} = 0.771 - 0.05916\ \log \frac{0.050}{0.103} = 0.790\ V$$

Note that the half-cell potentials at equilibrium are in between the values for the two half-cells before reaction.

(c) $E_{cell} = E_{Mn} - E_{Fe} = 1.52 - 0.671 = 0.85\ V$

(d) At equilibrium, $E_{Mn} = E_{Fe}$, and so E_{cell} is zero volts.

Note that if one of the species had not been initially present in a half-reaction, we could not have calculated an initial potential for that half-reaction.

19.4 Formal Potential—Use It for Defined Nonstandard Solution Conditions

The E^0 values listed in Table 19.1 refer to standard conditions as denoted by the superscript 0; this means they presume *all* species are at an activity of 1 M. However, the potential of a half-reaction may depend on the conditions of the solution. For example, the E^0 value for $Ce^{4+} + e^- \rightleftharpoons Ce^{3+}$ is 1.61 V. However, we can change this potential by changing the acid used to acidify the solution. (See Table C.5 in Appendix C.) This change in potential happens because the anions of the different acids differ in their ability to form complexes with one form of the cerium relative to the other, and the concentration ratio of the two forms of the free cerium ion is thereby affected.

If we know the form of the complex, we could write a new half-reaction involving the acid anion and determine an E^0 value for this reaction, keeping the acid and all other species at unit activity. However, the complexes are frequently of unknown composition. So we define the **formal potential** and designate this as $E^{0'}$. This is the standard potential of a redox couple with the oxidized and reduced forms at 1 M concentrations and *with the solution conditions specified*. For example, the formal potential of the Ce^{4+}/Ce^{3+} couple in 1 M HCl is 1.28 V. The Nernst equation is written as usual, using the formal potential in place of the standard potential. Table C.5 lists some formal potentials.

> The formal potential is used when not all species are known. In a sense, the formal potential provides the same convenience as available with the use of conditional complexation constants or solubility products.

Dependence of Potential on pH

Hydrogen or hydroxyl ions are involved in many redox half-reactions. We can change the potential of these redox couples by changing the pH of the solution. Consider the As(V)/As(III) couple:

> Many redox reactions involve protons, and their potentials are influenced greatly by pH.

$$H_3AsO_4 + 2H^+ + 2e^- \rightleftharpoons H_3AsO_3 + H_2O \tag{19.26}$$

$$E = E^0 - \frac{0.05916}{2} \log \frac{[H_3AsO_3]}{[H_3AsO_4][H^+]^2} \tag{19.27}$$

This can be rearranged to[4]

$$E = E^0 + 0.05916 \log [H^+] - \frac{0.05916}{2} \log \frac{[H_3AsO_3]}{[H_3AsO_4]} \tag{19.28}$$

or

$$E = E^0 - 0.05916 \, pH - \frac{0.05916}{2} \log \frac{[H_3AsO_3]}{[H_3AsO_4]} \tag{19.29}$$

The term $E^0 - 0.05916 \, pH$, where E^0 is the standard potential for the half-reaction, can be considered as equal to a formal potential $E^{0'}$, which can be calculated from the pH of the solution.[5] In 0.1 M HCl (pH 1), $E^{0'} = E^0 - 0.05916$. In neutral condition, it is $E^0 - 0.05916(7) = E^0 - 0.41$.

In strongly acid solution, H_3AsO_4 will oxidize I^- to I_2. But in neutral solution, the potential of the As(V)/As(III) couple ($E^{0'} = 0.146$ V) is less than that for I_2/I^-, and the reaction goes in the reverse; that is, I_2 will oxidize H_3AsO_3.

---

[4] The H^+ term in the log term can be separated as $(-0.05916/2) \log (1/[H^+]^2) = (+0.05916/2) \log [H^+]^2$. The squared term can be brought to the front of the log term to give $0.05916 \log [H^+]$.

[5] Actually, this is an oversimplification of the effect of pH in this particular case because H_3AsO_4 and H_3AsO_3 are also weak acids, and the effect of their ionization, that is, their K_a values, should be taken into account as well.

Dependence of Potential on Complexation

If an ion in a redox couple is complexed, the concentration of the free ion is reduced. This causes the potential of the couple to change. For example, E^0 for the Fe^{3+}/Fe^{2+} couple is 0.771 V. In HCl solution, the Fe^{3+} is complexed with the chloride ion. This reduces the concentration of Fe^{3+}, and so the potential is decreased. In 1 M HCl, the formal potential is 0.70 V. If we assume that the complex is $FeCl_4^-$, then the half-reaction would be

$$FeCl_4^- + e^- \rightleftharpoons Fe^{2+} + 4Cl^- \tag{19.30}$$

and if we assume that [HCl] is constant at 1 M,

$$E = 0.70 - 0.05916 \ \log \frac{[Fe^{2+}]}{[FeCl_4^-]} \tag{19.31}$$

In effect, we have stabilized the Fe^{3+} by complexing it, making it more difficult to reduce. So the reduction potential is decreased. If we complexed the Fe^{2+}, the reverse effect would be observed. So the presence of complexing agents that have different affinities for one form of the couple over another will affect the potential.

19.5 Limitations of Electrode Potentials

Electrode potentials (E^0 or $E^{0'}$) will predict whether a given reaction can occur, but they indicate nothing about the **rate** of the reaction. If a reaction is reversible, it will occur fast enough for a titration. But if the rate of the electron transfer step is slow, the reaction may be so slow that equilibrium will be reached only after a very long time. We say that such a reaction is **irreversible**.

Some reactions in which one half-reaction is irreversible do occur rapidly. Several oxidizing and reducing agents containing oxygen are reduced or oxidized irreversibly but may be speeded up by addition of an appropriate catalyst. The oxidation of arsenic(III) by cerium(IV) is slow, but it is catalyzed by a small amount of osmium tetroxide, OsO_4.

So, while electrode potentials are useful for predicting many reactions, they do not assure that a given reaction will actually occur. They are useful in that they will predict that a reaction will *not* occur if the potential differences are not sufficient.

Questions

1. What is an oxidizing agent? A reducing agent?
2. What is the Nernst equation?
3. What is the standard potential? The formal potential?
4. What is the function of a salt bridge in an electrochemical cell?
5. What are the NHE and SHE?
6. The standard potential for the half-reaction $M^{4+} + 2e^- = M^{2+}$ is +0.98 V. Is M^{2+} a good or a poor reducing agent?
7. What should be the minimum potential difference between two half-reactions so that a sharp end point will be obtained in a titration involving the two half-reactions?
8. Why cannot standard or formal electrode potentials always be used to predict whether a given titration will work?

9. Correlate the Brønsted-Lowry acid-base concept with redox reactions.

10. Do equilibrium electrode potential and cell potential behave similarly during redox titration?

11. How the complex formation affects the reduction potential?

12. Identify the oxidizing agent for the given reaction in the basic medium.

$$H_2O\ (l) + Zn\ (s) + NO_3^-\ (aq) + OH^-\ (aq) \rightarrow [Zn(OH)_4]^{2-}\ (aq) + NH_3\ (aq)$$

   (a) Zn (s)

   (b) $H_2O(l)$ (the oxygen)

   (c) NO_3^- (aq) (the nitrogen)

   (d) $NH_3(aq)$ (the nitrogen)

13. Which of the following statements is not true regarding the following observations?

$$Sn + 2AgBr \rightarrow 2Ag + SnBr_2$$

$$2Ag + SnBr_2 \rightarrow No\ reaction$$

   (a) Sn is a stronger reducing agent than Ag.

   (b) Ag^+ is a stronger reducing agent than Sn^{2+}.

   (c) The reducing potential for Ag^+ is more positive than that for Sn^{2+}.

   (d) Sn^{2+} is a stronger reducing agent than Ag^+.

14. In the modern periodic table where do you expect elements acting as good oxidizing agents?

   (a) On the right

   (b) At the bottom

   (c) In the top left

   (d) In the transition metals

15. Rank the following elements Ag, Au and Sn, according to their strength as reducing agent based on the following observations.

$$Sn + 2AgBr \rightarrow SnBr_2 + 2Ag$$

$$3Sn + 2AuBr_3 \rightarrow 3SnBr_2 + 2Au$$

$$3Ag + AuBr_3 \rightarrow 3AgBr + Au$$

   (a) Sn > Ag > Au

   (b) Sn > Au > Ag

   (c) Au > Ag > Sn

   (d) Ag > Au > Sn

Problems

Redox Strengths

16. Arrange the following substances in decreasing order of oxidizing strengths: H_2SeO_3, H_3AsO_4, Hg^{2+}, Cu^{2+}, Zn^{2+}, O_3, HClO, K^+, Co^{2+}.

17. Arrange the following substances in decreasing order of reducing strengths: I^-, V^{3+}, Sn^{2+}, Co^{2+}, Cl^-, Ag, H_2S, Ni, HF.

18. Which of the following pairs would be expected to give the largest end-point break in a titration of one component with the other in each pair?

   (a) $Fe^{2+} - MnO_4^-$ or $Fe^{2+} - Cr_2O_7^{2-}$

   (b) $Fe^{2+} - Ce^{4+}$ (H_2SO_4) or $Fe^{2+} - Ce^{4+}$ $(HClO_4)$

   (c) $H_3AsO_3 - MnO_4^-$ or $Fe^{2+} - MnO_4^-$

   (d) $Fe^{3+} - Ti^{2+}$ or $Sn^{2+} - I_3^-$

PROFESSOR'S FAVORITE PROBLEMS

(The following two problems contributed by Professor Bin Wang, Marshall University)

19. Identify the oxidizing agent and the reducing agent on the left side of the following reactions:
    (a) $2VO^{2+} + Sn^{2+} + 4H^+ \rightleftharpoons 2V^{3+} + Sn^{4+} + 2H_2O$
    (b) $Fe^{2+} + Fe(CN)_6^{3-} \rightleftharpoons Fe^{3+} + Fe(CN)_6^{4-}$
    (c) $Cu + 2Ag^+ \rightleftharpoons 2Ag + Cu^{2+}$
    (d) $I_2 + OH^- \rightleftharpoons HOI + I^-$
    (e) $2Fe^{2+} + H_2O_2 + 2H^+ \rightleftharpoons 2Fe^{3+} + 2H_2O$

20. The titration of 1.0512 g of an unknown iron sample required 28.75 mL of 0.1023 N $KMnO_4$. The iron was initially in the +2 oxidation state. The solution was strongly acidic. Write out the balanced reaction that takes place during this titration. What is the percentage of iron in the unknown sample?

Voltaic Cells

21. Construct the Galvanic cell and write down the corresponding expression for cell potential.
    (a) $2Cr\,(s) + 3Hg_2Cl_2\,(s) \rightarrow 2Cr^{3+}\,(aq) + 6Cl^-\,(aq) + 6Hg\,(l)$
    (b) $Ni^{2+}\,(aq) + H_2\,(g) + 2OH^-\,(aq) \rightarrow Ni\,(s) + 6H_2O\,(l)$
    (c) $AgCl\,(s) + I^-\,(aq) \rightarrow AgI\,(s) + Cl^-\,(aq)$
    (d) $Zn(s) + H_2SO_4(aq) \rightarrow ZnSO_4(aq) + H_2(g)$

22. Predict whether the following reactions are possible or not under the standard conditions.
    (a) Reduction of Fe^{3+} to Fe^{2+} by Fe.
    (b) Liberation of O_2 from water by permanganate ion (MnO_4^-) in the presence of an acid.
    (c) Oxygen (O_2) oxidize gold to $Au(CN)_2^-$ in the presence of CN^- and OH^- ions.
    (d) Silver spoon is used to stir a solution of $Cu(NO_3)_2$.

23. Write the equivalent voltaic cells for the following reactions (assume all concentrations are 1 M):
    (a) $6Fe^{2+} + Cr_2O_7^{2-} + 14H^+ \rightleftharpoons 6Fe^{3+} + 2Cr^{3+} + 7H_2O$
    (b) $IO_3^- + 5I^- + 6H^+ \rightleftharpoons 3I_2 + 3H_2O$
    (c) $Zn + Cu^{2+} \rightleftharpoons Zn^{2+} + Cu$
    (d) $Cl_2 + H_2SeO_3 + H_2O \rightleftharpoons 2Cl^- + SeO_4^{2-} + 4H^+$

24. For each of the following cells, write the cell reactions:
    (a) $Pt/V^{2+}, V^{3+}//PtCl_4^{2-}, PtCl_6^{2-}, Cl^-/Pt$
    (b) $Ag/AgCl(s)/Cl^-//Fe^{3+}, Fe^{2+}/Pt$
    (c) $Cd/Cd^{2+}//ClO_3^-, Cl^-, H^+/Pt$
    (d) $Pt/I^-, I_2//H_2O_2, H^+/Pt$

Potential Calculations

25. The methanol oxidation has a ΔG value of –937.9 kJ/mol. Calculate the standard cell potential for a methanol fuel cell.

$$2CH_3OH + 3O_2 \rightarrow 2CO_2 + 4H_2O$$

26. What is the value of E_{cell} for the given reaction when concertation of Cu^{2+} is 1.0 M and concentration of Zn^{2+} is 0.025 M?

$$Zn(s) + Cu^{2+}(aq) \rightarrow Cu(s) + Zn^{2+}(aq) \qquad E^0_{cell} = 1.10\ V$$

27. What is the electrode potential (vs. NHE) in a solution containing 0.50 M $KBrO_3$ and 0.20 M Br_2 at pH 2.5?

28. What is the electrode potential (vs. NHE) in the solution prepared by adding 90 mL of 5.0 M KI to 10 mL of 0.10 M H_2O_2 buffered at pH 2.0?

29. A solution of a mixture of Pt^{4+} and Pt^{2+} is 3.0 M in HCl, which produces the chloro complexes of the Pt ions (see Problem 20). If the solution is 0.015 M in Pt^{4+} and 0.025 M in Pt^{2+}, what is the potential of the half-reaction?

30. Equal volumes of 0.100 M UO_2^{2+} and 0.100 M V^{2+} in 0.10 M H_2SO_4 are mixed. What would the potential of a platinum electrode (vs. NHE) dipped in the solution be at equilibrium? Assume H_2SO_4 is completely ionized.

Cell Voltages

31. When a voltaic cell reaches equilibrium _____.

    (a) $E = 0$
    (b) $E_{cell} = 0$
    (c) $E = K$
    (d) $E_{cell} = Q$

32. What will be the molecular formula of mercurous nitrate, if the emf of the given cell is 0.0295 V at 25°C ?

33. From the standard potentials of the following half-reactions, determine the reaction that will occur, and calculate the cell voltage from the reaction:

$$PtCl_6^{2-} + 2e^- \rightleftharpoons PtCl_4^{2-} + 2Cl^-$$

$$E^0 = 0.68 \text{ V}$$

$$V^{3+} + e^- \rightleftharpoons V^{2+}$$

$$E^0 = -0.255 \text{ V}$$

34. Calculate the voltages of the following cells:

    (a) Pt/I^- (0.100 M), I_3^- (0.0100 M)//IO_3^- (0.100 M), I_2 (0.0100 M), H^+ (0.100 M)/Pt

    (b) $Ag/AgCl(s)/Cl^-$ (0.100 M)//UO_2^{2+} (0.200 M), U^{4+} (0.050 M), H^+ (1.00 M)/Pt

    (c) Pt/Tl^+ (0.100 M), Tl^{3+} (0.0100 M)//MnO_4^- (0.0100 M), Mn^{2+} (0.100 M), H^+ (pH 2.00)/Pt

35. From the standard potentials, determine the reaction between the following half-reactions, and calculate the corresponding standard cell voltage:

$$VO_2^+ + 2H^+ + e^- \rightleftharpoons VO^{2+} + H_2O$$
$$UO_2^{2+} + 4H^+ + 2e^- \rightleftharpoons U^{4+} + 2H_2O$$

PROFESSOR'S FAVORITE PROBLEM

Contributed by Professor Bin Wang, Marshall University

36. Use the Nernst equation to calculate the voltage of the following cell:

$$Ni\ (s)|NiSO_4\ (0.0020\ M)||CuCl_2\ (0.0030\ M)\ |Cu\ (s)$$

PROFESSOR'S FAVORITE PROBLEM

Contributed by Professor Yijun Tang, University of Wisconsin at Oshkosh

37. For a galvanic cell as shown below, assume that adding solid does not change the volume of the solution.

$$Cu(s)\ |\ 50.0\ mL\ 0.0167\ M\ Cu(NO_3)_2(aq)\ ||\ 50.0\ mL\ 0.100\ M\ AgNO_3\ |\ Ag(s)$$

    (a) Calculate the potentials of the Cu electrode and the Ag electrode. Is the Ag electrode anode or cathode?

(b) If 0.271 g of KCl is added to the original $AgNO_3$ solution, calculate the potential of the Ag electrode. Is the Ag electrode anode or cathode?

(c) If 0.812 g of KCl is added to the original $AgNO_3$ solution, calculate the potential of the Ag electrode. Is the Ag electrode anode or cathode?

Multiple Choice Questions

1. In an electrolytic cell, the electrode at which the electrons enter the solution is called the _____ and, the chemical change that occurs at this electrode is called _____.
   - (a) anode, oxidation
   - (b) anode, reduction
   - (c) cathode, reduction
   - (d) cathode, oxidation

2. Which of the following statements is false?
   - (a) All voltaic cells utilize electricity to initiate non-spontaneous chemical reactions.
   - (b) Oxidation occurs at the anode.
   - (c) Reduction occurs at the cathode.
   - (d) All electrochemical reactions involve the transfer of electrons.

3. The half reaction occurring at anode during the electrolysis of molten sodium bromide is
   - (a) $Br_2 + 2e^- \rightarrow 2Br^-$
   - (b) $2Br_2^- \rightarrow Br_2 + 2e^-$
   - (c) $Na^+ + e^- \rightarrow Na$
   - (d) $Na \rightarrow Na^+ + e^-$

4. During electrolysis,
   - (a) the cation gets oxidized.
   - (b) the anion gets oxidized.
   - (c) the anion gets reduced.
   - (d) both anions and cations get oxidized.

5. The process of reduction involves
   - (a) gain of one or more electrons by an atom, ion or molecule.
   - (b) removal of oxygen or addition of hydrogen.
   - (c) removal of an electronegative element or group or addition of an electropositive element or group.
   - (d) all of the above.

6. Which of the following statement is correct about electrochemical cell?
   - (a) Electrons are released at anode.
   - (b) Cathode is regarded as negative electrode.
   - (c) Chemical energy is convened into electrical energy.
   - (d) Salt bridge maintains the electrical neutrality of the solution.

7. For a spontaneous reaction emf of the cell is:
   - (a) Positive
   - (b) Negative
   - (c) Zero
   - (d) Fixed

8. Which of the following represents the relation between the emf and free energy change (ΔG) of a cell?
   - (a) $\Delta G = -nFE$
   - (b) $\Delta G = -FE/n$
   - (c) $\Delta G = -nRT/FE$
   - (d) $\Delta G = -nF^2E$

9. Given that $E^0_{Zn^{2+}/Zn} = -0.76\,V$ and $E^0_{Mg^{2+}/Mg} = -2.37\,V$, what will happen when zinc dust is added to $MgCl_2$ solution?
   - (a) Mg will be deposited.
   - (b) Zn will dissolve.
   - (c) $ZnCl_2$ will be formed.
   - (d) No reaction will occur.

10. An electrochemical cell is a device in which a redox reaction takes place indirectly and the decrease in _____ of the reaction appears largely in the form of electrical energy.
    - (a) chemical energy
    - (b) photochemical energy
    - (c) potential energy
    - (d) kinetic energy

11. The electrode $Pt/Fe^{2+} + (C_1)$, Fe^{3+} (C_2) belongs to the type:

    (a) inert-metal electrodes.

    (b) gas electrodes.

    (c) liquid electrodes.

    (d) metal-metal soluble salt electrode.

12. In the following cell, $Zn - Hg(c_1)|Zn^{2+}|Zn - Hg(c_2)$ are concentrating, i.e., 2 g, 0.5 g per 100 g of mercury, EMF of cell at 25° C is

    (a) −00178 V

    (b) +0.0168 V

    (c) +0.0178 V

    (d) −0.0168 V

13. The Nernst equation, $E = E^0 - \dfrac{RT}{nF} \ln Q$ indicates that the equilibrium constant K_c will be equal to Q when

    (a) $E = E^0$

    (b) $RT \ln F = 1$

    (c) $E = $ zero

    (d) $E^0 = 1$

14. Given E^0 for half-cell reaction:
    $E^0_{Zn^{2+}, Zn} = -0.7618$ vs NHE
    $E^0_{Hg_2Cl_2, Hg} = -0.2680$ vs NHE
    E^0 for the cell in which the reaction

$$Zn + Hg_2Cl_2 \rightarrow Zn^{2+} + 2Hg + 2Cl^-$$

    (a) (−0.7618 + 0.2680) V.

    (b) (0.7618 − 02680) V.

    (c) (−0.7618 − 02680) V.

    (d) (0.7618 + 0.2680) V.

15. Cu^+ (aq) is unstable in solution and undergoes simultaneous oxidation and reduction according to the reaction

$$2Cu^+ \text{ (aq)} \rightarrow Cu^{2+}\text{(aq)} + Cu\text{(s)}$$

    Choose correct E^0 for the above reaction if $E^0_{Cu^{2+}/Cu} = 0.34$ V and $E^0_{Cu^{2+}/Cu^+} = 0.15$ V

    (a) +0.49 V

    (b) 0.38 V

    (c) −0.19 V

    (d) −0.38 V

16. The Nernst equation giving dependence of electrode potential on concentration is

    (a) $E = E^0 + \dfrac{2.303RT}{nF} \log \dfrac{[M]}{[M^{n+}]}$

    (b) $E = E^0 + \dfrac{2.303RT}{nF} \log \dfrac{[M^{n+}]}{[M]}$

    (c) $E = E^0 - \dfrac{2.303RT}{nF} \log \dfrac{[M^{n+}]}{[M]}$

    (d) $E = E^0 - \dfrac{2.303RT}{nF} \log [M^{n+}]$

17. Given that

$$I_2 + 2e^- \rightarrow 2I^- \qquad E^0 = 0.54 \text{ V}$$

$$Br_2 + 2e^- \rightarrow 2Br^- \quad E^0 = 1.09 \text{ V}$$

    Predict which of the following is true.

    (a) I^- ions will be able to reduce bromine.

    (b) Br^- ions will be able to reduce iodine.

    (c) Iodine will be able to reduce bromine.

    (d) Bromine will be able to reduce iodide ions.

18. In a chemical cell represented by $Ag(s) \mid AgNO_3(aq) \parallel MCl(aq) \mid AgCl(s) \mid Ag(s)$

    (a) $[Ag^+]$ is larger in R.H.S.

    (b) emf depends on K_S of AgCl.

    (c) emf is independent of concentration of Cl^-.

    (d) emf depends on the concentration of NO_3^-.

Recommended References

Nernst Equation

1. L. Meites, "A 'Derivation' of the Nernst Equation for Elementary Quantitative Analysis," *J. Chem. Ed.*, **29** (1952) 142.

Electrode Sign Conventions

2. F. C. Anson, "Electrode Sign Convention," *J. Chem. Ed.*, **36** (1959) 394.
3. T. S. Light and A. J. de Bethune, "Recent Developments Concerning the Signs of Electrode Conventions," *J. Chem. Ed.*, **34** (1957) 433.

Standard Potentials

4. A. J. Bard, R. Parsons, and J. Jordan, eds., Standard Potentials in Aqueous Solution. New York: Marcel Dekker, 1985.
5. W. M. Latimer, *The Oxidation States of the Elements and Their Potentials in Aqueous Solutions*, 2nd ed. New York: Prentice Hall, 1952.

Chapter 20

POTENTIOMETRIC ELECTRODES AND POTENTIOMETRY

KEY THINGS TO LEARN FROM THIS CHAPTER

- Types of electrodes and electrode potentials from the Nernst equation (key Equations: 20.3, 20.10, 20.16)
- Liquid junctions and junction potentials
- Reference electrodes
- Accuracy of potentiometric measurements (key Equation: 20.36)
- The pH glass electrode (key Equation: 20.42)
- Standard buffers and the accuracy of pH measurements
- The pH meter
- Ion-selective electrodes
- The selectivity coefficient (key Equation: 20.46)

In Chapter 19, we mentioned measurement of the potential of a solution and described a platinum electrode whose potential was determined by the half-reaction of interest. This was a special case, and there are a number of electrodes available for measuring solution potentials. In this chapter, we list the various types of electrodes that can be used for measuring solution potentials and how to select the proper one for measuring a given analyte. The apparatus for making potentiometric measurements is described along with limitations and accuracies of potentiometric measurements. The important glass pH electrode is described, as well as standard buffers required for its calibration. The various kinds of ion-selective electrodes are discussed. The use of electrodes in potentiometric titrations is described in Chapter 21.

Potentiometric electrodes measure activity rather than concentration, a unique feature, and we will use activities in this chapter in describing electrode potentials. An understanding of activity and the factors that affect it are important for direct potentiometric measurements, as in pH or ion-selective electrode measurements. You should, therefore, review the material on activity and activity coefficients in Chapter 5.

Review activities in Chapter 5, for an understanding of potentiometric measurements.

Potentiometry is one of the oldest analytical methods, with foundations of electrode potentials and electrochemical equilibria laid down by J. Willard Gibbs (1839–1903) and Walther Nernst (1864–1941). Inert electrodes are used as indicating electrodes for redox titrations, and may be used in automatic titrators. The pH electrode is the most widely used potentiometric electrode. Ion selective electrodes are now more widely used than redox electrodes, for selectively measuring particular ions. The measurement of fluoride, for example in toothpaste, is one of the more important applications since fluoride is not easily measured otherwise. Clinical analyzers measure the electrolytes sodium, potassium, lithium (used in the treatment of manic depression), and calcium in blood using ion selective electrodes.

20.1 Metal Electrodes for Measuring the Metal Cation

An electrode of this type is a metal in contact with a solution containing the cation of the same metal. An example is a silver metal electrode dipping in a solution of silver nitrate.

For all electrode systems, an electrode half-reaction can be written from which the potential of the electrode is described. The electrode system can be represented by M/M^{n+}, in which the line represents an electrode–solution interface. For the silver electrode, we have

$$Ag|Ag^+ \tag{20.1}$$

and the half-reaction is

$$Ag^+ + e^- \rightleftharpoons Ag \tag{20.2}$$

The potential of the electrode is described by the Nernst equation:

$$E = E^0_{Ag^+,Ag} - \frac{2.303RT}{nF} \log \frac{1}{a_{Ag^+}} \tag{20.3}$$

where a_{Ag^+} represents the **activity** of the silver ion (see Chapter 5). The value of n here is 1. We will use the more correct unit of activity in discussions in this chapter because, in the interpretation of direct potentiometric measurements, significant errors would result if concentrations were used in calculations.

The potential calculated from Equation 20.3 is the potential *relative to the normal hydrogen electrode* (*NHE*—see Section 20.3). The potential becomes increasingly positive with increasing Ag^+ (the case for any electrode measuring a cation). That is, in a cell measurement using the NHE as the second half-cell, the voltage is

$$E_{measd.} = E_{cell} = E_{ind \text{ vs. NHE}} = E_{ind} - E_{NHE} \tag{20.4}$$

where E_{ind} is the potential of the **indicator electrode** (the one that responds to the test solution, Ag^+ ions in this case). Since E_{NHE} is zero,

$$E_{cell} = E_{ind} \tag{20.5}$$

corresponds to writing the cells as

$$E_{ref}|solution|E_{ind} \tag{20.6}$$

and

$$E_{cell} = E_{right} - E_{left} = E_{ind} - E_{ref} = E_{ind} - constant \tag{20.7}$$

> Increasing cation activity always causes the electrode potential to become more positive (if you write the Nernst equation properly).

> The indicator electrode is the one that responds to the analyte.

> The reference electrode completes the cell but does not respond to the analyte. It is usually separated from the test solution by a salt bridge.

> Any pure substance does not numerically appear in the Nernst equation (e.g., Cu, H₂O); their activities are taken as unity.

where E_{ref} is the potential of the **reference electrode**, whose potential is constant. Note that E_{cell} (or E_{ind}) may be positive or negative, depending on the activity of the silver ion or the relative potentials of the two electrodes. This is in contrast to the convention used in Chapter 19 for a voltaic cell, in which a cell was always set up to give a positive voltage and thereby indicate what the spontaneous cell reaction would be. In potentiometric measurements, we, in principle, measure the potential at zero current so as not to disturb the equilibrium, i.e., don't change the relative concentrations of the species being measured at the indicating electrode surface—which establishes the potential (see measurement of potential, below). We are interested in how the potential of the test electrode (indicating electrode) changes with analyte concentration, as measured against some constant reference electrode. Equation 20.7 is arranged so that changes in E_{cell} reflect the same changes in E_{ind}, *including sign*. This point is discussed further when we talk about cells and measurement of electrode potentials.

The activity of silver metal above, as with other pure substances, is taken as unity. So an electrode of this kind can be used to monitor the activity of a metal ion in solution. There are few reliable electrodes of this type because many metals tend to form an oxide coating that changes the potential.

20.2   Metal–Metal Salt Electrodes for Measuring the Salt Anion

The general form of this type of electrode is $M|MX|X^{n-}$, where MX is a slightly soluble salt. An example is the silver–silver chloride electrode:

$$Ag|AgCl_{(s)}|Cl^- \tag{20.8}$$

The (s) indicates a solid, (g) is used to indicate a gas, and (l) is used to indicate a pure liquid. A vertical line denotes a phase boundary between two different solids or a solid and a solution. The half-reaction is

$$\underline{AgCl} + e^- \rightleftharpoons \underline{Ag} + Cl^- \tag{20.9}$$

where the underline indicates a solid phase and the potential is defined by

$$\boxed{E = E^0_{AgCl,Ag} - \frac{2.303RT}{F} \log a_{Cl^-}} \tag{20.10}$$

The number of electrons, n, does not appear in the equation because here $n = 1$.

This electrode, then, can be used to measure the activity of chloride ion in solution. Note that, as the activity of chloride increases, the potential *decreases*. This is true of any electrode measuring an anion—the opposite for a cation electrode. A silver wire is coated with silver chloride precipitate (e.g., by electrically oxidizing it in a solution containing chloride ion, the reverse reaction of Equation 20.9). Actually, as soon as a silver wire is dipped in a chloride solution, a thin layer of silver chloride and is usually not required.

Note that this electrode can be used to monitor either a_{Cl^-} or a_{Ag^+}. It really senses only silver ion, and the activity of this is determined by the solubility of the slightly soluble AgCl. Since $a_{Cl^-} = K_{sp}/a_{Ag^+}$, Equation 20.10 can be rewritten:

> Increasing anion activity always causes the electrode potential to decrease.

> The Ag metal really responds to Ag^+, whose activity is determined by K°_{sp} and a_{Cl^-}.

$$E = E^0_{AgCl,Ag} - \frac{2.303RT}{F} \log \frac{K_{sp}}{a_{Ag^+}} \tag{20.11}$$

$$E = E^0_{AgCl,Ag} - \frac{2.303RT}{F} \log K_{sp} - \frac{2.303RT}{F} \log \frac{1}{a_{Ag^+}} \tag{20.12}$$

Comparing this with Equation 20.3, we see that

$$\boxed{E^0_{Ag^+,Ag} = E^0_{AgCl,Ag} - \frac{2.303RT}{F} \log K_{sp}} \tag{20.13}$$

K_{sp} here is the thermodynamic solubility product K°_{sp} (see Chapter 5), since activities, rather than concentrations, were used in arriving at it in these equations. We could have arrived at an alternative form of Equation 20.10 by substituting K_{sp}/a_{Cl^-} for a_{Ag^+} in Equation 20.3 (see Example 20.1).

In a solution containing a mixture of Ag^+ and Cl^- (e.g., a titration of Cl^- with Ag^+), the concentrations of each *at equilibrium* will be such that the potential of a silver wire dipping in the solution can be calculated by either Equation 20.3 or Equation 20.10. This is completely analogous to the statement in Chapter 19 that the potential of one half-reaction must be equal to the potential of the other in a chemical reaction at equilibrium. Equations 20.2 and 20.9 are the two half-reactions in this case, and when one is subtracted from the other, the result is the *overall chemical reaction*.

$$Ag^+ + Cl^- \rightleftharpoons AgCl \tag{20.14}$$

Note that as Cl^- is titrated with Ag^+, the former decreases and the latter increases. Equation 20.10 predicts an increase in potential as Cl^- decreases; and similarly, Equation 20.12 predicts the same increase as Ag^+ increases.

The silver electrode can also be used to monitor other anions that form slightly soluble salts with silver, such as I^-, Br^-, and S^{2-}. The E^0 in each case would be that for the particular half-reaction $\underline{AgX} + e^- \rightleftharpoons Ag + X^-$.

Another widely used electrode of this type is the **calomel electrode**, Hg, $Hg_2Cl_{2(s)}|Cl^-$. This will be described in more detail when we talk about reference electrodes.

Example 20.1

Given that the standard potential of the calomel electrode is 0.268 V and that of the Hg/Hg_2^{2+} electrode is 0.789 V, calculate K_{sp} for calomel (Hg_2Cl_2), for 298K.

Solution

For $Hg_2^{2+} + 2e^- \rightleftharpoons Hg$,

$$E = 0.789 - \frac{0.05916}{2} \log \frac{1}{a_{Hg_2^{2+}}} \tag{1}$$

For $Hg_2Cl_2 + 2e^- \rightleftharpoons 2Hg + 2Cl^-$,

$$E = 0.268 - \frac{0.05916}{2} \log(a_{Cl^-})^2 \tag{2}$$

Since $K_{sp} = a_{Hg_2^{2+}} \cdot (a_{Cl^-})^2$,

$$E = 0.268 - \frac{0.05916}{2} \log \frac{K_{sp}}{a_{Hg_2^{2+}}} \tag{3}$$

$$E = 0.268 - \frac{0.05916}{2} \log K_{sp} - \frac{0.05916}{2} \log \frac{1}{a_{Hg_2^{2+}}} \tag{4}$$

From (1) and (4),

$$0.789 = 0.268 - \frac{0.05916}{2} \log K_{sp}$$

$$K_{sp} = 2.4_4 \times 10^{-18}$$

20.3 Redox Electrodes—Inert Metals

In the redox electrode, an inert metal is in contact with a solution containing the soluble oxidized and reduced forms of the redox half-reaction. This type of electrode was mentioned in Chapter 19.

The inert metal used is usually platinum. The potential of such an inert electrode is determined by the ratio at the electrode surface of the reduced and oxidized species in the half-reaction:

$$M^{a+} + ne^- \rightleftharpoons M^{(a-n)+} \tag{20.15}$$

$$E = E^0_{M^{a+}, M^{(a-n)+}} - \frac{2.303RT}{nF} \log \frac{a_{M^{(a-n)+}}}{a_{M^{a+}}} \tag{20.16}$$

An example is the measurement of the ratio of MnO_4^-/Mn^{2+}:

$$MnO_4^- + 8H^+ + 5e^- \rightleftharpoons Mn^{2+} + 4H_2O \tag{20.17}$$

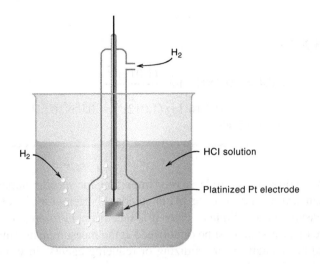

FIGURE 20.1 Hydrogen electrode.

$$E = E^0_{MnO_4^-,Mn^{2+}} - \frac{2.303RT}{5F} \log \frac{a_{Mn^{2+}}}{a_{MnO_4^-} \cdot (a_{H^+})^8} \qquad (20.18)$$

The pH is usually held constant, and so the ratio $a_{Mn^{2+}}/a_{MnO_4^-}$ is measured, as in a redox titration.

A very important example of this type of electrode is the **hydrogen electrode**, $Pt|H_2, H^+$:

$$H^+ + e^- \rightleftharpoons \frac{1}{2}H_2 \qquad (20.19)$$

$$\boxed{E = E^0_{H^+,H_2} - \frac{2.303RT}{F} \log \frac{(p_{H_2})^{1/2}}{a_{H^+}}} \qquad (20.20)$$

The construction of the hydrogen electrode is shown in Figure 20.1. A layer of platinum black must be placed on the surface of the platinum electrode by cathodically electrolyzing in a H_2PtCl_6 solution. The platinum black provides a larger surface area for adsorption of hydrogen molecules and catalyzes their oxidation. Too much platinum black, however, can adsorb traces of other substances such as organic molecules or H_2S, causing erratic behavior of the electrode.

For gases, we will use pressures, p (in atmospheres), in place of activity (or the thermodymic equivalent term for gases, fugacity).

The pressure of gases, in atmospheres, is used in place of activities. If the hydrogen pressure is held at 1 atm, then, since E^0 for Equation 20.19 is defined as zero,

$$\boxed{E = \frac{2.303RT}{F} \log \frac{1}{a_{H^+}} = -\frac{2.303RT}{F} pH} \qquad (20.21)$$

Example 20.2

Calculate the pH of a solution whose potential at 25°C measured with a hydrogen electrode at an atmospheric pressure of 1.012 atm (corrected for the vapor pressure of water at 25°C) is −0.324 V (relative to the NHE).

The vapor pressure of water above the solution must be subtracted from the measured gas pressure.

Solution

From Equation 20.20,

$$-0.324 = -0.05916 \log \frac{(1.012)^{1/2}}{a_{H^+}}$$
$$= -0.05916 \log(1.012)^{1/2} - 0.05916 \text{ pH}$$
$$\text{pH} = 5.48$$

While the hydrogen electrode is very important for specific applications (e.g., establishing standard potentials or the pH of standard buffers—see below), its use for routine pH measurements is limited. First, it is inconvenient to prepare and use. The partial pressure of hydrogen must be established at the measurement temperature. The solution should not contain other oxidizing or reducing agents since these will alter the potential of the electrode.

20.4 Voltaic Cells without Liquid Junction—For Maximum Accuracy

To make potential measurements, a complete cell consisting of two half-cells must be set up, as was described in Chapter 19. One half-cell usually is comprised of the test solution and an electrode whose potential is determined by the analyte we wish to measure. This electrode is the **indicator electrode**. The other half-cell is any arbitrary half-cell whose potential is not dependent on the analyte. This half-cell electrode is designated the **reference electrode**. Its potential is constant, and the measured cell voltage reflects the indicator electrode potential relative to that of the reference electrode. Since the reference electrode potential is constant, any *changes* in potential of the indicator electrode will be reflected by an equal change in the cell voltage.

It is possible to construct a cell without a salt bridge. For practical purposes, this is rare because of the tendency of the reference electrode potential to be influenced by the test solution.

There are two basic ways a cell may be set up, either without or with a salt bridge. The first is called a *cell without liquid junction*. An example of a cell of this type would be

$$\text{Pt}|\text{H}_2(\text{g}), \text{HCl(solution)}|\text{AgCl(s)}|\text{Ag} \tag{20.22}$$

The solid line represents an electrode–solution interface. An electrical cell such as this is a voltaic one, and the cell illustrated above is written for the *spontaneous reaction* by convention (positive E_{cell}—although we may actually measure a negative cell voltage if the indicator electrode potential is the more negative one; we haven't specified which of the half-reactions represents the indicator electrode). The hydrogen electrode is the anode, since its potential is the more negative (see Chapter 19 for a review of cell voltage conventions for voltaic cells). The potential of the left electrode would be given by Equation 20.20, and that for the right electrode would be given by Equation 20.10, and the cell voltage would be equal to the difference in these two potentials:

This cell is used to accurately measure the pH of "standard buffers." See Section 20.12.

$$E_{cell} = \left(E^0_{AgCl,Ag} - \frac{2.303RT}{F} \log a_{Cl^-}\right) - \left(E^0_{H^+,H_2} - \frac{2.303RT}{F} \log \frac{(p_{H_2})^{\frac{1}{2}}}{a_{H^+}}\right) \tag{20.23}$$

$$E_{cell} = E^0_{AgCl,Ag} - E^0_{H^+,H_2} - \frac{2.303RT}{F} \log \frac{a_{H^+}a_{Cl^-}}{(p_{H_2})^{\frac{1}{2}}} \tag{20.24}$$

The **cell reaction** would be (half-reaction)$_{right}$ − (half-reaction)$_{left}$ (to give a positive E_{cell} and the spontaneous reaction), or

$$\underline{AgCl} + e^- \rightleftharpoons \underline{Ag} + Cl^- \tag{20.25}$$

$$-\left(H^+ + e^- \rightleftharpoons \tfrac{1}{2}H_2\right) \tag{20.26}$$

$$\underline{AgCl} + \tfrac{1}{2}H_2 \rightleftharpoons \underline{Ag} + Cl^- + H^+ \tag{20.27}$$

Equation 20.23 would also represent the voltage if the right half-cell were used as an indicating electrode in a potentiometric measurement of chloride ion and the left cell were the reference electrode (see Equations 20.6 and 20.7). That is, the voltage (and hence the indicator electrode potential) would decrease with increasing chloride ion. If we were to use the hydrogen electrode as the indicating electrode to measure hydrogen ion activity or pH, we would reverse the cell setup in Equation 20.22 from left to right to indicate what is being measured. Equation 20.23 will be reversed as well, and the voltage (and indicator electrode potential) would increase with increasing acidity or decreasing pH ($E_{cell} = E_{ind} - E_{ref}$, Equation 20.7).

Cells without liquid junction are always used for the most accurate measurements because there are no uncertain potentials to account for and were used for measuring the pH of NIST standard buffers (see below). However, there are few examples of cells without liquid junction (sometimes called **cells without transference**), and they are inconvenient to use. Therefore, the more convenient (but less accurate) cells with liquid junction are commonly used.

20.5 Voltaic Cells with Liquid Junction—The Practical Kind

An example of this type of cell is

$$Hg|Hg_2Cl_2(s)|KCl(saturated)||HCl(solution), H_2(g)|Pt \tag{20.28}$$

The double line represents the **liquid junction** between two dissimilar solutions and is usually in the form of a **salt bridge**. The purpose of this is to prevent mixing of the two solutions. In this way, the potential of one of the electrodes will be constant, independent of the composition of the test solution, and determined by the solution in which it dips. The electrode on the left of cell 20.28 is the **saturated calomel electrode**, which is a commonly used reference electrode (see below). The cell is set up using the hydrogen electrode as the indicating electrode to measure pH.

Liquid-Junction Potential—We Can't Ignore This

The disadvantage of a cell of this type is that there is a potential associated with the liquid junction, called the **liquid-junction potential**. The potential of the above cells is

The presence of a liquid-junction potential limits the accuracy of potentiometric measurements.

$$E_{cell} = (E_{right} - E_{left}) + E_j \tag{20.29}$$

where E_j is the liquid-junction potential; E_j may be positive or negative. The liquid-junction potential results from the unequal diffusion of the ions on each side of the boundary. A careful choice of salt bridge (or reference electrode containing a suitable electrolyte) can minimize the liquid-junction potential and make it reasonably constant so that a calibration will account for it. The basis for such a selection is discussed as follows.

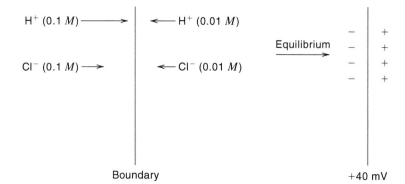

FIGURE 20.2 Representation of liquid-junction potential.

A typical boundary might be a fine-porosity sintered-glass frit with two different solutions on either side of it; the frit prevents appreciable mixing of the two solutions. The simplest type of liquid junction occurs between two solutions containing the same electrolyte at different concentrations. An example is HCl (0.1 M)||HCl (0.01 M), illustrated in Figure 20.2. Both hydrogen ions and chloride ions will migrate across the boundary in both directions, but the net migration will be from the more concentrated to the less concentrated side of the boundary, the driving force for this migration being proportional to the concentration difference. Hydrogen ions migrate about five times faster than chloride ions. Therefore, a net positive charge is built up on the right side of the boundary, leaving a net negative charge on the left side; that is, there is a separation of charge, and this represents a potential. A steady state is rapidly achieved by the action of this build-up positive charge in inhibiting the further migration of hydrogen ions; the converse applies to the negative charge on the left-hand side. Hence, a constant potential difference is quickly attained between the two solutions.

The E_j for this junction is +40 mV, and $E_{cell} = (E_{right} - E_{left}) + 40$ mV. This E_j is very large, owing to the rapid mobility of the hydrogen ion. As the concentration of HCl on the left side of the boundary is decreased, the net charge built up will be less, and the liquid-junction potential will be decreased until, at equal concentration, it will be zero, because equal amounts of HCl diffuse in each direction.

A second example of this type of liquid junction is 0.1 M KCl/0.01 M KCl. This situation is completely analogous to that above, except that in this case the K^+ and Cl^- ions migrate at nearly the same rate, with the chloride ion moving only about 4% faster. So a net *negative* charge is built up on the right side of the junction, but it will be relatively small. Thus, E_j will be negative and is equal to −1.0 mV.

How do We Minimize the Liquid-junction Potential?

The nearly equal migration of potassium and chloride ions makes it possible to significantly decrease the liquid-junction potential. This is possible because, if an electrolyte on one side of a boundary is in large excess over that on the other side, the flux of the migration of the ions of this electrolyte will be much greater than that of the more dilute electrolyte, and the liquid-junction potential will be determined largely by the migration of this more concentrated electrolyte. Thus, E_j of the junction KCl (3.5 M)||H_2SO_4 (0.05 M) is only −4 mV, even though the hydrogen ions diffuse at a much more rapid rate than sulfate.

Some examples of different liquid-junction potentials are given in Table 20.1. (The signs are for those as set up, and they would be the signs in a potentiometric measurement if the solution on the left were used for the salt bridge and the one on the right were the test solution. If solutions on each side of the junction were reversed, the signs of the junction potentials would be reversed.) It is apparent that the liquid junction

We minimize the liquid-junction potential by using a high concentration of a salt whose ions have nearly equal mobility, for example, KCl.

TABLE 20.1 Some Liquid-Junction Potentials at 25°C[a]

Boundary	E_j(mV)
0.1 M KCl‖0.1 M NaCl	+6.4
3.5 M KCl‖0.1 M NaCl	+0.2
3.5 M KCl‖1 M NaCl	+1.9
0.01 M KCl‖0.01 M HCl	−26
0.1 M KCl‖0.1 M HCl	−27
3.5 M KCl‖0.1 M HCl	+3.1
0.1 M KCl‖0.1 M NaOH	+18.9
3.5 M KCl‖0.1 M NaOH	+2.1
3.5 M KCl‖1 M NaOH	+10.5

[a]Adapted from G. Milazzo, *Electrochemie*. Vienna: Springer, 1952; and D. A. MacInnes and Y. L. Yeh, *J. Am. Chem. Soc.*, **43** (1921) 2563.

potential can be minimized by keeping a high concentration of a salt such as KCl, the ions of which have nearly the same mobility, on one side of the boundary. Ideally, the same high concentration of such a salt should be on both sides of the junction. This is generally not possible for the test solution side of a salt bridge. However, the solution in the other half-cell in which the other end of the salt bridge forms a junction can often be made high in KCl to minimize that junction potential. As noted before, this half-cell, which is connected via the salt bridge to form a complete cell, is the reference electrode. See the discussion of the saturated calomel electrode below.

As the concentration of the (dissimilar) electrolyte on the other side of the boundary (in the test solution) increases, or as the ions are made different, the liquid-junction potential will get larger. Very rarely can the liquid-junction potential be considered to be negligible. The liquid-junction potential with neutral salts is less than when a strong acid or base is involved. The variation is due to the unusually high mobilities of the hydrogen ion and the hydroxyl ion. Therefore, *the liquid-junction potential will vary with the pH of the solution*, an important fact to remember in potentiometric pH measurements. A potassium chloride salt bridge, at or near saturation, is usually employed, except when these ions may interfere in a determination. Ammonium chloride or potassium nitrate may be used if the potassium or chloride ion interferes.

Various types of electrolyte junctions or salt bridges have been designed, such as a ground-glass joint, a porous glass or ceramic plug, or a fine capillary tip. The reference electrode solution then contains saturated KCl solution, which slowly leaks through the bridge to create the liquid junction with the test solution.

Liquid-junction potentials are highly pH dependent because of the high mobilities of the proton and hydroxide ions.

20.6 Reference Electrodes: The Saturated Calomel Electrode

A requirement of a reference electrode is that its potential be fixed and stable, unaffected by the passage of small amounts of current required in making potentiometric measurements (ideally, the current in the measurement is zero, but in practice some small current must be passed—see below). Metal–metal salt electrodes generally possess the needed properties.

A commonly used reference electrode is the **saturated calomel electrode** (SCE). The term "saturated" refers to the concentration of potassium chloride; and at 25°C, the potential of the SCE is 0.242 V versus NHE. An SCE consists of a small amount of mercury mixed with some solid Hg_2Cl_2 (calomel), solid KCl, and enough saturated KCl solution to moisten the mixture. This is contacted with a saturated KCl solution containing some solid KCl to maintain saturation. A platinum electrode is immersed in

Reference electrodes are usually metal–metal salt types. The two most common are the Hg/Hg_2Cl_2 (calomel) and the Ag/AgCl electrodes.

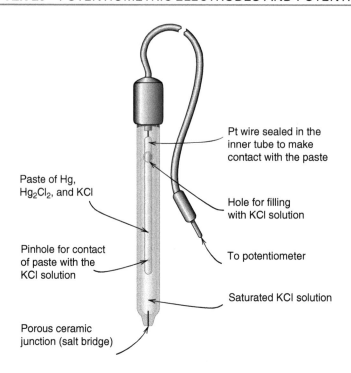

the paste to make contact with the small mercury pool formed, and the connecting wire from that goes to one terminal of the potential measuring device. A salt bridge serves as the contact between the KCl solution and the test solution and is usually a fiber or porous glass frit wetted with the saturated KCl solution. If a different salt bridge is needed to prevent contamination of the test solution (you can't use the SCE for chloride measurements!), then a double-junction reference electrode is used in which the KCl junction contacts a different salt solution that in turn contacts the test solution. This, of course, creates a second liquid-junction potential, but it is constant.

A commercial probe-type SCE is shown in Figure 20.3. This contains a porous fiber or frit as the salt bridge in the tip that allows very slow leakage of the saturated potassium chloride solution. It has a small mercury pool area and so the current it can pass without its potential being affected is limited (as will be seen below, a small current is usually drawn during potential measurements). The fiber salt bridge has a resistance of $\sim 2500 \ \Omega$ satisfactory for use with any modern high input impedance voltmeter, including a pH meter.

Example 20.3

Calculate the potential of the cell consisting of a silver electrode dipping in a silver nitrate solution with $a_{Ag^+} = 0.0100 \ M$ and an SCE reference electrode.

Solution

Neglecting the liquid-junction potential,

$$E_{cell} = E_{ind} - E_{ref}$$

$$E_{cell} = \left(E^0_{Ag^+,Ag} - 0.05916 \ \log \frac{1}{a_{Ag^+}} \right) - E_{SCE}$$

$$= 0.799 - 0.05916 \ \log \frac{1}{0.0100} - 0.242$$

$$= 0.439 \ V$$

Example 20.4

A cell voltage measured using an SCE reference electrode is −0.774 V. (The indicating electrode is the more negative half-cell.) What would the cell voltage be with a silver/silver chloride reference electrode (1 M KCl; $E = 0.228$ V) or with an NHE?

Solution

The potential of the Ag/AgCl electrode is more negative than that of the SCE by $0.242 − 0.228 = 0.014$ V. Hence, the cell voltage using the former electrode is less negative by this amount:

$$E_{\text{vs. Ag/AgCl}} = E_{\text{vs. SCE}} + 0.014$$
$$= −0.774 + 0.014 = −0.760 \text{ V}$$

Similarly, the cell voltage using the NHE is 0.242 V less negative:

$$E_{\text{vs. NHE}} = E_{\text{vs. SCE}} + 0.242 \text{ V}$$
$$= −0.774 + 0.242 = −0.532 \text{ V}$$

Reference electrode potentials are all relative. The measured cell potential depends on which one is used.

Potentials relative to different reference electrodes may be represented schematically on a scale on which the different electrode potentials are placed (see Reference 2). Figure 20.4 illustrates this for Example 20.4.

We should note that although calomel electrodes were once the gold standard, many laboratories now wish to limit the use of toxic mercury, and therefore use Ag/AgCl electrodes.

20.7 Measurement of Potential

We create a voltaic cell with the indicator and reference electrodes. We measure the voltage of the cell, giving a reading of the indicator electrode potential relative to the reference electrode. We can relate this to the analyte activity or concentration using the Nernst equation.

The pH Meter

pH measurements with a glass (or other) electrode involve the measurement of potentials (see Sections 20.11–20.16 below for detailed descriptions of how we measure pH). The **pH meter** is essentially a voltmeter.

The pH meter is a voltage measuring device designed for use with high-resistance glass electrodes and can be used to measure potential in both low- and high-resistance circuits. pH meters are typically built with very high input impedance operational amplifiers called **electrometers** as the front end.

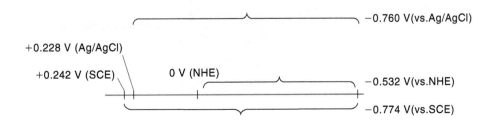

FIGURE 20.4 Schematic representation of electrode potential relative to different reference electrodes.

A pH meter or electrometer draws very small currents and are well suited for irreversible reactions that are slow to reestablish equilibrium. They are also required for high-resistance electrodes, like glass pH or ion-selective electrodes.

A pH meter is a high-input impedance voltmeter that senses the cell voltage, provides a digital readout (either in terms of voltage or pH) and often provides an amplified output to be acquired by an external data system device. Because very little current is drawn, chemical equilibrium is not perceptibly disturbed. This is vital for monitoring irreversible reactions that do not return to the prior state if an appreciable amount of current is drawn. The resistance of a typical glass pH electrode is of the order of $10^8 \Omega$.

Sufficiently sensitive pH meters are available that will measure the potential with a resolution of 0.1 mV. These are well suited for direct potentiometric measurements with both pH electrodes and other ion-selective electrodes.

The Cell for Potential Measurements

Impedance in an ac circuit is comparable to resistance in dc circuits, although aside from resistance it involves frequency dependent components. While no ac measurements are made by a pH meter, the term high input impedance adjective for the opeartional amplifiers simply indicate that they are compatible with ac measurements. Although in principle no current is (or should be) drawn by an ideal voltmeter, in practice even the high impedance voltmeters draw a finite current, albeit very small, in the 1-100 fA range.

In potentiometric measurements, a cell of the type shown in Figure 20.5 is set up. For direct potentiometric measurements in which the activity of one ion is to be calculated from the potential of the indicating electrode, the potential of the reference electrode will have to be known or determined. The voltage of the cell is described by Equation 20.7, and when a salt bridge is employed, the liquid-junction potential must be included. Then,

$$E_{cell} = (E_{ind} - E_{ref}) + E_j \tag{20.30}$$

The E_j can be combined with the other constants in Equation 20.30 into a single constant, assuming that the liquid-junction potential does not differ significantly from one solution to the next. We are forced to accept this assumption since E_j cannot be evaluated under most circumstances. E_{ref}, E_j, and E_{ind}^0 are lumped together into a constant k:

$$k = E_{ind}^0 - E_{ref} + E_j \tag{20.31}$$

Then (for a 1:1 reaction),

$$E_{cell} = k - \frac{2.303RT}{nF} \log \frac{a_{red}}{a_{ox}} \tag{20.32}$$

The constant k is determined by measuring the potential of a standard solution in which the activities are known.

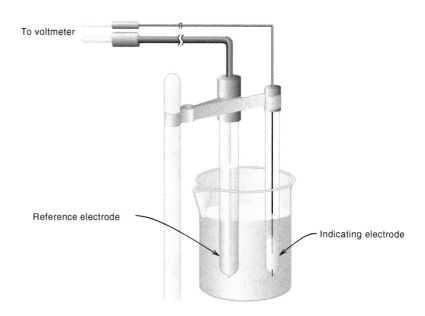

To voltmeter

Reference electrode

Indicating electrode

FIGURE 20.5 Cell for potentiometric measurements.

20.8 Determination of Concentrations from Potential Measurements

Usually, we are interested in determining the concentration of a test substance rather than its activity. Activity coefficients are not generally available, and it is inconvenient to calculate activities of solutions used to standardize the electrode.

If the ionic strength of all solutions is held constant at the same value, the activity coefficient of the test substance remains nearly constant for all concentrations of the substance. We can then write for the log term in the Nernst equation:

$$-\frac{2.303RT}{nF} \log f_i C_i = -\frac{2.303RT}{nF} \log f_i - \frac{2.303RT}{nF} \log C_i \qquad (20.33)$$

Under the prescribed conditions, the first term on the right-hand side of this equation is constant and can be included in k, (we will call this k'), so that at constant ionic strength,

$$\boxed{E_{cell} = k' - \frac{2.303RT}{nF} \log \frac{C_{red}}{C_{ox}}} \qquad (20.34)$$

If the ionic strength is maintained constant, activity coefficients are constant and can be included in k (to be called a new constant, k'). So concentrations can be determined from measured cell potentials.

In other words, the electrode potential changes by $\pm 2.303RT/nF$ volts for each 10-fold change in *concentration* of the oxidized or reduced form.

It is best to construct a **calibration curve** of potential versus log concentration; this should have a slope of $\pm 2.303RT/nF$. In this way, any deviation from this theoretical response will be accounted for in the calibration curve. Note that the intercept of the plot would represent the constant, k', which includes the standard potential, reference electrode potential, liquid junction potential, and the activity coefficient.

Since the ionic strength of an unknown solution is usually not known, a high concentration of an electrolyte is added both to the standards and to the samples to maintain about the same ionic strength. The standard solutions should have the same matrix as the test solutions, notably any species that will change the activity of the analyte, such as complexing agents. However, since the complete sample composition is often not known, this is frequently not possible.

20.9 Residual Liquid-Junction Potential—It Should Be Minimized

We have assumed above in Equations 20.32 and 20.34 that k or k' is the same in measurements of both standards and samples. This is so only if the liquid-junction potential at the reference electrode is the same in both solutions. But the test solution will usually have a somewhat different composition from the standard solution, and the magnitude of the liquid-junction potential will vary. The difference in the two liquid-junction potentials is called the **residual liquid-junction potential**, and it will remain unknown. The difference can be kept to a minimum by keeping the ionic strength of both solutions as close as possible, and especially by keeping the pH of the test solution and the pH of the standard solution as close as possible.

If the liquid-junction potentials of the calibrating and test solutions are identical, no error results (the residual $E_j = 0$). Our goal is to keep residual E_j as small as possible.

20.10 Accuracy of Direct Potentiometric Measurements—Voltage Error versus Activity Error

We can get an idea of the accuracy required in potentiometric measurements from the percent error caused by a 1-mV error in the reading at 25°C. For an electrode responsive to a monovalent ion such as silver,

$$E_{cell} = k - 0.05916 \, \log \frac{1}{a_{Ag^+}} \tag{20.35}$$

and

$$a_{Ag^+} = antilog \frac{E_{cell} - k}{0.05916} \tag{20.36}$$

A $\pm$1-mV error results in an error in a_{Ag^+} of $\pm$**4**, or in pAg units, an error of ~0.017. The absolute accuracy of most electrode based measurements is no better than 0.2 mV; this limits the maximum accuracy attainable in direct potentiometric measurements. The same percent error in activity will result for all activities of silver ion with a 1-mV error in the measurement. *The error is doubled when* n *is doubled* to 2. So, a 1-mV error for a copper/copper(II) electrode would result in an 8% error in the activity of copper(II). It is obvious, then, that the residual liquid junction potential can have an appreciable effect on the accuracy.

The accuracy and precision of potentiometric measurements are also limited by the **poising capacity** of the redox couple being measured. This is analogous to the buffering capacity in pH measurements. If the solution is very dilute, the solution is poorly poised and potential readings will be sluggish. That is, the solution has such a low ion concentration that it takes longer for the solution around the electrode to rearrange its ions and reach a steady state, when the equilibrium is disturbed during the measurement process. This is why a high input impedance voltmeter that draws very small current is preferred for potentiometric measurements in such solutions. To maintain a constant ionic strength, a relatively high concentration of an inert salt (ionic strength "buffer") can be added; this also helps reduce solution resistance, helpful when physically separate reference and indicator electrodes are used. Stirring helps speed up the equilibrium response.

In very dilute solutions, the potential of the electrode may be governed by other electrode reactions. In a very dilute silver solution, for example, $-\log(1/a_{Ag^+})$ becomes very negative and the potential of the electrode is very reducing. Under these conditions, an oxidizing agent in solution (such as dissolved oxygen) may be reduced *at the electrode surface*, setting up a second redox couple (O_2/OH^-); the potential will be a **mixed potential**.

Usually, the lower limit of concentration that can be measured with a degree of certainty is 10^{-5} to 10^{-6} M, although the actual range should be determined experimentally. As the solution becomes more dilute, a longer time should be allowed to establish the equilibrium potential reading because of slower approach to equilibrium. An exception to this limit is in pH measurements in which the hydrogen ion concentration of the solution is well poised, either by a buffer or by excess acid or base. At pH 10, the hydrogen ion concentration is 10^{-10} M, and this can be measured with a glass pH electrode (see Section 20.11). A neutral, unbuffered solution is poorly poised, however, and pH readings are sluggish. Pure water is a specially difficult sample to measure the pH of, both because of poor buffering and very high resistance. Potassium chloride is often deliberately added prior to pH measurement. The best choice is a refillable,

For a dilute or poorly poised solution, stirring the solution helps achieve an equilibrium reading. For a quantitative discussion of poising capacity, see E. R. Nightengale, "Poised Oxidation-Reduction Systems. A Quantitative Evaluation of Redox Poising Capacity and Its Relation to the feasibility of Redox Titrations," *Anal. Chem.*, **30**(2) (1958) 267–272.

liquid-filled electrode, ideally made of low-resistance glass. A flowing reference junction has a higher flow rate to minimize junction potentials. A fast leak rate is desirable with pure water so that the equilibrium potential can be established more quickly.

20.11   Glass pH Electrode—Workhorse of Chemists

The glass electrode, because of its convenience, is used almost universally for pH measurements today. Its potential is essentially not affected by the presence of oxidizing or reducing agents, and it is operative over a wide pH range. It is fast responding and functions well in physiological systems. No other pH-measuring electrode possesses all these properties.

Principle of the Glass Electrode

A typical construction of a pH glass electrode is shown in Figure 20.6. For measurement, only the bulb need be submerged. There is an internal reference electrode and electrolyte (Ag|AgCl|Cl$^-$) for making electrical contact with the glass membrane; its potential is necessarily constant and is set by the concentration of HCl. A complete cell, then, can be represented by

| reference electrode (external) || , | H$^+$ (unknown) | glass membrane | , | H$^+$ (internal) | reference electrode (internal) |

The double line represents the salt bridge of the reference electrode. The glass electrode is attached to the indicating electrode terminal of the pH meter while the external reference electrode (e.g., SCE) is attached to the reference terminal.

The potential of the glass membrane is given by

$$E_{glass} = \text{constant} - \frac{2.303RT}{F} \log \frac{a_{H^+ \text{ int}}}{a_{H^+ \text{ unk}}} \tag{20.37}$$

and the voltage of the cell is given by

$$E_{cell} = k + \frac{2.303RT}{F} \log a_{H^+ \text{ unk}} \tag{20.38}$$

where k is a constant that includes the potentials of the two reference electrodes, the liquid-junction potential, a potential at the glass membrane due to H$^+$ (internal), and a term known as the **asymmetry potential**.

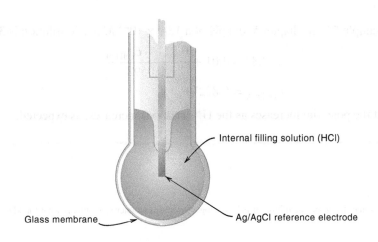

Internal filling solution (HCl)

Glass membrane

Ag/AgCl reference electrode

FIGURE 20.6 Glass pH electrode.

The glass pH electrode must be calibrated using "standard buffers." See Section 20.12.

The asymmetry potential is a small potential across the membrane that is present even when the solutions on both sides of the membrane are identical. It is associated with factors such as nonuniform composition of the membrane, strains within the membrane, mechanical and chemical attack of the external surface, and the degree of hydration of the membrane. It slowly changes with time, especially if the membrane is allowed to dry out, and is unknown. For this reason, *a glass pH electrode must be calibrated at least once a day*. The asymmetry potential will vary from one electrode to another, owing to differences in construction of the membrane.

Since $pH = -\log a_{H^+}$, Equation 20.38 can be rewritten[1]

$$E_{cell} = k - \frac{2.303RT}{F}pH_{unk} \tag{20.39}$$

or

$$pH_{unk} = \frac{k - E_{cell}}{2.303RT/F} \tag{20.40}$$

It is apparent that the glass electrode will undergo a $2.303RT/F$-volt response for each change of 1 pH unit (10-fold change in a_{H^+}); k must be determined by calibration with a **standard buffer** (see below) of known pH:

$$k = E_{cell} + \frac{2.303RT}{F}pH_{std} \tag{20.41}$$

Substitution of Equation 20.41 into Equation 20.39 yields

$$pH_{unk} = pH_{std} + \frac{E_{cell\,std} - E_{cell\,unk}}{2.303RT/F} \tag{20.42}$$

We usually don't resort to this calculation in pH measurements. Rather, the potential scale of the pH meter is calibrated in pH units (see Section 20.14).

Note that since the determination involves potential measurements with a very high-resistance glass membrane electrode (50 to 500 MΩ), it is essential to use a high input impedance voltmeter.

Example 20.5

A glass electrode–SCE pair is calibrated at 25°C with a pH 4.01 standard buffer, the measured voltage being 0.814 V. What voltage would be measured in a 1.00×10^{-3} *M* acetic acid solution? Assume $a_{H^+} = [H^+]$.

Solution

From Example 5.7 in Chapter 5, the pH of a 1.00×10^{-3} *M* acetic solution is 3.88;

$$\therefore 3.88 = 4.01 + \frac{0.814 - E_{cell\,unk}}{0.0592}$$

$$E_{cell\,unk} = 0.822V$$

Note that the potential increases as the H$^+$ (a cation) increases, as expected.

---

[1]We will assume the proper definition of pH as $-\log a_{H^+}$ in this chapter since this is what the glass electrode measures.

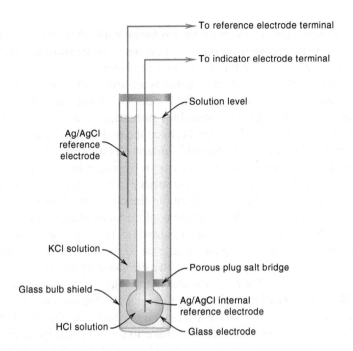

To reference electrode terminal

To indicator electrode terminal

Solution level

Ag/AgCl reference electrode

KCl solution

Glass bulb shield

HCl solution

Porous plug salt bridge

Ag/AgCl internal reference electrode

Glass electrode

FIGURE 20.7   Combination pH–reference electrode.

Combination pH Electrodes—A Complete Cell

Both an indicating and a reference electrode (with salt bridge) are required to make a complete cell so that potentiometric measurements can be made. It is convenient to combine the two electrodes into a single probe, so that only small volumes are needed for measurements. A typical construction of a combination pH–reference electrode is shown in Figure 20.7. It consists of a tube within a tube, the inner one housing the pH indicator electrode and the outer one housing the reference electrode (e.g., a Ag/AgCl electrode) and its salt bridge. There is one lead from the combination electrode, but it is split into two connectors at the end, one (the larger) going to the pH electrode terminal and the other going to the reference electrode terminal. It is important that the salt bridge be immersed in the test solution in order to complete the cell. The salt bridge is often a small plug in the outer ring rather than a complete ring as illustrated here. Combination electrodes are convenient, and therefore the most commonly used.

A combination electrode is a complete cell when dipped in a test solution.

What Determines the Glass Membrane Potential?

The pH glass electrode functions as a result of ion exchange on the surface of a hydrated layer. The membrane of a pH glass electrode consists of chemically bonded Na_2O and $-SiO_2$. The surface of a new glass electrode contains fixed silicate groups associated with sodium ions, $-SiO^-Na^+$. For the electrode to work properly, it must first be soaked in water. During this process, the outer surface of the membrane becomes *hydrated*. The inner surface is already hydrated. The glass membrane is usually 30 to 100 μm thick, and the hydrated layers are 10 to 100 nm thick.

When the outer layer becomes hydrated, the sodium ions are exchanged for protons in the solution:

$$\underset{\text{solid}}{-SiO^-Na^+} + \underset{\text{solution}}{H^+} \rightleftharpoons \underset{\text{solid}}{-SiO^-H^+} + \underset{\text{solution}}{Na^+} \qquad (20.43)$$

Other ions in the solution can exchange for the Na^+ (or H^+) ions, but the equilibrium constant for the above exchange is very large because of the affinity of the glass for protons. Thus, the surface of the glass is made up almost entirely of silicic acid, except in very alkaline solution, where the proton concentration is small. The $-SiO^-$ sites are fixed, but the protons are free to move and exchange with other ions. (By varying the glass composition, the exchange for other ions becomes more favorable, and this forms the basis of electrodes selective for other ions—see below.)

The potential of the membrane consists of two components, the boundary potential and the diffusion potential. The former is almost the sole hydrogen ion activity-determining potential. The **boundary potential** resides at the surface of the glass membrane, that is, at the interface between the hydrated gel layer and the external solution. When the electrode is dipped in an aqueous solution, a boundary potential is built up, which is determined by the activity of hydrogen ions in the external solution and the activity of hydrogen ions on the surface of the gel. One explanation of the potential is that the ions will tend to migrate in the direction of lesser activity, much as at a liquid junction. The result is a microscopic layer of charge built up on the surface of the membrane, which represents a potential. Hence, as the solution becomes more acidic (the pH decreases), protons migrate to the surface of the gel, building up a positive charge, and the potential of the electrode increases, as indicated by Equations 20.37 and 20.38. The reverse is true as the solution becomes more alkaline.

The **diffusion potential** results from a tendency of the protons in the inner part of the gel layer to diffuse toward the dry membrane, which contains $-SiO^-Na^+$, and a tendency of the sodium ions in the dry membrane to diffuse to the hydrated layer. The ions migrate at a different rate, creating a type of liquid-junction potential. But a similar phenomenon occurs on the other side of the membrane, only in the opposite direction. These in effect cancel each other, and so the net diffusion potential is very small, and the potential of the membrane is determined largely by the boundary potential. (Small differences in diffusion potentials may occur due to differences in the glass across the membrane—these represent a part of the asymmetry potential.)

Cremer described the first predecessor of the modern glass electrode [*Z. Biol.* **47** (1906) 56]. More than a hundred years later, exactly how a glass electrode works is still not eminently clear. Pungor has presented evidence that the establishment of an electrode potential is caused by charge separation, due to chemisorption of the primary ion (H^+) from the solution phase onto the electrode surface, that is, a surface chemical reaction. Counter ions of the opposite charge accumulate in the solution phase, and this charge separation represents a potential. A similar mechanism applies to other ion-selective electrodes (below). [See E. Pungor, "The New Theory of Ion-Selective Electrodes," *Sensors*, **1** (2001) 1–12 (this is an open access electronic journal: http://www.mdpi.com—Pungor author.)]

K. L. Cheng has proposed a theory of glass electrodes based on a capacitor model in which the electrode senses the hydroxide ion in alkaline solution (where a_{H^+} is very small), rather than sensing protons. [K. L. Cheng, "Capacitor Theory for Nonfaradaic Potentiometry," *Microchem. J.*, **42** (1990) 5.] Nonfaradaic here refers to reactions that do not involve a redox process. Cheng has performed isotope experiments that suggest the generally accepted ion exchange reaction between H^+ and Na^+ does not occur. He argues that the electrode actually responds to OH^- ions in alkaline solution (remember, [H^+] at pH 14 is only 10^{-14} *M*!) [C.-M. Huang et al., "Isotope Evidence Disproving Ion Exchange Reaction Between H^+ and Na^+ in pH Glass Electrode," *J. Electrochem. Soc.*, **142** (1995) L175]. While this theory is not generally accepted, Cheng et al. present some compelling arguments and experimental results that make this an interesting hypothesis. It has some commonality with Pungor's double-layer hypothesis.

Alkaline Error

Two types of error occur that result in non-Nernstian behavior (deviation from the theoretical response). The first is called the **alkaline error**. Such error is due to the capability of the membrane for responding to other cations besides the hydrogen ion.

The pH of the test solution determines the external boundary potential.

Does the glass electrode sense H^+ or OH^- in alkaline solutions?

As the hydrogen ion activity becomes very small, these other ions can compete successfully in the potential-determining mechanism. Although the hydrated gel layer prefers protons, sodium ions will exchange with the protons in the layer when the hydrogen ion activity in the external solution is very low (reverse of Equation 20.43). The potential then depends partially on the ratio of $a_{Na^+\ external}/a_{Na^+\ gel}$; that is, the electrode becomes a sodium ion electrode.

The error is negligible at pH less than about 9; but at pH values above this, the H^+ concentration is very small relative to that of other ions, and the electrode response to the other ions such as Na^+, K^+, and so on, becomes appreciable. In effect, the electrode appears to "see" more hydrogen ions than are present, and the pH reading is too low. The magnitude of this negative error is illustrated in Figure 20.8. Sodium ion causes the largest errors, which is unfortunate, because many samples, especially alkaline ones, contain significant amounts of sodium. Commercial general-purpose glass electrodes are usually supplied with a graphically represented alkaline error correction values, and if the sodium ion concentration is known, these electrodes are useful up to pH about 11.

By a change in the composition of the glass, the affinity of the glass for sodium ion can be reduced. If the Na_2O in the glass membrane is largely replaced by Li_2O, then the error due to sodium ions is markedly decreased. This is the so-called lithium glass electrode, high-pH electrode, or full-range electrode (0 to 14 pH range). Most pH electrodes in use today have glass membranes formulated to be capable of measurement up to pH 13.5 with reasonable accuracy if sodium error is corrected for. But if you need to make pH measurements in very alkaline solutions, the specially formulated electrodes are recommended. As mentioned before, it was the discovery that variation in the glass composition could change its affinity for different ions that led to the development of glasses selective for ions other than protons, that is, of ion-selective electrodes, that extended eventually to materials altogether different from glass.

> The glass electrode senses other cations besides H^+. This becomes appreciable only when a_{H^+} is very small, as in alkaline solution. We can't distinguish them from H^+, so the solution appears more acidic than it actually is.

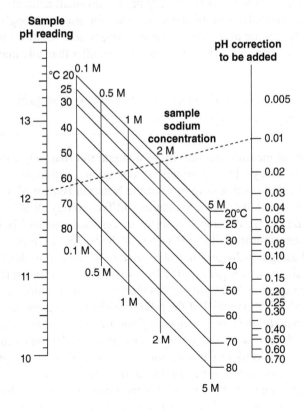

FIGURE 20.8 The sodium error of a good general purpose pH electrode. The example shows how to use this "nomogram" to correct the apparent measurement. Imagine that I have a solution, 0.5 M in sodium and the apparent pH read at 50°C is 12.10. We draw a line from the pH 12.10 point on the x-axis through the intersection point of the 50°C line and the 0.5 M line and find the line intersects the error axis at 0.01. The actual pH is therefore 12.10 + 0.01 = 12.11. (Courtesy Thermo Fisher Scientific Inc.)

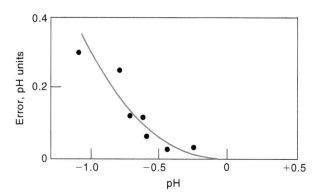

FIGURE 20.9 Error of glass electrode in hydrochloric acid solutions.(From L. Meites and L. C. Thomas, *Advanced Analytical Chemistry.* Copyright © 1958, McGraw-Hill, New York. Used with permission of McGraw-Hill Company.)

Acid Error

At very low pH values (pH < 1), the gel layer of the pH-sensitive glass membrane absorbs acid molecules. This absorption decreases the activity of hydrogen ions and results in a lower potential at the outer membrane phase boundary. The pH measurement therefore shows a higher pH value than the actual pH value of the sample solution. A second and possibly greater contributor to the **acid error** is more aptly described as the **water activity error**, is the second type causing non-Nernstian response. Such error occurs because the potential of the membrane depends on the activity of the water with which it is in contact. If the activity is unity, the response is Nernstian. In very acidic solutions, the activity of water is less than unity (an appreciable amount is used in solvating the protons), and a positive error in the pH reading results (Figure 20.9). A similar type of error will result if the activity of the water is decreased by a high concentration of dissolved salt or by addition of nonaqueous solvent such as ethanol. In these cases, a large liquid-junction potential may also be introduced and another error will thereby result, although this is not very large with small amounts of ethanol.

Similiar to specially formulated electrodes for use at strongly alkaline pH, specialized electrodes are available for use in strongly acid solutions that exhibit considerably less acid error. Acid error, in general, is smaller than alkaline error.

20.12 Standard Buffers—Reference for pH Measurements

Only the phosphate mixtures and borax are really buffers. Disodium tetraborate is effectively an equimolar mixture of orthoboric acid and its fully neutralized salt, and so is a buffer. The pH values change with temperature due to the temperature dependence of the K_a values.

Because we cannot measure the activity of a single ion (but only estimate it by calculation using the Debye–Hückel equation), operational definitions of pH have been proposed. One of these is that developed at the National Bureau of Standards (NBS), now called the National Institute of Standards and Technology (NIST), under the direction of Roger Bates. He developed a series of certified standard buffers for use in calibrating pH measurements. The pH values of the buffers were determined by measuring their pH using a hydrogen-indicating electrode in a cell without liquid junction (similar to the cell given by Equation 20.22). A silver/silver chloride reference electrode was used. From Equation 20.24, we see that the activity of the chloride ion must be calculated (to calculate the potential of the reference electrode) using the Debye–Hückel theory; *this ultimately limits the accuracy of the pH of the buffers to about ±0.01 pH unit.* Faced with the problem of choosing a convention for the ionic activity coefficient of a single species (chloride) to be used for the purpose of assigning pH values, Bates chose values for chloride that were similar to those for the mean activity coefficients (which can be measured) of HCl and NaCl in their mixtures. Hence, this is the basis for the operational definition of pH. This convention is known as the Bates–Guggenheim convention (Edward A. Guggenheim of Reading University in the U. K. and Bates were

TABLE 20.2 pH Values of NIST Buffer Solutions[a]

t (°C)	Buffer						
	Tetroxalate[b]	Tartrate[c]	Phthalate[d]	Phosphate[e]	Phosphate[f]	Borax[g]	Calcium Hydroxide[h]
0	1.666	—	4.003	6.984	7.534	9.464	13.423
5	1.668	—	3.999	6.951	7.500	9.395	13.207
10	1.670	—	3.998	6.923	7.472	9.332	13.003
15	1.672	—	3.999	6.900	7.448	9.276	12.810
20	1.675	—	4.002	6.881	7.429	9.225	12.627
25	1.679	3.557	4.008	6.865	7.413	9.180	12.454
30	1.683	3.552	4.015	6.853	7.400	9.139	12.289
35	1.688	3.549	4.024	6.844	7.389	9.102	12.133
38	1.691	3.549	4.030	6.840	7.384	9.081	12.043
40	1.694	3.547	4.035	6.838	7.380	9.068	11.984
45	1.700	3.547	4.047	6.834	7.373	9.038	11.841
50	1.707	3.549	4.060	6.833	7.367	9.011	11.705
55	1.715	3.554	4.075	6.834	—	8.985	11.574
60	1.723	3.560	4.091	6.836	—	8.962	11.449
70	1.743	3.580	4.126	6.845	—	8.921	—
80	1.766	3.609	4.164	6.859	—	8.885	—
90	1.792	3.650	4.205	6.877	—	8.850	—
95	1.806	3.674	4.227	6.886	—	8.833	—

[a] From R. G. Bates, *J. Res. Natl. Bur. Std.*, **A66** (1962) 179. (Reprinted by permission of the U.S. Government Printing Office.)

[b] 0.05 m potassium tetroxalate (m refers to molality, but only small errors result if molarity is used instead).

[c] Satd. (25°C) potassium hydrogen tartrate.

[d] 0.05 m potassium hydrogen phthalate.

[e] 0.025 m potassium dihydrogen phosphate, 0.025 m disodium monohydrogen phosphate.

[f] 0.008695 m potassium dihydrogen phosphate, 0.03043 m disodium hydrogen phosphate.

[g] 0.01 m borax.

[h] Satd. (25°C) calcium hydroxide.

charged by IUPAC to come up with a recommendation, and Guggenheim went along with Bates' suggestion). The partial pressure of hydrogen is determined from the atmospheric pressure at the time of the measurement (minus the vapor pressure of the water at the temperature of the solution).

The compositions and pH of NIST standard buffers are given in Table 20.2. The NIST pH scale is a "multi-standard" scale with several fixed points. The British Standards Institute, however, has developed an operational pH scale based on a single primary standard, and pH of other "standard" buffers are measured relative to this, these being secondary standards, rather than having a series of standard reference solutions. In practice, when accurate pH measurement is necessary, the pH meter is calibrated with two standard buffers that bracket the sample pH as closely as possible. Although the absolute value of the pH accuracy is no better than 0.01 unit, the buffers have been measured *relative to one another* to 0.001 pH. The potentials used in calculating the pH can be measured reproducibly this closely, and the discrimination of differences of thousandths of pH units is sometimes important (i.e., an electrode may have to be calibrated to a thousandth of a pH unit). The pH of the buffers is temperature dependent because of the dependence of the ionization constants of the parent acids or bases on temperature.

Note that several of these solutions are not really buffers, and they are actually standard pH solutions whose pH is stable since we do not add acid or base. They are resistant to pH change with minor dilutions (e.g., $H^+ \approx \sqrt{K_{a1}K_{a2}}$).

It should be pointed out that if a glass electrode–SCE cell is calibrated with one standard buffer and is used to measure the pH of another, the new reading will not correspond exactly to the standard value of the second because of the residual liquid-junction potential.

The $KH_2PO_4 - Na_2HPO_4$ buffer (pH 7.384 at 38°C) is particularly suited for calibration for blood pH measurements. Many blood pH measurements are made at 38°C, which is near body temperature; thus, the pH of the blood in the body is indicated.

For a discussion of the above NIST pH standard and other proposed definitions of pH, see the letters by W. F. Koch (*Anal. Chem.,* December 1, 1997, 700A; *Chem. & Eng. News,* October 20, 1997, 6).

20.13 Accuracy of pH Measurements

The residual liquid-junction potential limits the accuracy of pH measurement. Always calibrate at a pH close to that of the test solution.

The accuracy of pH measurements is governed by the accuracy to which the hydrogen ion activity of the standard buffer is known. As mentioned above, this accuracy is not better than ±0.01 pH unit because of several limitations. The first is in calculating the activity coefficient of a single ion.

A second limitation in the accuracy is the residual liquid-junction potential. The cell is standardized in one solution, and then the unknown pH is measured in a solution of a different composition. We have mentioned that this residual liquid-junction potential is minimized by keeping the pH and compositions of the solutions as near as possible. Because of this, *the cell should be standardized at a pH close to that of the unknown.* The error in standardizing at a pH far removed from that of the test solution is generally within 0.01 to 0.02 pH unit but can be as large as 0.05 pH unit for very alkaline solutions.

The residual liquid-junction potential, combined with the uncertainty in the standard buffers, limits the *absolute accuracy of measurement of pH of an unknown solution to about ±0.02 pH unit.* It may be possible, however, to *discriminate* between the pH of two similar solutions with differences as small as ±0.004 or even ±0.002 pH units, although their accuracy is no better than ±0.02 pH units. Such discrimination is possible because the liquid-junction potentials of the two solutions will be virtually identical in terms of true a_{H^+}. For example, if the pH values of two blood solutions are close, we can measure the difference between them accurately to ±0.004 pH. If the pH difference is fairly large, however, then the residual liquid-junction potential will increase and the difference cannot be measured as accurately. For discrimination of 0.02 pH unit, changes in the ionic strength may not cause serious errors, but for smaller pH changes than this, large changes in ionic strength will cause errors.

Potentiometric measurements of a_{H^+} are only about 5% accurate.

An error of ±0.02 pH unit corresponds to an error in a_{H^+} of $\pm4.8\%(\pm1.2$ mV),[2] and a discrimination of ±0.004 pH unit would correspond to a discrimination of $\pm1.0\%$ in a_{H^+} (±0.2 mV).

If pH measurements are made at a temperature other than that at which the standardization is made, other factors being equal, the liquid-junction potential will change with temperature. For example, in a rise from 25° to 38°C, a change of $+0.76$ mV has been reported for blood. Thus, for very accurate work, the cell must be standardized at the same temperature as the test solution.

---

[2]The electrode response is 59 mV/pH at 25°C.

20.14 Using the pH Meter—How Does It Work?

We have already mentioned that owing to the high resistance of the glass electrode, a high input impedance voltmeter (all pH meters qualify) must be used to make the potential measurements. If voltage is measured directly, Equation 20.40 or 20.42 is applied to calculate the pH. The value of $2.303RT/F$ at 298.16 K (25°C) is 0.05916; if a different temperature is used, this value should be corrected in direct proportion to the absolute temperature.

A digital pH meter is shown in Figure 20.10. The potential scale is calibrated in pH units, with each pH unit equal to 59.16 mV at 25°C (Equation 20.39). The pH meter is adjusted to indicate the pH of the standard buffer or the calibration function will cause it to calibrate itself with the known pH of the calibrant buffer. Then, the standard buffer is replaced by the unknown solution and the pH is read. This procedure, in effect, sets the constant k in Equation 20.40 and adjusts for the asymmetry potential as well as the other constants included in k.

Most pH meters contain a temperature adjustment dial, which changes the sensitivity response (mV/pH) so that it will be equal to $2.303RT/F$. For example, it is 54.1 mV at 0°C and 66.0 mV at 60°C.

> The temperature setting on the pH meter adjusts T in the RT/nF value, which determines the slope of the potential versus pH buffers.

Electrodes and meters are designed to have a point in calibrations lines, in the midrange of activity measurements, where the potential essentially has no variation with temperature. For pH glass electrodes, this is set at pH 7 (Figure 20.11). This is called the **isopotential point**, and the potential is zero. (pH meters actually measure potential which is converted to pH reading, and potentials can be recorded directly.) Any potential reading different from 0 mV for a pH 7.0 standard buffer is called the **offset** of that electrode. When the temperature is changed, the calibration slope changes, and the intersection of the curves establishes the actual isopotential point. If the isopotential point of the electrode differs from pH 7, then the temperature of the calibration buffer and the test solution should be the same for highest accuracy because a slight error will occur in the slope adjustment at different temperatures. For more details of

FIGURE 20.10 Typical PH meter. (Courtesy of Denver Instrument Company.)

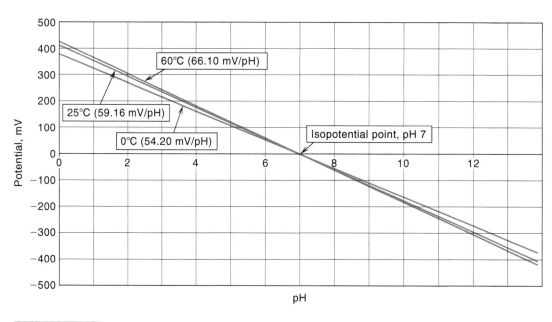

FIGURE 20.11 Isopotential point.

the isopotential point and its quantitative interpretation, see A. A. S. C. Machado, *Analyst*, **19** (1994) 2263.

In calibrating the pH meter, the electrodes are inserted in a pH 7.0 standard buffer. The temperature of the buffer is checked and the temperature adjustment knob is adjusted to that temperature. Using the standardized or calibration knob, the meter is adjusted to read 7.00. Many pH meters have microprocessors that recognize specific pH values, e.g., pH calibration standards of 4, 7, and 10. If you put an electrode in a pH 7 calibration standard for example, and press the calibration button (or equivalent), it will automatically calibrate the instrument to read the pH at 7.02, the pH of the NIST standard NaH_2PO_4-$Na_2H_2PO_4$ buffer. Next the slope is set by repeating the calibration with either a pH 4 (potassium hydrogen phthalate pH 4.01) or 10 (Na_2CO_3-$NaHCO_3$, pH 10.00) calibration standard, depending on the pH of the sample to be measured. Most pH meters now have temperature measurement capability and include a separate temperature probe so that temperature compensation can be automatic. The temperature probe may be incorporated in the electrode itself.

Most pH meters are precise to ±0.01 pH unit (±0.6 mV) with a full-meter scale of 14 pH units (about 840 mV). The meters can be set to read millivolts directly (usually with a sensitivity of 1400 mV full scale). Higher-resolution pH meters are capable of reading to ±0.001 pH unit; to accomplish this, the potential must be read to closer than 0.1 mV.

When the pH of an unbuffered solution near neutrality is measured, readings will be sluggish because the solution is poorly poised and a longer time will be required to reach a stable reading. The solution should be stirred because a small amount of the glass tends to dissolve, making the solution at the electrode surface alkaline (Equation 20.43, where H_2O—the source of H^+—is replaced by NaOH solution). See the discussion on measuring the pH of pure water at the end of Section 20.10.

20.15 pH Measurement of Blood—Temperature Is Important

Recall from Chapter 6 that, because the equilibrium constants of the blood buffer systems change with temperature, the pH of blood at the body temperature of 37°C is different than at room temperature. Hence, to obtain meaningful blood pH measurements that can be related to actual physiological conditions, the measurements should be made at 37°C and the samples should not be exposed to the atmosphere. (Also recall that the pH of a neutral aqueous solution at 37°C is 6.80, and so the acidity scale is changed by 0.20 pH unit.)

Some useful rules in making blood pH measurements are as follows:

1. Calibrate the electrodes using a standard buffer at 37°C, making sure to select the proper pH of the buffer at 37°C and to set the temperature on the pH meter at 37°C (slope = 61.5 mV/pH). It is a good idea to use two standards for calibration, narrowly bracketing the sample pH; this assures that the electrode is functioning properly. Also, the electrodes must be equilibrated at 37°C before calibration and measurement. The potential of the internal reference electrode inside the glass electrode is temperature dependent, as may be the potential-determining mechanism at the glass membrane interface; and the potentials of the SCE reference electrode and the liquid junction are temperature dependent. (We should note here that if pH or other potential measurements are made at less than room temperature, the salt bridge or the reference electrode should not contain saturated KCl, but somewhat less concentrated KCl, because solid KCl crystals will precipitate in the bridge and increase its resistance.)

2. Blood samples must be kept anaerobically to prevent loss or absorption of CO_2. Make pH measurements within 15 min after sample collection, if possible, or else keep the sample on ice and make the measurements within 2 h. The sample is equilibrated to 37°C before measuring. (If a pCO_2 measurement is to be performed also, do this within 30 min.)

3. To prevent coating of the electrode, flush the sample from the electrode with saline solution after each measurement. A residual blood film can be removed by dipping for *only* a few minutes in 0.1 *M* NaOH, followed by 0.1 *M* HCl and water or saline.

The pH measurement of blood samples must be made at body temperature to be meaningful.

Generally, venous blood is taken for pH measurement, although arterial blood may be required for special applications. The 95% confidence limit range (see Chapter 3) for arterial blood pH is 7.31 to 7.45 (mean 7.40) for all ages and sexes. A range of 7.37 to 7.42 has been suggested for subjects at rest. Venous blood may differ from arterial blood by up to 0.03 pH unit and may vary with the vein sampled. Intracellular erythrocyte pH is about 0.15 to 0.23 unit lower than that of the plasma.

20.16 pH Measurements in Nonaqueous Solvents

Measurement of pH in a nonaqueous solvent when the electrode is standardized with an aqueous solution has little significance in terms of possible hydrogen ion activity because of the unknown liquid-junction potential, which can be rather large, depending on the solvent. Measurements made in this way are usually referred to as "apparent pH." pH scales and standards for nonaqueous solvents have been suggested using an approach similar to the one for aqueous solutions. These scales have no rigorous relation to the aqueous pH scale, however. You are referred to the book by Bates (Reference 3) for a discussion of this topic. Some efforts have since been made to establish reference electrode potentials in mixed aqueous solvents at different temperatures [see, e.g., Bates et al., *J. Solut. Chem.* **8** (1979) 887–895. See also M. S. Frant, "How to Measure pH in Mixed & Nonaqueous Solutions," *Today's Chemist at Work,* American Chemical Society, June, 1995, p. 39]. Reference buffer solutions for use in 50% methanol:water solutions have been described (*J. Am. Chem. Soc.* **87** (1965) 415); Bates et al., have similiarly described pH standards in 10–40% ethanol:water (*Anal. Chem.* **52** (1980) 1598).

20.17 Ion-Selective Electrodes

See http://www.nico2000.net for an excellent tutorial (130-page beginners guide) on principles of pH and ion-selective electrodes, calibration, and measuring procedures.

It is important to know what other analytes an ISE responds to and the relative response compared to the analyte of interest. See Professor's Favorite Example at the end of this chapter - it is fortunate that that particular ISE produced a physically impossible result - had it produced a reasonable result, the presence of perchlorate in Martian soil would not have been so apparent.

The glass membrane pH electrode is the ultimate ion-selective electrode.

H$^+$ is a common interferent with ISEs, and so the pH must be above a limiting value, depending on the concentration of the primary ion (the one being measured).

The fluoride ion-selective electrode is one of the most successful and useful since the determination of fluoride is rather difficult by most other methods.

Various types of membrane electrodes have been developed in which the membrane potential is selective toward a given ion or ions, just as the potential of the glass membrane of a conventional glass electrode is selective toward hydrogen ions. These electrodes are important in the measurement of ions, especially in small concentrations. Generally, they are not "poisoned" by the presence of proteins, as some other electrodes are, and so they are ideally suited to measurements in biological media. This is especially true for the glass membrane ion-selective electrodes.

None of these electrodes is *specific* for a given ion, but each will possess a certain *selectivity* toward a given ion or ions. So they are properly referred to as **ion-selective electrodes** (ISEs).

Glass Membrane Electrodes

These are similar in construction to the pH glass electrode. Varying the composition of the glass membrane can cause the hydrated glass to acquire an increased affinity for various monovalent cations, with a much lower affinity for protons than the pH glass electrode has. The membrane potential becomes dependent on these cations, probably through an ion exchange mechanism similar to that presented for the glass pH electrode; that is, a boundary potential is produced, determined by the relative activities of the cations on the surface of the gel and in the external solution. Increased cation activity results in increased positive charge on the membrane and a positive increase in electrode potential.

The construction is similar to Figure 20.6. The internal filling solution will usually be the chloride salt of the cation to which the electrode is most responsive.

The sodium-sensitive type of electrode can be used to determine the activity of sodium ion in the presence of appreciable amounts of potassium ion. Its selectivity for sodium over potassium is on the order of 3000 or more.

Solid-State Electrodes

The construction of these electrodes is shown in Figure 20.12. The most successful example is the fluoride electrode. The membrane consists of a single crystal of lanthanum fluoride doped with some europium(II) fluoride to increase the conductivity of the crystal. Lanthanum fluoride is very insoluble, and this electrode exhibits Nerstian response to fluoride down to 10^{-5} M and non-Nerstian response down to 10^{-6} M (19 ppb!). This electrode has at least a 1000-fold selectivity for fluoride ion

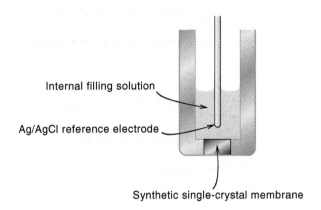

Internal filling solution

Ag/AgCl reference electrode

Synthetic single-crystal membrane

FIGURE 20.12 Crystal membrane electrode. (Reproduced by permission of Orion Research, Inc.)

over chloride, bromide, iodide, nitrate, sulfate, monohydrogen phosphate, and bicarbonate anions and a 10-fold selectivity over hydroxide ion. Hydroxide ion appears to be the only serious interference. The pH range is limited by the formation of hydrofluoric acid at the acid end and by hydroxide ion response at the alkaline end; the useful pH range is 4 to 9.

A useful solution for minimizing interferences with the fluoride electrode consists of a mixture of an acetate buffer at pH 5.0 to 5.5, 1 M NaCl, and cyclohexylenedinitrilo tetraacetic acid (CDTA). This solution is commercially available as TISAB (total ionic-strength adjustment buffer). A 1:1 dilution of samples and standards with the solution provides a high ionic-strength background, swamping out moderate variations in ionic strength between solutions. This keeps both the junction potential and the activity coefficient of the fluoride ion constant from solution to solution. The buffer provides a pH at which fluoride is almost entirely present as F⁻ and hydroxide concentration is very low. CDTA is a chelating agent, similar to EDTA, that complexes with polyvalent cations such as Al^{3+}, Fe^{3+}, and Si^{4+}, which would otherwise complex with F⁻ and reduce the fluoride activity.

> TISAB serves to adjust the ionic strength and the pH, and to prevent Al^{3+}, Fe^{3+}, and Si^{4+} from complexing the fluoride ion.

Liquid—Liquid Electrodes

The basic construction of these electrodes is shown in Figure 20.13. Here, the potential-determining "membrane" is a layer of a water-immiscible liquid ion exchanger held in place by an inert, porous membrane. The porous membrane allows contact between the test solution and the ion exchanger but minimizes mixing. It is either a synthetic, flexible membrane or a porous glass frit, depending on the manufacturer. The internal filling solution contains the ion for which the ion exchanger is specific plus a halide ion for the internal reference electrode.

> The filling solution for ISEs usually contains a chloride salt of the primary ion, for example, $CaCl_2$ for a Ca^{2+} electrode or KCl for a K^+ electrode. The chloride establishes the potential of the internal Ag/AgCl electrode.

An example of this electrode is the calcium-selective electrode. In one embodiment, the ion exchanger is an organophosphorus compound. The sensitivity of the electrode is governed by the solubility of the ion exchanger in the test solution. A Nernstian response is obtained down to about 5×10^{-5} M. The selectivity of this electrode is about 3000 for calcium over sodium or potassium, 200 over magnesium, and 70 over strontium. It can be used over the pH range of 5.5 to 11. Above pH 11, calcium hydroxide precipitates. A phosphate buffer should not be used for calcium measurements because the calcium activity will be lowered by complexation or precipitation. These and other liquid-membrane electrodes are often subject to fouling, for example by protein adsoption in biological fluids.

Table 20.3 summarizes the characteristics of some commercial ion-selective electrodes.

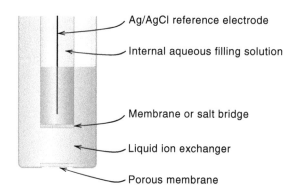

FIGURE 20.13 Liquid–membrane electrode.

Plastic Membrane—Ionophore Electrodes

A very versatile and relatively easy to prepare type of electrode is that in which a neutral lipophilic (organic loving) **ionophore** that selectively complexes with the ion of interest is dissolved in a soft plastic membrane. The ionophore should be lipophilic (as opposed to hydrophilic) so that it is not leached from the membrane upon exposure to aqueous solutions. The plastic membrane is usually polyvinylchloride (PVC) based and consists of about 33% PVC; about 65% plasticizer, for example, o-nitrophenyl ether (o-NPOE); about 1.5% ionophore. A modifier is generally added to increase electrical conductivity. For example, in a cation-selective ionophore based membrane, about 0.5% potassium tetrakis(p-chlorophenyl)borate (K(ϕCl$_4$B) to increase the conductivity and minimize interference from lipophilic anions such as SCN. The (ϕCl)$_4$B$^-$ ion is itself lipophilic and repels lipophilic anions that would otherwise penetrate the membrane and counter the metal ion response. A solution of these components is prepared in a solvent such as tetrahydrofuran (THF) and then is poured onto a glass plate to allow the THF to evaporate. The pliable membrane that results can then be mounted at the end of an electrode body.

TABLE 20.3 Typical Properties of Some Commercial Ion-Selective Electrodes

Electrode	Concentration Range (M)	Principal Interferences[a]
Liquid–liquid ion exchange electrodes		
Ca^{2+}	10^0– 10^{-5}	Zn^{2+}(3); Fe^{2+}(0.8); Pb^{2+}(0.6); Mg^{2+}(0.1); Na$^+$(0.003)
Cl$^-$	10^{-1}– 10^{-5}	I$^-$(17); NO$_3^-$(4); Br$^-$(2); HCO$_3^-$(0.2); SO$_4^{2-}$, F$^-$(0.1)
Divalent cation	10^0– 10^{-8}	Fe^{2+}, Zn^{2+}(3.5); Cu^{2+}(3.1); Ni^{2+}(1.3); Ca^{2+}, Mg^{2+}(1); Ba^{2+}(0.94); Sr^{2+}(0.54); Na$^+$(0.015)
BF$_4^-$	10^{-1}– 10^{-5}	NO$_3^-$(0.1); Br$^-$(0.04); OAc$^-$, HCO$_3^-$(0.004); Cl$^-$(0.001)
NO$_3^-$	10^{-1}– 10^{-5}	ClO$_4^-$(1000); I$^-$(20); Br$^-$(0.1); NO$_2^-$(0.04); Cl$^-$(0.004); CO$_3^{2-}$(0.0002); F$^-$(0.00006); SO$_4^{2-}$(0.00003)
ClO$_4^-$	10^{-1}– 10^{-5}	I$^-$(0.01); NO$_3^-$, OH$^-$(0.0015); Br$^-$(0.0006); F$^-$, Cl$^-$(0.0002)
K$^+$	10^0– 10^{-5}	Cs$^+$(1); NH$_4^+$(0.03); H$^+$(0.01); Na$^+$(0.002); Ag$^+$, Li$^+$(0.001)
Solid-state electrodes[b]		
F$^-$	10^0– 10^{-6}	Maximum OH$^-$ < 0.1 F$^-$
Ag$^+$ or S^{2-}	10^0– 10^{-7}	Hg^{2+} < $10^{-7}M$

[a] Number in parentheses is the relative selectivity for the interfering ion over the test ion (see The Selectivity Coefficient below).
[b] Interference concentrations given represent maximum tolerable concentrations.

Perhaps the most successful example of this type of electrode is the potassium ion-selective electrode incorporating the ionophore valinomycin. Valinomycin is a naturally occurring antibiotic with a cyclic polyether ring that has a cage of oxygens in the ring of just the right size for selectively complexing the potassium ion. Its selectivity for potassium is about 10^4 that for sodium.

Useful ionophores for a number of metal ions, especially alkali and alkaline earth ions, are **crown ethers**. These are synthetic neutral cyclic ether compounds that can be tailor-made to provide cages of the right size to selectively complex a given ion. A long hydrocarbon chain or phenyl group is usually attached to make the compound lipophilic. An example is the 14-crown-4 derivative illustrated in Figure 20.14, which is selective for lithium ion in the presence of sodium. The number 4 refers to the number of oxygens in the ring and the number 14 is the ring size. 14-Crown-4 compounds have the proper cage size to complex lithium. Placing bulky phenyl groups on the compound causes steric hindrance in the formation of the 2:1 crown ether:sodium complex and enhances the lithium selectivity (the lithium:crown complex is 1:1). The result is about 800-fold selectivity for lithium. Crown ether-based electrodes have been prepared for sodium, potassium, calcium, and other ions. Amide-based ionophores have been synthesized that selectively complex certain ions. Figure 20.15 shows some ionophores that have been used in PVC-based electrodes.

Pederson received the 1987 Nobel Prize for his pioneering work on crown ethers. See http://www.http://almaz.com/ nobel/chemistry/1987c.html.

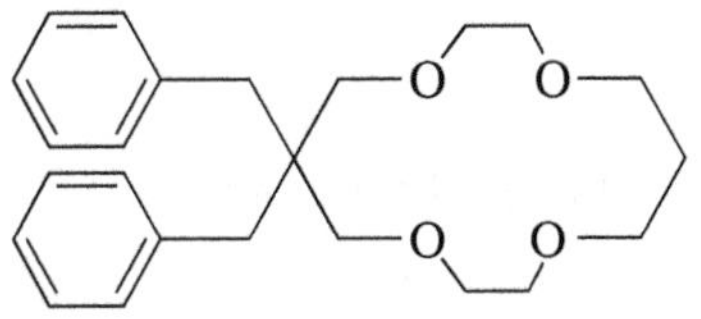

FIGURE 20.14 14-Crown-4 derivative that selectively binds lithium ion.

Mechanism of Membrane Response

The mechanisms of ion-selective electrode membrane response have not been as extensively studied as the glass pH electrode, and even less is known about how their potentials are determined. Undoubtedly, a similar mechanism is involved. The active membrane generally contains the ion of interest selectively bound to a reagent in the membrane, either as a precipitate or a complex. Otherwise, the electrode must be equilibrated in a solution of the test ion, in which case the ion also binds selectively to the membrane reagent. This can be compared to the—SiO^- H^+ sites on the glass pH electrode. When the ion-selective electrode is immersed in a solution of the test ion, a boundary potential is established at the interface of the membrane and the external solution. Again, the possible mechanism is due to the tendency of the ions to migrate in the direction of lesser activity to produce a liquid-junction-like potential. Positive ions will result in a positive charge and a change in the potential in the positive direction, while negative ions will result in a negative charge and a change in the potential in the negative direction.

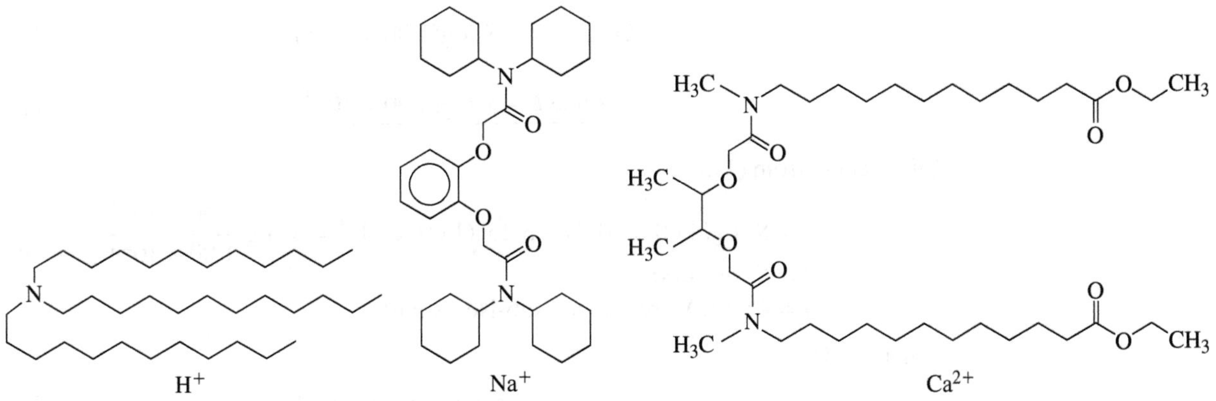

FIGURE 20.15 Ionophores for H^+, Na^+, and Ca^{2+}.

The secret in constructing ion-selective electrodes, then, is to find a material that has sites which show strong affinity for the ion of interest. Thus, the calcium liquid ion exchange electrode exhibits high selectivity for calcium over magnesium and sodium ions because the organophosphate cation exchanger has a high chemical affinity for calcium ions. The ion exchange equilibrium at the membrane–solution interface involves calcium ions, and the potential depends on the ratio of the activity of calcium ions in the external solution to that of calcium ion in the membrane phase.

Selectivity Coefficient

The potential of an ion-selective electrode in the presence of a single ion follows an equation similar to Equation 20.38 for the pH glass membrane electrode:

Don't forget the sign of z.

$$E_{ISE} = k + \frac{S}{z} \log a_{ion} \tag{20.44}$$

where S represents the slope (theoretically $2.303RT/F$) and z is the ion charge, *including sign*. Often, the slope is less than Nernstian; but for monovalent ion electrodes, it is usually close. The constant k depends on the nature of the internal reference electrode, the filling solution, and the construction of the membrane. It is determined by measuring the potential of a solution of the ion of known activity.

Example 20.6

A fluoride electrode is used to determine fluoride in a water sample. Standards and samples are diluted 1:10 with TISAB solution. For a 1.00×10^{-3} M (before dilution) standard, the potential reading relative to the reference electrode is -211.3 mV; and for a 4.00×10^{-3} M standard, it is -238.6 mV. The reading with the unknown is -226.5 mV. What is the concentration of fluoride in the sample?

Solution

Since the ionic strength remains constant due to dilution with the ionic-strength adjustment solution, the response is proportional to $\log[F^-]$:

$$E = k + \frac{S}{z} \log[F^-] = k - S \log[F^-]$$

where z is -1. First calculate S:

$$-211.3 = k - S \log(1.00 \times 10^{-3}) \tag{1}$$

$$\underline{-238.6 = k - S \log(4.00 \times 10^{-3})} \tag{2}$$

Subtract (2) from (1):

$$27.3 = S \log(4.00 \times 10^{-3}) - S \log(1.00 \times 10^{-3}) = S \log \frac{4.00 \times 10^{-3}}{1.00 \times 10^{-3}}$$
$$27.3 = S \log 4.00$$
$$S = 45.3 \text{ mV (somewhat sub-Nernstian)}$$

Calculate k:

$$-211.3 = k - 45.3 \log(1.00 \times 10^{-3})$$

$$k = -347.2 \text{ mV}$$

For the unknown:

$$-226.5 = -347.2 - 45.3 \, \log [F^-]$$

$$[F^-] = 2.16 \times 10^{-3} \, M$$

If the electrode is in a solution containing a mixture of cations (or anions, if it is an anion-responsive electrode), it may respond to the other cations (or anions). Suppose, for example, we have a mixture of sodium and potassium ions and an electrode that responds to both. The Nernst equation must include an additive term for the potassium activity:

$$E_{NaK} = k_{Na} + S \, \log(a_{Na^+} + K_{NaK}a_{K^+}) \tag{20.45}$$

The constant k_{Na} corresponds to that in the *Nernst* equation for the primary ion, sodium, alone. E_{NaK} is the potential of the electrode in a mixture of sodium and potassium. K_{NaK} is the **selectivity coefficient** of the electrode for *potassium over sodium*. It is equal to the reciprocal of K_{KNa}, which is the selectivity coefficient of *sodium over potassium*. Obviously, we want $a_{K^+}K_{NaK}$ to be small; this can be achieved by either a_{K^+} or K_{NaK} or both to be small.

K_{NaK} and k_{Na} are determined by measuring the potential of two different standard solutions containing sodium and potassium and then solving the two simultaneous equations for the two constants. Alternatively, one of the solutions may contain only sodium, and k_{Na} is determined from Equation 20.44.

A general equation, called the *Nikolsky* equation, can be written for mixtures of two ions of different charges:

$$\boxed{E_{AB} = k_A + \frac{S}{z_A} \log \, (a_A + K_{AB}a_B^{z_A/z_B})} \tag{20.46}$$

where z_A is the charge on ion A (the primary ion) and z_B is the charge on ion B. Thus, measurement of sodium in the presence of calcium using a sodium ion electrode would follow the expression:

$$E_{NaCa} = k_{Na} + S \, \log(a_{Na^+} + K_{NaCa}a_{Ca^{2+}}^{1/2}) \tag{20.47}$$

Since all electrodes respond more or less to hydrogen ions, the practice is to keep the activity of the hydrogen ion low enough that the product $K_{AH}a_{H^+}^{z_A}$ in the Nikolsky equation is negligible.

One problem with selectivity coefficients is that they often vary with the relative concentrations of ions and are not constant. For this reason, it is difficult to use them in calculations involving mixtures of ions. They are useful for predicting conditions under which interfering ions can be neglected. That is, in practice, conditions are generally adjusted so the product $K_{AB}a_B^{z_A/z_B}$ is negligible and a simple Nernst equation applies for the test ion. Usually, a calibration curve is prepared, and if an interfering ion is present, this can be added to standards at the same concentrations as in the unknowns; the result would be a nonlinear but corrected calibration curve. This technique can obviously only be used if the concentration of the interfering ion remains nearly constant in the samples.

The use of the selectivity coefficient in a calculation is illustrated in the following example.

No electrode is totally specific. Ideally, we can keep the product $K_{NaK}a_{Ka^+}$ negligible compared to a_{Na^+}.

$K_{NaK} = 1/K_{KNa}$.

$K_{NaH}a_{H^+}$ compared to a_{Na^+} determines the lower pH limit of the electrode.
Selectivity coefficients are generally not sufficiently constant to use in quantitative calculations.

Example 20.7

A cation-sensitive electrode is used to determine the activity of calcium in the presence of sodium. The potential of the electrode in 0.0100 M $CaCl_2$ measured against an SCE is 195.5 mV. In a solution containing 0.0100 M $CaCl_2$ and 0.0100 M NaCl, the potential is 201.8 mV. What is the activity of calcium ion in an unknown solution if the potential of the electrode in this is 215.6 mV versus SCE and the sodium ion activity has been determined with a sodium ion-selective electrode to be 0.0120 M? Assume Nernstian response.

Solution

The ionic strength of 0.0100 M $CaCl_2$ is 0.0300, and that of the mixture is 0.0400. Therefore, from Equation 5.20, the activity coefficient of calcium ion in the pure $CaCl_2$ solution is 0.55, and for calcium and sodium ions in the mixture, it is 0.51 and 0.83, respectively. Therefore,

$$k_{Ca} = E_{Ca} - 29.58 \; \log \; a_{Ca^{2+}}$$
$$= 195.5 - 29.58 \; \log(0.55 \times 0.0100)$$
$$= 262.3 \text{ mV}$$
$$E_{CaNa} = k_{Ca} + 29.58 \; \log(a_{Ca^{2+}} + K_{CaNa} a_{Na^+}^2)$$
$$201.8 = 262.3 + 29.58 \; \log[0.51 \times 0.0100 + K_{CaNa}(0.83 \times 0.0100)^2]$$
$$K_{CaNa} = 47$$
$$215.6 = 262.3 + 29.58 \; \log(a_{Ca^{2+}} + 47 \times 0.0120^2)$$
$$a_{Ca^{2+}} = 0.0196 \; M$$

Note that although the selectivity coefficient for Ca^{2+} is not very good (the electrode is a better sodium sensor!), the sodium contribution (0.0068) in the mixture is only about 0.3 that of the calcium (0.0196), due to the squared term for sodium.

Experimental Methods for Determining Selectivity Coefficients

There are various methods to determine selectivity coefficients. These include the separate solution method, the fixed interference method, and the matched potential method. See the text's website for details on applying these methods, as well as a discussion of the limitations of the Nikolsky equation. Also discussed is research on making ion-selective electrodes very sensitive.

Measurement with Ion-Selective Electrodes

Ion-selective electrodes are subject to the same accuracy limitations as pH electrodes. For $z_A = 2$, the errors per millivolt are doubled.

As with pH glass electrodes, most ion-selective electrodes have high resistance, and the measurement equipment must have high input impedance. A high resolution pH meter is generally used. It is often necessary to pretreat ion-selective electrodes by soaking in a solution of the ion to be determined.

The problems and accuracy limitations discussed under pH and other direct potentiometric measurements apply also to ion-selective electrodes.

Ion-selective electrodes measure only the *free* ion.

A calibration curve of potential versus log activity is usually prepared. If concentrations are to be measured, then the technique of maintaining a constant ionic strength as described earlier is used (Equation 20.34). For example, the concentration of unbound calcium ion in serum is determined by diluting samples and standards with 0.15 M NaCl. Only the *unbound* calcium is measured and not the fraction that is complexed.

The activity coefficient of sodium ion in normal human serum has been estimated, using ion-selective electrodes, to be 0.780 ± 0.001, and in serum water to be 0.747 (serum contains about 96% water by volume). Standard solutions of sodium chloride and potassium chloride are usually used to calibrate electrodes for the determination of sodium and potassium in serum. Concentrations of 1.0, 10.0, and 100.0 mmol/L sodium chloride can be prepared with respective activities of 0.965, 9.03, and 77.8 mmol/L for sodium ion, and the same concentrations of potassium chloride give potassium ion activities of 0.965, 9.02, and 77.0 mmol/L.

The response of many ion-selective electrodes is slow, and considerable time must be taken to establish an equilibrium reading. The response becomes even slower as the concentration is decreased. Some electrodes, on the other hand, respond sufficiently rapidly that they can be used to monitor reaction rates.

We can summarize the advantages and disadvantages of ion-selective electrodes and some precautions and limitations in their use as follows:

1. They measure activities rather than concentrations, a unique advantage but a factor that must be considered in obtaining concentrations from measurements. Errors in calculating concentrations can occur from ionic-strength effects.

2. They measure "free" ions (i.e., the portion that is not associated with other species). Chemical interference can occur from complexation, protonation, and the like.

3. They are not specific but merely more selective toward a particular ion. Hence, they are subject to interference from other ions. They respond to hydrogen ions and are, therefore, pH-limited.

4. They function in turbid or colored solutions, where photometric methods fail.

5. They have a logarithmic response, which results in a wide dynamic working range, generally from four to six orders of magnitude. This logarithmic response also results in an essentially constant, albeit relatively large, error over the working range where the Nernst relation holds.

6. In favorable cases, their response is fairly rapid, (except in dilute solution), often requiring less than 1 min for a measurement. Electrode response is frequently rapid enough to allow monitoring of flowing industrial process streams.

7. The response is temperature dependent, from the term RT/nF.

8. The required measuring equipment can be made portable for field operations, and small samples (e.g., 1 mL) can be analyzed.

9. The sample is not destroyed during measurement.

10. While certain electrodes may operate down to 10^{-6} M, many will not; commercially available electrodes rarely permit the lower applicable limit to reach those attainable by some other competing techniques.

11. Frequent calibration is required.

12. Few primary activity standards are available, as there are for pH measurements,[3] and the standard solutions that are used are not "buffered" in the ion being tested. Impurities, especially in dilute standards, may cause erroneous results.

The logarithmic response of potentiometric electrodes gives a wide dynamic range, but at some expense in precision.

---

[3]Activity standards similar to pH standards are available from the National Institute of Standards and Technology for some salts, such as sodium chloride.

In spite of some limitations, ion-selective electrodes have become important because they represent one approach to the analytical chemist's dream of an inexpensive portable probe that is specific for the test substance with perhaps minimal sample preparation for a solution phase sample.

20.18   Chemical Analysis on Mars Using Ion-Selective Electrodes

⌈Contributed by Professor Samuel P. Kounaves, Tufts University⌋ ## Professor's Favorite Example

In the summer of 2008, a NASA lander named Phoenix touched down on the arctic expanses of Mars. Onboard its deck were several analytical instruments, including four mug-sized wet chemistry laboratories (WCL) that contained an array of ISEs along with several other electrochemical sensors (Figure 20.16). Each WCL was designed to accept 1 cm^3 (~ 1 g) of soil from Phoenix's robotic arm (Figure 20.17), add it to 25 mL of an electrolyte solution, and then add five solid reagents stored in tiny crucibles. One crucible contained a small amount of dried salts to be used as calibration points for the ISEs, another contained a small amount of a solid nitrobenzoic acid to test the buffering capacity of the resulting soil/solution mixture, and three crucibles each containing 100 mg of $BaCl_2$ to titrate the solution for the soluble SO_4^{2-} in the soil.

After addition of the soil, a nitrate ISE that was to monitor the background level of the $LiNO_3$ electrolyte gave a response that would have required the 1 cm^3 of soil to contain 2 g of NO_3^-, a physical impossibility. After further analysis, it became clear that the ISE had not responded to NO_3^- but to the presence of ClO_4^-, an ion towards which this particular ISE is 1000 times more selective. The analysis was further confirmed by

FIGURE 20.16 The ion-selective electrode based wet chemistry laboratories (WCL) that went to Mars. The footprint is 6 6 cm; the mass of the WCL is 610 g. Credit: NASA Jet Propulsion Laboratory, Pasadena, CA.

FIGURE 20.17 Mosaic: A view of the four wet chemistry laboratory (WCL) units on the deck of the Phoenix lander and the scoop on the robot arm (RA) that delivers the soil. Credit: Samuel P. Kounaves, Tufts University and NASA/JPL.

the observation that the Ca^{2+} ISE had responded negatively to Ca^{2+}, a phenomena that is only observed when ClO_4^- is present in the solution. Three soil analyses confirmed that the results were reproducible.

The use of an ISE for determination of the soluble SO_4^{2-} in the soil was precluded since all known SO_4^{2-} ISEs show significant interference by other anionic species that were expected to be present in the soil. Thus, to measure the SO_4^{2-} it was necessary to titrate the SO_4^{2-} with $BaCl_2$ and monitor the concentration of Ba^{2+} using a barium ISE. When all the SO_4^{2-} had been precipitated, the rapid increase in the Ba^{2+} ISE signal indicated the end point and allowed determination of the SO_4^2.

For more details see S. P. Kounaves et al., *J. Geophys. Res.*, 114 (**2009**) E00A19; M. H. Hecht et al., *Science*, 325 (2009) 64; S. P. Kounaves et al., *J. Geophys. Res.*, 114 (**2010**) E00E10; S. P. Kounaves et al., *Geophys. Res. Lett.* 37 (**2010**), L09201.

Questions

1. Describe the working and limitations of standard hydrogen electrode.
2. Differentiate the boundary potential and diffusion potential for a glass pH electrode.
3. What are the combination electrodes and why they are used?
4. How potentiometric methods are different from voltammetry methods?
5. What is the liquid-junction potential? Residual liquid-junction potential? How can these be minimized?
6. Discuss the mechanism of the glass membrane electrode response for pH measurements.
7. What is the alkaline error and the acid error of a glass membrane pH electrode?
8. Incorporation of _____ into a polyvinylchloride membrane allows for the manufacture of an ion-selective electrode highly selective for calcium.

   (a) Di-*p*-octylphenyl phosphate   (b) Nonactin

   (c) Valinomycin   (d) Ertyhromycin

9. The ability of an indicator electrode to respond to a single ionic species and not any other is referred to as

   (a) sensitivity.   (b) accuracy.

   (c) precision.   (d) selectivity.

10. Why a saturated KCl solution is used in Calomel electrode?
11. Describe the different types of ion-selective electrodes. Include in your discussion the construction of the electrodes, differences in membranes, and their usefulness.
12. What is the selectivity coefficient? Discuss its significance and how you would determine its value. (See the text's website for descriptions of experimental methods to determine selectivity coefficients.)
13. What is the Nicolsky equation?

Problems

Standard Potentials

14. The potential for a X^{3+}/X^{2+} half cell is 0.755 V relative to the standard hydrogen electrode (SHE). What will be the potential when standard Calomel electrode or standard silver/silver chloride electrode is used?
15. The standard potential of the silver/silver bromide electrode is 0.073 V. Calculate the solubility product of silver bromide.

16. A sample of thiocyanate is titrated with silver solution. The potential at the end point of 0.202 V versus SCE. Calculate the standard potential for $Ag^+ + e^- = Ag$. The K_{sp} for silver thiocyanate is 1.00×10^{-12}.

Voltaic Cells

17. For each of the following reactions, (1) separate the reaction into its component half-reactions; (2) write a schematic representation of a cell in which the reaction would occur in the direction as written; (3) calculate the standard potential of the cell; (4) assign the polarity of each electrode under conditions that the reaction would occur as written.

    (a) $Ag + Fe^{3+} = Ag^+ + Fe^{2+}$

    (b) $VO_2^+ + V^{3+} = 2VO^{2+}$

    (c) $Ce^{4+} + Fe^{2+} = Ce^{3+} + Fe^{3+}$

18. For the following cells, write the half-reactions occurring at each electrode and the complete cell reaction, and calculate the cell potential:

    (a) $Pt, H_2(0.2\ atm)|HCl(0.5\ M)|Cl_2(0.2\ atm), Pt$

    (b) $Pt|Fe^{2+}(0.005\ M), Fe^{3+}(0.05\ M), HClO_4(0.1\ M)\|HClO_4(0.1\ M), VO_2^+(0.001\ M), VO^{2+}(0.002)\ M|Pt$

PROFESSOR'S FAVORITE QUESTION

Contributed by Professor Yijun Tang, University of Wisconsin, Oshkosh

19. A poly(vinyl ferrocene) or PVF film was coated on a gold electrode surface. The PVF film can be oxidized to PVF^+. The PVF^+/PVF redox pair had a standard reduction potential of 0.296 V *vs*. SCE at 25°C. If this PVF film coated electrode is immersed into a 2.0 mM $KAuCl_4$ solution, will $KAuCl_4$ be reduced to Au, given that the standard potential of the Au couple is 1.002 V. vs. NHE? Write the equation for the reaction above. Calculate the equilibrium constant for that equation. (See Tang, Y. and Zeng, X. *J. Electrochem. Soc.* **155** (2008) F82.

Redox Potentiometric Measurements

20. A copper electrode was placed in 3.25 m*M* $Cu(NO_3)_2$ at room temperature. What is the potential of this electrode relative to the standard hydrogen electrode (SHE) at equilibrium?

21. What would the potentials of the following half-cells at standard conditions be versus a saturated calomel electrode?

    (a) $Pt/Br_2(aq), Br^-$;

    (b) $Ag/AgCl/Cl^-$;

    (c) $Pt/V^{3+}, V^{2+}$.

22. What would be the observed potential if the following half-cell were connected with a saturated calomel electrode?

    $Fe^{3+}(0.00200\ M), Fe^{2+}(0.0500\ M)|Pt$ (numbers represent activities)

23. A 50-mL solution that is 0.10 *M* in both chloride and iodide ions is titrated with 0.10 *M* silver nitrate.

    (a) Calculate the percent iodide remaining unprecipitated when silver chloride begins to precipitate.

    (b) Calculate the potential of a silver electrode versus the SCE when silver chloride begins to precipitate and compare this with the theoretical potential corresponding to end point for the titration of iodide.

    (c) Calculate the potential at the end point for chloride. For simplicity, in lieu of activities, use concentrations in calculations.

24. The potential of the electrode Hg|Hg–EDTA, M–EDTA, M^{n+} is a function of the metal in M^{n+} and can be shown as

$$E = E^0_{Hg^{2+},Hg} - \frac{2.303RT}{2F} \log \frac{K_{f(HgEDTA)}}{K_{f(MEDTA)}}$$

$$- \frac{2.303RT}{2F} \log \frac{a_{MEDTA}}{a_{HgEDTA}} - \frac{2.303RT}{2F} \log \frac{1}{a_{M^{n+}}}$$

The stability of M–EDTA must be less than that of Hg–EDTA (a very stable chelate; $K_{f(Hg-EDTA)} = 10^{22}$). A Hg|Hg–EDTA electrode can be used to monitor M^{n+} during the course of a titration with EDTA. Starting with the Hg|Hg^{2+} electrode, derive the above equation. This represents a metal–metal chelate–metal ion electrode.

25. The potential of a hydrogen electrode in an acid solution is −0.465 V when measured against an SCE reference electrode. What would the potential be measured against a normal calomel electrode (1 M KCl)?

pH Measurements

26. For the given cell,

$$Pt|Q, QH_2, H^+ \,||\, 1\,M\,KCl|HgCl_2\,(s)\,|Hg\,(l)\,|Pt$$

What will be the pH, when $E_{cell} = 0$?

27. The reversible reduction potential of pure water is −0.413 V under 1 atm H2 pressure. If the reduction is considered to be $2H^+ + 2e^- \longrightarrow H_2$, calculate the pH of the pure water.

28. The pH of the solution in the cell,

$$Pt|H_2\,(g)\,|HCl\,(g)\,|AgCl\,(s)\,|Ag$$

is 0.65. Calculate the EMF of the cell. $\left(E_{Cl^-/AgCl,Ag} = 0.2224V\right)$

29. (a) How accurately can the pH of an unknown solution generally be measured? What limits this? What is this (calculate it) in terms of millivolts measured? In terms of percent error in the hydrogen ion activity?

    (b) How precisely can the pH of a solution be measured? How much is this in terms of millivolts measured? In terms of percent variation in the hydrogen ion activity?

30. A glass electrode was determined to have a potential of 0.395 V when measured against the SCE in a standard pH 7.00 buffer solution. Calculate the pH of the unknown solution for which the following potential readings were obtained (the potential decreases with increasing pH):

    (a) 0.467 V   (b) 0.209 V

    (c) 0.080 V   (d) −0.013 V

31. Calculate the potential of the cell consisting of a hydrogen electrode ($P_{H_2} = 1$ atm) and a saturated calomel reference electrode

    (a) in a solution of 0.00100 M HCl

    (b) in a solution of 0.00100 M acetic acid, and

    (c) in a solution containing equal volumes of 0.100 M acetic acid and 0.100 M sodium acetate. Assume that activities are the same as concentrations.

32. The quinhydrone electrode can be used for the potentiometric determination of pH. The solution to be measured is saturated with quinhydrone, an equimolar mixture of quinone (Q) and hydroquinone (HQ), and the potential of the solution is measured with a platinum electrode. The half-reaction and its standard potential are as follows:

quinone (Q)   hydroquinone (HQ)

What is the pH of a solution saturated with quinhydrone if the potential of a platinum electrode in the solution, measured against a saturated calomel electrode, is −0.205 V? Assume the liquid-junction potential to be zero.

Ion-Selective Electrode Measurements

33. It can be shown from Equations 20.44 and 20.45 that, for monovalent ions, $\log K_{AB} = (k_B - k_A)/S$. Derive this equation.

34. A potassium ion-selective electrode is used to measure the concentration of potassium ion in a solution that contains 6.0×10^{-3} M cesium (activity). From Table 20.3, the electrode responds equally to either ion ($K_{KCs} = 1$). If the potential versus a reference electrode is -18.3 mV for a 5.0×10^{-3} M KCl solution and $+20.9$ mV in the sample solution, what is the activity of K^+ in the sample? Assume Nernstian response.

35. The nitrate concentration in an industrial effluent is determined using a nitrate ion-selective electrode. Standards and samples are diluted 20-fold with 0.1 M K_2SO_4 to maintain constant ionic strength. Nitrate standards of 0.0050 and 0.0100 M give potential readings of -108.6 and -125.2 mV, respectively. The sample gives a reading of -119.6 mV. What is the concentration of nitrate in the sample?

36. The perchlorate concentration in a sample containing 0.015 M iodide is determined using a perchlorate ion-selective electrode. All samples and standards are diluted 1:10 with 0.2 M KCl to maintain constant ionic strength. A 0.00100 M $KClO_4$ standard gives a reading of -27.2 mV, and a 0.0100 M KI standard gives a reading of $+32.8$ mV. The sample solution gives a reading of -15.5 mV. Assuming Nernstian response, what is the concentration of perchlorate in the sample?

37. The potential of a glass cation-sensitive electrode is measured against an SCE. In a sodium chloride solution of activity 0.100 M, this potential is 113.0 mV, and in a potassium chloride solution of the same activity, it is 67.0 mV.

    (a) Calculate the selectivity coefficient of this electrode for potassium over sodium, using the relationship derived in Problem 33.

    (b) What would be the expected potential in a mixture of sodium ($a = 1.00 \times 10^{-3}$ M) and potassium ($a = 1.00 \times 10^{-2}$ M) chlorides? Assume Nernstian response, 59.2 mV/decade.

38. The selectivity coefficient for a cation-selective electrode for B^+ with respect to A^+ is determined from measurements of two solutions containing different activities of the two ions. The following potential readings were obtained: (1) 2.00×10^{-4} M $A^+ + 1.00 \times 10^{-3}$ M B^+, $+237.8$ mV; and (2) 4.00×10^{-4} M $A^+ + 1.00 \times 10^{-3}$ M B^+, $+248.2$ mV. Calculate K_{AB}. The electrode response is 56.7 mV/decade.

39. A sodium glass ion-selective electrode is calibrated using the separate solution method, for sodium response and potassium response. The two calibration curves have slopes of 58.1 mV per decade, and the sodium curve is 175.5 mV more positive than the potassium curve. What is K_{NaK} for the electrode? See the text website for a description of the separate solution method.

40. A valinomycin-based potassium ion-selective electrode is evaluated for sodium interference using the fixed interference method. A potassium calibration curve is prepared in the presence of 140 mM sodium. The straight line obtained from extrapolation of the linear portion deviates from the experimental curve by 17.4 mV at a potassium concentration corresponding to 1.5×10^{-5} M. If the linear slope is 57.8 mV per decade, what is K_{NaK} for the electrode? See the text website for a description of the fixed interference method.

Multiple Choice Questions

1. When a silver strip is dipped in aqueous Cu^{2+} ions, then what would happen?

    (a) Cu^{2+} ions would be reduced.

    (b) Ag would be oxidized.

    (c) Copper would be deposited on the silver strip.

    (d) No reaction would occur.

2. Which manufacturing process depends upon electrolysis?
   - (a) Haber process
   - (b) Down's process
   - (c) Czochralski process
   - (d) Constant process

3. A pH meter is an example of
   - (a) an ion-selective electrode.
   - (b) a fuel cell.
   - (c) an electrolytic cell.
   - (d) a reference electrode.

Recommended References

Activity Coefficients

1. J. Kielland, "Individual Activity Coefficients of Ions in Aqueous Solutions," *J. Am. Chem. Soc.*, **59** (1937) 1675.

Relative Potentials

2. W.-Y. Ng, "Conversion of Potentials in Voltammetry and Potentiometry," *J. Chem. Ed.*, **65** (1988) 727.

pH Electrodes and Measurements

3. R. G. Bates, *Determination of pH*. New York: Wiley, 1964.
4. H. B. Kristensen, A. Salomon, and G. Kokholm, "International pH Scales and Certification of pH," *Anal. Chem.*, **63** (1991) 885A.
5. J. V. Straumford, Jr., "Determination of Blood pH," in D. Seligson, ed., *Standard Methods of Clinical Chemistry*, Vol. **2**. New York: Academic, 1958, pp. 107–121.
6. C. C. Westcott, *pH Measurements*. New York: Academic, 1978.
7. H. Galster, *pH Measurement. Fundamentals, Methods, Applications, Instrumentation*. New York: VCH, 1991.

Ion-Selective Electrodes and Measurements

8. R. G. Bates, "Approach to Conventional Scales of Ionic Activity for the Standardization of Ion-Selective Electrodes," *Pure Appl. Chem.*, **37** (1974) 573.
9. E. Bakker, P. Buhlmann, and E. Pretsch, "Carrier-Based Ion-Selective Electrodes and Bulk Optodes. 1. General Characteristics," *Chem. Rev.*, **97** (1997) 3083.

Selectivity Coefficients

10. Y. Umezawa, *CRC Handbook of Ion Selective Electrodes: Selectivity Coefficients*. Boca Raton, FL: CRC Press, 1990.
11. V. P. Y. Gadzekpo and G. D. Christian, "Determination of Selectivity Coefficients of Ion-Selective Electrodes by a Matched-Potential Method." *Anal. Chim. Acta*, **166** (1984) 279.
12. Y. Umezawa, K. Umezawa, and H. Ito, "Selectivity Coefficients for Ion-Selective Electrodes: Recommended Methods for Reporting k_{ij}^{pot} Values," *Pure Appl. Chem.*, **67** (1995) 507.
13. G. Horvai, "The Matched Potential Method, a Generic Approach to Characterize the Differential Selectivity of Chemical Sensors," *Sensors and Actuators B*, **43** (1997) 94. See also G. Horvai, *Trends in Anal. Chem.*, **16** (1997) 260.
14. E. Bakker, R. K. Meruwa, E. Pretsch, and M. E. Meyerhoff, "Selectivity of Polymer Membrane-Based Ion-Selective Electrodes: Self-Consistent Model Describing the Potentiometric Response in Mixed Ion Solutions of Different Charge," *Anal. Chem.*, **66** (1994) 3021.
15. E. Bakker, "Determination of Improved Selectivity Coefficients of Polymer Membrane Ion-Electrodes by Conditioning with a Discriminating Ion," *J. Electrochem. Soc.*, **143**(4) (1996) L83.

16. E. Bakker, "Determination of Unbiased Selectivity Coefficients of Neutral Carrier-Based Cation-Selective Electrodes," *Anal. Chem.*, **69** (1997) 1061.

17. E. Bakker. "Review: Selectivity of Liquid Membrane Ion-Selective Electrodes," *Electroanalysis*, **9** (1997) 7.

Ultrasensitive Ion-Selective Electrodes

18. S. Mathison and E. Bakker, "Effect of Transmembrane Electrolyte Diffusion on the Detection Limit of Carrier-Based Potentiometric Sensors," *Anal. Chem.*, **70** (1998) 303.

19. Y. Mi, S. Mathison, R. Goines, A. Logue, and E. Bakker, "Detection Limit of Polymeric Membrane Potentiometric Ion Sensors: How Low Can We Go Down to Trace Levels?" *Anal. Chim. Acta*, **397** (1999) 103.

20. T. Sokalski, A. Ceresa, T. Zwickl, and E. Pretsch, "Large Improvement of the Lower Detection Limit of Ion-Selective Polymer Membrane Electrodes," *J. Am. Chem. Soc.*, **119** (1997) 11,347.

21. T. Sokalski, T. Zwickl, E. Bakker, and E. Pretsch, "Lowering the Detection Limit of Solvent Polymeric Ion-Selective Electrodes. 1. Modeling the Influence of Steady-State Ion Fluxes," *Anal. Chem.*, **71** (1999) 1204.

22. T. Sokalski, A. Cera, M. Fibbioli, T. Zwickl, E. Bakker, and E. Pretsch, "Lowering the Detection Limit of Solvent Polymeric Ion-Selective Electrodes. 2. Influence of Composition of Sample and Internal Electrolyte Solution," *Anal. Chem.* **71** (1999) 1210.

Chapter 21

REDOX AND POTENTIOMETRIC TITRATIONS

KEY THINGS TO LEARN FROM THIS CHAPTER

- Balancing redox reactions
- Calculating the reaction equilibrium constant from standard potentials (key Equation: 21.1; Example 21.3)
- Calculating redox titration curves
- Redox indicators

- Iodimetry and iodometry
- Preparing the analyte for titration
- Potentiometric titrations
- Derivative titrations—using spreadsheets for plotting
- Gran plots

Volumetric analyses based on titrations with reducing or oxidizing agents are very useful for many determinations. They may be performed using visual indicators or by measuring the potential with an appropriate indicating electrode to construct a potentiometric titration curve. In this chapter, we discuss redox titration curves based on half-reaction potentials and describe representative redox titrations and the necessary procedures to obtain the sample analyte in the correct oxidation state for titration. The construction of potentiometric titration curves is described, including derivative titration curves and Gran plots. You should first review the balancing of redox reactions since balanced reactions are required for volumetric calculations.

Common to all the redox procedures discussed below is the need to get the analyte to be titrated in the proper oxidation state, and removal of the conditioning agent, which is itself an oxidizing or reducing agent. See Section 21.8 below for typical procedures.

Some examples of useful redox titrations include measuring the ascorbic acid (a reducing agent) content of vitamin C tablets, or of sulfur dioxide in wines, by titration with iodine. The Karl Fisher titration of water in samples involves iodine. The measure of saturation in a fatty acid is determined as the iodine or bromine number, in which the grams of iodine or bromine absorbed by a 100-gram sample are measured. The iron content of an ore can be determined by titration of iron(II) with potassium permanganate.

21.1 First: Balance the Reduction–Oxidation Reaction

The calculations in volumetric analysis require a balanced reaction. The balancing of redox reactions is reviewed on the text website.

Various methods are used to balance redox reactions, and we shall use the **half-reaction method**. In this technique, the reaction is broken down into two parts: the oxidizing part and the reducing part. In every redox reaction, an oxidizing agent reacts with a reducing agent. The oxidizing agent is reduced in the reaction while the reducing agent is oxidized. Each of these constitutes a *half-reaction*, and the overall reaction can be broken down into these two half-reactions. Thus, in the reaction

$$Fe^{2+} + Ce^{4+} \rightarrow Fe^{3+} + Ce^{3+}$$

There are various ways of balancing redox reactions. Use the method you are most comfortable with. A more thorough review is available on the text **website**.

Fe^{2+} is the reducing agent and Ce^{4+} is the oxidizing agent. The corresponding half-reactions are:

$$Fe^{2+} \rightarrow Fe^{3+} + e^-$$

and

$$Ce^{4+} + e^- \rightarrow Ce^{3+}$$

To balance a reduction–oxidation reaction, each half-reaction is first balanced. There must be a net gain or loss of zero electrons in the overall reaction, and so the second step is multiplication of one or both of the half-reactions by an appropriate factor or factors so that, when they are added, the electrons cancel. The final step is addition of the half-reactions. The above simple 1:1 reaction just illustrates the half-reaction concept. You should review the methods in the text website for balancing more complex reactions in both acid and alkaline media.

21.2 Calculation of the Equilibrium Constant of a Reaction—Needed to Calculate Equivalence Point Potentials

At the equivalence point, we have unknown concentrations that must be calculated from K_{eq}. This is calculated by equating the two Nernst equations, combining the concentration terms to give K_{eq}, and then solving for K_{eq} from ΔE^0.

Before we discuss redox titration curves based on reduction–oxidation potentials, we need to learn how to calculate equilibrium constants for redox reactions from the half-reaction potentials. The reaction equilibrium constant is used in calculating equilibrium concentrations at the equivalence point, in order to calculate the equivalence point potential. Recall from Chapter 19 that since a cell voltage is zero at reaction equilibrium, the difference between the two half-reaction potentials is zero (or the two potentials are equal), and the Nernst equations for the half-reactions can be equated. When the equations are combined, the log term is that of the equilibrium constant expression for the reaction (see Equation 19.20), and a numerical value can be calculated for the equilibrium constant. This is a consequence of the relationship between the free energy and the equilibrium constant of a reaction. Recall from Equation 5.10 that $\Delta G^\circ = -RT \ln K$. Since $\Delta G^\circ = -nFE^0$ for the reaction, then

$$-RT \ln K = -nFE^0 \qquad (21.1)$$

or

$$E^0 = \frac{RT}{nF} \ln K$$

For the spontaneous reaction, ΔG° is negative and E^0 is positive.

Example 21.1

Calculate the potential in a solution at 298 K (vs. NHE) when 5.0 mL of 0.20 M Ce^{4+} solution is added to 5.0 mL of 0.40 M Fe^{2+} solution, using the cerium half-reaction. Compare with Example 19.4.

Solution

This is the same as Example 19.4 where we used the iron half-reaction to calculate the potential since both $[Fe^{2+}]$ and $[Fe^{3+}]$ were known. We begin with $0.40 \times 5.0 = 2.0$ mmol Fe^{2+} and add $0.20 \times 5.0 = 1.0$ mmol Ce^{4+}. So we form 1.0 mmol each of Fe^{3+} and Ce^{3+}, leaving 1.0 mmol Fe^{2+}:

$$
\begin{array}{ccccccc}
Fe^{2+} & + & Ce^{4+} & \rightleftharpoons & Fe^{3+} & + & Ce^{3+} \\
1.0 + x & & x & & 1.0 - x & & 1.0 - x
\end{array}
$$

where the numbers and x represent millimoles. In order to use the cerium half-reaction, we need to solve for x. This can only be done using the equilibrium constant, which is obtained by equating the two half-reaction potentials. The Ce^{4+}/Ce^{3+} half-reaction is

$$Ce^{4+} + e^- \rightleftharpoons Ce^{3+}$$

$$E = 1.61 - 0.05916 \ \log \ \frac{[Ce^{3+}]}{[Ce^{4+}]}$$

Therefore,

At equilibrium, the potentials of the two half-reactions are equal.

$$1.61 - 0.05916 \ \log \ \frac{[Ce^{3+}]}{[Ce^{4+}]} = 0.771 - 0.05916 \ \log \ \frac{[Fe^{2+}]}{[Fe^{3+}]}$$

$$0.84 = 0.05916 \ \log \ \frac{[Ce^{3+}][Fe^{3+}]}{[Ce^{4+}][Fe^{2+}]} = 0.05916 \ \log \ K_{eq}$$

$$\frac{[Ce^{3+}][Fe^{3+}]}{[Ce^{4+}][Fe^{2+}]} = 10^{0.84/0.05916} = 10^{14.2} = 1.6 \times 10^{14} = K_{eq}$$

Note that the large magnitude of K_{eq} indicates the reaction lies far to the right at equilibrium. Now, since the volumes cancel, we can use millimoles instead of millimoles/milliliter (molarity) and

$$[Ce^{3+}] = 1.0 - x \approx 1.0 \ mmol$$

$$[Ce^{4+}] = x \ mmol$$

$$[Fe^{3+}] = 1.0 - x \approx 1.0 \ mmol$$

$$[Fe^{2+}] = 1.0 + x \cong 1.0 \ mmol$$

Therefore,

$$\frac{(1.0 \ mmol)(1.0 \ mmol)}{(x \ mmol)(1.0 \ mmol)} = 1.6 \times 10^{14}$$

$$x = 1.6 \times 10^{-14} \ mmol$$

In a total volume of 10 mL, in terms of concentration, this will be $1.6 \times 10^{-15} \ M$.

We see how very small $[Ce^{4+}]$ is. Nevertheless, it is finite, and knowing its concentration, we calculate the potential from the Nernst equation, using millimoles:

$$E = 1.61 - 0.05916 \ \log \ \frac{[Ce^{3+}]}{[Ce^{4+}]} = 1.61 - 0.05916 \ \log \ \frac{1.0 \ mmol}{1.6 \times 10^{-14} \ mmol}$$

$$= 0.79 \ V$$

This compares with 0.753 V calculated in Example 19.4. See also Example 19.4 for a Goal Seek based solution.

Obviously, it is easier to make the computations using the half-reaction we have the most information about; in essence, the potential of that half-reaction must be calculated anyway during the calculation using the other half-reaction. This was already discussed and demonstrated in Example 19.4. The calculations illustrate that in a mixture, the concentrations of all species *at equilibrium* are such that the potential of each half-reaction is the same. Note that the potential will be close to the standard potential (E^0) of the half-reaction in which there is an excess of the reactant; in this case, there is an excess of Fe^{2+}.

The potential will approximate E^0 of the half-reaction for which the reactant is in excess.

It should be pointed out here that the n values in the two half-reactions do not have to be equal in order to equate the Nernst equations. For convenience, the two half-reactions are generally adjusted to the same n value before the Nernst relationships are equated.

When there are stoichiometric amounts of reactants, for example, at the equivalence point of a titration, the equilibrium concentrations of the species in neither half-reaction is known, and an approach similar to the calculation in Example 21.1 or 19.4 is required.

Example 21.2

Calculate the potential of a solution at 298 K obtained by reacting 10 mL each of $0.20\ M$ Fe^{2+} and $0.20\ M$ Ce^{4+}.

Solution

The reactants are essentially quantitatively converted to equivalent quantities of Fe^{3+} and Ce^{3+} and the concentration of each of the products is $0.10\ M$ (neglecting the amount of the reverse reaction):

$$
\begin{array}{ccccc}
Fe^{2+} & + & Ce^{4+} & \rightleftharpoons & Fe^{3+} & + & Ce^{3+} \\
x & & x & & 0.10 - x & & 0.10 - x
\end{array}
$$

where x represents the molar concentration of Fe^{2+} and Ce^{4+} after the reaction reaches equilibrium. We can solve for x as in Example 21.1 and then plug it in the Nernst equation for either half-reaction to calculate the potential (do this for practice). Another approach follows.

The potential is given by either Nernst equation:

$$E = E^0_{Fe^{3+},Fe^{2+}} - \frac{0.05916}{n_{Fe}}\ \log\frac{[Fe^{2+}]}{[Fe^{3+}]};$$

$$n_{Fe}E = n_{Fe}E^0_{Fe^{3+},Fe^{2+}} - 0.05916\ \log\frac{x\ mmol/mL}{0.10\ mmol/mL}$$

$$E = E^0_{Ce^{4+},Ce^{3+}} - \frac{0.05916}{n_{Ce}}\ \log\frac{[Ce^{3+}]}{[Ce^{4+}]};$$

$$n_{Ce}E = n_{Ce}E^0_{Ce^{4+},Ce^{3+}} - 0.05916\ \log\frac{0.10\ mmol/mL}{x\ mmol/mL}$$

Note that Nernst equations for both species are written for reductions even though one of the species, here Fe^{2+}, is actually being oxidized in the reaction. We can add these equations together and solve for E, the potential of each half-reaction, and hence the potential of the solution at equilibrium:

$$n_{Fe}E + n_{Ce}E = n_{Fe}E^0_{Fe^{3+},Fe^{2+}} + n_{Ce}E^0_{Ce^{4+},Ce^{3+}}$$

$$-0.05916\ \log\frac{x\ mmol/mL}{0.10\ mmol/mL}\cdot\frac{0.10\ mmol/mL}{x\ mmol/mL}$$

$$E = \frac{n_{Fe}E^0_{Fe^{3+},Fe^{2+}} + n_{Ce}E^0_{Ce^{4+},Ce^{3+}}}{n_{Fe} + n_{Ce}} = \frac{(1)0.77 + (1)1.61}{1 + 1} = 1.19\ V$$

The above approach is general, that is, E for stoichiometric quantities of reactants (E at the equivalence point of a titration) is given by

Use this equation to calculate the equivalence point potential, *if* there is no polyatomic species or proton dependence. See Problem 13 for those cases.

$$\boxed{E = \frac{n_1 E^0_1 + n_2 E^0_2}{n_1 + n_2}} \tag{21.2}$$

where n_1 and E_1^0 are respectively the n value (number of electrons involved) and the standard potential for one half-reaction and n_2 and E_2^0 are the values for the other half-reaction. In other words, E is the weighted average of the E^0 values. In the above example, it was the simple average since the n values were each unity. *This equation holds only for reactions in which there are no polyatomic species (e.g., $Cr_2O_7^{2-}$) and no hydrogen ion dependence (or when hydrogen ion activity is unity, i.e., the pH is zero).* If pH and concentration factors are involved, these must be considered (as in Problem 13), and the equation will contain additional terms. If **formal potentials** are used, that is, the potentials applicable for the specified conditions of acidity (see Chapter 19) are used, Equation 21.2 can still be used.

21.3 Calculating Redox Titration Curves

We can use our understanding of redox equilibria to describe titration curves for redox titrations. The shape of a titration curve can be predicted from the E^0 values of the analyte half-reaction and the titrant half-reaction. Roughly, the potential change in going from one side of the equivalence point to the other will be equal to the difference in the two E^0 values; the potential will be near E^0 for the analyte half-reaction before the equivalence point and near that of the titrant half-reaction beyond the equivalence point.

The potential change at the end point will be approximately ΔE^0 for the reactant and titrant half-reactions.

Consider the titration of 100 mL of 0.1 M Fe^{2+} with 0.1 M Ce^{4+} in 1 M HNO_3. Each millimole of Ce^{4+} will oxidize one millimole of Fe^{2+}, and so the end point will occur at 100 mL. The titration curve is shown in Figure 21.1. This is actually a plot of the potential of the solution being titrated relative to the NHE; the potential of the NHE by definition is zero.

In Chapter 20, we saw that the potential difference between redox half-cells can be measured with inert electrodes such as platinum in a cell similar to that in Figure 19.1. The electrode dipped in the titration or test solution is called the **indicator electrode**, and the other is called the **reference electrode**, whose potential remains constant. Hence, the potential of the indicator electrode will change relative to that of the reference electrode as indicated in Figure 21.1. The potential relative to the NHE is plotted against the volume of titrant. This is analogous to plotting the pH of a solution

The indicator electrode monitors potential changes throughout the titration.

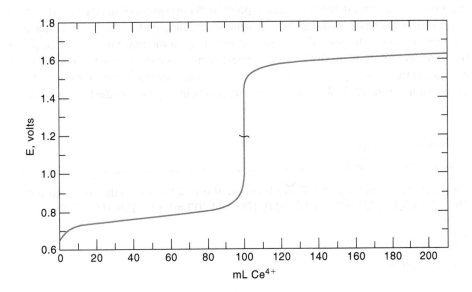

FIGURE 21.1 Titration curve for 100 mL 0.1 M Fe^{2+} versus 0.1 M Ce^{4+}.

versus the volume of titrant in an acid–base titration, or pM against volume of titrant in a precipitation or complexometric titration. In a redox titration, it is the redox potential rather than the pH that changes with concentration. At the beginning of the titration, we nominally have only a solution of Fe^{2+}, and so we cannot calculate the potential. As soon as the first drop of titrant is added, a known amount of Fe^{2+} is converted to Fe^{3+}, and we now know the ratio of $[Fe^{2+}]/[Fe^{3+}]$. So the potential can be determined from the Nernst equation of this couple. It will be near the E^0 value for this couple (the sample) *before* the end point.

Note that, since $[Fe^{2+}]/[Fe^{3+}]$ is equal to unity at the midpoint of the titration and $\log 1 = 0$, the potential is equal to E^0 at this point in the titration. This will only be true if the half-reaction is symmetrical. For example, in the half-reaction $I_2 + 2e^- \rightleftharpoons 2I^-$, the $[I^-]$ would be twice $[I_2]$ at midway in a titration, and the ratio would be $[I^-]^2/[I_2] = (2)^2/(1) = 4$. So the potential would be less than E^0 by $(-0.059/2) \log 4$, or -0.018 V.

At the equivalence point of our titration, we have the following conditions:

$$Fe^{2+} + Ce^{4+} \rightleftharpoons Fe^{3+} + Ce^{3+}$$
$$x \qquad x \qquad C-x \qquad C-x$$

where C is the concentration of Fe^{3+}, which we know since all the Fe^{2+} is converted to Fe^{3+} (x is negligible compared to C). Now, we have an unknown quantity in both half-reactions, and so we must solve for x by equating the two Nernst equations, as was done in Example 21.1. Then, we can calculate the potential from either half-reaction. Alternatively, Equation 21.2 could be used since this is a symmetrical reaction (no polyatomic species).

Beyond the equivalence point, we have an excess of Ce^{4+} and an unknown amount of Fe^{2+}. Since we now know the concentrations of both Ce^{3+} and Ce^{4+} in the Ce^{4+}/Ce^{3+} half-reaction, it is easier to calculate the potential from its Nernst equation. Note that here, with an excess of titrant, the potential is near the E^0 value of the *titrant*. At 200% of the titration, $[Ce^{4+}]/[Ce^{3+}] = 1$, and E is E^0 of the cerium couple.

Example 21.3 illustrates that the magnitude of the end-point break is directly related to the difference in the E^0 values of the sample and the titrant half-reactions. At least a 0.2-V difference is required for a sharp end point.

The equivalence point for this titration is indicated in Figure 21.1. Because the reaction is symmetrical, the equivalence point (*inflection point* of the curve—that point at which it is steepest) occurs at the midpoint of the rising part of the curve. In nonsymmetrical titrations, the inflection point will not occur at the midpoint. For example, in the titration of Fe^{2+} with MnO_4^-, the steepest portion occurs near the top of the break because of the consumption of protons in the reaction, causing it to be nonsymmetrical, as shown in Example 21.4 below, and the titration curve given on the text website.

> A potential change of 0.2 V is minimally needed for a sharp end point.

> Unlike the case for an acid-base titration where the initial pH is easily calculated, in a redox titration, the initial potential often cannot be computed because we know the concentration of only one form of the redox couple.

Example 21.3

Calculate the potential at 298 K as a function of titrant volume in the above titration of 100 mL of 0.100 M Fe^{2+} at 10.0, 50.0, 100, and 200 mL of 0.100 M Ce^{4+}.

Solution

The reaction is

$$Fe^{2+} + Ce^{4+} \rightleftharpoons Fe^{3+} + Ce^{3+}$$

10.0 mL : mmol Ce^{4+} added = 0.100 $M \times$ 10.0 mL = 1.00 mmol

10.0 mL : mmol Ce^{4+} added = 0.100 $M \times$ 10.0 mL = 1.00 mmol

mmol Fe^{2+} reacted = 1.00 mmol = mmol Fe^{3+} formed

mmol Fe^{2+} left = 0.100 $M \times$ 100 mL − 1.00 mmol = 9.0 mmol Fe^{2+}

$$E = 0.771 - 0.05916 \, \log \frac{9.0}{1.00} = 0.715 \text{ V}$$

50.0 mL: One-half the Fe^{2+} is converted to Fe^{3+} (5.00 mmol each)

$$E = 0.771 - 0.05916 \, \log \frac{5.00}{5.00} = 0.771 \text{ V}$$

100 mL: mmol Fe^{3+} = 10.0 − $x \approx$ 10.0

mmol Fe^{2+} = x

mmol Ce^{3+} = 10.0 − $x \approx$ 10.0

mmol Ce^{4+} = x

We must solve for x. Since our calculations are for when equilibrium is achieved between the two half-reactions, the two Nernst equations are equal:

$$0.771 - \frac{0.05916}{1} \, \log \frac{[Fe^{2+}]}{[Fe^{3+}]} = 1.61 - \frac{0.05916}{1} \, \log \frac{[Ce^{3+}]}{[Ce^{4+}]}$$

$$-0.84 = -0.05916 \, \log \frac{[Fe^{3+}][Ce^{3+}]}{[Fe^{2+}][Ce^{4+}]} = -0.05916 \, \log K_{eq}$$

$$K_{eq} = 1.7 \times 10^{14}$$

Substituting into K_{eq} and solving for x (use millimoles, since volumes cancel):

$$\frac{(10.0)(10.00)}{(x)(x)} = 1.7 \times 10^{14}$$

$$x = 7.7 \times 10^{-7} \text{ mmol } Fe^{2+} = \text{mmol } Ce^{4+}$$

Use either half-reaction to calculate the potential:

$$E = 0.771 - 0.05916 \, \log \frac{7.7 \times 10^{-7}}{10.0} = 1.19 \text{ V}$$

Compare this with the potential calculated in Example 21.2. Try calculating this potential using the Ce^{4+}/Ce^{3+} Nernst equation. Note that this potential is halfway between the two E^0 potentials.

200 mL: We have 100 mL excess titrant (Ce^{4+}). It is now easier to use the Ce^{4+}/Ce^{3+} half-reaction:

mmol Ce^{3+} = 10.0 − $x \approx$ 10.0

mmol Ce^{4+} = 0.100 $M \times$ 100 mL + $x \approx$ 10.0 mmol

$$E = 1.61 - 0.05916 \, \log \frac{10.0}{10.0} = 1.61 \text{ V}$$

[We could have used the Fe^{3+}/Fe^{2+} half-reaction to calculate this potential by calculating x ($[Fe^{2+}]$) as above from K_{eq}.]

Figure 21.1 can also be generated on the basis of an explicit equation. Consider that we are beginning with $V_{Fe,in}$ L of a $C_{Fe(II)}$ M solution of Fe^{2+} and we titrate this with a Ce^{4+} solution of concentration $C_{Ce(IV)}$ M, V_{Ce} L being added at any point. For convenience, we designate the total volume at any point as V_T L ($= V_{Fe,in} + V_{Ce}$). The total moles of Fe and Ce (irrespective of oxidation states) in the solution are respectively $V_{Fe,in} \times C_{Fe(II)}$ and $V_{Ce} \times C_{Ce(IV)}$. At any time, the amount of Ce^{3+} and Fe^{3+} formed are the same and let this be x moles.

The equilibrium constant as defined in Problems 21.1 through 21.3 can be expressed as

$$\frac{[Fe^{3+}][Ce^{3+}]}{[Fe^{2+}][Ce^{4+}]} = K_{eq} = 1.6 \times 10^{14} = \frac{x^2}{(V_{Fe,in}C_{Fe(II)} - x)\,(V_{Ce}C_{Ce\,(IV)} - x)}$$

Because all of the amounts in moles are divided by the same volume V_T, this cancels out. The above represents a quadratic equation in x, where using the usual notations, $a = (K_{eq} - 1)$, $b = -K_{eq}(V_{Fe,in}C_{Fe(II)} + V_{Ce}C_{Ce(IV)})$, $c = K_{eq}V_{Fe,in}C_{Fe(II)}V_{Ce}C_{Ce(IV)}$. Example 21.3.xlsx in the web supplement uses this single equation to calculate x (Ce^{3+} or Fe^{3+}) throughout. As in many other cases, the negative root of the quadratic equation provides the correct solution. The potential can then be calculated either from

$$E_{Ce} = 1.61 - 0.05916 \, \log \frac{x}{(V_{Ce}C_{Ce(IV)} - x)}$$

or

$$E_{Fe} = 1.61 - 0.05916 \, \log \frac{(V_{Fe,in}C_{Fe(II)} - x)}{x}$$

Both should lead to the same potential. Note that unlike acid–base, EDTA or precipitation titrations, the initial value of the measurand (e.g., pH, pM, or in this case cell potential) is undefined and cannot be measured with no titrant added. Even in calculations, when the equilibrium constant is high as in the present case, reliable results for the residual Ce^{4+} are not produced until several (in this case 15) mL of Ce^{4+} is added. See the plot in the Excel spreadsheet on the text **website**, both equations above provide the same potential.

For nonsymmetrical reactions, we must keep track of the ratio in which the chemicals react. Also, if protons are consumed or produced in the reaction, the change in $[H^+]$ must be calculated. Review balancing redox reactions on the text **website** for determining the numbers of protons involved in a reaction.

Example 21.4

Calculate the potential at the equivalence point in the titration of 100 mL of 0.100 M Fe^{2+} in 0.500 M H_2SO_4 with 100 mL of 0.0200 M $MnO_4{}^-$.

Solution

The reaction is

$$5Fe^{2+} + MnO_4{}^- + 8H^+ \rightleftharpoons 5Fe^{3+} + Mn^{2+} + 4H_2O$$

$$x \qquad \tfrac{1}{5}x \qquad\qquad C - x \quad \tfrac{1}{5}C - \tfrac{1}{5}x$$

Keep track of millimoles and the ratio in which things react. One millimole Fe^{2+} reacts with $\frac{1}{5}$ mmol $MnO_4{}^-$.

$$\text{mmol Fe}^{3+} = 0.100 \, M \times 100 \text{ mL} - x \approx 10.0$$
$$\text{mmol Fe}^{2+} = x$$
$$\text{mmol Mn}^{2+} = \tfrac{1}{5}(10.0) - \tfrac{1}{5}x \approx 2.00$$
$$\text{mmol MnO}_4^- = \tfrac{1}{5}x$$

Solve for x by equating the two Nernst equations (since we are at equilibrium, they are equal). Multiply the Fe^{2+}/Fe^{3+} half-reaction by 5 to equate the electrons:

$$0.771 - \frac{0.05916}{5} \log \frac{[Fe^{2+}]^5}{[Fe^{3+}]^5} = 1.51 - \frac{0.05916}{5} \log \frac{[Mn^{2+}]}{[MnO_4^-][H^+]^8}$$

$$-0.74 = \frac{0.05916}{5} \log \frac{[Mn^{2+}][Fe^{3+}]^5}{[MnO_4^-][Fe^{2+}]^5[H^+]^8} = -\frac{0.05916}{5} \log K_{eq}$$

$$K_{eq} = 5_{.0} \times 10^{62}$$

We started with $0.50 \, M \times 0.10 \text{ L} = 0.05 \text{ mol H}_2SO_4$. We consumed 8 mol H^+ or 4 mol H_2SO_4 per 5 mol Fe^{2+}. For 0.01 mol Fe^{2+} thus $(8 \text{ mol H}^+/5 \text{ mol Fe}^{2+}) \times 0.01 \text{ mol Fe}^{2+} = 0.016 \text{ mole H}^+$ or 0.008 mol H_2SO_4 was consumed, leaving 0.042 mol H_2SO_4. As the total volume is now 0.2 L, the H_2SO_4 concentration is 0.21 M. Although H_2SO_4 is a diprotic acid, the second dissociation constant is weak enough such that the second dissociation step does not proceed to a significant degree at these concentrations. We can thus assume $[H^+] = 0.21 \, M$. For the other species, we can use moles since the volumes all cancel:

> We must calculate the concentration of H^+ after reaction.

$$\frac{(0.002)(0.01)^5}{(1/5x)(x)^5(0.21)^8} = 5_{.0} \times 10^{62}$$

$$x = 2.8 \times 10^{-12} \text{ mol Fe}^{2+}; \quad \text{mol MnO}_4^- = \frac{1}{5}(2.8 \times 10^{-12})$$
$$= 5.6 \times 10^{-13}$$

Use either half-reaction to calculate the potential:

$$E = 0.771 - 0.05916 \log \frac{2.8 \times 10^{-12}}{0.01} = 1.34 \text{ V}$$

E^0 for the Mn^{2+}/MnO_4^- couple is 1.51 V. Note that the potential halfway between the two E^0 potentials is 1.14 V. The equivalence point (inflection point) for this unsymmetrical titration reaction is therefore closer to the titrant couple and the titration curve is asymmetric.

Unlike the case for the titration of Fe^{2+} with Ce^{4+}, the expressions involved in any reaction that involves the transfer of more than one electron are generally too complicated to solve with a single explicit equation through the whole titration. The present case of the titration of Fe^{2+} by acidic MnO_4^- where the Fe^{2+}, MnO_4^-, and H^+ are all consumed, and in different molar ratios, is no exception. Nevertheless, it is possible to numerically simulate the titration in three separate segments. The high equilibrium constant of the reaction makes it appropriate to assume that prior to the equivalence point, MnO_4^- stoichiometrically oxidizes Fe^{2+} and the cell potential can be determined by following the Fe^{2+}/Fe^{3+} couple. At the equivalence point, the potential is calculated as worked out above, and after the equivalence point, the potential is calculated from the $Mn^{2+}/MnO_4^- - H^+$ couple. Example 21.4.xlsx on the text **website** illustrates these calculations and the resulting redox titration plot.

21.4 Visual Detection of the End Point

Obviously, the end point can be determined by measuring potential with an indicating electrode (Chapter 20) relative to a reference electrode and plotting this against the volume of titrant. But as in other titrations, it is usually more convenient to use an indicator that can be observed with the naked eye. There are three methods used for visual indication.

Self-Indication

If the titrant is highly colored, this color may be used to detect the end point. For example, a 0.02 M solution of potassium permanganate is deep purple. A dilute solution of potassium permanganate is pink. The product of its reduction, Mn^{2+}, is extremely faint pink, nearly colorless. During a titration with potassium permanganate, the purple color of the MnO_4^- is removed as soon as it is added because it is reduced to Mn^{2+}. As soon as the titration is complete, a fraction of a drop of excess MnO_4^- solution imparts a definite pink color to the solution, indicating that the reaction is complete. Obviously, the end point does not occur at the equivalence point, but at a fraction of a drop beyond. The titration error is small and can be corrected for by running a blank titration, or it is accounted for in standardization.

Starch Indicator

This indicator is used for titrations involving iodine. Starch forms a complex with I_2 that is a dark-blue color. The color reaction is sensitive to very small amounts of iodine and is not very reversible. In titrations of reducing agents with iodine, the solution remains colorless up to the equivalence point. A fraction of a drop of excess titrant turns the solution a definite blue.

Redox Indicators

Compare redox indicators with acid–base indicators. Here, the potential determines the ratio of the two colors, rather than the pH.

The above two methods of end-point indication do not depend on the half-reaction potentials, although the completeness of the titration reaction and hence the sharpness of the end point do. Examples such as these first two methods of visual indication are few, and most other types of redox titrations are detected using **redox indicators**. These are highly colored dyes that are weak reducing or oxidizing agents that can be oxidized or reduced; the colors of the oxidized and reduced forms are different. The oxidation state of the indicator and hence its color will depend on the potential at a given point in the titration. A half-reaction and Nernst equation can be written for the indicator:

$$Ox_{ind} + ne^- \rightleftharpoons Red_{ind} \quad (21.3)$$

$$E_{ind} = E_{ind}^0 - \frac{0.05916}{n} \log \frac{[Red_{ind}]}{[Ox_{ind}]} \quad (21.4)$$

E_{In}^0 must be near the equivalence point potential. A potential change of at least 120 mV is needed for a color change for $n = 1$ (of the indicator half-reaction) and 60 mV for $n = 2$.

The half-reaction potentials during the titration determine E_{In} and hence the ratio of $[Red_{ind}]/[Ox_{ind}]$. This is analogous to the ratio of the different forms of a pH indicator being determined by the pH of the solution. So the ratio, and therefore the color, will change as the potential changes during the titration. If we assume, as with acid–base indicators, that the ratio must change from 10/1 to 1/10 in order that a sharp color change can be seen, then *a potential equal to 2 × (0.05916/n) V is required*. If n for the

TABLE 21.1 Redox Indicators

| | Color | | | |
Indicator	Reduced Form	Oxidized Form	Solution	E^0 (V)
Nitroferroin	Red	Pale blue	1 M H_2SO_4	1.25
Ferroin	Red	Pale blue	1 M H_2SO_4	1.06
Diphenylaminesulfonic acid	Colorless	Purple	Diluted acid	0.84
Diphenylamine	Colorless	Violet	1 M H_2SO_4	0.76
Methylene blue	Blue	Colorless	1 M acid	0.53
Indigo tetrasulfonate	Colorless	Blue	1 M acid	0.36

indicator is equal to 1, then a 0.12 V change is required. If E^0_{In} *is near the equivalence point potential* of the titration, where there is a rapid change in potential in excess of 0.12 V, then the color change occurs at the equivalence point. Again, this is analogous to the requirement that the pK_a value of an acid–base indicator be near the pH of the equivalence point.

If there is a hydrogen ion dependence in the indicator reaction, an appropriate hydrogen ion term will appear in the corresponding Nernst equation, and the potential at which the indicator changes color will depend on the hydrogen ion concentration.

Redox indicators for appropriate indication of the end point have a transition range over a certain potential, and this transition range must fall within the steep equivalence point break of the titration curve. The redox indicator reaction must be both *rapid*, and it must be *reversible*. If the reaction is slow or is *irreversible* (slow rate of electron transfer), the color change will be gradual and the end point will not be sharp.

While there are many useful acid-base indicators, good redox indicators are few. Table 21.1 lists some of the common indicators arranged in order of decreasing standard potentials. Ferroin [tris(1,10-phenanthroline)iron(II) sulfate] is one of the best indicators. It is useful for many titrations with cerium(IV). It is oxidized from red to a pale blue color at the equivalence point. Other useful indicators are listed in Table 21.1. Diphenylaminesulfonic acid is used as an indicator for titrations with dichromate in acid solution. The potential of the $Cr_2O_7^{2-}/Cr^{3+}$ couple is lower than that of the cerium couple, and so with this oxidant an indicator with a lower E^0 is required. The color at the end point is purple. However, dichromate titrations are no longer commonly used; because of the carcinogenicity of Cr(VI), special disposal precautions are needed. The choice of the indicator may depend on the sample titrated since the magnitude of the end-point break is also dependent on the potential of the sample half-reaction.

21.5 Titrations Involving Iodine: Iodimetry and Iodometry

Redox titrations are important in many application areas, for example, in food, pharmaceutical, and general industrial analyses. Titration of sulfite in wine using iodine is a common example. Alcohol can be determined based on its oxidation by potassium dichromate. Examples in clinical analysis are rare since most analyses involve trace determinations, but these titrations are still extremely useful for standardizing reagents. You should be familiar with some of the more commonly used titrants.

Iodine is an oxidizing agent that can be used to titrate fairly strong reducing agents. On the other hand, iodide ion is a mild reducing agent and serves as the basis for determining strong oxidizing agents.

Iodimetry

In iodimetry, the titrant is I_2 and the analyte is a reducing agent. The end point is detected by the appearance of the blue starch–iodine color.

Iodine is a moderately strong oxidizing agent and can be used to titrate reducing agents. Titrations with I_2 are called **iodimetric methods**. These titrations are usually performed in neutral or mildly alkaline (pH 8) to weakly acidic solutions. If the pH is too alkaline. I_2 will disproportionate to hypoiodate and iodide:

$$I_2 + 2OH^- \rightleftharpoons IO^- + I^- + H_2O \qquad (21.5)$$

There are three reasons for keeping the solution from becoming strongly acidic. First, the starch used for the end-point detection tends to hydrolyze or decompose in strong acid, and so the end point may be affected. Second, the reducing power of several reducing agents is increased in neutral solution. For example, consider the reaction of I_2 with As(III):

$$H_3AsO_3 + I_2 + H_2O \rightleftharpoons H_3AsO_4 + 2I^- + 2H^+ \qquad (21.6)$$

This equilibrium is affected by the hydrogen ion concentration. At low hydrogen ion concentration, the equilibrium is shifted to the right. We have already seen in Equation 19.25 that in neutral solution the potential of the As(V)/As(III) couple is decreased sufficiently that arsenic(III) will reduce I_2. But in acidic solution, the equilibrium is shifted the other way, and the reverse reaction occurs. The third reason for avoiding acid solutions is that the I^- produced in the reaction tends to be oxidized by dissolved oxygen in acidic solution:

$$4I^- + O_2 + 4H^+ \rightarrow 2I_2 + 2H_2O \qquad (21.7)$$

The pH for the titration of arsenic(III) with I_2 can be maintained neutral by adding $NaHCO_3$. The bubbling action of the CO_2 formed also helps remove the dissolved oxygen and maintains a blanket of CO_2 over the solution to prevent air oxidation of the I^-.

Iodine is a more selective oxidizing titrant than stronger oxidants like cerium (IV), permanganate, or dichromate.

Because I_2 is not a strong oxidizing agent, the number of reducing agents that can be titrated is limited. Nevertheless, several important determinations are possible, and the moderate oxidizing power of I_2 makes it a more selective titrant than other, stronger oxidants. Some commonly determined substances are listed in Table 21.2. Antimony behaves similarly to arsenic, and the pH is critical for the same reasons. Tartrate is added to complex the antimony and keep it in solution to prevent hydrolysis.

Although high-purity I_2 can be obtained by sublimation, iodine solutions are usually standardized against a primary standard reducing agent such as As_2O_3 (As_4O_6). Arsenious oxide is dissolved in dilute HCl or NaOH. The solution is neutralized after dissolution is complete. If arsenic(III) solutions are to be kept for any length of time, they should be neutralized or acidified because arsenic(III) is slowly oxidized in alkaline solution.

TABLE 21.2 Some Substances Determined by Iodimetry

Substance Determined	Reaction with Iodine	Solution Conditions
H_2S	$H_2S + I_2 \rightarrow S + 2I^- + 2H^+$	Acid solution
SO_3^{2-}	$SO_3^{2-} + I_2 + H_2O \rightarrow SO_4^{2-} + 2I^- + 2H^+$	
Sn^{2+}	$Sn^{2+} + I_2 \rightarrow Sn^{4+} + 2I^-$	Acid solution
As(III	$H_2AsO_3^- + I_2 + H_2O \rightarrow$ $HAsO_4^{2-} + 2I^- + 3H^+$	pH 8
N_2H_4	$N_2H_4 + 2I_2 \rightarrow N_2 + 4H^+ + 4I^-$	

Iodine has a low solubility in water but the complex I_3^- is very soluble. So iodine solutions are prepared by dissolving I_2 in a solution of potassium iodide where iodide is present in a large excess:

$$I_2 + I^- \rightarrow I_3^- \tag{21.8}$$

Therefore, I_3^- is the actual species used in the titration.

Example 21.5

The purity of a hydrazine (N_2H_4) sample is determined by titration with iodine. A sample of the oily liquid weighing 1.4286 g is dissolved in water and diluted to 1 L in a volumetric flask. A 50.00-mL aliquot is taken with a pipet and titrated with standard iodine solution, requiring 42.41 mL. The iodine was standardized against 0.4123 g primary standard As_2O_3 by dissolving the As_2O_3 in a small amount of NaOH solution, adjusting the pH to 8, and titrating, requiring 40.28 mL iodine solution. What is the percent purity by weight of the hydrazine?

Solution

Standardization:

$$H_2AsO_3^- + I_2 + H_2O \rightarrow HAsO_4^{2-} + 2I^- + 3H^+$$

Each As_2O_3 gives $2H_2AsO_3^-$, so mmol $I_2 = 2 \times$ mmol As_2O_3.

$$M_{I_2} \times 40.28 \text{ mL } I_2 = \frac{412.3 \text{ mg As}_2\text{O}_3}{197.85 \text{ mg As}_2\text{O}_3/\text{mmol}} \times 2 \text{ mmol } I_2/\text{mmol As}_2\text{O}_3$$

$$M_{I_2} = 0.1034_7 \text{ mmol/mL}$$

With molarity, keep track of millimoles and the ratios in which things react.

Analysis:

$$N_2H_4 + 2I_2 \rightarrow N_2 + 4H^+ + 4I^-$$

$$\text{mmol } N_2H_4 = \frac{1}{2} \times \text{mmol } I_2$$

$$\text{weight of } N_2H_4 \text{ titrated} = 1.4286 \text{ g} \times \frac{50.00 \text{ mL}}{1000.0 \text{ mL/L}} = 0.07143 \text{ g/L}$$
$$= 71.43 \text{ mg sample}$$

$$\text{weight } N_2H_4 \text{ present} = [0.1034_7 \, M \, I_2 \times 42.41 \text{ mL } I_2 \times \frac{1}{2} \, (\text{mmol } N_2H_4/\text{mmol } I_2)$$
$$\times 32.045 \text{ mg } N_2H_4/\text{mmol}] = 70.31 \text{ mg}$$

$$\% \text{ Purity} = 70.31 \text{mg}/71.43 \text{ mg} \times 100\% = 98.43\%$$

Note that because of the low molecular weight of hydrazine, it would have been difficult to weigh out the required sample to four significant figures, and by titrating an accurately measured aliquot, a larger sample could be weighed.

Iodometry

Iodide ion is a weak reducing agent and will reduce strong oxidizing agents. It is not used, however, as a titrant mainly because of the lack of a convenient visual indicator system, as well as other factors such as speed of the reaction.

When an excess of iodide is added to a solution of an oxidizing agent, I_2 *is produced in an amount equivalent to the oxidizing agent present.* This I_2 can, therefore, be

In iodometry, the analyte is an oxidizing agent that reacts with I^- to form I_2. The I_2 is titrated with thiosulfate, using disappearance of the starch–iodine color for the end point. In iodometry, we titrate liberated iodine, in iodimetry, we titrate with iodine.

titrated with a reducing agent, and the result will be the same as if the oxidizing agent were titrated directly. The titrating agent used is sodium thiosulfate.

Analysis of an oxidizing agent in this way is called an **iodometric method**. Consider, for example, the determination of dichromate:

$$Cr_2O_7^{2-} + 6I^- \text{ (excess)} + 14H^+ \rightarrow 2Cr^{3+} + 3I_2 + 7H_2O \tag{21.9}$$

$$\boxed{I_2 + 2S_2O_3^{2-} \rightarrow 2I^- + S_4O_6^{2-}} \tag{21.10}$$

The millimoles thiosulfate per millimole analyte is needed for calculations. 2 mmol thiosulfate is needed for each mmol I_2 produced.

Each $Cr_2O_7^{2-}$ produces $3I_2$, which in turn react with $6S_2O_3^{2-}$. The millimoles of $Cr_2O_7^{2-}$ are equal to one-sixth the millimoles of $S_2O_3^{2-}$ used in the titration.

Iodate can be determined iodometrically:

$$IO_3^- + 5I^- + 6H^+ \rightarrow 3I_2 + 3H_2O \tag{21.11}$$

Each IO_3^- produces $3I_2$, which again react with $6S_2O_3^{2-}$, and the millimoles of IO_3^- are obtained by dividing the millimoles of $S_2O_3^{2-}$ used in the titration by 6.

Example 21.6

A 0.200-g sample containing copper is analyzed iodometrically. Copper(II) is reduced to copper(I) by iodide; CuI precipitates:

$$2Cu^{2+} + 4I^- \rightarrow \underline{2CuI} + I_2$$

What is the percent copper in the sample if 20.0 mL of 0.100 M $Na_2S_2O_3$ is required for titration of the liberated I_2?

Solution

One-half mole of I_2 is liberated per mole of Cu^{2+}, and since each I_2 reacts with $2S_2O_3^{2-}$, each Cu^{2+} is equivalent to one $S_2O_3^{2-}$, and mmol Cu^{2+} = mmol $S_2O_3^{2-}$.

$$\%Cu = \frac{0.100 \text{ mmol } S_2O_3^{2-}/mL \times 20.0 \text{ mL } S_2O_3^{2-} \times Cu}{200 \text{ mg sample}} \times 100\%$$

$$= \frac{0.100 \text{ mmol/mL} \times 20.0 \text{ mL} \times 63.54 \text{ mg Cu/mmol}}{200 \text{ mg sample}} \times 100\% = 63.5\%$$

Note that we are not adding a known excess of a reagent and back titrating; rather, we are converting a stronger oxidant, which cannot be directly titrated with thiosulfate, to an equivalent amount of iodine that can be titrated with thiosulfate with a visual indicator. It is therefore classified as a direct titration.

Why not titrate the oxidizing agents directly with the thiosulfate? Because strong oxidizing agents oxidize thiosulfate to oxidation states higher than that of tetrathionate (e.g., to SO_4^{2-}), but the reaction is generally not stoichiometric. Also, several oxidizing agents form mixed complexes with thiosulfate (e.g., Fe^{3+}). By reaction with iodide, the strong oxidizing agent is destroyed and an equivalent amount of I_2 is produced, which will react stoichiometrically with thiosulfate and for which a satisfactory indicator exists. The titration can be considered a direct titration.

The end point for iodometric titrations is detected with starch. The disappearance of the blue starch $-I_2$ color indicates the end of the titration. The starch is not added at the beginning of the titration when the iodine concentration is high. Instead, it is added just before the end point when the dilute iodine color becomes pale yellow. There are two reasons for such timing. One is that the iodine–starch complex is only slowly dissociated, and a diffuse end point would result if a large amount of the iodine were adsorbed on the starch. The second reason is that most iodometric titrations are performed in strongly acid medium and the starch has a tendency to hydrolyze in acid solution. The reason for using acid solutions is that reactions between many oxidizing agents and iodide are promoted by high acidity. Thus,

$$2MnO_4^- + 10I^- + 16H^+ \rightarrow 5I_2 + 2Mn^{2+} + 8H_2O \tag{21.12}$$

$$H_2O_2 + 2I^- + 2H^+ \rightarrow I_2 + 2H_2O \tag{21.13}$$

as examples.

The starch is added near the end point.

The titration should be performed rapidly to minimize air oxidation of the iodide. Stirring should be efficient to prevent local excesses of thiosulfate because it is decomposed in acid solution:

$$S_2O_3^{2-} + 2H^+ \rightarrow H_2SO_3 + S \tag{21.14}$$

Indications of such excess is the presence of colloidal sulfur, which makes the solution cloudy. In iodometric methods, a large excess of iodide is added to promote the reaction (common ion effect). The unreacted iodide does not interfere, but it may be air-oxidized if the titration is not performed immediately or too much time is taken for the titration.

Sodium thiosulfate solution is standardized iodometrically against a pure oxidizing agent such as $K_2Cr_2O_7$, KIO_3, $KBrO_3$, or metallic copper (dissolved to give Cu^{2+}). With potassium dichromate, the deep green color of the resulting chromic ion makes it a little more difficult to determine the iodine–starch end point. When copper(II) is titrated iodometrically, the end point is diffuse unless thiocyanate ion is added. The primary reaction is given in Example 21.6. But iodine is adsorbed on the surface of the cuprous iodide precipitate and only slowly reacts with the thiosulfate titrant. The thiocyanate coats the precipitate with CuSCN and displaces the iodine from the surface. The potassium thiocyanate should be added near the end point since it is slowly oxidized by iodine to sulfate. The pH must be buffered to around 3. If it is too high, copper(II) hydrolyzes and cupric hydroxide will precipitate. If it is too low, air oxidation of iodide becomes appreciable because it is catalyzed in the presence of copper. Copper metal is dissolved in nitric acid, with oxides of nitrogen being produced. These oxides will oxidize iodide, and they are removed by addition of urea. Some examples of iodometric determinations are listed in Table 21.3.

Example 21.7

A solution of $Na_2S_2O_3$ is standardized iodometrically against 0.1262 g of high-purity $KBrO_3$, requiring 44.97 mL $Na_2S_2O_3$. What is the molarity of the $Na_2S_2O_3$?

Solution

The reactions are

$$BrO_3^- + 6I^- + 6H^+ \rightarrow Br^- + 3I_2 + 3H_2O$$
$$3I_2 + 6S_2O_3^{2-} \rightarrow 6I^- + 3S_4O_6^{2-}$$

So mmol $S_2O_3^{2-} = 6 \times$ mmol BrO_3^-:

$$M_{S_2O_3^{2-}} \times 44.97 \text{ mL} = \frac{126.2 \text{ mg KBrO}_3}{167.01 \text{ (mg/mmol KBrO}_3)} \times 6 \left(\text{mmol } S_2O_3^{2-}/\text{mmol BrO}_3^-\right)$$

$$M_{S_2O_3^{2-}} = 0.1008_2 \text{ mmol/mL}$$

TABLE 21.3 Iodometric Determinations

Substance Determined	Reaction with Iodine
MnO_4^-	$2MnO_4^- + 10I^- + 16H^+ \rightleftharpoons 2Mn^{2+} + 5I_2 + 8H_2O$
$Cr_2O_7^{2-}$	$Cr_2O_7^{2-} + 6I^- + 14H^+ \rightleftharpoons 2Cr^{3+} + 3I_2 + 7H_2O$
IO_3^-	$IO_3^- + 5I^- + 6H^+ \rightleftharpoons 3I_2 + 3H_2O$
BrO_3^-	$BrO_3^- + 6I^- + 6H^+ \rightleftharpoons Br^- + 3I_2 + 3H_2O$
Ce^{4+}	$2Ce^{4+} + 2I^- \rightleftharpoons 2Ce^{3+} + I_2$
Fe^{3+}	$2Fe^{3+} + 2I^- \rightleftharpoons 2Fe^{2+} + I_2$
H_2O_2	$H_2O_2 + 2I^- + 2H^+ \underset{\text{[Mo(VI) catalyst]}}{\rightleftharpoons} 2H_2O + I_2$
As(V)	$H_3AsO_4 + 2I^- + 2H^+ \rightleftharpoons H_3AsO_3 + I_2 + H_2O$
Cu^{2+}	$2Cu^{2+} + 4I^- \rightleftharpoons \underline{2CuI} + I_2$
HNO_2	$2HNO_2 + 2I^- \rightleftharpoons I_2 + 2NO + H_2O$
SeO_3^{2-}	$SeO_3^{2-} + 4I^- + 6H^+ \rightleftharpoons \underline{Se} + 2I_2 + 3H_2O$
O_3	$O_3 + 2I^- + 2H^+ \rightleftharpoons O_2 + I_2 + H_2O$ (can determine in presence of O_2 at pH 7–8:5)
Cl_2	$Cl_2 + 2I^- \rightleftharpoons 2Cl^- + I_2$
Br_2	$Br_2 + 2I^- \rightleftharpoons 2Br^- + I_2$
HClO	$HClO + 2I^- + H^+ \rightleftharpoons Cl^- + I_2 + H_2O$

21.6 Titrations with Other Oxidizing Agents

We have already mentioned some oxidizing agents that can be used as titrants. The titrant should be fairly stable and should be convenient to prepare and to handle. If the oxidizing agent is too strong, it is not likely to be stable as a reagent. Thus, fluorine is one of the strongest oxidizing agents known, but it is certainly not convenient to use in the analytical laboratory ($E^0 = 3.06$ V) as a reagent. Chlorine would make a good titrant, except that it is volatile from aqueous solution, and to prepare and maintain a standard solution would be difficult.

Potassium permanganate is a widely used oxidizing titrant. It acts as a self-indicator for end-point detection and is a very strong oxidizing agent ($E^0 = 1.51$ V). The solution is stable if precautions are taken in its preparation. When the solution is first prepared, small amounts of reducing impurities in the solution reduce a small amount of the MnO_4^-. In neutral solution, the reduction product of this permanganate is MnO_2, rather than Mn^{2+} produced in acid medium. The MnO_2 acts as a catalyst for further decomposition of the permanganate, which produces more MnO_2, and so on. This is called **autocatalytic decomposition**. The solution can be stabilized by removing the MnO_2. So, before standardizing, the solution is boiled to hasten oxidation of all impurities and is allowed to stand overnight. The MnO_2 is then removed by filtering through a sintered-glass filter. Potassium permanganate can be standardized by titrating primary standard sodium oxalate, $Na_2C_2O_4$, which, dissolved in acid, forms oxalic acid:

$$5H_2C_2O_4 + 2MnO_4^- + 6H^+ \rightleftharpoons 10CO_2 + 2Mn^{2+} + 8H_2O \qquad (21.15)$$

The solution must be heated for rapid reaction. The reaction is catalyzed by the Mn^{2+} product and it goes very slowly at first until some Mn^{2+} is formed. Pure electrolytic iron metal can also be used as the primary standard. It is dissolved in acid and reduced to Fe^{2+} for titration (see Section 21.8).

A difficulty arises when permanganate titrations of iron(II) are performed in the presence of chloride ion. The oxidation of chloride ion to chlorine by permanganate at room temperature is normally slow. However, the oxidation is catalyzed by the presence of iron. If an iron sample has been dissolved in hydrochloric acid, or if stannous chloride has been used to reduce it to iron(II) (see below), the titration can be performed by adding the **Zimmermann–Reinhardt reagent**. This contains manganese(II) and phosphoric acid. The manganese(II) reduces the potential of the MnO_4^-/Mn^{2+} couple sufficiently so that permanganate will not oxidize chloride ion; the formal potential is less than E^0, due to the large concentration of Mn^{2+}. This decrease in the potential decreases the magnitude of the end-point break. Therefore, phosphoric acid is added to complex the iron(III) and decrease the potential of the Fe^{3+}/Fe^{2+} couple also; the iron(II) is not complexed. In other words, iron(III) is removed from the solution as it is formed to shift the equilibrium of the titration reaction to the right and give a sharp end point. The overall effect is still a large potential break in the titration curve, but the entire curve has been shifted to a lower potential.

The Z–R reagent prevents oxidation of Cl^- by MnO_4^- and sharpens the end point.

An added effect of complexing the iron(III) is that the phosphate complex is nearly colorless, while the chloro complex (normally present in chloride medium) is deep yellow. A sharper end-point color change results.

Potassium dichromate, $K_2Cr_2O_7$, is a slightly weaker oxidizing agent than potassium permanganate. The great advantage of this reagent is its availability as a primary standard, and thus, generally the solution need not be standardized. However, in the titration of iron(II), standardizing potassium dichromate against electrolytic iron is preferable for most accurate results since the green color of the chromic ion can introduce a small error in the end point (diphenylamine sulfonate indicator), and the standardization corrects for this. Admittedly, this is necessary only for the most accurate work.

Oxidation of chloride ion is not a problem with dichromate. However, the formal potential of the $Cr_2O_7^{2-}/Cr^{3+}$ couple is reduced from 1.33 to 1.00 V in 1 M hydrochloric acid, and phosphoric acid must be added to reduce the potential of the Fe^{3+}/Fe^{2+} couple. Such addition is also necessary because it decreases the equivalence point potential to near the standard potential for the diphenylamine sulfonate indicator (0.84 V). Otherwise, the end point would occur too soon.

Cerium(IV) is a powerful oxidizing agent. Its formal potential depends on the acid used to keep it in solution (it must be kept acidic; it hydrolyzes to form ceric hydroxide otherwise). Titrations are usually performed in sulfuric acid or perchloric acid. In H_2SO_4, the formal potential is 1.44 V, and in $HClO_4$, it is 1.70 V. So cerium(IV) is a stronger oxidizing agent in $HClO_4$. Cerium(IV) can be used for most titrations in which permanganate is used, and it possesses a number of advantages. It is a very strong oxidizing agent and its potential can be varied by choice of the acid used. The rate of oxidation of chloride ion is slow, even in the presence of iron, and titrations can be carried out in the presence of moderate amounts of chloride without the use of a Zimmermann–Reinhardt type of preventive solution. The solution can be heated but should not be boiled, or chloride ion will be oxidized. Sulfuric acid solutions of cerium(IV) are stable indefinitely. Nitric acid and perchloric acid solutions, however, do decompose, but only slowly. An added advantage of cerium is that a salt of cerium(IV), ammonium hexanitratocerate, $(NH_4)_2Ce(NO_3)_6$, can be obtained as a primary standard, and the solution does not have to be standardized. The main disadvantage of cerium(IV) is its increased cost over potassium permanganate, although this is not usually a prohibitive factor as time is saved. Ferroin is a suitable indicator for many cerium(IV) titrations.

Cerium(IV) solutions can be standardized against primary standard As_2O_3, $Na_2C_2O_4$, or electrolytic iron. The reaction with arsenic(III) is slow, and it must be catalyzed by adding either osmium tetroxide (OsO_4) or iodine monochloride (ICl). Ferroin is used as the indicator. The reaction with oxalate is also slow at room temperature, and the same catalysts can be used. The reaction is rapid, however, at room temperature in the presence of 2 M perchloric acid. Nitroferroin is used as the indicator.

Cerium(IV) solutions to be standardized are usually prepared from diammonium ceric sulfate, $(NH_4)_4Ce(SO_4)_4 \cdot 2H_2O$; diammonium ceric nitrate, $(NH_4)_2Ce(NO_3)_6$ (*not* the high-purity primary standard variety, though); or hydrous ceric oxide, $CeO_2 \cdot 4H_2O$. Primary standard $(NH_4)_2Ce(NO_3)_6$ is expensive and used when the savings in time is justifiable.

21.7 Titrations with Other Reducing Agents

Standard solutions of reducing agents are not used as widely as oxidizing agents are because most of them are oxidized by dissolved oxygen. They are, therefore, less convenient to prepare and use. **Thiosulfate** is the only common reducing agent that is stable to air oxidation and that can be kept for long periods of time. This is the reason that iodometric titrations are attractive for determining oxidizing agents. However, stronger reducing agents than iodide ion are sometimes required.

Iron(II) is only slowly oxidized by air in sulfuric acid solution and is a common titrating agent. It is not a strong reducing agent ($E^0 = 0.771$ V) and can be used to titrate strong oxidizing agents such as cerium(IV), chromium(VI) (dichromate), and vanadium(V) (vanadate). Ferroin is a good indicator for the first two titrations, and oxidized diphenylamine sulfonate is used for the last titration. Iron(II) oxidizes slowly; standardization should be checked daily.

Chromium(II) and **titanium(III)** are very powerful reducing agents, but they are readily air-oxidized and difficult to handle. The standard potential of the former is -0.41 V (Cr^{3+}/Cr^{2+}) and that of the latter is 0.04 V (TiO^{2+}/Ti^{3+}). The oxidized forms of copper, iron, silver, gold, bismuth, uranium, tungsten, and other metals have been titrated with chromium(II). The principal use of Ti^{3+} is in the titration of iron(III) as well as copper(II), tin(IV), chromate, vanadate, and chlorate.

21.8 Preparing the Solution—Getting the Analyte in the Right Oxidation State before Titration

When samples are dissolved, the element to be analyzed is usually in a mixed oxidation state or is in an oxidation state other than that required for titration. There are various oxidizing and reducing agents that can be used to convert different metals to certain oxidation states prior to titration. The excess preoxidant or prereductant must generally be removed before the metal ion is titrated.

Reduction of the Sample Prior to Titration

Reducing agents that can be readily removed are used to reduce the analyte, prior to titration with an oxidizing agent.

The reducing agent should not interfere in the titration or, if it does, unreacted reagent should be readily removable. Most reducing agents will, of course, react with oxidizing titrants, and they must be removable. **Sodium sulfite**, Na_2SO_3, and **sulfur dioxide** are good reducing agents in acid solution ($E^0 = 0.17$ V), and the excess can be removed by bubbling with CO_2 or in some cases by boiling. If SO_2 is not available, sodium sulfite or bisulfite can be added to an acidified solution. Thallium(III) is reduced to the +1 state, arsenic(V) and antimony(V) to the +3 state, vanadium(V) to the +4 state, and selenium and tellurium to the elements. Iron(III) and copper(II) can be reduced to the +2 and +1 states, respectively, if thiocyanate is added to catalyze the reaction.

Stannous chloride, $SnCl_2$, is usually used for the reduction of iron(III) to iron(II) for titrating with cerium(IV) or dichromate. The reaction is rapid in the presence of chloride (hot HCl). When iron samples (e.g., ores) are dissolved (usually in hydrochloric acid), part or all of the iron is in the +3 oxidation state and must be reduced. The reaction with stannous chloride is

$$2Fe^{3+} + SnCl_2 \text{ (aq)} + 2Cl^- \rightarrow 2Fe^{2+} + SnCl_4 \text{ (aq)} \tag{21.16}$$

The reaction is complete when the yellow color of the iron(III)–chloro complex disappears. The excess tin(II) is removed by addition of mercuric chloride:

$$SnCl_2 + 2HgCl_2 \text{ (excess)} \rightarrow SnCl_4 + \underline{Hg_2Cl_2} \text{ (calomel)} \tag{21.17}$$

A large excess of cold $HgCl_2$ must be added rapidly with stirring. If too little is added, or if it is added slowly, some of the mercury will be reduced, by local excesses of $SnCl_2$, to elemental mercury, a gray precipitate. The calomel, Hg_2Cl_2, which is a milky-white precipitate, does not react at an appreciable rate with dichromate or cerate, but mercury will. In order to prevent a large excess of tin(II) and subsequent danger of formation of mercury, the stannous chloride is added dropwise until the yellow color of iron(III) just disappears. If a gray precipitate is noted after the $HgCl_2$ is added, the sample must be discarded. Stannous chloride can also be used to reduce As(V) to As(III), Mo(VI) to Mo(V), and, with $FeCl_3$ catalyst, U(VI) to U(IV). However, because of its toxicity, the use of mercury and its compounds in the laboratory is gradually disappearing; this hampers the use of $SnCl_2$ as reductant for titrations as well.

Metallic reductors are widely used for preparing samples. These are usually used in a granular form in a column through which the sample solution is passed. The sample is eluted from the column by slowly passing dilute acid through it. The oxidized metal ion product does not interfere in the titration and no excess reductant is present since the metal is insoluble. For example, lead can be used to reduce tin(IV):

$$\underline{Pb} + Sn^{4+} \rightarrow Pb^{2+} + Sn^{2+} \tag{21.18}$$

The solution eluted from the column will contain Pb^{2+} and Sn^{2+}, but no metallic Pb. Table 21.4 lists several commonly used metallic reductors and some elements they will reduce. The reductions are carried out in acid solution. In the case of zinc, metallic zinc is amalgamated with mercury to prevent attack by acid to form hydrogen:

$$Zn + 2H^+ \rightarrow Zn^{2+} + H_2 \tag{21.19}$$

Sometimes, the reduced sample is rapidly air-oxidized and the sample must be titrated under an atmosphere of CO_2, by the addition of sodium bicarbonate to an acid solution. Air must be excluded from tin(II) and titanium(III) solutions. Sometimes, elements

TABLE 21.4 Metallic Reductors

Reductor	Species Reduced
Zn (Hg) (Jones reductor)	Fe(III) → Fe(II), Cr(VI) → Cr(II), Cr(III) → Cr(II), Ti(IV) → Ti(III), V(V) → V(II), Mo(VI) → Mo(III), Ce(IV) → Ce(III), Cu(II) → Cu
Ag (1 M HCl) (Walden reductor)	Fe(III) → Fe(II), U(VI) → U(IV), Mo(VI) → Mo(V) (2 M HCl), Mo(VI) → Mo(III) (4 M HCl), V(V) → V(IV), Cu(II) → Cu(I)
Al	Ti(IV) → Ti(III)
Pb	Sn(IV) → Sn(II), U(VI) → U(IV)
Cd	ClO_3^- → Cl^-

rapidly air-oxidized are eluted from the column into an iron(III) solution, with the end of the column immersed in the solution. The iron(III) is reduced by the sample to give an equivalent amount of iron(II), which can be titrated with permanganate or dichromate. Molybdenum(III), which is oxidized to molybdenum(VI) by the iron, and copper(I) are determined in this way.

Oxidation of the Sample Prior to Titration

Very strong oxidizing agents are required to oxidize most elements. Hot anhydrous **perchloric acid** is a strong oxidizing agent. It can be used to oxidize chromium(III) to dichromate. The mixture must be diluted and cooled very quickly to prevent reduction. Dilute perchloric acid is not a strong oxidizing agent, and the solution needs only to be diluted following the oxidation. Chlorine is a product of perchloric acid reduction, and this must be removed by boiling the diluted solution. See Chapter 2 for precautions in using perchloric acid.

Potassium persulfate, $K_2S_2O_8$, is a powerful oxidizing agent that can be used to oxidize chromium(III) to dichromate, vanadium(IV) to vanadium(V), cerium(III) to cerium(IV), and manganese(II) to permanganate. The oxidations are carried out in hot acid solution, and a small amount of silver(I) catalyst must be added. The excess persulfate is destroyed by boiling. This boiling will always reduce some permanganate.

Bromine can be used to oxidize several elements, such as Tl(I) to Tl(III) and iodide to iodate. The excess is removed by adding phenol, which is brominated. **Chlorine** is an even stronger oxidizing agent. **Permanganate** oxidizes V(IV) to V(V) and Cr(III) to Cr(VI). The latter reaction is rapid only in alkaline solution. It has been used to oxidize trace quantities of Cr(III) in hot acid solution, however. Excess permanganate is destroyed by adding hydrazine, the excess of which is destroyed by boiling. **Hydrogen peroxide** will oxidize Fe(II) to Fe(III), Co(II) to Co(III) in mildly alkaline solution, and Cr(II) to Cr(VI) in strongly alkaline solution. The last oxidation can also be accomplished directly by adding granular sodium peroxide. Excess H_2O_2 can be destroyed by many catalysts.

For most redox determinations, specified procedures have been described for the preparation of different elements in various types of samples. You should be able to recognize the reasoning behind the operations from the discussions in this chapter.

The only common redox titration applied in the clinical laboratory is for the analysis of calcium in biological fluids. Calcium oxalate is precipitated and filtered, the precipitate is dissolved in acid, and the oxalate, which is equivalent to the calcium present, is titrated with standard potassium permanganate solution. This method is largely replaced now by more convenient techniques such as complexometric titration with EDTA (Chapter 8) or measurement by atomic absorption spectrophotometry (Chapter 12).

21.9 Potentiometric Titrations (Indirect Potentiometry)

Volumetric titrations are usually most conveniently performed with a visual indicator. In cases where a visual indicator is unavailable, potentiometric indication of the end point can often be used. Potentiometric titrations are among the most accurate known because the potential follows the actual change in activity and, therefore, the end point will often coincide directly with the equivalence point. And, as we have mentioned in our discussions of volumetric titrations, end-points are more easily sensed than with visual indicators. Potentiometry is, therefore, often employed for dilute solutions.

Potentiometric titrations are straightforward. They involve measurement of an indicating electrode potential against a convenient reference electrode and plotting the change of this potential difference against volume of titrant. See Figure 20.5 for a potentiometric titration setup. A large potential break will occur at the equivalence point. Since we are interested only in locating when a large potential *change*, signifying the end-point, occurs, the *correct* potential of the indicating electrode need not be known. For example, in pH titrations, the glass electrode does not have to be calibrated with a standard buffer; it will still give the same *shape* of titration curve that may be shifted up or down on the potential axis. It is a good idea, however, to have some indication of the correct value so the end point can be anticipated and any anomalous difficulties can be detected.

Because we are not interested in "absolute" potentials, the liquid-junction potential becomes unimportant. It will remain somewhat constant throughout the titration, and small changes will be negligible compared to the change in potential at the end point. Also, the potential need not be read very closely, and so any pH meter can be used for such titrations.

pH Titrations—Using pH Electrodes

We have shown in Chapter 7 that in acid–base titrations the pH of the solution exhibits a large break at the equivalence point. Instead of a visual indicator, this pH change can be easily monitored with a glass pH electrode. By plotting the measured pH against volume of titrant, one can obtain titration curves similar to those shown in Chapter 7. The end point is taken as the **inflection point** of the large pH break occurring at the equivalence point; this is the steepest part of the curve.

A glass pH electrode is used to follow acid–base titrations.

Precipitation Titrations—Using Silver Electrodes

The indicating electrode in precipitation titrations is used to follow the change in pM or pA, where M is the cation of the precipitate and A is the anion. In the titration of chloride ion with silver ion, for example, either Equation 20.3 or 20.10 will hold. In the former equation, the term $\log (1/a_{Ag^+})$ is equal to pAg; and in the latter equation, the term $\log a_{Cl^-}$ is equal to $-pCl$. Therefore, the potential of the silver electrode will vary in direct proportion to pAg or pCl, changing $2.30 \, RT/F$ V (ca. 59 mV) for each 10-fold change in a_{Ag^+} or a_{Cl^-}. A plot of the potential versus volume of titrant will give a curve identical in shape to that in Figure 10.1. (Note that since $a_{Ag^+}a_{Cl^-} = $ constant, a_{Cl^-} is proportional to $1/a_{Ag^+}$ and pCl is proportional to $-pAg$, so the same shape curve results if we plot or measure either pCl or pAg.)

A silver electrode is used to follow titrations with silver ion.

Redox Titrations—Using Platinum Electrodes

Because there is generally no difficulty in finding a suitable indicator electrode, redox titrations are widely used; an inert metal such as platinum is usually satisfactory for the electrode. Both the oxidized and reduced forms are usually soluble and their ratio varies throughout the titration. The potential of the indicating electrode will vary in direct proportion to $\log (a_{red}/a_{ox})$, as in the calculated potential for the titration curves shown in Figure 21.1 for the titration of Fe^{2+} with Ce^{4+}. As pointed out, the potential is determined by either half-reaction. Generally, the pH in these titrations is held nearly constant, and any H^+ term in the Nernst equation will drop out of the log term.

A potentiometric plot, as in Figure 21.1, is useful for evaluating or selecting a suitable visual indicator for the titration, particularly for a new titration. From a knowledge of the transition potential, it is often possible to select an indicator whose color

An inert electrode (e.g., Pt) is used to follow redox titrations.

A potentiometric titration curve is used to select the appropriate redox indicator (with E_{In}^0 near $E_{eq.pt.}$).

transition occurs within this potential range. Or the potential can actually be measured during the visual titration and the color transition range on the potentiometric curve noted to see whether it corresponds to the equivalence point.

Ion-selective Electrodes in Titrations—Measuring pM

The term log a_{ion} in Equation 20.44 is equal to $-pIon$, and so ion-selective electrodes (ISE) can be used to monitor changes in pM during a titration. For example, a cation-selective glass electrode that is sensitive to silver ion can be used to follow changes in pAg in titrations with silver nitrate. A calcium-sensitive electrode can be used for the titration of calcium with EDTA. The electrode should not respond to sodium ion since the disodium salt of EDTA is usually used. If the electrode responds to a second ion in the solution whose activity remains approximately constant throughout the titration, then Equation 20.46 will hold, and the titration curve will be distorted; this is so because the electrode potential is determined by $log(a_{ion} + constant)$ and not $log\ a_{ion}$. If the contribution from the second ion is not too large, then the distortion will not be too great and a good break will still occur at the end point. Titrations involving anions can also be monitored with anion-selective electrodes. For example, fluoride ion can be precipitated with lanthanum(III), and a fluoride electrode can be used to mark the end point of the titration.

Potentiometric titrations are always more accurate than direct potentiometry because of the uncertainties involved in direct potential measurements. Whereas accuracies of better than a few percent are rarely possible in direct potentiometry, accuracies of a few tenths of a percent are common by potentiometric titration. We can make some general statements concerning potentiometric titrations:

> *Potentiometric titrations are more accurate than direct ISE measurements because the liquid-junction potential is not important.*

1. The potential readings are usually sluggish in dilute solutions and near the end point because the solution is poorly poised.

2. It is necessary to plot the potential only near the end point. Small increments of titrant are added near the end point, 0.1 or 0.05 mL, for example. The exact end point volume need not be added, but it is determined by interpolation of the E versus volume plot.

3. The polarity of the indicating electrode relative to the reference electrode may change during the titration. That is, the potential difference may go from one polarity to zero and then to the reverse polarity; most voltmeters/pH meters today can read either polarity but if not, the polarity of the potential-measuring device may have to be changed.

Derivative Titrations

Plotting or recording the first or second derivative of a titration curve can more accurately pinpoint the end point. The use of such plots for determining end points has been discussed extensively in Chapter 7 and **web** supplements in Chapter 7 (see 7.11 derivative titrations easy method). Instead of $\Delta pH/\Delta V$ and $\Delta^2 pH/\Delta V^2$, we plot here $\Delta E/\Delta V$ and $\Delta^2 E/\Delta V^2$.

See also the text website for Chapter 21 for an alternate approach for end point selection by derivative titirations by entering signal and volume data near the end point in the program. It is simple to use, but for more accurate plots, use the 8.11 derivative titrations made easy program. See Reference 4 by Carter and Huff for a discussion of second derivative plots.

In these methods, the volume increment should not be too large or there will not be sufficient points near the end point. If the increments are small enough, then extrapolation of the second derivative plots may not be necessary at all because there will be two or more points on the straight-line portion of the plot that passes through zero. On the other hand, the increments should not be small enough to be tedious and to fall within experimental error of the volume measurement. Of course, these small volume increments are taken only near the end point. In some titrations, the potential break is sufficiently large that the magnitude of potential change can be noted with equal added volume increments and the end point taken as that point where the change is largest. Also, it is convenient sometimes merely to titrate to an end-point potential, which has been determined by calculations or empirically from a measured titration curve.

A word of caution should be mentioned with respect to derivative methods. The derivatives tend to emphasize noise or scatter in the data points, being worse for the second derivative. Hence, if a particular titration is subject to noise or potential drift, a derivative plot is not advised.

Each time a derivative is taken, the noise is amplified.

Gran's Plots for End-Point Detection

Gunnar Gran (References 5, 6) suggested an unique way of predicting where the end point will be well before the end point is actually reached. Assume that instead of plotting the electrode potential (which is a logarithmic function of concentration) against volume of titrant, we plotted the concentration of analyte remaining at each point in the titration. A straight-line plot would in principle result (neglecting volume changes) in which the extrapolated concentration would decrease to zero at the equivalence point (assuming the equilibrium for the titration reaction lies far to the right). This is because at 20% titrated, 80% of the sample will remain, at 50% titrated, 50% will remain, at 80% titrated, 20% will remain, and so on. (In practice, a plot in the region closer to the end point is made.) Similarly, a plot of titrant concentration beyond the equivalence point would be a linear plot of increasing concentration that would extrapolate to zero concentration at the end point.

Consider the titration of chloride ion with silver nitrate solution. Except near the equivalence point where the solubility becomes appreciable compared to the unreacted chloride, the concentration of chloride in solution at any point in the titration is calculated from the initial moles less the moles reacted with $AgNO_3$:

$$[Cl^-] = \frac{C_{Cl}V_{Cl} - C_{Ag}V_{Ag}}{V_T} \tag{21.20}$$

where C_{Cl} and V_{Cl} are respectively the molar concentration and volume of the chloride sample taken and C_{Ag} and V_{Ag} are respectively the molar concentration and volume of the silver titrant and volume added, and V_T is the total volume at any point ($V_{Cl} + V_{Ag}$).

The potential of a chloride ion-selective electrode (neglecting activity coefficients) is

$$E_{cell} = k - S \log[Cl^-] \tag{21.21}$$

or

$$\log[Cl^-] = \frac{k - E_{cell}}{S} \tag{21.22}$$

where S is the empirical potentiometric slope in the Nernst equation (theoretically 0.059) and k the empirical potentiometric cell constant (theoretically the difference in E^0 values for the indicating and reference electrodes). Substituting Equation 21.20 in Equation 21.22:

$$\log \left(\frac{C_{Cl}V_{Cl} - C_{Ag}V_{Ag}}{V_T} \right) = \frac{k - E_{cell}}{S} \tag{21.23}$$

$$V_T \, \text{antilog} \left(\frac{k - E_{\text{cell}}}{S} \right) = C_{Cl}V_{Cl} - C_{Ag}V_{Ag} \qquad (21.24)$$

A plot of V_{Ag} (the independent variable) versus the left-hand side of the equation will give a straight line (the concentrations are corrected for volume changes in the above calculations). This is called a **Gran plot** (see References 5–7). The equivalence point occurs when mmol Cl = mmol Ag; that is, when the left-hand term (*y*-axis) is zero. The plot would be as illustrated in Figure 21.2. There is curvature near the end point because of the finite solubility of the silver chloride; that is, the antilog term does not go to zero (the potential would have to go to infinity), so extrapolation is made over several points before the end point.

The application of Equation 21.24 to a Gran plot implies knowledge of the constant *k* in the Nernst equation in order to construct the zero intercept on the *y*-axis. This (and the slope) can be determined from standards.

The Gran plot can also be performed empirically in a number of ways. A calibration curve of potential versus analyte concentration can be constructed and used to convert potential readings directly into concentration readings: the end-point intercept would then correspond to zero concentration on the *y* axis. The antilogarithm of the potential or pH reading can be calculated and plotted against volume of titrant ($E \propto \log C$, antilog $E \propto C$). The intercept would then correspond to the potential determined for zero analyte concentration. Electronic antilog amplifiers are also readily available for signal transformation.

A Gran plot can also be obtained for the titrant beyond the end point (in which the antilog term increases linearly from zero at the end point). In this case, the intercept potential is best determined from a blank titration and extrapolation of the linear portion of the *y*-axis to zero milliliters.

The antilog values, which are directly proportional to concentration, must be corrected for volume changes. Corrected values are obtained by multiplying the observed values by $(V_T)/V^0$, where V^0 is the initial volume and V_T is the total volume at any point. A Gran plot for a weak acid–strong base titration and a strong acid–strong base titration as monitored by a pH electrode is discussed in the **website** supplement.

In addition to the advantage of a linear plot, Gran plots do not require measurements precisely near the end point, where the potential tends to drift because of the low level of the ion being sensed and where very small increments of titrant must be added. Only a few points are needed on the straight line at a distance away from the end point.

A typical Gran plot is shown in Figure 21.3 for the titration of small amounts of chloride with silver ion. The excess titrant is monitored with a Ag/Ag_2S electrode. A plot proportional to the titrant concentration is shown (right-hand ordinate) along with the usual S-shaped potentiometric plot (left-hand ordinate); the potentiometric inflection point is barely discernible due to the small concentrations involved. In contrast, the straight-line Gran plot is readily extrapolated back to the horizontal axis to determine the end point (a blank titration is performed and the linear blank plot is extrapolated to zero milliliters to accurately determine the horizontal axis). Curvature of the straight line around the end point generally indicates appreciable solubility of a precipitate, dissociation of a complex, and so on.

Several advantages accrue from the linear plots. It is only necessary to obtain a few points to define the straight line, and the end point is easily identified by extrapolating the line to the horizontal axis. Points only need be accurately determined a bit away from the equivalence point, where the titrant is in sufficient excess to suppress dissociation of the titration product and where electrode response is rapid because one of the ions is at relatively high levels compared to the levels at the equivalence point. In case of small inflection points (Figure 21.3), the end point is more readily defined by a Gran plot.

A Gran plot converts a logarithmic response to a linear plot.

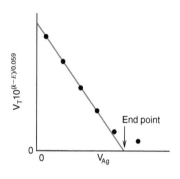

FIGURE 21.2 Gran plot as given for Equation 21.24.

With a Gran plot, we do not have to actually measure where the end point in the titrations is.

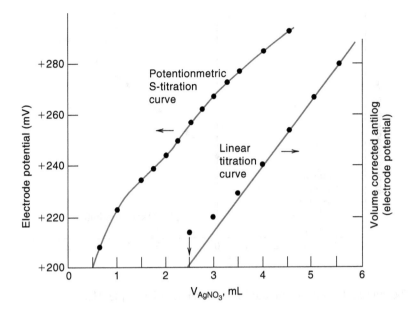

FIGURE 21.3 Gran plot for titration of 100 mL 5×10^{-5} M Cl$^-$ with 2×10^{-3} M AgNO$_3$ using Ag$_2$S electrode. (Courtesy of Orion Research, Inc.)

A Gran-type plot can also be obtained by plotting the reciprocal of a first derivative curve, that is, $\Delta E/\Delta V$ versus V. Since in a derivative titration $\Delta E/\Delta V$ goes to infinity at the equivalence point, the reciprocal will go to zero where the intersection of the two lines occurs, and a V- or inverted V-shaped plot results. In this application, the average volume between the two increments is plotted, as in the first derivative plot. The $\Delta E/\Delta V$ values must be corrected for volume changes to obtain straight lines ($\Delta E/\Delta V$ is linearly dependent on volume changes). See Section 7.11 and Figures 7.13 and 7.14.

A Gran-type plot is convenient in **standard additions** or **known additions** procedures. Standard additions methods are useful ways of calibration when the sample matrix affects the analyte signal. In these methods, a signal is recorded for the sample, and then a known amount of standard is added to the sample and the change in signal is measured. This latter measurement provides calibration in the same matrix as the unknown analyte, and the matrix should have the same effect on both unknown and standard. In this case, it is the electrode response that is calibrated. Most analytical methods give a linear response to analyte; but in potentiometry, the response is logarithmic. With the Gran approach, a linear plot can be obtained, simplifying calculation. The potential of the sample is initially recorded and then known amounts of standard are added to the sample. The antilog values of the potential are plotted as a function of the amount of standard added, and the best straight line is drawn through them (e.g., by least-squares analysis). Extrapolation to the horizontal axis (determined from similar measurements on a "blank" with extrapolation to zero concentration) gives the equivalent amount of analyte in the sample (Figure 21.4).

In applying the standard additions method, it is most convenient to add small volumes of concentrated standard to the sample solution in order to minimize volume change and thereby make volume corrections unnecessary. For example, 100 µL of a 1000-ppm standard might be added to 10 mL of sample to increase the concentration by 10 ppm. The volume change is only 1% and can generally be ignored. The concentration increments should be comparable to the unknown concentration.

An added advantage to adding small volumes is that it avoids significant dilution of the matrix, keeping matrix effects constant.

A first derivative titration can be used to prepare a Gran plot.

Standard additions calibration corrects for sample matrix effects. The standard is added to the sample.

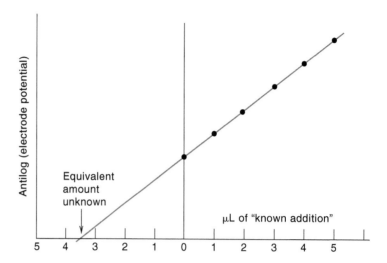

FIGURE 21.4 Standard addition antilog plot for a potentiometric sensor.

The standard additions method is illustrated in Example 21.8.

Example 21.8

The calcium ion concentration in serum is determined using an ion-selective electrode by the standard addition method. The potential measured with the electrode in the sample is +217.6 mV. Addition of 100 μL of a 2000-ppm standard to 2.00 mL of sample and measurement of the potential gives +226.8 mV. Assuming a Nernstian response (59.2/2 mV per 10-fold change in activity), what is the concentration of calcium in the sample?

Solution

Since the analyte and the standard are subjected to the same matrix and ionic strength, the electrode responds in a Nernstian fashion to concentration (see Section 20.8). We can write

$$E = k + 29.6 \ \log[Ca^{2+}]$$

The standard (0.100 mL) is diluted in the sample (2.00 mL) about 1:20 to give an added concentration of 100 ppm or, more precisely, correcting for the 5% volume change:

$$C = 2000 \ \text{ppm} \times \frac{0.100 \ \text{mL}}{2.10 \ \text{mL}} = 95.2 \ \text{ppm}$$

Let x equal the unknown concentration in parts per million:

$$217.6 \ \text{mV} = k + 29.6 \ \log x \tag{1}$$

$$\underline{226.8 \ \text{mV} = k + 29.6 \ \log(x + 95.2)} \tag{2}$$

Subtracting (2) from (1):

$$-9.2 \ \text{mV} = 29.6 \ \log x - 29.6 \ \log(x + 95.2)$$

$$-9.2 \ \text{mV} = 29.6 \ \log \frac{x}{x + 95.2}$$

$$\log \frac{x}{x + 95.2} = -0.31_1$$

$$\frac{x}{x + 95.2} = 0.467$$

$$x = 83.5 \ \text{ppm}$$

If the actual slope for the electrode is not known, then multiple additions of the standard should be made to determine the actual shape.

Standard addition provides a powerful means of correcting for matrix dependence of analyte response, whether the response is linear or logarithmic. It cannot, however, correct for the presence of non-analyte species in the sample to which the sensor responds.

Automatic Titrators

There are numerous automatic titrators that employ potentiometric end-point detection. They usually can automatically record the first or second derivative of the titration curve and read out the end-point volume. The sample is placed in the titration vessel, and the titrant, drawn from a reservoir, is placed in a syringe-driven buret. The volume is digitally read from the displacement of the syringe plunger by the electronic driver. The derivative value, especially with a pH or other potentiometric sensor, is often used to control the buret delivery speed, i.e., as the rate of change of pH increases, the delivery is slowed down to allow more accurate end point detection. Titrators may also employ photometric detection of indicator color changes. An automatic titrator is shown in Figure 21.5. Automatic titrators make volumetric analyses rapid, reproducible, and convenient. While instrumental methods provide many advantages, classical volumetric analyses are still widely used and are very useful, especially for major constituents, for example, in the pharmaceutical, chemical, and petrochemical industries.

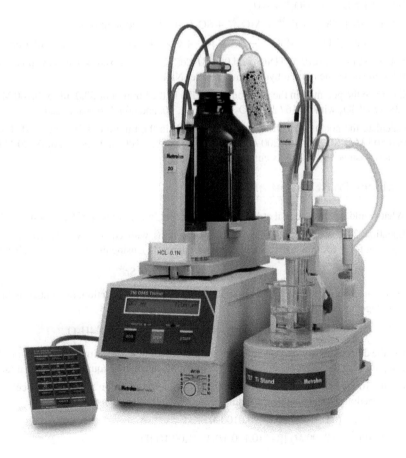

FIGURE 21.5 Automatic potentiometric titrator. (Courtesy of Brinkmann Instruments, Inc.)

Questions

1. Describe the ways in which the end points of redox titrations may be detected visually.
2. Distinguish between iodimetry and iodometry.
3. Why are iodimetric titrations usually done in neutral solution and iodometric titrations in acid solution?
4. Does the end point in a permanganate titration coincide with the equivalence point? Explain and suggest how any discrepancies might be corrected.
5. Explain the function of the Zimmermann–Reinhardt reagent in the titration of iron(II) with permanganate.

Problems

Balancing Redox Reactions

6. Balance the following aqueous reactions:
    (a) $ClO^- (aq) + Cr(OH)_4^- (aq) \rightarrow CrO_4^{2-} (aq) + Cl^- (aq)$ (basic solution)
    (b) $NO_3^- (aq) + Cu (s) \rightarrow NO (g) + Cu^{2+} (aq)$
    (c) $H_3PO_3 + HgCl_2 \rightarrow \underline{Hg_2Cl_2} + H_3PO_4 + HCl$

7. Balance the following aqueous reactions:
    (a) $Fe(OH)_2 (s) + H_2O (l) + O_2 (g) \rightarrow Fe(OH)_3 (s)$ (basic solution)
    (b) $MnO_4^- + H_2S \rightarrow Mn^{2+} + S$
    (c) $SbH_3 + Cl_2O \rightarrow H_4Sb_2O_7 + HCl$
    (d) $PbO_2 (s) + Cl^- (aq) \rightarrow PbCl_2 (s) + O_2 (g)$ (acidic solution)
    (e) $Al + NO_3^- \rightarrow AlO_2^- + NH_3$
    (f) $FeAsS + ClO_2 \rightarrow Fe^{3+} + AsO_4^{3-} + SO_4^{2-} + Cl^-$ (acid solution)
    (g) $K_2NaCo(NO_2)_6 + MnO_4^- \rightarrow K^+ + Na^+ + Co^{3+} + NO_3^- + Mn^{2+}$ (acid solution)

8. Equal volumes of $0.10\ M$ $TlNO_3$ and $0.20\ M$ $Co(NO_3)_3$ are mixed. What is the potential in the solution versus the normal hydrogen electrode (NHE)?

9. Calculate the potential in the solution (vs. NHE) in the titration of 50.0 mL of 0.100 M Fe^{2+} in $1.00\ M$ $HClO_4$ with $0.0167\ M$ $Cr_2O_7^{2-}$ at 10, 25, 50, and 60 mL titrant added.

10. Calculate the potential of the solution (vs. NHE) in the titration of 100 mL of $0.100\ M$ Fe^{2+} in $0.500\ M$ H_2SO_4 with $0.0200\ M$ $KMnO_4$ at 10.0, 50.0, 100, and 200 mL of 0.020 M $KMnO_4$ titrant. Assume the H_2SO_4 is completely ionized.

Equivalence Point Potentials

11. What would be the potential at the equivalence point in the titration of Fe^{3+} with Sn^{2+}?

12. Equation 21.2 was derived using two half-reactions with equal n values. Derive a similar equation for the following reaction (used in Problem 11), using the numerical n values:

$$2Fe^{3+} + Sn^{2+} \rightleftharpoons 2Fe^{2+} + Sn^{4+}$$

13. Derive an equation similar to Equation 21.2 for the following reaction, remembering to include a hydrogen ion term:

$$5Fe^{2+} + MnO_4^- + 8H^+ \rightleftharpoons 5Fe^{3+} + Mn^{2+} + 4H_2O$$

Use the derived equation to calculate the end-point potential in the titration in Example 21.4 and compare with the value obtained in that example by calculating equilibrium concentrations.

14. For the following cells, calculate the cell voltage before reaction and each half-cell potential after reaction. Also calculate the equilibrium constants of the reactions:
    (a) $Zn|Zn^{2+}$ (0.250 M) $||Cd^{2+}$ (0.0100 M) $|Cd$
    (b) $Pb|Pb^{2+}$ (0.0100 M) $||I_3^-$ (0.100 M), I^- (1.00 M) $|Pt$

15. From standard reduction potentials, calculate the equilibrium constant at 25°C for the reaction:

$$2MnO_4^- \text{ (aq)} + 10Cl^- \text{ (aq)} + 16H^+ \text{ (aq)} \longrightarrow 2Mn^{2+} \text{ (aq)} + 5Cl_2 \text{ (g)} + 8H_2O \text{ (l)}$$

Quantitative Calculations

16. Selenium in a 10.0-g soil sample is distilled as the tetrabromide, which is collected in aqueous solution where it is hydrolyzed to SeO_3^{2-}. The SeO_3^{2-} is determined iodometrically, requiring 4.5 mL of standard thiosulfate solution for the titration. If the thiosulfate titer is 0.049 mg $K_2Cr_2O_7$/mL, what is the concentration of selenium in the soil in ppm?

17. The calcium in a 5.00-mL serum sample is precipitated as CaC_2O_4 with ammonium oxalate. The filtered precipitate is dissolved in acid, the solution is heated, and the oxalate is titrated with 0.00100 M $KMnO_4$, requiring 4.94 mL. Calculate the concentration of calcium in the serum in meq/L (equivalents based on charge).

18. A 2.50-g sample containing As_2O_5, Na_2HAsO_3 and inert material is dissolved and the pH is adjusted to neutral with excess $NaHCO_3$. The As(III) is titrated with 0.150 M I_2 solution, requiring 11.3 mL to just reach the end point. Then, the solution (all the arsenic in the +5 state now) is acidified with HCl, excess KI is added, and the liberated I_2 is titrated with 0.120 M $Na_2S_2O_3$, requiring 41.2 mL. Calculate the percent As_2O_5 and Na_2HAsO_3 in the sample.

19. If 1.00 mL $KMnO_4$ solution will react with 0.125 g Fe^{2+} and if 1.00 mL $KHC_2O_4 \cdot H_2C_2O_4$ (tetroxalate) solution will react with 0.175 mL of the $KMnO_4$ solution, how many milliliters of 0.200 M NaOH will react with 1.00 mL of the tetroxalate solution? (All three protons on the tetroxalate are titratable.)

20. The sulfide content in a pulp plant effluent is determined with a sulfide ion-selective electrode using the method of standard additions for calibration. A 10.0-mL sample is diluted to 25.0 mL with water and gives a potential reading of −216.4 mV. A similar 10.0-mL sample plus 1.00 mL of 0.030 M sulfide standard diluted to 25.0 mL gives a reading of −224.0 mV. Calculate the concentration of sulfide in the sample.

Gran Plots

21. Starting with the K_a expression for a weak acid HA and substituting the titrant volumes in [HA] and [A^-], show that the following expression holds up to the equivalence point for the titration of HA with a strong base B:

$$V_B[H^+] = K_a(V_{eq.pt.} - V_B) = V_B 10^{-pH}$$

where V_B is the volume of added base and $V_{eq.pt.}$ the volume added at the equivalence point. A plot of V_B versus $V_B 10^{-pH}$ gives a straight line with a slope of $-K_a$ and an intercept corresponding to the equivalence point.

Multiple Choice Questions

1. The standard redox potential of water oxidation to dioxygen is −1.23 V.

$$2H_2O \longrightarrow O_2 + 4H^+ + 4e^-$$

The redox potential of the same reaction at pH = 7 would be:

(a) −0.41 V   (b) −1 V

(c) −0.82 V   (d) −1.64 V

2. The corrosion of iron in contact with an acidic aqueous solution undergoes the following reaction in the **anaerobic** condition,

$$Fe + 2H^+ \rightleftharpoons Fe^{2+} + H_2 \tag{1}$$

and the following reaction in the **aerobic** condition.

$$2Fe + O_2 + 4H^+ \rightleftharpoons Fe^{2+} + 2H_2O \tag{2}$$

During the corrosion, Fe(II) ions are formed in both conditions. If the water is polluted with Cr(IV), the following reaction may take place.

$$7H^+ + 3Fe^{2+}\,(aq) + HCrO_4^-\,(aq) \rightleftharpoons 3Fe^{3+} + Cr^{3+}\,(aq) + 4H_2O\,(l) \qquad (3)$$

Reaction (3) be broken down to the following redox half-reactions:

$$3Fe^{3+} + 3e^- \rightleftharpoons 3Fe^{2+}; \quad E^0 = +0.77\text{ V} \qquad (4)$$

$$7H^+ + HCrO_4^- + 3e^- \rightleftharpoons Cr^{3+} + 4H_2O; \quad E^0 = +1.38\text{ V} \qquad (5)$$

The standard potentials of these reactions are with respect to the normal hydrogen electrode. What would be the approximate value of the equilibrium constant of reaction 3 at 298 K?

(a) 10^{11} (b) 10^{31}
(c) 10^{-31} (d) 10^{-11}

3. The cell potential for the following electrochemical system at 25°C is

$$\text{Al (s)} \,|\, Al^{3+}\,(0.01M) \,||\, Fe^{2+}\,(0.1M) \,|\, \text{Fe (s)}$$

(a) 1.23 V (b) 1.21 V
(c) 1.22 V (d) −2.10 V

4. FAD is a redox-active molecule which takes part in many important biological reactions. The redox potential of FAD at pH 7.0 is given below.

$$FAD + 2H^+ + 2e^- \rightleftharpoons FADH_2; \quad E^0_{FADH_2/FAD} = -0.180\text{ V}$$

Calculate the redox potential when the media is acidified to pH = 0

(a) 0 V (b) 0.24 V
(c) 0.12 V (d) None of the above

5. Given that $E^0\,(Cl_2/Cl^-) = 1.35$ V and $K_{sp}\,(AgCl) = 10^{-10}$ at 25°C, E^0 corresponding to the electrode reaction

$$\frac{1}{2}Cl_2 + Ag^+ + e^- \rightarrow AgCl$$

is
$$\left[\frac{2.303\ RT}{F} = 0.06\text{V} \right]$$

(a) 0.75 V (b) 1.05 V
(c) 1.65 V (d) 1.95 V

6. Given

(i) $Zn + 4NH_3 \rightarrow Zn\,(NH_3)_4^{\,2+} + 2e^-; \quad E^0 = 1.03\text{V}$

(ii) $Zn \rightarrow Zn^{2+} + 2e^-; \quad E^0 = 0.763\text{V}$

the formation constant of the complex $Zn\,(NH_3)_4^{\,2+}$ is approximately

$$\left(\frac{2.303\ RT}{F} = 0.0591 \right)$$

(a) 1.1×10^5 (b) 1×10^7
(c) 1×10^9 (d) 1×10^{12}

7. For the cell reaction,

$$Sn + Sn^{4+} \rightleftharpoons 2Sn^{2+}$$

separate electrode reactions could be written with the respective standard electrode potential data at 25°C as

$$Sn^{4+} + 2e^- \rightarrow Sn^{2+}; \quad E^0 = +0.15\text{ V}$$

$$Sn^{2+} + 2e^- \rightarrow Sn; \quad E^0 = -0.14\text{ V}$$

When RT/F is given as 25.7 mV, logarithm of the equilibrium constant ($\ln K$) is obtained as

(a) 22.6 (b) 226
(c) 2.26 (d) 2.26×10^{-1}

8. The electrochemical cell potential (E), after the reactants and products reach equilibrium, is (E^0 is the standard cell potential and n is the number of electrons involved)

(a) $E = E^0 + nF/RF$

(b) $E = E^0$

(c) $E = E^0 - RT/nF$

(d) $E = 0$

Recommended References

Redox Equations

1. C. A. Vanderhoff, "A Consistent Treatment of Oxidation–Reduction," *J. Chem. Ed.*, **25** (1948) 547.
2. R. G. Yolman, "Writing Oxidation–Reduction Equations," *J. Chem. Ed.*, **36** (1959) 215.

Equivalence Point Potential

3. A. J. Bard and S. H. Simpsonsen, "The General Equation for the Equivalence Point Potential in Oxidation–Reduction Titrations," *J. Chem. Educ.*, **37** (1960) 364.

Derivative Titrations

4. K. N. Carter and R. B. Huff, "Second Derivative Curves and End-Point Determination," *J. Chem. Ed.*, **56** (1979) 26.

Gran's Plots

5. G. Gran, "Determination of Equivalent Point in Potentiometric Titrations," *Acta Chem. Scand.*, **4** (1950) 559.
6. G. Gran, "Determination of the Equivalence Point in Potentiometric Titrations. Part II," *Analyst*, **77** (1952) 661.
7. C. C. Westcott, "Ion-Selective Measurements by Gran Plots with a Gran Ruler," *Anal. Chim. Acta*, **86** (1976) 269.
8. H. Li, "Improvement of Gran's Plot Method in Standard Addition and Subtraction Methods by a New Plot Method," *Anal. Lett.*, **24** (1991) 473.

Chapter 22

VOLTAMMETRY AND ELECTROCHEMICAL SENSORS

KEY THINGS TO LEARN FROM THIS CHAPTER

- Electrolytic cells
- Current–voltage curves
- Supporting electrolytes

- Amperometric electrodes
- Chemically modified electrodes
- Ultramicroelectrodes

Electrolytic methods include some of the most accurate, as well as most sensitive, instrumental techniques. In these methods, an analyte is oxidized or reduced at an appropriate electrode in an electrolytic cell by application of a voltage (Chapter 19), and the amount of electricity (quantity or current) involved in the electrolysis is related to the amount of analyte. The fraction of analyte electrolyzed may be very small, in fact negligible, in the current–voltage techniques of voltammetry. Micromolar or smaller concentrations can be measured. Since the potential at which a given analyte will be oxidized or reduced is dependent on the particular substance, selectivity can be achieved in electrolytic methods by appropriate choice of the electrolysis potential. Owing to the specificity of the methods, prior separations are often unnecessary. These methods can therefore be rapid.

As stated in Chapter 19, electrochemical methods are generally not as widely used as spectrochemical or chromatographic methods. But voltammetric techniques, for example, may be used for trace analyses at the parts per billion level, and respond differently to different forms of the same element (e.g., As(III) vs. As(V)) in a mixture. Electrochemically active organic compounds can also be measured. In all but ultratrace analyses, consumption of analyte is negligible and so the analysis can be considered non-destructive. The amperometric oxygen electrode is widely used to measure oxygen in water, or to follow chemical reactions that consume or produce oxygen.

In this chapter, we discuss voltammetric methods and associated electrochemical sensors, including chemically modified electrodes. Voltammetric techniques use a microelectrode for microelectrolysis. Here, the potential is scanned and a dilute solution of the analyte produces, at a given potential, a limiting current (microamperes or less), which is proportional to the analyte concentration. Amperometry is a subclass of voltammetry where the current measured at a fixed potential is proportional to the concentration of a given species, and it can be used for direct measurement, or, for example, be used to follow the course of a titration. Amperometry is the basis of many electrochemical sensors.

We describe in detail each of these techniques in this chapter. It will be helpful to review Chapter 20 on potentiometry before reading this material.

In voltammetry, the potential is scanned at a microelectrode, and at a certain potential, an electroactive analyte is reduced or oxidized. The current increases in proportion to the analyte concentration.

22.1 Voltammetry

Voltammetry is essentially electrolysis on a microscale, using a micro working electrode (e.g., a platinum wire). As the name implies, it is a current–voltage technique. The potential of the micro working electrode is varied (scanned slowly) and the resulting current is recorded as a function of applied potential. The recording is called a

voltammogram. If an electroactive (reducible or oxidizable) species is present, a current will be recorded when the applied potential becomes sufficiently negative or positive for it to be reduced or oxidized. [By convention, a cathodic (reduction) current is taken to be positive and an anodic (oxidation) current is taken to be negative]. If the solution is dilute, the current will reach a limiting value directly proportional to its concentration because the analyte must be transported to the electrode in order to be electrolyzed and this mass transport rate becomes diffusion controlled. We will see below that the limiting current is proportional to the concentration of the species. The microelectrode restricts the current to a few microamperes or less, and so in most applications the concentration of the test substance in solution remains essentially unchanged after the voltammogram is recorded.

The Voltammetric Cell—An Electrolytic Cell

A voltammetric cell consists of the micro **working electrode**, the **auxiliary electrode**, and a **reference electrode**, usually an Ag/Agcl electrode or a saturated calomel electrode (SCE). A **potentiostat** varies the potential of the working electrode, relative to the reference electrode, which has a fixed potential. The current passing through the working electrode is recorded as a function of its potential measured against the reference electrode, but the voltage is applied and current passes between the working and auxiliary electrodes, as in Figure 22.1. In this manner, the current–voltage curve is not disturbed by an appreciable solution resistance, which creates an iR drop (voltage drop) between the working and auxiliary electrodes, as in nonaqueous solvents. Ohm's law states that voltage is equal to the product of current and resistance: $E = iR$, where i is the current in amperes and R the resistance in ohms. When current flows, the recorded potential is distorted—shifted—by an amount equal to $i \times R$, where R is the solution resistance. If appreciable, this causes the current–potential curve (see Figure 22.2) to be distorted and drawn out over a larger potential range. With a three-electrode system, the recorded potential is that between the working electrode and the reference electrode, with essentially no flow of current between these two electrodes and no iR drop.

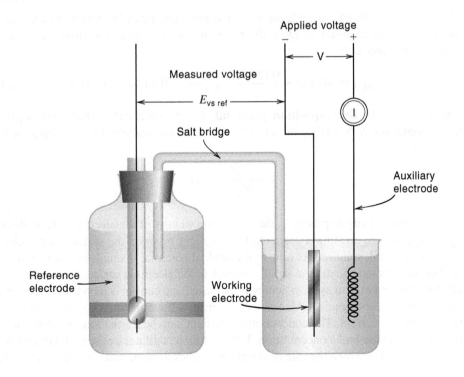

FIGURE 22.1 Three-electrode setup for voltammetric measurements.

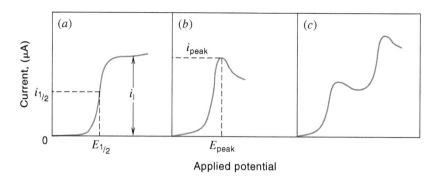

FIGURE 22.2 Different types of voltammetric curves. (*a*) Stirred solution or rotated electrode, (*b*) unstirred solution, and (*c*) stepwise reduction (or oxidation) of analyte or of a mixture of two electroactive substances (unstirred solution). In (*a*), we measure the limiting current, i_1, which is proportional to the concentration. $E_{1/2}$ (where the current is half the limiting current) is characteristic of the analyte, being related to E^0 of the redox couple. In (*b*), the peak current is proportional to the concentration. E_{peak} is also a characteristic of the analyte, but is not directly related to E^0.

The Basis of Voltammetry: The Current-Voltage Curve

See Example 20.4 and Figure 20.4 for conversion of potentials from one reference electrode to another.

Potentials in voltammetry are by convention referred to the SCE. The following relationship can be used to convert potentials versus SCE to the corresponding potentials versus normal hydrogen electrode (NHE), and vice versa:

$$E_{\text{vs. SCE}} = E_{\text{vs. NHE}} - 0.242 \qquad (22.1)$$

This relationship can be used to calculate the applied potential required for the electrolysis of the test ion at the microelectrode. Suppose, for example, we place a 10^{-3} M solution of cadmium nitrate in the test cell with a carbon microelectrode and impress a voltage difference between the working and auxiliary electrodes, making the microelectrode negative relative to the SCE. The electrode reaction will be

The voltammetric cell is really an electrolytic cell in which the electrochemical reaction as a result of the applied potential is the reverse of the spontaneous reaction (as in a voltaic cell). See Section 19.2.

$$\text{Cd}^{2+} + 2e^- \rightarrow \underline{\text{Cd}} \quad E^0 = -0.403 \text{ V} \qquad (22.2)$$

The minimum working electrode potential to begin reducing cadmium [back-emf (electromotive force) required to force the reaction] can be calculated from the Nernst equation (Chapter 20):

In Equation 22.3 the activity of solid Cd is unity.

$$E_{\text{vs. SCE}} = -0.403 - \frac{0.05916}{2} \log \frac{1}{10^{-3}} - 0.242 = -0.556 \text{ V} \qquad (22.3)$$

This is called the **decomposition potential**. As the potential is increased beyond the decomposition potential, the current will increase linearly in accordance with Ohm's law,

$$i = \frac{E}{R_{\text{circuit}}} = kE \qquad (22.4)$$

A limiting current is reached because the analyte is being electrolyzed as fast as it can diffuse to the electrode.

As the electrolysis proceeds, the ions in the vicinity of the electrode are depleted by being reduced, creating a concentration gradient between the surface of the electrode and the bulk solution. As long as the applied potential is small, the ions from the bulk of the solution can diffuse rapidly enough to the electrode surface to maintain the electrolysis current. But as the potential is increased, the current is increased, creating a larger concentration gradient. Hence, the ions must diffuse at a more rapid rate in order to maintain the current. The concentration gradient, and hence the rate of diffusion, is proportional to the bulk concentration. If the solution is dilute, a potential will eventually be reached at which the rate of diffusion reaches a maximum and all the ions are reduced as fast as they can diffuse to the electrode surface. Hence, a **limiting current** value, i_l, is reached, and further increased potential will not result in increased current.

A typical recorded current–voltage curve is illustrated in Figure 22.2. If the solution is stirred or the electrode is rotated, an S-shaped plot [curve (a)] is obtained; that is, the limiting current remains constant once it is established. This occurs because the **diffusion layer**, or thickness of the concentration gradient across which the analyte must diffuse, remains small and constant since the analyte is continually brought near to the electrode surface by mass transfer (stirring). But if the electrode is unstirred and in a quiet solution, the diffusion layer will extend farther out into the solution with time, with the result that the limiting current decreases exponentially with time and a "peaked" wave is recorded [curve (b)]. For this reason and others, the voltage scan rate using stationary microelectrodes is usually fairly rapid, for example, 50 mV/second. (In reality, even with stirred solutions, the waves tend to be "peaked" to some extent.)

Although the decomposition potential required to initiate the electrolysis will vary slightly with concentration, the potential at which the current is one-half the limiting current is independent of concentration. This is called the **half-wave potential** $E_{1/2}$. It is a constant related to the standard or formal potential of the redox couple, and so voltammetry serves as a qualitative tool for identifying the reducible or oxidizable species.

An electrode whose potential is dependent on the magnitude of the current flowing is called a **polarizable electrode**. If the electrode area is small and a limiting current is reached, then the electrode is said to be **depolarized**. Therefore, a substance that is reduced or oxidized at a microelectrode is referred to as a **depolarizer** and the redox process is said to take place under depolarized conditions.

If a depolarizer is reduced at the working electrode, a **cathodic current** is recorded at potentials more negative than the decomposition potential. If the depolarizer is oxidized, then an **anodic current** is recorded at potentials more positive than the decomposition potential.

> The solution can be recovered essentially unchanged following a voltammetric measurement due to the small currents passed.

> The potential at which the analyte is electrolyzed is a qualitative identifier of the analyte.

> According to IUPAC definitions, "depolarizer" is simply a synonym for an electractive substance, i.e., a sustance that can be oxidized or reduced at an electrode.

Stepwise Reduction or Oxidation

An electroactive substance may be reduced to a lower oxidation state at a certain potential and then be reduced to a still lower oxidation state when the potential is decreased to an even lower value. For example, copper(II) in ammonia solution is reduced at a graphite electrode to a stable Cu(I)–ammine complex at −0.2 V versus SCE, which is then reduced to the metal at −0.5 V, each a one-electron reduction step. In such cases, two successive voltammetric waves will be recorded as in Figure 22.2c. The relative heights of the waves will be proportional to the number of electrons involved in the reduction or oxidation. In this case, the two waves would be of equal height.

When a solution contains two or more electroactive substances that are reduced at different potentials, then a similar stepwise reduction will occur. For example, lead is reduced at potentials more negative than −0.4 V versus SCE ($Pb^{2+} + 2e^- \rightarrow Pb$), and cadmium is reduced at potentials more negative than −0.6 V ($Cd^{2+} + 2e^- \rightarrow Cd$). So a solution containing a mixture of these would exhibit two voltammetric waves at a graphite electrode, one for lead at −0.4 V, followed by another stepwise wave for cadmium at −0.6 V. The relative heights will be proportional to the relative concentrations of the two substances, as well as the relative n values in their reduction or oxidation.

Mixtures of electroactive substances can be determined by their stepwise voltammetric waves. There should be at least 0.2 V between the $E_{1/2}$ values for good resolution between the successive reduction or oxidation steps. If the $E_{1/2}$ values are equal, then a single composite wave will be seen, equal in height to the sum of the individual waves. If a major component in much higher concentration is reduced (or oxidized for anodic

> The height of a voltammetric wave is proportional to the number of electrons in the electrolysis reaction, and the concentration of the electroactive spices.

scans) before the test substance(s) of interest, its response is likely to mask succeeding waves and a limiting current might not even be reached. In such cases, most of this interfering substance will have to be removed before the analysis can be performed. A common procedure is to preelectrolyze it at a macroelectrode using a setup similar to Figure 22.1, at a potential corresponding to its limiting current plateau but not sufficient to electrolyze the test substance.

Similiar to stepwise reductions, stepwise oxidations may also occur to give stepwise anodic waves.

Why Supporting Electrolyte is Needed for Voltammetric Measurements?

It was assumed above that when a concentration gradient existed in a quiescent solution, the only way the reducible ion could get to the electrode surface was by diffusion. It can also get to the electrode surface by electrical (coulombic) attraction or repulsion.

The supporting electrolyte is an "inert" electrolyte in high concentration that "swamps out" the attraction or repulsion of the analyte ion at the charged electrode.

The electrode surface will be either positively or negatively charged, depending on the applied potential, and this surface charge will either repel or attract the ion diffusing to the electrode surface. This will cause an increase or decrease in the limiting current, which is called the **migration current**. The migration current can be prevented by adding a high concentration, at least 100-fold greater than the test substance, of an inert **supporting electrolyte** such as potassium nitrate. Potassium ion is reduced only at a very negative potential (not attainable in aqueous solution, hydrogen ion is reduced before this) and will not interfere. The high concentration of inert ions essentially eliminates the attraction or repelled forces between the electrode and the analyte, and the inert ions are attracted or repulsed instead. The inert ions are not electrolyzed, however.

A second reason for adding a supporting electrolyte is to decrease the iR drop of the cell. For this reason, at least 0.1 M supporting electrolyte is commonly added. This is true for nearly all electrochemical techniques, potentiometry being an exception. The choice of the supporting electrolyte is influenced by the need to provide optimum conditions for the particular analysis, such as buffering at the proper pH or elimination of interferences by selective complexation of some species. With some exceptions, such as the Cu(II)-tetrammine complex mentioned previously, when a metal ion is complexed, it is generally stabilized against electrolysis. In all such cases, the reduction half-wave is shifted to more negative reduction potentials; it may even become nonelectroactive. Commonly used complexing agents include tartrate, citrate, cyanide, ammonia, and EDTA.

Irreversible Reduction or Oxidation

If a substance is reduced or oxidized reversibly, then its half-wave potential will be near the standard potential for the redox reaction. If it is reduced or oxidized irreversibly, the mechanism of electron transfer at the electrode surface involves a slow step with a high energy of activation. Therefore, extra energy, in the form of increased applied potential, must be applied to the electrode for the electrolysis to occur at an appreciable rate. This is called the **activation overpotential**. Therefore, $E_{1/2}$ will be more negative than the standard potential in the case of a reduction, or more positive in the case of an oxidation. An irreversible wave is more drawn out than a reversible wave. Nevertheless, an S-shaped wave is still obtained, and its diffusion current is still the same as it would be if it were reversible because i_l is limited only by the rate of diffusion of the substance to the electrode surface.

Dependence of Working Potential Range on the Electrode

The potential range over which voltammetric techniques can be used will depend on the electrode material, the solvent, the supporting electrolyte, and the pH. If a platinum electrode is used in aqueous solution, the limiting positive potential would be oxidation of water ($H_2O \rightarrow \frac{1}{2}O_2 + 2H^+ + 2e^-$), unless the supporting electrolyte contains a more easily oxidizable ion (e.g., Cl^-). E^0 for the water half-reaction is $+1.0$ V versus SCE, and so the limiting positive potential is about $+1$ V versus SCE, depending on the pH. The negative limiting potential will be from the reduction of hydrogen ions. Platinum has a low hydrogen overvoltage at low current densities, and so this will occur at about -0.1 V versus SCE. Because oxygen is not reduced at these potentials, it need not be removed from the solution, unless oxygen interferes chemically.

Carbon electrodes are frequently used for voltammetry. Their positive potential limit is essentially the same as with platinum electrodes, but more negative potentials can be reached because hydrogen has a rather high overvoltage on carbon. Potentials of about -1 V versus SCE or more can be used, again depending on the pH. With potentials more negative than -0.1 versus SCE, oxygen must be removed from solutions because it is electrochemically reduced. This is conveniently done by bubbling nitrogen through the solution for 10 to 15 min through a small tube. The nitrogen is passed through water first to saturate it with water vapor so the test solution is not evaporated. Following deaeration, the tube is withdrawn and nitrogen is passed above the surface of the solution to prevent air from being absorbed.

An advantage of carbon electrodes is that unlike platinum electrodes, surface oxide coatings are not formed. While carbon electrodes can be used at fairly negative potentials, a dropping mercury electrode (DME) has been traditionally preferred because better reproducibility could be achieved. This is because the electrode surface is constantly renewed (small mercury drops fall from a capillary attached to a mercury reservoir). Voltammetric techniques using a dropping mercury electrode are called **polarography**, developed in 1922 by Jaroslav Heyrovský; polarography is the precursor to most voltammetric techniques.

Solid electrode voltammetry is used largely for the oxidation of substances at fairly positive potentials, although it is also useful for very easily reducible substances. However, reproducibility frequently suffers because the surface characteristics of the electrodes are not reproducible and the surface becomes contaminated. For this reason, the related technique of polarography had been preferred. However, because of the toxicity of mercury, alternative electrode materials, including exotic compositions, are continuously being investigated. Many have unique properties: Indium tin oxide (ITO) electrodes are of great utility in spectroelectrochemical studies as they are optically transparent. Many unusual electrooxidations can be carried out on a boron doped diamond (BDD) anode, not possible on virtually any other electrode. Single-walled carbon nanotubes (SWCNTs) and multiwalled carbon nanotubes (MWCNTs) are being incorporated in many electrode formulations because of their unusual ability to carry out many electrooxidation/reduction processes not otherwise possible, and their increased surface areas enhance sensitivity. Screen-printed electrodes are used in many disposable applications. Renewable bismuth films coated on other electrodes share many advantageous features of a mercury electrode. Alteration of surface properties of conventional electrodes affecting analytical performance is greatly ameliorated by multiple step pulsed amperometry that electrochemically cleans the electrode surface before each measurement; this is discussed in Chapter 16 in the context of amperometric detectors for HPLC.

Water or protons are easily reduced at a platinum electrode, limiting the available negative potential range to about -0.1 V versus SCE.

Potentials of -1 V versus SCE can be reached with a carbon electrode and -2 V with a mercury electrode. Oxygen must be removed for measurements more negative than -0.1 V. This is done by bubbling with nitrogen.

22.2 Amperometric Electrodes—Measurement of Oxygen

Amperometry involves the measurement of current at a fixed potential; the magnitude of the current is linearly related to the concentration of the electroactive species. There are many amperometric sensors; an important example is an oxygen sensor, often called the oxygen electrode.

The oxygen electrode consists of a thin polymer film such as that of Teflon stretched over a platinum or gold cathode. The polymer film allows diffusion of gases but is impermeable to ions in solution (Figure 22.3). Oxygen diffuses through the membrane and is reduced at the cathode, producing a current. A potential suitable to cause reduction of the oxygen is applied between the oxygen-indicating electrode and the reference electrode, usually a silver–silver chloride electrode built into the probe. An electrolyte in solution or a gel is usually placed between the membrane and the glass insulator to provide electrical contact between the reference electrode and the indicating electrode.

The rate of diffusion of oxygen to the cathode is proportional to the partial pressure of the oxygen in the sample [for water, the dissolved oxygen (DO) content] to which the electrode is exposed, and the measured current is proportional to this. Measurements are made at atmospheric pressure. Gases (e.g., Cl_2) that are also reduced at the potential used will interfere. Hydrogen sulfide poisons the electrode.

The meter is precalibrated by exposure of the probe to samples with known oxygen content, for example, air with assumed 20.9% O_2 content or water saturated with either oxygen or air. At 37°C and sea level (P_{O_2} of 159 torr), air-saturated water contains 5.6 μL O_2 per milliliter and oxygen-saturated water contains 28 μL O_2 per milliliter. See Reference 7 for a discussion of the calibration of electrodes and the calculation of P_{O_2} and oxygen concentration.

Amperometric oxygen sensors are often used in biochemistry to follow the consumption or release of oxygen in biochemical and enzymatic reactions. In the clinical laboratory, the concentrations of enzymes or substrates whose reactions involve consumption of oxygen are often determined by measuring the rate of oxygen consumption.

The oxygen electrode is widely used to measure DO in natural water, and the terms DO probes and DO meters are widely used. DO is usually expressed in the units of mg O_2/L H_2O; for air-saturated water at sea level, DO will range from 8.56 mg/L at 0°C to 5.95 mg/L at 22°C. The DO levels are critical to aquatic life. Saturation DO levels as a function of temperature are widely tabulated, see, for example, http://www.dnr.mo.gov/env/esp/wqm/DOSaturationTable.htm.

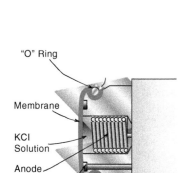

FIGURE 22.3   Details of an oxygen electrode. (Courtesy of Arthur H. Thomas Company.)

"O" Ring
Membrane
KCl Solution
Anode coil
Cathode ring

22.3 Electrochemical Sensors: Chemically Modified Electrodes

Like potentiometric electrodes discussed in Chapter 20, amperometric devices constitute another class of popular electrochemical sensors. In recent years there has been a great deal of interest in the development of various types of electrochemical sensors that exhibit increased selectivity or sensitivity. These enhanced measurement capabilities of amperometric sensors are achieved by chemical modification of the electrode surface to produce **chemically modified electrodes** (CMEs).

An enzyme layer can impart some chemical selectivity to the electrode.

All chemical sensors consist of a **transducer**, which transforms the response to the analyte into a signal that can be detected (a current in the case of amperometric sensors). The response must be chemically selective; this may be accomplished by a

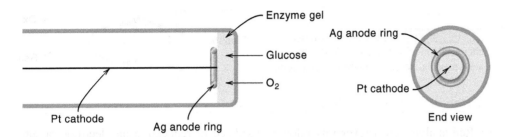

FIGURE 22.4 Amperometric glucose electrode

chemically selective layer. The transducer may be optical (e.g., a fiber-optic cable sensor), electrical (potentiometric, amperometric), thermal, and so on. We are concerned here with amperometric transducers.

Enzyme-Based Electrodes for Measuring Substrates

Enzymes are often utilized in a chemically modified electrode layer to impart the selectivity needed. We discussed potentiometric enzyme electrodes in Chapter 20. An example of an amperometric enzyme electrode is the glucose sensor, illustrated in Figure 22.4. The enzyme glucose oxidase is immobilized in a gel (e.g., acrylamide) and coated on the surface of a platinum wire cathode. The gel also contains a chloride salt and makes contact with a silver–silver chloride ring to complete the electrochemical cell. The enzyme *glucose oxidase* catalyzes the aerobic oxidation of glucose as follows:

$$\text{glucose} + O_2 + H_2O \xrightarrow{\text{glucose oxidase}} \text{gluconic acid} + H_2O_2 \qquad (22.5)$$

(see Chapter 23). A potential (ca. +0.6 V vs. Ag/AgCl) is applied to the platinum electrode at which H_2O_2 is electrochemically oxidized:

$$H_2O_2 \rightarrow O_2 + 2H^+ + 2e^- \qquad (22.6)$$

Glucose and oxygen from the test solution diffuse into the gel where their reaction is catalyzed to produce H_2O_2; part of this diffuses to the platinum cathode where it is oxidized to give a current that is proportional to the glucose concentration. The remainder eventually diffuses back out of the membrane. Reducing agents like ascorbic acid (vitamin C) can interfere in oxidative measurements. In one embodiment, the glucose sensor is covered with a thin cation exchange membrane that will not allow a negative ion like ascorbate to pass through but neutral glucose is easily transported. Yet another alternative design of a glucose electrode is to coat the membrane of an oxygen electrode with the glucose oxidase gel. Then the depletion of oxygen due to the reaction is measured.

Other examples of amperometric enzyme sensors based on the measurement of oxygen or hydrogen peroxide include electrodes for the measurement of galactose in blood (using the enzyme galactose oxidase), oxalate in urine (using the enzyme oxalate oxidase), and cholesterol in blood serum (using the enzyme cholesterol oxidase). Ethanol is determined by reacting with a cofactor, nicotinamide adenine dinucleotide (NAD^+) in the presence of the enzyme alcohol dehydrogenase to produce the reduced form of NAD^+, NADH, which is then electrochemically oxidized as the measurand. Lactate in blood is similarly determined (using the enzyme lactate dehydrogenase).

A redox mediator layer catalyzes the electrochemical reaction, so smaller potentials are required.

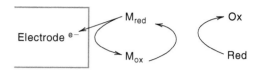

FIGURE 22.5 Redox mediator-based chemically modified elecrode. Red is the analyte being measured, in the reduced form.

Catalytic Electrodes—Redox Mediators

There are standard chemical equivalents of redox electrocatalysis. The oxidation of As(III) by Ce(IV) is very slow but proceeds rapidly if iodide is added - Ce(IV) rapidly oxidizes iodide to iodine - iodine is then rapidly reduced back to iodide by As(III).

Often analytes are irreversibly (slowly) oxidized or reduced at an electrode; a substantial overpotential beyond the thermodynamic redox potential (E^0) is needed for electrolysis to occur at a reasonable rate. This problem of slow electron transfer kinetics has spawned much research in the development of *electrocatalysts*, which may be covalently attached to the electrode, chemisorbed, or trapped in a polymer layer. The basis of electrocatalytic CMEs is illustrated in Figure 22.5. Red represents the analyte in the reduced form, which is irreversibly oxidized, and Ox is its oxidized form. The redox mediator is electrochemically reversible and is oxidized at a lower potential. The analyte reacts rapidly with the oxidized form of the mediator, M_{ox}, to produce M_{red}, which is immediately oxidized at the electrode surface. The electrochemical reaction takes place near the thermodynamic E^0 value of the mediator. If a lower potential is applied, there is less chance for interference from other electrochemically active species. Electrochemical mediators include ruthenium complexes, ferrocene derivatives, and *o*-hydroxybenzene derivatives. Mediators such as methylene blue catalyze the oxidation of H_2O_2 so that a potential of only about +0.2 V versus Ag/AgCl needs to be applied, instead of the usual +0.6 V.

A protective layer enhances selectivity and reduces chemical fouling of the electrode.

Electrodes are sometimes coated with protective layers to prevent fouling from larger molecules (e.g., proteins). A layer of cellulose triacetate, for example, will allow the small H_2O_2 molecule to pass but not the larger ascorbic acid molecule present in biological fluids, which is oxidized at the same potential. Anionic Nafion membranes allow cations but not anions to pass through.

Ultramicroelectrode responses are independent of the diffusion layer thickness and of flow. They exhibit increased signal-to-noise ratio.

22.4 Ultramicroelectrodes

Amperometric electrodes made on a microscale, on the order of 5 to 30 μm diameter, possess a number of advantages. The electrode is smaller than the diffusion layer thickness. This results in enhanced mass transport that is independent of flow, increased signal-to-noise ratio, and electrochemical measurements can be made in high-resistance media, such as nonaqueous solvents. An S-shaped sigmoid current–voltage curve is recorded in a quiet solution instead of a peak-shaped curve because of the independence from the diffusion layer. The limiting current, i_l, of such microelectrodes is given by

$$i_l = 2nFDCd \tag{22.7}$$

where n is the electron change, F is the Faraday constant, D is the diffusion coefficient, C is the concentration, and d is the electrode diameter.

There are various ways of constructing ultramicroelectrodes. A typical construction is shown in Figure 22.6. The microdisk is the electrode. Electrodes are typically made from carbon fiber, and they can be used for electrooxidation measurements at positive potentials in various microenvironments. Wang and Angnes, for example, have described co-deposition of rhodium and glucose oxidase on carbon fiber ultramicroelectrodes to construct miniature glucose sensors (*Anal. Chem.* **64** (1992) 456). Ultramicroelectrodes generate currents that are a few nanoamperes or less, and sensitive and well-shielded equipment is required for measurements. The

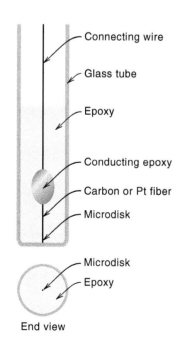

- Connecting wire
- Glass tube
- Epoxy
- Conducting epoxy
- Carbon or Pt fiber
- Microdisk

- Microdisk
- Epoxy

End view

FIGURE 22.6

Ultramicroelectrode construction.

photograph of a commercially available carbon fiber ultramicroelectrode is shown in Figure 22.7.

22.5 Microfabricated Electrochemical Sensors

Microfabrication technologies have had a radical impact on many areas of our life. Microfabrication of electrochemical sensors has made possible mass production of such devices and they are now widely used in health care applications, especially for the measurement of chemical constituents of interest in blood. Potentiometric pH sensors form the basis of measurement of blood pH, pCO_2, and urea [blood urea nitrogen (BUN), via the enzyme urease that generates ammonia from urea], and ion-selective electrode sensors are used for the measurement of sodium, potassium, chloride, and ionic calcium. Amperometric sensors are used for the measurement of glucose, lactate, cholesterol, and creatinine (via the production of H_2O_2 with appropriate oxidase enzymes, often more than one enzyme is needed for the desired reaction chain) and pO_2 by reduction to OH^-. The measurement of solution electrical conductivity (is discussed in Chapter 16; although it is a relatively simple electrical measurement of a bulk solution property, it is not an *electrochemical* measurement–no chemistry is carried out at the electrodes. However, conductometric sensors are frequently discussed in conjunction with electrochemical sensors; conductometric measurements are used for measuring blood hematocrit levels. The structures of some of these sensors are given in the text **website**; see Microfabricated Sensors for Blood Chemistry.pdf in the supplement to Chapter 22.

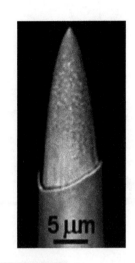

FIGURE 22.7 Carbostar 1, voltammetric carbon fiber ultramicroelectrode. Courtesy of Kation Scientific USA (www.Kationscientific.com).

22.6 Micro and Ultramicroelectrode Arrays

Communication between nerves (neural signaling) occurs via release and reception of molecules called neurotransmitters; biogenic amines such as dopamine constitute a typical example. Many neurotransmitter molecules are electroactive and can be electrochemically sensed. To map the status of nerve cells (which neuron is "firing," i.e., releasing signaling molecules) that grow in close proximity, isolated individual electrodes in close proximity are needed. Arrays of electrodes were originally constructed using bundles of fine wires to serve the needs of such electrophysiological studies. The ability to fabricate very closely spaced individually addressable electrodes in almost any imaginable pattern by photolithographic techniques has revolutionized such studies. Often the electrodes are patterned on glass or ITO (indium tin oxide) so that they can be simultaneously observed through an inverted microscope. See http://www.multichannelsystems.com/products-mea/microelectrode-arrays for examples of complex multielectrode array patterns.

The ability to have closely spaced electrodes that can be made of a variety of materials (virtually every noble metal in various combination has been used) is of utility in other areas. With closely spaced electrode arrays, for reversible redox systems an analyte can be oxidized at one electrode and the oxidized form can be reduced on the next and the signal can be considerably amplified. See http://abtechsci.com/labproducts.html to see many types of commercially available interdigitated microelectrodes (IMEs) and individually addressable microelectrodes, both in an array format. The URL http://www.abtechsci.com/ime-animated.html shows an animation of such an electrode array with successively increasing magnification steps. Closely spaced microelectrode arrays have also expanded the power of stripping voltammetry techniques. In anodic stripping voltammetry, for example, the working electrode is first held at a considerably reducing potential with a stirred sample such that many metal ions present in the sample that can be reduced at that potential are deposited and thus concentrated on the electrode surface. After the deposition step, the electrode potential is slowly scanned to anodic values. The most easily oxidized metal first comes off the electrode and then the next most easily oxidizable metal and so on. In such anodic stripping voltammetry (ASV) techniques, the potential at which the current peak occurs is characteristic of the particular metal and the magnitude of the current peak is related to the concentration of the metal present. Ultramicroelectrode arrays make such techniques highly sensitive and portable. A description of a field-usable instrument can be found in www.tracedetect.com, with a demonstrative video clip in http://www.tracedetect.com/video/tracedetect_yt.mp4.

Questions

1. Define back-emf, overpotential, and iR drop.
2. Define half-wave potential, depolarizer, DME, residual current, and voltammetry.
3. Why do we use cyclic voltammetry?
4. Distinguish between voltammetry and amperometry.
5. Give two reasons for using a supporting electrolyte in voltammetry.
6. Why the reference electrode is placed near the working electrode in a three-electrode cell?
7. A solution contains about 10^{-2} M Fe^{3+} and 10^{-5} M Pb^{2+}. It is desired to analyze for the lead content polarographically. Fe^{3+} is reduced to Fe^{2+} at all potentials accessible with the DME up to -1.5 V versus SCE, and is reduced along with Fe^{2+} to the metal at potentials more negative than -1.5 V. Pb^{2+} is reduced at -0.4 V. Suggest a scheme for measuring the lead polarographically.
8. What effect does complexation have on the voltammetric reduction of a metal ion?
9. What is a chemically modified electrode?
10. What is the function of an electrocatalyst in a chemically modified electrode?
11. What are the advantages of an ultramicroelectrode?

Problems

Voltammetry/Amperometry

12. The limiting current of lead in an unknown solution is 6.60 μA. One milliliter of a 2.00×10^{-3} M lead solution is added to 15.0 mL of the unknown solution and the limiting current of the lead is increased to 13.5 μA. What is the concentration of lead in the unknown solution?

13. Iron(III) is polarographically reduced to iron(II) at potentials more negative than about $+0.4$ V versus SCE and is further reduced to iron(0) at -1.5 V versus SCE. Iron(II) is also reduced to the metal at -1.5 V. A polarogram is run (using a DME) on a solution containing Fe^{3+} and/or Fe^{2+}. A current is recorded at zero applied volts, and its magnitude is 12.5 μA. A wave is also recorded with $E_{1/2}$ equal to -1.5 V versus SCE, and its height is 30.0 μA. Identify the iron species in solution (3+ and/or 2+) and calculate the relative concentration of each.

PROFESSOR'S FAVORITE QUESTION

Contributed by Professor Charles S. Henry, Colorado State University

14. Substance A is electrolyzed at a constant current. During such an experiment, what happens to the potential at the working electrode? Assume that the reaction consists of a reduction of A to B where the initial solution consists of all A.

   (a) The potential remains constant throughout the entire experiment.

   (b) The potential becomes more positive as more and more of A is reduced to B.

   (c) The potential becomes more negative as more and more of A is reduced to B.

   (d) The potential does not change in a constant current experiment.

1. Which of the following does not affect the polarization at an electrode?

   (a) Diffusion of the analyte to the electrode surface

   (b) Diffusion of the product from the electrode surface

   (c) Convection

   (d) The standard cell potential for the redox couple

   (e) A significant activation barrier for the reaction

2. The condition of full polarization is observed in which of the following forms of electrochemistry?

   (a) Potentiometry   (b) Voltammetry

   (c) Coulometry   (d) Electrogravimetry

   (e) Ohmetry

3. A large excess of an inert electrolyte is used in voltammetry, this ensures that a redox active analyte arrives at the electrode surface primarily by

   (a) diffusion in a concentration gradient at the electrode surface.

   (b) migration in the electrical field at the electrode surface.

   (c) convective transport in a thermal gradient at the electrode surface.

   (d) electrostatic attraction to an image charge in the electrode.

4. The double vertical line "‖" in the standard notation for a voltaic cell represents:

   (a) a phase boundary.   (b) gas electrode.

   (c) a wire (metal) connection.   (d) a salt bridge.

   (e) a standard hydrogen electrode

5. Amongst the following the false statement is

   (a) Oxidation and reduction half-reactions occur at electrodes in electrochemical cells.

   (b) All electrochemical reactions involve the transfer of electrons.

   (c) Reduction occurs at the cathode.

   (d) Oxidation occurs at the anode.

   (e) All voltaic (galvanic) cells involve the use of electricity to initiate nonspontaneous chemical reactions.

6. Which among the following is not a feature of the immobilized enzymes?

   (a) They cannot be re-used.

   (b) It produces reproducible results.

   (c) Stability exists.

   (d) Same catalytic activity is present for number of analysis.

7. The following principle is used for the transducers employed in the bulk of enzyme electrodes

   (a) Amperometric   (b) Optical

   (c) Magnetic   (d) Colorimetric

Recommended References

General

1. J. J. Lingane, *Electroanalytical Chemistry*, 2nd ed. New York: Interscience, 1958. A classic. Excellent general text.

2. J. Wang, *Analytical Electrochemistry*, 2nd ed. New York: John Wiley, 2000. Good introduction to modern electroanalytical techniques.

3. A. J. Bard and L. R. Faulkner, *Electrochemical Methods: Fundamentals and Applications*, 2nd ed. New York: Wiley, 2001

4. A. J. Bard and M. Stratmann, editors-in-chief, *Encyclopedia of Electrochemistry*, 10 volumes + index volume, New York: Wiley.Vol. 3, *Instrumentation and Electroanalytical Chemistry*, P. Unwin, ed., 2003;Vol. 9, *Bioelectrochemistry*, G. S. Wilson, ed., 2002; Vol. 10, *Modified Electrodes*, I. Rubinstein and M. Fujihira, eds., 2003.

Voltammetry/Amperometry

5. M. R. Smyth and J. G. Vos, *Analytical Voltammetry*. Amsterdam: Elsevier Science, 1992.
6. R. N. Adams, *Electrochemistry at Solid Electrodes*. New York: Marcel Dekker, 1969. A classic, in its 5th printing.
7. M. A. Lessler and G. P. Brierley, in D. Glick, ed., *Methods of Biochemical Analysis*, Vol. 17. New York: Interscience, 1969, p. 1. Describes the oxygen electrode.
8. Web: www.pineinst.com, (Go to Books, Manuals and Documents, Educational Materials for Educator's Reference Guide for Electrochemistry). An excellent 66-page tutorial on the principles of voltammetry and instrumentation. You can download a sample voltammetry experiment from the Pine Instrument Company website.

Chapter 23

KINETIC METHODS OF ANALYSIS

"There is nothing permanent but change."
—Heraclitus

KEY THINGS TO LEARN FROM THIS CHAPTER

- First-order reactions, half-life (key Equations: 23.3, 23.4)
- Second-order reactions, half-life (key Equations: 23.7, 23.8)
- Enzyme catalysis—Michaelis constant (key Equation: 23.14); using spreadsheets to calculate
- Substrate determinations (Table 23.1), enzyme determinations (Table 23.2)

We mentioned in Chapters 5 and 21 the use of catalysts to alter the reaction rate in certain redox reactions; the titration reaction of As(III) with Ce(IV) is catalyzed by OsO_4. The catalyst is added in sufficiently high concentration to make the reaction occur immediately. If the catalyst concentration is low and the reaction slow, then we can measure the rate of the reaction and relate it to the catalyst concentration. In this chapter, we describe the basic kinetics of rate-limited reactions. Then we discuss reactions catalyzed by specific catalysts called enzymes and the measurement of the reaction rate to either determine the enzyme activity (concentration) or the concentration of the catalyst's substrate by adding a fixed amount of enzyme to the solution.

Kinetic methods are particularly useful in clinical chemistry. Many analyses are based on enzymatic reactions. The rate of the reaction may be measured, or it may be allowed to go to completion before measurement. For example, glucose reacts with oxygen in the presence of the enzyme, glucose oxidase, to form gluconic acid and hydrogen peroxide. The latter can be measured spectrophotometrically or electrochemically and related to the glucose concentration. This, in fact, is probably the most widely performed chemical analysis in the world. Many dehydrogenase enzyme reactions are used in which the analyte substrate reacts with nicotinamide adenine dinucleotide (NAD^+) to produce NADH, which is then measured. An example is the determination of lactic acid, using the enzyme lactic acid dehydrogenase. The determination of enzymes themselves are important, many reacting with a substrate to form a product that then reacts with NADH, for example, glutamic-oxaloacetic transaminase (GOT) determination.

23.1 Basics of Kinetics

Kinetics is the description of **reaction rates**. The **order** of a reaction defines the dependence of reaction rates on the concentrations of reacting species. Order is determined empirically and is not necessarily related to the stoichiometry of the reaction. Rather, it is governed by the **mechanism** of the reaction, that is, by the number of species that must collide for the reaction to occur.

The order of a reaction defines the number of species that must react, not the ratio in which they react (stoichiometry).

First-Order Reactions

Reactions in which the rate of the reaction is directly proportional to the concentration of a single substance are known as first-order reactions. Consider the reaction

$$A \rightarrow P \tag{23.1}$$

Substance A might be a compound that is decomposing to one or more products. The rate of the reaction is equal to the rate of disappearance of A, and it is proportional to the concentration of A:

$$-\frac{dA}{dt} = k[A] \tag{23.2}$$

This is a **rate expression**, or **rate law**. The minus sign is placed in front of the term on the left side of the equation to indicate that A is decreasing in concentration as a function of time. The constant k is the **specific rate constant** at the specified temperature and has the dimensions of reciprocal time, for example, s^{-1}. The **order of a reaction** is the sum of the exponents to which the concentration terms in its rate expression are raised. Thus, this is a first-order reaction and its rate depends only on the concentration of A.

Equation 23.2 is known as the **differential form** of the first-order rate law. The **integrated form** of the equation is

$$\log[A] = \log[A]_0 - \frac{kt}{2.303} \tag{23.3}$$

where $[A]_0$ is the initial concentration of $A(t = 0)$ and $[A]$ is its concentration at time t after the reaction is started. This equation gives the amount of A that has reacted after a given time interval. It is a straight-line equation, and if t is plotted versus $\log[A]$ (which can be measured at different times), a straight line with slope $-k/2.303$ and intercept $\log[A]_0$ is obtained. Thus, the rate constant can be determined.

Note that, from Equation 23.2, the rate of the reaction (*not* the rate constant) will decrease as the reaction proceeds because the concentration of A decreases. Since $[A]$ decreases logarithmically with time (see Equation 23.3), it follows that the rate of the reaction will decrease exponentially with time. The time required for one-half of a substance to react is called the **half-life** of the reaction, $t_{1/2}$. The ratio of $[A]/[A]_0$ at this time is $\frac{1}{2}$. By inserting this in Equation 23.3 and solving for $t_{1/2}$, we see that for a first-order reaction

$$t_{1/2} = \frac{0.693}{k} \tag{23.4}$$

After the reaction is half complete, then one-half of the remaining reacting substances will react in the same time $t_{1/2}$, and so on. This is the exponential decrease we mentioned. Theoretically, it would take an infinitely long time for the reaction to go to completion, but for all practical purposes, it is complete (99.9%) after 10 half-lives. It is important to note that the half-life, and hence, the time for the reaction to go to completion is independent of the concentration for first-order reactions.

Radioactive decay is an important example of a first-order reaction.

Second-Order Reactions

Suppose we have the following reaction:

$$A + B \rightarrow P \tag{23.5}$$

The rate of the reaction is equal to the rate of disappearance of either A or B. If it is empirically found to be

$$-\frac{dA}{dt} = -\frac{dB}{dt} = k[A][B] \tag{23.6}$$

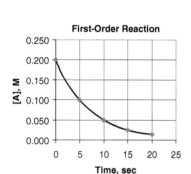

First-Order Reaction

The reaction is half-complete every 5 seconds. Shown here are four half-reactions.

The rate of reaction slows down with time.

Consider a reaction complete after 10 half-lives. It really takes an infinite time for completion.

then the reaction is first order with respect to [A] and to [B] and second order overall (the sum of the exponents of the concentration terms is 2). The specific reaction rate constant has the dimensions of reciprocal molarity and time, for example, $M^{-1}s^{-1}$.

The integrated form of Equation 23.6 depends on whether the initial concentrations of A and B ($[A]_0$ and $[B]_0$) are equal. *If they are equal*, the equation is

$$kt = \frac{[A]_0 - [A]}{[A]_0[A]} \qquad (23.7)$$

If $[A]_0$ and $[B]_0$ *are not equal*, then

$$kt = \frac{2.303}{[B]_0 - [A]_0} \log \frac{[A]_0[B]}{[B]_0[A]} \qquad (23.8)$$

If the concentration of one species, say B, is very large compared with the other and its concentration remains essentially constant during the reaction, then Equation 23.6 reduces to that of a first-order rate law:

$$-\frac{dA}{dt} = k'[A] \qquad (23.9)$$

By making the concentration of one species large compared to the other, a second-order reaction behaves as a (pseudo) first-order reaction.

where k' is equal to $k[B]$; the integrated form becomes

$$kt = \frac{2.303}{[B]_0} \log \frac{[A]_0}{[A]} \qquad (23.10)$$

Since $[B]_0$ is constant, Equation 23.10 is identical in form to Equation 23.3. This is a **pseudo first-order reaction**.

The half-life of a second-order reaction in which $[A]_0 = [B]_0$ is given by

$$t_{1/2} = \frac{1}{k[A]_0} \qquad (23.11)$$

The half-life of a second-order reaction depends on the concentrations.

Thus, unlike the half-life of a first-order reaction, the half-life here depends on the initial concentration.

A reaction between A and B is not necessarily second order. Reactions of a fraction rate order are common. A reaction such as $2A + B \rightarrow P$ may be third order (rate $\propto [A]^2[B]$), or it may be second order (rate $\propto [A][B]$), or a more complicated order (even a fractional order).

Reaction Time

The time for a reaction to go to completion will depend on the rate constant k and, in the case of second-order reactions, the initial concentrations. A first-order reaction is essentially instantaneous if k is greater than 10 s^{-1} (99.9% complete in less than 1 s). When k is less than $10^{-3}s^{-1}$, the time for 99.9% reaction exceeds 100 min. Although it is more difficult to predict the time for second-order reactions, they can generally be considered to be instantaneous if k is greater than about 10^3 or $10^4 M^{-1}s^{-1}$. If k is less than $10^{-1}M^{-1}s^{-1}$, the reaction requires hours for completion.

23.2 Catalysis

The rates of some reactions can be accelerated by the presence of a **catalyst**. A catalyst can be defined as a substance that alters the rate of a reaction without shifting the equilibrium and without itself undergoing any net change. We are interested in catalysis in which the rate of the reaction is proportional to the concentration of the catalyst. This forms the basis for the quantitative analysis of a catalyst. The rate of decrease of a reactant or increase of a product is measured and related to the catalyst concentration. These techniques are extremely sensitive for many catalysts.

Many analyses carried out by redox methods utilize catalysis, since the reactions may be slow. These same reactions can be used to measure the concentration of catalysts. The most commonly applied catalytic method is for the determination of traces of iodide from its catalytic action on the Ce(IV)-As(III) reaction, in which As(III) is oxidized to As(V), and yellow Ce(IV) is reduced to colorless Ce(III). This is called the Sandell–Kolthoff reaction, developed in 1937: E. B. Sandell and I. M. Kolthoff, "Microdetermination of iodide by catalytic method," *Mikochim. Acta*, **1** (1937) 9–25.

The Sandell–Kolthoff reaction is widely used for the determination of urinary iodine (UI). UI is used as the measure of iodine sufficiency. A subject population is considered iodine deficient if the median UI is < 100 μg/L. In affluent countries, UI is measured by spiking urine samples with a known concentration of ^{129}I (this isotope of iodine does not occur naturally), diluting and ICP-MS analysis. In developing countries, the urine sample is digested to oxidize organics (in earlier years perchloric acid digestion was used, presently the preferred method is to add 1 M ammonium persulfate (200 μL/mL urine), incubate for 30 min at 95°C), allow to cool and add Ce(IV) and As(III) solutions. The absorbance is then measured at 410 nm (where Ce(IV) absorbs) after a preset period, typically 10 or 15 min. A greater disappearance of Ce(IV) (less absorbance) indicates a greater amount of iodine. UI levels down to 10 μg/L can be measured. The method is easily automated by continuous flow methods or using microplate readers (the microplate readers can provide measurements as a function of time, increasing data reliability) allowing for many samples to be analyzed rapidly. The U.S. Food and Drug Administration uses this as the reference method for determining iodine in foods, after perchloric acid digestion; good throughput is achieved using automated segmented flow analyzers.

23.3 Enzyme Catalysis

Enzymes are proteins that are nature's catalysts in the body.

Enzymes are remarkable naturally occurring proteins that catalyze *specific* reactions with a high degree of efficiency. Enzymes range in formula weight from 10,000 to 2,000,000. They are, of course, intimately involved in biochemical reactions in the body, that is, the life process itself. The determination of certain enzymes in the body is, therefore, important in the diagnosis of diseases. But aside from this, enzymes have proved extremely useful for the determination of **substrates**, the substances whose reaction the enzymes catalyze.

Enzyme: Kinetics

We can describe the rate equation for enzyme reactions from a simple reaction model. The typical enzyme-catalyzed reaction can be represented as follows:

$$E + S \underset{k_2}{\overset{k_1}{\rightleftharpoons}} ES \overset{k_3}{\longrightarrow} P + E \tag{23.12}$$

The reaction rate is first order with respect to substrate and enzyme. If [S] is large, the reaction becomes zero order with respect to S.

where E is the enzyme, S is the substrate, ES is the **activated complex** that imparts a lower energy barrier to the reaction, P is the product(s), and the k's are the rate constants for each step. That is, the enzyme forms a complex with the substrate, which then dissociates to form product (Figure 23.1). The rate of reaction, R, is proportional to the complex concentration and, therefore, to the substrate and enzyme concentrations:

$$R = k_3[ES] = k[S][E] \tag{23.13}$$

Assuming that k_1 and k_2 are much larger than k_3, we find the rate of the reaction to be limited by the rate of dissociation of the activated complex.

The dependence of an enzyme-catalyzed reaction rate on substrate concentration is illustrated in Figure 23.2. An enzyme is characterized by the number of molecules of substrate it can complex per unit time and convert to product, that is, the **turnover number**. As long as the substrate concentration is small enough with respect to the enzyme concentration that the turnover number is not exceeded, the reaction rate is directly proportional to substrate concentration, that is, it is first order with respect to substrate (Equation 23.13). If the enzyme concentration is held constant, then the overall reaction is first order and directly proportional to substrate concentration ($k[E]$ = constant in Equation 23.13). This serves as the basis for substrate determination.[1] However, if the amount of substrate exceeds the turnover number for the amount of enzyme present, the enzyme becomes *saturated* with respect to the number of molecules it can complex (saturated with respect to substrate), and the reaction rate reaches a maximum value. At this point, the reaction becomes independent of further substrate concentration increases, that is, becomes **pseudo zero order** if the enzyme concentration is constant (Figure 23.2); in Equation 23.13, [ES] becomes constant and R = constant.

When the enzyme is saturated with respect to substrate, then the overall reaction is first order with respect to enzyme concentration ($k[S]$ = constant in Equation 23.13). This becomes the basis for enzyme determination since a linear relationship between reaction rate and enzyme concentration will exist. Since substrate is consumed in the reaction, however, it must be kept at a high enough concentration that the reaction remains zero order with respect to substrate during the time of the reaction (i.e., the enzyme remains saturated). Eventually, at high enzyme concentrations, insufficient substrate will be available for saturation, and a plot similar to Figure 23.2 will result.

Enzymes: Properties

The rate of an enzymatic reaction depends on a number of factors, including the temperature, pH, ionic strength, and so forth. The rate of the reaction will increase as the temperature is increased, up to a point. Above a certain temperature, the activity of the enzyme is decreased because, being a protein, it becomes *denatured*, that is, the tertiary structure of the enzyme is destroyed as hydrogen bonds are broken. Protein tertiary structure refers to the 3-D structure, determined by bonding interactions between amino acids, including hydrogen bonds, salt bridges, dissulfide bonds, and nonpolar hydrophobic interactions. The steric nature of an enzyme is critical in its catalytic mechanism. Most animal enzymes become denatured at temperatures above about 40°C.

As with other catalytic reactions, temperature changes as small as 1 or 2°C may result in changes as high as 10 to 20% in the reaction rate under analytical conditions. So it is important that the temperature be controlled during the measurements of enzyme reactions.

Enzymes should be stored at 5°C or less since they are eventually deactivated over a period of time, even at moderate temperatures. Some enzymes lose activity when frozen.

The reaction rate will be at a maximum at a certain pH, owing to complex acid–base equilibria such as acid dissociation between the substrate, the activated complex, and the products. Also, the maximum rate may depend on the ionic strength and on the type of buffer used. For example, the rate of aerobic oxidation of glucose in the presence of the enzyme glucose oxidase is maximum in an acetate buffer at pH 5.1, but in a phosphate buffer of the same pH, it is decreased.

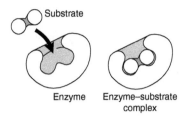

FIGURE 23.1 Mechanism of enzyme activity. [D. Leja, National Human Genome Research Institute. Reproduced by permission.]

When the enzyme is saturated with substrate, the reaction rate is proportional to the enzyme concentration.

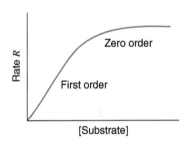

FIGURE 23.2 Dependence of enzyme-catalyzed reaction rate on substrate concentration. At high concentrations, the enzyme becomes saturated with substrate and the reaction rate becomes maximum and constant since [ES] becomes constant (Equation 23.13).

Above the optimum temperature, the enzyme becomes denatured. When you cook an egg, the protein is denatured.

There is also an optimum pH for enzyme reactions.

---

[1] Substrates need not be determined from the reaction rate. Instead, the reaction may be allowed to proceed until the substrate is completely converted to product. The concentration of the product is measured before (blank) and after the reaction. Each of these techniques is discussed in more detail below under the determination of enzymes and of enzyme substrates.

The **activity** of an enzyme preparation will vary from one source to another because the enzymes are usually not purified to 100% enzyme. That is, the percent enzyme will vary from one preparation to another. The activity of a given preparation is expressed in **international units** (I.U.). An international unit has been defined by the International Union of Biochemistry as "the amount that will catalyze the transformation of one micromole of substrate per minute under defined conditions." The defined conditions will include temperature and pH. For example, a certain commercial preparation of glucose oxidase may have an activity of 30 units per milligram. Thus, for the determination of a substrate, a certain number of units of enzyme is taken. The **specific activity** is the units of enzyme per milligram of *protein*. **Molecular activity** is defined as units per molecule of enzyme, that is, it is the number of molecules of substrate transformed per minute per molecule of enzyme. The **concentration** of an enzyme in solution should be expressed as international units per milliliter or liter.

> Enzyme concentrations are usually expressed in activity and not molar units.

Enzyme: Inhibitors and Activators

Although enzymes catalyze only certain reactions or certain types of reaction, they are still subject to interference. When the activated complex is formed, the substrate is adsorbed at an *active site* on the enzyme. Other substances of similar size and shape may be adsorbed at the active site. Although adsorbed, they will not undergo any transformation. However, they do compete with the substrate for the active sites and slow down the rate of the catalyzed reaction. This is called **competitive inhibition**. For example, the enzyme succinic dehydrogenase will specifically catalyze the dehydrogenation of succinic acid to form fumaric acid. But other compounds similar to succinic acid will competitively inhibit the reaction. Examples are *other* diprotic acids such as malonic and oxalic acids. Competitive inhibition can be reduced by increasing the concentration of the substrate relative to that of the interferent so that the majority of enzyme molecules combine with the substrate.

Noncompetitive inhibition occurs when the inhibition depends only on the concentration of the inhibitor. This is usually caused by adsorption of the inhibitor at a site other than the active site but one that is necessary for activation. In other words, an inactive derivative of the enzyme is formed. Examples are the reaction of the heavy metals mercury, silver, and lead with sulfhydryl groups (–SH) on the enzyme. The sulfhydryl group is tied up by the heavy metal ($ESH + Ag^+ \rightarrow ESAg + H^+$), and this reaction is irreversible. This is why heavy metals are poisons; they inactivate enzymes in the body.

Some enzymes require the presence of a certain metal for activation, perhaps to form a complex of the proper stereochemistry. Any substance that will complex with the metal ion may then become an inhibitor. For example, magnesium ion is required as an activator for a number of enzymes. Oxalate and fluoride will complex the magnesium, and they are inhibitors. Activators of enzymes are sometimes called **coenzymes**.

Substrate inhibition sometimes occurs when excessive amounts of substrate are present. In cases such as this, the reaction rate actually decreases after the maximum velocity has been reached. This is believed to be due to the fact that there are so many substrate molecules competing for the active sites on the enzyme surfaces that they block the sites, preventing other substrate molecules from occupying them.

The Michaelis Constant

As explained previously, an enzyme at a given concentration eventually becomes saturated with respect to substrate as the substrate concentration is increased, and the reaction rate becomes maximum, R_{max}. The **Lineweaver–Burk equation** describes the relationship between the enzyme effectiveness as a catalyst and the maximum rate:

$$\frac{1}{R} = \frac{1}{R_{max}} + \frac{K_m}{R_{max}[S]}$$

(23.14)

where K_m is the **Michaelis constant**. The Michaelis constant is a measure of the enzyme activity and can be shown to be equal to $(k_2 + k_3)/k_1$ in Equation 23.12. It is also equal to the substrate concentration at one-half the maximum rate, $R_{max}/2$, as derived from Equation 23.14 by setting $R = R_{max}/2$. A plot of 1/[S] versus 1/R gives a straight line whose **intercept** is $1/R_{max}$ and whose **slope** is K_m/R_{max}. Thus, the Michaelis constant, which is characteristic for an enzyme with a substrate, can be determined.

Example 23.1

An enzyme reaction gave the following reaction rate absorbance data as a function of substrate concentration:

[S](M)	$R(\Delta A/\text{min})$
0.00400	0.093
0.0100	0.231
0.0400	0.569
0.0800	0.758
0.120	0.923
0.160	0.995
0.240	1.032

Prepare a spreadsheet for a Lineweaver–Burk plot and calculate the maximum rate and the Michaelis constant.

Solution

We will use the Excel regression program in Data Analysis, as illustrated in Chapter 11 at the end of Section 11.7, Excel Exercise, with the Excel result shown in the text website for that. Review this material as necessary.

	A	B	C	D	E	F	G	H	I
1	Lineweaver-Burk Plot								
2	[S], M	R, ΔA/min	1/[S], M⁻¹	1/R, ΔA⁻¹min					
3	0.004	0.093	250	10.753					
4	0.01	0.231	100	4.329					
5	0.04	0.569	25	1.757					
6	0.08	0.758	12.5	1.319					
7	0.12	0.923	8.333	1.083					
8	0.16	0.995	6.25	1.005					
9	0.24	1.032	4.167	0.969					
10									
11	SUMMARY OUTPUT		Input Y range D3:D9	. Input X range C3:C9					
12	Regression Statistics and ANOVA included in the text website version								
13									
14		Coefficients	andard Err	t Stat	P-value	Lower 95%	Upper 95%	ower 95.0%	Upper 95.0%
15	Intercept	0.736935925	0.082605	8.921256	0.000295	0.524594	0.949278	0.524594	0.94927755
16	X Variable	0.039525692	0.000807	48.99997	6.69E-08	0.037452	0.041599	0.037452	0.04159924
17									
18	Intercept = 1/R_max								
19	K_m (M) = slope x R_max =B16/B15=			0.053635					
20									
21	Uncertainty:								
22	K_m =		0.053635 ±		0.006111				
23	K_m = 0.054 ± 0.006M			[±B22*SQRT((0.000807/0.03955)^2+(0.0826/0.7369)^2)]					

A small K_m means a fast reaction rate and easy substrate saturation.

The significance of K_m is illustrated in Figure 23.3 (similar to Figure 23.2). When the reaction rate increases rapidly with substrate concentration, K_m is small (curve 1). The substrate that gives the lowest K_m for a given enzyme is often (but not necessarily) the enzyme's natural substrate, hence the reason for the rapid increase in rate with increased substrate concentration. A small K_m indicates that the enzyme becomes saturated at small concentrations of substrate. Conversely, a large K_m indicates that high concentrations of substrate are required to achieve maximum reaction velocity. In such a case, it would be difficult to achieve zero-order rate with respect to substrate, and the substrate would not be a good one for determining the enzyme.

Specificity of Enzyme

In general, there are four types of enzyme specificity: (1) **absolute specificity**, in which the enzyme will catalyze only one reaction; (2) **group specificity**, in which the enzyme will act on molecules with certain functional groups, such as amino, phosphate, or methyl groups; (3) **linkage specificity**, in which the enzyme will act on a particular type of chemical bond; and (4) **stereochemical specificity**, in which the enzyme will act on a particular steric or optical isomer.

In addition to the substrate being acted on, many enzymes require a second cosubstrate. Such a cosubstrate may activate many enzymes and is an example of a **cofactor** or **coenzyme** (described previously). An example is nicotinamide adenine dinucleotide (NAD^+), which is a cofactor for many dehydrogenase reactions by acting as a hydrogen acceptor:

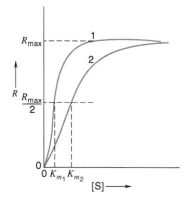

FIGURE 23.3 Illustration of relationship of Michaelis constant ($K_m = [S]$ at $R_{max}/2$), to reaction rate. Curve 1. Small K_m. Curve 2. Large K_m.

NAD$^+$ is a common cofactor in clinical chemistry measurements. The reactions are monitored by measuring the NADH concentration.

$$SH_2 + NAD^+ \xrightleftharpoons{enzyme} S + NADH + H^+ \qquad (23.15)$$

where SH_2 is the reduced form of the substrate, S is its oxidized (dehydrogenated) form, and NADH is the reduced form of NAD^+.

Nomenclature of Enzyme

Enzymes are classified according to the type of reaction and the substrate, that is, according to their reaction specificity and their substrate specificity. Most enzyme names end in "ase." Enzymes can be divided into four groups based on the kind of chemical reaction catalyzed: (1) Those that catalyze addition (*hydrolases*) or removal (*hydrases*) of water. Hydrolases include esterases, carbohydrases, nucleases, and deaminases, while hydrases include enzymes such as carbonic anhydrase and fumarase. (2) Those that catalyze the transfer of electrons: *oxidases* and *dehydrogenases*. (3) Those that catalyze transfer of a radical, such as *transaminases* (amino groups), *transmethylases* (methyl groups), or *transphosphorylases* (phosphate groups). (4) Those that catalyze splitting or forming of a C—C bond: *desmolases*. For example, α-glucosidase acts on any α-glucoside. The rate of reaction may be different for different glucosides. More generally, however, enzymes show absolute specificity to one particular substrate. Thus, glucose oxidase catalyzes the aerobic (oxygen) oxidation of glucose to gluconic acid plus hydrogen peroxide:

Glucose oxidase is used for determining glucose.

$$C_6H_{12}O_6 + O_2 + H_2O \xrightarrow{glucose\ oxidase} C_6H_{12}O_7 + H_2O_2 \qquad (23.16)$$

Actually, this enzyme shows almost complete specificity for β-D-glucose; α-D-Glucose reacts at a rate of 0.64 relative to 100 for the β form. In the latter form, all the hydrogens are axial and the hydroxyl groups are equatorial, allowing the molecule to lie down flat on the enzyme active site and form the enzyme–substrate complex. The α

form does not have the same arrangement of hydrogens and hydroxyls and cannot lie flat on the enzyme. Thus, the aerobic conversion of glucose (usually 36% α and 64% β) depends on the mutarotation of the α form to the β form. The mutarotation (equilibrium) is shifted as the β form is removed. Another enzyme, mutarotase, will affect the mutarotation, but this is usually not necessary. There is one other substance, 2-deoxy-D-glucose, that is affected by glucose oxidase. The relative rate of its reaction is about 10% of that of β-D-glucose, and it is usually not present in blood samples being analyzed for glucose.

There are thousands of enzymes in nature, and most of these exhibit absolute specificity.

Determination of Enzymes

Enzymes themselves can be analyzed by measuring the amount of substrate transformed in a given time or the product that is produced in a given time. The substrate should be in excess so that the reaction rate depends only on the enzyme concentration. The results are expressed as international units of enzyme. For example, the activity of a glucose oxidase preparation can be determined by measuring manometrically or amperometrically the number of micromoles of oxygen consumed per minute. On the other hand, the use of enzymes to develop specific procedures for the determination of substrates, particularly in clinical chemistry, has proved to be extremely useful. In this case, the enzyme concentration is in excess so the reaction rate is dependent on the substrate concentration.

> Enzyme activities are measured by determining the rate of substrate conversion, under pseudo zero-order substrate conditions.

Determination of Enzyme Substrates

Two general techniques may be used for measuring enzyme substrates. First, **complete conversion** of the substrate may be utilized. Before and following completion of the enzymatically catalyzed reaction, a product is analyzed or the depletion of a reactant that was originally in excess is measured. The analyzed substance (net change) is then related to the original substrate concentration. These reactions are often not stoichiometric with respect to the substrate concentration because of possible side reactions or instability of products or reactants. Also, the reaction may require extraordinarily long times for completion. For these reasons, the analytical procedure is usually standardized by preparing a calibration curve of some type in which the measured quantity is related to known concentrations or quantities of the substrate.

The second technique employed for substrate determination is the measurement of the **rate** of an enzymatically catalyzed reaction, as is used to determine enzyme activity. This may take one of three forms. First, the time required for the reaction to produce a preset amount of product or to consume a preset amount of substrate may be measured. Second, the amount of product formed or substrate consumed in a given time may be measured. These are single-point measurements (called *end-point measurements*) and require well-defined reaction conditions. They are easy to automate or may be performed manually. A third procedure is continuous measurement of a product or substrate concentration as a function of time to give the slope of the reaction rate curve, $\Delta c/\Delta t$. These are the so-called true rate measurements. The measurements must generally be made during the early portion of the reaction where the rate is pseudo first order.

> The enzyme is not consumed, so its concentration just needs to be held constant for rate measurements.

Rate methods are generally more rapid than end-point methods or complete conversion reactions. Complete conversion reactions, on the other hand, are less subject to interference from enzyme inhibitors or activators as long as sufficient time is allowed for complete conversion. See also the discussion of enzyme electrodes in Chapter 20 for a different approach to measuring substrates.

Example 23.2

The blood alcohol content of an individual is determined enzymatically by reacting ethanol with NAD^+ in the presence of the enzyme alcohol dehydrogenase to produce NADH (Table 23.1). The rate of formation of NADH is measured at 340 nm (Figure 23.4). The following absorbances are recorded for a 0.100% (wt/vol) alcohol standard and the unknown, treated in the same way. Use a spreadsheet to calculate the unknown concentration and its uncertainty, from the slopes of the absorbance changes.

$T(s)$	A_{std}	A_{unk}
0	0.004	0.003
20	0.052	0.036
40	0.099	0.070
60	0.147	0.098
80	0.201	0.132
100	0.245	0.165

We will use the Excel regression program in Data Analysis, as illustrated in Chapter 11 at the end of Section 11.7, Excel Exercise, with the Excel result shown in the text website for that. Review this material as necessary. You will prepare SUMMARY OUTPUTS for both the standard and the unknown.

Solution

	A	B	C	D	E	F	G	H	I
1	T (s)	A_{std}	A_{unk}						
2	0	0.004	0.003						
3	20	0.052	0.036						
4	40	0.099	0.07						
5	60	0.147	0.098						
6	80	0.201	0.132						
7	100	0.245	0.165						
8									
9	(Regression Statistics and ANOVA for each included in text website version)								
10	Regression for standard:								
11	SUMMARY OUTPUT (0.100 wt% standard); Input Y range B2:B7, Input X range A2:A7:								
12									
13		Coefficient	andard Err	t Stat	P-value	Lower 95%	Upper 95%	ower 95.0%	Jpper 95.0%
14	Intercept	0.003238	0.001603	2.0202	0.11347	-0.00121	0.007688	-0.00121	0.0076883
15	X Variable 1	0.002429	2.65E-05	91.7468	8.5E-08	0.002355	0.002502	0.002355	0.0025021
16									
17	Regression for unknown:								
18	SUMMARY OUTPUT; Input Y range C2:C7, Input X range A2:A7:								
19									
20		Coefficient	andard Err	t Stat	P-value	Lower 95%	Upper 95%	ower 95.0%	Jpper 95.0%
21	Intercept	0.003571	0.001115	3.20443	0.03276	0.000477	0.006666	0.000477	0.0066659
22	X Variable 1	0.001609	1.84E-05	87.3945	1E-07	0.001557	0.00166	0.001557	0.0016597
23									
24	$C_{unk} = C_{std}$*slope unk/slope std = 0.100*B22/B15 =				0.066235 wt%				
25									
26	For calculating uncertainies, let's multiply the variables by 1,000 to simplify equations:								
27	$C_{unk} = 0.100$*(1.609±0.018)/(2.429±0.026)								
28		= 0.100*	0.662353		± 0.01026				
29					[± C28*SQRT((0.026/2.429)^2 +(0.018/1.609)^2)]				
30	C_{unk} = 0.0662 ± 0.0010 wt%								

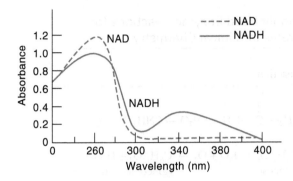

Example of Enzymatic Analyses

Spectrophotometric methods are widely used to measure enzyme reactions. The reaction product may have an absorption spectrum quite different from the substrate, allowing simple measurement of the product or substrate. In other cases, a dye-forming reagent is employed that will react with the product or the substrate and the increase or decrease in color is measured. Frequently, the chromogen is enzymatically coupled with the product using a second enzyme.

1. Dehydrogenase Reactions. The reduced (NADH) and oxidized (NAD^+) forms of nicotinamide adenine dinucleotide exhibit marked differences in their ultraviolet absorption spectra and are, therefore, widely used for following the course of dehydrogenase reactions. The ultraviolet absorption spectra for NAD^+ and NADH are given in Figure 23.4. NAD has negligible absorption at 340 nm while NADH has an absorption maximum, and so it is a simple matter to monitor the increase or decrease in NADH concentration.

An example using NADH for measurement is the determination of the enzyme **lactic acid dehydrogenase** (LDH), which is important in confirming myocardial infarction (heart attack). NAD^+ is required in the LDH-catalyzed oxidation of lactic acid to pyruvic acid:

$$
\begin{array}{ccc}
CH_3 & & CH_3 \\
| & LDH & | \\
CHOH + NAD^+ & \rightleftharpoons & C{=}O + NADH + H^+ \\
| & & | \\
CO_2H & & CO_2H \\
\text{lactic acid} & & \text{pyruvic acid}
\end{array}
\tag{23.17}
$$

The reaction is reversible and can be employed in either direction. In the forward reaction, serum containing an unknown amount of LDH would be added to a solution containing enzyme saturating concentrations of lactic acid and NAD, and the increase in absorbance at 340 nm would be measured as a function of time.

NADH can sometimes be used to follow enzyme reactions in which it is not directly involved by using it as a coupling agent in a secondary reaction with the product. For example, serum glutamic-oxaloacetic transaminase (GOT) catalyzes the reaction of α-ketoglutarate and asparate, and the product is reduced by NADH in the presence of another enzyme, malic acid dehydrogenase (MDH):

$$\alpha\text{-ketoglutarate} + \text{asparate} \xrightleftharpoons{\text{GOT}} \text{glutamate} + \text{oxaloacetate} \tag{23.18a}$$

$$\text{oxaloacetate} + \text{NADH} \xrightleftharpoons{\text{MDH}} \text{malate} + \text{NAD}^+ \tag{23.18b}$$

The second reaction is fast compared to the first in the presence of an excess of MDH, and so the rate of decrease of NADH concentration is directly proportional to the GOT activity.

Dehydrogenase reactions are monitored by measuring the UV absorbance of NADH.

TABLE 23.1　Examples of Commonly Used Enzyme Reactions for Determining Substrates in Clinical Chemistry

Substrate Determined	Enzyme	Reaction
Urea	Urease	$$\underset{\text{NH}_2-\overset{\overset{\displaystyle O}{\|\|}}{C}-\text{NH}_2 + \text{H}_2\text{O} \rightarrow 2\text{NH}_3 + \text{CO}_2}{}$$
Glucose	Glucose oxidase	$\underset{\text{(glucose)}}{C_6H_{12}O_6} + H_2O + O_2 \rightarrow \underset{\text{(gluconic acid)}}{C_6H_{12}O_7} + H_2O_2$
Uric acid	Uricase	$\underset{\text{(uric acid)}}{C_5H_4O_3} + 2H_2O + O_2 \rightarrow \underset{\text{(allantoin)}}{C_4H_6O_3N_4} + CO_2 + H_2O_2$
Galactose	Galactose oxidase	D-Galactose + $O_2 \rightarrow$ D-galactohexodialdose + H_2O_2
Blood alcohol	Alcohol dehydrogenase	Ethanol + $NAD^+ \rightarrow$ acetaldehyde + NADH + H^+

2. Commonly Determined Substrates. A list of some substrates determined in blood or urine is given in Table 23.1. They are discussed in order in the paragraphs following.

Urease was the first enzyme to be isolated and crystallized. It quantitatively converts urea to ammonia and carbon dioxide. The amount of urea is calculated from a determination of either the ammonia or the carbon dioxide produced, usually the former. This can be done spectrophotometrically by reacting the ammonia with a color reagent.

Glucose is usually determined by measuring the hydrogen peroxide produced upon addition of glucose oxidase. This is done spectrophotometrically by coupling the hydrogen peroxide as it is produced with a reagent such as o-toluidine. This coupling occurs in the presence of a second enzyme, horseradish peroxidase, and a colored product results. Commercial preparations of glucose oxidase usually contain impurities that react with and consume part of the hydrogen peroxide and so the conversion is not stoichiometric. Catalase is an enzyme impurity, for example, that is specific for the decomposition of hydrogen peroxide. Nevertheless, the fraction of hydrogen peroxide converted to the dye product is constant, and a calibration curve can be prepared using different concentrations of glucose.

There are a number of possible inhibitors in the glucose determination. Most of them, however, occur in the second enzymatic reaction. The glucose oxidase method would be more specific, then, if the hydrogen peroxide were measured directly without the need for a second enzyme. For example, added iodide ion, in the presence of a molybdenum(VI) catalyst, is rapidly oxidized to iodine. The iodine concentration can be followed amperometrically (Chapter 22). An alternative is to measure the depletion of oxygen amperometrically.

Uric acid is usually determined by measuring its ultraviolet absorption at 292 nm. However, the amount of uric acid in blood is small and the absorption is not specific. So, after the measurement, the uric acid is destroyed by adding the enzyme uricase. The absorbance is measured again. The *difference* in absorbance is due to the uric acid present. Since only uric acid will be decomposed by uricase, the method becomes specific. A similar procedure for the colorimetric determination of uric acid involves the oxidation of uric acid with molybdate to form molybdenum blue, a molybdenum(V) compound. In principle, uric acid could be determined as glucose was, but impurities in uricase preparations usually rapidly destroy the very small amount of hydrogen peroxide produced.

Oxidase reactions are monitored by measuring O_2 depletion or H_2O_2 production.

TABLE 23.2 Examples of Commonly Determined Enzymes in Clinical Chemistry

Enzyme	Abbreviation	Reaction
Glutamic–pyruvic transaminase	GPT	α-Ketoglutarate + L-alanine $\overset{GOT}{\rightleftharpoons}$ glutamate + pyruvate
		Pyruvate + NADH + H$^+$ $\overset{LDH}{\rightleftharpoons}$ lactate + NAD$^+$
Glutamic–oxaloacetic transaminase	GOT	α-Ketoglutarate + aspartate $\overset{GOT}{\rightleftharpoons}$ glutamate + oxaloacetate
		Oxaloacetate + NADH + H$^+$ $\overset{MDH}{\rightleftharpoons}$ malate + NAD$^+$
Creatinine phosphokinase	CK	Creatinine phosphate + ADP $\overset{CK}{\rightleftharpoons}$ creatine + ATP
		ATP + glucose $\overset{hexokinase}{\rightleftharpoons}$ ADP + glucose 6-phosphate
		Glucose 6-phosphate + NAD$^+$ $\overset{G\text{-}6PDH}{\longrightarrow}$ 6-phosphogluconate + NADH + H$^+$
Lactate dehydrogenase	LDH	L-Lactate + NAD$^+$ $\overset{LDH}{\rightleftharpoons}$ pyruvate + NADH + H$^+$
α-Hydroxybutyrate dehydrogenase	HBD	α-Ketobutyrate + NADH + H$^+$ $\overset{HBD}{\rightleftharpoons}$ α-hydroxybutyrate + NAD$^+$
Alkaline phosphatase		Na thymolphthalein monophosphate $\overset{pH\ 10.1}{\longrightarrow}$ Na thymolphthalein + phosphate
Acid phosphatase		Same as alkaline phosphatase, except pH 6.0

Galactose is determined in the same manner as glucose, by oxidizing the chromogen by H_2O_2 in the presence of peroxidase. Blood alcohol can be determined by UV measurement of the NADH produced.

3. Commonly Determined Enzymes. Table 23.2 summarizes reactions used to determine the activity of some enzymes frequently determined in the clinical laboratory. The pyruvate formed in the GPT reaction is coupled with NADH in the presence of added LDH for measurement. CK catalyzes the transfer of a phosphate group from creatinine phosphate to the nucleotide adenosine diphosphate (ADP) to produce adenosine triphosphate (ATP). The ATP is reacted with glucose in the presence of the enzyme hexokinase to form glucose-6-phosphate, which can then be reacted with NAD in the presence of glucose-6-phosphodehydrogenase. CK is also determined now by high-performance liquid chromatography, size exclusion chromatography, ion exchange chromatography, or electrophoresis.

> Coupled enzyme reactions are often used for detection reactions.

Natural LDH consists of five components called **isozymes**, or **isoenzymes**, the ratio of which varies with the tissue source. The two electrophoretically fastest components occur in high percentage in LDH from heart muscle, and the level of these is preferentially increased in blood after heart muscle damage. The LDH method measures total LDH isozymes, which will usually indicate the heart damage. The two heart muscle isozymes mentioned, however, more readily catalyze the reduction of α-ketobutyrate than the slower moving components of hepatic origin and are referred to as α-hydroxybutyrate dehydrogenase (HBD) active. Elevated HBD can be determined by the reaction given and is more specific for myocardial infarction than LDH; it also remains elevated for longer periods after infarct.

The **phosphatases** are determined by measuring the blue color of thymolphthalein in highly alkaline solution after a specific time; the high alkalinity stops the enzyme reactions.

Enzyme reactions can, of course, be measured by other techniques than spectrophotometry. Techniques that have been used include amperometry, conductivity, coulometry, and ion-selective electrodes. Certain enzyme reactions have been measured using enzyme electrodes whose response is specific for the particular reaction. These are described in Chapters 20 and 22.

Enzyme inhibitors and **activators** may be determined by employing enzyme reactions. The easiest technique is to measure the decrease or increase in the rate of the enzymatic reaction. Or the enzyme may be "titrated" with an inhibitor (or vice versa), and the amount of inhibitor required to completely inhibit the reaction measured. Trace elements, for example, have been determined by their inhibition or activation of enzyme reactions.

Many enzymes ideally represent the analytical chemist's dream of an absolutely specific reactant, but because of inhibitor effects, as well as problems associated with pH and ionic strength control, they must be used with some caution.

Most of the procedures discussed above can be adapted to automatic rate monitoring systems for enzymatic analysis.

Questions

General

1. Distinguish between a first-order and a second-order reaction.
2. What is the half-life of a reaction? How many half-lives does it take for a reaction to go to completion?
3. What is a pseudo-first-order reaction?
4. What is turnover number?
5. Why is the rate of an enzyme-catalyzed reaction proportional to the amount of E.S complex?
6. Suggest a way to determine whether a particular reaction between two substances A and B is first order or second order.

Enzymes

7. What is an international unit?
8. What is the difference between competitive inhibition and noncompetitive inhibition of an enzyme?
9. What is the chemical basis of enzyme specificity?
10. Under what conditions can a bisubstrate enzyme-catalyzed reaction with a double-displacement mechanism be treated kinetically as if it is a single-substrate reaction? (i.e., under what conditions does its Michaelis-Menten equation approximate that for a single-substrate reaction?)
11. Why are heavy metals often poisons in the body?
12. What are coenzymes?
13. Suggest a way to test whether an enzyme inhibitor is competitive or noncompetitive.
14. Suggest how a Lineweaver–Burk plot can be used to determine whether an inhibitor is competitive or noncompetitive.

Problems

Kinetics

15. A first-order reaction requires 15.0 min for 50% conversion to products. How much time is required for 90% conversion? For 99% conversion?
16. A first-order reaction required 30.0 s for 40% conversion to products. What is the half-life of the reaction?

17. A solution is 0.100 M in substances A and B, which react by a second-order reaction. If the reaction is 15.0% complete in 6.75 min, what is its half-life under these conditions? What would be the half-life if A and B were each at 0.200 M, and how long would it take for 15.0% completion of the reaction?

18. Sucrose is hydrolyzed to glucose and fructose:

$$C_{12}H_{22}O_{11} + H_2O \rightarrow C_6H_{12}O_6 + C_6H_{12}O_6$$

In dilute aqueous solution, the water concentration remains essentially constant, and so the reaction is pseudo first order and follows first-order kinetics. If 25.0% of a 0.500 M sucrose solution is hydrolyzed in 9.00 h, in how much time will the glucose and fructose concentration be equal to one-half the concentration of the remaining sucrose?

19. Hydrogen peroxide decomposes by a second-order reaction,

$$2H_2O_2 \xrightarrow{\text{catalyst}} 2H_2O + O_2$$

If 45.0% of a 0.1000 M solution decomposes in 9.60 min, how much time is required for the evolution of 100 mL O_2 from 100.0 mL of a 0.1000 M solution of H_2O_2 at standard temperature and pressure?

Enzyme

20. The activity of a glucose oxidase preparation is determined by measuring the volume of oxygen gas consumed as a function of time. A 10.0-mg fraction of the preparation is added to a solution containing 0.01 M glucose and saturated in oxygen. After 20.0 min, it is determined that 10.5 mL oxygen is consumed at standard temperature and pressure (STP). What is the activity of the enzyme preparation expressed in enzyme units per milligram? If the purified enzyme has an activity of 61.3 units/mg, what is the percent purity of this enzyme preparation?

Spreadsheet Problem

21. When an apple is sliced, it turns brown on exposure to air due to catalysis of the oxidation of phenols in the apple by o-diphenyl oxidase enzyme. An experiment is performed to determine the Michaelis constant of o-diphenyl oxidase in which fresh pieces of apple are ground up and then centrifuged to produce a supernatant that will serve as the enzyme source (see http://users.rcn.com/jkimball.ma.ultranet/BiologyPages/E/EnzymeKinetics.html). Catechol is used as the substrate. A fixed amount of the enzyme preparation is added to a tube containing 0.300 mM catechol, and the change in absorbance is measured at 540 nm at 1-min intervals for several minutes. The experiment is repeated with three other tubes containing 0.600, 1.20, and 4.80 mM catechol. The following results are obtained (mM catechol/ΔA/min): 0.30/0.020; 0.60/0.035; 1.20/0.048; 4.80/0.081. Prepare a spreadsheet to determine K_m and its uncertainty. (See the text website for a solution.)

Multiple Choice Questions

1. If the concept of half-life is generalized to quarter-life of a first order chemical reaction, it will be equal to
   - (a) $\ln 2/k$
   - (b) $\ln 4/k$
   - (c) $4/k$
   - (d) $1/4k$

2. For a first order reaction $A \rightarrow$ products, the plot of $\ln\left(\dfrac{[A]_t}{[A]_0}\right)$ vs. time, where $[A]_0$ and $[A]_t$ refer to concentrations at time $t = 0$ and t respectively, is
   - (a) a straight line with a positive slope passing through origin.
   - (b) a straight line with a negative slope passing through origin.
   - (c) an exponential curve asymptotic to the time axis.
   - (d) a curve asymptotic to the $\ln\left(\dfrac{[A]_t}{[A]_0}\right)$ axis.

3. The concertation of a reactant R varies with time for two different reactions as shown in the following plots:

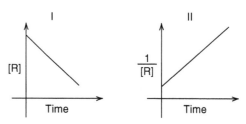

The orders of these two reactions I and II, respectively are

(a) zero and one.   (b) one and zero.

(c) zero and two.   (d) two and zero.

4. In an enzyme-catalyzed reaction,

$$E + S \underset{k_{-1}}{\overset{k_1}{\rightleftharpoons}} ES \overset{k_2}{\longrightarrow} E + P$$

$k_2 = 3.42 \times 10^4 \, s^{-1}$. If $[E]_0 = 1 \times 10^{-2} \, mol \, dm^{-3}$, the magnitude of maximum velocity and turn over number using Michaelis-Menten kinetics are

(a) $3.42 \times 10^2 \, mol \, dm^{-3} s^{-1}$; $3.42 \times 10^4 \, s^{-1}$

(b) $3.42 \times 10^6 \, mol \, dm^{-3} s^{-1}$; $3.42 \times 10^4 \, s^{-1}$

(c) $3.42 \times 10^4 \, mol \, dm^{-3} s^{-1}$; $3.42 \times 10^6 \, s^{-1}$

(d) $3.42 \times 10^4 \, mol \, dm^{-3} s^{-1}$; $3.42 \times 10^2 \, s^{-1}$

5. Which one of the following statements is not correct?

(a) Catalyst does not initiate any reaction.

(b) The value of equilibrium constant is changed in the presence of a catalyst in the reaction at equilibrium.

(c) Enzymes catalyze mainly biochemical reactions.

(d) Coenzymes increase the catalytic activity of enzyme.

6. A catalyst is a substance:

(a) which changes the equilibrium of a reaction so that the concentration of product increases.

(b) which increases the rate of reaction and increases the equilibrium concentration of the product.

(c) both (a) and (b).

(d) None of these.

7. In chemical reaction, catalyst:

(a) alters the amount of the products.   (b) lowers the activation energy.

(c) decreases the ΔH of forward reaction. (d) increases the ΔH of forward reaction.

8. A catalytic poison:

(a) decomposes the catalyst.   (b) decomposes the reactants.

(c) occupies the active centers.   (d) none of these.

9. The order of a reaction depends on:

(a) the number of molecules whose concentration changes.

(b) the number of molecules whose concentration is constant.

(c) the number of molecules whose concentration is minimum.

(d) all of the above.

10. A reaction is 20% completed in 20 minutes. How long will it take for 80% completion, if the reaction is a zero-order reaction?

(a) 40 minutes   (b) 70 minutes

(c) 60 minutes   (d) 80 minutes

11. The half-life period of a zero-order reaction of the type A → Product, is 100 minutes. How long will it take for 80% of the reaction to complete?

   (a) 80 minutes
   (b) 100 minutes
   (c) 120 minutes
   (d) 160 minutes

12. Which of the following can be unit of rate of a reaction?

   (a) Moles liter^{-1} sec^{-1}
   (b) Moles sec^{-1}
   (c) Moles liter^{-1}
   (d) None of these

13. Half-life period of any first order reaction is _____ of the initial concentration.

   (a) variable
   (b) independent
   (c) infinite
   (d) zero

14. If the half-life period of N_2O_5 decomposition is 1117.7 sec and it is a first order reaction, then the rate-constant for the decomposition of N_2O_5 is

   (a) 6.2×10^{-4} sec^{-1}
   (b) 6.2×10^{-5} sec^{-1}
   (c) 6.2×10^{-3} sec^{-1}
   (d) 6.2×10^{-6} sec^{-1}

Recommended References

Kinetics

1. H. H. Bauer, G. D. Christian, and J. E. O'Reilly, eds., *Instrumental Analysis*. Boston: Allyn and Bacon, 1978, Chapter 18, "Kinetic Methods," by H. B. Mark, Jr.
2. R. A. Greinke and H. B. Mark, Jr., "Kinetic Aspects of Analytical Chemistry," *Anal. Chem.*, **46** (1974) 413R.
3. H. L. Pardue, "A Comprehensive Classification of Kinetic Methods of Analysis Used in Clinical Chemistry," *Clin. Chem.*, **23** (1977) 2189.
4. D. Perez-Bendito and M. Silva, *Kinetic Methods in Analytical Chemistry*. New York: Wiley, 1988.
5. H. A. Mottola, *Kinetic Aspects of Analytical Chemistry*. New York: Wiley-Interscience, 1988.

Enzymes

6. H. U. Bergmeyer, *Methods of Enzymatic Analysis*, 3rd ed. New York: Wiley. A series of 12 volumes plus index volume, 1983–1987.

Websites

7. www.worthington-biochem.com. Link to the Worthington enzyme manual that gives the properties of enzymes, their assay, and references.
8. www.chem.qmul.ac.uk/iubmb/enzyme. Enzyme nomenclature from the International Union of Biochemistry (IUB) and the International Union of Pure and Applied Chemistry (IUPAC).
9. www.chem.qmw.ac.uk/iubmb/kinetics. Symbolism and terminology in kinetics.

Chapter 24

AUTOMATION IN MEASUREMENTS

"If you don't know how to do something, you don't know how to do it with a computer."
—Anonymous (From J. F. Ryan, *Today's Chemist at Work*, November, 1999, p. 7)

KEY THINGS TO LEARN FROM THIS CHAPTER

- Process control—continuous and discrete analyzers
- Automatic instruments
- Flow injection analysis
- Sequential injection analysis

The services of the analytical chemist are constantly increasing as more and better analytical tests are developed, particularly in the environmental and clinical laboratories. The analyst often must handle a large number of samples and/or process vast amounts of data. Instruments are available that will automatically perform many or all of the steps of an analysis, greatly increasing the load capacity of the laboratory. The data generated are typically processed by computers interfaced to the analytical instruments. An important type of automation is in process control whereby the progress of an industrial plant process is monitored in real time (i.e., online), and continuous analytical information is fed to control systems that maintain the process at preset conditions.

Billions of tests are run annually in clinical chemistry laboratories, and automation plays a key role in processing large numbers of samples. Clinical chemistry analyzers are mainly automatic systems, for example, based on discrete or continuous analyzers for the rapid measurement of many samples in a day. Often, multiple analytes are simultaneously measured in a sample. Online measurements are used in the chemical industry to monitor chemical production, to assure reactions are proceeding as designed, and to minimize bad production lots, as well as to to use real time measurements in feedback loops to improve efficiency and purity of product; an example might be to control pH during the reaction.

In this chapter, we briefly consider the types of automated instruments and devices commonly used and the principles behind their operation. Their application to process control is discussed. And we describe the techniques of flow injection analysis and sequential injection analysis that allow most common analytical measurements to be performed automatically using microliter volumes of samples and reagents.

24.1 Automation: Principles

There are two basic types of automation equipment. **Automatic devices** perform specific operations at given points in an analysis, frequently the *measurement step*. Thus, an *automatic titrator* will stop a titration at the end point, either mechanically or electrically, upon sensing a change in the property of the solution. **Automated devices**, on the other hand, control and regulate a *process* without human intervention. They do so through mechanical and electronic devices that are regulated by means of *feedback information* from a sensor or sensors. Hence, an *automated titrator* may maintain a sample pH at a preset level by addition of acid or base as drift from the set pH is sensed by a pH electrode. Such an instrument is called a *pH-stat* and may be used, for example, in maintaining the pH during an enzyme reaction, which releases or consumes protons. The rate of the reaction can actually be determined from a recording of the rate of addition of acid or base to keep the solution pH constant.

Automated devices are widely used in process control systems, whereas automatic instruments of various sophistication are used in the analytical laboratory for performing analyses. The latter may perform all steps of an analysis, from sample pickup and measurement through data reduction and display.

24.2 Automated Instruments: Process Control

In process analysis, analytical measurements are performed on chemical processes to provide information about the progress of the process or the quality of product. There are various ways in which process analysis may be performed, as illustrated in Figure 24.1. Samples may be taken intermittently and transported to the laboratory for measurement. This allows access to the usual laboratory instruments and the ability to perform a variety of measurements. But this procedure is relatively slow and the chemical process is usually complete before the analytical information is available. Hence, laboratory analysis is more suited for *quality control*, to ascertain the quality of a product. More efficient measurements can be made if the instrument is transferred to the chemical plant. But for true real-time analysis, instruments should be interfaced directly to the chemical process, with automatic sampling and analysis. A more idealized approach would be to place a sensor directly inline so that measurements are continuous and no chemical treatment of the sample is needed. Such sensors are more limited in scope and availability since they must be selective for a given analyte and must withstand the chemical environment, not be poisoned, and remain in calibration. An even more idealized approach is a noninvasive measurement. For example, an analyte may exhibit an absorbance spectrum that allows selective measurement by the passage of light through the chemical system. Again, these types of process measurements will be limited.

An important aspect of real-time process analysis is the use of the analytical data to control the chemical process via feedback of the information to a controller that can alter the addition of chemical reactants to maintain an intermediate, for example, at a preset level. The application of online measurements with feedback control of the chemical process can save a chemical company millions or even hundreds of millions of dollars in a year by providing optimization of the process for achieving maximum reaction efficiency and product formation and avoiding failed reactions, detecting contaminants, and the like. This application of analytical chemistry has become a critical part of industrial production. The University of Washington Center for Process Analysis & Control (CPAC) is supported by industrial sponsors for conducting research on state-of-the-art process measurement technology (see http://cpac.apl.washington.edu/).

The measurement devices may be classed as continuous or discrete (batch) instruments. The **continuous instrument** constantly measures some physical or chemical property of the sample and yields an output that is a continuous (smooth) function of time. A **discrete**, or **batch**, **instrument** analyzes a discrete or batch-loaded sample, and information is supplied only in discrete steps. In either case, information on the measured variable is fed back to monitoring or control equipment. Each technique utilizes conventional analytical measurement procedures and must be capable of continuous unattended operation.

Automatic instruments improve the analyst's efficiency by performing some of the operations done manually. Automated instruments control a system based on the analysis results.

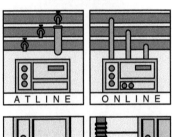

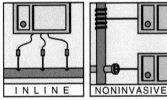

FIGURE 24.1 Methods for process analysis. [From J. B. Callis, D. Illman, and B. R. Kowalski, *Anal. Chem.*, **59** (1987) 624A. Copyright 1987 by the American Chemical Society. Reprinted by permission of the copyright owner.]

The use of analytical data for automated process control can save millions of dollars in improved production efficiencies and product quality.

Continuous Analyzers

Continuous-process control instruments may make measurements directly in a flowing stream or a batch process reactor such as a fermentor. This generally precludes any analytical operation on the sample, and direct sensing devices such as electrodes must be used. If a sample dilution, temperature control, or reagent addition is required, or measurements are made with nonprobe-type instruments, then a small fraction of the stream is diverted into a test stream where reagents may be mixed continuously and automatically with the sample, and the test measurement is made. The sample may be passed through a filter prior to measurement.

Process control instruments operate by means of a **control loop**, which consists of three primary parts:

1. A **sensor** or measuring device that monitors the variable being controlled.
2. A **controller** that compares the measured variable against a reference value (set point) and feeds the information to an operator.
3. An **operator** that activates some device such as a valve to bring the variable back to the set point.

The control loop operates by means of a **feedback mechanism**, illustrated in Figure 24.2. The process may be any industrial process that produces some desired product. It has one or more inputs that can be controlled to provide the desired product (output). The sensor measures the variable to be controlled (e.g., pH, temperature, reactant). The information is fed to the controller, which compares the measured variable against a reference set point. The difference is fed to the operator that actuates the valve (opens or closes it) or some other appropriate device to adjust the variable back to the set point.

These devices are characterized by their **dead time**. This is equal to the time interval, after alteration of the variable at the input, during which no change in the variable is sensed by the detector at the output. It includes the analytical dead time. It can be minimized by keeping the detector as near the input as possible and by high flow rate. The sensor may actually be placed at the input, before the process. In this manner, corrective action on the manipulated variable (if it changes erroneously) is taken before the process occurs, and an error need not occur in the output prior to corrective action. However, the result is not detected at the output. Such systems may be called **feedforward systems**, and are **open-loop controls**, as opposed to **closed-loop controls** in feedback systems. Sophisticated algorithms, using computers, are used for

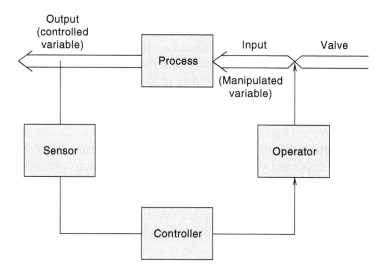

FIGURE 24.2 Feedback control loop.

control, based on chemometric methods and multivariate analysis. These are beyond the scope of this text.

Discrete Analyzers

In discrete analyzers, a batch sample is taken at selected intervals and then analyzed, with the information being fed to the controller and operator in the usual fashion. Obviously, the sampling and analytical dead times are increased over continuous analyzers, and the manipulated variable is held at a fixed value between measurements. If a transient error occurs between measurements, it may not be detected and corrected for. On the other hand, a short transient error may be detected during the measurement interval and a correction for this applied during the entire interval between measurements.

Discrete measurements must be made when the sensing instrument requires discrete samples, as in a chromatograph, or a flow injection analyzer (see below).

Instruments Used in Automated Process Control

In principle, any conventional measuring device or technique can be used in process control. But laboratory instruments normally are not suited for online measurement. The instruments must be more robust and are designed for unattended operation, with minimal skill required for operation, since plant personnel are often not trained analytical chemists. (But you, as the resident analytical chemist will be in charge of selecting the measurement technique, the instrument, and making sure reliable data are obtained!)

The choice is dictated by cost and applicability to the problem. The most widely used methods include spectrophotometry to measure color, ultraviolet or infrared absorption, turbidity, film thickness, and the like; electrochemistry, primarily potentiometry, for the measurement of pH and cation or anion activity; and gas and liquid chromatography, especially in the petrochemical industry where complex mixtures from distillation towers are monitored. Spectrophotometric and other measurements are often rapidly made using flow injection analysis.

24.3 Automatic Instruments

Automatic instruments, as mentioned before, are not feedback control devices but rather are designed to automate one or more steps in an analysis. They are generally intended to analyze multiple samples, either for a single analyte or for several analytes.

Automatic instruments will perform one or more of the following operations:

Automatic instruments relieve the analyst from several operations. The precise nature of automatic operations improves the analytical precision.

1. Sample pickup (e.g., from a small cup on a turntable or assembly line)
2. Sample dispensing
3. Dilution and reagent addition
4. Incubation
5. Placing of the reacted sample in the detection system
6. Reading and recording the data
7. Processing of the data (correct for blanks, correct for nonlinear calibration curves, calculate averages or precisions, correlate with the sample number, etc.).

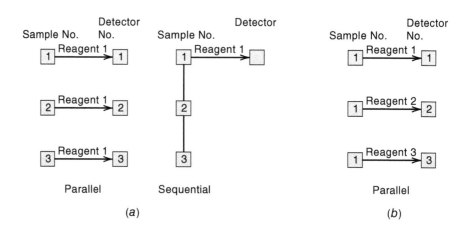

FIGURE 24.3 Discrete analyzers.
(*a*) Single channel (batch).
(*b*) Multichannel.

Clinical instruments may also contain a feature for deproteinizing the sample. Instruments that provide only a few of these steps, primarily electronic data processing, are called **semiautomatic instruments**.

While all automatic instruments are discrete in the terminology of automated process analyzers, in that they analyze individual discrete samples, they may be classified as follows:

1. **Discrete sampling instruments.** In discrete sampling, each sample undergoes reaction (and usually measurement) in a separate cuvet or chamber. These samples may be analyzed sequentially or in parallel (see below).

2. **Continuous-flow sampling instruments.** In continuous-flow sampling, the samples flow sequentially and continuously in a tube, perhaps being separated by air bubbles. They are each sequentially mixed with reagents in the same tube at the same point downstream and then flow sequentially into a detector.

Discrete samplers have the advantage of minimizing or avoiding cross contamination between samples. But continuous-flow instruments require fewer mechanical manipulations and can provide very precise measurements.

Discrete instruments may be designed to analyze samples for one analyte at a time. These are the so-called **batch instruments**, or **single-channel analyzers**. See Figure 24.3. However, those that analyze the samples in parallel, that is, simultaneously rather than sequentially one at a time, can analyze a large number very quickly; and they can readily be changed to perform different analyses. Discrete analyzers may also analyze separate aliquots of the same sample (one in each cup) in parallel for several different analytes. These are **multichannel analyzers**.

Continuous-flow instruments may also be single-channel (batch) instruments that analyze a continuous series of samples sequentially for a single analyte (Figure 24.4). Or they may be multichannel instruments in which the samples are split at one or more

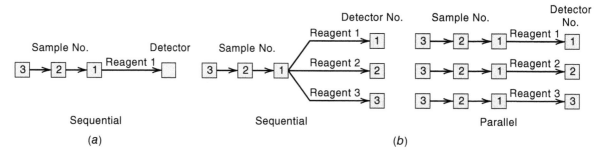

FIGURE 24.4 Continuous-flow analyzers. (*a*) Single channel (batch). (*b*) Multichannel.

points downstream into separate streams for different analyte analyses, or separate aliquots of samples may be taken with separate streams in parallel.

Modern-day instruments are so sophisticated that they actually possess *automated* features whether they perform an analysis automatically or not. For example, they may monitor sample chamber temperature and by feedback to a regulator maintain it constant (important in enzyme reactions).

24.4 Flow Injection Analysis

Flow injection analysis (FIA) is based on the injection of a liquid sample into a moving, nonsegmented continuous carrier stream of a suitable liquid. The injected sample forms a zone, which is then transported toward a detector. Mixing with the reagent in the flowing stream occurs mainly by diffusion-controlled processes, and a chemical reaction occurs. The detector continuously records the absorbance, electrode potential, or other physical parameter as it changes as a result of the passage of the sample material through the flow cell.

An example of one of the simplest FIA methods, the spectrophotometric determination of chloride in a single-channel system, is shown in Figure 24.5. This is based on the release of thiocyanate ions from mercury(II) thiocyanate and its subsequent reaction with iron(III) and measurement of the resulting red color (for details, see Experiment 39). The samples, with chloride contents in the range 5 to 75 ppm chloride, are injected (S) through a 30-μL valve into the carrier solution containing the mixed reagent, pumped at a rate of 0.8 mL/min. The iron(III) thiocyanate is formed on the way to the detector (D) in a mixing coil (50 cm long, 0.5 mm i.d.), as the injected sample zone disperses in the carrier stream of reagent. The mixing coil minimizes band broadening (of the sample zone) due to centrifugal forces, resulting in sharper recorded peaks. The absorbance A of the carrier stream is continuously monitored at 480 nm in a micro-flow-through cell (volume of 10 μL) and recorded (Figure 24.5*b*). To demonstrate the reproducibility of the analytical readout, each sample in this experiment was

FIA is like HPLC without a column. It is low pressure, and there is no separation. The injected sample mixes and reacts with the flowing stream. A transient signal (peak) is recorded.

FIA measurements are very rapid.

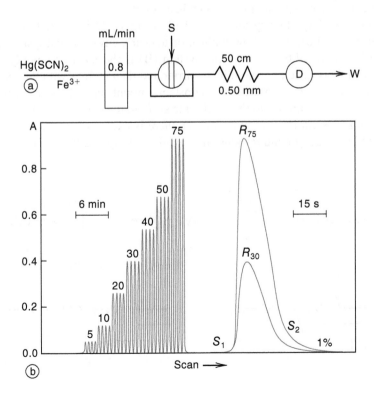

FIGURE 24.5 (*a*) Flow injection diagram for the spectrophotometric determination of chloride: S is the point of sample injection, D is the detector, and W is the waste.
(*b*) Analog output showing chloride analysis in the range of 5 to 75 ppm Cl with the system depicted in (*a*).

injected in quadruplicate, so that 28 samples were analyzed at seven different concentrations of chloride. As this took 14 min, the average sampling rate was 120 samples per hour. The fast scan of the 75- and 30-ppm sample peaks (shown on the right in Figure 24.5*b*) confirms that there was less than 1% of the solution left in the flow cell at the time when the next sample (injected at S2) would reach it, and that there was no carryover when the samples were injected at 30-sec intervals.

A key feature of FIA is that since all conditions are reproduced, dispersion is very controlled and reproducible. That is, all samples are sequentially processed in exactly the same way during passage through the analytical channel, or, in other words, what happens to one sample happens in exactly the same way to any other sample.

A peristaltic pump is typically used to propel the stream. For process analysis, these are not suitable because the pump tubing must be frequently changed, and more rugged pumps are used, such as syringe pumps, or pumping is by means of air displacement in a reservoir. The injector may be a loop injector valve as used in high-performance liquid chromatography. A bypass loop allows passage of carrier when the injection valve is in the load position. The injected sample volumes may be between 1 and 200 µL (typically 25 µL), which in turn requires no more than 0.5 mL of reagent per sampling cycle. This makes FIA a simple, microchemical technique that is capable of having a high sampling rate and minimum sample and reagent consumption. The pump, valve, and detector may be computer controlled for automated operation.

FIA is a general solution-handling technique, applicable to a variety of tasks ranging from pH or conductivity measurement to colorimetry and enzymatic assays. To design any FIA system properly, one must consider the desired function to be performed. For pH measurement, or in conductimetry, or for simple atomic absorption, when the original sample composition is to be measured, the sample must be transported through the FIA channel and into the flow cell in an undiluted form in a highly reproducible manner. For other types of determinations, such as spectrophotometry, the analyte has to be converted to a compound measurable by a given detector. The prerequisite for performing such an assay is that during the transport through the FIA channel, the sample zone is mixed with reagents and sufficient time is allowed for production of a desired compound in a detectable amount.

Besides the single-line system, described in Figure 24.5, a variety of manifold configurations may be used to allow application to nearly any chemical system. Several are shown in Figure 24.6. The two-line system (B) is the most commonly used, in which the sample is injected into an inert carrier and then merges with the reagent. In this manner, the reagent is diluted by a constant amount throughout, even when the sample is injected, in contrast to the single-line system; reagent dilution by the sample in a single-line system is all right as long as there is excess reagent and the reagent does not exhibit a background response that would shift upon dilution. If two reagents

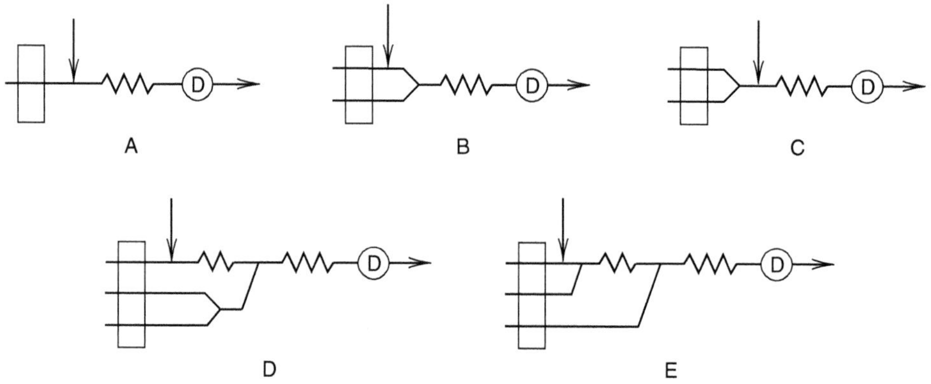

Only a few microliters of sample are required.

A two-line system is most often used.

FIGURE 24.6 Types of FIA manifolds. A, single line; B, two-line with a single confluence point; C, reagent premixed into a single line; D, two-line with a single confluence point and reagent premix; E, three-line with two confluence points.

are unstable when mixed, then they may be mixed online (C or D), or they may merge with the sample following injection (E). Mixing coils may be interspersed between confluence points to allow dispersion before merging.

24.5 Sequential Injection Analysis

Sequential injection analysis (SIA) is a computer-controlled, single-line, injection technique that simplifies the manifold and is more robust for unattended operation. The flow is intermittent, and only a few microliters of reagent are used. Instead of an injection valve for sample introduction, SIA uses a multiport selection valve as shown in Figure 24.7. The common port of the valve (in the center) can access any of the other ports by electrical rotation of the valve. It is connected to a reversible piston pump via a two-position valve; carrier is drawn into the pump in one position and delivered toward the valve in the other. A holding coil is placed between the pump and the valve to prevent aspirated (injected) solution from entering the pump. The different ports of the valve are connected to sample, reagent, standards, waste reservoir, and the detector. In operation, the pump is first filled with carrier. The valve is switched to the detector position, and the piston pushes carrier through the system until it exits at the waste end. The valve is then switched to the sample position, and a few microliters of sample are drawn in by reversing the flow of the pump using precise timing. (Before beginning the analysis, the sample and reagent tubes are filled by aspirating some excess into the holding coil and expelling it to waste via the auxiliary waste port.) The pump is stopped during rotation of the valve to avoid pressure surges. After the sample is introduced, the reagent is aspirated, next to the sample. Thus, the sample and reagent solutions are sequentially injected into the holding coil, and hence the name sequential injection analysis. Finally, the valve is switched to the detector port, the pump is changed to forward flow, and the injected solutions are propelled through the reaction coil to the detector flow cell. The solutions merge via diffusion and secondary forces and react to form a product that is detected, resulting in a transient signal as in conventional FIA.

Only one pump and valve are required in SIA. The entire operation is computer controlled for the precise timing required, and commercial systems have programmable

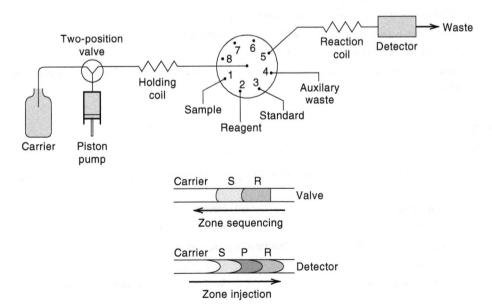

FIGURE 24.7 SIA manifold and sequential injection. S, P, and R are sample, product, and reagent, respectively.

software to set sequences and aspiration times. The software also collects peak height data, prepares calibration curves, and calculates concentrations, with precision data. Different reagents can be introduced to perform a different analysis, simply by programming from the keyboard to aspirate from a different reagent port, rather than changing the plumbing as required in FIA. Two-reagent chemistries may be used by sandwiching a small volume of sample between them so that all three zones overlap before reaching the detector.

See www.flowinjection.com for examples of commercial SIA and FIA systems. This includes a detailed tutorial on FIA and SIA. You may also request a free CD of the e-book by Professor Jaromir Ruzicka, which includes Professor Elo Hansen's extensive database of FIA lierature (Reference 11 below).

24.6 Laboratory Information Management Systems

Analytical laboratories can generate large amounts of information, particularly with automated systems. It is a challenge to manage samples, measurements, and data and meet good laboratory practice requirements. Laboratory information management systems (LIMs) utilize dedicated software to process, store, and manage data. There are a variety of commercial software packages to perform multiple functions. Examples include logging in samples, with sample identification, bar coding, electronic signatures for approvals and records, reporting, delivery of results to end users via web access, e-mail, fax, etc. They may include standard operating procedures (SOPs), maintenance procedures, calibration, traceability, and audit trails. Automated calculation of results, with statistical analysis, is often included. Such systems are critical for large laboratory operations.

Questions

1. Distinguish between an automatic instrument and an automated instrument.
2. Distinguish between discrete and continuous automated devices.
3. Distinguish between discrete and continuous-flow sampling automatic instruments.
4. What is a feedback loop?
5. What are the three primary parts of a control loop?
6. Distinguish between single and multiple channel analyzers?
7. Describe the principles of flow injection analysis.
8. Describe the principles of sequential injection analysis and how it differs from conventional flow injection analysis.

Multiple Choice Questions

1. The overall gain of the system in a closed loop control system with positive value of feedback gain is _____.

   (a) decrease.   (b) increase.

   (c) be unaffected.   (d) any of the above.

2. Which among the following is not a feature of a good control system?

   (a) Good stability.   (b) Slow response.

   (c) Good accuracy.   (d) Sufficient power handling capacity.

3. With a feedback system, the transient response,

   (a) rises slowly. (b) rises quickly.

   (c) decays slowly. (d) decays quickly.

4. Which of the following elements is not used in an automatic control system?

   (a) Error detector (b) Final control element

   (c) Sensor (d) Oscillator

5. The automatic control system with output as a variable is called

   (a) closed loop system. (b) servomechanism.

   (c) automatic regulating system. (d) process control system.

6. The following represents an illustration of closed loop system:

   (a) Automatic washing machine (b) Automatic electric iron

   (c) Bread toaster (d) Electric hand drier

7. The time responsible for introducing an error in the temperature regulation of applications associated with ON-OFF controllers is called:

   (a) rise time. (b) dead time.

   (c) switching time. (d) decay time.

Recommended References

General

1. G. D. Christian and J. E. O'Reilley, eds., *Instrumental Analysis*, 2nd ed. Boston: Allyn and Bacon, 1986. Chapter 25, "Automation in Analytical Chemistry," by K. S. Fletcher and N. C. Alpert. Provides an excellent detailed but brief description of various types and operations of automated and automatic instruments.

Flow Injection Analysis

2. J. Ruzicka and E. H. Hansen, *Flow Injection Analysis*, 2nd ed. New York: Wiley, 1988.
3. M. Valcarcel and M. D. Luque de Castro, *Flow Injection Analysis, Principles and Applications*. Chichester: Ellis Horwood, 1987.
4. Z. Fang, *Flow Injection Separation and Preconcentration*. Weinheim: VCH, 1993.
5. Z. Fang, *Flow Injection Atomic Absorption Spectrometry*. New York: Wiley, 1995.
6. J. L. Burguera, ed., *Flow Injection Atomic Spectroscopy*. New York: Marcel Dekker, 1989.
7. S. D. Kolev and I. D. McKelvies, eds., *Advances in Flow Injection Analysis and Related Techniques*. Amsterdam: Elsevier, 2008. (Vol. 54 of Wilson & Wilson's Comprehensive Analytical Chemistry series.)
8. M. Trajanowicz, ed., *Advances in Flow Injection Analysis*. Weinheim: Wiley-VCH, 2008.

Sequential Injection Analysis

9. G. D. Christian, "Sequential Injection Analysis for Electrochemical Measurements and Process Analysis," *Analyst*, **119** (1994) 2309.
10. P. J. Baxter and G. D. Christian, "Sequential Injection Analysis: A Versatile Technique for Bioprocess Monitoring," *Accounts Chem. Res.*, **29** (1996) 515.

Web Databases

11. Hansen's FI Bibliography, http://www.flowinjection.com/Elos_Database/FIAdatabase.aspx, or go to www.flowinjection.com and click on the DATABASE link. You can download the database. To search, hit Ctrl F. Search for analyte, matrix, author, etc. Almost 20,000 references on flow injection analysis.

12. Chalk's Flow Analysis Database, https://www.fia.unf.edu. Over 17,000 references on flow methods, covering 1954 to 2007. You can search by analyte, author, technique, etc.

FIA E-book

13. Jarda Ruzicka, *Flow Injection Analysis*, 4th ed. Free CD tutorial with much historical information and photos of pioneers. Available from www.fialab.com. Click on the TUTORIAL link.

Process Analysis

14. G. D. Christian and E. D. Yalvac, "Process Analytical Chemistry," in R. Kellner, J.-M. Mermot, M. Otto, and H. M. Widner, eds. *Analytical Chemistry*, Weinheim: Wiley-VCH, 1998.
15. J. B. Callis, D. L. Illman, and B. R. Kowalski, "Process Analytical Chemistry," *Anal. Chem.*, **59** (1987) 624A.
16. M. T. Riebe and D. J. Eustace, "Process Analytical Chemistry. An Industrial Perspective," *Anal. Chem.*, **62** (1990) 65A.

LITERATURE OF ANALYTICAL CHEMISTRY

"To reinvent the wheel is a waste of time and talent."

—Anonymous

When defining a problem, one of the first things the analyst does is to go to the scientific literature and see if the particular problem has already been solved in a manner that can be employed. There are many reference books in selected areas of analytical chemistry that describe the commonly employed analytical procedures in a particular discipline and also some of the not-so-common ones. These usually give reference to the original chemical journals. For many routine or specific analyses, prescribed standard procedures have been adopted by the various professional societies.

If you do not find a solution to the problem in reference books, then you must resort to the scientific journals. *Chemical Abstracts* is the logical place to begin a literature search. This journal contains abstracts of all papers appearing in the major chemical publications of the world. Yearly and cumulative indexes are available to aid in the literature search. The element or compound to be determined as well as the type of sample to be analyzed can be looked up to obtain a survey of the methods available. Author indexes are also available. Your library may subscribe to *SciFinder Scholar*, the online access to *Chemical Abstracts*. You can search by chemical substance, topic, author, company name, and access abstracted journals. Once you have an article online, you can link to referenced articles in the paper.

You can also locate many relevant references for a specific problem by using Web search engines. Following is a selected list of some references in analytical chemistry. References to the various methods of measurements are included at the ends of the chapters covering these methods throughout the text.

A.1 Journals[1]

1. *American Lab*oratory
2. *Analytical Biochem*istry
3. *Analytical Chim*ica *Acta* (P. K. Dasgupta is an editor – take a look!)
4. *Analytical Abstracts*
5. *Analytical Chem*istry[2]
6. *Analytical Instrum*entation
7. *Analytical Lett*ers
8. *Analyst*, The
9. *Appl*ied *Spectros*copy
10. *Clin*ica *Chim*ica *Acta*
11. *Clin*ical *Chem*istry

---

[1] The *Chemical Abstracts* abbreviation of each title is indicated by italics.

[2] This journal has a biannual volume every April that reviews in alternate years the literature of various analytical techniques and their applications in different areas of analysis.

12. *Electroanalysis*
13. Journal of *AOAC In*ternational
14. Journal of *Chromatog*raphic *Science*
15. Journal of *Chromatog*raphy
16. Journal of *Electroanal*ytical *Chem*istry and *Interfac*ial *Electrochem*istry
17. *Microchem*ical Journal
18. *Spectrochim*ica *Acta*
19. *Talanta* (G. D. Christian, Editor-in-Chief—take a look!)
20. Zeitschrift für *analy*tische *Chem*ie

A.2　General References

Some general references are given in Chapter 1, including encyclopedias, and more specific references and useful websites are given in the chapters throughout the text. Following are several classical references, general and substance specific, that provide much useful information for the analyst. They are available in many libraries.

1. *Annual Book of ASTM Book of Standards*, Multiple volumes for many different industrial materials. Philadelphia: American Society for Testing and Materials.
2. R. Belcher and L. Gordon, eds., *International Series of Monographs on Analytical Chemistry*. New York: Pergamon. A multivolume series.
3. N. H. Furman and F. J. Welcher, eds., *Scott's Standard Methods of Chemical Analysis*, 6th ed., 5 vols. New York: Van Nostrand, 1962–1966.
4. I. M. Kolthoff and P. J. Elving, eds., *Treatise on Analytical Chemistry*. New York: Interscience. A multivolume series.
5. I. M. Kolthoff, E. B. Sandell, E. J. Meehan, and S. Bruckenstein, *Quantitative Chemical Analysis*, 4th ed. London: Macmillan, 1969.
6. C. N. Reilly, ed., *Advances in Analytical Chemistry and Instrumentation*. New York: Interscience. A multivolume series.
7. C. L. Wilson and D. W. Wilson, *Comprehensive Analytical Chemistry*, G. Sveha, ed., New York: Elsevier. A multivolume series.
8. *Official Methods of Analysis of AOAC International*, 18th ed., Revision 3, G. W. Lewis and W. Horwitz, eds. Gaithersburg, MD: AOAC International, 2010. Available on CD-ROM.

A.3　Inorganic Substances

1. *ASTM Methods for Chemical Analysis of Metals*. Philadelphia: American Society for Testing and Materials, 1956.
2. F. E. Beamish and J. C. Van Loon, *Analysis of Noble Metals*. New York: Academic, 1977.
3. T. R. Crompton, *Determination of Anions: A Guide for the Analytical Chemist*. Berlin: Springer, 1996.

A.4 Organic Substances

1. J. S. Fritz and G. S. Hammond, *Quantitative Organic Analysis*. New York: Wiley, 1957.
2. T. S. Ma and R. C. Rittner, *Modern Organic Elemental Analysis*. New York: Marcel Dekker, 1979.
3. J. Mitchell, Jr., I. M. Kolthoff, E. S. Proskauer, and A. W. Weissberger, eds., *Organic Analysis*, 4 vols. New York: Interscience, 1953–1960.
4. S. Siggia, Jr., and J. G. Hanna, *Quantitative Organic Analysis via Functional Group Analysis*, 4th ed. New York: Wiley, 1979.
5. A. Steyermarch, *Quantitative Organic Microanalysis*, 2nd ed. New York: Academic, 1961.

A.5 Biological and Clinical Substances

1. M. L. Bishop, E. P. Fody, and L. E. Schoeff, eds, *Clinical Chemistry: Techniques, Principles, Correlations*, 6th ed. Baltimore: Lippincot Williams & Wilkins, 2010.
2. G. D. Christian and F. J. Feldman, *Atomic Absorption Spectroscopy. Applications in Agriculture, Biology, and Medicine*. New York: Wiley-Interscience, 1970.
3. D. Glick, ed., *Methods of Biochemical Analysis*. New York: Interscience, A multivolume series.
4. R. J. Henry, D. C. Cannon, and J. W. Winkelman, eds., *Clinical Chemistry. Principles and Techniques*, 2nd ed. Hagerstown, MD: Harper & Row, 1974.
5. M. Reiner and D. Seligson, eds., *Standard Methods of Clinical Chemistry*. New York: Academic. A multivolume series from 1953.
6. C. A. Burtis, E. R. Ashwood, and D. E. Bruns, eds., *Tietz Fundamentals of Clinical Chemistry*, 6th ed., St. Louis, MO: Saunders: Elsevier, 2008. 976 pages.

A.6 Gases

1. C. J. Cooper and A. J. DeRose, *The Analysis of Gases by Gas Chromatography*. New York: Pergamon, 1983.

A.7 Water and Air Pollutants

1. *Quality Assurance Handbook for Air Pollution Measurement Systems*, U.S.E.P.A., Office of Research and Development, Environmental Monitoring and Support Laboratory, Research Triangle, NC 27711. Vol. I, *Principles*. Vol. II, *Ambient Air Specific Methods*.
2. *Standard Methods for the Examination of Water and Wastewater*. New York: American Public Health Association.

A.8 Occupational Health and Safety

1. National Institute of Occupational Health and Safety (NIOSH), P. F. O'Connor, ed., *Manual of Analytical Methods*, 4th ed. Washington, DC: DHHS (NIOSH) Publication No. 94–113 (August 1994).

REVIEW OF MATHEMATICAL OPERATIONS: EXPONENTS, LOGARITHMS, AND THE QUADRATIC FORMULA

B.1 Exponents

It is convenient in mathematical operations, even when working with logarithms, to express numbers semiexponentially. Mathematical operations with exponents are summarized as follows:

$$N^a N^b = N^{a+b} \qquad \text{e.g., } 10^2 \times 10^5 = 10^7$$

$$\frac{N^a}{N^b} = N^{a-b} \qquad \text{e.g., } \frac{10^5}{10^2} = 10^3$$

$$(N^a)^b = N^{ab} \qquad \text{e.g., } (10^2)^5 = 10^{10}$$

$$\sqrt[a]{N^b} = N^{b/a} \qquad \text{e.g., } \sqrt{10^6} = 10^{6/2} = 10^3$$

$$\sqrt[3]{10^9} = 10^{9/3} = 10^3$$

The decimal point of a number is conveniently placed using the **semiexponential form**. The number is written with the decimal placed in the units position, and it is multiplied by 10 raised to an integral exponent equal to the number of spaces the decimal was moved to bring it to the units position. The exponent is negative if the decimal is moved to the right (the number is less than 1) and it is positive if the decimal is moved to the left (the number if 10 or greater). Some examples are:

Number	Semiexponential Form
0.00267	2.67×10^{-3}
0.48	4.8×10^{-1}
52	5.2×10^{1}
6027	6.027×10^{3}

Any number raised to the zero power is equal to unity. Thus, $10^0 = 1$, and 2.3 in semiexponential form is 2.3×10^0. Numbers between 1 and 10 do not require the semiexponential form to place the decimal.

B.2 Taking Logarithms of Numbers

It is convenient to place numbers in the semiexponential form for taking logarithms or for finding a number from its logarithm. The following rules apply:

$$N = b^a$$

$$\log_b N = a$$

or
$$N = 10^a$$
$$\log_{10} N = a$$

For example
$$\log 10^2 = 2$$
$$\log 10^{-3} = -3$$

Also,
$$\log ab = \log a + \log b$$

for example
$$\log 2.3 \times 10^{-3} = \log 2.3 + \log 10^{-3}$$
$$= 0.36 - 3$$
$$= -2.64$$
$$\log 5.67 \times 10^7 = \log 5.67 + \log 10^7$$
$$= 0.754 + 7$$
$$= 7.754$$

The exponent is actually the **characteristic** of the logarithm of a number, and the logarithm of the number between 1 and 10 is the **mantissa**. Thus, in the example $\log 2.3 \times 10^{-3}$, -3 is the characteristic and 0.36 is the mantissa.

B.3   Finding Numbers from Their Logarithms

The following relationships hold:
$$\log_{10} N = a$$
$$N = 10^a = \text{antilog } a$$

For example,
$$\log N = 0.371$$
$$N = 10^{0.371} = \text{antilog } 0.371 = 2.35$$

In general, to find a number from its logarithm, write the number in exponent form, and then break the exponent into the mantissa (m) and the characteristic (c). Then, take the antilog of the mantissa and multiply by the exponential form of the characteristic:
$$\log_{10} N = mc$$
$$N = 10^{mc} = 10^m \times 10^c$$
$$N = (\text{antilog } m) \times 10^c$$

For example,
$$\log N = 2.671$$
$$N = 10^{2.671} = 10^{.671} \times 10^2$$
$$= 4.69 \times 10^2 = 469$$
$$\log N = 0.326$$
$$N = 10^{0.326} = 2.12$$
$$\log N = -0.326$$
$$N = 10^{-0.326} = 10^{0.674} \times 10^{-1}$$
$$= 4.72 \times 10^{-1} = 0.472$$

Whenever the logarithm is a negative number, the exponent is broken into a negative integer (the characteristic) and a positive noninteger less than 1 (the mantissa), as in the last example. Note that in the example, the sum of the two exponents is equal to the original exponent (-0.326). Another example is

$$\log N = -4.723$$
$$N = 10^{-4.723} = 10^{0.277} \times 10^{-5}$$
$$= 1.89 \times 10^{-5} = 0.0000189$$

B.4 Finding Roots with Logarithms

It is a simple matter to find a given root of a number using logarithms. Suppose, for example, you wish to find the cube root of 325. Let N represent the cube root:

$$N = 325^{1/3}$$

Taking the logarithm of both sides,

$$\log N = \log 325^{1/3}$$

This $\dfrac{1}{3}$ can be brought out front:

$$\log N = \frac{1}{3} \log 325 = \frac{1}{3}(2.512) = 0.837$$
$$N = 10^{0.837} = \text{antilog } 0.837 = 6.87$$

B.5 The Quadratic Formula

A quadratic equation of the general form

$$ax^2 + bx + c = 0$$

can be solved by use of the quadratic formula:

$$x = \frac{-b \pm \sqrt{b^2 - 4ac}}{2a}$$

Quadratic equations are frequently encountered in calculation of equilibrium concentrations of ionized species using equilibrium constant expressions. Hence, an equation of the following type might require solving:

$$\frac{x^2}{1.0 \times 10^{-3} - x} = 8.0 \times 10^{-4}$$

or

$$x^2 = 8.0 \times 10^{-7} - 8.0 \times 10^{-4}x$$

Arranging in the quadratic equation form above, we have

$$x^2 + 8.0 \times 10^{-4}x - 8.0 \times 10^{-7} = 0$$

or

$$a = 1 \qquad b = 8.0 \times 10^{-4} \qquad c = -8.0 \times 10^{-7}$$

Therefore,

$$x = \frac{-8.0 \times 10^{-4} \pm \sqrt{(8.0 \times 10^{-4})^2 - 4(1)(-8.0 \times 10^{-7})}}{2(1)}$$

$$= \frac{-8.0 \times 10^{-4} \pm \sqrt{0.64 \times 10^{-6} + 3.2_0 \times 10^{-6}}}{2}$$

$$= \frac{-8.0 \times 10^{-4} \pm \sqrt{3.8_4 \times 10^{-6}}}{2}$$

$$= \frac{-8.0 \times 10^{-4} \pm 1.9_6 \times 10^{-3}}{2} = \frac{1.1_6 \times 10^{-3}}{2}$$

$$x = 5.8_0 \times 10^{-4}$$

A concentration can only be positive, and so the negative value of x is not a solution.

You can use Excel Solver for solving quadratic equations. See Chapter 5.

Appendix C

TABLES OF CONSTANTS

TABLE C.1 Dissociation Constants for Acids

Name	Formula	Dissociation Constant, at 25°C			
		K_{a1}	K_{a2}	K_{a3}	K_{a4}
Acetic	CH_3COOH	1.75×10^{-5}			
Alanine	$CH_3CH(NH_2)COOH^a$	4.5×10^{-3}	1.3×10^{-10}		
Arsenic	H_3AsO_4	6.0×10^{-3}	1.0×10^{-7}	3.0×10^{-12}	
Arsenious	H_3AsO_3	6.0×10^{-10}	3.0×10^{-14}		
Benzoic	C_6H_5COOH	6.3×10^{-5}			
Boric	H_3BO_3	6.4×10^{-10}			
Carbonic	H_2CO_3	4.3×10^{-7}	4.8×10^{-11}		
Chloroacetic	$ClCH_2COOH$	1.51×10^{-3}			
Citric	$HOOC(OH)C(CH_2COOH)_2$	7.4×10^{-4}	1.7×10^{-5}	4.0×10^{-7}	
Ethylenediaminetetraacetic	$(CO_2^-)_2NH^+CH_2CH_2NH^+(CO_2^-)_2{}^a$	1.0×10^{-2}	2.2×10^{-3}	6.9×10^{-7}	5.5×10^{-11}
Formic	$HCOOH$	1.76×10^{-4}			
Glycine	$H_2NCH_2COOH^b$	4.5×10^{-3}	1.7×10^{-10}		
Hydrocyanic	HCN	7.2×10^{-10}			
Hydrofluoric	HF	6.7×10^{-4}			
Hydrogen sulfide	H_2S	9.1×10^{-8}	1.2×10^{-15}		
Hypochlorous	$HOCl$	1.1×10^{-8}			
Iodic	HIO_3	2×10^{-1}			
Lactic	$CH_3CHOHCOOH$	1.4×10^{-4}			
Leucine	$(CH_3)_2CHCH_2CH(NH_2)COOH^b$	4.7×10^{-3}	1.8×10^{-10}		
Maleic	cis-HOOCCH : CHCOOH	1.5×10^{-2}	2.6×10^{-7}		
Malic	$HOOCCHOHCH_2COOH$	4.0×10^{-4}	8.9×10^{-6}		
Nitrous	HNO_2	5.1×10^{-4}			
Oxalic	$HOOCCOOH$	6.5×10^{-2}	6.1×10^{-5}		
Phenol	C_6H_5OH	1.1×10^{-10}			
Phosphoric	H_3PO_4	1.1×10^{-2}	7.5×10^{-8}	4.8×10^{-13}	
Phosphorous	H_3PO_3	5×10^{-2}	2.6×10^{-7}		
o-Phthalic	$C_6H_4(COOH)_2$	1.12×10^{-3}	3.90×10^{-6}		
Picric	$(NO_2)_3C_6H_5OH$	4.2×10^{-1}			
Propanoic	CH_3CH_2COOH	1.3×10^{-5}			

(continued)

TABLE C.1 Dissociation Constants for Acids(*Continued*)

Name	Formula	Dissociation Constant, at 25°C			
		K_{a1}	K_{a2}	K_{a3}	K_{a4}
Salicyclic	$C_6H_4(OH)COOH$	1.07×10^{-3}	1.82×10^{-14}		
Sulfamic	NH_2SO_3H	1.0×10^{-1}			
Sulfuric	H_2SO_4	$\gg 1$	1.2×10^{-2}		
Sulfurous	H_2SO_3	1.3×10^{-2}	1.23×10^{-7}		
Trichloroacetic	Cl_3COOH	1.29×10^{-1}			

[a] The first two carbonyl protons are most readily dissociated, with K_a values of 1.0 and 0.032, respectively.
The protons on the more basic nitrogens are most tightly held (K_{a3} and K_{a4}).
[b] K_{a1} and K_{a2} for stepwise dissociation of the protonated form $R—CH—CO_2H$
$$|$$
$$NH_3^+$$

TABLE C.2a Dissociation Constants for Basic Species

Name	Formula	Dissociation Constant, at 25°C	
		K_{b1}	K_{b2}
Ammonia	NH_3	1.75×10^{-5}	
Aniline	$C_6H_5NH_2$	4.0×10^{-10}	
1-Butylamine	$CH_3(CH_2)_2CH_2NH_2$	4.1×10^{-4}	
Diethylamine	$(CH_3CH_2)_2NH$	8.5×10^{-4}	
Dimethylamine	$(CH_3)_2NH$	5.9×10^{-4}	
Ethanolamine	$HOC_2H_4NH_2$	3.2×10^{-5}	
Ethylamine	$CH_3CH_2NH_2$	4.3×10^{-4}	
Ethylenediamine	$NH_2C_2H_4NH_2$	8.5×10^{-5}	7.1×10^{-8}
Glycine	$HOOCCH_2NH_2$	2.3×10^{-12}	
Hydrazine	H_2NNH_2	1.3×10^{-6}	1.0×10^{-15}
Hydroxylamine	$HONH_2$	9.1×10^{-9}	
Methylamine	CH_3NH_2	4.8×10^{-4}	
Piperidine	$C_5H_{11}N$	1.3×10^{-3}	
Pyridine	C_5H_5N	1.7×10^{-9}	
Triethylamine	$(CH_3CH_2)_3N$	5.3×10^{-4}	
Trimethylamine	$(CH_3)_3N$	6.3×10^{-5}	
Tris(hydroxymethyl)aminomethane	$(HOCH_2)_3CNH_2$	1.2×10^{-6}	
Zinc hydroxide	$Zn(OH)_2$		4.4×10^{-5}

TABLE C.2b Acid Dissociation Constants for Basic Species[a]

Name	Formula	Dissociation Constant, at 25°C	
		K_{a1}	K_{a2}
Ammonia	NH_4^+	5.71×10^{-10}	
Aniline	$C_6H_5NH_3^+$	2.50×10^{-5}	
1-Butylamine	$CH_2(CH_2)_2CH_2NH_3^+$	2.44×10^{-11}	
Diethylamine	$(CH_3CH_2)_2NH_2^+$	1.18×10^{-11}	
Dimethylamine	$(CH_3)_2NH_2^+$	1.69×10^{-11}	

(*continued*)

TABLE C.2b Acid Dissociation Constants for Basic Species[a] (*Continued*)

Name	Formula	Dissociation Constant, at 25°C	
		K_{a1}	K_{a2}
Ethanolamine	$HOC_2H_4NH_3^+$	3.1×10^{-10}	
Ethylamine	$CH_3CH_2NH_3^+$	2.33×10^{-11}	
Ethylenediamine	$NH_3C_2H_4NH_3^{2+}$	1.41×10^{-7}	1.18×10^{-10}
Glycine	$HOOCCH_2NH_3^+$	4.4×10^{-3}	
Hydrazine	$H_3NNH_3^{2+}$	10.0	1.0×10^{-15}
Methylamine	$CH_3NH_3^+$	2.08×10^{-11}	
Piperidine	$C_5H_{11}NH^+$	7.7×10^{-12}	
Pyridine	$C_5H_5HN^+$	5.9×10^{-6}	
Triethylamine	$(CH_3CH_2)_3NH^+$	1.89×10^{-11}	
Trimethylyamine	$(CH_3)_3NH^+$	1.59×10^{-10}	
Tris(hydroxmethyl) aminomethane	$(HOCH_2)_3CNH_3^+$	8.3×10^{-9}	
Zinc hydroxide	$Zn(OH)_2H^+$	2.27×10^{-10}	

[a] Some tabulations list only acid dissociation constants for all acidic and basic species. We may represent basic compounds as protonated acids (the conjugate acids) and tabulate K_a values. K_b for the conjugate base is K_w/K_a (for diprotic species such as ethylenediamine, $K_{b1} = K_w/K_{a2}$; $K_{b2} = K_w/K_{a1}$). In the case of $Zn(OH)_2$ base, the first OH^- is essentially all ionized, and so only K_{a1} (and hence conjugate base K_{b2}) is listed.

TABLE C.3 Solubility Product Constants

Substance	Formula	K_{sp}
Aluminum hydroxide	$Al(OH)_3$	2×10^{-32}
Barium carbonate	$BaCO_3$	8.1×10^{-9}
Barium chromate	$BaCrO_4$	2.4×10^{-10}
Barium fluoride	BaF_2	1.7×10^{-6}
Barium iodate	$Ba(IO_3)_2$	1.5×10^{-9}
Barium manganate	$BaMnO_4$	2.5×10^{-10}
Barium oxalate	BaC_2O_4	2.3×10^{-8}
Barium sulfate	$BaSO_4$	1.0×10^{-10}
Beryllium hydroxide	$Be(OH)_2$	7×10^{-22}
Bismuth oxide chloride	$BiOCl$	7×10^{-9}
Bismuth oxide hydroxide	$BiOOH$	4×10^{-10}
Bismuth sulfide	Bi_2S_3	1×10^{-97}
Cadmium carbonate	$CdCO_3$	2.5×10^{-14}
Cadmium oxalate	CdC_2O_4	1.5×10^{-8}
Cadmium sulfide	CdS	1×10^{-28}
Calcium carbonate	$CaCO_3$	8.7×10^{-9}
Calcium fluoride	CaF_2	4.0×10^{-11}
Calcium hydroxide	$Ca(OH)_2$	5.5×10^{-6}
Calcium oxalate	CaC_2O_4	2.6×10^{-9}
Calcium sulfate	$CaSO_4$	1.9×10^{-4}
Copper(I) bromide	$CuBr$	5.2×10^{-9}
Copper(I) chloride	$CuCl$	1.2×10^{-6}
Copper(I) iodide	CuI	5.1×10^{-12}
Copper(I) thiocyanate	$CuSCN$	4.8×10^{-15}
Copper(II) hydroxide	$Cu(OH)_2$	1.6×10^{-19}
Copper(II) sulfide	CuS	9×10^{-36}

(*continued*)

TABLE C.3 Solubility Product Constants(*Continued*)

Substance	Formula	K_{sp}
Iron(II) hydroxide	$Fe(OH)_2$	8×10^{-16}
Iron(III) hydroxide	$Fe(OH)_3$	4×10^{-38}
Lanthanum iodate	$La(IO_3)_3$	6×10^{-10}
Lead chloride	$PbCl_2$	1.6×10^{-5}
Lead chromate	$PbCrO_4$	1.8×10^{-14}
Lead iodide	PbI_2	7.1×10^{-9}
Lead oxalate	PbC_2O_4	4.8×10^{-10}
Lead sulfate	$PbSO_4$	1.6×10^{-8}
Lead sulfide	PbS	8×10^{-28}
Magnesium ammonium phosphate	$MgNH_2PO_4$	2.5×10^{-13}
Magnesium carbonate	$MgCO_3$	1×10^{-5}
Magnesium hydroxide	$Mg(OH)_2$	1.2×10^{-11}
Magnesium oxalate	MgC_2O_4	9×10^{-5}
Manganese(II) hydroxide	$Mn(OH)_2$	4×10^{-14}
Manganese(II) sulfide	MnS	1.4×10^{-15}
Mercury(I) bromide	Hg_2Br_2	5.8×10^{-23}
Mercury(I) chloride	Hg_2Cl_2	1.3×10^{-18}
Mercury(I) iodide	Hg_2I_2	4.5×10^{-29}
Mercury(II) sulfide	HgS	4×10^{-53}
Silver arsenate	Ag_3AsO_4	1.0×10^{-22}
Silver bromide	$AgBr$	4×10^{-13}
Silver carbonate	Ag_2CO_3	8.2×10^{-12}
Silver chloride	$AgCl$	1.0×10^{-10}
Silver chromate	Ag_2CrO_4	1.1×10^{-12}
Silver cyanide	$Ag[Ag(CN)_2]$	5.0×10^{-12}
Silver iodate	$AgIO_3$	3.1×10^{-8}
Silver iodide	AgI	1×10^{-16}
Silver phosphate	Ag_3PO_4	1.3×10^{-20}
Silver sulfide	Ag_2S	2×10^{-49}
Silver thiocyanate	$AgSCN$	1.0×10^{-12}
Strontium oxalate	SrC_2O_4	1.6×10^{-7}
Strontium sulfate	$SrSO_4$	3.8×10^{-7}
Thallium(I) chloride	$TlCl$	2×10^{-4}
Thallium(I) sulfide	Tl_2S	5×10^{-22}
Zinc ferrocyanide	$Zn_2Fe(CN)_6$	4.1×10^{-16}
Zinc oxalate	ZnC_2O_4	2.8×10^{-8}
Zinc sulfide	ZnS	1×10^{-21}

TABLE C.4 Formation Constants for Some EDTA Metal Chelates ($M^{n+} + Y^{4-} \rightleftharpoons MY^{n-4}$)

Element	Formula	K_f
Aluminum	AlY^-	1.35×10^{16}
Bismuth	BiY^-	1×10^{23}
Barium	BaY^{2-}	5.75×10^7
Cadmium	CdY^{2-}	2.88×10^{16}
Calcium	CaY^{2-}	5.01×10^{10}

(continued)

TABLE C.4 Formation Constants for Some EDTA Metal Chelates ($M^{n+} + Y^{4-} \rightleftharpoons MY^{n-4}$)(*Continued*)

Element	Formula	K_f
Cobalt (Co^{2+})	CoY^{2-}	2.04×10^{16}
(Co^{3+})	CoY^{-}	1×10^{36}
Copper	CuY^{2-}	6.30×10^{18}
Gallium	GaY^{-}	1.86×10^{20}
Indium	InY^{-}	8.91×10^{24}
Iron (Fe^{2+})	FeY^{2-}	2.14×10^{14}
(Fe^{3+})	FeY^{-}	1.3×10^{25}
Lead	PbY^{2-}	1.10×10^{18}
Magnesium	MgY^{2-}	4.90×10^{8}
Manganese	MnY^{2-}	1.10×10^{14}
Mercury	HgY^{2-}	6.30×10^{21}
Nickel	NiY^{2-}	4.16×10^{18}
Scandium	ScY^{-}	1.3×10^{23}
Silver	AgY^{3-}	2.09×10^{7}
Strontium	SrY^{2-}	4.26×10^{8}
Thorium	ThY	1.6×10^{23}
Titanium (Ti^{3+})	TiY^{-}	2.0×10^{21}
(TiO^{2+})	$TiOY^{2-}$	2.0×10^{17}
Vanadium (V^{2+})	VY^{2-}	5.01×10^{12}
(V^{3+})	VY^{-}	8.0×10^{25}
(VO^{2+})	VOY^{2-}	1.23×10^{18}
Yttrium	YY^{-}	1.23×10^{18}
Zinc	ZnY^{2-}	3.16×10^{16}

TABLE C.5 Some Standard and Formal Reduction Electrode Potentials

Half-Reaction	$E^0(V)$	Formal Potential (V)
$F_2 + 2H^+ + 2e^- \rightleftharpoons 2HF$	3.06	
$O_3 + 2H^+ + 2e^- \rightleftharpoons O_2 + H_2O$	2.07	
$S_2O_8^{2-} + 2e^- \rightleftharpoons 2SO_4^{2-}$	2.01	
$Co^{3+} + e^- \rightleftharpoons Co^{2+}$	1.842	
$H_2O_2 + 2H^+ + 2e^- \rightleftharpoons 2H_2O$	1.77	
$MnO_4^- + 4H^+ + 3e^- \rightleftharpoons MnO_2 + 2H_2O$	1.695	
$Ce^{4+} + e^- \rightleftharpoons Ce^{3+}$		1.70 (1 M $HClO_4$); 1.61 (1 M HNO_3); 1.44 (1 M H_2SO_4)
$HClO + H^+ + e^- \rightleftharpoons \frac{1}{2}Cl_2 + H_2O$	1.63	
$H_5IO_6 + H^+ + 2e^- \rightleftharpoons IO_3^- + 3H_2O$	1.6	
$BrO_3^- + 6H^+ + 5e^- \rightleftharpoons \frac{1}{2}Br_2 + 3H_2O$	1.52	
$MnO_4^- + 8H^+ + 5e^- \rightleftharpoons Mn^{2+} + 4H_2O$	1.51	
$Mn^{3+} + e^- \rightleftharpoons Mn^{2+}$		1.51 (8 M H_2SO_4)
$ClO_3^- + 6H^+ + 5e^- \rightleftharpoons \frac{1}{2}Cl_2 + 3H_2O$	1.47	
$PbO_2 + 4H^+ + 2e^- \rightleftharpoons Pb^{2+} \rightleftharpoons 2H_2O$	1.455	
$Cl_2 + 2e^- \rightleftharpoons 2Cl^-$	1.359	

(*continued*)

TABLE C.5 Some Standard and Formal Reduction Electrode Potentials(*Continued*)

Half-Reaction	E^0(V)	Formal Potential (V)
$Cr_2O_7^{2-} + 14H^+ + 6e^- \rightleftharpoons 2Cr^{3+} + 7H_2O$	1.33	
$Tl^{3+} \rightleftharpoons 2e^- \rightleftharpoons Tl^+$	1.25	0.77 (1 M HCl)
$IO_3^- + 2Cl^- + 6H^+ + 4e^- \rightleftharpoons ICl_2^- + 3H_2O$	1.24	
$MnO_2 + 4H^+ + 2e^- \rightleftharpoons Mn^{2+} + 2H_2O$	1.23	
$O_2 + 4H^+ + 4e^- \rightleftharpoons 2H_2O$	1.229	
$2IO_3^- + 12H^+ + 10e^- \rightleftharpoons I_2 + 6H_2O$	1.20	
$SeO_4^{2-} + 4H^+ + 2e^- \rightleftharpoons H_2SeO_3 + H_2O$	1.15	
$Br_2(aq) + 2e^- \rightleftharpoons 2Br^-$	1.087[a]	
$Br_2(l) + 2e^- \rightleftharpoons 2Br^-$	1.065[a]	
$ICl_2^- + e^- \rightleftharpoons \frac{1}{2}I_2 + 2Cl^-$	1.06	
$VO_2^+ + 2H^+ + e^- \rightleftharpoons VO^{2+} + H_2O$	1.000	
$HNO_2 + H^+ + e^- \rightleftharpoons NO + H_2O$	1.00	
$Pd^{2+} + 2e^- \rightleftharpoons Pd$	0.987	
$NO_3^- + 3H^+ + 2e^- \rightleftharpoons HNO_2 + H_2O$	0.94	
$2Hg^{2+} + 2e^- \rightleftharpoons Hg_2^{2+}$	0.920	
$H_2O_2 + 2e^- \rightleftharpoons 2OH^-$	0.88	
$Cu^{2+} + I^- + e^- \rightleftharpoons CuI$	0.86	
$Hg^{2+} + 2e^- \rightleftharpoons Hg$	0.854	
$Ag^+ + e^- \rightleftharpoons Ag$	0.799	0.228 (1 M HCl); 0.792 (1 M HClO$_4$)
$Hg_2^{2+} + 2e^- \rightleftharpoons 2Hg$	0.789	0.274 (1 M HCl)
$Fe^{3+} + e^- \rightleftharpoons Fe^{2+}$	0.771	
$H_2SeO_3 + 4H^+ + 4e^- \rightleftharpoons Se + 3H_2O$	0.740	
$PtCl_4^{2-} + 2e^- \rightleftharpoons Pt + 4Cl^-$	0.73	
$C_6H_4O_2$ (quinone) $+ 2H^+ + 2e^- \rightleftharpoons C_6H_4(OH)_2$	0.699	0.696 (1 M HCl, H$_2$SO$_4$, HClO$_4$)
$O_2 + 2H^+ + 2e^- \rightleftharpoons H_2O_2$	0.682	
$PtCl_6^{2-} + 2e^- \rightleftharpoons PtCl_4^{2-} + 2Cl^-$	0.68	
$I_2(aq) + 2e^- \rightleftharpoons 2I^-$	0.6197[b]	
$Hg_2SO_4 + 2e^- \rightleftharpoons 2Hg + SO_4^{2-}$	0.615	
$Sb_2O_5 + 6H^+ + 4e^- \rightleftharpoons 2SbO^+ + 3H_2O$	0.581	
$MnO_4^- + e^- \rightleftharpoons MnO_4^{2-}$	0.564	
$H_3AsO_4 + 2H^+ + 2e^- \rightleftharpoons H_3AsO_3 + H_2O$	0.559	0.577 (1 M HCl, HClO$_4$)
$I_3^- + 2e^- \rightleftharpoons 3I^-$	0.5355	
$I_2(s) + 2e^- \rightleftharpoons 2I^-$	0.5345[b]	
$Mo^{6+} + e^- \rightleftharpoons Mo^{5+}$		0.53 (2 M HCl)
$Cu^+ + e^- \rightleftharpoons Cu$	0.521	
$H_2SO_3 + 4H^+ + 4e^- \rightleftharpoons S + 3H_2O$	0.45	
$Ag_2CrO_4 + 2e^- \rightleftharpoons 2Ag + CrO_4^{2-}$	0.446	
$VO^{2+} + 2H^+ + e^- \rightleftharpoons V^{3+} + H_2O$	0.361	
$Fe(CN)_6^{3-} + e^- \rightleftharpoons Fe(CN)_6^{4-}$	0.36	0.72 (1 M HClO$_4$, H$_2$SO$_4$)
$Cu^{2+} \rightleftharpoons 2e^- \rightleftharpoons Cu$	0.337	
$UO_2^{2+} + 4H^+ + 2e^- \rightleftharpoons U^{4+} + 2H_2O$	0.334	
$BiO^+ + 2H^+ + 3e^- \rightleftharpoons Bi + H_2O$	0.32	
$Hg_2Cl_2(s) + 2e^- \rightleftharpoons 2Hg + 2Cl^-$	0.268	0.242 (sat'd KCl—SCE); 0.282 (1 M KCl)
$AgCl + e^- \rightleftharpoons Ag + Cl^-$	0.222	0.228 (1 M KCl)

(continued)

TABLE C.5 Some Standard and Formal Reduction Electrode Potentials(*Continued*)

Half-Reaction	E^0(V)	Formal Potential (V)
$SO_4^{2-} + 4H^+ + 2e^- \rightleftharpoons H_2SO_3 + H_2O$	0.17	
$BiCl_4^- + 3e^- \rightleftharpoons Bi + 4Cl^-$	0.16	
$Sn^{4+} + 2e^- \rightleftharpoons Sn^{2+}$	0.154	0.14 (1 M HCl)
$Cu^{2+} + e^- \rightleftharpoons Cu^+$	0.153	
$S + 2H^+ + 2e^- \rightleftharpoons H_2S$	0.141	
$TiO^{2+} + 2H^+ + e^- \rightleftharpoons Ti^{3+} + H_2O$	0.1	
$Mo^{4+} + e^- \rightleftharpoons Mo^{3+}$		0.1 (4 M H_2SO_4)
$S_4O_6^{2-} + 2e^- \rightleftharpoons 2S_2O_3^{2-}$	0.08	
$AgBr + e^- \rightleftharpoons Ag + Br^-$	0.071	
$Ag(S_2O_3)_2^{3-} + e^- \rightleftharpoons Ag + 2S_2O_3^{2-}$	0.01	
$2H^+ + 2e^- \rightleftharpoons H_2$	0.000	
$Pb^{2+} + 2e^- \rightleftharpoons Pb$	−0.126	
$CrO_4^{2-} + 4H_2O + 3e^- \rightleftharpoons Cr(OH)_3 + 5OH^-$	−0.13	
$Sn^{2+} + 2e^- \rightleftharpoons Sn$	−0.136	
$AgI + e^- \rightleftharpoons Ag + I^-$	−0.151	
$CuI + e^- \rightleftharpoons Cu + I^-$	−0.185	
$N_2 + 5H^+ + 4e^- \rightleftharpoons N_2H_5^+$	−0.23	
$Ni^{2+} + 2e^- \rightleftharpoons Ni$	−0.250	
$V^{3+} + e^- \rightleftharpoons V^{2+}$	−0.255	
$Co^{2+} + 2e^- \rightleftharpoons Co$	−0.277	
$Ag(CN)_2^- + e^- \rightleftharpoons Ag + 2CN^-$	−0.31	
$Tl^+ + e^- \rightleftharpoons Tl$	−0.336	−0.551 (1 M HCl)
$PbSO_4 + 2e^- \rightleftharpoons Pb + SO_4^{2-}$	−0.356	
$Ti^{3+} + e^- \rightleftharpoons Ti^{2+}$	−0.37	
$Cd^{2+} + 2e^- \rightleftharpoons Cd$	−0.403	
$Cr^{3+} + e^- \rightleftharpoons Cr^{2+}$	−0.41	
$Fe^{2+} + 2e^- \rightleftharpoons Fe$	−0.440	
$2CO_2(g) + 2H^+ + 2e^- \rightleftharpoons H_2C_2O_4$	−0.49	
$Cr^{3+} + 3e^- \rightleftharpoons Cr$	−0.74	
$Zn^{2+} + 2e^- \rightleftharpoons Zn$	−0.763	
$2H_2O + 2e^- \rightleftharpoons H_2 + 2OH^-$	−0.828	
$Mn^{2+} + 2e^- \rightleftharpoons Mn$	−1.18	
$Al^{3+} + 3e^- \rightleftharpoons Al$	−1.66	
$Mg^{2+} + 2e^- \rightleftharpoons Mg$	−2.37	
$Na^+ + e^- \rightleftharpoons Na$	−2.714	
$Ca^{2+} + 2e^- \rightleftharpoons Ca$	−2.87	
$Ba^- + 2e^- \rightleftharpoons Ba$	−2.90	
$K^+ + e^- \rightleftharpoons K$	−2.925	
$Li^+ + e^- \rightleftharpoons Li$	−3.045	

[a] E^0 for Br_2(l) is used for saturated solutions of Br_2 while E^0 for Br_2(aq) is used for unsaturated solutions.
[b] E^0 for I_2(s) is used for saturated solutions of I_2, while E^0 for I_2(aq) is used for unsaturated solutions.

APPENDIX D

    Safety in the Laboratory

    See the Text Website

APPENDIX E

    Periodic Tables on the Web

    See the Text Website

Appendix F

ANSWERS TO PROBLEMS

CHAPTER 1

Multiple Choice Questions
1. (d)
2. (b)
3. (a)
4. (d)

CHAPTER 2
19. 24.920 mL
20. $V_{25}^0 = 25.071$ mL; $V_{20}^0 = 24.041$ mL
21. 10 mL: +0.05; 20 mL: +0.08; 30 mL: +0.08; 40 mL: +0.06; 50 mL: +0.11
22. 0.05140 M
23. (b)
24. 15.920 g

Multiple Choice Questions
1. (b)
2. (d)
3. (b)

CHAPTER 3
6. (iii)
7. (c)
8. (a) 87009 (Five significant figures) (b) 0.0009 (One significant figures) (c) 9.000,000 (Seven significant figures)
9. (a) 5.060×10^{-3} (for sig. fig.) (b) 9.00×10^{-2} (Three sig. fig.) (c) 2000 (One three sig. fig.)
10. 68.9466
11. 177.3
12. (a) 322.000 g, (b) 5.0 g, (c) 1.08 g
13. 0.008_2
14. To the nearest 0.01 g for three significant figures
15. (a) 100.0 [both have four significant figures, but uncertainty here is 1 part per thousand versus about 1 part in 3600] (b) 0.935 [have three significant figures] (c) 0.10 [both have two significant figures, but uncertainty here is 1 part is 10 versus about 1 part in 60]
16. (a) 100 meq/L, (b) –2 meq/L, (c) –2%
17. (a) 126.2 g (b) 126.2 g (c) 1.8 g

18. (a) –4.3979, (b) 3×10^{12}
19. (a) 0.052% S.D., 0.16% c.v., (b) 0.0021% S.S., 8.8% c.v.
20. (a) 0.42 ppm, (b) 0.41%, (c) 0.21 ppm, (d) 0.20%
21. (a) 0.052% ppm, (b) 0.026% ppm, (c) 0.027%
22. (a) 1.07% (b) 10,700
23. (a) $s_a = \pm 9.1; 517 \pm 9$, (b) $s_a = \pm 0.067; 6.82 \pm 0.07$, (c) $s_a = \pm 1.1; 981 \pm 1$
24. (a) ±0.0003, (b) ±0.032 (c) ±42
25. $s_a = \pm 0.023; 1.22 \pm 0.03$
26. (a) 0.88 W/m^2, The terms for the aerosols seem to contribute most to the uncertainty (b) 1.13 ± 0.89 K (c) Yes
27. 0.5024 – 0.5030 M
28. Brand B
29. (a) 139.6 ± 0.47 meq/L, or 139.1 – 140.1 meq/L
    (b) 139.6 ± 0.64 meq/L, or 139.0 – 140.2 meq/L
    (c) 139.6 ± 1.17 meq/L, or 138.4 – 140.8 meq/L
30. ±3.9 ppm
31. $\mu = 5.174 \pm 0.24$; A cell count of 4.5 (million) falls outside the range of 95% confidence interval, and therefore it is too low.
32. $\pm 1._2$ meq/L
33. Ave. = 95.4, S.D. = 3.0_3. Analysis is within 95% C.L.
34. 0.1064 – 0.1072 M
35. (a) Q for 3.16 = 0.5; < 0.64, so can't reject. (b) Mean = 3.04_4; S.D. = 0.07_4; C.L. = $\pm 0.09_2$. t is 2.776 at 95% C.L. So, results agree.
36. $t_{calc} = 1.8_6$. This is smaller than the tabulated t value for 16 degrees of freedom at the 95% confidence level, but not at the 90% confidence level. It appears there is a fair probability the differences between the two populations is real. More studies are indicated.
37. $t_{calc} = 4.9258; t_{table} = 3.355$. So, differences are significant.
38. $t_{calc} = 0.8_7; t_{table} = 2.365$. So, no significant difference.
39. $s_p = 0.0258$ absorbance units.
40. $F_{calc} = 2.79; F_{table} = 4.88$. So, no significant difference.
41. $t_{calc} = 0.713, t_{table} = 2.31$, so no difference.
42. $t_{calc} = 5._9$. Exceeds t_{calc}, even at 99% C.L. So is significant difference.
43. $t_{calc} = 3._6$ (> t_{table}), 95% probability significant difference.
44. (a) $t_{calc} = 0.2857, t_{table} = 3.182$. (b) $t_{calc} = 0.1624, t_{table} = 2.228$. No significant differences.

45. $Q = 0.52$. Since the calculated Q is less than the tabulated Q, the value 5.81 ppm must not be rejected.

46. Q for 0.1050 is 0.76. Q_{table} is 0.829. So 95% certain suspected value not due to accidental error.

47. Q for 33.27 is 0.70. Q_{table} is 0.970. So 95% certain suspected value not due to accidental error.

48. $Q = 0.50$; $Q_{table} = 0.710$. So 22.09 is valid.

49. (a) $Q_{calc} = 0.31$ (for 41.99) and 0.51 (for 45.71). $Q_{table} = 0.56$. So both valid numbers. (b) $(43.58 \pm 1.22)\%$ (c) $t_{calc} = 1.82$, $t_{table} = 2.365$. So, yes.

50. 0.44 ppm

51. Value is 0.5027 ± 0.00026 for both.

52. $139.0 - 140.2$; $138.4 - 140.8$

53. $m = 53.7_5$ for both.

54. 3.05 ppm

55. $m = 0.205 \pm 0.004$; $b = 0.00_0 \pm 0.01_2$; 3.05 ± 0.08 ppm

56. The equations will fail if the errors are not randomly distributed, if there are more outliers, for example. They will also fail if the functional form that the data should exhibit is not a straight line.

57. $r_2 = 0.84$

58. $r_2 = 0.978$

59. $t_{calc} = 1.8_6$, $t_{table} = 2.262$, no significant difference. $r_2 = 0.998$

60. D.L. = 0.17 ppm; total reading 0.36

61. (a) 0.0028_5 (b) $m = 0.7493$ (c) 0.0023 ppb

62. 1.6 g

63. (a) 800 (b) $s = 13$

64. $n = 3.84$

Multiple Choice Questions

1. (c)
2. (b)
3. (c)
4. (d)

CHAPTER 4

2. (c)
3. (b)
4. (d)
12. (a)
13. (b)
14. (a) 12.5 g (b) 5.0 g (c) 100 g
15. (a) 5.23% (b) 55.0% (c) 1.82%
16. (a) 244.27 (b) 218.16 (c) 431.73 (d) 310.18
17. (a) 1.98 (b) 4.20 (c) 1.28 (d) 3.65 (e) 2.24 (f) 1.31

18. (a) 5.06 g (b) 2.38 g (c) 7.80 g (d) 2.74 g (e) 4.46 g (f) 7.64 g

19. (a) 5.84×10^4 mg, (b) 3.42×10^4 mg, (c) 1.71×10^3 mg, (d) 284 mg, (e) 7.01×10^3 mg, (f) 2.25×10^3 mg

20. 1.216 g/mL

21. 25%

22. 23.077 %

23. 0.0333, 0.100, 0.001, 0.0333 mmol/mL of Mn^{2+}, NO_3^-, K^+, SO_4^{2-}

24. 0.00147 g/mL

25. (a) 0.408 M (b) 0.300 M (c) 0.147 M

26. (a) 7.10 g, (b) 49.0 g, (c) 109 g

27. (a) 1.40 g, (b) 8.08 g, (c) 4.00 g

28. 8.06 mL

29. (a) 11.6 M, (b) 15.4 M, (c) 14.6 M, (d) 17.4 M, (e) 14.8 M

30. (b)

31. (c)

32. (a)

33. (b)

34. (b)

35. (c)

36. (b)

37. 1.1 mg/L Na^+; 2.3 mg/L SO_4^{2-}

38. 1.06×10^3 mg/L

39. (a) 5.88×10^{-6} mol/L, (b) 2.92×10^{-6} mol/L, (c) 2.27×10^{-5} mol/L, (d) 1.58×10^{-6} mol/L, (e) 2.73×10^{-5} mol/L, (f) 1.00×10^{-5} mol/L

40. (a) 10.0 mg/L, (b) 27.8 mg/L, (c) 15.8 mg/L, (d) 16.3 mg/L, (e) 13.7 mg/L, (f) 29.8 mg/L

41. 0.00702 g; 1.79×10^{-5} mol/L

42. (a) 0.123%, (b) 1.23% (c) 1.23×10^3 ppm

43. (a) 0.254 g/L (b) 0.165 g/L

44. 156 mL

45. 0.160 g

46. (a)

47. (b)

48. (b)

49. 5.00 M

50. 65 mL

51. 100 mL

53. 0.172%

54. Stock solution 4.809×10^{-3} M; internal standard 5.962×10^{-3} M

55. (a) 865 ± 65 mg/L, (b) Regular coffee $0.264 - 0.529$ mg/mL brewed, (c) 588.5 mg intake.

56. 0.0220 M

57. 5.218×10^{-4} moles

58. 0.171 M

59. 390 mg

60. 0.007 M

61. 99.57%

62. 9.25 mg

63. (a) 26.0% (b) 89.2%

64. 15.3%

65. 4.99%

66. 7.53 mL

67. 1.04×10^3 mg

68. 22.7 mL

69. 143.9

70. 88.4%

71. 12.3 mL

72. 6.47 %

73. 47.1%

74. 6.13%

75. 1.52 mg

76. 15.3 mg BaO per mL EDTA

77. 20.0 mg Fe_2O_3/mL $KMnO_4$

78. 51.2 mg Br/mL

79. (a) 36.46 g/eq, (b) 85.57 g/eq, (c) 389.91 g/eq, (d) 41.04 g/eq, (e) 60.05 g/eq

80. (a) 0.250 mol/L, (b) 0.125 mol/L, (c) 0.250 mol/L, (d) 0.125 mol/L, (e) 0.250 mol/L

81. 16 g equiv^{-1}

82. (a) 128.1 g/eq, (b) 64.05 g/eq

83. 108.3 g/eq

84. (a) 151.91 g/eq, (b) 17.04 g/eq, (c) 17.00 g/eq, (d) 17.00 g/eq

85. 4.093 meq

86. 0.608 eq/L

87. 4.945 g/L

88. 0.180 eq/L

89. 0.267 N

90. 0.474 g $KHC_2O_4.H_2C_2O_4$/g Na_2CO_4

91. 4.903 g

92. 84.5 meq/L

93. 10.0 mg/dL

94. 8.76 g/L

95. 0.610_7 g

96. (a) 0.132_8 g (b) 1.31_0 g

97. 0.7203_1 g Mn/g Mn_3O_4; 2.0700 g Mn_2O_3/g Mn_3O_4; 1.0617_0 g Ag_2S/g $BaSO_4$; 0.46906 g $CuCl_2$/g AgCl; 1.0656 g MgI_2/g PbI_2

98. 45.20%

Multiple Choice Questions

1. (a)

2. (d)

3. (b)

4. (a)

CHAPTER 5

1. (a)

2. (d)

3. (c)

4. $[A] = 9.0 \times 10^{-5}M$, $[B] = 0.50\ M$, $[C] = [D] = 0.30 - x \approx 0.30\ M$

5. $[A] = 4.3 \times 10^{-7}M$, $[B] = 0.30\ M$, $[C] = 0.80\ M$

6. 60%

7. 0.085%

8. 10%

9. 1.1×10^{-22}

10. $[Cr_2O_7{}^{2-}] = 5 \times 10^{-11}M$, $[Fe^{2+}] = 3 \times 10^{-10}M$

11. $1 - (c);\ 2 - (a);\ 3 - (e);\ 4 - (b);\ 5 - (d)$

12. (a) $3\left[Bi^{3+}\right] + \left[H^{+}\right] = 2\left[S^{2-}\right] + \left[HS^{-}\right] + \left[OH^{-}\right]$
    (b) $\left[Na^{+}\right] + \left[H^{+}\right] = 2\left[S^{2-}\right] + \left[HS^{-}\right] + \left[OH^{-}\right]$

15. $\left[F^{-}\right] + \left[HF\right] + 2\left[HF_2{}^{-}\right] = 2\left[Ba^{2+}\right]$

16. $2\left[Ba^{2+}\right] = 3\left(\left[PO_4{}^{3-}\right] + \left[HPO_4{}^{2-}\right] + \left[H_2PO_4{}^{-}\right] + \left[H_3PO_4\right]\right)$

17. 2.88

18. $\left[Ag^{+}\right] + \left[Ag\left(NH_3\right)^{+}\right] + \left[Ag\left(NH_3\right)_2^{+}\right] + \left[H_3O^{+}\right] + \left[NH_4^{+}\right] = \left[OH^{-}\right] + \left[Br^{-}\right]$

20. $\left[Na^{+}\right] + \left[H_3O^{+}\right] + 2\left[Ba^{+}\right] + 2\left[Al^{3+}\right] = \left[ClO_4^{-}\right] + \left[NO_3^{-}\right] + 2\left[SO_4^{2-}\right] + \left[HSO_4^{-}\right] + \left[OH^{-}\right]$

21. (a) 0.1 M, (b) 0.3 M

22. 0.35 M

23. (a) 0.30 (b) 0.90 (c) 0.90 (d) 3.3

24. (a) 0.08 (b) 1.20 (c) 3.0 (d) 3.0 (e) 9.0

25. 0.40

26. 0.96_5

27. $f_{Na^+} = 0.867, f_{SO_4{}^{2-}} = 0.56, f_{Al^{3+}} = 0.35$

28. 0.0019 M

29. 0.0084 M

30. $f_{\pm} = 0.925$

31. (a) 3.25 (b) pH = 3.18; pH = 3.28

Multiple Choice Questions

1. (d)

2. (d)

3. (d)

CHAPTER 6

8. (a) pH = 1.30, pOH = 12.70, (b) pH = 1.92, pOH = 12.08, (c) pH = −0.08, pOH = 14.08, (d) pH = 4.92, pOH = 9.08, (e) pH = 4.70, pOH = 9.30

9. (a) pH = 1.52, pOH = 12.48, (b) pH = 0.92, pOH = 13.08, (c) pH = −0.38, pOH = 14.34, (d) pH = 4.30, pOH = 9.70, (e) pH = 4.30, pOH = 9.40

10. (a) $3.8 \times 10^{-10}M$, (b) $5.0 \times 10^{-14}M$, (c) $1.0 \times 10^{-7}M$, (d) $5.3 \times 10^{-15}M$

11. (a) $3.4 \times 10^{-4}M$ (b) $0.63\ M$ (c) $2.5 \times 10^{-9}M$, (d) $4.0\ M$, (e) $4.5 \times 10^{-15}M$, (f) $18\ M$

12. pH = 12.70, pOH = 1.30

13. 11.65

14. $V_a = 0.10\ V_b$

15. $2.3 \times 10^{-7}M$; pH = 6.64

16. 6.20

17. 3.2%

18. (a) 5.71×10^{-10} (b) 8.88

19. 10.47

20. 18_2 g/mol

21. 2.74

22. 8.80

23. 1.2

24. $0.014_5\ M$

25. $0.0661\ M$

26. 11.54

27. $0.019\ M$

28. Four-fold

29. 11.12

30. 10.57

31. 8.45

32. 2.92

33. 4.85

34. 8.72

35. 3.33

36. 7.00

37. 9.49

38. (i) 2.54, (ii) 8.71, (iii) 4.16, (iv) 8.21, (v) 13.74, (vi) 12.96, (vii) 8.34, (viii) 10.98

39. (a) (i) A^{2-}; (ii) HA^-; (iii) H_2A (b) (i) HA^{2-} (ii) A^{3-} (iii) H_2A^-

40. 4.05

41. 9.42

42. 5.12

43. 4.58

44. 9.24

45. 0.50_8 g

46. 7.3 g

47. $0.46_7\ M$

48. (a) 16 (b) 0.16

49. 2.62

50. 5.41

51. $2.0 \times 10^{-3}M\ HPO_4{}^{2-}$, $9.6 \times 10^{-4}M\ H_2PO_4{}^-$

52. 7.53

53. $\beta = 0.095$ mol/L per pH (or 10.5 pH per mol/L acid or base); ΔpH HCl = −0.011 pH; ΔpH NaOH = +0.011 pH

54. $0.868\ M = [HOAc] = [OAc^-]$

55. 1.62 g Na_2HPO_4; 0.82 g KH_2PO_4

56. 0.24 mL, 5.4 g

57. $[H_2SO_3] = 7.63 \times 10^{-5}M$; $[HSO_3] = 9.90 \times 10^{-3}M$; $[SO_3{}^{2-}] = 1.22 \times 10^{-4}M$

59. $4.8 \times 10^{-6}M$

62. pH = 3.88, $[OAc^-] = 10^{-3.88}M$

63. $[OAc^-] = 1.7 \times 10^{-6}M$

65. (a) $[H_2A] = 5.6 \times 10^{-4}M$; $[HA^-] = 4.5 \times 10^{-4}M$; $[A^{2-}] = 1.0 \times 10^{-5}M$ (b) $[H_2A] = 2.0 \times 10^{-8}M$; $[HA^-] = 1.1 \times 10^{-6}M$; $[A^{2-}] = 1.0 \times 10^{-3}M$

69. 6.8246 vs. 6.833 NIST

70. 8.209

71. 8.345

73. (i) E; (ii) C; (iii) A; (iv) B; (v) D; (vi) C; (vii) E

74. $n = 3$, (c) only an equivalence point at $V = 30.0$ mL

Multiple Choice Questions

1. (d)

2. (b)

3. (b)

CHAPTER 7

15. (c)

16. (b)

17. (b)

18. (d)

19. (a)

20. (c)

21. (d)

22. $0.022\ M$

23. 0.0261 M

24. 0.1060 M

25. 0.1025 M

26. 0.1288 M

27. 0.1174 M

29. ΔpH = 2, transition around pH = 14 – pK_b

30. pH 13.00, 12.70, 7.00, 1.90

31. pH 2.70, 4.10, 4.70, 8.76, 12.07

32. pH 11.12, 9.84, 9.24, 5.27, 2.04

33. pH 2.00, 3.00, 5.00, 7.00, 9.76, 12.16

34. pH 9.72, 7.60, 7.12, 6.64, 4.54, 1.96

35. (a) 1.0,   (b) 1.477,   (c) 7,   (d) 12.30

36. (a) 1.0,   (b) 1.511,   (c) 7,   (d) 12.523

37. 79.2%

38. 35.5 mL

39. 201 mg/g, 838 g/mol

40. 62.8%

41. 9.85 g %

42. 0.0200 M H_3PO_4, 0.0300 M HCl

43. 34.4 % Na_2CO_3, 21.5 % $NaHCO_3$

44. 403 mg Na_2CO_3, 110 mg NaOH

45. 4.50 mmol Na_2CO_3, 6.00 mmol $NaHCO_3$

46. 3.00 mmol Na_2CO_3, 1.5_0 mmol NaOH

47. 41.7%

48. $25._2$%

49. 1 hr, pH = 11.94, 5 hr, pH = 11.59

50. (c)

51. (b)

52. (b)

54. A. 0.02 M HCO_3^- and 0.02 M CO_3^{2-} B. 0.015 M CO_3^{2-} and 0.01 M OH^-

55. (b) 167.7

56. (e) Tartaric acid

57. (c) Histidine

58. 60.56%

59. 86.00%

60. 2.80%

61. (a) pH 3.917   (b) 47.04 mg   (c) 4.80 mg

62. 120 mmol Na_2HPO_4 $(0.12M)$ and 72.9 mmol NaOH (2.915 g)

63. 0.379 mL

64. (a) pH = 6.50   (b) Lower

Multiple Choice Questions

1. (c)

2. (d)

3. (c)

4. (c)

CHAPTER 8

5. (d)

12. [Ca^{2+}] = 0.0017 M, [$Ca(NO_3)^+$] = 0.0033 M

13. 1.4×10^{-3} M

14. 2.0×10^{-5} M

15. [Ag^+] = 3.4×10^{-16} M; [$Ag(S_2O_3)^-$] = 2.2×10^{-7} M; [$Ag(S_2O_3)_2^{3-}$] = 9.9×10^{-3} M

16. (a) $2.7_8 \times 10^7$ (b) $3.9_0 \times 10^{17}$

17. (a) (1) 1.60   (2) 2.12   (3) 4.80   (4) 7.22,   (b) (1) 1.60   (2) 2.12   (3) 9.87   (4) 17.37

18. 1.2_6

19. 2.10×10^{14}

20. 9.306 g

21. $9.543 \times 10^{-3} M$

22. $7.00 \times 10^{-3} M$

23. $0.1039_8 M$

24. 10.01 mg $CaCO_3$/mL EDTA

25. 10.01 (mg $CaCO_3$/L.H_2O)/mL EDTA

26. 2.89×10^3 ppm

27. 9.93 mg/dL; 4.95 meq/L

28. 98.79%

29. 3.04 ppm

30. 119 mg/L

31. 20.4 %

Multiple Choice Questions

1. (b)

2. (b)

3. (c)

CHAPTER 9

12. (c)

13. (d)

16. (b)

17. (c)

18. (a)

19. (b)

20. (a)

21. (d)

22. (c)

23. (c)

24. 16.2 g

25. 82.2 g

26. 0.2138, 1.902, 0.1314, 0.6474

27. 0.586 g

28. 98.68%

29. 0.6888 g

30. 1.071%

31. 636 mg

32. 26 mL

33. 1.75%

34. 24.74 g

35. 2.571 g

36. 42.5% Ba, 37.5% Ca

37. 79.98%

38. 0.846 g AgCl, 1.154 g AgBr

39. 55.444 %

40. 0.04045 g/100 mL

41. (a) $K_{sp} = [Ag^+][SCN^-]$,   (b) $K_{sp} = [La^{3+}][IO_3^-]^3$,
(c) $K_{sp} = [[Hg_2^{2+}][Br^-]^2$,   (d) $K_{sp} = [Ag^+][[Ag(CN)_2^-]$,
(e) $K_{sp} = [Zn^{2+}]^2[Fe(CN)_6^{4-}]$,   (f) $K_{sp} = [Bi^{3+}]^2[S^{2-}]^3$

42. 8.20×10^{-19}

43. $1.3 \times 10^{-4} M$ Ag$^+$, $6.5 \times 10^{-5} M$ CrO$_4^{2-}$

44. $1.9 \times 10^{-8} M$

45. $1.3 \times 10^{-17} M$

46. $5.1 \times 10^{-7} M$, $1.0 \times 10^{-9} M$

47. 3.8

48. 1.6×10^{-4} g

49. $1.7 \times 10^{-6} M$

50. AB: $s = 2 \times 10^{-9} M$, AC$_2$: $s = 1 \times 10^{-6} M$

51. Bi$_2$S$_3$ is 4×10^7 times more soluble than HgS.

52. 0.33 M excess F$^-$ needed. Is feasible.

53. (a) $K_{sp}f_{Ba2+}f_{SO_4^{2-}}$ (b) $K_{sp}f_{Ag^+}{}^2f_{CrO_4^{2-}}$

54. $2.0 \times 10^{-5} M$

55. 4.1×10^{-4} g

Multiple Choice Questions

1. (a)

2. (c)

3. (c)

CHAPTER 10

2. **(c)**

3. **(c)**

5. 1–(b); 2–(c); 3–(d); 4–(a)

8. $s = 2.1 \times 10^{-4} M$; $[IO_3^-] = 1.5 \times 10^{-4} M$; $[HIO_3] = 6.9 \times 10^{-5} M$

9. $s = 6.1 \times 10^{-3} M$; $[HF] = 1.20 \times 10^{-2} M$; $[F^-] = 8.01 \times 10^{-5} M$

10. $s = 2.7 \times 10^{-5} M$; $[H_2S] = 2.7 \times 10^{-5} M$; $[HS^-] = 2.7 \times 10^{-10} M$; $[S_2^-] = 2.9 \times 10^{-18} M$

11. $s = 8.4 \times 10^{-3} M$; $[Ag(en)^+] = 5.96 \times 10^{-5} M$; $[Ag(en)_2^{2+}] = 8.4 \times 10^{-3} M$

12. $[Ag^+] = 1.8 \times 10^{-4} M$ ($2.1 \times 10^{-4} M$ if include HIO$_3$ formation)

13. $[Pb^{2+}] = 2.7 \times 10^{-5} M$

14. $8.4 \times 10^{-3} M = s$ ($7.0 \times 10^{-3} M$ if correct for en consumed by reiteration)

15. 29.77 %

16. $1.0 \times 10^{-6} M$

17. 5.434 g/L

18. 0.029 mL excess titrant

20. Yes

Multiple Choice Questions

1. (b)

2. (d)

3. (d)

4. (b)

5. (d)

CHAPTER 11

38. 0.25 μm, 250 nm

39. 7.5×10^{14} Hz, 25,000 cm^{-1}

40. $20,000 - 150,000$ Å; $5,000 - 670$ cm^{-1}

41. 9.5×10^4 cal einstein^{-1}

42. 0.70 A, 0.10 A, 0.56 T, 0.10 T

43. 20

44. 4.25×10^3 cm^{-1} mol^{-1} L

45. (a) 0.492   (b) 67.8%

46. 5.22×10^4 cm^{-1} mol^{-1} L

47. 2.05

48. 0.528 g

49. 79.0%

50. (a) 0.00387 g/day   (b) 0.0985 mmol/L   (c) 0.25

51. 0.054 mg

52. 0.405 ppm N

53. 1.59$_3$

57. (a) $1.3 \times 10^{-7} M$ (b) 200   (c) 120

58. 551 nm

59. (b) slope = εb_2 as c $\to$ 0 and εb_1 as c$\to\infty$

60. (a) 256 nm,   (b) Deuterium or xenon-arc. LEDs also available, (c) Quartz or fused silica cells

Multiple Choice Questions

1. (a)
2. (a)
3. (a)
4. (c)
5. (d)
6. (a)
7. (b)
8. (b)
9. (d)
10. (b)
11. (d)
12. (b)
13. (c)
14. (d)
15. (c)
16. (d)
17. (a)
18. (b)
19. (a)
20. (d)
21. (b)
22. (a)
23. (d)
24. (a)
25. (d)
26. (a)
27. (d)
28. (a)
29. (d)
30. (a)
31. (c)
32. (d)
33. (b)
34. (d)
35. (c)
36. (b)

CHAPTER 12

25. 1.82 ppm per 0.0044 A
26. 13%
27. 4.5×10^{-9} %
28. 190 ppm
29. 5.9 ppm

30. 84 ppm
31. 79 M&Ms in the bag

Multiple Choice Questions

1. (a)
2. (d)
3. (a)
4. (d)
5. (a)
6. (d)
7. (d)
8. (a)
9. (a)
10. (b)
11. (a)
12. (d)

CHAPTER 13

12. $D = K_D \left(1 + 2K_p K_D [HBz]_a\right) / \left(1 + K_a / [H^+]_a\right)$
13. 99.8
14. $9.04\ 10^{-6}$ mol
15. 5.7
16. 56.5%
17. 95%
18. 0.016 M
19. 96.2% extracted with 10 mL; 92.6% extracted with 5 mL
20. 2.7%
21. (a) 1.3×10^{-7} (b) 200   (c) 12_3

Multiple Choice Questions

1. (a)
2. (d)
3. (c)
4. (b)
5. (d)
6. (a)

CHAPTER 14

10. 1.47
11. 0.052 cm/plate
12. $1.0_6 \times 10^3$ cm
13. mL/min, N, H: 120.2, 2420, 0.123; 90.3, 2500, 0.120; 71.8, 2550, 0s.118; 62.7, 2380, 0.126, 50.2, 2230, 0.135; 39.9, 1830, 0.164; 31.7, 1640, 0.183. Optimum near 75 mL/min.
14. $R_s = 0.96$. Not quite resolved.

Multiple Choice Questions

1. (a)
2. (d)
3. (c)
4. (c)
5. (c)
6. (c)
7. (c)
8. (b)
9. (a)

CHAPTER 15

14. 3315 torr inlet pressure
15. 20.9 ppm

Multiple Choice Questions

1. (d)
2. (a)
3. (b)
4. (c)
5. (d)
6. (a)
7. (b)
8. (b)
9. (a)
10. (c)
11. (a)
12. (c)

CHAPTER 16

86. (a) 2058 plates   (b) 0.014 cm   (c) 1.09   (d) 57 cm   (e) t_{R_A} = 29.86 min; t_{R_B} = 32.56 min
87. 270 mmol/L
88. 6.7 g resin
89. (a) HCl,   (b) H_2SO_4,   (c) $HClO_4$,   (d) H_2SO_4
92. 5Hz
93. 61 μg/mL
94. 277.5 nm
96. 120 nL; C_{Cl}/C_{IO_3} = 0.53
97. C_{Cl}/C_{IO_3} = 0.225
100. MeOH: H_2O: 141 psi; acetonitrile: H_2O: 73.5 psi
101. For 0.02 μM LiOH, 54.4 S/cm; For H_2O, 54.8 nS/cm
102. 149.85 μS/cm
103. (a) D_{K+} = 1.95 × 10^{-9} m^2s^{-1}; D_{Cl-} = 2.03 × 10^{-9} m^2s^{-1}, (b) r_{K+} = 0.125 nm; r_{Cl-} = 0.120 nm
104. κ = 6.37 × 10^3 cm^{-1}

105. G = 23.5 nS
109. pH 9.00: μ_{eo} = 2.3 × 10^{-8} m^2/ Vs Net flow is towards the negative electrode.

 pH 3.00, μ_{eo} = −2.0 × 10^{-8} m^2/ Vs Net flow is towards the positive electrode.

110. R_s = 28
112. 5.3
113. 0.119 mM
115. 35.22, 34.92, 32.41 ng/mL

Multiple Choice Questions

1. (c)
2. (c)
3. (b)
4. (b)
5. (c)
6. (a)
7. (c)
8. (a)
9. (c)
10. (c)

CHAPTER 17

20. 217.07017 Da
21. [M − H]−, m/z 155 = 100%; M + 1, m/z 156 = 7.7%; M + 2, m/z 157 = 33%
22. R = 69, 692
23. R_{FWHM} = 10, 665
24. R = 34
25. −2.03 ppm
26. v = 6.67 × 10^5 ms^{-1}, R = 32.1 cm
27. 14 ns

Multiple Choice Questions

1. (c)
2. (a)
3. (b)
4. (b)
5. (a)
6. (a)
7. (a)
8. (b)
9. (a)
10. (d)
11. (a)
12. (a)

13. (a)

14. (a)

15. (c)

16. (a)

17. (c)

18. (d)

CHAPTER 18

Multiple Choice Questions

1. (b)

2. (b)

3. (c)

4. (d)

5. (c)

6. (c)

7. (a)

8. (d)

9. (d)

CHAPTER 19

12. (c)

13. (d)

14. (a)

15. (a)

16. O_3, $HClO$, Hg^{2+}, H_2SeO_3, H_3AsO_4, Cu^{2+}, Co^{2+}, Zn^{2+}, K^+

17. Ni, H_2S, Sn^{2+}, V^{3+}, I^-, Ag, Cl^-, Co^{2+}, HF

18. (a) $Fe^{2+} - MnO_4^-$　(b) $Fe^{2+} - Ce^{4+}(HClO_4)$,　(c) $H_3AsO_3^- MnO_4^-$　(d) $Fe^{3+} - Ti^{2+}$

19. (a) VO^{2+}/Sn^{2+},　(b) $Fe(CN)_6^{3-}/Fe^{2+}$,　(c) Ag^+/Cu, (d) I_2/I_2,　(e) H_2O_2/Fe^{2+}

20. 15.62%

23. (a) Pt/Fe^{2+}, $Fe^{3+}//Cr_2O_7^{2-}$, Cr^{3+}, H^+/Pt,　(b) Pt/I^-, $I_2//IO_3^-$, I_2, H^+/Pt,　(c) $Zn/Zn^{2+}//Cu^{2+}/Cu$,　(d) Pt/H_2SeO_3, SeO_4^{2-}, $H^+//Cl_2$, Cl^-/Pt

24. (a) $2V^{2+} + PtCl_6^{2-} = 2V^{3+} + PtCl_4^{2-} + 2Cl^-$　(b) $Ag + Fe^{3+} + Cl^- = AgCl + Fe^{2+}$　(c) $3Cd + ClO_3^- + 6H^+ = 3Cd^{2+} + Cl^- + 3H_2O$　(d) $2I^- + H_2O_2 + 2H^+ = I_2 + H_2O$

25. 0.810V

26. 1.134V

27. 1.24V

28. 0.419V

29. 0.65V

30. 0.216V

31. (c)

32. $Hg_2(NO_3)_2$

33. $PtCl_6^{2-} + 2V^{2+} = PtCl_4^{2-} + 2V^{3+} + 2Cl^-$, 0.94V

34. (a) 0.57V　(b) 0.071V　(c) 0.09V

35. $2VO_2^+ + U^{4+} = 2VO^{2+} + UO_2^{2+}$, 0.67V

36. 0.592V

37. (a) Cu: 0.286 V; Ag: 0.740 V; cathode,　(b) 0.706 V; cathode,　(c) 0.277 V; anode

Multiple Choice Questions

1. (c)

2. (a)

3. (b)

4. (b)

5. (d)

6. (b)

7. (a)

8. (a)

9. (d)

10. (c)

11. (a)

12. (b)

13. (c)

14. (d)

15. (b)

16. (b)

17. (a)

18. (b)

CHAPTER 20

8. (a)

9. (d)

14. 0.558 V

15. 5.4×10^{-13}

16. 0.799V

17. (a) −0.028 V,　(b) 0.639V,　(c) 0.84V

18. (a) 1.353V,　(b) 0.034

19. $3PVF + AuCl_4^- = 3PVF^+ + Au + Cl^-$, $K = 6.9 \times 10^{70}$

20. 0.32 V

21. (a) 0.845 V,　(b) −0.020 V,　(c) −0.497 V

22. 0.446 V

23. (a) 1×10^{-4} %　(b) 0.084 V　(c) 0.261V

25. −0.505 V

26. 7.1

27. 7

28. 0.2992 V

29. (a) 0.02 pH unit, 1.2 mV, 4.8%, (b) 0.002 pH unit, 0.12 mV, 0.48%

30. (a) 5.78, (b) 10.14, (c) 12.32, (d) 13.89

31. (a) −0.419 V, (b) −0.472 V, (d) −0.524 V

32. 11.2

34. 0.015 M

35. 0.0079 M

36. $4.8 \times 10^{-4} M$

37. (a) 0.16, (b) 18.9 mV

38. 0.020

39. 9.5×10^{-4}

40. $1.0_7 \times 10^4$

Multiple Choice Questions

1. (d)

2. (b)

3. (a)

CHAPTER 21

6. (a) $2OH^- + 2Cr(OH)_4^- + 3ClO^- \rightarrow 2CrO_4^{2-} + 3Cl^- + 3H_2O +$
$2H^+ + 2OH^- + 2Cr(OH)_4^- + 3ClO^- + 2OH^- \rightarrow 2CrO_4^{2-}$
$+3Cl^- + 5H_2O$

(b) $2NO_3^- + 3Cu + 8H^+ \rightarrow 3Cu^{2+} + 2NO + 4H_2O$

(c) $H_3PO_3 + 2HgCl_2 + H_2O = H_3PO_4 + Hg_2Cl_2 + 2H^+ + 2Cl^-$

7. (a) $4Fe(OH)_2 + 2H_2O + O_2 \rightarrow 4Fe(OH)_3$

(b) $2MnO_4^- + 5H_2S + 6H^+ = 2Mn^{2+} + 5S + 8H_2O$

(c) $2SbH_3 + 4Cl_2O + 3H_2O = H_4Sb_2O_7 + 8Cl^- + 8H^+$

(d) $2PbO_2 + 4H^+ + 4Cl^- \rightarrow 2PbCl_2 + 2H_2O + O_2$

(e) $8Al + 3NO_3^- + 5OH^- + 2H_2O = 8AlO_2^- + 3NH_3$

(f) $5FeAsS + 14ClO_2 + 12H_2O = 5Fe^{3+} + 5AsO_4^{3-} +$
$5SO_4^{2-} + 14Cl^- + 24H^+$

(g) $5K_2NaCo(NO_2)_6 + 12MnO_4^- + 36H^+ = 10K^+ + 5Na^+ +$
$5Co^{3+} + 30NO_3^- + 12Mn^{2+} + 18H_2O$

8. 1.127 V

9. 0.735 V, 0.771 V, 1.218 V, 1.28 V

10. 0.715 V, 0.771 V, 1.35 V, 1.46 V

11. 0.360 V

12. $E = \left[(1) E^0_{Fe^{3+},Fe^{2+}} + (2) E^0_{Sn^{4+},Sn^{2+}} \right] / [(1) + (2)]$

13. $E = \left(nFe E^0_{Fe^{3+},Fe^{2+}} + nMn E^0_{MnO_4^-,Mn^{2+}} \right) / \left(n_{Fe} + n_{Mn} \right)$ pH, 1.33V (for pH 0.68)

14. (a) 0.319 V, −0.780 V, 1.6×10^{12} (b) 0.691 V, −0.154 V, 2.6×10^{22}

15. 2×10^{25}

16. 8.9 ppm

17. 4.96 meq/L

18. 11.5% Na_2HAsO_3, 3.5_4 % As_2O_5

19. 1.47 mL

20. $3.8 \times 10^{-3} M$

Multiple Choice Questions

1. (c)

2. (b)

3. (a)

4. (b)

5. (d)

6. (c)

7. (a)

8. (d)

CHAPTER 22

12. $11.9 \times 10^{-5} M$

13. $[Fe^{3+}]/[Fe^{2+}] = 5:1$

14. (c)

Multiple Choice Questions

1. (d)

2. (b)

3. (a)

4. (d)

5. (e)

6. (a)

7. (a)

CHAPTER 23

15. 49.8 min, 99.7 min

16. 40.8 s

17. 38 min; 19 min, 3.4 min

18. 12.7 h

19. 165 h

20. 3.82%

21. $K_m = 1.20 \pm 0.05$ mM

Multiple Choice Questions

1. (b)

2. (b)

3. (c)

4. (a)

5. (b)

6. (b)

7. (b)

8. (c)

9. (a)

10. (d)

11. (d)

12. (a)

13. (b)

14. (a)

CHAPTER 24

Multiple Choice Questions

1. (a)

2. (b)

3. (d)

4. (d)

5. (d)

6. (b)

7. (b)

CHAPTER 25

6. $4.1 \times 10^{-6} g/L_{air}$

7. 380 ppt

Multiple Choice Questions

1. (a)

2. (b)

3. (b)

4. (a)

EXPERIMENTS

"Theory guides, experiment decides."
—I. M. Kolthoff

"I hear and I forget.
I see and I remember.
I do and I understand."
—Confucius

The following experiments follow the order of coverage in the text, after introductory experiments on the use of the analytical balance and volumetric glassware. Before beginning experiments, you should review the material in Chapter 2 on the basic tools of analytical chemistry and their use and the material in Appendix D on the text's website on safety in the laboratory. It will also be helpful to review data handling in Chapter 3, particularly significant figures and propagation of errors, so you will know how accurately to make each required measurement. You will also use the spreadsheets described in that chapter and Chapter 11 for calibration plots and unknown calculations, including precisions.

Video Resource. Professor Christopher Harrison from San Diego State University has a YouTube "Channel" of videos of different types of experiments, some illustrating laboratory and titration technique: http://www.youtube.com/user/crharrison. You are encouraged to look at the ones dealing with buret rinsing, pipetting, and aliquoting a sample, before you begin experiments. Also, you will find useful the examples of acid–base titrations illustrating methyl red or phenolphthalein indicator change at end points. There are a few specific experiments that may be related to ones you may perform from your textbook, for example, EDTA titration of calcium or Fajan's titration of chloride. If you perform an iodometric determination in which iodine is titrated with thiosulfate using starch indicator, the glucose analysis using titration video gives a good illustration of the starch end point.

Use of Apparatus

EXPERIMENT 2   USE OF THE PIPET AND BURET AND STATISTICAL ANALYSIS

Principle

Practice in the use of pipets is checked by weighing the amount of water delivered in successive pipettings. Precision of delivery of different volumes is determined. Similar experiments may be done with a buret. The experiments may also be used for calculation (calibration) of the volumes of the glassware if the temperature of the water is measured, as described in Table 2.4.

Solutions and Chemicals Required

Cleaning solution; distilled water.

Procedure

1. **Cleaning the glassware.** Check your buret and pipets for cleanliness by rinsing with distilled water and allowing to drain. Clean glassware will retain a continuous, unbroken film of water when emptied. If necessary, clean the buret with a detergent and a buret brush. For pipets, try warm water, then detergent plus warm water. If these do not work, use a commercial cleaning solution. Rinse the buret or pipet several times with tap water and finally with distilled water. Leave the buret filled with distilled water when not in use. Always check your volumetric glassware for cleanliness before use, and clean whenever necessary. Burets if thoroughly rinsed and filled with distilled water immediately after use should remain clean for weeks; pipets, however, become contaminated easily and must be cleaned frequently.

2. **Pipetting.** Practice filling a 25-mL pipet and adjusting the meniscus to the calibration mark until you are proficient. The forefinger used to adjust the level should be only slightly moistened. If it is too wet, it will be impossible to obtain an even flow.

   Weigh a clean, dry 50-mL Erlenmeyer flask and a rubber stopper to the nearest milligram (0.001 g).

   Transfer 25 mL water from the pipet to the flask; allow the water to run out with the tip of the pipet touching the side of the flask, being careful that no water splashes or is otherwise deposited on the neck of the flask. It is important that the neck of the flask and the stopper remain dry throughout the experiment. The pipet must be held vertically. Allow to drain for 10 s before removing the pipet. Do not blow out the last drop. Replace the stopper and weigh.

   Perform the procedure at least two times. Only the outside and the neck of the flask need be dry for subsequent weighings. Calculate the standard deviation and the range of delivery in parts per thousand as described in Chapter 3.

   Repeat, using 1-, 5-, 10-, and 50-mL pipets. Compare the precision of delivery for the different pipets.

3. **Use of the buret.** Check the stopcock (if ground glass) of your buret for proper lubrication, and be sure the bore and tip are clear. Fill with distilled water and place in the buret clamp; the water should be at room temperature. Displace any air bubbles from the tip of the buret, and adjust to near the zero mark. Allow to stand for a few minutes to check for leakage and read the initial volume to the nearest 0.01 mL. Use a meniscus reader, and be careful to avoid parallax.

   Using a procedure similar to that for the pipets, draw off about 5 mL into the weighed flask. Touch the tip of the buret to the wall of the flask to remove the adhering drop, and read the volume to the nearest 0.01 mL. Insert the stopper and weigh to the nearest milligram. Repeat the operation, adding to the flask the next 5 mL water, and weigh again. Continue in this way until the entire 50 mL have been weighed.[1]

   Subtract the weight of the empty flask from each of the succeeding weights to get the weight of 5, 10, 15, and so forth, milliliters of water at the temperature observed.

   Repeat the entire procedure. At each approximate volume, the weight, compared to the exact measured volume, should be within 0.03 g of that predicted from the first measurement. Recheck any points that are not.

4. **Calibration of glassware.** If your instructor wishes you to use the data from these experiments to calibrate the volumes of the glassware, use Table 2.4 to convert the weight in air to the volume. The temperature of the water delivered must be

---

[1]To minimize evaporation errors, your instructor may direct you to refill the buret for each volume delivery (0–5, 0–10, ... 0–50 mL) and deliver it into an empty (dry) flask.

determined; the Table 2.4 spreadsheet is on your text website, and you can enter data in that by copying it to your computer desktop.

5. **Statistics: The normal error curve.** In this experiment, you will make 20 measurements of the weight of water delivered by a 1-mL pipet, as described above, by adding successive pipet volumes to a small Erlenmeyer flask and calculating the weight increase each time. Be sure to rapidly stopper the flask each time. Plot the frequencies of the weights on graph paper to prepare a normal error curve. Calculate the standard deviation and relate this to the plotted curve. Do about 68% of the readings fall within one standard deviation?

Gravimetry

EXPERIMENT 5   GRAVIMETRIC DETERMINATION OF SO$_3$ IN A SOLUBLE SULFATE

Principle

Sulfate is precipitated as barium sulfate with barium chloride. After the filtering with filter paper, the paper is charred off, and the precipitate is ignited to constant weight. The SO$_3$ content is calculated from the weight of BaSO$_4$.

Equation

$$SO_4^{2-} + Ba^{2+} \rightarrow \underline{BaSO_4}$$

Solutions and Chemicals Required

1. **Provided.** Dil. (0.1 M) AgNO$_3$, conc. HNO$_3$, conc. HCl.
2. **To prepare.** 0.25 M BaCl$_2$. Dissolve about 5.2 g BaCl$_2$ (need not be dried) in 100 mL distilled water.

Things to Do before the Experiment

1. **Obtain your unknown and dry.** Check out a sample of a soluble sulfate from the instructor and dry it in the oven at 110 to 120°C for at least 2 h. Allow it to cool in a desiccator for at least 0.5 h.
2. **Prepare crucibles.** Clean three porcelain crucibles and covers. Place them over Tirrill burners and heat at the maximum temperature of the burner for 10 to 15 min; then place in the desiccator to cool for at least 1 h. Weigh accurately each crucible with its cover. Between heating and weighing, the crucibles and covers should be handled only with a pair of tongs.
3. Prepare the 0.25 M BaCl$_2$ solution.

Procedure

1. **Preparation of the sample.** Weigh accurately to four significant figures, using the direct method, three samples of about 0.5 to 0.7 g each. Transfer to 400-mL beakers, dissolve in 200 to 250 mL distilled water, and add 0.5 mL concentrated hydrochloric acid to each.
2. **Precipitation.** Assume the sample to be pure sodium sulfate and calculate the millimoles barium chloride required to precipitate the sulfate in the largest sample. Heat the solutions nearly to boiling on a wire gauze over a Tirrill burner and adjust the burner to keep the solution just below the boiling point. Add slowly from a buret,

drop by drop, 0.25 M barium chloride solution until 10% more than the above calculated amount is added; stir vigorously throughout the addition. Let the precipitate settle, then test for complete precipitation by adding a few drops of barium chloride without stirring. If additional precipitate forms, add slowly, with stirring, 5 mL more barium chloride; let settle, and test again. Repeat this operation until precipitation is complete. Leave the stirring rods in the beakers, cover with watch glasses, and digest on the steam bath until the supernatant liquid is clear. (The initial precipitate is fine particles. During digestion, the particles grow to filterable size.) This will require 30 to 60 min or longer. Add more distilled water if the volume falls below 200 mL.

3. **Filtration and washing of the precipitate.** Prepare three 11-cm No. 42 Whatman filter papers or equivalent for filtration; the paper should be well fitted to the funnel so that the stem of the funnel remains filled with liquid, or the filtration will be very slow. See the discussion in Chapter 2 for the proper preparation of filters. Filter the solutions while hot; be careful not to fill the paper too full, as the barium sulfate has a tendency to "creep" above the edge of the paper. Wash the precipitate into the filter with hot distilled water, clean the adhering precipitate from the stirring rod and beaker with the rubber policeman, and again rinse the contents of the beaker into the filter. Examine the beaker very carefully for particles of precipitate that may have escaped transfer. Wash the precipitate and the filter paper with hot distilled water until no turbidity appears when a few milliliters of the washings acidified with a few drops of nitric acid are tested for chloride with silver nitrate solution. During the washing, rinse the precipitate down into the cone of the filter as much as possible. Examine the filtrate for any precipitate that may have run through the filter.

4. **Ignition and weighing of the precipitate.** Loosen the filter paper in the funnels and allow to drain for a few minutes. Fold each filter into a package enclosing the precipitate, with the triple thickness of paper on top. Place in the weighed porcelain crucibles and gently press down into the bottom. Inspect the funnels for traces of precipitate; if any precipitate is found, wipe it off with a small piece of moist ashless filter paper and add to the proper crucible. Place each crucible on a triangle on a tripod or the ring of a ring stand, in an inclined position with the cover displaced slightly. Heat gently with a small flame (Tirrill burner) until all the moisture has been driven off and the paper begins to smoke and char. Adjust the burner so that the paper continues to char without catching fire. If the paper inflames, cover the crucible to smother the fire, and lower the burner flame. When the paper has completely carbonized and no smoke is given off, gradually raise the temperature enough to burn off the carbon completely. A red glowing of the carbon as it burns is normal, but there should be no flame. The precipitate should finally be white with no black particles. Allow to cool. Place the crucible in a vertical position in the triangle, and moisten the precipitate with three or four drops of dilute (1:4) sulfuric acid. Heat *very gently* until the acid has fumed off. (This treatment converts any precipitate that has been reduced to barium sulfide by the hot carbon black to barium sulfate.) Then, cover the crucibles and heat to dull redness in the full flame of the Tirrill burner for 15 min.

Allow the covered crucibles to cool in a desiccator for at least 1 h and then weigh them. Heat again to redness for 10 to 15 min, cool in the desiccator, and weigh again. Repeat until two successive weighings agree within 0.3 to 0.4 mg.

Calculation

Calculate and report the percent SO_3 in your unknown for each portion analyzed. Report also the mean and the relative standard deviation.

Acid–Base Titrations

EXPERIMENT 7   DETERMINATION OF REPLACEABLE HYDROGEN IN ACID BY TITRATION WITH SODIUM HYDROXIDE

Principle

One-tenth molar sodium hydroxide is prepared and standardized against primary standard potassium acid phthalate (KHP). The unknown is titrated with the standardized sodium hydroxide.

Equations

$$HA^- + OH^- \rightarrow A^{2-} + H_2O$$

Solutions and Chemicals Required

1. **Provided.** KHP, 0.2% phenolphthalein solution in 90% ethanol, 50% NaOH solution.

2. **To prepare.**[1] 0.1 M NaOH solution. This solution must be carbonate free. The Na_2CO_3 in NaOH pellets is insoluble in 50% NaOH solution. The precipitated Na_2CO_3 will settle to the bottom of the container after several days to a week, and the clear, syrupy supernatant (approximately 19 M in NaOH) can be carefully withdrawn or decanted from the container to obtain carbonate-free NaOH. The solution is prepared in a borosilicate (Kimax or Pyrex) beaker (it gets very warm) and after cooling is stored in a polyethylene bottle. Your instructor will provide the carbonate-free 50% solution to you.

    Distilled water free from carbon dioxide will be needed in this experiment. To prepare this, fill a 1000-mL Florence flask nearly to the shoulder with distilled water, insert a boiling rod with the cupped end down (the cup must be empty), heat to boiling on a wire gauze over a Meker burner, and boil for 5 min. Cover the flask with an inverted beaker and allow to cool overnight, or cool under the cold water tap. Prepare an additional 800 mL the same way.

    Fill a 500-mL bottle, with rubber stopper, half full of CO_2-free water, add 2.5 mL clear 50% NaOH solution, stopper, and swirl to mix.[2] Avoid exposing the solutions to the atmosphere as much as possible. Finally, add more CO_2-free water to nearly fill the bottle, stopper, and shake thoroughly.

    The sodium hydroxide thus prepared is approximately 0.1 M. It will be standardized against primary standard potassium acid phthalate.

---

[1] If you are to use this solution in Experiment 8 to standardize HCl, prepare 1 L instead of 500 mL.

[2] If the NaOH were pure, only 2 mL (2 g NaOH) would be required for 500 mL of 0.1 M NaOH. Two and one-half milliliters is required to compensate for the water and Na_2CO_3 content of the pellets.

Things to Do before the Experiment

1. **Obtain KHP and dry.** Dry about 4 g primary standard potassium acid phthalate in a weighing bottle at 110 to 120°C for 1 to 2 h. Cool in a desiccator for at least 30 to 40 min before weighing.

2. **Obtain your unknown and dry.** The unknown may be either a liquid or a solid. If it is a solid, obtain it in a weighing bottle before the day of the experiment and dry for 2 h at 110 to 120°C. Store in the desiccator. If the unknown is a liquid, obtain it in a clean 250-mL volumetric flask and dilute to volume with CO_2-free water. Mix well.

Procedure

1. **Standardization of the 0.1 M NaOH solution against potassium acid phthalate.** Weigh accurately three portions of the dried potassium acid phthalate of about 0.7 to 0.9 g each, and transfer to clean 250-mL wide-mouth Erlenmeyer flasks. (These quantities are designed for titrations using 50-mL burets. If you are supplied a 25-mL buret, your instructor will direct you to adjust the quantities of KHP and the unknown accordingly.) The direct method of weighing, using a tared weighing dish, may be used with this material. Dissolve each sample in about 50 mL CO_2-free distilled water. Rinse your buret with three small portions of the 0.1 M NaOH solution, fill, and adjust to near zero. Record the initial volume reading to the nearest 0.02 mL. Add 2 to 3 drops phenolphthalein indicator to each KHP sample and titrate with the 0.1 M NaOH to a faint pink end point. The color should persist at least 30 s. Split drops at the end of the titration. Estimate the buret reading to the nearest 0.02 mL.

   Calculate the molarity of the NaOH to four significant figures from the weight of KHP used (three significant figures if the molarity is slightly less than 0.1 M). Use the average of the results.

   Keep the NaOH bottle stoppered with a rubber stopper, and protect the solution from the air as much as possible to minimize absorption of CO_2. Proceed as soon as possible to the determination of the unknown acid; standard sodium hydroxide should be used within one week of standardization. After that time, changes in molarity may have occurred, and restandardization will be necessary.[1]

2. **Determination of replaceable hydrogen in an unknown acid.** If the unknown is a weak acid (e.g., KHP, acetic acid, or vinegar) you will be instructed to use phenolphthalein indicator (color change same as above). If it is a strong acid, you may use another indicator, such as chlorophenol red (color change from yellow to violet).

   If the unknown is a solid, dry it for 2 h at 110°C. Cool for 30 min. Your instructor will inform you of the approximate weight of sample to take so that about 30 to 40 mL NaOH will be used. If it is a KHP unknown, approximately 1 g may be taken. Weigh three samples to the nearest 0.1 mg and transfer them into numbered 250-mL wide-mouth Erlenmeyer flasks. Dissolve in 50 mL CO_2-free water, warming if necessary. Add two drops of indicator and titrate with 0.1 M NaOH until the color change persists 30 s.

   If the unknown is a liquid, transfer with a pipet three 50-mL aliquots from the 250-mL volumetric flask and titrate as above.

---

[1] If you are to use this solution in Experiment 8 to standardize the HCl, your instructor will advise you to save the remaining solution after completing this experiment. In that event, you should prepare and standardize the HCl within one week of standardizing the NaOH.

Calculation

1. **Solids.** Calculate and report the percent of replaceable hydrogen or percent KHP in your unknown for each portion analyzed.

2. **Liquid.** Calculate and report the milligrams replaceable hydrogen in your unknown. Since one-fifth of the sample was taken for each determination (50 mL out of 250 mL), the weight determined in each aliquot must be multiplied by 5.

Complexometric Titration

EXPERIMENT 10   DETERMINATION OF WATER HARDNESS WITH EDTA

Principle

Water hardness, due to Ca^{2+} and Mg^{2+}, is expressed as mg/L $CaCO_3$ (ppm). The total of Ca^{2+} and Mg^{2+} is titrated with standard EDTA using an Eriochrome Black T indicator.[1] A standard EDTA solution is prepared from dried (do not exceed 80°C) $Na_2H_2Y \cdot 2H_2O$ (purity $100.0 \pm 0.5\%$). If the sample does not contain magnesium, Mg–EDTA is added to the titration flask to provide a sharp end point with the Eriochrome Black T, since calcium does not form a sufficiently strong chelate with the indicator to give a sharp end point. See the discussion in Chapter 8 for a more complete description.

Equations

Titration:

$$Ca^{2+} + H_2Y^{2-} \rightarrow CaY^{2-} + 2H^+$$
$$Ca^{2+} + MgY^{2-} \rightarrow CaY^{2-} + Mg^{2+}$$

End point:

$$Mg^{2+} + HIn^{2-} \rightarrow MgIn^- + H^+$$
$$\underset{\text{(red)}}{MgIn^-} + \underset{\text{(colorless)}}{H_2Y^{2-}} \rightarrow \underset{\text{(colorless)}}{MgY^{2-}} + \underset{\text{(blue)}}{HIn^{2-}} + H^+$$

The free acid parent of the indicator is H_3In, and that of the titrant EDTA is H_4Y.

Solutions and Chemicals Required

1. **Provided.** 0.5% (wt/vol) Eriochrome Black T indicator solution in ethanol, 0.005 M Mg–EDTA (prepared by adding stoichiometric amounts of 0.01 M EDTA and 0.01 M $MgCl_2$). The indicator solution should be prepared fresh every few days, as it is unstable. A portion of the Mg–EDTA solution, when treated with pH 10 buffer and Eriochrome Black T, should turn a dull violet color; one drop 0.01 M EDTA should change this to blue and one drop 0.01 M $MgCl_2$ should change it to red.

2. **To prepare**

   (a) **NH_3–NH_4Cl buffer solution, pH 10.** Dissolve 3.2 g NH_4Cl in water, add 29 mL conc. NH_3, and dilute to about 50 mL. The buffer solution is best stored for long periods of time in a polyethylene bottle to prevent leaching of metal ions from glass.

   (b) **Standard 0.01 M EDTA solution.** Dry about 1.5 g reagent-grade $Na_2H_2Y \cdot 2H_2O$ in a weighing bottle at 80°C for 2 h. Cool in a desiccator for 30 min

---

[1] If it is desired to titrate only Ca^{2+}, this can be done at pH 12 (use NaOH), where $Mg(OH)_2$ is precipitated and does not titrate. Hydroxynaphthol blue indicator is used.

and accurately weigh (to the nearest milligram) approximately 1.0 g. Transfer to a 250-mL volumetric flask. (This is the disodium salt of EDTA; the free acid is insoluble). Add about 2.00 mL distilled, *deionized* water and shake or swirl periodically until the EDTA has dissolved. EDTA dissolves slowly and may take 0.5 h or longer. If possible, the solution should be allowed to stand overnight before using. If any undissolved particles remain, addition of three pellets of NaOH may aid dissolution, but there is danger of adding metallic impurities. After the EDTA is dissolved, dilute to 250.0 mL and shake thoroughly to prepare a homogeneous solution. Then, rinse a clean polyethylene bottle with three small portions of the EDTA solution and transfer the remainder of the solution to the bottle for storage. (Polyethylene is preferable to glass for storage because EDTA solutions gradually leach metal ions from glass containers, resulting in a change in the concentration of free EDTA.) Calculate the molarity of the EDTA solution.

Things to Do before the Experiment

Dry the $Na_2H_2Y \cdot 2H_2O$ and prepare the standard EDTA solution.

Procedure

Obtain a water sample from your instructor. Add with a pipet or a buret a 50-mL aliquot of the sample to a 250-mL wide-mouth Erlenmeyer flask, add 2 mL of the buffer solution, 0.5 mL of the Mg–EDTA solution, and five drops of the indicator solution. (If the unknown contains magnesium, addition of Mg–EDTA is not necessary—consult your instructor.) Avoid adding too much indicator with dilute solutions or the end-point change may be too gradual. The indicator will not become wine red until magnesium is added. *The buffer should be added before the indicator*, so that small amounts of iron present in the water will not react with the indicator. [An end-point color change from wine red to violet indicates a high level of iron in the water. After the titration, add 1 g $FeSO_4 \cdot 7H_2O$ to convert CN^- to harmless $Fe(CN)_6^{4-}$.] If the water sample is likely to contain copper, add a few crystals of hydroxylamine hydrochloride. This reduces copper(II) to copper(I), which does not interfere.

Titrate with 0.01 M EDTA until the color changes from wine red through purple to a pure blue. The reaction (color change) is slow at the end point, and titrant must be added slowly and the solution stirred thoroughly in the vicinity of the end point. A comparison solution for the proper color at the end point may be prepared by adding 2 mL of pH 10 buffer to 50 mL distilled water, five drops indicator, a few drops Mg–EDTA, and a few drops EDTA.

If the end point for the first titration is less than 10 mL, double the volume of sample for the remaining two titrations.

Calculation

Calculate and report the hardness of the water as ppm $CaCO_3$ for each portion analyzed.

EXPERIMENT 11 DETERMINATION OF THE CONCENTRATION OF UNKNOWN ZINC IONS IN A SOLUTION WITH EDTA

Contributed by Dr. Sandeep Kaur and Dr. Dhanraj Masram, University of Delhi

Objective

To perform complexometric volumetric titrations with EDTA (ethylenediaminetetraacetic acid).

1. Using primary-standard zinc salt (ZnO) a primary-standard zinc ion solution will be prepared.

2. Using the primary-standard zinc ion solution, a supplied EDTA solution will be standardized.

3. The standardized EDTA solution (secondary standard) will be used for determination of zinc in an unknown solution.

Principle

EDTA, belongs to the class of compounds known as aminopolycarboxylic acids and undergoes successive acid dissociations to form negatively charged ions. The ion formed has the capability to "wrap" itself around positively charged metal ions (except for the alkali metal ions of charge +1) in aqueous solution (1 mol EDTA to 1 mol metal ion) (see Eq below). The species so formed is called a complex and this complex formation reaction between EDTA and many metal ions has a very large equilibrium/-formation constant. Because of this property of EDTA it is used for various purposes like chemical analysis of metals, medical removal of heavy metals in accidental poisonings, boiler water softening by chelation, removal of hard water scale by cleaning agents and so on.

For zinc ions in water the reaction can be written as:

$$Zn^{2+} \text{ (aq)} + EDTA^{4-} \text{ (aq)} \rightarrow Zn(EDTA)^{2-} \text{ (aq)} \qquad K_f = 3.2 \times 10^{16}$$

For consistent results of titrations, the pH of the solutions should be controlled by using buffer solutions. This is because EDTA is an acid substance with four weak acid dissociations, and hence, the reactions with metal ions are pH dependent. The metal ions that react strongly with EDTA should be titrated in acidic solution (e.g. zinc), whereas the metals that react weakly with EDTA must be titrated in alkaline solution (e.g. Ca and Mg).

Procedure

I. **Preparation of glassware and apparatus**. The following clean glassware and laboratory apparatus is required for the experiment for each student:

A 50 mL burette and its stand, a burette funnel, a weighing bottle and its lid (use a clean, dry weighing boat), a 10 or 25 mL transfer pipet, a glass stirring rod, a small funnel, two small beakers a spatula, all available Erlenmeyer flasks, a 250 mL volumetric flask with stopper, a small watch glass to fit a small beaker, a rubber pipet squeeze bulb.

Instructions

(a) The fume hood fan should be in operating condition. The glassware and apparatus should be cleaned thoroughly if necessary, with a 1% solution of detergent in warm water. The cleaned glassware and apparatus should then be rinsed with tap water followed by distilled water.

(b) The glassware should be used carefully and to avoid breakage should not be left standing in an unstable position.

(c) The spatula (and the weighing bottle and its lid if used) should be dried in the oven at 110 or 120°C for 15 minutes and the bottle and lid should not be allowed to vacuum seal.

(d) The spatula and weighing bottle with its lid should be carefully removed from the oven (on a heat proof surface), taken to the bench position and allowed to cool to room temperature before being used.

II. Preparation of a primary standard solution of Zn^{2+}.

1. A small clean beaker is taken and labelled for identification. Then using a top-loading balance ~0.25–0.30 g of primary-standard zinc oxide is weighed in the clean, dry weighing container. First the weight of the empty container is taken followed by weight of the container plus the zinc sample. The initial and final mass values are recorded in a **data table**. The difference gives the weight of the zinc sample. One should be careful as to not transfer any solid over the balance.

2. The weighed ZnO is then transferred to the labelled beaker and distilled water is added to about the 20 mL mark. In the fume hood, with the help of a graduated cylinder about 5–10 mL of 6 M hydrochloric acid (dilute HCl) is measured and carefully added to the ZnO in the beaker (Caution: hazardous). The beaker should be kept in the fume hood and covered with a clean watch glass due to the reaction that occurs. If required, the covered beaker can be heated gently on a hot plate inside the fume hood. One should be careful as to boil away all of the liquid.

3. The resulting solution of zinc ions is then removed from the fume hood (Caution: hot beaker). Using a clean funnel and a wash bottle the solution in the beaker is transferred quantitatively from the beaker to a clean 250 mL volumetric flask. The beaker and the funnel are rinsed with distilled water and the washed water is also added to the volumetric flask. The flask is filled with distilled water to the mark line of the volumetric flask. The flask is then stoppered with a clean stopper and turned several times by holding the stopper in place with one hand so that the solution is completely uniform. The small beaker is cleaned so that it can be used for the next part of the experiment.

III. Burette and pipet preparation and pipetting of portions of the standard zinc solution.

1. The burette stand, 50 mL burette, 10 mL volumetric transfer pipet, a burette funnel if needed, two small beakers and at least three Erlenmeyer flasks are arranged on the workstation. The outside of all the glassware is cleaned and dried. The burette is set-up and the burette tap are checked for any leaking. The inner walls of the burette and the transfer pipet should be clean, and the capillary tips should not be broken/plugged as dirty glassware leads to error in the analysis.

2. ~20 mL supplied EDTA is poured into a labelled, clean small beaker and used to rinse the inside walls of the beaker and then the same EDTA solution is poured into the burette for rinsing its inner walls. This EDTA is then drained out through the tip of the burette into a waster beaker. The burette funnel is also rinsed with EDTA. This rinsing process can be repeated again. All the rinse solution portions are collected in the waste beaker and are discarded into a sink with the cold-water tap running.

3. Next a larger volume of EDTA is taken in the rinsed beaker and the rinsed burette is filled near to the 0.00 mL mark, clearing the tip of air bubbles.

4. A second clean small beaker is labelled and used for the standard zinc ion solution. The beaker is rinsed with about a 20 mL volume of zinc ion solution from the volumetric flask. The same portion is used to rinse out the 10- or 25-mL transfer pipet as well and collected in a waste beaker. This is rinsing process is repeated again followed by addition of 40–50 mL of the zinc ion solution into the rinsed beaker. All the rinse solution portions are collected in the waste beaker and are discarded into a sink with the cold-water tap running.

5. The Erlenmeyer flasks taken for the titrations should be clean from the insides but need not be dry. Using a clean and dry squeeze bulb 10- or 25-mL portion of the standard zinc ion solution is carefully transferred from the beaker into each of the three clean Erlenmeyer flasks. The tip of the pipet should be wiped off before the transfer. The pipet should always be used to transfer from a beaker or any other vessel and never directly from a volumetric flask or a storage bottle.

6. Then to each Erlenmeyer flask add distilled water approximately to the 20 mL mark, 5 mL of pH 10 buffer solution, and three (3) drops of Eriochrome black T (EBT) indicator solution. The solution should be mixed well. The indicator color should be violet at this point due to the formation of Zn-EBT complex and changes to blue at the endpoint of the titration when the zinc is displaced from the indicator by EDTA.

IV. **Standardization of the supplied EDTA solution.**

1. The starting volume of the EDTA in the burette is read and the value is recorded in the **data table**. The first sample flask containing the known zinc ion solution is titrated slowly with small addition volumes of the EDTA solution with swirling of the flask to ensure mixing. The tip of the burette is placed 1 or 2 cm down into the opening of the flask to avoid any loss of the solution.

2. When a blue color begins to appear in the flask, the volumes of the EDTA addition is decreased and only one or two drops are added slowly at a time. The inside walls of the flask and the burette tip should be washed with a stream of distilled water using the wash bottle from time to time. The end-point color of the titration is when the violet color changes to blue.

3. The final volume reading is recorded in the **data table** and the titration volume is determined by taking the difference of the initial and final readings. On repeating the experiment with the zinc ion solutions in the other two Erlenmeyer flasks same titration volume to the endpoint should be obtained. Hence, based on the first titration in the repeat experiments all but the final 1 mL can be added rapidly. All the volumes are recorded in the **data table**. The repeat experiments are performed till one obtains three acceptable titration volumes with a difference of only 0.20 mL.

V. **Endpoint indicator correction.**

1. The endpoint indicator volume correction (indicator blank) is determined by adding only 25 mL of distilled water, 3 drops of indicator and 5 mL of buffer solution to a clean Erlenmeyer flask. The indicator blank is zero, 0.00 mL, if the solution is blue (endpoint color). However, if the solution is violet (possibly due to impurity of metal ions), the volume of EDTA solution (should only be a very small volume) required to produce the endpoint is determined by titration and recorded in the **data table**.

2. The data collected for the known zinc ion solution using the supplied EDTA solution should be verified with the instructor before proceeding to the next part of the experiment. Also, as the EDTA will be used for the titration of the unknown zinc solution it should be left in the burette.

VI. **Analysis of the unknown zinc ion solution.**

1. Using a clean funnel, the supplied (unknown) zinc ion solution is transferred quantitatively from the supplied vessel provided to a clean 250 mL volumetric flask. The supplied vessel and the funnel are rinsed with distilled water and the washed water is also added to the volumetric flask. The flask is filled with

distilled water to the mark line of the volumetric flask. The flask is then stoppered with a clean stopper and turned several times by holding the stopper in
place with one hand so that the solution is completely uniform.

2. Similar procedure is followed for the titration of the unknown solution with
   EDTA as discussed earlier for the known primary standard zinc ion solution.
   This is repeated several times till acceptable readings for the volume of EDTA
   are obtained and recorded in the **data table**.

3. The end-point indicator volume correction (indicator blank) is also determined
   for the unknown zinc ion solution.

4. The data table is verified with the instructor. All the glassware and apparatus
   used are cleaned up and all solutions may be discarded into the sink with the
   cold-water tap running.

VII. Report and calculations.

1. The mass data for weighing the zinc oxide sample is recorded and molarity
   (mol/L) of the zinc ion solution (5 places after the decimal point) in the volumetric flask is calculated.

2. For each titration, the titration volume determined, and the corrected titration
   volume are entered in the **data table**. The mean corrected volume is calculated
   for the readings taken. This is then used to calculate the molarity (mol/L) of
   the supplied unknown EDTA solution (5 places after the decimal point).

3. Using the EDTA whose molarity has now been calculated, titrate the unknown
   supplied zinc ion solution and record the data for the titrations performed in
   the **data table**. Calculate the mean corrected volume in this case as discussed
   in point 2. Using the mean corrected volume, the molarity of the unknown
   zinc ion solution is calculated.

Molarity (M) = solute concentration in moles per liter (mol/L)

$$= \frac{\text{Amount of solute (mol)}}{\text{Total solution volume (L)}}$$

$$\text{Molarity } (M) = X \text{g ZnO} \times \frac{1 \text{mol ZnO}}{81.38 \text{ g ZnO}} \times \frac{1}{0.2500 \text{ L}}$$

Molarity of Primary Standard Zinc Ion = Y mol/L zinc ion = Y *M*

Molarity of supplied EDTA (mol/L)

$$\text{Volume zinc (L)} \times \text{Molarity zinc (mol/L)} \times \frac{1 \text{ mol EDTA}}{1 \text{ mol zinc}} \times \frac{1}{\text{Volume EDTA } (L)}$$

Volume zinc (L) × Molarity zinc (mol/L) = Calculates mol of Zinc

$$\frac{1 \text{ mol EDTA}}{1 \text{ mol zinc}} = \text{Converts to mol of EDTA}$$

$$\frac{1}{\text{Volume EDTA (L)}} = \text{Converts to mol/L EDTA}$$

Molarity of supplied EDTA (mol/L)

$$= 0.02500 \text{ or } 0.01000 \text{ L} \times \text{Y mol/L} \times \frac{1 \text{ mol EDTA}}{1 \text{ mol zinc}} \times \frac{1}{Z \text{ L}}$$

Molarity of supplied EDTA (mol/L) = XXX *M*
For unknown zinc ion solution:
The molarity of the supplied EDTA solution and the volume of EDTA determined for titration of 10 or 25 mL of the unknown zinc solution are used to
calculate the molarity of the unknown zinc ion solution.

Precautions to be Taken During the Experiment

1. The burette should be cleared and well-cleaned for any impurity.

2. Air bubble present in the nozzle of the burette must be removed before taking the initial reading.

3. There should be no leakage from the burette during titration.

4. While taking the burette reading, reading the pipette or measuring flask the eye should be kept in level with the liquid surface or while etc.

5. While taking readings, read the lower meniscus in case of colorless solution and upper meniscus in case of colored solutions. Accurate and correct readings of the volume of solutions used should be taken.

6. Simply touch the inner surface of the titration flask with the nozzle of the pipette to expel the last drop of solution from it, do not blow through the pipette for this purpose.

7. Titration should not be carried out with the funnel on top of the burette as it may change the initial reading due to drops falling from the funnel.

8. The shaking of the titration flask should be continuous during the addition of the solution from the burette.

9. Time should not be wasted in bringing the burette reading to zero before each titration.

10. Addition of Indicator must be according to the procedure as less or more quantity of indicator can lead to different set of readings.

Potentiometric Measurements

EXPERIMENT 14   POTENTIOMETRIC DETERMINATION OF FLUORIDE IN DRINKING WATER USING A FLUORIDE ION-SELECTIVE ELECTRODE[1]

Principle

Fluoride in a water sample is determined by measurement with a fluoride ion-selective combination electrode (contains reference electrode built in). First, you will determine whether the electrode response is Nernstian over a wide range of concentrations. Then, you will determine fluoride in the unknown by comparing potential measurements with standards over a narrower range, bracketing the unknown; a calibration curve will be prepared.

Equations

$$E = k - \frac{2.303RT}{F} \log a_{F^-}$$
$$= k' - \frac{2.303RT}{F} \log[F^-] \text{ (if ionic strength is held constant)}$$

Solutions and Chemicals Required

1. **Provided.** TISAB (total ionic-strength adjustment buffer) solution, which is prepared with 57 mL glacial acetic acid, 58 g sodium chloride, and 4 g CDTA (cyclohexylenedinitrilotetraacetic acid) in about 500 mL water, adjusted to pH 5.0 to 5.5

---

[1] This experiment can also be used to determine % NaF or % SnF$_2$ in toothpaste by preparing a suspension of the paste in water and pipetting the supernatant, or to determine the fluoride content of children's fluoride tablets or drops. See T. Light and C. Cappucino, *J. Chem. Ed.*, 52 (1975) 247.

with 5 M NaOH and diluted to a total volume of 1 L. A 1:1 dilution of *all* samples with this solution serves the following:

(a) It provides a high total ionic strength background, swamping out variations in ionic strength between samples.

(b) It buffers the samples at pH 5 to 6. (In acid media, HF forms; while in alkaline media, OH^- ion interferes in the electrode response.)

(c) The CDTA preferentially complexes with polyvalent cations present in water (e.g., Si^{4+}, Al^{3+}, Fe^{3+}), which otherwise would complex with F^- and change its concentration.

2. To prepare

(a) **Stock standard 0.1 M fluoride solution.** Dry about 1 g NaF at 110°C for 1 h, cool in a desiccator for 30 min. Weigh out 0.45 to 0.50 g of the dried NaF (to the nearest milligram), transfer to a 100-mL volumetric flask, dissolve, and dilute to volume with distilled deionized water. Shake thoroughly and transfer to a polyethylene bottle (rinse with a few small portions first). Fluoride tends to adsorb on glass and should be stored in plastic containers. **Caution:** Fluoride is poisonous. Handle with care. Commercially prepared fluoride solutions may be used.

(b) **Linearity standards.** By serial dilution of the stock solution with distilled deionized water, prepare a series of solutions of about 10^{-2}, 10^{-3}, 10^{-4}, 10^{-5}, and 10^{-6} M fluoride (calculate accurate concentrations). (Do *not* pipet by mouth!) For example, dilute the stock solution 10:100, 1:100, and 1:1000 mL to prepare the first three solutions. Then, dilute the 10^{-4} M solution 10:100 and 1:100 mL to prepare the last two. Transfer to polyethylene bottles. These solutions should be prepared on the day of use.

(c) **Calibration standards.** The unknown concentration will be within 1×10^{-3} and 1×10^{-2} M fluoride, and a calibration curve will be prepared using concentrations of fluoride to bracket the unknown. Using the above procedure, prepare additional standards of 2×10^{-3} and 4×10^{-3} M fluoride (calculate the accurate concentrations). Prepare on the day of use.

Things to Do before the Experiment

Prepare the stock NaF solution. This will require drying of NaF.

Procedure

1. Determination of range of response and range of linearity. Connect the electrode leads to an expanded-scale pH meter. Add with a pipet 10 mL of the 10^{-6} M standard solution and 10 mL TISAB solution to the small plastic beaker provided. Place the electrode in the beaker. Stir the solution with a magnetic stirrer and small stirring bar during measurement. You may make readings in pH units. (1 pH = 59.2 mV at 25°C; arbitrarily take pH 0 as 0 mV). The meter should also allow reading in millivolts. When a steady reading is obtained, record the value. Rinse and blot the electrode and repeat, going from dilute to concentrated standard solutions. Prepare a spreadsheet of the data, and chart mV vs. log C. Report the slope in mV/decade, the intercept, and the r^2 value. Report also the range of linearity.

2. Standardization for unknown. Record the readings of the standard solution from 1×10^{-3} M to 1×10^{-2} M, and plot a calibration curve as above over this range. Enter the formula in one cell for calculating an unknown concentration from the millivolt reading and the slope and intercept of the calibration curve.

3. **Analysis of unknown.** After preparing the calibration curve, obtain an unknown fluoride sample. This may be a synthetic solution, in which case obtain the unknown in a 250-mL volumetric flask. Immediately dilute to volume with distilled deionized water and transfer to a polyethylene bottle. Add 10 mL of the unknown with a pipet to a small plastic beaker followed by 10 mL TISAB. Record the mV reading as above. Make at least three separate runs (separate additions and potential readings). **Note**: The unknown should be measured at the same time the calibration curve is prepared. The mV scale should not be adjusted between calibration and sample measurements. If it is, take a new reading of one of the standards, and readjust to the original reading.

Calculations

From the spreadsheet calibration curve, determine the concentration of fluoride in the unknown solution. Report the results in parts per million fluoride, along with the standard deviation for the three measured samples.

Reduction–Oxidation Titrations

EXPERIMENT 15 ANALYSIS OF AN IRON ALLOY OR ORE BY TITRATION WITH POTASSIUM DICHROMATE

Principle

An iron alloy or ore is dissolved in HCl and the iron is then reduced from Fe(III) to Fe(II) with stannous chloride ($SnCl_2$). The excess $SnCl_2$ is oxidized by addition of $HgCl_2$. The calomel formed (insoluble Hg_2Cl_2) does not react at an appreciable rate with the titrant. The Fe(II) is then titrated with a standard $K_2Cr_2O_7$ solution to a diphenylamine sulfonate end point.

Equations

Reduction:

$$2Fe^{3+} + Sn^{2+} \text{ (slight excess)} \rightarrow 2Fe^{2+} + Sn^{4+} + Sn^{2+} \text{ (due to excess)}$$
$$Sn^{2+} + 2Hg^{2+} + 2Cl^- \rightarrow Sn^{4+} + \underline{Hg_2Cl_2} \text{ (white precipitate)}$$

If too much Sn^{2+} is added, then

$$Sn^{2+} + \underline{Hg_2Cl_2} \rightarrow Sn^{4+} + 2Hg^0 + 2Cl^-$$

(Black Hg^0 precipitate makes end-point determination impossible. The sample must be discarded because Hg^0 reacts with $Cr_2O_7{}^{2-}$.)

Titration:

$$6Fe^{2+} + Cr_2O_7{}^{2-} + 14H^+ \rightarrow 6Fe^{3+} + 2Cr^{3+} + 7H_2O$$

Solutions and Chemicals Required

1. **Provided.** 0.28% (wt/vol) of the sodium salt of *p*-diphenylamine sulfonate indicator, 0.5 *M* $SnCl_2$ in 3.5 *M* HCl (with mossy tin added to stabilize against air oxidation: $Sn^{4+} + Sn^0 \rightarrow 2Sn^{2+}$), saturated $HgCl_2$ solution, conc. HCl, 6 *M* HCl, conc. H_3PO_4-conc. H_2SO_4 mixture (15 mL each added to 600 mL water and cooled to room temperature), 0.1 *M* $FeCl_3$ in 6 *M* HCl.

2. **To prepare.**[1] Standard $1/60 \, M \, K_2Cr_2O_7$ solution (approx. 0.017 M). Dry about 3 g primary standard $K_2Cr_2O_7$ in a weighing bottle at 120°C for at least 2 h. Drying in the oven for longer periods of time (i.e., until the next lab period) will do no harm. Place the $K_2Cr_2O_7$ in the desiccator to cool for 30 to 40 min. Weigh accurately to the nearest milligram, about 1.3 g in a weighing dish, and transfer quantitatively to a 200-mL beaker. Dissolve in about 200 mL water and transfer quantitatively to a 250-mL volumetric flask. Dilute to volume and mix thoroughly. Rinse a clean 1-L bottle with three small portions of the solution and transfer the remainder of the solution to the bottle for storage. Calculate the molarity of the solution. $K_2Cr_2O_7$ may also be prepared approximately and standardized against electrolytic iron wire (primary standard) using the same procedure as given below for an alloy sample.

Things to Do before the Experiment

1. **Dry the necessary amount of $K_2Cr_2O_7$.**

2. **Obtain and dry or dissolve your unknown as required.**

    (a) **Alloy sample.** This may be in the form of a wire. Your instructor will provide you with three separate (weighed) pieces of the unknown, each to be placed in separate, labeled 500-mL Erlenmeyer flasks containing 10 mL conc. HCl. Cover with inverted 100-mL beakers and store these dissolving samples in your desk overnight or longer. Alternatively, the samples can be dissolved on the day of the experiment by heating on a hot plate or steam bath in 400-mL beakers covered with ribbed watch glasses in the hood to hasten dissolution. (After dissolution, rinse the cover glass and the sides of the beaker, catching all the rinsings in the beaker. Use as little water as possible. The final volume should not be more than 50 mL.)

    (b) **Ore sample.** Check out a sample of an iron ore from the instructor. Dry in the oven at 110 to 120°C for at least 2 h; longer drying will do no harm.

Procedure

1. **Reduction of the iron and trial titration.** Before titrating your unknown, it is advisable to perform one or two trial titrations. This can be done while the (ore) samples are dissolving. Add approximately 10 mL 0.1 M $FeCl_3$ solution in 6 M HCl to a 600-mL beaker and add about 50 mL water. (This dilutes the sample sufficiently that when all the Fe^{3+} is reduced, the solution will be nearly colorless. If the volume is less, then the pale green of the Fe^{2+} makes detection of complete reduction more difficult). Place a ribbed watch glass on the beaker and heat nearly to boiling on a hot plate in the hood; the solutions must be very close to the boiling point, perhaps simmering gently, but not boiling violently since $FeCl_3$ can be lost due to volatilization. Add 0.5 M stannous chloride solution with a dropper through the lip of the beaker until the color begins to fade; then, continue the addition drop by drop, swirling the beaker and allowing each drop to react before adding the next, until the solution is colorless. It will first become pale yellow and then will gradually turn more clear. It may never get completely colorless but may instead go to a pale green due to the ferrous ion (this will depend on the amount of iron). Whichever you get (colorless or pale green) stop addition and allow the solution to heat for two more minutes. If the yellow color returns, add a few more drops of $SnCl_2$ until it becomes colorless or pale green again. Repeat dropwise addition

---

[1] If you are to use this solution also for Experiment 16 and/or 26, prepare 500 mL, taking accurately about 2.5 g $K_2Cr_2O_7$, dissolving in 200 mL water, and transferring to a 500-mL volumetric flask. You need to save at least 150 mL for Experiment 16 and 100 mL for Experiment 26. The 250 mL you prepare for this experiment will be enough if you are careful not to waste any.

of $SnCl_2$ until the solution does not return to the yellow color. At this point, add two drops excess, no more. (If more than two drops are added, the stannous chloride can be oxidized with a few drops of potassium permanganate solution and the above reduction process repeated.) Remove from the hot plate, rinse down the cover glass and sides of the beaker, and cool quickly to room temperature by immersing the bottom of the beaker in cold water. Two to three samples may be taken this far together; *the remainder of the procedure must be carried out with each sample individually without interruption through the titration.* If any sample turns yellow again while awaiting its turn, it must be reheated and sufficient stannous chloride added to discharge the color, with two drops excess. Fill your 50-mL buret with the standard $K_2Cr_2O_7$ and have it ready for titration.

To one sample, which must be at room temperature, add 100 mL water and then add rapidly 15 mL saturated mercuric chloride solution, previously measured out, while stirring and immediately mix thoroughly. A slight, white precipitate should form. If either a heavy gray or black precipitate (elemental mercury) or no precipitate forms, too much or not enough (to reduce all the Fe^{3+}) stannous chloride has been added; in either case, the sample must be discarded. Mix for 2 min, then add 100 mL of the H_3PO_4–H_2SO_4 mixture and six to eight drops diphenylamine sulfonate indicator. Titrate immediately with the $K_2Cr_2O_7$ solution, stirring constantly, until the green color changes to a purple or violet blue that remains for at least 1 min. (The acid mixture provides the protons consumed in the titration reaction and forms a nearly colorless phosphate complex with the Fe^{3+} titration product, which sharpens the end point.)

2. **Alloy sample.** The sample should by now be dissolved in the Erlenmeyer flasks (or in the heated 600-mL beakers). Adjust the volume to 40 to 60 mL with distilled water. Heat nearly to boiling and follow the same procedure as used in the trial titration, starting with addition of stannous chloride. All three samples can be taken up to the point to just before the addition of mercuric chloride and then must be treated one at a time up through the titration.

3. **Ore sample.** After cooling the dried sample in a desiccator for 30 to 40 min, weigh out by difference (the iron ore may be hygroscopic) three samples; consult the instructor as to the size of the samples. Transfer to 600-mL beakers. Add 10 mL water and swirl until the sample is completely moistened and in suspension; then cover with ribbed watch glasses and add 10 mL concentrated HCl, pouring it through the lip of the beaker. Heat on a hot-plate in the hood until the iron ore has dissolved to give a clear, red-brown solution; with some samples there may be an insoluble sandy residue, which may be disregarded. [Silica or insoluble sulfides (black) or silicates may remain.] The hot plate should be adjusted to keep the solutions just barely at the boiling point; vigorous boiling should be avoided since it may cause loss of material and excessive evaporation of acid. If necessary, add 6 M HCl to keep the volume about 20 mL. When all the iron has been dissolved, the insoluble residue (if any) will be gray or white, with no black or reddish particles, after adding stannous chloride to reduce the iron. When the solution appears clear, add distilled water to bring the volume to about 50 mL and follow the same procedure used in the trial titration starting with addition of stannous chloride to the hot solution. All three samples can be carried up to the point just before addition of mercuric chloride and then must be treated one at a time up through the titration.

Calculations

1. **Alloy sample.** Calculate and report the milligrams iron in each portion of the unknown analyzed, along with the mean and the precision.

2. **Ore sample.** Calculate and report the percent iron in each portion of the unknown analyzed, along with the mean and the precision.

EXPERIMENT 17   IODOMETRIC DETERMINATION OF COPPER

Principle

A copper metal sample is dissolved in nitric acid to produce Cu(II), and the oxides of nitrogen are removed by adding H_2SO_4 and boiling to SO_3 fumes. The solution is neutralized with NH_3 and then slightly acidified with H_3PO_4. [The H_3PO_4 also complexes any iron(III) that might be present and prevents its reaction with I^-.] Finally, the solution is treated with excess KI to produce CuI and an equivalent amount of I_3^-, which is titrated with standard $Na_2S_2O_3$ solution, using a starch indicator. KSCN is added near the end point to displace absorbed I_2 on the CuI by forming a layer of CuSCN. For best accuracy, the $Na_2S_2O_3$ is standardized against high-purity copper wire since some error occurs from reduction of copper(II) by thiocyanate.

Equations

$$2Cu^{2+} + 5I^- \rightarrow 2\underline{CuI} + I_3^-$$
$$I_3^- + 2S_2O_3^{2-} \rightarrow 3I^- + S_4O_6^{2-}$$

Solutions and Chemicals Required

1. **Provided.** 6 M HNO_3, conc. H_2SO_4, 3 M H_2SO_4, conc. H_3PO_4, 6 M NH_3, KSCN.
2. **To prepare.**
   (a) **Starch solution.** Prepare the day of the experiment as described in Experiment 16, or use Thiodene indicator.
   (b) **0.1 M $Na_2S_2O_3$ solution.** Prepare as described in Experiment 16.

Things to Do before the Experiment

Prepare the 0.1 M $Na_2S_2O_3$ solution. Although this can be prepared on the day of the experiment, it is preferable to prepare it at least a day before it is standardized. The solution tends to lose some of its titer right after preparing.

Procedure

1. **Standardization of the 0.1 M $Na_2S_2O_3$ solution.** The same procedure is used as will be used for analyzing the sample. Weigh out three 0.20- to 0.25-g samples of pure electrolytic copper foil and add to 250-mL Erlenmeyer flasks. In a hood, dissolve in 10 mL 6 M HNO_3, heating on a steam bath if necessary. Do in a fume hood. Add 10 mL conc. H_2SO_4 and evaporate to copious white SO_3 fumes. Cool and add carefully 20 mL water. Boil 1 to 2 min and cool. Add 6 M NH_3 dropwise with swirling of the sample solution until the first dark blue of the $Cu(NH_3)_4^{2+}$ complex appears. Then, add 3 M H_2SO_4 until the blue color just disappears, followed by 2.0 mL conc. H_3PO_4. Cool to room temperature.

   From this point, each sample must be treated separately. Dissolve about 2 g KI in 10 mL water and add to one of the flasks. Titrate *immediately* with the $Na_2S_2O_3$ solution until the yellow color of I_3^- *almost* disappears. Add 2 to 3 mL of the starch solution or approximately 0.4 g Thiodene indicator, and titrate until the blue color begins to fade (should be less than 0.5 mL). Finally, add about 2 g KSCN and continue the titration until the blue color just disappears.

Repeat with the other two samples. Calculate the molarity of the $Na_2S_2O_3$.

2. **Determination of copper in an unknown.** Add 10 mL of 6 M HNO_3 to each of three clean 250-mL Erlenmeyer flasks and take these to your instructor, who will add an unknown sample to each. (This unknown may be copper foil as used in the standardization.) Dissolve each and titrate as described for the standardization of $Na_2S_2O_3$. **Note**: In place of H_3PO_4, approximately 2 g ammonium bifluoride, NH_4HF_2 or $NH_4F \cdot HF$, may be added to complex any iron and at the same time adjust the solution to the proper acidity. This experiment is also suitable for determining copper in about 0.3 g brass.

Calculation

Calculate the grams copper in each unknown sample and report the values of each and the relative standard deviation. (If a weighed brass sample is analyzed, report the percent copper.)

EXPERIMENT 19   MICROSCALE QUANTITATIVE ANALYSIS OF HARD-WATER SAMPLES USING AN INDIRECT POTASSIUM PERMANGANATE REDOX TITRATION[1]

Principle

The calcium in an unknown hard-water sample is precipitated as calcium oxalate in ammonia solution, and the precipitate is quantitatively filtered and washed, and is then dissolved in dilute sulfuric acid. The oxalic acid is titrated with standardized potassium permanganate solution.

Note: This experiment may serve as a template for microscale versions of some other experiments in the text. Statistical comparison of the microscale experiment with a similar conventional macroscale experiment using a 50-mL buret shows comparable precision. But there is a slight negative determinate error in the microscale experiment, averaging about 50 ppm for ca. 500-ppm samples, probably due to physical loss of some of the precipitate, using a large funnel. Richardson suggests using a microfunnel to minimize this problem. For additional information on microscale titrations, see M. M. Singh, C. B. McGown, Z. Szafran, and R. M. Pike, *J. Chem. Educ.*, 77 (2000) 625.

Equations

$$Ca^{2+} + C_2O_4^{2-} \rightarrow CaC_2O_4$$
$$CaC_2O_4 + 2H^+ \rightarrow Ca^{2+} + H_2C_2O_4$$
$$5H_2C_2O_4 + 2MnO_4^- + 6H^+ \rightarrow 10CO_2 + 2Mn^{2+} + 8H_2O$$

Solutions and Chemicals Required

1. **Provided.** Conc. H_2SO_4, conc. HCl, conc. HNO_3, primary standard $Na_2C_2O_4$ (dried at 120°C for 1 h), 0.35 M $(NH_4)_2C_2O_4$, 0.10 M $AgNO_3$, dil. (1:10) NH_3 solution, methyl red (0.02% in 60% ethanol—dissolve first in the ethanol portion). Provide concentrated acids and $AgNO_3$ in dropper bottles.

---

[1]Courtesy of Professor J. N. Richardson, M. T. Stauffer, and J. L. Henry Shippensburg University. Details of statistical analysis of the experiment in *J. Chem. Educ.*, 80(1) (2003) 65.

2. **Approx. 0.2 *M* KMnO₄.** One liter is enough for 30 students. Prepare as follows (see Section 21.6). The preparation may be scaled up for more students. Calculate the weight of potassium permanganate required to make 1 L of 0.02 *M* solution. Weigh out in a weighing dish about 0.05 g more than this amount. Transfer to two 600-mL beakers, placing about half of the permanganate in each, add 500 mL of distilled water to each, cover with watch glasses, heat to boiling, and boil *gently* for 1 to 2 min, *not longer*. (Longer boiling, 0.5 to 1 h, is desirable but will evaporate the solution and change its concentration. The solution may be heated for a longer period at a temperature just below its boiling point.) Allow to stand for at least 24 h before proceeding with the next step. Keep covered with watch glasses at all times to exclude dust and vapors, and to retard evaporation.

Mount a sintered-glass filter in a filter flask, and filter the permanganate solution through the crucible. Do not stir or swirl the solution; a sediment will have settled to the bottom that would clog the filter and make filtration very slow. For this reason, discard the last few milliliters of the solution from the first beaker. Use a filter trap; if any tap water backs up from the suction pump, the permanganate will be contaminated. Pour the filtered solution into a clean brown glass bottle with glass stopper and shake until homogeneous. The solution must never come into contact with organic material, including corks and rubber stoppers.

3. **Unknown hard-water solution.** Prepared by dissolving Ca 20 g of dried $CaCO_3$ in a minimum volume of 1 *M* HCl. To this solution is added a few drops of methyl red indicator, followed by dropwise addition of 1 *M* NaOH until the methyl red end point is noted (red to yellow). The resulting solution is then quantitatively transferred to a 2.000-L volumetric flask and diluted to the mark followed by thorough mixing. Student unknowns may then be dispensed from a buret into individual 100.00-mL volumetric flasks. Typical aliquots range from 10.00 to 20.00 mL, resulting in diluted unknowns with concentrations ranging from ca. 400 to 800 ppm Ca^{2+}.

Microburet

Constructed using a 2.000-mL graduated pipet, 2-cm length of latex tubing, 10-mL plastic syringe barrel, and an automatic delivery pipet tip. A detailed illustration of the microburet, as well as documentation regarding its construction, are provided in M. M. Singh, C. B. McGown, Z. Szafran, and R. M. Pike, *J. Chem. Educ.*, 75 (1998) 371.

Procedure

1. **Standardization of KMnO₄.** Each bench prepares 200 mL of dilute (1:20 v:v) sulfuric acid using concentrated H_2SO_4. (**Caution:** Acid should be added *slowly* to water with stirring.) Remove dissolved gases from this solution by boiling for 5 to 10 min with a glass stirring rod in the beaker to keep solution from bumping. Cool the solution to room temperature using an ice bath and store in a tightly capped, appropriately labeled polypropylene bottle. All students at a bench will share this solution (with typically three or four students occupying a lab bench). *Do not proceed past this point unless the standardization titration will be attempted the same day.*

To a 100.00-mL volumetric flask, each bench adds 0.5000 g of primary standard $Na_2C_2O_4$, dissolves it in the dilute H_2SO_4 prepared previously, and dilutes to the mark, mixing thoroughly. Again, this solution is shared among all members of a lab bench. From this point on, each student works individually. Rinse a 2-mL microburet with the primary standard solution and then fill it with the solution. Transfer 1.500 to 2.000 mL of the solution into a clean 30-mL beaker, recording the volume delivered exactly. Dilute to a total volume of about 5 mL using the dilute H_2SO_4. Prepare three more samples this way, using a different volume of $Na_2C_2O_4$ solution each time.

Clean a 2-mL microburet, rinse with and then fill with the 0.02 M $KMnO_4$ solution. Calculate the approximate volume of this titrant needed to reach an end point for the primary standard sample with the smallest volume. Add rapidly all but about 0.2 mL of this amount from the buret, with constant but gentle stirring with a stirring rod. Let this solution stand until the color disappears, which may take a minute or two. Heat the solution to 55 to 60°C, and complete the titration at this temperature (the temperature can be monitored by a thermometer, or one can watch for initial formation of steam). The remaining titrant is added dropwise, allowing each drop to react before adding the next. The end point is the first perceptible pink that persists for at least 30 s. Repeat this procedure for each of the remaining samples of $Na_2C_2O_4$. Calculate the molarity of the $KMnO_4$ for each titration, along with the average molarity and standard deviation.

2. **Analysis of an Unknown Hard-Water Sample.** Each student is assigned a labeled 100.00-mL volumetric flask containing an unknown hard-water sample. Add distilled water to the mark and invert the flask to mix thoroughly. Use a microburet to transfer an aliquot of hard water of about 1 mL into a clean 30 mL beaker. The volume transferred should be recorded to the precision of the buret. To this sample add 10 mL of distilled water, 7 drops of concentrated HCl, and a drop or two of methyl red indicating solution. Heat the resulting solution to nearly boiling, and add 1 mL of 0.35 M $(NH_4)_2C_2O_4$. (**Note:** students typically obtain this solution from a community macroburet or from a small graduated cylinder because they are only concerned with having an excess of oxalate ion in the solution.)

Prepare and fill a microburet with dilute (1:10 v:v) NH_3 solution. Add this solution dropwise to the unknown sample in the beaker, noting that the solution should become cloudy as the precipitate (CaC_2O_4) starts to form. Continue the addition of NH_3 solution until the solution becomes alkaline to methyl red as indicated by a change from pink to pale yellow. Addition of excess NH_3 solution should be avoided. Cover the beaker with Parafilm, and allow the precipitate to digest overnight. If the pink returns, add more NH_3 solution to obtain the methyl red end point. Repeat this procedure for two more hard-water samples, using a slightly different aliquot size each time.

Gravity filter the precipitated CaC_2O_4 in each sample through 42.5-mm diameter No. 42 or equivalent Whatman filter paper seated in a long-stem filter funnel.[2] Note that transfer of the precipitate must be quantitative. Wash the filtered precipitate with cold distilled water until the filtrate is clear upon addition of $HNO_3/AgNO_3$. This is done by collecting a few drops of filtrate in a test tube and adding a drop or two of the acid, followed by a drop or two of $AgNO_3$ solution.

Remove the filter paper from the funnel and place in a clean 30-mL beaker. Add about 3 mL of 1:10 (v:v) sulfuric acid (previously boiled as in step). Then stir the mixture until the solid CaC_2O_4 is dissolved and the filter paper is torn apart. Add about 10 mL of distilled water to the mixture, and heat to just below the boiling point. Titrate the mixture with the standardized $KMnO_4$ solution as follows: (1) add 0.10 mL of titrant and allow the mixture to stand until the color fades, and (2) continue the titration as normal until an end point is reached. The temperature of the solution should be maintained above 55°C for the duration of the titration. Repeat this procedure for each of the other unknown samples.

---

[2]Improved recovery may be achieved by using a smaller funnel to fit the paper.

Calculation

Using the data obtained, along with a balanced chemical reaction equation, calculate the concentration of calcium in parts per million for each trial. Report the mean and standard deviation.

Potentiometric Titrations

EXPERIMENT 20   pH TITRATION OF UNKNOWN SODA ASH

Principle

The unknown soda ash is titrated with standard HCl using a potentiometric (pH) end point measured with a pH meter using a pH glass electrode–saturated calomel reference electrode combination. The end-point breaks are compared with indicator color changes.

Equations

$$CO_3{}^{2-} + H^+ \rightarrow HCO_3{}^- \text{ (phenolphtalein end point)}$$
$$HCO_3{}^- + H^+ \rightarrow H_2O + CO_2 \text{ (methyl purple end point)}$$

Note that between the first and second end points, a gradual decrease in pH due to the $HCO_3{}^-/CO_2$ buffer system will occur. This will give a poor visual end point, unless the buffer couple is destroyed. In practice, the visual titration used for standardization is continued until the methyl purple end point is reached, at which time the solution is gently boiled to remove the CO_2, leaving only the remaining $HCO_3{}^-$, which is then titrated to completion (see Chapter 7 for a more detailed discussion).

Solutions and Chemicals Required

1. **Provided.** 0.2% phenolphthalein in 95% ethanol, 0.1% methyl purple in water, primary standard Na_2CO_3, standard pH 7 buffer.

2. **To prepare.** Standard 0.1 M HCl solution. Use the solution prepared in Experiment 8. If this solution is not available, prepare and standardize 500 mL as described in Experiment 7. Alternatively, the acid may be standardized against the primary standard Na_2CO_3 by pH titration as described below for the unknown soda ash.

Things to Do before the Experiment

Prepare and standardize the HCl solution. This will require prior drying of primary standard Na_2CO_3.

   Obtain the unknown soda ash from your instructor and dry for at least 2 h at 160°C. Cool at least 30 min in a desiccator before weighing.

Procedure

The glass electrode to be used for pH measurements should have been soaked and stored in 0.1 M KCl for at least one day prior to its use. Always store the electrode in KCl solution when not in use. Calibrate the pH meter as described by your instructor, using the pH 7 standard buffer. This will consist essentially of adjusting the meter to read pH 7.00 with the electrodes immersed in the buffer solution.

1. **Trial titration.** The purpose of this titration is to locate quickly and approximately the two end points. Weigh accurately by difference a dried sample of unknown soda ash (0.2 to 0.3 g) and add it to a 400-mL beaker containing a magnetic stirring bar.

Add approximately 50 mL water and a few drops phenolphthalein indicator. The indicators are for the purpose of making a comparison between the potentiometric end points and the indicator color changes. Place the beaker on a magnetic stirrer, immerse the electrodes, and start the stirrer, being careful not to touch the electrodes to the stirring bar. Titrate with standard HCl, taking readings about every 2 mL. After the phenolphthalein color disappears, add a few drops methyl purple indicator and titrate at 2-mL increments until the second end point is reached. Add a few increments beyond the end point. The correct color for the second end point can be determined by comparison with the color of a few drops of the indicator in a solution of 0.20 g potassium acid phthalate in 100 mL water. Prepare a spreadsheet to plot your titration curve of pH (ordinate) versus volume of HCl (abscissa), and also the first derivative plot, and locate the approximate end points. See Chapters 7 and 21 and your text website for those chapters for the preparation of a spreadsheet for derivative titrations.

2. **Final titration.** Weigh accurately another sample of the unknown and titrate as before, but make pH readings every 5 mL to within 3 mL of each end point (both sides of end point). Then, make readings of 0.50- to 1-mL intervals within 1 mL of the end point. Near the end point, *take readings as quickly as possible because the pH will tend to drift as CO_2 escapes from the solution.* Note and record the points at which the indicators change color. Use the spreadsheet you prepared above to plot your titration curve and its first derivative. Print out the titration curve and indicate on this curve the range in which the indicators change color. Determine the end point from the second inflection point of the curve. Repeat the titration on two more portions of the unknown. Be sure to rinse the electrodes between titrations.

Calculations

Calculate and report the percent Na_2CO_3 and Na_2O in your unknown for each portion analyzed. Hand in the plots of the titration curves with your report. Report also the mean percent value and the precision.

Spectrochemical Measurements

Contributed by Dr. Sandeep Kaur and Dr. Dhanraj Masram, University of Delhi

EXPERIMENT 21   SPECTROCHEMICAL MEASUREMENTS: VERIFICATION OF LAMBERT-BEER'S LAW FOR $K_2Cr_2O_7$ COLORIMETRICALLY

Objective

To verify Lambert-beer's law for $K_2Cr_2O_7$ colorimetrically.

Theory

Determining the concentration of an unknown $K_2Cr_2O_7$ solution is the primary objective of this experiment. In this experiment, the $K_2Cr_2O_7$ solution that is used has a blue color, so colorimeter users are instructed to use the red LED. More light is absorbed (and less transmitted) by a colored solution of higher concentration than a solution of lower concentration.

Procedure

1. Five samples of known concentration (standard solutions) are to be prepared.

2. Then each solution is transferred to a small, rectangular cuvette which is placed into the colorimeter or spectrometer. The absorbance of each solution is computed by the amount of light that penetrates the solution and strikes the photocell. When a graph is plotted for absorbance *vs.* concentration for the standard solutions, a direct

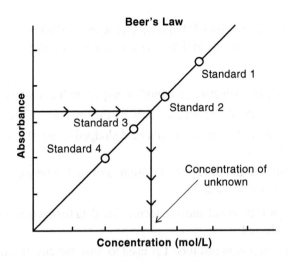

FIGURE Beer's law

relationship should be displayed. This direct relationship between absorbance and concentration for a solution is known as *Beer's law*.

3. This is followed by determination of the concentration of an unknown $K_2Cr_2O_7$ solution by measuring its absorbance. The corresponding concentration on the horizontal axis can be found by locating the absorbance of the unknown on the vertical axis of the graph. Slope of the Beer's law curve can also be used to find the concentration of the unknown solution.

Requirements

- Colorimeter cuvette
- Six test tubes
- Test tube rack
- Stirring rod
- Two 10 mL pipets or measuring cylinders
- Two 100 mL beakers
- 0.01 M $K_2Cr_2O_7$ solution
- Distilled water
- Tissues

Procedure

1. Small volumes of 0. 01 M $K_2Cr_2O_7$ solution and distilled water are taken in two separate beakers.

2. Five standard solutions are to be prepared according to the chart given below in clean, dry, test tubes that are labelled **1–5**. The solutions are prepared using pipets or measuring cylinders. Each solution is thoroughly mixed with a stirring rod. The stirring rod is cleaned and dried between uses.

Test Tube	0.01 M $K_2Cr_2O_7$ (mL)	Distilled H_2O (mL)	Concentration (M)
1	2	8	0.002
2	4	6	0.004
3	6	4	0.006
4	8	2	0.008
5	~10	0	0.0100

3. A blank is prepared by filling 3/4 of a cuvette with distilled water.

    For proper measurements and correct use of cuvettes, the following should be kept in mind:

    - The outside of each cuvette should be wiped with a lint-free tissue.
    - The cuvettes should be handled only by the top edge of the ribbed sides.
    - Any bubbles in the cuvette should be dislodged by gently tapping the cuvette on a hard surface.
    - The cuvette should always be positioned in such a way that the light passes through the clear sides.

    After **steps 1–3**, absorbance-concentration data for the five standard solutions can be collected.

    (a) The solution in test tube **1** is used to rinse the cuvette twice with ~1 mL amount and then filled upto 3/4. The outside of the cuvette is wiped with a tissue and then placed in the device (Colorimeter or Spectrometer). This is followed by closing the lid on the Colorimeter.

    (b) The procedure is repeated for test tubes **2, 3** and **4**. The measurement from test tube **5** is the original 0.01 M $K_2Cr_2O_7$ solution. The unknown solution is measured only after the testing of the standard solutions is completed.

4. The absorbance value of the unknown $K_2Cr_2O_7$ solution is determined after the testing of the standard solutions is completed.

    (a) About 5 mL of the *unknown* $K_2Cr_2O_7$ is taken in a clean, dry, test tube. The number of the unknown is recorded in the **data table**.

    (b) The unknown is then taken in the cuvette. The absorbance value displayed in the meter is read and recorded only when the displayed absorbance value stabilizes. The value is recorded as measurement **6** in the **data table**.

    (c) The absorbance and concentration values of the known solutions are then plotted in a graph and analyzed. The absorbance value that is closest to the absorbance reading obtained for the unknown is marked on the graph. The concentration of the unknown $K_2Cr_2O_7$ solution is calculated and the value obtained for the unknown concentration is recorded in the **data table**.

    (d) The remaining solutions are disposed off.

TABLE Data table

Trial	Concentration (mol/L)	Absorbance
1	0.002	
2	0.004	
3	0.006	
4	0.008	
5	0.010	
6	Unknown number	

EXPERIMENT 22   SPECTROPHOTOMETRIC DETERMINATION OF IRON

Principle

A complex of iron(II) is formed with 1,10-phenanthroline, $Fe(C_{12}H_8N_2)_3^{2+}$, and the absorbance of this colored solution is measured with a spectrophotometer. The spectrum is plotted to determine the absorption maximum. Hydroxylamine (as the hydrochloride salt to increase solubility) is added to reduce any Fe^{3+} to Fe^{2+} and to maintain it in that state.

Equations

$$4Fe^{3+} + 2NH_2OH \rightarrow 4Fe^{2+} + N_2O + 4H^+ + H_2O$$

$\frac{1}{3}Fe^{2+} +$ 1,10-phenanthroline $\rightarrow$ tris(1,10-phenanthroline) iron(II)

Solutions and Chemicals Required

1. **Standard iron(II) solution.** Prepare a standard iron solution by weighing 0.0176 g ferrous ammonium sulfate, $Fe(NH_4)_2(SO_4)_2 \cdot 6H_2O$. Quantitatively transfer the weighed sample to a 250-mL volumetric flask and add sufficient water to dissolve the salt. Add 0.7 mL conc. sulfuric acid, dilute exactly to the mark with distilled water, and mix thoroughly. This solution contains 10.0 mg iron per liter (10 ppm); if the amount weighed is other than specified above, calculate the concentration.

2. **1,10-Phenanthroline solution.** Dissolve 25 mg 1,10-phenanthroline monohydrate in 25 mL water. Store in a plastic bottle.

3. **Hydroxylammonium chloride solution.** Dissolve 10 g hydroxylammonium chloride in 100 mL water.

4. **Sodium acetate solution.** Dissolve 10 g sodium acetate in 100 mL water.

Procedure

Into a series of 100-mL volumetric flasks, add with pipets 1.00, 2.00, 5.00, 10.00, and 25.00 mL of the standard iron solution. Into another 100-mL volumetric flask, place 50 mL distilled water for a blank. The unknown sample will be furnished in another 100-mL volumetric flask. To each of the flasks (including the unknown) add 1.0 mL of the hydroxylammonium chloride solution and 5.0 mL of the 1,10-phenanthroline solution. Buffer each solution by the addition of 8.0 mL of the sodium acetate solution to produce the red color of ferrous 1,10-phenanthroline. [The iron(II)–phenanthroline complex forms at pH 2 to 9. The sodium acetate neutralizes the acid present and adjusts the pH to a value at which the complex forms.] Allow at least 15 min after adding the reagents before making absorbance measurements so that the color of the complex can fully develop. Once developed, the color is stable for hours. Dilute each solution to exactly 100 mL. The standards will correspond to 0.1, 0.2, 0.5, 1, and 2.5 ppm iron, respectively.

Obtain the absorption spectrum of the 2.5-ppm solution by measuring the absorbance from about 400 to 700 nm (or the range of your instrument). Take readings at 25-nm intervals except near the vicinity of the absorption maximum, where you should take readings at 5- or 10-nm intervals. Follow your instructor's directions for the operation of your spectrophotometer. The blank solution should be used as the reference solution. Plot the absorbance against the wavelength and select the wavelength of the absorption maximum. This may be done with a spreadsheet. From the molar concentration of the iron solution and the cell pathlength, calculate the molar absorptivity of the iron(II)–phenanthroline complex at the absorption maximum.

Prepare a calibration curve by measuring the absorbance of each of the standard solutions of the wavelength of maximum absorbance. Measure the unknown in the same way. Using a spreadsheet, prepare a calibration curve by plotting the absorbance of the standards against concentration in ppm. From this plot and the unknown's absorbance, determine the final concentration of iron in your unknown solution. Perform the calculations of the measured concentration by entering the formula in a cell of the spreadsheet using the slope and intercept values and the measured absorbance. (See Chapters 3 and 11 for preparing the spreadsheet.) Report the number of micrograms of iron in your unknown along with the molar absorptivity and the spectrum of the iron(II)–phenanthroline complex.

EXPERIMENT 25 SPECTROPHOTOMETRIC DETERMINATION OF LEAD ON LEAVES USING SOLVENT EXTRACTION

Principle

Lead on the surfaces of leaves is dissolved by shaking with nitric acid solution. The lead is extracted as the dithizone complex into methylene chloride at pH above 9. The intensity of the color of the complex is measured spectrophotometrically and compared to a calibration curve prepared similarly from lead standards to calculate the amount of lead. Sulfite is been added as a masking agent to eliminate most interference from other metals[1].

Equation

Solutions and Chemicals Required

1. **Provided.** 1 M HNO_3, 0.1 M HNO_3, thymol blue indicator solution (0.1% in water), 2 M NH_3 solution, ammonia sulfite solution (350 mL conc. NH_3 solution, and 1.5 g Na_2SO_3 diluted to 1 L; the pH is about 11).

---

[1]Normally, cyanide is also added to mask other metals and provide maximum selectivity. But for safety reasons, this is omitted from the masking solution for this experiment. For illustrative purposes, you can assume the measurements are due soley to lead.

2. **To prepare**

   (a) **Stock 1000-ppm standard lead solution.** Dissolve 0.160 g $Pb(NO_3)_2$ and dilute to 100 mL in a volumetric flask.

   (b) **Standard 10-ppm working solution.** On the day of the experiment, dilute 1 mL of the stock solution to 100 mL in a volumetric flask.

   (c) **Dithizone solution.** Dissolve 7.5 mg dithizone in 300 mL methylene chloride. This should be prepared fresh on the day of the experiment.

Things to Do before the Experiment

Collect leaf samples. These can be from trees near a road or highway and from some that are more isolated for comparison. Collect at least two large leaves from each tree and place each in a clean plastic bag and seal. The leaves selected should be reasonably free from dirt or other visible contamination. Record the location of the leaves. Your instructor will advise you of the number of trees you should sample.

Procedure

1. **Preparation of calibration curve.** This should be done at the time the samples are to be analyzed. The chelate formation and solvent extraction are to be performed in clean 6-oz vials with caps. To each of six labeled vials add with pipets 0 (blank), 2, 4, 6, or 8 mL of the 10-ppm lead standard and sufficient water to bring the volume to about 20 mL. Add about 60 mL of the ammonia–sulfite solution using a graduated cylinder and 25 mL of the CH_2Cl_2–dithizone solution using a pipet (*not by mouth*). Stopper the vial and shake for about a minute. Using a pipet, withdraw most of the heavier methylene chloride layer and filter through dry filter paper (Whatman No. 40) into a dry Bausch and Lomb Spectronic 20 measuring tube or the equivalent, or centrifuge before transferring. (The samples should be prepared for measurement now so they can be measured when the standards are.)

   Using one of the standards, measure the absorbance from 400 to 600 nm in 20-nm increments to determine the wavelength of maximum absorption. Using this wavelength, measure the absorbance of each standard, using the blank to zero the instrument. Plot absorbance against micrograms of lead taken to prepare a calibration curve, using a spreadsheet (Chapters 3 and 11).

2. **Determination of lead on leaves.** For each plastic bag containing a leaf sample, heat 20 mL of 0.1 M HNO_3 to about 70°C. Add 20 mL to each bag, close, and shake for about 2 min. Pour into clean 100-ml- beakers. Add one drop thymol blue indicator solution to each, followed by dropwise addition of 2 M NH_3 until the indicator color change is complete (to blue) and add a couple of extra drops. *The solution should smell of ammonia.* Then, add 60 mL of the ammonia–sulfite solution—and then add with a pipet 25 mL of the CH_2Cl_2–dithizone solution and proceed with the extraction measurement as with the standards.

Calculation

Blot each leaf dry, place on a sheet of paper, and trace the outline of the leaf. Cut out the leaf outline and weigh on an analytical balance to three figures. Cut out a 10-cm × 10-cm (100 cm^2) square from the same paper and weigh. Calculate the area of the leaf in cm^2.

From the measured absorbance of each sample and the calibration curve, calculate the micrograms of lead on the leaf and report the amount of lead in μg Pb/100 cm^2 leaf. Is there any correlation of lead content with proximity of the tree to a roadway?

EXPERIMENT 26   SPECTROPHOTOMETRIC DETERMINATION OF MANGANESE AND CHROMIUM IN MIXTURE

Principle

Manganese and chromium concentrations may be determined simultaneously by measurement of the absorbance of light at two wavelengths, after the metals have been oxidized to $Cr_2O_7^{2-}$ and MnO_4^-. Beer's law has been shown to apply closely if the solutions are at least 0.5 M in H_2SO_4. $Cr_2O_7^{2-}$ has an absorption maximum at 440 nm and MnO_4^- has one at 545 nm. (A somewhat more intense maximum is at 525 nm, but there is less interference from $Cr_2O_7^{2-}$ at 545 nm.) Equations similar to Equations 11.16 and 11.17 are solved for the unknown concentrations from the measured absorbances at the two wavelengths. The four constants ($\epsilon b = k$) are determined by measurements of absorbance at the two wavelengths using pure solutions of known concentration; a calibration curve is prepared at each wavelength for both $Cr_2O_7^{2-}$ and MnO_4^- and the slopes of the curves (A versus C) are used to obtain an average k value.

Equations

The unknown contains Cr^{3+} and Mn^{2+}. The former is oxidized to $Cr_2O_7^{2-}$ by heating with peroxydisulfate (persulfate) in the presence of a silver catalyst:

$$2Cr^{3+} + 3S_2O_8^{2-} + 7H_2O \xrightarrow[\Delta]{Ag^+} Cr_2O_7^{2-} + 6SO_4^{2-} + 14H^+$$

Mn^{2+} is oxidized in part by peroxydisulfate, but also by periodate:

$$2Mn^{2+} + 5S_2O_8^{2-} + 8H_2O \xrightarrow[\Delta]{Ag^+} 2MnO_4^- + 10SO_4^{2-} + 16H^+$$
$$2Mn^{2+} + 5IO_4^- + 3H_2O \longrightarrow 2MnO_4^- + 5IO_3^- + 6H^+$$

For the mixture,

$$A_{440} = k_{Cr,440}C_{Cr} + k_{Mn,440}C_{Mn}$$
$$A_{545} = k_{Cr,545}C_{Cr} + k_{Mn,545}C_{Mn}$$

The k values are determined from the slopes of the calibration curves of the pure solutions:

$$k_{Cr,440} = A_{440}/C_{Cr} \qquad k_{Cr,545} = A_{545}/C_{Cr}$$
$$k_{Mn,440} = A_{440}/C_{Mn} \qquad k_{Mn,545} = A_{545}/C_{Mn}$$

Solutions and Chemicals Required

1. **Provided.** 18 M H_2SO_4, $K_2S_2O_8$, KIO_4, $AgNO_3$.

2. **To prepare**

   (a) **Standard 0.002 M $MnSO_4$ solution.** Dry about 1 g $MnSO_4$ at 110°C for 1 h, cool for 30 min, and weigh out about 0.08 g (to the nearest tenth milligram). Transfer to a 250-mL volumetric flask, dissolve, and dilute to volume. Calculate the molarity of the solution and the concentration of Mn in mg/L (ppm).

   (b) **Standard 0.0178 M $K_2Cr_2O_7$ solution.** Use the solution prepared in Experiment 15 or prepare 100 mL as directed there (weigh to the nearest milligram). Calculate the molarity of the solution and the concentration of Cr in mg/L (remember there are two Cr atoms per molecule of $K_2Cr_2O_7$).

   (c) **0.1 M $AgNO_3$.** Dissolve about 0.2 g $AgNO_3$ in about 12 mL water.

Things to Do before the Experiment

Prepare the standard $MnSO_4$ and $K_2Cr_2O_7$ solutions. This will require drying $MnSO_4$ and $K_2Cr_2O_7$.

Procedure

1. **Calibration (determination of k values). Note:** The absorbance of the calibration solutions and the unknown should be read at the same time. Therefore, get all solutions prepared before making any readings. They are all sufficiently stable that they could be allowed to set until another laboratory period but it is best not to.

   (a) **Manganese.** Add with pipets aliquots of 10, 15, and 25 mL of the standard $MnSO_4$ solution into three different 250-mL Erlenmeyer flasks. Add distilled water to bring the volume in each flask to about 50 mL. To each flask add 10 mL conc. H_2SO_4 (CAREFULLY, using a graduated cylinder) and 0.5 g solid KIO_4 (potassium periodate or metaperiodate, depending on the manufacturer). Heat each to boiling for about 10 min, cool, transfer quantitatively to 250-mL volumetric flasks, and dilute to the mark with distilled water. Determine the absorbance of each solution at 440 and 545 nm, using $0.5\ M\ H_2SO_4$ as a blank solution. Permanganate solutions containing periodate are stable. The absorbance at 440 nm will be less than 0.1 and, hence, the spectrophotometric error (precision) will be large (see Figure 11.29). But this is acceptable, in fact, desirable, because the correction for manganese absorption at this wavelength is small; that is, a relatively large error in determining a small correction results in only a small error.

   (b) **Chromium.** Add 10-, 15-, and 25-mL aliquots of the standard $K_2Cr_2O_7$ solution to 250-mL volumetric flasks, add about 100 mL distilled water and 10 mL conc. H_2SO_4, mix thoroughly, and dilute to 250 mL with distilled water. Determine the absorbance of each solution at 440 and 545 nm, using $0.5\ M\ H_2SO_4$ as the blank solution. The absorbance in this case will be small (< 0.1) at 545 nm.

   (c) **Determination of k values.** Using a spreadsheet, plot absorbance versus concentration in units of mg/L for each solution at each wavelength, and plot the least-squares straight line through each set of data points. The lines should intercept at zero absorbance and zero concentration; under Chart Options, you can instruct Excel to plot the intercept at zero. The slopes of these lines are the coefficients ($k = A/C$) to be used in determining the concentrations of chromium and manganese in the unknown. These slopes relate absorbance and concentration for the instrumental parameters used. Therefore, one should use the same instrument, cuvets, cuvet position, and volumes of solutions for all determinations in this experiment.

2. **Analysis of unknown.** Obtain a mixture of Mn^{2+} and Cr^{3+} or $Cr_2O_7{}^{2-}$ unknown in a 250-mL volumetric flask and dilute to volume. Transfer with a pipet three 50-mL aliquots into 250-mL Erlenmeyer flasks. The procedure may be stopped up to this point. But once the peroxydisulfate is added, the oxidation should be completed.

   To each flask add 5 mL conc. H_2SO_4 (BE CAREFUL!), then mix well. Add about 1 or 2 mL of $0.1\ M\ AgNO_3$ solution and 1.0 g solid potassium peroxydisulfate ($K_2S_2O_8$). BE CAREFUL! PEROXYDISULFATE IS A STRONG OXIDIZING AGENT THAT CAN REACT VIOLENTLY WITH REDUCING AGENTS! USE ONLY AS DIRECTED. Do not spill. Dissolve peroxydisulfate and heat the solution to boiling and boil gently for about 5 min. Cool the solution and then add 0.5 g KIO_4. Again heat to boiling for 5 min.

   Cool each solution to room temperature, quantitatively transfer to 250-mL volumetric flasks, and dilute to volume. The solutions at this point (or before dilution) are stable and can be saved until another laboratory period if necessary. Also save the serial standards to calibrate the instrument at the same time.

From the unknown absorbances at the two wavelengths, calculate the parts per million of Cr and Mn in your unknown using Beer's law for the mixture and the determined constants. Use a spreadsheet similar to the one given in Chapter 11 and in your text **website** to perform the mixture calculation. The calculated results will have the same units as used in determining the constants. Keep in mind the dilutions made. Report the results for each portion analyzed.

EXPERIMENT 29   INFRARED DETERMINATION OF A MIXTURE OF XYLENE ISOMERS

Principle

Meta- and *para*-xylene are determined in mixtures using *ortho*-xylene as an internal standard, to compensate for variation in cell length between runs. The infrared spectrum of the unknown mixture is recorded and the relative height of peaks of the two compounds are compared with those of standard mixtures, using the baseline technique.

Solutions and Chemicals Required

1. **Provided.** *Ortho-, meta-,* and *para*-xylene.

2. **To prepare.** *Meta*-xylene/*para*-xylene standards. Prepare a series of standards (use available burets), all containing 30% (vol/vol) *o*-xylene as internal standard, by mixing the appropriate volumes of the three isomers to give 25, 35, and 45 vol % of *m*-xylene. The corresponding concentrations of *p*-xylene will be 45, 35, and 25%, respectively.

Procedure

Consult your instructor on the proper operation of your instrument. Handle the infrared cell carefully, avoiding contact with water and the fingers. Fill the cell with pure *m*-xylene and obtain a spectrum on this from 2 to 15 μm, being sure to record the last peak just before 15 μm (692 cm^{-1}). Each time you run a sample, be sure to check 0% T by placing a card in the sample beam and adjust the pen to 0% T. Empty the cell, rinse and fill with *p*-xylene, and run a spectrum on this. Repeat for *o*-xylene. Run spectra on each of the standard mixtures. From the spectra of the pure substances, choose a peak of each isomer to measure. Using the baseline method (see Figure 11.11), measure P_0/P for the peak for each compound. Prepare a calibration curve of the ratio of $\log \left(P_0/P\right)_{\text{meta}} / \log \left(P_0/P\right)_{\text{ortho}}$ and of $\log \left(P_0/P\right)_{\text{para}} / \log \left(P_0/P\right)_{\text{ortho}}$ versus concentration for the meta and para isomers, respectively. See Chapter 15 and your text **website** for spreadsheet preparation using an internal standard.

Obtain an unknown mixture of meta and para isomers from your instructor. Prepare a mixture of this with *o*-xylene by adding 70 parts of the unknown to 30 parts *o*-xylene. Run the spectrum on this mixture and, using the baseline method and the same peaks as before, measure P_0/P for the three compounds and calculate $\log(P_0/P)/ \log \left(P_0/P\right)_{\text{ortho}}$ for the two unknown isomers. Compare with the calibration curve to determine the percent concentrations of the meta and para isomers; use the spreadsheet for calculations. Remember to divide by 0.7 to convert to initial concentrations.

Atomic Spectrometry Measurements

EXPERIMENT 30   DETERMINATION OF CALCIUM BY ATOMIC ABSORPTION SPECTROPHOTOMETRY

Principle

The effects of instrumental parameters and of phosphate and aluminum on calcium absorption are studied [see, e.g., W. Hoskins et al., *J. Chem. Ed.*, **54** (1977) 128]. Calcium in an unknown synthetic or serum sample is determined by comparing the absorbance with that of standards.

Solutions and Chemicals Required

500 ppm Ca, 4% $SrCl_2$, 2000 ppm NaCl, 100 ppm phosphate, ethanol.

1. **Provided.** Ethanol, 4%, $SrCl_2$ solution (wt/vol), stock solution of 140 meq/L Na and 4.1 meq/L K (see footnote below).

2. **To prepare**

   (a) **500 ppm Ca stock solution.** Dissolve 1.834 g (accurately weighed) $CaCl_2 \cdot 2H_2O$ in water and dilute to 1 L in a volumetric flask. Dilute this 1:10 to prepare a 50-ppm stock solution. Use this to prepare the solutions required below. (Commercial 1000-ppm Ca^{2+} solutions may be used.)

   (b) **2000-ppmNa stock solution.** Dissolve 0.51 g NaCl in 100 mL water.

   (c) **100 ppm phosphate.** Dissolve 0.15 g Na_2HPO_4 in 1 L water.

   (d) **100-ppmAl stock solution.** Dissolve 0.18 g $Al_2(SO_4)_3 \cdot K_2SO_4 \cdot 24H_2O$ in 100 mL water. ($AlCl_3$ may be used, but take care when adding water.)

Study of Instrumental Parameters

Follow your instructor's directions for the operation of the instrument. If you have a single-beam instrument, the hollow-cathode lamp should be allowed to warm up for 30 min before the experiment. A few minutes should be adequate with a double-beam instrument. An air–acetylene flame should be used with a premix burner.

1. **Burner height.** Adjust the fuel and support gas pressures until the flame is near stoichiometric (just a slight yellow color to the flame). Then, turn up the fuel pressure to impart a strong yellow glow to the flame (fuel rich). The yellow glow is due to unburned carbon particles in the rich flame. In a lean flame, an excess of oxidant is present and the flame appears blue. Prepare and aspirate a 5-ppm calcium solution and note its absorbance at 422.67 nm. Adjust the wavelength setting to obtain maximum absorbance. The monochromator is now set exactly at the calcium line. With the burner height adjusting knob, raise the burner so that the light beam just passes over the tip of it (base of the flame). Use distilled water to zero the instrument, and then measure the absorbance of the 5-ppm calcium solution. Lower the burner in increments (six to eight steps) and record the absorbance at each height.

   Plot the absorbance against heights of observation in the flame and select the optimum height.

2. **Fuel/air ratio.** Hold the air pressure constant and adjust the fuel pressure in increments from a very fuel-rich to a lean flame. Record the absorbance of 5 ppm Ca at each increment.

   Select the optimum fuel pressure and vary the air pressure in a similar manner. Plot absorbance against gas pressure for both the fuel and the air, noting the pressure setting of the one held constant. Select the optimum fuel, and air settings. Is this a rich, stoichiometric, or lean flame?

Interference Studies

1. **Effect of phosphate.** Prepare a solution containing 5 ppm Ca and 10 ppm phosphate. Record the absorbance of this solution, using the optimum conditions determined above, and compare to that of 5 ppm Ca. Explain the results.

   Prepare a solution containing 5 ppm Ca, 10 ppm phosphate, and 1% $SrCl_2$. Prepare also a solution of 5 ppm Ca and 1% $SrCl_2$. Record the absorbance of the solutions. Compare the absorption of the first solution with that of the phosphate-containing solution above and with that of the 1% $SrCl_2$-containing solution. Explain.

2. **Effect of sodium.** Prepare a solution containing 5 ppm Ca and 1000 ppm Na. Record the absorbance and compare with that of 5 ppm Ca. Explain any difference.

3. **Effect of aluminum.** Prepare a solution containing 5 ppm Ca and 10 ppm Al. Record the absorbance and compare with that of 5 ppm Ca by itself. Suggest a possible reaction for the results.

Determination of Calcium in an Unknown

(The method of standard additions below may be used instead of the following procedures.)

1. **Synthetic unknown.** Obtain an unknown from your instructor and dilute to give a concentration of 5 to 15 ppm Ca. Prepare a series of calcium standards of 0, 2.5, 5, 7.5, 10, and 15 ppm from the 50-ppm stock solution. If the unknown contains phosphate, add $SrCl_2$ to standards and the unknown to give a final concentration of 1%. Record the absorbance (or % absorption and convert to absorbance) of these and prepare a calibration curve of absorbance versus concentration. Determine the concentration of the unknown in the usual manner.

2. **Serum.** Calcium in serum or an "artificial serum" as described in the footnote to Experiment 31 is determined by diluting 1:20 with 1% $SrCl_2$ solution. The normal calcium content of serum is about 100 ppm, and so that analyzed solution contains about 5 ppm Ca. Sodium and potassium equal to that in the sample are added to the standards.

   Add 0.5 mL of the unknown serum or the "artificial serum" to a 10-mL volumetric flask and dilute to volume with 1% $SrCl_2$. (If the method of standard additions is to be used also for comparison, dilute 2.5 mL of unknown to 50 mL with 1% $SrCl_2$). Prepare standards of 0, 3, 4, 5, 6, and 8 ppm Ca, each also containing 1% $SrCl_2$, 6.9 meq/L Na, and 0.21 meq/L K.[1]

   Prepare a calibration curve from the absorbance of the standards and from this determine the concentration of calcium in the unknown. Use a spreadsheet.

Method of Standard Additions

This procedure may be used, instead of the one above, to analyze the unknowns, and it illustrates the usefulness of the method of standard additions for compensating for matrix effects. Dilute your unknown as described above, using distilled water to give a

---

[1] A stock solution of 140 meq/L Na and 4.1 meq/L K (20 times the concentration in the standard solution) can be prepared by dissolving 8.1 g NaCl and 0.21 KCl in 1 L water. This contains the normal levels of Na and K in serum and compensates for ionization interference due to these elements in the serum. Dilute this solution 1:20 in the standards.

concentration of about 5 ppm Ca (1:20 for serum, e.g., 2.5 mL diluted to 50 mL with 1% $SrCl_2$). Transfer with a pipet separate 10.0-mL aliquots of the diluted unknown to three separate clean and dry test tubes or flasks. Add to these 50.0, 100, and 150 μL, respectively, of the 500-ppm standard calcium solution (or 25.0, 50.0, and 75.0 μL of a 1000-ppm solution if available). This results in an increase in the calcium concentration in the diluted unknown of about 2.5, 5.0, and 7.5 ppm, respectively, depending on the exact concentration of the standard, and brackets the unknown. The volume changes can be considered negligible. Use an appropriate syringe microliter pipet if available (e.g., a 50-μL Eppendorf pipet or Finn-pipette) or else a 0.1-mL graduated measuring pipet. For best accuracy, the pipet should be calibrated (see Chapter 2).

Zero the instrument with distilled water and aspirate the diluted unknown and the standard addition samples. The absorbance increases in the latter are due to the added calcium. Using a spreadsheet, prepare a plot of absorbance against added concentration of calcium (starting at zero added, i.e., the sample). From the x-axis intercept of the plot, determine the concentration of calcium in the diluted unknown. See the spreadsheet in Chapter 12 for preparation of a standard additions plot and unknown calculation. Calculate the concentration in the original sample. How does this method account for phosphate interference?

EXPERIMENT 31   FLAME EMISSION SPECTROMETRIC DETERMINATION OF SODIUM

Principle

The intensity of sodium emission in a flame at 589.0 nm is compared with that of standards. If an internal standard instrument is available, the ratio of sodium to lithium emission is measured.

Solutions and Chemicals Required

1. **Stock standard NaCl solution (1000 ppm Na).** Dry about 1 g NaCl at 120°C for 1 h and cool for 30 min. Weigh and dissolve 0.254 g NaCl in water and dilute to 1 L. Care must be taken to avoid sodium contamination, especially from the water and glassware. A blank must be run to correct for sodium in the water.

2. **$LiNO_3$ internal standard solution (1000 ppm Li).** (Lithium is not required if a direct-intensity instrument, rather than an internal standard instrument, is used.) Dissolve 0.99 g $LiNO_3$ in water and dilute to 100 mL.

3. **Working standard solutions**[1]

---

[1]The unknown may be simply a sodium chloride solution, or it may be serum. If serum is analyzed, then the standards should be prepared over a narrower concentration range to better bracket the unknown. Sodium in serum (approximately 140 meq/L, or 3200 ppm Na) may be determined by simple 1:100 dilution (e.g., 0.1 mL diluted to 10 mL) or 1:500 if required by the instrument. An alternative unknown is an "artificial serum" prepared by dissolving the following salts in water and diluting to 1 L:

NaCl	8.072 g
KCl	0.21 g
KH_2PO_4	0.18 g
$CaCl_2 \cdot 2H_2O$	0.37 g
$MgSO_4 \cdot 7H_2O$	0.25 g

This contains 138.1 meq/L Na, which can be varied from unknown to unknown. Bovine serum albumin (60 g) may be added as a source of protein. Serum contains about 6% (wt/wt) protein.

(a) **Direct-intensity instrument.** Prepare standards of 0, 10, 20, 30, and 40 ppm Na by diluting 0, 5, 10, 15, 20, and 25 mL of the stock NaCl solution to 50 mL. (It may be better with some instruments to prepare solutions five times more dilute than this by adding 0, 1, 2, 3, 4, and 5 mL of solution to the flasks. This may result in a more linear calibration curve. Follow your instructor's directions.)

(b) **Internal standard instrument.** Prepare the same solutions as above for the direct-reading instrument, but add 5 mL of the stock lithium solution to each flask. (This results in 100 ppm Li in each solution. The recommended concentration may vary from one manufacturer to another, and so your instructor may direct you to add a different amount.)

Things to Do before the Experiment

Dry the NaCl at 120°C for 1 hour and cool in a desiccator.

Procedure

Have your instructor add your unknown to a 100-mL volumetric flask.

1. **Direct-reading instrument.** Dilute your unknown to volume with water. Follow your instructor's directions for the operation of the instrument. Several atomic absorption instruments can be used for measuring emission. Set the zero reading while aspirating distilled water (the blank). Aspirate each standard and the unknown and record their emission intensity readings. With some instruments, the 100% reading is set with the most concentrated standard.

    Using a spreadsheet, plot the emission readings for the standards against concentration and determine the concentration in the unknown solution from the calibration curve. From this, calculate the micrograms of solution in your unknown if it is water, and ppm or meq/L if it is serum (see footnote).

2. **Internal standard instrument.** Add the same amount of lithium solution to your unknown as was added to your standard, and dilute to volume with water. Follow your instructor's directions for the operation of the instrument. The lithium emission line is 670.8 nm. Prepare a calibration curve as above for the direct-reading instrument, but record the ratio of the Na/Li emission intensities. Determine the concentration of sodium in the unknown solution and report micrograms of sodium in the unknown if it is water, and ppm or meq/L if it is serum (see footnote 1). Use a spreadsheet as described in Chapter 15 for the internal standard plot and calculation.

Solid-Phase Extraction and Chromatography

Contributed by University of Washington: Thomas C. Leach and Professor Robert E. Synovec

EXPERIMENT 32   SOLID-PHASE EXTRACTION WITH PRECONCENTRATION, ELUTION, AND SPECTROPHOTOMETRIC ANALYSIS

Introduction and Theory

The purpose of this laboratory experience is to introduce the student to the concepts and methodology involving solid phase extraction (SPE) of an analyte of interest from an initial solution, preconcentration and elution of that analyte, and then absorbance spectrophotometric analysis of the preconcentrated analyte solution. **The basic idea is to develop the skill and knowledge in order to take a sample in which an analyte of interest is initially at too dilute of a concentration to analyze directly, and**

then to preconcentrate the analyte, to obtain a significantly more concentrated sample that is amenable to analysis. In this experiment, the test analyte is 2-((4-(Dimethylamino)phenyl)azo)benzoic acid or Methyl Red (*F.W.* = 269.31), which is a weak acid, so it is necessary to work in acidic solution conditions to keep Methyl Red protonated (in the neutral form), thus it will behave like a typical organic compound. Also, in the acidic form the Methyl Red solution is red, and is easily observed if at sufficient concentration.

The structure of the Methyl Red indicator is shown below.

Methyl Red ($C_{15}H_{15}N_3O_2$)

This compound is minimally soluble in pure water. It is supplied to students sufficiently diluted for the purposes of preparing a standard curve and spiking the sample.

The experiment proceeds is as follows. The initial sample solution containing Methyl Red is at a sufficiently low concentration so it is barely perceptible that there is Methyl Red present, and absorbance measurements of this initial sample will confirm that this initial sample would benefit from SPE-based preconcentration prior to analysis via absorbance. Additionally, in a separate measurement, the initial sample will be spiked with Methyl Red to perform quantitative analysis following the Standard Addition Method (SAM) procedure. Preconcentration of both the initial sample and spiked sample is performed using separation SPE cartridges (300 mg Hi-Cap C18). These SPE column cartridges contain C18, a very non-polar stationary phase on a solid substrate (like a liquid chromatography stationary phase). The basic principle involved with SPE-based preconcentration is to first extract the analyte from a large volume of an initial sample solvent onto the stationary phase, and second, to extract (i.e., remove) the analyte from the stationary phase using a much smaller volume of an eluting solvent. Then, the preconcentration factor (P), i.e., the factor by which the initial sample has had it's concentration amplified, is the ratio of the initial sample volume (V_i), divided by the eluting solvent volume (V_f), or $P = V_i/V_f$. This theory is summarized as follows:

Goal of preconcentration:

$$[MR]_f \gg [MR]_i \tag{1}$$

where $[MR]_f$ is the concentration of Methyl Red following preconcentration and $[MR]_i$ is the concentration of Methyl Red in an initial sample. In terms of the preconcentration factor,

$$[MR]_f = P[MR]_i \tag{2}$$

Substituting in for concentration,

$$mole_f/V_f = P\,(mole_i/V_i) \tag{3}$$

Since ideally,

$$mole_f = mole_i \tag{4}$$

then the preconcentration factor is,

$$P = V_i/V_f \tag{5}$$

Therefore, selecting a $V_i \gg V_f$ will ideally provide a large preconcentration factor.

Experimentally, it is very important to reproduce the same P from one sample to another in order to obtain accurate quantification.

In this experiment, Methyl Red in the acidified neutral form is sufficiently non-polar so it is quantitatively extracted from the initial aqueous sample matrix (polar and acidified solvent) since the SPE stationary is also non-polar (C18). Then, the Methyl Red is quantitatively eluted using ethanol (a solvent that is sufficiently non-polar relative to the initial aqueous sample matrix). This eluted sample is then analyzed (following pH adjustment) using absorbance spectrophotometry. The preconcentration factor P ideally will be 50 since the initial volume, V_i, is 500 mL and the final volume, V_f, of the eluted sample is 10 mL. The spiked sample is analyzed similarly, and the SAM is used for quantification. Additionally, a traditional calibration plot of Methyl Red is prepared using standard solutions. The slope of the calibration plot (for the standards) is compared to the slope for the SAM plot. If the sample matrix does not interfere with detection of the analyte, agreement between the slopes should be obtained. A discrepancy between the slopes of the two plots indicates that matrix interferences may be present and the SAM would be the preferred method of analysis.

Prelab Preparation

At least one day before requesting a SPE sample, clean and drain and label a 2 L bottle. Attach an approximately two foot length of flexible PVC tubing to the air line nozzle in one of the hoods and blow air into the bottle with the air valve turned to a fully opened position until all of the water in the bottle has evaporated. Weigh the dry bottle on a balance with at least a 10 kg capability and record this tare weight.

Clean and dry two vials capable of holding 30 mL to give to the lab assistant at the time the SPE sample is requested. Label the vials with the name and account number when requesting the SPE sample.

One lab period after submitting the vial, the SPE sample will be issued in a plastic test tube. This sample represents two liters of original sample that has been issued in concentrated form to facilitate transport and handling. (This is a similar to buying frozen orange juice. It is shipped to the store in small containers to facilitate transportation and storage then diluted to a gallon to reconstitute it to its original strength.)

Students will also receive a 5.57×10^{-6} *M* pink reference solution of Methyl Red in one of the vials that was turned in to the lab assistant, and a clear water/ethanol/acetic acid make-up solution in the other vial.

When getting the unknown sample also obtain two High Capacity SPE columns from the lab assistant. 5.0 mL syringes should be available in the lab.

Sample Preparation

The density of the sample after dilution will be considered to be exactly 1 kg/L. After receiving the unknown sample from the lab assistant, transfer it quantitatively to the two liter bottle. Add deionized water to the bottle to bring the final weight of the bottle and sample to a weight of approximately 2,000 g plus the tare weight of the bottle. [Note: the weight of the water with sample does not need to be exactly 2,000 g, but the exact weight to the nearest gram must be measured and recorded so that the concentration of methyl red can be calculated based on a sample volume of 2 L (2,000 g). In the Calculating and Reporting Results section, the final calculations will adjust for the difference in actual weight of the sample from 2,000 g.]

Sample Aliquot Preparation

Rinse a clean 100 mL beaker three times with small portions of the sample.

Fill the beaker with the sample.

Use the sample in the beaker to rinse a clean 500 mL volumetric flask and long stem funnel three times. Using the long stem funnel fill a 500 mL volumetric flasks almost to the mark with sample from the 2 L bottle. Use a plastic pipet to fill the 500 mL flask to the mark with the sample in the 100 mL beaker.

Pour the 500 mL aliquot of sample in the volumetric flask into a 500 mL Erlenmeyer flask. You will now rinse the flask to obtain quantitative transfer.

Quantitatively pipet 10 mL of deionized water into a clean 100 mL beaker. While turning the 500 mL volumetric flask pour the deionized water from the beaker down the neck of the flask so water coats the entire surface of the flask. Transfer all of this rinse water into the Erlenmeyer flask. Repeat this process twice more so that the volumetric flask gets rinsed out quantitatively with a total of 30 mL of deionized water that is transferred into the Erlenmeyer flask.

Quantitatively pipet 10 mL of the water/ethanol/acetic acid make-up solution into the Erlenmeyer flask to result in a total volume of 540 mL in the Erlenmeyer flask.

Spiked Aliquot Preparation

Fill the 500 mL volumetric flask with sample as was done for the sample preparation step above, being sure to first rinse the flask three times with the sample solution in the 100 mL beaker before filling it. Transfer the entire volume to a second 500 mL Erlenmeyer flask. As with the sample aliquot, quantitatively pipet 10 mL of deionized water into a clean 100 mL beaker. While turning the 500 mL volumetric flask pour the deionized water from the beaker down the neck of the flask so water coats the entire surface of the flask. Transfer all of this rinse water into the Erlenmeyer flask. Repeat this process twice more so that the volumetric flask is rinsed out quantitatively with a total of 30 mL of deionized water that is transferred into the Erlenmeyer flask.

Quantitatively transfer 10 mL of the $5.57 \times 10^{-6} M$ reference solution to the Erlenmeyer flask to result in a total volume of 540 mL in the Erlenmeyer flask.

Solid Phase Extraction

In the hood place a few milliliters of concentrated acetic acid in a small (~ 30 mL) beaker or vial. Using an adjustable pipet add 1 mL of concentrated glacial acetic acid to the Erlenmeyer flasks containing the sample and spiked sample.

Set up one of the 500 mL Erlenmeyer flasks containing the sample using clamps and a ring stand so the flask is secure and tips at an angle so, as it empties, the last few milliliters of liquid can be easily sucked out of a single location.

Plug in vacuum tubing to an aspirator and attach Tygon tubing to the free end.

Place a two inch piece of Tygon tubing on the end of a graduated 1 mL pipet.

Mix 1 mL of acetic acid with 500 mL of deionized water (DIW). (This makes a 0.2% acetic acid solution, which will equal the acetic acid concentration of the sample when it is extracted.)

Prior to connecting them to the rest of the experimental apparatus, condition two SPE cartridge columns by pushing 5 mL of ethanol through each of them with a syringe, then rinse the syringe with the 0.2% acetic acid solution and push 5 mL of 0.2% acetic acid through each one. The syringe will only lock into one end of the SPE column, which is referred to as the large port (top), with the small port (bottom) of the SPE column serving as the drain to waste for the conditioning step.

To finish setting up the experimental apparatus, insert the large port (top) of one of the conditioned SPE columns into the Tygon tubing leading to the vacuum and the small end into the Tygon tubing leading to the 1 mL graduated pipet. The small port (bottom) of the SPE column is connected to the larger Tygon tubing leading to the vacuum. Note that the flow of liquid in the conditioning step is reversed from the preconcentration step. See the picture below.

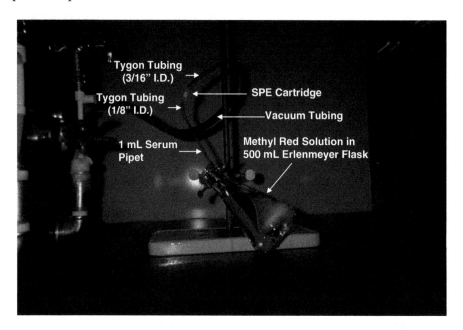

Place the 1 mL plastic graduated pipet into the Erlenmeyer flask so the tip of the pipet goes into the deepest portion of the flask. Stir the solution well with the 1 mL graduated pipet to mix.

Turn on the aspirator to draw the sample through the SPE column. After drawing the sample through the SPE column the aspirated solution goes directly into the aspirator and down the drain. The Methyl Red is trapped directly on the column and no aspirator trap is necessary. *While the sample is being extracted onto the column prepare the standards as described in the section: **Standard Calibration Plot for Absorption Coefficient Determination***

When only one or two milliliters of sample is left in the Erlenmeyer flask, rinse the sides of the flask down with 10 mL of DIW quantitatively pipetted into a 100 mL beaker as was done with the rinsing of the 500 mL volumetric flasks. Aspirate this rinse through the SPE column until no liquid remains in the flask. Repeat this process two more times. At this point 571 mL have been passed through the SPE column. [500 mL original sample + 30 mL rinse for the volumetric flask + a 10 mL aliquot of water/ethanol/acetic acid make-up solution + 1 mL acetic acid + 30 mL rinse for the Erlenmeyer flask].

After the sample aliquot is collected on one column, remove that column and insert the second column in between the vacuum tubing and graduated pipet. Place the 1mL graduated pipet into the Erlenmeyer flask containing the spiked aliquot so the tip of the pipet goes into the deepest portion of the flask. Stir the solution with the graduated pipet to mix.

As before with the sample, when only one or two milliliters of sample is left in the Erlenmeyer flask, rinse the sides of the flask down with 10 mL of DIW quantitatively pipetted into a 100 mL beaker. Aspirate through the SPE column until no liquid remains in the flask. Repeat this process two more times. At this point 571 mL have been passed through the SPE column. [500 mL original sample + 30 mL rinse for the volumetric flask + a 10 mL aliquot of the $5.57 \times 10^{-6}\ M$ reference solution + 1 mL acetic acid + 30 mL rinse for the Erlenmeyer flask].

Removal of sample and spike solution from the SPE column

After removing both columns from the vacuum system, draw exactly 4 mL ethanol into a 5 mL syringe. Connect the syringe to the column containing the sample and place the other end of the column into the mouth of a 10 mL volumetric flask. Slowly push the ethanol through the column into the 10 mL volumetric flask. ***Repeat this process with an additional aliquot of exactly 3 mL of ethanol***. Some methyl red analyte may appear to remain on the column. Do not attempt to remove it using any additional ethanol rinses because this will bias the analysis, and indeed, application of the SAM provides correction for this effect.

In the hood place a few milliliters of concentrated HCl in a small beaker or vial ($\sim$ 30 mL). Use an adjustable pipet to add exactly 0.5 mL of concentrated HCl to the 10 mL volumetric flask. Using a plastic dropper add deionized water to the flask to bring the contents to the mark. Stopper the flask and invert ten times to mix the sample. Transfer the sample to a glass vial and cap it.

Rinse the 10 mL volumetric flask with DIW and repeat the removal process with the spiked solution, placing it in a separate vial that will then be capped.

Standard Calibration Plot for Absorption Coefficient Determination

The presence of matrix effects may alter the absorption coefficient of the analyte in a sample. One benefit of extracting analyte from the sample and then rediluting it in pure ethanol and water is to minimize matrix interferences. To measure the extent to which matrix effects may still be present after the sample (and spiked sample) have been extracted then rediluted to 10 mL, a plot using the SAM is used and compared with a standard curve made with dilutions of the reference solution with deionized water.

Prepare five reference standard dilutions of the 5.57×10^{-6} M reference solution with:

1. 4 mL reference
2. 3 mL reference with 1 mL DIW
3. 2 mL reference with 2 mL DIW
4. 1 mL reference with 3 mL DIW
5. 4 mL DIW (use as a blank)

Spectrophotometric Analysis

Use your spectrometer to measure the absorbance of the standards, sample and spike at 520 nm or select a wavelength near that value where the maximum absorbance is observed. Use the prepared blank in the dilution 5) above to calibrate the spectrometer.

The original volume of the spike added and the final volume of the extract are the same (10 mL each), so the increase in concentration due to the addition of Methyl Red in the spiked extract will equal the concentration of Methyl Red in the reference, that is 5.57×10^{-6} M. Plot a two point SAM line for the absorbance vs. molar concentration for the sample and spiked sample extract solutions. Also, plot the absorbance vs. molar concentration for the five reference standards (including the blank) prepared. From the slope of the five point standard line and the slope of the SAM line, determine and compare the molar absorptivity obtained by each method. Were any significant matrix effects seen between the extracted samples and the diluted reference solution in this experiment?

Calculating and Reporting Results

Divide the y intercept of the two point line (SAM plot) by its slope to determine the molar concentration of Methyl Red in the 10 mL of extract. Multiply this value by the volume of the extract (10 mL) to determine the total number of moles of Methyl Red in 500 mL of the sample. Assuming the density of the sample is exactly 1 g/mL, divide the 500 g weight of sample analyzed by the weight of sample you measured in the sample bottle in grams to determine the fraction of total sample used for analysis. Divide the total moles of Methyl Red in 500 mL by this fraction to determine the total moles of Methyl Red in the sample.

Using the total moles of Methyl Red in the sample **calculate and report the molarity, M, of Methyl Red in exactly 2 liters of sample**.

Waste Disposal

All of the waste generated in this experiment can go down the drain after it is neutralized to a pH above 5. Add some sodium bicarbonate to 600 mL beaker and fill it half way with water. Add the waste slowly to the beaker and allow it to react. If excess bicarbonate is left in the bottom of the beaker the waste can go down the drain. If all of the bicarbonate has been dissolved add more until it falls to the bottom of the beaker in excess.

Supplemental Information for Instructors for the SPE Experiment

0.3 g Methyl Red (F.W. 269.309) to the nearest 0.1 mg is dissolved in 100 mL ethanol for a 0.01114 M solution (Solution A).

A 1/10 dilution of Solution A in ethanol is made for a 0.03% or 1.114×10^{-3} M solution (Solution B).

A 1/200 dilution of Solution B in 0.5% Acetic acid is made for a 5.57×10^{-6} M (Working Solution).

The Working Solution is used for making unknowns and spiking solutions.

5 to 45 grams of the Working Solution is issued to students as an unknown sample. See Note Below.

Students dilute their unknown to 2000 grams for the SPE analysis. Scoring of results is less stringent than the error that is introduced due to the deviation of the density of water from 1 g/mL at normal temperatures.

A 0.5% Ethanol—0.5% Acetic acid solution is issued to students as a Make-Up solution.

Note: On 11/3/2011 the temperature in the laboratory was 24.5 deg C. Water density at this temperature is 0.997171 g/mL.

Four times a Class A 200 mL ± 0.1 mL volumetric flask was tarred, filled to the mark with Methyl Red working solution and weighed with the Methyl Red.

The weights and densities were

199.411 g	0.997055 g/mL
199.382 g	0.996910 g/mL
199.351 g	0.996755 g/mL
199.372 g	0.996860 g/mL
Avg. density	0.996895 g/mL

The density of water is 0.997171 g/mL at 24.5 deg C.
The difference in the average density between the solution and water was:

$$0.996895 - 0.997171 = -0.000276$$
$$(-0.000276/0.997171) \times 100 = -0.0277\%$$

The possible percent error in the class A 200 mL volumetric flask is:

$$(0.1 \text{ mL}/200 \text{ mL}) \times 100 = 0.05\%$$

Since the per cent different in density between the Methyl Red working solution and water is less than the uncertainty of the volumetric glassware, used the density of the Methyl Red working solution is assumed to be the same as that of water.

EXPERIMENT 34 GAS CHROMATOGRAPHIC ANALYSIS OF A TERTIARY MIXTURE

Principle

A mixture of pentane, hexane, and heptane is separated by gas chromatography. A number of different types of columns can be used, and a simple thermal conductivity or hot wire detector is satisfactory. The instrument response for each compound is calibrated by running a standard mixture of the compounds. The order of separation is determined by running the individual compounds.

An alternative experiment is to analyze a two-component mixture of *n*-hexane and *n*-heptane, and use *n*-pentane as an internal standard.

Solutions and Chemicals Required

1. **Provided.** Acetone.
2. **To prepare.** Standard mixture. Prepare the following standard mixture: *n*-pentane (5.00 mL), *n*-hexane (10.00 mL), *n*-heptane (15.00 mL). All mixtures should be fresh the day of use and kept in plastic-stoppered vials. Alternatively, the standards may be weighed and results reported on a weight/weight basis (or use density to calculate weights from volumes).

   Caution. Exercise great care not to damage the Hamilton syringes. Syringes require proper cleaning and handling to give consistent results. After each use, remove the plunger, rinse thoroughly with acetone, and allow to dry. Insert the syringe through a septum on one end of a glass tube and insert a clean syringe containing acetone into the other end. Force acetone through the tubing and the dirty syringe until clean. Dry by drawing air through the barrel of the syringe with a water aspirator. The glass will appear frosted when completely dry.

Procedure

Obtain an unknown mixture from the instructor. Check the instrument instructions and your instructor regarding the operation of the chromatograph. Do not make any temperature adjustments. Using the appropriate syringe for the instrument, obtain chromatograms of each separate component of the mixture, three chromatograms of your standard mixture, and three chromatograms of your unknown.

Having obtained the necessary data, leave the instrument in the manner prescribed by the instructor. Especially important is the decrease in gas flow and adjustment of filament conditions to a "stand-by" state. Also, clean, rinse, and dry with acetone all syringes and vials.

Analysis of Data

The peak areas on the chromatographic curves are equal to height times width at half-height.[1] Make measurements with a millimeter ruler. Your instrument may automatically print peak areas for each compound, in which case, use these.

Individual components are recognized by the positions of their peaks with respect to the origin (the retention time).

The standard mixture is used for quantitative calibration. Since the detector does not exhibit equal response to all compounds, then a calibration factor must be determined for each, using the standard mixture. One simple way of calibration is just to make a direct comparison of the absolute peak area for each compound with its percent in the mixture. Thus, if 25% compound A gives a peak area of 40, then a peak area of 80 for that compound would correspond to 50% A. Obviously, greatest accuracy would be obtained by preparing a calibration curve of area versus percent compound over a range of percentages for each compound. This would compensate for the fact that the net volume of a given total volume of the individual compounds may vary with the composition.

From the average of the measured chromatographic areas for each compound of the unknown, calculate the average volume percentage of each compound.

Internal Standard Calibration

Your instructor may give you a two-component unknown of *n*-hexane and *n*-heptane, and direct you to add *n*-pentane to all solutions as an internal standard. In this case, prepare the standards as before, but for the unknown, take 25.00 mL and add 5.00 mL of *n*-pentane. Calculate the ratios of the analyte to *n*-pentane areas for the calibration. See Chapter 15 for a description of the use of internal standards.

Mass Spectrometry

Adapted from experiment courtesy of Thomas Leach, University of Washington

EXPERIMENT 37   CAPILLARY GAS CHROMATOGRAPHY-MASS SPECTROMETRY

Introduction

The most powerful tool in the analytical laboratory for the characterization of complex mixtures of organic molecules is the capillary gas chromatograph-mass spectrometer or GC-MS. Its extraordinary capabilities are attained by separating the components of a mixture using capillary gas chromatography, a high resolution chromatographic technique, combined with detection by mass spectrometry to obtain molecular information about the eluted compounds by comparing mass spectra with MS databases.

A typical column is 30 meters long with an internal diameter of 0.25 mm. It has a 0.25 micron film on the inside wall containing the stationary phase made up of a polysiloxane polymer shown below. The two silicon bonds are linked by oxygen to form a chain, leaving two silicon bonds (the vertical bonds) free to attach to

---

[1] In place of determining the areas under the peaks by measurement and using these for the computations, a somewhat more accurate method is to photocopy the chromatograms first and then cut out from the original chromatograms carefully with scissors each peak area. These pieces are then weighed on the analytical balance and the weights are used in the calculations. If you use this method, hand in the "cutouts" with your report. The method of choice is electronic integration.

organic functional groups. In one common variation encountered, 95% of these bonds are methyl groups and 5% of them are phenyl groups. The name for the polymer is 5% Phenyl/95% Dimethyl Polysiloxane (manufacturer names for these include DB-5, HP-5, and Rtx-5).

Some material injected into the column may, over time, contaminate the inner layer of the stationary phase near the head of the column. In that case, the first few inches of the column may sometimes be cut to improve its performance. Therefore, the column may be shorter than 30 meters in practice.

The column can operate at temperatures between −60 deg C and 325 to 350 deg C.

As each component leaves the chromatography column it enters a rapid scan mass spectrometer where it is fragmented by electron ionization. The mass-to-charge (m/z) ratio of each fragment is measured. Differing compounds will break into different patterns of fragments with different m/z ratios.

An essential element of this instrument is a computer with a graphics display terminal. In addition to controlling the operating parameters of the gas chromatograph and the mass spectrometer, the data system allows the progress of the analysis to be monitored as it takes place, writes all of the data to permanent storage on disk, and provides a variety of routines for post run analysis of the data.

Experiment

In this experiment, you will separate two or more compounds, along with an internal standard to correct for experimental variations such as injection volume. You will inject air or butane to determine the time of a non-retained substance (t_0). Your instructor will give directions on how to operate your particular GC-MS system, including the software for controlling the system and for collecting and processing data. You will plot a conventional chromatogram in the "total ion current" mode, i.e., record a "total ion chromatogram". And you will take the mass spectrum at the peaks of interest and identify the compounds. The NIST (National Institute of Standards and Technology) database created in 2008 of ion fragmentation patterns is available to identify compounds from their fragmentation patterns in the mass spectrometer by comparing them to fragmentation patterns collected from numerous other mass spectrometers by the National Institute of Standards and Technology.

You should be able to print the chromatogram and the mass spectra for your report. You will be given standards marked, e.g., A and B, to prepare a calibration curve for the two peaks identified as A and B. The program will quantitate the peaks that appear at the retention times within specified windows, adjust responses to their relative response to the internal standard, match the compounds selected with the qualifier ions and display a report on the screen.

Report

1. Turn in a TIC for the air or butane injection. Report the retention time measured. This represents the time of a non-retained substance in the column (t_0).

2. Report the retention time and identity of each compound quantitated.

3. Provide the total ion chromatograms (TIC) for the standards and the sample runs.

4. Provide a calibration curve plot, along with the equation of the best-fit line, that is used to determine the concentration of each unknown.

5. Circle the parent ion peaks and qualitative ion peaks in the fragmentation patterns that appear with each analyte in the sample.

6. Report the concentrations of the unkowns in the sample.

7. Print out the summary report.

Flow Injection Analysis

EXPERIMENT 38   CHARACTERIZATION OF PHYSICAL PARAMETERS OF A FLOW INJECTION ANALYSIS SYSTEM

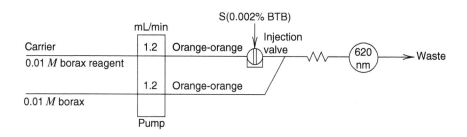

Principles

The purpose of this experiment is to learn the operation of your flow injection apparatus and to estimate (a) the flow rates of the carrier and reagent streams, (b) the volumes of the sample injection loop and of the flow channels, (c) the dispersion, D^{max}, of the sample plug at the peak maximum (two-line and single line), and (d) the maximum sampling frequency. The two-line system is used for all experiments, except for a single-line dispersion experiment.

Equations

$$D = C^0/C = H^0/H$$
$$D^{max} = C^0/C^{max} = H^0/H^{max}$$

H^0 is the recorded steady-state signal for pure sample, and H^{max} is the recorded peak maximum for the injected sample.

Solutions and Chemicals Required

1. **Provided.** Stock 0.1 M borax solution ($Na_2B_4O_7$), stock 0.4% (wt/vol) bromothymol blue (BTB) solution (0.4 g dissolved in 25 mL 96% ethanol and made to 100 mL with 0.01 M borax). **NOTE:** The bromothymol blue should not be injected in acid solution since the acid form of the indicator adsorbs on the plastic tubing.

2. **To prepare.** Dilute 100 mL of 0.1 M borax solution to 1 L to prepare 0.01 M working solution. Dilute 1 mL of the 0.4% bromothymol blue solution to 200 mL with the 0.01 M borax solution. This 0.002% BMT solution will serve as the working solution. The absorbance of this solution in a 1-cm-long cell is about 1.2.

Assembly of Apparatus

Assemble the peristaltic pump tubes, injector, reactor, and detector as described in the instrument operation manual or by your instructor. There will be three peristaltic pump tubes, one for the carrier (0.89 mm i.d., orange–orange color-coded stops), one for the reagent (0.89 mm i.d., orange–orange), and one for the sample (0.89 mm i.d., orange–orange).

    Connect the sampling tube to the inlet of the injection loop and the sample pump tube to the outlet of the loop (the sample will be drawn into the loop by means of the pump).

Procedure

1. **Checking the flow system.** Turn on the detector and recorder and allow to warm up. Fill the carrier bottle and the reagent bottle with the 0.01 *M* sodium borate solution. Fill a small beaker approximately half full with the 0.002% bromothymol blue solution. Place the appropriate tubes in the bottles. Place the injector in the load position and turn on the peristaltic pump. The channels will fill and solution will exit to the waste bottle. The blue sample solution will fill the injector sample loop. Allow to flow until all air bubbles are gone. **NOTE:** Occasionally an air bubble may get stuck in the detector, as indicated by a deflection on the recorder. This is most conveniently dislodged by introducing a large air bubble in the carrier stream by momentarily pulling the carrier tube out of the carrier solution.

    Set the detector at 620 nm. The injected sample should provide a peak maximum of about 0.15 absorbance unit with a 1-cm path flow cell. Adjust your recording device (strip chart recorder or computer—use a chart speed of 0.5 cm/s if a recorder) to accommodate this range. With the carrier being pumped, turn the injection valve to the inject position. Note the blue sample plug as it passes through the channels. If sample continues to be aspirated (to waste) in the inject position of your valve, in order to conserve sample, the sampling tube may be removed from the sample solution, causing air to be aspirated. When the sample reaches the detector, there will be a deflection followed by return to baseline. If necessary, adjust the recording sensitivity so that the deflection will be about two-thirds full scale. After the sample has passed the detector, turn the valve to the load position to refill the sample loop. Continue the flow until the sample liquid reaches the end of the injector waste tube (this assures proper flushing of the sample loop with the new sample), then inject the sample and record the peak. Continue several injections in this manner until you are satisfied the system is operating properly, the detector is not drifting, and reproducible peaks are being obtained. You are now ready to perform the experiment.

2. **Determination of flow rates.** Fill a 10-mL graduated cylinder with distilled water, record the volume, and insert the carrier tube in the cylinder. Turn on the pump and simultaneously start a stop watch (or begin timing with a clock). Pump for 5 min, remove the tube, and turn off the pump. Record the volume of water remaining in the cylinder. Calculate the flow rate in milliliters per minute.

    Perform a similar experiment for the reagent flow and then for the sample flow (injector in the load position).

    Finally, measure the flow rate of the waste stream from the detector, by collecting the waste for 5 min.

    If pump tubing of equal internal diameter is used for all channels, the flow rates should be similar. Also, the flow rate of the waste stream should total the combined flow rates of the carrier and reagent streams.

Note that flow rate is directly proportional to the square of the internal radius of a pump tube, so rates can be appropriately adjusted by changing the pump tubes (i.e., flow is proportional to cross-section area $= \pi R^2$).

3. **Estimation of sample loop volume.** With the pump turned on, place the valve in the load position and remove the sampling tube from the sample. This will allow the loop to fill with air. Then, insert the tube in the sample solution and, with a stopwatch, measure the time from when the sample just enters the loop to when it leaves the loop. Perform the measurement several times and take the average. From a knowledge of the sample flow rate determined above and the measured time to fill the loop, calculate the sample loop volume in microliters. **NOTE:** This is an estimate and will not include the dead volume of the holes in the injector rotor where the loop is connected.

4. **Estimation of the flow system volume.** Fill the sample loop with air and inject the air into the carrier. Measure the time to reach the merging point of the carrier and reagent streams. From a knowledge of the carrier flow rate determined above and the measured time, calculate the volume from the injection valve to the merging point, in microliters. Perform a similar experiment, but measure the time for the air to go from the merging point to the detector. From a knowledge of the combined flow of the carrier and reagent streams in the reactor module determined above and the measured time, calculate the volume from the reactor module inlet to the detector, in microliters.

5. **Determination of total dead time.** With the recording device on, simultaneously inject a sample and start a stopwatch. Measure the time for the sample to reach the detector, that is, the time for the recorder to begin deflection. This represents the dead time from injection to initial measurement.

6. **Time to flush sampling tube and sample loop.** When injecting different samples, it will be necessary to flush the previous sample completely from the sampling tube and the injection loop. Determine the time required to just flush the previous sample by introducing air into the tube and then reinserting the tube in the sample solution with the pump running. Measure the time for the sample solution to reach and fill the sample loop. Use at least three times this time for introduction of each sample in all future experiments, to allow adequate washing of the loop by the new sample.

From a knowledge of the sample flow rate, also calculate the volume of solution required to flush the line.

7. **Determination of dispersion: Two-line system.** Adjust the recording sensitivity so the peak deflection for an injected sample is about one-fifth full scale. Make several injections of the sample and determine the average peak height. This represents H^{max}. Turn the pump off and insert both the carrier tube and the reagent tube in the sample solution. Turn the pump on and record the signal until a steady-state signal is obtained. This represents H^0. **NOTE:** Alternatively, H^0 may be determined by inserting the exit waste tube in the sample solution and reversing the pump flow to fill the cell with sample solution. Calculate the dispersion at the peak height, D^{max}. **NOTE:** If transmittance is recorded rather than absorbance, convert the readings to absorbance for calculating D^{max}.

8. **Determination of dispersion: Single-line system.** Also determine the dispersion by clamping the reagent tube to convert to a single-line system; remove the tube from the pump. Determine the dispersion as above and compare with that of the two-line system.

9. **Determination of maximum sampling frequency.** Expand the recording time axis 10-fold. Inject a sample and record the rise and fall of the peak. Measure the distance from baseline to baseline on the peak and convert this to seconds. This will represent

the minimum time between injections. Report the maximum sampling frequency in samples per hour.

At the end of the experiment, wash the system thoroughly by pumping distilled water for a few minutes. Include flushing the sampling tube and valve loops. Then release the pump cassettes and pump tubes. If the system is not to be used for an extended time, your instructor will instruct you to empty it.

In your report, list all instrument parameters and solutions employed.

EXPERIMENT 40   THREE-LINE FIA: SPECTROPHOTOMETRIC DETERMINATION OF PHOSPHATE

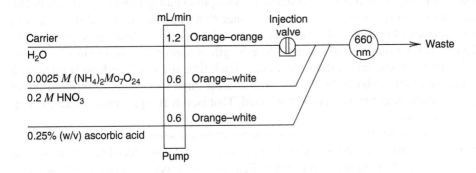

Principle

The analytical procedure is based upon the following reactions:

$$7H_3PO_4 + 12(NH_4)_6Mo_7O_{24} + 51H^+ \rightarrow 7(NH_4)_3PO_4 \cdot 12MoO_3 + 51NH_4^+ + 36H_2O$$

Mo(VI) – YELLOW COMPLEX + ascorbic acid → Mo (V) – BLUE COMPLEX

The carrier is distilled water. Reagent 1 is a 0.0025 M solution of heptamolybdate. Reagent 2 is 5% ascorbic acid (wt/vol) in a 10% aqueous solution of glycerine. The glycerine aids in preventing the colored complex from adhering to the walls of the flow-through cell upon injection of the sample into the carrier stream. The sample merges successively with the reagents and passes through the reaction coils. The phosphate in the sample combines with the heptamolybdate, forming a yellow-colored complex. This yellow complex then reacts with the ascorbic acid, which reduces the molybdenum from the +6 state to the +5 state, forming a blue-colored complex, which has an extremely high absorptivity and is measured spectrophotometrically at 660 nm. The height of the recorded peak is proportional to the concentration of phosphate. The calibration curve linearity and slope depend upon the extent of reaction, that is, how much of the blue complex has been formed. This is a function of the kinetic nature of the FIA procedure in that the reaction may not reach its steady-state value, but rather some fraction of the steady-state value, since it is a slow reaction. The extent of reaction is dependent upon the particular reaction system. Addition of antimony(III) catalyzes the reduction by ascorbic acid.

Perform Experiment 38 first to characterize and learn how to use the system, or else read through it.

Solutions and Chemicals Required

Reagent 1. In a 100-mL volumetric flask, place approximately 50 mL of distilled water and carefully add 1.3 mL of concentrated nitric acid. To this add 0.309 g of ammonium heptamolybdate, $(NH_4)_6Mo_7O_{24} \cdot 4H_2O$, and dilute to the mark with distilled water. This results in a solution that is 0.2 M in nitric acid and 0.0025 M in heptamolybdate.

Reagent 2. In a 100-mL volumetric flask place 5 g of ascorbic acid, approximately 50 mL of distilled water, and 10 mL of glycerine. Mix thoroughly prior to diluting to the mark with distilled water. This results in a solution that is 5% (wt/vol) in ascorbic acid.

Standard solutions. Standard solutions of phosphate (as P) in the 10- to 100-ppm P range are made by suitable dilution of a stock solution containing 100 ppm P (0.440 g of anhydrous KH_2PO_4 per liter).

Procedure

Assemble the injector, reactor, and detector as described by the manufacturer or your instructor. Orange–white color-coded peristaltic pump tubing (0.64 mm i.d.) should be used for reagents 1 and 2, while orange–orange (0.89-mm i.d.) pump tubing should be used for the H_2O carrier and sample. The sample injection size should be approximately 25 μL. This corresponds to a sample loop length of approximately 13 cm when using 0.5-mm inside diameter Micro-line tubing. Each time the injection loop is filled with a new solution, it should be flushed with at least three loop volumes of the solution before filling with the aliquot to be injected. That is, run 100 μL through a 25-μL loop and then stop.

Turn on the detector and allow it to warm up for several minutes to stabilize. If a monochromator is used, set to 660 nm. (Since the color produced is specific for the analyte product, a simple visible white light source detector may be used without a monochromator.) Use a chart speed of approximately 0.5 cm/s if a strip chart recorder is used. With the water carrier and the two reagents being pumped, inject the highest concentration standard solution. Adjust the recording sensitivity to give about 75% deflection with this standard.

Inject the individual phosphate standards successively in triplicate, thereby yielding a series of peaks for each standard. The precision of the procedure is determined by injecting the 50-ppm standard 10 times. Report the percent relative standard deviation.

Obtain an unknown from your instructor and inject it at least three times to obtain an average peak height. Using a spreadsheet, prepare a calibration curve from the peaks recorded for the standard solutions. From this calibration curve, calculate the concentration of phosphate in your unknown solution.

Your report should include all instrument operation parameters, as well as reagents and solutions used in performing the experiment.

Catalyzed Reaction

The sensitivity of this reaction can be enhanced through the use of a catalyst or by using stopped-flow techniques. Antimony(III) catalyzes the reaction, and phosphorus concentrations in the 1- to 10-ppm range can be analyzed with this system without the need for elevated temperatures for increasing reaction rate (the uncatalyzed reaction is actually quite slow, requiring up to 10 min for complete reaction to occur). To perform the catalyzed reaction, a stock antimony(III) solution is prepared by placing 1.10 g of potassium antimony(III) tartrate, $KSbOC_4H_4O_6 \cdot \frac{1}{2}H_2O$, in a 100-mL volumetric flask and diluting to the mark with distilled water. This makes a 0.033 *M* or 4000-ppm solution in antimony(III). A 2.5-mL aliquot of this solution is then added during the preparation of reagent 2, prior to diluting to the mark, yielding a 5% ascorbic acid solution, which is also 100 ppm in antimony(III). Appropriate dilutions of the 100-ppm phosphate standard solution are then made to yield concentrations from 1 to 10 ppm. The experiment is then run in the same manner as before, substituting the new ascorbic acid solution containing antimony and the new phosphate standards. The recorder sensitivity is adjusted using the 10-ppm standard.

Shutdown Procedure

At the end of the experiment, flush the system thoroughly by pumping distilled water through *all pump tubes* for several minutes. After this has been done, release the clamps that hold the pump tubes and release the pump tubes from around the pump rollers. If the system is not to be used for an extended time, your instructor will advise you as to the proper procedure for emptying it.

Compare with Manual Spectrophotometry

You made triplicate measurements of each standard and the unknown in this experiment. About how long did it take to make these measurements (once the reagents were prepared)? You will have probably performed some spectrophotometric experiments. How long do you think the same number of measurements would take using conventional spectrophotometry, considering the cuvet has to be rinsed for each measurement. How much reagent would be consumed for each measurement compared with the amount used in this experiment?

pH Indicators			pH Transition Intervals		
Cresol red	pink	0.2		1.8	yellow
m-Cresol purple	red	1.2		2.8	yellow
Thymol blue	red	1.2		2.8	yellow
p-Xylenol blue	red	1.2		2.8	yellow
2,2′,2″,4,4′-Pentamethoxy-triphenylcarbinol	red	1.2		3.2	colorless
2,4-Dinitrophenol	colorless	2.8		4.7	yellow
4-Dimethylaminoazobenzene	red	2.9		4.0	yellow–orange
Bromochlorophenol blue	yellow	3.0		4.6	purple
Bromophenol blue	yellow	3.0		4.6	purple
Methyl orange	red	3.1		4.4	yellow–orange
Bromocresol green	yellow	3.8		5.4	blue
2,5-Dinitrophenol	colorless	4.0		5.8	yellow
Alizarinsulfonic acid sodium salt	yellow	4.3		6.3	violet
Methyl red	red	4.4		6.2	yellow–orange
Methyl red sodium salt	red	4.4		6.2	yellow–orange
Chlorophenol red	yellow	4.8		6.4	purple
Hematoxylin	yellow	5.0		7.2	violet
Litmus extra pure	red	5.0		8.0	blue
Bromophenol red	orange–yellow	5.2		6.8	purple
Bromocresol purple	yellow	5.2		6.8	purple
4-Nitrophenol	colorless	5.4		7.5	yellow
Bromoxylenol blue	yellow	5.7		7.4	blue
Alizarin	yellow	5.8		7.2	red
Bromothymol blue	yellow	6.0		7.6	blue
Phenol red	yellow	6.4		8.2	red
3-Nitrophenol	colorless	6.6		8.6	yellow–orange
Neutral red	bluish-red	6.8		8.0	orange–yellow
4,5,6,7-Tetrabromophenolphthalein	colorless	7.0		8.0	purple
Cresol red	orange	7.0		8.8	purple
1-Naphtholphthalein	brownish	7.1		8.3	blue–green
m-Cresol purple	yellow	7.4		9.0	purple
Thymol blue	yellow	8.0		9.6	blue
p-Xylenol blue	yellow	8.0		9.6	blue
Phenolphthalein	colorless	8.2		9.8	red–violet
Thymolphthalein	colorless	9.3		10.5	blue
Alizarin yellow GG	light yellow	10.2		12.1	brownish-yellow
Epsilon blue	orange	11.6		13.0	violet

[a]Adapted from *pH Indicators*, E. Merck and Co.

Grades of Chemicals

Grade	Purity	Notes
Technical or commercial	Indeterminate quality	May be used as a cleaning agent or in very preliminary exploratory work. Should not be used in analytical work.
C.P. (Chemically pure)	More refined, but still unknown quality	
U.S.P.	Meets minimum purity standards	Conforms to tolerance set by the U.S. Pharmacopoeia for contaminants dangerous to health
A.C.S. reagent	High purity	Conforms to minimum specifications set by the Reagent Chemicals Committee of the American Chemical Society
Primary standard	Highest purity	Required for accurate volumetric analysis (for standard solutions)

Concentrations of Commercial Reagent-Grade Acids and Bases[a]

Reagent	F. Wt.[b]	M[c]	% by Wt	Density (20°) (g/cm³)
H_2SO_4	98.08	17.6	94.0	1.831
$HClO_4$	100.5	11.6	70.0	1.668
HCl	36.46	12.4	38.0	1.188
HNO_3	63.01	15.4	69.0	1.409
H_3PO_4	98.00	14.7	85.0	1.689
$HC_2H_3O_2$	60.05	17.4	99.5	1.051
NH_3	17.03	14.8	28.0	0.898

[a] These are approximate concentrations and cannot be used for preparing standard solutions.
[b] Formula weight.
[c] Molarity.

NIST Tolerances for Volumetric Glassware, Class A[a]

Capacity (mL) (Less Than and Including)	Tolerances (mL)		
	Volumetric Flasks	Transfer Pipets	Burets
1000	±0.30		
500	±0.15		
100	±0.08	±0.08	±0.10
50	±0.05	±0.05	±0.05
25	±0.03	±0.03	±0.03
10	±0.02	±0.02	±0.02
5	±0.02	±0.01	±0.01
2		±0.006	

[a] Corning Pyrex glassware and Kimball KIMAX, Class A, conform to these tolerances.

Fundamental Physical Constants

Avogadro's number (N) = 6.022×10^{23} atoms/g-at wt = mol^{-1}
Boltzmann constant (k) = 1.38065×10^{-23} J/K = 8.6173×10^{-5} eV/K
Gas constant (R) = 8.3145 J/(mol · K)
Planck constant (h) = 6.6261×10^{-34} J · s = 4.1357×10^{-15} eV · s
Velocity of light (c) = 2.99792×10^{8} m/s

International Atomic Weights Based on $^{12}C = 12$

Element	Symbol	Atomic Number	Atomic Weight[a]	Element	Symbol	Atomic Number	Atomic Weight[a]
Actinium	Ac	89	(227)	Mendelevium	Md	101	(258)
Aluminum	Al	13	26.9815	Mercury	Hg	80	200.59
Americium	Am	95	(243)	Molybdenum	Mo	42	95.94
Antimony	Sb	51	121.76	Neodymium	Nd	60	144.24
Argon	Ar	18	39.948	Neon	Ne	10	20.180
Arsenic	As	33	74.9216	Neptunium	Np	93	(237)
Astatine	At	85	(210)	Nickel	Ni	28	58.69
Barium	Ba	56	137.33	Niobium	Nb	41	92.906
Berkelium	Bk	97	(247)	Nitrogen	N	7	14.0067
Beryllium	Be	4	9.0122	Nobelium	No	102	(259)
Bismuth	Bi	83	208.980	Osmium	Os	76	190.2
Bohrium	Bh	107	(264)	Oxygen	O	8	15.9994
Boron	B	5	10.811	Palladium	Pd	46	106.4
Bromine	Br	35	79.904	Phosphorus	P	15	30.9738
Cadmium	Cd	48	112.41	Platinum	Pt	78	195.08
Calcium	Ca	20	40.08	Plutonium	Pu	94	(244)
Californium	Cf	98	(251)	Polonium	Po	84	(209)
Carbon	C	6	12.011	Potassium	K	19	39.098
Cerium	Ce	58	140.12	Praseodymium	Pr	59	140.907
Cesium	Cs	55	132.905	Promethium	Pm	61	(145)
Chlorine	Cl	17	35.453	Protactinium	Pa	91	(231)
Chromium	Cr	24	51.996	Radium	Ra	88	(266)
Cobalt	Co	27	58.9332	Radon	Rn	86	(222)
Copper	Cu	29	63.546	Rhenium	Re	75	186.2
Curium	Cm	96	(247)	Rhodium	Rh	45	102.905
Dubrium	Db	105	(262)	Rubidium	Rb	37	85.47
Dysprosium	Dy	66	162.50	Ruthenium	Ru	44	101.07
Einsteinium	Es	99	(252)	Rutherfordium	Rf	104	(261)
Erbium	Er	68	167.26	Samarium	Sm	62	150.35
Europium	Eu	63	151.96	Scandium	Sc	21	44.956
Fermium	Fm	100	(257)	Seaborgium	Sg	106	(266)
Fluorine	F	9	18.9984	Selenium	Se	34	78.96
Francium	Fr	87	(223)	Silicon	Si	14	28.086
Gadolinium	Gd	64	157.25	Silver	Ag	47	107.870
Gallium	Ga	31	69.72	Sodium	Na	11	22.9898
Germanium	Ge	32	72.61	Strontium	Sr	38	87.62
Gold	Au	79	196.967	Sulfur	S	16	32.066
Hafnium	Hf	72	178.49	Tantalum	Ta	73	180.948
Hassium	Hs	108	(265)	Technetium	Tc	43	(98)
Helium	He	2	4.0026	Tellurium	Te	52	127.60
Holmium	Ho	67	164.930	Terbium	Tb	65	158.925
Hydrogen	H	1	1.00794	Thallium	Tl	81	204.38
Indium	In	49	114.82	Thorium	Th	90	232.038
Iodine	I	53	126.9045	Thulium	Tm	69	163.934
Iridium	Ir	77	192.2	Tin	Sn	50	118.71
Iron	Fe	26	55.845	Titanium	Ti	22	47.87
Krypton	Kr	36	83.80	Tungsten	W	74	183.85
Lanthanum	La	57	138.91	Uranium	U	92	238.03
Lawrencium	Lw	103	(262)	Vanadium	V	23	50.9415
Lead	Pb	82	207.2	Xenon	Xe	54	131.30
Lithium	Li	3	6.9417	Ytterbium	Yb	70	173.04
Lutetium	Lu	71	174.967	Yttrium	Y	39	88.905
Magnesium	Mg	12	24.312	Zinc	Zn	30	65.37
Manganese	Mn	25	54.9380	Zirconium	Zr	40	91.22
Meitnerium	Mt	109	(268)				

[a] Numbers in parentheses indicate mass of most stable known isotope. For elements 110 to 118, see the periodic tables listed in Appendix E. Elements 110, 111, 112, 114, and 116 have been given the names Darmstadium (110), Roentgenium (111), Copernicum (112), Flerovium (114), and Livermorium (116). Others are awaiting official names by IUPAC. Element 118 is temporarily named Unuoctium. Element 113 has recently been claimed to have been created.

Formula Weights

AgBr	187.78	K_2CrO_7	294.19
AgCl	143.32	KHC_2O_4	128.13
Ag_2CrO_4	331.73	$KHC_2O_4 \cdot H_2C_2O_4$	218.16
AgI	234.77	$KHC_8H_2O_4 (KHP)$	204.23
$AgNO_3$	169.87	$KH(IO_3)_2$	389.92
AgSCN	165.95	K_2HPO_4	174.18
Ag_2SO_4	311.80	KH_2PO_4	136.09
$Al(C_9H_6ON)_3 (AlOx_3)$	459.46	$KHSO_4$	136.17
Al_2O_3	101.96	KI	166.01
$Al_2(SO_4)_3$	342.14	KIO_3	214.00
As_2O_3	197.85	KIO_4	230.00
$BaCO_3$	197.35	$KMnO_4$	158.04
$BaCl_2$	208.25	KNO_3	101.11
$BaCl_2 \cdot 2H_2O$	244.27	KOH	56.11
$BaCrO_4$	253.33	KSCN	97.18
BaO	153.34	K_2SO_4	174.27
$BaSO_4$	233.40	$MgCl_2$	95.22
Bi_2O_3	466.0	$Mg(C_9H_6ON)_2 (MgOx_2)$	312.60
$C_6H_{12}O_6$ (glucose)	180.16	$MgNH_4PO_4$	137.35
CO_2	44.01	MgO	40.31
$CaCl_2$	110.99	$Mg_2P_2O_7$	222.57
$CaCO_3$	100.09	$MgSO_4$	120.37
CaC_2O_4	128.10	MnO_2	86.94
CaF_2	78.08	Mn_2O_3	157.88
CaO	56.08	Mn_3O_4	228.81
$CaSO_4$	136.14	$MnSO_4$	151.00
CeO_2	172.12	$Na_2B_4O_7 \cdot 10H_2O$	381.37
$Ce(SO_4)_2$	332.25	NaBr	102.90
$(NH_4)_2Ce(NO_3)_6$	548.23	$Na(C_6H_5)_4B$	342.20
$(NH_4)_4Ce(SO_4)_4 \cdot 2H_2O$	632.6	$NaC_2H_3O_2$	82.03
Cr_2O_3	151.99	$Na_2C_2O_4$	134.00
CuO	79.54	NaCl	58.44
Cu_2O	143.08	NaClO	74.44
$CuSO_4$	159.60	NaCN	49.01
$Fe(NH_4)_2(SO_4)_2 \cdot 6H_2O$	392.14	Na_2CO_3	105.99
FeO	71.85	$NaHCO_3$	84.01
Fe_2O_3	159.69	$Na_2H_2EDTA \cdot 2H_2O$	372.23
Fe_3O_4	231.54	NaOH	40.00
HBr	80.92	NaSCN	81.07
$HC_2H_3O_2$ (acetic acid)	60.05	Na_2SO_4	142.04
HCl	36.46	$Na_2S_2O_3 \cdot 5H_2O$	248.18
$HClO_4$	100.46	$Ni(C_4H_7O_2N_2)_2 (Ni-DMG_2)$	288.94
$H_2C_2O_4$	90.04	NH_3	17.03
$H_2C_2O_4 \cdot 2H_2O$	126.07	NH_4Cl	53.49
H_5IO_6	227.94	$(HOCH_2)_3CNH_2$ (THAM; Tris)	121.14
HNO_3	63.01	NH_2CONH_2 (Urea)	60.06
H_2O	18.015	$(NH_4)_2C_2O_4 \cdot H_2O$	142.11
H_2O_2	34.01	NH_4NO_3	80.04
H_3PO_4	98.00	$(NH_4)_2SO_4$	132.14
H_2S	34.08	$(NH_4)_2S_2O_8$	228.18
H_2SO_3	82.08	NH_2SO_3H	97.09
H_2SO_4	98.08	$PbCrO_4$	323.18
HSO_3NH_2 (sulfamic acid)	97.09	$PbSO_4$	303.25
HgO	216.59	P_2O_5	141.94
Hg_2Cl_2	472.09	Sb_2O_3	291.50
$HgCl_2$	271.50	SiO_2	60.08
$Hg(NO_3)_2$	324.61	$SnCl_2$	189.60
KBr	119.01	SnO_2	150.69
$KBrO_3$	167.01	$SrSO_4$	183.68
$K(C_6H_5)_4B$	358.31	SO_2	64.06
KCl	74.56	SO_3	80.06
$KClO_3$	122.55	TiO_2	79.90
KCN	65.12	V_2O_5	181.88
K_2CrO_4	194.20	$Zn_2P_2O_7$	304.68